The Maths of
Life and Death

KIT YATES

The Maths of Life and Death

Why Maths Is (Almost) Everything

Quercus

First published in Great Britain in 2019 by

Quercus Editions Ltd
Carmelite House
50 Victoria Embankment
London EC4Y 0DZ

An Hachette UK company

A CIP catalogue record for this book is available
from the British Library.

HB ISBN 978 1 78747 542 7
TPB ISBN 978 1 78747 541 0
Ebook ISBN 978 1 78747 539 7

Certain names and identifying details have been changed
whether or not so noted in the text.

10 9 8 7 6 5 4 3 2 1

Designed and typeset by EM&EN
Printed and bound in Great Britain by Clays Ltd, Elcograf S.p.A.

Paper: ts

For my parents,
Tim, Nancy and Mary,
who taught me how to read,
and my sister, Lucy,
who taught me how to write.

Contents

Introduction

ALMOST EVERYTHING

My four-year-old son loves playing out in the garden. His favourite activity is digging up and inspecting creepy crawlies, especially snails. If he is patient enough, after the initial shock of being uprooted, they will emerge cautiously from the safety of their shells and start to glide over his little hands leaving viscid trails of mucus. Eventually, when he tires of them, he will discard them, somewhat callously, in the compost heap or on the woodpile behind the shed.

Late last September, after a particularly busy session in which he had unearthed and disposed of five or six large specimens, he came to me as I was sawing up wood for the fire and asked 'Daddy, how many snails is [*sic*] there in the garden?' A deceptively simple question for which I had no good answer. It could have been 100 or it could have been 1000. To be quite honest, he would not have comprehended the difference. Nevertheless, his question piqued an interest in me. How could we figure this out together?

We decided to conduct an experiment. The next weekend, on Saturday morning, we went out to collect snails. After ten minutes, we had a total of 23 of the gastropods. I took a sharpie from my back pocket and proceeded to place a subtle cross on the back of each one. Once they were all marked up, we tipped up the bucket and released the snails back into the garden.

A week later we went back out for another round. This time, our ten-minute scavenge brought us just 18 snails. When we inspected them closely we found that three of them had the cross on their shells, while the other 15 were unblemished. This was all the information we needed to make the calculation.

The idea is as follows: the number of snails we captured on the first day, 23, is a given proportion of the total population of the garden, which we want to get a handle on. If we can work out this proportion then we can scale up the number of snails we caught to find the total population of the garden. So we use a second sample (the one we took the following Saturday). The proportion

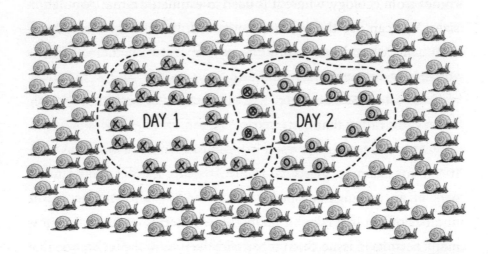

Figure 1: The ratio (3:18) of the number of snails recaptured (marked ⊗) to the total number captured on the second day (marked O) should be the same as the ratio (23:138) of the number captured on the first day (marked X) to the total number of snails in the garden (marked and unmarked).

of marked individuals in this sample, 3/18, should be representative of the proportion of marked individuals in the garden as a whole. When we simplify this proportion, we find that the marked snails make up one in every six individuals in the population at large (you can see this illustrated in Figure 1). Thus we scale up the number of marked individuals caught on the first day, 23, by a factor of 6 to find an estimate for the total number of snails in the garden, which is 138.

After finishing this mental calculation I turned to my son, who had been 'looking after' the snails we had collected. What did he make of it when I told him that we had roughly 138 snails living in our garden? 'Daddy,' he said, looking down at the fragments of shell still clinging to his fingers, 'I made it dead.' Make that 137.

This simple mathematical method, known as capture–recapture, comes from ecology, where it is used to estimate animal population sizes. You can use the technique yourself, by taking two independent samples and comparing the overlap between them. Perhaps you want to estimate the number of raffle tickets that were sold at the local fair or to estimate the attendance at a football match using ticket stubs rather than having to do an arduous head count.

Capture–recapture is used in serious scientific projects as well. It can, for example, give vital information on the fluctuating numbers of an endangered species. By providing an estimate of the number of fish in a lake,[1] it might allow fisheries to determine how many permits to issue. Such is the effectiveness of the technique that its use has evolved beyond ecology to provide accurate estimates on everything from numbers of drug addicts[2] in a population to the number of war dead in Kosovo.[3] This is the pragmatic power that simple mathematical ideas can wield. These are the sorts of concepts that we will explore throughout this book and that I use routinely in my day job as a mathematical biologist.

•

When I tell people I am a mathematical biologist, the reaction I get is usually a polite nodding of the head accompanied by an awkward silence, as if I was about to test them on their recall of the quadratic formula or Pythagoras' theorem. More than simply being daunted, people struggle to understand how a subject like maths, which they perceive as being abstract, pure and ethereal, can have anything to do with a subject like biology, which is typically thought of as being practical, messy and pragmatic. This artificial dichotomy is often first encountered at school: if you liked science but you weren't so hot on algebra, then you were pushed down the life sciences route. If, like me, you enjoyed science but you weren't into cutting dead things up (I fainted once, at the start of a dissection class, when I walked into the lab to see a fish head sitting at my bench space) then you were guided towards the physical sciences. Never the twain shall meet.

This happened to me. I dropped biology at sixth-form and took A-levels in maths, further maths, physics and chemistry. When it came to university, I had to further streamline my subjects, and felt sad that I had to leave biology behind forever; a subject I thought had incredible power to change lives for the better. I was hugely excited about the opportunity to plunge myself into the world of mathematics, but I couldn't help worrying that I was taking on a subject that seemed to have very few practical applications. I couldn't have been more wrong.

Whilst I plodded through the pure maths we were taught at

university, memorising the proof of the intermediate value theorem or the definition of a vector space, I lived for the applied maths courses. I listened to lecturers as they demonstrated the maths that engineers use to build bridges so that they don't resonate and collapse in the wind, or to design wings that ensure planes don't fall out of the sky. I learned the quantum mechanics that physicists use to understand the strange goings-on at subatomic scales and the theory of special relativity that explores the strange consequences of the invariance of the speed of light. I took courses explaining the ways in which we use mathematics in chemistry, in finance and in economics. I read about how we use mathematics in sport to enhance the performance of our top athletes and how we use mathematics in the movies to create computer-generated images of scenes that couldn't exist in reality. In short, I learned that mathematics can be used to describe almost everything.

In the third year of my degree I was fortunate enough to take a course in mathematical biology. The lecturer was Philip Maini, an engaging Northern Irish professor in his forties. Not only was he the pre-eminent figure in his field (he would later go on to be elected to the Fellowship of the Royal Society), but he clearly loved his subject, and his enthusiasm spread to the students in his lecture theatre.

More than just mathematical biology, Philip taught me that mathematicians are human beings with feelings, not the one-dimensional automata that they are often portrayed to be. A mathematician is more than just, as the Hungarian probabilist, Alfréd Rényi, once put it, 'a machine for turning coffee into theorems'. As I sat in Philip's office awaiting the start of the interview for a PhD place, I saw, framed on the walls, the numerous rejection letters he had received from the Premier League clubs to whom he had jokingly applied for vacant managerial positions. We ended up talking more about football than we did about maths.

Crucially at this point in my academic studies, Philip helped me to become fully reacquainted with biology. During my PhD under his supervision, I worked on everything, from understanding the way locusts swarm and how to stop them, to predicting the complex choreography that is the development of the mammalian embryo and the devastating consequences when the steps get out of sync. I built models explaining how birds' eggs get their beautiful pigmentation patterns and wrote algorithms to track the movement of free-swimming bacteria. I simulated parasites evading our immune systems and modelled the way in which deadly diseases spread through a population. The work I started during my PhD has been the bedrock for the rest of my career. I still work on these fascinating areas of biology, and others, with PhD students of my own, in my current position as an associate professor (senior lecturer) in applied mathematics at the University of Bath.

•

As an applied mathematician, I see mathematics as, first and foremost, a practical tool to make sense of our complex world. Mathematical modelling can give us an advantage in everyday situations and it doesn't have to involve hundreds of tedious equations or lines of computer code to do so. Mathematics, at its most fundamental, is pattern. Every time you look at the world you are building your own model of the patterns you observe. If you spot a motif in the fractal branches of a tree, or in the multi-fold symmetry of a snowflake, then you are seeing maths. When you tap your foot in time to a piece of music, or when your voice reverberates and resonates as you sing in the shower, you are hearing maths. If you bend a shot into the back of the net or catch a cricket ball on its parabolic trajectory, then you are doing maths. With every new experience, every piece of sensory information, the models you've made of your environment are refined, reconfigured and rendered ever more detailed and complex. Building mathematical models, designed to capture our intricate reality, is the best way we have of making sense of the rules that govern the world around us.

I believe that the simplest, most important models are stories and analogies. The key to exemplifying the influence of the unseen undercurrent of maths is to demonstrate its effects on people's lives: from the extraordinary to the everyday. When viewed through the correct lens we can start to tease out the hidden mathematical rules that underlie our common experiences.

The seven chapters of this book explore the true stories of life-changing events in which the application (or misapplication) of mathematics has played a critical role: patients crippled by faulty genes and entrepreneurs bankrupted by faulty algorithms; innocent victims of miscarriages of justice and the unwitting victims of software glitches. We follow stories of investors who have lost fortunes and parents who have lost children, all because of mathematical misunderstanding. We wrestle with ethical dilemmas from screening to statistical subterfuge and examine pertinent societal issues such as political referenda, disease prevention, criminal justice and artificial intelligence. In this book we will see that mathematics has something profound or significant to say on all of these subjects, and more.

Rather than just pointing out the places in which maths might

crop up, throughout these pages I will arm you with simple mathematical rules and tools that can help you in your everyday life: from
getting the best seat on the train, to keeping your head when you
get an unexpected test result from the doctor. I will suggest simple
ways to avoid making numerical mistakes and we will get our hands
dirty with newsprint when untangling the figures behind the headlines. We will also get up close and personal with the maths behind
consumer genetics and observe maths in action as we highlight the
steps we can take to help halt the spread of a deadly disease.

As you'll hopefully have worked out by now, this is not a maths
book. Nor is it a book for mathematicians. You will not find a single
equation in these pages. The point of the book is not to bring back
memories of the school mathematics lessons you might have given
up years ago. Quite the opposite. If you've ever been disenfran-
chised and made to feel that you can't take part in mathematics or
aren't good at it, consider this book an emancipation.

I genuinely believe that maths is for everyone and that we can all
appreciate the beautiful mathematics at the heart of the complicated
phenomena we experience daily. As we will see in the following
chapters, maths is the false alarms that play on our minds and the
false confidence that helps us sleep at night; the stories pushed at us
on social media and the memes that spread through it. Maths is the
loopholes in the law and the needle that closes them; the technology
that saves lives and the mistakes that put them at risk; the outbreak
of a deadly disease and the strategies to control it. It is the best
hope we have of answering the most fundamental questions about
the enigmas of the Cosmos and the mysteries of our own species.
It leads us on the myriad paths of our lives and lies in wait, just
beyond the veil, to stare back at us as we draw our final breaths.

1

THINKING EXPONENTIALLY

Exploring the Awesome Power and Sobering Limits of Exponential Behaviour

Darren Caddick is a driving instructor from Caldicot, a small town in South Wales. In 2009 he was approached by a friend with a lucrative offer. By contributing just £3,000 to a local investment syndicate and recruiting two more people to do the same, he would see a return of £23,000 in just a couple of weeks. Initially, thinking it was too good to be true, Caddick resisted the temptation. Eventually though, his friends convinced him that 'nobody would lose, because the scheme would just keep going and going and going', so he decided to throw in his lot. He lost everything and, ten years on, is still living with the consequences.

Unwittingly, Caddick had found himself at the bottom of a pyramid scheme that couldn't 'just keep going'. Initiated in 2008, the 'Give and Take' scheme ran out of new investors and collapsed in less than a year, but not before sucking in £21 million from over 10,000 investors across the UK, 90% of whom lost their £3,000 stake. Investment schemes that rely on investors recruiting multiple others in order to realise their payout are doomed to failure. The number of new investors needed at each level increases in proportion to the number of people in the scheme. After fifteen rounds of recruitment, there would be over 10,000 people in a pyramid scheme of this sort. Although that sounds like a large number, it was easily achieved by Give and Take. Fifteen rounds further on, however, and one in every seven people on the planet would need to invest to keep the scheme going. This rapid growth phenomenon, which led to an inevitable lack of new recruits and the eventual collapse of the scheme, is known as exponential growth.

No use crying over spoilt milk

Something grows exponentially when it increases in proportion to its current size. Imagine, when you open your pint of milk in the morning, a single cell of the bacteria *Streptococcus faecalis* finds its way into the bottle before you put the lid back on. *Strep f.* is one of the bacteria responsible for the souring and curdling of milk, but one cell is no big deal, right?[4] Maybe it's more worrying when you find out that, in milk, *Strep f.* cells can divide to produce two daughter cells every hour.[5] At each generation, the number of cells increases in proportion to the current number of cells, so their numbers grow exponentially.

The curve that describes how an exponentially growing quantity increases is reminiscent of a quarter-pipe ramp used by skaters, skateboarders and BMXers. Initially, the gradient of the ramp is very low – the curve is extremely shallow and gains height only gradually (as you can see from the first curve in Figure 2). After two

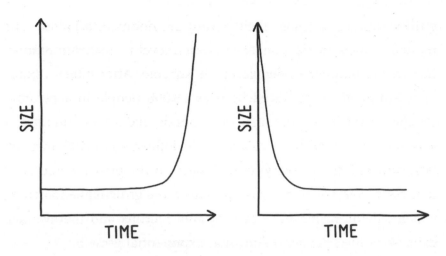

Figure 2: J-shaped exponential growth (*left*) and decay (*right*) curves.

hours there are four *Strep f.* cells in your milk and after four hours there are still only 16, which doesn't sound like too much of a problem. As with the quarter-pipe though, the height of the exponential curve and its steepness rapidly increase. Quantities that grow exponentially might appear to grow slowly at first, but they can take off quickly in a way that seems unexpected. If you leave your milk out on the side for 48 hours, and the exponential increase of *Strep f.* cells continues, when you come to pouring it on your cereal again there could be almost a thousand trillion cells in the bottle – enough to make your blood curdle, let alone the milk. At this point the cells would outnumber the people on our planet 40,000 to one. Exponential curves are sometimes referred to as 'J-shaped' as they almost mimic the letter J's steep curve. Of course, as the bacteria use up the nutrients in the milk and change its pH, the growth conditions deteriorate and the exponential increase is only sustained for a relatively short period of time. Indeed, in almost every real-world scenario, long-term exponential growth is unsustainable, and in many cases pathological, as the subject of the growth uses up resources in an unviable manner. Sustained exponential growth of cells in the body, for example, is a typical hallmark of cancer.

Another example of an exponential curve is a free fall waterslide, so called because the slide is initially so steep that the rider feels the sensation of free fall. This time, as we travel down the slide, we are surfing an exponential *decay* curve, rather than a *growth* curve (you can see an example of such a graph in the second image of Figure 2). Exponential decay occurs when a quantity *decreases* in proportion to its current size. Imagine opening a huge bag of M&Ms, pouring them out onto the table and eating all the sweets that land with the M-side facing upwards. Put the rest back in the bag for tomorrow. The next day, give the bag a shake and pour out the M&Ms. Again, eat the M-up sweets and put the rest back in the bag. Each time you pour the sweets out of the bag you get to eat roughly half of those that remain, irrespective of the number you start with. The number of sweets decreases in proportion to the number left in the bag, leading to exponential decay in the number of sweets. In the same way, the exponential water slide starts high up and almost vertical, so that the height of the rider decreases very rapidly; when we have large numbers of sweets the number we get to eat is also large. But the curve ever-so-gradually gets less and less steep until it is almost horizontal towards the end of the slide; the fewer sweets we have left, the fewer we get to eat each day. Although an individual sweet landing M-up or M-down is random and unforeseeable, the predictable waterslide curve of exponential decay emerges in the number of sweets we have left over time.

Throughout this chapter we will uncover the hidden connections between exponential behaviour and everyday phenomena:

the spread of a disease through a population or a meme through the internet; the rapid growth of an embryo or the all-too-slow growth of money in our bank accounts; the way in which we perceive time and even the explosion of a nuclear bomb. As we progress we will carefully unearth the full tragedy of the Give and Take pyramid scheme. The stories of the people whose money was sucked in and swallowed serve to illustrate just how important it is to be able to think exponentially, which in turn will help us anticipate the sometimes surprising pace of change in the modern world.

A matter of great interest

On the all-too-rare occasions when I get to make a deposit into my bank account, I take solace from the fact that no matter how little I have in there, it is always growing exponentially. Indeed, a bank account is one place where there are genuinely no limitations on exponential growth, at least on paper. Provided that the interest is compounded (i.e. interest is added to our initial amount and earns interest itself) then the total amount in the account increases in proportion to its current size – the hallmark of exponential growth. As Benjamin Franklin put it: 'Money makes money, and the money that money makes, makes more money.' If you could wait long enough, even the smallest investment would become a fortune. But don't go and lock up your rainy-day fund just yet. If you invested £100 at 1% per year it would take you over 900 years to become a millionaire. Although exponential growth is often associated with rapid increases, if the rate of growth and the initial investment are small, exponential growth can feel very slow indeed.

The flip side of this is that, because you are charged a fixed rate of interest on the outstanding amount (often at a large rate), debts on credit cards can also grow exponentially. As with mortgages, the earlier you pay your credit cards off and the more you pay early on, the less you end up paying overall, as exponential growth never gets a chance to take off.

•

Paying off mortgages and sorting out other debts was one of the main reasons given by victims of the Give and Take scheme for getting involved in the first place. The temptation of quick and easy money to reduce financial pressures was too much for many to resist, despite the nagging suspicion that something wasn't quite right. As Caddick admits, 'The old adage of "If something looks too good to be true, then it probably is" is really, really true here.'

The scheme's initiators, pensioners Laura Fox and Carol Chalmers, had been friends since their days at a Catholic convent school. The pair, both pillars of their local community – one vice-president of her local Rotary club, the other a well-respected grandmother – knew exactly what they were doing when they set up their fraudulent investment scheme. Give and Take was cleverly designed to ensnare potential investors, whilst hiding the pitfalls. Unlike the traditional two-level pyramid scheme, in which the person at the top of the chain takes money directly from the investors they have recruited, Give and Take operated as a four-level 'aeroplane' scheme. In an aeroplane scheme, the person at the top of the chain is known as the 'pilot'. The pilot recruits two 'co-pilots', who each recruit two 'crew members', who finally each recruit two 'passengers'. In Fox and Chalmers' scheme, once the hierarchy of fifteen people was complete, the eight passengers paid their £3,000 to the organisers, who passed a huge £23,000 payout to the initial investor with £1,000 skimmed off the top. Part of this money was donated to charity, with letters of thanks from the likes of the NSPCC adding legitimacy to the scheme. Part was kept by the organisers to ensure the continued smooth operation of the scheme.

Having received their payout, the pilot then drops out of the scheme and the two co-pilots are promoted to pilot, awaiting the recruitment of eight new passengers at the bottom of their trees. Aeroplane schemes are particularly seductive for investors, as new participants need only recruit two other people in order to multiply their investment by a factor of 8 (although, of course, these two are required to recruit two more and so on). Other, flatter, schemes require far more recruitment effort per individual for the same returns. The steep four-level structure of Give and Take meant that crew members never took money directly from the passengers they recruited. Since new recruits are likely to be friends and relatives of the crew members, this ensures that money never travels directly between close acquaintances. This separation of the passengers from the pilots, whose payouts they fund, renders recruitment easier and reprisals less likely, making for a more attractive investment opportunity and thus facilitating the recruitment of thousands of investors to the scheme.

In the same way, many investors in the Give and Take pyramid scheme were given the confidence to invest by stories of successful payouts that had been made previously, and in some cases, even witnessing these payouts at first hand. The scheme's organisers, Fox and Chalmers, hosted lavish private parties at the Somerset hotel owned by Chalmers. Flyers handed out at the parties included pictures of the scheme's members, sprawled on cash-covered beds or waving fists of fifty-pound notes at the camera. To each of these parties the organisers also invited some of the scheme's 'brides' – those people (mainly women) who had made it to the position of pilot of their pyramid cell and were due to receive their payouts. The brides would be asked a series of four simple questions such as 'What part of Pinocchio grows when he lies?' in front of an audience of 200 to 300 potential investors.

This 'quiz' aspect of the scheme was supposed to exploit a loop-hole in the law, which Fox and Chalmers believed allowed for such investments if an element of 'skill' was involved. In mobile-phone footage of one such event, Fox can be heard shouting 'We are gambling in our own homes and that's what makes it legal.' She was wrong. Miles Bennet, the lawyer prosecuting the case, explained, 'The quiz was so easy that there were never any people in the payout position who didn't get their money. They could even get a friend or a committee member to help with the questions and the committee knew what the answers were!'

This didn't stop Fox and Chalmers using these prize-giving parties as inoculants in their low-tech viral marketing campaign. Upon seeing the brides presented with their £23,000 cheques, many of the invited guests would invest and encourage their friends and family to do the same, forming the pyramid beneath them. Pro-viding each new investor passed the baton to two or more others, the scheme would continue indefinitely. When Fox and Chalmers started the scheme, back in the spring of 2008, they were the only two pilots. By recruiting friends to invest and indeed help organise the scheme, the pair quickly brought four more people on board. These four recruited eight more and then 16 and so on. This exponential doubling of the numbers of new recruits in the scheme closely mimics the doubling of the number of cells in a growing embryo.

The exponential embryo

When my wife was pregnant with our first child we were obsessed, like many first-time parents-to-be, by trying to find out what was going on inside my wife's midriff. We borrowed an ultrasound heart monitor in order to listen in to our baby's heartbeat; we signed up for clinical trials in order to get extra scans; and we read website after website describing what was going on with our daughter as she grew and continued to make my wife sick on a daily basis. Amongst our 'favourites' were the 'How big is your baby?'-type websites, which compare, for each week of gestation, the size of an unborn baby to a common fruit, vegetable or other appropriately sized foodstuff. They give substance to prospective parents' unborn foetuses with epigrams such as 'Weighing about one and a half ounces and measuring about three and a half inches, your little angel is roughly the size of a lemon' or 'Your precious little turnip now weighs about five ounces and is approximately five inches long from head to bottom'.

What really struck me about these websites' comparisons was how quickly the sizes changed from week to week. At week four, your baby is roughly the size of a poppy seed, but by week five, she has ballooned to the size of a sesame seed! This represents an increase in volume of roughly 16 times over the course of a week.

Perhaps, though, this rapid increase in size shouldn't be so surprising. When the egg is initially fertilised by the sperm, the resulting zygote undergoes sequential rounds of cell division, called 'cleavage', which allow the number of cells in the developing embryo to increase rapidly. First, it divides into two. Eight hours later these two further subdivide into four and, after eight more hours, four become eight, which soon turn into 16, and so on – just like the number of new investors at each level of the pyramid scheme. Subsequent divisions occur almost synchronously every eight hours. Thus, the number of cells grows in proportion to the quantity of cells comprising the embryo at a given moment in time: the more cells there are, the more new cells are created at the subsequent division. In this case, since each cell creates exactly one daughter cell at each division event, the factor by which the number of cells in the embryo increases is two; in other words, the size of the embryo doubles every generation.

During human gestation the period in which the embryo grows exponentially is, thankfully, relatively short. If the embryo were to carry on growing at the same exponential rate for the whole pregnancy, the 840 synchronous cell divisions would result in a super-baby comprising roughly 10^{253} cells. To put that into context, if every atom in the universe were itself a copy of our universe then the total number of atoms in all these universes would be roughly equivalent to the number of super-baby's cells. Naturally, cell division becomes less rapid as more complex events in the life of the embryo are choreographed. In reality the number of cells comprising an average new-born baby can be approximated at a relatively modest two trillion. This number of cells could be achieved in fewer than 41 synchronous division events.

The destroyer of worlds

Exponential growth is vital for the rapid expansion in the number of cells necessary for the creation of a new life. However, it was also the astonishing and terrifying power of exponential growth that led nuclear physicist J. Robert Oppenheimer to proclaim 'Now I am become Death, the destroyer of worlds.' This growth was not the growth of cells, nor even of individual organisms, but of energy created by the splitting of atomic nuclei.

During the Second World War, Oppenheimer was the head of the Los Alamos laboratory where the Manhattan Project – to develop the atomic bomb – was based. The splitting of the nucleus (tightly bound protons and neutrons) of a heavy atom into smaller constitutive parts had been discovered by German chemists in 1938. It was named 'nuclear fission' in analogy to the binary fission, or splitting, of one living cell into two, as occurs to such great effect in the developing embryo. Fission was found either to occur naturally, as the radioactive decay of unstable chemical isotopes, or to be induced artificially by bombarding the nucleus of one atom with subatomic particles in a so-called 'nuclear reaction'. In either case, the splitting of the nucleus into two smaller nuclei, or fission products, was concurrent with the release of large amounts of energy in the form of electromagnetic radiation, as well as the energy associated with the movement of the fission products. It was quickly recognised that these moving fission products, created by a first nuclear reaction, could be used to impact on further nuclei, splitting more atoms and releasing yet more energy: a so-called 'nuclear chain reaction'. If each nuclear fission produced, on average, more than one product that could be used to split subsequent atoms, then, in theory, each fission could trigger multiple other splitting events. Continuing this process, the number of reaction events would increase exponentially, producing energy on an unprecedented scale. If a material could be found that would permit this unchecked nuclear chain reaction, the exponential increase in energy emitted over the short timescale of the reactions would potentially allow such a *fissile* material to be weaponised.

In April 1939, on the eve of the outbreak of war across Europe, French physicist Frédéric Joliot-Curie (son-in-law of Marie and Pierre and also a Nobel Prize winner in collaboration with his wife), made a crucial discovery. He published in the journal *Nature* evidence that, upon fission caused by a single neutron, atoms of the uranium isotope, U-235, emitted on average 3.5 (later revised down to 2.5) high energy neutrons.[6] This was precisely the material required to drive the exponentially growing chain of nuclear reactions. The 'race for the bomb' was on.

With Nobel Prize winner Werner Heisenberg and other celebrated German physicists working for the Nazis' parallel bomb project, Oppenheimer knew he had his work cut out at Los Alamos. His main challenge was to create the conditions that would facilitate an exponentially growing nuclear chain reaction allowing the almost instantaneous release of the huge amounts of energy required for an atom bomb. To produce this self-sustaining and sufficiently rapid chain reaction, he needed to ensure that enough of the neutrons emitted by a fissioning U-235 atom were reabsorbed by the nuclei of other U-235 atoms, causing them to split in turn. He found that, in naturally occurring uranium, too many of the emitted neutrons are absorbed by U-238 atoms (the other significant isotope, which makes up 99.3% of naturally occurring uranium)[7] meaning that any chain reaction dies out exponentially instead of growing. In order to produce an exponentially growing chain reaction, Oppenheimer needed to refine extremely pure U-235 by removing as much of the U-238 in the ore as possible.

These considerations gave rise to the idea of the so-called *critical mass* of the fissile material. The critical mass of uranium is the amount of material required to generate a self-sustaining nuclear chain reaction. It depends on a variety of factors. Perhaps most crucial is the purity of the U-235. Even with 20% U-235 (compared to the naturally occurring 0.7%), the critical mass is still over 400 kilograms, making high purity essential for a feasible bomb. Even when he had refined sufficiently pure uranium to achieve supercriticality, Oppenheimer was left with the challenge of the delivery of the bomb itself. Clearly he couldn't just package up a critical mass of uranium in a bomb and hope it didn't explode. A single, naturally occurring, decay in the material would trigger the chain reaction and initiate the exponential explosion.

With the spectre of the Nazi bomb-developers constantly at their backs, Oppenheimer and his team came up with a hastily developed idea for the delivery of the atomic bomb. The 'gun-type' method involved firing one subcritical mass of uranium into another, using conventional explosives, to create a single supercritical mass. The chain reaction would then be kicked off by a spontaneous fission event emitting the initiating neutrons. The separation of the two subcritical masses ensured that the bomb would not detonate until required. The high levels of uranium enrichment achieved (around 80%) meant that only 20 to 25 kilograms were required for critical-ity. But Oppenheimer couldn't risk the failure of his project ceding the advantage to his German rivals, so he insisted on much larger quantities.

In the event, by the time enough pure uranium was finally ready, the war in Europe was already over. However, the war in the Pacific region raged on, with Japan showing little sign of surrender despite significant military disadvantages. Understanding that a land invasion of Japan would significantly increase the Americans' already heavy casualties, General Leslie Groves, director of the Manhattan Project, issued the directive authorising the use of the atomic bomb on Japan as soon as weather conditions permitted.

After several days of poor weather, caused by the tail end of a typhoon, on the 6th of August 1945 the sun rose in blue skies above Hiroshima. At 07:09 in the morning an American plane was spotted in the skies above Hiroshima and the air-raid warning sounded loudly across the city. Seventeen-year-old Akiko Takakura had recently taken up a job as a bank clerk. On her way to work as the siren sounded, she took refuge with other commuters in the public air-raid shelters strategically positioned around the city.

Air-raid warnings were not an uncommon experience in Hiroshima; the city was a strategic military base, housing the headquarters of Japan's Second General Army. So far, though, Hiroshima had largely been spared from the fire-bombing that rained down on so many other Japanese cities. Little did Akiko and her fellow commuters know that Hiroshima was being artificially preserved in order that the Americans might measure the full scale of the destruction caused by their new weapon.

At half past seven the all-clear was sounded. The B-29 flying overhead was nothing more sinister than a weather plane. As Akiko emerged from her air-raid shelter, along with many of the others, she breathed a sigh of relief: there would be no air-raid this morning.

Unbeknownst to Akiko and Hiroshima's other citizens, as they continued on their journeys to work, the B-29 radioed in reports of clear skies above Hiroshima to the *Enola Gay* – the plane carrying the gun-type fission bomb known as the 'Little Boy'. As children made their way to school and workers continued on their everyday routines, heading for offices and factories, Akiko arrived at the bank in central Hiroshima where she worked. Female clerks were supposed to arrive 30 minutes before the men in order to clean their offices ready for the day to begin, so by ten past eight Akiko was already inside the largely empty building and hard at work.

At 08:14, the cross-hairs of the T-shaped Aioi Bridge came into the sights of Colonel Paul Tibbets, piloting the *Enola Gay*. The 4400-kilogram Little Boy was released and began its 6-mile descent towards Hiroshima. After free-falling for around 45 seconds, the bomb was triggered at a height of about a mile above the ground. One subcritical mass of uranium was fired into another, creating a supercritical mass ready to explode. Almost instantaneously the spontaneous fissioning of an atom released neutrons, at least one of which was absorbed by a U-235 atom. This atom in turn fissioned and released more neutrons, which were absorbed in their turn by more atoms. The process rapidly accelerated, leading to an exponentially growing chain reaction and the simultaneous release of huge amounts of energy.

As she wiped the desktops of her male colleagues, Akiko looked out of her window and saw a bright white flash, like a strip of burning magnesium. What she couldn't know was that exponential growth had allowed the bomb to release energy equivalent to 30 million sticks of dynamite in an instant. The bomb's temperature increased to several million degrees, hotter than the surface of the sun. A tenth of a second later, ionising radiation reached the ground, causing devastating radiological damage to all living creatures exposed to it. A second further on and a fireball, 300 metres across and with a temperature of thousands of degrees Celsius, ballooned above the city. Eye witnesses describe the sun rising for a second time over Hiroshima that day. The blast wave, travelling at the speed of sound, levelled buildings across the city, throwing Akiko across the room and knocking her unconscious. Infrared radiation burned exposed skin for miles in every direction. People on the ground close to the bomb's hypocentre were instantly vaporised or charred to cinders.

Akiko was sheltered from the worst of the bomb's blast by the bank's earthquake-proof building. When she regained conscious-ness, she staggered out onto the street. As she emerged, she found that the clear blue morning skies had gone. The second sun over Hiroshima had set almost as quickly as it had risen. The streets were dark and choked with dust and smoke. Bodies lay where they had fallen for as far as the eye could see. At only 260 metres away, Akiko was one of the closest to the hypocentre of the bomb to survive the terrible exponential blast.

The bomb itself and the resulting firestorms that spread across the city are estimated to have killed around 70,000 people, 50,000 of whom were civilians. The majority of the city's buildings were also completely destroyed. Oppenheimer's prophetic musings had come true. The justification for the bombings of both Hiroshima and, three days later, Nagasaki, in the context of ending the Second World War are still debated to this day.

The nuclear option

Whatever the rights and wrongs of the atomic bomb itself, the greater understanding of the exponential chain reactions caused by nuclear fission that was developed as part of the Manhattan Project gave us the technology required to generate clean, safe, low-carbon energy through nuclear power. One kilogram of U-235 can release roughly three million times more energy than burning the same amount of coal.[8] Despite evidence to the contrary, nuclear energy suffers from a poor reputation for safety and environmental impact. In part, exponential growth is to blame.

On the evening of the 25th of April 1986, Alexander Akimov checked in for the night shift at the power plant in which he was shift supervisor. An experiment designed to stress-test the cooling pump system was due to get underway in a couple of hours. As he initiated the experiment, he could have been forgiven for thinking how lucky he was – at a time when the Soviet Union was collapsing and 20% of its citizens were living in poverty – to have a stable job at the Chernobyl nuclear power station.

At around 11 p.m., in order to reduce the power output to around 20% of normal operating capacity for the purposes of the test, Akimov remotely inserted a number of control rods between the uranium fuel rods in the reactor core. The control rods acted to absorb some of the neutrons released by atomic fission, so that these neutrons couldn't cause too many other atoms to split. This put a break on the rapid growth of the chain reaction that would be allowed to run exponentially out of control in a nuclear bomb. However, Akimov accidentally inserted too many rods, causing the power output of the plant to drop significantly. He knew that this would cause reactor poisoning – the creation of material, like the control rods, that would further slow the reactor and decrease the temperature, which would lead to more poisoning and further cooling in a self-reinforcing feedback loop. Panicking now, he overrode the safety systems, placing over 90% of the control rods under manual supervision and removing them from the core in order to prevent the debilitating total shutdown of the reactor.

As he watched the needles on the indicator gauges rise as the power output slowly increased, Akimov's heart-rate gradually returned to normal. Having averted the crisis, he moved to the next stage of the test, shutting down the pumps. Unbeknownst to Akimov, back-up systems were not pumping coolant water as fast as they should have been. Although initially undetectable, the slow-flowing coolant water had vaporised, impairing its ability both to absorb neutrons and to reduce the heat of the core. Increased heat and power output led to more water flash-boiling into steam, allowing more power to be produced: another, altogether more deadly positive feedback loop. The few remaining control rods that Akimov did not have under his manual supervision were automatically reinserted in order to rein in the increased heat generation, but they weren't enough. Upon realising the power output was increasing too rapidly, Akimov pressed the emergency shutdown button designed to insert all the control rods and power down the core, but it was too late. As the rods plunged into the reactor they caused a short but significant spike in power output leading to an overheated core, fracturing some of the fuel rods and blocking further insertion of the control rods. As the heat energy rose exponentially, the power output increased to over ten times the usual operating level. Coolant water rapidly turned to steam, causing two massive pressure explosions, destroying the core and spreading the fissile radioactive material far and wide.

Refusing to believe reports of the core's explosion, Akimov relayed incorrect information about the reactor's state, delaying vital containment efforts. Upon eventually realising the full extent of the destruction, he worked, unprotected, with his crew to pump water into the shattered reactor. As they worked, crew members received doses of 200 grays per hour. A typical fatal dose is around ten grays, meaning that these unprotected workers received fatal doses in less than five minutes. Akimov died two weeks after the accident from acute radiation poisoning.

The official Soviet death toll from the Chernobyl disaster was just 31, although some estimates, including individuals involved in the large-scale clean-up, are significantly higher. This is not to mention the deaths caused by the dispersal of radioactive material outside the immediate vicinity of the power plant. A fire that ignited in the shattered reactor core burned for nine days. The fire drew into the atmosphere hundreds of times more radioactive material than had been released during the bombing of Hiroshima, causing widespread environmental consequences for almost all of Europe.[9]

On the weekend of the 2nd of May 1986, for example, unseasonably heavy rainfall lashed the highlands of the UK. Contained within the falling rain droplets were the radioactive products of the fallout from the explosion – strontium-90, caesium-137 and iodine-131. In total, around 1% of the radiation released from the Chernobyl reactor fell on the UK. These radio-isotopes were absorbed by the soil, incorporated by the growing grass and then eaten by the sheep that grazed the land. The result – radioactive meat.

The Ministry of Agriculture immediately placed restrictions on the sale and movement of sheep in the affected areas, with implications for nearly 9000 farms and over four million sheep. Lake District sheep farmer, David Elwood, struggled to believe was what happening. The cloud carrying the invisible, almost undetectable, radio-isotopes cast a long shadow over his livelihood. Every time he wanted to sell sheep he had to isolate them and call in a government inspector to check their radiation levels. Each time the inspectors came they would tell him restrictions would only last another year or so. Elwood lived under this cloud for over 25 years until the restrictions were finally lifted in 2012.

It should, however, have been much easier for the government to inform Elwood and other farmers when radiation levels would be safe enough for them to sell their sheep freely. Radiation levels are remarkably predictable, thanks to the phenomenon of exponential *decay*.

The science of dating

Exponential decay, in direct analogy to exponential growth, describes any quantity that *decreases* with a rate proportional to its current value – remember the reduction in the number of M&Ms each day and the waterslide curve that described their decline. Exponential decay describes phenomena as diverse as the elimination of drugs in the body[10] and the rate of decrease of the head on a pint of beer.[11] In particular, it does an excellent job of describing the rate at which the levels of radiation emitted by a radioactive substance decrease over time.[12]

Unstable atoms of radioactive materials will spontaneously emit energy in the form of radiation, even without an external trigger, in a process known as radioactive decay. At the level of an individual atom, the decay process is random – quantum theory implies that it is impossible to predict when a given atom will decay. However, at the level of a material comprising huge numbers of atoms, the decrease in radioactivity is a predictable exponential decay. The number of atoms decreases in proportion to the number remaining. Each atom decays independently of the others. The rate of decay can be characterised by the half-life of a material – the time it takes for half of the unstable atoms to decay. Because the decay is exponential, no matter how much of the radioactive material is present to start with, the time for its radioactivity to decrease by half will always be the same. Pouring M&Ms out on the table each day and eating the M-up sweets leads to a half-life of one day – we expect to eat half of the sweets each time we pour them out of the bag.

The phenomenon of exponential decay of radioactive atoms is the basis of radiometric dating, the method used to date materials by their levels of radioactivity. By comparing the abundance of radioactive atoms to that of their known decay products, we can theoretically establish the age of any material emitting atomic radiation. Radiometric dating has well-known uses, including approximating the age of the Earth and determining the age of ancient artefacts like the Dead Sea Scrolls.[13] If you ever wondered how on Earth they knew that archaeopteryx was 150 million years old[14] or that Ötzi the iceman died 5300 years ago,[15] the chances are that radiometric dating was involved.

Recently, more accurate measurement techniques have facilitated the use of radiometric dating in 'forensic archaeology' – the use of exponential decay of radio-isotopes (amongst other archaeological techniques) to solve crimes. In November 2017, radio-carbon dating was used to expose the world's most expensive whisky as a fraud. The bottle, labelled as a 130-year-old Macallan single malt, was proved to be a cheap blend from the 1970s, much to the chagrin of the Swiss hotel that sold a single shot of it for $10,000. In December 2018, in a follow-up investigation, the same lab found that over a third of 'vintage' Scotch whiskies they tested were also fakes. But perhaps the most high-profile use of radiometric dating concerns verification of the age of historical art works.

·

Before the Second World War, only 35 paintings by Dutch Old Master Johannes Vermeer were known to exist. In 1937 a remarkable new work was discovered in France. Lauded by art critics as one of Vermeer's greatest works, *The Supper at Emmaus* was quickly procured at great expense for the Museum Boijmans Van Beuningen in Rotterdam. Over the next few years several more, hitherto unknown Vermeers surfaced. These were quickly appropriated by wealthy Dutchmen, in part in an attempt to prevent the loss of important cultural property to the Nazis. Nevertheless, one of these Vermeers, *Christ with the Adulteress*, ended up with Hermann Göring, Hitler's designated successor.

After the war, when this lost Vermeer was discovered in an Austrian salt mine, along with much of the Nazis' looted artwork, a great search was undertaken to find out who was responsible for the sale of the paintings. The Vermeer was eventually traced back to Han van Meegeren, himself a failed artist whose work was derided by many art critics as derivative of the Old Masters. Unsurprisingly, immediately after his arrest, Van Meegeren was incredibly unpopular with the Dutch public. Not only was he suspected of selling Dutch cultural property to the Nazis – a crime punishable by death – but he had also made huge sums of money through the sale and lived lavishly in Amsterdam throughout the war, when many of the city's residents were starving. In a desperate attempt at self-preservation, Van Meegeren claimed that the painting he sold to Göring was not a genuine Vermeer, but one that he himself had forged. He also confessed to the forgeries of the other new 'Vermeers', as well as recently discovered works by Frans Hals and Pieter de Hooch.

A special commission set up to investigate the forgeries appeared to verify Van Meegeren's claims, in part based on a new forgery, *Christ and the Doctors*, which the commission had him paint. By the time Van Meegeren's trial started in 1947, he was hailed a national hero, having tricked the elitist art critics who had so derided him, and fooled the Nazi high command into buying a worthless fake. He was cleared of collaboration with the Nazis and given a sentence of just a year in prison for forgery and fraud, but died of a heart attack before his sentence began. Despite the verdict, many (especially those who had bought the 'Van Meegeren Vermeers') still believed the paintings to be genuine and continued to contest the findings.

In 1967, *The Supper at Emmaus* was re-examined using lead-210 radiometric dating. Despite Van Meegeren being meticulous in his forgeries, using many of the materials Vermeer would originally have employed, he could not control the method by which these materials were created. For authenticity he used genuine

17th-century canvases and mixed his paints according to original formulae, but the lead he used for his white lead paint was only recently extracted from its ore. Naturally occurring lead contains radioactive isotope lead-210 and its parent radioactive species (from which lead is created by decay), radium-226. When the lead is extracted from its ore, most of the radium-226 is removed, leaving only small amounts, meaning relatively little new lead-210 is created in the extracted material. By comparing the concentration of lead-210 and radium-226 in samples, it is possible to date the lead paint accurately using the fact that the radioactivity of lead-210 decreases exponentially with a known half-life. A far higher proportion of lead-210 was found in *The Supper at Emmaus* than there would have been there if it were genuinely painted 300 years earlier. This established for certain that Van Meegeren's forgeries couldn't have been painted by Vermeer in the 17th century, as the lead which Van Meegeren used for his paints had not yet been mined.[16]

Ice bucket flu

Had Van Meegeren been around today, it's likely that his work would have been neatly parcelled up into a convenient click-bait article entitled something like 'Nine paintings you won't believe aren't the real thing', and spread around the internet. Modern-day fakes, such as the doctored photo of multimillionaire presidential candidate Mitt Romney appearing to line up six letter-adorned supporters to read 'RMONEY' instead of 'ROMNEY', or the Photoshopped snap of 'Tourist Guy' posing on the viewing deck of the South Tower of the World Trade Centre seemingly unaware of the low-flying plane approaching in the background, have achieved the global exposure that viral marketers' dreams are made of.

Viral marketing is the phenomenon by which advertising objectives are achieved through a self-replicating process akin to the spread of a viral disease (the mathematics of which we will look into more deeply in Chapter 7). One individual in a network infects others, who in turn infect others. As long as each newly infected individual infects at least one other, the viral message will grow exponentially. Viral marketing is a subfield of an area known as memetics, in which a 'meme' – a style, behaviour or, crucially, an idea – spreads between people through a social network, just like a virus. Richard Dawkins coined the word 'meme' in his 1976 book, *The Selfish Gene*, in order to explain the way in which cultural information spreads. He defined memes as units of cultural transmission. In analogy to genes, the units of heritable transmission, he proposed that memes could self-replicate and mutate. The examples he gave of memes included tunes, catch-phrases and, in a wonderfully innocent indication of the times in which he wrote the book, ways of making pots or building arches. Of course, in 1976, Dawkins had not come across the internet in its current form, which has allowed the spread of once unimaginable (and arguably pointless) memes including #thedress, rickrolling and Lolcats.

One of the most successful, and perhaps genuinely organic, examples of a viral marketing campaign was the ALS ice bucket challenge. During the summer of 2014, videoing yourself having a bucket of cold water thrown over your head and then nominating others to do the same, whilst possibly donating to charity, was *the* thing to do in the northern hemisphere. Even I caught the bug.

Adhering to the classic format of the ice bucket challenge, after being thoroughly soaked I nominated two other people in my video, whom I later tagged when I uploaded it to social media. As with the neutrons in a nuclear reactor, as long as, on average, at least one person takes up the challenge for every video posted, the meme becomes self-sustaining, leading to an exponentially increasing chain reaction.

In some variants of the meme, those nominated could either undertake the challenge and donate a small amount to the amyotrophic lateral sclerosis (ALS) association or another charity of their choice, or choose to shirk the challenge and donate significantly more in reparation. In addition to increasing the pressure on nominated individuals to participate in the meme, the association with charity had the added bonus of making people feel good about themselves by raising awareness, and promoting a positive image of themselves as altruistic. This self-congratulatory aspect acted to increase the infectiousness of the meme. By the start of September 2014, the ALS association reported receiving over 100 million dollars in additional funding from over three million donors. As a result of the funding received during the challenge, researchers discovered a third gene responsible for ALS, demonstrating the viral campaign's far-reaching impact.[17]

In common with some extremely infective viruses like flu, the ice bucket challenge was also highly seasonal (an important phenomenon, in which the rate of disease spread varies throughout the year, and that we will meet again in Chapter 7). As autumn approached and colder weather hit the northern hemisphere, getting doused in ice-cold water suddenly seemed like less fun, even for a good cause. By the time September arrived, the craze had largely died off. Just like the seasonal flu, though, it returned the next summer and the summer after in similar formats, but to a largely saturated population. In 2015, the challenge raised less than 1% of the previous year's total for the ALS association. People exposed to the virus in 2014 had typically built up a strong immunity, even to slightly mutated strains (different substances in the bucket, for example). Tempered by the immunity of apathy, each new outbreak soon died out as each new participant failed, on average, to pass on the virus to at least one other.

Is the future exponential?

There is a parable involving exponential growth that is told to French children to illustrate the dangers of procrastination. One day, it is noted that an extremely small algal colony has formed on the surface of the local lake. Over the next few days, the colony is found to be doubling its coverage of the surface of the lake each day. It will continue to grow like this until it covers the lake, unless something is done. If left unchecked, it will take 60 days to cover the surface of the lake, poisoning its waters. Since the algal coverage is initially so small and there is no immediate threat, it is decided to leave the algae to grow until it covers half the surface of the lake, when it will be more easily removed. The question is then asked, 'On which day will the algae cover half of the lake?'

A common answer that many people give to this riddle, without thinking, is 30 days. But, since the colony doubles in size each day, if the lake is half-covered one day it will be completely covered the next day. The perhaps surprising answer, therefore, is that the algae will cover half the surface of the lake on the 59th day, leaving only one day to save the lake. At 30 days the algae take up less than a billionth of the capacity of the lake. If you were an algal cell in the lake, when would you realise you were running out of space? Without understanding exponential growth, if someone told you on the 55th day, when the algae covered only 3% of the surface, that the lake would be completely choked in five days' time, would you believe them? Probably not.

This serves to highlight the way in which we, as humans, have been conditioned to think. Typically, for our forebears, the experiences of one generation were very much like the last: they did the same jobs, used the same tools and lived in the same places as their ancestors. They expected their descendants to do the same. However, the growth of technology and social change is now occurring so rapidly that noticeable differences occur within single generations. Some theoreticians believe that the rate of technological advancement is itself increasing exponentially.

Computer scientist Vernor Vinge encapsulated just such ideas in a series of science-fiction novels and essays,[18] in which successive technological advancements arrive with increasing frequency until a point at which new technology outstrips human comprehension. The explosion in artificial intelligence ultimately leads to a 'technological singularity' and the emergence of an omnipotent all-powerful superintelligence. American futurist Ray Kurzweil attempted to take Vinge's ideas out of the realm of science fiction and apply them to the real world. In 1999, in his book *The Age of Spiritual Machines*, Kurzweil hypothesised the 'law of accelerating returns'.[19] He suggested that the evolution of a wide range of systems – including our own biological evolution – occurs at an exponential pace. He even went so far as to pin the date of Vinge's 'technological singularity'

– the point at which we will experience, as Kurzweil describes it, 'technological change so rapid and profound it represents a rupture in the fabric of human history' – to around 2045.[20] Amongst the implications of the singularity, Kurzweil lists 'the merger of biological and nonbiological intelligence, immortal software-based humans, and ultra-high levels of intelligence that expand outward in the universe at the speed of light.' While these extreme, outlandish predictions should probably have been confined to the realms of science fiction, there are examples of technological advances which really have sustained exponential growth over long periods.

Moore's law – the observation that the number of components on computer circuits seems to double every two years – is a well-cited example of exponential growth of technology. Unlike Newton's laws of motion, Moore's law is not a physical or natural law, so there is no reason to suppose it will continue to hold forever. However, between 1970 and 2016 the law has held remarkably steady. Moore's law is implicated in the wider acceleration of digital technology, which in turn contributed significantly to economic growth in the years surrounding the turn of the last century.

In 1990, when scientists undertook to map all three billion letters of the human genome, critics scoffed at the scale of the project, suggesting that it would take thousands of years to complete at the current rate. But sequencing technology improved at an exponential pace. The complete 'Book of Life' was delivered in 2003, ahead of schedule and within its one-billion-dollar budget.[21] Today, sequencing an individual's whole genetic code takes under an hour and costs less than a thousand dollars.

•

Population explosion

The story of the algae in the lake highlights that our failure to think exponentially can be responsible for the collapse of ecosystems and populations. One species on the endangered list, despite clear and persistent warning signs, is, of course, our own.

Between 1346 and 1353, the Black Death, one of the most devastating pandemics in human history (infectious disease spread being a subject which we will investigate in more detail in Chapter 7), swept through Europe, killing 60% of its population. At this point the total population of the world was reduced to around 370 million. Since then the global population has increased constantly without abating. By 1800, the human population had almost reached its first billion. The perceived rapid increase in population at that time prompted the English mathematician, Thomas Malthus, to suggest that the human population grows at a rate that is proportional to its current size.[22] As with the cells in the early embryo or the money left untouched in a bank account, this simple rule suggests exponential growth of the human population on an already crowded planet.

A favoured trope of many science-fiction novels and films (take the recent blockbusters, *Interstellar* and *Passengers*, for example), is to solve the problems of the world's growing population through space exploration. Typically, a suitable Earth-like planet is discovered and prepared for habitation for the overspilling human race. Far from being a purely fictional fix, in 2017 eminent scientist Stephen Hawking gave credibility to the proposition of extraterrestrial colonisation. He warned that humans should start leaving the Earth within the next 30 years, in order to colonise Mars or the

Moon, if our species is to survive the threat of extinction presented by overpopulation and associated climate change. Disappointingly, though, if our growth rate continued unchecked, even shipping half of the Earth's population over to a new Earth-like planet would only buy us another 63 years until the human population doubled again and both planets reached saturation point. Malthus forecast that exponential growth would render the idea of interplanetary colonisation futile when he wrote: 'The germs of existence contained in this spot of earth, with ample food, and ample room to expand in, would fill millions of worlds in the course of a few thousand years.'

However, as we have already found (remember the bacteria *Strep f.* growing in the milk bottle at the start of this chapter), exponential growth cannot be sustained forever. Typically, as a population increases, the resources of the environment that sustains it become more sparsely distributed and the net rate of growth (the difference between the birth rate and the death rate) naturally drops. The environment is said to have a 'carrying capacity' for a particular species – an inherent maximum sustainable population limit. Darwin recognised that environmental limitations would cause a 'struggle for existence' as individuals 'compete for their places in the economy of nature'. The simplest mathematical model to capture the effects of competition for limited resources, within or between species, is known as the logistic growth model.

In Figure 3, logistic growth looks exponential initially as the population grows freely in proportion to its current size, unrestricted by environmental concerns. However, as the population increases, resource scarcity brings the death rate ever closer to the birth rate. The net population growth rate eventually decreases to zero: new births in the population are only sufficient to replace those that

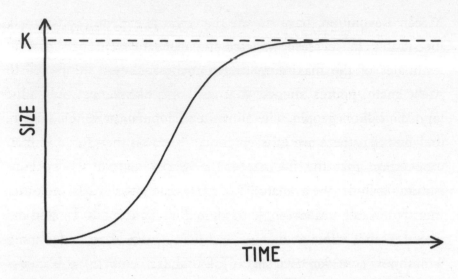

Figure 3: The logistic growth curve increases almost exponentially at first, but then growth slows as resources become a limiting factor and the population approaches the carrying capacity, K.

have died and no more, meaning that the numbers plateau at the carrying capacity. Scottish scientist Anderson McKendrick (one of the earliest mathematical biologists, with whom we will become better acquainted in Chapter 7 in the context of his work on modelling the spread of infectious disease) was the first to demonstrate that logistic growth occurred in bacterial populations.[23] The logistic model has since been shown to be an excellent representation of a population introduced into a new environment, capturing the growth of animal populations as diverse as sheep,[24] seals[25] and cranes.[26]

The carrying capacities of many animal species remain roughly constant, as they depend on the resources available in their environments. For humans, however, a variety of factors, amongst them the Industrial Revolution, the mechanisation of agriculture and the

Green Revolution, have meant that our species has consistently been able to increase its carrying capacity. Although current estimates of the maximum sustainable population of the Earth vary, many figures suggest that it is somewhere between nine and ten billion people. The eminent sociobiologist, E. O. Wilson, believes that there are inherent, hard limits on the size of human population that the Earth's biosphere can support.[27] The constraints include: the availability of fresh water, fossil fuels and other non-renewable resources, environmental conditions (including, most notably, climate change), and living space. One of the more commonly considered factors is food availability. Wilson estimates that, even if everyone were to become vegetarian, eating food produced directly rather than feeding it to livestock (since eating animals is an inefficient way to convert plant energy into food energy), the present 1.4 billion hectares of arable land would only produce enough food to support ten billion people.

If the (near seven and a half billion) human population continues to grow at its current rate of 1.1% per year, then we will reach ten billion inside 30 years. Malthus expressed his fears of overpopulation way back in 1798, when he warned: 'The power of population is so superior to the power of the Earth to produce subsistence for man, that premature death must in some shape or other visit the human race.' In the context of human history, we are now well within the last day left to save the lake.

There are, however, reasons for optimism. Although the human population is still increasing in number, effective birth control and the reduction in infant mortality (leading to lower reproduction rates) mean we are doing so at a slower rate than in previous generations. Our growth rate reached a peak of around

2% per year in the late 1960s, but is projected to fall below 1% per year by 2023.[28] To put that into context, if growth rates had stayed at 1960s rates it would have taken only 35 years for the population size to double. In fact, we only reached the 7.3 billion mark (double the 3.65 billion world population of 1969) in 2016 – nearly 50 years later. At a rate of just 1% per year we can expect the doubling time to increase to 69.7 years, almost twice as long as the doubling period based on 1969 rates. A small drop in the rate of increase makes a huge difference when exponential growth is concerned. It seems that, by slowing our population growth as we head towards the planet's carrying capacity, we are naturally beginning to buy ourselves some more time. There are, however, reasons why exponential behaviour may make us, as individuals, feel like we have less time left than we think.

Time flies when you're getting old

Do you remember, when you were younger, that summer holidays seemed to last an eternity? For my children, who are four and six, the wait between consecutive Christmases seems like an inconceivable stretch of time. In contrast, as I get older, time appears to pass at an alarming rate, with days blending into weeks and then into months, all disappearing into the bottomless sinkhole of the past. When I chat weekly to my septuagenarian parents, they give me the impression that they barely have time to take my call, so busy are they with the other activities in their packed schedules. When I ask them how they fill their week, however, it often seems like their unrelenting travails might comprise the work of just a single day for me. But then what would I know about competing time pressures: I just have two kids, a full-time job and a book to write.

I should remember not to be too caustic with my parents, though, because it seems that perceived time really does run more quickly the older we get, fuelling our increasing feelings of overburdened time-poverty.[29] In an experiment carried out in 1996, a group of younger people (19–24) and a group of older people (60–80) were asked to count out three minutes in their heads. On average the younger group clocked an almost-perfect three minutes and three seconds of real time, but the older group didn't call a halt until a staggering three minutes and 40 seconds, on average.[30] In other related experiments, participants were asked to estimate the length of a fixed period of time during which they had been undertaking a task.[31] Older participants consistently gave shorter estimates for the length of the time period they had experienced than younger groups. For example, by the point at which two minutes of real time had elapsed, the older group, on average, had clocked less than 50 seconds in their heads, leading them to question where the remaining minute and ten seconds had gone.

This acceleration in our perception of the passage of time has little to do with leaving behind those carefree days of youth and filling our calendars with adult responsibilities. In fact, there are a number of competing ideas that provide explanations for why, as we age, our perception of time accelerates. One theory is related to the fact that our metabolism slows as we get older, matching the slowing of our heartbeats and our breathing.[32] Just as with a stop watch that is set to run fast, children's versions of these 'biological clocks' tick more quickly. In a fixed period of time they experience more beats of these biological pacemakers (breaths or heartbeats, for example), making them feel like a longer period of time has elapsed.

A competing theory suggests that our perception of time's passage depends upon the amount of new perceptual information we are subjected to from our environment.[33] The more novel stimuli there are, the longer our brains take to process the information. The corresponding period of time seems, at least in retrospect, to last longer. This argument can be used to explain the movie-like perception of events playing out in slow-motion in the moments immediately preceding an accident. So unfamiliar is the situation for the accident victim in these scenarios that the amount of novel perceptual information is correspondingly huge. It might be that rather than time actually slowing down during the event, our recollection of the events is decelerated in hindsight, as our brain records more detailed memories based on the flood of data it experiences. Experiments on subjects experiencing the unfamiliar sensation of free fall have demonstrated this to be the case.[34]

This theory ties in nicely with the acceleration of perceived time. As we age, we tend to become more familiar with our environments and with life experiences more generally. Our daily commutes, which might initially have appeared long and challenging journeys full of new sights and opportunities for wrong turns, now flash by as we navigate their familiar routes on autopilot.

It is different for children. Their worlds are often surprising places filled with unfamiliar experiences. Youngsters are constantly reconfiguring their models of the world around them, which takes mental effort and seems to make the sand run more slowly through their hour-glasses than for routine-bound adults. The greater our acquaintance with the routines of everyday life, the quicker we perceive time to pass and, generally, as we age, this familiarity increases. This theory suggests that, in order to make our time last longer, we should fill our lives with new and varied experiences, eschewing the time-sapping routine of the everyday.

Neither of the above ideas manages to explain the almost perfectly regular rate at which our perception of time seems to accelerate. That the length of a fixed period of time appears to reduce continually as we age suggests an 'exponential scale' to time. We employ exponential scales instead of traditional linear scales when measuring quantities that vary over a huge range of different values. The most well-known examples are scales for energy waves like sound (measured in decibels) or seismic activity. On the exponential Richter scale (for earthquakes), an increase from magnitude 10 to magnitude 11 would correspond to a ten-fold increase in ground movement, rather than a 10% increase as it would do on a linear scale. At one end, the Richter scale was able to capture the low-level tremor felt in Mexico City in June 2018 when Mexican football fans in the city celebrated their goal against Germany at the World Cup. At the other extreme, the scale recorded the 1960 Valdivia earthquake in Chile. The magnitude 9.6 quake released energy equivalent to over a quarter of a million of the atomic bombs dropped on Hiroshima.

If the length of a period of time is judged in proportion to the time we have already been alive, then an exponential model of perceived time makes sense. As a 34-year-old, a year accounts for just under 3% of my life. My birthdays seem to come around all too quickly these days. But to a ten-year-old, waiting 10% of their life for the next round of presents requires almost-saintly patience. To my four-year-old son, the idea of having to wait a quarter of his life until he is the birthday boy again is almost intolerable. Under this exponential model, the proportional increase in age that a four-year-old experiences between birthdays is equivalent to a 40-year-old waiting until they turn 50. When looked at from this relative perspective, it makes sense that time seems only to accelerate as we age.

It's not uncommon for us to categorise our lives into decades – our carefree twenties, our serious thirties and so on – which suggests that each period should be afforded an equal weighting. However, if time really does appear to speed up exponentially, chapters of our life spanning different lengths of time might feel like they are of the same duration. Under the exponential model, the ages from 5 to 10, 10 to 20, 20 to 40 and even 40 to 80 might all seem equally long (or short). Not to precipitate the frantic scribbling of too many bucket lists, but under this model the 40-year period between 40 and 80, encompassing much of middle and old age, might flash by as quickly as the five years between your fifth and tenth birthdays.

It should be some small compensation, then, for pensioners Fox and Chalmers, jailed for running the Give and Take pyramid scheme, that the routine of prison life, or just the exponentially increasing passage of perceived time, should make their sentences seem to pass very quickly indeed.

In total, nine women were sentenced for their part in the scheme. Although some were forced to pay back part of the money they made during the scheme's operation, very little of the millions of pounds invested in the scheme was recovered. None of this money made its way to the scheme's defrauded investors – the unsuspecting victims who lost everything because they underestimated the power of exponential growth.

From the explosion of a nuclear reactor to the explosion of the human population and from the spread of a virus to the spread of a viral marketing campaign, exponential growth and decay can play an unseen, but often critical, role in the lives of normal people like you and me. The exploitation of exponential behaviour has spawned branches of science that can convict criminals and others that can now, quite literally, destroy worlds. Failing to think exponentially means our decisions, like uncontrolled nuclear chain reactions, can have unexpected and exponentially far-reaching consequences. Amongst other innovations, the exponential pace of technological advancements has hastened in the era of personalised medicine, in which anyone can have their DNA sequenced for a relatively modest sum. This genomics revolution has the potential to lend unprecedented insight into our own health traits, but only, as we will examine in the next chapter, if the mathematics that underpins modern medicine is able to keep pace.

2

SENSITIVITY, SPECIFICITY AND SECOND OPINIONS

Why Maths Makes Medicine Matter

When I saw the unread email sitting in my inbox I immediately felt a surge of adrenalin. It started in my stomach and moved down through my arms, causing my fingers to tingle. My pulse throbbed behind my ears as I unconsciously held my breath. I opened the message and, skimming past the preamble of text, clicked immediately on the 'View your reports' link. A browser window opened up, I logged in and clicked on the section headed 'Genetic Health Risk'. As I scanned the list I was relieved to see 'Parkinson's Disease: Variants not detected', 'BRCA1/BRCA2: Variants not detected', 'Age-Related Macular Degeneration: Variants not detected'. My anxiety subsided as I scrolled past more and more diseases to which I was not genetically predisposed. When I reached the bottom of the list of all-clears, my eyes flicked back to one outlying entry I had missed: 'Late-Onset Alzheimer's Disease: Increased risk'.

When I started writing this book, I thought it would be interesting to investigate the mathematics behind take-at-home genetic tests. So I signed up to 23andMe, probably the best-known personal genomics company out there. How better to understand such results than by taking the test for myself? For a not inconsiderable fee, they sent me a tube in which to collect two millilitres of saliva, which I then sealed and sent back to them; 23andMe promised over 90 reports on my traits, health and even my ancestry. Over the next few months I didn't give it a second thought, never really believing that anything significant would come to light. When the email arrived, however, it suddenly hit me that a comprehensive indication of my future health lay just a couple of clicks away. And then there I was, sitting in front of my computer screen, facing what seemed to be quite serious health implications.

In order to get a better understanding of what 'increased risk' meant, I downloaded the full 14-page report on my Alzheimer's risk. I only had a superficial knowledge of what Alzheimer's was, and wanted to find out more. The first sentence of the report did little to assuage my anxiety: 'Alzheimer's disease is characterised by memory loss, cognitive decline, and personality changes.' As I read on, I found that 23andMe had detected the epsilon-4 (ε4) variant in one of the two copies of the *apolipoprotein E* (*APOE*) gene. The first quantitative information in the report informed me that '... on average, a man of European descent with this variant has a 4–7% chance of developing late-onset Alzheimer's disease by age 75, and a 20–23% chance by age 85.'

Although these figures meant something in an abstract way, I found them hard to parse. There were three things I really wanted to know. Firstly, what, if anything, could I do about my new-found predicament? Secondly, how much worse off was I compared to the average individual in the population? Finally, how much could I trust the figures that 23andMe had supplied me with? As I scrolled down, the next information I hit on answered my first question: 'There is currently no known prevention or cure for Alzheimer's disease.' To find the answers to my other questions I would have to dig deeper into the report. My interest in the mathematical interpretation of genetic tests had suddenly become much more urgent and personal.

•

As medicine becomes a progressively quantitative discipline, mathematical formulae often provide the dispassionate basis for key decisions, whether with regards to the availability of a particular treatment or, at a more personal level, concerning our lifestyle choices. In this chapter, we will explore these formulae to find out whether they have a solid grounding in science or whether they are outmoded numerology that need to be discredited and discarded. Ironically, we will draw on centuries-old mathematics to suggest more refined replacements.

As diagnostic technology advances, we are being subjected to more medical evaluations than ever before. We will investigate the surprising effects of false positive results on the most ubiquitous medical screening programmes, and get to grips with how tests can be both highly accurate and yet very imprecise at the same time. We will encounter the dilemmas introduced by tools like pregnancy tests, which give both false positives and false negatives, and see how, in different diagnostic contexts, these incorrect results can be put to good use.

Whole genome sequencing, wearable technology and advances in data science have delivered us into the infancy of the personalised medicine era. As we take our first tentative steps in this new age for healthcare, I will reinterpret the results of my own DNA screening in order to understand what my disease risk profile really looks like and determine whether the mathematical methodology currently used to interpret personalised genetic screens stands up to scrutiny.

What are the odds?

In 2007, 23andMe, named after the 23 pairs of chromosomes that comprise typical human DNA, became the first company to offer personal DNA testing for the purposes of establishing a person's ancestry. The following year, thanks to a $4 million investment from Google, 23andMe marketed a saliva test that could estimate how likely you were to suffer from nearly 100 different conditions, from alcohol intolerance to atrial fibrillation. So comprehensive was their list of traits and so potentially transformative the power of the results that *Time Magazine* named the test 'Invention of the Year'.

But the good times didn't last for 23andMe. In 2010 the US Food and Drug Administration (FDA) notified the personal genomics company that their tests were considered to be medical devices and therefore required federal approval. In 2013, with 23andMe still lacking this approval, the FDA ordered them to stop providing disease risk factors until the accuracy of their tests had been verified. 23andMe customers filed a class-action law suit, alleging that they had been misled about what the personal profiling company could deliver. At the height of these troubles, in December 2014, 23andMe launched their health-related services in the UK. Given the controversies, I wondered about the reliability of the tests they might run on my own DNA, were I to send them a sample.

Reading about the experiences of 33-year-old web developer, Matt Fender, in the *New York Times* didn't do anything to allay my concerns. As a self-confessed nerd and member of the increasingly large community of the 'worried well', Fender is 23andMe's ideal customer. After receiving his profile data and running it through a third party interpreter, Fender discovered he was positive for the PSEN1 mutation. PSEN1 is an indicator of early-onset Alzheimer's with 'complete penetrance', meaning that everyone who has the mutation gets the disease: no ifs, no buts. Unsurprisingly, Fender was alarmed by the thought of losing his ability to think abstractly, to problem solve, and to recall coherent memories. The diagnosis reduced his meaningful life expectancy by at least 30 years.

Unable to take his mind off the implications of the mutation, he sought reassurance. Without any family history of Alzheimer's, Fender struggled to convince geneticists to order a confirmatory follow-up test. Instead, he resorted to taking a second do-it-yourself genetic test. He sent off another spit-kit, this time from Ancestry.com, and awaited the results. They came back five weeks later: negative for PSEN1. Slightly relieved, but even more puzzled than before, Fender eventually managed to convince a doctor to grant him a clinical assessment, which confirmed Ancestry.com's negative result.

The sequencing technology employed by 23andMe and Ancestry.com, with an error rate of just 0.1%, seems incredibly reliable. However, when testing nearly a million genetic variants, it's worth remembering that, even with this low error rate, around 1000 mistakes should be expected. It is worrying, but maybe not surprising, that there can be discord between the results of two independent companies. Perhaps more concerning is the evident lack of post-result support provided. Patients requesting at-home genetic profiles are left to deal with their results in almost complete medical isolation.

After 23andMe gradually gained approval from the FDA for a significantly reduced range of genetic tests, the company relaunched in the US in 2017 and their home DNA test kit became one of Amazon's top-selling items on Black Friday that year. Despite (or perhaps because of) my own misgivings, I ordered a kit and sent off my saliva sample for testing.

In almost every cell of the human body is a nucleus that contains a copy of our DNA – the so-called 'Book of Life'. We inherit these long, twisted ladders of amino acids in 23 pairs of chromosomes, one of each pair coming from each of our parents. Each chromosome in a pair contains copies of the same genes as its partner, whose sequences are similar, but not necessarily exactly the same. For example, there are two main variants of the Alzheimer's-associated APOE gene that 23andMe test for, called ε3 and ε4. The ε4 variant is associated with increased risk of late-onset Alzheimer's. Because there are two chromosomes, you can either have one copy of ε4 (and one copy of ε3), two copies of ε4 (and no copies of ε3) or no copies of ε4 (and two copies of ε3) – the number of copies is known as your genotype. Two copies of ε3 is the most common genotype and is the baseline against which the likelihood of Alzheimer's is judged. The more copies of the ε4 variant you have, the higher the associated risk of developing Alzheimer's.

But how high is high? Given that 23andMe had found that I had a particular genotype, what was my 'predicted risk' – the probability of developing the disease? To be confident of the risks they had predicted for me I needed to make sure their mathematical analysis was on a sound footing before jumping to any conclusions.

•

The best way to get a handle on the predicted Alzheimer's risk would be to select a huge number of individuals, representative of the population at large, ascertain their genotype and then check up on them regularly to see who develops Alzheimer's. With these representative data it would then be easy to compare the risk of getting Alzheimer's given a particular genotype, with the risk in the general population – the so-called 'relative risk'. Usually though, this sort of longitudinal study is prohibitively expensive due to the large number of individuals required (especially for rare diseases) and the long time frame over which they must be observed.

More common, but less powerful, is a case-controlled trial in which a number of individuals already suffering from Alzheimer's are selected, along with a number of 'controls' – individuals with similar backgrounds, but without the disease. (We will see in Chapter 3 why controlling carefully for the background of individuals is of great importance.) Unlike the longitudinal study, in which participants are selected independently of their disease status, the subjects in the case-controlled trial are skewed towards disease carriers, so we aren't able to extract an estimate for the incidence of the disease in the population at large. This means we get a biased prediction of relative risk of the disease. However, these trials do allow us to accurately calculate something called 'odds ratios', which don't require knowledge of the total incidence in the population.

If you've ever been to a greyhound stadium or had a flutter on a horse race you might remember that the probability of a particular animal winning the race is often expressed in terms of odds. In a given race, an outsider might have odds of 5-to-1 against. This means that if the same race was run a total of six times we would expect to see this outsider lose five times and win once. The probability of the outsider winning, then, is 1-in-6, or 1/6. The natural way to think about 'odds against' is as the ratio of the probability that an event doesn't happen to the probability that it does happen (5/6-to-1/6 in this case, or more simply 5-to-1). Conversely, the favourite for the race might have odds of 2-to-1 on. In sports betting it's traditional to always put the larger number first, so we need to distinguish between odds on and odds against. 'Odds on', the converse of odds against, expresses the ratio of the probability that an event happens to the probability that it doesn't happen. With odds of 2-to-1 on, if the same race was run three times in total, we might expect the favourite to win twice and lose once. The probability of the favourite winning, then, is 2-in-3 or 2/3, and the probability of it losing is 1/3, which is how we get back to odds on of 2/3-to-1/3 or more simply 2-to-1.

When you hear commentators or book-keepers describing an 'odds on favourite' it is usually in races with a small number of horses. The phrase, however, is a tautology. Any horse that is odds on must be the favourite, because there can only be one horse in any race which is more likely to win than lose. In a race with a larger number of horses it is unusual for a single horse to win more races than it loses. For example, in the UK's most famous horse race, the Grand National, a total of 40 horses compete against each other. Even the 2018 winner, Tiger Roll, who started as favourite for (and eventually went on to win) the 2019 race as well, had odds of 4-to-1 against. Because most horses will not be likely to win most of their races, unless explicitly stated otherwise, odds at the race course, with the largest number first, are usually the odds against.

In medical scenarios the opposite is true. Odds are typically expressed as odds on – the probability that an event happens vs. the probability that it doesn't – and because we are usually talking about rare diseases (with less than a 50% prevalence in the population), the smaller number usually comes first.

To see how to calculate medical odds and the desired odds ratio, let's consider a hypothetical case-controlled study into the effects of having a single ε4 variant (as picked up in my DNA profile) on the incidence of Alzheimer's by age 85, presented in Table 1. The odds of developing Alzheimer's by the age of 85, given that you have one copy of the ε4 variant (like me), is the number of people with the disease (100) divided by the number of people without the disease (335): 100-to-335 or, expressed as a fraction, 100/335. By the same logic, drawing figures from the second row of the table, the odds of developing the disease by age 85 if you have two copies of the common ε3 variant are 79-to-956 or 79/956. The odds *ratio*, then, is a comparison of the odds of having the disease given one genotype (one copy of the ε4 variant and one copy of the ε3 variant, for example) versus the odds of having the disease given the most common genotype (two copies of the ε3 variant). For the hypothetical figures given in Table 1 the odds ratio is 100/335 divided by 79/956 which works out to be 3.61. Crucially, odds ratios don't require us to know the incidence in the population at large and so can easily be calculated from case control studies.

Although the odds ratios themselves don't provide the relative risk (the ratio of the risk of getting the disease with the ε3/ε4 genotype to the risk of getting the disease with the ε3/ε3 genotype),

	ALZHEIMER'S BY 85	NO ALZHEIMER'S BY 85
ε3/ε4	100	335
ε3/ε3	79	956

Table 1: Results from a hypothetical case-controlled study on the impact of a single ε4 variant on the occurrence of Alzheimer's by age 85.

they can be combined with the overall population risk of the disease and the known genotype frequencies to find the disease probability for a given genotype. This calculation is not trivial. Indeed, there is not even a unique way to do the calculation. I tried to replicate the late-onset Alzheimer's risks in my genetic report using the same method as 23andMe and data taken directly from the report or from papers they cited.[35] (In case you're interested, the calculation I undertook to find the disease probabilities involved using a non-linear solver in order to resolve a system of three coupled equations for three unknown conditional probabilities – the sort of thing I enjoy dirtying my hands with in my day job.) I found small, but potentially significant, discrepancies between my figures and theirs. My calculations seemed to suggest that I should view the precision of 23andMe's figures with a degree of scepticism.

My conclusion was reinforced when I came across the findings of a 2014 study that investigated the risk-calculation methods of three of the leading personal genomic companies, including 23andMe.[36] The authors found that differences in the overall population risk, the genotype frequencies, and the mathematical formulae used all contributed to significantly different predicted risks between the companies. When predicted risks were used to categorise individuals into elevated, decreased or unchanged risk categories, the discrepancies became even more stark. The study found that 65% of all individuals tested for prostate cancer were placed in contrasting risk categories (elevated or decreased) by at least two of the three companies. In almost two-thirds of cases, one company might have been telling the individual they were healthy whilst another company told them they had a significantly increased risk of prostate cancer.

Setting aside the potential for error of the genetic tests themselves, I had an answer to my third question: inconsistencies in the mathematical approach mean that numerical risk calculations presented in personal genomics health reports should be viewed with some scepticism.

A Eureka moment

Personalised DNA testing is by no means the only area in which health-related tools are being put into our own hands. There are now phone apps that can monitor heart rates or estimate aerobic fitness, and at-home tests that claim to be able to diagnose anything from allergies and blood pressure problems to thyroid problems or even HIV infection. But long before the advent of pricey personalised DNA testing and phone apps that measure your mindfulness or keep tabs on your abs came the cheapest, most easily calculable and decidedly low-tech personal diagnostic tool: the body mass index (BMI). An individual's BMI is calculated by measuring their mass in kilograms and then dividing that by the square of their height in metres.

For recording and diagnostic purposes, anyone with a BMI below 18.5 is classified as 'underweight'. The 'normal weight' range extends from 18.5 to 24.5 and the 'overweight' classification spans 24.5 to 30. 'Obesity' is defined as having a BMI above 30. Although it is difficult to estimate exactly, obesity may be implicated in around 23% of US deaths. The trend is mirrored, to a slightly less extreme extent, throughout the world. In Europe, obesity is second only to smoking as a cause of premature death. Obesity in adults and children is on the rise in almost every country and its prevalence has doubled over the past 30 years. People who are found to have an obese BMI are warned of the dangers of potentially life-threatening conditions like type-2 diabetes, strokes, coronary heart disease and some types of cancer, as well as the increased risks of psychological problems such as depression. Today, more people in the world die from being overweight than from being underweight.

Given the health implications related to a diagnosis of obesity, or even of being overweight, you might have assumed that the metric used to diagnose these conditions, the BMI, would have a strong theoretical and experimental basis. Sadly this is far from the truth. In fact, BMI was first cooked up in 1835 by Belgian Adolphe Quetelet, a renowned astronomer, statistician, sociologist and mathematician but, notably, not a physician.[37] Using some decidedly shaky mathematics, Quetelet concluded that 'The weight of developed persons, of different heights, is nearly as the square of the stature'. Notably, though, Quetelet derived this statistic from average population-level data and did not suggest that this ratio would hold true for any given individual. Neither did Quetelet suggest that his ratio, which would become known as the 'Quetelet index', could be used for making inferences on how over- or underweight an individual was, less still about their health. This development would not come until 1972. In response to unprecedented levels of obesity, American physiologist Ancel Keys (who would later make the link between saturated fat and cardiovascular disease) undertook a study to find the best indicator of excess weight.[38] He came up with the same ratio of mass to height squared as Quetelet, and argued that the measure would be a good indicator of obesity in the population.

Theoretically, individuals who are overweight have a higher mass than their height would suggest and hence a higher BMI. Underweight people would have a correspondingly lower BMI. Keys' BMI formula gained popularity because it was so simple. As we became more overweight as a species and detrimental health outcomes began to be definitively associated with obesity, epidemiologists began to use BMI as a way to track risk factors associated with being overweight. In the 1980s, the World Health Organization, the UK's National Health Service (NHS) and the United States' National Institutes of Health (NIH) all officially adopted the single-value BMI to define obesity for all individuals. Insurance companies on both sides of the Atlantic now routinely use BMI in order to determine premiums and even whether or not they will insure an individual at all.

Whilst it's true that fatter people typically have a higher BMI, it is perhaps unsurprising that this phenomenological catch-all does not work for everybody. The main problem with BMI is that it can't distinguish between muscle and fat. This is important because excess body fat is a good predictor of cardiometabolic risk. BMI is not. If the definition of obesity were instead based on high percentage body fat, between 15 and 35% of men with non-obese BMIs would be reclassified as obese.[39] For example, 'skinny-fat' individuals, with low muscle but high levels of body fat and consequently normal BMI, fall into the undetected 'normal-weight obesity' category. A recent cross-population study of 40,000 individuals found that 30% of people with BMI in the normal range were cardiometabolically unhealthy. It seems that the obesity crisis may be much worse than our BMI-based figures suggest. However, it turns out that BMI both under- and over-diagnoses obesity. The same study found that up to half of the individuals that BMI classified as overweight and over a quarter of BMI-obese individuals were metabolically healthy.

These incorrect classifications have implications for the way in which we measure and record obesity at a population level. Perhaps more worryingly though, diagnosing healthy individuals as overweight or obese based on their BMI can also have detrimental effects on their mental health.[40] As a teenager, journalist and author Rebecca Reid battled with eating disorders. She cites a biology lesson in which she was taught how to measure BMI as a major trigger point for her struggles. Despite previously being content with her body, when Rebecca measured her BMI she was placed in the overweight category. She became obsessed with the metric, to the extent that she began a strict diet and exercise programme which saw her lose ten pounds in just a few weeks. At one point, she passed out alone in her bedroom while trying to restrict herself to just 400 calories a day. When not dieting, she was punishing herself by overeating and then making herself sick to compensate. Rather than a gentle reminder to encourage her to take more exercise, Rebecca describes being placed in the overweight category as a 'confidence shattering claxon'. Ironically, irrespective of their body shape and size, individuals recovering from eating disorders are routinely classified as 'recovered' when their BMI reaches 19 – just inside the 'healthy' range. After taking the incredibly difficult step of admitting to themselves they have a problem and seeking help, some eating disorder sufferers have even been denied support based on their 'healthy' BMIs.

BMI is clearly not an accurate indicator of health at either end of the scale. Instead, it would be useful to access a direct measure of the body fat percentage that is so closely linked to cardiometabolic health outcomes. To do that we need to borrow a 2000-year-old idea from the ancient city-state of Syracuse on the island of Sicily.

·

Around 250 BCE Archimedes, the pre-eminent mathematician of antiquity (and conveniently a local) was asked by Hiero II, king of Syracuse, to help resolve a contentious issue. The king had commissioned a metalsmith to make him a crown out of pure gold. After receiving the finished crown and hearing rumours of the metalsmith's less-than-honest reputation, the king worried that he had been cheated and that the metalsmith had used an alloy of gold and some other cheaper, lighter metal to cut costs. Archimedes was charged with figuring out if the crown was a dud without taking a sample from it or otherwise disfiguring it.

The illustrious mathematician realised that to resolve the issue he would need to calculate the density of the crown. If the crown were less dense than pure gold then he would know that the metalsmith had cheated. The density of pure gold was easily calculated by taking a regularly shaped gold block, working out the volume and then weighing it to find its mass. Dividing the mass by the volume gave the density. So far so good. If Archimedes could just repeat the procedure with the crown, he'd be able to compare the two densities. Weighing the crown was easy enough, but the problem came when trying to work out its volume, because of its irregular shape. This problem stumped Archimedes for some time, until one day he decided to take a bath. As he got into his extremely full tub, he noticed that some of the water overflowed. As he wallowed, he realised that the volume of water that overflowed from a completely full bath would be equal to the submersed volume of his irregularly shaped body. Immediately he had a method for determining the volume, and hence the density, of the crown. Vitruvius tells us that Archimedes was so happy with his discovery that he jumped straight out of the bath and ran naked and dripping down the street shouting 'Eureka!' ('I have found it!') – the original eureka moment.

Even today, Archimedes' 'displacement' method is used to calculate the volume of irregularly shaped objects. If you were thinking of starting a health drive, you might use it to work out how much smoothie a combination of irregularly shaped fruit and vegetables will make when blended. Alternatively, by blowing as much air as you can into an empty airtight bag and then sealing and submersing it in water, you can use Archimedes' principle to estimate your lung capacity a few weeks into your new exercise programme.

Sadly, despite the utility of the displacement method described in the common retelling of the story, it is unlikely that this is actually how Archimedes solved the problem. Archimedes' measurements of the volume of water displaced by the crown would have had to have been infeasibly accurate. Instead, it is more likely that Archimedes used a related idea from hydrostatics, which would later become known as Archimedes' principle.

The principle states that an object placed in a fluid (a liquid or a gas) experiences a buoyant force equal to the weight of fluid it displaces. That is, the larger the submersed object, the more fluid it displaces and, consequently, the larger the upwards force it experiences to counteract its weight. This explains why extremely large cargo ships float, providing the weight of the ship and its cargo is less than the weight of water they displace. The principle is also closely related to the property of density – the mass of an object divided by its volume. An object whose density is greater than that of water weighs more than the water it displaces, so the buoyant force is not enough to counteract the object's weight. Consequently, the object sinks.

Using this idea, all Archimedes needed to do was to take a pan balance with the crown on one side and an equal mass of pure gold on the other. In air, the pans would balance. However, when the scales were placed underwater, a fake crown (which would be larger in volume than the same mass of denser gold) would experience a larger buoyant force as it displaced more water, and its pan would consequently rise.

It is precisely this principle from Archimedes that is used when accurately calculating body fat percentage. A subject is first weighed in normal conditions and then reweighed whilst sitting completely submerged on an underwater chair attached to a set of scales. The differences in the dry and underwater weight measurements can then be used to calculate the buoyant force acting on the individual whilst underwater, which in turn can be used to determine their volume, given the known density of water. Their volume can then be used, in conjunction with figures for the density of fat and lean components of the human body, to estimate the body fat percentage and provide more accurate assessments of health risks.

The God equation

BMI is just one of a huge number of different mathematical tools that are used routinely throughout the practice of modern medicine. Others range from simple fractions for calculating drug doses to complex algorithms for reconstructing images from CAT scans. In UK healthcare, there is perhaps one formula that stands out above all others in its contentiousness, importance and wide-ranging implications. The 'God equation' dictates which new drugs will be paid for by the NHS: it literally determines who lives and who dies. If you have a child who is terminally ill, you might argue that no price is too high to pay in order to buy you some more time with your little one. The 'God equation' says otherwise.

In November 2016, Daniella and John Else's 14-month-old son, Rudi, was rushed to the Sheffield Children's Hospital. He was hooked up to a ventilator to keep him breathing, with doctors telling Daniella and John that Rudi might not last the night. The cause of the alarm was a common chest infection that most children would fight off. Most children, however, do not have Spinal Muscular Atrophy (SMA).

When Rudi was six months old, after doctors had failed to figure out what was wrong with him, Daniella and John helped to diagnose their son with SMA after finding out that John's cousin had been afflicted by the same disorder. Rudi's type of the progressive muscle-wasting disease comes with a life expectancy of just two years. Miraculously, there is a drug, Spinraza, developed by a company called Biogen, which can halt and even reverse some of the debilitating effects of SMA. This drug has the potential to improve and extend the lives of SMA sufferers like Rudi, but in England in 2016, when Rudi was fighting for his life in hospital, it was not available free of charge.

In theory, in the US, as soon as the FDA approves a drug for sale it is made available to patients. Spinraza was approved by the FDA in December 2016. In practice, most insurance companies have a 'prior authorization' list for expensive or potentially risky drugs. For each treatment, the list stipulates a range of conditions that must be met before it is released for a particular patient. Spinraza is on every insurance company's prior authorization list. Of course, access to healthcare in the US also depends on being able to afford medical insurance. In 2017, 12.2% of Americans were uninsured, and the US remains the only industrialised nation without universal healthcare coverage.

By contrast, in England, healthcare is available to everyone, free at the point of use, and largely paid for by general taxation. The European Medicines Agency (EMA) and the Medicines and Healthcare Products Regulatory Agency are responsible for approving the safety and efficacy of drugs in England. In May 2017, the EMA approved Spinraza for use. However, since the NHS has only a limited budget, it is not able to sanction every new treatment that comes on the market. Decisions taken one way or another might, for example, lead to cuts in social care provision, lack of diagnostic or treatment equipment for cancer patients, or understaffing of neonatal care units. The National Institute for Health and Care Excellence (NICE) is the body responsible for making these tough choices. When it comes to drugs, there is a well-established formula by which NICE ensures that its decisions are objective.

The God equation attempts to balance how much extra 'health

benefit' a drug gives to a patient against how much extra the NHS is being asked to pay for it. Assessing the former is a difficult task. How can one compare the advantages of a drug that reduces the incidence of heart disease, for example, against the benefits of a drug that prolongs the life of a cancer patient?

NICE uses a common benchmark known as a quality-adjusted life year or QALY. When comparing a new treatment to the existing therapy, the QALY accounts not just for how much a drug might extend life, but also the quality of life that it affords. A single QALY might result from a cancer drug that extends life by two years, but leaves patients in only 50% health, or it might result from knee-replacement surgery, which does nothing to extend a patient's remaining ten-year life expectancy but improves their quality of life by 10%. Successful treatment of testicular cancer might garner a large number of QALYs, as the typically young patients have a dramatically extended life expectancy without a reduction in their quality of life.

Once a reliable figure for QALYs has been established, the difference in QALYs and the change in costs between the new and old treatments can be compared. If the QALYs decrease, then the new treatment will be rejected out of hand. If the QALYs increase and the cost decreases then clearly funding a more effective, cheaper new treatment is a no-brainer. However, if, as is most often the case, both the QALYs and the cost increase, NICE is left with a decision to make. In these cases the incremental cost-effectiveness ratio (ICER) is calculated by dividing the increase in QALYs by the increase in cost. The ICER tells us the increased cost per QALY gained. Typically, NICE set their threshold for the maximum ICER they will fund to be between £20,000 and £30,000 per QALY.

In August 2018, SMA sufferers and their families, including Daniella, John and Rudi, waited anxiously to find out whether NICE would sanction Spinraza for use on the NHS. NICE recognised that Spinraza 'provides important health benefits' for patients with SMA. The results of quality of life improvement were also extremely positive. Spinraza was expected to generate an additional 5.29 QALYs. The additional cost, however, ran to an enormous £2,160,048, giving an ICER of over £400,000 per QALY gained: way, way above NICE's threshold. Despite compelling testimony from SMA sufferers and the carers of SMA patients, the God equation meant that the only option was to prohibit the use of Spinraza on the NHS.

Fortunately for the Else family, Rudi is enrolled in an expanded access programme run by manufacturer Biogen, allowing infants with type 1 SMA to receive the drug. In February 2019, he received his tenth injection and is now a thriving three-year-old, far exceeding the Spinraza-less life expectancy of type 1 SMA sufferers. However, Spinraza, a life-saving and life-extending drug, remains unapproved by NICE for SMA sufferers in England.

False alarms

The application of the 'God equation' can be seen as an attempt to take difficult life-and-death decisions out of our subjective hands and place them under the control of an objective mathematical formula. This point of view plays on the seeming impartiality and objectivity of mathematics, but neglects to recognise that the subjective decisions are simply being diverted out of sight in the form of judgements on quality of life and cost-effectiveness thresholds at earlier stages of the decision-making process. We will look more closely at the theme of mathematics' seeming impartiality in Chapter 6, when we consider the applications of algorithmic optimisation to our everyday lives.

Far away from the behind-the-scenes bureaucracy that informs the often unseen decisions taken in our healthcare systems, maths is being used on the front lines in hospitals in order to save lives. As we will see shortly, one particularly important place where maths is starting to have an impact is in the reduction of false alarms in the intensive care unit (ICU).

False alarms typically refer to an alarm triggered by something other than the expected stimulus. A staggering 98% of all burglar-alarm activations in the US are thought to be false alarms. This prompts the question, 'Why have an alarm at all?' As we get used to incorrect alerts we become more reluctant to investigate their causes.

Burglar alarms are by no means the only warnings with which we have become over-familiar. When the smoke detector goes off, we are usually already opening a window and scraping the soot off our toast. When we hear a car alarm outside, very few of us will even get off the sofa and stick our heads outside to investigate. When alarms become an inconvenience rather than an aid, and when we no longer trust their output, we are said to be suffering alarm fatigue. This is a problem because situations in which alarms become so routine that we ignore them, or disable them completely, can be less sensible than not having the alarm in the first place, as the Williams family found out to their great cost.

Michaela Williams spent much of her junior year at high school dreaming of becoming a fashion designer. For some time, she had been suffering with long-lasting, frequent, and painful sore throats. Despite tonsillectomies being more prone to complications in adolescents than in children, Michaela and her family took the decision to undergo the surgery to improve her quality of life. Three days after her 17th birthday, Michaela checked in as an outpatient to her local surgical centre. After a routine procedure, which took less than hour, Michaela was taken to the recovery room while her mother was told that the operation had been a success and that she would be able to take her daughter home later that day. In order to ease her discomfort while in the recovery room, Michaela was given fentanyl, a powerful opioid painkiller. Amongst the known, but relatively infrequent, side effects of fentanyl is respiratory depression. To be on the safe side, the nurse hooked Michaela up to a monitor, measuring her vital signs, before going to check on other patients. Despite having the curtains drawn around her, the monitor would rapidly alert the nurse to any deterioration in Michaela's condition.

Or it would have done, had the monitor not been muted.

While looking after several patients simultaneously in the recovery room, persistent false alarms had been a nuisance preventing the nurses from doing their jobs efficiently. Having to stop a procedure on one patient to reset an alarm for another was not only costing nurses vital time but also disrupting their concentration. So the nurses had devised a simple solution to allow them to continue their tasks uninterrupted. It became routine practice in the recovery room to turn down, or even completely mute the monitors to avoid the persistent false alarms.

Shortly after the curtains were drawn around her, the fentanyl caused Michaela's breathing to drop drastically. The alarm highlighting hypoventilation was triggered, but no one could see the flashing light through the curtain and, certainly, no one could hear it. As Michaela's oxygen levels continued to fall, her neurons began to fire uncontrollably, setting off a chaotic electrical storm that caused irreparable damage to her brain. By the time she was next checked, 25 minutes after the administration of the fentanyl, she was so severely brain-damaged that all chances of survival were gone. She died 15 days later.

•

For patients like Michaela, who are recovering from operations, or who have to spend time in intensive care, having their vital signs monitored with automated alarms that detect everything from heart rate and blood pressure to blood oxygenation and inter-cranial pressure, has obvious benefits. Typically, these monitors are rigged so that when the detected signal moves above or below a given threshold, the alarm is triggered. However, approximately 85% of automated warnings in ICUs are false alarms.[41]

Two factors cause these high rates of false alarms. Firstly, for obvious reasons, alarms in ICUs are set to be extremely sensitive: thresholds for alarms are set deliberately close to normal physiological levels in order to ensure that even the slightest abnormalities are highlighted. Secondly, rather than requiring a sustained abnormal signal, alarms are triggered the instant a signal crosses a threshold. When combined, the slightest rise in blood pressure, for example, even for an instant, is enough to trigger the alarm. While this spike could be indicative of dangerous hypertension, it is far more likely to be caused by natural variation or noise in the measuring equipment. However, if blood pressure were to stay high for a sustained period of time, we would be less likely to attribute this to a measuring error. Fortunately, mathematics has a simple way to solve this problem.

The solution is known as filtering. This is the process by which a signal at a given point is replaced by the average over its neighbouring points. This sounds complicated, but we encounter filtered data all the time. When climate scientists claim that 'We have just experienced the warmest year since records began' they are not comparing temperature data on a day-by-day basis. Instead, they might average across all the days of the year, smoothing over fluctuating daily temperatures, giving a result that makes for easier comparison.

Filtering tends to smooth signals out, making spikes less pronounced. When you take a photo with a digital camera in low light conditions, the long exposures required often result in grainy images. Occasional bright pixels appear in dark areas of the image and vice versa. Since the intensity of pixels in a digital photo is represented numerically, filtering can be used to replace the value of each pixel by the average value of its neighbouring pixels, filtering out the noise and giving a smoother resulting image.

We can also use different sorts of averages when filtering. The average we are most familiar with is the *mean*. To find the mean, we add up all the values in a data set and divide by how many values there are. If, for example, we wanted to find the mean height of Snow White and the seven dwarves, we would add together their heights and divide by eight. This average would be skewed by Snow White, whose relative tallness makes her an outlier in the data set. A more representative average would be the *median*. To find the median height of the crew we line up the dwarves and Snow White in height order (Snow White at the front, Dopey at the back) and take the height of the middle person. Since we have eight (an even number) people in the line, we don't have a single middle person. Instead we take the mean height of the middle two (Grumpy and Sleepy) as our median. By using the median we have successfully removed the outlying height of Snow White which so biased the mean. For the same reason, the median is often used when presenting data on average income. As evident in Figure 4, the high wages of the very well-off individuals in our societies tends to distort the mean – an idea we will encounter again in the context of misleading mathematics in the courtroom in the next chapter. The median gives us a better idea than the mean of what to expect of a 'typical' household's disposable income. Of course, it could be argued that Snow White's height or the income of high earners should not be

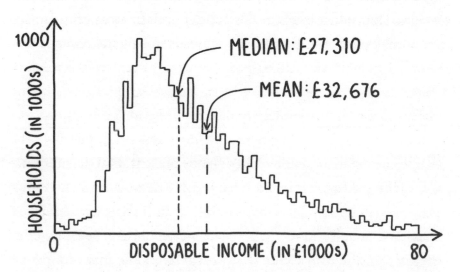

Figure 4: The frequency of UK households with a given disposable (after tax) income (in £1000 blocks) in 2017. The median (£27,310) might be considered a better representation of the 'typical' household disposable income than the mean (£32,676).

neglected in these statistics, as they are as valid as any other data points in the set. While this may be the case, the point is that neither mean nor median is correct in any objective sense. The different averages are simply useful for different applications.

When filtering a grainy digital image, we want to remove the effects of spurious pixel values. When averaging over neighbouring pixel values, mean filtering would modulate, but not completely remove, these extreme values. Conversely, median filtering ignores the values of extremely noisy pixels with impunity.

For the same reason, median filtering is beginning to be used in our ICU monitors to prevent false alarms.[42] Taking the median over a number of sequential readings, alarms are triggered only if thresholds are breached for a sustained (although still short) period of time, rather than by a one-off spike or dip in a monitor's read-out. Median filtering can reduce the occurrence of false alarms in ICU monitors by as much as 60% without jeopardising patient safety.[43]

•

False alarms are a subcategory of errors known as false positives. A false positive, as the name suggests, is a test result that indicates that a particular condition or attribute is present when it isn't. Typically, false positives occur in binary tests. These are tests with two possible outcomes – positive or negative. In the context of medical tests, false positives result in people who are not sick being told that they are. In the courtroom, false positives are the innocent people convicted of crimes they didn't commit. (We will meet many such victims in the next chapter.)

A binary test can be wrong in two ways. The four possible outcomes of a binary test (two correct and two incorrect) can be read

PREDICTED CONDITION	TRUE CONDITION	
	POSITIVE	NEGATIVE
POSITIVE	TRUE POSITIVE	FALSE POSITIVE
NEGATIVE	FALSE NEGATIVE	TRUE NEGATIVE

Table 2: The four possible outcomes of a binary test.

off from Table 2 above. As well as false positives, there are also false negatives.

In the context of disease diagnostics, you might assume that false negatives are potentially more damaging, since they tell patients that they do not have the disease for which they are being tested, when in fact they do have it. We will meet some of the unsuspecting victims of false negatives later in this chapter. False positives can also have surprising and serious implications, but for completely different reasons.

The big screen

Take disease screening, for example. Screening is the mass testing, for a particular disease, of asymptomatic people belonging to a high-risk group. For example, in the UK, women over 50 are invited to routine breast screens, as they are at increased risk of developing breast cancer. The occurrence of false positives in medical screening programmes is currently a subject of intense debate.

The prevalence of undiagnosed breast cancer amongst women in the UK may be around 0.2%. This means that, at any given time, for every 10,000 women in the UK who have not been diagnosed, we would expect 20 of them to have breast cancer. This doesn't sound very high, but this is because, in the majority of cases, breast cancer is detected quickly. In fact, one in eight women will be diagnosed with breast cancer during her lifetime. In the UK, around one in ten of these women is diagnosed late (at stage three or four). Late diagnosis significantly reduces the chances of long-term survival, supporting the argument that regular mammograms are of vital importance, particularly for women in vulnerable age categories. However, there is a mathematical problem with our breast cancer screens, of which most people are unaware.

Kaz Daniels is a mother of three from Northampton. In 2010, she went for her first routine mammogram at the age of 50. A week after her appointment she received a letter in the post asking her to go back for further tests in two days' time. Understandably, given the urgency of the recall, she was petrified. She spent the next two days feeling too worried to eat and too anxious to sleep, ruminating on the possible consequences of a positive diagnosis.

Most patients who undergo mammograms perceive them to be a fairly accurate way of screening for breast cancer. Indeed, for people who do have breast cancer, the test will pick this up roughly nine times out of ten. For people who don't have the disease, the results of the test will tell you this correctly nine out of ten times.[44] Knowing these statistics and having received a positive mammography result, Kaz considered it likely that she had the disease. However, a simple mathematical argument demonstrates that, in fact, the opposite is true.

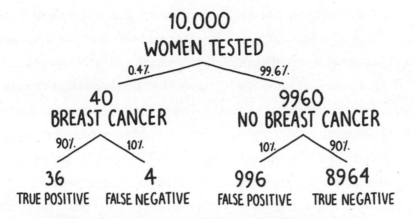

PROPORTION OF CORRECT POSITIVES: 36/(36+996)

Figure 5: Of 10,000 women aged over 50 tested, 36 will be correctly identified as positive, whereas 996 will be told they are positive despite not having the disease.

The prevalence of undiagnosed breast cancer in women over 50 – those who are invited for routine screening – is slightly higher than in the general female population, and can be estimated at around 0.4%. The fates of 10,000 such women are broken down in Figure 5. We can see that, on average, only 40 of them will have breast cancer, so 9960 will not. However, one in ten, or 996, of the women who are free of the disease will be given an incorrect positive diagnosis. Compared to the 36 women who are correctly diagnosed as having the disease, this means that a positive test result is correct only in 36 of 1032 cases or 3.48% of the time. The proportion of positive test results that are true positives is known as the precision of the test. Of the 1032 women to receive a positive result, only 36 of them will actually have breast cancer. To put it another way, if your mammogram comes back positive, the overwhelming likelihood is still that you don't have breast cancer. Despite appearing to be quite an accurate test, the low prevalence of the disease in the population makes it extremely imprecise.

Poor Kaz didn't know this, and neither do many of the women who undergo such tests. Indeed, many doctors are unable to interpret positive mammograms. In 2007, a group of 160 gynaecologists were given the following information about the accuracy of mammograms and the prevalence of breast cancer in the population:[45]

— The probability that a woman has breast cancer is 1% (prevalence).
— If a woman has breast cancer, the probability that she tests positive is 90%.
— If a woman does not have breast cancer, the probability that she nevertheless tests positive is 9%.

The physicians were then faced with a multiple-choice question asking them to identify which of the following statements best characterised the chances that a patient with a positive mammogram actually has breast cancer:

A. The probability that she has breast cancer is about 81%.
B. Out of 10 women with a positive mammogram, about 9 have breast cancer.
C. Out of 10 women with a positive mammogram, about 1 has breast cancer.
D. The probability that she has breast cancer is about 1%.

The most popular answer amongst the gynaecologists was A – that a positive result in the mammogram will be correct 81% of the time (around eight times out of ten). Are they right? Well, we

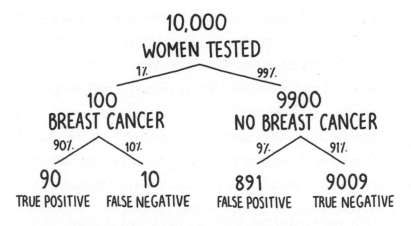

PROPORTION OF CORRECT POSITIVES: 90/(90+891)

Figure 6: Of 10,000 hypothetical women in the multiple-choice question, 90 will be correctly identified as positive, whereas 891 will be told they are positive despite not having the disease.

can work out the correct answer by considering the updated decision tree shown in Figure 6. With a 1% background prevalence, of 10,000 randomly selected women, on average 100 will have breast cancer. Ninety of these will be told correctly that they have the disease by the mammogram. Of the 9900 women who don't have breast cancer, 891 will be incorrectly told that they do have breast cancer. Of a total of 981 women with a positive result, only 90 of them – or roughly 9% – will actually have the disease. Worryingly, the gynaecologists massively overestimated the true value. The correct answer, C, was chosen by around one fifth of respondents, a worse result than if all the doctors had just selected from the four answers at random.

In the event, Kaz's follow-up scans gave her the all-clear, as was likely to be the case. Her travails, though, are typical of the majority of women who receive a positive mammography result. With repeated mammograms, as directed by most screening programmes, the chances of receiving a false positive go up. Assuming false positives occur with equal probability of 10% (or 0.1) in each test, the correct diagnosis of a true negative occurs with a probability of 90% (or 0.9). After seven independent tests the probability of never having received a false positive (0.9 multiplied by itself six times, or 0.9^7) drops to less than a half (approximately 0.47). In other words, it only takes seven mammograms before an individual free from breast cancer is more likely to have received a false positive than not. With mammograms ordered every three years after the age of 50, women in the screening programme might expect at least one false positive over the course of their lifetime.

The illusion of certainty

Of course, these high frequency false positives raise questions about the cost-benefit balance of screening programmes. High rates of false positives can have damaging psychological effects and lead patients to delay or cancel future mammograms. However, the problems with screening go beyond simple false positives. Writing in the *British Medical Journal*,[46] Muir Gray, former director of the UK National Screening Programme, admitted, 'All screening programmes do harm; some do good as well, and, of these, some do more good than harm at reasonable cost.'

In particular, screening can lead to the phenomenon of over-diagnosis. Although more cancers are detected through breast screening, many of these are cancers that are so small or slow-growing that they would never be a threat to a woman's health, causing no problems if left undetected. Nevertheless, the C-word induces such mortal fear in most ordinary people that many will, often on medical advice, undergo painful treatment or invasive surgery unnecessarily.

Similar debates surround other mass screening programmes, including the smear test for cervical cancer (a disease that we will revisit in Chapter 7, when we reconsider the cost-effectiveness and equality of vaccination programmes), the PSA test for prostate cancer and screens for lung cancer. It is important, therefore, that we understand the difference between screens and diagnostic tests. Screening processes can be thought of in analogy to searching for a job. The initial application for a job allows the employer to shortlist people for interview in an efficient way based on a few desirable characteristics. In the same way, screens are designed to cast a wide, less-discriminating net across a broad population to identify people who have not yet developed clear symptoms. They are typically less accurate tests, but can be applied in a cost-effective manner to large numbers of people. Employers use more resource-intensive and informative methods, like assessment centres and interviews, to decide which candidates they will hire. Similarly, once a population of potentially unwell people has been identified through a screen, they can be followed up with more expensive, but more discerning, diagnostic tests to confirm or dismiss the initial screen results. You wouldn't assume you had got the job just because you had been invited to interview. Equally, you shouldn't assume you have a disease on the basis of a positive screen result. When the prevalence of a disease is low, screenings will produce many more false positives than true positives.

The problems caused by false positives in medical screens are, in part, due to our unquestioning attitude towards the accuracy of medical tests. The phenomenon is often known as *the illusion of certainty*. We are so desperate for a definitive answer, one way or another, particularly in medical matters, that we forget to treat our results with the required degree of scepticism.

In 2006, 1000 adults in Germany were asked whether a series of tests gave results that were 100% certain.[47] Although 56% correctly identified mammograms as having some inaccuracy, the vast majority believed DNA tests, fingerprint analysis and HIV tests to be 100% conclusive, which they demonstrably are not.

In January 2013, journalist Mark Stern spent a week in bed with a fever. He booked an appointment with his new doctor, who decided that the best course of action would be to take a blood sample and run a batch of tests. A few weeks later, now feeling better after having taken a course of antibiotics, Mark was alone in his flat in Washington DC when the phone rang. He answered to find it was his doctor on the other end of the line with his test results. Mark was completely unprepared for the conversation that would unfold.

'Your ELISA test went up to positive,' his doctor said, cutting to the chase. 'You should go ahead and assume that you have HIV.' Despite being unaware that his doctor had even run the ELISA test for HIV (or the follow-up Western blot test), when faced with this evidence and the advice of his doctor, Mark had little choice but to reconcile himself to his shock HIV-positive diagnosis. Before ending the call, Mark's doctor suggested he come in the next day for confirmatory tests.

That night, Mark and his boyfriend reviewed their previous negative HIV tests from recent months and tried to think of all the events in the intervening period that could have led to HIV infection. Being in a committed monogamous relationship and practising safe sex, they found it hard to think of any possibilities. They found it even harder to get to sleep that night.

The next morning, panicked, confused and exhausted from lack of sleep, Mark reported to the surgery. As his doctor drew blood from his arm to send away for a confirmatory RNA test, he reiterated his conviction that Mark was HIV-positive and suggested a rapid immunoassay test be taken in the surgery to confirm his belief. As Mark waited out the longest 20 minutes of his life for the test to resolve itself, he considered what his life with HIV would be like. Although no longer the stigmatised death sentence it once was, he knew a diagnosis would lead him to re-evaluate and question many aspects of his life, not least how he had come to be HIV-positive in the first place.

At the end of the agonising wait, no red line had appeared in the results window. Instead, the window allowed a small ray of hope to shine through the clouds onto the troubled landscape of Mark's mind. The test was negative. Two weeks later Mark received the results of the more accurate RNA test – also negative. With a further immunoassay coming back negative, the clouds lifted as his doctor was finally convinced that Mark was negative for HIV.

In truth, Mark's original ELISA and Western blot tests were ambiguous. His ELISA test did come back with raised levels of antibodies, indicating a positive test. However, at the time he took the test, ELISA had reported false positive rates of around 0.3%.[48] His Western blot test – a more accurate test designed to catch such false positives – came back with results that were indicative of a lab error. However, Mark's doctor, having never seen this error before, misinterpreted the results. His diagnosis may have been biased by his knowledge that Mark was a gay man, placing him in a high-risk category for HIV. In turn, Mark, blinkered by the illusion of certainty, trusted in the judgement of his doctor and the accuracy of the tests.

Two tests are better than one

The concept of accuracy for two-outcome binary tests is poorly understood by many. From the point of view of the proportion of the population who don't have the disease (which will typically be the vast majority) we could define the 'accuracy' of the test as the proportion of these people who are correctly identified as disease-free – the 'true negatives'. The higher the proportion of true negatives (and thus the lower the rate of false positives), the more accurate the test. In fact, the proportion of true negatives is known as the 'specificity' of a test. If a test is 100% specific, then only people who genuinely have the disease will test positive – there are no false positives.

Even completely specific tests are not guaranteed to identify *everyone* who has the disease. Perhaps we should classify accuracy based on the point of view of the people who actually have the disease. If you were in their shoes, wouldn't you consider it a priority for your disease to be picked up first time by the test? So perhaps the 'accuracy' of a test could be the proportion of 'true positives' – the people who have the disease and are identified correctly as such. In fact, this proportion is known as the test's 'sensitivity'. A test with 100% sensitivity would correctly alert all affected patients to their condition.

A test's precision is found by calculating the number of true positives and dividing by the total number of positives, both true and false. The low precision of breast cancer screens, at just 3.48%, is what so surprised us earlier in the chapter. The term 'accuracy', however, is typically reserved for the number of true positives and true negatives divided by the total number of people taking the test. This makes sense, as it is the proportion of times the test is giving the correct result, one way or another.

Definitive error rates for the ELISA test for HIV that failed Mark Stern are hard to determine. However, most studies agree on a specificity of around 99.7% and a sensitivity closely approximating 100%. A negative test result implies the recipient is almost certainly HIV-free, but, on average, 3 of every 1000 people negative for HIV will be given an incorrect HIV-positive diagnosis. The UK has an HIV prevalence of just 0.16%. Considering the 1,000,000 randomly selected UK citizens depicted in Figure 7, on average 1600 will be HIV-positive and 998,400 will not. Of the 998,400 HIV-negative patients undergoing an ELISA test, even with a specificity as high as 99.7%, 2995 will be given an incorrect positive diagnosis. These false positive numbers outweigh the 1600 true positives by almost two to one. As with breast cancer screening, since the prevalence of HIV is low, and because the ELISA test lacks a tiny amount of specificity, the proportion of people with positive diagnoses who are genuinely positive (the precision of the test) is low, at just over one third. The accuracy of the test, however, is extremely high. It gives 997,005 correct results (positive or negative) for every 1,000,000

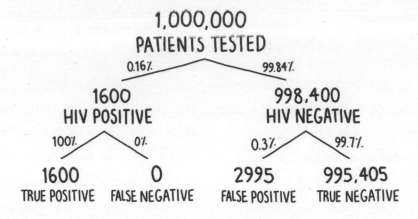

PRECISION: 1600/(1600+2995)

Figure 7: Of 1,000,000 UK citizens undergoing the ELISA test, 1600 will be correctly identified as HIV-positive, whereas 2995 will be told they are HIV-positive despite not having the disease.

people tested – an accuracy of over 99.7%. Even extremely accurate tests can be alarmingly imprecise.

One simple way to improve the precision of a test is simply to run a second test. This is why the first test for many diseases (as we have seen is the case when detecting breast cancer) is a low-specificity screen. It is designed to inexpensively highlight as many potential cases as possible while missing as few as possible. The second test is usually diagnostic and will have a much higher specificity, ruling out the majority of the false positives. Even if a more specific test is not available, a re-run of the same test on all the positive-testing patients can dramatically improve the precision. For the ELISA test, the first attempt effectively increases the prevalence of HIV-positive individuals in the retested population from 0.16% to around 34.8%: the value of the first test's precision. When we

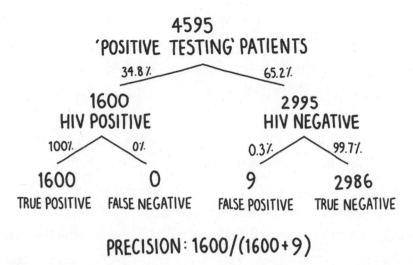

Figure 8: Of 4595 people who originally tested positive, the 1600 true positives will still be identified as such, but the number of false positives will reduce to just 9.

run the test again, as depicted in the decision tree in Figure 8, the majority of the original false positives are rooted out by the test's high precision, whereas the true HIV-positive individuals are still correctly identified as such. The precision improves to 1600/1609, which equates to roughly 99.4%.

•

In theory, it is possible to have a test that is both completely sensitive and completely specific: a test that identifies all the people, and only those people, who have the disease. Such a test could genuinely be claimed to be 100% accurate.

Completely accurate tests are not without precedent. In December 2016, a global team of researchers developed a blood test for Creutzfeldt-Jakob disease (CJD).[49] In a controlled trial, the fatal degenerative brain disorder, thought to be caused by eating beef from animals infected with mad cow disease, correctly identified all 32 patients who had the disease (complete sensitivity) with no false positives (complete specificity) amongst the 391 control patients.

Although there doesn't necessarily have to be a trade-off between sensitivity and specificity, in practice there usually is. False positives and negatives are typically negatively correlated: the fewer false positives the more false negatives and vice versa. In practice, effective tests will find a threshold level at which to draw the line between complete specificity and complete sensitivity; a balance struck somewhere in between the two extremes, as close as possible to both.

The reason this trade-off exists is that we are typically testing for proxies rather than the phenomena themselves. The test that misdiagnosed Mark Stern as HIV-positive does not test for the HIV virus. Rather, it tests for antibodies that the body's immune system raises in an attempt to fight off the virus. However, high HIV-associated antibody loads can be raised by something as innocuous as the flu vaccination. Similarly, most home pregnancy tests do not look for the presence of a viable embryo implanted in the woman's womb. Typically, these tests look for elevated levels of the hormone HCG, produced after implantation of the embryo. Such proxy indicators are often called surrogate markers. Tests can be wrong because markers similar to the surrogate can trigger a positive result.

CJD diagnostic tests, for example, have typically been based on brain scans and biopsies measuring the potential effect on the brain of the faulty proteins that are the root cause of the condition. Unfortunately, the characteristics evaluated by these tests are similar to the characteristics of people with dementia, making clear diagnoses difficult. Rather than looking for subtly different symptoms that could be confused with those of other diseases, the new CJD blood test detects the infectious proteins that always give rise to the disease. This is why the test can be so conclusive: if the malformed proteins are found then that person has the disease, if not, then they haven't. When testing for the root cause of a disease, rather than a proxy, it really is that simple.

•

Another common reason for the failure of proxy tests occurs when the surrogate marker itself is produced by something other than the phenomenon for which we were hoping to test. Anna Howard was just 20 when she woke up feeling sick one morning in June 2016. Despite the fact that she and her boyfriend of nine months, Colin, were not actively trying for a baby, she decided to take a pregnancy test just in case. She was surprised when the little blue line slowly appeared, as if by magic, as she watched the wand. This was not something either of them had planned, but having convinced themselves they would make good parents, Colin and Anna decided to keep the baby and even started to try out some names.

Eight weeks into her pregnancy, Anna started to bleed. Her GP referred her to hospital for a scan to check that the baby was OK. After the scan, doctors informed Anna that she was miscarrying. They told her to come back the next day for further confirmatory tests. The next day, however, a hormone test, not dissimilar to a home pregnancy test, showed that Anna's levels of HCG, the 'pregnancy hormone', were still high enough to indicate a viable pregnancy. Consequently, the doctors informed her that the miscarriage diagnosis was a false alarm.

A week later Anna was bleeding again, and in extreme pain, so she returned to the hospital. This time, fearing an ectopic pregnancy, the doctors carried out an inspection of Anna's reproductive tract with a fibre optic camera. Thankfully, they found no evidence of a foetus growing in the wrong place, but the growth in Anna's womb was no foetus either. Rather than a healthy baby, Anna had a gestational trophoblastic neoplasia (GTN) – a cancerous tumour – growing in her uterus. The tumour was growing at roughly the same rate as a foetus and producing HCG, the proxy indicator for pregnancy, deceiving the pregnancy tests, Anna, and the medics alike into thinking her life-threatening cancer was a normal healthy baby.

Although tumours like Anna's GTN are rare, other types of tumour are also capable of fooling pregnancy tests into giving false positives by producing the surrogate indicator HCG. Indeed, the teenage cancer trust state that pregnancy tests have been used to aid the diagnoses of testicular cancer for at least the last decade. In fact, only a small minority of testicular tumours will give a positive result. But in these cases, the fact that any positive results are known to be false positives for pregnancy means that raised HCG levels are extremely likely to be the result of a tumour.

Pregnancy tests are manifestly capable of giving (in some cases quite useful) false positives. However, the levels of HCG in the urine can be so low that these tests are also capable of false negatives. False negative pregnancy tests, although less common than false positives, can have significant adverse effects on mothers-to-be. In one case, a woman miscarried after she was sanctioned to undergo a surgical procedure that would never have been undertaken had she known she was pregnant.[50] Another woman's ectopic pregnancy was missed by urine tests, leading to a ruptured fallopian tube and life-threatening blood loss.[51]

•

In most cases, once pregnancy is well established, typically around 12 weeks in the UK, we abandon the proxy hormonal markers in favour of ultrasound scans, which directly demonstrate the presence of a developing foetus in the womb. However, the purpose of the ultrasound scans themselves is rarely to establish pregnancy, but rather to check that the foetus is developing normally. One of the tests that is run at this stage is the nuchal scan. The scan is designed to detect cardiovascular abnormalities in the developing foetus, which are typically associated with chromosomal abnormalities like Patau's syndrome, Edwards' syndrome and Down's syndrome. For most people, our DNA is bunched into 23 numbered pairs of chromosomes. In the three conditions tested for by the nuchal scan, one of the numbered pairs has an extra chromosome, meaning it is actually a chromosome triple or a 'trisomy'.

The nuchal scan is not quite as simple as a binary test. It doesn't predict absolutely whether an unborn child has Down's syndrome. Rather, it presents the parents-to-be with an assessment of the risk of the condition. Nevertheless, based on the scan, pregnancies are clearly categorised into high-risk and low-risk, and this distinction is used when relaying the test results to the parents. If an unborn child is categorised as low-risk (less than a 1 in 150 chance) of having Down's, then no further testing is offered, but if the child is in the high-risk category then the more accurate amniocentesis is often offered. Fluid containing foetal skin cells is removed from the amniotic sac surrounding the foetus using a needle. Piercing the womb and the amniotic sac comes with a risk: between 5 and 10 in every 1000 pregnancies that are tested with amniocentesis are subsequently miscarried. However, the increased specificity of the test makes the risk of amniocentesis acceptable for many parents-to-be. The test is able to be more accurate than a scan because it explicitly detects the extra chromosome in the baby's DNA (extracted from the foetal skin cells) rather than a proxy marker. It picks out the false positives from the first test and provides the parents of the true positives time to make an informed decision about whether to continue with their pregnancy. The cases that slip through the cracks are the false negatives – the parents who are told incorrectly that their child is at low risk of Down's and are offered no further testing.

Flora Watson and Andy Burrell were one such pair of parents. Back in 2002, having had a scare four weeks into her second pregnancy, Flora decided to pay for the relatively new nuchal test to be carried out privately at ten weeks gestation. After the ultrasound scan, Flora was told that she had an extremely low chance of having a baby with Down's syndrome. In fact, the likelihood of having a baby with Down's was compared to that of winning the lottery – about 1 in 14 million. This is more reassurance than most parents can expect from these sorts of screens. Flora was satisfied that she need not go through the potentially risky amniocentesis procedure to confirm what the nuchal test had already told her. Instead, she could get on with making excited preparations for the birth of her second child.

Five weeks before her due date, however, Flora noticed that something was wrong. Her unborn baby had started to move less and less. Three weeks later she was in hospital delivering Christopher. He came quickly, just half an hour after her arrival in hospital. When he emerged, he was so purple and contorted that Flora thought he was dead. The nurses assured her and Andy that he was very much alive but the news they delivered next would alter the future of their family.

Christopher had Down's syndrome. On hearing the news, Andy rushed out of the room and Flora began to cry. What should have been a celebration turned, for them, almost to a wake for the loss of their 'healthy baby'. For the next 24 hours, Flora recalls, 'I just couldn't touch him or have him anywhere near me.' So Christopher lay alone on the first night of his life, cared for only by the nurses on the ward. When the rest of the family arrived to meet the new arrival things got worse. Having brought up another son with learning difficulties, Andy's father urged them to leave Christopher in the hospital. Flora's mother wouldn't even look at the baby.

The life that awaited Flora and Andy when they brought Christopher home was very different from the one they had eagerly anticipated all those months before, when given the results of their nuchal scan. The whole family eventually reconciled themselves with Christopher's condition, but the pressures of looking after a disabled child took their toll. With the time pressures and exhaustion placing too much of a strain on their relationship, Flora and Andy separated. Flora insists she wouldn't have terminated her pregnancy had Christopher's Down's syndrome been diagnosed earlier. She still gets angry, though, that she was denied time to

adjust and make preparations for her son's condition – a complaint
we will hear again in Chapter 6 when we discover the perils of
automated algorithmic diagnosis. Perhaps the ensuing family heart-
break that followed Christopher's birth might have been avoided if
it weren't for the false negative test result.

•

Whether we like it or not, false positives and false negatives are inescapable. Mathematics and modern technology can help to deal with
some of these issues, with tools like filtering at the forefront of the
battle, but other problems we must learn to deal with by ourselves.
We should remember that screens are not diagnostic tests and that
their results should be taken with a pinch of salt. That's not to say
we should completely ignore a positive screen result, but we should
wait for the results of a more accurate follow-up before we lose too
much sleep. The same is true of personalised genetic tests. The risk
categories we are placed in may vary from company to company
and they can't all be right. As Matt Fender found when faced with
a potentially life-limiting Alzheimer's diagnosis, a second test might
help to give a more definitive answer.

For some tests, a more accurate version is not available. In these
cases, we should remember that even a second run of the same test
can dramatically improve the precision of its results. We should
never be afraid to ask for a second opinion. It is clear that even doctors – the perceived experts – don't always have the firmest grasp
of the figures, despite the illusion of confidence they exude. Before
you start to worry yourself unduly, based on assertions of a single
test, find out its sensitivity and specificity, and work out the likelihood of an incorrect result. Question the illusion of certainty and

take the power of interpretation back into your own hands. As we'll see in the next chapter, not stopping to question authority figures, especially those exploiting the laws of mathematics, has landed more than one person on the right side of the law but on the wrong side of the prison-cell door.

3

THE LAWS OF MATHEMATICS

Investigating the Role of Mathematics in the Law

Sally Clark walked into the bedroom of her cottage where, minutes before, her husband, Steve, had left their eight-week-old infant son, Harry, asleep. She screamed. Harry lay slumped in his bouncy chair, blue in the face and not breathing. Despite her husband's resuscitation attempts and those of the ambulance crew, Harry was pronounced dead a little over an hour later. A horrible tragedy to befall any new mother. But this was the second time it had happened to Sally Clark.

A little over a year earlier, Steve left their home in the leafy Manchester suburb of Wilmslow to enjoy his department's Christmas dinner. Sally put their 11-week-old son, Christopher, to bed in his Moses basket on her own that evening. Around two hours later, upon finding Christopher unconscious and grey, she called an ambulance. Despite their efforts, Christopher never woke up. A post-mortem carried out three days later would attribute his death to a lower-respiratory-tract infection.

After Harry's death however, Christopher's post-mortem results were re-examined. A cut to the lip and bruising to the legs, which were originally attributed to resuscitation attempts, were given a more sinister interpretation. When Christopher's preserved tissue samples were re-analysed, evidence of recent ante-mortem (pre-death) bleeding in the lungs, missed by the first examination, led the pathologist to suggest smothering.

Harry's post-mortem indicated retinal haemorrhaging, spinal damage and tears in the brain tissue: key indications that Harry might have been shaken to death. Taking both post-mortems together, police felt they had enough evidence to arrest Sally and Steve Clark. The crown prosecution service decided not to litigate against Steve (since he had not been present when Christopher died), but Sally was charged with the murders of both of her sons.

The trial that followed would see not one, but four mathematical mistakes, which would contribute to what is often referred to as Britain's worst miscarriage of justice. By telling Sally's story, in this chapter we will investigate the sometimes tragic, but all-too-common, courtroom mistakes that can result from mathematical errors. Along the way, we will meet the participants in similar calamities: the criminal whose conviction was quashed on a mathematical technicality; the judge whose mathematical misunderstanding may have helped to free Amanda Knox, the infamous US student accused of murder. But first, let us hear the case of the French military officer exiled to a brutal prison camp for a crime he didn't commit.

The Dreyfus affair

Maths in the courtroom has a long and not so distinguished history. The first notable (mis)use was in a political scandal that would divide the French Republic and that would become known worldwide as 'the Dreyfus affair'. In 1894, a French cleaning lady, working undercover at the German embassy in Paris, recovered a discarded memo. The discovery of the handwritten message, offering French military secrets to the Germans, led to a witch hunt for a possible German spy in the midst of the French Army. The search culminated in the arrest of the French Jewish artillery officer, Captain Alfred Dreyfus.

At the resulting court martial, disliking the opinion of the genuine handwriting expert who suspected Dreyfus was innocent, the French government drafted in the unqualified Alphonse Bertillon, the head of the 'Bureau of Identification' in Paris. Confusingly, Bertillon claimed that Dreyfus had written the note to make the writing look like a forgery of his own handwriting – a practice known as autoforgery. Bertillon proceeded to concoct an abstruse mathematical analysis based on a series of similarities in the pen strokes of repeated polysyllabic words in the memo. He claimed that the probability of a similarity between pen strokes at the starts or ends of any pair of repeated words was 1/5. He went on to calculate that the probability of the four coincidences he had detected amongst the 26 starts and ends of the 13 repeated polysyllabic words was 1/5 multiplied by itself three times – which works out to be a tiny 16 in 10,000 – making their occurrence by chance seem extremely unlikely. Bertillon suggested that the similarities were not coincidences, but 'must have been done carefully on purpose, and must denote a purposeful intention, probably a secret code'.[52] His argument was enough to persuade, or at least baffle, a seven-man jury. Dreyfus was convicted and condemned to life imprisonment in solitary confinement on the lonely penal colony of Devil's Island, several miles off the coast of French Guiana.

Such was the opacity of Bertillon's mathematical argument at the time, that neither Dreyfus' defence team, nor the government commissioner present at the court, understood any of the argument. It is likely that the presiding judges were equally confused, but were too intimidated by the pseudo-mathematical arguments presented to do anything about it. It took Henri Poincaré, one of the most prodigious mathematicians of the 19th century (and whom we will meet again in Chapter 6 when we come across his million-dollar problem) to unpick Bertillon's mystifying calculations. Drafted in over a decade after the original conviction, Poincaré quickly spotted the error in Bertillon's calculations. Instead of calculating the probability of four coincidences in the list of 26 starts and ends of the 13 repeated words, Bertillon had calculated the probability of four coincidences in four words, which naturally is far less likely.

As an analogy, imagine inspecting the person-shaped silhouettes at the end of a target practice at a shooting range. Upon finding ten shots to either the head or chest you might assume that the marksman was a sharp shooter. When you discover that the session involved firing 100 or even 1000 shots, you might be less impressed. The same was true for Bertillon's analysis. Four coincidences from four possibilities is indeed very unlikely, but there are 14,950 different ways of choosing four options from the 26 starts and ends of the words that Bertillon analysed. The real probability of the four coincidences Bertillon had spotted was roughly 18 in 100, over 100 times larger than the figure he used to convince the jury. When you take into account the fact that Bertillon would have been equally as happy to find five, six, seven or more coincidences, we can recalculate the probability of finding four *or more* coincidences as roughly eight in ten. Finding what Bertillon considered an 'unusual' number of coincidences is far more likely than not finding them. By exposing Bertillon's miscalculation and arguing that even attempting to apply probability theory to such a question was not legitimate, Poincaré was able to debunk the aberrant handwriting analysis and in so doing to exonerate Dreyfus.[53] After suffering four years of intolerable conditions on Devil's Island and a further seven years living in disgrace back in France, Dreyfus was finally released in 1906 and promoted to a major in the French Army. His honour restored, with great magnanimity he went on to serve his country in the First World War, distinguishing himself on the front-line at Verdun.

Dreyfus' case demonstrates both the power of mathematically backed arguments and the ease with which they can be abused. We will revisit this theme several times in the coming chapters: the tendency, when a mathematical formulation is presented, for heads to nod sagely without asking for further explanation, in deference to the savant who has conjured it into being. The mystery surrounding many mathematical arguments is, in part, what makes them both so impenetrable and, often undeservedly, so impressive. Very little is ever challenged. A mathematical form of the illusion of certainty (the phenomenon we met in the previous chapter, that leads people to accept unquestioningly the results of medical tests) strikes would-be doubters dumb. The tragedy is that we have failed to learn the lessons from Dreyfus' trial and from numerous other mathematical miscarriages of justice throughout history. As a result, innocent victims have suffered the same fate over and over again.

Guilty until proven innocent?

Just as we saw with medical tests in the previous chapter, the law is full of instances in which binary judgments have to be made: right or wrong, true or false, innocent or guilty. The courtrooms of many Western democracies abide by the maxim 'innocent until proven guilty' – that the burden of proof should rest with the accuser, not the accused. Almost all countries have done away with the converse presumption 'guilty until proven innocent', a practice bound to result in more false positives and fewer false negatives. However, there are some modern-day countries in which the balance leans towards the presumption of guilt and away from innocence. The Japanese criminal justice system, for example, has a conviction rate of 99.9%, with most of these convictions backed up with a confession.[54] For comparison, in 2017/18 the UK's crown court had a conviction rate of 80%. Japan's high conviction rate sounds like an impressive statistic, but is it likely that the Japanese police get the right person in over 999 out of every 1000 cases?

This high conviction rate is due, in part, to the tough interrogative techniques practised by Japanese detectives. As routine, they are allowed to detain suspects for up to three days without charge, can interrogate suspects without a lawyer present, and are not required to record interviews. In turn, these uncompromising techniques are a result of the Japanese legal system, in which establishing motive through confession is a hugely important part of obtaining a guilty verdict. This is compounded by the pressure, applied to the interrogators by their superiors, to extract confessions before physically investigating the evidence relating to the case. The interrogator's task is made easier by the seeming willingness of many Japanese suspects to confess, in order to avoid the shame brought upon their families by a high-profile trial. The prevalence of false confessions in the Japanese justice system was highlighted recently by the arrests of four innocent people for malicious internet threats. Before the genuine perpetrator eventually owned up to his crimes, two of the accused had already been coerced into giving false confessions.

Japan's preference for the assumption of guilt is a notable exception. So strong is the 'innocent until proven guilty' sentiment in most of the rest of the world, that it is enshrined as an international human right in the United Nations' Universal Declaration of Human Rights. Eighteenth-century English judge and politician William Blackstone even went as far as to quantify the sentiment, stating, 'It is better that ten guilty persons escape than that one innocent suffer.' This view places us firmly in the camp of the false negative, acquitting people who may well have committed a crime, but cannot be proven guilty. Even if there is evidence for the accused's guilt, unless that evidence can convince jurors or judges beyond reasonable doubt, the accused often walks away. In Scottish courts, a third verdict exists which reduces the false negative rate, if only in name. The 'not proven' verdict can be applied to acquittals in which the judge or jury is not sufficiently convinced of the accused's innocence to declare them not guilty. In these cases, although the accused is still acquitted, the verdict itself is not incorrect.

73 million to one

At Sally Clark's trial in an English courtroom, the conflicting evidence made it difficult for the jury to arrive at a clear-cut guilty or not guilty. Sally was adamant that she had not killed her children. Home Office pathologist and expert witness for the prosecution, Dr Alan Williams, contended otherwise. The medical forensic evidence he presented was intricate and confusing for the jury. In the lead-up to the trial, the brain tears, spinal injuries and retinal haemorrhages that Williams had originally 'found' in Harry's post-mortem were readily discredited by independent experts. Consequently, the prosecution changed tack and tried to persuade the jury that Harry had been smothered to death, not shaken as was originally claimed. Even Williams changed his mind. Nothing in the medical evidence was clear-cut.

On top of this, the fierce contest between defence and prosecution over the circumstantial evidence surrounding the two deaths served to foment the storm of confusion further. The prosecution painted a picture of Sally as a vain and selfish career woman who resented the changes that having children had brought to her lifestyle and her body. A woman so desperate to get back to her pre-maternal life, that she had killed her infant sons. Why then, argued the defence, did she have a second child so soon after the first? And why had she become pregnant with and then given birth to a third child while the trial was being prepared? The defence argued Sally was clearly distraught about her first son's death. The prosecution twisted the argument, intimating that there was something suspicious in her overt grief. The doctor who first saw Christopher when he arrived at the hospital countered that there was nothing unusual in Sally's distress after having lost her firstborn child. The arguments went back and forth, adding to the mist clouding the jurors' vision of the truth.

It was into this confusion that expert witness, Professor Sir Roy Meadow, swept. While the pathologists argued over the extent of 'pulmonary haemorrhage' and 'subdural haematomas', Meadow guided the jurors away from the rocks of confusion and towards the safety of a verdict, with a clear beacon of light: a single statistic. He testified that the chance of two children from an affluent family suffering sudden infant death syndrome (SIDS – often referred to as cot death) was one in 73 million. For many of the jurors, this was the most important piece of information they took from the trial: 73 million was too huge a number to ignore.

In 1989, Meadow, at the time an eminent British paediatrician, had edited a book, *ABC of Child Abuse,* in which was contained the aphorism that came to be known as Meadow's law: 'One sudden infant death is a tragedy, two is suspicious and three is murder until proved otherwise.'[55] This glib maxim is, however, based on a fundamental misunderstanding of probability. The same misunderstanding with which Meadow would mislead the jury in the case of Sally Clark: the simple difference between dependent and independent events.

The independence mistake

Two events are dependent if knowledge of one event influences the probability of the other. Otherwise they are independent. When

IQ	SEX		TOTAL
	MALE	FEMALE	
>110	125	125	250
<110	375	375	750
TOTAL	500	500	1000

Table 3: 1000 people broken down by IQ and sex.

presented with the probabilities of individual events, it is common practice to multiply these probabilities together to find the probability of the combination of events occurring. For example, the probability that a randomly chosen person from the population is female is 1/2. As illustrated in Table 3, of 1000 people, on average, 500 of them will be female. The probability that a randomly chosen person in the population scores above 110 on a particular IQ test is 1/4. This corresponds to a total of 250 out of the 1000 people considered in Table 3. To find out the probability that someone is female *and* has an IQ over 110, we multiply the probabilities 1/2 and 1/4 together to give a probability of 1/8. This agrees with the 125 (1000/8) people in the female high IQ entry of Table 3. Multiplying the two probabilities together to find the joint probability of being female and having a high IQ is perfectly acceptable because IQ and sex are independent: having a particular IQ says nothing about your sex and being of a particular sex says nothing about your IQ.

The prevalence of autism in the UK is roughly 1 per 100,[56] or equivalently, 10 per 1000. We might assume that to find the probability of being female and autistic we can simply multiply the

AUTISTIC	SEX		TOTAL
	MALE	FEMALE	
YES	8	2	10
NO	492	498	990
TOTAL	500	500	1000

Table 4: 1000 people broken down by sex and whether or not they have autism.

two probabilities (1/2 and 1/100) together to give a probability of 1/200, or equivalently, a prevalence of 5 in every 1000 people. However, autism and sex are not independent. When we analyse 1000 randomly chosen people in the population, as in Table 4, we see that autism in males (8 per 500) is four times more likely than in females (2 per 500). Only 1 in 5 of those on the autistic spectrum are female.[57] We need this extra piece of information to calculate the probability that a randomly chosen person from the population is both female and autistic is 2/1000, not 5/1000 as we would have erroneously computed by assuming independence. This illustrates how easy it is to make significant mistakes when we use incorrect assumptions about the independence of events.

The events that Meadow was considering in his testimony were the deaths of each of Sally Clark's children by SIDS. For his figures, Meadow used a – then unpublished – report on SIDS for which he had been asked to write the preface.[58] The UK-based report studied 363 SIDS cases from a total of 473,000 live births over a three-year period. As well as providing an overall population rate of SIDS occurrence, the report stratified the data by mother's age, household income, and whether anyone in the household smoked.

For an affluent, non-smoking family such as the Clarks, in which the mother was over the age of 26, there was just one SIDS case for every 8543 live births.

Meadow's first mistake was to assume that the incidence of SIDS cases were entirely independent events. In doing so he felt justified in calculating the probability of two SIDS deaths in the family by multiplying the figure 8543 by itself, to arrive at a probability of approximately one occurrence in every 73 million pairs of live births. To justify his assumptions he even went so far as to state 'There is no evidence that cot deaths run in families, but there is plenty of evidence that child abuse does.' With this figure in hand, he suggested that with a birth rate in the UK of around 700,000 a year, such a pair of cot deaths could be expected roughly once every 100 years.

His assumption was wildly off the mark. There are many known risk factors associated with SIDS, including smoking, premature birth, and bed-sharing. In 2001, researchers at the University of Manchester also identified markers in genes related to the regulation of the immune system that put children at increased risk of SIDS.[59] Many more genetic risk factors have since been identified.[60] Children who share the same parents are likely to share many of the same genes and, potentially, the increased risk of SIDS. If one child dies from SIDS, then it is likely that the family has some of the attendant risk factors. Hence the probability of subsequent deaths is greater than the background population average. In reality, it is thought that around one family a year in Britain suffers a second SIDS death.

An analogy for the probability of SIDS deaths is to imagine ten bags of marbles. Nine of these bags each contains ten white marbles.

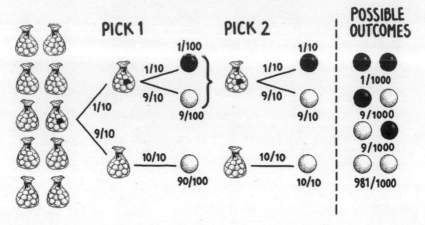

Figure 9: A decision tree for finding the probability of picking black or white marbles. To calculate the probability of picking a black or white marble at each attempt, follow the appropriate branches of the tree and multiply the probabilities on each arm. For example, picking a black marble on the first attempt happens with probability 1/100. Once we have chosen a bag on the first attempt, we pick from the same bag on the second attempt. The probabilities of each of the two-pick combinations are illustrated at the right of the dashed line.

The final bag contains nine white marbles and one black one. This initial state is illustrated on the left of Figure 9. On your first go, you choose a bag at random and pick a marble at random from this bag. Since there are 100 marbles and they are all equally likely to be selected, the probability of choosing the black marble on this first go is one in 100. For your second pick you put the first marble you chose back in its bag and draw another one from the same bag, completely ignoring the other nine bags. If your first pick was the black marble, then you know you are choosing from the bag containing the black marble for your second pick. This makes the probability of choosing the black marble much higher, at one in ten,

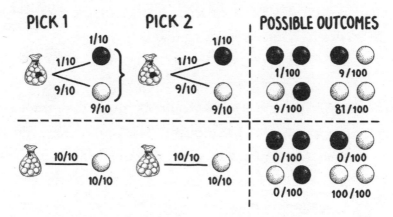

Figure 10: Two alternative decision trees in which the bag you choose from is pre-ordained, but still the same bag for both picks. For each tree, the probabilities of each of the two-pick combinations are illustrated at the right of the dashed line. Clearly, if we are picking from a bag with no black marbles then the only possibility is picking two white marbles.

rather than one in 100. In this scenario, choosing two black marbles (with a probability of one in 1000), is much more likely than simply multiplying the original probability of choosing one black marble by itself (to give a probability of one in 10,000). In the same way, once you have had one child die from SIDS, the probability that the second will also die from SIDS is known to increase.

In fact, with SIDS, the risk factors for your family are not chosen randomly when your first child is born, they are pre-existing – it could be argued that, right from the start, you are either choosing from the bag with the black marble in it or not. This alternative interpretation is illustrated as the pair of decision trees in Figure 10. If you are choosing from the bag with the black marble on both occasions then the probability of choosing two black marbles increases to one in 100. Certainly, simply multiplying the background population risk of SIDS by itself to work out the probability of two SIDS deaths is the wrong thing to do.

•

There were further issues with Meadow's use of the stratified rate of one SIDS case in 8543 live births. The report from which he cherry-picked this figure also gave a significantly larger overall population risk – just one in 1303 – calculated without stratifying the data by socio-economic indicators. Meadow chose not to use this alternative figure. Instead, by specifically taking account of the Clarks' background, Meadow produced a number that made a single SIDS case appear far less likely (and, because of his mistake of ignoring dependence between the deaths, a double SIDS case less likely still), whilst neglecting those factors that made it appear more likely. For example, he chose to ignore the fact that both of Sally's children were boys and that SIDS is almost twice as likely in boys as it is in girls. Taking this into account would have undermined the prosecution's argument by making a double SIDS death seem more likely. The prospect that Sally killed both of her children would then appear commensurately less likely.

Although the prosecution's biasing of the statistical evidence by selectively choosing only detrimental background traits might in itself have been considered unethical or misleading, there is a deeper problem associated with this practice. The stratification of the data in the original report, from which Meadow drew the statistics, was implemented in order to identify high-risk demographics, so as to more efficiently deploy stretched healthcare resources. It was never meant to be used to infer the risk of SIDS for a given individual in these groups. The report was a broad-brush investigation into nearly half a million births in the UK, which meant the individual circumstances of each birth could not be investigated in detail. In contrast, the examination of Sally Clark was an extremely detailed investigation into a particular allegation. The prosecution chose only those aspects of Sally and Steve's background that fitted with the report and assumed that they could use this to characterise the SIDS risk of the Clarks' children. But this falsely supposes that the characteristics of the individual are the same as those of the population. This is a classic example of what is known as an ecological fallacy.

The ecological fallacy

One type of ecological fallacy occurs when we make the unsophisticated assumption that a single statistic can characterise a diverse population. As an example, in the UK in 2010, women had an average life expectancy of 83 years. Men, however, had a life expectancy of just 79 years. The overall population life expectancy was 81 years. A simple example of an ecological fallacy would be to state that because the average life expectancy of females is higher than that of males, any randomly chosen female will live longer than any randomly chosen male. This fallacy has a special (and appropriate) name – 'a sweeping generalization'. Another commonly seen and unsophisticated ecological fallacy based on *increasing* life expectancy is the statement 'We are all living longer', which is so often reached for by lazy journalists. It is not the case that everyone will live longer than they might have expected to previously. Clearly these are naive suggestions at best.

However, ecological fallacies can be more subtle than this. Perhaps it would surprise you to know that despite having a mean life expectancy of just 78.8 years, the majority of British males will live longer than the *overall* population life expectancy of 81 years. At first this statement seems contradictory, but in fact it is due to a discrepancy in the statistics we use to summarise the data. The small, but significant, number of people who die young brings down the mean age of death (the typically quoted life expectancy in which everyone's age at death is added together and then divided by the total number of people). Surprisingly, these early deaths take the mean well below the median (the age that falls exactly in the middle: as many people die before this age as after). The median age of death for UK males is 82, meaning that half of them will be at least this age when they die. In this case, the summary statistic presented – the mean age at death of 78.8 years – is a particularly misleading descriptor of the population of British men.

The bell curve, or normal distribution, which can be used to characterise many everyday data sets, from heights to IQ scores, is a beautifully symmetrical curve in which half of the data lies on one side of the mean and half on the other. This implies that the mean and the median – the middle-most data value – tend to coincide for characteristics that follow this distribution. Because we are familiar with the idea that this prominent curve can describe real-life information, many of us assume that the mean is a good marker of the 'middle' of a data set. It surprises us when we come across distributions in which the mean is skewed away from the median. The distribution of ages at death for British males, displayed in Figure 11, is clearly far from symmetrical. We typically refer to such distributions themselves as 'skewed'.

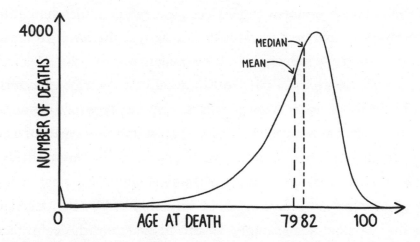

Figure 11: The age-dependence of the number of deaths per year for males in Great Britain follows a skewed distribution. The mean age at death is just under 79, while the median age is 82.

As we saw in the previous chapter (where we first introduced the median for the purposes of avoiding false alarms) household income distribution is another statistic for which the median paints a very different picture to the mean. The UK household income distribution shown in Figure 4, for example, also has a very skewed distribution, much like a slightly messier, flipped version of Figure 11. The majority of UK households have a low disposable income, but there are a small, but significant, number of high earners that skew the distribution. In the UK in 2014, two-thirds of the population had a weekly income below the 'average'.

An initially even more surprising example is the old riddle: 'What is the probability that the next person you meet when walking down the street will have more than the average number of legs?' The answer is 'Almost certain'. The very few people who have no legs or one leg are responsible for a small reduction in the mean so that everyone with two legs has more than the average. In this case it would be ridiculous to assume that the mean correctly characterised any individual in the population.

Clearly, using the wrong sort of average to describe a population can cause an ecological fallacy. Another type of ecological fallacy, known as Simpson's paradox, occurs when we try to take an average of averages. Simpson's paradox has ramifications in diverse areas from measuring the health of the economy,[61] to understanding voter profiles[62] and, perhaps most importantly, in drug development.[63] Imagine, for example, we are in charge of the controlled trial for a new drug, Fantasticol, which is designed to help people reduce their blood pressure. There are 2000 people signed up to the trial, with even numbers of men and women. For control purposes we split them into two groups of 1000. The patients in group A will receive Fantasticol and those in group B will receive a placebo. At the end of the trial 56% (560 out of 1000) people given the drug are found to have reduced blood pressure, while just 35% (350 out of 1000) in the placebo group have lower blood pressure (see Table 5). It seems that Fantasticol really does make a difference.

In order to target the drug properly, it's important to know whether there are any sex specific effects. Consequently, we break

TREATMENT	A: FANTASTICOL	B: PLACEBO
IMPROVEMENT	560	350
NO IMPROVEMENT	440	650
IMPROVEMENT PERCENTAGE	56%	35%

Table 5: Fantasticol appears to give a better overall improvement rate than the placebo.

SEX	MALE		FEMALE	
TREATMENT	A: FANTASTICOL	B: PLACEBO	A: FANTASTICOL	B: PLACEBO
IMPROVEMENT	40	200	520	150
NO IMPROVEMENT	160	600	280	50
TOTAL	200	800	800	200
IMPROVEMENT RATE	20%	25%	65%	75%

Table 6: When subjects are broken down by sex, patients of both sexes who are taking the placebo do better than patients taking Fantasticol.

the figures down to discover how the drug affects males and females separately. This more detailed breakdown is given in Table 6. When we analyse the stratified results we get a bit of a shock. Amongst the men in the trial, 25% (200 out of 800 in group B) who took the placebo had improved blood pressure readings, but only 20% (40 out of 200 in group A) who took Fantasticol improved. Amongst the women the same trend was apparent; 75% (150 out of 200 in group B) of women who were taking the placebo improved, in comparison to just 65% (520 out of 800 in group A) of those taking Fantasticol. For both sexes, a higher proportion of patients taking the placebo improved than of those taking the real drug. Looking at the data this way, it seems that Fantasticol is less effective than the placebo. How can it be that when the data are stratified the trial tells one story, but when amalgamated they tell the opposite story, and which is correct?

The answer lies in something called a 'confounding' or a 'lurking' variable. In this case the variable is sex. It turns out that one's sex is very important to the results. Throughout the course of the trial, women's blood pressures improved naturally more often than

SEX	MALE		FEMALE	
TREATMENT	A: FANTASTICOL	B: PLACEBO	A: FANTASTICOL	B: PLACEBO
IMPROVEMENT	100	125	325	375
NO IMPROVEMENT	400	375	175	125
TOTAL	500	500	500	500
IMPROVEMENT RATE	20%	25%	65%	75%

Table 7: The proportions of men and women who improve under each treatment stay the same as in Table 6 when males and females are distributed evenly between the two groups.

men's did. Because the split of participants' sexes was different in the two groups (800 females and 200 males in drug group A and 200 females and 800 males in placebo group B), group A benefitted significantly from many more women improving naturally, making Fantasticol appear to be more effective than the placebo. Although there were even numbers of men and women in the trial, because they were not distributed evenly throughout the two groups, taking the average of the drug's separate success rates for the two sexes (20% for men and 65% for women) doesn't give the overall 56% success rate for Fantasticol that was originally observed in Table 5. You can't just average an average.

It's only acceptable to average an average if we are sure we have controlled for the confounding variables. If we had known in advance that sex was one such variable, then we would have known it was necessary to stratify the results by sex to get the true picture of Fantasticol's efficacy. Alternatively, we could have controlled for sex by ensuring we had equal numbers of men and women in each group, as in Table 7. The improvement rates for men and women

TREATMENT	A: FANTASTICOL	B: PLACEBO
IMPROVEMENT	425	500
NO IMPROVEMENT	575	500
IMPROVEMENT RATE	42.5%	50%

Table 8: Now we have controlled for the confounding variable of sex, it is clear that Fantasticol does worse than the placebo.

taking Fantasticol or the placebo remain the same as in Table 6. However, when the results are amalgamated in Table 8, and we look at the improvement rates for Fantasticol (42.5% improvement rate) it is clear that the drug does worse, not better, than the placebo (50% improvement rate). There may, of course, be other confounding variables, such as age or social demographic, for example, which we haven't accounted for.

Ecological fallacies and well-regulated controls are serious considerations for those who design clinical trials (as we saw in Chapter 2 and will see again, but for different reasons, in Chapter 4), but they have been known to confound other areas of medicine too. In the 1960s and 1970s a curious phenomenon was observed in children whose mothers had smoked while pregnant. Children with low birth-weights born to smoking mothers were significantly less likely to die in their first year of life than those born to non-smoking mothers. Low birth-weight had long been associated with higher infant mortality, but it seemed that smoking during pregnancy was providing some protection to low birth-weight babies.[64] In reality, it was nothing of the sort.[65] The solution to the paradox lay in a confounding variable.

Although lower birth weight is *correlated* with higher infant mortality, it does not *cause* higher infant mortality. Typically, both can be caused by some other adverse condition: a confounding variable. Both smoking and other adverse health conditions can reduce birth-weight and increase infant mortality, but they do so to different extents. Smoking causes many, otherwise healthy, children to be born underweight. Other causes of low birth-weight are typically more detrimental to a child's health, leading to higher infant mortality rates in these cases. The much larger proportion of low birth-weight children born to smoking mothers, combined with their only slightly increased infant mortality rate, means that a smaller proportion of these children die in their first year than children born with a low birth-weight due to some more life-threatening condition.

The ecological fallacy Meadow committed, by pigeonholing the Clarks into the low-risk SIDS category, made their two children's deaths seem far more suspicious than if the higher, population rate of SIDS had been used. Even to use the overall population rate of SIDS would be to commit an ecological fallacy. Arguably though, the population-level assumption is less partisan, and therefore more appropriate to a situation in which a woman's liberty is on the line. The erroneous assumption of independent SIDS deaths made matters worse.

The prosecutor's fallacy

Meadow was not yet done with his statistical blundering. He was allowed to make an even graver statistical error. This mistake is so common in courtrooms that it is known as the 'prosecutor's fallacy'. The argument starts by showing that, if the suspect is innocent, seeing a particular piece of evidence is extremely unlikely. For Sally Clark this was the assertion that, if she was innocent of killing her two children, the probability of the two infant deaths occurring was as low as one in 73 million. The prosecutor then deduces, incorrectly, that an alternative explanation – the guilt of the suspect – is extremely likely indeed. The argument neglects to take into account any possible alternative explanations, in which the suspect is innocent: the death of Sally's children by natural causes, for example. It also neglects the possibility that the explanation that the prosecution is proposing, in which the suspect is guilty (double infant murder in Sally's case), may be just as unlikely, if not more so, than the innocent explanation.

To explain the problems with the prosecutor's fallacy, let's imagine we are investigating a crime. The one piece of evidence we have is the partial registration number of a car, which must have been that of the perpetrator, seen driving away from the scene. Let's pretend, for the purposes of this example, that all number plates are composed of seven numbers, each chosen from the digits 0 to 9. There are ten possibilities for each of the seven numbers, which means there are $10 \times 10 \times 10 \times 10 \times 10 \times 10 \times 10$ or 10,000,000 (ten million) such number plates out there on the road. The eye-witness who reported the number plate remembered the first five numbers, but couldn't read the last two. Once these first five digits are specified, we are choosing from a much smaller pool of cars with just two unknown numbers. There are ten choices for each of these two unknown digits, meaning that there are only 100 (10×10) possible plates out there with the prescribed first five digits.

A suspect is found whose number plate matches the five digits remembered by the witness. If the suspect is innocent, then there are only 99 other cars out there, out of the ten million cars on the road, whose plates match the first five digits. Therefore, the probability that the witness observed such a number plate if the suspect is innocent is 99/10,000,000 (just under a hundred in ten million), less than one in one hundred thousand (1/100,000). This tiny probability of seeing the evidence if the suspect is innocent seems to overwhelmingly indicate the suspect's guilt. However, to assume so is to commit the prosecutor's fallacy.

The probability of seeing the evidence if the suspect is innocent is not the same as the probability of the suspect *being* innocent, once that piece of evidence has been observed. Recall that 99 of the 100 cars that match the witness's description do not belong to the suspect. The suspect is just 1 in 100 people who drive such a car. The probability of the suspect's guilt given their number plate, therefore, is just 1/100 – exceedingly unlikely. Of course, other evidence tying the suspect to the area of the crime or eliminating the other cars from being in the area would increase the probability of the suspect's guilt. However, based on the single piece of evidence, the overwhelmingly likely conclusion should be that the suspect is innocent.

The prosecutor's fallacy is only truly effective when the chance of the innocent explanation is extremely small, otherwise it is too easy to see through the fallacious argument. For example, imagine investigating a London burglary. Blood belonging to the perpetrator, discovered at the crime scene, is found to be of the same type as that of a suspect, but no other evidence is retrieved. Only 10% of the population share this blood type. So the probability of finding blood of this type at the scene if the accused is innocent (i.e. someone else in the population committed the crime) is 10%. The prosecutor's fallacy would be to infer that the probability that the

suspect is innocent in light of the blood evidence is also only 10%
– that the probability of guilt is 90%. Clearly, in a city like London,
with a population of ten million, there will be roughly a million
other people (10% of the total population) who match the blood
type found at the crime scene. This makes the probability of the
suspect's guilt, based on the blood evidence alone, literally one in
a million. Even though finding that blood type was relatively rare
(one in ten), because there are so many people who share that blood
type, that piece of evidence on its own says very little about the guilt
or innocence of a suspect whose blood matches.

•

In the example above the fallacy was relatively obvious. To assume
that the probability of innocence could be as low as one in ten
based purely on the blood type of an individual in a large population seems absurd. However, in Sally Clark's case, the figures were
small enough to make the fallacy quite opaque to a jury untrained
in statistics. It is doubtful whether Meadow himself even knew he
was committing the fallacy when he stated: '. . . the chance of the
children dying naturally in these circumstances is very, very long
odds indeed: one in 73 million.'

The inference that an untrained jury might draw from this statement runs something along these lines: 'The deaths of two infants
from natural causes is extremely rare; so for a family in which two
infants have died, the odds that these deaths were *unnatural* is
correspondingly extremely high.'

Meadow reinforced this misconception by putting the figure
of one in 73 million into a more colourful, but spurious, context.
He claimed that the chance of two SIDS deaths in one family was

equivalent to betting on the 80-to-1 outsider in the Grand National four years running and winning each time. This made the chance of an innocent explanation for the two infant deaths seem very unlikely, with the jury left to assume that the alternative, that Sally had murdered her two children, was therefore extremely likely indeed.

Two children dying from SIDS *is* an extremely unlikely event. This in itself, though, does not provide us with useful information about how likely it is that Sally murdered her children. In fact, the alternative explanation proposed by the prosecution is even more unlikely. Double infant murder has been calculated to be between 10 and 100 times less frequent than double SIDS death.[66] Assuming the latter figure suggests a probability of guilt of just 1 in 100, even before any other mitigating evidence has even been considered. However, the probability of double murder was never presented to the jury for comparison. Sally's defence never critically questioned Meadow's statistic, leaving it to stand unchallenged.

•

After deliberating for two days, on the 9th of November 1999, the jury found Sally guilty, convicting her by a 10–2 majority. One of the jurors was reported to have confided to a friend that Meadow's statistic was the piece of evidence that swayed the majority of the jury in their verdicts. Sally was sentenced to life imprisonment. As her sentence was read out, Sally looked over to her husband, Steve, who mouthed the words, 'I love you.' He was her biggest supporter and he would not stop fighting for her throughout the time she lived in prison, in what she called her 'living hell'. As she was taken down she looked back across the gallery and silently spoke his words back to him. 'I love you.'

The media wasted no time sticking the knife in. The *Daily Mail's* headline ran 'Driven by drink and despair, the solicitor who killed her babies' while the *Daily Telegraph* proposed that 'Baby killer was "lonely drunk"'. Sally's reputation in the outside world was in tatters, but as a convicted child-killer and the daughter of a policeman, life on the inside looked like it would be hell for her.

Sally spent a year in prison, locked away from her husband and her young son. Her only comforts were the letters she received from strangers who believed she was innocent. On the outside, Steve also maintained his belief in Sally's innocence. After nearly 12 months of hard work, they were finally ready to face the judges again in the court of appeal. The primary basis for the appeal was the inaccuracy of the statistics. Expert statisticians explained to the judges the ecological fallacy in pigeonholing the Clarks into a low SIDS risk category, the erroneous assumption of independence Meadow had made by squaring the probability of a single SIDS death, and the prosecutor's fallacy to which the jury had been subjected.

The presiding judges appeared to understand all of these arguments and took them into consideration. In their summary, they accepted that Meadow's statistics were not accurate but argued that they were only ever supposed to be ball-park figures. The judges believed the prosecutor's fallacy to be so obvious that it should have been objected to by Sally's defence lawyer. The judges took the fact that no objections were raised as evidence that the fallacy was abundantly clear to everyone:

> It is stating the obvious to say that the statement 'In families with two infants, the chance that both will suffer true SIDS deaths is 1 in 73 million' is not the same as saying 'If in a

family there have been two infant deaths, then the chance that they were both unexplained deaths with no suspicious circumstances is 1 in 73 million'. You do not need the label 'the prosecutor's fallacy' for that to be clear.

The judges concluded that the role of the statistical evidence in the trial was so insignificant that there was no possibility that the jury was misled. Far from being the rock that the jury were offered to cling to in the storm of contradictory medical evidence, the statistics, it seemed, were no more than a drop in the ocean, 'a sideshow' as the judges dismissed them. Sally's original conviction was upheld and that same evening she was taken back to prison.

•

Sally Clark's is by no means the only trial in which probability has been misused and misunderstood. In 1990, Andrew Deen was maligned by the same prosecutor's fallacy in his trial for the rape of three women in his native Manchester in the north-west of England. He was convicted and sentenced to 16 years in prison. At the trial, the prosecuting barrister, Howard Bentham, presented DNA evidence from semen found on one of the victims. Bentham claimed that DNA from a sample of Deen's blood matched the DNA of the semen sample. When he asked the expert witness 'So the likelihood of this being any other man but Andrew Deen is one in three million?' the expert answered 'Yes'. The expert went on to add 'My conclusion is that the semen originated from Andrew Deen.' Even the judge in his summing up claimed the one in three-million figure 'approximated pretty well to certainty'.

In fact, the one in three-million figure should be interpreted as the probability that a randomly chosen individual from the population at large had a DNA profile matching that of the semen found at the crime scene. Given the roughly 30 million males in the UK at the time, we might expect ten of them to match the profile, dramatically increasing the probability of Deen's innocence from an improbable one in three million to a very likely nine in ten. Of course not all of the 30 million males in the UK are possible suspects. However, even if we restrict ourselves to the seven million people who live within an hour's drive of Manchester city centre, we would still expect at least one other male to match the profile, leaving the odds of Deen's innocence at evens: one-to-one. The prosecutor's fallacy had led the jury to believe Deen was millions of times more likely to be guilty than the evidence really suggested.

In fact, even the DNA evidence that linked Deen to the crimes was not as convincing as the expert witness had claimed. It was shown on appeal that Deen's DNA and the DNA found at the crime scene were nowhere near as similar as originally thought. Instead of one in three million, the probability of a random match with someone other than Deen was actually around 1 in 2500, making Deen's innocence dramatically more likely. Combined with the more than three million males in the vicinity of the crime scene, giving over 1000 other potentially-matching individuals, the probability of Deen's guilt based on DNA dropped to less than one in a thousand. The revised interpretation of the forensic evidence, and the recognition that both the original judge and expert witness had committed the prosecutor's fallacy, caused Deen's conviction to be quashed.

Knox and the knife

Another case in which the understanding of DNA evidence and probability combined to play a pivotal role was that of murdered British student Meredith Kercher. In 2007, Kercher was stabbed to death in the apartment she shared with fellow exchange student Amanda Knox, in Perugia, Italy. Two years later, in 2009, Knox and her Italian former boyfriend Raffaele Sollecito were convicted unanimously of Kercher's murder. One piece of evidence presented by the prosecution, which proved crucial to Knox and Sollecito's convictions, was a knife, of a size and shape consistent with some of the wounds inflicted on Kercher. The knife was found in Sollecito's kitchen and had Knox's DNA on the handle, tying both Sollecito and Knox to the weapon. There was also a second sample of DNA on the blade of the knife, although it was small, just a few cells in fact. When a DNA profile was produced from the cells it was found to be a positive match for the victim, Kercher.

In 2011, Knox and Sollecito appealed against their lengthy prison sentences. Lawyers for the defence primarily focussed on discrediting the only evidence that physically tied Knox and Sollecito to the murder – the DNA evidence on the knife.

Almost everyone (the only exceptions being identical siblings) has a genome – the read-out of all the As, Ts, Cs and Gs that characterise the long strings of DNA in each of their cells – that is unique to them. If each of the roughly three billion base pairs in a person's genome were read off and stored, the resulting sequence would constitute a genuinely unique identifier for that person. A DNA profile used in court, or stored in a DNA database, however, is not an exact read-out of an individual's whole genome. When DNA profiles were first conceived of, generating a full-genome profile would have comprised too much data, taken too long and been too expensive. The comparison of two profiles would also have taken an unfeasible amount of time.

Instead, a DNA profile is produced by analysing 13 specific regions, known as loci, of a person's DNA. Because we inherit one chromosome in each pair from each of our parents, there are two regions of DNA associated with each locus. Each of these regions comprises, in part, a 'short tandem repeat': a small segment of DNA repeated many times. The number of repeats at a given locus varies significantly between individuals. Indeed, these 13 loci are specifically chosen because of the diversity in the number of repeated segments, meaning there are astronomically large numbers of different combinations of repeat numbers across the 13 loci. The DNA profile, then, is just the list of the numbers of repeats at each locus, which can be read off from a chart known as an electropherogram. The electropherogram represents the raw DNA sequence and looks a little like the read-out of a seismometer (used for measuring earthquakes) with low-level background noise interspersed with peaks at particular positions, corresponding to each of the loci used in the profile. The electropherogram for the sample extracted from the blade of the knife is displayed in Figure 12.

Creating an individual electropherogram can be likened to recording the results of two rolls of each of 13 separate dice, in order, each die having up to 18 faces. Two randomly chosen individuals' profiles being a perfect match can be thought of as rolling exactly the same sequence twice. Under ideal conditions, the probability of two randomly selected, unrelated individuals' profiles matching

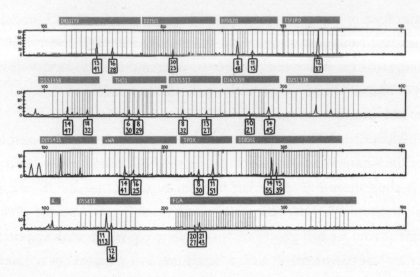

Figure 12: The electropherogram from the DNA sample on the blade of the knife, allegedly belonging to Meredith Kercher. The peaks corresponding to the 13 loci used in a standard DNA profile are labelled. In some instances only one peak is visible, indicating that the sample's owner inherited the same number of repeats for that locus from both of their parents. The top number in each box gives the number of DNA segment repeats. The bottom number gives the strength of the signal, corresponding to the height of the peak. Most peaks' signal strengths are lower than the desired minimum of 50.

is lower than one in a hundred trillion – effectively making the DNA profile a unique identifier. If the positions of the peaks on two electropherograms match exactly then it can reasonably be assumed they come from the same person.

Sometimes DNA matches can be ambiguous because the age or quality of the DNA sample can lead to only partial profiles being recovered – the signal is not attainable at every locus. Partial profiles cannot give such a definitive match. It is also possible, especially for small samples, that the signal that comes through on the electropherogram is drowned out by the background noise produced during the analysis. For that reason there are accepted standards on the strength of the signals in the DNA profile. This was the only hope left to Knox's defence.

At the time of the original trial, Dr Patrizia Stefanoni, the chief technical director of the forensic genetic investigation section of the Rome Police, decided that, based on its tiny size, instead of dividing the knife DNA sample into two, she needed to use all of the available DNA to create a sufficiently strong profile. (This was strictly against good practice: with two samples a low-strength or ambiguous profile can be re-validated using the second sample. But because of her action, no reserve sample would be available for a potential second test.) As noted in the original trial, the electropherogram possessed clear peaks in all the right places and was an incredibly close match for Kercher's profile. However, as can be read off from the numbered boxes in Figure 12, most of the peak heights in the profile fell well short of even the most relaxed of standards. Because Stefanoni had not followed the proper procedures for generating the profile, the defence team at the appeal were able to discredit the knife DNA evidence.

In response, the prosecution asked for a small number of cells, missed by the original swab but discovered by independent forensic experts, to be retested to confirm the results of the first test. Presiding judge Claudio Hellmann rejected the prosecution's requests to have the minuscule sample retested.

On the 3rd of October 2011, the mixed jury of judges and laypeople retired to consider their verdict. They returned, later than expected, to a courtroom whose atmosphere had been slowly building and that was now extremely tense and thick with pent-up emotion. Despite all the evidence that had been reviewed, no one really knew which way the pendulum would swing. As the verdicts were read out, Knox collapsed in her chair and began to cry – mixed tears of joy and relief. The jury cleared her of Kercher's murder. In his summarising 'motivations' document, in justifying his refusal to allow the second knife DNA sample to be tested, Judge Hellmann stated that 'The sum of two results, both unreliable due to not having been obtained by a correct scientific procedure, cannot give a reliable result.' But Leila Schneps and Coralie Colmez, authors of the 2013 book, *Math on Trial: How Numbers Get Used and Abused in the Courtroom*, suggest that Judge Hellmann was wrong; sometimes two unreliable tests *are* better than one.[67]

To understand their argument, imagine, rather than testing DNA for a match, we have a dice to roll. We would like to determine whether the dice is fair, in which case a six should come up one sixth of the time, or if it is weighted, in which case we are told a six should appear 50% of the time. Because we don't want to presuppose anything about the situation, assume that, before we carry out our tests, each of these scenarios is equally likely.

We begin by doing a test where we roll the dice 60 times. If the dice is unbiased, then we would expect to see a six come up ten times on average. If it were weighted, we would expect to see 30 sixes on average. If we find 30 sixes or more in the trial then we will be extremely confident that the dice is weighted because it would be exceedingly unlikely for this to happen by chance using an unweighted dice. Similarly, if we find 10 or fewer sixes then we will be confident that the dice is fair. If the number of sixes falls somewhere in between 10 and 30 then we will be able to calculate the probability of the dice being weighted by comparing the

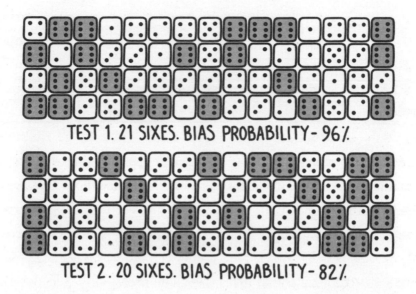

TEST 1. 21 SIXES. BIAS PROBABILITY - 96%.

TEST 2. 20 SIXES. BIAS PROBABILITY - 82%.

Figure 13: Two separate tests of the dice. We roll 21 sixes from 60 rolls on the first test, but only 20 sixes on the second test. The second test seems to undermine the first.

probability of that number of sixes coming up with the weighted dice to the probability of seeing that number of sixes with the unweighted dice.

In the experiment, we record the rolls seen in the top half of Figure 13 – comprising 21 sixes in total. The probability of seeing this many sixes with an unweighted dice is low, at just 0.000297. With a weighted dice the probability of seeing 21 sixes is still quite small, but at 0.00693 it is over 20 times more likely than if the dice is unweighted. The 21 sixes are far more likely to have come from a weighted dice than an unweighted one. We can find the combined probability of seeing 21 sixes under both of these scenarios by adding their respective probabilities together to give 0.00722. The proportion of this probability that the weighted dice accounts for

	PROBABILITY GIVEN UNWEIGHTED DICE	PROBABILITY GIVEN WEIGHTED	TOTAL PROBABILITY FOR BOTH SCENARIOS	PROBABILITY DICE IS WEIGHTED
TEST 1	0.000297	0.00693	0.00722	96%
TEST 2	0.000780	0.00364	0.00442	82%
COMBINED	0.00000155	0.000168	0.000170	99%

Table 9: The probability of seeing the different numbers of sixes in each of the tests if the dice is fair (columns 1) or weighted towards sixes (column 2). The total probability with the two dice (column 3) and the probability that the dice is weighted (column 4).

is 0.00693/0.00722, which gives 0.96. The probability that the dice is weighted, therefore, is 96%. Fairly convincing, but perhaps not convincing enough to convict a murderer.

In order to be sure, we undertake a second test in which we roll the dice another 60 times. This time, if we count up the sixes in the bottom half of Figure 13 we find only 20. As summarised in Table 9, the probability of seeing this number of sixes if the dice is unweighted is 0.000780 and if the dice is weighted is 0.00364 – only around five times more likely now. Although not hugely different to the results of the first test, applying the same calculation gives a slightly less convincing 82% chance of the dice being weighted. It seems that undertaking this second test has cast doubt on the results of the first. The second test certainly doesn't seem to confirm our conviction that the dice is weighted beyond reasonable doubt.

However, when we combine the results, as in Figure 14, we find we have rolled the dice 120 times. For an unbiased dice we would expect six to come up 20 times on average. Instead it came up 41 times. The probability of seeing 41 sixes from 120 rolls is

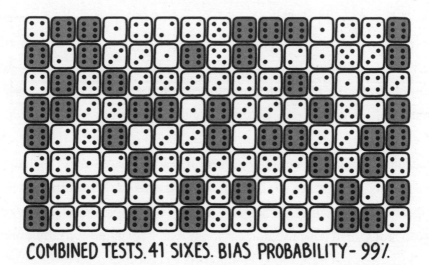

COMBINED TESTS. 41 SIXES. BIAS PROBABILITY – 99%.

Figure 14: When the tests are combined we find 41 sixes from a total of 120 rolls. This suggests the overwhelming likelihood that the dice is weighted.

just 0.00000155 if the dice is unweighted, whereas, if the dice is weighted, 41 sixes are over 100 times more likely, at 0.000168. The probability that the dice is weighted given these 41 sixes is, therefore, over 99%.

Surprisingly, combining the two less persuasive investigations makes for a much more convincing result than either of the individual tests alone. A similar technique is often employed in the scientific practice of systematic reviews. Systematic reviews in medicine, for example, consider multiple clinical trials, which may not, in themselves, be conclusive about the effectiveness of a given treatment due to the small number of trial participants. When the results of multiple independent trials are combined, however, it is often possible to draw statistically significant conclusions about the effectiveness or otherwise of the intervention. Perhaps the best-known use of systematic reviews is in the analysis of alternative medicines (the seeming 'positive effects' of which we will explain, in the next chapter, as being primarily caused by mathematical artefacts), for which little funding is available to conduct large-scale clinical trials. By combining multiple seemingly inconclusive tests, systematic reviews have debunked alternative therapies from the use of cranberries to treat urinary tract infections[68] to the use of vitamin C for preventing the common cold.[69]

Similarly, Schneps and Colmez argue that the combination of a pair of potentially inconclusive DNA tests might have provided stronger evidence for the link between Kercher's DNA and the knife in Sollecito's kitchen. Judge Hellmann's decision deprived the court of the opportunity to hear such evidence and consequently denied the world the opportunity to see the effect that evidence might have had on the outcome of the trial.

Blinded by maths

The astronomically small probabilities generated by a complete DNA sample appear to be very convincing statistics, but we should remember not to be blindsided by these very big or small numbers in the courtroom. We should always be careful to consider the circumstances that led to their generation and remember that, without proper interpretation, simply quoting an extremely small number out of context does not in itself demonstrate the guilt or innocence of a suspect.

The 'one in 73 million' figure concocted by Meadow in Sally Clark's case is one such cautionary example. Due to a combination of erroneous independence assumptions (assuming that having had one baby die due to SIDS doesn't alter the probability of having a second baby die from SIDS) and ecological fallacies (pigeonholing the Clarks into a lower-risk category based on some cherry-picked demographic details), the figure was far smaller than it should have been. To compound these problems, the figure was also presented in such a way that any reasonable jury might have assumed that one in 73 million was the probability of Sally's innocence, rather than it being the probability of one possible alternative explanation for the infants' deaths – the prosecutor's fallacy. Indeed, a jury did find her guilty, based, in no small part, on Meadow's presentation of his incorrect figure.

If we should be careful of being too readily convinced of someone's guilt by extremely small probabilities, we should not simply accept the refutation of these figures as evidence of someone's innocence. Andrew Deen was defamed by the prosecutor's fallacy, making the probability of his guilt, based on the DNA evidence alone, seem much more likely than it was. At his appeal, Deen's defence argued for a revised figure of one in 2500 for the probability of a DNA match, making him one of thousands of potentially matching suspects in the vicinity of the crime. One might argue that this makes the DNA evidence effectively worthless. This argument, however, is equally erroneous and is known as the 'defence attorney's fallacy'. The DNA evidence should not be discarded, but assimilated alongside the other evidence implicating or exonerating a suspect. Deen's conviction was ruled unsafe, in part because of the misleading effect the prosecutor's fallacy had on the jury. At his retrial, however, Deen pleaded guilty and was convicted of rape.

In the same way, Schneps and Colmez present a compelling mathematical argument that by denying a retest of the DNA, Judge

Hellmann, presiding at Amanda Knox's appeal, may have helped secure her freedom. In 2013, Knox's appeal acquittal was quashed and a judge ordered that the later-found DNA sample be retested. The DNA was found to belong, in fact, to Knox herself and not to Kercher. At her final appeal, in 2015, judges heard evidence that the collection and examination of the knife had been severely compromised. Errors ranged from the knife being collected and stored in an unsealed envelope and then an unsterilised cardboard box, to police officers not wearing the correct protective clothing and even one officer who had been at Kercher's apartment before handling the knife later in the day. Contamination in the lab was also difficult to rule out, with at least 20 of Kercher's samples having previously been tested in the lab before the alleged murder weapon was examined. If the original DNA found on the knife had genuinely arrived there by contamination, no number of retests would change the fact that the DNA belonged to Kercher or answer how it came to be on the knife. Indeed, had more of the contaminated DNA sample been available, the suggestion of retesting could have erroneously provided more misplaced confidence in Knox's guilt.

By getting hung up on the details of a neat mathematical argument, a complex computation or a memorable figure, we often neglect to ask the most pertinent question: is the calculation in question even relevant?

•

In Sally Clark's case, the statistic that most influenced the jurors was Meadow's estimate of two SIDS deaths occurring in a single family. Upon closer analysis we might question why this figure was even calculated at all. No one at the trial was arguing that both of the

Clarks' children died of SIDS. At the time of Christopher's death, the pathologist conducting the post-mortem certified that Christopher had died of a lower-respiratory-tract infection. This is not the same as a diagnosis of SIDS, which is effectively the diagnosis reached when everything else has been ruled out. The defence claimed natural causes, the prosecution claimed murder, but no-one suggested that SIDS should be considered the cause of both infants' deaths. Meadow's figure purporting to describe the probability of two SIDS deaths in the same family had no business being anywhere near the courtroom. Yet this figure seems to have been a significant factor in the minds of the jurors when they came to their conclusion that Sally was guilty of murdering her two infant sons.

At her second appeal, in January 2003, Sally's lawyers presented new evidence that had been discovered since her original conviction. The evidence from the post-mortem of Sally's second son, Harry, clearly indicated the presence of the bacteria *Staphylococcus aureus* in his cerebrospinal fluid. According to experts, this infection was extremely likely to have led to some form of bacterial meningitis that caused Harry's death. Although the new microbiology evidence was enough for Sally's conviction to be considered unsafe, the appeal judges stated that the misuse of statistics in the original trial would have been enough to uphold the appeal.

On the 29th of January 2003, Sally was set free. She returned to Steve and her third son, who was by then four years old. In a statement given upon her release she talked of being allowed finally to grieve for the death of her babies, the importance of returning to her husband, of being a mother to her little boy and of becoming a 'proper family again'. Despite her overwhelming delight at being reunited with her family, even this reprieve was not enough to make up for the years she had spent wrongfully incarcerated, blamed for killing two of the people she loved most. In March 2007, she was found dead at her home from alcohol poisoning, never having fully recovered from the effects of her wrongful conviction.

•

We can extend the lessons learned in the courtroom to other areas of our lives. As we will see in the next chapter, it is prudent to bring a questioning attitude to the figures that catch our eyes in newspaper headlines, the claims pushed at us in adverts or the Chinese whispers we catch from our friends and colleagues. In fact, in any area where someone has a vested interest in manipulating the figures, which is almost everywhere figures occur, we should treat claims sceptically and ask for more explanation. Anyone who is confident in the veracity of their figures will be happy to provide it. Maths and statistics can be difficult to understand, even for trained mathematicians; this is why we have experts in those areas. If needs be, ask for help from a professional, a Poincaré, who can lend an expert opinion. Any mathematician worth their salt will be happy to oblige. Even more importantly, before a mathematical smokescreen is whipped up in front of us, we must vigorously question whether mathematics is even an appropriate tool to use.

There is no doubt that, with the increasing prevalence of quantifiable forms of evidence, mathematical arguments have an irreplaceable role in some parts of our modern justice system, but in the wrong hands mathematics can act as a tool that impedes justice, costing innocent people their livelihoods and, in extreme cases, their lives as well.

4

DON'T BELIEVE THE TRUTH

Debunking Media Statistics

Don't Believe the Truth was the title of Mancunian rock band Oasis' sixth album. Growing up in Manchester in the 1990s, I was mad for the band. I'd been to see them at a number of venues around the city and just after this album came out, in 2005, I went to see them play again, at the City of Manchester Stadium, home of my beloved Manchester City Football Club. As a teenager I used to go to gigs fairly regularly in a range of venues around Manchester: the Apollo, Night and Day, the Roadhouse, and for bigger bands, the Manchester Arena.

By 2017, Oasis had long since broken up and I hadn't lived in Manchester or been to a gig there for over ten years, but many of the music venues I used to frequent were still going strong. On the 22nd of May that year, at around half past ten in the evening, an Ariana Grande concert had just finished at the Manchester Arena. The audience, many of whom were teenagers or younger, were streaming into the foyer to meet their waiting parents. Motionless, in the middle of the crowd, stood 23-year-old Salman Abedi. On his shoulders he carried a rucksack filled with the nuts and bolts that enveloped his home-made bomb. At 22:31 he detonated it. He killed 22 innocent victims and injured hundreds more. It was the worst terrorist attack on UK soil since the 2005 bombings, which targeted the London transport network, killing 52 members of the public.

At the time of the attack I was not in Manchester; I was not even in the country. I was visiting Mexico City for work. Because of the six-hour time delay, I watched reports of the attack come in one after the other as my afternoon progressed and most of the UK slept, as yet unaware. Despite being over 5000 miles away, having traversed that very foyer after a gig myself, I felt somehow more connected to the incident: more shocked and more appalled than by many recent terrorist incidents. Over the next few days I read as much as I could about the attack and how the people of my hometown had reacted. One article, in the *Daily Star*, particularly caught my attention. It was entitled '"Dates matter to jihadis" Manchester Arena attack on Lee Rigby anniversary'. The article's author highlighted a tweet by Sebastian Gorka, then deputy assistant to US president Donald Trump, which read: 'Manchester explosion happens on 4th anniversary of the public murder of Fusilier Lee Rigby. Dates matter to jihadi terrorists.'

Gorka had noticed a coincidence between the dates of two Islamist terrorist attacks. The first, on the 22nd of May 2013, a butcherous knife attack on a British Army soldier by two Christianity-to-Islam converts of Nigerian descent. The second, on the 22nd of May 2017, a suicide bombing on a non-political target by a life-long Muslim of Libyan ancestry. Gorka was suggesting, in his tweet, that the Manchester Arena attack was meticulously planned so that it could be carried out on the anniversary of Lee Rigby's murder. Obviously, if this were true it would lend credibility to the idea that Islamist terrorists are a well-organised and coherent group, capable of striking at will on a chosen date. This, however, is somewhat at odds with the 'lone wolf' picture that has since been painted of Abedi.

Organisation and order in a terrorist group seems to make them more threatening than if attacks are seen to be carried out at random with no central control or coherence. The purpose of Gorka's tweet seemed to be to heighten the fear of Islamist terrorism, perhaps with the aim of supporting President Trump's embattled executive order: 'Protecting the Nation from Foreign Terrorist Entry into the United States', banning many Muslims from travelling to the US, which was, at the time, undergoing several legal challenges. But is this really the case? I wondered. Should we really believe Gorka's assertion, given credence by the *Daily Star*? Isn't this the sort of unfounded, hyped-up rhetoric that serves the terrorists' purposes perfectly? How likely is it, I wondered, that two terrorist incidents would happen on the same day of the year, purely by chance?

•

We are constantly bombarded by numbers and figures: in what we read, what we watch and what we hear. Large-cohort studies into the ways in which 21st-century lifestyles impact on our health, for example, are accruing faster than ever. Simultaneously, there is a concomitant increase in the numerical skills required to interpret their findings. In many cases there is no hidden agenda, the statistics are just difficult to interpret. There are, however, many reasons why it might benefit one party or another to put a spin on a particular finding.

In the era of fake news, it's difficult to know who to trust. Believe it or not, most mainstream media outlets base most of their stories on facts. Truthfulness and accuracy are near the top (if not the top) of the list on almost all codes of journalistic ethics and integrity.[70] In addition to moral obligations to tell the truth, libel cases can be extremely damaging and expensive, so there is a financial incentive to get the facts right.

Where many media organisations differ in their reporting of the facts, however, is in the slant they place on a story. When, for example, President Trump's tax reform bill (its title 'Tax Cuts and Jobs Act' not without spin itself) was passed in December 2017, journalist Ed Henry on Fox reported it as 'a major victory' and a 'desperately needed win for the president'. Lawrence O'Donnell on MSNBC, however, referred to Republican senators who voted for the bill as 'The ugliest display of pigs at the trough that I have ever seen in Congress'. Jack Tapper on CNN led with the question 'Has there ever been a piece of major legislation passed by the Congress with less [popular] support?'

You will have had no trouble detecting the different verbal bias applied to the above story and making inferences about the political agendas promoted by the three news outlets. Partisanship is easy to detect through people's words. Numbers, on the other hand, are easier to spin surreptitiously. Statistics can be cherry-picked to present a particular angle on a story. Other figures are ignored altogether and misrepresentative stories created purely by omission. Sometimes it is the studies themselves that are unreliable. Small, unrepresentative or biased samples, in conjunction with leading questions and selective reporting can all make for unreliable statistics. More subtle still are the statistics used out of context so that we have no way to judge whether, for example, a 300% increase in cases of a disease represents an increase from one patient to four or from 500,000 patients to two million. Context is important. It's not that these different interpretations of numbers are lies – each one is a small piece of the true story on which someone has shone a light from their preferred direction – it's just that they are not the whole truth. We are left to try to piece together the true story behind the hyperbole.

In this chapter, we will analyse and demystify the tricks, traps and transformations in newspaper headlines, advertising hoardings and political soundbites. We will expose similar mathematical manipulations employed in places where we might expect better: in patient advice publications and even in scientific articles. We will provide simple ways of recognising when we are not being told the whole story and tools to help us reverse the spin applied to a statistic, as we try to find out when we should believe the 'truth'.

The birthday problem

The most subtle, and often effective, mathematical misdirections are the ones in which it doesn't even appear that there is a number at play. By stating 'Dates matter to jihadi terrorists', Gorka was implicitly asking us to evaluate the probability that two terrorist incidents could have occurred on the same day by chance, making it clear that he didn't think that this was very likely. The way to find out the real answer lies in a mathematical thought-experiment known as the 'birthday problem'.

The birthday problem asks 'How many people do you need to have at a gathering before the probability of at least two people sharing a birthday rises above 50%?' Typically, when first posed this problem, people plump for a number like 180, which is roughly half the number of days in the year. This is because we tend to put ourselves in the room and think about the probability of someone else matching our own birthday. 180 is, in fact, way, way too many. Making the reasonable assumption that birthdays are roughly evenly distributed throughout the year, the answer is just 23 people. This is because we are not concerned about the particular day on which the birthday falls, just that there is a match.

To gain an insight into why the number required is so low, we can start by considering the number of pairs of people in the room – pairs of birthdays falling on the same day, after all, being what the question is about. To calculate the number of pairs with 23 people in the room, imagine lining everyone up and asking them to shake hands with each other. The first person shakes hands with the other 22 people, the second with the 21 people she has not shaken hands with yet, the third with 20 others and so on. Finally, the penultimate person shakes hands with the last and we are left to add up 22 + 21 + 20 + . . . + 1. This is arduous, though easy enough for 23 people, but borders on tedious when the number of people in the room gets above 50. Sums like this – of consecutive whole numbers, starting from one – are called triangular numbers, since we can lay out these numbers of objects in orderly triangular arrays, as we have done in Figure 15. Fortunately, there is a neat formula for triangular numbers. For a general number of people, N, in the room, the number of handshakes is given by $N \times (N-1)/2$. For 23 people, this gives $23 \times 22/2$ or 253 pairs. Perhaps it's not surprising then that the

Figure 15: The number of handshakes between 23 people. The first person shakes hands with 22 others, the second with 21, and so until the penultimate person is left with only one person to shake hands with. The total number of handshakes between 23 people is the sum of the first 22 whole numbers. The formula for the triangular numbers tells us that there are 253 pairs of people with just 23 people in the room.

probability of at least one pair of people with the same birthday rises to above 50% with so many pairs of people in the room.

In order to put a figure on this probability, it's easier to think first about the probability that no one shares a birthday. This is exactly the same mathematical technique that we employed in Chapter 2 when calculating how many mammograms a woman might attend before the probability of her receiving a false positive diagnosis increased to above 1/2. With any single pair of people we can easily find the probability they don't share a birthday. The first person can have their birthday on any of the 365 days of the year, and the second on any of the remaining 364 days. So the probability of a single pair of people not sharing a birthday is pretty close to certain, at 364/365 (or 99.73%). However, since there are 253 pairs of people, and we are interested in finding out the probability that none of them share a birthday, we need each of the other 252 pairs to also have distinct birthdays. If all of these pairings were independent of each other, the probability that none of the 253 pairs of people share a birthday would be given by the probability that one pair don't share a birthday, 364/365, multiplied by itself 252 more times or $(364/365)^{253}$. Although 364/365 is quite close to 1, when multiplied by itself hundreds of times, the probability of no matching pairs of birthdays turns out to be 0.4995, just under 1/2. Since no one sharing a birthday or two-or-more people sharing a birthday are the only two possibilities (in mathematical parlance they are 'collectively exhaustive') the probabilities of the two events must add up to one. Therefore, the probability that two or more people share a birthday is 0.5005, just more than 1/2.

In reality, not all the pairs of birthdays will be independent of each other. If person A shares a birthday with person B and person B shares a birthday with person C, then we know something about the pairing A–C: they must also share a birthday – they are no longer independent. If they were independent they would have only a 1/365 chance of sharing a birthday. The exact calculation of the probability of a match, taking these dependencies into account, is only slightly more involved than when we assumed independence in the previous paragraph. It relies on thinking about adding people to the room one at a time. For two people we established that the probability of not sharing a birthday is 364/365. Adding a third person to the mix, they can have their birthday on any of the remaining 363 days of the year, if they are not to share a birthday with either of the others. So the probability of three people not sharing a birthday is (364/365) × (363/365). A fourth person can only have their birthday on one of the 362 remaining days, so the probability of all four not sharing a birthday drops slightly to (364/365) × (363/365) × (362/365). The pattern continues until we add the 23rd person to the party. They can have their birthday on any of the remaining 343 days of the year. The probability that 23 people don't share a birthday is given by the protracted multiplication

$$\frac{364}{365} \times \frac{363}{365} \times \frac{362}{365} \times \dots \times \frac{343}{365}.$$

This expression tells us that the exact probability that two people in a group of 23 don't share a birthday (accounting for possible dependencies) is 0.4927, just under 1/2. Using the idea of

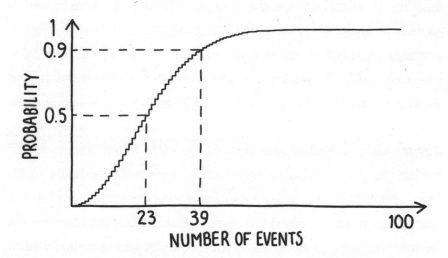

Figure 16: The probability of two or more events falling on the same day increases with the number of events. When there are 23 events, the probability of a match is just above 1/2. When there are 39 independent events the probability of at least two of them falling on the same day rises to nearly 0.9.

collective exhaustion (that either no shared birthdays or at least one shared birthday are the only two options) again, the only other possibility – that at least two people do share a birthday – has a probability just over 1/2, at 0.5073. By the time there are 70 people in the group, there are 2415 pairs of people. The exact calculation tells us that the probability of a match is overwhelmingly likely, at 0.999. Figure 16 shows how the probability of two events falling on the same day of the year changes as the number of independent events we consider increases from one to 100.

I used the surprising results of the birthday problem to impress my literary agent when we met up for the first time to discuss writing this book. I bet him the next round of drinks that I would be able to find two people, in the relatively quiet pub, who shared a birthday. After a quick scan of the room, he readily took me on and indeed offered to buy the next *two* rounds if I could find such a pair, so unlikely did he think the prospect of a match. Twenty minutes and a lot of baffled looks and superficial explanations later ('It's OK,' a slightly-worse-for-wear version of myself told the people I accosted, 'I'm a mathematician') I had found my pair of birthday-sharers and the drinks were on Chris. It probably wasn't quite fair of me; I'd already done my own head count when I went to the bar for the previous round and counted about 40 customers. With that number of punters, there was only ever an 11% chance that I would lose the bet. I should have been staking the next *two* rounds against Chris's one, not the other way around. More than just a facile maths hustle to exploit unsuspecting victims in the pub, though, the high probability of a match for surprisingly low numbers of events has some far deeper implications. In particular, it can help us test Gorka's implication of jihadis' ability to strike at will.

Over the five-year period between April 2013 and April 2018, there were at least 39 terrorist attacks against Western (European Union, North American or Australian) nations committed by Islamist terrorists. At first glance it seems unlikely that two of these would fall on the same day if they simply occurred at random throughout the year. However, because there are 741 possible pairs of events, the probability of two falling on the same day is very likely indeed, at around 88%, as captured in Figure 16. With this high probability we should be extremely surprised if two of these attacks did *not* fall on the same day. Of course, this says nothing about the likelihood of future terrorist attacks, but it seems Gorka was giving more credit to the organisational skills of Islamist terrorists than they deserve.

•

The same 'birthday problem' reasoning tells us that we have to be careful how we interpret the DNA evidence that is so instrumental in many modern criminal trials (as exemplified in the previous chapter). In 2001, while searching through Arizona's state DNA database of 65,493 samples, a scientist discovered a partial match between two unrelated profiles. Nine out of thirteen loci matched between the samples. To put that in perspective, for two given unrelated individuals we would only expect a match of this calibre roughly once in every 31 million profiles sampled. This shock finding prompted a search for more possible matches. By the time all the profiles in the database were compared, 122 pairs of profiles matching at nine or more loci had been found.

Based on this study[71] and now doubting the uniqueness of DNA identifiers, lawyers across the US argued for similar comparisons to be made in other DNA databases, including the national DNA database containing 11 million samples. If 122 matches had turned up in a database as small as 65,000 people, could DNA really be relied upon to uniquely identify suspects in a country with a population of 300 million?[72] Were the probabilities associated with DNA profiles incorrect and therefore risking the safety of DNA-based convictions across the nation? Some lawyers believed so and even submitted the Arizona findings as evidence in order to cast doubt on the reliability of the DNA evidence in their defendants' trials.

In fact, using the formula for the triangular numbers, we can work out that comparing each of the 65,493 samples in the Arizona database with every other gives a total of over two billion unique pairs of samples. With a probability of one match per 31 million pairs of unrelated profiles we should expect 68 partial (i.e. matching at nine loci) matches. The difference between the expected 68 matches and the 122 that were found might easily be explained by the profiles of close relatives in the database. These profiles are significantly more likely to throw up a partial match than those of unrelated individuals. Rather than shaking our confidence in DNA evidence, in the light of the insight gained from triangular numbers, the database findings agree nicely with the mathematics.

Authority figures

In the original *Daily Star* article highlighting the coincidence in the dates of the murder of Fusilier Lee Rigby and the Manchester Arena attack, the probability we needed to evaluate in order to assess Gorka's claim was hidden out of sight. This is in direct contrast to the way in which most advertisers use figures. If flattering enough figures can be found, then they are generally presented prominently. Advertisers know that numbers are widely perceived as being indisputable pieces of hard fact. Adding a figure to an advert can be extremely persuasive and lend power to the promoter's argument. The apparent objectivity of statistics seems to say 'Don't just trust what we're saying, trust this piece of indisputable evidence.'

Between 2009 and 2013, L'Oréal advertised and sold the Lancôme Génifique line of 'anti-aging' products. Alongside the usual advertising pseudo-science ('Youth is in your genes. Reactivate it', 'Now, boost genes' activity and stimulate the production of "youth proteins"') was a bar graph purporting to show that 85% of consumers had 'perfectly luminous' skin, 82% had 'astonishingly even' skin, 91% had 'cushiony soft' skin, and 82% found skin's 'overall appearance improved' after just seven days. The hopelessly nebulous description of the improvements aside, these figures sound extremely impressive, a ringing endorsement of the product.

Burrow down a little further into the study behind the figures, however, and we find a quite different story. Women who took part in the study were asked to apply Génifique twice a day and consider how they felt about statements including: 'Skin appears more radiant/luminous'; 'Skin tone/complexion appears more even'; and 'Skin feels softer'. They were then asked to rate their agreement with the statements on a nine-point scale whose extremes ranged from one 'disagree completely' to nine 'agree completely'. The subjects were not asked to rate the degree of radiance, softness or evenness of their skin; just how much they agreed or disagreed that there was an improvement at all. They were certainly not asked to provide adverbs like 'perfectly' or 'astonishingly'.

The results of the survey showed that, although 82% of women did agree (giving a score between six and nine on the nine-point scale) that their skin appeared more even after seven days, fewer than 30% 'agree[d] completely'. Similarly, although 85% agreed that their skin looked more radiant/luminous, only 35.5% agreed completely. L'Oréal had been massaging the results of their own survey to make them appear more impressive than they were.

Perhaps of more concern was the size of the study. With just 34 participants, it's hard to be sure that the results are reliable, because of an effect known as 'small sample fluctuations'. Small sample sizes will typically show greater deviations from the true population mean than large samples. To illustrate this, imagine I have a fair coin – one that comes down heads 50% of the time and tails 50% of the time. For some reason, I want to convince people that the coin is biased in favour of tails. Let's say they will be convinced if I can show them that the coin comes up tails at least 75% of the time. How do my chances of winning them over change as the sample size – the number of flips of the coin – increases?

I might try to get away with just flipping the coin once. If it comes up tails I'm happy; one tail in one flip exceeds the 75% threshold. This happens on half of the occasions that I flip the coin once. A single flip gives me the best chance of convincing someone that the coin is biased, but they would be right to object that they need to see more data to be convinced, and to ask me to flip the coin again. With two flips I need to achieve two tails to convince people that the coin is biased. One tail and one head won't cut it as the number of tails is only 50% in this case. As we can see in Figure 17, two tails is just one of the four equally likely outcomes from two flips of an unbiased coin, so I win over only a quarter of

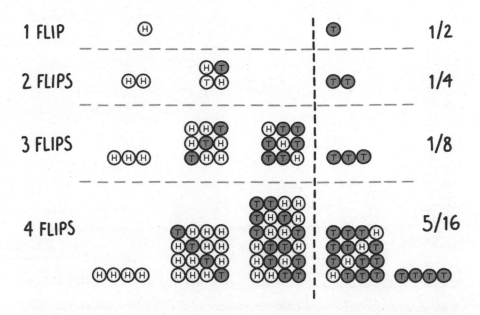

Figure 17: The possible combinations of heads and tails that can arise with different numbers of coin flips, for up to four flips. The dividing line separates outcomes for which the proportion of tails that occur is at least 75% from those where it is less.

the people I try to convince. The probability of seeing at least 75% tails diminishes rapidly as the size of the sample is increased, as seen in Figure 18. By the time I'm asked to increase the sample size to 100 flips my chances of convincing someone that the coin is biased drop to 0.00000009.

As the sample size increases the variation around the mean (in this case the mean would be 50% tails) decreases: it becomes harder and harder to convince someone of something that isn't true. That's why, with only 34 participants in the study, we should be sceptical about the reliability of the results presented in L'Oréal's advert.

Typically, adverts with small sample sizes report their findings

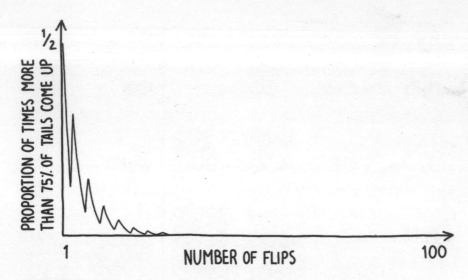

Figure 18: The chances of convincing someone that a genuinely fair coin is biased in favour of heads diminishes rapidly as the number of flips increases.

in percentages (82% had astonishingly even skin) rather than ratios (28 of 34 had astonishingly even skin) to hide the embarrassingly small sample sizes. The tell-tale sign of the small sample size, however, is if, as in the Génifique advert, you find two percentages the same (82% also found the skin's overall appearance improved). There are relatively few options to choose from in a small sample size if you want to convince your audience that your product is good, but not too good (figures between 95% and 100%, for example, might look suspicious). With a larger sample size it's far less likely that exactly the same number of people will give positive answers to two different questions.

In 2014, the Federal Trade Commission (FTC) wrote to L'Oréal charging them with deceptive advertising over the Génifique range.[73] The FTC claimed that the numbers in the adverts' charts were 'false or misleading' and not proven by scientific studies. In response, L'Oréal agreed to stop 'making claims about these products that misrepresent the results of any test or study'.

As well as small sample bias, it's possible that the Génifique study also suffered from sampling biases like 'voluntary response bias' or 'selection bias'. If L'Oréal recruited participants for the study by placing an advert on their website, for example, then they would likely recruit women who were already susceptible to the perceived benefits of the product and likely to give it a good review (voluntary response bias). Alternatively, they might have hand-picked the women themselves specifically because they had given good reviews on L'Oréal products in the past (selection bias).

There are other, even more dubious ways in which favourable figures for a study, poll or political soundbite could be arrived at. If the first study of 34 participants doesn't come back with a favourable result, then why not do another one? Sooner or later the large variation will give you the impressive responses you need. Alternatively, why not just run a larger trial and cherry-pick the participants who give you the best responses. This is known as data manipulation or, less technically, 'fudging the data'. A common example of this phenomenon is reporting bias. Scientists who investigate pseudo-scientific phenomena such as alternative medicine or extrasensory perception (psychic abilities) often bemoan what they perceive as reporting bias amongst investigators sympathetic to the cause. Unscrupulous researchers present only the 'positive results' (participants who report a benefit from a treatment, for example, or runs of a 'psychic' choosing the correct colour of the next card in a shuffled deck), while the majority of 'negative results' are discarded, making findings seems more favourable than they actually are. When two or more types of biases combine, they can give wildly different results to those expected in an unbiased sample, as the editors of the *Literary Digest* magazine discovered.

Hard to Digest

In the lead-up to the 1936 election to determine the president of the United States, the editors of the well-respected monthly magazine, *Literary Digest*, took it upon themselves to conduct a poll to predict the winner. The candidates were the incumbent president, Franklin D. Roosevelt, and his Republican challenger, Alf Landon. The *Digest* had a proud history of correctly predicting the next president, going all the way back to 1916. Four years earlier, in 1932, they had predicted Roosevelt's victory margin to within a percentage point.[74] In 1936, their poll was to be as ambitious and expensive as any poll ever previously undertaken. The *Digest* created a list of around ten million names (roughly one quarter of the voting population) based on records of automobile registration and names in telephone directories. In August, they sent out straw polls to everyone they had identified, and trumpeted in the magazine[75] '. . . if past experience is a criterion, the country will know to within a fraction of 1 percent the actual popular vote of forty million [voters]'.

By the 31st of October, over 2.4 million votes had been returned and counted. The *Digest* were ready to announce their result. 'Landon, 1,293,669; Roosevelt, 972,897' was the headline of the article.[76] The *Digest* had Landon to win by a wide margin: 55% to 41% of the popular vote (with a third candidate, William Lemke, polling 4%) and taking 370 of the 531 electoral votes. Only four days later, when the genuine election results were announced, the editors of the *Digest* were shocked to find that Roosevelt had been re-elected to the White House. It wasn't a narrow victory either, it was a landslide. Roosevelt won 60.8% of the popular vote – the largest share since 1820. He took 523 of the electoral votes to Landon's eight. The *Digest* was out by nearly 20 percentage points in their prediction of the popular vote. We might expect a large variation in the results with a small sample size, but the *Literary Digest* had polled 2.4 million people. With such a large sample, how did they get it so wrong?

The answer is sampling bias. The first issue the poll suffered from was selection bias. In 1936 America was still in the grip of the Great Depression. Those people who owned cars and telephones were likely amongst the better-off in society. Consequently, the list that the *Digest* compiled was skewed towards the upper- and middle-class voters among whom, leaning further towards the right in their political opinions, support for Roosevelt was less strong. Many of the poorer people, who comprised Roosevelt's core support, were completely unaccounted for in the *Digest* poll.

Perhaps even more significant for the results of the poll was a phenomenon known as non-response bias. Of the ten million names on the original list, fewer than a quarter responded. The survey was no longer sampling the population it originally intended to. Even if the initial demographic selected had been representative of the population as a whole (which it wasn't), the people who responded to the survey tended to have different political attitudes to the people who didn't. The typically wealthier and better-educated people who did answer tended to be supporters of Landon, rather than Roosevelt. Together, these two sampling biases combined to give embarrassingly incorrect results, leaving the *Digest* a laughing stock.

That same year, using just 4500 participants,[77] *Fortune* magazine was able to predict the margin of Roosevelt's victory to within 1%. The *Literary Digest* did not come well out of the comparison. The dent that their previously impeccable credibility sustained on the back of the results is cited as a significant factor in hastening the magazine's demise less than two years later.[78]

Do the math

While political pollsters have found they need to be increasingly statistically aware in order to get accurate results, politicians themselves are finding that they can get away with more statistical manipulation, misappropriation and malpractice than ever before. When running for the Republican nomination in November 2015 Donald Trump tweeted an image with the following statistics:

Blacks killed by whites — 2%

Blacks killed by police — 1%

Whites killed by police — 3%

Whites killed by whites — 16%

Whites killed by blacks — 81%

Blacks killed by blacks — 97%

The source of the figures was attributed to the 'Crime Statistics Bureau – San Francisco'. As it turned out, the Crime Statistics Bureau didn't exist and the statistics were wildly off the mark. Some real comparative statistics for 2015 (with raw figures given in Table 10, page 153) from the FBI read as follows:

Blacks killed by whites — 9%

Whites killed by whites — 81%

Whites killed by blacks — 16%

Blacks killed by blacks — 89%

Evidently, Trump's tweet massively overplayed the number of homicides committed by black people, effectively transposing the figures for 'white-on-white' and 'black-on-white' murders. Nevertheless it was retweeted over 7000 times and 'Liked' over 9000 times. This is a classic example of confirmation bias. People retweeted the false message because it came from a source they respected and chimed with their pre-existing prejudices. They didn't stop to check if it were true or not, and neither did Trump. When he was questioned by journalist Bill O'Reilly on Fox News about his motivations for disseminating the image, after first claiming, in typical hyperbolic style, 'I'm probably the least racist person on earth,' he followed up with '. . . am I gonna check every statistic?'

•

Trump's tweet in 2015 came at the height of the national debate about police brutality, particularly brutality targeting black victims. Such cases, including most notably the deaths of the unarmed black teenagers, Trayvon Martin and Michael Brown, were the catalyst for the formation and rapid expansion of the 'Black Lives Matter' movement. Between 2014 and 2016 Black Lives Matter held mass protests, including marches and sit-ins, across the United States. By September 2016 the movement had chapters in the UK whose protests drew the ire of right-leaning journalist Rod Liddle. A mathematically oriented blog-post[79] drew my attention to Liddle's comments in the British tabloid newspaper, the *Sun*, on the foundation of the original Black Lives Matter movement in the US:

> It was set up to protest about American cops shooting black suspects instead of just arresting them.
>
> There's no doubt US cops are a bit trigger-happy. And maybe especially when a black suspect hoves into view.
>
> There's also no doubt whatsoever that the greatest danger to black people in the USA is ... er ... other black people.
>
> Black-on-black murders average more than 4000 each year. The number of black men killed by US cops — rightly or wrongly — is little more than 100 each year.
>
> Go on, do the math.

So here it is – I did the math.

Let's consider the statistics for 2015, the latest full calendar year for which Liddle could have accessed data. According to FBI statistics[80] summarised in Table 10, 3167 white people and 2664 black people were murdered in 2015. Of the homicides in which the victim was white, 2574 (81.3%) were perpetrated by white offenders and 500 (15.8%) by black offenders. Of the homicides in which the victim was black, 229 (8.6%) were perpetrated by white offenders and 2380 (89.3%) by black offenders. So Liddle's claim of 4000 'black-on-black' homicides a year is a significant over-exaggeration; by approximately 70%. Given that black people comprised just 12.6% of the US population in 2015 and white people 73.6%, it is alarming that black individuals make up 45.6% of the homicide victims.[81]

Although a far more prominently debated issue, figures for the number of people killed by police are harder to obtain. The fatal

RACE/ETHNICITY OF VICTIM	TOTAL	RACE/ETHNICITY OF OFFENDER	
		WHITE	BLACK
WHITE	3167	2574 (81.3%)	500 (15.8%)
BLACK	2664	229 (8.6%)	2380 (89.3%)

Table 10: Homicide statistics for 2015 broken down by the race/ethnicity of the victim and perpetrator. The disparities between the total column and the sum of white and black victim columns are due to cases in which the ethnicity of the victim is different or unknown.

shooting of black teenager Michael Brown by white police officer Darren Wilson, and the subsequent protests that took hold of Ferguson, Missouri, marked a tipping point for the Black Lives Matter movement. The protests also served to shine a spotlight onto the FBI's 'annual count of homicides by police'. The FBI was found to be recording fewer than half of all killings by police in the US.[82] In response, in 2014, the *Guardian* began its campaign, 'The Counted', to compile more accurate figures. So successful was the project that in October 2015, the then FBI-director, James Comey, called it 'embarrassing and ridiculous' that the *Guardian* had better data on civilian deaths at the hands of the police than the FBI.[83]

The *Guardian*'s figures show that, of the 1146 people 'rightly or wrongly' (to echo Liddle) killed by police in 2015, 307 (26.8%) were black and 584 (51.0%) were white (while the remaining victims were of different or undetermined ethnicities). Again, Liddle's figure is way off the mark. His suggestion of 100 police killings of black citizens a year is less than a third of the true value.

If Liddle was trying to answer the question 'If a black person in the US is killed, is it more likely to have been another black person or a police officer who killed them?', then using the correct figures it's clear that black people kill almost eight times (2380 vs 307) more black people than police do. But this question seems disingenuous. Would you believe that dogs were more murderous than bears if I told you that, in 2017, dogs killed 40 US citizens while bears killed just two? Of course not. Dogs are not inherently more dangerous than bears, there are just more of them in the US. To put it another way, if you were to be left alone in a room with a bear or a dog, which would you prefer it to be? I don't know about you, but I'd probably opt for the dog.

For the same reason, given that there are over 40.2 million black US citizens and only 635,781 full-time 'law enforcement officers' (those who carry a firearm and a badge)[84] it's not surprising that more killings are perpetrated by black people than law enforcement officers. A more appropriate question for Liddle to ask might have been 'If a black US citizen comes across someone while out walking alone, who should they be more scared will kill them: another black person, or a law enforcement officer?'

To find out the answer we need to compare the 'per capita' rates of black-victim killings perpetrated by black people and by police officers. We find the per capita rates, as presented in Table 11, by dividing the total number of black victims killed by a particular group (black people or police officers) by the size of the group. Black people were responsible for 2380 killings of other black people in 2015, but with over 40.2 million black US citizens, the per capita rate is relatively small – around 1 in 17,000. Police officers were 'rightly or wrongly' responsible for killing 307 black people in 2015. With 635,781 police officers, this amounts to a per capita killing rate that is just below 1 killing per 2000 police officers – over

KILLER	NUMBER OF BLACK-VICTIM KILLINGS	SIZE OF POPULATION	PER CAPITA KILLING RATE
BLACK CITIZENS	2380	40,241.818	1/16908
LAW ENFORCEMENT OFFICERS	307	635,781	1/2071

Table 11: The number of killings in which a black citizen was the victim, stratified by whether the killing was committed by another black person or a law enforcement officer. The sizes of the two populations are also presented and used to work out the per capita killing rate.

eight times higher than the rate for black US citizens. It seems that a black person walking down the street should be more alarmed to see a police officer approaching them than another black person.

Of course we have not accounted for the fact that encounters with the police are often confrontational and that US police are typically armed. It's perhaps not surprising that those authorised to wield lethal force do so more frequently than the general population at large. By exactly the same mathematics, we can show that white people should also be more scared of law enforcement officers (per capita white killing rate of 1 per 1000 officers) than other white people (per capita white killing rate of 1 per 90,000 white people), despite more white people killing other white people than police officers killing white people. The fact that police officers have twice as high a per capita rate of killing white people than black people is because there are more white people in the country. Again, it is perhaps unsettling that the rate is only twice as high, given that there are almost six times as many white people as black people in the US.

So while Liddle's statistics are incorrect, perhaps more importantly, by asking 'Who kills the most?' rather than 'Who is killed the most?', his *Sun* article diverts attention away from a statistic that is at the heart of the Black Lives Matter movement: that the 12.6% of the population who are black account for 26.8% of police killings while the 73.6% who are white account for just 51.0%. Are there hidden links (the type of 'lurking variables' we met in the previous chapter, explaining the supposed benefits of smoking to low birth-weight babies) that might explain this disparity? Almost certainly there are. For example, poorer people are more likely to commit crime and, in the US, black people are more likely to be poor. Whether or not these factors account for the huge over-representation of black people in the police homicide figures remains to be seen.

Careless pork costs lives

Liddle's piece wasn't the first or the last time that the *Sun* newspaper was to be embroiled in statistical controversy. In 2009, under the admittedly inspired headline 'Careless pork costs lives', the *Sun* reported just one of many hundreds of results from a 500-page study by the World Cancer Research Fund, on the effect of consuming 50 grams of processed meat per day.[85] The tabloid newspaper shocked readers with the 'fact' that eating a bacon sandwich every day would increase the risk of colorectal cancer by 20%.

But the figure was sensationalised. When stated in terms of 'absolute risks' – the proportion of people exposed or unexposed to a particular risk factor (e.g. eating bacon sandwiches or not eating

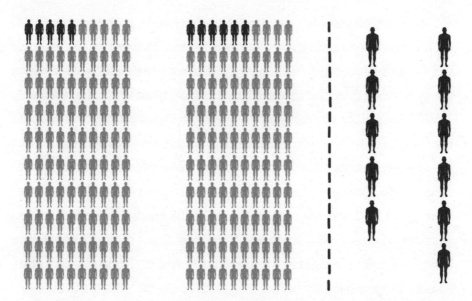

Figure 19: A comparison of the absolute figures (5 in 100 vs 6 in 100) (*left*) makes it appear that the increase in risk of eating 50g of processed meat per day is small. By focussing on the relatively small number of people who have the disease (*right*) the relative risk increase of 20% (1 in 5) seems very large.

bacon sandwiches) who are expected to develop a given outcome (e.g. cancer) in each case – it turns out that 50 grams of processed meat per day increases the absolute lifetime risk of developing colorectal cancer from 5% to 6%. On the left of Figure 19 we consider the fates of two groups of 100 individuals. Of 100 people who eat a bacon sandwich every day, only one more of them will develop colorectal cancer than in a group of 100 people who abstain.

Instead of using the more objective absolute risk, the *Sun* chose to focus on the 'relative risk' – the risk of a particular outcome (e.g. developing cancer) for people exposed to a given risk factor (e.g. eating bacon sandwiches) as a proportion of the risk for the general population. If the relative risk is above one then an exposed individual is more likely to develop the disease when compared to someone without the exposure. If it is below one then the risk is decreased. On the right-hand side of Figure 19, by neglecting the people who are not affected by the disease, the increase in relative risk (6/5 or, equivalently, 1.2) seems much more dramatic. Although it is true that the relative risk for those eating 50 grams of processed meat per day represents an increase of 20%, the absolute risk increased by only 1%. But an increased risk of 1% doesn't sell many papers. Sure enough, the article's headline was sufficiently inflammatory to spark the 'Save our Bacon' media firestorm. Over the next few days, the furore over the figure saw scientists branded 'health Nazis' who had declared a 'war on bacon'.

·

Another attention-grabbing media trick is deliberately to change what we consider and accept to be the 'normal' population. The most honest way to report relative risk is to present an increased or decreased risk for a particular subgroup in comparison to the background risk in the general population. Sometimes the disease risk levels of the largest subpopulation are used as the baseline and any deviations in risk are reported relative to that population. When the disease is rare, the disease-free cohort makes up almost the whole population anyway, so the disease-free subpopulation risk is a good approximation to the general population risk. Consider reporting the risks of breast cancer for women with BRCA1 or BRCA2 genetic mutations, for example. It seems sensible to talk about the increase in absolute risk for the 0.2% of women with these mutations relative to the general population rather than a decreased risk for the 99.8% of women without these mutations. Unfortunately, this type of candid transparent reporting doesn't always make for the best headlines, so we see many of the big news outlets manipulating the way statistics are presented again and again in order sell stories.

In a 2009 story headlined 'Nine in ten people carry gene which increases chance of high blood pressure', the *Daily Telegraph* ran a story containing this sentence: 'Scientists found that one gene variant carried by almost 90 per cent of the population increased the chance of developing high blood pressure by 18 per cent.' The figures actually reported in the journal *Nature Genetics* were that 10% of individuals possessed genetic variants that left them at a 15% lower risk than the 90% of the population with a different variant.[86] The 18% figure didn't appear in the journal article. Although technically correct, the *Telegraph*'s story had mischievously changed the reference population to the smaller one – the 10% of people with the lowered risk. Since a 15% decrease from the reference value of one takes us to 0.85, the article's author recognised that the increase required to get back up to one is approximately 18% of this smaller figure. With one mathematical sleight of hand, the *Telegraph* not only increased the size of the relative risk but managed to turn what might have been a good news story for 10% of the population into a bad news story for 90% of the population. The *Telegraph* was by no means alone in its manipulation of the figures – many other papers spun the story in the same dubious way in order to entice their audiences.

Often, after reading a sensationalist article, you will find you haven't been presented with the absolute risks – usually two small figures (certainly never more than 100%), one for those subject to the focal condition or intervention and the other for the remaining population. On other occasions it may be claimed that the risk is increased or decreased for more than half of the population. In these cases you should think carefully about whether you choose to accept the article's argument. If you want to find out the truth behind the headlines, consider following it up in a publication that gives you access to the absolute stats, or even in the original scientific article itself, which can increasingly be accessed online and for free.

A different frame of mind

Newspapers are by no means alone in their dubious reporting of risks and probabilities. In the medical arena, when communicating the risks of treatments or when reporting the efficacy of drugs and their side effects, there are more statistical games that the presenter can play in order to advance their agenda. One simple way to suggest a particular interpretation involves framing figures in a positive or a negative light. In one study from 2010, participants were presented with a number of numerical statements about medical procedures and asked to rank the risk they associated with each on a scale from one (not risky at all) to four (very risky).[87] In amongst them were the statements 'Mr Roe needs surgery: 9 in 1000 people die from this surgery.' and 'Mr Smythe needs surgery: 991 in 1000 people survive this surgery.' Take a second to think about whose shoes you'd rather be in: Mr Roe's or Mr Smythe's?

Of course, the two statements frame the same statistic in two contrasting ways – the first using mortality rates and the second survival rates. For participants with low numeracy skills, the positively framed statement about survival was perceived to be almost a whole point less risky on the four-point scale. Even people with greater numeracy skills perceived the risk attached to the negatively framed statement as higher.

When examining the results of medical trials it is not uncommon to see positive outcomes reported in relative terms, in order to maximise their perceived benefit, whilst side effects are reported in absolute terms in an attempt to minimise the appearance of their risk. This practice is known as 'mismatched framing' and was found to occur in roughly a third of the articles reporting the harms and benefits of medical treatments in three of the world's leading medical journals.[88]

Perhaps even more worryingly, this phenomenon also appears prevalently in patient advice literature. In the late 1990s, the US National Cancer Institute (NCI) created the 'Breast Cancer Risk Tool' in order to educate and inform the public of their risks of the disease. Along with many other studies, the online app reported the results of a recent clinical trial, on over 13,000 women at increased risk of breast cancer, in which the benefits and potential side effects of the drug tamoxifen were assessed.[89] In the trial, the women were divided into two, roughly even, groups (known as the two 'arms' of the trial). Women in the first arm were administered tamoxifen, whilst women in the second arm were given a placebo treatment as a control.

At the end of the five-year study, in order to assess the effect of the drug, the numbers of women with invasive breast cancer in each group were compared, as were the numbers of women with other types of cancer. In its Breast Cancer Risk Tool the NCI reported the relative risk reduction: 'Women [taking tamoxifen] had about 49% fewer diagnoses of invasive breast cancer.' The big figure of 49% seems quite impressive. However, when it came to quantifying the possible side effects, an absolute risk was presented: '. . . annual rate of uterine cancer in the tamoxifen arm [of the trial] was 23 per 10,000 compared to 9.1 per 10,000 in the placebo arm'. These tiny fractions seem to indicate that the risk of uterine cancer from the tamoxifen treatment doesn't change much at all. Consciously or unconsciously, while gathering data for their online risk information tool, the NCI's researchers emphasised the benefits of tamoxifen for reducing breast cancer incidence while simultaneously minimising the perception of the increased risk of uterine cancer. If these figures had been used to calculate a relative risk, in order to put the two statistics on a level playing field, it would have been reasonable to present a figure of 153% increased risk of uterine cancer to balance the 49% reduced risk of breast cancer.

Even in the abstract of the original article, the 49% figure for the reduction in breast cancer is used, but the increase in uterine cancer is presented as a relative risk ratio of 2.53. Using percentages instead of decimals to highlight perceived benefits is one of another family of tricks referred to as 'ratio bias'.[90] Our susceptibility to ratio bias has been confirmed in simple experiments in which blindfolded subjects are asked to choose a jelly bean from a tray at random.[91] Drawing a red jelly bean represented a $1 win for the picker. When given the choice between a tray with nine white jelly beans and a single red one or a tray with 91 white jelly beans and nine red ones the second tray is chosen more often, despite giving the smaller chance of picking out a winning sweet. Presumably this is because the higher number of red jelly beans on the tray leads to the perception of a higher chance of picking one irrespective of the number of the other beans. One subject commented 'I picked the ones with the more red jelly beans because it looked like there were more ways to get a winner.'

The absolute figures from the tamoxifen study showed that cases of invasive breast cancer were reduced from 261 per 10,000 without the treatment down to 133 per 10,000 with it. Ironically, had the ratio bias and mismatched framing been excluded in favour of the absolute numbers, the Breast Cancer Risk Tool's users would easily have been able to see that the total breast cancer cases prevented (128 per 10,000) hugely outweighed the uterine cancers caused by the treatment (14 per 10,000), with no manipulation of the original clinical data needed.

Regressive attitudes

It is likely that the majority of statistical misrepresentation in a medical context is done unconsciously by researchers who are unaware of some common statistical pitfalls. In clinical trials, for example, it is typical to take a group of people who are unwell, to give them a proposed treatment for their ailment, and to monitor them for improvement in order to understand the effect of the medicine. If the symptoms are alleviated, it seems natural to give credit to the treatment.

Imagine, for example, recruiting a large number of people suffering from joint pain and asking them to sit still while you sting them with live bees. (Although this sounds absurd, it is a genuine alternative therapy known as apipuncture. Apipuncture has recently grown in popularity, in part due to its promotion by Gwyneth Paltrow on her Goop lifestyle website.) Now imagine that, miraculously, some of the sufferers' joint pain goes away – on average, they start to feel better after the therapy. Can we conclude that apipuncture is actually an effective therapy for joint pain? Probably not. In reality, there is no scientific evidence to support apipuncture's efficacy for treating any disorder. Indeed, adverse reactions to bee-venom therapy are common and are known to have killed at least one patient. So how can we explain the positive results of our hypothetical trial? What causes the patients' improvement?

Conditions like joint pain fluctuate in their severity over time. It's likely that the sufferers recruited for the trial, especially for something as extreme and alternative as apipuncture, are at a particularly low ebb and are desperate for some resolution to their ailments when they sign up. If they receive treatment when their pain is at its worst, then it is highly likely that some time later they will be feeling better, irrespective of the benefits of the treatment. This phenomenon is known, ostentatiously, as 'regression to the mean'. It affects many trials in which there is an element of randomness to the results.

To understand better how regression to the mean works, consider the results of an exam. Take an extreme case in which students are asked to answer 50 'yes/no' multiple-choice questions on a subject of which they know nothing. With the students effectively guessing entirely at random, the scores from the test could range from zero all the way up to 50, but there will be very few people who get only a small number of answers right and very few that get almost none of the answers wrong. From the distribution of scores

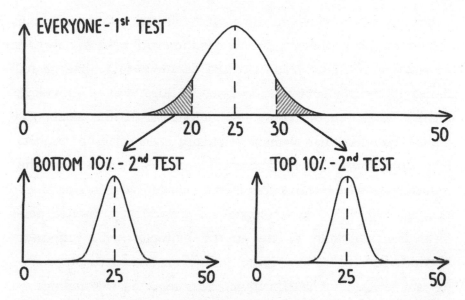

Figure 20: The spread of marks on a 50-question 'yes/no' multiple-choice questionnaire. When those with the top 10% of scores (*right shaded region*) are retested, their mean score is the same as the mean overall score. The same is true for the bottom 10% (*left shaded region*). Both high-scoring and low-scoring populations have regressed towards the mean.

given in Figure 20, it is clear that there will be more people who get middling scores nearer to the average of 25. If we analyse the students in the top 10%, their scores will be, by definition, significantly higher than the whole population average. Should we therefore expect these students to perform significantly above average when we retest them with fresh questions? Of course not. We would again expect their scores to be evenly distributed around a mean score of 25. The same would be true if we retested the bottom 10%. The individuals we picked out on the basis of their extreme scores in the first test will have, on average, reverted towards the mean in the second.

In real exams, skill and work ethic will play a significant role in determining a student's results, but there will probably also be an element of chance relating to the questions that come up on the exam and the subjects prioritised during revision. Providing there is some random component, regression to the mean will register its effect. The element of chance is especially prominent in multiple-choice exams, for which even a student without the requisite knowledge can guess at the right answer. In one study conducted in 1987, 25 test-anxious US students, who had performed unexpectedly poorly on the multiple-choice Scholastic Aptitude Test (SAT), were given the hypertension drug propranolol and retested.[92] The *New York Times* reported the findings of the study as 'A drug used to control high blood pressure has dramatically improved Scholastic Aptitude Test scores for students suffering unusually severe anxiety . . .' Students taking propranolol improved their scores, on average, by a remarkable 130 points on a scale from 400 to 1600. It appears, at first, that propranolol has a significant effect. It turns out, however, that even non-anxious students retaking the test improve their scores by around 40 points. When we consider that the students selected for the trial were chosen precisely because they performed worse than their IQ or other academic indicators had suggested they should, it wouldn't be surprising to see them significantly increase their scores, even without propranolol, as a result of their regression to the mean.

In the absence of a similar set of poorly performing students who undertook the retest without taking the drug – the so-called 'control cohort' – it is impossible to determine the effects of the intervention. Based only on the treated cohort, it is tempting to attribute the improved performance to the effects of the drug.

However, the results of the purely random multiple-choice test demonstrate that the regression of an extreme cohort towards the mean is a purely statistical phenomenon.

•

Avoiding the spurious inference of causality is of great importance in medical trials. One way to do this (as we have already seen in Chapters 2 and 3) is to conduct a randomised controlled trial in which patients are allocated to one of two groups at random. As in the tamoxifen breast cancer trial, patients in the 'treatment arm' receive the genuine treatment and patients in the 'control arm' receive a dummy or 'placebo' treatment. If both the patients and the administrators of the treatment are kept in the dark about which arm of the trial the patient is in, the trial is known as double-blind – widely considered the gold standard for clinical trials. In a double-blind randomised controlled trial any difference between improvement in the control group and improvement in the treatment group can be attributed solely to the treatment, ruling out regression to the mean.

Historically, any improvement of patients in the control arm of the trial has been termed the placebo effect – the benefit of receiving what is perceived as a treatment, even if it's just a sugar pill. However, it is becoming increasingly clear that this effect is composed of two quite different phenomena. One, perhaps smaller part is the genuine psychosomatic effect that makes patients feel better just because they believe they are being treated. This 'true placebo' effect confers some genuine change in the patient's judgement of their symptoms. The psychosomatic benefit is larger if the patient knows they are receiving the real treatment and, interestingly, even if the person administering the treatment knows, hence the reason for double-blinding.

The other, perhaps more significant, reason for the improvement of patients in the control arm is regression to the mean. This simple statistical effect confers no benefit to the patients at all. The only way to determine which is the more important of the two placebo components is to compare the effects of the dummy treatment with the effects of no treatment at all. These types of trial are often considered unethical, but enough studies have been done in the past to indicate that the majority of the so-called placebo effect is actually a result of regression to the mean – from which patients derive no benefit.[93]

Many proponents of alternative medicine argue that, even if their treatment is nothing more than a placebo effect, the benefit of the placebo can be significant and is worth having. However, if the majority of the placebo effect is caused by regression to the mean, which provides no benefit to the patient, this argument doesn't hold water. Other alternative medicine gurus argue that, rather than putting stock in 'artificial clinical trials', it is important to consider 'real world' results – or, to paraphrase, 'uncontrolled trial outcomes that focus solely on how patients' conditions change after treatment'. Unsurprisingly, these 'quacks' clutch at any argument that allows them to wilfully misinterpret the effects of regression to the mean as a genuine causative benefit of their unscientific treatments. As Pulitzer Prize winner, Upton Sinclair, put it: 'It is difficult to get a man to understand something, when his salary depends upon his not understanding it.'

•

Away from medicine, regression to the mean also has far-reaching consequences for the interpretation of cause and effect in the context of law-making. On the 16th of October 1991, 32-year-old Suzanna Gratia sat down to eat with her parents in Luby's Cafeteria in Killeen, Texas. At its peak-time lunch hour the restaurant was unusually busy, with over 150 people packed around the square tables. At 12:39, George Hennard, an unemployed merchant seaman, accelerated his blue Ford Ranger pick-up truck towards the restaurant and drove straight through the front window into the dining area. He immediately jumped out of the driver-side door and, holding his Glock 17 pistol in one hand and his Ruger P89 in the other, started to shoot.

Initially suspecting an armed robbery, Gratia and her parents dropped to the floor and upended the table as a makeshift barrier between them and the gunman. As shot after shot rang out, however, it became chillingly clear to Gratia that this man was not there to rob the restaurant; he was there to kill indiscriminately, and to kill as many people as possible.

The gunman approached within a few metres of their table and Gratia went for her purse. In it she kept concealed a .38 calibre Smith & Wesson she had been given for self-defence a number of years earlier. As she reached for the gun, however, her blood froze. She remembered that she had taken the cautious decision to leave the revolver under the passenger seat of her car so as not to fall foul of Texas' concealed-weapons law. She refers to it as 'the stupidest decision of my life'.

Gratia's father heroically decided that he would have to tackle the gunman before everyone in the restaurant was murdered. He sprang up from behind his table and rushed at Hennard. He didn't make it more than a few feet. Shot in the chest, he fell to the floor, mortally wounded. In his search for more victims, Hennard veered away from the table behind which Gratia and her mother were still concealed. At the same time another customer, Tommy Vaughan, threw himself through a window at the back of the restaurant in a desperate bid to get away. Seeing the broken window as a potential escape route, Gratia grabbed her mother, Ursula, and insisted: 'Come on, we gotta run, we gotta get outta here.' Running as hard as she could, Gratia got herself quickly through the window and outside the restaurant unscathed. She turned back to ensure her mother had followed her, but found herself alone. Instead, Ursula had crawled to where her husband's body lay prone on the floor and cradled his head as he lay dying. Slowly, methodically, but surely enough, Hennard made his way back to where she sat and shot her in the head.

Gratia's parents were just two of the 23 victims Hennard murdered that day. Twenty-seven others were wounded. At the time it was the worst mass shooting in US history. Gratia went on to give her powerful testimony across the country, in support of legalising the bearing of concealed weapons. Prior to the Luby's massacre in 1991, a total of ten states had 'shall-issue' concealed-carry laws. These laws require that, provided an applicant satisfied a set of objective criteria, they must be issued with a permit to carry a concealed weapon – with no discretion on the part of the issuer. Between 1991 and 1995, 11 more states passed similar laws and, on the 1st of September 1995, George W. Bush signed a law making Texas the 12th.

Understandably, given that gun control is such a contentious issue in the United States, there was great interest in understanding the effects that these concealed-carry laws had on violent crime.

Advocates of gun control suggested that more concealed weapons might lead to the fatal escalation of relatively minor disputes as well as an increase in the number of guns available to criminal factions. The gun-rights lobby suggested that the increased probability of a perpetrator's victim being armed might deter potential criminals, or at the very least allow citizens to attempt to end mass shootings sooner. The first studies comparing crime rates pre-introduction of the laws to those post-introduction seemed to indicate that rates of murder and violent crime had reduced in the immediate aftermath of the issuing of these concealed-carry laws.[94]

However, two factors were typically neglected in these studies. The first was the decrease in violent crime across the whole country at the time when large numbers of concealed-carry laws were introduced. Between 1990 and 2001 increases in policing, rising numbers of incarcerations and the receding crack cocaine epidemic all contributed to a fall in murders across the US from around 10 per 100,000 per year to around 6 per 100,000 per year.[95] The prevalence of homicide decreased by almost exactly the same amount in states with and without concealed-carry laws. When murder rates in concealed-carry states are considered relative to the overall US murder rate, the suggested impact of concealed-carry laws diminishes significantly. Perhaps even more important is one study's finding that once regression to the mean was accounted for, the data '. . . gives no support to the hypothesis that shall-issue laws have beneficial effects in reducing murder rates'.[96] It was common for states to issue concealed-carry laws in response to increasing levels of violent crime. That relative murder rates seemed to dip subsequent to their introduction was seemingly unrelated to the concealed-carry laws. Instead, the laws were found to be associated with increased relative murder rates *prior* to their introduction. This gave a false impression of the laws' effectiveness as crime rates naturally dipped from their abnormally high levels.

Spotting the spin

The debate on gun legislation in the US still rages today. In the wake of the Las Vegas shooting, in October 2017, in which 58 people were killed and hundreds more injured, Sebastian Gorka, recently relieved of his duties at the White House, participated in a round-table debate on gun control. Gorka, who, as we saw at the start of the chapter, is no stranger to making bold, unsubstantiated claims, waded into a discussion about restricting the sales of firearms and their accessories and took the debate in an unexpected direction:

> . . . It's not about inanimate objects. The biggest problem we have is not mass shootings, they are the anomaly. You do not make legislation out of outliers. Our big issue is black African gun crime against black Africans . . . black young men are murdering each other by the bushel.

Assuming Gorka was referring to African Americans, this sounds very much like a rehash of the bad statistics that were discredited earlier in this chapter. Gorka's repeated transgression highlights one of the situations in which you should be most on your guard against bad statistics: the serial offender. People who have shown a disregard for the accuracy of their figures once are unlikely to be more scrupulous in the future. The *Washington Post*'s Glenn Kessler, one of the pioneers of political fact-checking, regularly analyses and rates the statements of politicians on a scale from one to four 'Pinocchios', depending on the degree to which they have bent the truth. The same names crop up over and over again in his reports.

There are other more subtle signs that indicate a manipulated statistic. If presenters are confident of the veracity of their figures then they won't be afraid to give the context and the source for others to check. As with Gorka's terrorism tweet, a contextual vacuum is a red flag when it comes to believability. Lack of details on survey results, including the sample size, the questions asked, and the source of the sample – as we saw in L'Oréal's advertising campaign – is another warning sign. Mismatched framing, percentages, indexes and relative figures without the absolutes, as in the NCI's 'Breast Cancer Risk Tool', should set alarm bells ringing. The spurious inference of a causative effect from uncontrolled studies or sub-sampled data – as often seen in the conclusions drawn from trials of alternative medicine – are yet more tricks to watch out for. If an initially extreme statistic suddenly rises or falls – as with gun crime in the US – be on the look-out for regression to the mean.

More generally, when a statistic is pushed your way, ask yourself the questions 'What's the comparison?', 'What's the motivation?' and 'Is this the whole story?' Finding the answers to these three questions should take you a long way towards determining the veracity of the figures. Not being able to find the answers tells its own story.

•

There are multiple ways to be economical with the truth using mathematics. The stats proclaimed in newspapers, championed in adverts or spouted by politicians are frequently misleading, occasionally disingenuous, but rarely completely incorrect. The seeds of the truth are usually contained within their figures, but very rarely the whole fruit. Sometimes these distortions are a result of wilful misrepresentation, while on other occasions the perpetrator is genuinely unaware of the bias they are imposing or the errors in their calculations. We will explore the catastrophic consequences of such genuine mathematical mistakes in more portentous contexts in the following chapter.

In his classic book, *How to Lie with Statistics*, Darrell Huff suggests that 'despite its mathematical base, statistics is as much an art as it is a science'. Ultimately, the degree to which we believe the stats we come across should depend on how complete a picture the artist paints for us. If it is a richly detailed, realist landscape with context, a trusted source, clear expositions and chains of reasoning, then we should be confident in the veracity of the numbers. If, however, it is a dubiously inferred claim, supported by a minimalist single statistic on an otherwise empty canvas, we should think hard about whether we believe this 'truth'.

5

WRONG PLACE, WRONG TIME

The Evolution of Our Number Systems and
How They Let Us Down

Alex Rossetto and Luke Parkin were in their second year of sports science degrees at Northumbria University. In March 2015, they signed up for a trial designed to investigate the effects of caffeine on exercise. The students were to be given 0.3 grams of caffeine, and then put through their paces. Instead, because of a simple mathematical error, they found themselves in intensive care fighting for their lives.

After drinking the caffeine, dissolved in a mixture of orange juice and water, Rossetto and Parkin agreed to take part in a commonly administered exercise performance trial, known as the Wingate test. The students were asked to get on an exercise bike and pedal as hard as they could to see how the caffeine affected their anaerobic power output. But, shortly after ingesting the caffeine cocktail, before even getting near to the bikes, the students started to feel dizzy, reporting blurred vision and heart palpitations. They were immediate rushed to the Accident and Emergency department and put on dialysis machines. Over the coming days Rossetto and Parkin each lost nearly two stone in weight.

Instead of 0.3 grams of powdered caffeine, the researchers administering the test had made a mistake when calculating the dose, stirring in an astonishing 30 grams of powdered caffeine. The students had ingested the equivalent of around 300 cups of regular coffee in a few seconds. Ten grams has been known to be fatal in adults. Fortunately for Parkin and Rossetto, both were young and healthy enough to tolerate the massive overdose with few long-term effects.

The error occurred because the researchers administering the test had typed a decimal point into their mobile phones two spaces too far to the right, turning 0.30 grams into 30 grams. This is not the first time a misplaced decimal point has had dramatic effects. Other similar mistakes have had consequences ranging from funny to farcical, and even fatal.

•

In the spring of 2016, construction worker Michael Sergeant sent out an invoice for £446.60 after completing a week's work. A few days later he was surprised and excited to find £44,660 credited to his bank account after the director of the company he billed put his decimal point in the wrong place. For a few days Sergeant lived the life of a rock star. He spent thousands of pounds on a new car, drugs, drink, gambling, designer clothes, watches and jewellery before the police finally caught up with him. Sergeant was forced to pay back the remaining money and to complete community service for his small-scale opportunism.

On a much larger scale, in the run-up to the 2010 general

election in the UK, the Conservative Party published a document highlighting the disparities between rich and poor areas of the UK under the incumbent Labour government. The document claimed that 54% of girls in Britain's most deprived areas became pregnant before the age of 18, compared to 19% in Britain's most affluent areas. Rather than acting as a stinging rebuke, highlighting the supposed social inequality fostered under 13 years of Labour rule, the figures were turned on their heads when Labour commentators and politicians pointed out that in fact the figures were only 5.4% and 1.9%. Quite apart from making a glaring error with a decimal point, the unquestioning attitude with which the Conservatives suggested that over half of girls in some areas were pregnant as teenagers was seized upon as an illustration of just how out of touch the Conservatives were with their electorate. Despite the great embarrassment to the Conservatives caused by the misplaced decimal points, they went on to win the 2010 general election, their mistake proving not to be fatal.

However, it was exactly that for 85-year-old pensioner Mary Williams. On the 2nd of June 2007, community nurse, Joanne Evans, visited Mrs Williams, as a favour to a colleague. Evans was charged with delivering her diabetic patient's insulin shot for the day. She filled up her first insulin injecting pen with the required 36 'units' of insulin, but as she tried to inject it, the pen jammed. She tried again with the two other pens she had brought, but each one failed. Worried about what would happen to Mrs Williams if she did not get her insulin, the nurse returned to her car to procure a regular syringe. Although the pens were marked simply in 'units' of insulin and the syringe in millilitres, Evans knew that each 'unit' corresponded to 0.01 millilitres. She filled up the 1-millilitre syringe and injected it into Mrs Williams' arm. She repeated the process three more times to complete the dosage, not stopping to question why she had had to deliver multiple injections when a single dose had sufficed for her other patients. With the job finally completed, she left Mrs Williams and continued on her round. It was only later in the day that she realised her terrible mistake: instead of injecting 0.36 millilitres of insulin, she had given Mrs Williams 3.6 millilitres – ten times too much. She immediately called a doctor, but by that time Mrs Williams had already suffered a fatal, insulin-induced heart attack.

Although it is easy to lampoon the mistaken protagonists in these stories for their obvious errors, the prevalence of such stories demonstrates that simple mistakes can and do happen, often with serious consequences. In part, the gravity of the repercussions of these mistakes is the fault of our decimal place value system. In a number like 222 each of the 2s represents a different number: 2, 20 and 200, with each being ten times bigger than the last. It is the scaling factor, 10, which makes placing a decimal point in the wrong place so serious. Perhaps, if we used the binary system – the system on which all our modern computerised technology is based, in which each place is only a factor of two larger than the last – we could avoid these errors. Injecting twice as much insulin or even prescribing four times as much caffeine might not have had such serious ramifications.

In this chapter we explore more of the costly errors that result from the systems that currently enumerate our everyday lives. We uncover the often hidden influence of seemingly long-disused numerical systems that provide a window on our human history and shine a light on our biology. We discover the flaws that afflict them, and look at the alternative systems being advocated that are helping to avoid common mistakes. We follow the natural selection of our counting systems down dead ends and along convergent paths that parallel the very evolution of our human cultures. Just as with cultural biases, we unmask the mathematical thinking that is so deeply ingrained in our subconscious that we don't even recognise its restriction of our perspectives.

The place

The current number system we adhere to is known as the 'decimal place value system'. 'Place value', because the same digit in a different position can represent a different numerical value. 'Decimal', because the same digit in an adjacent position represents a number ten times bigger or smaller than its neighbour. The multiplication factor between places, 10, is known as the base. Why we use base 10 rather than some other base is more of an accident of our biology than any well-thought-out plan. Although some of our forebears chose a different base, the vast majority of cultures that developed numerical systems (the Armenians, the Egyptians, the Greeks, the Romans, the Indians, and the Chinese, amongst others) chose decimal. The simple reason is that when we realised the need to enumerate, in much the same way as we teach our children today, we counted using our ten fingers.

Although base 10 was the most common system adopted by our ancestors, some cultures chose other bases constructed from different aspects of our biology. The native Yuki people of California counted in base eight, using the spaces between their fingers as markers, rather than the fingers themselves. The Sumerians used base 60, pointing to the 12 finger joints of the four fingers of the right hand, using the right thumb as a pointer, and keeping track of up to five sets of 12 (60) with the five fingers of the left hand. The Oksapmin people of Papua New Guinea use a system based on the number 27: starting with the thumb on one hand (1), travelling up and down the arms, taking in the nose (14) and finishing with the little finger on the other hand (27). So, whilst the ten fingers are by no means the only body parts to inspire a number system, they are the most obvious and hence the most commonly used by our ancestors when first developing mathematics.

Once a culture had established a system of counting, it opened up the possibility of developing higher mathematics that could be put to use for practical purposes. Indeed, many of the oldest human civilisations were well versed in sophisticated maths. By the third millennium BCE, the Egyptians, for example, could add, subtract, multiply, and use simple fractions. Aptly, they knew the formula for the volume of a pyramid and there is evidence that they had stumbled across right-angled triangles with sides of length 3, 4 and 5, a so-called Pythagorean triple, long before Pythagoras. The Egyptians used the common base 10, but had no place value system. Instead they had separate hieroglyphs for different powers of ten. These pictorial representations of numbers were not written in any particular order – the Egyptians knew how much each was worth by looking at the picture. The number 1 was just a single stroke, much as we have today, 10 was a cattle yoke, 100 was a coil of rope and 1000 an ornate water lily. 10,000 was a bent finger, 100,000 was a tadpole and 1,000,000 was the god Heh – the personification of infinity or eternity. A million was about as big as it got for the ancient Egyptians. If they wanted to represent the number 1999, they would draw one water lily, nine coiled ropes, nine yokes and nine vertical dashes. Although awkward, the system works well enough for numbers under around a billion. However, if the Egyptians had been able to fathom the number of stars in the universe (estimated at a huge 1,000,000,000,000,000,000,000,000 in our decimal place value system) they would have had to draw the god Heh a billion, billion times – not really feasible.

The Romans were, in many ways, a far more advanced civilisation than the Egyptians. Famously, they popularised, on a large scale, inventions like books, concrete, roads, indoor plumbing, and the concept of public health. However, their number system was more primitive. They used a system of seven symbols: I, V, X, L, C, D and M to represent the numbers 1, 5, 10, 50, 100, 500 and 1000 respectively. Appreciating that their numeral system was somewhat cumbersome, the Romans ensured that numbers were always written from left to right, largest to smallest, so the characters could then simply be added together. MMXV for example would represent 1000 + 1000 + 10 + 5 or 2015.

Because writing long numbers was so unwieldy, an exception to the rule was introduced. If a smaller number was ever found to the left of a larger number, it meant that it was to be subtracted from the larger number. The number 2019, for example, would be written MMXIX instead of MMXVIIII, the I subtracted from the final X to give 9, saving crucial characters. If this didn't complicate matters enough, it is probably the case that the standardised rules and symbols of Roman numerals as we think of them today were not the same as those that the Romans actually used. The Etruscans, for example, may have used symbols like I, Λ, X, ↑ and X instead of I, V, X, L and C, although even these are debated. The regulated symbols and rules for writing Roman numerals described above may have evolved over many centuries in post-Roman Europe. The systems used by real Romans were likely far less uniform.

Nevertheless Roman numerals did not suffer the same annihilation as Egyptian hieroglyphs once the Roman Empire crumbled. Today, Roman numerals adorn many buildings to denote the year in which they were completed, allowing architects to lend an air of antiquity to a recently completed project. For that reason, the late 1800s were a particularly tough time for stone masons. The Boston Public Library bears the inscription MDCCCLXXXVIII – at 13 characters, the longest Roman numeral of the last millennium – for the year 1888 in which it was completed. It's not just architects who feel that writing a number in Roman numerals gives it more gravitas. Fashion style guides propose that by using Roman numerals on your watch you are indicating you are more sophisticated than your average Joe. It's true that Elizabeth II, the name of the longest serving British monarch, does read less like a film sequel than Elizabeth 2. Films and television shows also make use of Roman numerals to denote the date of their production, but for different reasons. In the early days of film, since Roman numerals were harder to read quickly, the practice prevented most people from deducing too readily that they were watching recycled material, while still satisfying the film-maker's copyright requirements.

Despite their niche longevity, Roman numerals never took over the world because their notational intricacies actively hindered the development of higher mathematics. Indeed, the Roman Empire is notable for its lack of eminent mathematicians and contributions to mathematics. As we have seen, each number in the Roman system is a potentially complex equation instructing the reader to add or subtract a series of symbols to come up with a result. This makes even the simple addition of two of these Roman numerals difficult. It wasn't possible, for example, to write down two numbers, one on top of the other, and add up the digits in each column, as we are all taught nowadays in our earliest maths lessons. Two identical symbols in the same position in two different Roman numbers didn't necessarily mean the same thing. One can't simply subtract the digits of MMXV from the digits of MMXIX from right to left (X – V gives 5, I – X gives –9, etc.) in order to find that the gap between 2019 and 2015 is four years. Crucially, the Romans lacked the concept of a positional or place value number system.

•

Long before the Romans and the Egyptians, the people of Sumer, located in what is now modern Iraq, had a far more advanced number system. The Sumerians, often referred to as the originators of civilisation, developed a wide range of technologies and tools for agricultural purposes, including irrigation, the plough and possibly even the wheel. With the burgeoning of their agricultural society, it became necessary, for bureaucratic purposes, to be able to measure plots of land accurately and to determine and record taxation. So around five thousand years ago the Sumerians invented the first place value system – a system whose most fundamental concepts would, eventually, spread across the globe. Numbers were written in a prescribed order. A symbol further to the left represented a larger value than the same symbol placed further to the right. In our modern place value system, in the number 2019, the 9 represents nine ones, the 1, one ten, the 0, no hundreds and the 2, two thousands.

Each time we move to the left, the same digit represents a number 10 times larger. Although the Sumerians chose base 60, they employed exactly the same principle. The furthest-right column represented units, the next column to the left 60s, the next 3600s and so on. In the Sumerians' sexagesimal system, the digits 2019 would represent 9 ones, one 60, no 3600s and two 216,000s, giving a total, in decimal, of 432,069. Conversely, if the Sumerians had wanted to write the year 2019 in sexagesimal it would have looked something like 33 39, where the symbol 33 represents 33 lots of 60 (1980) and the symbol 39 represents the remaining 39 units.

The development of place value is arguably the most important scientific revelation of all time. It is no coincidence that the widespread European adoption of the base 10 Hindu-Arabic place value system (the system we still use today) in the 15th century closely preceded the Scientific Revolution. Place value systems allow any number, no matter how large, to be tamed with just a few simple symbols. In the Egyptian and Roman systems, the position of a symbol had no global meaning. Instead the value was determined by the symbol itself, which meant that both cultures were hamstrung by the finite number of figures they could reasonably represent. The Sumerians, however, could represent any number they chose with their set of 60 symbols. Their sophisticated positional system allowed them to carry out advanced mathematics such as solving quadratic equations (which arise naturally in an agricultural context when apportioning land) and trigonometry.

Perhaps the primary reason the Sumerians used the sexagesimal system was because it significantly eased working with fractions and division. Sixty has lots of factors: the numbers 1, 2, 3, 4, 5, 6, 10, 12, 15, 20, 30 and 60 all divide into 60 exactly with no remainders.

Trying to divide a pound (made up of 100 pence) or a dollar or a Euro (made up of 100 cents) between six people will cause a disagreement over who gets the four remaining pennies. A Sumerian mina, made of 60 shekels, however, could be divided neatly between 2, 3, 4, 5, 6, 10, 12, 15, 20 or even 30 people without causing a fight. Using the Sumerian's base 60, it also becomes easy for us to measure and divide up a cake exactly and evenly between, for example, 12 people. One twelfth, in the sexagesimal place value system, is just five-sixtieths. They would write this neatly as 0.5 (the first place after the point representing sixtieths in sexagesimal instead of tenths in decimal), as opposed to the ugly 0.083333 ... (8 hundredths, 3 thousandths, 3 ten-thousandths etc.) in our decimal place value system. For that reason, just like a circular cake, Sumerian astronomers divided the arc of the night sky into 360 (that is, 6 × 60) degrees, helping them to make astronomical predictions.

The Ancient Greeks, building on the Sumerian tradition, divided each degree into 60 minutes (denoted ') and each minute into 60 seconds (denoted "). Indeed, the words 'minute' (think 'mine-yute' not 'min-it') just means an extremely small division (in this case of the circle) and 'second' simply refers to the second level of division of the degree. The sexagesimal place value system is still used in astronomy today and allows astronomers to capture the size of objects that are vastly different in the night sky. Because of its astronomical connections, the circular symbol for degrees, as in 360°, which is now also used for temperature, is thought originally to have symbolised the sun. Less romantically (and more mathematically), it's possible that it was natural to use a superscript ° for degrees after ' and " had been used for its subdivisions, minutes and seconds, completing the sequence O, I, II.

The time

Although we may be less familiar with the minutes and seconds used in astronomy, there is a far better known sexagesimal system which governs the rhythms of our everyday lives: time. From the moment we wake up to the moment we fall asleep, whether we know it or not, we are frequently thinking in sexagesimal. It is no coincidence that hours, the temporal divisions of our cyclic days, are also broken down into 60 minutes and each minute further into 60 seconds.

Hours themselves, however, are grouped into sets of 12. Despite primarily using base 10, it was the ancient Egyptians who divided the day into 24 segments: 12 day-hours and 12 night-hours, mimicking the number of months in the solar calendar. During daylight, time was recorded using sundials with 10 divisions. Two twilight hours were added, one at either end of the day, for periods that were not yet dark, but for which the sundial was of no use. The night was similarly divided into 12, based on the rising of particular stars in the night sky.

Because the Egyptians prescribed 12 hours in every daylight period, the length of their hours changed throughout the year as the amount of daylight changed with the seasons: longer in the summer and shorter in the winter. The Ancient Greeks realised that to make significant progress with their astronomical calculations, equal segments of time would be a necessity, so they introduced the idea of dividing the day into 24 equal-length hours. However, it wasn't until the advent of the first mechanical clocks in 14th-century Europe that this idea really caught on. By the early 1800s, reliable mechanical clocks were widespread. Most cities in Europe divided their day into two sets of 12 equal hours.

The divisions of the day into two 12-hour periods is still standard throughout much of the English speaking world. Most countries, however, use the 24-hour clock, which distinguishes, for example, eight o'clock in the morning (08:00) from 8 o'clock in the evening (20:00) with numbers that are unambiguously 12 hours apart. The US, Mexico, the UK and much of the Commonwealth (Australia, Canada, Egypt, India, etc.), however, still make use of the abbreviations a.m. (*ante meridiem*) and p.m. (*post meridiem*) or simply 'before noon' and 'after noon' to distinguish 8:00 in the morning from 8:00 in the evening. This discrepancy has occasionally been known to cause problems, especially to me.

When I was a graduate student I was offered the opportunity to visit collaborators in Princeton. I am a bit of a nervous traveller, something I have inherited from my dad. Every time I set off from the house on an international trip I hear him listing 'Money, tickets, passports' in an anxious voice in the back of my head. In much the same way, my recollection of the Pythagorean theorem for right-angled triangles: 'the square of hypotenuse is equal to the sum of the squares of the other two sides' still plays in the Irish accent of my high-school maths teacher, Mr Reid.

Unsurprisingly, on my outwards trip from Heathrow I arrived an excessive four hours early for my flight. I bumped into my more relaxed and experienced supervisor, who was catching a slightly earlier flight than mine, over two and a half hours later. My academic visit was a productive one, but my travel paranoia meant that I cut short my sightseeing trip to New York on my last day in the US to make sure I got back to Princeton in enough time to get in a good night's sleep. That evening, with bags packed, room scoured, money, tickets, and passports checked and double checked, I set my alarm for four a.m. in order to guarantee I wasn't late for my nine o'clock flight.

I duly woke up at four in the morning and boarded a train from Princeton. I arrived at Newark International Airport two and a half hours later. When I looked for my flight on the departure board, however, I couldn't find it. I scanned again and again, but the list skipped straight from the 8:59 to Saint Lucia to the 9:01 to Jacksonville. I went to the information desk and asked the lady sitting behind it about the flight. 'I'm afraid the only flight we have going to London today leaves this evening, sir.' I couldn't believe it. How had I made this mistake? I'd been so careful in my preparations, yet it seemed I'd overlooked the fact that the flight I thought I was getting didn't even exist. Then it dawned on me. I asked the assistant what time the flight was leaving this evening. 'Why, that would be the nine p.m. flight this evening, sir,' she replied.

I had mixed up a.m. and p.m., a mistake that is simply not possible to make in the 24-hour system. Fortunately, I had got it wrong in the right direction. My punishment was a 14-hour wait to board my flight, but the internet abounds with stories of people who have made this mistake in the opposite direction, completely missing their flight by 12 hours and having to fork out for a new ticket. Needless to say, this experience has done little to reduce my travel anxiety.

I found it difficult enough to arrive at the airport on time in the 21st century, but just imagine how challenging long-distance travel would have been with the confused and asynchronous time system of the early 19th century. By the 1820s, although most countries in Europe had broken their day into 24 equally sized hours, comparing the time between countries was so difficult as to be almost meaningless. Few nations had managed to enforce a single time across the whole of their dominion, let alone coordinate with their neighbours. Bristol, in the west of the UK, could be as much as 20 minutes *behind* Paris, while London found itself 6 minutes *ahead* of Nantes, in the west of France. The reason for the discrepancies was, typically, because each city used a local time based on the position of the sun in the sky. Since Oxford is one and a quarter degrees further west than London, the sun is at its peak there around five minutes later, setting local Oxford time five minutes behind London time. The 24 hours corresponding to one 360° rotation of the Earth on its axis means that every degree of longitude is worth four minutes of time. Bristol, two and a half degrees west of London, was a further five minutes behind Oxford.

In the end, it was the problems that local time posed for long-distance travel on the burgeoning railway network that led to the coordination of time across the UK. Using local time in the different cities of the UK led to timetabling disarray, and to several near misses due to confusion between drivers and signalmen. In 1840, the Great Western Railway adopted Greenwich Mean Time (GMT) across its network. The industrial cities of Liverpool and Manchester took up the cause in 1846. With the advent of telegraphy, times could be transmitted around the country almost instantaneously from the Royal Observatory in Greenwich, allowing cities to synchronise their clocks. Although the vast majority of the country quickly came on board with 'railway' time, some cities, particularly those with strong religious traditions, refused to give up their 'God-given' solar time in favour of the soulless pragmatism imposed by the railways.

Not until 1880, when Britain's parliament finally passed legislation, did the majority of solar-time stalwarts finally fall into line. Having said that, the bells of Tom Tower in Christ Church, a constituent college of the University of Oxford, still chime at five past the hour.

Italy, France, Ireland and Germany all followed quickly in adopting a uniform time across their countries, with Paris time nine minutes ahead of GMT and Dublin 25 minutes behind. But the situation was not so simple in the United States. A single time across the 58 degrees of longitude of the contiguous US mainland would not be practical for regions nearly four solar hours apart. In the winter, when the sun was setting in Maine, it would only just be lunchtime on the west coast. Clearly local time had some part to play, but in the middle of the 1800s the situation was extreme, with every major city holding to its own local time. Consequently, most of the railway companies operating across New England in 1850 had their own time, typically based on the location of their head offices or one of their more popular stations. At some busy junctions, up to five different times were respected. It is thought that the confusion caused by this lack of uniformity contributed to numerous accidents. After one particularly troubling incident in 1853, which lead to the deaths of 14 passengers, plans were put in place to standardise time on the railroads of New England. Eventually, it was proposed that the whole of the US be divided into a series of time zones, each one hour behind the next, running east to west. On the 18th of November 1883, known to many across the country as 'The Day of Two Noons', station clocks were reset across the continent. The US was divided into five time zones: Intercolonial, Eastern, Central, Mountain and Pacific.

Inspired by the US's subdivisions, in October 1884, at the

International Meridian Conference in Washington D.C., Canadian Sir Sandford Fleming proposed to carve the whole Earth up into a series of 24 time zones, creating a globally standardised clock. The globe was divided by 24 imaginary lines, known as meridians, running from the South to the North Pole. The universal day was to begin at midnight at the *prime* meridian in Greenwich. By the year 1900 almost everywhere on Earth was part of some standard time zone, but it wasn't until 1986 that all countries measured their times with reference to the prime meridian, when Nepal finally set its clocks five hours and 45 minutes ahead of Greenwich Mean Time. Having time zones that were offset from each other by regular portions of an hour saved a great deal of trouble and confusion, greatly simplifying timetabling and trade between neighbouring countries. However, the introduction of time zones didn't completely eradicate confusion. Typically it meant that when mistakes were made, time calculations were not delayed by just a few minutes, but sometimes by up to an hour – a lag with the potential to cause disaster.

•

As the leader of the 26th of July Movement, in 1959, Fidel Castro, along with his brother Raúl and comrade Che Guevara, had overthrown the US-backed Cuban dictator, Fulgencio Batista. Acting on his Marxist-Leninist philosophy, Castro quickly turned Cuba into a one-party state, nationalising industries and businesses as part of sweeping social reforms. The US government could not countenance a Soviet-sympathising communist state on its doorstep. By 1961, as the Cold War neared its zenith, the US hierarchy had come up with a plan to overthrow Castro. Fearing Soviet reprisal in

Berlin, the American president, John F. Kennedy, insisted that the US be seen to have no involvement in the coup. Consequently, a group of over 1000 Cuban dissidents, known as Brigade 2506, were trained up for the invasion of Cuba in secret camps in Guatemala. The US also stationed ten B26 bombers (the type of plane with which the US had armed Castro's predecessor) in nearby Nicaragua to assist with the invasion. On the 17th of April, the exile brigade was to stage an invasion at the Bay of Pigs, on Cuba's southern coast. The idea was to spark an uprising in the hope that huge numbers of oppressed Cuban natives would take up the exiles' cause.

The plan was in trouble even before it could be enacted. On the 7th of April, a full ten days before the anticipated attack, the *New York Times* got wind of the plans and ran a front-page story alleging that the US had been training anti-Castro dissidents. Castro, now alert to the potential of invasion, took stringent precautions, imprisoning known dissidents who might assist in the uprising and readying his military. Nevertheless, on Saturday the 15th of April, two days before the invasion, the US B26s flew to Cuba in an attempt to destroy Castro's air force. Their mission was an almost complete failure, destroying very few of Castro's operational aircraft and ditching at least one of their own, enemy-strafed B26s in the sea north of Cuba.

The botched mission had the additional effect of sending Cuban Foreign Minister Raúl Roa storming to the United Nations. At an emergency session of the General Assembly, Roa claimed, correctly, that the US had bombed Cuba. With the world's spotlight focussed on the issue, Kennedy refused to risk providing further evidence of US involvement and cancelled the air strike planned for the morning of the 17th to assist the exiles when they disembarked.

Since Brigade 2506 was composed entirely of Cuban dissidents with no obvious link to the US, Kennedy had plausible deniability with regard to their actions. On the morning of the 17th of April, he sanctioned their landing on the beaches of the Bay of Pigs. They were confronted by 20,000 well-prepared Cuban troops. Kennedy, again fearing international reprisals, refused to issue orders for Castro's army to be shelled or for planes to assist from above. By the evening of the 18th of April, the exiles' invasion was on the ropes. In a last-ditch rescue attempt, Kennedy issued an order for the Nicaragua-based B26s to strike at the Cuban military. The bombers would be protected by jets from the US aircraft carrier situated just over the horizon to the east of Cuba. The air strike was scheduled for 06:30 on the morning of the 19th.

As the allotted time approached, the jets set off to meet the B26s, only to find that they had not arrived. In fact, the B26s, working on Nicaraguan Central Time, arrived a whole hour later, at 07:30 Cuban Eastern time. With no protection from the jets that had long since given up on their mission, Castro's planes were able to shoot down two B26s bearing American insignia, proving beyond doubt the US's involvement in the attempted coup. The political ramifications of the simple time-zone mistake were huge, driving Cuba firmly into the arms of the Soviets and precipitating the Cuban Missile Crisis a year later.

By the dozenal

The failure of the Bay of Pigs invasion is attributable, in part, to the division of the day, and hence the world, into two sets of 12-hour time zones. However, the mistake would have been similarly disastrous had the Earth been broken up using a different base. With 60 or even just 10 segments, Nicaragua's time zone would still have lagged behind Cuba's by roughly the same amount of time. In fact, there are many people who think the base 12 ('duodecimal' or 'dozenal') system is far superior to our predominant decimal system. Both the Dozenal Society of Great Britain and the Dozenal Society of America argue that the dozenal system's six factors: 1, 2, 3, 4, 6 and 12, in comparison to just four for decimal (1, 2, 5 and 10), give it an advantage – and I think they have a point.

My two children have taught me, through painful experience, that it's important to divide things equally. I'm sure they would prefer that they both had just one sweet each rather than one having five if the other had six. When we stopped at a service station on the way up to their grandparents' I bought a packet of Starburst. I passed the packet into the back for the kids to share. Little did I know that there were 11 starburst in the packet and that I had passed back an odd number of sweets for the kids to share between them. The fall-out that characterised the remainder of that long journey north is the reason why I am careful now to buy only even numbers of sweets. Similarly, I have friends with three children who will only buy treats in triplicate. If you're the manufacturer of such child-focussed products, you can maximise your audience and minimise the potential for an upset by selling in sets of 12, catering for families of 1, 2, 3, 4, 6 or even 12 children. Similarly, next time you are dividing something up and it's important that everyone gets the same amount (cutting a cake at a children's party, for example) dividing into twelves will allow you increased flexibility in the number of people you can equitably accommodate. That said, if it isn't the sweets or the cake I'm sure the kids will manage to find something else to argue about.

The main basis for favouring dozenal over decimal is that, just as with the Sumerians' base 60, more fractions have a 'nice' closed representation in base 12 than in base 10. For example, in decimal, 1/3 has to be represented by the messy non-terminating decimal 0.33333 . . ., whereas in dozenal 1/3 can simply be thought of as four twelfths and written as 0.4 (the first place after the point representing twelfths in duodecimal). But why does this matter? Well, not having an exact representation of a number can make a difference when making repeated measurements. As an example, consider a metre-length of wood that you would like to divide into three equally long pieces to form the legs of a stool. Using a coarse decimal ruler, you approximate the first third as 33cm and the second third again as 33cm. But this leaves the final third at 34cm. The resulting stool, with mismatched legs, might not be so comfortable to sit on. On a dozenal ruler, one third, or equivalently four-twelfths, of a metre would be an exact marking, allowing you to divide the wood exactly into three equally sized legs.

Advocates of the dozenal system claim it would reduce the necessity for rounding-off and hence mitigate a number of common problems. To some extent they're right. Although a wonky stool is perhaps of only minor inconvenience, the simple rounding errors that result from having to truncate the representation of numbers in our current decimal system can have more serious implications.

For example, a simple rounding error in a German election in 1992 nearly led to the leader of the victorious Social Democrat Party being denied a seat in the parliament, when the Green Party's share of the vote was reported as 5.0% instead of 4.97%.[97] In a completely different context, in 1982, a newly created Vancouver Stock Exchange index plummeted continuously over a period of nearly two years, despite the market's bullish performance.[98] Every time a transaction took place the value of the index was rounded down to three decimal places, consistently knocking value off the index. With 3000 transactions per day the index lost around 20 points a month, undermining market confidence.

Imperial rule

Despite its tendency to reduce errors associated with rounding, the upheaval and consternation it would cause means that the conversion to dozenal looks unlikely to be implemented by an industrialised nation any time soon. However, many of the burgeoning industrialised nations of the past made extensive use of imperial measurement systems, which lean heavily on base 12. There are 12 inches in a foot and 12 lines to an inch. Originally, there were also 12 ounces to the imperial pound. The word ounce is derived from the same Latin word as inch, *uncia*, meaning a twelfth part. Indeed, the imperial troy system, used for measuring precious metals and gemstones, still divides the troy pound into 12 troy ounces. The old British monetary pound comprised 20 shillings each made up of 12 pence. This meant that the 240-pence pound could be divided up evenly in 20 different ways.

Although there are some perceived advantages of the imperial system (the most commonly cited is forcing children to become familiar with obscure times tables), its non-uniformity (16 ounces to the pound, 14 pounds to the stone, 11 cubits to the rod, 4 poppy-seeds to the barleycorn, etc.) has largely been abandoned in favour of the decimal metric system. Today the US remains in the company of Liberia and Myanmar as one of only three countries worldwide not to make extensive use of a metric system. Myanmar is currently trying to switch to metric. The US's lack of conformity is based largely on scepticism and traditionalist stubbornness on the part of many of its citizens. In one episode of *The Simpsons*, so often a window into contemporary American life, Grampa Simpson rants that, 'The metric system is the tool of the devil. My car gets 40 rods to the hogs head and that's the way I likes it.'

The UK began its transition to metric in 1965 and is now a nominally metric country. Nevertheless, the UK has never quite relinquished the imperial measurements that it fathered. It still clings strongly to miles, feet and inches used for height and distance, pints for milk and beer, and stones, pounds and ounces used collo-quially to measure weight. In February 2017, UK Secretary of State for Environment, Food and Rural Affairs and two-time Conservative leadership candidate, Andrea Leadsom, even suggested that British manufacturers might be allowed to sell goods using the old impe-rial system after leaving the European Union. Although appealing to a small minority of Grampa Simpsons overwhelmed by nostalgia for a by-gone 'golden era', switching back to imperial would leave the UK almost completely isolated in international trade. Much like the switch to dozenal, it would be incredibly expensive and time-consuming to implement, as well as creating mountains of needless bureaucracy. Bureaucracy and expense, combined with the reticence of the people who live in the few remaining non-metric countries, are also the primary reasons why the metric system has not yet been universally adopted. But while the US remains the last industrial nation to use imperial units[99] almost ubiquitously, it will continue to experience episodes in which it finds itself lost in translation.

•

On the 11th of December 1998, NASA launched its $125 million Mars Climate Orbiter: a robot designed to investigate the Martian climate and to act as a communications relay to the Mars Polar Lander. In contrast to the Polar Lander, the Orbiter was never designed to reach the surface of Mars. In fact, any approach closer than 85 km would cause it to break up due to atmospheric buffeting. On the 15th September 1999, after successfully negotiating its nine-month journey through the solar system, a final series of manoeuvres was initiated to bring the Orbiter to the ideal altitude of around 140 km above the surface of Mars. On the morning of the 23rd of September, the Orbiter fired its main thruster and then disappeared out of sight, 49 seconds earlier than expected, behind the Red Planet. It never came back into view. A post-accident investigation board concluded that the Orbiter had been on an incorrect trajectory that would have taken it to within 57 km of the surface, low enough for the atmosphere to destroy the fragile probe. When the board further investigated the reason for the discrepancy, they found that a piece of software, supplied by US aerospace and defence contractor Lockheed Martin, had been sending data about the Orbiter's thrust in imperial units. NASA, one of the foremost scientific institutions of the world, was, unsurprisingly, expecting those measurements in standard international metric units. The error meant that the Orbiter fired its thrusters too vigorously and consequently became just another 338 kilograms (or, if you prefer, 745 pounds) of space junk as it fell apart deep in the Martian atmosphere.

·

Realising that most of the rest of the world had converted to the metric system, and anticipating the sorts of mistakes that would afflict NASA, in 1970, Canada decided to make the move to metric. By the mid-1970s, products were labelled in metric units, temperature was reported in Celsius rather than Fahrenheit and snowfall was measured in centimetres. By 1977 road signs had all been converted to metric and speed limits were measured in kilometres per hour instead of miles per hour. For practical reasons, some industries took longer to convert to metric than others. In 1983, Air Canada's new Boeing 767s were the first to be calibrated in metric units. Fuel was measured in litres and kilograms rather than gallons and pounds.

On the 23rd of July 1983, one of the newly revamped 767s landed in Montreal after a routine flight from Edmonton. Following a brief turnaround, which included refuelling and a crew change, at 17:48 Flight 143 took off from Montreal on the return trip with 61 passengers and eight crew on board.

Cruising at an altitude of 41,000 feet or, as the electronic metric gauge read, 12,500 metres, Captain Robert Pearson set the plane to autopilot and relaxed. Around an hour into the flight, Pearson was startled by a loud beeping accompanied by flashing lights on the control panel. The warnings were indicating low fuel pressure to the aircraft's left engine. Assuming a fuel pump had failed, an unperturbed Pearson switched off the alarm. Even without the pump, gravity should have continued to draw fuel into the engine. Seconds later the same alarm sounded and warning lights again flashed on the dashboard. This time it was the right-hand engine. Again Pearson turned off the alarm.

He realised, however, that with both engines potentially faulty, he would need to divert to nearby Winnipeg to get the plane checked out. As he had this thought, the left engine spluttered and failed. Pearson radioed Winnipeg, urgently informing them that he would need to undertake a single-engine emergency landing. As he desperately tried to restart the left engine Pearson heard a noise from the control panel that neither he, nor his First Officer, Maurice Quintal, had ever heard before. The second engine gave out and the electronic flight instruments, powered by engine-generated electricity, went blank. The reason neither Pearson nor Quintal had ever heard the alarm before was because none of their training had dealt with the loss of both engines. It was assumed that the chances of both engines failing simultaneously were so small as to be negligible.

These engine failures were not even the first malfunctions the plane had experienced that day. When Pearson took charge of the plane earlier in the day, he had been informed that the fuel gauge was not working properly. Rather than grounding the flight and waiting 24 hours for the replacement part, Pearson decided that the amount of fuel required for the journey should be calculated by hand. Being a veteran pilot with over 15 years' experience, this was nothing new to him. Based on average fuel efficiencies and leaving some margin for error, the ground crew worked out that to make the trip to Edmonton the plane would need 22,300 kilograms of fuel. Upon landing at Montreal a dipstick was used to find that the plane still contained 7682 litres. This volume was multiplied by the density of the fuel, 1.77 kilograms per litre, to find that the plane already had 13,597 kilograms of fuel on board. This meant that the ground crew needed to add a further 8703 kilograms to reach the total of 22,300 kilograms. Using the fuel density of 1.77 kilograms per litre told them that adding the extra 8703 kilograms could be achieved by topping up the fuel tanks by 4917 litres. Perhaps Pearson should have noticed there was a problem at this point rather than later in the flight. When checking the ground crew's calculations he might have remembered that the density of jet fuel is less than the density of water at one kilogram per litre, but, then again, Canada had only recently gone metric. Unfortunately, during Air Canada's protracted switch to metric, the figure 1.77 that the plane's documentation gave for the density of fuel was wrong: 1.77 converts litres of jet fuel to pounds, not kilograms. The correct figure should have been less than half that, 0.803, in order to convert litres to kilograms. Because of this error, Pearson really had only 6169 kilograms of fuel already on board at Montreal. This meant that the ground crew should have added 20,088 litres, four times more than the 4917 litres they had calculated. Instead of the required 22,300 kilograms of fuel, Flight 143 took off with less than half that amount. The engines hadn't failed because of a mechanical fault. The 767 had simply run out of fuel.

The stricken plane continued to glide towards Winnipeg, with the only hope being that they could make a 'deadstick', or unpowered landing if their timing was just right. As luck would have it Pearson was also an experienced glider pilot, so he set to calculating the plane's optimal glide speed in order to maximise their chances

of making it to Winnipeg. However, as Flight 143 emerged from
the clouds, the limited instruments available, powered by back
up batteries, told Pearson that they would never make it. Pearson
radioed air-traffic control at Winnipeg with their situation. He
was informed that the only airstrip which might be in range was
at Gimli, approximately 12 miles away from their current position.
In what seemed another stroke of luck, Quintal had been stationed
at Gimli when he was a pilot in the Royal Canadian Air Force, so
knew the airfield well. What neither he, nor anyone in the control
tower at Winnipeg, knew was that Gimli had since become a public
airport and that part of the airport had been converted into a motor
sports arena. At that very moment, the track was hosting a car race,
with thousands of people in cars and campervans spectating from
the close surrounds of the runway.

   As the plane approached the runway, Quintal attempted to lower
the landing gear, but the hydraulic systems had given out when the
engines stopped working. Gravity was enough to pull the rear landing gear into place. Although the front landing gear also descended,
it wouldn't lock in place: a piece of serendipity which would shortly
be instrumental in saving many lives. With the engines silent, the
race spectators on the ground had no idea that the free-gliding, 100-
ton tin can was approaching until it was almost upon them. As the
plane hit the tarmac, Pearson braked as hard as he could, bursting
two of the rear tyres. Simultaneously the unlocked front landing
gear collapsed back into the plane, unable to support its weight. The
nose hit the ground, causing fountains of sparks to spray from the
undercarriage. The increased friction brought the plane to a swift
halt, just a few hundred metres from the dumbfounded onlookers.
Quick-thinking race stewards rushed onto the track to extinguish

small friction-induced fires that had started in the nose and all 69 passengers and crew descended the emergency slides safely.

The Millennium Bug bites

That Pearson managed to land the plane with barely any instruments or on-board computers is a hugely impressive feat. As we move further into the 21st century, many modern technologies continue to experience the exponential acceleration of their development and propagation that we encountered in Chapter 1. In particular, computers are progressively pervading our modern lives and we are consequently becoming increasingly vulnerable to their failure. In the years preceding the turn of the new millennium, the 'Millennium Bug' loomed large over companies which relied on computer software for their operation. The software glitch was the legacy of an almost preposterously simple computer programming oversight of the 1970s and 1980s.

If someone asks you for your date of birth, for brevity it's not unusual to give a six-digit answer. There may be some ambiguity when a 10-year-old and a 110-year-old are asked to write down their birthdays, but the correct year for each can usually be inferred from the context. Computers, however, often operate without such context. In an attempt to be as economical as possible with memory (which was expensive in the early days of computing), most programmers employed a six-digit date format. Typically, they allowed their programs to assume the date belonged in the 1900s. This left scope for error if the date genuinely belonged in the next century. As the dawn of the new millennium approached, computer experts began to warn that many computer programs might be unable to distinguish between 2000 and 1900, or the first year of any other century for that matter.

When the clock eventually ticked over to midnight on the 1st of January 2000, very little appeared to change. No planes fell out of the sky, no funds were wiped out and no nuclear missiles were launched. The lack of dramatic and immediate consequences led to the widespread belief that fears of the effects of the Millennium Bug were blown way out of proportion. Some cynics even suggested that the computer industry may have deliberately over-exaggerated the scale of the problem in order to bring in a big pay day. The opposing view is that the stringent preparation prior to the event helped to avert many potential disasters. There are many frivolous accounts of unremedied systems. Amusingly, the website of the US Naval Observatory, the organisation responsible for keeping the nation's official time, showed the date '1 Jan 19100'. However, some of the symptoms of the Millennium Bug were not so easy to laugh off.

In 1999, the pathology laboratory at the Northern General Hospital in Sheffield was a regional hub for Down's syndrome testing. Test results from pregnant women across the east of the UK were sent to Sheffield to be analysed by their sophisticated computer model, run on the NHS computer system, PathLAN. The model took in a range of data about the women, including date of birth, weight and the results of a blood test, in order to calculate a risk of their baby having Down's syndrome. This assessment of risk helped the women to decide how to proceed with their pregnancy, with high-risk mothers-to-be offered more definitive testing.

Throughout January 2000, staff in Sheffield found a number of isolated minor errors (relating to dates) in the PathLAN system, but these were quickly and easily rectified and not worried about. Later in the month, a midwife based at one of the hospitals served by the Northern General reported seeing far fewer high-risk Down's syndrome cases than she would have expected. She reported the same findings three months later, but on both occasions was assured by staff in the lab that nothing was amiss. In May, a midwife from a different hospital reported similarly infrequent high-risk test results. Eventually, the manager of the pathology laboratory was convinced to look into the results. He quickly realised that something *was* amiss. The Millennium Bug had bitten with full force.

In the pathology lab's computer model, the mother's date of birth was used, with reference to the current date, to calculate her age. The mother's age is an important risk factor, with older mothers being significantly more likely to conceive a child with Down's syndrome. After the 1st of January 2000, instead of a year of birth like 1965 being subtracted from 2000 to give a mother's age of 35, 65 was subtracted from 0, giving a negative age that the computer couldn't make sense of. Instead of triggering a warning, the nonsensical ages dramatically skewed the risk calculation, placing many older mothers in a lower-risk category than they should have been. As a consequence (in a similar misfortune to that which befell Flora Watson, the mother of baby Christopher, whose heart-wrenching 'false negative' story we heard in Chapter 2) over 150 women were sent letters incorrectly categorising their unborn babies as being at low risk of having Down's: false negatives. Of these, four women, who might otherwise have been offered further testing, gave birth to children with Down's syndrome and two other women had traumatic late abortions.

·

Binary thinking

The computers on which we have increasingly come to rely work with the most primitive base possible – base-two or binary. With decimal base 10, we require nine digits and a zero to represent any number. In the base two binary system, we require only one digit other than zero. All binary numbers are strings of just ones and zeros. Indeed, the word 'binary' comes from the Latin *binarius* meaning 'consisting of two parts'. In the binary place value system, the same digit one place to the left of a neighbour represents a number larger by a factor of 2, as opposed to a factor of 10, as we are used to in decimal. The first column on the right represents units, the second from the right 2s, the third 4s, the fourth 8s and so on. To build a number like 11 we need a one, a two and an eight, but no fours, so 11 has the binary representation 1011. There is an old mathematical joke 'There are only 10 types of people: those who understand binary and those who don't', 10, of course, representing the number two in binary.

Binary is the base of choice for computers, not because there is something inherently nice about doing mathematics in binary, but because of the way computers are built. Every modern computer comprises billions of tiny electronic components called transistors, which communicate with each other, transferring and storing data. The voltage flow though a transistor is a good way to represent a numerical value. Rather than having ten reliably distinguishable voltage options for each transistor and working in decimal, it makes more sense to have just two voltage options: on and off. This 'true or false' system means a small voltage can be used to deliver a reliable signal that is not mistaken if it fluctuates slightly. By combining the true or false outputs of these transistors with logical operations like 'and', 'or' and 'not', mathematicians have shown that it is possible, in theory, to compute the answer to any mathematical calculation that has an answer, no matter how complex. Modern day computers have gone a long way to realising this theory practically. They are capable of performing incredibly complicated tasks by converting our requests into a series of ones and zeros and applying cold, hard logic to flip these bits back and forth until they provide a lucid answer. Despite the everyday miracles we are able to achieve by enslaving the binary place value system to the machines that live on our desks and in our pockets, there are times when this most primitive base has let its masters down.

·

Christine Lynn Mayes was just 17 when she enrolled in the US army in 1986. She spent three years serving abroad in Germany as a cook before retiring from active duty, after which she returned home to study business at Indiana University of Pennsylvania, where she met her boyfriend David Fairbanks. In October of 1990, in need of money to support her studies, Mayes re-enlisted in the army reservists. She joined the 14th Quartermaster Detachment, a unit that would be charged with water purification. On Valentine's Day 1991, the unit was called to combat as part of operation Desert Storm. Three days later, Mayes shipped out to the Middle East. On the day she left the US, Fairbanks got down on one knee and proposed. Mayes willingly accepted his offer but, worried that she would lose it, declined to take the ring with her. 'All right, then, it'll be here when you get back,' were Fairbanks' last words to his fiancée before she left for Saudi Arabia. Fairbanks took the ring home with him and placed it atop a photo of Christine next to his stereo. He would never have the chance to place the ring on her finger.

When the 14th Quartermaster Detachment disembarked from the airbase in the oil-rich city of Dhahran in Saudi Arabia, they were transported a short distance to their temporary barracks in the city of Al Khobar on the Gulf Coast. The temporary building that housed Mayes' unit, as well as other American and British units, was little more than a corrugated-metal warehouse recently converted for human habitation. Six days after her arrival, on Sunday the 24th of February, Mayes rang home to tell her mother she had arrived safely and that her unit would soon be moving 40 miles further north towards the Kuwaiti border. The next day, having completed her shift, while others in her unit relaxed or worked out, Mayes took the chance to sleep, little suspecting that the events that would decide her fate had already been set in motion.

Despite having launched over 40 scud missiles at Saudi Arabia during the course of the Gulf War, fewer than ten of the Iraqis' attacks caused any significant damage. Most missiles that did reach Saudi Arabia were off course and landed in civilian areas, rather than their intended military targets. In part, the Iraqis' lack of success was due to the Americans' Patriot missile system. The system was designed to detect incoming missiles and launch an 'interception' in order to destroy the offending projectile mid-flight. The system relied upon an initial radar detection, followed by a more detailed confirmatory detection, designed to ensure the target was a genuine missile rather than spurious noise detected by an overactive first radar. In order to make the more detailed detection, the secondary radar was sent the time and location of the first sighting together with an estimation of the projectile's velocity. These could then be used to produce a narrow window to search for the potential positions of the missile, allowing a more detailed verification.

For accuracy, the Patriot system counted time in tenths of a second. Unfortunately, although it has the nice short decimal representation '0.1', in binary one tenth has an infinitely repeating expansion that looks something like 0.00011001100110011001100 . . . After the initial '0.0', the four digits 0011 just repeat over and over again. No computer can store infinitely many numbers, so instead the Patriot missile approximated one tenth using 24 binary digits. Because this number is a truncated representation, it is different from the true value of one tenth by around one ten-millionth of a second. The programmers who wrote the code that governed the Patriot system assumed that such a small discrepancy would make no practical difference. However, when the system had been up and running for a long period of time, the error in the Patriot's internal clock accumulated to something significant. After about 12 days, the total error in the Patriot's recording time would be almost a second.

At 20:35 on the 25th of February, the Patriot system had been running for over four days in a row. As Mayes slept, the Iraqi army launched a warhead atop a Scud missile towards the eastern coast of Saudi Arabia. Minutes later, as the missile crossed into Saudi Arabian air space, the Patriot's first radar detected the missile and fed its data to the second radar for verification. When data was passed from one radar to the next, the time of detection was out by almost a third of a second. With the incoming Scud travelling at over 1600 metres per second, its position was miscalculated by over

500 metres. When the second radar searched the region in which it expected to find the missile, it drew a blank. The missile alert was assumed to be a false alarm and removed from the system.[100]

At 20:40 the missile hit the barracks where Mayes was sleeping, killing her and 27 of her colleagues and injuring almost 100 others. This single attack, three days before the end of hostilities, was responsible for the deaths of one third of all the US soldiers killed during the First Gulf War and could perhaps have been averted, if computers spoke in a different language – with a different base.

No base, however, is capable of representing every number exactly with just a finite set of digits. With a different base, the Patriot missile detection error might have been avoided, but other errors would undoubtedly have occurred instead. So, despite the infrequent errors that it engenders, the energy and reliability advantages afforded by binary make it the most sensible choice of base for our current computers. However, these advantages quickly evaporate if we try to employ binary in a societal context.

•

Picture yourself chatting to an attractive stranger you are pressed up against on the crowded bus. As your stop nears, you ask them for their mobile phone number and they obligingly reel off some combination of 11 digits of the form 07XXX-XXX-XXX, the format common to all mobile phone numbers in the UK. To achieve the same variety of numbers in binary, each mobile phone number would have to be at least 30 digits long. Imagine trying to take down 111011100110101100100111111111 before the bus reaches the stop and you have to get off. 'Was that a one after the seventh zero or a zero?'

Of more immediate relevance is the potentially damaging binary thinking that pervades our society. From time immemorial, quick yes or no decisions have meant the difference between life and death. Our primitive brains had no time to calculate the probability that a falling rock would land on our heads. Coming face to face with a dangerous animal required a snap decision: fight or flight. More often than not a quick (and overcautious) binary decision was better than a slow measured one that weighed up all the options. As we evolved into more complex societies, we retained these binary judgements. We fell back on stereotypes of our fellow human beings as good or bad, saints or sinners, friends or enemies. These classifications are crude, but they provided us with an easy shortcut that dictated how to react when faced with each individual. Over time, these stereotypes have become further entrenched by the binary caricatures that are a pre-requisite in many popular dualist religions. There is no room for followers of these religions to doubt what the characteristics of good and evil look like.

But for most of us nowadays, such quick decisions and absolute caricatures have little relevance. We have time to meditate more deeply on important life choices. People are too complex to be classified by a single binary descriptor, too ambiguous, too subtle. Binary thinking would leave no space on the page for some of our favourite characters: the Snapes, the Gatsbys or the Hamlets of the literary world. The reason we like these mixed personas, entrenched in moral ambiguity, is precisely because they reflect our own complex, flawed personalities. But still we reach for the comfortable certainty of binary labels to show the outside world who we are as people: we are red or blue, we are left or right, we are theists or atheists. We trick ourselves into self-defining as one of two options, when in reality there are so many more colours in the spectrum.

•

In my own subject, mathematics, our biggest struggle is with such self-imposed false dichotomies: those who believe that they can do maths and those who think they can't. There are far too many of the latter. But there is almost no one who understands no maths at all, no one who cannot count. At the other extreme, for hundreds of years there have been no mathematicians who understand all of known mathematics. We all sit somewhere on this spectrum; how far we travel to the left or to the right depends on how much we think this knowledge can be useful to us.

Understanding the number systems around us, for example, gives us an insight into the history and culture of our species. These seemingly strange and often unfamiliar systems are not to be feared but to be celebrated. They tell us how our forebears thought, and reflect aspects of their traditions. They also hold up a more tangible mirror to our most basic biology, demonstrating that mathematics is as intrinsic to us as the fingers on our hands or the toes on our feet. They teach us the language of our modern technology and help us to avoid simple mathematical mistakes. Indeed, as we will see in the next chapter, by dissecting the mistakes we have made in the past, modern mathematics-based technology is (sometimes with dubious success) providing ways to avoid the same miscalculations in the future.

6

RELENTLESS OPTIMISATION

The Unconstrained Potential of Algorithms,
from Evolution to E-commerce

'In 100 metres, turn right . . . Turn right,' instructed the disembodied voice from the sat nav. With his wife and two of his children in the car, learner driver Roberto Farhat did just that. He had taken over the driving from his wife – a confident driver with 15 years' experience – only minutes before. As he turned off the A6, a 2-ton Audi travelling on the opposite carriageway careered into the passenger side of the car at 45 miles per hour. In paying close attention to his sat nav, Farhat had missed the road signs warning him *not* to turn right. Remarkably he walked away from the crash unhurt. His four-year-old daughter, Amelia, was not so lucky. She died in hospital three hours later.

We have all come to rely on devices like satellite navigation systems to simplify our increasingly fraught lives. In determining the quickest route from A to B, sat navs have an intricate job to perform. On-demand calculation in the form of an algorithm is the only feasible option to achieve this task. It would be challenging for one device to hold all possible routes between a distant pair of start and finish points. The vast number of beginnings and ends that might be requested magnifies the difficulty of the task astronomically. Given the difficulty of the problem, it is impressive that sat nav algorithms are very rarely wrong. But when they do make mistakes they can be disastrous.

An algorithm is a sequence of instructions that exactly specifies a job. The task could be anything from organising your record collection to cooking a meal. The earliest recorded algorithms, though, were strictly mathematical in nature. The ancient Egyptians had a simple algorithm for multiplying two numbers together and the Babylonians had rules for finding square roots. In the third century BCE, Ancient Greek mathematician Eratosthenes invented his 'sieve' – a simple algorithm for sifting out the primes from a range of numbers – and Archimedes had his 'method of exhaustion' for finding the digits of pi.

In pre-Enlightenment Europe, increased skill in mechanical manipulation allowed for the physical manifestation of algorithms in tools like clocks and, later, gear-based calculators. By the mid-19th century this skill had progressed to such a degree that the polymath Charles Babbage was able to build the first mechanical computer, for which pioneering mathematician Ada Lovelace wrote the first computer programs. Indeed, it was Lovelace who recognised that Babbage's invention had applications far beyond the purely mathematical calculations for which it had originally been designed: that entities like musical notes or, perhaps most importantly, letters could be encoded and manipulated with the machine. First electro-mechanical, and then purely electrical, computers were harnessed for exactly this purpose by the allies during the

Second World War to run algorithms that broke German ciphers. Although, in principle, the algorithms could have been implemented by hand, the prototype computers executed their commands with a speed and accuracy unmatchable by an army of humans.

The increasingly complex algorithms that computers now execute have become a vital part of the efficient handling of our day-to-day routines, from typing a query into a search engine or taking a photo on your phone to playing a computer game or asking your digital personal assistant what the weather will be like this afternoon. We won't accept just any old solution either: we want the search engine to bring up the most relevant answer to our questions, not just the first one it finds; we want to know with accuracy the probability of rain at five p.m., so we can decide whether we need to take our coats with us to work; we want the sat nav to guide us along the fastest route from A to B, not the first route it discovers.

Conspicuously absent from most definitions of an algorithm – a list of instructions to achieve a task – are the inputs and outputs, the data that give algorithms relevance. For example, in a recipe, the inputs are the ingredients, while the meal you serve up on the table is the output. For a sat nav, the inputs are the start and finish points you specify together with the map that the device holds in its memory. The output is the route the machine decides to take you on. Without these tethers to the real world, algorithms are just abstract sets of rules. As often as not, when the malfunction of an algorithm makes the news, it is incorrect inputs or unexpected outputs that are the real story, not the rules themselves.

In this chapter we discover the maths behind the unrelenting algorithmic optimisation in our everyday lives: from the way our search results are ordered in Google to the stories pushed at us on

Facebook. We expose the deceptively simple algorithms that solve difficult problems and upon which our modern-day tech giants rely: from Google Maps' navigation system to Amazon's delivery routes. We also step back from the computerised world of modern-day technology and deliver some algorithms directly into *your* control: the simple optimisation algorithms you can use to get the best seat on the train or to choose the shortest queue at the supermarket.

Although some algorithms can perform tasks of unimaginable complexity, there are sometimes aspects of their performance that are, to put it mildly, sub-optimal. Tragically for the Farhat family, an out-of-date map caused their sat nav to provide the wrong directions. The route-finding rules themselves were not at fault and, had the map been up to date, it is likely the accident would never have happened. Their story illustrates the awesome power of modern algorithms. These incredible tools, which have pervaded and simplified many aspects of our daily lives, are not to be feared. At the same time, they must be treated with due reverence and their inputs and outputs kept under close surveillance. With human supervision, however, comes the potential for censorship and bias. By considering what can happen when, for the sake of impartiality, manual control is curbed, we discover that prejudice may lie hidden, hard-coded within the algorithm itself: an imprint of the inclinations of its creator. No matter how useful algorithms can be, a little understanding of their inner workings, rather than blind faith in their error-free operation, can save time, money and even lives.

The million-dollar questions

In 2000, the Clay Mathematics Institute announced a list of seven 'Millennium Prize Problems', considered to be the most important unresolved problems in mathematics.[101] The list includes: the Hodge conjecture; the Poincaré conjecture; the Riemann hypothesis; the Yang–Mills existence and mass gap problem; the Navier–Stokes existence and smoothness problem; the Birch and Swinnerton-Dyer conjecture and the P vs NP problem. Although their names mean little to many outside some relatively small sub-fields of mathematics, Landon Clay, the institute's eponymous major donor, indicated just how significant he believed each one of these problems to be when he stumped up $1 million for the proof or disproof of each.

At the time of writing, only the Poincaré conjecture has been solved. The Poincaré conjecture is a problem in the mathematical field of topology. You can think of topology as geometry (the maths of shapes) with dough. In topology the actual shapes of objects themselves are not important; instead objects are grouped together by the number of holes they possess. For example, a topologist sees no difference between a tennis ball, a rugby ball or even a Frisbee. If they were all made of dough, they could theoretically be squashed, stretched or otherwise manipulated to look like each other without making or closing any holes in the dough. However, to a topologist, these objects are fundamentally different to a rubber ring, a bike's inner tube or a basketball hoop which, bagel-like, each have a hole through the middle of them. A figure of 8 with two holes and a pretzel with three are different topological objects again.

In 1904, French mathematician Henri Poincaré (the very same Poincaré who intervened to set right a mathematical travesty and exonerate Captain Alfred Dreyfus in Chapter 3) suggested that the simplest possible shape in four dimensions was the four-dimensional version of a sphere. To explain what Poincaré meant by 'simple', imagine making a loop of string around an object. If you can keep the string on the surface and pull it tight so that the loop disappears, then the article is, topologically, the same as a sphere. This idea is known as 'simple connectivity'. If you can't always do this trick with the string then you have a more complex topological object. Imagine threading the string, from underneath, through the centre of a bagel and over the top. Now when you pull the string the bagel gets in the way so that the loop never disappears. The bagel, with one hole, is fundamentally more complex than the football, with none. The result in three dimensions was already well known, but Poincaré suggested that the same idea would hold in four dimensions of space. His conjecture was later generalised to state that the same idea should hold in any dimension. However, by the time the Millennium prizes were announced, the conjecture had been proved to be true in every other dimension, leaving only Poincaré's original four-dimensional conjecture unproved.

In 2002 and 2003, reclusive Russian mathematician Grigori Perelman shared three dense mathematical papers with the topology community.[102] These papers purported to solve the problem in four dimensions. It took several groups of mathematicians three years to be sure of the veracity of his proof. In 2006, the year Perelman turned 40 – the cut-off age for the prize – he was awarded the Fields Medal, mathematics' equivalent of the Nobel Prize. Although the awarding of the prize made a small ripple of news outside mathematics, it was nothing compared to the stories that started to circulate when Perelman became the first person ever to turn down the Fields Medal. In his rejection statement, Perelman said: 'I'm not interested in money or fame. I don't want to be on display like an animal in a zoo.' When the Clay Mathematics Institute was finally satisfied, in 2010, that he had done enough to merit the $1 million for solving one of their Millennium Prize Problems, he turned their money down too.

P vs NP

Although undoubtedly a hugely important piece of work in the field of pure mathematics, Perelman's proof of the Poincaré conjecture has few practical applications. The same is true for the majority of the other six Millennium Prize Problems that, at the time of writing, remain unsolved. The proof or disproof of problem number seven, however – known succinctly, and somewhat cryptically, in the mathematical community as 'P vs NP' – has the potential for wide-ranging implications, in areas as diverse as internet security and biotechnology.

At the heart of the P vs NP challenge is the idea that it is often easier to verify a correct solution to a problem than it is to produce the solution in the first place. This most important of open mathematical questions asks whether every problem that can be checked efficiently by a computer can also be solved efficiently.

To draw an analogy, imagine you are putting together a jigsaw puzzle of a featureless image, like a picture of clear blue sky. To try all the possible combinations of pieces to see if they fit together is a difficult job; to say it would take a long time is an understatement. However, once the jigsaw is complete it is easy to check that it has been done correctly. More rigorous definitions of what efficiency means are expressed mathematically in terms of how quickly the algorithm works as the problem gets more complicated – when more pieces are added to the jigsaw. The set of problems that can be solved quickly (in what is known as 'Polynomial time') is called P. A bigger group of problems that can be checked quickly, but not necessarily solved quickly, is known as NP (which stands for 'Nondeterministic Polynomial time'). P problems are a subset of NP problems, since by solving the problem quickly, we have automatically verified the solution we find.

Now imagine building an algorithm to complete a *generic* jigsaw puzzle. If the algorithm is in P, then the time taken to solve it might depend on the number of pieces, that number's square, that number's cube or even high powers of the number of pieces. For example, if the algorithm depends on the square of the number of pieces, then it might take 4 (2^2) seconds to complete a 2-piece jigsaw, 100 (10^2) seconds for a 10-piece jigsaw, and 10,000 (100^2) seconds for a 100-piece jigsaw. This sounds like a relatively long time, but it is still in the realm of only a few hours. However, if the algorithm is in NP, then the time taken to solve it might grow exponentially with the number of pieces. A 2-piece jigsaw might still take 4 (2^2) seconds to solve, but a 10-piece jigsaw could take 1024 (2^{10}) seconds and a 100-piece jigsaw 1,267,650,600,228,229,401,496,703,205,376 (2^{100}) seconds – vastly outstripping the time elapsed since the Big Bang. Both algorithms take longer to complete with more pieces, but algorithms to solve generic NP problems quickly become unserviceable as the problem size increases. For all intents and purposes, P might as well stand for those problems that can be solved **P**ractically and NP for **N**ot **P**ractically.

The P vs NP challenge asks whether all the problems in the NP class, which can be checked quickly but for which there is no known quick solution algorithm, are in fact also members of the P class. Could it be that the NP problems have a practical solution algorithm, but we just haven't found it yet? In mathematical shorthand, does P equal NP? If it does, then, as we shall see, the potential implications, even for everyday tasks, are huge.

·

Rob Fleming, the protagonist of Nick Hornby's classic 90s novel, *High Fidelity*, is the music-obsessed owner of the second-hand record store, Championship Vinyl. Periodically, Rob reorganises his enormous record collection according to different classifications: alphabetical, chronological and even autobiographical (telling his life story through the order in which he bought his records). Quite apart from being a cathartic exercise for music lovers, sorting enables data to be interrogated quickly and reordered to display its different nuances. When you click the button that allows you to toggle between ordering your emails by date, sender or subject, your email client is implementing an efficient sorting algorithm. eBay implements a sorting algorithm when you choose to look at the items matching your search term by 'best match', 'lowest price' or 'ending soonest'. Once Google has decided how well webpages match the search terms you entered, the pages need to be quickly sorted and presented to you in the right order. Efficient algorithms that achieve this goal are highly sought after.

One way to sort a number of items might be to make lists with the records in every possible permutation, then check each list to see if the order is correct. Imagine we have an incredibly small record collection comprising one album by each of Led Zeppelin, Queen, Coldplay, Oasis and Abba. With just these five albums there are already 120 possible orderings. With six there are 720 and with ten there are already over three million permutations. The number of different orderings grows so rapidly with the number of records that any self-respecting record fan's collection would easily prohibit them from being able to consider all the possible lists: it's just not feasible.

Fortunately, as you probably know from experience, sorting out your record collection, books or DVDs is a P problem – one of those for which there is a practical solution. The simplest such algorithm is known as 'bubble sort' and works as follows. We abbreviate the artists in our meagre record collection to L, Q, C, O and A, and decide to organise them alphabetically. Bubble sort looks along the shelf from left to right and swaps any neighbouring pairs of records that it finds out of order. It keeps passing over the records until no pairs are out of order, meaning that the whole list is sorted. On the first pass, L stays where it is as it comes before Q in the alphabet, but when Q and C are compared they are found to be in the wrong order and swapped. The bubble sort carries on by swapping Q with O and then with A as it completes its first pass, leaving the list as L, C, O, A, Q. By the end of this pass, Q has been 'bubbled' to its rightful place at the end of the list. On the second pass, C is swapped with L and A is promoted in favour of O, so that O now resides in the correct place: C, L, A, O, Q. We need two more passes before A makes its way to the front and the list is ordered alphabetically.

With five records to sort we had to trawl though the unsorted list four times, doing four comparisons each time. With ten records we would have to undertake nine passes, with nine comparisons each time. This means that the amount of work we have to do during the sort grows almost like the square of the number of objects we are sorting. If you have a large collection then this is still a lot of work, but 30 records would take hundreds of comparisons instead of the trillions upon trillions of possible permutations we might have to check with a brute-force algorithm listing all the possible orders. Despite this huge improvement, the bubble sort is typically derided as inefficient by computer scientists. In practical applications, like Facebook's news feed or Instagram's photo feed, where billions of posts have to be sorted and displayed according to the tech giants' latest priorities, simple bubble sorts are eschewed in favour of more recent and more efficient cousins. The 'merge sort', for example, works by dividing the posts into small groups which it then sorts quickly and merges together in the right order.

In the build-up to the 2008 US presidential election, shortly after declaring his candidacy, John McCain was invited to speak at Google to discuss his policies. Eric Schmidt, Google's then CEO, joked with McCain that running for the presidency was very much like interviewing at Google. Schmidt then proceeded to ask McCain a genuine Google interview question: 'How do you determine good ways of sorting one million 32-bit integers in two megabytes of RAM?' McCain looked flummoxed, and Schmidt, having had his fun, quickly moved on to his next serious question. Six months later, when Barack Obama was in the hot seat at Google, Schmidt threw him the same question. Obama looked at the audience, wiped his eye and started off, 'Well, er . . .' Sensing Obama's awkwardness,

Schmidt tried to step in, only for Obama to finish, looking Schmidt straight in the eye, '. . . no, no, no, no, I think the bubble sort would be the wrong way to go', to huge applause and cheers from the assembled computer scientists in the crowd. Obama's unexpectedly erudite response – sharing an in-joke about the inefficiency of a sorting algorithm – was characteristic of the seemingly effortless charisma (afforded by meticulous preparation) that characterised his whole campaign, eventually bubbling him all the way up to the White House.

•

With efficient sorting algorithms to hand, it's great to know that next time you want to re-order your books or reshuffle your DVD collection it won't take you longer than the lifetime of the universe.

In contrast to this, there are problems that are simple to state, but which might require an astronomical amount of time to solve. Imagine that you work for a big delivery company like DHL or UPS and you have a number of packages you need to deliver during your shift, before dropping your van back at the depot. Since you get paid by the number of packages you deliver and not the time you spend doing the delivery, you want to find the quickest route to visit all of your drop-off points. This is the essence of an old and important mathematical conundrum known as the 'travelling salesman problem'. As the number of locations to which you have to deliver increases, the problem gets extremely tough extremely quickly in a so-called 'combinatorial explosion'. The rate at which possible solutions increase as you add new locations grows faster even than exponential growth. If you start with 30 places to deliver to, then you have 30 choices for your first drop-off, 29 for your second, 28 for your third and so on. This gives a total of $30 \times 29 \times 28 \times \ldots \times 3 \times 2$ different routes to check. In real terms the number of routes with just 30 destinations is roughly 265 nonillion – that's 265 followed by 30 zeros. This time though, unlike the sorting problem, there is no shortcut – no practical algorithm that runs in polynomial time to find the answer. Verifying a correct solution is just as hard as finding one in the first place, since all the other possible solutions need to be checked as well.

Back at delivery headquarters there might be a logistics manager attempting to assign the deliveries to be made each day to a number of different drivers, whilst also planning their optimal routes. This related task is known as the vehicle-routing problem and is even harder than the travelling salesman one. These two challenges appear everywhere – from planning bus routes across a city, collecting the mail from postboxes and picking items off warehouse shelves, to drilling holes in circuit boards, making microchips and wiring up computers.

The only redeeming feature of all of these problems is that, for certain tasks, we can recognise good solutions when they are put in front of us. If we ask for a delivery route which is shorter than 1000 miles, then we can easily check if a given solution fits the bill, even if it is not easy to find such a path in the first place. This is known as the 'decision version' of the travelling salesman problem, for which we have a yes or no answer. It is one of the NP class of problems for which finding solutions is hard, but checking them is easy.

Despite their difficulty, it is possible to find exact solutions for some specific sets of destinations even if it is not possible more generally. Bill Cook, a professor of combinatorics and optimisation at the University of Waterloo in Ontario, spent nearly 250 years of computer time on a parallel super-computer calculating the shortest route between all the pubs in the United Kingdom. The giant pub crawl takes in 49,687 establishments and is just 40,000 miles long – on average one pub every 0.8 miles. Long before Cook started on his calculations, Bruce Masters from Bedfordshire in England was doing his own practical version of the problem. He holds the aptly titled Guinness world record for visiting the most pubs. By 2014 the 69-year-old had had a drink in 46,495 different watering holes. Starting in 1960, Bruce estimates that he has travelled over one million miles in his quest to visit all of the UK's pubs – over 25 times longer than Bill Cook's most efficient route. If you're planning a similar odyssey for yourself, or even just a local pub crawl, it's probably worth consulting Cook's algorithm first.[103]

•

The vast majority of mathematicians believe that P and NP are fundamentally different classes of problems – that we will never have fast algorithms to dispatch salespeople or route vehicles. Perhaps that's a good thing. The yes-no 'decision version' of the travelling salesman problem is the canonical example of a subgroup of problems known as NP-complete. There is a powerful theorem which tells us that if we were able to come up with a practical algorithm that solved one NP-complete problem then we would be able to transmute this algorithm to solve any other NP problem, proving that P equals NP – that P and NP are in fact the same class of problems. Since almost all internet cryptography relies on the difficulty of solving certain NP problems, proving P equals NP could be disastrous for our online security.

On the plus side, though, we might be able to develop fast

algorithms to solve all sorts of logistical problems. Factories could schedule tasks to operate at maximum efficiency and delivery companies could find efficient routes to transport their packages, potentially bringing down the price of goods – even if we could no longer safely order them online! In the scientific realm, proving P equals NP might provide efficient methods for computer vision, genetic sequencing and even the prediction of natural disasters.

Ironically, although science might be a big winner if P equals NP, scientists themselves might be the biggest losers. Some of the most astounding scientific discoveries have relied enormously on the creative thinking of highly trained and dedicated individuals, embedded deeply in their fields: Darwin's theory of evolution by natural selection, Andrew Wiles' proof of Fermat's Last Theorem, Einstein's theory of general relativity, Newton's equations of motion. If P equals NP, then computers would be able to find formal proofs of any mathematical theorem that is provable – many of the greatest intellectual achievements of humankind might be reproduced and superseded by the work of a robot. A lot of mathematicians would be out of a job. At its heart, it seems, P vs NP is the battle to discover whether human creativity can be automated.

Greedy algorithms

Optimisation problems, like the travelling salesman problem, are so difficult because we are trying to find the very best solution from an inconceivably large set of possibilities. Sometimes, though, we might be prepared to accept a quick, good solution rather than a slow, perfect one. Maybe I don't need to find the optimal way to minimise the space taken up by the things I put into my bag before I set off to work. Perhaps I just need to find a way that gets everything in. If this is the case then we can start to take shortcuts in how we solve problems. We can use heuristic algorithms (commonsense approximations or rules of thumb) that are designed to get us close to the best solution for a wide range of variants of a problem.

One such family of solution techniques are known as greedy algorithms. These short-sighted procedures work by making the best local choice in an attempt to find globally optimal solutions. While they work quickly and efficiently, they're not guaranteed to produce the optimal solution or even a good one. Imagine you're visiting somewhere for the first time and you want to climb the highest hill around to see the lay of the land. A greedy algorithm to get to the top might first find the steepest incline at your current position, followed by taking a step in that direction. Repeating this procedure for every step, will eventually lead you to a point where you face a decline in every direction. This means you've made it to the top of a hill, but not necessarily the highest hill. If you want to climb the highest peak to get the best view, then this greedy algorithm isn't guaranteed to get you there. It might be that the route to the top of the small hillock you've just scaled started off more steeply than the route that takes you up to the local mountain range, so you followed the hillock path erroneously, based on your heuristic myopia. Greedy algorithms can find solutions, but they're not always guaranteed to be the best solutions. There are particular problems, however, for which greedy algorithms are known to give the optimal solution.

A map held by a sat nav can be thought of as a set of junctions that are connected by lengths of road. The problem that faces sat navs, to find the shortest route between two locations through a maze of roads and junctions, sounds as difficult as the travelling salesman problem. Indeed, the number of possible routes blows up astronomically quickly as the number of roads and junctions increases. Just a smattering of roads and a sprinkling of junctions is enough to push the number of possible routes into the trillions. If the only way to find the solution were to calculate all the possible routes and compare the total distance traversed for each one, then this would be an NP problem. Fortunately for everyone who uses a sat nav, it turns out that there is an efficient method – Dijkstra's algorithm – which finds the solution to the 'shortest path problem' in polynomial time.[104]

For example, when trying to find the shortest route from home to the cinema, Dijkstra's algorithm works backwards from the cinema. If the shortest distance from home to all the junctions that are connected to the cinema by a single stretch of road is known, then the job becomes simple. We can simply calculate the shortest trip to the cinema by adding the lengths of the paths from home to the nearby junctions to the lengths of the roads connecting the junctions to the cinema. Of course, at the start of the process, the distances from home to the nearby junctions are not known. However, by applying the same idea again, we can find the shortest paths to these penultimate junctions by using the shortest paths from home to the junctions that connect to them. Applying this logic recursively, junction by junction, takes us all the way back to the house where we start our journey. Finding the shortest route through the road network simply requires us to repeatedly make good local choices – a greedy algorithm. To reconstruct the route we just keep track of the junctions we had to pass through to achieve this shortest distance. It is likely that some variation of Dijkstra's algorithm is crunching numbers under the hood when you ask Google Maps to find you the best route to the cinema.

When you arrive at the cinema and go to pay for parking at the meter, it's more than likely that the ticket machine won't provide change. If there are enough coins in your pocket then you probably want to try to make up the exact price as quickly as possible. One greedy algorithm, which many of us will reach for intuitively, involves inserting coins sequentially, each time adding the largest-value coin that is less than the remaining total.

Most currencies, including those of Britain, Australia, New Zealand, South Africa, and Europe, share the 1-2-5 structure, with coins or notes increasing repeatedly in this pattern as the denominations of the currency get larger. The British system, for example, has 1, 2 and 5 pence coins. Next come the 10, 20 and 50 pence coins, then £1 and £2 coins followed by a £5 note and finally £10, £20 and £50 notes. So to make up 58 pence in change in this system using the greedy algorithm you would select the 50 pence coin, leaving 8 pence to make up; 20 and 10 pence would take you over the total so next add a 5 pence, followed by a 2 pence, and finally a 1 pence, or a penny. It turns out that for all the 1-2-5 currencies, as well as the US coinage system, the greedy algorithm described above actually does make up the total using the smallest number of coins.

The same algorithm isn't guaranteed to work in every currency. If, for some reason, there was a 4 pence coin as well, then the last 8 pence of the 58 might have been made up more simply using two 4 pence coins instead of a 5, a 2 and a 1. Any currency for which each coin or note is at least twice as valuable as the next smallest denomination will satisfy the greedy property. This explains the prevalence of the 1-2-5 structure – the ratios of 2 or 2.5 between denominations guarantee that the greedy algorithm will work, while the simple decimal system is preserved. Because making change is such a common procedure almost all of the world's currencies have been converted so that they satisfy the greedy property. Tajikistan, with the 5, 10, 20, 25 and 50 diram coins, is the only country whose coinage doesn't satisfy the greedy property. It's quicker to make 40 diram with two 20s than with the 25, 10 and 5 diram coins that the greedy algorithm would suggest.

On the subject of being greedy, have you ever tried asking for 43 Chicken McNuggets at McDonald's? It may sound unlikely, but these battered, deep-fried poultry morsels have given rise to some interesting mathematics. In the UK, McNuggets were originally served in boxes of 6, 9 or 20. While having lunch with his son in McDonald's, mathematician Henri Picciotto wondered exactly which numbers of Chicken McNuggets he would not be able to order with combinations of the three boxes. His list contained 1, 2, 3, 4, 5, 7, 8, 10, 11, 13, 14, 16, 17, 19, 22, 23, 25, 28, 31, 34, 37, and 43. All the other numbers of McNuggets were achievable and known, from that day forth, as the McNugget numbers. The largest number you can't make with multiples of a given set of numbers is called the Frobenius number. So 43 was the Frobenius number for Chicken McNuggets. Sadly, when McDonald's started selling packs of four Chicken McNuggets, the Frobenius number tumbled to just 11. Ironically, even with the new box of 4, the greedy algorithm fails when trying to make 43 Chicken McNuggets (two boxes of 20 take you straight to 40, but there's no box of 3), so even though it's now possible, asking for 43 Chicken McNuggets at the drive-thru might still prove to be a difficult problem.

Highly evolved

When they work, greedy algorithms are highly efficient methods for solving problems. When they fail, though, they can be worse than useless. If you're keen to venture into the great outdoors to commune with nature by climbing the highest mountain around, getting stuck on top of a mole hill in your back garden because you followed an inflexible greedy algorithm is less than optimal. Fortunately, there are a number of algorithms inspired by nature itself which help to get us over both the proverbial and literal hump.

One procedure, known as ant colony optimisation, sends out armies of computer-generated ants to explore a virtual environment inspired by a real-world problem. When tackling the travelling salesman problem, for example, the ants walk between nearby destinations, reflecting real ants' ability to perceive only their local environment. If the ants find a short route around all the points then they retrospectively lay pheromone on that route to guide other ants. The more popular, and correspondingly shorter, routes are reinforced and attract more ant traffic. As in the real world, the deposited pheromone evaporates, allowing the ants the flexibility to remodel the fastest route if the destinations change. Ant colony optimisation is used to find efficient solutions to NP challenges like the vehicle-routing problem and to answer some of the toughest questions in biology, including understanding the way in which proteins fold from simple one-dimensional chains of amino acids into intricate three-dimensional structures.

Ant colony optimisation is just one of a family of nature-inspired tools known as swarm intelligence algorithms. Despite communicating locally with only a small number of neighbours, flocks of starlings or schools of fish exhibit extremely rapid, but coherent, changes in direction. Information about a predator at one edge of a school of fish, for example, propagates quickly through to the other side of the group. By borrowing these local interaction rules, algorithm designers can dispatch huge flocks of well-connected artificial agents to explore an environment. Their rapid, swarm-like communication allows them to stay in touch with discoveries made by other individuals in their search for optimal surroundings.

Far and away nature's most famous algorithm is evolution. In its simplest form, evolution works by combining the traits of parents to produce children. The children that are better equipped to survive and reproduce in their environment pass on their characteristics to more offspring in the next generation. Sometimes there are mutations between generations, allowing new traits to be introduced, which may fare better or worse than those already in the population. Just three simple rules – select, combine and mutate – are enough to generate the biodiversity that solves some of the planet's most difficult problems.

Before we get carried away by this eulogy to the panacea of biological evolution, it's important to recognise that evolutionary solutions are often good but rarely, if ever, perfect. On wildlife documentaries, or in articles about the natural world, it's not uncommon to hear of animals being 'perfectly' adapted to their environment. From the desert-dwelling kangaroo rat that has evolved to go its whole life without ever drinking water, extracting all the moisture it needs from its food, to the notothenioid fish that have developed 'antifreeze' proteins so they can survive in sub-zero oceans, evolution has produced animals that are brilliantly adapted to their challenging environments.

The search for perfection, however, should not be confused with evolution's blind exploration of the possibilities. Evolution typically finds a solution that works better than any previous solution for that environment, but it doesn't always come up with the best way to solve a problem.

The UK's red squirrel population provides a classic example. With its sharp claws, flexible hind feet and long tail essential for balance, it is well adapted to climbing trees in the search for food. Its teeth grow continuously throughout its life, allowing squirrels to crack open the hard outer shells of nuts with abandon and without their teeth ever wearing out. It was seemingly perfectly adapted to its environment, until an even better-suited relative arrived. The significantly larger grey squirrel finds and eats more food, as well as digesting and storing it more efficiently. Although the grey never fought or killed the red, its superior adaptation meant that it quickly came to dominate the broad-leafed woodlands of England and Wales, outcompeting the red and taking over its ecological niche. Our perception of many species' exemplary adaptation perhaps owes more to our limited imagination of what a genuinely 'perfect' solution could look like, than it does to evolution finding a true optimum.

Despite evolution not necessarily finding the best-possible solution, the central tenets of this best-known of natural problem-solving algorithms have been plagiarised by computer scientists many times over, most notably in the so-called 'genetic' algorithms. These tools are employed to solve scheduling problems (including designing fixture lists for major sporting leagues) and to provide good, if not perfect, solutions to difficult NP problems like the 'knapsack problem'.

The knapsack problem imagines a trader who has a lot of goods to take to market in a rucksack which has only a fixed capacity. She can't take everything with her, so she has to make a choice. The different items each have different sizes and different profits associated with them. A good solution to the knapsack problem is a selection of goods that fit into the bag and give a high potential profit. Variations of the knapsack problem crop up when cutting shapes out of pastry or trying to be economical with your use of wrapping paper at Christmas. It appears when loading cargo ships and packing transport trucks. When download managers determine which data chunks to download and in what order to maximise the use of limited internet bandwidth, they are trying to solve the knapsack problem.

A genetic algorithm starts by generating a given number of potential solutions to a problem. These solutions are the 'parent' generation. For the knapsack problem, this parent generation comprises lists of packages that could fit in the knapsack. The algorithm ranks the solutions by how well they solve the problem. For the knapsack problem the ranking is based on the potential profit generated by that list of packages. Two of the best solutions – lists generating the most profit – are then *selected*. Some of the packages from one good knapsack solution are thrown out and the remainder are *combined* with some of the packages from the other good solution. There is also the possibility of *mutation* – that a randomly chosen package might be removed from the knapsack and replaced with another. Once the first 'child' solution in the new generation is produced, two more top-performing 'parent' solutions are chosen and allowed to reproduce. That way the better solutions in the parent generation pass their characteristics on to more child solutions in the next generation. This combination process is repeated until there are enough children to replace all the original solutions in the parent generation. Having had their turn, the parent solutions are then killed off, the new child solutions promoted to parent status and the whole selection, combination, mutation cycle starts again.

Because of the randomness inherent in the way the child solutions are created, the algorithm doesn't guarantee that *all* the offspring it produces will be better than their parents. In fact, many will be worse. However, by being selective about which of these children are allowed to reproduce – a virtual survival of the fittest – the algorithm dispenses with the dud solutions and allows only the best ones to pass on their characteristics to the next generation. As with other optimisation algorithms, it's possible that the solutions can hit a local maximum, for which any change will cause a decrease in fitness even though we have not yet reached the best-possible solution. Fortunately, the random processes of combination and mutation allow us to move away from these local peaks and to push on towards even better solutions.

The randomness that is such an important feature of genetic algorithms also has a role to play in our everyday lives. When you find yourself stuck in a rut, listening to the same songs by the same bands over and over again, you might hit the shuffle button. In its purest form, shuffle will just pick a random song for you. It's like a genetic algorithm without the selection and combination stages, but with a high degree of mutation. It might be one way to find a new band you like, but you may have to wade through a heap of Justin Bieber or One Direction songs to get there.

Many music-streaming services now provide far more algorithmically sophisticated ways to mix up your listening. If you've been playing a lot of the Beatles and Bob Dylan recently, a genetic algorithm might suggest you try a band that combines certain characteristics of the two – the Traveling Wilburys (the Bob Dylan–George Harrison supergroup) for example. By skipping songs or listening to them the whole way through, you signal a measure of their fitness, so the algorithm knows which 'solutions' to work from in the future.

There are also Netflix plug-ins that select random films or box-sets for you to watch, working from your previous preferences. Similarly, rafts of companies have recently sprung up offering to help relieve your food fatigue by sending you random selections of their products. From cheeses and wines to fruit and vegetables, you can start to optimise your gastronomic experience, exploring tastes you might not have even known were out there, while the caterers learn, based on your feedback, what to send you next. From fashion to fiction, companies are using tools from the evolutionary algorithms stable in an attempt to reinvigorate our everyday consumer experience.

Optimal stopping

The mathematical underpinning of some of the optimisation algorithms discussed above seems to suggest that they are solely the preserve of the tech giants who exploit them on such a huge scale for commercial gain. There are, however, some more straightforward algorithms – albeit supported by sophisticated mathematics

– that can be employed to make small but significant improvements to our everyday lives. One such family is known as 'optimal stopping strategies' and provides a way of choosing the best time to take action in order to optimise the outcome of a decision-making processes.

Pretend, for example, that you are looking for somewhere to take your partner out to dinner. You're both quite hungry, but you'd like to find somewhere nice. You don't want to dive into the first place you see. You consider yourself a good judge, so you'll be able to rank the quality of each restaurant relative to the others. You figure you'll have time to check out up to ten restaurants before your partner gets fed up with traipsing around. Because you don't want to look indecisive, you decide that you won't go back to a restaurant once you have rejected it.

The best strategy for this sort of problem is to look at and reject some restaurants out of hand in order to get a feel for what's out there. You could just choose the first restaurant you come to, but given you have absolutely no information about what's out there, there's only a one in ten chance you'll choose the best one at random. So you're better off waiting until you've judged a number of restaurants first, before choosing the first one you see that's better than all the others you've looked at so far. This restaurant-picking strategy is illustrated in Figure 21. The first three restaurants are judged for quality, but then rejected. The seventh restaurant is better than all the others so far and so that's where you stop and eat. But is three the right number to reject? The optimal stopping question asks, how many restaurants should you look at and reject, just to get a sense of what's on offer? If you don't look at enough then you won't get a good feel for what's

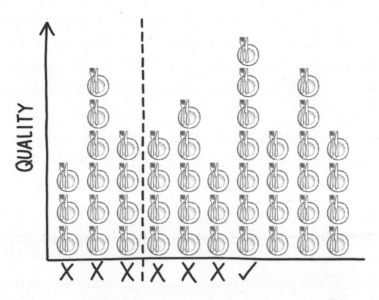

Figure 21: The optimal strategy is to assess, but reject, every option up to a given cut-off point (dashed line) and then accept the next option you evaluate which is better than all the previous options.

available, but if you rule out too many before taking the plunge then your remaining choice is limited.

The maths behind the problem is complicated, but it turns out you should judge and reject roughly the first 37% of the restaurants (rounded down to three if there are only 10) before accepting the next one that is better than all the previous ones. More precisely, you should reject the fraction $1/e$ of the available options, where e is a mathematical shorthand for a number known as Euler's number.[105] Euler's number is approximately 2.718, so the fraction $1/e$ is roughly 0.368 or, as a percentage, approximately 37%. Figure 22 illustrates how the probability of choosing the best of 100 restaurants changes as you vary the number of restaurants you reject out of hand. Unsurprisingly, when you jump in and make a decision too

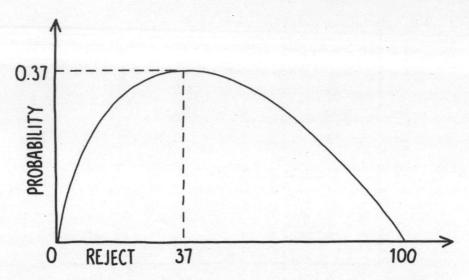

Figure 22: The probability of choosing the best option is maximised when we judge and reject 37% of the options before accepting the next one we assess that is better than all those we have previously seen. In this scenario, the probability you choose the best restaurant is 0.37 or 37%.

soon, you are effectively guessing blindly so the probability is low. Similarly, when you wait too long, it's likely you've already missed the best option. The probability of choosing the best option is maximised when you reject the first 37 options.

But what if the best restaurant was in the first 37%? In this case you miss out. The 37%-rule doesn't work every time: it's a probabilistic rule. In fact, this algorithm is only guaranteed to work 37% of the time. That's the best you can do given the circumstances, but it's better than the 10% of times you would have chosen the best restaurant if you had just picked the first of ten at random, and way better than the 1% success rate if you had to choose at random between 100 restaurants. The relative success rate improves the more options you have to choose from.

The optimal stopping rule doesn't just work for restaurants. In fact, the problem first came to the attention of mathematicians as the 'hiring problem'.[106] If you have to interview a set number of candidates for a job one after the other and at the end of each interview you have to tell the candidate whether they've got the job or not, then use the 37% rule. Interview 37% of the candidates and use those as your benchmark. Take the first interviewee after that who is better than all of the others you've seen so far and reject the rest.

When I reach the checkout at my local supermarket, I walk past the first 37% (4 of the 11), passively noting how long they are and then join the first queue thereafter that is shorter than all the others I've seen. If I am running, with a group of friends, for the packed last train after a night out and we want to find the carriage with the most spare seats so we can all sit together, we use the 37% rule. We head past the first three carriages on an eight-carriage train, remembering how empty they are, and then take the first carriage after that that has more free seats than any of the first three.

Although rooted in realism, some of the above scenarios are a bit contrived, but they can be made more pragmatic. What happens if half of the restaurants you try don't have a table free? Then, understandably, it turns out you should spend less time rejecting restaurants out of hand. Instead of checking out the first 37%, look at just the first 25% before choosing the next one that's better than the ones you've encountered.

What if you decide you've got enough time to risk going back to an earlier carriage on the train, but the probability that it has filled up in the meantime is 50%? Because you're broadening your options by going back you can afford to look for a bit longer – rejecting the first 61% of the carriages before choosing the next most empty carriage. Do make sure you get on the train before it starts to move off though.

There are optimal stopping algorithms that can tell you when to sell your house, or how far away from the cinema you should park to maximise your chances of getting a spot while minimising the distance you have to walk. The caveat is that, as the situation becomes more realistic, the maths becomes much more difficult and we lose the easy percentage rules.

There are even optimal stopping algorithms that can tell you how many people to date before you decide to settle down. You first need to decide how many partners you think you might get through by the time you'd like to settle down. Perhaps you might have one partner a year between your 18th and your 35th birthdays, making a total of 17 potential partners to choose from. Optimal stopping suggests that you play the field for about six or seven years (roughly 37% of 17 years) trying to gauge who's out there for you. After that you should stick with the first person who comes along who's better than all of the others you've dated so far.

Not many people are comfortable with letting a predefined set of rules dictate their love life. What if you find someone you're truly happy with in the first 37%? Can you really cold-heartedly reject them because you're on an algorithmic love mission? What if you follow all the rules and the person you decide is best for you doesn't think that you're the best for them? What if your priorities change halfway through? Fortunately, in matters of the heart, as with other more obviously mathematical optimisation problems, we don't always need to look for the very best solution, the single person who is the perfect fit – the One. There are likely to be multiple people out there who will be a good match and with whom we can be happy. Optimal stopping doesn't hold the answers to all of life's problems.

Indeed, despite the tremendous potential for algorithms to facilitate many aspects of our daily lives, they are far from the best solution to every challenge. Although an algorithm might be employed to simplify and accelerate a monotonous task, there are often risks associated with their use. Their tripartite nature – comprising inputs, rules and outputs – means there are three areas where things can go wrong. Even if the user is confident that the rules of the procedure are specified to their requirements, incautious inputs and unregulated outputs can lead to disastrous consequences, as online salesman Michael Fowler found out to his detriment. The American's algorithmically inspired retail master-plan, which unravelled so abruptly in 2013, has its roots in Britain at the beginning of the Second World War.

Keep calm and check your algorithm

Towards the end of July 1939 the dark clouds of war loomed ominously over Great Britain. In the air hung the possibility of heavy bombing, poison gas or even a Nazi occupation. Worried about public morale, the British government resurrected a shadowy organisation first instituted in the last year of the First World War to influence news reporting at home and abroad: the Ministry for Information. Pre-empting an Orwellian amalgam of the Ministries for Truth and Peace, the new Ministry for Information would be responsible for propaganda and censorship during the war.

In August of 1939 the Ministry designed three posters. Topped by a Tudor crown, the first read 'Freedom is in peril, defend it with all your might'. The second read 'Your courage, your cheerfulness, your resolution will bring us victory'. By late August, hundreds of thousands of copies of these two posters had been printed, ready for use should war break out. They were distributed widely during the early months of the war to an audience, in the Great British public, who were largely left feeling either apathetic or patronised.

The third poster, which was printed at the same time, was held back for the severe and potentially demoralising aerial bombardment that was expected. But, by the time the Blitz actually began in September 1940, over a year after the start of the war, paper shortages, combined with the perceived condescension of the first two posters, led to the mass pulping of all three. The third poster was seen by almost no one outside the Ministry of Information.

In 2000, in the quiet market town of Alnwick, secondhand booksellers Mary and Stuart Manley took delivery of a crate of used books they had recently purchased at auction. After emptying it out, they found a sheet of creased red paper at the bottom. As they unfolded it they read the five words of the 'lost' Ministry of Information poster: 'Keep calm and carry on'.

The Manleys liked the poster so much that they framed it and hung it on the wall of their shop, where it attracted the attention of their customers. By 2005 they were selling 3000 copies a week. But it was in 2008 that the meme really exploded into the global public consciousness. As recession gripped around the world, many sought to evoke the indomitable, stiff-upper-lipped demeanour that had previously seen Brits struggle through tough times. 'Keep calm and carry on' sold just the ticket. The message was transcribed to mugs, mouse mats, keyrings and every other piece of merchandise you can imagine. Even toilet paper did not escape the treatment. The message was transmuted in advertising campaigns for products as diverse as Indian restaurants ('Keep calm and curry on') and condoms ('Keep calm and carry one'). Almost any combination of 'Keep calm and [insert verb] [insert noun]' seemed to resonate. Almost any.

This simple idea was harnessed by online merchandiser Michael Fowler. In 2010, Fowler's company, Solid Gold Bomb, was selling pre-printed T-shirts with around 1000 different designs when Fowler hit on an idea to increase the efficiency of his workflow. Instead of paying to store huge numbers of printed T-shirts, he would move to printing on demand. This would allow him to advertise many more designs, that would be printed only when an order was placed. Once the printing process was streamlined, he set to writing computer programs that would automatically generate the designs. Almost overnight, the number of Solid Gold Bomb's offerings jumped from 1000 to over ten million. One such algorithm, created in 2012, took in a list of verbs and a list of nouns, combining them following the simple 'Keep calm and [word from verb list] [word from noun list]' formula. The phrases generated by the procedure were then screened automatically for syntactical errors, superimposed on the image of a T-shirt and listed for sale on Amazon for around $20 each. At Solid Gold Bomb's sales peak Fowler was selling 400 T-shirts a day with phrases like 'Keep calm and kick ass' or 'Keep calm and laugh a lot'. The problem was that he had also automatically listed several T-shirts on the world's biggest online retailer with phrases like 'Keep calm and kick her' or 'Keep calm and rape a lot'.

Surprisingly, the phrases went almost unnoticed for a year. Then one day in March 2013, Fowler's Facebook page was suddenly inundated with death-threats and allegations of misogyny. Despite acting quickly to pull down the designs, the damage was already done. Amazon suspended Solid Gold Bomb's pages, sales dropped to nearly nothing and, despite staggering on for three months, the company eventually went under. The algorithm Fowler designed, which seemed like such a good idea at the time, eventually ended up costing him and his employees their livelihoods.

Amazon did not escape the episode unscathed either. The day after Solid Gold Bomb had issued their formal apology for the debacle, Amazon was still listing T-shirts with the slogans 'Keep calm and grope a lot' and 'Keep calm and knife her'. A boycott of the retail giant was organised, with Lord Prescott, formerly the Deputy Prime Minister of the United Kingdom, even joining in the twiticism: 'First Amazon avoids paying UK tax. Now they're make money from domestic violence.' Unsurprisingly, given the tech giant's heavy reliance on computer-automated procedures, this was just one of the many pitfalls of unsupervised algorithmic activity that the world's most valuable retailer has stumbled into.

•

In 2011 Amazon had already found itself the subject of algorithmic controversy as a result of automated pricing strategies. On the 8th of April that year, Michael Eisen, a computational biologist at Berkeley, asked one of his researchers to get the lab a new copy of *The Making of a Fly*, a classic, but out-of-print work in evolutionary developmental biology. When the researcher went to Amazon he was pleased to find two new copies of the book for sale. When he looked more closely, though, he found that one of the books, sold by *profnath*, was on sale for $1,730,045.91. The other book, sold by *bordeebook*, was on sale for over $2 million. No matter how much he needed the book, Eisen couldn't justify the price, so instead he decided to watch the items to see if the price would come down. The next day when he checked the prices he found that things were worse: both books were now priced at nearly $2.8 million. The day after they had risen to over $3.5 million.

Eisen quickly figured out a method to the madness. Each day *profnath* would set their price to be 0.9983 of *bordeebook*'s offering. Later in the day *bordeebook* would scan *profnath*'s listing and set its price to be roughly 1.27 times as large. Day after day *bordeebook*'s price tag was inflated in proportion to its current size, causing it to grow exponentially; *profnath*'s listing lagging just behind. A human vendor controlling the prices would quickly have realised when the fees being asked for the books stretched beyond a sensible level. Unfortunately, this dynamic repricing wasn't being handled by a human, but by one of the range of repricing algorithms that are on offer to Amazon sellers. Apparently, no one thought to include a price-cap option on these algorithms, or, if they did, the sellers decided against using it.

All the same, *profnath*'s marginal under-pricing strategy made some sense. It ensured their book was the cheapest available and consequently that it appeared at the top of the search list, whilst not giving away too much in terms of profit. But why would *bordeebook* choose an algorithm that continually priced their book out of the market while it lay unordered, taking up space in their warehouse? It doesn't seem to make any sense, unless of course *bordeebook* never actually owned the book at all. Eisen suspects that *bordeebook* were trading on the trust and reliability indicated by their strong user ratings. If someone decided to buy the book from them, they would quickly buy the real copy from *profnath* and send it on to their buyer. The price hike allowed them to cover the postage they would have had to pay and still make a profit on the item.

Ten days after Eisen first spotted the exorbitant prices, they had spiralled further and reached the $23 million mark. Sadly, on the 19th of April, someone at *profnath* noticed the ridiculous price they were asking for a 20-year-old textbook and spoiled Eisen's fun by dropping the price back down to $106.23. The next day *bordeebook*'s price was sitting at $134.97, roughly 1.27 times *profnath*'s price, ready for the cycle to begin again. The price peaked again in August 2011, but this time to a mere $500,000, where it stayed unnoticed for the next three months. Apparently, someone learned their lesson and introduced a cap, although not a very realistic one. At the time of writing you can find around 40 listings for the book starting at the more reasonable price of around seven dollars.

Despite the extortionate price tag, *The Making of a Fly* is not the most expensive item ever listed or sold on Amazon. In January 2010, engineer Brian Klug found a copy of a Windows 98 CD-ROM called 'Cells' for sale on Amazon for nearly $3 billion (plus $3.99 postage and packing). The high price was presumably the result of another price spiral, with a second copy of the same CD-ROM listed by another seller who had topped out at a comparatively modest $250,000. Klug proceeded to enter his credit card details and purchase the item. A few days later Amazon emailed him to apologise that they couldn't fulfil his order. Disappointed, but probably equally relieved, Klug emailed back on the off chance that Amazon would honour the 1% credit against purchases made on the site using his Amazon credit card.

Flash crash

Algorithmic price spirals, like those that affected Amazon, don't always coil upwards. If you've ever invested in the stock market, or even just put some savings in an account linked to it, then you will have heard the familiar refrain: 'The value of your investment can go down as well as up'. Transactions on the stock market are increasingly being implemented as so-called algo-trading. Computers can sense and respond to changes in the market in a fraction of the time it would take their human counterparts. If a big order to sell a particular financial product flashes up on the screen, it may indicate that the price of that product is dropping and that traders are hoping to get rid of their assets at a good price before it drops further. In the time it takes a human to read the message and to click the button to sell their own assets, high-frequency algorithmic traders will already have sold theirs and the price will have dropped significantly. Human traders just can't compete. It is estimated that 70% of trading on Wall Street is now dealt with by these so-called black-box machines. That's why the big city traders and banks are increasingly turning to maths and physics graduates, rather than brokers, to assist with writing and, perhaps more importantly, understanding these algorithmic traders.

On the 6th of May 2010, after an already poor morning for the markets, bit-part trader, Navinder Sarao, operating from his London bedroom, turned on the bespoke algorithm he had recently finished modifying. The algorithm was designed to make a lot of money very quickly by spoofing the market – having other traders believe and act on a trend in the market that wasn't actually there.

His program was designed to rapidly place orders to sell a financial product known as E-mini futures contracts, but, before anyone could buy them, to cancel his order to sell.

Offering to sell his contracts at a price that was a little higher than the current best price ensured that no one, not even a fast-acting algorithm, would be tempted to take him up on his offer before his algorithm had a chance to cancel. When he executed the program it worked like a charm. High-frequency algorithmic traders recognised huge numbers of sell orders coming in and decided to sell their own E-mini futures contracts before the price dropped – as it inevitably would if the market became saturated by so many sales. Once the price of the futures crashed to a point at which Sarao was happy, he proceeded to turn off his program and buy up the now cheap contracts. Sensing the lack of sales, algorithmic traders rapidly regained confidence in and bought up the futures contracts once more allowing the price to recover. Sarao made a killing.

It is estimated that his spoofing made him $40 million. His algorithm was hugely successful – perhaps too successful. High-frequency trading algorithms reacted to huge sales volumes in the futures market. In just 14 seconds, the algorithms traded over 27,000 E-mini contracts, accounting for 50% of the day's total trading volume. They then began to sell other types of futures contracts to mitigate further losses. The fire sale then bled through into equities and out into the wider market. In the five minutes between 14:42 and 14:47 the Dow Jones dropped almost 700 points, bringing the total deficit for the day to nearly 1000 points – the biggest single-day drop in the index's history – and wiping $1 trillion off the stock market. The high frequency trading algorithms may not have caused the crash, but their unscrutinised, rapid trading certainly exacerbated it. Once the market bottomed out, however, and algorithmic confidence returned, they were also responsible for the rapid readjustment of most stocks back to near their opening values.

Sarao escaped justice for nearly five years while US financial regulators blamed a whole raft of other factors for the flash crash. In 2015, however, he was arrested and extradited to the US for the part his scheme played in the 2010 crash. He pleaded guilty to illegally manipulating the market and faces up to 30 years in prison, as well as having to pay back the money he earned through his illegal trading. Crime, it seems, even algorithmically assisted crime, doesn't pay.

Bang on trend

Sarao's bedroom market manipulation illustrates just how straightforward it can be to employ algorithms for malign purposes. Too often we picture them simply as impartial sequences of instructions that can be followed dispassionately, forgetting that all algorithms are developed for a reason. Just because the rules themselves are predefined and can be executed impassively, it doesn't follow that the purpose for which they are being employed is unbiased, even if impartiality is the algorithm designer's original intention.

Twitter, often heralded as the bastion of transparency amongst social media platforms, uses a relatively straightforward algorithm to determine which topics are trending. The algorithm looks for sharp spikes in hashtag usage rather than promoting topics purely on the basis of high volume. This seems sensible: looking at the acceleration rather than just the rate of usage allows brief but important events, such as the request for blood donors (#dondusang – blood donation) or the offer of shelter for the night (#porteouverte – open door) in the aftermath of the coordinated 2015 terrorist attacks in Paris to rise quickly to prominence. If high volume were the only trending criterion then we would never hear of anything but Harry Styles (#harrystyles) and Game of Thrones (#GoT).

Unfortunately, this same set of rules means that social topics that build slowly are rarely catapulted to the prominence they might warrant. In September and October of 2011, throughout the 'Occupy movement', the hashtag #occupywallstreet never trended in the movement's native New York City despite it being Twitter's most popular hashtag during that period. Although they had less volume overall, more transient stories during that period, like Steve Job's death (#ThankYouSteve) or Kim Kardashian's marriage (#KimKWedding), piqued attention in the right way to climb up Twitter's trending rankings. It is worth remembering that even genuinely pragmatic algorithms can have biases hard-coded within them that influence the direction in which the spotlight is shone on the global stage.

Of perhaps greater concern are situations in which the results of seemingly independent algorithms are subject to human intervention. In May 2016, Facebook's 'trending' news section was accused of anti-conservative bias in an exposé article on the technology news website Gizmodo. Gizmodo heard testimony from a former Facebook news curator who claimed that right-wing stories on US political figures like Mitt Romney and Rand Paul, amongst others, were being kept from Facebook's list of trending topics by human intervention. Even when conservative stories were organically trending on Facebook it was alleged that they were not making it onto the trending list. In other cases, stories were purported to have been artificially 'injected' into the trending list even if they weren't popular enough to merit inclusion.

In response to the accusations of political bias, Facebook decided to fire its trending editorial team and 'make the product more automated'. By divesting more power to the algorithm and removing a degree of human control, Facebook hoped to play on the perception of algorithmic objectivity. Just hours after their decision, the trending topics section was promoting a fake news story from the right wing, reporting that 'closet liberal' Fox News anchor Megyn Kelly had been fired for her alleged support of Hillary Clinton. This would be just the first of a whole barrage of predominantly right-leaning fake news stories that would come to characterise Facebook's trending section over the next two years, making the allegations of anti-conservative bias seem bland in comparison. The issue of reliability eventually led Facebook to pull the plug on the trending platform altogether in June 2018.

•

We place trust in supposedly impartial algorithms because we are wary of obvious human inconsistencies and inclinations. But although computers may implement algorithms in an objective manner following a predefined set of rules, the rules themselves are written by humans. These programmers might hard-code their biases, conscious or unconscious, directly into the algorithm itself, obfuscating their prejudices by translating them into computer code. The idea that we should be reassured about the neutrality of their trending news stories because Facebook, one of the world's foremost technology companies, had ceded power to one of its own algorithms doesn't really hold water.

In common with Solid Gold Bomb's offensive T-shirts and Amazon's spiralling prices, Facebook's travails highlight the need for more, not less, human supervision. As algorithms become increasingly complicated, their outputs can become commensurately unpredictable and need to be policed with greater scrutiny. This scrutiny is not the sole responsibility of the tech giants, however. As optimisation algorithms come to pervade more and more facets of our everyday lives, we – the front-line users of such shortcuts – need to shoulder part of the responsibility for ensuring the veracity of the outputs we are fed. Do we trust the source of the news stories we read? Does the route the sat nav is suggesting make sense? Do we think the automated price we are being asked to pay represents value for money? Although algorithms can provide us with information that facilitates vital decisions, in the end they are no substitute for our own subtle, biased, irrational, inscrutable, but ultimately human judgements.

When we investigate the tools at the forefront of the battle against infectious disease in the next chapter, we will find that exactly the same thesis holds true: although advances in modern medicine have gone a long way towards halting the spread of infectious disease, mathematics shows that amongst the most effective ways we have to keep the lid on an epidemic are the simple actions and choices we make as individuals.

7

SUSCEPTIBLE, INFECTIVE, REMOVED

Containing Disease Is in Our Own Hands

During the Christmas holidays towards the end of 2014, the 'Happiest Place on Earth' became a place of abject misery for many families. Hundreds of thousands of parents and children visited California's Disneyland over the holiday period, hoping to take home magical memories that would last a lifetime. Instead, some of them left with a memento that they hadn't bargained for: a highly infectious disease.

One of those visitors was four-month-old Mobius Loop. His mother, Ariel, and father, Chris, were self-confessed Disneyland fanatics, to the extent that they got married there in 2013. As a trained nurse, Ariel was acutely aware of the dangers of exposing her prematurely born son's developing immune system to infectious disease. She confined her new-born almost exclusively to the house. She also insisted that anyone who wanted to visit Mobius before his first round of vaccinations, at two months, was up-to-date with their own jabs for seasonal flu, tetanus, diphtheria and whooping cough.

In mid-January 2015, with his first round of vaccinations completed and their annual passes burning a hole in their pockets, Ariel and Chris decided to take Mobius to 'experience the magic' at Disneyland. After a day watching parades and meeting super-sized cartoon characters, the Loops returned home, delighted by how much Mobius enjoyed his first adventure in the land of Disney.

Two weeks later, after a night spent struggling to get her son to sleep, Ariel noticed raised red marks on Mobius' chest and the back of his head. She took his temperature and found he was running a fever of 102° Fahrenheit (39° Celsius). Unable to control his temperature, Ariel rang her doctor, who instructed her to take the baby straight to the emergency room. When they arrived, they were met outside the hospital by an infection-control team wearing full protective clothing. Ariel and Chris were given their own masks and gowns and were rushed through a back entrance into a negative-pressure isolation room. Once inside, medical professionals inspected Mobius carefully, before asking Ariel to restrain him as they drew blood for a definitive test. Despite never having seen a case of the disease before, the emergency room staff all suspected the same thing: measles.

Due to the effectiveness of vaccination programmes begun in the 1960s, few citizens of Western countries, including many medical professionals, have ever witnessed at first hand how severe the symptoms of measles can be. But travel to less developed countries like Nigeria, where annual measles cases in the tens of thousands are routinely reported, and you might get a better picture of the disease. Complications can include pneumonia, encephalitis, blindness and even death.

In the year 2000, measles was officially declared eliminated across the whole of the United States.[107] Eliminated status meant that measles was no longer continuously circulating in the country and that any new cases were the result of outbreaks triggered by individuals returning from abroad. In the nine years from 2000 to 2008 there were just 557 confirmed cases of measles in the US. But in 2014 alone there were 667 cases. As 2015 approached, the outbreak emanating from Disneyland, which affected the Loops and dozens of other families, spread rapidly across the country. By the time it died out it had infected over 170 people in 21 states. The Disneyland flare-up is part of a trend of increasingly common large outbreaks. Measles is on the rise again in the United States and Europe, placing vulnerable people at risk.

•

Disease has afflicted humans ever since our hominin lineage first diverged from that of chimps and bonobos. Much of the story of our history has the often unwritten subplot of contagious disease running through it. Malaria and tuberculosis, for example, have recently been discovered to have affected significant swathes of the Ancient Egyptian population over 5000 years ago. From 541 to 542 CE the global pandemic known as 'the Plague of Justinian' is estimated to have killed 15 to 25% of the world's 200-million-strong population. Following Cortés' invasion of Mexico, the native population dropped from around 30 million in 1519 to just 3 million 50 years later; the Aztec physicians had no power to resist the previously unseen diseases brought over by the Western conquistadors. The list goes on.

Even today, in our increasingly medically advanced civilisation, disease-causing pathogens are still sufficiently sophisticated that modern medicine has not been able to eliminate them from our daily lives. Most people experience the banality of the common cold almost every year. If you yourself have not had flu, then you surely know several people who have. Fewer in the developed world will have experienced cholera or tuberculosis, but these pandemic diseases are not uncommon in much of Africa and Asia. Interestingly, though, even within communities in which disease prevalence is high, succumbing is not a certainty. Part of our morbid fascination with diseases is their seemingly random occurrence, visiting untold horrors on some while leaving others in the same community completely untouched.

There is, however, a little known, but highly successful, field of science working in the background to unpick the mysteries of infectious disease. By suggesting preventative measures to halt the spread of HIV, and bringing the Ebola crisis to heel, mathematical epidemiology is playing a crucial role in the fight against large-scale infection. From highlighting the risks to which the growing anti-vaccination movement is exposing us, to fighting global pandemics, maths is at the heart of the crucial life-and-death interventions that allow us to wipe diseases clean off the face of the Earth.

The scourge of smallpox

By the middle of the 18th century, smallpox was endemic throughout the world. In Europe alone it is estimated that 400,000 people a year died from the disease – up to 20% of all deaths on the continent. Half of those who survived were left blind and disfigured. Working as a doctor in rural Gloucestershire, Edward Jenner had been witness to the truly held belief of his patients: that becoming a milkmaid could protect you from smallpox. Jenner deduced that the mild disease cowpox, which most milkmaids were exposed to, provided some immunity against smallpox.

To investigate his hypothesis, in 1796, Jenner carried out a pioneering experiment into disease prevention that would be considered wildly unethical today.[108] He extracted pus from a lesion on the arm of a milkmaid infected with cowpox, and smeared it into a cut on the arm of an eight-year-old boy, James Phipps. The boy rapidly developed lesions and a fever, but within ten days was back on his feet, as fit and healthy as before the inoculation. As if it wasn't enough to have been infected by Jenner once, two months later Phipps submitted to Jenner inoculating him again, this time with the more dangerous smallpox. After several days, when Phipps failed to develop symptoms of smallpox, Jenner concluded that he was immune to the disease. Jenner named his protective process a 'vaccination', after the Latin word *vaccas,* meaning cow. In 1801, Jenner recorded his hopes for the discovery '. . . that the annihilation of the Small Pox, the most dreadful scourge of the human species, must be the final result of this practice.' Eventually, after a concerted vaccination effort by the World Health Organization nearly 200 years later, in 1977, his dream became a reality.

The story of Jenner's development of vaccinations provides an indelible link between smallpox and the history of modern disease prevention. Mathematical epidemiology also finds its roots in the attempt to diminish smallpox, but the subject's origins go back even further than Jenner.

Long before Jenner developed his idea of vaccination, in a desperate attempt to save themselves from the ever increasing incidence of smallpox, the people of India and China practised variolation. In contrast to vaccination, variolation involved exposing oneself to a small amount of material associated with the disease itself. In the case of smallpox, powdered scabs of previous victims were often blown up the nose or pus introduced into a cut in the arm. The aim was to induce a milder form of smallpox, which although still unpleasant, was far less dangerous and would provide the patient with lifelong immunity from the severe symptoms of the full-blown disease. The practice quickly spread to the Middle East and thence into Europe in the early 1700s, where smallpox was rife.

Despite its seeming effectiveness, variolation had its detractors. In some cases, the practice failed to protect patients from a second, more serious attack of smallpox as their immunity waned. Perhaps even more damaging to variolation's reputation were the 2% of cases in which patients died as a result of their treatment. The death of Octavius, the four-year-old son of the English monarch King George III, was one such high-profile case, which did little to improve the public's perception of the practice. Although a 2% mortality rate was still significantly lower than the 20–30% associated with the natural spread of the diseases, critics argued that many variolated patients might never have been exposed to smallpox naturally and that widespread treatment was an unnecessary risk. It was also observed that variolated patients could spread the full-blown disease just as effectively as naturally infected smallpox victims. In the absence of controlled medical trials, however, quantifying the effect of variolation and removing the lingering shadow over the procedure was not easily achieved.

This was exactly the sort of public health issue that piqued the interest of Swiss mathematician Daniel Bernoulli, one of the great unsung scientific heroes of the 18th century. Amongst his many mathematical achievements, Bernoulli's studies in fluid dynamics led to him creating equations that provide an explanation for how wings can create the lift required to allow planes to fly. Before he mastered advanced mathematics, however, Bernoulli's first degree was in medicine. His later studies into fluid flow combined with his medical knowledge led him to discover the first procedure that could be used to measure blood pressure. By puncturing the wall of a pipe with a hollow tube, Bernoulli could determine the pressure of the fluid running through the pipe by looking at how high it rose up the tube. The uncomfortable practice that developed from his findings involved inserting a glass tube directly into a patient's artery. This method was not supplanted by a less invasive alternative for over 170 years.[109] Bernoulli's broad academic background also led him to apply a mathematical approach to determine the overall efficacy of variolation, a question to which traditional medical practitioners could only guess the answer.

Bernoulli suggested an equation to describe the proportion of people of a given age who had never had smallpox, and were hence still susceptible to the disease.[110] He calibrated his equation with a life table, collated by Edmund Halley (of comet-spotting fame), which described the proportion of live births surviving to any given age. From this he was able to calculate the proportion of people who had had the disease and recovered, as well as the proportion who had succumbed. With a second equation, Bernoulli was able to account for the number of lives that would be saved if variolation were practiced routinely on everyone in the population. He concluded that with universal variolation, nearly 50% of infants born would survive to the age of 25, which, although depressing by today's standards, was a marked improvement on the 43% if small-pox were allowed to rage freely in the population. Perhaps even more remarkably, he showed that this one simple medical interven-tion had the power to raise average life expectancy by over three years. For Bernoulli, the case for medical intervention by the state was clear. In concluding his paper he wrote, 'I simply wish that, in a matter which so closely concerns the wellbeing of the human race, no decision shall be made without all the knowledge which a little analysis and calculation can provide'.

Today, the purpose of mathematical epidemiology has not strayed far from Bernoulli's original aims. With basic mathematical models we can begin to forecast the progression of diseases and understand the effect of potential interventions on disease spread. With more complex models we can start to answer questions relat-ing to the most efficient allocation of limited resources or tease out the unexpected consequences of public health interventions.

The S-I-R model

At the end of the 19th century, poor sanitation and crowded living environments in colonial India led to a series of deadly epidem-ics including cholera, leprosy and malaria sweeping through the country and killing millions.[111] The outbreak of a fourth disease, one whose very name has inspired terror for hundreds of years, would give rise to one of the most important developments in the history of epidemiology.

No one is entirely sure how the disease reached Bombay in August 1896, but there is no doubt about the devastation it caused.[112] The most likely explanation seems to be that a merchant ship, harbouring several highly undesirable stowaways, set sail from the British colony of Hong Kong. Two weeks later it docked at Port Trust in Bombay (now Mumbai). As the sweating longshoremen busied themselves unloading the ship's cargo in 30°C heat, several of the stowaways disembarked unnoticed and hurried off towards the city's slums. These free-riders were harbouring an unwanted cargo of their own, which would throw first Bombay, and then the rest of India, into chaos. The stowaways were rats carrying the fleas responsible for the spread of the bacterium *Yersinia pestis*: the plague.

The first cases of plague amongst Bombayites were detected in the Mandvi region encompassing the port. The disease spread unbridled through the city and, by the end of 1896, was killing 8000 people a month. By the beginning of 1897, the plague had spread to nearby Poona (now Pune) and would subsequently spread across India. By May 1897, strict containment measures had seemingly caused the plague to die out. However, the disease would return periodically to haunt India for the next 30 years, killing over 12 million of its citizens.

•

It was into one such plague outbreak that a young Scottish military physician, named Anderson McKendrick, arrived in 1901. He would spend almost 20 years in India, carrying out research (remember, in Chapter 1, we saw that McKendrick was the first scientist to show that bacteria increased to a carrying capacity according to the logistic growth model), public health interventions and gaining a deeper understanding of zoonotic diseases – those diseases, like swine flu, that can spread between animals and humans. Eventually his prowess in both research and practice would see McKendrick rise to be the head of the Pasteur Institute in Kasauli. Ironically, whilst in Kasauli he contracted brucellosis – a debilitating disease caused by drinking unpasteurised milk. As a result he was sent on several periods of medical leave back home to Scotland.

It was on one of these leave periods, inspired by an earlier meeting with fellow Indian Medical Service physician and Nobel Prize winner, Sir Ronald Ross, that he decided to take up the study of mathematics. Mathematical study and research would dominate McKendrick's final years in India, before he was eventually invalided home permanently in 1920 after contracting a tropical bowel disease.

Back in Scotland, McKendrick took up the position of superintendent of the laboratory of the Royal College of Physicians of Edinburgh. It was here that he met a young and talented biochemist by the name of William Kermack. Not long after meeting McKendrick, Kermack was caught up in a devastating explosion which left him instantly and permanently blind. Despite this set-back, his partnership with McKendrick flourished. Inspired by data on the plague outbreaks in Bombay, collected while McKendrick was in India, they conducted the single most influential study in the history of mathematical epidemiology.[113]

Together they derived one of the earliest and most prominent mathematical models of disease spread. To make their model work, they split the population into three basic categories according to disease status. People who had not yet had the disease were labelled, somewhat ominously, as 'susceptibles'. Everyone was assumed to be born susceptible and capable of being infected. Those who had contracted the disease and were capable of passing it to susceptibles were the 'infectives'. The third group were euphemistically referred to as the 'removed' class. Typically, these were the people who had had the disease and recovered with immunity or those who had succumbed to the disease and died. These 'removed' individuals no longer contributed to the spread of the disease. This classic mathematical representation of disease spread is referred to as the S-I-R model.

In their paper, Kermack and McKendrick demonstrated the utility of their S-I-R model by showing it was able to accurately recreate the rise and fall of the number of cases of plague in the 1905 outbreak in Bombay. In the 90 years since its inception, the S-I-R model (and its variants) have had great success in describing all sorts of other diseases. From dengue fever in Latin America to swine fever in the Netherlands and norovirus in Belgium, the S-I-R model can provide vital lessons for disease prevention.

Presenteeism, predictions and the plague problem

In recent years, the advent of zero-hour contracts and the increase in temporary employment – a hallmark of the burgeoning 'gig' economy – have contributed to the rise in people coming to work while ill. While absenteeism has been the subject of extensive research, the costs of 'presenteeism' have only recently begun to be understood. Studies combining mathematical modelling and workplace attendance data have drawn some surprising conclusions. Measures implemented to reduce employee absence, including reducing paid sick leave, are causing a marked rise in people coming to work regardless of how bad they might be feeling, leading unintentionally to more illness and lowered rates of efficiency overall.

The problem of presenteeism is particularly prevalent in health-care and teaching. Ironically, nurses, doctors and teachers feel so obligated to the large numbers of people they safeguard that they often put them at risk by coming into work whilst under the weather. It is the hospitality industry, however, which has perhaps the most acute presenteeism problem. One study found that in the US alone over 1000 outbreaks of the vomiting bug, norovirus, were linked to contaminated food in the four years from 2009 to 2012.[114] Over 21,000 people fell sick as a result and 70% of the outbreaks were linked to unwell food service workers.

Five years after that study concluded, Chipotle Mexican Grill became a high-profile victim of the detrimental consequences of presenteeism. From 2013 to 2015 Chipotle was ranked as the strongest Mexican restaurant brand in the US. Despite having a paid sick leave policy in place, workers at many Chipotle branches across the US reported managers requiring them to turn up to work when ill, under the threat of losing their jobs.

On the 14th of July 2017, Paul Cornell went out to enjoy a bur-rito in the Sterling, Virginia branch of Chipotle. Despite suffering from stomach cramps and nausea, that same evening an unnamed food worker clocked in for work. Twenty-four hours later, Cornell was in hospital, hooked up to an intravenous drip and suffering the extreme stomach pain, nausea, diarrhoea and vomiting consistent with a full-blown norovirus infection. One hundred and thirty-five other staff and customers also contracted the virus after visiting the restaurant. In the five days following the outbreak Chipotle's share price slumped, wiping over a billion dollars off the company's market value and leading its own shareholders to file a class-action law suit against it. By the end of 2017, Chipotle didn't even make it into the top half of America's favourite Mexican restaurant chains.

The S-I-R model illustrates the importance of not coming in to work when unwell. By staying at home until fully recovered, you effectively take yourself from the infected class straight to the removed class. The model demonstrates that this simple action can reduce the size of an outbreak by diminishing the opportunities for the disease to pass to susceptible individuals. Not only that, but you also give yourself a better chance of a speedy recovery by not 'working through the pain'. The S-I-R model describes how, if everyone with an infectious disease followed this practice, we would all benefit through fewer preventable closures of restaurants, schools and hospital wards.

•

Perhaps more than for its descriptive ability, however, the S-I-R model is vaunted for its predictive power. Instead of always looking back at past epidemics, the S-I-R model allowed Kermack and McKendrick to look forward – to predict the explosive dynamics of disease outbreaks and understand the sometimes mysterious patterns of disease progression. Indeed, they used their model to tackle some of the most hotly contested questions in epidemiology at the time. One such debate centred on the question 'What causes a disease to die out?' Is it the case that a disease simply infects everyone in the population? That once the susceptible population is exhausted, perhaps the disease just has nowhere to go? Alternatively, perhaps the disease-causing pathogen becomes less potent over time, to the point at which it is unable to infect healthy individuals any more.

In their influential paper, the two Scottish scientists were able to show that neither of these is necessarily the case. When looking at the status of their population at the end of a simulated outbreak, they found that there were always some susceptible individuals remaining. This is, perhaps, in direct contrast to our intuition (fostered by movies and media scare stories) which would suggest that a disease dies out because there are no more people left to infect. In reality, as infective people recover or die, contact between the remaining infectives and susceptibles becomes so infrequent that the infectives never get a chance to pass on the disease before being removed (recovering with immunity or dying) themselves. The S-I-R model predicts that, ultimately, outbreaks die out from of a lack of infective people, not a lack of susceptibles.[115]

In the small community of 1920s epidemic modellers, Kermack and McKendrick's S-I-R model was a towering contribution. It lifted the study of disease progression high above the rooftops of the purely descriptive studies that had gone before and allowed far off glimpses into the future. However, the windows of insight it provided were restricted by the narrow foundations on which the model was built: the numerous assumptions that limited the situations in which it was able to make useful predictions. These assumptions included: a constant rate of human-to-human disease transmission; that infected people were also instantaneously infectious; and that population numbers didn't change. While useful for describing some diseases some of the time, these assumptions don't hold for the majority.

For example, ironically, the Bombay plague data that Kermack and McKendrick used to 'validate' their model breaks many of these assumptions. For starters, the Bombay plague was not primarily passed from human to human, but spread by rats carrying fleas that were, in turn, carrying the plague bacterium. Their model also assumed a constant rate of transmission between infective carriers and their susceptible victims. In fact (as with the viral spread of the more trivial ice bucket challenge that you will recall from Chapter 1) there was a strong seasonal component to the plague in Bombay, with flea density and bacterial abundance at dramatically higher levels from January to March, leading to a consequent increase in the transmission rate.

Nevertheless, future generations of mathematicians would adapt the seminal S-I-R model, loosening its restrictive assumptions and expanding the range of diseases to which mathematics could lend its insight.

·

One of the first adaptations made to the original S-I-R model was to represent diseases that confer no immunity on their victims. One such disease progression, typical of some sexually transmitted diseases, like gonorrhoea, has no removed population at all. As soon as someone recovers from gonorrhoea, they are capable of being infected again. Since no one dies from the symptoms of gonorrhoea, no one is ever 'removed' by the disease. Such models are typically labelled S-I-S, mimicking the progression pattern of an individual from susceptible to infective and back to susceptible again. Since the population of susceptible people is never exhausted, but renewed as people recover, the S-I-S model predicts that diseases can become self-sustaining or 'endemic', even within an isolated population with no births or deaths. In England, gonorrhoea's endemic status has contributed to it becoming the second most common sexually transmitted infection, with over 44,000 reported cases in 2017.

In fact, even more adaptations to the basic model are needed to properly represent sexually transmitted diseases like gonorrhoea. Their progression pattern is not as simple as diseases like the common cold in which everyone can infect everyone else. With sexually transmitted diseases, infectives typically only infect people corresponding to their preferred sexual orientation. Since the majority of sexual encounters are heterosexual, the most obvious mathematical model divides the population into males and females and allows infection only between these two groups rather than amongst everyone. Models in which the bipartite nature of heterosexual interactions is taken into account produce slower disease spread than models in which it is assumed that everyone can transmit the disease to everyone else irrespective of gender and sexual orientation. Such models of sexually transmitted diseases are, however, full of potential pitfalls.

HPV – More than just the cancer virus

The memories of my fifth birthday were still fresh when my mother was diagnosed with cervical cancer at the age of 40. She endured round after round of arduous and debilitating chemo- and radiotherapy. Thankfully, at the end of the gruelling process, she was told that she was in complete remission. I was surprised to learn, in later life, that cervical cancer is one of the few cancers caused primarily by a virus – a cancer you can catch, typically through

sexual intercourse. I find the idea that my father might have harboured the virus that caused my mother's cancer hard to take. He
cared so devotedly for her when the cancer came back. It was only
his strength of will that held our family together when she died a
few weeks before her 45th birthday. Even unknowingly, how could
it have been him?

As it turns out, the vast majority of cases of cervical-cancer-
causing human papillomavirus (HPV) infections *are* transmitted
through sexual intercourse. Over 60% of allcervical cancers are
caused by two strains of HPV.[116] Indeed, HPV is the most frequent
sexually transmitted disease in the world.[117] Men can carry the
virus asymptomatically and pass it to their sexual partners, contributing to cervical cancer's status as the fourth most common cancer
in women, with around half a million new cases and quarter of a
million deaths reported worldwide each year.

In 2006, the first revolutionary vaccines against HPV were
approved by the US Food and Drug Administration. Unsurprisingly,
given the high incidence rates, there was great hope surrounding
the licensing of the vaccine. Studies undertaken in the UK around
the time of the vaccine's deployment indicated that the most cost-
effective strategy would be to immunise adolescent girls between
the ages of 12 and 13, the likely future sufferers of cervical cancer.[118]
Related studies in other countries, considering mathematical models
of the heterosexual transmission of the disease, confirmed that
vaccinating females only was the best course of action.[119]

However, these preliminary studies ultimately demonstrated
that any mathematical model is only as good as the assumptions
underpinning it and the data parameterising it. The majority of
these analyses neglected to include an important feature of HPV

in their modelling assumptions: that the strains of HPV guarded against by the vaccine can also cause a range of non-cervical diseases in both women and men.[120]

If you've ever had a wart or a verruca then you will have harboured at least one of five types of HPV. Eighty percent of people in the UK will be infected with one strain of HPV at some point during their life. As well as causing cervical cancer, HPV types 16 and 18 contribute to 50% of penile cancer, 80% of anal cancers, 20% of mouth cancers, and 30% of throat cancers.[121] Famously, when in recovery from throat cancer, actor Michael Douglas was asked if he regretted his lifetime of smoking and drinking. He candidly told *Guardian* reporters that he had no regrets, because his cancer had been caused by HPV that he contracted through oral sex. In both the US and the UK, the majority of cancers caused by HPV are not cervical.[122] Significantly, HPV types 6 and 11 also cause nine out of ten cases of anogenital warts.[123] In the US approximately 60% of the healthcare costs associated with all non-cervical HPV infections are spent on the treatment of these warts.[124] Cervical cancer is an important part of the HPV narrative, but it is not the whole story.

In 2008, at the time the vaccine was first being rolled out, German virologist Harald zur Hausen was awarded the Nobel Prize in medicine 'for his discovery of human papilloma viruses causing cervical cancer'. The link to other cancers and diseases was somewhat ignored by the prize committee and most of the rest of the world. The one UK study that did account for non-cervical cancers was not able to do so with any certainty because, at the time, the burden of the diseases and the impact of the vaccination against them was not properly understood. Most models suggested that by vaccinating a sufficiently high proportion of females, the prevalence of HPV-related diseases in unprotected males would also decline. The general public, perhaps aware only of HPV's role in cervical cancer – the common cancer that spreads like an infectious disease – accepted without question the decision only to vaccinate girls. Why should boys be vaccinated if they don't suffer from the head-line HPV-cancer?

But imagine the public outcry if a vaccination for the AIDS-causing human immunodeficiency virus (HIV) was developed and it was ordained that only women would be given the vaccination for free, in the hope that men would be protected through women's immunity. Quite aside from the issues associated with partial vac-cination coverage and vaccine inefficiency, perhaps the first point that critics would make would be about the protection of gay men – should they be left with no defence against the deadly virus? The same argument holds true in the case of HPV. By neglecting homo-sexual relationships in their mathematical models, early studies had ignored the effects of same-sex couplings. Models based on sexual networks including homosexual relationships have a higher rate of disease transmission than those that consider only heterosexual relationships.[125] The prevalence of HPV in men who have sex with men is significantly higher than in the general population.[126] In the US, the incidence rate of anal cancer in this group is over 15 times higher. At 35 per 100,000, it is comparable to the rates of cervical cancer in women *before* cervical screening was introduced and sig-nificantly higher than current rates of cervical cancer in the US.[127] When models were recalibrated, taking into account homosexual relationships, new knowledge on the protection afforded against non-cervical cancers, and up-to-date information on the length of protection that the vaccinations provide, it was found that vaccinating boys as well as girls becomes a cost-effective option.

In April 2018, the UK's National Health Service eventually offered the HPV vaccination to homosexual men between the ages of 15 and 45. In July of the same year, advice based on a new cost-effectiveness study recommended that all boys in the UK be given the HPV vaccination at the same age as girls.[128] Thankfully, my daughter and my son will be afforded equal protection against catching and disseminating the virus that killed their grandmother. It just goes to show that the conclusions drawn from even the most sophisticated of mathematical models are only as strong as their weakest assumptions.

The next pandemic

Another confounding factor with HPV infection is asymptomatic disease-carrying. People can harbour the virus, infecting others without showing symptoms themselves. For this reason, another adaptation that is commonly made to the basic S-I-R model, in order to represent diseases more realistically, is to include a class of people who, once infected, are capable of passing on the disease whilst remaining asymptomatic. The so-called 'carrier' class changes the S-I-R model into an S-C-I-R model and is vital for representing the transmission of many diseases, including some of the most deadly of our time.

Some patients experience a short period of flu-like symptoms a few weeks after contracting HIV. The severity of the symptoms varies broadly and some carriers will not even notice anything wrong. Despite no outwardly obvious symptoms, the virus slowly damages the patient's immune system, leaving them open to opportunistic infections like tuberculosis or cancers to which people with healthy immune systems might not succumb. Patients at the later stages of HIV infection are said to have acquired immune deficiency syndrome (AIDS). One of the main reasons why HIV/AIDS has become pandemic, meaning it has spread throughout the world and is still spreading, is this long incubation period. Carriers who are unaware that they even have the virus spread the disease far more rapidly than people who know they are HIV-positive. Each year, for the last 30 years or more, HIV has been one of the top causes of death by infectious disease worldwide.

HIV is thought to have emerged from non-human primates in Central Africa early in the 20th century. Possibly as a result of humans handling infected primates caught for bush-meat, a mutated form of the simian immunodeficiency virus (SIV) jumped species into humans and was able to spread from human to human through the exchange of bodily fluids. Zoonotic diseases which jump between species, like the original strains of HIV, present one of the biggest potential threats to public health.

In 2018, England's deputy chief medical officer, Professor Jonathan Van-Tam, singled out one such disease, the H7N9 virus – a new strain of bird flu – to be the most probable cause of the next global flu pandemic. The virus is currently highly prevalent in Chinese bird populations and has infected over 1500 people. To put this into perspective, the Spanish flu, the most deadly pandemic of the 20th century, infected roughly 500 million people worldwide. However, the mortality rate of the Spanish flu was only around 10%. H7N9 kills roughly 40% of those it infects. Fortunately, so far,

H7N9 has not acquired the crucial ability to pass between humans, which would allow it to mushroom to the scale of the Spanish flu. Despite animal experiments suggesting that it is just three mutations away from being able to do so, perhaps, like its predecessor, the bird-flu strain H5N1, it never will. It is quite possible that the next global pandemic will not be from an emerging disease at all, but perhaps from one we have seen many times before.

Patient zero

One afternoon in late 2013, two-year-old Emile Ouamouno was playing with some of the other children in the remote Guinean village of Meliandou. One of the children's favourite haunts was a huge, hollow cola tree on the outskirts of the village: perfect for hiding in. The tree's deep and dark cavity also provided the ideal roost for a population of insect-eating free-tailed bats. Whilst playing in the bat-infested tree, Emile came into contact either with fresh bat guano, or perhaps even face to face with a bat itself.

On the 2nd of December, Emile's mother noticed that her usually energetic young toddler was tired and lethargic. After feeling the heat of the fever emanating from his forehead, she took him to bed to recover. However, he soon started vomiting and excreting black diarrhoea. He died four days later.

Having cared diligently for her son, Emile's mother also contracted the disease and died a week later. Emile's sister Philomène succumbed next, followed by their grandmother on the first day of the new year. The village midwife, who had cared for the family during their illnesses, unknowingly took the disease with her to neighbouring villages and thence to hospital in the nearest town of Guéckédou, where she sought treatment for her affliction. From there, one of many conduits for the onward spread of the disease was through a health worker who treated the midwife. She spread the virus to a hospital in Macenta, about 50 miles to the east, where she infected the doctor who attended her. In turn, he infected his brother in the city of Kissidougou, 80 miles to the north-west, and so the spread continued.

On the 18th of March, the number of cases and their extent were becoming a serious worry. Health officials publicly announced the outbreak of an, as yet unidentified haemorrhagic fever 'which strikes like lightning.' Two weeks later, upon identification of the disease, Médecins Sans Frontières called the scale of its spread 'unprecedented'. From this point onwards, Emile Ouamouno, an otherwise unremarkable child, would be transformed into someone the world would never forget. Tragically, he would become infamous as 'Patient Zero': the victim of the first animal-to-human transmission of what had become the biggest, most uncontrolled Ebola outbreak of all time.

That we even know the progression of the disease is a tribute to the huge detail in which the epidemic was analysed by scientists and healthcare professionals placing themselves directly in its path. A method known as 'contact tracing' allows epidemic experts to work their way backwards through many generations of infected individuals, all the way down to the originating case – Patient Zero – hence Emile's sobriquet. By asking infected individuals to list all the people they have had contact with during and after the incubation period of the disease – when they are infected but not necessarily symptomatic – scientists can build up a picture of their contact network. By iterating the process many times on the individuals in the network, disease spread can often be pinned down to a single source. As well as allowing us to learn about the complex pattern of disease spread in order to suggest methods to prevent future outbreaks, contact tracing also allows us to take real-time measures to control the spread of disease. It can inform effective strategies to contain a disease in its early stages. Everyone who has had direct contact with an infected individual within the time-frame of incubation is quarantined until they have been shown to be free of the disease or to be infected. If infected, they can be kept in isolation until they are no longer likely to pass on the disease.

•

In practice, though, contact networks are often incomplete and there are many disease carriers who are not known to the authorities. In fact, there are many individuals who don't even know that they have the disease themselves due to the incubation period – the window of time, post infection, before symptoms occur. With Ebola, the incubation period can be up to 21 days, but on average it is around 12 days. In October 2014, it became clear that the epidemic in West Africa could potentially take on global proportions. Ostensibly to protect its citizens, the UK government announced that enhanced Ebola screening would take place for passengers entering the UK from high-risk countries at five major UK airports and the Eurostar terminal in London.

A similar programme in Canada during the time of the SARS (severe acute respiratory syndrome) epidemic in 2004 screened nearly half a million travellers, none of whom were found to have the raised temperature indicative of SARS. The programme cost the

Canadian government $15 million. In hindsight, the SARS screening programme was a futile measure, which may have gone some way to reassuring the Canadian public that they were safe but was ineffective as an intervention strategy.

With that expense in mind, as well as what smacked of a needlessly fraught reaction, a team of mathematicians from the London School of Hygiene & Tropical Medicine developed a simple mathematical model incorporating an incubation period.[129] Considering the average 12-day incubation period for Ebola and the six-and-a-half-hour flight time from Freetown in Sierra Leone to London, the mathematicians calculated that only about 7% of Ebola-carrying individuals boarding planes would be detected by the expensive new measures. They suggested that the money might be better spent on the developing humanitarian crisis in West Africa, which would strike at the source of the problem and consequently reduce the risk of transmission to the UK. This is an example of mathematical intervention at its best – simple, decisive and evidence-based. Rather than speculating on how effective the screening measures could be, a simple mathematical representation of the situation can give powerful insights and help to direct policy.

R-nought and the exponential explosion

The transmission pathway used to identify Emile Ouamouno as Ebola's Patient Zero was far from unique. The disease radiated from its epicentre in Mellandou through multiple distinct pathways. In fact, in its early stages, the disease replicated in an exponential fashion through multiple independent channels, much like the memes or viral marketing campaigns described in Chapter 1. One person infected three others, who went on to infect others, who infected more still, and so the outbreak exploded. Whether or not an outbreak promulgates itself to infamy or dies out into obscurity can be determined by a single number, unique to that outbreak – the basic reproduction number.

Think of a population completely susceptible to a particular disease, much like the native inhabitants of Mesoamerica in the 1500s before the arrival of the conquistadors. The average number of previously unexposed individuals infected by a single, freshly introduced disease carrier is known as the 'basic reproduction number' and often denoted R_0 (pronounced 'R-nought' or 'R-zero'). If a disease has an R_0 of less than 1, then the infection will die out quickly as each infectious person passes on the disease, on average, to less than one other individual. The outbreak cannot sustain its own spread. If the R_0 is larger than 1 then the outbreak will grow exponentially.

Take a disease like SARS, for example, with a basic reproduction number of 2. The first person with the disease is Patient Zero. They spread the disease to two others, who each spread the disease to two others and thence to two others each. Just as we saw in Chapter 1, Figure 23 illustrates the exponential growth that characterises the initial phase of the infection. If the spread was able to continue like this, ten generations down the chain of progression, over 1000 people would be infected. Ten steps further on, the toll would rise to over one million.

In practice, just as with the spread of a viral idea, the expansion of a pyramid scheme, the growth of a bacterial colony or the proliferation of a population, the exponential growth predicted

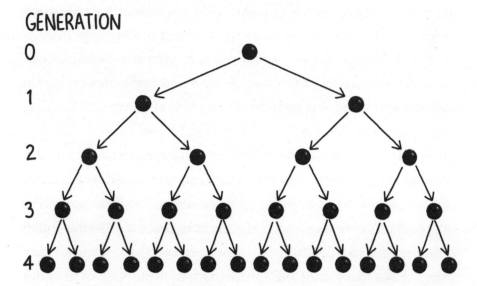

Figure 23: Exponential spread of a disease with basic reproduction number, R_0 of 2. The initial infected individual is assumed to be in Generation Zero. As we enter the fourth generation 16 people become newly infected.

by the basic reproduction number is rarely sustained beyond a few generations. Outbreaks eventually peak and then decline due to the decreasing frequency of infected–susceptible contacts. Ultimately, even when there are no infectives left and the outbreak is officially over, some susceptibles will remain. Way back in the 1920s, Kermack and McKendrick came up with a formula that used the basic reproduction number to predict how many susceptible individuals would be left untouched at the end of an outbreak. With an estimated R_0 of around 1.5, Kermack and McKendrick's formula predicts that the Ebola outbreak of 2013– 2016 would have afflicted 58% of the population if no intervention had been taken. In contrast, polio outbreaks have been found to have an R_0 of around 6, which, under Kermack and McKendrick's prediction, means only a quarter of 1% would survive unscathed without interposition.

The basic reproduction number is a ubiquitously useful descriptor of an outbreak because it wraps up all the subtleties of disease transmission into a single figure. From the way in which the infection develops in the body, to the mode of transmission, and even the structure of the societies within which it spreads, it captures all the outbreak's key features and allows us to react accordingly. R_0 can typically be broken down into three components: the size of the population; the rate at which susceptibles become infected (often known as the force of infection); and the rate of recovery or death from the disease. Increasing the first two of these factors increases R_0, while increasing the recovery rate reduces it. The bigger the population and the faster the disease spreads between individuals, the more likely an outbreak is to occur. The more quickly that individuals recover, the less time they have to pass on the disease to others and, consequently, the smaller the likelihood of an outbreak occurring. For many human diseases, it is only the first two factors that we can control. Although antibiotic or antiviral medicines may shorten the course of some diseases, the rate of recovery or fatality is often a property inherent to the disease-causing pathogen itself. A quantity closely related to R_0 is the *effective* reproduction number (often denoted R_e) – the average number of secondary infections caused by an infectious individual at a *given point* in the outbreak's progression. If, by intervention, R_e can be brought below 1, then the disease will die out.

Although crucial for disease control, R_0 does not tell us how serious a disease is for an infected individual. An extremely infectious disease like measles, for example, with an R_0 of between 12 and 18, is typically considered less serious for an individual than a disease like Ebola with a smaller R_0 of around 1.5. While measles spreads quickly, its case fatality rate is small compared to the 50–70% of Ebola patients who will eventually die of the disease.

Perhaps surprisingly, diseases with high case fatality rates tend to be less infectious. If a disease kills too many of its victims too quickly then it reduces its chances of being passed on. Diseases that kill most of the people they infect and also spread efficiently are very rare, and are usually confined to disaster movies. Although a high case fatality rate significantly raises the fear associated with an outbreak, diseases with high R_0 but lower case fatality may end up killing more people by virtue of the larger numbers they infect.

The maths dictates that once we have decided that a disease needs to be controlled, case fatality rates don't provide useful information on how to decelerate the spread. The three factors that comprise R_0, however, suggest important interventions that can bring a halt to deadly disease outbreaks before they run their unencumbered courses.

Taking control

One of the most effective options for reducing disease spread is vaccination. By taking people directly from susceptible to removed, bypassing the infective state, it effectively reduces the size of the susceptible population. Vaccination, however, is typically a precautionary measure applied in an attempt to reduce the probability of outbreaks. Once outbreaks are in full swing, it is often impractical to develop and test an effective vaccine within a useful time frame.

An alternative strategy, employed for animal diseases, and that has the same diminishing effect on R_e, the effective reproduction number, is culling. In 2001, when Britain was in the grip of the foot-and-mouth crisis, the decision was taken to cull. By slaughtering infected individuals the infectious period was lowered from up to three weeks to a matter of days, dramatically reducing the effective reproduction number. In this outbreak, however, culling only the infected animals was not enough to control the disease. Some infectives inevitably slipped through the net, causing infection in others nearby. In response, the government implemented a 'ring culling' strategy, slaughtering animals (infected or not) within a 3-kilometre radius of affected farms. At first glance killing uninfected individuals seems a pointless exercise. However, because it reduces the population of susceptible animals in the local area – one of the factors that contributes to the reproduction number – the maths dictates that it slows the spread of the disease.

For active outbreaks of human diseases in unvaccinated populations, culling is clearly not an option. Quarantine and isolation, however, can prove extremely efficient ways to reduce the transmission rate and, consequently, the effective reproduction number. Isolating infective patients reduces the rate of spread, whilst quarantining healthy individuals reduces the effective susceptible population. Both actions contribute to decreasing the effective reproduction number. Indeed, the last smallpox outbreak in Europe, in Yugoslavia in 1972, was rapidly brought under control by extreme quarantine measures. Up to 10,000 potentially infectious individuals were held under armed guard in hotels commandeered for that express purpose, until the threat of new cases had passed.

In less extreme cases, simple applications of mathematical

modelling are able to suggest the most effective duration to isolate infected patients.[130] They can also determine whether or not to quarantine a proportion of the uninfected population, weighing up the economic costs of quarantining healthy individuals against the risk of an enlarged disease outbreak. This sort of mathematical modelling really comes into its own for situations in which carrying out field studies on disease progression is impractical for logistical or ethical reasons. For example, it's inhumane, during a disease out-break, to deprive a fraction of a population of an intervention that might save lives for the purposes of a study. Similarly, it's impractical in the real world to quarantine a high proportion of the population for a long period of time. There are no such concerns when run-ning a mathematical model. We can test models in which everyone is quarantined, or no one, or anywhere in between, in an attempt to balance the economic impact of this enforced isolation with the effect it has on the progression of the disease.

This is the real beauty of mathematical epidemiology – the ability to test out scenarios that are infeasible in the real world, sometimes with surprising and counter-intuitive results. Maths has shown, for example, that for diseases like chickenpox (varicella), isolation and quarantine may be the wrong strategy. Trying to seg-regate children with and without the disease will undoubtedly lead to numerous missed school and work days in order to avoid what is widely considered to be a relatively mild disease. Perhaps more significantly, though, mathematical models prove that quarantining healthy children can defer the date at which they catch the disease to a point when they are older, at which age the complications aris-ing from chickenpox can be far more serious. Such counter-intuitive effects of a seemingly sensible strategy like isolation might never have been fully understood if it weren't for mathematical interventions.

If quarantine and isolation have unexpected consequences for some diseases, they simply don't work at all for others. Mathematical models of disease spread have identified that the degree to which a quarantining strategy succeeds depends on the *timing* of peak infectiousness.[131] If a disease is especially infectious in the early stages, when patients are asymptomatic, then they may spread the disease to the majority of their expected victims before they can be isolated. Fortunately, in the case of Ebola, for which many other potential control avenues are blocked, the majority of transmissions occur after patients are symptomatic and can be isolated.

In fact, the infectious period of Ebola extends to such an extreme that – even after their death – victims' viral loads remain high. The dead may still infect individuals who come into contact with their corpses. Notably, the funeral of one traditional healer in Sierra Leone was one of the major flashpoints in the early spread of the outbreak. With the cases increasing rapidly throughout Guinea, people started to become more desperate. Apprised of her powers, Ebola patients from Guinea crossed the border into Sierra Leone to consult the well-renowned healer, who believed she could cure the disease. Unsurprisingly, the healer herself quickly fell sick and died. Her burial attracted hundreds of mourners over several days, all observing traditional funeral practices including washing and touching the corpse. That single event was directly linked to over 350 Ebola deaths and facilitated the full-blown introduction of the disease into Sierra Leone.

In 2014, at around the peak of the Ebola outbreak, a mathematical study concluded that, approximately 22% of new Ebola cases were attributable to deceased Ebola victims.[132] The same study suggested that by limiting traditional practices, including burial rituals, the basic reproduction number might be reduced to a level at which the outbreak would become unsustainable. One of the most important interventions enforced by the governments of West Africa and humanitarian organisations working in the area was to restrict traditional funeral procedures and ensure that all Ebola victims were given safe and dignified burials. In combination with education campaigns providing alternatives to unsafe traditional practices and imposing travel restrictions even on seemingly healthy individuals, the Ebola outbreak was eventually brought to heel. On the 9th of June 2016, almost two and a half years after Emile Ouamouno's infection, the West African Ebola outbreak was declared over.

Herd immunity

As well as actively helping to tackle infectious disease, mathematical models of epidemics can also help us to understand the unusual features of different disease landscapes. For example, a number of interesting questions surround childhood diseases like mumps and rubella: why do these diseases sweep over us periodically, affecting only children? Perhaps they have a particular predilection towards some elusive childhood quality? And why have they persisted for so long in our society? Perhaps they lie dormant for several years, biding their time between major outbreaks only to strike our most defenceless?

The reason childhood diseases show these typical periodic outbreak patterns in young people is because the effective reproduction number varies over time with the population of susceptible individuals. After a big outbreak has affected large swathes of the unprotected child population, a disease like scarlet fever doesn't just disappear. It persists in the population, but with an effective reproduction number that hovers around 1. The disease only just sustains itself. As time goes by, the population ages and new, unprotected children are born. As the unguarded fraction of the population grows, the effective reproduction number becomes higher and higher, making new outbreaks increasingly likely. When it finally takes off, the victims to whom the disease spreads are usually at the unprotected younger end of the demographic, because most of the older populace are already immune through experiencing the disease. Those people who didn't get the disease as children are typically afforded some protection by the fact that they fraternise with fewer of the infected age group.

The idea that a large population of immune individuals can slow or even halt the spread of a disease, as with the dormant periods between outbreaks of childhood diseases, is a mathematical concept known as 'herd immunity'. Surprisingly, this community effect does not require everyone to be immune to the disease for the whole population to be protected. By reducing the effective reproduction number to less than 1, the chain of transmission can be broken and the disease stopped in its tracks. Crucially, herd immunity means that those with immune systems too weak to tolerate vaccination, including the elderly, new-borns, pregnant women and people with HIV, can still benefit from the protection of vaccinations. The threshold for the immune fraction of the population required to

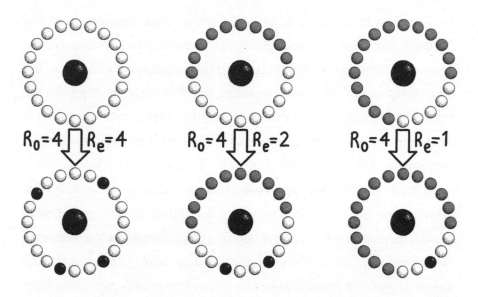

Figure 24: A single infectious individual (black) meets 20 susceptible (white) or vaccinated (grey) individuals, during her week-long infectious period. When no one is vaccinated (*left*) the single infectious individual infects four others, meaning the basic reproduction number, R_0, is 4. When half of the population is vaccinated (*middle*) only two susceptible individuals become infected. The effective reproduction number, R_e, is reduced to 2. Finally (*right*) when 3/4 of the population is vaccinated, only one other person becomes infected on average. The effective reproduction number is reduced to the critical value of 1.

protect the susceptible portion varies depending on how infectious the disease is. The basic reproduction number, R_0, holds the key to how large that proportion is.

Take, for example, a person infected with a virulent strain of flu. If they meet 20 susceptible people during the week in which they are infectious and four of them become infected then the basic reproduction number of the disease, R_0, is 4. Each susceptible person has a one in five chance of being infected. This illustrates how the reproduction number depends on the size of the susceptible population. If our flu patient had only met ten susceptible people during their infectious week (as in the middle panel of Figure 24), with the probability of transmission remaining the same, then they would have infected only two of them, on average, halving the effective reproduction number from 4 to 2.

The most effective way to reduce the size of the susceptible population is through vaccination. The question of how many to vaccinate in order to achieve herd immunity relies on reducing the effective reproduction number to below 1. If we could vaccinate 3/4 of the population then (as in the situation on the right of Figure 24), of the original 20 contacts our flu patient made in a week, only 1/4 (i.e. five) would still be susceptible. On average, only one of them would become infected. It's no coincidence that this critical vaccination threshold, for achieving herd immunity for a disease with basic reproduction number 4, requires 3/4 (which is $1 - 1/4$) of the population to be vaccinated. In general we can only afford to leave $1/R_0$ of the population unvaccinated and must protect the remaining fraction ($1 - 1/R_0$ of the population) if we are to achieve the herd immunity threshold. For smallpox, whose basic reproduction number is around 4, we can afford to leave one 1/4 (or equivalently 25%) of the population unprotected. Vaccinating only 80% (5% above the 75% critical immunisation threshold to provide a buffer) of the susceptible population against smallpox was enough, in 1977, to complete one of the greatest accomplishments of our species – to wipe a human disease clean off the face of the Earth. The feat has never been repeated.

The debilitating and dangerous implications of smallpox

infection alone made it a suitable target for eradication. Its low critical immunisation threshold also made it a relatively easy target. Many diseases are harder to protect against because they spread more easily. Chickenpox, with an estimated R_0 of around 10, would require 9/10 of the population to be immune before the rest would be effectively protected and the disease wiped out. Measles, by far the most infectious human-to-human disease on Earth, with an R_0 estimated to be between 12 and 18, would require between 92% and 95% of the population to be vaccinated. A study that modelled the spread of the 2015 Disneyland measles outbreak – in which Mobius Loop was infected – suggested vaccination rates amongst those exposed to the disease may have been as low as 50%, way below the threshold required for herd immunity.[133]

Mr MMR

Since its introduction in 1988, England's rate of vaccination against measles, through the combined measles, mumps and rubella (MMR) injection, had been steadily rising. In 1996 the vaccination rate hit a record high at 91.8% – close to the critical immunisation threshold for eliminating measles. Then, in 1998, something happened that would derail the vaccination process for years to come.

This public health disaster was not caused by disease-ridden animals, poor sanitation or even failures of government policy, but instead by a sombre five-page publication in the well-respected medical journal, the *Lancet*.[134] In the study, lead author Andrew Wakefield proposed a link between the MMR vaccine and autism spectrum disorders. On the back of his 'findings' Wakefield launched his own personal anti-MMR campaign, stating in a press conference, 'I can't support the continued use of these three vaccines given in combination until this issue has been resolved.' Most of the mainstream media couldn't resist the bait.

Amongst the *Daily Mail*'s headlines covering the story were 'MMR killed my daughter', 'MMR fears gain support' and 'MMR safe? Baloney. This is one scandal that's getting worse.' In the years that followed Wakefield's article, the story snowballed, becoming the biggest UK science story of 2002. Whilst indulging the fears of many fretful parents, the media's coverage of the story typically failed to mention that Wakefield's study was conducted on just 12 children, an extremely small cohort from which to draw meaningful large-scale conclusions. Any coverage that did sound a note of caution about the study was drowned out by the warning sirens emanating from most news outlets. As a result, parents started to withdraw permission for their children to be vaccinated. In the ten years that followed the publication of the infamous *Lancet* paper, the MMR uptake rate would drop from above 90% to below 80%. Confirmed cases of measles would increase from 56 in 1998 to over 1300 ten years later. Cases of mumps, which had been becoming less prevalent throughout the 1990s, suddenly skyrocketed.

In 2004, as instances of measles, mumps and rubella continued to increase, one investigative journalist, Brian Deer, sought to expose Wakefield's work as fraudulent. Deer reported that, prior to submitting his paper, Wakefield had received over £400,000 from lawyers looking for evidence against the pharmaceutical companies that manufacture vaccines. Deer also uncovered documents that, he claimed, showed that Wakefield had submitted patents for a rival vaccine to MMR. Crucially, Deer claimed to have evidence that Wakefield had manipulated the data in his paper to give the false impression of a link to autism. Deer's evidence of Wakefield's scientific fraud and extreme conflicts of interest eventually led to the offending paper's retraction by the *Lancet*'s editors. In 2010, Wakefield was struck off the medical register by the General Medical Council. In the 20 years since Wakefield's original paper, at least 14 comprehensive studies on hundreds of thousands of children across the world have found no evidence of a link between MMR and autism. Sadly, though, Wakefield's influence lives on.

•

Although MMR vaccination in the UK has returned to pre-scare levels, vaccination rates across the developed world as a whole are dropping and measles cases are increasing. In Europe, in 2018, there were more than 60,000 cases of measles, with 72 proving fatal – double the number from the previous year. In the main this is a result of the emergence of the growing anti-vaccination movement. The World Health Organization lists what it calls 'vaccine hesitancy' as one of 2019's top ten global health threats. The *Washington Post*, amongst other media outlets, attributes the rise of the 'anti-vaxxers' directly to Wakefield, describing him as 'the founder of the modern anti-vaccination movement'. The doctrines of the movement, how-ever, have expanded far beyond Wakefield's now debunked findings. They range from assertions that vaccines contain dangerous levels of toxic chemicals to allegations that vaccines actually infect chil-dren with the diseases they are trying to prevent. In reality, toxic chemicals like formaldehyde are produced in higher amounts by our own metabolic system than the trace amounts found in vaccines. Similarly, vaccinations causing the disease they are designed to protect against is an extremely rare occurrence, especially so in otherwise healthy individuals.

Despite many convincing refutations of their claims, 'anti-vax' rhetoric has risen to prominence as a result of support from high-profile celebrities including Jim Carrey, Charlie Sheen and Donald Trump. In a twist almost too implausible to be believed, in 2018, Wakefield confirmed his own elevation to celebrity status when he started dating former supermodel Elle Macpherson.

Alongside the rise of the celebrity activist has come the emergence of social media, allowing these personalities to promulgate their views directly to their fans on their own terms. With the erosion of trust in the mainstream media, people are increasingly turning to these echo chambers for reassurance. The rise of these alternative platforms has provided a space for the anti-vaccination movement to grow, unthreatened and unchallenged by evidence-based science. Wakefield himself even described the emergence of social media as having 'evolved beautifully' – for his purposes, perhaps.

•

We all have choices to make that affect our likelihood of contracting infectious disease: whether to holiday in exotic countries; whom to let our children play with; whether we travel on crowded public transport. When we are ill, other choices we make affect our likelihood of transmitting disease to others: whether we cancel the much-anticipated catch-up with our friends; whether we keep our children home from school; whether we cover our mouths when we cough. The crucial decision as to whether we vaccinate ourselves and our dependents can only be taken ahead of time. It affects our chances, not only of catching, but also of transmitting diseases.

Some of these decisions are inexpensive, making their adoption straightforward. It costs nothing to sneeze into a tissue or a handkerchief. The simple act of washing your hands frequently and carefully has been shown to reduce the effective reproduction numbers of respiratory illnesses like flu by as much as 3/4. For some diseases, this might be enough to take us below the threshold value of R_0, beneath which an infectious disease cannot break out.

Other decisions provide us with more of a dilemma. It is always tempting to send the kids to school even if we know it increases the number of potentially infectious contacts they will make and thus increases the possibility of an epidemic. At the heart of all our choices should be an understanding of their risks and consequences.

Mathematical epidemiology provides a way to assess and understand these decisions. It explains why it is better for everyone if you stay away from work or school if you are ill. It tells us how and why washing our hands can help prevent outbreaks by reducing the force of infection. Sometimes, counter-intuitively, it can highlight that the most terror-inducing diseases are not always the ones we should worry about most.

On a broader scale it suggests strategies to tackle disease outbreaks and the preventative measures we can take to avoid them. In conjunction with reliable scientific evidence, mathematical epidemiology demonstrates that vaccination is a no-brainer. Not only does it protect you, it protects your family, your friends, your neighbours and your colleagues. World Health Organization figures show that vaccines prevent millions of deaths every year and could prevent millions more if we could improve global coverage.[135] They are the best way we have of preventing outbreaks of deadly diseases and the only chance we have of terminating their devastating impacts for good. Mathematical epidemiology is a glimmer of hope for the future, a key that can unlock the secrets of how to achieve these monumental tasks.

Epilogue

MATHEMATICAL EMANCIPATION

Mathematics has shaped our history: through the ancestors who won evolution's numbers game and the diseases that strained and filtered our species. Our biology reflects mathematics' constant unchanging rules. At the same time, our mathematical aesthetics have morphed to reflect our very physiology, and our mathematical understanding has co-evolved with us over millions of years to its current state.

In today's society, mathematics underpins almost everything we do. It is vital to the ways in which we communicate with each other and the methods we use to navigate from place to place. It has completely altered how we buy and sell and it has revolutionised the manner in which we work and relax. Its influence can be felt in almost every courtroom and every hospital ward, in every office and every home.

Maths is being used daily to achieve previously unimaginable tasks. Sophisticated mathematical algorithms allow us to find the answer to almost any question in a matter of seconds. People across the world are linked in an instant by the mathematical power of the internet. The guardians of justice use mathematics as a force for good when detecting criminals through forensic archaeology.

We must remember, however, that mathematics is only as benign as the person or people who wield it. After all, the same mathematics that implicated the art forger, Han van Meegeren, as a criminal also gave us the atomic bomb. It is clear that we should strive to understand the full implications of the mathematical tools we so frequently submit ourselves to. What starts with friend-recommendations and personalised advertising might end with trending fake news or the erosion of our privacy.

As mathematics becomes an increasingly prevalent part of our everyday lives, the opportunities for unexpected disaster multiply. As many times as we have appreciated the wonderful uses of mathematics to achieve hitherto unthinkable feats, we have seen the disastrous consequences of mathematical mistakes. Careful mathematics may have put man on the moon, but careless mathematics destroyed the multi-million-dollar Mars Orbiter. When handled appropriately, maths can be a powerful tool for criminal analysis, but when abused by unscrupulous shysters it can cost innocents their liberty. At its best, maths is the state-of-the-art medical technology that can save lives, but at its worst it is the miscalculated doses that end them. It is our duty to learn from mathematical mistakes so they are not revisited in the future or, better still, are rendered completely unrepeatable.

Mathematical modelling can provide us with some access to what that future will look like. Mathematical models don't just describe the world as it is – the data they are calibrated against – but rather they provide some degree of clairvoyance. Mathematical epidemiology allows us to peer into the future of disease progression and to take proactive preventative measures, rather than always playing reactive games of catch-up. Optimal stopping can give us the best chance of making the best choice when we are not allowed to see all the options beforehand. Personal genomics may revolutionise the understanding of our future disease risk, but only if we can standardise the mathematics with which we interpret the results.

Mathematics has been, is, and ever will be a current running almost unseen beneath the surface of our affairs. We should, however, be careful not to get carried away with the flow by attempting to extend its application beyond its remit. There are places where mathematics is completely the wrong tool for the job, activities in which human supervision is unquestionably necessary. Even if some of the most complex mental tasks can be farmed out to an algorithm, matters of the heart can never be broken down into a simple set of rules. No code or equation will ever imitate the true complexities of the human condition.

Nevertheless, a little mathematical knowledge in our increasingly quantitative society can help us to harness the power of numbers for ourselves. Simple rules allow us to make the best choices and avoid the worst mistakes. Small alterations in the way we think about our rapidly evolving environments help us to 'keep calm' in the face of rapidly accelerating change, or adapt to our increasingly automated realities. Basic models of our actions, reactions and interactions can prepare us for the future before it arrives. The stories relating other people's experiences are, in my view, the simplest and most powerful models of all. They allow us to learn from the mistakes of our predecessors so that, before we embark on any numerical expedition, we ensure we are all speaking the same language, have synchronised our watches, and checked we've got enough fuel in the tank.

Half the battle for mathematical empowerment is daring to question the perceived authority of those who wield the weapons – shattering the illusion of certainty. Appreciating absolute and relative risks, ratio biases, mismatched framing and sampling bias gives us the power to be sceptical of the statistics screamed from newspaper headlines, the 'studies' pushed at us in adverts, or the half-truths that come tumbling from the mouths of our politicians. Recognising ecological fallacies and dependent events allows us to disperse obfuscating smoke screens, making it harder to fool us with mathematical arguments, be they in the courtroom, the classroom or the clinic.

We must ensure that the person with the most shocking statistics doesn't always win the argument, by demanding an explanation of the maths behind the figures. We shouldn't let medical charlatans delay us from receiving potentially life-saving treatment when their alternative therapies are just a regression to the mean. We mustn't let the anti-vaxxers make us doubt the efficacy of vaccinations, when mathematics proves that they can save vulnerable lives and wipe out disease.

It is time for us to take the power back into our own hands, because sometimes maths really is a matter of life and death.

Acknowledgements

The title and the main thrust for *The Maths of Life and Death*, as a book about the hidden places in which maths affects our everyday lives, came from a drunken conversation in the pub the first time I met my agent, Chris Wellbelove, face to face. Chris has looked at every draft of every pitch and chapter that I have sent him and done so much more besides. I owe him a great debt for taking a chance on me and for steering me successfully through the process of pitching and writing my first book.

From the day I signed with Quercus, my editor, Katy Follain, has had my back. She has looked at numerous drafts of the book and made suggestions that have immeasurably improved it. Similarly, Sarah Goldberg, my US editor, has had a huge influence on the direction that the book has taken. The fact that Katy and Sarah took the time to get their heads together and give me coherent feedback makes me all the more grateful. Thanks to them and to all the other people who have worked tirelessly behind the scenes at Quercus and Scribner to make this book happen.

I owe a huge debt of gratitude to all the people whom I contacted when writing this book and who so kindly agreed to share their stories with me. Your tales of mathematical catastrophes and triumphs are the fabric of the book. It simply wouldn't have happened without the time and generosity you put in to answering my long lists of seemingly irrelevant questions.

I am grateful to the Institute for Mathematical Innovation at the University of Bath, who have supported me through an internal secondment. This has been hugely influential in allowing me to deliver the book I wanted with the care it deserved. More broadly, many of my colleagues across the wider university to whom I have spoken about the book have given me great encouragement and support. My *alma mater* Somerville College, Oxford also provided me a place to work when I needed to be out of the house, and I am grateful for that.

At the start of the writing process, when I realised I needed sound quantitative minds to critique my work, I looked to my former PhD colleagues and close friends Gabriel Rosser and Aaron Smith. Not quite knowing what they were letting themselves in for, they agreed to look over early drafts of the manuscript for me despite both having small babies and many other life complications to contend with. My sincerest gratitude goes to them for the improvements their comments have made to the book.

My great friend and colleague Chris Guiver has been so kind as to let me lodge at his house once a week for over a year while I have been writing this book. He has acted as an excellent sounding board for my ideas, discussing and debating with me long into the night about the book, about science and about life more generally. Chris, you probably don't realise how much your generosity has meant to me. Thank you.

My parents, Tim and Mary, have been my most staunch supporters throughout this process. They have read the whole book through *twice*. They are my audience of intelligent laypeople. More than just their insightful comments and thorough proof-reading, however, I owe them my education and my values. You have supported me through highs and lows. I will never be able to thank you enough.

My sister, Lucy, was responsible for helping me to weave the threads of my early ideas together into something resembling a coherent pitch. It is no exaggeration to say the book would not exist without the time and effort she put into sensitively critiquing my writing and setting me on the right path at the start.

I owe a, perhaps less tangible, debt of thanks to my wider family for, amongst other things, never complaining when I went off to sleep in the middle of a family gathering because I'd stayed up late the previous night to work on the book. The importance of that respite cannot be overestimated.

Finally, to the people who have probably endured most for this book: my family. My wife, Caroline, has been incredibly supportive of the project, even fiddling with the genetics sections of the book when I let her. Not only has she been supporting a fledgling author, but she's also been a bad-ass mum and full-time CEO to boot. My admiration for you is unwavering. Lastly, to Em and Will. Thanks for keeping my feet on the ground. Every worry goes out of my head when I come home; there isn't space for anything but you two. Even if this book doesn't sell a single copy, I know it won't make a difference to you.

References

Introduction

1 *page xi* 'By providing an estimate of the number of fish in a lake'

Pollock, K. H. (1991). Modeling capture, recapture, and removal statistics for estimation of demographic parameters for fish and wildlife populations: past, present, and future. *Journal of the American Statistical Association, 86*(413), 225. https://doi.org/10.2307/2289733

2 *page xi* 'numbers of drug addicts'

Doscher, M. L., & Woodward, J. A. (1983). Estimating the size of subpopulations of heroin users: applications of log-linear models to capture/recapture sampling. *The International Journal of the Addictions, 18*(2), 167–82.

Hartnoll, R., Mitcheson, M., Lewis, R., & Bryer, S. (1985). Estimating the prevalence of opioid dependence. *Lancet, 325*(8422), 203–5. https://doi.org/10.1016/S0140-6736(85)92036-7

Woodward, J. A., Retka, R. L., & Ng, L. (1984). Construct validity of heroin abuse estimators. *International Journal of the Addictions, 19*(1), 93–117. https://doi.org/10.3109/10826088409055819

3 *page xi* 'the number of war dead in Kosovo'

Spagat, M. (2012). *Estimating the Human Costs of War: The Sample Survey Approach.* Oxford University Press. https://doi.org/10.1093/oxfordhb/9780195392777.013.0014

Chapter 1

4 *page 2* '*Strep f.* is one of the bacteria responsible for the souring and curdling of milk, but one cell is no big deal, right?'

Botina, S. G., Lysenko, A. M., & Sukhodolets, V. V. (2005). Elucidation of the taxonomic status of industrial strains of thermophilic lactic acid bacteria by sequencing of 16S rRNA genes. *Microbiology*, *74*(4), 448–52. https://doi.org/10.1007/s11021-005-0087-7

5 *page 2* 'Maybe it's more worrying when you find out that, in milk, *Strep f.* cells can divide to produce two daughter cells every hour'

Cárdenas, A. M., Andreacchio, K. A., & Edelstein, P. H. (2014). Prevalence and detection of mixed-population enterococcal bacteremia. *Journal of Clinical Microbiology*, *52*(7), 2604–8. https://doi.org/10.1128/JCM.00802-14

Lam, M. M. C., Seemann, T., Tobias, N. J., Chen, H., Haring, V., Moore, R. J., . . . Stinear, T. P. (2013). Comparative analysis of the complete genome of an epidemic hospital sequence type 203 clone of vancomycin-resistant Enterococcus faecium. *BMC Genomics*, *14*, 595. https://doi.org/10.1186/1471-2164-14-595

6 *page 12* 'He published in the journal *Nature* evidence that, upon fission caused by a single neutron, atoms of the uranium isotope, U-235, emitted on average 3.5 (later revised to 2.5) high energy neutrons'

Von Halban, H., Joliot, F., & Kowarski, L. (1939). Number of neutrons liberated in the nuclear fission of uranium. *Nature*, *143*(3625), 680. https://doi.org/10.1038/143680a0

7 *page 13* 'the other isotope, which makes up 99.3% of naturally occurring uranium'

Webb, J. (2003). Are the laws of nature changing with time? *Physics World*, *16*(4), 33–8. https://doi.org/10.1088/2058-7058/16/4/38

8 *page 17* 'One kilogram of uranium can release roughly three million times more energy than burning the same amount of coal.'

Bernstein, J. (2008). *Nuclear Weapons: What You Need to Know*. Cambridge University Press.

9 *page 19* 'The fire drew into the atmosphere hundreds of times more radioactive material than had been released during the bombing of Hiroshima, causing widespread environmental consequences for almost all of Europe.'

International Atomic Energy Agency. (1996). Ten years after Chernobyl: what do we really know? In *Proceedings of the IAEA/WHO/EC International Conference: One Decade after Chernobyl: Summing Up the Consequences*. Vienna: International Atomic Energy Agency.

10 *page 20* 'the elimination of drugs in the body'

Greenblatt, D. J. (1985). Elimination half-life of drugs: value and limitations. *Annual Review of Medicine, 36*(1), 421–7. https://doi.org/10.1146/annurev.me.36.020185.002225

Hastings, I. M., Watkins, W. M., & White, N. J. (2002). The evolution of drug-resistant malaria: the role of drug elimination half-life. *Philosophical Transactions of the Royal Society of London.* Series B: Biological Sciences, 357(1420), 505–19. https://doi.org/10.1098/rstb.2001.1036

11 *page 20* 'the rate of decrease of the head on a pint of beer'

Leike, A. (2002). Demonstration of the exponential decay law using beer froth. *European Journal of Physics, 23*(1), 21–6. https://doi.org/10.1088/0143-0807/23/1/304

Fisher, N. (2004). The physics of your pint: head of beer exhibits exponential decay. *Physics Education, 39*(1), 34–5. https://doi.org/10.1088/0031-9120/39/1/F11

12 *page 21* 'In particular, it does an excellent job of describing the rate at which the levels of radiation emitted by a radioactive substance decrease over time.'

Rutherford, E., & Soddy, F. (1902). LXIV. The cause and nature of radioactivity. Part II. *The London, Edinburgh, and Dublin Philosophical Magazine and Journal of Science, 4*(23), 569–85. https://doi.org/10.1080/14786440209462881

Rutherford, E., & Soddy, F. (1902). XLI. The cause and nature of radioactivity. Part I. *The London, Edinburgh, and Dublin Philosophical Magazine and Journal of Science, 4*(21), 370–96. https://doi.org/10.1080/14786440209462856

13 *page 21* 'determining the age of ancient artefacts like the Dead Sea scrolls'

Bonani, G., Ivy, S., Wölfli, W., Broshi, M., Carmi, I., & Strugnell, J. (1992). Radiocarbon dating of Fourteen Dead Sea Scrolls. *Radiocarbon, 34*(03), 843–9. https://doi.org/10.1017/S0033822200064158

Carmi, I. (2000). Radiocarbon dating of the Dead Sea Scrolls. In L. Schiffman, E. Tov, & J. VanderKam (eds.), *The Dead Sea Scrolls: Fifty Years After Their Discovery. 1947–1997* (p. 881).

Bonani, G., Broshi, M., & Carmi, I. (1991). 14 Radiocarbon dating of the Dead Sea scrolls. *'Atiqot,* Israel Antiquities Authority.

14 *page 21* 'archaeopteryx was 150 million years old'

Starr, C., Taggart, R., Evers, C. A., & Starr, L. (2019). *Biology: The Unity and Diversity of Life,* Cengage Learning.

15 *page 21* 'Ötzi the iceman died 5300 years ago'

Bonani, G., Ivy, S. D., Hajdas, I., Niklaus, T. R., & Suter, M. (1994). Ams 14C age determinations of tissue, bone and grass samples from the ötztal ice man. *Radiocarbon*, *36*(02), 247–250. https://doi.org/10.1017/S0033822200040534

16 *page 24* 'This established for certain that Van Meegeren's forgeries couldn't have been painted by Vermeer in the 17th century as the lead which Van Meegeren used for his paints had not yet been mined.'

Keisch, B., Feller, R. L., Levine, A. S., & Edwards, R. R. (1967). Dating and authenticating works of art by measurement of natural alpha emitters. *Science*, *155*(3767), 1238–42. https://doi.org/10.1126/science.155.3767.1238

17 *page 26* 'As a result of the funding received during the challenge, researchers discovered a third gene responsible for ALS, demonstrating the viral campaign's far-reaching impact.'

Kenna, K. P., van Doormaal, P. T. C., Dekker, A. M., Ticozzi, N., Kenna, B. J., Diekstra, F. P., . . . Landers, J. E. (2016). NEK1 variants confer susceptibility to amyotrophic lateral sclerosis. *Nature Genetics*, *48*(9), 1037–42. https://doi.org/10.1038/ng.3626

18 *page 28* 'Computer scientist Vernor Vinge encapsulated just such ideas in a series of science fiction novels'

Vinge, V. (1986). *Marooned in Realtime*. Bluejay Books/ St. Martin's Press.

Vinge, V. (1992). *A Fire Upon the Deep*. Tor Books.

Vinge, V. (1993). The coming technological singularity: how to survive in the post-human era. In *NASA. Lewis Research Center, Vision 21: Interdisciplinary Science and Engineering in the Era of Cyberspace* (pp. 11–22). Retrieved from https://ntrs.nasa.gov/search.jsp?R=19940022856

19 *page 28* 'In 1999, in his book *The Age of Spiritual Machines*, Kurzweil hypothesised the "law of accelerating returns".'

Kurzweil, R. (1999). *The Age of Spiritual Machines: When Computers Exceed Human Intelligence*. Viking.

20 *page 29* 'He even went so far as to pin the date of Vinge's "technological singularity" – the point at which we will experience, as Kurzweil describes it, "technological change so rapid and profound it represents a rupture in the fabric of human history" – to around 2045'

Kurzweil, R. (2004). The law of accelerating returns. In *Alan Turing: Life and Legacy of a Great Thinker* (pp. 381–416). Springer Berlin Heidelberg. https://doi.org/10.1007/978-3-662-05642-4_16

21 *page 29* 'The complete "Book of Life" was delivered in 2003, ahead of schedule and within its one-billion-dollar budget.'

Gregory, S. G., Barlow, K. F., McLay, K. E., Kaul, R., Swarbreck, D., Dunham, A., ... Bentley, D. R. (2006). The DNA sequence and biological annotation of human chromosome 1. *Nature, 441*(7091), 315–21. https://doi.org/10.1038/nature04727

International Human Genome Sequencing Consortium. (2001). Initial sequencing and analysis of the human genome. *Nature, 409*(6822), 860–921. https://doi.org/10.1038/35057062

Pennisi, E. (2001). The human genome. *Science, 291*(5507), 1177–80. https://doi.org/10.1126/SCIENCE.291.5507.1177

22 *page 30* 'The perceived rapid increase in population at that time prompted the English mathematician, Thomas Malthus, to suggest that the human population grows at a rate that is proportional to its current size.'

Malthus, T. R. (2008). *An Essay on the Principle of Population.* (Ed. R. Thomas and G. Gilbert) Oxford University Press.

23 *page 32* 'was the first to demonstrate that logistic growth occurred in bacterial populations'

McKendrick, A. G., & Pai, M. K. (1912). The rate of multiplication of micro-organisms: a mathematical study. *Proceedings of the Royal Society of Edinburgh, 31*, 649–53. https://doi.org/10.1017/S0370164600025426

24 *page 32* 'sheep'

Davidson, J. (1938). On the ecology of the growth of the sheep population in South Australia. *Trans. Roy. Soc. S. A., 62*(1), 11–148.

Davidson, J. (1938). On the growth of the sheep population in Tasmania. *Trans. Roy. Soc. S. A., 62*(2), 342–6.

25 *page 32* 'seals'

Jeffries, S., Huber, H., Calambokidis, J., & Laake, J. (2003). Trends and status of harbor seals in Washington State: 1978–1999. *The Journal of Wildlife Management, 67*(1), 207. https://doi.org/10.2307/3803076

26 *page 32* 'cranes'

Flynn, M. N., & Pereira, W. R. L. S. (2013). Ecotoxicology and environmental contamination. *Ecotoxicology and Environmental Contamination, 8*(1), 75–85.

27 *page 33* 'The eminent sociobiologist, E. O. Wilson, believes that there are inherent, hard limits on the size of human population that the Earth's biosphere can support.'

Wilson, E. O. (2002). *The Future of Life* (1st ed.). Alfred A. Knopf.

28 *page 34* 'Our growth rate reached a peak of around 2% per year in the late 1960s, but is projected to fall below 1% per year by 2023.'

Raftery, A. E., Alkema, L., & Gerland, P. (2014). Bayesian Population Projections for the United Nations. *Statistical Science: A Review Journal of the Institute of Mathematical Statistics*, *29*(1), 58–68. https://doi.org/10.1214/13-STS419

Raftery, A. E., Li, N., Ševčíková, H., Gerland, P., & Heilig, G. K. (2012). Bayesian probabilistic population projections for all countries. *Proceedings of the National Academy of Sciences of the United States of America*, *109*(35), 13915–21. https://doi.org/10.1073/pnas.1211452109

United Nations Department of Economic and Social Affairs Population Division. (2017). World population prospects: the 2017 revision, key findings and advance tables, *ESA/P/WP/2*.

29 *page 35* 'I should remember not to be too caustic with my parents, though, because it seems that perceived time really does run more quickly the older we get, fuelling our increasing feelings of overburdened time-poverty.'

Block, R. A., Zakay, D., & Hancock, P. A. (1999). Developmental changes in human duration judgments: a meta-analytic review. *Developmental Review*, *19*(1), 183–211. https://doi.org/10.1006/DREV.1998.0475

30 *page 35* 'On average the younger group clocked an almost-perfect three minutes and three seconds of real time, but the older group didn't call a halt until a staggering three minutes and 40 seconds, on average.'

Mangan, P., Bolinskey, P., & Rutherford, A. (1997). Underestimation of time during aging: the result of age-related dopaminergic changes. In *Annual Meeting of the Society for Neuroscience*.

31 *page 35* 'In other related experiments, participants were asked to estimate the length of a fixed period of time during which they had been undertaking a task. Older participants consistently gave shorter estimates for the length the time period they had experienced than younger groups.'

Craik, F. I. M., & Hay, J. F. (1999). Aging and judgments of duration: Effects of task complexity and method of estimation. *Perception & Psychophysics*, *61*(3), 549–60. https://doi.org/10.3758/BF03211972

32 *page 35* 'One theory is related to the fact that our metabolism slows as we get older, matching the slowing of our heartbeats and our breathing.'

Church, R. M. (1984). Properties of the Internal Clock. *Annals*

of the New York Academy of Sciences, 423(1), 566–82. https://doi.
org/10.1111/j.1749-6632.1984.tb23459.x

Craik, F. I. M., & Hay, J. F. (1999). Aging and judgments of duration:
effects of task complexity and method of estimation. *Perception &
Psychophysics, 61*(3), 549–60. https://doi.org/10.3758/BF03211972

Gibbon, J., Church, R. M., & Meck, W. H. (1984). Scalar timing in
memory. *Annals of the New York Academy of Sciences, 423*(1 Timing and Ti),
52–77. https://doi.org/10.1111/j.1749-6632.1984.tb23417.x

33 *page 36* 'A competing theory suggests that our perception of time's passage
depends upon the amount of new perceptual information we are subjected
to from our environment.'

Pennisi, E. (2001). The human genome. *Science, 291*(5507), 1177–80.
https://doi.org/10.1126/SCIENCE.291.5507.1177

34 *page 36* 'Experiments on subjects experiencing the unfamiliar sensation of
free fall have demonstrated this to be the case.'

Stetson, C., Fiesta, M. P., & Eagleman, D. M. (2007). Does time really
slow down during a frightening event? *PLoS ONE, 2*(12), e1295. https://doi.
org/10.1371/journal.pone.0001295

Chapter 2

35 *page 49* 'I tried to replicate the late onset Alzheimer's risks in my genetic
report using the same method as 23andMe and data taken directly from the
report or from papers they cited.'

Farrer, L. A., Cupples, L. A., Haines, J. L., Hyman, B., Kukull, W. A.,
Mayeux, R., . . . Duijn, C. M. van. (1997). Effects of age, sex, and ethnicity on
the association between apolipoprotein E genotype and Alzheimer disease.
JAMA, 278(16), 1349. https://doi.org/10.1001/jama.1997.03550160069041

Gaugler, J., James, B., Johnson, T., Scholz, K., & Weuve, J. (2016).
2016 Alzheimer's disease facts and figures. *Alzheimer's & Dementia, 12*(4),
459–509. https://doi.org/10.1016/J.JALZ.2016.03.001

Genin, E., Hannequin, D., Wallon, D., Sleegers, K., Hiltunen, M.,
Combarros, O., . . . Campion, D. (2011). APOE and Alzheimer disease: a
major gene with semi-dominant inheritance. *Molecular Psychiatry, 16*(9),
903–7. https://doi.org/10.1038/mp.2011.52

Jewell, N. P. (2004). *Statistics for Epidemiology.* Chapman & Hall/CRC.

Macpherson, M., Naughton, B., Hsu, A. and Mountain, J. (2007), *Estimating Genotype-Specific Incidence for One or Several Loci*, 23andMe.

Risch, N. (1990). Linkage strategies for genetically complex traits. I. Multilocus models. *American Journal of Human Genetics*, 46(2), 222–8.

36 *page 49* 'My conclusion was reinforced when I came across the findings of a 2014 study which investigated the risk-calculation methods of three of the leading personal genomic companies, including 23andMe.'

Kalf, R. R. J., Mihaescu, R., Kundu, S., de Knijff, P., Green, R. C., & Janssens, A. C. J. W. (2014). Variations in predicted risks in personal genome testing for common complex diseases. *Genetics in Medicine*, 16(1), 85–91. https://doi.org/10.1038/gim.2013.80

37 *page 51* 'In fact, BMI was first cooked up in 1835 by Belgian Adolphe Quetelet, a renowned astronomer, statistician, sociologist and mathematician but, notably, not a physician.'

Quetelet, L. A. J. (1994). A treatise on man and the development of his faculties. *Obesity Research*, 2(1), 72–85. https://doi.org/10.1002/j.1550-8528.1994.tb00047.x

38 *page 52* 'In response to unprecedented levels of obesity, American physiologist Ancel Keys (who would later make the link between saturated fat and cardiovascular disease) undertook a study to find the best indicator of excess weight.'

Keys, A., Fidanza, F., Karvonen, M. J., Kimura, N., & Taylor, H. L. (1972). Indices of relative weight and obesity. *Journal of Chronic Diseases*, 25(6–7), 329–43. https://doi.org/10.1016/0021-9681(72)90027-6

39 *page 52* 'If the definition of obesity were instead based on high percentage body fat, between 15 and 35% of men with non-obese BMIs would be reclassified as obese.'

Tomiyama, A. J., Hunger, J. M., Nguyen-Cuu, J., & Wells, C. (2016). Misclassification of cardiometabolic health when using body mass index categories in NHANES 2005–2012. *International Journal of Obesity*, 40(5), 883–6. https://doi.org/10.1038/ijo.2016.17

40 *page 53* 'These incorrect classifications have implications for the way in which we measure and record obesity at a population level. Perhaps more worryingly though, diagnosing healthy individuals as overweight or obese based on their BMI can also have detrimental effects on their mental health.'

McCrea, R. L., Berger, Y. G., & King, M. B. (2012). Body mass index and common mental disorders: exploring the shape of the association and its moderation by age, gender and education. *International Journal of Obesity,* *36*(3), 414–21. https://doi.org/10.1038/ijo.2011.65

41 *page 63* 'However, approximately 85% of automated warnings in ICUs)are false alarms.'

Sendelbach, S., & Funk, M. (2013). Alarm fatigue: a patient safety concern. *AACN Advanced Critical Care, 24*(4), 378–86; quiz 387-8. https://doi.org/10.1097/NCI.0b013e3182a903f9

Lawless, S. T. (1994). Crying wolf: false alarms in a pediatric intensive care unit. *Critical Care Medicine, 22*(6), 981–85.

42 *page 66* 'For the same reason, median filtering is beginning to be used in our ICU monitors to prevent false alarms.'

Mäkivirta, A., Koski, E., Kari, A., & Sukuvaara, T. (1991). The median filter as a preprocessor for a patient monitor limit alarm system in intensive care. *Computer Methods and Programs in Biomedicine, 34*(2–3), 139–44. https://doi.org/10.1016/0169-2607(91)90039-V

43 *page 66* 'Median filtering can reduce the occurrence of false alarms in ICU monitors by as much as 60% without jeopardising patient safety.'

Imhoff, M., Kuhls, S., Gather, U., & Fried, R. (2009). Smart alarms from medical devices in the OR and ICU. *Best Practice & Research Clinical Anaesthesiology, 23*(1), 39–50. https://doi.org/10.1016/J.BPA.2008.07.008

44 *page 68* 'Indeed, for people who have breast cancer, the test will pick this up roughly nine times out of ten. For people who don't have the disease, the results of the test will tell you this correctly nine out of ten times.'

Hofvind, S., Geller, B. M., Skelly, J., & Vacek, P. M. (2012). Sensitivity and specificity of mammographic screening as practised in Vermont and Norway. *The British Journal of Radiology, 85*(1020), e1226–32. https://doi.org/10.1259/bjr/15168178

45 *page 70* 'In 2007, a group of 160 gynaecologists were given the following information about the accuracy of mammograms and the prevalence of breast cancer in the population'

Gigerenzer, G., Gaissmaier, W., Kurz-Milcke, E., Schwartz, L. M., & Woloshin, S. (2007). Helping doctors and patients make sense of health statistics. *Psychological Science in the Public Interest, 8*(2), 53–96. https://doi.org/10.1111/j.1539-6053.2008.00033.x

46 *page 72* 'Writing in the *British Medical Journal*, Muir Gray, former director of the UK National Screening Programme, admitted'

Gray, J. A. M., Patnick, J., & Blanks, R. G. (2008). Maximising benefit and minimising harm of screening. *BMJ (Clinical Research Ed.)*, 336(7642), 480–83. https://doi.org/10.1136/bmj.39470.643218.94

47 *page 74* 'In 2006, 1000 adults in Germany were asked whether a series of tests gave results that were 100% certain.'

Gigerenzer, G., Gaissmaier, W., Kurz-Milcke, E., Schwartz, L. M., & Woloshin, S. (2007). Helping doctors and patients make sense of health statistics. *Psychological Science in the Public Interest*, 8(2), 53–96. https://doi.org/10.1111/j.1539-6053.2008.00033.x

48 *page 75* 'However, at the time he took the test, ELISA had reported false positive rates of around 0.3%.'

Cornett, J. K., & Kirn, T. J. (2013). Laboratory diagnosis of HIV in adults: a review of current methods. *Clinical Infectious Diseases*, 57(5), 712–18. https://doi.org/10.1093/cid/cit281

49 *page 79* 'In December 2016, a global team of researchers developed a blood test for Creutzfeldt-Jakob disease (CJD).'

Bougard, D., Brandel, J.-P., Bélondrade, M., Béringue, V., Segarra, C., Fleury, H., . . . Coste, J. (2016). Detection of prions in the plasma of presymptomatic and symptomatic patients with variant Creutzfeldt-Jakob disease. *Science Translational Medicine*, 8(370), 370ra182. https://doi.org/10.1126/scitranslmed.aag1257

50 *page 83* 'In one case, a woman miscarried after she was sanctioned to undergo a surgical procedure that would never have been undertaken had she known she was pregnant.'

Sigel, C. S., & Grenache, D. G. (2007). Detection of unexpected isoforms of human chorionic gonadotropin by qualitative tests. *Clinical Chemistry*, 53(5), 989–90. https://doi.org/10.1373/clinchem.2007.085399

51 *page 83* 'Another woman's ectopic pregnancy was missed by urine tests, leading to a ruptured fallopian tube and life-threatening blood loss.'

Daniilidis, A., Pantelis, A., Makris, V., Balaouras, D., & Vrachnis, N. (2014). A unique case of ruptured ectopic pregnancy in a patient with negative pregnancy test – a case report and brief review of the literature. *Hippokratia*, 18(3), 282–84.

Chapter 3

52 *page 91* 'Bertillon suggested that the similarities were not coincidences, but 'must have been done carefully on purpose, and must denote a purposeful intention, probably a secret code".'

    Schneps, L., & Colmez, C. (2013). *Math on trial : how numbers get used and abused in the courtroom*, Basic Books (New York).

53 *page 92* 'By exposing Bertillon's miscalculation and arguing that even attempting to apply probability theory to such a question was not legitimate, Poincaré was able to debunk the aberrant handwriting analysis and in so doing to exonerate Dreyfus.'

    Jean Mawhin. (2005). Henri Poincaré. A life in the service of science. *Notices of the American Mathematical Society, 52*(9), 1036–44.

54 *page 93* 'The Japanese criminal justice system, for example, has a conviction rate of 99.9%, with most of these convictions backed up with a confession.'

    Ramseyer, J. M., & Rasmusen, E. B. (2001). Why is the Japanese conviction rate so high? *The Journal of Legal Studies, 30*(1), 53–88. https://doi.org/10.1086/468111

55 *page 96* 'In 1989, Meadow, at the time an eminent British paediatrician, had edited a book, *ABC of Child Abuse*, in which was contained the aphorism that came to be known as Meadow's law: 'One sudden infant death is a tragedy, two is suspicious and three is murder until proved otherwise'.

    Meadow, R. (Ed.) (1989). *ABC of Child Abuse* (First edition). British Medical Journal Publishing Group.

56 *page 97* 'The prevalence of autism in the UK is roughly 1 per 100'

    Brugha, T., Cooper, S., McManus, S., Purdon, S., Smith, J., Scott, F., . . . Tyrer, F. (2012). *Estimating the Prevalence of Autism Spectrum Conditions in Adults – Extending the 2007 Adult Psychiatric Morbidity Survey – NHS Digital.*

57 *page 98* 'Only one in five of those on the autistic spectrum are female'

    Ehlers, S., & Gillberg, C. (1993). The Epidemiology of Asperger Syndrome. *Journal of Child Psychology and Psychiatry, 34*(8), 1327–50. https://doi.org/10.1111/j.1469-7610.1993.tb02094.x

58 *page 98* 'For his figures, Meadow used a - then unpublished - report on SIDS for which he had been asked to write the preface.'

    Fleming, P. J., Blair, P. S. P., Bacon, C., & Berry, P. J. (2000). *Sudden*

unexpected deaths in infancy: the CESDI SUDI studies 1993–1996. The
Stationery Office.

Leach, C. E. A., Blair, P. S., Fleming, P. J., Smith, I. J., Platt, M. W., Berry,
P. J., ... Group, the C. S. R. (1999). Epidemiology of SIDS and explained
sudden infant deaths. *Pediatrics, 104*(4), e43.

59 *page 99* 'In 2001, researchers at the University of Manchester also identified
markers in genes related to the regulation of the immune system which put
children at increased risk of SIDS.'

Summers, A. M., Summers, C. W., Drucker, D. B., Hajeer, A. H., Barson,
A., & Hutchinson, I. V. (2000). Association of IL-10 genotype with sudden
infant death syndrome. *Human Immunology, 61*(12), 1270–73. https://doi.
org/10.1016/S0198-8859(00)00183-X

60 *page 99* 'Many more genetic risk factors have since been identified.'

Brownstein, C. A., Poduri, A., Goldstein, R. D., & Holm, I. A. (2018).
The genetics of Sudden Infant Death Syndrome. In *SIDS: Sudden Infant and
Early Childhood Death: The Past, the Present and the Future.*

Dashash, M., Pravica, V., Hutchinson, I. V., Barson, A. J., & Drucker, D.
B. (2006). Association of Sudden Infant Death Syndrome with VEGF and
IL-6 Gene polymorphisms. *Human Immunology, 67*(8), 627–33. https://doi.
org/10.1016/J.HUMIMM.2006.05.002

61 *page 106* 'measuring the health of the economy'

Ma, Y. Z. (2015). Simpson's paradox in GDP and per capita GDP
growths. *Empirical Economics, 49*(4), 1301–15. https://doi.org/10.1007/
s00181-015-0921-3

62 *page 106* 'understanding voter profiles'

Nurmi, H. (1998). Voting paradoxes and referenda. *Social Choice and
Welfare, 15*(3), 333–50. https://doi.org/10.1007/s003550050109

63 *page 106* 'drug development'

Abramson, N. S., Kelsey, S. F., Safar, P., & Sutton-Tyrrell, K. (1992).
Simpson's paradox and clinical trials: What you find is not necessarily what
you prove. *Annals of Emergency Medicine, 21*(12), 1480–82. https://doi.
org/10.1016/S0196-0644(05)80066-6

64 *page 109* 'Low birth-weight had long been associated with higher infant
mortality, but it seemed that smoking during pregnancy was providing some
protection to low birth-weight babies.'

Yerushalmy, J. (1971). The relationship of parents' cigarette smoking
to outcome of pregnancy – implications as to the problem of inferring

causation from observed associations. *American Journal of Epidemiology,* *93*(6), 443–56. https://doi.org/10.1093/oxfordjournals.aje.a121278

65   *page 109* 'In reality, it was nothing of the sort.'

Wilcox, A. J. (2001). On the importance – and the unimportance – of birthweight. *International Journal of Epidemiology, 30*(6), 1233–41. https://doi.org/10.1093/ije/30.6.1233

66   *page 114* 'Double infant murder has been calculated to be between ten and 100 times less frequent than double SIDS death.'

Dawid, A. P. (2005). Bayes's theorem and weighing evidence by juries. In Richard Swinburne (ed.), *Bayes's Theorem.* British Academy. https://doi.org/10.5871/bacad/9780197263419.003.0004

Hill, R. (2004). Multiple sudden infant deaths – coincidence or beyond coincidence? *Paediatric and Perinatal Epidemiology, 18*(5), 320–26. https://doi.org/10.1111/j.1365-3016.2004.00560.x

67   *page 122* 'But Leila Schneps and Coralie Colmez, authors of the 2013 book, *Math on Trial: How Numbers Get Used and Abused in the Courtroom,* suggest that Judge Hellmann was wrong, sometimes two unreliable tests are better than one.'

Schneps, L., & Colmez, C. (2013). *Math on Trial: How Numbers Get Used and Abused in the Courtroom.*

68   *page 126* 'the use of cranberries to treat urinary tract infections'

Jepson, R. G., Williams, G., & Craig, J. C. (2012). Cranberries for preventing urinary tract infections. *Cochrane Database of Systematic Reviews,* (10). https://doi.org/10.1002/14651858.CD001321.pub5

69   *page 126* 'the use of vitamin C for preventing the common cold.'

Hemilä, H., Chalker, E., & Douglas, B. (2007). Vitamin C for preventing and treating the common cold. *Cochrane Database of Systematic Reviews,* (3). https://doi.org/10.1002/14651858.CD000980.pub3

Chapter 4

70   *page 134* 'Truthfulness and accuracy are near the top (if not the top) of the list on almost all codes of journalistic ethics and integrity.'

American Society of News Editors (2019). ASNE Statement of Principles. Retrieved March 16, 2019, from https://www.asne.org/content.asp?pl=24&sl=171&contentid=171

International Federation of Journalists. (2019). Principles on Conduct of Journalism – IFJ. Retrieved March 16, 2019, from https://www.ifj.org/who/rules-and-policy/principles-on-conduct-of journalism.html

Associated Press Media Editors. (2019). Statement of Ethical Principles – APME. Retrieved March 16, 2019, from https://www.apme.com/page/EthicsStatement?&hhsearchterms=%22ethics%22

Society of Professional Journalists. (2019). SPJ Code of Ethics. Retrieved March 16, 2019, from https://www.spj.org/ethicscode.asp

71 *page 141* 'Based on this study'

Troyer, K., Gilboy, T., & Koeneman, B. (2001). A nine STR locus match between two apparently unrelated individuals using AmpFlSTR® Profiler Plus and Cofiler. In *Genetic Identity Conference Proceedings, 12th International Symposium on Human Identification*. Retrieved from https://www.promega.ee/~/media/files/resources/conference proceedings/ishi 12/poster abstracts/troyer.pdf

72 *page 142* 'If 122 matches had turned up in a database as small as 65,000 people, could DNA really be relied upon to uniquely identify suspects in a country with a population of 300 million?'

Curran, J. (2010). Are DNA profiles as rare as we think? Or can we trust DNA statistics? *Significance, 7*(2), 62–6. https://doi.org/10.1111/j.1740-9713.2010.00420.x

73 *page 146* 'In 2014, the Federal Trade Commission (FTC) wrote to L'Oréal charging them with deceptive advertising over the Génifique range.'

Ramirez, E., Brill, J., Ohlhausen, M. K., Wright, J. D., Terrell, M., & Clark, D. S. (2014). In the matter of L'Oréal USA, Inc., a corporation. Docket No. C. Retrieved from https://www.ftc.gov/system/files/documents/cases/140627lorealcmpt.pdf

74 *page 148* 'Four years earlier, in 1932, they had predicted Roosevelt's victory margin to within a percentage point.'

Squire, P. (1988). Why the 1936 *Literary Digest* poll failed. *Public Opinion Quarterly, 52*(1), 125. https://doi.org/10.1086/269085

75 *page 148* 'In August, they sent out straw polls to everyone they had identified, and trumpeted in the magazine'

Simon, J. L. (2003). *The Art of Empirical Investigation*. Transaction Publishers.

76 *page 148* 'The *Digest* were ready to announce their result. "Landon, 1,293,669; Roosevelt, 972,897" was the headline of the article.'

Literary Digest. (1936). Landon, 1,293,669; Roosevelt, 972,897: Final Returns in 'The Digest's' Poll of Ten Million Voters. *Literary Digest, 122,* 5–6.

77 *page 150* 'That same year, using just 4,500 participants, *Fortune* magazine was able to predict the margin of Roosevelt's victory to within 1%.'

Cantril, H. (1937). How accurate were the polls? *Public Opinion Quarterly, 1*(1), 97. https://doi.org/10.1086/265040

Lusinchi, D. (2012). 'President' Landon and the 1936 *Literary Digest* poll. *Social Science History, 36*(01), 23–54. https://doi.org/10.1017/S014555320001035X

78 *page 150* 'The dent that their previously impeccable credibility sustained on the back of the results is cited as a significant factor in hastening the magazine's demise less than two years later.'

Squire, P. (1988). Why the 1936 *Literary Digest* poll failed. *Public Opinion Quarterly, 52*(1), 125. https://doi.org/10.1086/269085

79 *page 151* 'A mathematically oriented blog-post'

'Rod Liddle said, "Do the math". So I did.' Blog post from polarizingthevacuum, 8 September 2016. Retrieved 21 March, 2019, from https://polarizingthevacuum.wordpress.com/2016/09/08/rod-liddle-said-do-the-math-so-i-did/#comments

80 *page 152* 'According to FBI statistics'

Federal Bureau of Investigation. (2015). *Crime in the United States: FBI — Expanded Homicide Data Table 6.* Retrieved from https://ucr.fbi.gov/crime-in-the-u.s/2015/crime-in-the-u.s.-2015/tables/expanded_homicide_data_table_6_murder_race_and_sex_of_vicitm_by_race_and_sex_of_offender_2015.xls

81 *page 152* 'Given that black people comprised just 12.6% of the US population in 2015 and white people 73.3%, it is alarming that black individuals make up 45.6% of the homicide victims.'

U.S. Census Bureau. (2015). *American FactFinder – Results.* Retrieved from https://factfinder.census.gov/bkmk/table/1.0/en/ACS/15_5YR/DP05/0100000US

82 *page 153* 'The FBI was found to be recording fewer than half of all killings by police in the US.'

Swaine, J., Laughland, O., Lartey, J., & McCarthy, C. (2016). The counted: people killed by police in the US. Retrieved from https://www.theguardian.com/us-news/series/counted-us-police-killings

83 *page 153* 'So successful was the project, that in October 2015, the then

FBI director, James Comey, called it "embarrassing and ridiculous" that the *Guardian* had better data on civilian deaths at the hands of the police than the FBI.'

    Tran, M. (2015, October 8). FBI chief: 'unacceptable' that *Guardian* has better data on police violence. *The Guardian*. Retrieved from https://www. theguardian.com/us-news/2015/oct/08/fbi-chief-says-ridiculous-guardian-washington-post-better-information-police-shootings

84 *page 154* 'only 635,781 full-time "law enforcement officers" (those who carry a firearm and a badge)'

    Federal Bureau of Investigation. (2015). Crime in the United States: Full-time Law Enforcement Employees. Retrieved from https://ucr.fbi.gov/ crime-in-the-u.s/2015/crime-in-the-u.s.-2015/tables/table-74

85 *page 156* 'Liddle's piece wasn't the first or the last time that the *Sun* newspaper was to be embroiled in statistical controversy. In 2009, under the, admittedly inspired headline "Careless pork costs lives", the *Sun* reported just one of many hundreds of results from a 500-page study by the World Cancer Research Fund, on the effect of consuming 50 grams of processed meat per day.'

    World Cancer Research Fund, & American Institute for Cancer Research. (2007). Second Expert Report | World Cancer Research Fund International. http://discovery.ucl.ac.uk/4841/1/4841.pdf

86 *page 159* 'The figures actually reported in the journal *Nature Genetics* were that 10% of individuals possessed genetic variants which left them at a 15% lower risk than the 90% of the population with a different variant.'

    Newton-Cheh, C., Larson, M. G., Vasan, R. S., Levy, D., Bloch, K. D., Surti, A., . . . Wang, T. J. (2009). Association of common variants in NPPA and NPPB with circulating natriuretic peptides and blood pressure. *Nature Genetics, 41*(3), 348–53. https://doi.org/10.1038/ng.328

87 *page 160* 'In one study from 2010, participants were presented with a number of numerical statements about medical procedures and asked to rank the risk they associated with each on a scale from one (not risky at all) to four (very risky).'

    Garcia-Retamero, R., & Galesic, M. (2010). How to reduce the effect of framing on messages about health. *Journal of General Internal Medicine, 25*(12), 1323–29. https://doi.org/10.1007/s11606-010-1484-9

88 *page 161* 'This practice is known as "mismatched framing" and was found to occur in roughly a third of journal articles reporting the harms

and benefits of medical treatments in three of the world's leading medical journals.'

Sedrakyan, A., & Shih, C. (2007). Improving depiction of benefits and harms. *Medical Care*, *45*(10 Suppl 2), S23–S28. https://doi.org/10.1097/MLR.0b013e3180642f69

89 *page 161* 'Along with many other studies, the online app reported the results of a recent clinical trial, on over 13,000 women at increased risk of breast cancer, in which the benefits and potential side effects of the drug Tamoxifen were assessed.'

Fisher, B., Costantino, J. P., Wickerham, D. L., Redmond, C. K., Kavanah, M., Cronin, W. M., . . . Wolmark, N. (1998). Tamoxifen for prevention of breast cancer: report of the National Surgical Adjuvant Breast and Bowel Project P-1 Study. *JNCI: Journal of the National Cancer Institute*, *90*(18), 1371–88. https://doi.org/10.1093/jnci/90.18.1371

90 *page 162* 'Using percentages instead of decimals to highlight perceived benefits is one of another family of tricks referred to as "ratio bias".'

Passerini, G. and Macchi, L. and Bagassi, M. (2012). A methodological approach to ratio bias. *Judgment and Decision Making*, *7*(5).

91 *page 162* 'Our susceptibility to ratio bias has been confirmed in simple experiments in which blindfolded subjects are asked to choose a jelly bean from a tray at random.'

Denes-Raj, V., & Epstein, S. (1994). Conflict between intuitive and rational processing: When people behave against their better judgment. *Journal of Personality and Social Psychology*, *66*(5), 819–29. https://doi.org/10.1037/0022-3514.66.5.819

92 *page 166* 'In one study conducted in 1987, 25 test-anxious US students, who had performed unexpectedly poorly on the multiple choice Scholastic Aptitude Test (SAT), were given the hypertension drug propranolol and retested.'

Faigel, H. C. (1991). The effect of beta blockade on stress-induced cognitive dysfunction in adolescents. *Clinical Pediatrics*, *30*(7), 441–5. https://doi.org/10.1177/000992289103000706

93 *page 168* 'These types of trial are often considered unethical, but enough studies have been done in the past to indicate that the majority of the so-called placebo effect is actually a result of regression to the mean – from which patients derive no benefit.'

Hróbjartsson, A., & Gøtzsche, P. C. (2010). Placebo interventions for all clinical conditions. *Cochrane Database of Systematic Reviews*, (1). https://doi.org/10.1002/14651858.CD003974.pub3

94 *page 171* 'The first studies comparing crime rates pre-introduction of the laws to those post-introduction seemed to indicate that rates of murder and violent crime had reduced in the immediate aftermath of the issuing of these concealed-carry laws.'

Lott, J. R. (2000). *More Guns, Less Crime: Understanding Crime and Gun Control Laws* (2nd edn). University of Chicago Press.

Lott, Jr., J. R., & Mustard, D. B. (1997). Crime, deterrence, and right-to-carry concealed handguns. *The Journal of Legal Studies*, 26(1), 1–68. https://doi.org/10.1086/467988

Plassmann, F., & Tideman, T. N. (2001). Does the right to carry concealed handguns deter countable crimes? Only a count analysis can say. *The Journal of Law and Economics*, 44(S2), 771–98. https://doi.org/10.1086/323311

Bartley, W. A., & Cohen, M. A. (1998). The effect of concealed weapons laws: an extreme bound analysis. *Economic Inquiry*, 36(2), 258–65. https://doi.org/10.1111/j.1465-7295.1998.tb01711.x

Moody, C. E. (2001). Testing for the effects of concealed weapons laws: specification errors and robustness. *The Journal of Law and Economics*, 44(S2), 799–813. https://doi.org/10.1086/323313

95 *page 171* 'Between 1990 and 2001 increases in policing, rising numbers of incarcerations and the receding crack cocaine epidemic all contributed to a fall in murders across the US from around ten per 100,000 per year to around six per 100,000 per year.'

Levitt, S. D. (2004). Understanding why crime fell in the 1990s: four factors that explain the decline and six that do not. *Journal of Economic Perspectives*, 18(1), 163–90. https://doi.org/10.1257/089533004773563485

96 *page 171* 'Perhaps even more important is one study's finding that once regression to the mean was accounted for, the data "... gives no support to the hypothesis that shall-issue laws have beneficial effects in reducing murder rates".'

Grambsch, P. (2008). Regression to the mean, murder rates, and shall-issue laws. *The American Statistician*, 62(4), 289–95. https://doi.org/10.1198/000313008X362446

Chapter 5

97 *page 196* 'For example, a simple rounding error in a German election in 1992 nearly led to the leader of the victorious Social Democrat Party being denied a seat in the parliament, when the Green Party's share of the vote was reported as 5.0% instead of 4.97%.'

Weber-Wulff, D. (1992). Rounding error changes parliament makeup. *The Risks Digest, 13*(37).

98 *page 196* 'In a completely different context, in 1982, a newly created Vancouver Stock Exchange index plummeted continuously over a period of nearly two years, despite the market's bullish performance.'

McCullough, B. D., & Vinod, H. D. (1999). The numerical reliability of econometric software. *Journal of Economic Literature, 37*(2), 633–65. https://doi.org/10.1257/jel.37.2.633

99 *page 198* 'But, while the US remains the last industrial nation to use imperial units'

Technically, United States customary units are slightly different to their close relatives of the British imperial system. The differences, however, are not important for the purposes of this book, so we will refer to both measurement systems as 'imperial'.

100 *page 210* 'The missile alert was assumed to be a false alarm and removed from the system.'

Wolpe, H. (1992). *Patriot missile defense: software problem led to system failure at Dhahran, Saudi Arabia*, United States General Accounting Office, Washington D.C. Retrieved from https://www.gao.gov/products/IMTEC-92-26.

Chapter 6

101 *page 217* 'In 2000, the Clay Mathematics Institute announced a list of seven "Millennium Prize Problems", considered to be the most important unresolved problems in mathematics'

Jaffe, A. M. (2006). The millennium grand challenge in mathematics. *Notices of the AMS* 53.6.

102 *page 218* 'In 2002 and 2003 reclusive Russian mathematician Gregori

Perelman shared three dense mathematical papers with the topology community.'

Perelman, G. (2002). The entropy formula for the Ricci flow and its geometric applications. Retrieved from http://arxiv.org/abs/math/0211159

Perelman, G. (2003). Finite extinction time for the solutions to the Ricci flow on certain three-manifolds. Retrieved from http://arxiv.org/abs/math/0307245

Perelman, G. (2003). Ricci flow with surgery on three-manifolds. Retrieved from http://arxiv.org/abs/math/0303109

103 *page 226* 'If you're planning a similar odyssey for yourself, or even just a local pub crawl, it's probably worth consulting Cook's algorithm first.'

Cook, W. (2012). *In Pursuit of the Traveling Salesman: Mathematics at the Limits of Computation.* Princeton University Press.

104 *page 229* 'Fortunately for everyone who uses a sat nav, it turns out that there is an efficient method – Dijkstra's algorithm – which finds the solution to the "shortest path problem" in polynomial time.'

Dijkstra, E. W. (1959). A note on two problems in connexion with graphs. *Numerische Mathematik, 1*(1), 269–71.

105 *page 239* 'More precisely you should reject the fraction 1/e of the available options, where *e* is a mathematical shorthand for a number known as Euler's number.'

Euler's number first appeared in the 17th century, when Swiss mathematician Jacob Bernoulli (uncle of the early mathematical biologist, Daniel Bernoulli, whose epidemiological exploits are relayed in Chapter 7) was investigating compound interest. In Chapter 1 we encountered compound interest, which means that interest is paid into the account so that it can accrue interest itself. Bernoulli wanted to know how the amount of interest accrued at the end of a year depends on how often the interest is compounded.

Imagine, for simplicity, that the bank pays a special rate of 100% a year on an initial investment of £1. Interest is added to the account at the end of each fixed period and interest can then be paid on that interest in the next period. What happens if the bank decides to pay interest only once a year? At the end of the year, we receive £1 in interest, but there is no time left to accrue further interest on the interest, so we are left with £2. Alternatively, if the bank decides to pay us every six months, then after half a year the bank calculates the interest owed using half the yearly rate (i.e. 50%) leaving us with £1.50 in the account. The same procedure is repeated at the end of the year, giving 50% interest on the £1.50 in the account, and leaving a total of £2.25 at the end of the year.

By compounding more often, the money in the account by the end of the year increases. Compounding quarterly, for example, gives £2.44, monthly compounding yields £2.61. Bernoulli was able to show that by using continuous compounding (i.e. calculating and accruing interest infinitely often, but with an infinitely small rate), the amount of money at year-end would peak at approximately £2.72. To be more precise, we would have precisely e (Euler's number) pounds at the end of the year.

106 *page 241* 'In fact, the problem first came to the attention of mathematicians as the "hiring problem".'

Ferguson, T. S. (1989). Who solved the secretary problem? *Statistical Science, 4*(3), 282–89. https://doi.org/10.1214/ss/1177012493

Gilbert, J. P., & Mosteller, F. (1966). Recognizing the maximum of a sequence. *Journal of the American Statistical Association, 61*(313), 35. https://doi.org/10.2307/2283044

Chapter 7

107 *page 257* 'In the year 2000, measles was officially declared eliminated across the whole of the United States.'

Fiebelkorn, A. P., Redd, S. B., Gastañaduy, P. A., Clemmons, N., Rota, P. A., Rota, J. S., . . . Wallace, G. S. (2017). A comparison of postelimination measles epidemiology in the United States, 2009–2014 versus 2001–2008. *Journal of the Pediatric Infectious Diseases Society, 6*(1), 40–48. https://doi. org/10.1093/jpids/piv080

108 *page 259* 'To investigate his hypothesis, in 1796, Jenner carried out a pioneering experiment into disease prevention that would be considered wildly unethical today.'

Jenner, E. (1798). *An inquiry into the causes and effects of the variolae vaccinae, a disease discovered in some of the western counties of England, particularly Gloucestershire, and known by the name of the cow pox.* (Ed. S. Low).

109 *page 261* 'This method was not supplanted by a less invasive alternative for over 170 years.'

Booth, J. (1977). A short history of blood pressure measurement. *Proceedings of the Royal Society of Medicine, 70*(11), 793–9.

110 *page 261* 'Bernoulli suggested an equation to describe the proportion of people of a given age who had never had smallpox, and were hence still susceptible to the disease.'

Bernoulli, D., & Blower, S. (2004). An attempt at a new analysis of the mortality caused by smallpox and of the advantages of inoculation to prevent it. *Reviews in Medical Virology, 14*(5), 275–88. https://doi.org/10.1002/rmv.443

111 *page 262* 'At the end of the 19th century, poor sanitation and crowded living environments in colonial India led to a series of deadly epidemics including cholera, leprosy and malaria sweeping through the country and killing millions.'

Hays, J. N. (2005). *Epidemics and Pandemics: Their Impacts on Human History.* ABC-CLIO.

Watts, S. (1999). British development policies and malaria in India 1897–c.1929. *Past & Present, 165*(1), 141–81. https://doi.org/10.1093/past/165.1.141

Harrison, M. (1998). 'Hot beds of disease': malaria and civilization in nineteenth-century British India. *Parassitologia, 40*(1–2), 11–18. Retrieved from http://www.ncbi.nlm.nih.gov/pubmed/9653727

Mushtaq, M. U. (2009). Public health in British India: a brief account of the history of medical services and disease prevention in colonial India. *Indian Journal of Community Medicine: Official Publication of Indian Association of Preventive & Social Medicine, 34*(1), 6–14. https://doi.org/10.4103/0970-0218.45369

112 *page 263* 'No one is entirely sure how the disease reached Bombay in August 1896, but there is no doubt about the devastation it caused.'

Simpson, W. J. (2010). *A Treatise on Plague Dealing with the Historical, Epidemiological, Clinical, Therapeutic and Preventive Aspects of the Disease.* Cambridge University Press. https://doi.org/10.1017/CBO9780511710773

113 *page 264* 'Inspired by data on the plague outbreaks in Bombay, collected while McKendrick was in India, they conducted the single-most influential study in the history of mathematical epidemiology.'

Kermack, W. O., & McKendrick, A. G. (1927). A contribution to the mathematical theory of epidemics. *Proceedings of the Royal Society A: Mathematical, Physical and Engineering Sciences, 115*(772), 700–721. https://doi.org/10.1098/rspa.1927.0118

114 *page 266* 'One study found that in the US alone over 1000 outbreaks of the vomiting bug, norovirus, were linked to contaminated food in the four years from 2009 to 2012.'

Hall, A. J., Wikswo, M. E., Pringle, K., Gould, L. H., Parashar, U. D. (2014). Vital signs: food-borne norovirus outbreaks – United States, 2009–2012. *MMWR. Morbidity and Mortality Weekly Report, 63*(22), 491–5.

115 *page 268* 'The S-I-R model predicts that, ultimately, outbreaks die out from of a lack of infective people, not a lack of susceptibles.'

Murray, J. D. (2002). *Mathematical Biology I: An Introduction.* Springer.

116 *page 271* 'Over 60% of all cervical cancers are caused by two strains of the human papillomavirus (HPV).'

Bosch, F. X., Manos, M. M., Muñoz, N., Sherman, M., Jansen, A. M., Peto, J., . . . Shah, K. V. (1995). Prevalence of human papillomavirus in cervical cancer: a worldwide perspective. International Biological Study on Cervical Cancer (IBSCC) Study Group. *Journal of the National Cancer Institute, 87*(11), 796–802.

117 *page 271* 'Indeed, HPV is the most frequent sexually transmitted disease in the world.'

Gavillon, N., Vervaet, H., Derniaux, E., Terrosi, P., Graesslin, O., & Quereux, C. (2010). Papillomavirus humain (HPV): comment ai-je attrapé ça ? *Gynécologie Obstétrique & Fertilité, 38*(3), 199–204. https://doi.org/10.1016/J.GYOBFE.2010.01.003

118 *page 271* 'Studies undertaken in the UK around the time of the vaccine's deployment indicated that the most cost-effective strategy would be to immunise adolescent girls between the ages of 12 and 13, the likely future sufferers of cervical cancer.'

Jit, M., Choi, Y. H., & Edmunds, W. J. (2008). Economic evaluation of human papillomavirus vaccination in the United Kingdom. *BMJ (Clinical Research Ed.), 337*, a769. https://doi.org/10.1136/bmj.a769

119 *page 271* 'Related studies in other countries, considering mathematical models of the heterosexual transmission of the disease, confirmed that vaccinating females only was the best course of action.'

Zechmeister, I., Blasio, B. F. de, Garnett, G., Neilson, A. R., & Siebert,

U. (2009). Cost effectiveness analysis of human papillomavirus-vaccination programs to prevent cervical cancer in Austria. *Vaccine, 27*(37), 5133–41. https://doi.org/10.1016/J.VACCINE.2009.06.039

120 *page 272* 'that the strains of HPV guarded against by the vaccine can also cause a range of non-cervical diseases in both women and men.'

Kohli, M., Ferko, N., Martin, A., Franco, E. L., Jenkins, D., Gallivan, S., . . . Drummond, M. (2007). Estimating the long-term impact of a prophylactic human papillomavirus 16/18 vaccine on the burden of cervical cancer in the UK. *British Journal of Cancer, 96*(1), 143–50. https://doi.org/10.1038/sj.bjc.6603501

Kulasingam, S. L., Benard, S., Barnabas, R. V, Largeron, N., & Myers, E. R. (2008). Adding a quadrivalent human papillomavirus vaccine to the UK cervical cancer screening programme: a cost-effectiveness analysis. *Cost Effectiveness and Resource Allocation, 6*(1), 4. https://doi.org/10.1186/1478-7547-6-4

Dasbach, E., Insinga, R., & Elbasha, E. (2008). The epidemiological and economic impact of a quadrivalent human papillomavirus vaccine (6/11/16/18) in the UK. *BJOG: An International Journal of Obstetrics & Gynaecology, 115*(8), 947–56. https://doi.org/10.1111/j.1471-0528.2008.01743.x

121 *page 272* 'As well as causing cervical cancer, HPV types 16 and 18 contribute to 50% of penile cancer, 80% of anal cancers, 20% of mouth, and 30% of throat cancers.'

Hibbitts, S. (2009). Should boys receive the human papillomavirus vaccine? Yes. *BMJ, 339*, b4928. https://doi.org/10.1136/BMJ.B4928

parkin, D. M., & Bray, F. (2006). Chapter 2: The burden of HPV-related cancers. *Vaccine, 24*, S11–S25. https://doi.org/10.1016/J.VACCINE.2006.05.111

Watson, M., Saraiya, M., Ahmed, F., Cardinez, C. J., Reichman, M. E., Weir, H. K., & Richards, T. B. (2008). Using population-based cancer registry data to assess the burden of human papillomavirus-associated cancers in the United States: Overview of methods. *Cancer, 113*(S10), 2841–54. https://doi.org/10.1002/cncr.23758

122 *page 272* 'In both the US and UK, the majority of cancers caused by HPV are not cervical.'

Hibbitts, S. (2009). Should boys receive the human papillomavirus vaccine? Yes. *BMJ, 339*, b4928. https://doi.org/10.1136/BMJ.B4928

ICO/IARC Information Centre on HPV and Cancer. (2018). United Kingdom Human Papillomavirus and Related Cancers, Fact Sheet 2018.

Watson, M., Saraiya, M., Ahmed, F., Cardinez, C. J., Reichman, M. E., Weir, H. K., & Richards, T. B. (2008). Using population-based cancer registry data to assess the burden of human papillomavirus-associated cancers in the United States: Overview of methods. *Cancer, 113*(S10), 2841–2854. https://doi.org/10.1002/cncr.23758

123 *page 272* 'Significantly, HPV types 6 and 11 also cause nine out of ten cases of anogenital warts'

Yanofsky, V. R., Patel, R. V, & Goldenberg, G. (2012). Genital warts: a comprehensive review. *The Journal of Clinical and Aesthetic Dermatology, 5*(6), 25–36.

124 *page 272* 'In the US approximately 60% of the healthcare costs associated with all non-cervical HPV infections are spent on the treatment of these warts.'

Hu, D., & Goldie, S. (2008). The economic burden of noncervical human papillomavirus disease in the United States. *American Journal of Obstetrics and Gynecology, 198*(5), 500.e1–500.e7. https://doi.org/10.1016/J.AJOG.2008.03.064

125 *page 273* 'Models based on sexual networks including homosexual relationships have a higher rate of disease transmission than those which only consider heterosexual relationships.'

Gómez-Gardeñes, J., Latora, V., Moreno, Y., & Profumo, E. (2008). Spreading of sexually transmitted diseases in heterosexual populations. *Proceedings of the National Academy of Sciences of the United States of America, 105*(5), 1399–404. https://doi.org/10.1073/pnas.0707332105

126 *page 273* 'The prevalence of HPV in men who have sex with men is significantly higher than in the general population.'

Blas, M. M., Brown, B., Menacho, L., Alva, I. E., Silva-Santisteban, A., & Carcamo, C. (2015). HPV Prevalence in multiple anatomical sites among men who have sex with men in Peru. *PLOS ONE, 10*(10), e0139524. https://doi.org/10.1371/journal.pone.0139524

McQuillan, G., Kruszon-Moran, D., Markowitz, L. E., Unger, E. R., & Paulose-Ram, R. (2017). Prevalence of HPV in Adults aged 18–69: United States, 2011–2014 *NCHS Data Brief*, (280), 1–8. Retrieved from http://www.ncbi.nlm.nih.gov/pubmed/28463105

127 *page 273* 'In the US, the incidence rate of anal cancer in this group is over

15 times higher. At 35 per 100,000 it is comparable to the rates of cervical cancer in women before cervical screening was introduced and significantly higher than current rates of cervical cancer in the US.'

D'Souza, G., Wiley, D. J., Li, X., Chmiel, J. S., Margolick, J. B., Cranston, R. D., & Jacobson, L. P. (2008). Incidence and epidemiology of anal cancer in the multicenter AIDS cohort study. *Journal of Acquired Immune Deficiency Syndromes (1999), 48*(4), 491–99. https://doi.org/10.1097/QAI.0b013e31817aebfe

Johnson, L. G., Madeleine, M. M., Newcomer, L. M., Schwartz, S. M., & Daling, J. R. (2004). Anal cancer incidence and survival: the surveillance, epidemiology, and end results experience, 1973–2000. *Cancer, 101*(2), 281–8. https://doi.org/10.1002/cncr.20364

Qualters, J. R., Lee, N. C., Smith, R. A., & Aubert, R. E. (1987). Breast and cervical cancer surveillance, United States, 1973–1987. *Morbidity and Mortality Weekly Report: Surveillance Summaries.* Centers for Disease Control & Prevention (CDC).

U.S. Cancer Statistics Working Group. U.S. Cancer Statistics Data Visualizations Tool, based on November 2017 submission data (1999–2015): U.S. Department of Health and Human Services, Centers for Disease Control and Prevention and National Cancer Institute; www.cdc.gov/cancer/dataviz, June 2018.

Noone, A. M., Howlader, N., Krapcho, M., Miller, D., Brest, A., Yu, M., Ruhl, J., Tatalovich, Z., Mariotto, A., Lewis, D. R., Chen, H. S., Feuer, E. J., Cronin, K. A. (eds). SEER Cancer Statistics Review, 1975–2015, National Cancer Institute. Bethesda, MD, https://seer.cancer.gov/csr/1975_2015/, External based on November 2017 SEER data submission, posted to the SEER website, April 2018.

Chin-Hong, P. V., Vittinghoff, E., Cranston, R. D., Buchbinder, S., Cohen, D., Colfax, G., . . . Palefsky, J. M. (2004). Age-specific prevalence of anal human papillomavirus infection in HIV-negative sexually active men who have sex with men: The EXPLORE Study. *The Journal of Infectious Diseases, 190*(12), 2070–76. https://doi.org/10.1086/425906

128 *page 274* 'In July of the same year, advice based on a new cost-effectiveness study recommended that all boys in the UK be given the HPV vaccination at the same age as girls.'

Brisson, M., Bénard, É., Drolet, M., Bogaards, J. A., Baussano, I., Vänskä, S., . . . Walsh, C. (2016). Population-level impact, herd immunity, and elimination after human papillomavirus vaccination: a systematic review and meta-analysis of predictions from transmission-dynamic models. *Lancet. Public Health*, *1*(1), e8–e17. https://doi.org/10.1016/S2468-2667(16)30001-9

Keeling, M. J., Broadfoot, K. A., & Datta, S. (2017). The impact of current infection levels on the cost-benefit of vaccination. *Epidemics*, *21*, 56–62. https://doi.org/10.1016/J.EPIDEM.2017.06.004

Joint Committee on Vaccination and Immunisation. (2018). Statement on HPV vaccination. Retrieved from https://www.gov.uk/government/publications/jcvi-statement-extending-the-hpv-vaccination-programme-conclusions

Joint Committee on Vaccination and Immunisation. (2018). Interim statement on extending the HPV vaccination programme. Retrieved March 7, 2019, from https://www.gov.uk/government/publications/jcvi-statement-extending-the-hpv-vaccination-programme

129 *page 279* 'With that expense in mind, as well as what smacked of needlessly fraught reaction, a team of mathematicians from the London School of Hygiene and Tropical Medicine developed a simple mathematical model incorporating an incubation period.'

Mabey, D., Flasche, S., & Edmunds, W. J. (2014). Airport screening for Ebola. *BMJ (Clinical Research Ed.)*, *349*, g6202. https://doi.org/10.1136/bmj.g6202

130 *page 285* 'In less extreme cases, simple applications of mathematical modelling are able to suggest the most effective duration to isolate infected patients.'

Castillo-Chavez, C., Castillo-Garsow, C. W., & Yakubu, A.-A. (2003). Mathematical Models of Isolation and Quarantine. *JAMA: The Journal of the American Medical Association*, *290*(21), 2876–77. https://doi.org/10.1001/jama.290.21.2876

131 *page 286* 'Mathematical models of disease spread have identified the degree to which the effectiveness of a quarantining strategy depends on the *timing* of peak infectiousness.'

Day, T., Park, A., Madras, N., Gumel, A., & Wu, J. (2006). When is quarantine a useful control strategy for emerging infectious diseases? *American Journal of Epidemiology*, *163*(5), 479–85. https://doi.org/10.1093/aje/kwj056

Peak, C. M., Childs, L. M., Grad, Y. H., & Buckee, C. O. (2017).

Comparing nonpharmaceutical interventions for containing emerging epidemics. *Proceedings of the National Academy of Sciences of the United States of America, 114*(15), 4023–8. https://doi.org/10.1073/pnas.1616438114

132 *page 287* 'In 2014, around the peak of the Ebola outbreak, a mathematical study concluded that, approximately 22% of new Ebola cases were attributable to deceased Ebola victims.'

Agusto, F. B., Teboh-Ewungkem, M. I., & Gumel, A. B. (2015). Mathematical assessment of the effect of traditional beliefs and customs on the transmission dynamics of the 2014 Ebola outbreaks. *BMC Medicine, 13*(1), 96. https://doi.org/10.1186/s12916-015-0318-3

133 *page 291* 'A study which modelled the spread of the 2015 Disneyland measles outbreak – in which Mobius Loop was infected – suggested vaccination rates amongst those exposed to the disease may have been as low as 50% – way below the threshold required for herd immunity.'

Majumder, M. S., Cohn, E. L., Mekaru, S. R., Huston, J. E., & Brownstein, J. S. (2015). Substandard vaccination compliance and the 2015 measles outbreak. *JAMA Pediatrics, 169*(5), 494. https://doi.org/10.1001/jamapediatrics.2015.0384

134 *page 291* 'This public health disaster was not caused by disease-ridden animals, poor sanitation or even failures of government policy, but instead by a sombre five-page publication in the well-respected medical journal, the *Lancet*.'

Wakefield, A., Murch, S., Anthony, A., Linnell, J., Casson, D., Malik, M., . . . Walker-Smith, J. (1998). RETRACTED: Ileal-lymphoid-nodular hyperplasia, non-specific colitis, and pervasive developmental disorder in children. *Lancet, 351*(9103), 637–41. https://doi.org/10.1016/S0140-6736(97)11096-0

135 *page 296* 'World Health Organization figures show that vaccines prevent millions of deaths every year and could prevent millions more if we could improve global coverage.'

World Health Organisation: strategic advisory group of experts on immunization. (2018). *SAGE DoV GVAP Assessment report 2018. WHO.* World Health Organization. Retrieved from https://www.who.int/immunization/global_vaccine_action_plan/sage_assessment_reports/en/

A HISTORY OF MODERN ART

Painting • Sculpture • Architecture • Photography

FOURTH EDITION

A HISTORY OF MODERN ART

Painting · Sculpture · Architecture · Photography

H. H. ARNASON

MARLA F. PRATHER
Revising author, Fourth Edition

DANIEL WHEELER
Revising author, Third Edition

Fourth Edition

1,438 illustrations, 504 in colour

THAMES AND HUDSON

This fourth edition published in Great Britain in 1998 by
Thames and Hudson Ltd, London

First edition 1969
Second Edition 1977
Third edition 1986
Revised and enlarged 1988

British Library Cataloguing-in-Publication Data
A catalogue record for this book is available from the
British Library

ISBN 0-500-23757-3

Publisher's note: Marla F. Prather created the structure
for revision of the entire fourth edition and revised
chapters 1 to 21. Chapters 22 to the Epilogue are the
collaborative efforts of the following scholars and writers,
working from Marla Prather's plans for these chapters:
Anne F. Collins, Saundra Goldman, Anna Hammond,
Gloria Kury, Christine Liotta, Roxana Marcoci and Lynn
Matheny. Daniel Wheeler was the reviser of the previous,
third edition.

Printed and bound in Singapore

TABLE OF CONTENTS

PREFACES / ACKNOWLEDGMENTS
FROM PREVIOUS EDITIONS

FIRST AND SECOND EDITIONS

The arts of painting, sculpture, and architecture are arts of space. For this reason it is essential to approach these arts in the twentieth century, or in any other period, through an analysis of the artist's attitude toward spatial organization. Since space is normally defined as extension in all directions, this attitude can be seen relatively easily in architecture and sculpture, which traditionally are three-dimensional masses or volumes surrounded by space and, in the case of architecture, enclosing space. Architecture is frequently defined as "the art of enclosing space," a definition which gives primary importance to the interior space, despite the fact that many architectural styles throughout history have been largely concerned with the appearance of the exterior, or the organization of outdoor space.

The Classic and Romantic eclecticism of the nineteenth-century academic styles in architecture added nothing new to the history of architectural spatial experiment. The development of reinforced concrete and structural steel, however, provided the basis for a series of new experiments in twentieth-century architecture. The most significant product of steel construction is, of course, the skyscraper, which in its characteristic spatial development is perhaps more interesting as an exterior form than as an interior space (if one excepts the new interior space effects resultant from the all-glass sheathing). It is in the flexible material of reinforced concrete that many of the most impressive twentieth-century architectural-spatial experiments have been realized, as well as in the use of new structural principles such as those embodied in Buckminster Fuller's geodesic domes.

Until the twentieth century a work of sculpture has characteristically been a three-dimensional object existing in surrounding space. The formal problems of the sculptor have thus involved the exploitation of his material (bronze, clay, stone, etc.), the integration of the sculptural elements with their environment, and the relation of these elements to surrounding space. In the various broad cycles of sculptural history—ancient, medieval, Renaissance, Baroque—the development of sculpture has tended to follow a pattern in which the early stages emphasized frontality and mass, and the later stages, openness and spatial existence.

In the twentieth century, sculpture has continued in one way or another most of the sculptural-spatial tendencies of the past. Modern experimental sculptures have also made fundamental new departures, particularly in the exploration of sculpture as construction or as assemblage; in experimentation with new materials; and in sculpture as enclosing space. Partly as a result of influence from primitive or archaic art, there has also been a contrasting abandonment of the developed systems of full spatial organization in favor of a return to frontality achieved through simplified masses.

It is probably more difficult for the spectator to comprehend the element of space in painting than in either architecture or sculpture. A painting is physically a two-dimensional surface to which pigments, usually without appreciable bulk, have been applied. Except insofar as the painter may have applied his paint thickly, in impasto, the painting traditionally has no projecting mass, and any suggestion of depth on the surface of the canvas is an illusion created through various technical means. The instant a painter draws a line on the blank surface he introduces an illusion of the third dimension. This illusion of depth may be furthered by overlapping or spacing of color shapes, by the different visual impacts of colors—red, yellow, blue, black, or white—by different intensities or values, and by many other devices known throughout history. The most important of these, before the twentieth century, were linear and atmospheric perspective (discussed in Chapter 1).

Perspective, although known in antiquity, became for the Renaissance a means for creating paintings that were "imitations of nature"—visual illusions that made the spectator think he was looking at a man, a still life, or a landscape rather than at a canvas covered with paint. Perhaps the greatest revolution of early modern art lay in the abandonment of this attitude and the perspective technique that made it possible. As a consequence, the painting—and the sculpture—became a reality in itself, not an imitation of anything else; it had its own laws and its own reason for existence. As will be seen, after the initial experiments carried out by the Impressionists, Post-Impressionists, Fauves, and Cubists, there has been no logical progression. The call for a "return to nature" recurs continually, but there is no question that the efforts of the pioneers of modern art changed modern artists' way of seeing, and in some degree, modern man's.

The principal emphasis of this book revolves around

this problem of "seeing" modern art. It is recognized that this involves two not necessarily compatible elements: the visual and the verbal. Any work of art history and/or criticism is inevitably an attempt to translate a visual into a verbal experience. Since the mind is involved in both experiences, there are some points of contact between them. Nevertheless, the two experiences are essentially different and it must always be recognized that the words of the interpreter are at best only an approximation of the visual work of art. The thesis of this book, insofar as it has a thesis, is that in the study of art the only primary evidence is the work of art itself. Everything that has been said about it, even by the artist himself, may be important, but it remains secondary evidence. Everything that we can learn about the environment that produced it—historically, socially, culturally—is important, but again is only secondary or tertiary evidence. It is for this reason that an effort has been made to reproduce most of the works discussed. For the same reason a large part of the text is concerned with a close analysis of these works of art—and with detailed descriptions of them as well. This has been done in the conviction that simple description has an effect in forcing the attention of the spectator on the painting, sculpture, or building itself. If, after studying the object, he disagrees with the commentator, all the better. In the process he has learned something about visual perception.

This book is intended for the general reader and the student of modern art. As the title suggests, the emphasis is on the development of modern painting, sculpture, and architecture, although there is reference to the graphic arts—drawing and printmaking—and to the arts of design, when these are of significance in the work of an artist or of a period. The book deals predominantly with twentieth-century art, but attention has been paid to the nineteenth-century origins of modern painting. Whereas modern sculpture and architecture saw their beginnings at the end of the nineteenth century with a small number of pioneers, such as Auguste Rodin in sculpture and Louis Sullivan in architecture, the sources of modern painting must be traced much further back in the century.

A discussion of some changes in attitudes toward pictorial space and subject matter around 1800 is followed by a brief summary of Romanticism, aspects of the academic, Classic tradition, and an account of Courbet's approach to reality, both in subject and in materials. With Édouard Manet and the Impressionists the account becomes more explicit in the examination of individual artists, their works, and the movements with which they were associated. It is, nevertheless, still necessarily selective, since only the major figures of Impressionism, Post-Impressionism, and Neo-Impresionism can be discussed in any depth. Although there were many other excellent Impressionist and Post-Impressionist painters, and some

of these are mentioned, the main emphasis is on the artists of greatest stature. Another feature of the book should be mentioned at this point: every effort has been made to place the black-and-white illustrations as close as possible to the accompanying text. The colorplates are grouped with or near the chapters they illustrate.

Emphasis is placed on these matters of physical makeup because they are obviously important to any student, and because the approach used makes them important in a very particular sense. The main body of the book deals with the twentieth-century, for the earlier part of which—up to 1940—the works and the attitudes of the pioneers of modernism are examined at length. In discussing their works, the approach is primarily the close examination of the works as aesthetic objects in which problems of space, color, line, and the total organization of the surface are studied. Iconography—the meaning of and attitude toward subject matter—significant in many phases of modern art such as Expressionism, Surrealism, and Pop Art, is given appropriate attention. Also, it is recognized that a work of art or architecture cannot exist in a vacuum. It is the product of a total environment—a social and cultural system—with parallels in literature, music, and the other arts, and relations to the philosophy and science of the period. Aside from being a broad, analytical survey of modern art and architecture, the book is conceived as a dictionary in which much factual information is included. This is particularly true of the latter part, art since 1950. As we approach our own time it is possible to discuss at length only a minority of the talented artists who are creating the art and architecture of the future. Thus, the accounts of these, with a few exceptions, are summary and factual, intended to suggest the main directions of contemporary art and architecture rather than to pretend completeness.

Since this is a history of international modern art and architecture, the criterion for inclusion has been art and architecture which have in some way had international implications. In the late nineteenth and earlier twentieth centuries the emphasis is largely on French, German, and Italian developments. In architecture, further accent is placed on English and American experiments, since these did have some international impact. Other national movements—De Stijl in Holland, Suprematism and Constructivism in Russia—are also discussed in detail for the same reasons. However, painting and sculpture in the United States and in England before World War II are discussed only briefly, since, despite the presence of many talented artists in these countries, they cannot be said to have affected the international stream of modern art until mid-century. For the same reason, Abstract Expressionism in the United States in the 1940s and 1950s, as well as British and American Pop Art, are treated at some length, as a consequence of their world-wide dispersion.

My primary acknowledgment is to Charles Rufus Morey (1875-1955), even though he might not have approved of the subject matter. I am indebted to Robert Goldwater for reading the manuscript and correcting numerous errors of fact. Needless to say he is in no sense responsible for the opinions expressed by the author.

At the Solomon R. Guggenheim Museum, my thanks are due to: Anna Golfinopoulos, my research assistant, who achieved miracles of efficient organization under extreme pressure. Also to Susan Earl, research assistant during the first stages of the book; Joan Hall, librarian; Robert Mates and Paul Katz, photographers; Darrie Hammer, and Ward Jackson—all for aid beyond the call of duty.

At Harry N. Abrams, Inc., my thanks are due to: the late Milton S. Fox, whose sensitivity, perception, and humor were a constant delight and whose premature death was to me a great personal tragedy. Also to the late Sam Cauman, who read the manuscript; Barbara Lyons, Director, Photo Department, Rights and Reproductions; Mary Lea Bandy, Mr. Fox's assistant; and to all the other members of the Abrams firm who worked so hard on this project.

A particular note of appreciation goes to all the collectors, museums, art galleries, and other institutions who so kindly granted permission to reproduce works of art and provided photographs or color transparencies.

In this context it will be obvious that I have drawn heavily on the resources of a few museums and collectors;

first, The Museum of Modern Art, New York; then other New York museums such as the Solomon R. Guggenheim Museum and the Whitney Museum of American Art. For sculpture I have drawn heavily on the collection of the Hirshhorn Museum and Sculpture Garden in Washington, D.C. The reasons I have used these collections so intensively are: 1) They are among the greatest collections of modern art in existence. Taken together (and with the modern and American collections of The Metropolitan Museum of Art) they constitute the most comprehensive collections of modern painting and sculpture to be found in any one area. 2) I am most intimately familiar with the paintings and sculpture in these museums. 3) For the student who may wish to secure slides or additional research material, these institutions have extensive facilities. Again, my thanks to all who have helped me so generously.

For the 1976 edition, my particular appreciation, aside from that already expressed, goes to my assistant Dorothy Anderson. Thanks are also due to Francis Naumann, who gave me valuable assistance, particularly in the fields of Conceptualism and other recent movements. The list of those at Harry N. Abrams, Inc., who have worked untiringly on the many tasks involved in revising this work and making it current has been greatly enlarged from those already mentioned. My gratitude goes particularly to: Edith M. Pavese for her capable editing and thoughtful supervision; Margaret L. Kaplan, Executive Editor, for skillful coordination of the various phases of what proved to be a complicated undertaking.

—H. H. A.

THIRD EDITION

In the Third Edition presented here, the story of contemporary art and architecture has been brought up-to-date right through the mid-1980s. Since so much of the new art produced during the 1970s and thereafter represents some form of reaction against the whole modernist tradition that went before—or at least against its culminating phase in 1960s Minimalism—the earlier chapters have received almost as much fresh attention as the new chapters addressed to the various forms of Conceptual Art, to Photorealism, Pattern, Decoration, and the New Imagery, to Neo-Expressionism and, finally, a resurgent, albeit radically altered, abstraction. This was done not only to incorporate recent research and to realize a clearer, more cohesive organization, but also to provide firm antecedents for the numerous references, positive as well as negative, made to earlier modernism by artists of the so-called "Post-Minimal" age, a *fin-de-siècle* period of broad revisionist and retrospective thinking. Among the older chap-

ters, those that have been rather substantially rewritten are Cubism and Fauvism. Throughout the pre-1970 material, however, readers familiar with the old editions will find some new writing, as well as fresh illustrations, even in the section devoted to Minimalism. They will also discover that the history of photography has been introduced and interwoven with the ongoing history of painting and sculpture. This particular expansion of the book has been undertaken not only to acknowledge the aesthetic importance of the work realized in the medium of photography, but also to provide a background for photography's current ubiquitous presence among the mix of media increasingly favored, by all manner of artists, for their endless capacity to yield new figurative, formal, and expressive effects. In architecture, too, the recent trend has been revisionist, and so, along with a very modest amount of new text and reorganization in the original material on architecture, there is an additional chapter entitled "Post-

Modernism," touching on the various forms of reaction manifested against the International Style dominant during the century's third quarter.

For the 1986 edition of *History of Modern Art*, the first acknowledgment must go to H. H. Arnason, who not only created the book in the first place, but then generously allowed another person to revise the text once he found it impossible, for reasons of health, to carry out the task himself. As the writer to whom this privilege passed, I met with Arnason on several occasions, and was fortunate enough to benefit from his general advice before circumstances forced him to withdraw from all further participation in the development of the present volume. Still, he must have been very much with us in spirit, if not in fact, for just as these acknowledgments were being assembled—always the final event before a book goes on press—the publisher and I learned that Arnason had died. It was as if he had held fast just long enough to *will* us safely clear of a thousand hurdles. And so in gratitude for his confidence and aid, both tangible and intangible, I want to dedicate my own sections in the new edition to the memory of Harvey Arnason, scholar, teacher, critic, curator, and administrator, an art historian as much at ease with archival research as with artists in their studios, with collectors as much as with students, with specialized monographic writing as much as with the broader statements needed in a general survey.

In preparing the 1986 edition of Arnason's *History of Modern Art*, I have attempted not so much to "present" new material as to discover it, as the curious viewer/reader himself might in the act of following contemporary art through its present and ongoing processes of unforeseeable, yet ultimately logical, development. Consequently, I write not as an authority but rather as a student of modern art, albeit a serious—even passionate—one of some years standing. Moreover, in discovering, studying, or revealing works of art and architecture, I have carefully avoided the role of critic, as well as that of theorist. Rather than engage in speculative value judgments, or some narrowly defined ideological approach, I invariably find it more rewarding and instructive—for myself and I hope for readers—to remain open to the distinctive qualities of each work or trend, to learn its special language, and to allow the particular truth this conveys to unfold. And since art, like poetry, almost always seeks a deeper, less obvious reality, the challenge of that reality can be further enriched by the findings of those who have interviewed the artist, observed his work at length, and commented upon its possibilities in a knowing, constructive manner. Once the art has been experienced from these several perspectives, criticality emerges, as problems are posed, resolved, and assessed, like a cat's cradle of checks and bal-

ances, woven by the competing interests of time, place, personal circumstance, and sociocultural context. Thus, what I have wanted to offer is a balanced view of recent events in the visual arts—a view balanced, that is, between the subjective vision of the artist himself, an objective, informed consideration of what that vision produced, and my own keen sense of identification with both.

To obtain such a synthesis, I have, of course, incurred a very great many debts, all of which I am eager to acknowledge with the warmest feelings of gratitude. For present purposes, my most immediate recognition must go to the nearly countless writers cited in the bibliography, whose books, articles, reviews, monographs, and catalogues provided not only information, concepts, and artists' statements, but even the essential nomenclature and style of address without which the sort of broad, integrated survey I have attempted would not be possible. However, a still more primary debt of gratitude is the one I owe to the distinguished faculty who first taught me art history and how to think as well as write about it. At New York University's Institute of Fine Arts these were Professors Colin Eisler, Robert Goldwater, Jim Jordan, Theodore Reff, Robert Rosenblum, William Rubin, and Gert Schiff. Here, however, I must reserve a special place for Robert Rosenblum and Gert Schiff, to acknowledge the gentle persistence of their urging that I cease editing other people's books and undertake one of my own. In venturing this partial, if not altogether first, step in that direction, I have been long about it, long enough certainly for my mentors to be absolved of all blame for the present consequences of their well-meant encouragement. Quite particular to the enterprise at hand has been the spirited interest and active support offered by Dr. Ruth Kaufmann, the kind of friend that only the lucky can boast and a specialist whose knowledge of current art is peerless. Another informed guide who tried to set me off in the right direction across the largely uncharted terrain of the new is Charles Stuckey, now of the National Gallery in Washington, D.C. I have also been the beneficiary of advice from Dr. Anthony F. Janson of the John and Mabel Ringling Museum of Art in Sarasota, Florida, on how to cut or shrink the old text in order to permit the inclusion of another fifteen years of art, in addition to the history of photography, without making the book an object that only a weight-lifter could handle. If I have managed to find my way through the maze of fresh knowledge concerning the Russian avant-garde, it is thanks to the generously shared scholarship of Dr. Marian Burleigh Motley, a friend at Princeton University. Professor Ursula Meyer interrupted her own writing and quite literally went out of her way to clarify ideas and lend illustrations for the section on Conceptualism, as did Sidney Geist when I approached him about Brancusi.

While Sabine Rewald of New York's Metropolitan Museum of Art helped me with Balthus, Dr. Madeleine Fidell Beaufort in Paris answered my frantic call for assistance in obtaining documents on a key Matisse in the Soviet Union. In the course of adding photography to the text, I sought aid and received it from many able and selfless individuals, among them Peter Galassi of The Museum of Modern Art in New York, Sam Wagstaff, John Waddell, Pierre Apraxine, Frau Monika Schmela in Düsseldorf, Christine Faltermeier, Maria Morris Hambourg at The Metropolitan Museum, Beth Gates-Warren of Sotheby's in New York City, and Gerd Sander of the Sander Gallery, also in New York. And in speaking of museums and galleries, the author of a book like this could scarcely be too extravagant in voicing thanks for the patience with which such institutions—here cited either in the captions or in the Photo Credits—endure what must seem interminable requests for facts and images. At Harry N. Abrams, I have been remarkably fortunate in my sponsoring editor, Margaret Kaplan, a professional par excellence and graceful about it. May her trust in me not have been misplaced! Other members of Abrams's capable staff who overcame terrible difficulties to bring off this complex project are Barbara Lyons, director of the Photo Department, Pamela Phillips, photo researcher, and Pamela Bass, photo assistant. They all have my sincere thanks. And so does my "other" publisher, Alexis Gregory of The Vendome Press, without whose tolerance I could not have stayed so long at a different well.

—D. W.

ACKNOWLEDGMENTS FOR THE FOURTH EDITION

A book of such scope and complexity naturally involves the efforts of many people, and I am very grateful to the editors, writers, and colleagues who have helped bring this project to fruition. Eve Sinaiko, Senior Editor at Abrams, contributed significantly to this volume in the early stages, helping to shape its content. Anna Hammond skillfully edited my texts and wrote several entries on photography. Editor Katherine Rangoon Doyle expertly managed the project in its final stages. Catherine Ruello, formerly of Abrams, took on the daunting task of photo research for the book with unflagging commitment. Executive Editor Julia Moore oversaw the project with great patience throughout the many years it took to complete it, supported by Assistant Editor Monica Mehta. I would also like to thank Anne F. Collins, a doctoral candidate at the University of Texas, who is responsible for the revisions in chapter 22; Christine Liotta, who revised chapters 23 and 26 on architecture; and Anne F. Collins, Saundra Goldman, Anna Hammond, Gloria Kury, and Roxana Marcoci, who provided the texts for the Epilogue.

At the National Gallery, I am grateful to a number of colleagues for providing advice in their areas of expertise, including Frank Kelly, Curator of American Art, and Nancy Anderson, an Associate Curator in the same department. Jeffrey Weiss, Associate Curator of Twentieth-Century Art and a specialist in the art of Picasso, provided many helpful suggestions for chapter 10, on Cubism, which was written entirely anew for this edition. Mark Rosenthal, a curator at the Solomon R. Guggenheim Museum and Nan Rosenthal, Curator of Twentieth-Century Art at The Metropolitan Museum of Art kindly fielded many of my questions, as did art historians David Sylvester and Angelica Zander Rudenstine. At The Museum of Modern Art, I am grateful to Peter Galassi, Curator of Photography, and Beatrice Kernan, Associate Curator, Department of Drawings. Lynn Matheny, a doctoral candidate in art history at UCLA, worked closely with me in the final stages of the book, handling numerous details of research with remarkable speed and skill and writing occasional sections of the text.

—M.P.

Unless otherwise noted, all paintings are oil on canvas. Measurements are not given for objects that are inherently large (architecture, architectural sculpture, wall paintings) or small (drawings, prints, photographs). Height precedes width. A probable measuring error of more than one percent is indicated by "c." A list of credits for the illustrations appears at the end of the book.

1 The Prehistory of Modern Painting

Various dates are used to mark the point at which modern art supposedly began. The most commonly chosen, perhaps, is 1863, the year of the Salon des Refusés in Paris, where Édouard Manet first showed his scandalous painting *Déjeuner sur l'herbe* (color plate 6, page 35). But other and even earlier dates may be considered: 1855, the year of the first Paris Exposition Universelle (a kind of world's fair), in which **GUSTAVE COURBET** (1819–77) built a separate pavilion to show *The Painter's Studio* (fig. 1); 1824, when the English landscapists John Constable and Richard Parkes Bonington exhibited their brilliant, direct-color studies from nature at the Paris Salon (an annual exhibition of contemporary art juried by members of the French Academy); or even 1784, when **JACQUES-LOUIS DAVID** (1748–1825) finished his *Oath of the Horatii* (fig. 2) and the Neoclassical movement had assumed a position of dominance in Europe and the United States.

Each of these dates has significance for the development of modern art, but none categorically marks a completely new beginning. For what happened was not that a new outlook suddenly appeared; rather, a gradual metamorphosis took place in the course of a hundred years. It embodied a number of separate developments: shifts in patterns of patronage, in the role of the French Academy, in the system of art instruction, in the artist's position in society. The period under discussion was one of profound social and political upheaval, with bloody revolutions in the United States and France and industrial revolution in England. Artists are, like everyone else, affected by changes in society—sometimes, as in the case of David or Courbet, quite directly. Social changes lead inevitably to changes in attitudes toward artistic means and issues—toward subject matter and expression, toward the use of color and line, and toward the nature and purpose of a work of art and the role it plays vis-à-vis its diverse audience.

A major aspect of modern art is the challenge it posed to traditional methods of representing three-dimensional space. Neoclassicism, which dominated the arts in Europe and America in the second half of the eighteenth century, has at times been called an eclectic and derivative style that perpet-

1. GUSTAVE COURBET.
The Painter's Studio (Atelier).
1854–55. Oil on canvas,
11'10" x 19'7" (3.6 x 6 m).
Musée d'Orsay, Paris

2. JACQUES-LOUIS DAVID. *The Oath of the Horatii.* 1784. Oil on canvas, 10'10" x 14' (3.4 x 4.3 m). Musée du Louvre, Paris

3. Detail of the *Ara Pacis,* scene of an imperial procession. 13–9 B.C.E., marble frieze. Rome

uated the classicism of Renaissance and Baroque art, a classicism that might otherwise have expired. Yet in Neoclassical art a fundamental Renaissance visual tradition was seriously opposed for the first time—the use of perspective recession to govern the organization of pictorial space. Indeed, it may be argued that David's work was crucial in shaping the attitudes that led, ultimately, to twentieth-century abstract art. David and his followers did not actually abandon the tradition of a pictorial structure based on linear and atmospheric perspective. They were fully wedded to the idea that a painting was an adaptation of classical relief sculpture (fig. 3): they subordinated atmospheric effects; emphasized linear contours; arranged their figures as a frieze across the picture plane and accentuated that plane by closing off pictorial depth through the use of such devices as a solid wall, a back area of neutral color, or an impenetrable shadow. The result,

as seen in *The Oath of the Horatii,* is an effect of figures composed along a narrow stage behind a proscenium, figures that exist in space more by the illusion of sculptural modeling than by their location within a pictorial space that has been constructed according to the principles of linear and atmospheric perspective (fig. 2).

Perspective space, the method of representing depth against which David was reacting, had governed European art for the preceding four hundred years. Its basis was single-point perspective, perfected in early fifteenth-century Florence and the logical outcome of the naturalism of fourteenth-century art: a single viewpoint was assumed, and all lines at right angles to the visual plane were made to converge toward a single point on the horizon, the vanishing point. These mathematical-optical principles were discovered by the architect Filippo Brunelleschi before 1420, first applied by the painter Masaccio in 1425, and then written down by the artist and theorist Leon Battista Alberti about 1435. Fifteenth-century artists used perspective to produce the illusion of organized depth by the converging lines of roof beams and checkered floors, to establish a scale for the size of figures in architectural space, and to give objects a diminishing size as they recede from the eye.

Nowhere in the Early Renaissance did this science receive a more lucid or poetic expression than in the painting of PIERO DELLA FRANCESCA (c. 1406–1492). Mystically convinced that a divine order underlay the surface irregularities of natural phenomena, this mathematician-artist endowed his forms with ovoid, cylindrical, or cubic perfection, fixed their relationships in exact proportions, and further clarified all these geometries with a suffusion of cool, silvery light (fig. 4). Such is the abstract wonder produced by this conceptual approach to perceptual reality that it would prove irresistible to a whole line of modern artists, beginning with David and Jean-Auguste-Dominique Ingres and continuing through Georges Seurat and Pablo Picasso. Piero and his fif-

4. PIERO DELLA FRANCESCA. *Flagellation of Christ.* c. 1450. Tempera on panel, 23¼ x 32" (59.1 x 81.3 cm). Galleria Nazionale delle Marche, Palazzo Ducale, Urbino, Italy

6. RAPHAEL SANZIO. *The School of Athens*. 1510–11. Fresco. Stanza della Segnatura, Vatican Palace, Rome

5. JAN VAN EYCK. *Madonna with Chancellor Nicolas Rolin*. c. 1435. Oil on panel, 26 x 24⅜" (66 x 61.9 cm). Musée du Louvre, Paris

teenth-century peers in Italy elaborated the one-point perspective system by shifting the viewpoint to right or left along the horizon line, or above or below it. Atmospheric perspective, another fifteenth-century technique, developed in Flanders rather than Italy, added to the illusion of depth by progressively diminishing color and value contrasts relative to the presumed distance from the viewer. Thus, a distant background landscape might be painted with less saturated colors and soft contours, in contrast to strongly colored and sharply defined foreground figures (fig. 5). With these as their means, Renaissance painters attained control over naturalistic representation of the human figure and environment.

Early in the sixteenth century, the Renaissance conception of space reached a second major climax with **RAPHAEL SANZIO** (1483–1520), in such works as *The School of Athens* (fig. 6), where the nobility of the theme is established by the grandeur of the architectural space. For this supreme masterpiece of the High Renaissance, Raphael employed both linear and atmospheric perspective not only as devices for unifying a vast and complex pictorial space but also as a metaphor of the longing for harmonious unification. The divergent philosophies of Plato and Aristotle, the human and the divine, the ancient and the modern, and, perhaps most of all, the searing divisions that afflicted Christendom on the eve of the Protestant Reformation, coalesced in the converging lines of Raphael.

So exquisite was the balance struck by Raphael and his contemporaries, among them Leonardo da Vinci, Mi-

chelangelo, and Titian, that the succeeding generation of painters gave up studying nature in favor of basing their art on that of the High Renaissance, an art that had already conquered nature and refined away its dross element. Thus, where the older masters had looked to nature and found their grand harmonious style, the new artists discovered what the sixteenth century called *maniera* (a manner), which in the early twentieth century was defined as Mannerism. This stylistic designation is used to describe the work of certain artists after the High Renaissance and includes the late work of Michelangelo. Mannerist paintings may include figural grace exaggerated into extreme attenuation and twisting, choreographic poses, jammed, irrational spaces or distorted perspectives, shrill colors, scattered compositions, cropped images, polished surfaces, and an atmosphere of glacial coolness, even in scenes of high, often erotic emotion or violence. All of these qualities apply to the work of one of the finest Italian Mannerists, **AGNOLO BRONZINO** (1502–72), whose supreme technical skills created compositions of sensuous beauty and dazzling complexity (fig. 7). In their preference for allowing the ideal to prevail over the real, the Mannerists can be seen as forerunners of the bent toward abstraction that would become a dominant trend in modernist art, whether pursued for reasons of objective formal analysis or to express some inner, subjective state of emotional or spiritual necessity.

Emotional or spiritual necessity lay at the core of the *Isenheim Altarpiece,* painted by Raphael's German contemporary **MATTHIAS GRÜNEWALD** (d. 1528), the creator of a harrowing and moving Crucifixion (fig. 8). In this dreadful scene, set against a darkening wilderness and floodlit with a harsh, glaring light, the tortured body of Christ hangs so heavy on the rude cross that the arms seem all ·but wrenched from their sockets, while the hands strain upward like claws frozen in rigor mortis. Under the fearsome diadem of brambles, the head slumps, with eyes closed and the

7. AGNOLO BRONZINO. *Allegory of Time and Love.* c. 1546.
Oil on panel, 61 x 56¼" (154.9 x 143 cm).
The National Gallery, London

8. MATTHIAS GRÜNEWALD. *Crucifixion,* center panel of the *Isenheim Altarpiece* (closed). 1512–15. Oil on panel, 9'9½" x 10'9"
(3 x 3.3 m). Musée d'Unterlinden, Colmar, France

mouth twisted in agony. Supporting the lacerated, gangrenous body are the feet, crushed together by an immense, heavy spike. As the ghostly pale Virgin swoons in the arms of Saint John, Mary Magdalene falls to the ground wringing her hands in grief, while John the Baptist points to the martyred Christ with a stabbing gesture, as if to reenact the violence wreaked upon him. Here the unity of the human and the divine has been sought not in harmonious perfection of outward form, such as that realized by Raphael, but rather in an appalling image of spiritual and physical suffering. Placed upon the high altar of a hospital chapel in the monastery of Saint Anthony at Isenheim in Alsace, Grünewald's *Crucifixion* was part of an elaborate polyptych designed to offer example and solace to the sick and the dying. With its contorted forms, dissonant colors, passionate content, and spiritual purpose, the *Isenheim Altarpiece* stands near the apex of a long tradition that later resurfaced in the Expressionist art of twentieth-century Europe and even more explicitly in the art of the contemporary American painter Jasper Johns (see fig. 613).

In the seventeenth century, along with the religious Counter-Reformation, came an aesthetic reform designed to cleanse art of Mannerism's self-referential excesses and

9. CARAVAGGIO. *The Entombment of Christ.* 1602–4. Oil on canvas, 9'10¼" x 6'8" (3 x 2 m). Pinacoteca Vaticana, Rome

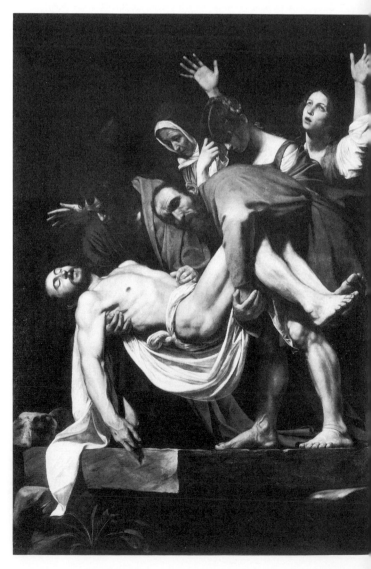

reinvigorate it with something of the broad, dramatically communicative naturalism realized in the High Renaissance. It began with one of the most revolutionary and influential painters in all of history, the Italian Michelangelo Merisi, known as CARAVAGGIO (1572–1610). Singlehandedly and still in his twenties, Caravaggio introduced a blunt, warts-and-all kind of naturalism that sought to make the greatest mysteries of the Christian faith seem present and palpable (fig. 9). Once enhanced by the artist's sense of authentic gesture and his bold light-dark contrasts, full-bodied illusionistic painting assumed an optical and emotive power never before seen in European art. Its impact was felt with stunning force among countless Italian followers, in the Spain of Diego Velázquez, the Netherlands of Rembrandt van Rijn, and the France of Georges de la Tour.

The innovations of Caravaggio ushered in the age of the Baroque, which gave rise during the eighteenth century to such decorators as GIOVANNI BATTISTA TIEPOLO (1696–1770), in whose hands perspective painting became an expressive instrument to be merged with architecture and sculpture in the creation of gigantic symphonies of space, a dynamic world of illusion alive with sweeping, rhythmic movement (fig. 10). Tiepolo resorted to every trompe l'oeil trick of perspective and foreshortening learned in two centuries of experimentation in order to create the illusion of seemingly infinite space. Only with the advent of Neoclassicism at the end of the eighteenth century was there a halt in representing the expansion of space—and indeed, a desire to limit it severely.

David's followers in the first half of the nineteenth century were increasingly concerned with placing an emphasis on an art of ideal subject matter. Both Neoclassicism and Romanticism, it could be argued, were flights from the immediate world to a reality evoked from impressions of the Orient, Africa, or the South Seas; or to a fictional world derived from the art and literature of classical antiquity, the Middle Ages, and the Renaissance. The differences between the two schools lay partly in the particular subjects selected, the Neoclassicists obviously leaning to antiquity and the Romantics to the Middle Ages or what they considered exotic, the East. Even this distinction was blurred as the century wore on, since Ingres, the classicist, made rather a speciality of Oriental odalisques, and Eugène Delacroix, the Romantic, at various times turned to Greek mythology.

The clearest formal distinction between Neoclassical and Romantic painting in the nineteenth century may be seen in the approaches to plastic form and techniques of applying paint. The Neoclassicists continued the Renaissance tradition of glaze painting to attain a uniform surface unmarred by the evidence of active brushwork, whereas the Romantics revived the textured surface of Peter Paul Rubens, Rembrandt, and the Rococo period. Neoclassicism in painting established the principle of balanced frontality to a degree that transcended even the High Renaissance or the classical Baroque of NICOLAS POUSSIN (1594–1665) (fig. 11). Ro-

10. GIOVANNI BATTISTA TIEPOLO. *Translation of the Holy House to Loreto*. 1744. Ceiling fresco (destroyed 1915). Formerly Church of the Scalzi, Venice

mantic painters relied on diagonal recession in depth and indefinite atmospheric-coloristic effects more appropriate to the expression of the inner imagination than the clear light of reason. During the Romantic era there developed an increasingly high regard for artists' sketches, which were thought to capture the individual touch of the artist, thereby communicating deep and authentic emotion. Such attitudes were later crucial for much abstract painting in America and France following World War II.

In analyzing classical, Romantic, and Realist painting in the first part of the nineteenth century, a number of factors other than attitudes toward technique or spatial organization must be kept in mind. The Neoclassicism of David and his followers involved specifically moralistic subject matter related to the philosophic ideals of the French Revolution and based on the presumed stoic and republican virtues of early Rome. Yet painters were hampered in their pursuits of a truly classical art by the lack of adequate prototypes in ancient painting. There was, however, a profusion of ancient sculpture. Thus, it is not surprising that classical paintings such as *The Oath of the Horatii* (see fig. 2) should emulate sculptured figures in high relief within a restricted stage, as in the ancient Roman *Ara Pacis*, (Altar of Peace), which David saw in Rome, where he painted *The Oath*. The "moralizing" atti-

left: 11. Nicolas Poussin.
Mars and Venus. c. 1630.
Oil on canvas,
61" x 7' (155 cm x 2.1 m).
Museum of Fine Arts, Boston
Augustus Hemenway Fund, Arthur Wheelwright Fund

below left: 12. Jacques-Louis David.
The Death of Marat. 1793.
Oil on canvas, 63¾ x 50⅜"
(161.9 x 127.9 cm).
Musées Royaux des Beaux-Arts
de Belgique, Brussels

tudes of his figures make the stage analogy particularly apt, for David's radically distilled composition results from his attitude toward the subject—a deliberate attempt to replace eighteenth-century royalist elaboration with republican simplification and austerity. Though commissioned for Louis XVI, whom David later helped send to the guillotine, this rigorous composition of brothers heroically swearing allegiance to Rome came to be seen as a manifesto of revolutionary sentiment.

One of David's greatest paintings, *The Death of Marat* (fig. 12), contains all the elements referred to—spatial compression, sculptural figuration, highly dramatized subject. But it also reminds us of David's power of realistic presentation, a power brought to bear not on a scene from classical antiquity but, significantly, on a contemporary event. Murdered by a counterrevolutionary in his bath (where he sought relief from a painful skin disease), the revolutionist Marat becomes in David's hands a secular martyr, and a means to highly effective political propaganda. By virtue of its convincing verisimilitude, this painting forms a link between the French portraitists of the eighteenth century and the nineteenth-century Realist tradition of Courbet and his followers.

Benjamin West (1738–1820). Like David, the American painter Benjamin West turned to the *Ara Pacis* when composing his scene from antiquity (fig. 13). West and his contemporary John Singleton Copley, a portraitist best known for his painting of *Paul Revere*, were the first artists from colonial America to achieve real international distinction. Pennsylvanian by birth, West studied in Rome and settled permanently in London, where he was a founding member and eventually president of the Royal Academy, and painter to King George III, a unique distinction for an American artist. West's art was securely rooted in a European tradition that elevated history painting, depictions in the grand manner of historical or religious subjects, above all other genres. He translated his Roman sketches of the *Ara Pacis* into a sober funerary procession, the quintessence of classical dignity and repose.

13. BENJAMIN WEST. *Agrippina Landing at Brundisium with the Ashes of Germanicus.* 1768. Oil on canvas, 64½" x 7'10½" (164 cm x 2.4 m). Yale University Art Gallery, New Haven

GIFT OF LOUIS M. RABINOWITZ

JEAN-AUGUSTE-DOMINIQUE INGRES (1780–1867). Another classicist of paramount importance to the development of modern painting was Jean-Auguste-Dominique Ingres, a pupil of David who during his long life remained the exponent and defender of the Davidian classical tradition. Ingres's style was essentially formed by 1800 and cannot be said to have changed radically in works painted at the end of his life. Although he was a vociferous opponent of most of the new doctrines of Romanticism and Realism, he did introduce certain factors that affected the younger artists who opposed in spirit everything he stood for. Ingres represented to an even greater degree than did David the influence of Renaissance classicism, particularly that of Raphael. Although David was a superb colorist, he tended to subordinate his color to the classical ideal except when he was carried away by the pageantry of the Napoleonic style. Ingres, on the contrary, used a palette both brilliant and delicate, combining classical clarity with Romantic sensuousness, often in liberated, even atonal harmonies of startling boldness (color plate 1, page 33).

The sovereign quality that Ingres brought to the classical tradition was that of drawing, and it was his drawing, his expression of line as an abstract entity—coiling and uncoiling in self-perpetuating complications that seem as much autonomous as descriptive—which provided the link between his art and that of Edgar Degas and Picasso.

FRANCISCO DE GOYA (1746–1828). One of the major figures of eighteenth- and nineteenth-century art, who had a demonstrable influence on what occurred subsequently, was the Spaniard Francisco José de Goya y Lucientes. In a lengthy career Goya carried his art through many stages, from penetrating portraits of the Spanish royal family to a particular concern with the human propensity for barbarity in his middle and late periods. The artist expressed this bleak vision in monstrous, even fantastic images that were the result of penetrating observation. His brilliant cycle of prints, The Disasters of War (fig. 14), depicts the devastating results of Spain's popular uprisings against Napoléon's armies during the Peninsular War. In one of the most searing indictments of war in the history of art, Goya described atrocities committed on both sides of the conflict with reportorial vividness and personal outrage. While sympathetic to the modern ideas espoused by the great thinkers of the Enlightenment, or the Age of Reason, Goya was deeply cynical about the irrational side of human nature and its capacity for the most grotesque cruelty. Because of their inflammatory and ambivalent message, his etchings were not published until 1863, well after his death. During his lifetime

14. FRANCISCO DE GOYA. Plate 30 from The Disasters of War, 1810–11. Etching, 1863 edition, image 5 x 6⅛" (12.8 x 15.5 cm). Hispanic Society of America, New York

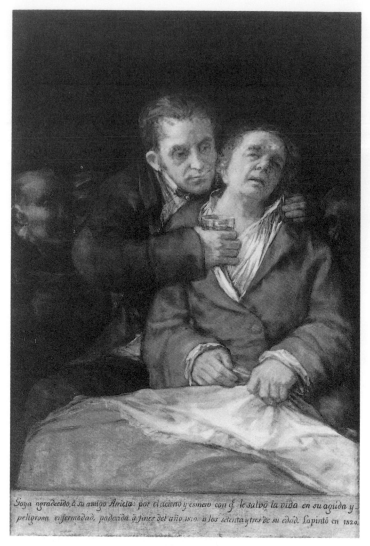

Goya ngradecido, á su amigo Arrieta: por el acierto y esmero con q. le salvó la vida en su aguda y
peligrosa enfermedad, padecida á fines del año 1819. á los setenta y tres de su edad. Lo pintó en 1820.

15. FRANCISCO DE GOYA. *Self-Portrait with Dr. Arrieta.* 1820.
Oil on canvas, 45½ x 31⅛" (117 x 79 cm).
The Minneapolis Institute of Arts
THE ETHEL MORRISON VAN DERLIP FUND

Goya was not very well known outside Spain, despite the
final years he spent in voluntary exile in the French city of
Bordeaux, but once his work had been rediscovered by
Édouard Manet at mid-century it made a strong impact on
the mainstream of modern painting.

Like David's *Death of Marat,* Goya's depiction of himself
being tended by his physician (fig. 15) is a secular event transformed into a tragedy of religious proportions. While Goya's
thinly brushed surface contrasts with David's starkly delineated forms, both works share spatial devices calculated to draw
the viewer directly into the pictorial realm. Goya, who had
already been left permanently deaf by another near-fatal illness, is surrounded by a phantomlike chorus of onlookers. As
a vehicle for the investigation of a personal psyche, rather than
a means to create a heroic message for public consumption,
Goya's stunning and eerie self-portrait anticipates countless
images of modern alienation in twentieth-century art.

EUGÈNE DELACROIX (1798–1863). The French Romantic movement really came into its own with Eugène
Delacroix—through his exploration of exotic themes, his

accent on violent movement and intense emotion, and,
above all, through his reassertion of Baroque color and
emancipated brushwork (color plate 2, page 33). The intensive study that Delacroix made of the nature and capabili-
ties of full color derived not only from the Baroque but
also from his contact with English color painters such as
John Constable, Richard Bonington, and J. M. W. Turner.
His greatest originality, however, may lie less in the actual
freedom and breadth of his touch than in the way he juxtaposed colors in blocks of mutually intensifying complementaries, such as vermilion and blue-green or violet and gold,
arranged in large sonorous chords or, sometimes, in small,
independent, "divided" strokes. These techniques and their
effects had a profound influence on the Impressionists and
Post-Impressionists, particularly Vincent van Gogh (who
made several copies after Delacroix) and Paul Cézanne.

Northern Romanticism

If one characteristic of Romanticism is the supremacy of
emotion over reason, the movement found its most characteristic manifestation in Germany rather than in France.
Indeed, although the main lines of twentieth-century painting are traditionally traced to French Neoclassical, Romantic,
and Realist art, there were critical contemporary developments in Germany, England, Scandinavia, and the Low
Countries throughout much of the nineteenth century. One
may, in fact, trace an almost unbroken Romantic tradition in
Germany and Scandinavia—a legacy that extends from the
late eighteenth century through the entire nineteenth century to Edvard Munch, the Norwegian forerunner of Expressionism, and the later German artists who admired him.

A number of Romantic painters were active in these countries at the beginning of the nineteenth century: the Germans Caspar David Friedrich and Philipp Otto Runge, the
Danish-German Asmus Jacob Carstens, and two Englishmen, William Blake and Joseph Mallord William Turner.
These artists developed very individualized styles but based
their common visionary expression on the vastness and mystery of nature rather than on the religious sources traditional
to much art from the medieval through the Baroque periods.
Comparable attitudes are manifest in the work of European
and American landscapists, from Constable or Thomas Cole
to the early landscapes of Piet Mondrian. Implicit in this
Romantic vision is a sense that the natural world can communicate spiritual and cultural values, at times formally religious, at times broadly pantheistic.

With CASPAR DAVID FRIEDRICH (1774–1840), the leading German painter of Romantic landscape, the image of
nature was by definition a statement of the sublime, of the
infinite and the immeasurable. His landscapes are filled with
mysterious light and vast distances, and the human beings,
when they appear, occupy a subordinate or largely contemplative place (fig. 16). In his ghostly procession of monks
into the ruined apse of a Gothic church, Friedrich clearly
draws formal parallels between the towering forms of the
apse and the framing "architecture" of nature.

Romantic Landscape and the Barbizon School

Although landscape painting in France during the first part of the nineteenth century was a relatively minor genre, by mid-century certain close connections with the English landscapists of the period began to have crucial effects. The painter Richard Parkes Bonington (1802–28), known chiefly for his watercolors, lived most of his brief life in France, where, for a short time, he shared a studio with his friend Delacroix. Bonington's direct studies from nature exerted considerable influence on several artists of the Romantic school, including Delacroix, as well as Camille Corot and his fellow Barbizon painters. Although he painted cityscapes as well as genre and historical subjects, it was the spectacular effects of Bonington's luminous marine landscapes that directly affected artists like Johan Barthold Jongkind and Eugène Boudin, both important precursors of Impressionism (see fig. 48). Indeed, many of the English landscapists visited France frequently and exhibited in the Paris Salons, while Delacroix spent time in England and learned from the direct nature studies of the English artists. Foremost among these were **JOHN CONSTABLE** (1776–1837) and **JOSEPH MALLORD WILLIAM TURNER** (1775–1851). Constable spent a lifetime recording in paint those locales in the English countryside with which he was intimately familiar (color plate 3, page 34). Though his paintings and the sketches he made from nature were the product of intensely felt emotion, Constable never favored the dramatic historical landscapes, with their sublime vision of nature, for

16. CASPAR DAVID FRIEDRICH. *Cloister Graveyard in the Snow.* 1819. Oil on canvas, 48 × 67" (121.9 × 170.2 cm). Formerly Nationalgalerie, Berlin (destroyed in World War II)

which Turner was justifiably famous in his own day. Ambitious, prolific, and equipped with virtuoso technical skills, Turner was determined to make landscapes in the grand tradition of Claude Lorrain and Poussin. His first trip to Italy in 1819 was an experience with profound consequences for his art. In his watercolors and oils Turner explored his fascination with the forces of nature, often

17. THÉODORE ROUSSEAU. *Under the Birches, Evening,* called *Le Curé.* 1842–43. Oil on panel, 16⅝ × 25⅜" (42.3 × 64.4 cm) The Toledo Museum of Art GIFT OF ARTHUR J. SECOR

18. JEAN-FRANÇOIS MILLET. *The Gleaners*. 1857. Oil on canvas, 33 x 44" (83.8 x 111.8 cm). Musée d'Orsay, Paris

tainebleau, rather than the brilliant sunlight of the seashore, that appealed to them. This in itself could be considered a Romantic interpretation of nature, as the expression of intangibles through effects of atmosphere. The Romantic landscape of the Barbizon School merged into a kind of Romantic Realism in the paintings of JEAN-FRANÇOIS MILLET (1814–75), who paralleled Courbet in his passion for the subject of peasants at work but whose interpretation, with primary emphasis on the simplicity and nobility of agrarian experience, was entirely different in manner (fig. 18). As if to redeem the grinding poverty of unpropertied farm life, he shaped his field laborers with the monumentality of Michelangelo and integrated them into landscape compositions of Poussinesque grandeur and calm. Because of this reverence for peasant subjects Millet exerted great influence on Van Gogh.

Another painter who worked outdoors in the Fontainebleau forest was ROSA BONHEUR (1822–99), who is best known for her skillful and sympathetic depictions of animals in the landscape. To prepare her most celebrated picture, *The Horse Fair* (fig. 19), she visited the horse market dressed as a man (for her skirts would have been a "great hindrance") and studied the animals' anatomy and movement. As was often the case with popular works of art in the nineteenth century, this painting was reproduced as a lithograph and circulated widely in Europe and the United States.

CAMILLE COROT (1796–1875). An influential French landscapist of the nineteenth century before Impressionism was Jean-Baptiste-Camille Corot. Only peripherally associated with the Barbizon School's, Corot's work cannot easily

destructive ones, and the ever-changing conditions of light and atmosphere in the landscape. His dazzling light effects could include the delicate reflections of crepuscular light on the Venetian canals or a dramatic view across the Thames of the Houses of Parliament in flames (color plate 4, page 34). Turner's painterly style could sometimes verge on the abstract, and his paintings are especially relevant to developments in twentieth-century art.

The principal French movement in Romantic landscape is known as the Barbizon School, a loose group named for a village in the heart of the forest of Fontainebleau, southeast of Paris. The painters who went there to work drew more directly on the seventeenth-century Dutch landscape tradition than on that of England (fig. 17). In this, the emphasis continued to be on unified, tonal painting rather than on free and direct color. It was the interior of the forest of Fon-

19. ROSA BONHEUR. *The Horse Fair*. 1853; retouched 1855. Oil on canvas, 8' ¼" x 16'7" (2.4 x 5.1 m). The Metropolitan Museum of Art, New York GIFT OF CORNELIUS VANDERBILT, 1887

20. Camille Corot. *Island and Bridge of San Bartolomeo*. 1826–28. Oil on paper mounted on canvas, 10⅝ x 16⅞" (27 x 16.9 cm). Private collection

be categorized. His studies of Roman scenes have a classical purity of organization comparable to that of Poussin and a clarity of light and color similar to that of the English water-colorists. Like his English contemporaries, Corot spent a good deal of his time drawing and painting directly from nature. Beginning in 1825, he spent three years in Italy making open-air studies that in their delicate tonalities capture the special character of southern light. One of his best-known works from this Italian sojourn, *Island and Bridge of San Bartolomeo* (fig. 20), possesses a classical balance and clarity while it demonstrates Corot's striking approach to form. His structures are tightly interlocking horizontals and verticals, all harmoniously defined within a narrow range of ochers and browns. The strong sense of architectural geometry, of contrasting masses and planes, emerges not by way of conventional modeling, but through a regard for form as a series of nearly abstract volumes. These small landscapes exerted a great influence on the development of the Impressionism and Post-Impressionism of Claude Monet and Cézanne. From the landscapes of the Roman period, Corot turned to a more Romantic mode in delicate woodland scenes in tones of silvery gray, Arcadian landscapes sometimes populated by diminutive figures of nymphs and satyrs (see fig. 46). His late portraits and figure studies, on the contrary, are solidly realized and beautifully composed, works that are closely related to the studio scenes and figures of Post-Impressionist, Fauve, and Cubist tradition.

The Academic School and the Salon

Since a large part of this book is concerned with revolts by experimental artists against the academic system, a brief summary of the official academy of art, particularly as it existed in mid-nineteenth-century France, is in order.

The term "academy," in the sense of a school of arts, letters, philosophy, or science, may be traced back to Plato and Athens in the fourth century B.C.E. It was revived in the latter fifteenth century C.E. with the recrudescence of Platonism in Italy. During the Middle Ages and much of the Renaissance and Baroque periods, guilds were organized chiefly to protect artists' rights as craftspeople rather than as creative artists. The origin of the modern academy of art is associated with Leonardo da Vinci at the end of the fifteenth century, after which the idea gained strength in Italy during the sixteenth century as painters and sculptors sought to elevate their position from the practical to the liberal arts. The academy in its modern sense really began in the seventeenth century, when academies of arts and letters and of science were established in many countries of Europe. The French Académie des Beaux-Arts (Academy of Fine Arts) was founded in 1648 and, in one form or another, dominated the production of French art until the 1880s. An artist's survival could depend on his or her acceptance in the annual Salons, large public exhibitions that were open only to Academy members and held for many years in the Louvre. In 1791, during the Revolution, the jury that judged the Salon entries was dis-

21. WILLIAM-ADOLPHE BOUGUEREAU. *Return of Spring (Le Printemps).* 1886. Oil on canvas, 7'½" x 50" (2.1 m x 127 cm). Joslyn Art Museum, Omaha GIFT OF FRANCIS T. B. MARTIN, 1951

men exhibiting at the Salon exceeded twenty percent. Women, however, were not admitted to the prestigious École des Beaux-Arts until the late nineteenth century, nor were they allowed to compete for the most coveted of academic honors, the Prix de Rome, an award that funded the winner's classical education in Rome.

During the nineteenth century, the Salons occupied an even more influential place. In contrast to previous centuries, they now became vast public affairs in which thousands of paintings were hung and thousands rejected, for the revolutionary attempt to "democratize" the academic Salon resulted in a very eclectic mélange that contrasted sharply with the relatively small invitational exhibitions of the eighteenth century. Although the new Salons were selected by juries, presumably competent and occasionally distinguished, they reflected the taste or lack of taste of the new middle-class buyers of art. This might not have been pernicious in itself, except insofar as the authority of the academic tradition persisted, but the reputation, and even the livelihood, of artists continued to be dependent upon acceptance in these official exhibitions.

Typical Salon paintings ranged from pseudoclassical compositions, whose large scale tended to attention-gaining vastness, to the nearly photographic history paintings of Jean-Louis-Ernest Meissonier (1815–91). Among the many other genres on which the largely second- and third-rate artists of the Salon depended, particularly popular were works of extreme sentimentality combined with extreme realism and superficially classical erotic compositions like *Return of Spring* (fig. 21) by **WILLIAM-ADOLPHE BOUGUEREAU** (1825–1905).

Although the revolution of modern art was, as we have seen, in large degree the revolt against this cumbersome academic Salon system, it must be remembered that the leading artists of the nineteenth century participated gladly in the Salons—indeed, they could rarely afford not to. The famous Salon of 1855 devoted a room each to Delacroix and Ingres, and both were awarded grand medals of honor. Courbet may have built a separate pavilion to show *The Painter's Studio* (see fig. 1), which had been rejected, but he also showed other works in the official Salon. Not only did many of the Romantics, Realists, and Impressionists whom we now regard as pioneers of modern art exhibit regularly or occasionally in the Salon, but there were certainly a number of "Salon painters" of distinction, among them Henri Fantin-Latour and Pierre Puvis de Chavannes (see fig. 75). It is also true that Édouard Manet, Odilon Redon, and Edgar Degas, and many other artists highly regarded today for their stylistic innovations, were, to one degree or another, successful Salon painters.

banded and the exhibition was opened to all artists. The results proved so disastrous that the jury was reinstated. Similar exhibitions were staged in London by the Royal Academy, founded in 1768. In the eighteenth century it is difficult to find a painter or sculptor who is well recognized today who was not an academician and did not exhibit in the Salons, though revisionist histories of the period have carefully searched out the developments that took place outside this mainstream.

Until 1790, membership to women in the French Academy was limited to four at any given time. In fact, two of David's contemporaries, Elisabeth Vigée-Lebrun and Adélaïde Labille-Guiard, served there with distinction, virtually dominating the genre of portraiture in the years before the Revolution. It was thanks to the efforts of Labille-Guiard that after the Revolution the Academy ceased to impose limits on female enrollment, and by 1835, the number of wo-

2 Realism, Impressionism, and Early Photography

In the industrial expansion of the nineteenth century, the new and prosperous middle class assumed a larger and larger part in every phase of European and American life. Although governmental patronage of the arts still persisted, official commissions and state purchases were a feasible income source for a limited number of artists, and by 1880 the state relinquished its control over the ever-popular Salon exhibitions. The new patrons were predominantly the newly rich bourgeoisie. Fundamentally materialistic in its values, this increasingly dominant segment of society had less interest in the fantasy of Romantic art than in a kind of pictorial verisimilitude that could convey meticulous visual facts verifiable in the external world of here and now. Such developments in the market and in the attitudes of the public consumers of art helped bring about the eventual demise of history painting. For private collectors did not covet large, cumbersome scenes from antiquity, but preferred the genres of portraiture, landscape, and scenes from daily life.

The Invention of Photography and the Rise of Realism

The public taste for visual fact served as an all-important stimulant to the research that finally brought about the invention of photography. In early 1839 it was publicly demonstrated that a new mechanical technology had been discovered for resolving and permanently fixing upon a flat surface and in minute detail the exact tonal, if not full-color, image of the three-dimensional world. From that time, Western painters would hardly be able to create new imagery without some consciousness of the special conditions introduced by the medium of photography. Nor, it should be added, would photographers ever feel altogether free to work without some awareness of the rival aesthetic standards, qualities, and prestige imposed by the age-old, handwrought image-making processes of painting and drawing.

Long before anyone had ever seen a daguerreotype (an early form of the photograph), painters had already anticipated many of the salient characteristics of photographic form: a snapshotlike cropping of figures by the framing edges (color plate 12, page 37); motion arrested or frozen in full flight or mid-step to reveal its physiognomy in a way seldom perceptible to the naked eye; a related "stop-action" informality of human pose and gesture; ghostly residual or afterimages left in the wake of speeding objects; imagery defined purely in terms of tone, free of contour lines; wide-angle views and exaggerated foreshortenings; and, perhaps most of all, a generalized flatness of image even where there is depth illusion, an effect produced in normal photography by the camera's relative consistency of focus throughout the field and its one-eyed or monocular view of things.

None of these characteristics, however, became a commonplace of painting style until the mass proliferation of photographic imagery made them a ubiquitous and unavoidable feature of modern life. The most immediate and obvious impact on painting can be seen in the work of those artists, eager to achieve a special kind of optical veracity unknown until the advent of photography, a trend often thought to have culminated in the "instantaneity" of Impressionism, and resurgent in the Photorealism of the 1970s.

Other artists, however, or even the same ones, took the scrupulous fidelity of the photographic image as good reason to work imaginatively or conceptually and thus liberate their art from the dictum of perceptual truth. Some indeed saw the aberrations and irregularities peculiar to photography as a source of fresh ideas for creating a whole new language of form. And among painters of the period were some accomplished photographers, such as Edgar Degas and Thomas Eakins (see fig. 64). Quite apart from its role as a model of pictorial mimesis, photography could serve painters as a shortcut substitute for closely observed preparatory drawings and as a vastly expanded repertoire of reliable imagery, drawn from whatever remote, exotic corner of the globe into which adventurous photographers had been able to lug their cumbersome nineteenth-century equipment.

Meanwhile, early photographers found themselves no less symbiotically involved with painting than the painters were bound up with photography. Buoyed by pride in their tech-

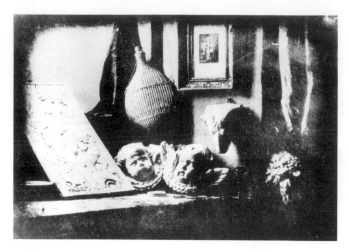

22. LOUIS-JACQUES-MANDÉ DAGUERRE. An early daguerreotype,
taken in the artist's studio. 1837.
Société Française de Photographie, Paris

nical superiority, photographers even felt compelled to match painting in its artistic achievements. For painters enjoyed the freedom to select, synthesize, and emphasize at will and thus attain not only the poetry and expression thought to be essential to art but also a higher order of visual, if not optical, truth. In one of the earliest known photographs, LOUIS-JACQUES-MANDÉ DAGUERRE (1789–1851), an artist already famous for his immense, sensationally illusionistic painted dioramas, had sought to emulate art as much as nature: the reality he chose to record was that of a still life arranged in a manner conventional to painting since at least the seventeenth century (fig. 22). This was a daguerreotype, a metal plate coated with a light-sensitive silver solution, which, once developed in a chemical bath, resulted in a unique likeness that could not

below: 23. OSCAR G. REJLANDER. *The Two Ways of Life*. 1857.
Combination albumen print.
Royal Photographic Society, Bath, England

be replicated, but that recorded the desired image with astonishing clarity.

Simultaneously in England, the photographic process became considerably more versatile when WILLIAM HENRY FOX TALBOT (1800–1877) discovered a way of fixing a light-reflected image on the silver treated surface of paper, producing what its creator called a calotype. Talbot invented a technique that converted a negative into a positive, a procedure that remains fundamental to photography even now. Because the negative made it possible for the image to be replicated an infinite number of times, unlike the impractical daguerreotype, painting-conscious photographers such as OSCAR G. REJLANDER (1813–75) could assemble elaborate, multifigure compositions in stagelike settings. They pieced the total image together from a variety of negatives and then orchestrated them, rather in the way contemporary history painters prepared their grand narrative scenes for the Salon (figs. 23, 24). And indeed Rejlander emulated not only the artistic methods and ambitions of academic masters like

below: 24. THOMAS COUTURE. *Decadence of the Romans*. 1847.
Oil on canvas, 15'3" x 25'3" (4.6 x 7.7 m).
Musée d'Orsay, Paris

25. HENRY PEACH ROBINSON. Sketch for *Carrolling*. 1886–87.
Pencil on paper, 13 x 25" (33 x 65.5 cm).
Royal Photographic Society, Bath, England

26. HENRY PEACH ROBINSON. *Carrolling*. 1887.
Combination albumen print.
Royal Photographic Society, Bath, England

THOMAS COUTURE (1815–79), but also their pretensions to high moral purpose. The British photographer HENRY PEACH ROBINSON (1830–1901) used sketches (fig. 25) to prepare his elaborate exhibition photographs, which he compiled from multiple negatives (fig. 26).

But just as the Realist spirit inspired some progressive painters to seek truth in a more direct and simplifying approach to subject and medium, many enlightened photographers sought to purge their work of the artificial, academic devices employed by Rejlander and concentrate on what photography then did best—report the world and its life candidly. The practitioners of this kind of direct photography thus allowed art and expression to occur naturally, through the choice of subject, view, framing, light, and the constantly improving means for controlling the latter—lenses, shutter speeds, plates, and processing chemicals. Even the work of Daguerre and Talbot (figs. 27, 28) shows this taste for creating a straightforward record of the everyday world, devoid of theatrics or sentiment. Their first street scenes already have the oblique angles and random cropping of snapshots, if not the bustling life, which moved in and out of range too rapidly to be caught by the slow photosensitive materials of the 1830s

and 1840s. Contemporaries called Daguerre's Paris a "city of the dead," since the only human presence it registered was that of a man who stood still long enough for his shoes to be shined and, coincidentally, his picture to be taken.

Among the first great successes in photography were the portraitists, foremost among them France's NADAR (Gaspard-Félix Tournachon; 1820–1910) and England's JULIA MARGARET CAMERON (1815–79), whose own powerful personalities gave them access to many of the most illustrious people of the age and the vision necessary to render these with unforgettable forthrightness and penetration (fig. 29). A prolific writer, as well as caricaturist, hot-air balloon photographer, and dynamic man about Paris, Nadar wrote in 1856:

Photography is a marvellous discovery, a science that has attracted the greatest intellects, an art that excites the most astute minds—and one that can be practiced by any imbecile. . . . Photographic theory can be taught in an hour, the basic technique in a day. But what cannot be taught is the feeling for light. . . . It is how light lies on the face that you as artist must capture. Nor can one be taught how to grasp the personality of the sitter. To produce an intimate likeness rather than a banal portrait, the result of mere chance, you must put yourself at once in communion with the sitter, size up his thoughts and his very character.

27. LOUIS-JACQUES-MANDÉ DAGUERRE.
Boulevard du Temple, Paris. c. 1838.
Daguerreotype.
Bayerisches Nationalmuseum, Munich

left: 28. WILLIAM HENRY FOX TALBOT. *Trafalgar Square, London, During the Erection of the Nelson Column*. London, England. 1843. Salt print from a paper negative, 6¾ x 8⅜" (17.1 x 21.3 cm). The Museum of Modern Art, New York
GIFT OF WARNER COMMUNICATIONS, INC.
COPY PRINT © 1995 THE MUSEUM OF MODERN ART, NEW YORK

below, left: 29. NADAR (GASPARD-FÉLIX TOURNACHON). *Sarah Bernhardt*. 1864. Photograph, 9⅞ x 9⅝" (24.8 x 20.5 cm). International Museum of Photography at George Eastman House, Rochester, New York

below, right: 30. JULIA MARGARET CAMERON. *Mrs. Herbert Duckworth as Julia Jackson*. c. 1867. Albumen print, 13⅜ x 9⅝" (34 x 24.4 cm). International Museum of Photography at George Eastman House, Rochester, New York

At the same time that Cameron liked to dress up her friends and family and reenact scenes from the Bible and Tennyson's *Idylls of the King* as photographed costume dramas, styled in a late-Romantic Pre-Raphaelite manner (color plate 5, page 34), she also approached portraiture with an intensity that simply swept away all staginess but the drama of character and accentuated chiaroscuro (fig. 30). Throughout her work, however, she maintained, like the good

Victorian she was, only the loftiest aims. In answer to the complaint that her pictures always appeared to be out of focus, Cameron stated:

What is focus—who has a right to say what focus is the legitimate focus—My aspirations are to ennoble Photography and to secure for it the character and uses of High Art by combining the real & ideal & sacrificing nothing of Truth by all possible devotion to Poetry & Beauty—

Color plate 1. Jean-Auguste-Dominique Ingres. *Odalisque with Slave.* 1842. Oil on canvas, 30 x 41½"
(76.2 x 105.4 cm). Walters Art Gallery, Baltimore

Color plate 2. Eugène Delacroix. *The Lion Hunt.* 1861. Oil on canvas, 30⅛ x 38¾" (76.5 x 98.4 cm).
The Art Institute of Chicago Potter Palmer Collection

Color plate 3. JOHN CONSTABLE. *The Hay Wain*. 1819–21. Oil on canvas, 51¼" x 6'1" (130.2 cm x 1.9 m). The National Gallery, London

Color plate 4. JOSEPH MALLORD WILLIAM TURNER. *The Burning of the Houses of Parliament*. 1834–35. Oil on canvas, 36½ x 48½" (92.7 x 123.2 cm). The Cleveland Museum of Art

JOHN L. SEVERANCE COLLECTION

Color plate 5. WILLIAM HOLMAN HUNT. *The Hireling Shepherd*. 1851. Oil on canvas, 30¹⁄₁₆ x 43⅛" (76.4 x 109.5 cm). Manchester City Art Galleries, England

Color plate 6. ÉDOUARD MANET. *Déjeuner sur l'herbe (Luncheon on the Grass)*. 1863. Oil on canvas, 6'9⅛" x 8'10¼" (2.1 x 2.7 m). Musée d'Orsay, Paris

left: Color plate 7. ANDO HIROSHIGE. *Moon Pine at Ueno* from *One Hundred Views of Famous Places in Edo*. 1857. Color woodcut, 13¾ x 8⅝" (34.9 x 21.9 cm). The Brooklyn Museum, New York

right: Color plate 8. JAMES ABBOTT MCNEILL WHISTLER. *Symphony in White No. II: The Little White Girl*. 1864. Oil on canvas, 30⅛ x 20⅛" (76.5 x 51.1 cm). Tate Gallery, London

Color plate 9. CLAUDE MONET. *Impression: Sunrise.* 1872. Oil on canvas, 17¾ x 21¾" (45.1 x 55.2 cm).
Musée Marmottan, Paris

Color plate 10. CLAUDE MONET. *The Bridge at Argenteuil.* 1874. Oil on canvas, 23⅝ x 31½" (60 x 80 cm).
Musée d'Orsay, Paris

Color plate 11. AUGUSTE RENOIR. *Moulin de la Galette*. 1876. Oil on canvas, 51½ x 69" (130.8 x 175.3 cm).
Musée d'Orsay, Paris

left: Color plate 12. EDGAR DEGAS. *Ballerina and Lady with a Fan*. 1885.
Pastel on paper, 25⅛ x 19¼" (63.8 x 48.9 cm). Philadelphia Museum of Art
THE JOHN G. JOHNSON COLLECTION

Color plate 13. MARY CASSATT. *Little Girl in a Blue Armchair*. 1878. Oil on canvas, 35 x 51" (88.9 x 129.5 cm). National Gallery of Art, Washington, D.C.
COLLECTION MR. AND MRS. PAUL MELLON, 1983

Color plate 14. GEORGE CATLIN. *Buffalo Bull's Back Fat, Head Chief, Blood Tribe*. 1832. Oil on canvas, 29 x 24" (73.7 x 61 cm). National Museum of American Art, Smithsonian Institution, Washington, D.C. GIFT OF MRS. JOSEPH HARRISON, JR.

Color plate 15. REMBRANDT PEALE. *Rubens Peale with a Geranium*. 1801. Oil on canvas, 28¼ x 24" (71.7 x 61 cm). National Gallery of Art, Washington, D.C. PATRON'S PERMANENT FUND

Color plate 16. FREDERIC E. CHURCH. *Twilight in the Wilderness*. 1860. Oil on canvas, 40 x 64" (101.6 x 162.6 cm). The Cleveland Museum of Art

Color plate 17. WINSLOW HOMER. *The Fox Hunt.* 1893.
Oil on canvas, 38 x 68½" (96.5 x 174 cm).
The Pennsylvania Academy of the Fine Arts, Philadelphia
JOSEPH E. TEMPLE FUND

Right: Color plate 18. THOMAS EAKINS. *The Gross Clinic.* 1875.
Oil on canvas, 8' x 6'6" (2.4 x 2 m). Jefferson Medical College of
Thomas Jefferson University, Philadelphia

Color plate 19. GEORGES SEURAT. *A Sunday Afternoon on the Island of La Grande Jatte.* 1884–86. Oil on canvas, 6'9½" x 10'1¼"
(2.1 x 3.1 m). The Art Institute of Chicago HELEN BIRCH BARTLETT MEMORIAL COLLECTION

Color plate 20. PAUL SIGNAC. *Opus 217. Against the Enamel of a Background Rhythmic with Beats and Angles, Tones and Tints, Portrait of M. Félix Fénéon in 1890. 1890.* Oil on canvas, 29⅛ x 36½" (73.5 x 92.5 cm). Private Collection
PROMISED GIFT OF DAVID ROCKEFELLER © 1997 ADAGP, PARIS/ARTISTS RIGHTS SOCIETY (ARS), NEW YORK

Color plate 21. PAUL CÉZANNE. *The Bay from L'Estaque.* c. 1885. Oil on canvas, 31½ x 39⅝" (80 x 100.6 cm). The Art Institute of Chicago MR. AND MRS. MARTIN A. RYERSON COLLECTION

31. UNKNOWN PHOTOGRAPHER. *Frederick Douglass.* 1847.
Daguerreotype, 3¼ x 2¾" (8.3 x 7 cm).
Collection William Rubel

32. CHARLES NÈGRE. *The Gargoyle of Notre Dame.* Original negative 1840s or 1850s, reprinted 1960. Calotype, 13⅛ x 9¼"
(33.3 x 23.5 cm). The Metropolitan Museum of Art, New York
ROGERS FUND, 1962

In America the daguerreotype experienced wild popularity almost from the very moment instruction manuals and
equipment arrived on the East Coast. The many commercial
studios that grew up around the new medium, especially in
New York and Philadelphia, gave rise to hundreds of amateur photographers. These provided inexpensive portrait services to sitters who could have never undertaken the expense
of a commissioned painted portrait. An American propensity for visual fact guaranteed the particular success of the
daguerreotype, at least on an economic if not on an aesthetic level. From the countless efforts of anonymous photographers emerge forthright records of ordinary citizens and
valuable likenesses of extraordinary personalities, such as a
portrait of the eminent abolitionist writer and orator Frederick Douglass (fig. 31). An escaped slave, Douglass bolstered his eloquent public protests against slavery with firsthand experience. He was an adviser to President Abraham
Lincoln during the Civil War and afterward became the first
African-American to hold high office in the United States
government. The photograph of Douglass as a young man
was taken at the end of a lecture tour he had given in Great
Britain and Ireland in order to avoid recapture in America.
Like the best daguerreotypes, it possesses tremendous appeal
by virtue of its unaffected, straightforward presentation of
the sitter.

Documentary pursuits were obviously well served by the
representational integrity available in straight photography.
When the French photographer **CHARLES NÈGRE**
(1820–79) dragged his heavy paraphernalia high up among
the gargoyles of Paris's Notre Dame Cathedral (fig. 32), he
created an image almost surreal in its visual juxtapositions
and one of significant architectural interest as well. Documentary photographers could provide eager audiences with
imagery of the relatively unknown environments of the western American landscape (fig. 33), the wonders of ancient
Egypt, or the exotic backstreets of old China (fig. 34).

When it came to war, photography robbed armed conflict
of the operatic glamour often given it by traditional painting.
This first became evident in the photographs taken by Roger
Fenton (1819–69) in the Crimea, but it had even more shattering effect in the work of **MATHEW BRADY** (1823–96) and
such associates as Alexander Gardner (1821–82), who bore
cameras onto the body-strewn fields in the wake of battle
during the American Civil War (fig. 35). (Because the required length of exposure was still several seconds, the battles
themselves could not be photographed.) Now that beautifully particularized, mutely objective—or indifferent—pictures
disclosed all too clearly the harrowing calm brought by violence, or the almost indistinguishable likeness of the dead on
either side of the conflict, questions of war and peace took on
a whole new meaning. When Brady's photograph is compared to a contemporary Civil War scene by one of the greatest American painters of the era, Winslow Homer (see fig.
63), it becomes clear that photography brought home the
graphic reality of war in a way no painter could.

As a medium of communication, as well as expression,

left: 33. TIMOTHY O'SULLIVAN. *Ancient Ruins in the Cañon de Chelle.* 1873. Collodion print, 10⅞ x 8" (27.6 x 20.3 cm). National Archives, Washington, D.C.

photography was both the product and the agent of the powerful social changes that had been taking place in Western civilization since the revolutions of the late eighteenth century. Along with photography, and its popularization of pictorial imagery, came the immense expansion of journalism, as newspapers and periodicals were in some measure liberated from the tight and often punitive censorship of governments. A new kind of journalism and a new kind of critical writing were everywhere evident. Many of the leading authors and critics of France and England wrote regularly for the new popular reviews.

HONORÉ DAUMIER (1808–79). The changes in attitude are perhaps best seen in the emergence of the caricature—the satirical comment on life and politics, which became a standard feature of journals and newspapers (fig. 36). Although visual attacks on individuals or corrupt economic or political systems have always existed, and artists such as Goya have been motivated by a spirit of passionate invective, drawings that ranged from gentle amusement to biting satire to vicious attack now became a commonplace in every part of Europe and America. These images could be cheaply disseminated through the popular medium of lithography. Just as the rise of Realism in art and the seemingly opposed expansion of the inflated Salon picture were closely allied to the emergence of large-scale bourgeois patronage, so were the expansion of the newspaper and journal and

34. JOHN THOMPSON. *Physic Street, Canton.* c. 1868. Carbon print from *Illustration of China and Its People* (London, 1873), 10½ x 8" (26.7 x 20.3 cm). Peabody Essex Museum, Salem, Massachusetts

35. MATHEW BRADY. *Dead Soldier, Civil War.* c. 1863. Gelatin-silver print. Library of Congress, Washington, D.C.

36. HONORÉ DAUMIER. *Battle of the Schools: Idealism vs. Realism.* 1855. Lithograph, 6⁵⁄₁₆ x 8⅛" (16 x 18 cm). Bibliothèque Nationale, Paris

37. HONORÉ DAUMIER. *The Third-Class Carriage.* c. 1863–65. Oil on canvas, 25¼ x 35½" (64.1 x 90.2 cm). The Metropolitan Museum of Art, New York
H. O. HAVEMEYER COLLECTION, BEQUEST OF MRS. H. O. HAVEMEYER, 1929

the popularity of the daily or weekly caricatures of the problems—and sometimes the tragedies—of everyday life.

The greatest of the satirical artists to emerge in this new environment was Honoré Daumier, who, although best known for some four thousand lithographic drawings he contributed to French journals such as *La Caricature* and *Le Charivari,* was also a major painter and sculptor. But while Daumier was concerned with all the details of everyday life in Paris, and his satire is thus rooted in realistic observation, he cannot be called only a Realist. His paintings dealing with themes from Cervantes and Molière place him partly in the Romantic camp, not only for his use of literary subjects but also for the dramatic chiaroscuro he employed to obtain his effects. This drama of light and shade transforms even mundane subjects, such as *The Third-Class Carriage* (fig. 37), from simple illustration to a scene of pathos. The painting is a crucial example of French Realist art in its sympathetic depiction of contemporary working-class life.

Daumier's studies of the theater, of the art world, and of politics and the law courts run the gamut from simple commentary to perceptive observation to bitter satire. His political caricatures were at times so biting in their attacks on the establishment that they were censored. For a caricature he did of King Louis Philippe, the artist was imprisoned for six months. The profundity of Daumier's horror at injustice and brutality is perhaps best illustrated in the famous lithograph *Rue Transnonain* (fig. 38), which shows the dead bodies of a family of innocent workers who were murdered by the civil guard during the republican revolt of 1834. With journalistic bluntness, the title refers to the street where the killings occurred. In its powerful exploitation of a black-and-white medium, as well as the graphic description of its gruesome subject, this is a work comparable to Goya's *Disasters of War* (see fig. 14).

GUSTAVE COURBET (1819–77). Like Daumier, Gustave Courbet inherited certain traits from his Romantic predecessors. Throughout his life, he produced both erotic subjects that border on academicism and Romantic interpretations of the artist's world in many self-portraits. But most important, he created unsentimental records of contemporary life more persuasive than any other artist's since the seventeenth century, works that today secure his place as the leading exponent of modern Realism in painting. These works possess a compelling physicality and an authenticity of vision that the artist derived from the observed world. Most notably, Courbet insisted upon the very material with which a painting is constructed: oil paint. He understood that a painting is in itself a reality—a two-dimensional world of stretched canvas defined by the nature and textural consistency of paint—and that the artist's function is to define this world.

A Burial at Ornans (fig. 39) is significant in the history of modern painting for its denial of illusionistic depth, its apparent lack of a formal composition, and an approach to subject matter so radical that it was seen as an insult to everything for which the venerable French Academy stood. In his assertion

38. HONORÉ DAUMIER. *Rue Transnonain, April 15, 1834.* Published in *L'Association Mensuelle,* July 1834. Lithograph, 11½ x 17½" (29.2 x 44.5 cm). Graphische Sammlung Albertina, Vienna

39. GUSTAVE COURBET. *A Burial at Ornans.* 1849–50. Oil on canvas, 10'3½" x 21'9" (3.1 x 6.6 m). Musée d'Orsay, Paris

of paint texture, the artist drew on and amplified the intervening experiments of the Romantics, although with a different end in view. Whereas the Romantics used broken paint surfaces to emphasize the intangible elements of the spirit, Courbet employed them as a means to make the ordinary world more tangible. While the *Burial* documents a commonplace event in Courbet's small provincial hometown in eastern France, its ambitious size conforms to the dimensions of history painting, a category reserved for the noblest events from the past. Courbet's casual, friezelike arrangement of village inhabitants, made up mostly of his friends and relatives, is an accurate depiction of local funerary customs (men and women are segregated, for example), but it imports no particular narrative message. It was not only this lack of compositional hierarchy and pictorial rhetoric that outraged French critics. Courbet's representation of ordinary residents of the countryside did not coincide with established Parisian norms for the depiction of rural folk. These are not the ennobled peasants of Millet or the pretty, happy maids of traditional academic painting. Rather, they are highly individualized, mostly unattractive, and aggressively real.

Combined with his unorthodox approach to formal matters, Courbet's unpolished treatment of working-class subjects was associated, as was the mid-century battle cry of Realism, with left-wing dissidence. Any art-historical consideration of Realism is inextricably tied to the tumultuous political and social history during this period in France, the country where Realism developed most coherently. Following the devastating defeat of France's imperial government by the Germans in the Franco-Prussian War in 1871, Courbet took part in the popular uprising that resulted in the short-lived, radically republican Paris Commune. As head of the Federation of Artists, he proposed the peaceful dismantling of a Napoleonic monument in Paris, the Vendôme column. When the column was

actually toppled and destroyed by the Communards, the conservative regime that brutally suppressed the Commune imprisoned Courbet and held him financially responsible for the damage. To avoid confiscation of his work, the artist eventually fled to Switzerland, where he died in 1877.

In a group of landscapes of the 1850s and 1860s, *The Source of the Loue* among them (fig. 40), Courbet experimented with a form of extreme close-up of rocks abruptly cut off at the edges of the painting. *The Source*, again a view from the artist's native province of Franche-Comté, may owe something to the emerging art of photography, both in its fragmentary impression and in its subdued, almost monochromatic color. Courbet's landscape, however, combined a sense of observed reality with an even greater sense of the ele-

40. GUSTAVE COURBET. *The Source of the Loue.* 1864.
Oil on canvas, 39¼ x 56" (99.7 x 142.2 cm).
The Metropolitan Museum of Art, New York

ments and materials with which the artist was working: the rectangle of the picture plane and the emphatic texture of the oil paint, which asserted its own nature at the same time that it was simulating the rough-hewn, sculptural appearance of exposed rocks.

Courbet, in his later career, experimented with many different approaches to the subjects of landscape, portrait, still life, and the figure. His fascination with a type of realism in which precise observation of nature is combined with the aggressive, expressive statement of the pictorial means culminated in a group of seascapes of the late 1860s. In *The Waves* (fig. 41) he appears to have selected an unrelieved expanse of open sea in order to demonstrate his ability to translate this subject into a primarily two-dimensional organization in which the third dimension is realized as the relief texture of the projecting oil paint. Like *A Burial at Ornans*, this seascape is not compositionally structured around a central pictorial focus. Indeed, it is devoid of all human presence and concentrates, to a degree previously achieved only by Turner, on sea and sky, all tangibly evoked with vigorous strokes from Courbet's palette knife.

Courbet's contribution to painting is revealed when such landscapes as *The Source of the Loue* and *The Waves* are compared with a much earlier landscape by John Constable, *The Hay Wain* (color plate 3, page 34). The paintings of the English colorists Constable, Turner (color plate 4, page 34), and Bonington had a profound influence on the French Romantics, particularly on the Barbizon School of landscapists and, eventually, the Impressionists. Constable's textural statement of brush gesture was well adapted to Romantic narrative and landscape, out of which Courbet's painting emerged. The radiant color used by the Englishman was a revelation. When *The Hay Wain* was shown at the 1824 Salon in Paris it caused a sensation, even sending Delacroix back to his studio to rework one of his paintings. Its sun-filled naturalism, however, was premature for the French Realists and Romantics. With the exception of Delacroix, they continued to prefer the more subdued color harmonies and the less vivid light of the studio when painting the out-of-doors. From early in his career, Constable dedicated himself to the close observation of the life and landscape around his father's property in Suffolk. Yet his bucolic visions of the English countryside rarely make reference to the harsher realities of agricultural life during a period of economic strife and civil unrest among agrarian workers. That task would be left to later English painters such as Ford Madox Brown (see fig. 42). Constable could produce accurate, unaffected nature studies composed entirely in the open air (a half-century before the Impressionists), while remaining committed to the study of the seventeenth-century landscapes of the French classical painter Claude Lorrain. The paint texture of his *Hay Wain* is direct and unconcealed to a degree rarely observed in painting before his time, but his landscapes are still seen with the eyes of the Renaissance-Baroque tradition. Space opens up into depth, and paint is converted into fluffy, distant clouds, richly leaved trees, and reflecting pools of water.

41. GUSTAVE COURBET. *The Waves*. c. 1870. Oil on canvas, 12¾ x 19" (32.4 x 48.3 cm). Philadelphia Museum of Art
THE LOUIS E. STERN COLLECTION

Realism in England: The Pre-Raphaelites

On the English side of the Channel, the social and artistic revolution that fueled the Realist movement in France assumed virtually religious zeal, as a group of young painters rebelled against what they considered the decadence of current English art and called for renewal modeled on the piously direct, primitive naturalism practiced in Florence and Flanders prior to the late work of Raphael and the High Renaissance. **WILLIAM HOLMAN HUNT** (1827–1910), John Everett Millais (1829–96), and Dante Gabriel Rossetti (1828–82) joined with several other sympathetic spirits to form a secret, reform-minded artistic society, which they called the Pre-Raphaelite Brotherhood. The ardent, if not monkish, PRB determined to eschew all inherited Mannerist and Baroque artifice and to search for truth in worthwhile, often Christian subject matter presented with naive, literal fidelity to the exact textures, colors, light, and, above all, the outlines of nature.

For *The Hireling Shepherd* (color plate 5, page 34) Hunt pursued "truth to nature" in the countryside of Surrey, where he spent months throughout the summer and autumn working en plein air, or out-of-doors, to set down the dazzling effect of morning sunlight. Reviving the techniques of the fifteenth-century Flemish masters, Hunt laid—over a white wet ground—glaze after glaze of jewellike, translucent color and used the finest brushes to limn every last wildflower blossom and blade of grass. Late in life, he asserted that his "first object was to portray a real Shepherd and Shepherdess . . . sheep and absolute fields and trees and sky and clouds instead of the painted dolls with pattern backgrounds called by such names in the pictures of the period." But it seems that he was also motivated to create a private allegory of the need for mid-Victorian spiritual leaders to cease their sectarian disputes and become true shepherds tending their distracted flocks. Thus, while the shepherd, ruddy-faced from beer in his hip keg, pays court to the shepherdess, the lamb on her lap makes a

42. FORD MADOX BROWN. *Work.* 1852, 1855–63. Oil on canvas, 53" x 6'5¹¹⁄₁₆" (135 cm x 2 m). Manchester City Art Galleries, England

deadly meal of green apples, just as the sheep in the background succumb from being allowed to feed on corn. Gone are the smooth chiaroscuro elisions, rhythmic sweep, and gestural grace of earlier art, replaced by a fresh-eyed naturalism.

FORD MADOX BROWN (1821–93). Never an official member of the PRB, Brown was introduced to the group by the precocious young poet and painter Rossetti. The large canvas *Work* (fig. 42), which preoccupied him for several years, is Brown's monumental testament to the edifying and redemptive power of hard toil. Like Hunt, he relies on a microscopic accuracy of detail to deliver his moralizing message. At the center of the sun-drenched composition are the muscular navvies who dig trenches for new waterworks in the London suburb of Hampstead. A proliferation of genre detail orbits around this central action of the painting like satellite vignettes; these reinforce Brown's theme with emblems of the various social strata that made up contemporary Victorian society. At the far left, for example, is a ragged flower seller followed by two women distributing temperance tracts. In the shade of a tree at the right, an indigent couple tends to their infant while in the background an elegantly dressed couple rides horseback. The two men at the far right are portraits: the philosopher Thomas Carlyle and a leading Christian Socialist, Frederick Denison Maurice, "brain-

workers" whose relatively conservative ideas about social reform dovetailed with Brown's own genuine attempts to correct social injustice. His painting is the most complete illustration of the familiar Victorian ethic that promoted work as the foundation of material advancement, national progress, and spiritual salvation.

Impressionism

ÉDOUARD MANET (1832–83). Although he probably would not have found himself at odds with Brown's left-wing politics, Manet was a realist with an artistic temperament far from the Englishman's accumulation of exacting detail. Despite his commitment to depict subjects of modern life in a modern style, he sought recognition through the time-honored venue of the Salon. He never exhibited with his fellow Impressionists in the eight independent exhibitions they organized between 1874 and 1886. Manet's *Déjeuner sur l'herbe* (color plate 6, page 35), rejected by the Salon of 1863, created a major scandal when it was shown in the Salon des Refusés, a now-infamous exhibition mounted by the state that year to house the large number of works barred by jurors from the main Salon. Although the subject was based on respectable academic precedents from the Renaissance— Giorgione's *Pastoral Concert* (1510) and a detail from an engraving by Marcantonio Raimondi of *The Judgment of Paris* (c. 1520), after a lost cartoon by Raphael—popular indignation arose because the classical, pastoral subject had been translated into contemporary terms. Raphael's goddesses and Giorgione's nymphs had become models, or even prostitutes, one naked, one partially disrobed, relaxing in the woods with two men whom Salon visitors would have recognized by their dress as artists or students. The woman in the foreground is Victorine Meurent, the same model who posed for *Olympia;* the men are Manet's relations. Manet clearly abided by the convictions of his supporter Charles Baudelaire, the great poet and art critic who exhorted artists to paint scenes of modern life and to invent contemporary contexts for old subjects such as the nude. But in his ironic quotations from Renaissance sources, Manet created what seemed to be a conscious affront to tradition, and by failing to encode his contemporary scene as idealized allegory, he seemed deliberately to flout accepted social convention.

Although Manet ostensibly adopted the structure of Raphael's original composition, with the bending woman in the middle distance forming the apex of the classical receding triangle of which the three foreground figures serve as the base and sides, his broad, abrupt handling of paint and his treatment of the figures as silhouettes rather than carefully modeled volumes tend to collapse space, flatten forms, and assert the nature of the canvas as a two-dimensional surface. Manet's subject, as well as the sketchiness of technique, infu-

riated the professional critics, as though they sensed in this work the prophecy of a revolution that was to destroy the comfortable world of secure values of which they felt themselves to be the guardians.

Manet's *Olympia* (fig. 43), painted in 1863 and exhibited in the Salon of 1865, created an even greater furor than the *Déjeuner.* Here, again, the artist's source was exemplary: Titian's *Venus of Urbino,* which Manet had copied as a young man visiting Florence. *Olympia,* however, was a slap in the face to all the complacent academic painters who regularly presented their exercises in scarcely disguised eroticism as virtuous homage to the gods and goddesses of classical antiquity. Venus reclining has become an unidealized, modern woman who, by details such as ribbon and slippers, would have been recognized as a common prostitute. She stares out at the spectator with an impassive boldness that bridges the gap between the real and the painted world in a startling manner, and effectively shatters any illusion of sentimental idealism in the presentation of the nude figure. *Olympia* is a brilliant design in black and white, with the nude figure silhouetted against a dark wall enriched by red underpainting that creates a Rembrandtesque glow of color in shadow. The black maidservant brings flowers offered by a client; the cat clearly reacts to her. The volume of the figure is created entirely through its contours—again, the light-and-dark pattern and the strong linear emphasis annoyed the critics as much as the shocking subject matter. Even Courbet found *Olympia* so unsettling in

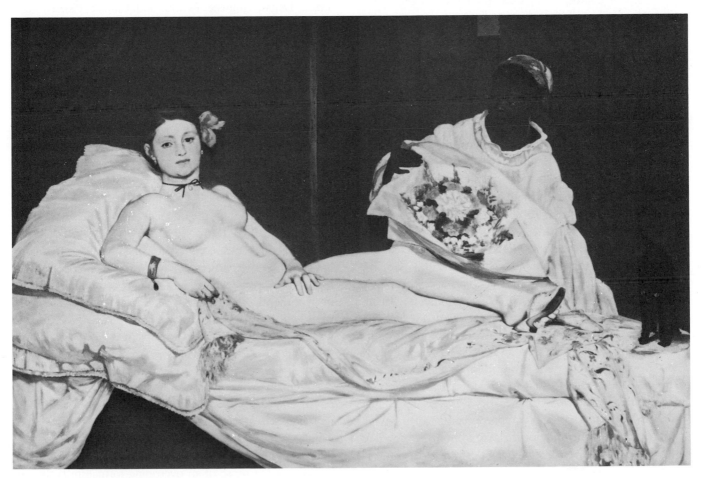

43. ÉDOUARD MANET. *Olympia.* 1863. Oil on canvas, 51" x 6'2¾" (130 cm x 1.9 m). Musée d'Orsay, Paris

44. ÉDOUARD MANET. *Portrait of Émile Zola*. 1868.
Oil on canvas, 57⅛ x 44⅞" (145.1 x 114 cm).
Musée d'Orsay, Paris

the playing-card flatness of its forms, all pressed hard against the frontal plane, that he likened the picture to "a Queen of Spades getting out of a bath."

Portrait of Émile Zola (fig. 44), an informal genre portrait, continues the Realists' critique of the academic concept of the grand and noble subject, carrying the attack several steps further in its emphasis on the plastic means of painting itself. Zola, the great Realist novelist and proponent of social reform, was Manet's friend and supporter. Manet painted into the work what are perhaps more revealing credos of his own than those of the writer who had so brilliantly defended him against his detractors. On the desk, for example, is the pamphlet Zola wrote on Manet. The bulletin board at the right holds a small print of *Olympia*, behind which is a reproduction of Velázquez's *Bacchus*. Beside *Olympia* is a Japanese print, and behind Zola's figure is seen part of a Japanese screen. The Japanese print and screen are symptomatic of the growing enthusiasm for Asian and specifically Japanese art then spreading among the artists of France and throughout Europe. In the art of Asia experimental French painters found some of the formal qualities they were seeking in their own work. Manet's conscious "appropriation" of the work of older artists, sometimes for the sake of homage as well as irony, has become a major strategy of Postmodern artists (see fig. 863).

After photography, the most powerful and pervasive influ-

ence on the nineteenth-century advance of modernist form may have come from Asian art, especially the color woodcut prints that began blowing west shortly after the American naval officer Matthew Perry had forced open the ports of Japan in 1853. With their steep and sharply angled views, their bold, snapshotlike croppings or near-far juxtapositions, and their flat-pattern design of brilliant, solid colors set off by the purest of contour drawing, the unfamiliar works of such masters as Kitagawa Utamaro, Katsushika Hokusai, and Andō Hiroshige (color plate 7, page 35) struck susceptible eyes as amazingly new and fresh. These prints reinforced the growing belief in the West that pictorial truth lay not in illusion but in the intrinsic qualities of the artist's means and the opaque planarity of the picture surface.

The attraction of this art for Manet is evident in every aspect of the *Portrait of Zola*. The critic Henri Fouquier denounced the picture in the newspaper *Le National*, declaring: "The accessories are not in perspective, and the trousers are not made of cloth." The criticism contained an obvious but remarkable truth: the trousers were not made of cloth; they were made of paint.

By the mid-1870s Manet had turned his attention to plein-air painting in which he sought to apply his brush-sketch technique to large figure compositions and to render these in full color that caught all the lightness and brilliance of natural sunlight. Although he did not show in their exhibitions, he had been in close contact with the Impressionists Claude Monet, Edgar Degas, Auguste Renoir, and Camille Pissarro. Like Degas, Manet was predominantly a painter of the human figure, not a landscapist like his friend Monet. But in the 1870s he moved gradually away from his modernizations of past art to contemporary genre scenes, which he composed in the heightened palette and loose brushwork of the Impressionists.

JAMES ABBOTT MCNEILL WHISTLER (1834–1903). Closely allied to his friend Manet in the search for a new pictorial reality suitable for "a painter of modern life" was the American expatriate artist James Abbott McNeill Whistler. Whistler, who trained in Paris and resided in London, exhibited a painting at the 1863 Salon des Refusés that caused a critical scandal rivaling the one incited by Manet's *Déjeuner sur l'herbe*. Much more than Manet, however, Whistler transformed Japanese sources into an emphatic aestheticism, in which Japanese-inspired style (*japonisme*) became the means to express a lyrical, atmospheric sensibility. Even so, in an authentic Realist manner, he consistently painted only what he had observed. The shift in emphasis was evident in the early 1860s, when he made pictures like *Symphony in White No. II: The Little White Girl* (color plate 8, page 35), where *japonisme* makes its presence felt not only in the obvious décor of an Asian blue-and-white porcelain vase, painted fan, and azalea blossoms, but also in the more subtle, and more important, flattening effects of the nuanced, white-on-white rendering of the model's dress and the rectilinear, screenlike divisions of the architectural framework. Further reinforcing the sense of an overriding formalism is the off-center composition, with its resulting crop of the figure

45. JAMES ABBOTT MCNEILL WHISTLER. *Nocturne in Black and Gold: The Falling Rocket.* 1875. Oil on oak panel, 23¾ x 18⅜" (60.3 x 46.7 cm). The Detroit Institute of Arts GIFT OF DEXTER M. FERRY, JR.

analogizing them to a more immaterial, evocative, and non-descriptive medium. For his aim was to create not a duplicate of nature, but rather what he called "an arrangement of line, form, and colour first." By the mid-1870s, when the artist had settled permanently in London, his painting *Nocturne in Black and Gold: The Falling Rocket* (fig. 45) seemed so purely abstract that it prompted the art critic John Ruskin to accuse Whistler openly of "flinging a pot of paint in the public's face." Whistler brought a libel suit against the critic and won a token judgment, for indeed he had worked from an observed subject, the fiery bursts and glittering sparks of a nocturnal fireworks sprayed over a midnight sky.

A great part of the struggle of nineteenth-century experimental painters was an attempt to recapture the color, light, and changeability of nature that had been submerged in the rigid stasis and studio gloom of academic tonal formulas. The color of the Neoclassicists had been defined in local areas but modified with large passages of neutral shadows to create effects of sculptural modeling. The color of Delacroix and the Romantics flashed forth passages of brilliant blues or vermilions from an atmospheric ground. As Corot and the Barbizon School of landscape painters sought to approximate more closely the effects of the phenomenal world with its natural light, they were almost inevitably bent in the direction of the subdued light of the interior of the forest, of dawn, and of twilight. Color, in their hands, thus became more naturalistic, although they did choose to work with those effects of nature most closely approximating the tradition of studio light. However, the increasingly delicate and abstract, almost monochromatic tonality of the post-1850, proto-Impressionist landscapes painted indoors by Corot (fig. 46), so different from the clear colors and structural firmness of his earlier Italian work (see fig. 20), may well have been an attempt to imitate

along two edges of the canvas. To emphasize his commitment to the principle of art for art's sake, an art committed to purely formal values, Whistler gave his pictures musical titles—arrangements, symphonies, and nocturnes—thereby

46. CAMILLE COROT. *Ville d'Avray.* 1871–74. Oil on canvas, 21⅝ x 31½" (54.9 x 80 cm). The Metropolitan Museum of Art, New York

BEQUEST OF CATHERINE LORILLARD WOLFE, 1887

47. EUGÈNE CUVELIER. *Untitled (View of Fontainebleau Forest in the Mist).* 1859–62. Salted-paper print from wet-collodion glass negative, 7¹³⁄₁₆ x 10⅛" (19.8 x 25.7 cm). The Metropolitan Museum of Art, New York

THE HOWARD GILMAN FOUNDATION AND JOYCE AND ROBERT MENSCHEL GIFTS, 1988

the seamless, line-free value range of contemporary photography, a medium that greatly interested him (fig. 47). In their smoky tissues of feathery leaves and branches, the moody, autumnal, parklike scenes painted by the aging Corot seem to be painterly counterparts of the blurred and halated landscape photographs of the period.

One of the few French landscapists of the mid-century determined to work away from the forest in full sunlight was **EUGÈNE BOUDIN** (1824–98), who painted the fashionable vacationers at seaside resorts around Le Havre with a direct, on-the-spot brilliance that fascinated both Courbet and Baudelaire (fig. 48). It inspired Claude Monet, Boudin's student and friend, and, through Monet, the movement of Impressionism.

Artists had always sketched in the open air and, as we have seen, some such as Corot and the Barbizon painters made compositions entirely en plein air, though even these

artists usually completed paintings in the studio. The Impressionists were the most consistently devoted to plein-air painting, even for some of their most ambitious works, and to capturing on canvas as faithfully as possible the optical realities of the natural world. However, the more the Realist artists of the mid-nineteenth century attempted to reproduce the world as they saw it, the more they understood that reality rested not so much in the simple objective world of natural phenomena as in the eye of the spectator. Landscape in actuality could never be static and fixed. It was a continuously changing panorama of light and shadow, of moving clouds and reflections on the water. The same scene, observed in the morning, at noon, and at twilight, was actually three different realities. Further, as painters moved out of doors from the studio they came to realize that nature, even in its dark shadows, was not composed of black or muddy browns. The colors of nature were the full spectrum—blues,

48. EUGÈNE BOUDIN. *On the Beach at Trouville.* 1863. Oil on panel, 10 x 18" (25.4 x 45.7 cm). The Metropolitan Museum of Art, New York

BEQUEST OF AMELIA B. LAZARUS, 1907

greens, yellows, and whites, accented with notes of red and orange, and often characterized by an absence of black.

In the sense that the artists were attempting to represent specific aspects of the observed world more precisely than any before them, Impressionism offered a specialized view of the world, based on intense observation as well as new discoveries in the science of optics. But Impressionism was much more than this. Behind the landscapes of Monet and the figure studies of Renoir lay not only the social realism of Courbet and the romantic realism of the Barbizon painters but also the significant sense of abstract structure that had appeared with David and Ingres and developed into a new aesthetic with Courbet and Manet.

CLAUDE MONET (1840–1926). Louis Leroy, the satirical critic for *Le Charivari*, was the first to speak of a school of "Impressionists," a term he derisively based on the title of a painting by Monet (color plate 9, page 36) that was included in the first Impressionist exhibition at the studio of the photographer Nadar. The show marked a significant historical occasion, for it was organized by a group of artists outside the official apparatus of the Salon and its juries, the first of many independent, secessionist group exhibitions that punctuate the history of vanguard art in the modern period.

The work in question offers a harmony of sky and water, an example of the type of atmospheric dissolution later explored by Whistler in his Nocturnes (see fig. 45). It is a thin veil of light gray-blue shot through with the rose-pink of the rising sun. Reflections on the water are suggested by short, broken brushstrokes, and the ghostly forms of boats and smokestacks are described with loose patches of color, rather than gradations of light and dark tones. *Impression: Sunrise*, for Monet, was an attempt to capture the ephemeral aspects of a changing moment, more so, perhaps, than any of his paintings until the late Venetian scenes or *Waterlilies* of 1905. *The Bridge at Argenteuil* (color plate 10, page 36) might be considered a more classic version of developed Impressionism. It represents a popular sailing spot in the Parisian suburb of Argenteuil, where Monet settled in 1871. Scenes of contemporary leisure were a mainstay of Impressionist subject matter. The painting glistens and vibrates, giving the effect of brilliant, hot sunlight shimmering on the water in a scene of contemplative stillness. Monet employed no uniform pattern of brushstroke to define the surface: the trees are treated as a fluffy and relatively homogeneous mass; the foreground blue of the water and the blue of the sky are painted quite smoothly and evenly, while the reflections of the tollhouse in the water are conveyed by dabs of thick yellow impasto; the boats and the bridge are drawn in with firm, linear architectural strokes. The complete avoidance of blacks and dark browns and the assertion of modulated hues in every part of the picture introduce a new world of light-and-color sensation. And even though distinctions are made among the textures of various elements of the landscape—sky, water, trees, architecture—all elements are united in their common statement as paint. The overt declaration of the actual physical texture of the paint itself, heavily brushed or laid on with

a palette knife, derives from the broken paint of the Romantics, the heavily modeled impasto of Courbet, and the brush gestures of Manet. But now it has become an overriding concern with this group of young Impressionists. The point has been reached at which the painting ceases to be solely or even primarily an imitation of the elements of nature.

In this work, then, Impressionism can be seen not simply as the persistent surface appearance of natural objects but as a never-ending metamorphosis of sunlight and shadow, reflections on water, and patterns of clouds moving across the sky. This is the world as we actually see it: not a fixed, absolute perspective illusion in the eye of a frozen spectator within the limited frame of the picture window, but a thousand different glimpses of a constantly changing scene caught by a constantly moving eye.

Monet's intuitive grasp of the reality of visual experience becomes particularly evident when a work like *Boulevard des Capucines, Paris* (fig. 49)—painted from an upstairs window in Nadar's former studio, where the Impressionists later held their first exhibition—is compared with contemporary photographs of a similar scene (fig. 50; see fig. 27). Whereas both camera-made images show "cities of the dead," the one

49. CLAUDE MONET. *Boulevard des Capucines, Paris (Les Grands Boulevards)*. 1873–74. Oil on canvas, 31¼ x 23¼" (79.4 x 59.1 cm). The Nelson-Atkins Museum of Art, Kansas City, Missouri
KENNETH A. AND HELEN F. SPENCER FOUNDATION ACQUISITIONS FUND

50. EDWARD ANTHONY. *A Rainy Day on Broadway, New York,*
from the series Anthony's Instantaneous Views.
1859. One-half of a stereograph.
International Museum of Photography at George Eastman
House, Rochester, New York

because its slow emulsion could record virtually no sign of life and the other because its speed froze every horse, wheel, and human in mid-movement, Monet used his rapidly executed color spotting to express the dynamism of the bustling crowd and the flickering, light- and mist-suffused atmosphere of a wintry day, the whole perceived within an instant of time. Still, far from allowing his broken brushwork to dissolve all form, he so deliberated his strokes that simultaneously every patch of relatively pure hue represents a ray of light, a moment of perception, a molecule of atmosphere or form in space, and a tile within the mosaic structure of a surface design. With its decorative clustering of color touches, its firm orthogonals, and its oblique Japanese-style or photographic view, the picture emerges as a statement of the artist's sovereign strength as a pictorial architect. Monet's is a view of modern Paris, for throughout the Second Empire Louis-Napoléon undertook a major program of urban renewal that called for the construction of wide boulevards and new sewers, parks, and bridges. While the renovations generally improved circumstances for commerce, traffic, and tourism, they destroyed many of the narrow streets and small houses of old Paris. Life in these bustling new *grands boulevards,* and in the cafés and theaters that lined them, provided fertile subjects for the painters of modern life.

AUGUSTE RENOIR (1841–1919). The artists associated with the Impressionist movement were a group of diverse individuals, all united by a common interest, but each intent on the exploration of his or her own separate path. Auguste Renoir was essentially a figure painter who applied the principles of Impressionism in the creation of a lovely dream

world that he then transported to the Paris of the late nineteenth century. *Moulin de la Galette* (color plate 11, page 37), painted in 1876, epitomizes his most Impressionist moment, but it is an ethereal fairyland in contemporary dress. In *Moulin de la Galette* the commonplace scene of an afternoon Sunday dance in the picturesque Montmartre district of Paris is transformed into a color- and light-filled reverie of fair women and gallant men, who, in actuality, were the painters and writers of Renoir's acquaintance and the working-class women with whom they chatted and danced. The lights flicker and sway over the color shapes of the figures—blue, rose, and yellow—and details are blurred in a romantic haze of velvety brushwork that softens their forms and enhances their beauty. This painting and his many other comparable paintings of the period are so saturated with the sheer joy of a carefree life that it is difficult to recall the financial problems that Renoir, like Monet, was suffering during these years.

The art of the Impressionists was largely urban. Even the landscapes of Monet had the character of a Sunday in the country or a brief summer vacation. But this urban art commemorated a pattern of life in which the qualities of insouciance, charm, and good living were extolled in a manner equaled only by certain aspects of the eighteenth century. Now, however, the good life was no longer the prerogative of a limited aristocracy; even the most poverty-stricken artist could enjoy it. But the moments of relaxation were relatively rare and spaced by long intervals of hard work and privation for those artists who, unlike Cézanne or Degas, were not blessed with a private income.

EDGAR DEGAS (1834–1917). Though born into a family belonging to the *grande bourgeoisie,* Degas was far from financially secure, especially after the death of his father in 1874. He thought of himself as a draftsman in the tradition of Ingres, yet he seems to have had little interest in exhibiting at the Salon after the 1860s or in selling his works. On the contrary, he took an active part in the Impressionist exhibitions between 1874 and 1886, in which he showed his paintings regularly. But while associated with the Impressionists, Degas did not share their enthusiasm for the world of the out-of-doors. His racehorse scenes of the late 1860s and early 1870s, which were open-air subjects, however, were painted in his studio. Though he produced extraordinary landscapes in his color monotypes, drawing horses and people remained Degas's primary interest, and his draftsmanship continued to be more precise than that of Manet's comparable racetrack subjects. This was particularly true in later works, where Degas became one of the first artists to exploit the new knowledge of animal movement recorded in the 1870s and 1880s by **EADWEARD MUYBRIDGE** (1830–1904) and others in a series of stop-action photographs (figs. 51, 52; see fig. 64), images that made forever obsolete the "hobbyhorse" pose—legs stretched forward and backward—long conventional in paintings of running beasts.

When Degas became really interested in light and color-light organization, an interest that arose from his enthusiasm for the world of the theater, the concert hall, and the ballet,

51. EDGAR DEGAS. *The Jockey.* 1889.
Pastel on paper,
12½ x 19¼" (31.8 x 48.9 cm).
Philadelphia Museum of Art
W. P. WILSTACH COLLECTION

52. EADWEARD MUYBRIDGE.
Horse in Motion. 1878.
Wet-plate photographs

he expressed it in terms of dramatic indoor effects. Theater and ballet gave him unequaled opportunities for the exercise of his brilliant skill as a draftsman, for the exploration of artificial light, and for creating an essentially abstract organization from the costumes and motions of ballet dancers. Degas was a student of Japanese prints and was strongly influenced by their patterns. He was also an accomplished photographer, and earlier than almost any other painter, he discovered how to use photography as a means for developing a fresh, original, even oblique vision of the world. The pastel *Ballerina and Lady with a Fan* (color plate 12, page 37), for instance, could be a photographic fragment vividly translated into color, but what appears casually conceived is in fact precisely constructed. This dance scene, like many others created by Degas during the 1880s and 1890s, is presented from above, with the woman in the loge acting as the observer who, when our vantage point as viewers is taken into account, also becomes part of what is observed. The performers in blue in the background are abruptly decapitated and the bowing ballerina in the foreground is truncated from the knees down. The rhymes created between forms—the fan and the tutus, for example—and the juxtaposition of forms, such as the faces of the spectator and the foreground dancer, which are actually separated by considerable distance, are typical of the ingenious visual puns Degas made in these works to reinforce the

essentially two-dimensional nature of the image. By concentrating on the members of the audience as well as the performers in many of his ballet pictures, Degas could exploit the social and psychological nuances of theater life, where seeing was as important as being seen.

In addition to being a skilled draftsman, painter, and photographer, Degas was also an accomplished printmaker and sculptor (color plate 45, page 115). Increasingly, during the 1880s and 1890s, he employed these media to explore commonplace subjects: milliners in their shops; exhausted laundresses; women occupied with the everyday details of their toilet: bathing, combing their hair, dressing, or drying themselves after the bath. In a work like *The Tub* (fig. 53) Degas goes beyond any of his predecessors in presenting the nude figure as part of an environment in which she fits with unconscious ease. Here the figure is beautiful in the easy, compact sculptural organization, but the routine character of her pose takes away any element of the erotic, while the bird's-eye view and the cutoff edges translate the photographic fragment into an abstract arrangement reminiscent of Japanese prints. When Degas exhibited this work among a group of similar pastels in the last Impressionist exhibition in 1886, many critics responded to them with descriptions, such as comparisons to animals, that betrayed a virulent but commonplace misogyny. In his unflattering, naturalistic de-

53. Edgar Degas. *The Tub*. 1886. Pastel on cardboard, 23⅝ x 32⅝" (60 x 82.9 cm). Musée d'Orsay, Paris

the American **Mary Cassatt** (1845–1926). Unlike her brother-in-law Manet, Morisot exhibited in all but one of the eight exhibitions organized by the Impressionists. Despite considerable success in her lifetime, the first retrospective of her work in this century did not take place until 1987. Morisot's subject of choice, like that of Cassatt, was modern domestic life, one arena of Parisian life to which these upper-class women could gain unrestricted access. They brought to their subjects an unsentimental detachment that male artists reserved for the street or the café. Morisot frequently employed her family members as models, as in *Hide and Seek* (fig. 54), which portrays her sister Edma during a stroll in the country with her daughter. Morisot joined Monet and Renoir in banishing black from her palette and developed a unique style of sketchy, deftly applied strokes to realize scenes and figures as materializations of light and color. The Philadelphia-born Cassatt, who arrived in Paris with her wealthy family in 1866, had an almost Japanese love of line, but often used it with the painterly freedom of Manet and the high-keyed color of the plein-air Impressionists. In her skewed perspectives, intimate views, and photographic cropping, however, she was at one with Degas, her closest artistic associate (color plate 13, page 37). Like Degas, she was a keen observer of the theatrical world, and she possessed a strong graphic sensibility which made her a highly successful printmaker. Like Morisot, Cassatt generally kept to mother-and-child subjects, but the sensuous immediacy and directness with which she treated them betrayed the daring modernity of her whole approach to art. Cassatt's elite social status permitted her to play the all-important role of tastemaker to American millionaires, whose acquisitions of Impressionist masterpieces

pictions of women, Degas had failed to create the eroticized objects of desire that inhabited academic paintings of the nude and that were, as feminist art historians have pointed out, designed for consumption by a predominantly male audience. While Degas's pastels were destined for the same viewers, they transgress accepted norms of "femininity" and, in their monumental simplicity, are among the most moving treatments of the nude in modern art.

Two painters who showed regularly with the Impressionists and who identified strongly with the Manet-Degas orbit were the Frenchwoman **Berthe Morisot** (1841–95) and

54. Berthe Morisot. *Hide and Seek*. 1873. Oil on canvas, 17¾ x 21¾" (45.1 x 55.2 cm). Collection Mrs. John Hay Whitney

55. CLAUDE MONET. *Les Nuages (Clouds)*. 1916–26. Oil on canvas, left panel of three, each panel
6'6¾" x 13'11⅜" (2 x 4.3 m). Musée de l'Orangerie, Paris

helped to elevate standards of art appreciation in the United States, as well as establishing the foundations of the great collections of modern art now accessible in American museums.

By the mid-1880s, the Impressionists were undergoing a period of self-evaluation, which prompted Monet and his circle to move in various directions to renew their art once again, some by reconsidering aspects of traditional drawing and composition, others by enlarging or refining the premises on which they had been working. Foremost among the latter stood Monet, who remained true to direct visual experience, but with such intensity, concentration, and selection as to push his art toward antinaturalistic subjectivity and pure decorative abstraction. In series after series he withdrew from the urban and industrial world, once so excitingly consistent with the nineteenth century's materialist concept of progress, and looked to unspoiled nature, as if to record its scenes quickly before science and technology could destroy them forever. Thus motivated, Monet so magnified his sensations of the individual detail perceived in an instant of time that instead of fragmenting large, complex views like that in *Boulevard des Capucines, Paris* (see fig. 49), he broke up simple, unified subjects—poplars, haystacks, the facade of Rouen Cathedral—into representations of successive moments of experience, each of which, by its very nature, assumed a uniform tonality and texture that tended to reverse objective analysis into its opposite: subjective synthesis. In a long, final series, painted at his home at Giverny, in the French countryside, Monet retired to an environment of his own creation, a water garden, where he found a piece of the real world—a sheet of clear liquid afloat with lily pads, lotus blossoms, and reflections from the sky above—perfectly at one with his conception of the canvas as a flat, mirrorlike surface shimmering with an image of the world as a dynamic materi-

alization of light and atmosphere (fig. 55). In these late, deeply pondered masterpieces, Monet's empirical interest in luminary phenomena had become a near-mystical obsession, its lyrical poetry often expressed in monochrome blue and mauve. As single, all-over, indivisible images, they also looked forward, from 1926, when the long-lived Monet died, to the late 1940s and 1950s, the time of the New York Abstract Expressionists, who would produce a similarly "holistic" kind of painting, environmental in its scale but now entirely abstract in its freedom from direct reference to the world outside (color plate 238, page 453).

Nineteenth-Century Art in the United States

At the opening of the twentieth century, American painting and sculpture, like their counterparts in Great Britain, lagged considerably behind the most progressive developments in continental Europe. American artists historically lacked the critical support systems of established art academies (therefore many studied abroad) and a tradition of government patronage. Despite the shortcomings of its art apparatus, American culture had sporadically generated major figures, such as Benjamin West and John Singleton Copley in colonial times and the late nineteenth-century expatriates Whistler, Cassatt, and John Singer Sargent (see fig. 60), capable of holding their own in the European art capitals. After the Civil War Americans had proved vigorously original in architecture and through this medium (see chapter 4) had made a decisive contribution to the visual arts. But that promising start was brought to a halt by the same academic bias that had held so many painters and sculptors in thrall to outmoded aesthetics inherited from the classical and Romantic past. While the more self-assured among the innately gifted

Europeans grew strong through resistance to the tyrannical Academy, Americans generally felt the need to master tradition rather than innovate against it. However, they also knew that to progress in their art it would be necessary to achieve their own artistic identity, assimilating influences from the generative centers in Europe until these had been transformed by authentic native sensibility into something independent and distinctive.

Many American artists of the nineteenth century saw art as a means to define the American national character as a distinct quality. American history and culture seemed at this time to be curiously polarized between the material and the mystical. In the colonial period, and for some time after, Americans most consistently expressed their artistic personality in some form of Realism.

GEORGE CATLIN (1796–1872) was an artist particularly concerned with national identity, but in a manner quite apart from the landscapists of the Hudson River School. He abandoned a successful practice as a portraitist for wealthy white patrons in Philadelphia to paint the individuals and culture of America's indigenous populations. For six years Catlin traveled west of the Mississippi, executing portraits and scenes of daily life among more than fifty Native American tribes. Then thought to be on the verge of extinction (due to aggressive government policies to "remove" them from United States territories), Native Americans were a source of tremendous curiosity for Easterners. Catlin presented his paintings, together with actual costumes and artifacts, in a famous "Indian Gallery" that toured the United States and Europe, bringing images of an exotic world to audiences who felt nostalgia for the disappearance of America's native inhabitants while for the most part sanctioning their obliteration. Among Catlin's many paintings of tribal leaders is a striking portrait of the chief of the Blackfoot tribe, *Buffalo Bull's Back Fat* (color plate 14, page 38), which was shown in the Paris Salon of 1846 and admired by the poet and critic Charles Baudelaire. While Catlin's paintings stand as important ethnographic sources because of their accurate depictions of tribal dress and custom, they are also powerful examples of the art of portraiture.

The American taste for realism occasionally grew so obsessive in its concern for the integrity of palpable things that reality, once scientifically measured, delineated, or actually dissolved, entered the realm of the ideal, thereupon becoming a vehicle of feeling, intuition, or metaphysical meaning.

Certainly this penchant for realism is evident in the rivetingly precise portraiture of John Singleton Copley and in the almost magical illusionism of *Rubens Peale with a Geranium* (color plate 15, page 38), painted by REMBRANDT PEALE (1778–1860). Peale was the son of Charles Willson Peale, an enlightened man of many talents, chiefly interested in natural history and painting. Charles helped to found the Pennsylvania Academy of the Fine Arts in Philadelphia and headed an extraordinary family dynasty of artists. Like his father, Rembrandt (most of the seventeen children were named after old-master painters or famous scientists) excelled

at portraiture, though he also tried his hand at history painting. His touching and meticulous portrayal of his farsighted younger brother, a botanist who posed with what was purportedly the first geranium specimen to enter the country, is his greatest work. The two interests of the Peale family—science and art—are brought together in this freshly conceived composition, just as the two genres of still life and portraiture are seamlessly merged.

The urge to see not only the actual but also the essential lodged within it could lead to an obsessive exactitude in American art. This tendency emerges dramatically in the trompe-l'oeil painters of the late nineteenth century, most notably John Frederick Peto (1854–1907) and WILLIAM MICHAEL HARNETT (1848–92). The latter's *The Old Violin* (fig. 56) is a tour de force of verisimilitude, visual deception, and classical simplicity. While Harnett's still lifes delighted the public with their artistic legerdemain, nineteenth-century critics often condemned the work as old-fashioned illusionism bereft of elevating moral content. It was not until the 1930s that Harnett's reputation was resurrected, when a

56. WILLIAM M. HARNETT. *The Old Violin.* 1886. Oil on canvas, 38 x 24" (96.5 x 61 cm).
National Gallery of Art, Washington, D.C.
GIFT OF MR. AND MRS. RICHARD MELLON SCAIFE IN HONOR OF PAUL MELLON

57. THOMAS COLE. *The Oxbow (The Connecticut River near Northampton)*. 1836. Oil on canvas, 51½" x 6'4" (130.8 cm x 1.9 m). The Metropolitan Museum of Art, New York

GIFT OF MRS. RUSSELL SAGE

taste for Surrealism and abstraction revised attitudes about his technical wizardry and highly structured formal organization. Harnett's examination of the nature of visual perception reemerged in the later twentieth century in the Superrealist art of Richard Estes and Duane Hanson (color plate 397, page 676; color plate 399, page 676).

America's greatest artistic achievements in the middle of the nineteenth century came in the development of a kind of landscape painting that merged elements of scientific Realism with idealized, even religious interpretations of nature. This occurred particularly in the works created by members of the so-called Hudson River School, a diverse group of artists (not a literal school or association) who worked in New England and upstate New York. For the most part, their landscapes presented the young American republic as a land of pristine wilderness which, by virtue of its majestic scale and untouched beauty, seemed suffused with divine presence. Artists thus invested landscape painting with moral meaning and used it as a vehicle for defining America's national identity as a new "Promised Land." *The Oxbow (The Connecticut River near Northampton)* (fig. 57), an icon of American art that established a virtual formula for landscapes of the Hudson River School, was painted by its founder, THOMAS COLE (1801–48). *The Oxbow* offers a vast panoramic view of a famous bend in the Connecticut River from a stormy promontory where, amid the rocks and tree trunks, Cole has depicted himself at his easel. Symbolically, the storm clouds at left have passed over the fertile, cultivated fields on the banks of the river, signs of a benign human presence in nature where dwell, according to Cole, "freedom's offspring—peace, security and happiness." More than the artists who followed him, Cole expressed through his paintings serious doubts about the country's expansionist tendencies, which resulted in the destruction of wilderness as new frontiers were settled.

Cole's greatest protégé was FREDERIC EDWIN CHURCH

(1826–1900), who reinforced his fine-art education by traveling extensively, mostly in North and South America, and by studying the theories of Alexander von Humboldt, a famous German naturalist who called for a "heroic landscape painting" that would bring scientific accuracy to the depiction of nature's grandest scenes. Church was a master of virtuoso, highly theatrical interpretations of those grand scenes, from volcanos in the Ecuadorian Andes to icebergs in the Arctic north. In *Twilight in the Wilderness* (color plate 16, page 38), he translated a blazing sunset into a sublime vision of biblical proportions, implying, as had Cole, a spiritual, pantheistic view of nature. It has also been suggested, however, that the fiery skies of the painting could portend the imminent conflagration of the Civil War. Though it is difficult to imagine in today's art world, *Twilight* caused a sensation when it, like Church's other masterpieces, was exhibited by itself in a New York gallery, where visitors paid an entrance fee to see the artist's newest creation.

ROBERT SCOTT DUNCANSON (1821–72), Church's contemporary, was from Cincinnati, where a thriving landscape tradition in the Hudson River School style had developed. Born into a poor family of free African-American tradespeople, Duncanson was a serious advocate of the abolitionist movement, frequently donating his work in support of antislavery causes. In the art of painting he was self-taught, though in nineteenth-century America this was often the case with white artists as well (Cole, for example). Like his fellow painters in the East, Duncanson took the valleys and rivers of his own region as the subjects for his landscapes (fig. 58). The tranquillity of mood in *Blue Hole, Flood Waters, Little Miami River* and the symmetrical organization of its composition suggest a classical vision of landscape in the manner of the seventeenth-century French paintings of Claude Lorrain.

By the end of the century, the tight, descriptive style of the Hudson River School had been transformed in the hands of

58. ROBERT S. DUNCANSON. *Blue Hole, Flood Waters, Little Miami River.* 1851. Oil on canvas, 29¼ x 42¼" (74.3 x 107.3 cm). Cincinnati Art Museum

GIFT OF NORBERT HEERMAN AND ARTHUR HELBIG

artists like **GEORGE INNESS** (1825–94) into something more subjective and less dependent on the phenomena of the natural world (fig. 59). An admirer of the French Barbizon painters, Inness painted unpretentious subjects in a highly poetic style far from the grandiose statements of Frederic Church. For him, nature is the domesticated landscape, rather than untamed wilderness, and it is filled with evocative atmosphere and soft harmonies of green and gold. The way in which formal concerns supersede the need for literal transcription in Inness's work has parallels with many American and European artists at the end of the 1800s, who were creating the foundation for abstraction in the next century.

Whatever the developments in painting and sculpture back home, no artist had sensed the coming of the new more presciently than glamorous expatriates such as Whistler and Cassatt, whose histories belong within the context of their European contemporaries. Born to American parents living abroad, **JOHN SINGER SARGENT** (1856–1925) also spent most of his career in Paris and London, though he received many portrait commissions from leading American families. His flashing, liquid stroke and flattering touch in portraiture made him one of the most famous and wealthy artists of his time. Moreover, in a major work like *Madame X* (fig. 60) Sargent revealed himself almost mesmerized, like a latter-day Ingres, by the abstract qualities of pure line and flat silhouette. At the same time, he so caught the explicit qualities of surface and inner character that the painting created a scandal when publicly shown at the Salon, for in addition to the figure's already décolleté dress, Sargent had placed the left-hand strap off her shoulder. To placate an offended public, he adjusted the strap as seen here after the Salon closed. The experience prompted the artist to leave Paris eventually and establish his practice in London. In his most painterly works, meanwhile, Sargent carried gestural virtuosity, inspired by Frans Hals and Diego Velázquez, to levels of pictorial autonomy not exceeded before the advent of the Abstract Expressionists after World War II.

Americans who chose photography as their medium of expression stood out in the international Salons organized for exhibiting the new camera-made art, and indeed often won the major prizes and set the standards for both technical mastery and aesthetic vision. This will be seen in the images of Man Ray, Edward Steichen, Alfred Stieglitz, Paul Strand, and Edward Weston, reproduced in later chapters as examples of the best work of its kind. Meanwhile, Whistler, Cassatt, and Sargent also had their American peers in photography, principally **GERTRUDE KÄSEBIER** (1852–1934) and **CLARENCE H. WHITE** (1871–1925), both of whom worked in the Impressionist manner favored by many of the so-called

59. GEORGE INNESS. *Sundown.* 1894. Oil on fabric, 45 x 70" (114.3 x 177.8 cm). National Museum of American Art, Smithsonian Institution, Washington, D.C.

GIFT OF WILLIAM T. EVANS

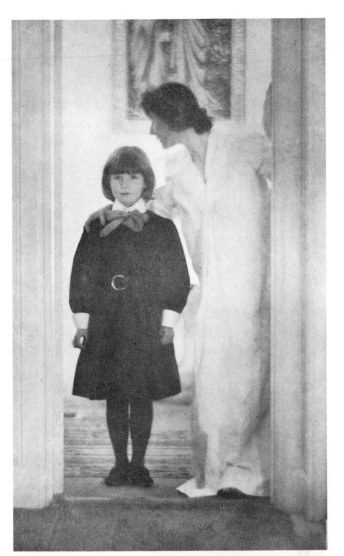

above: 60. JOHN SINGER SARGENT. *Madame X (Mme Gautreau).*
1884. Oil on canvas, 6'10⅛" x 43¼"
(2.1 m x 109.2 cm).
The Metropolitan Museum of Art, New York
ARTHUR H. HEARN FUND, 1916

above right: 61. GERTRUDE KÄSEBIER.
Blessed Art Thou among Women. c. 1900.
Platinum print on Japanese tissue,
9⅜ x 5½" (23.8 x 13.9 cm).
The Museum of Modern Art, New York
GIFT OF MRS. HERMINE M. TURNER
COPY PRINT © 1995 THE MUSEUM OF MODERN ART, NEW YORK

right: 62. CLARENCE H. WHITE. *Ring Toss.* 1899. Platinum print.
Library of Congress, Washington, D.C.

63. WINSLOW HOMER. *Prisoners from the Front*. 1866. Oil on canvas, 24 x 38" (61 x 96.5 cm). The Metropolitan Museum of Art, New York GIFT OF MRS. FRANK B. PORTER, 1922

Pictorialist photographers during the later nineteenth century (figs. 61, 62). Like Cassatt, Käsebier specialized in the mother-and-child theme, while White aligned himself with the elite social content of Sargent's portraiture and the Orientalizing aestheticism of Whistler's musical "arrangements." The camera offered an unparalleled capacity to render the reality so beloved by Americans, at the same time that it also could easily transform visual facts, as Käsebier and White did with their soft focus, and thus endow them with the special meaning characteristically sought by American artists.

WINSLOW HOMER (1836–1910). More exclusively rooted in American experience than Whistler, Cassatt, and Sargent, and thus more representative of the point of departure for American art in the twentieth century, were Winslow Homer and Thomas Eakins. As an illustrator for *Harper's Weekly* magazine in New York, Homer, a virtually self-taught artist, was assigned to the front during the Civil War. In his illustrations and in the oil paintings he began to make during the war, Homer tended to focus on the quiet moments of camp life rather than the high drama of battle. A landscape laid waste by those battles is the setting for the artist's greatest early work, *Prisoners from the Front* (fig. 63). At the left, three Confederate soldiers—a disheveled youngster, an old man, and a defiant young officer—surrender to a Union officer at the right. Although Homer's painting represents a fairly unremarkable occurrence in the war, it achieves the impact of history painting in the significance of its theme. His subtle characterization of the varying classes and types of the

participants in the tragic conflict alludes to the tremendous difficulties to be faced during Reconstruction between these warring cultures.

In 1866 Homer traveled to Paris, where *Prisoners from the Front* was being exhibited. While he shared in the practice of plein-air painting and some of the subject matter of the Impressionists, Homer always insisted on the physical substance of things, rarely allowing paint to disintegrate form. Much of his mature work centered on the ocean, either breezy, sun-drenched watercolors made in the Caribbean or dramatic views of the human struggle with the high seas off the coast of Maine, where he settled in 1884. In *The Fox Hunt* (color plate 17, page 39), the struggle between the forces of nature plays itself out in a haunting drama featuring a fox, weakened by the deprivations of winter, descended upon by hungry crows. The spareness of this striking composition, with its graceful silhouettes, cropped forms, and slanted perspectives, attests to Homer's sophisticated knowledge of *japonisme* as well as his ability to extract great emotional intensity from the simplest of scenes.

THOMAS EAKINS (1844–1916). In his determination to fuse art and science for the sake of an uncompromising Realism in painting, the Philadelphia painter Thomas Eakins all but revived the Renaissance tenets of Leonardo da Vinci. Not only did Eakins dissect cadavers right along with medical students (a traditional method of artistic training) and join Eadweard Muybridge in his studies of motion with stop-action photography, especially those devoted to human

left: 64. THOMAS EAKINS. *The Pole Vaulter.* 1884–85. Multiple-exposure gelatin-silver print. The Metropolitan Museum of Art, New York
GIFT OF CHARLES BREGLER

below: 65. THOMAS EAKINS. *Max Schmitt in a Single Scull.* 1871. Oil on canvas, 32½ x 46¼" (82.6 x 117.5 cm). The Metropolitan Museum of Art, New York
PURCHASE, ALFRED N. PUNNETT ENDOWMENT FUND AND GEORGE D. PRATT GIFT, 1934

movement (figs. 64, 52), but he even had an assistant pose on a cross in full sunlight as the model for a Crucifixion scene and provided a nude male model for his female drawing students, a step that forced his resignation from the august Pennsylvania Academy of the Fine Arts. Eakins's extraordinary early painting *Max Schmitt in a Single Scull* (fig. 65) was his first treatment of an outdoor subject. In the foreground the artist's friend, a celebrated oarsman, pauses momentarily while rowing on Philadelphia's Schuylkill River and looks toward the viewer; in the middle distance, Eakins has depicted himself midstroke, also looking at us. A crystalline light and carefully ordered composition lead us into this rational pictorial space, in which each detail is keenly observed and convincingly rendered. Eakins has here produced a Realism that transcends mere illusionism by way of a magical clarity, as if time were suspended in a single instant.

Eakins was arguably the greatest American portraitist of the nineteenth century and his large painting *The Gross Clinic*

66. HENRY OSSAWA TANNER.
The Banjo Lesson. 1893.
Oil on canvas, 49 x 35½"
(124.5 x 90.2 cm).
Hampton University
Museum, Hampton, Virginia

(color plate 18, page 39) is a masterpiece of the genre. The artist looked to the seventeenth-century precedent of Rembrandt's *Anatomy Lesson of Dr. Tulp* for his heroic portrayal of a distinguished surgeon performing an operation before his class. Like Max Schmitt, Dr. Gross looks up during a break in the action. In the midst of the dark, richly hued painting, light falls on his forehead, the intellectual focal point of the scene, and, even more dramatically, on the details of the surgery. The work is instructive for the ways in which artists can signify gender difference. The only female observer of the event, the mother of the patient, turns away in horror. Unlike the dispassionate male participants, she is overcome with emotion and reduced to a passive role, just as were the female members of the Horatii family in David's painting (see fig. 2).

HENRY OSSAWA TANNER (1859–1937) was an African-American artist who, by the end of the nineteenth century, had achieved significant international distinction. During the so-called Harlem Renaissance of the 1920s, he was recognized as the most important black artist of his generation.

Though he studied in Philadelphia with Eakins, whose portrait style profoundly influenced his own, Tanner found a more accepting atmosphere in his adopted city of Paris. He exhibited widely during his lifetime, including at the Paris Salon, and eventually befriended members of the avant-garde circle around Paul Gauguin in the rural artists' communities of Brittany. The French government awarded him the prestigious Legion d'Honneur. Tanner's best-known work, *The Banjo Lesson* (fig. 66), was probably made during a trip to Philadelphia, when he said he painted "mostly Negro subjects," a genre he felt had been stereotypically cast by white artists. With a loose weave of elongated strokes, Tanner softly defines the central pair of figures, bathing them in a light that imparts a spiritual stillness to the scene, a light not unlike that used in the many compositions of religious subjects that make up the bulk of Tanner's work.

ALBERT PINKHAM RYDER (1847–1917). In the art of Albert Pinkham Ryder the sense of dream is utterly dominant, even in paintings so overloaded with the material sub-

67. ALBERT PINKHAM RYDER. *Moonlight Marine.* c. 1890s. Oil on panel, 11⅜ x 12" (28.9 x 30.5 cm).
The Metropolitan Museum of Art, New York
SAMUEL D. LEE FUND, 1934

commissions. The expanding economy led to innumerable sculptural monuments and architectural decorations, though these mostly followed the academic traditions of Rome or Paris. AUGUSTUS SAINT-GAUDENS (1848–1907) received academic training in New York, Paris, and Rome, though, contrary to most nineteenth-century academic sculptors, European or American, he took fresh inspiration from sculptures of the Renaissance, notably those of Donatello. By infusing the classical tradition with his own brand of naturalism, Saint-Gaudens produced a number of first-rate portraits and could enliven the most banal of commemorative or allegorical sculptural monuments. But the request he received from the author Henry Adams to create a memorial to his late wife, who had committed suicide, presented challenges quite apart from the usual equestrian statue or portrait made from life (fig. 68). The sculptor's inspired solution, stimulated by Adams's interest in Buddhism, was an austere, mysteriously draped figure that seems to personify a state of spiritual withdrawal from the physical world. With eyes downcast and a face shrouded in shadow, the sculpture constitutes an unforgettable image of eternal repose.

stance of thick paint that, as the critic Lloyd Goodrich has written, "a tiny canvas weighs heavy in the hand." This was the product of years spent in a trial-and-error process of working and reworking a single picture, carried out by a painter who declared that "the artist should fear to become the slave of detail. He should strive to express his thought and not the surface of it. What avails a storm cloud accurate in form and color if the storm is not therein." The storm Ryder wanted was of the sort stirred up by the German composer Richard Wagner, whose sublimely Romantic music deeply touched the artist, as it did many of his contemporaries. Ryder, a solitary figure, was unfortunately a dangerously experimental technician, applying wet paint on wet paint and mixing his oils with what was probably bitumen, so that his pictures did not dry properly, but have gone on "ripening" until they have actually darkened and decayed, ironically destroying many of the exquisite nuances he had sought in endless modifications. What remains, however, tends to dramatize the extraordinary reductiveness of the final image (fig. 67). With all detail refined away and the whole simplified to an arrangement of broad, dramatically contrasted shapes, a painting by Ryder often evokes the convoluted, emotional rhythms of Rubens and Delacroix as well as the Gothic, visionary world of German Expressionism. Little wonder that Ryder was featured in the famous 1913 Armory Show in New York as a progenitor of modernism, as well as the painter whom the young Jackson Pollock most admired among all his American pictorial forebears.

As for American sculpture during the later nineteenth century, it was nothing if not prolific, especially in public

68. AUGUSTUS SAINT-GAUDENS. *Adams Memorial.* 1886–91.
Bronze and granite, height 70" (177.8 cm).
Rock Creek Cemetery, Washington, D.C.

3 Post-Impressionism

So various and distinctive were the reactions against the largely sensory nature and supposed formlessness of Impressionism that the English critic Roger Fry, who was among the first to take a comprehensive look at these developments, could do no better than characterize them, in 1910, with the essentially non-defining, umbrella designation *Post-Impressionism*. The inadequacy of the term is borne out by the fact that, unlike *Impressionism*, it did not enter the language until a quarter-century after the first appearance of the artistic phenomena it purported to describe. Among other things, it implies that Impressionism was itself a homogeneous style, when in fact it collected a highly diverse range of artistic sensibilities under a single rubric. In fact, the Post-Impressionists were schooled in Impressionism and many of them continued to respect its exponents and share much of their outlook. Cézanne, for example, who could not abide the paintings of Paul Gauguin, always admired the work of Monet. Gauguin, on the other hand, learned to paint in part from the Impressionist Pissarro and, like most of those in his circle, came to regard Cézanne as an almost legendary figure, even collecting his paintings. Yet both artists are considered key Post-Impressionists. Despite its shortcomings as a term, *Post-Impressionism* has become a common and convenient label for an innovative group of artists working in France in the late nineteenth and early twentieth centuries. And it does specify the one element that Seurat, Cézanne, Gauguin, and Van Gogh all had in common—their determination to move beyond what they regarded as the relatively passive registration of perceptual experience practiced by the Impressionists and give formal expression to the conceptual realm of ideas and intuitions.

By the 1880s numerous schisms and ideological differences had surfaced within the Impressionist ranks as the artists set about organizing their independent exhibitions. Renoir, Monet, and Degas, as we have seen, were all beginning to reconsider their painting in the light of their own altered consciousness. The Impressionists were far from mindless or uncritical portrayers of bourgeois ease and pastoral pleasures, as they have sometimes been called; these artists were all vibrantly in tune with their times and capable of highly subjective interpretations of their world. By the 1880s, that world had begun to change rapidly in ways that could only distress a Europe imbued with the nineteenth-century positivist belief in ever-accelerating progress under the impetus of science and industry. Change was rapid indeed at this time, but not always progressive. As industrial waste combined with expanding commerce to destroy the unspoiled rivers and meadows so often painted by Monet, the master withdrew ever deeper into his "harem of flowers" at Giverny, a synthetically natural world of his own creation and so bound up with his late painting as to be virtually inseparable from it. As rising socioeconomic expectations throughout industrialized Europe met with conservative backlash, the clash of ideologies brought anarchist violence. As scientists broke down the theories of classical science, a new physics, chemistry, and psychology emerged. As the radical philosopher Friedrich Nietzsche declared God dead and found truth less in reason than in the irrational, new debates arose concerning the nature and role of religion. The most advanced artists of the period found little to satisfy their needs in the optimistic utilitarianism that had dominated Western civilization since the Enlightenment in the eighteenth century. Gradually, therefore, they abandoned the Realist tradition of Daumier, Courbet, and Manet that had opened such a rich vein of aesthetic exploration, climaxing in the Impressionism of the 1870s. Instead, they sought to discover, or recover, a new and more complete reality, one that would encompass the inner world of mind and spirit as well as the outer world of physical substance and sensation.

Inevitably, this produced paintings that seemed even more shocking than Impressionism in their violation of both academic principles and the polished illusionism desired by eyes now thoroughly under the spell of photography. And so, just as Monet and Renoir were beginning to enjoy a measure of critical and financial success, the emerging painters who inspired the term *Post-Impressionist* reopened the gap—wider now than ever before—that separated the world of advanced art from that of society at large. Whatever their divergences, the Post-Impressionists all brought heightened tension and excitement to art by reweighting their values in favor of the ideal or romantic over the real, of symbol over sight, of concept over percept. Less and less were perception and its translation into art seen as an end unto themselves; rather, they became a means toward the knowledge of form in the service of expressive content. In arriving at the antinaturalism that followed, artists depended upon and counterbalanced the dualities of mind and spirit, thought and emotion.

GEORGES SEURAT (1859–91). Trained in the academic tradition of the Paris École des Beaux-Arts, Georges Seurat was a devotee of classical Greek sculpture and of such classical masters as Piero della Francesca, Poussin, and Ingres. He also studied the drawings of Hans Holbein, Rembrandt, and Millet, and learned principles of mural design from the academic Symbolist Pierre Puvis de Chavannes (fig. 75). He early became fascinated by theories and principles of color organization, which he studied in the writings and paintings of Delacroix, the texts of the critics Charles Blanc and David Sutter, and the scientific treatises of Michel-Eugène Chevreul, Ogden Rood, and Charles Henry. These researches in optics, aesthetics, and color theory studied, among other phenomena, the ways in which juxtaposed colors affect one another. According to Chevreul, for example, when two colors are placed side by side, each will impose its own complementary on its neighbor. Thus, if red is placed next to blue, the red will cast a green tint on the blue, altering it to a greenish blue. Simultaneously, the blue imposes its own opposite, a pale orange, on the red. Although the rational, scientific basis for such theories appealed to Seurat, he was no cold, methodical theorist. Like any great artist, his highly personal form of expression evolved through careful experiments with his craft. Working with his younger friend Paul Signac he sought to synthesize the color experiments of the Impressionists and the classical structure inherited from the Renaissance, combining the latest concepts of pictorial space, traditional illusionistic perspective space, and the newest scientific discoveries in the perception of color and light.

It is astonishing that Seurat produced such a body of masterpieces in so short a life (he died at thirty-one in a diphtheria epidemic). His greatest work, and one of the landmarks of modern art, is *A Sunday Afternoon on the Island of La Grande Jatte* (color plate 19, page 39). More briefly known as *La Grande Jatte*, it was the most notorious painting shown in the last of the eight Impressionist exhibitions, in 1886. Seurat worked for over a year on this monumental painting, preparing it with twenty-seven preliminary drawings and thirty color sketches (fig. 69). Seurat's hauntingly beautiful conté crayon drawings reveal his interest in masters of black and white from Rembrandt to Goya. He avoided lines to define contours and depended instead on shading to achieve soft, penumbral effects. In the preparatory works for *La Grande Jatte*, ranging from studies of individual figures to oils that laid in most of the final composition, Seurat analyzed, in meticulous detail, every color relationship and every aspect of pictorial space. His unique color system was based on the

69. GEORGES SEURAT.
The Couple, study for *La Grande Jatte.* 1884.
Conté crayon on paper, 12¼ x 9¼" (31.1 x 23.5 cm).
The British Museum, London

Impressionists' intuitive realization that all nature was color, not neutral tone. But the Impressionist painters for the most part had made no organized, scientific effort to achieve their remarkable optical effects. Their technique involved the placement of pure, unblended colors in close conjunction on the canvas and their fusion on the retina of the eye as glowing, vibrating patterns of mixed color. This effect appears in many paintings by Monet, as anyone realizes who has stood close to one of his works, first experiencing it simply as a pattern of discrete color strokes, and then moving gradually away from it to observe all the elements of the scene come into focus.

Seurat, building on Impressionism and scientific studies of optical phenomena, constructed his canvases with an overall pattern of small, dotlike brushstrokes of generally complementary colors—red and green, violet and yellow, blue and orange—and with white. Various names have been given to Seurat's method, including *Divisionism, Pointillism,* and *Neo-Impressionism.* Despite their apparent uniformity overall, on close inspection the "dots" vary greatly in size and shape and are laid down over areas where color has been brushed in more broadly. The intricate mosaic produced by Seurat's painstaking and elaborate technique is somewhat analogous to the actual medium of mosaic; its glowing depth of luminosity gives an added dimension to color experience. Unfortunately, some of the pigments he used were unstable, and in less than a decade oranges had turned to brown and brilliant emerald green had dulled to olive, reducing somewhat the original chromatic impact of the painting.

In *La Grande Jatte* Seurat began with a simple contemporary scene of middle-class Parisians relaxing along the banks of the Seine River. What drew him most strongly to the particular scene, perhaps, was the manner in which the figures could be arranged in diminishing perspective along the banks of the river, although delightful inconsistencies in scale abound throughout the composition. At this point in the picture's evolution the artist was concerned as much with the recreation of a fifteenth-century exercise in linear perspective as with the creation of a unifying pattern of surface color dots. Broad contrasting areas of shadow and light, each built of a thousand minute strokes of juxtaposed color-dot complementaries, carry the eye from the foreground into the beautifully realized background. Certainly, Seurat was here attempting to reconcile the classical tradition of Renaissance perspective painting with a modern interest in light, color, and pattern. The immense size of the canvas gives it something of the impact of a Renaissance wall fresco, and is especially dramatic since it is composed of brushstrokes smaller than a pea.

More important in the painting than depth or surface pattern is the magical atmosphere that the artist was able to create from the abstract patterns of the figures. The art historian Meyer Schapiro notes that Seurat depicts "a society at rest and, in accord with his own art, it is a society that enjoys the world in pure contemplation and calm." Seurat's figures are like those of a mural by Piero della Francesca in their quality of mystery and of isolation one from another. But by placing the figures in strict profile or frontal views, he flattens their forms and endows them with the immobility of stage flats, producing a static atmosphere critics have often described as "Egyptian." In the accumulation of commonplace details, which are filled with amusing vignettes of social caricature, there is an effect of poetry that makes of the whole a profoundly moving experience. Seurat's paintings anticipate not only the abstraction of Piet Mondrian but also the strange Surrealist stillness and space of Giorgio de Chirico and René Magritte. The most scientific and objective of all the painters of his time is, curiously, also one of the most poetic and mysterious.

In a preparatory oil for a later painting, such as the study for *Le Chahut* (fig. 70), the color dots are so large that the figures and their spatial environment are dissolved in color patterns that cling to the surface of the picture. The painting depicts a provocative and acrobatic dance, then popular in Parisian cafés, which Seurat beautifully orchestrates into a series of decorative rhythms, giving form to a then-current theory that ascending lines induce feelings of gaiety in the viewer. For his imagery Seurat was clearly inspired by the colorful posters that were placed around Paris to advertise the dance halls. Around the study for *Le Chahut* he painted a border of multicolored strokes (as he had for *La Grande Jatte*) and he probably painted the actual wooden frame around the work as well. Van Gogh and Gauguin also expressed interest in the way their works were framed, preferring solutions much simpler than the ornate, gilded frames usually placed on their paintings.

70. GEORGES SEURAT. Study for *Le Chahut*. 1889–90.
Oil on canvas, 21⅞ x 18⅜" (55.6 x 46.7 cm).
Albright-Knox Art Gallery, Buffalo

Seurat's chief colleague and disciple was **PAUL SIGNAC** (1863–1935), who strictly followed the precepts of Neo-Impressionism in his landscapes and portraits. Like his colleagues, Signac regarded his revolutionary artistic style as a form of expression parallel to his radical political anarchism. This movement was closely associated with other Neo-Impressionists in France and Belgium who believed that only through absolute creative freedom could art help to bring about social change. In his later, highly coloristic seascapes, Signac provided a transition between Neo-Impressionism and the Fauvism of Henri Matisse and his followers. And through his book, *D'Eugène Delacroix au Néo-impressionisme,* he was also the chief propagandist and historian of the movement. His portrait of the critic Félix Fénéon (color plate 20, page 40), who had been the first to use the term *Neo-Impressionism,* is a fascinating example of the decorative formalism that the artists around Seurat favored. The painting, with its spectacular spiral background inspired by patterns Signac had found in a Japanese print, contains private references to the critic's ideas and to those of the color theorist Charles Henry, with which Signac was very familiar.

Seurat's Neo-Impressionist technique in the hands of other painters too often declined into a decorative formula for use in narrative exposition. This was also the fate of Impressionism; as it gradually became accepted and then, much later, immensely popular throughout the world, its color and light, quotidian subjects, radical cropping, and photographic immediacy—initially so shocking—were all too quickly discovered to be charming, endearing, and capable of every kind of vulgarization. Seurat's Neo-Impressionism, on the other hand, affected not only Fauvism and certain aspects of Cubism in the early twentieth century, but also some Art Nouveau painters and designers.

PAUL CÉZANNE (1839–1906). Of all the nineteenth-century painters who might be considered prophets of twentieth-century experiment, the most significant for both achievement and influence was Paul Cézanne. Cézanne struggled throughout his life to express in paint his ideas about the nature of art, ideas that were among the most revolutionary in the history of art. Son of a well-to-do merchant turned banker in Aix-en-Provence, he had to resist parental disapproval to embark on his career, but once having won the battle, unlike Monet or Gauguin, he did not need to struggle for mere financial survival. As we view the splendid control and serenity of his mature paintings, it is difficult to realize that he was an isolated, socially awkward man of a sometimes violent disposition. This aspect of his character is evident in some of his early mythological figure scenes, which were baroque in their movement and excitement. At the same time he was a serious student of the art of the past, one who studied the masters in the Louvre, from Paolo Veronese to Poussin to Delacroix.

Cézanne's unusual combination of logic and emotion, of reason and unreason, represented the synthesis that he sought in his paintings. Outside of Degas, the Impressionists in the 1870s largely abandoned traditional drawing in an effort to communicate a key visual phenomenon—that objects in nature are not seen as isolated phenomena separated from one another by defined contours. In their desire to realize the observed world through spectrum colors in terms of which the spaces between objects are actually color intervals like the objects themselves, they tended to destroy the objects as three-dimensional entities existing in three-dimensional space, and to re-create solids and voids as color shapes functioning within a limited depth. It was this destruction or at least subordination of the distinction between the illusion of three-dimensional mass and that of three-dimensional depth that led to the criticism of Impressionism as formless and insubstantial. Even Renoir felt compelled to restudy the drawing of the Renaissance in order to recapture some of that "form" which Impressionism was said to have lost. And Seurat, Gauguin, Van Gogh, each in his own way, consciously or unconsciously, was seeking a kind of expression based on, but different from, that of the Impressionists.

Certainly, Cézanne's conviction that Impressionism had taken a position excluding qualities of Western painting important since the Renaissance prompted his oft-quoted and frequently misunderstood remark that he wanted to "make of Impressionism something solid like the art of the museums." By this, obviously—as his paintings document—he did not mean to imitate the old masters. He realized, quite correctly, that in their paintings artists like Veronese or Poussin had created a world similar to but quite distinct from

the world in which they lived, and that painting, resulting from the artist's various experience, emerged as a separate reality in itself. This kind of reality, the reality of the painting, Cézanne sought in his own work. And he felt progressively that his sources must be nature and the objects of the world in which he lived, rather than stories or myths from the past. For this reason he expressed his desire "to do Poussin over again from nature."

Cézanne's mature position was arrived at only after long and painful thought, study, and struggle with his medium. The theoretical position that, late in his life, he tried to express in his idiosyncratic language was probably achieved more through the discoveries he made on his canvas with a fragment of nature before him than through his studies in museums. In the art of the museum he found corroboration of what he already instinctively knew. In fact, what he knew in his mature landscapes is in many ways evident in such an early work as *Uncle Dominic as a Monk* (fig. 71). Uncle Dominic, his mother's brother, is painted in the white habit of the Dominican order, modeled out from the blue-gray ground in sculptured paint laid on in heavy strokes with the palette knife. The sources in Manet and Courbet are evident, but the personality revealed by the portrait—the forceful temperament of the artist himself—is different from that in the work of either predecessor. The painting, which is not so

71. PAUL CÉZANNE. *Uncle Dominic as a Monk.* c. 1866.
Oil on canvas, 25½ x 21¼" (64.7 x 54 cm).
The Metropolitan Museum of Art, New York

72. PAUL CÉZANNE. *Battle of Love*. c. 1880. Oil on canvas, 14⅞ x 18¼" (37.8 x 46.2 cm). National Gallery of Art, Washington, D.C.
GIFT OF THE W. AVERELL HARRIMAN FOUNDATION IN MEMORY OF MARIE N. HARRIMAN

much a specific portrait as a study of passion held in restraint, exemplifies Cézanne's use of the matrix of paint to create a unified pictorial structure, a characteristic of all his work. Throughout the 1860s, Cézanne exorcised his own inner conflict in scenes of murder and rape and, around 1870, he attempted his own *Déjeuner sur l'herbe* in a strange, heavy-handed variant on Manet's painting. After his exposure to the Impressionists, he returned to the violence of these essays in his *Battle of Love* (fig. 72), a curious recasting of the classic theme of the bacchanal. Here he has moved away from the somber tonality and sculptural modeling of the 1860s into an approximation of the light blue, green, and white of the Impressionists. Nevertheless, this is a painting remote from Impressionism in its subject and effect. The artist subdues his greens with grays and blacks, and expresses an obsession with the sexual ferocity he is portraying that is notably different from the stately bacchanals of Titian and even more intense than the comparable scenes of Rubens. At this stage he is still exploring the problem of integrating figures or objects and surrounding space. The figures are clearly outlined, and thus

exist sculpturally in space. However, their broken contours, sometimes seemingly independent of the cream-color area of their flesh, begin to dissolve solids and integrate figures with the shaped clouds that, in an advancing background, reiterate the carnal struggles of the bacchanal.

In the 1880s all of Cézanne's ideas of nature and painting came into focus in a magnificent series of landscapes, still lifes, and portraits. During most of this period he was living at Aix, in southern France, largely isolated from the Parisian art world. *The Bay from L'Estaque* (color plate 21, page 40) today is a work so familiar from countless reproductions that it is difficult to realize how revolutionary it was at the time of its execution. Viewed from the hills above the red-tiled houses of the village of L'Estaque and looking toward the Bay of Marseilles, the scene does not recede into a perspective of infinity in the Renaissance or Baroque manner. Buildings in the foreground are massed close to the spectator and presented as simplified cubes with the side elevation brightly lighted, prominently asserting their fluctuating identity both as frontalized color shapes parallel to the picture plane and as the walls of buildings at right angles to it. The buildings and intervening trees are composed in ocher, yellow, orange-red, and green, with little differentiation as objects recede from the eye. Cézanne once explained to a friend that sunlight cannot be "reproduced," but that it must be "represented" by some other means. That means was color. He wished to re-create nature with color, feeling that drawing was a consequence of the correct use of color. In *The Bay from L'Estaque*, contours are the meeting of two areas of color. Since these colors vary substantially in value contrasts or in hue, their edges are perfectly defined. However, the nature of the definition tends to allow color planes to slide or "pass" into one another, thus to join and unify surface and depth, rather than to separate them in the manner of traditional outline drawing. Apparent in the composition of this painting is Cézanne's intuitive realization that the eye took in a scene both consecutively and simultaneously, with profound implications for construction of the painting and for the future of modern painting.

The middle distance of *The Bay from L'Estaque* is the bay itself, an intense area of dense but varied blue stretching from one side of the canvas to the other and built up of meticulously blended brushstrokes. Behind this is the curving horizontal range of the hills and, above these, the lighter, softer blue of the sky, with only the faintest touches of rose. The abrupt manner in which the artist cuts off his space at the sides of the painting has the effect of denying the illusion of recession in depth. The blue of the bay asserts itself even more strongly than the ochers and reds of the foreground, with the result that space becomes both ambiguous and homogeneous. The painting must be read simultaneously as a panorama in depth and as an arrangement of color shapes on the surface.

Cézanne never, even at the end of his life, had any desire to abandon nature entirely. When he spoke of "the cone, the cylinder, and the sphere" he was not thinking of these geometric shapes as the end result, the final abstraction into which he wanted to translate the landscape or the still life, as has sometimes been assumed. Abstraction for him was a method of stripping off the visual irrelevancies of nature in order to begin rebuilding the natural scene as an independent painting. Thus the end result was easily recognizable from the original motif, as photographs of Cézanne's sites have proved, but it is essentially different: the painting is a parallel but distinct and unique reality.

Few if any artists in history have devoted themselves as intently as Cézanne to the separate themes of landscape, portrait, and still life. It is important to understand that to him these subjects involved similar problems. In all of them he was concerned with the re-creation, the *realization*, of the scene, the object, or the person. The fascination of the still life for Cézanne, as for generations of painters before him and after, lies in the fact that it involves a subject delimited and controllable as no landscape or portrait sitter can possibly be. In *Still Life with Basket of Apples* (color plate 22, page 73), he carefully arranged the wine bottle and tilted basket of apples, scattered the other apples casually over the tablecloth, and placed the plate of biscuits at the back of the table. He had only to

look until all these elements began to transform themselves into the relationships on which the final painting would be based. The apple obsessed Cézanne as a three-dimensional form that was difficult to control as a distinct object and to assimilate into the larger unity of the canvas. To attain this goal and at the same time to preserve the nature of the individual object, he modulated the circular forms with small, flat brushstrokes, distorted the shapes, and loosened or broke the contours to set up spatial tensions among the objects and thereby unify them as color areas. By tilting the wine bottle out of the vertical, flattening and distorting the perspective of the plate, or changing the direction of the table edge as it moved under the cloth, Cézanne was able, while maintaining the illusion of real appearance, to concentrate on the relations and tensions existing among objects as the significant visual experience. To try to express in words all the subtleties through which Cézanne gained his final result is to confront how his paintings, as Roger Fry observed, "still outrange our pictorial apprehension." However, we can comprehend that he was one of the great constructors and colorists in the history of painting, one of the most penetrating observers, and one of the most subtle minds.

Cézanne's figure studies, such as the various versions of *The Card Players* (fig. 73), remind one in their massive, closed architecture of his paintings of quarries. The artist brings great solemnity to this genre subject by his grandly balanced composition, by the amplitude of the figures, who solidly occupy their space, and by discarding the usual narratives included in older paintings on this theme, such as the card-shark subjects of Caravaggio. Although the sitters were local farmhands with whom the artist was probably long acquainted, there is little sense of particularized portraiture in *The Card Players*.

Cézanne usually relied on friends or family members as models, most frequently his patient wife, Hortense Fiquet, but for *Boy in a Red Waistcoat* (color plate 23, page 73) he hired a young Italian boy as a model. During a residence in Paris between 1888 and 1890, he painted four portraits of

73. PAUL CÉZANNE. *The Card Players*. 1890–92. Oil on canvas, 25¾ x 32¼" (65.4 x 81.9 cm). The Metropolitan Museum of Art, New York. BEQUEST OF STEPHEN C. CLARK, 1960

the seemingly melancholy boy in the same white shirt, blue cravat, and brilliant red waistcoat. Only in this composition, however, did the artist situate his subject frontally, to maximize the relaxed contrapposto of the boy's lanky torso. He is centered against a large swag of drapery so variegated by patches of color that its local hue is indeterminate. The painting is constructed as much by its network of broken lines as by the all-over tessellation of its flat, planar strokes. At the same time that the lines contour forms and endow them with an unmistakable fullness, they also intersect and continue from edge to edge, across figure and ground alike, thereby integrating the whole of the field into a kind of asymmetrical armature. A similar equilibrium of volume and plane occurs in the brushwork, where the squarish, "constructive" strokes build up forms while articulating the surface as a mosaic pattern of translucent, painterly slabs.

After 1890 Cézanne's brushstrokes became larger and more abstractly expressive, the contours more broken and dissolved, with color floating across objects, sustaining its own identity independent of the object. These tendencies were to lead to the wonderfully free paintings done at the very end of the artist's life, of which *Mont Sainte-Victoire Seen from Les Lauves* of 1902–6 is one of the supreme examples (color plate 24, page 73). Here the brushwork acts the part of the individual musician in a superbly integrated symphony orchestra. Each stroke exists fully in its own right, but each is nevertheless subordinated to the harmony of the whole. This is both a structured and a lyrical painting, one in which the artist has achieved the integration of structure and color, of nature and painting. It belongs to the great tradition of Renaissance and Baroque landscape, seen, however, as an infinite accumulation of individual perceptions. These are analyzed by the painter into their abstract components and then reconstructed into the new reality of the painting.

To the end of his life, Cézanne held fast to his dream of remaking Poussin after nature, continuously struggling with the problem of how to pose classically or heroically nude figures in an open landscape. For the largest canvas he ever worked, Cézanne painted as Poussin had done—that is, from his imagination—and relied on his years of plein-air experience to evoke Impressionist freshness. *The Large Bathers* (color plate 25, page 74), in many ways the culmination of thirty years of experiment with the subject of female bathers in an outdoor setting, was painted the year of Cézanne's death and is unfinished. In it the artist subsumed the fierce emotions expressed in the earlier *Battle of Love* within the grandeur of his total conception. Thus, while the figures have been so formalized as to seem little more than part of the overall pictorial architecture, their erotic potential now charges the scene. It can be sensed as much in the high-vaulted trees as in the sensuous, shimmering beauty of the brushwork, with its unifying touches of rosy flesh tones and earthy ochers scattered throughout the delicate blue-green haze of sky and foliage. *The Large Bathers,* in its miraculous integration of linear structure and painterly freedom, of form and color, of eye and idea, provided the touchstone model

for Fauves and Cubists alike, as well as the immediate antecedent for such landmark, yet disparate paintings as Matisse's *Bonheur de vivre* (color plate 52, page 118) and Picasso's *Les Demoiselles d'Avignon* (color plate 89, page 166).

Symbolism

The formulation of modern art at the end of the nineteenth century involved not only a search for new forms, a new plastic reality, but also a search for new content and new principles of synthesis. This search was meaningful in itself and not simply a continuation of outworn clichés taken from antiquity, French history, or mythology. It became, as we shall see, a conscious program in Gauguin's paintings, first in the small artists' colony of Pont-Aven in Brittany and later in the South Seas. In Brittany it was the peculiar individuality and religious devotion of the inhabitants which intrigued him and which he wanted to capture in such works as *Vision after the Sermon* (color plate 29, page 77). In Tahiti it was the mystery of a beautiful people living in an earthly paradise haunted by the gods and spirits of their religion (color plate 30, page 77). In both cases the subjects he pursued were those of the Romantic tradition: the exotic, the otherworldly, the mystical. In this, Gauguin and the artists who were associated with him at Pont-Aven and later at Le Pouldu in Brittany, as well as very different painters like Odilon Redon and Gustave Moreau, were affected by the Symbolist spirit.

Symbolism in literature and the visual arts was a popular—if radical—movement of the late 1800s, a direct descendant of the Romanticism of the eighteenth and the early nineteenth centuries that stemmed from a reaction against Realism in art and against materialism in life. In literature, its founders were mainly poets: Charles Baudelaire and Gérard de Nerval; the leaders at the close of the century were Jean Moréas, Stéphane Mallarmé, and Paul Verlaine. In music, the German Richard Wagner was a great force and influence and Claude Debussy the outstanding French master. Baudelaire had defined Romanticism as "neither a choice of subjects nor exact truth, but a mode of feeling"—something found within rather than outside the individual—"intimacy, spirituality, color, aspiration toward the infinite." Symbolism arose from the intuitive ideas of the Romantics; it was an approach to an ultimate reality, a pure essence that transcended particular physical experience. The credo that the work of art is a consequence of the emotions, of the inner spirit of the artist rather than of observed nature, dominated the attitudes of the Symbolist artists and was to recur continually in the philosophies of twentieth-century Expressionists, Dadaists, Surrealists, and even of Mondrian and other makers of abstract art.

For the Symbolists the reality of the inner idea, of the dream or symbol, was paramount, but could be expressed only obliquely, as a series of images or analogies out of which the final revelation might emerge. Symbolism led some poets and painters back to organized religion, some to mysticism and arcane religious cults, and others to aesthetic creeds that were essentially antireligious. During the 1880s, at the moment when artists were pursuing the idea of the dream, Sigmund

Freud was beginning the studies that were to lead to his theories of the significance of dreams and the subconscious.

Symbolism in the visual arts was manifested in a diverse range of styles. Paul Gauguin, one of the movement's leading exponents, sought in his paintings what he termed a "synthesis of form and color derived from the observation of the dominant element." He advised a fellow painter not to "copy nature too much. Art is an abstraction; derive this abstraction from nature while dreaming before it, but think more of creating than the actual result."

In these statements may be found many of the concepts of twentieth-century experimental painting, from the idea of color used arbitrarily rather than to describe an object visually, to the primacy of the creative act, to painting as abstraction. Gauguin's ideas, which he called "Synthetism," involved a synthesis of subject and idea with form and color, so that *Vision after the Sermon* is given its mystery, its visionary quality, by its abstract color patterns.

Symbolism in painting—the search for new forms, anti-naturalistic if necessary, to express a new content based on emotion (rather than intellect or objective observation), intuition, and the idea beyond appearance—may be interpreted broadly to include most of the experimental artists who succeeded the Impressionists and opposed their artistic goals. The Symbolist movement, while centered in France, was international in scope and had adherents in America, Belgium, England, and elsewhere in Europe. These artists had a common concern with problems of personal expression and pictorial structure. They were anticipated, inspired, and abetted by two older academic masters—Gustave Moreau and Pierre Puvis de Chavannes—and also by Odilon Redon, an artist born into the generation of the Impressionists who attained his artistic maturity much later than his exact contemporaries.

GUSTAVE MOREAU (1826–98). Appointed professor at the École des Beaux-Arts in 1892, Gustave Moreau displayed remarkable talents as a teacher, numbering among his students Henri Matisse and Georges Rouault, as well as others who were to be dubbed the Fauves, or "wild beasts," of early modern painting. Inspired by Romantic painters such as Delacroix, Moreau's art exemplified the spirit of the *mal-du-siècle,* an end-of-the-century tendency toward profound melancholy or soul sickness, often expressed in art and literature through decadent and morbid subject matter. In several compositions around 1876, Moreau interpreted the biblical subject of Salomé, the young princess who danced for her stepfather Herod, demanding in return the execution of Saint John the Baptist. The bloody head of the saint appears in Moreau's painting as if called forth in a grotesque hallucination (fig. 74). He conveys his subject with meticulous draftsmanship and an obsessive profusion of exotic detail combined with jewellike color and rich paint texture. Moreau's spangled, languidly voluptuary art did much to glamorize decadence in the form of the femme fatale, or fatal woman, the image of woman as an erotic and destructive force, a concept fostered by Baudelaire's great poems *Les Fleurs du Mal*

74. GUSTAVE MOREAU. *The Apparition.* c. 1876. Oil on canvas, 21¼ x 17½" (54 x 44.5 cm). Fogg Art Museum, Harvard University Art Museums, Cambridge, Massachusetts
GRENVILLE L. WINTHROP BEQUEST

(The Flowers of Evil) (1857) and the mid-century pessimistic philosophy of Arthur Schopenhauer. Among scores of male artists in the last decades of the nineteenth century, Salomé frequently served as the archetype of a fictitious, castrating female who embodied all that is debased, sexually predatory, and perverse. The femme fatale also played a central role in the work of such composers as Wagner and Richard Strauss, writers on the order of Gustave Flaubert, Joris-Karl Huysmans (who admired Moreau's painting enormously), Stéphane Mallarmé, Oscar Wilde, and Marcel Proust, and a great number of artists, ranging from Rossetti and Moreau to such fin-de-siècle figures as Redon, Aubrey Beardsley, Edvard Munch, and Gustav Klimt. Moreover, the deadly temptress thrived well into the twentieth century, as she resurfaced in Picasso's *Les Demoiselles d'Avignon,* (color plate 89, page 166), the paintings and drawings of Egon Schiele, and the *amour fou* art of the Surrealists.

PIERRE PUVIS DE CHAVANNES (1824–98). Although Puvis de Chavannes, with his simple, naive spirit, bleached colors, and near-archaic handling, would seem to have stood at some polar extreme from the elaborate, hothouse art of Moreau, the two painters were alike in a certain academicism and in the curious attraction this held for the younger generation. In Puvis, the reasons for such an anomaly can be readily discerned in the mural painting seen here (fig. 75).

75. PIERRE PUVIS DE CHAVANNES. *Summer*. 1891.
Oil on canvas, 59" x 7'7½" (149.9 cm x 2.3 m).
The Cleveland Museum of Art
GIFT OF MR. AND MRS. J. H. WADE, 1916

The allegorical subject and its narrative treatment are sufficiently traditional, but the organization in large, flat, subdued color areas, and the manner in which the plane of the wall is respected and even asserted, embodied a compelling truth in the minds of artists who were searching for a new idealism in painting. Although the abstract qualities of the murals are particularly apparent to us today, the classical withdrawal of the figures—as still and quiet as those in Piero della Francesca and Seurat—transforms them into emblems of that inner light which the Symbolists extolled.

ODILON REDON (1840–1916). Symbolism in painting found one of its earliest and most characteristic exponents in

Odilon Redon, called the "prince of mysterious dreams" by the critic and novelist Huysmans. His artistic roots were in nineteenth-century Romanticism, while his progeny was the twentieth-century Surrealists. Redon felt Impressionism lacked the ambiguity that he sought in his art, and though he frequently found his subjects in the study of nature, at times observed under the microscope, they were transformed in his hands into beautiful or monstrous fantasies.

Redon studied for a time in Paris with the painter Jean-Léon Gérôme, but he was not temperamentally suited to the rigors of academic training. He suffered from periodic depression—what he called his "habitual state of melancholy"—and much of his early work stems from memories of an unhappy childhood in and around Bordeaux, in southwest France. Like Seurat, Redon was a great colorist who also excelled at composing in black and white, and the first twenty years of his career were devoted almost exclusively to monochrome drawing, etching, and lithography, works he referred to as his *noirs*, or works in black. His activity in printmaking contributed to the rise in popularity of etching in the 1860s, prompted in part by a new appreciation in France of Rembrandt's prints, which Redon especially admired for their expressive effects of light and shadow. Redon studied etching with Rodolphe Bresdin, a strange and solitary artist who created a graphic world of meticulously detailed fantasy based on the work of Dürer and Rembrandt. Through him, Redon was attracted to these artists, but later he found a closer affinity with the graphics of Francisco Goya, the greatest printmaker of the early nineteenth century (see fig. 14). Redon himself became one of the modern masters of the medium of lithography, in which he invented a world of dreams and nightmares based not only on the examples of past artists but also on his close and scientific study of anatomy.

Redon's own predilection for fantasy and the macabre drew him naturally into the orbit of Delacroix, Baudelaire, and the Romantics. In his drawings and lithographs, where he pushed his rich, velvety blacks to extreme limits of expression, he developed and refined the fantasies of Bresdin in nightmare visions of monsters, or tragic-romantic themes taken from mythology or literature. Redon was close to the Symbolist poets and was almost the only artist who was successful in translating their words into visual equivalences. He dedicated a portfolio of his lithographs to Edgar Allan Poe, whose works had been translated by Baudelaire and Mallarmé, and he created three different series of lithographs inspired by Gustave Flaubert's *Temptation of St. Anthony* (fig. 76), a novel that achieved cult status among Symbolist writers and artists in the 1880s. Huysmans reviewed Redon's exhibitions enthusiastically at the same time that he was him-

76. ODILON REDON. *Anthony: What Is the Object of All This?
The Devil: There Is No Object.* Plate 18 from The Temptation of
St. Anthony, third series. 1886. Lithograph, printed in black,
composition 12¼ x 9⅞" (31.1 x 25.1 cm).
The Museum of Modern Art, New York

Color plate 22. PAUL CÉZANNE. *Still Life with Basket of Apples.* c. 1895.
Oil on canvas, 24⅜ x 31" (61.9 x 78.7 cm).
The Art Institute of Chicago HELEN BIRCH-BARTLETT MEMORIAL COLLECTION

right: Color plate 23. PAUL CÉZANNE. *Boy in a Red Waistcoat.* 1888–95.
Oil on canvas, 35¼ x 28¾" (89.5 x 73 cm).
National Gallery of Art, Washington, D.C. COLLECTION MR. AND MRS. PAUL MELLON

Color plate 24. PAUL CÉZANNE. *Mont Sainte-Victoire Seen from Les Lauves.* 1902–6. Oil on canvas, 25½ x 32"
(64.8 x 81.3 cm). Collection Mrs. Louis C. Madeira, Gladwyne, Pennsylvania

Color plate 25. PAUL CÉZANNE. *The Large Bathers*. 1906. Oil on canvas, 6'10" x 8'2" (2.1 x 2.5 m).
Philadelphia Museum of Art WILSTACH COLLECTION

Color plate 26.
ODILON REDON. *Roger and Angelica*. c. 1910. Pastel on paper on canvas,
36½ x 28¾" (92.7 x 73 cm).
The Museum of Modern Art, New York

LILLIE P. BLISS COLLECTION

Color plate 27. HENRI ROUSSEAU. *Carnival Evening.* 1886. Oil on canvas, 46 x 35⅛" (116.8 x 89.2 cm).
Philadelphia Museum of Art LOUIS E. STERN COLLECTION

Color plate 28. Henri Rousseau. *The Dream*. 1910. Oil on canvas, 6'8½" x 9'9½" (2.04 x 2.98 m).
The Museum of Modern Art Gift of Nelson A. Rockefeller

Color plate 29. PAUL GAUGUIN. *Vision after the Sermon.* 1888. Oil on canvas, 28¾ x 36¼" (73 x 92.1 cm).
National Gallery of Scotland, Edinburgh

Color plate 30.
PAUL GAUGUIN. *Where Do We
Come From? What Are We?
Where Are We Going?*
1897–98.
Oil on canvas, 4'6¾" x 12'3"
(1.4 x 3.7 m).
Museum of Fine Arts, Boston
TOMPKINS COLLECTION

Color plate 31.
Vincent van
Gogh.
The Night Café.
1888.
Oil on canvas,
27½ x 35"
(69.9 x 88.9 cm).
Yale University Art
Gallery, New
Haven
Bequest of Stephen C. Clark

Color plate 32.
Vincent van
Gogh.
The Starry Night.
1889. Oil on canvas, 29 x 36¼"
(73.7 x 92.1 cm).
The Museum of
Modern Art,
New York
Acquired through the Lillie P.
Bliss Bequest

above, left: Color plate 33. PAUL SÉRUSIER.
The Talisman (Landscape of the Bois d'Amour).
1888. Oil on wood cigar-box cover, 10½ x 8⅜"
(26.7 x 21.3 cm). Private collection

above, right: Color plate 34. ÉDOUARD VUILLARD.
Self-Portrait. 1892. Oil on board, 14 x 11"
(35.6 x 27.9 cm).
Private collection
© 1997 ADAGP, PARIS/ARTISTS RIGHTS SOCIETY (ARS), NEW YORK

Color plate 35. PIERRE BONNARD. *Nu à contre jour*
(Nude against the Light). 1908. Oil on canvas,
49 x 42½" (124.5 x 108 cm).
Musées Royaux des
Beaux-Arts de Belgique, Brussels
© 1997 ADAGP, PARIS/ARTISTS RIGHTS SOCIETY (ARS), NEW YORK

Color plate 36.
PIERRE BONNARD. *Dining Room on the Garden.* 1934–35. Oil on canvas, 50⅛ x 53½" (127.3 x 135.9 cm). Solomon R. Guggenheim Museum, New York
© 1997 ADAGP, PARIS/ARTISTS RIGHTS SOCIETY (ARS), NEW YORK

Color plate 37.
HENRI DE TOULOUSE-LAUTREC. *Moulin Rouge—La Goulue.* 1891. Color lithograph, 6'3¼" x 4' (1.9 x 1.2 m). Victoria and Albert Museum, London

self moving away from Émile Zola's naturalism and into the Symbolist stream with his novel of artistic decadence *Against the Grain*, in which he discussed at length the art of Redon and of Gustave Moreau.

After 1890 Redon began to work seriously in color, almost immediately demonstrating a capacity for exquisite, original harmonies, in both oil and pastel, that changed the character of his art from the macabre and the somber to the joyous and brilliant. Gone are the nightmarish visions of previous decades, replaced by a new enthusiasm for religious and mythological subjects. Yet in *Roger and Angelica* (color plate 26, page 74), the subject—the rescue of Angelica by the hero Roger, based on a Renaissance poem—seems almost incidental. We can hardly decipher the two tiny figures, Roger on a winged horselike hippogriff, at left, and Angelica at right, naked and chained to a rock; they are lost amid clouds of brilliant, amorphous color, where nothing is fixed in space and everything is subsumed in an atmosphere of overpowering sensuality.

HENRI ROUSSEAU (1844–1910). Although Henri Rousseau was actually four years older than Paul Gauguin, his name tends to be linked with a younger generation of artists, including Picasso and Robert Delaunay, because of the remarkable artistic circles with which he associated, sometimes unwittingly, in Paris. Though Rousseau shared many formal concerns with his Post-Impressionist contemporaries, such as Gauguin and Seurat, he is often regarded as a phenomenon apart from them, for he worked in isolation, admired the academic painters they abhorred, and was "discovered" by several younger vanguard artists. Yet, in its rejection of traditional illusionism and search for a poetic, highly personalized imagery, his work partakes of concerns held by many Symbolist artists at the end of the century.

In 1893, Rousseau retired from a full-time position as an inspector at a municipal toll station in Paris. He is known as the "Douanier" because he was thought to have been a customs inspector, one of many myths about this self-taught painter. Once he turned his full energies to painting, Rousseau seems to have developed a native ability into a sophisticated technique and an artistic vision that, despite its ingenuousness, was not without aesthetic sincerity or self-awareness.

Rousseau was a fascinating mixture of naïveté, innocence, and wisdom. Seemingly humorless, he combined a strange imagination with a way of seeing that was magical, sharp, and direct. In 1886, he exhibited his first painting, *Carnival Evening* (color plate 27, page 75), at the Salon des Indépendants. The black silhouettes of trees and house are drawn in painstaking detail—the accretive approach typically used by self-taught painters. Color throughout is low-keyed, as befits a night scene, but the lighted bank of clouds beneath the dark sky creates a sense of clear, wintry moonlight, as do the two tiny figures in carnival costumes in the foreground, who glow with an inner light. Despite the laboriously worked surface and the naïveté of his drawing, the artist fills his painting with poetry and a sense of dreamy unreality.

By 1890 Rousseau had exhibited some twenty paintings at

77. HENRI ROUSSEAU. *Myself: Portrait Landscape.* 1890. Oil on canvas, 56¼ x 43¼" (142.9 x 109.9 cm). Národni Muzeum, Prague

the Salon des Indépendants. His work had achieved a certain public notoriety, due to the constant mockery it endured in the press, but it also increasingly excited the interest of serious artists and such important writers as the poet and playwright Alfred Jarry and, somewhat later, Guillaume Apollinaire. Redon had already seen something unique in the art of Rousseau, and during the 1890s his admirers included Degas, Henri de Toulouse-Lautrec, and even Renoir. Despite such recognition, Rousseau never made a significant living from his painting and spent most of his post-retirement years in desperate poverty. His self-portrait, entitled *Myself: Portrait Landscape* (fig. 77), extols the great 1889 Exposition Universelle in Paris and shows the Eiffel Tower, a hot-air balloon, a flag-decked ship, and an iron bridge over the Seine, the Pont des Arts. In the foreground stands the majestic figure of the bearded artist, dressed in black and with a black beret. He holds his brush and palette, inscribed with the names of his past and present wives. Rousseau, whose confidence in his own talent bordered on the delusional, unabashedly presents himself as a master of modern painting.

With painstaking care and deliberation, Rousseau continually painted and drew scenes of Paris, still lifes, portraits of his friends and neighbors, and details of plants, leaves, and animals. His romantic passion for far-off jungles filled with strange terror and beauty recurred in several of his best-known canvases. While such subjects had precedents in Salon

78. HENRI ROUSSEAU. *The Sleeping Gypsy*. 1897. Oil on canvas, 51" x 6'7" (129.5 cm x 2 m). The Museum of Modern Art, New York GIFT OF MRS. SIMON GUGGENHEIM

PHOTOGRAPH © 1995 THE MUSEUM OF MODERN ART, NEW YORK

painting, Rousseau's tropical paintings are worlds away from the clichéd scenes of North Africa or other exotic locales of the academicians. Yet to the end of his life Rousseau continued his uncritical admiration for the Salon painters, particularly for the technical finish of their works. *The Sleeping Gypsy* (fig. 78) is one of the most entrancing and magical paintings in modern art. By this time he could create mood through a few elements, broadly conceived but meticulously rendered. The composition has a curiously abstract quality (the mandolin and bottle foretell Cubist studio props), but the tone is overpoweringly strange and eerie—a vast and lonely landscape framing a mysterious scene. In this work, as in others, Rousseau expressed qualities of strangeness that were unvoiced, inexpressible rather than apparent, anticipating the standard vocabulary of the Surrealists. This painting was lost for years until it was rediscovered in 1923 in a coal dealer's shop by Louis Vauxcelles, the art critic who had created the term "Fauves." In fact, it has been suggested that the term, meaning "wild beast," occurred to Vauxcelles because one of Rousseau's jungle paintings was exhibited very near the works of Matisse and his colleagues at the infamous 1905 Salon d'Automne.

The last series painted by Rousseau, and the acme of his vision, also present scenes of the jungle. The animals and plants he observed in the Paris Zoo and Botanical Garden were transformed in his paintings into scenes of tropical mystery. The setting in each is a mass of broad-leaved jungle foliage, gorgeous orchids, exotic fruits, and curious animals peering out of the underbrush. The climax of such scenes is

The Dream of 1910 (color plate 28, page 76), Rousseau's last great work, painted not long before he died. In the midst of a wildly abundant jungle, with its wide-eyed lions, brilliant birds, pale moon, and dark-skinned flute player, a nude woman rests incongruously on a splendid Victorian sofa. Rousseau's charmingly simple explanation of this improbable creation was that the woman on the couch dreamed herself transported to the midst of the jungle.

During his later years, Rousseau became something of a celebrity for the new generation of artists and critics. He held regular musical soirées at his studio. In 1908 Picasso, discovering a portrait of a woman by Rousseau at a junk shop (though he no doubt knew Rousseau's work already), purchased it for five francs and gave a large party for him in his studio, a party that has gone down in the history of modern painting. It was half intended as a joke on the innocent old painter, who took himself so seriously as an artist, but more an affectionate tribute to a naive man recognized to be a genius. In 1910, the year of his death, Rousseau's first solo exhibition was held in New York at Alfred Stieglitz's Gallery 291, thanks to the efforts of the young American painter Max Weber (see fig. 473), who had befriended Rousseau in Paris.

PAUL GAUGUIN (1848–1903) was unquestionably the most powerful and influential artist to be associated with the Symbolist movement. But while he became intimate with Symbolist poets, especially Mallarmé, and enjoyed their support, Gauguin deplored the literary content and traditional form of such artists as Moreau and Puvis. He also rejected the optical naturalism of the Impressionists, while initially retaining their rainbow palette and then vastly extending its potential for purely decorative effects. For Gauguin, art was an abstraction to be dreamed in the presence of nature, not an illustration but a synthesis of natural forms reinvented imaginatively. His purpose in creating such an anti-Realist art was to express invisible, subjective meanings and emotions experienced by a spirit he attempted to free from the corrupting sophistication of the modern industrial world, and renew by contact with the innocence and sense of mystery he sought in nonindustrial societies. He constantly used for painting the analogy of music, of color harmonies, of color and lines as forms of abstract expression. In his search he was attracted, to a greater degree even than most of his generation, to Asian and so-called "primitive" art. In his art we find the origins of modern primitivism, the desire to attain a kind of expression that discards the accumulation of Western culture and the forces of industrialization and urbanization. Gauguin turned away from these corrupt forces to what were regarded as the elemental verities of "primitive" societies, whether the religious art of Breton peasants in northwest France or the traditions of the Maori peoples in Polynesia. By establishing contact with supposedly "uncivilized" cultures, so the myth went, the enlightened artist could be in touch with the primitive side of his or her own nature and express it through art. Of course, such notions were forged at a time when European countries were aggressively colonizing the very societies Western artists sought to emulate.

In searching for the life of a "savage," no French artist traveled farther from the reaches of Western urban life than Gauguin. He has become, sometimes to the detriment of a serious consideration of his art, a romantic symbol, the personification of the artist as rebel against society. After years of wandering, first in the merchant marine, then in the French navy, he settled down, in 1871, to a prosaic but successful life as a stockbroker in Paris, married a Danish woman, and had five children. For the next twelve years the only oddity in his respectable, bourgeois existence was the fact that he began painting, first as a hobby and then with increasing seriousness. He even managed to show a painting in the Salon of 1876 and to exhibit in four Impressionist exhibitions from 1879 to 1882. He lost his job, probably due to a stock-market crash, and by 1886, after several years of family conflict and attempts at new starts in Rouen and Copenhagen, he had largely severed his family ties, isolated himself, and become involved with the Impressionists.

Gauguin, who had spent the early years of his childhood in Peru, seems almost always to have had a nostalgia for far-off, exotic places. This feeling ultimately crystallized in the conviction that his salvation and perhaps that of all contemporary artists lay in abandoning modern civilization and its encumbrance of classical Western culture to return to some simpler, more elemental pattern of life. From 1886 to 1891 he moved between Paris and the Breton villages of Pont-Aven and Le Pouldu, with a seven-month interlude in Panama and Martinique in 1887 and a tempestuous but productive visit with Van Gogh in Arles the following year. In 1891 he sailed for Tahiti, returning to France for two years in 1893 before settling in the South Seas for good. His final trip, in the wake of years of illness and suffering, was to the island of Hivaoa in the Marquesas Islands, where he died in 1903.

The earliest picture in which Gauguin brought his revolutionary ideas to full realization is *Vision after the Sermon*, painted in 1888 in Pont-Aven (color plate 29, page 77). This is a startling and pivotal work, a pattern of red, blue, black, and white tied together by curving, sinuous lines and depicting a Breton peasant's biblical vision of Jacob wrestling with the Angel. The innovations used here were destined to affect the ideas of younger groups such as the Nabis and the Fauves. Perhaps the greatest single departure is the arbitrary use of color in the dominating red field within which the protagonists struggle, their forms borrowed from a Japanese print. Gauguin has here constricted his space to such an extent that the dominant red of the background visually thrusts itself forward beyond the closely viewed heads of the peasants in the foreground. Though the brilliant red hue may have been stimulated by his memory of a local religious celebration which included fireworks and bonfires in the fields, this painting was one of the first complete statements of color as an expressive end in itself.

From the beginning of his life as an artist, Gauguin did not restrict himself to painting. He carved sculptures in mar-

79. PAUL GAUGUIN. *Be in Love and You Will Be Happy.* 1889.
Painted wood relief, 37½ x 28½" (95.3 x 72.4 cm).
Museum of Fine Arts, Boston ARTHUR TRACY CABOT FUND

80. VINCENT VAN GOGH.
The Garden of the Presbytery at Nuenen, Winter.
1884. Pen and pencil on paper,
15¼ x 20¾" (38.7 x 52.7 cm).
National Museum Vincent van Gogh,
Amsterdam

ble and wood and learned the rudiments of ceramics to become one of the most innovative ceramists of the century. By the late 1880s he had ventured into printmaking, another medium to which he brought all his experimental genius. One of Gauguin's most enigmatic works, made in Brittany during an especially despondent period for the artist, is a wooden panel that he carved in low relief, painted, and titled with the cynical admonition *Be in Love and You Will Be Happy* (fig. 79). Here a woman, "whom a demon takes by the hand," faces the forces of temptation, symbolized in part by a small fox. Such themes, the struggle between knowledge and innocence, good and evil, life and death, recur in Gauguin's Tahitian compositions. Among the faces in the relief, the one with its thumb in its mouth, at the upper right, is the artist himself.

Gauguin was disappointed, upon his arrival in Tahiti, to discover how extensively Western missionaries and colonials had encroached upon native life. The capital of Papeete was filled with French government officials, and the beautiful Tahitian women often covered themselves in ankle-length missionary dresses. Nor was the island filled with the indigenous carvings of ancient gods Gauguin had hoped for, so he set about making his own idols, based in part on Egyptian and Buddhist sculptures he knew from photographs. One such invention can be seen at the left in his grand, philosophical painting *Where Do We Come from? What Are We? Where Are We Going?* (color plate 30, page 77). Over twelve feet long, this is the most ambitious painting of Gauguin's career. It presents a summation of his Polynesian imagery, filled with Tahitians of all ages situated at ease within a terrestrial paradise devoid of any sign of European civilization. "It's not a canvas done like a Puvis de Chavannes," Gauguin said, "with studies from nature, then preliminary cartoons etc. It is all dashed off with the tip of the brush, on burlap full of knots and wrinkles, so that its appearance is terribly rough." The multiple forms and deep spaces of this complex composition are tied together by its overall tonalities in green and blue. It was this element—color—that the artist called "a mysterious language, a language of the dream."

VINCENT VAN GOGH (1853–90). Whereas Gauguin was

an iconoclast, caustic in speech, cynical, indifferent, and at times brutal to others, Van Gogh was filled with a spirit of enthusiasm for his fellow artists and overwhelming love for humanity. This love had led him, after a short-lived experience as an art dealer and an attempt to follow theological studies, to become a lay preacher in a Belgian coal-mining area. There he first began to draw in 1880. After study in Brussels, the Hague, and Antwerp, he went to Paris in 1886, where he met Toulouse-Lautrec, Seurat, Signac, and Gauguin, as well as members of the original Impressionist group.

In his early drawings, meanwhile, Van Gogh revealed his roots in traditional Dutch landscape, portrait, and genre painting (fig. 80). Works like this one extol, through the same perspective structures, the broad fields and low-hanging skies that the seventeenth-century artists had loved. Van Gogh never abandoned perspective even in later years, when he developed a style with great emphasis on the linear movement of paint over the surface of the canvas. For him—and this is already apparent in the early drawings—landscape itself had an expressive, an emotional significance.

After his exposure to the Impressionists in Paris, Van Gogh changed and lightened his palette. Indeed, he discovered his deepest single love in color—brilliant, unmodulated color—which in his hands took on a character radically different from the color of the Impressionists. Even when he used Impressionist techniques, the peculiar intensity of his vision gave the result a specific and individual quality that could never be mistaken.

The passion in Van Gogh's art arose from his intense, overpowering response to the world in which he lived and to the people whom he knew. His mental troubles are well known, indeed confusion about them, prompted by the incident in which he sliced off part of his ear during Gauguin's visit, has overshadowed a reasoned understanding of his work. Van Gogh may have suffered from a neurological disorder, perhaps a severe form of epilepsy, that was no doubt exacerbated by physical ailments and excessive drinking. He was prone to depression and suffered acutely during seizures, but he painted during long periods of lucidity, bringing tremendous intelligence and imagination to his work. His letters to his brother Theo, an art dealer who tried in vain to find a market for Vincent's work, are among the most moving and informative narratives by an artist that we have. They reveal a highly sensitive perception that is fully equal to his emotional response. He was sharply aware of the extraordinary effects he was achieving through his expressive use of color. "Instead of trying to reproduce exactly what I have before my eyes, I use color more arbitrarily," he wrote, "in order to express myself forcibly." Echoing the Symbolist ideas of Gauguin, Van Gogh told Theo that he "was trying to exaggerate the essential and to leave the obvious vague."

Van Gogh could also present the darker side of existence. Thus, of *The Night Café* (color plate 31, page 78) he says: "I have tried to express the terrible passions of humanity by means of red and green." *The Night Café* is a nightmare of deep-green ceiling, blood-red walls, and discordant greens of furniture. The perspective of the brilliant yellow floor is tilted so precipitously that the contents of the room threaten to slide toward the viewer. The result is a terrifying experience of claustrophobic compression that anticipates the Surrealist explorations of fantastic perspective, none of which has ever quite matched it in emotive force.

Vincent van Gogh was one of the few artists of his generation who carried on the great tradition of portraiture, from his first essays in drawing to his last self-portraits, painted a few months before his suicide in 1890. The intense *Self-Portrait* from 1888 was made in Arles and was dedicated to Gauguin (fig. 81). It formed part of an exchange of self-portraits

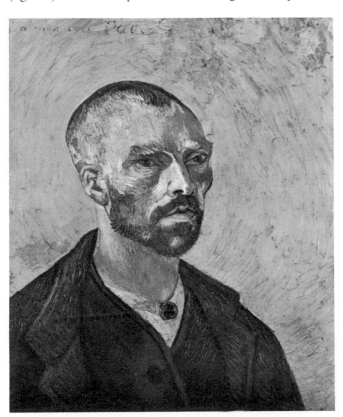

81. VINCENT VAN GOGH. *Self-Portrait*. 1888. Oil on canvas,
24½ x 20½" (62.2 x 52.1 cm). Fogg Art Museum,
Harvard University Art Museums, Cambridge, Massachusetts
BEQUEST FROM THE MAURICE WERTHEIM COLLECTION, CLASS OF 1906

among Van Gogh's artist friends to support his notion of an ideal brotherhood of painters. The beautifully sculptured head (which Van Gogh said resembled that of a Buddhist monk) and the solidly modeled torso are silhouetted against a vibrant field of linear rhythms painted, according to the artist, in "pale malachite." The coloristic and rhythmic integration of all parts, the careful progression of emphases, from head to torso to background, all demonstrate an artist in superb control of his plastic means. "In a picture," he wrote to Theo, "I want to say something comforting as music. I want to paint men and women with that something of the eternal which the halo used to symbolize, and which we seek to give by the actual radiance and vibration of our colorings."

The universe of Van Gogh is forever stated in *The Starry Night* (color plate 32, page 78). This work was painted in June 1889 at the sanatorium of Saint-Rémy, where he had been taken after his second breakdown. The color is predominantly blue and violet, pulsating with the scintillating yellow of the stars. *The Starry Night* is both an intimate and a vast landscape, seen from a high vantage point in the manner of the sixteenth-century landscapist Pieter Brueghel the Elder. In fact, the peaceful village, with its prominent church spire, is a remembrance of a Dutch rather than a French town. The great poplar tree in the foreground shudders before our eyes, while above whirl and explode all the stars and planets of the universe. Van Gogh was intrigued by the idea of painting a nocturnal landscape from his imagination. Scholars have tried to explain the content of the painting through literature, astronomy, and religion. Though their studies have shed light on Van Gogh's interests, none has tapped a definitive source that accounts for the astonishing impact of this painting, which today ranks among the most famous works of art ever made. When we think of Expressionism in painting, we tend to associate with it a bravura brush gesture, arising from the spontaneous or intuitive act of expression and independent of rational processes of thought or precise technique. The anomaly of Van Gogh's paintings is that they are supernatural or at least extrasensory experiences evoked with a touch as meticulous as though the artist were painfully and exactly copying what he was observing before his eyes.

The Nabis

Neo-Impressionism, created by Seurat and Signac, made its appearance in 1884, when a number of artists who were to be associated with the movement exhibited together at the Groupe des Artistes Indépendants in Paris. Later that year the Société des Artistes Indépendants was organized through the efforts of Seurat, Henri-Edmond Cross, Redon, and others, and was to become important to the advancement of early twentieth-century art as an exhibition forum. Also important were the exhibitions of *Les XX* (*Les Vingt*, or "The Twenty") in Brussels. Van Gogh, Gauguin, Toulouse-Lautrec, and Cézanne exhibited at both the Indépendants and Les XX. James Ensor, Henry van de Velde, and Jan Toorop exhibited regularly at Les XX and its successor, La Libre Esthétique, whose shows became increasingly dominated first by the attitudes of

82. ÉMILE BERNARD. *Buckwheat Harvesters, Pont-Aven.* 1888. Oil on canvas, 29¼ x 36" (74.3 x 91.4 cm). Private collection

the Neo-Impressionists and then by the Nabis.

The Nabis, who took their name from the Hebrew word for "prophet," were a somewhat eclectic group of artists whose principal contributions—with some outstanding exceptions—lay in a synthesizing approach to masters of the earlier generation, not only to Seurat but also to Cézanne, Redon, and Gauguin, particularly the last, for his art theory as well as the direct example of his painting.

Gauguin had been in some degree affected by the ideas of his young friend ÉMILE BERNARD (1868–1941) when the two were working together in Pont-Aven in the summer of 1888. He may well have derived important elements of his style from Bernard's notion of *cloisonnisme,* a style based on medieval enamel and stained-glass techniques, in which flat areas of color are bounded by dark, emphatic contours (fig. 82). Certainly, the arbitrary, nondescriptive color, the flat areas bounded by linear patterns, the denial of depth and sculptural modeling—all stated by Gauguin in *Vision after the Sermon* (color plate 29, page 77)—were congenial to and influential for the Nabis as well as for Art Nouveau decoration.

PAUL SÉRUSIER (1863–1927), one of the young artists under Gauguin's spell at Pont-Aven, experienced something of an epiphany when the older master undertook to demonstrate his method during a painting session in a picturesque wood known as the Bois d'Amour: "How do you see these trees?" Gauguin asked. "They are yellow. Well then, put down yellow. And that shadow is rather blue. Render it with pure ultramarine. Those red leaves? Use vermilion." This permitted the mesmerized Sérusier to paint a tiny work on a cigar-box lid (color plate 33, page 79), which proved so daring in form, even verging on pure abstraction, that the artist and his friends thought it virtually alive with supernatural power. And so they entitled the painting *The Talisman* and dubbed themselves the "Nabis." The group included Sérusier, Maurice Denis, Pierre Bonnard, Paul Ranson, and later Aristide Maillol, Édouard Vuillard, Félix Vallotton, Ker-Xavier Roussel, and Armand Séguin. The Nabis were artists of varying abilities, but included three outstanding talents: Bonnard, Vuillard, and Maillol.

The Nabis were symptomatic of the various interests and enthusiasms of the end of the century. Among these were literary tendencies toward organized theory and elaborate celebrations of mystical rituals. Denis and Sérusier wrote extensively on the theory of modern painting; and Denis was responsible for the formulation of the famous phrase, "a picture—before being a warhorse, a female nude, or some anecdote—is essentially a flat surface covered with colors assembled in a particular order." The Nabis sought a synthesis of all the arts through continual activity in architectural painting, the design of glass and decorative screens, book illustration, poster design, and stage design for the advanced theater of Ibsen, Maurice Maeterlinck, Strindberg, Wilde, and notably for Alfred Jarry's shocking play *Ubu Roi* (King Ubu).

La Revue Blanche, a magazine founded in 1891, became one of the chief organs of expression for Symbolist writers and painters, Nabis, and other artists of the avant-garde. Bonnard, Vuillard, Denis, Vallotton, and Toulouse-Lautrec (who was never officially a Nabi, although associated with the group) all made posters and illustrations for *La Revue Blanche.* The magazine was a meeting ground for experimental artists and writers from every part of Europe, including Van de Velde, Edvard Munch, Marcel Proust, André Gide, Ibsen, Strindberg, Wilde, Maxim Gorky, and Filippo Marinetti.

ÉDOUARD VUILLARD (1868–1940). The Nabis produced two painters of genius, Édouard Vuillard and Pierre Bonnard, whose long working lives remind us how short a time actually separated us from the worlds of two centuries. Both were much admired; their reputations, however, were for a long time private rather than public. Their world is an intimate one, consisting of corners of the studio, the living room, the familiar view from the window, and portraits of family and close friends. Vuillard's *Self-Portrait* of 1892 (color plate 34, page 79) shows him at the moment when his style was closest to the theories of Gauguin and the Nabis, using color arbitrarily for expressive rather than naturalistic ends. He produced in this early work an image so bold in palette and reductive in design that it rivals the Fauve paintings that Matisse, his exact contemporary, made in the following decade (color plate 51, page 117).

In his early works Vuillard used the broken paint and small brushstroke of Seurat or Signac, but without their rigorous scientific methods. In *Woman in Blue with Child* (fig. 83) he portrayed the Parisian apartment of Thadée Natanson, cofounder of the *Revue Blanche,* and his famously beautiful and talented wife, Misia, who is depicted in the painting playing with her niece. As was often his practice, Vuillard probably used his own photograph of the apartment as an *aide-mémoire* while working up his composition. It is a typical turn-of-the-century interior, sumptuously chock-a-block with flowered wallpaper, figured upholstery, and decorative objects. In Vuillard's hands, the interior became a dazzling surface pattern of muted blues, reds, and yellows, comparable to a Persian painting in its harmonious richness. Space may be indicated by the tilted perspective of the chaise longue and the angled folds of the standing screen, but the forms of the woman and child are flattened so as to be virtually indistinguishable from the surrounding profusion of patterns. Such quiet scenes of Parisian middle-class domesticity have been

called "intimist"; in them, the flat jigsaw puzzle of conflicting patterns generates shimmering after-images that seem to draw from everyday life an ineffable sense of strangeness and magic.

PIERRE BONNARD (1867–1947). Of all the Nabis, Pierre Bonnard was the closest to Vuillard, and the two men remained friends until the latter's death. Like Vuillard, Bonnard lived a quiet and unobtrusive life, but whereas Vuillard stayed a bachelor, Bonnard early became attached to a young woman whom he ultimately married in 1925. It is she who appears in so many of his paintings, as a nude bathing or combing her hair, or as a shadowy but ever-present figure seated at the breakfast table, appearing at the window, or boating on the Seine.

After receiving training both in the law and in the fine arts, Bonnard soon gained a reputation making lithographs, posters, and illustrated books. His most important early influences were the work of Gauguin and Japanese prints. The impact of the latter can be seen in his adaptation of the *japoniste* approach to the tilted spaces and decorative linear rhythms of his paintings. But from the beginning Bonnard also evinced a love of paint texture. This led him from the relatively subdued palette of his early works to the full luminosity of high-keyed color rendered in fragmented brushstrokes, a development that may well owe something to both the late works of Monet and the Fauve paintings of Matisse.

The large folding screen *Promenade of the Nursemaids, Frieze of Fiacres* (fig. 84) is made of four lithographs, based

83. ÉDOUARD VUILLARD. *Woman in Blue with Child.* c. 1899. Oil on cardboard, 19⅛ x 22¼" (48.6 x 56.5 cm). Glasgow Art Gallery and Museum, Kelvingrove PRESENTED BY SIR JOHN RICHMOND, 1948

84. PIERRE BONNARD. *Promenade of the Nursemaids, Frieze of Fiacres.* 1899. Color lithograph on four panels, each 54 x 18¾" (137.2 x 47.6 cm). The Museum of Modern Art, New York

on a similarly painted screen. With its tilted perspectives and abbreviated, silhouetted forms, it shows Bonnard at his most *japoniste* and decorative. At the same time, the figures of mother and children, the three heavily caped nurses, and the marching line of fiacres, or carriages, reveal a touch of gentle satire that well characterizes the penetrating observation Bonnard could combine with a brilliant simplicity of design. Like his fellow Nabis, Bonnard believed in eliminating barriers between the realms of popular decorative arts and the high-art traditions of painting and sculpture. He envisioned an art of "everyday application" that could extend to fans, prints, furniture, or, in this case, color lithographs adapted to the format of a four-part Japanese screen.

In *Nude against the Light* (color plate 35, page 79), Bonnard has moved from the public sphere of Parisian streets to the intimate world of the nude in a domestic interior, a subject he exploited throughout his career as a means to investigate light and color. Bonnard silhouetted the model, his ever-youthful wife, Marthe, against the sun-drenched surfaces of her boudoir. Light falls through the tall French windows, strongly illuminating the side of the woman turned from our view but visible in the mirror at left. This use of reflections to enlarge and enrich the pictorial space, to stand as a picture within a picture, became a common strategy of Bonnard's as well as Matisse's interiors (color plate 62, page 154). But in its quiet solemnity and complete absence of self-consciousness, Bonnard's nude is deeply indebted to the precedent of Degas's bathers, even to the detail of the round tub (see fig. 53). Like those of the older artist, Bonnard's composition is disciplined and complex, carefully structured to return the eye to the solid form of the nude, which he surrounds with a multitude of textures, shapes, and colors. But Bonnard creates an expressive mood all his own. As she douses herself with perfume, the model seems almost transfixed by the warm, radiant light that permeates the scene.

Bonnard's color became brighter and gayer; until in a painting like *Dining Room on the Garden,* executed in 1934–35 (color plate 36, page 80), he had long since recovered the entire spectrum of luminous color, and had learned from Cézanne that color could function constructively as well as sensually. In this ambitious canvas Bonnard tackled the difficult problem of depicting an interior scene with a view through the window to a garden beyond, setting the isolated, geometric forms of a tabletop still life against a lush exterior landscape. Now the model, his wife Marthe, is positioned to one side, an incidental and ghostly presence in this sumptuous display. By the time of *Dining Room on the Garden,* virtually all the great primary revolutions of twentieth-century painting had already occurred, including Fauvism, with its arbitrary, expressive color, and Cubism, with its reorganization of Renaissance pictorial space. Moreover, painting had found its way to pure abstraction in various forms. Perfectly aware of all this, Bonnard was nonetheless content to go his own way. In the work seen here, for instance, there is evidence that he had looked closely at Fauve and Cubist paintings, particularly the works of Matisse—who was a devoted admirer of his—and had used what he wanted of the new approaches without at any time changing his basic attitudes.

Although **HENRI DE TOULOUSE-LAUTREC** (1864–1901) may be seen as the heir of Daumier in the field of printmaking, he also served, along with his contemporaries Gauguin and Van Gogh, as one of the principal bridges between nineteenth-century vanguard painting and early twentieth-century experiments of Edvard Munch, Pablo Picasso, and Henri Matisse. Lautrec was interested in Goya and the line drawings of Ingres, but he was above all a passionate disciple of Degas, both in his admiration of Degas's draftsmanship and in the disengaged attitude and calculated formal strategies he brought to the depiction of his favorite subjects—the theaters, brothels, and bohemian cabarets of Paris.

Due to years of inbreeding in his old, aristocratic family, Lautrec was permanently disfigured from a congenital disease that weakened his bones. Against his family's wishes, he pursued art as a profession after receiving an education in the private Parisian studios of Léon Bonnat and Fernand Cormon, painters who provided students with a more open and tolerant atmosphere than that found in the École des Beaux-Arts. In Cormon's studio Lautrec met Bernard and Van Gogh, both of whom he rendered in early portraits.

Lautrec is best known for his color lithographs in the 1890s of performers in Montmartre dance halls, but in the previous decade he had proved himself to be a sensitive portraitist with paintings and drawings of a colorful cast of characters, including Carmen Gaudin, the woman portrayed in *"À Montrouge"—Rosa La Rouge* (fig. 85). The artist was drawn to the simple clothes, unruly red hair, and tough look of this young working-class woman, who, arms dangling informally, averts her face as she is momentarily silhouetted against the lighted window. Lautrec creates this simplified composition out of his characteristically long strokes of color in warm, subdued tonalities. But the somber mood of the painting has also to do with its subject. Lautrec's painting was inspired by a gruesome song written by his bohemian friend, the famous cabaret singer Aristide Bruant, about a prostitute who conspires to kill her clients.

The naturalism of Lautrec's early portraits gave way in the 1890s to the brightly colored and stylized works that make his name synonymous with turn-of-the-century Paris. His earliest lithographic poster, designed for the notorious dance hall called the Moulin Rouge (color plate 37, page 80), features the scandalous talents of La Goulue (the greedy one), a dancer renowned for her gymnastic and erotic interpretations of the *chahut,* the dance that had attracted Seurat in 1889 (see fig. 70). Lautrec's superb graphic sensibility is apparent in the eye-catching shapes that, albeit abbreviated, were the result of long observation. Their snappy curves and crisp silhouettes were born of an Art Nouveau aesthetic that dominated the arts across Europe at the time and that is examined in chapter five.

85. Henri de Toulouse-Lautrec. *"À Montrouge"—Rosa La Rouge.* 1886–87. Oil on canvas, 28½ x 19¼" (72.4 x 48.9 cm). The Barnes Foundation, Merion, Pennsylvania

4 The Origins of Modern Architecture and Design

In architecture, a pattern of stylistic revivals had predominated in Europe since the fifteenth century and in the Americas since the sixteenth. Throughout the nineteenth century, architects continued to expand this tendency. By the first decades of the 1800s, the Neoclassical movement was well established in both Europe and America as the dominant style. In America, this Neoclassical style, or classical revival, coincided with the aims of a young republic that regarded itself as the new Arcadia. The Neoclassical style in architecture, which consisted of endless recombinations of ancient Greek and Roman motifs, symbolized the ideal public virtues of democracy, liberty, and reason. During the later eighteenth century, particularly in England, an opposing trend, the Gothic revival, had also gained considerable momentum and continued to make headway during the first half of the nineteenth; it was accompanied by revivals of Renaissance and Baroque classicism. All these movements produced much of the monumental architecture of the late nineteenth century and the early twentieth.

Nineteenth-Century Industry and Engineering

New ideas in architecture emerged during the nineteenth century in the context of engineering, with the spread of industrialization and the use of novel building materials, notably iron, glass, and hollow ceramic tile. The use of iron for structural elements became fairly common in English industrial building only after 1800, although there are sporadic instances during the eighteenth century. These new materials—developed or improved during the Industrial Revolution—permitted greater flexibility and experimentation in design, as well as larger scale. During the first half of the new century, a concealed iron-skeleton structure was used in buildings of all types, and exposed columns and decorative ironwork appeared, particularly in nontraditional structures and notably in the Royal Pavilion at Brighton (fig. 86), designed by JOHN NASH (1752–1835). Iron was also used extensively in the many new railway bridges built throughout Europe, as the railway system expanded (fig. 87). GUSTAVE

EIFFEL (1832–1923), whose famous 984-foot tower, built for the Paris Exposition Universelle of 1889, was the most dramatic demonstration of the possibilities of exposed-metal construction, using iron in a variety of ways. Roofs of iron and glass over commercial galleries or bazaars became popular and ambitious in design at this time, especially in Paris. In greenhouses, iron gradually replaced wood as a frame for glass panes, with a consequent enlargement of scale. The Jardin d'Hiver (Winter Garden) in Paris, designed by Hector Horeau in 1847 and since destroyed, measured 300 by 180 feet in floor plan and rose to a height of 60 feet—a quite remarkable size at the time. In works such as this originated the concept of the monumental glass-and-metal structure, of which the Crystal Palace, created for the London Exposition

86. JOHN NASH. Royal Pavilion, Brighton, England. 1815–23, as remodeled

87. GUSTAVE EIFFEL. Truyère Bridge, Garabit, France. 1880–84

most daring and experimental buildings in iron and glass. The Gare Saint-Lazare in Paris supplied the painter Monet with a new experience in light, space, and movement, which he painted several times. The use of exposed iron columns supporting iron-and-glass arches penetrated a few traditional structures, such as the Bibliothèque Sainte-Geneviève (fig. 89) by **HENRI LABROUSTE** (1801–75). Stamped, cast-iron details were very popular, and a substantial industry in their prefabrication developed before 1850. For decades thereafter, however, the structural and decorative use of cast iron declined, in large part because of the conservative tastes of revival-oriented theorists in architecture. Only at the very end of the nineteenth century, with the new Art Nouveau style, did metal emerge once more as the architectural element on which much of twentieth-century architecture was based.

In architecture, as in the decorative arts, the latter half of the nineteenth century often represented a highly eclectic period in the history of taste. The classical and Gothic revivals of the early century had been largely superseded by elaborate buildings in the Renaissance- and Baroque-revival styles; of these the Paris Opera (fig. 90), designed by **CHARLES GARNIER** (1825–98), is perhaps the supreme example. These remained, far into the twentieth century, the official styles for public and monumental buildings. At the turn of the century, America, with its growing population, burgeoning cities, and expanding economy, was emerging as a world power. The expression of this in public buildings showed a heightened consciousness of historical revivalism, which could encompass Gothic, classical, or colonial styles, among others. More and more American architects received their training at the École des Beaux-Arts in Paris, the international heart of academicism in architecture. For American banks, civic monuments, libraries, capitols, and other public buildings coast to coast, the preferred mode was academic classicism on a grandiose scale. For example, the enormous central mall in Washington, D.C., which runs between the Capitol and the Lincoln Memorial, and which is lined with mostly classical

of 1851 by **SIR JOSEPH PAXTON** (1801–65), was the great exemplar (fig. 88). The Crystal Palace was erected through modern building methods that are today taken for granted—prefabricated modules that allow for easy assembly. Paxton developed the ridge-and-furrow roof construction while designing greenhouses. Yet the building, an airy barrel-vaulted gallery, also recalls the space and shape of a gigantic, secular cathedral. The Crystal Palace was dismantled and reassembled in Sydenham, in south London, where it remained until it burned to the ground in 1936.

The sheds of the new railway stations, precisely because they were structures without tradition, provided some of the

88. SIR JOSEPH PAXTON. Crystal Palace, London. 1851

89. HENRI LABROUSTE. Reading room, Bibliothèque Sainte-Geneviève, Paris. 1843–50

90. CHARLES GARNIER. The Opera, Paris. 1861–74

façades, was a product of this period.

The tremendous expansion of machine production during the nineteenth century had as its corollary the development of mass consumption of new, more cheaply manufactured products. This growing mass market tended to purchase products with less attention to their quality or beauty than to their price. Thus there arose early in the nineteenth century the profession of modern advertising, whose credo it was that the most ornate, the most unoriginal products sell best. By the time of the London Exposition of 1851, the vast array of manufactured objects exhibited in Paxton's daring glass-and-metal Crystal Palace represented and even flaunted the degeneration of quality brought on by the emergence of a machine economy.

To a romantic, progressive nineteenth-century thinker and artist like William Morris (see fig. 103), friend and collaborator of the Pre-Raphaelites, the machine was destroying the values of individual craftsmanship on which the high level of past artistic achievement had rested. Poet, painter, designer, and social reformer, Morris fought against the rising tide of what he saw as commercial vulgarization, for he passion-

91. PHILIP WEBB and WILLIAM MORRIS. Red House, Bexley Heath, Kent, England. 1859–60

92. HENRY HOBSON RICHARDSON. Stoughton House, Cambridge, Massachusetts. 1883

93. MᴄKɪᴍ, Mᴇᴀᴅ & Wʜɪᴛᴇ. William G. Low House, Bristol, Rhode Island. 1886–87. Demolished 1960s

ately believed that the industrial worker was alienated both from the customer and the craft itself and was held hostage to the dictates of fashion and profit. Morris founded an artists' cooperative, called the Firm, which made furniture, fabrics, and wallpapers. He turned to the Middle Ages for design inspiration as well as for the medieval model of guild organization for skilled laborers. His friend, the artist and designer Walter Crane, summarized their common goals by calling for "a return to simplicity, to sincerity; to good materials and sound workmanship; to rich and suggestive surface decoration, and simple constructive forms." Morris's designs were ideally intended to raise the dignity of craftsperson and owner alike and to enhance the beauty of ordinary homes. The Firm's progeny was the Arts and Crafts movement in the British Isles, which exerted tremendous influence in the United States, especially on the oak Mission Style furniture of Gustav Stickley.

Domestic Architecture in England and the United States

Morris asked the architect **PHILIP WEBB** (1831–1915) to design his own home, called the Red House, in Kent, in 1859 (fig. 91). In accordance with Morris's desires, Red House was generally Gothic in design, but Webb interpreted his mandate freely. Into a simplified, traditional, red-brick Tudor Gothic manor house, with a steeply pitched roof, he introduced exterior details of classic Queen Anne windows and oculi. The effect of the exterior is of harmony and compact unity, despite the disparate elements. The plan is an open L-shape with commodious, well-lit rooms arranged in an easy and efficient asymmetrical pattern. The interior, from the furniture and stained-glass windows down to the fire irons, was designed by Webb and Morris in collaboration with Morris's wife, Jane, and the Pre-Raphaelite artists Rossetti and Edward Burne-Jones.

Aware of such English experiments, **HENRY HOBSON RICHARDSON** (1838–86) nonetheless developed his own style in the United States, which fused native American traditions with those of medieval Europe. The Stoughton House,

in Cambridge, Massachusetts (fig. 92), is one of his most familiar structures in the shingle style then rising to prominence on the eastern seaboard. The main outlines are reminiscent of Romanesque- and Gothic-revival traditions, but these are architecturally less significant than the spacious, screened entry and the extensive windows, which suggest his concern with the livability of the interior. The unornamented shingled exterior wraps around the main structure of the house to create a sense of simplicity and unity.

While the firm of **MCKIM, MEAD & WHITE** was in some degree responsible for the hold that academic classicism had on American public architecture for half a century, the firm's historicism was based less on generic Greek and Roman prototypes than on American eighteenth-century architecture. These prolific architects produced some of the most successful classical public buildings in America, including the Boston Public Library, New York's magnificent Pennsylvania Railroad Station (see fig. 272), sadly demolished in 1966, and the influential Rhode Island State Capitol in Providence, which established the model for many state capitols in the early years of the twentieth century. In their domestic architecture, McKim, Mead & White drew on the qualities of American colonial building and, like Webb and Morris, were able to select the historical or geographic style that best suited a particular site or commission. The William G. Low House of 1886–87 (fig. 93), a superb example of the shingle style in Bristol, Rhode Island, is revolutionary when viewed in contrast to the enormous mansions in revivalist styles that were then fashionable in nearby, wealthy Newport. The shingle-covered structure, with its bay windows and extended side porch, is formed of a single triangular gable, and has a striking geometric integrity.

Twentieth-century architecture is rooted in nineteenth-century technological advances and in the invention of new types of structures—railway stations, department stores, and exhibition halls, buildings without tradition that lent themselves to experimental design. However, many of modern architecture's most revolutionary concepts—attention to utility and comfort, the notion of beauty in undecorated forms,

94. CHARLES F. VOYSEY. Forster House, Bedford Park, London. 1891

and of forms that follow function—were first conceived in modest individual houses and small industrial buildings. The pioneer efforts of Webb in England and of Richardson and McKim, Mead & White in the United States were followed by the important experimental housing of Charles F. Voysey, Arthur H. Mackmurdo, and Charles Rennie Mackintosh in

95. CHARLES RENNIE MACKINTOSH. Library, Glasgow School of Art. 1907–9

the British Isles, and in the United States by the revolutionary house designs of Frank Lloyd Wright.

CHARLES F. VOYSEY (1857–1941), through his wallpapers and textiles, had an even more immediate influence on European Art Nouveau than did the Arts and Crafts movement of William Morris. The furniture he designed had a rectangular, medievalizing form with a simplicity of decoration and a lightness of proportion that some historians have claimed anticipated later Bauhaus furniture, though Voysey was perplexed by such assertions. Voysey's architecture was greatly influenced by Arthur H. Mackmurdo (1851–1942), who, in structures such as his own house and his Liverpool exhibition hall, went beyond Webb in the creation of a style that eliminated almost all reminiscences of English eighteenth-century architecture. A house by Voysey in Bedford Park, London, dated 1891 (fig. 94), is astonishingly original in its rhythmic groups of windows and door openings against broad white areas of unadorned, starkly vertical walls. While Voysey was a product of the English Gothic revival, he looked not to the great cathedrals for inspiration, but to the domestic architecture of rural cottages. In his later country houses he returned, perhaps at the insistence of clients, to suggestions of more traditional Tudor and Georgian forms. He continued to treat these, however, in a ruggedly simple manner in which plain wall masses were lightened and refined by articulated rows of windows.

Glasgow, principally in the person of CHARLES RENNIE MACKINTOSH (1868–1928), was one of the most remarkable centers of architectural experiment at the end of the nineteenth century. Mackintosh's most considerable work of architecture, created at the beginning of his career, is the Glasgow School of Art of 1898–99 and its library addition, built between 1907 and 1909 (fig. 95). The essence of this building is simplicity, clarity, monumentality, and, above all, an organization of interior space that is not only functional but highly expressive of its function. The huge, rectangular studio windows are imbedded in the massive stone façade, creating a balance of solids and voids. The rectangular heaviness of the walls, softened only by an occasional curved masonry element, is lightened by details of fantasy, particularly in the ironwork, that show a relationship to Art Nouveau, but derive in part from the curvilinear forms of medieval Scottish and Celtic art. The library addition is a large, high-ceilinged room with surrounding balconies. Rectangular beams and columns are used to create a series of views that become three-dimensional, geometric abstractions.

The same intricate play with interior vistas seen in this library is apparent in a series of four public tea rooms, also in Glasgow, commissioned by Catherine Cranston. For the last of these, the Willow Tea Rooms (fig. 96), Mackintosh was able to design the entire building. Most details of the interior design, furniture, and wall painting are the products of both Mackintosh and his wife, Margaret Macdonald. They shared with Morris and the followers of the Arts and Crafts movement the notion that all architectural elements, no matter how small, should be integrated into a total aesthetic experi-

96. CHARLES RENNIE MACKINTOSH and MARGARET MACDONALD.
Front salon of Willow Tea Rooms, Glasgow. 1902–4

ence. The stripped-down rectangular partitions and furniture, including Mackintosh's famous ladder-back chairs, and the curvilinear figure designs on the walls illustrate the contrasts that form the spirit of Art Nouveau. The decorations are close to those in paintings of Gustav Klimt and works of the Vienna Secession movement (see figs. 107, 116), and in fact may have influenced them. Mackintosh exhibited with the Vienna Secession in 1900, and his architectural designs, furniture, glass, and enamels were received more favorably there than they had been in London. He died virtually penniless in 1928, and it was not until the 1960s that his reputation was fully revived. Today his furniture and decorative arts are among the most coveted of the modern era.

After 1900, the architectural leadership that England had assumed, particularly through designs for houses and smaller public buildings by Webb, Voysey, and Mackintosh, declined rapidly. In the United States, Frank Lloyd Wright and a few followers carried on their spectacular experiments in detached domestic architecture. However, for many years even Wright himself was given almost no large-scale building commissions. The pioneer efforts of Louis Sullivan and the Chicago School of architects were challenged by the massive triumph of the Beaux-Arts tradition of academic eclecticism, which was to dominate American public building for the next thirty years.

The Chicago School and the Origins of the Skyscraper

Modern architecture may be said to have emerged in the United States with the groundbreaking commercial buildings of H. H. Richardson, an architect who, though eclectic, did not turn to the conventional historicist styles of Gothic pastiche and classical revival. In his Marshall Field Wholesale Store in Chicago (fig. 97), now destroyed, Richardson achieved an effect of monumental mass and stability through the use of graduated rough blocks of reddish Massachusetts sandstone and Missouri granite in a heavily rusticated struc-

ture inspired by Romanesque architecture and fifteenth-century Italian Renaissance palaces. The windows, arranged in diminishing arcaded rows that mirror the gradual narrowing of the masonry wall from ground to roof, are integrated with the interior space rather than being simply holes punched at intervals into the exterior wall. The landmark Marshall Field building was constructed with traditional load-bearing walls, columns, and girders (rather than the steel frames of later skyscrapers), but its elemental design, free of picturesque ornament, influenced the later work of the Chicago School, specifically that of Louis Sullivan and, ultimately (though he would never acknowledge it), Frank Lloyd Wright.

The progressive influence of Richardson's proto-American architecture was counteracted after his death as more and more young American architects studied in Paris in the academic environment of the École des Beaux-Arts (where, in fact, Richardson had trained). The 1893 World's Columbian Exposition in Chicago, a vast and highly organized example of quasi-Roman city planning (fig. 98), was a collaboration among many architects, including McKim, Mead & White, the Chicago firm of BURNHAM AND ROOT, and RICHARD MORRIS HUNT (1827–95), the doyen of American academic architecture. This world's fair celebrated the four-hundredth anniversary of Columbus's voyage. Though modeled on European precedents, the Columbian Exposition set out to assert American ascendancy in industry and particularly the visual arts. Ironically, while international expositions in London and Paris had unveiled such futuristic architectural marvels as the Crystal Palace and the Eiffel Tower, America looked to the classical European past for inspiration. The gleaming white colonnades of the buildings, conceived on an awesome scale and arranged around an immense reflecting pool, formed a model for the future American dream city, one that affected a generation of architects and their clients.

Throughout the nineteenth century, continual expansion and improvement in the production of structural iron and

97. HENRY HOBSON RICHARDSON. Marshall Field Wholesale Store, Chicago. 1885–87. Demolished 1931

98. RICHARD MORRIS HUNT; McKIM, MEAD & WHITE; BURNHAM AND ROOT; and other architects.
World's Columbian Exposition, Chicago. 1893. Demolished

steel permitted raising the height of commercial buildings, a necessity born of growing urban congestion and rising real-estate costs. At the same time, there were sporadic experiments in increasing the scale of windows beyond the dictates of Renaissance palace façades. A vast proportion of Chicago was destroyed in the great fire of 1871—an event that cleared the way for a new type of metropolis fully utilizing the new architectural and urban-planning techniques and materials developed in Europe and America over the previous hundred years.

The Home Insurance Building in Chicago (fig. 99), designed and constructed between 1883 and 1885 by the architect **WILLIAM LE BARON JENNEY** (1832–1907), was only ten stories high, no higher than other proto-skyscrapers already built in New York (the illustration shown here includes the two stories that were added in 1890–91). Its importance rested in the fact that it embodied true skyscraper construction, in which the internal metal skeleton carried the weight of the external masonry shell. This innovation, which was not Jenney's alone, together with the development of the elevator, permitted buildings to rise to great heights and led to the creation of the modern urban landscape of skyscrapers. With metal-frame construction, architects could eliminate load-bearing walls and open up façades, so that a building became

99. WILLIAM LE BARON JENNEY. Home Insurance Building, Chicago. 1884–85. Demolished 1929

Among the many architects attracted to Chicago from the East or from Europe by the opportunities engendered by the great fire was the Dane **DANKMAR ADLER** (1844–1900), an engineering specialist. He was joined in 1879 by a young Boston architect, **LOUIS SULLIVAN** (1856–1924), and in 1887 the firm of Adler and Sullivan hired a twenty-year-old draftsman, Frank Lloyd Wright. The 1889 dedication of the Adler and Sullivan Auditorium helped stimulate a resurgence of architecture in Chicago after the devastations of the fire. And, like Jenney's Home Insurance Building, it emerged under the influence of Richardson's Marshall Field store. The Auditorium included offices, a hotel, and, most significantly, a spectacular new concert hall (color plate 38, page 113), then the largest in the country, with a highly successful acoustical system designed by Adler. The shallow concentric arches in the hall made for a majestic yet intimate space, and Sullivan's use of a sumptuous decoration of natural and geometric forms, richly colored mosaics, and painted panels proved that ornament was one of the architect's great strengths as a designer. By the late 1920s, the Chicago Opera Company had moved to new quarters, but fortunately demolishing the Auditorium during the ensuing years of national depression proved too expensive to carry out.

Adler and Sullivan entered the field of skyscraper construction in 1890 with their Wainwright Building in St. Louis. Then, in the Guaranty Trust Building (now the Prudential Building) in Buffalo, New York, Sullivan designed the first masterpiece of the early skyscrapers (fig. 100). Here, as in the Wainwright Building, the architect attacked the problem of skyscraper design by emphasizing verticality, with the result that the piers separating the windows of the top ten stories are uninterrupted through most of the building's

100. LOUIS SULLIVAN. Guaranty Trust Building (now Prudential Building), Buffalo. 1894–95

a glass box in which solid, supporting elements were, for that time, reduced to a minimum. The concept of the twentieth-century skyscraper as a glass-encased shell framed in a metal grid was here stated for the first time.

101. JOHN AUGUSTUS ROEBLING. Brooklyn Bridge, New York. 1869–83

height. At the same time, he seemed well aware of the peculiar design problem involved in the skyscraper, which is basically a tall building consisting of a large number of superimposed horizontal layers. Thus, Sullivan accentuated the individual layers with ornamented bands under the windows, as well as throughout the attic storey, and crowned the building with a projecting cornice that brings the structure back to the horizontal. This tripartite division of the façade—base, piers, and attic—has frequently been compared to the form of a classical column. Meanwhile, the treatment of the columns on the ground floor emphasizes the openness of the interior space. Above them, the slender piers between the windows soar aloft and then join under the attic storey in graceful arches that tie the main façades together. The oval, recessed windows of the attic, being light and open, blend into the elegant curve of the summit cornice. Sullivan's famed ornament covers the upper part of the structure in a light, overall pattern, helping to unify the façade while also emphasizing the nature of the terra-cotta sheathing over the metal skeleton as a weightless, decorative surface, rather than a weight-bearing element. "It must be every inch a proud and soaring thing," said Sullivan of the tall office building, "rising in sheer exaltation that from bottom to top as a unit without a single dissenting line." In theory, Sullivan felt that a building's interior function should determine its exterior form, hence his famous phrase, "form follows function." But practical experience taught him that function and structure did not always generate the most appealing forms.

Although the Chicago School of architecture maintained its vitality into the first decade of the twentieth century, neo-academic styles were also firmly established, with New York architects leading the way. The Gothic-revival style could be applied to structures other than buildings, such as the Brooklyn Bridge (1869–83), which, as one of the country's most spectacular suspension bridges, represented a triumph of American technology (fig. 101). It was designed by **JOHN AUGUSTUS ROEBLING** (1806–69), who barely lived to see work on the project begun (his son Washington completed the work with his wife, Emily). Before construction of the bridge, the East River was crossed by up to a thousand ferry trips a day. With its soaring Gothic towers and graceful steel suspension cables, the Brooklyn Bridge forms a dramatic gateway to New York City. No wonder numbers of American painters and photographers were drawn to its grandeur as a

102. CASS GILBERT. Woolworth Building, New York. 1911–13

pictorial motif. By 1913 the Woolworth Building (fig. 102), designed by **CASS GILBERT** (1859–1934), loomed 792 feet above the streets of New York—fifty-two stories of Late Gothic-style stone sheathing serving mainly to add weight, disguise the metal-case construction frame, and create a sense of soaring verticality. Until 1931, the Woolworth Building was New York's tallest building, a Gothic cathedral of the modern age and a shining symbol of American capitalism. It has since been overshadowed by later and taller buildings, but remains one of the most distinguished structures on the New York skyline.

5 Art Nouveau

As seen in the section on Post-Impressionism, and the trend toward functional design in late nineteenth-century architecture, the fin de siècle was a period of synthesis in the arts, a time when artists sought new directions that in themselves constituted a reaction against the tide of "progress" represented by industrialization. At the same time, the architects discussed in the last chapter recognized that new patterns of life in the industrial age called for new types of buildings—bridges, railroad stations, and skyscrapers. The great innovators, such as Richardson, Sullivan, or Mackintosh, were able to draw selectively upon older styles without resorting to academic pastiche, and to create from the past something new and authentic for the present. While these architects did not abandon tradition, their work constituted a powerful countercurrent to the hegemony of Beaux-Arts historicism. At the same time, the search of the Symbolist poets, painters, and musicians for spiritual values was part of this reaction. Gauguin used the term "synthetism" to characterize the liberating color and linear explorations that he pursued and transmitted to disciples. The phenomenal thing about this synthesizing spirit is that in the last decades of the nineteenth century and the first of the twentieth it became a great popular movement that affected the taste of every part of the population in both Europe and the United States. This was the movement called Art Nouveau, a French term meaning simply "new art." Art Nouveau was a definable style that emerged from the experiments of painters, architects, craftspeople, and designers and for a decade permeated not only painting, sculpture, and architecture, but also graphic design, magazine and book illustration, furniture, textiles, glass and ceramic wares, jewelry, and even clothing. As the name implies, Art Nouveau represented a rejection of historical revivalism and a conscious search for new and genuine forms that could express the modern age—which is not to say that it was without formal precedent.

Art Nouveau grew out of the English Arts and Crafts movement, whose chief exponent and propagandist, as we saw earlier, was the artist and poet **WILLIAM MORRIS** (1834–96). The Arts and Crafts movement came forth as a revolt against

the new age of mechanization, a Romantic effort on the part of Morris and others to implement the philosophy of the influential critic John Ruskin, who stated that true art should be both beautiful and useful and should base its forms on those found in nature (fig. 103). They saw the world of the artist-craftsperson in the process of destruction by mass production, and they fought for a return to some of the standards of simplicity, beauty, and craftsmanship that they associated with earlier centuries, notably the Middle Ages. Ironically for these social reformers, it was only the wealthy, in the end, who could afford their handcrafted goods. The ideas of the Arts and Crafts movement spread rapidly throughout Europe and

103. WILLIAM MORRIS. Detail of *Pimpernel* wallpaper. 1876. Victoria and Albert Museum, London

104. WILLIAM BLAKE. *The Great Red Dragon and the Woman Clothed with the Sun.* 1805–10. Watercolor, 16⅛ x 13⅛" (41 x 33.5 cm). National Gallery of Art, Washington, D.C.
ROSENWALD COLLECTION

ered most of the sources for their decoration in nature, especially in plant forms, often given a symbolic or sensuous overtone, or in microorganisms that the new explorations in botany and zoology were making familiar.

In the later works of Seurat and his followers, as we have seen, line took on an abstract, expressive function. The formal innovations of Gauguin and of the Nabis were rooted in linear as well as color pattern. Toulouse-Lautrec reflected (and influenced) the Art Nouveau spirit in his paintings and prints, particularly in his subtle use of descriptive, expressive line. Through his highly popular posters (color plate 37, page 80), aspects of Art Nouveau graphic design were spread throughout the Western world.

While Art Nouveau in architecture was largely a phenomenon on the continent, there were many precursors of the movement in England, particularly in painting. The mystical visions of **WILLIAM BLAKE** (1757–1827; fig. 104), expressed in fantastic rhythms of line and color, were an obvious ancestor, as were the late, highly decorative paintings of the Pre-Raphaelite artist Dante Gabriel Rossetti. The Aesthetic Movement of the 1870s and 1880s, with its emphasis on beautiful but useful objects and a decorative style largely shaped by the reigning craze for *japonisme,* was an important forerunner of Art Nouveau. A masterpiece of the period is James Abbott McNeill Whistler's famous Peacock Room from 1876–77 (color plate 39, page 113), made in collaboration with the architect Thomas Jeckyll (1827–81) for the London house of the shipping magnate Frederick Richards Leyland. Designed to showcase Leyland's Chinese porcelain collection, the Peacock Room was eclectic in style, as was typical of the Aesthetic Movement, but it is generally suffused with an exotic, Oriental flavor. Like his exact contemporary William Morris (and perhaps in response to him), Whistler, known for his paintings and prints (color plate 8, page 35; see fig. 45), wished to prove his strength as a designer of an integrated decorative environment. The result is a shimmering interior of dark greenish-blue walls embellished with circular abstract patterns and golden peacock motifs, which Whistler called *Harmony in Blue and Gold.* On the south wall of the room he depicted two fighting peacocks, one rich and one poor, that slyly alluded to a protracted quarrel between the artist and his patron. Leyland, who refused to pay Whistler for a great deal of work on the room that he had never commissioned, finally banned the irascible artist from the house after he invited the public and press to view his creation without Leyland's permission. The room was later acquired intact by an American collector and is now housed in the Freer Gallery of Art in Washington, D.C.

The drawings of the young English artist **AUBREY BEARDSLEY** (1872–98) were immensely elaborate in their black-and-white stylization and decorative richness (fig. 105). Associated with the so-termed "aesthetic" or "decadent" literature of Oscar Wilde and other fin-de-siècle writers, they were admired for their beauty and condemned for their sexual content. They probably constitute the most significant contribution of the English to Art Nouveau graphic art.

found support in the comparable theories of French, German, Belgian, and Austrian artists and writers. Implicit in the movement was the concept of the synthesis of the arts based on an aesthetic of dynamic linear movement. Many names were given to this phenomenon in its various manifestations, but ultimately Art Nouveau became the most generally accepted, probably through its use by the Parisian shop and gallery called the Maison de l'Art Nouveau, which, with its exhibitions and extensive commissions to artists and artisans, was influential in propagating the style. In Germany the term *Jugendstil,* or "youth style," after the periodical *Jugend,* became the name.

In 1890, it must be remembered, most accepted painting, sculpture, and architecture was still being produced in enormous quantity under the aegis of the academies. In contrast to the Arts and Crafts movement, Art Nouveau made use of new materials and machine technologies both in buildings and in decoration. In actuality, Art Nouveau artists could not avoid the influence of past styles, but they explored those that were less well known and out of fashion with the current academicians. They derived from medieval or Asian art forms or devices congenial to their search for an abstraction based on linear rhythms. Thus the linear qualities and decorative synthesis of eighteenth-century Rococo; the wonderful linear interlace of Celtic and Saxon illumination and jewelry; the bold, flat patterns of Japanese paintings and prints (color plate 7, page 35); the ornate motifs of Chinese and Japanese ceramics and jades—all were plumbed for ideas. The Art Nouveau artists, while seeking a kind of abstraction, discov-

105. AUBREY BEARDSLEY. *Salome with the Head of John the Baptist.* 1893. India ink and watercolor, 10⅞ x 5¾" (27.6 x 14.6 cm). Princeton University Library, Special Collection/Department of Rare Books, New Jersey

Beardsley, like Edgar Allan Poe and certain of the French Symbolists, was haunted by Romantic visions of evil, of the erotic, and the decadent. In watered-down imitations, his style in drawing appeared all over Europe and America in popularizations of Art Nouveau book illustrations and posters. Beardsley based one of his illustrations for Wilde's 1894 poem *Salome* on this drawing, in which the stepdaughter of Herod holds up the severed head of Saint John the Baptist in a gesture of grotesque eroticism.

106. HENRY VAN DE VELDE. *Tropon.* c. 1899. Advertising poster, 47⅝ x 29⅛" (121 x 74 cm). Museum für Kunst und Gewerbe, Hamburg

One of the most influential figures in all the phases of Art Nouveau was the Belgian HENRY VAN DE VELDE (1863–1957). Trained as a painter, he eventually produced abstract compositions in typical Art Nouveau formulas of color patterns and sinuous lines (fig. 106). For a time he was interested in Impressionism and read widely in the scientific theory of color and perception. He soon abandoned this direction in favor of the Symbolism of Gauguin and his school and attempted to push his experiments in symbolic statement through abstract color expression further than had any of his contemporaries. Ultimately, Van de Velde came to believe that easel painting was a dead end and that the solution for contemporary society was to be found in the industrial arts. Though significantly influenced by Morris's theories, unlike the Englishman, he regarded the machine as a potentially positive agent that could "one day bring forth beautiful products." It was finally as an architect and designer that he made his major contributions to Art Nouveau and the origins of twentieth-century art.

GUSTAV KLIMT (1862–1918). In Austria, the new ideas of Art Nouveau were given expression in the founding in 1897 of the movement known as the Vienna Secession and, shortly thereafter, in its publication, *Ver Sacrum*. This diverse group was so named because it seceded from Vienna's conservative exhibiting society, the Künstlerhaus, and opposed the intolerance for new, antinaturalist styles at the Academy of Fine Arts. The major figure of the Secession was Gustav

107. GUSTAV KLIMT. Detail of dining-room mural, Palais Stoclet, Brussels. c. 1905–8. Mosaic and enamel on marble

Klimt, in many ways the most complete and most talented exponent of pure Art Nouveau style in painting. Klimt was well established as a successful decorative painter and fashionable portraitist, noted for the brilliance of his draftsmanship, when he began in the 1890s to be drawn into the stream of new European experiments. He became conscious of Dutch Symbolists such as Jan Toorop, of the Swiss Symbolist Ferdinand Hodler (fig. 122), of the English Pre-Raphaelites and of Aubrey Beardsley (color plate 5, page 34; see figs. 42, 105). His style was also formed on a study of Byzantine mosaics. A passion for erotic themes led him not only to the creation of innumerable drawings, sensitive and explicit, but also to the development of a painting style that integrated sensuous nude figures with brilliantly colored decorative patterns of a richness rarely equaled in the history of modern art. He was drawn more and more to mural painting, and his murals for the University of Vienna involved elaborate and complicated symbolic statements composed of voluptuous figures floating through an amorphous, atmospheric limbo. Those he created for the Palais Stoclet, designed by Josef Hoffmann, in Brussels between 1905 and 1910 (figs. 107, 116, 117) are executed in glass, enamel, metal, and semiprecious stones. They combined figures conceived as flat patterns (except for modeled heads and hands) with an overall pattern of abstract spirals—a glittering complex of volumetric forms embedded in a mosaic of jeweled and gilded pattern (color plate 40, page 113). Although essentially decorative in their total effect, the Stoclet murals mark the moment when modern painting was on the very edge of nonrepresentation. Through his relationship with the younger Austrian artist Egon Schiele, as well as through his own obsessive, somewhat morbid and febrile eroticism, Klimt also occupies a central place in the story of Austrian Expressionism.

Art Nouveau Architecture and Design

In architecture it is difficult to speak of a single, unified Art Nouveau style except in the realms of surface ornament and interior decoration. Despite this fact, certain aspects of architecture derive from Art Nouveau graphic art and decorative or applied arts, notably the use of the whiplash line in ornament and a generally curvilinear emphasis in decorative and even structural elements. The Art Nouveau spirit of imaginative invention, linear and spatial flow, and nontraditionalism fed the inspiration of a number of architects in continental Europe and the United States and enabled them to experiment more freely with ideas opened up to them by the use of metal, glass, and reinforced-concrete construction. Since the concept of Art Nouveau involved a high degree of specialized design and craftsmanship, it did not lend itself to the developing field of large-scale mass construction. However, it did contribute substantially to the outlook that was to lead to the rise of a new and experimental architecture in the early twentieth century.

The architectural ornament of Louis Sullivan in the Guaranty Trust Building (see fig. 100) or the Carson Pirie Scott store (fig. 108) was the principal American manifestation of the Art Nouveau spirit in architecture, and a comparable spirit permeated the early work of Sullivan's great disciple, Frank Lloyd Wright (see fig. 264). The outstanding American name in Art Nouveau was that of Louis Comfort Tiffany, but his expression lay in the fields of interior design and decorative arts (color plate 41, page 114). In these he was not only in close touch with the European movements but himself exercised a considerable influence on them. One of Art Nouveau's greatest contributions lies in the field of graphic design—posters, advertisements, and book illustration—and in the design of lamps, textiles, furniture, ceramics, glass, and wallpaper.

ANTONI GAUDÍ (1852–1926). In Spain, Antoni Gaudí was influenced as a student by a Romantic and Symbolist concept of the Middle Ages as a Golden Age, which for him and other Spanish artists became a symbol for the rising nationalism of Catalonia. Also implicit in Gaudí's architecture was his early study of natural forms as a spiritual basis for architecture. He was drawn to the work of the influential French architect and theorist Eugène Emmanuel Viollet-le-Duc (1814–79). The latter was a leading proponent of the

108. LOUIS SULLIVAN. Detail of cast-iron ornament on the facade of the Carson Pirie Scott and Company department store, Chicago. c. 1903–4

109. ANTONI GAUDÍ. Church of the Sagrada Familia, Barcelona.
1883–1926

Gothic revival and a passionate restorer of medieval buildings who analyzed Gothic architectural structure in the light of modern technical advances. His writings on architecture, notably his *Entretiens,* which appeared in French, English, and American editions in the 1860s and 1870s, were widely read by architects. His bold recommendations on the use of direct metal construction influenced not only Gaudí, but a host of other experimental architects at the end of the nineteenth century. While Gaudí's early architecture belonged in part to the main current of Gothic revival, it involved a highly idiosyncratic use of materials, particularly in textural and coloristic arrangement, and an even more imaginative personal style in ornamental ironwork. His wrought-iron designs were arrived at independently and frequently in advance of the comparable experiments of mainstream Art Nouveau. Throughout his later career, from the late 1880s until his death in 1926, Gaudí followed his own direction, which at first was parallel to Art Nouveau and later was independent of anything that was being done in the world or would be done until the middle of the twentieth century, when his work was reassessed.

Gaudí's first major commission was to complete the Church of the Sagrada Familia in Barcelona, already begun

as a neo-Gothic structure by the architect Francisco de Villar (fig. 109). Gaudí worked intermittently on this church from 1883 until his death, leaving it far from complete even then. The main parts of the completed church, particularly the four great spires of one transept, are only remotely Gothic, and although influences of Moorish and other architectural styles may be traced, the principal effect is of a building without historic style—or rather one that expresses the imagination of the architect in the most personal and powerful sense. In the decoration of the church emerges a profusion of fantasy in biological ornament flowing into naturalistic figuration and abstract decoration. Brightly colored mosaic embellishes the finials of the spires. These forms were not arbitrary; they were tied to Gaudí's structural principles and his often hermetic language of symbolic form, informed by his spiritual beliefs. To complete the church according to the architect's plans, an enormous amount of construction has been undertaken since his death and continues today.

At the Güell Park in Barcelona (also unfinished), Gaudí produced a demonstration of sheer fantasy and engineering ingenuity, a Surrealist combination of landscaping and urban planning. This gigantic descendant of the Romantic, Gothic-style English gardens of the eighteenth century is a mélange of sinuously curving walls and benches, grottos, porticos, arcades, all covered with brilliant mosaics of broken pottery and glass (color plate 42, page 114). In it can be observed the strongly sculptural quality of Gaudí's architecture, a quality that differentiates it from most aspects of Art Nouveau. He composed in terms not of lines but of twisted masses of sculpturally conceived masonry. Even his ironwork has a sculptural heaviness that transcends the usually attenuated elegance of Art Nouveau line.

Gaudí designed the Casa Milá apartment house, a large structure freestanding around open courts, as a whole continuous movement of sculptural volumes. The façade, flowing around the two main elevations, is an alternation of void and sculptural mass. The undulating roof lines and the elaborately twisted chimney pots carry through the unified sculptural theme. Ironwork grows over the balconies like luscious exotic vegetation. The sense of organic growth continues in

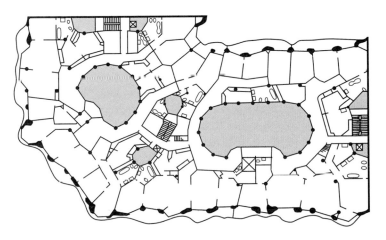

110. ANTONI GAUDÍ. Typical floor plan,
Casa Milá apartment house, Barcelona. 1905–7

111. VICTOR HORTA. Stairwell of interior, Tassel House, Brussels. 1892–93

112. VICTOR HORTA. Auditorium, Maison du Peuple, Brussels. 1897–99. Demolished 1964

the floor plan (fig. 110), where one room or corridor flows without interruption into another. Walls are curved or angled throughout, to create a feeling of everlasting change, of space without end.

In his concept of architecture as dynamic space joining the interior and external worlds and as living organism growing in a natural environment, in his daring engineering experiments, in his imaginative use of materials—from stone that looks like a natural rock formation to the most wildly abstract color organizations of ceramic mosaic—Gaudí was a visionary and a great pioneer.

VICTOR HORTA (1861–1947). If any architect might claim to be the founder of Art Nouveau architecture, it is the Belgian Victor Horta. Trained as an academician, he was inspired by Baroque and Rococo concepts of linear movement in space, by his study of plant growth and of Viollet-le-Duc's structural theories, and by the engineer Gustave Eiffel.

The first important commission carried out by Horta was the house of a Professor Tassel in Brussels, where he substantially advanced Viollet-le-Duc's structural theories. The stair hall (fig. 111) is an integrated harmony of linear rhythms, established in the balustrades of ornamental iron, the whiplash curves atop the capitals, the arabesque designs

on the walls and floor, and the winding steps. Line triumphs over sculptured mass as a multitude of fanciful, tendrillike elements blend into an organic whole that boldly exposes the supporting metal structure. In Horta's Maison du Peuple, built in 1897–99 as the headquarters of the Belgian Socialist Party, and now destroyed, the architect worked for a clientele that departed from the usual upper-middle-class patrons of his houses. The façade was a curtain wall of glass supported on a minimum of metal structural frames and wrapped around the irregularly curving edge of an open plaza. The auditorium interior (fig. 112) became an effective glass enclosure in which the angled, exposed frame girders formed articulated supports for the double curve of the ceiling. The interior was given variety and interest by the combination of vertical glass walls, angled metal struts, and an open, curving ceiling, and in general achieved a harmonious unity between ornament and articulated structure.

Many of the chief examples of Art Nouveau architecture are to be found in the designs of department stores and similar commercial buildings. The large-scale department store was a characteristic development of the later nineteenth century, superseding the older type of small shop and enclosing the still older form of the bazaar. Thus it was a form of building without traditions, and its functions lent themselves to an architecture that emphasized openness and spatial flow as well as ornate decorative backgrounds. Horta's department store in Brussels, À l'Innovation (fig. 113), made of the façade a display piece of glass and curvilinear metal supports. Such department stores, like Sullivan's Carson Pirie Scott store in Chicago (1899–1904), sprang up all over Europe and America at the beginning of the twentieth century. Their utilitarian purpose made them appropriate embodiments of the new discoveries in mass construction: exposed-metal and glass structure and decorative tracery. Horta's design was notable for its expression of the interior on the glass skin of the exterior, so that the three floors, with a tall central "nave" and flanking, four-story side aisles, are articulated in the façade.

113. VICTOR HORTA. À l'Innovation department store, Brussels.
1901. Destroyed by fire 1967

114. HECTOR GUIMARD. Entrance to the Porte Dauphine
Métropolitain station, Paris. 1901

Many distinguished architects were associated with Art Nouveau in one context or another, but few of their works can be identified with the style to the same degree as Horta's. The stations designed by **HECTOR GUIMARD** (1867–1925) for the Paris Métropolitain (subway) can be considered pure Art Nouveau, perhaps because they were less architectural structures than decorative signs or symbols. Guimard's Métro stations were constructed for the enormous Exposition Universelle in 1900 out of prefabricated parts of cast iron and glass. At odds with the prevailing taste for classicism, they created a sensation in Paris, causing one critic to compare them, perhaps not inappropriately, to a dragonfly's wings. This entrance for the Porte Dauphine station (fig. 114) is a rare example of an original Guimard canopy still intact.

If one were to remove the elaborate "sea horse" ornament from the façade of the Atelier Elvira by **AUGUST ENDELL** (1871–1925) in Munich (fig. 115), a relatively simple and austere structure would remain. However, details of the interior, notably the stair hall, did continue the delicate undulations of Art Nouveau. The Palais Stoclet in Brussels (figs. 116, 117) by the Austrian architect **JOSEF HOFFMANN** (1870–1955), is a flat-walled, rectangular structure, although the dining-room murals by Klimt (fig. 107) and the interior furnishings and decorations represent a typical Art Nouveau synthesis of decorative accessories. Hoffmann's starkly ab-

stract language is closer to the elegant geometry of the Viennese Jugendstil school than the flourishes of the Belgian Art Nouveau, and really belongs in the context of works such as Otto Wagner's Postal Savings Bank in Vienna and the architecture of Peter Behrens, Adolf Loos, Joseph Maria Olbrich, and van de Velde, all discussed in chapter twelve (see figs. 277, 281, 278, 279, 282).

115. AUGUST ENDELL. Atelier Elvira, Munich. 1897

116. JOSEF HOFFMANN. Dining room, Palais Stoclet, Brussels. 1905–11

117. JOSEF HOFFMANN. Palais Stoclet, Brussels. 1905–11

Toward Expressionism: Late Nineteenth-Century Painting beyond France

In retrospect it is apparent that Jugendstil was as important for the new ideas it evoked in painters of the era as for the immediate impetus given architects and designers working directly in an Art Nouveau style. The Norwegian Edvard Munch, who had a sizable impact on German Expressionism; the Swiss Ferdinand Hodler; the Belgian James Ensor, a progenitor of Surrealism; and the Russian Vasily Kandinsky, one of the first abstract artists—all grew up in the environment of Jugendstil or Art Nouveau. Although some of the artists in this section formed their highly individual styles in response to advanced French art, they also drew extensively from their own local artistic traditions.

EDVARD MUNCH (1863–1944). By 1880, at the time Edvard Munch began to study painting seriously, Oslo (then Kristiania), Norway, had a number of accomplished painters and a degree of patronage for their works. But the tradition was largely academic, rooted in the French Romantic Realism of the Barbizon School and in German lyrical naturalism, in

part because Norwegian painters usually trained in Germany. In Kristiania, Munch was part of a radical group of bohemian writers and painters who worked in a naturalist mode.

Thanks to scholarships granted by the Norwegian government, Munch lived intermittently in Paris between 1889 and 1892, and he had already spent three weeks in the French capital in 1885. Though it is not clear what art he saw while in France, the Post-Impressionist works he was bound to have encountered surely struck a sympathetic chord with his incipient Symbolist tendencies. In 1892 his reputation had grown to the point where he was invited to exhibit at the Verein Berliner Künstler (Society of Berlin Artists). His retrospective drew such a storm of criticism that the members of the Society voted to close it after less than a week. Sympathetic artists, led by Max Liebermann, left the Society and formed the Berlin Secession. This recognition—and controversy—encouraged Munch to settle in Germany, where he spent most of his time until 1908.

In evaluating Munch's place in European painting it could be argued that he was formed not so much by his Norwegian origins as by his exposure first to Paris during one of the most exciting periods in the history of French painting, and then to Germany at the moment when a new and dynamic art was emerging. The singular personal quality of his paintings and prints, however, is unquestionably a result of an intensely literary and even mystical approach that was peculiarly Scandinavian, intensified by his own tortured psyche. The painter moved in literary circles in Norway as well as in Paris and Berlin, was a friend of noted writers, among them the playwright August Strindberg, and in 1906 made stage designs for Ibsen's plays *Ghosts* and *Hedda Gabler*.

Although he lived to be eighty years old, the specters of

118. EDVARD MUNCH. *The Sick Child*. 1885–86. Oil on canvas, 47 x 46⅝" (119.4 x 118.4 cm). Nasjonalgalleriet, Oslo

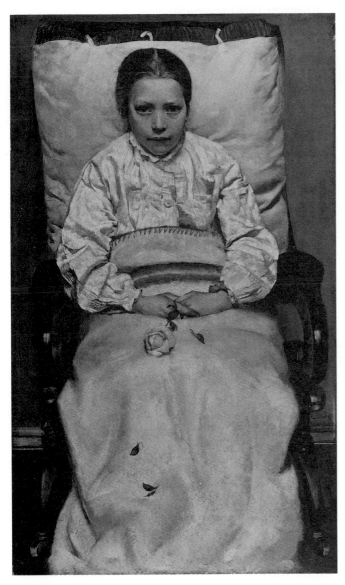

119. CHRISTIAN KROHG. *The Sick Girl.* 1880–81. Oil on wood, 40⅛ x 22⅞" (101.9 x 58.1 cm). Nasjonalgalleriet, Oslo

sickness and death hovered over Munch through much of his life. His mother and sister died of tuberculosis while he was still young. His younger brother succumbed in 1895, five years after the death of their father, and Munch himself suffered from serious illnesses. Sickness and death began to appear early in his painting and recurred continually. The subject of *The Sick Child,* related to the illness and death of his sister (fig. 118), haunted him for years. He reworked the painting over and over, and made prints of the subject as well as two later versions of the painting.

Munch's mentor, the leading Norwegian naturalist painter **CHRISTIAN KROHG** (1852–1925), had earlier explored a similar theme, common in the art and literature of the period, but did so with an astonishingly direct and exacting style (fig. 119). This sitter confronts the viewer head-on, her pale skin and white gown stark against the white pillow. Krohg, who had also watched his sister die of tuberculosis, avoided the usual sentimentality or latent eroticism that normally pervaded contemporary treatments of such subjects. While his version could not have predicted the melodrama or painter-

ly handling of Munch's painting, the connection between the two canvases is obvious, though Munch, typically, denied the possibility of influence.

Munch was enormously prolific, and throughout his life experimented with many different themes, palettes, and styles of drawing. The works of the 1890s and the early 1900s, in which the symbolic content is most explicit, are characterized by the sinuous, constantly moving, curving line of Art Nouveau, combined with color dark in hue but brilliant in intensity. Influences from Gauguin and the Nabis are present in his work of this period. The anxieties that plagued Munch were frequently given a more general and ambiguous, though no less frightening, expression. *The Scream* (fig. 120) is an agonized shriek translated into bands of color that echo like sound waves across the landscape and through the blood-red sky. The image has its source in Munch's experience. As he walked across a bridge with friends at sunset, he was seized with despair and "felt a great, infinite scream pass through nature." *The Scream,* the artist's best-known and most exploited image, has become a familiar symbol of modern anxiety and alienation.

Munch was emerging as a painter at the moment when Sigmund Freud was developing his theories of psychoanalysis, and the painter, in his obsessions with sex, death, and woman as a destructive force, at times seemed almost a classic example of the problems Freud was exploring. In painting after painting, *The Dance of Life* among them (color plate

120. EDVARD MUNCH. *The Scream.* 1893. Oil and tempera on board, 35¾ x 29" (90.8 x 73.7 cm). Nasjonalgalleriet, Oslo

121. EDVARD MUNCH.
Vampire. 1902.
Woodcut and lithograph,
from an 1895 woodcut,
14⅞ x 21½" (38 x 54.6
cm). Munch-Museet, Oslo

43, page 114), Munch transformed the moon and its long reflection on the water into a phallic symbol. *The Dance of Life* belongs to a large series Munch called The Frieze of Life, which contained most of his major paintings (including *The Scream)*, addressing the central themes of love, sexual anxiety, and death. The subject of this painting, based on Munch's troubled personal history of love and rejection, was in step with international trends in Symbolist art.

Munch was one of the major graphic artists of the twentieth century, and like Gauguin, he took a very experimental approach to printmaking and contributed to a revival of the woodcut medium. He began making etchings and lithographs in 1894 and for a time was principally interested in using the media to restudy subjects he had painted earlier. The prints, however, were never merely reproductions of the paintings. In each case he reworked the theme in terms of the new medium, sometimes executing the same subject

in drypoint, woodcut, and lithography, and varying each of these as each was modified from the painting. For *Vampire* (fig. 121), Munch used an ingenious method he had invented for bypassing the tedious printmaking process of color registration, which involves inking separate areas of the woodblock for each color and painstakingly aligning the sheet for every pass (one for each color) through the press. Munch simply sawed his block into sections, inked each section separately, and assembled them like a puzzle for a single printing. In *Vampire,* he actually combined this technique with lithography to make a "combination print." Through these innovative techniques, Munch gained rich graphic effects, such as subtly textured patterns (by exploiting the rough grain of the woodblock), and translucent passages of color. His imagery, already explored in several drawings and paintings made in Berlin, is typically ambiguous. While the redheaded woman could be embracing her lover, the omi-

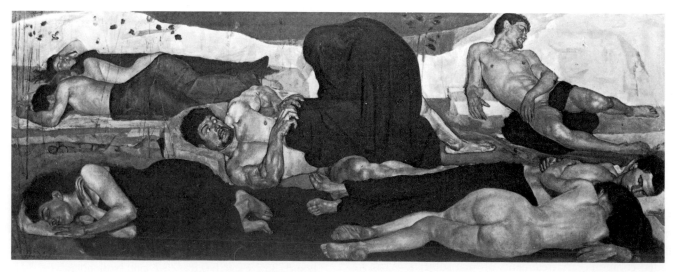

122. FERDINAND HODLER. *Night*. 1889–90. Oil on canvas, 45¾" x 9'9¾" (116.2 cm x 3 m). Kunstmuseum, Bern

nous shadow that looms above them and the title of the work imply something far more sinister.

FERDINAND HODLER (1853–1918). The Swiss artist Ferdinand Hodler too knew the ravages of sickness and death, and, like Van Gogh, had undergone a moral crisis in the wake of deep religious commitment. In his paintings, he avoided the turbulent drama of earlier Romantic expressions and sought instead frozen stillness, spareness, and purity, the better to evoke some sense of the unity within the apparent confusion and complexity of the empirical world. This can be discerned in the famous *Night* (fig. 122), where the surfaces and volumes of nature have been described in exacting detail, controlled by an equally intense concern for abstract pattern, a kind of rigorously balanced, friezelike organization the artist called "parallelism." As in Munch's *The Scream*, the imagery in *Night* was an expression of Hodler's own obsessive fears. The central figure, who awakens to find a shrouded phantom crouched above him, is a self-portrait, as is the man at the upper right, while the women in the foreground are depictions of his mistress and ex-wife. The painting was banned from an exhibition in Geneva in 1891 as indecent, but encountered a better reception when it was later shown in Paris.

JAMES ENSOR (1860–1949), whose long life was almost exactly contemporary with Munch's, differed from the Norwegian in being the inheritor of one of the great traditions in European·art—that of Flemish and Dutch painting. The consciousness of this heritage even compelled him to assume from time to time, though half satirically, the appearance of Rubens (fig. 124), and it is difficult to believe that he could have conceived his pictorial fantasies without knowledge of the late-medieval works of Hieronymus Bosch (despite his denials) and Pieter Brueghel the Elder. Outside of a three-year period of study at the Royal Academy of Fine Arts in Brussels, Ensor spent his entire life in his native Ostend, on the coast of Belgium, where his Flemish mother and her relatives kept a souvenir shop filled with toys, seashells, model ships in bottles, crockery, and, above all, the grotesque masks popular at Flemish carnivals.

By 1881, Ensor could produce such accomplished traditional Realist-Romantic paintings that he was accepted in the Brussels Salon and by 1882 in the Paris Salon. His approach already was changing, however, with the introduction of brutal or mocking subject matter, harshly authentic pictures of miserable, drunken tramps and inexplicable meetings of masked figures. *Scandalized Masks* (fig. 123), for instance, shows simply a familiar setting, the corner of a room, with a masked figure sitting at a table. Another masked figure enters through the open door. The masked figures obviously derive from the Flemish carnivals of popular tradition, but here, isolated in their grim surroundings, they cease to be merely masked mummers. The mask becomes the reality, and we feel involved in some communion of monsters. This troubling and fantastic picture reveals intensification of the artist's inner mood of disturbance. In any event, the new works did not please the judges of the Brussels Salon, and in 1884 all Ensor's entries were rejected.

123. JAMES ENSOR. *Scandalized Masks.* 1883. Oil on canvas, 53 x 45" (134.6 x 114.3 cm). Musées Royaux des Beaux-Arts de Belgique, Brussels

Ensor's mature paintings still, decades later, have the capacity to shock. He was something of an eccentric and much affected by nineteenth-century Romanticism and Symbolism, evinced in his passionate devotion to the tales of Poe and Balzac, the poems of Baudelaire, and the work of several contemporary Belgian writers. He searched the tradition of fantastic painting and graphic art for inspiration—from the tormented visions of Matthias Grünewald and Bosch to Goya. Ensor was further affected by Daumier and his social satire, the book illustrator Gustave Doré, Redon, and the decadent-erotic imagery of the Belgian artist Félicien Rops.

During the late 1880s Ensor turned to religious subjects, frequently the torments of Christ. They are not interpreted in a narrowly religious sense, but are rather a personal revulsion for a world of inhumanity that nauseated him. This feeling was given its fullest expression in the most important painting of his career, *The Entry of Christ into Brussels in 1889* (color plate 11, page 115), a work from 1888–89 that depicts the Passion of Christ as the center of a contemporary Flemish kermis, or carnival, symptomatic of the indifference, stupidity, and venality of the modern world. Ensor describes this grotesque tumult of humanity with dissonant color, compressing the crowd into a vast yet claustrophobic space. Given the enormous size of *The Entry of Christ* (it is over twelve feet wide), it has been suggested that Ensor painted it in response to Seurat's *La Grande Jatte* (color plate 19, page 39), which had been much heralded at an exhibition in Brussels in 1887. While the French painting was a modern celebration of middle-class life, Ensor's was an indictment of it.

The figure of Christ, barely visible in the back of the crowd, was probably based on the artist's own likeness. This conceit—the artist as persecuted martyr—was being exploited by Gauguin in Brittany at virtually the same time. By extension, Ensor's hideous crowd refers to the ignorant citizens of Ostend, who greeted his work with incomprehension.

Even members of a more liberal exhibiting group of Brussels, called L'Essor, were somewhat uneasy about Ensor's new paintings. This led in 1884 to a withdrawal of some members, including Ensor, to form the progressive society *Les XX* (The Twenty). For many years the society was to support aspects of the new art in Brussels. Although it exhibited Manet, Seurat, Van Gogh, Gauguin, and Toulouse-Lautrec, and exerted an influence in the spread of Neo-Impressionism and Art Nouveau, hostility to Ensor's increasingly fantastic paintings grew even there. *The Entry of Christ* was refused admission and was never publicly exhibited before 1929. The artist himself was saved from expulsion from the group only by his own vote.

During the 1890s Ensor focused much of his talent for invective on the antagonists of his paintings, frequently with devastating results. This decade saw some of his most brilliant and harrowing works. The masks reappeared at regular intervals, occasionally becoming particularly personal, as in *Portrait of the Artist Surrounded by Masks* (fig. 124), a painting in which he portrayed himself in the manner of Rubens's self-portraits—with debonair mustache and beard, gay, plumed hat, and piercing glance directed at the spectator. The work is reminiscent of an earlier 1883 self-portrait by Ensor and, in a

126. MAX KLINGER. *The Abduction,* from the portfolio *A Glove.* 1881. Etching, printed in black plate,
4¹¹⁄₁₆ x 10⁹⁄₁₆" (11.9 x 26.82 cm). The Museum of Modern Art, New York

curious way, of a painting by Hieronymus Bosch of Christ surrounded by his tormentors (here, Ensor's critics): the personification of Good, isolated by Evil but never overwhelmed.

The ancestors of twentieth-century Surrealism can also be found in a trio of artists who emerged in German-speaking Europe during the Symbolist era: Böcklin, Klinger, and Kubin. The eldest of these was **ARNOLD BÖCKLIN** (1827–1901), born in Basel, educated in Düsseldorf and Geneva, and then resident in Italy until the end of his life. Like so many others within contemporary German culture, Böcklin fell prey to the classical dream, which prompted him to paint, with brutal, almost grotesque academic Realism, such mythical beings as centaurs and mermaids writhing in battle, or in some sexual contest between a hapless, depleted male and an exultant femme fatale. One of his most haunting images, especially for later painters like Giorgio de Chirico and the Surrealists, came to be called *Island of the Dead* (fig. 125), a scene depicting no known reality but universally appealing to the late Romantic and Symbolist imagination. With its uncanny stillness, its ghostly white-cowled figure, and its eerie moonlight illuminating rocks and the entrances to tombs against the deep, mysteriously resonant blues and greens of sky, water, and tall, melancholy cypresses, the picture provided inspiration not only for subsequent painters but also for numerous poets and composers.

The German artist **MAX KLINGER** (1857–1920) paid homage to Böcklin by making prints after several of his paintings, including *Island of the Dead.* Klinger is best known for a ten-part cycle of prints, entitled *A Glove,* whose sense of fantasy many a Surrealist might envy. As the etching seen here suggests (fig. 126), Klinger may have found a kindred spirit in the dark, sinister imaginings of Goya. The print in question, entitled *Abduction,* represents an episode in the nightmare of a man—the artist himself—fetishistically obsessed with a glove. After the Czech artist and illustrator **ALFRED KUBIN** (1877–1959) saw Klinger's highly influential prints in 1898, he wrote of the effect they had on him: "I looked and quivered with delight. Here a new art was thrown open to me, which offered free play for the imaginative expression

of every conceivable world of feeling. Before putting the engravings away I swore that I would dedicate my life to the creation of similar works." Actually, Kubin proved capable of phantasms well in excess of those found in the art of either Böcklin or Klinger (fig. 126). Indeed, he may be compared with Redon, though his perverse obsession with woman-as-destroyer themes places him alongside decadent fin-de-siècle artists like Rops—as well as far into the future, in the Freudian world of Surrealism. A decade after he drew *Butcher's Feast* (fig. 127), Kubin became a member of Kandinsky's Munich circle, and in 1911 he joined a set of brilliant and experimental artists in Germany called Der Blaue Reiter (the Blue Rider). Meanwhile, in 1909, he published his illustrated novel entitled *The Other Side,* often regarded as a progenitor of the hallucinatory fictions of Franz Kafka.

127. ALFRED KUBIN. *Butcher's Feast.* c. 1900. India-ink wash,
9⅛ x 9" (23.2 x 22.9 cm). Graphische Sammlung Albertina, Vienna

6 The Origins of Modern Sculpture

In the twentieth century, sculpture has emerged as a major art for the first time in four hundred years. Its modern development has been even more remarkable than that of painting. The painting revolution was achieved against the background of an unbroken great tradition extending back to the fourteenth century. In the nineteenth century, despite the prevalence of lesser academicians, painting was the principal visual art, producing during the first seventy-five years such artists as Goya, Blake, Friedrich, David, Ingres, Géricault, Delacroix, Constable, Turner, Corot, Courbet, and Manet. The leading names in sculpture during this same period were Bertel Thorvaldsen, François Rude, Pierre-Jean David d'Angers, Antoine-Louis Barye, Jean-Baptiste Carpeaux, Jules Dalou, Alexandre Falguière, and Constantin Meunier, none of whom, despite their accomplishments, have the world reputations of the leading painters.

The eighteenth century also was an age of painting rather than sculpture. During that century, only the French Neoclassical sculptor Jean-Antoine Houdon and the Italian Antonio Canova may be compared with the most important painters. Thus, the seventeenth century was the last great age of sculpture before the twentieth century and then principally in the person of the Baroque sculptor Gianlorenzo Bernini. In the United States, with the exception of one or two artists of originality, such as Augustus Saint-Gaudens (see fig. 68), sculptors were secondary figures until well into the twentieth century.

When we consider the dominant place that sculpture has held in the history of art from ancient Egypt until the seventeenth century of our era, this decline appears all the more remarkable. It was not for want of patronage. Although the 1700s provided fewer large public commissions than the Renaissance or the Baroque, the 1800s saw mountains of sculptural monuments crowding the parks and public squares or adorning architecture. By this time, however, academic classicism had achieved such a rigid grip on the sculptural tradition that it was virtually impossible for a sculptor to gain a commission or even to survive without conforming. The experimental painter could usually find a small group of enlightened private patrons, but for the sculptor the very nature of the medium and the tradition of nineteenth-century sculpture as a monumental and public art made this difficult. After the eighteenth century, the center for sculpture in Western Europe shifted from Italy to Paris, due in part to technological innovations and new sources for patronage, and the French Romantic period produced several sculptors of note, such as Carpeaux and Barye. This geographical shift was accompanied by a tendency of artists to turn away from the preferred Neoclassical technique of carving marble to the practice of modeling plaster and wax. These materials could carry the inflection of the artist's hand and serve as a means to make the work reveal the process of its own creation. The models could then be cast in permanent form, and as multiples in edition, in bronze.

The basic subject for sculpture from the beginning of time until the twentieth century had been the human figure. It is in terms of the figure, presented in isolation or in combination, in action or in repose, that sculptors have explored the elements of sculpture—space, mass, volume, line, texture, light, and movement. Among these elements, volume and space and their interaction have traditionally been a primary concern. Within the development of classical, medieval, or Renaissance and Baroque sculpture we may observe progressions from static frontality to later solutions in which the figure is stated as a complex mass revolving in surrounding space, and interpenetrated by it. The sense of activated space in Hellenistic, Late Gothic, and Baroque sculpture was often created through increased movement, expressed in a twisting pose, extended gesture, or by a broken, variegated surface texture whose light and shadow accentuated the feeling of transition or change. The Baroque propensity for space, movement, and spontaneity was part of the nineteenth-century sculptural tradition.

One of the most conclusive symptoms of the revival of sculpture at the turn of the century was the large number of important painters who practiced sculpture—Gauguin, Degas, Renoir, and Bonnard among them. These were followed a little later by Picasso, Matisse, Amadeo Modigliani, André Derain, Umberto Boccioni, and others.

Color plate 38. Dankmar Adler and Louis Sullivan. Concert hall of the Auditorium Building, Chicago. 1887–89

Color plate 39.
James Abbott McNeill Whistler. *The Peacock Room*. 1876–77. Panels: oil color and gold on tooled leather and wood. Installed in the Freer Gallery of Art, Washington, D.C.

Color plate 40. Gustav Klimt. *Death and Life*. 1908–11. Oil on canvas, 70⅛" x 6'6" (178.1 cm x 2 m). Private collection

Color plate 41.
LOUIS COMFORT
TIFFANY. Table Lamp.
c. 1900. Bronze and
Favrile glass, height
27" (68.6 cm),
shade diameter 18"
(45.7 cm). Lilian
Nassau Ltd.,
New York

Color plate 42.
ANTONI GAUDÍ. Detail
of surrounding wall,
Güell Park,
Barcelona. 1900–14

Color plate 43. EDVARD
MUNCH.
The Dance of Life.
1900. Oil on canvas,
49½" x 6'3½" (125.7
cm x 1.9 m).
Nasjonalgalleriet,
Oslo

above: Color plate 44. JAMES ENSOR.
*The Entry of Christ into Brussels in
1889.* 1888–89. Oil on canvas,
8'5" x 12'5" (2.6 x 3.8 m).
J. Paul Getty Museum, Malibu
© 1997 ESTATE OF JAMES ENSOR/VAGA, NEW YORK, NY

Color plate 45. EDGAR DEGAS.
Little Dancer Fourteen Years Old.
1878–81. Wax, hair, linen bodice,
satin ribbon and shoes, muslin tutu,
wood base, height 39" (99.1 cm)
without base.
Collection Mr. and Mrs. Paul
Mellon, Upperville, Virginia

below: Color plate 46. CAMILLE CLAUDEL. *Chatting Women.* 1897. Onyx and bronze, 17¾ x 16⅝ x 15⅜" (45 x 42.2 x 39 cm). Musée Rodin, Paris

Color plate 47. HENRI MATISSE. *La Desserte (Dinner Table).* 1896–97. Oil on canvas, 39¼ x 51½" (99.7 x 130.8 cm). Private collection, Paris

Color plate 48. HENRI MATISSE. *Luxe, calme et volupté.* 1904–5. Oil on canvas, 37 x 46" (94 x 116.8 cm). Musée d'Orsay, Paris

right: Color plate 50. HENRI MATISSE. *Woman with the Hat*.
1905. Oil on canvas, 32 x 23¾" (81.3 x 60.3 cm).
San Francisco Museum of Modern Art

above: Color plate 49. HENRI MATISSE. *The Open Window*.
1905. Oil on canvas, 21¾ x 18⅛" (55.2 x 46 cm).
Collection Mrs. John Hay Whitney, New York

Color plate 51. HENRI MATISSE. *Portrait of Mme Matisse/The
Green Line*. 1905. Oil and tempera on canvas, 15⅞ x 12⅞"
(40.3 x 32.7 cm).
Statens Museum for Kunst, Copenhagen

Color plate 52. HENRI MATISSE. *Le Bonheur de vivre (The Joy of Life)*. 1905–6. Oil on canvas, 68½" x 7'9¾" (174 cm x 2.4 m).
The Barnes Foundation, Merion, Pennsylvania

Color plate 53. HENRI MATISSE. *Landscape at Collioure*/Study for *Le Bonheur de vivre*. 1905. Oil on canvas, 18 x 21½" (45.7 x 54.6 cm).
Statens Museum for Kunst, Copenhagen

Color plate 54. HENRI MATISSE. *Blue Nude: Memory of Biskra*. 1907. Oil on canvas, 36¼ x 56⅛" (92.1 x 142.5 cm).
The Baltimore Museum of Art

Color plate 55. HENRI MATISSE. *Le Luxe II.*
1907–8. Casein on canvas, 6'10½" x 54"
(2.1 m x 138 cm).
Statens Museum for Kunst, Copenhagen
J. RUMP COLLECTION © 1997 SUCCESSION H. MATISSE/ARTISTS RIGHTS
SOCIETY (ARS), NEW YORK

Color plate 56. ANDRÉ DERAIN. *London Bridge.* 1906. Oil on canvas, 26 x 39"
(66 x 99.1 cm). The Museum of Modern Art, New York
GIFT OF MR. AND MRS. CHARLES ZADOK © 1997 ADAGP, PARIS/ARTISTS RIGHTS SOCIETY (ARS), NEW YORK

Color plate 57. ANDRÉ DERAIN. *Turning Road, L'Estaque.* 1906. Oil on canvas, 51" x 6'4¾" (129.5 cm x 1.9 m). The Museum of
Fine Arts, Houston THE JOHN A. AND AUDREY JONES BECK COLLECTION © 1997 ADAGP, PARIS/ARTISTS RIGHTS SOCIETY (ARS), NEW YORK

right: 128. HONORÉ DAUMIER.
Comte de Kératry (The Obsequious One). c. 1833. Bronze,
4⅞ x 5¼ x 3⅝" (12.4 x 13.3 x 8.9 cm).
Hirshhorn Museum and Sculpture Garden,
Smithsonian Institution, Washington, D.C.

HONORÉ DAUMIER (1810–79). The pioneer painter-sculptor of the nineteenth century was Honoré Daumier. His small caricatural busts were created between 1830 and 1832 to lampoon the eminent politicians of Louis Philippe's regime. Daumier could not resist mocking the apelike features of the Comte de Kératry (fig. 128), for he was not only a government official, but an art critic as well. These satirical sculptures anticipate the late works of Auguste Rodin in their expressive power and the directness of their deeply modeled surfaces, as does the later *Ratapoil* (fig. 129), a caustic caricature of neo-Bonapartism. Unlike Kératry, *Ratapoil* (hairy rat) is not based on an actual person but is a personification of all the unscrupulous agents of Louis-Napoléon (later Napoléon III). The arrogance and tawdry elegance of this fellow is subtly communicated through his twisting pose and the fluttering, light-filled flow of clothing, which is a counterpoise to the bony armature of the figure itself. He leans on his club, the source of his "persuasive" powers. Out of fear that it would be destroyed by the very forces it set out to indict, the clay figurine was kept hidden until after Daumier's death. Like the earlier busts, it was posthumously cast in bronze. The sculpture of Daumier, now much admired, was a private art, little known or appreciated until its relatively recent "discovery."

EDGAR DEGAS (1834–1917). Like Daumier's, Degas's sculpture was little known to the sculptors of the first modern generation, for most of it was never publicly exhibited during his lifetime. Degas was concerned primarily with the traditional formal problems of sculpture, such as the continual experiments in space and movement represented by his dancers and horses. His posthumous bronze casts retain the immediacy of his original modeling material, a pigmented wax, which he built up over an armature, layer by layer, to a richly articulated surface in which the touch of his hand is directly recorded. While most of his sculptures have the quality of sketches, *Little Dancer Fourteen Years Old* (color plate 45, page 115) was a fully realized work that was exhibited at the 1881 Impressionist exhibition, the only time the artist showed one of his sculptures. By combining an actual tutu, satin slippers, and real hair with more traditional media, Degas created an astonishingly modern object that foreshadowed developments in twentieth-century sculpture (color plate 399, page 677).

PAUL GAUGUIN (1848–1903), like Degas, was a painter with an often unorthodox approach to materials. Throughout his career he worked in a diverse range of media. In addition to making paintings, drawings, woodcuts, etchings, and monotypes, he carved in marble and wood (see fig. 79) and was one of the most innovative ceramic artists of the nineteenth century. His largest work in this medium,

129. HONORÉ DAUMIER. *Ratapoil*. 1851 (cast 1925). Bronze,
17⅜ x 6½ x 7¼" (44.1 x 16.5 x 18.4 cm).
Hirshhorn Museum and Sculpture Garden,
Smithsonian Institution, Washington, D.C.

130. PAUL GAUGUIN. *Oviri*. 1894. Partially glazed stoneware, 29½ x 7½ x 10⅝" (74.9 x 19.1 x 27 cm). Musée d'Orsay, Paris

131. AUGUSTE RODIN. *Man with the Broken Nose*. 1864. Bronze, 12¼ x 7½ x 6½" (31.1 x 19 x 16.5 cm). Hirshhorn Museum and Sculpture Garden, Smithsonian Institution, Washington, D.C.

Oviri (fig. 130), was executed during a visit to Paris between two Tahitian sojourns. The title means "savage" in Tahitian, a term with which Gauguin personally identified, for he later inscribed "Oviri" on a self-portrait. The mysterious, bug-eyed woman crushes a wolf beneath her feet and clutches a wolf cub to her side. Whether she embraces the cub or suffocates it is unclear, though Gauguin did refer to her as a "murderess" and a "cruel enigma." The head was perhaps inspired by the mummified skulls of Marquesan chiefs, while the torso derives from the voluptuous figures, symbols of fecundity, on the ancient Javanese reliefs at Borobudur, in Southeast Asia, of which Gauguin owned photographs. In his reliance upon non-Western sources and his determination to create an image of raw, primitive power, Gauguin's work

was directly linked to the concerns of European and American artists in the twentieth century. When it was exhibited in Paris after his death in 1906, *Oviri* made a significant impact on, among others, the young Pablo Picasso.

AUGUSTE RODIN (1840–1917). It was Auguste Rodin's achievement to rechart the course of sculpture almost single-handedly and to give the art an impetus that led to a major renaissance. There is no one painter who occupies quite the place in modern painting that Rodin can claim in modern sculpture. He began his revolution, as had Courbet in painting, with a reaction against the sentimental idealism of the academicians. His *Man with the Broken Nose* (fig. 131) was rejected by the official Salon as being offensively realistic on the one hand and, on the other, as an unfinished fragment, a head with its rear portion broken away. Rodin sought the likeness rather than the character of his model, a poor old man who frequented his neighborhood. This early work was already mature and accomplished, suggesting the intensity of the artist's approach to his subject as well as his uncanny ability to exploit simultaneously the malleable properties of the original clay and the light-saturated tensile strength of the final bronze material.

Rodin's examination of nature was coupled with a re-examination of the art of the Middle Ages and the Renaissance—most specifically, of Donatello and Michelangelo. Although much academic sculpture paid lip service to the

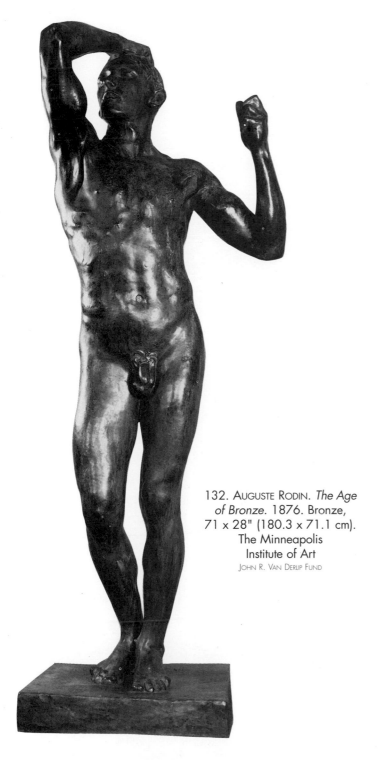

132. AUGUSTE RODIN. *The Age of Bronze*. 1876. Bronze, 71 x 28" (180.3 x 71.1 cm). The Minneapolis Institute of Art
JOHN R. VAN DERLIP FUND

tionably observed the model closely from all angles and was concerned with capturing and unifying the essential quality of the living form, he was also concerned with the expression of a tragic theme, inspired perhaps by his reaction to the Franco-Prussian War of 1870, so disastrous for the French. The figure, which once held a spear, was originally titled *Le Vaincu* (the vanquished one) and was based firmly on the *Dying Slave* of Michelangelo. By removing the spear, Rodin created a more ambivalent and ultimately more daring sculpture. The suggestion of action, frequently violent and varied, had been an essential part of the repertory of academic sculpture since the Late Renaissance, but its expression in nineteenth-century sculpture normally took the form of a sort of tableau of frozen movement. Rodin, however, studied his models in constant motion, in shifting positions and attitudes, so that every gesture and transitory change of pose became part of his vocabulary. As he passed from this concentrated study of nature to the later Expressionist works, even his most melodramatic subjects and most violently distorted poses were given credibility by their secure basis in observed nature.

In 1880 Rodin received a commission, the most important one of his career, for a portal to the proposed Museum of Decorative Arts in Paris. He conceived the idea of a free interpretation of scenes from Dante's *Inferno*, within a design scheme based on Lorenzo Ghiberti's great fifteenth-century gilded-bronze portals for the Florence Baptistry, popularly known as the Gates of Paradise. Out of his ideas emerged *The Gates of Hell* (fig. 133), which remains one of the masterpieces of nineteenth- and twentieth-century sculpture. It is not clear whether the original plaster sculpture of *The Gates* that Rodin exhibited at the Exposition Universelle in 1900 was complete, though it is clear that the idea of academic "finish" was abhorrent to the artist. *The Gates* were never installed in the Museum of Decorative Arts, and the five bronze casts that exist today were all made posthumously. Basing his ideas loosely on individual themes from Dante but also utilizing ideas from the poems of Baudelaire, which he greatly admired, Rodin created isolated figures, groups, or episodes, experimenting with different configurations over more than thirty years. *The Gates* contain a vast repertory of forms and images that the sculptor developed in this context and then adapted to other uses, and sometimes vice versa. For example, he executed the central figure of Dante, a brooding nude seated in the upper panel with his right elbow perched on his left knee, as an independent sculpture and exhibited it with the title *The Thinker; The Poet; Fragment of the Gate*. It became the artist's best-known work, in part because of the many casts that exist of it in several sizes. *The Gates* are crammed with a dizzying variety of forms on many different scales, above and across the lintel, throughout the architectural frieze and framework, and rising and falling restlessly within the plastic turmoil of the main panels. To convey the turbulence of his subjects, Rodin depicted the human figure bent and twisted to the limits of endurance, although with remarkably little actual distortion of anatomy. Imprisoned within the drama of their agitated and anguished state, the teeming fig-

High Renaissance, it was viewed through centuries of imitative accretions that tended to obscure the unique expression of the old masters. Rodin was in possession of the full range of historical sculptural forms and techniques by the time he returned in 1875 from a brief but formative visit to Italy, where he studied firsthand the work of Michelangelo, who, he said, "liberated me from academicism."

The Age of Bronze (fig. 132), Rodin's first major signed work, was accepted in the Salon of 1877, but its seeming scrupulous realism led to the suspicion that it might have been in part cast from the living model—which became a legitimate technique for making sculpture only in the 1960s (see color plate 303, page 529). Although Rodin had unques-

133. AUGUSTE RODIN. *The Gates of Hell.* 1880–1917. Bronze, 20'10" x 13'1" x 33½"
(6.4 m x 4 m x 85 cm). Musée Rodin, Paris

134. AUGUSTE RODIN. *Thought (Camille Claudel).*
1886. Marble, 29½ x 17 x 18"
(74.9 x 43.2 x 45.7 cm).
Musée d'Orsay, Paris

135. AUGUSTE RODIN. *The Crouching Woman.* 1880–82 (cast
1962). Bronze, 37½ x 26 x 21¾" (95.3 x 66 x 55.2 cm).
Hirshhorn Museum and Sculpture Garden, Smithsonian
Institution, Washington, D.C.

ures—coupled, clustered, or isolated—seem a vast and melancholy meditation on the tragedy of the human condition, on the plight of souls trapped in eternal longing, and on the torment of guilt and frustration.

The violent play on the human figure seen here was a forerunner of the Expressionist distortions of the figure developed in the twentieth century. An even more suggestive and modern phenomenon is to be found in the basic concept of the *Gates*—that of flux or metamorphosis, in which the figures emerge from or sink into the matrix of the bronze itself, are in the process of birth from, or death and decay into, a quagmire that both liberates and threatens to engulf them. Essential to the suggestion of change in the *Gates* and in most of the mature works of Rodin is the exploitation of light. A play of light and shadow moves over the peaks and crevasses of bronze or marble, becoming analogous to color in its evocation of movement, of growth and dissolution.

Although Rodin, in the sculptural tradition that had persisted since the Renaissance, was first of all a modeler, starting with clay and then casting the clay model in plaster and bronze, many of his most admired works are in marble. These also were normally based on clay and plaster originals, with much of the carving, as was customary, done by assistants. Rodin closely supervised the work and finished the marbles with his own hand. The marbles were handled with none of the expressive roughness of the bronzes (except in deliberately unfinished areas) and without deep undercutting. He paid close attention to the delicate, translucent, sen-

suous qualities of the marble, which in his later works he increasingly emphasized—inspired by unfinished works of Michelangelo—by contrasting highly polished flesh areas with masses of rough, unfinished stone. In his utilization of the raw material of stone for expressive ends, as in his use of the partial figure (the latter suggested no doubt by fragments of ancient sculptures), Rodin was initiating the movement away from the human figure as the prime medium of sculptural expression. The remarkable portrait of his lover and fellow sculptor, Camille Claudel (fig. 134), does away with the torso altogether to personify meditation through the abrupt juxtaposition of a smoothly polished, disembodied head with a coarsely hewn block of marble.

Like *The Thinker,* many of the figures made for *The Gates of Hell* are equally famous as individual sculptures, including *The Three Shades,* who stand hunched atop the gates. The tragedy and despair of the work are perhaps best summarized in *The Crouching Woman* (fig. 135), which looks at first glance like an extreme of anatomical distortion. In Rodin's hands, the contorted figure, a compact, twisted mass, attains a beauty that is rooted in intense suffering. The powerful diagonals, the enveloping arms, the broken twist of the head, all serve both an expressive and a formal purpose, emphasizing the agony of the figure and carrying the eye around the mass in a series of integrated views.

In *Iris, Messenger of the Gods* (fig. 136) the artist achieved an even more extreme pose, compounding the sensuality that characterizes so many of the later figures. The headless,

one-armed torso, by its maimed, truncated form and contorted pose, embodies tremendous vitality and displays a sexual boldness that far exceeds the thinly veiled, clichéd eroticism of academic sculptures of the female nude. Though the partial figure seems the sculptural equivalent of a sketch, it is a fragment that reaches an expressive completeness.

As a somewhat indirect memorial to French losses in the Franco-Prussian War, the city of Calais began formulating plans in 1884 for a public monument in memory of Eustache de Saint-Pierre, who in 1347, during the Hundred Years' War, had offered himself, along with five other prominent citizens, as hostage to the English in the hope of raising the long siege of the city. The commission held particular appeal for Rodin. In the old account of the event, the medieval historian Froissart describes the hostages as delivered barefoot and clad in

left: 136. AUGUSTE RODIN. *Iris, Messenger of the Gods.* 1890–91. Bronze, 32 x 36 x 24" (81.3 x 91.4 x 61 cm). Hirshhorn Museum and Sculpture Garden, Smithsonian Institution, Washington, D.C.

below: 137. AUGUSTE RODIN. *The Burghers of Calais.* 1886. Six figures, bronze, 6'10½" x 7'11" x 6'6" at base (2.1 x 2.4 x 2 m). Hirshhorn Museum and Sculpture Garden, Smithsonian Institution, Washington, D.C.

sackcloth, with ropes around their necks, bearing the keys to the city and fortress; this tale permitted the artist to be historically accurate and yet avoid the problem of period costumes without resorting to nudity. After winning the competition with a relatively conventional heroic design set on a tall base, Rodin created a group of greater psychological complexity, in which individual figures, bound together by common sacrifice, respond to it in varied ways (fig. 137). So he studied each figure separately and then assembled them all on a low rectangular platform, a nonheroic arrangement that allows the viewer to approach the figures directly. It took some doing for Rodin to persuade Calais to accept the work, for rather than an image of a readily recognizable historical event, he presented six particularized variations on the theme of human courage, deeply moving in their emotional range and thus very much a private monument as well as a public one.

The debt of *The Burghers of Calais* to the fourteenth- and fifteenth-century sculptures of Claus Sluter and Claus de Werve is immediately apparent, but this influence has been combined with an assertion of the dignity of common humanity analogous to the nineteenth-century sculptures of the Belgian artist Constantin Meunier. The rough-hewn faces, powerful bodies, and the enormous hands and feet transform these burghers into laborers and peasants and at the same time enhance their physical presence, as do the outsize keys. Rodin's tendency to dramatic gesture is apparent here, and the theatrical element is emphasized by the unorthodox organization, with the figures scattered about the base like a group of stragglers wandering across a stage. The informal, open arrangement of the figures, none of whom touch, is one of the most daring and original aspects of the sculpture. It is a direct attack on the classical tradition of closed, balanced groupings for monumental sculpture. The detached placing of the masses gives the intervening spaces an importance that, for almost the first time in modern sculpture, reverses the traditional roles of solid and void, of mass and space. Space not only surrounds the figures but interpenetrates the group, creating a dynamic, asymmetrical sense of balance. Rodin's wish that the work be placed not on a pedestal but at street level, situating it directly in the viewer's space, anticipated some of the most revolutionary innovations of twentieth-century sculpture.

Rodin's *Monument to Balzac* (fig. 138), a work he called "the sum of my whole life," was commissioned in 1891 by the French Writers' Association, at the urging of the novelist Émile Zola. Here the artist plunged so deeply into the privacy of individual psychology—his own perhaps even more than the subject's—that he failed to gain acceptance by the patrons. What these and the public perceived was a tall, shapeless mass crowned by a shaggy head so full of Rodin's "lumps and hollows" that it seemed more a desecrating caricature than a trib-

ute to the great novelist Honoré de Balzac, who had died in 1850. Little were they prepared to appreciate what even the artist characterized as a symbolism of a kind "yet unknown." The work provoked a critical uproar, with some proclaiming it a masterpiece and others reviling it as a monstrosity, whereupon it was withdrawn from the Paris Salon in 1898.

Rodin struggled to realize a portrait that would transcend any mere likeness and be a sculptural equivalent of a famously volcanic creative force. He made many different studies over a period of seven years, some of an almost academic exactness, others more emblematic. In the end, Rodin cloaked Balzac in the voluminous "monk's robe" he had habitually worn during his all-night writing sessions. The anatomy has virtually disappeared beneath the draperies, which are gathered up as if to muster and concentrate the whole of some prodigious inspiration, all reflected like a tragic imprint on the deep-set features of the colossal head. The figure leans back dramatically beneath the robe, a nearly abstract icon of generative power.

This symbolism of a kind "yet unknown" may have had its greatest impact in the photographs that Rodin commissioned **EDWARD STEICHEN** (1879–1973) to make of the original plaster cast of *Balzac* (fig. 139). With a typical sureness of aesthetic instinct, Rodin proposed that Steichen try working by moonlight. Not only did this technique avoid the flattening effect that direct sunlight would have had on the white material, but the long exposure that the dim light required invested the pictures with a sense of timelessness totally unlike the stop-action instantaneity normally associated with the camera. Rodin detested such imagery as a treacherous distortion of reality. "It is the artist," he insisted, "who is truthful, and it is photography that lies, for in reality time does not stop, and if the artist succeeds in producing the impression of a movement which takes several moments for accomplishment, his work is certainly less conventional than the scientific image, where time is abruptly suspended." Rodin spent most of his creative life studying motion and improvising ways to express its drama and emotive effect in sculptural form. Often he did this in drawings made from a moving model, as in a famous series based on the dancer Isadora Duncan (fig. 140), but always for the purpose of capturing in line the impression of a flowing, continuous process, not an arrested action.

CAMILLE CLAUDEL (1864–1943). Until the 1980s, when long overdue retrospectives of her work were held, Camille Claudel was known largely as Rodin's assistant and mistress. Tragically, she never recovered from her failed relationship with the sculptor and was confined to mental hospitals for the last thirty years of her life. Claudel's work shares many formal characteristics with that of her mentor, but hers was a highly original talent in a century that produced relatively few women sculptors of note. By the time she entered Rodin's studio at age twenty, she was already exhibiting at the Salon. Though she was an accomplished portraitist—she produced one of the most memorable images of Rodin—Claudel's particular strength lay with inventive solutions to multifigure compositions. An unusual and decidedly unheroic subject for

140. AUGUSTE RODIN. *Study of Isadora Duncan.* Pencil and watercolor on paper, 12¼ x 9" (31.1 x 22.9 cm).
LAURA GREEN AND TOBIAS BASKIN COLLECTION

sculpture, *Chatting Women* (color plate 46, page 116) presents four seated figures, rapt in discussion, who form a tightly knit group in the corner of a room. An immediacy of expression and the quotidian nature of the subject seem at odds with the inexplicable nudity of the figures and the use of strongly veined and colored onyx, which serves to underscore the abstract nature of the representation.

ARISTIDE MAILLOL (1861–1944). French artists who chose to work in a more traditional mode—Aristide Maillol and Antoine Bourdelle—represent opposition to Rodin in their attempts to preserve but also modernize the classical ideal. Maillol focused his attention on this restatement of classicism, stripped of all the academic accretions of sentimental or erotic idealism. Concentrating almost exclusively on the single nude female figure, standing, sitting, or reclining, and usually in repose, he stated over and over again the fundamental thesis of sculpture as integrated volume, as mass surrounded by tangible space.

Maillol began as a painter and tapestry designer; he was almost forty when the onset of a dangerous eye disease made the meticulous practice of weaving difficult, and he decided to change to sculpture. He began with wood carvings that had a definite relation to his paintings and to the Nabis and the Art Nouveau environment in which he had been working at the turn of the century. However, he soon moved to clay model-

left: 141. ARISTIDE MAILLOL. *The Mediterranean.* 1902–5. Bronze, 41 x 45 x 29¾" at base (104.1 x 114.3 x 75.6 cm), including base. The Museum of Modern Art, New York
GIFT OF STEPHEN C. CLARK
PHOTOGRAPH © 1995 THE MUSEUM OF MODERN ART, NEW YORK
© 1997 ADAGP, PARIS/ARTISTS RIGHTS SOCIETY (ARS), NEW YORK

below: 142. ARISTIDE MAILLOL. *The River.* Begun 1938–39; completed 1943. Lead (cast 1948), 53¾" x 7'6" x 66" (136.5 cm x 2.3 m x 167.7 cm), on lead base designed by the artist, 9¾ x 67 x 27¾" (24.3 x 170.1 x 70.4 cm). The Museum of Modern Art, New York
MRS. SIMON GUGGENHEIM FUND
PHOTOGRAPH © 1995 THE MUSEUM OF MODERN ART, NEW YORK
© 1997 ADAGP, PARIS/ARTISTS RIGHTS SOCIETY (ARS), NEW YORK

143. ANTOINE BOURDELLE. *Hercules the Archer*. 1909.
Bronze, 25 x 24 x 10¼" (63.5 x 61 x 26 cm).
The Art Institute of Chicago
A. A. MUNGER COLLECTION © 1997 ADAGP, PARIS/ARTISTS RIGHTS SOCIETY (ARS), NEW YORK

144. MEDARDO ROSSO. *The Concierge (La Portinaia)*. 1883.
Wax over plaster, 14½ x 12⅝" (36.8 x 32 cm).
The Museum of Modern Art, New York
MRS. WENDELL T. BUSH FUND
PHOTOGRAPH © 1995 THE MUSEUM OF MODERN ART, NEW YORK

ing and developed a mature style that changed little throughout his life. That style is summarized in one of his very first sculptures, *The Mediterranean* (fig. 141), a massive, seated female nude, integrated as a set of curving volumes in space. This work typifies Maillol's personal brand of classicism in its simplification of the body into idealized, geometric forms and in its quality of psychological withdrawal and composed reserve. Maillol could also achieve an art of dynamic movement, as in *The River* (fig. 142), a work that may have been affected by the late paintings and the sculptures of Renoir.

ANTOINE BOURDELLE (1861–1929) was a student and assistant of Rodin. Like Maillol, he sought to revitalize the classical tradition. His eclectic, somewhat archaeological approach drew on archaic and early fifth-century B.C.E. Greek sculpture, as well as the Gothic. *Hercules the Archer* (fig. 143) discloses Bourdelle's indebtedness to Rodin; its scarred modeling is an echo of Rodin's *Thinker* translated into violent action. The figure, braced against the rocks in a pose of fantastic effort, is essentially a profile, two-dimensional alternation of large areas of solid and void, possibly inspired by the famous Roman sculpture of the *Discobolus*, itself based on a Greek original.

MEDARDO ROSSO (1858–1928), born in Turin, Italy, in 1858, worked as a painter until 1880. After being dismissed from the Accademia di Brera in Milan in 1883, he lived in Paris for two years, worked in Dalou's atelier, and met Rodin. After 1889 Rosso spent most of his active career in Paris, until his death in 1928, and thus was associated more with French sculpture than with Italian. In a sense, he always remained a painter. Rosso extended the formal experiments of Rodin, deliberately dissolving sculptural forms until only an impression remained. His favorite medium, wax, allowed the most imperceptible transitions, so that it becomes difficult to tell at exactly what point a face or figure emerges from an amorphous shape and light-filled, many-textured surface. "Nothing is material in space," Rosso said. The primacy he gave to the play of light and shadow is evident in the soft *sfumato* that envelops the portrait of his old doorkeeper (fig. 144), a work once owned by Émile Zola. This quality has sometimes led critics to call him an "Impressionist" sculptor. In his freshness of vision, his ability to catch and record the significant moment, Rosso added a new dimension to sculpture, in works that are invariably intimist, small-scale, and antiheroic, and anticipated the search for immediacy that characterizes so much of the sculpture that followed.

7 Fauvism

Donatello au milieu des fauves! "Donatello among the wild beasts!" was the ready quip of Louis Vauxcelles, art critic for the review *Gil Blas,* when he entered Gallery VII at Paris's 1905 Salon d'Automne and found himself surrounded by blazingly colored, vehemently brushed canvases in the midst of which stood a small neo-Renaissance sculpture. With this witticism Vauxcelles gave the first French avant-garde style to emerge in the twentieth century its name, thereby provoking one of modernism's classic "scandals." It should be noted, however, that Vauxcelles was generally sympathetic to the work presented by the group of young painters, as were other liberal critics. The starting point of Fauvism was later identified by Henri Matisse, the sober and rather professorial leader of the Fauves, as "the courage to return to the purity of means." Matisse and his fellow painters—André Derain, Maurice de Vlaminck, Georges Rouault, Raoul Dufy, and others—committed the solecism of allowing their search for immediacy and clarity to show forth with bold, almost unbearable candor. While divesting themselves of Symbolist literary aesthetics, along with fin-de-siècle morbidity, the Fauves reclaimed Impressionism's direct, joyous embrace of nature and combined it with Post-Impressionism's heightened color contrasts and emotional, expressive depth. They emancipated color from its role of describing external reality and concentrated on the medium's ability to communicate directly the artist's experience before that reality by exploiting the pure chromatic intensity of paint.

Inevitably, artists so intent on achieving personal authenticity would never form a coherent movement or issue the kind of joint theoretical manifestos that proselytized the cause of many subsequent avant-garde movements. But before drifting apart as early as 1907, the Fauves made certain definite and unique contributions. Though none of them attempted complete abstraction, as did their contemporaries Vasily Kandinsky or Robert Delaunay, for example, they extended the boundaries of representation, based in part on their exposure to non-Western sources, such as the art of black Africa. For subject matter they turned to portraiture, still life, and landscape, but in the latter, at least in the art of

Matisse, they revisualized Impressionism's culture of leisure as a pagan ideal of *bonheur de vivre,* the joy of life. Most important of all, the Fauvist painters committed themselves to pictorial autonomy, which yielded an art delicately poised between expression derived from emotional, subjective experience and expression stimulated by pure optical sensation.

HENRI MATISSE (1869–1954) was born only two years after Bonnard and outlived him by seven years. These two were not only contemporaries but had much in common as artists. They were among the greatest colorists of the twentieth century, and each learned something from the other. Yet Matisse, one of the pioneers of twentieth-century experiment in painting, seems to belong to a later generation, to a different world. Like Bonnard, he was destined for a career in the law, but by 1891 he had enrolled in the Académie Julian, studying briefly with the academic painter Bouguereau (see fig. 21), who came to represent everything he rejected in art. The following year he entered the École des Beaux-Arts and was fortunately able to study in the class of Gustave Moreau (see fig. 74), a dedicated teacher who encouraged his students to find their own directions through constant study in the museums, as well as through individual experiment. In his class, Matisse met Georges Rouault, Albert Marquet, Henri-Charles Manguin, Charles Camoin, and Charles Guérin, all of whom were later associated with the Fauves.

Matisse's work developed slowly from the dark tonalities and literal subjects he first explored. By the late 1890s he had discovered Neo-Impressionism and artists such as Toulouse-Lautrec and, most importantly, Cézanne. In about 1898 he began to experiment with figures and still lifes painted in bright, nondescriptive color. In 1900, Matisse entered the atelier of Eugène Carrière, a maker of dreamily romantic figure paintings. There he met André Derain, who introduced him to Vlaminck the following year, completing the principal Fauve trio. Around that time, Matisse also worked in the studio of Antoine Bourdelle (see fig. 143), making his first attempts at sculpture and demonstrating the abilities that were to make him one of the great painter-sculptors of the twentieth century.

Among the paintings that Matisse copied in the Louvre

145. JAN DAVIDSZ. DE HEEM. *The Dessert*. 1640. Oil on canvas, 58⅝" x 6'8" (148.9 cm x 2 m). Musée du Louvre, Paris

during his student days under Moreau was a still life by the seventeenth-century Dutch painter Jan Davidsz. de Heem (fig. 145). Matisse's version (fig. 146) is a free copy, considerably smaller than the original. In 1915, he was to make a Cubist variation of the work (see fig. 332). At the Salon de la Société Nationale des Beaux-Arts (known as the Salon de la Nationale) of 1897, he exhibited his own composition of a still life, *Dinner Table* (color plate 47, page 116), which was not favorably received by the conservatives. Though highly traditional on the face of it, this work was one of Matisse's most complicated and carefully constructed compositions to date, and it was his first truly modern work. While it still depended on locally descriptive color, this painting revealed in its luminosity an interest in the Impressionists, and in the abruptly tilted table that crowds and contracts the space of the picture it anticipated the artist's subsequent move toward radical simplification in his later treatment of similar subjects (color plate 150, page 271). *Male Model* of about 1900 (fig. 147) carried this process of simplification and contraction several stages further, even to the point of some distortion of perspective, to achieve a sense of delimited space. The mod-

eling of the figure in abrupt facets of color was a direct response to the paintings of Cézanne, whose influence can also be seen in *Carmelina* (fig. 148), an arresting, frontalized arrangement in which the model projects sculpturally from the rectangular design of the background. Though more traditional in its use of light and dark shading to model the figure, *Carmelina* also contains occasional nondescriptive flourishes of brilliant red paint. With his own visage reflected in the mirror, Matisse introduced two themes—the studio interior and the artist with his model—to which he often returned in the course of his life, as if to restate the modernist's abandonment of direct attachment to the outside world in favor of an ever greater absorption in the world of art.

Between 1902 and 1905 Matisse exhibited at the Salon des Indépendants and at the galleries of Berthe Weill and Ambroise Vollard. The latter was rapidly becoming the principal dealer for the avant-garde artists of Paris. When the more liberal Salon d'Automne was established in 1903, Matisse showed there, along with Bonnard and Marquet. But most notorious was the Salon d'Automne of 1905, in which a room of paintings by Matisse, Vlaminck, Derain, and Rouault, among others, is supposed to have occasioned the remark that gave the group its permanent name.

The word *fauves* made particular reference to these artists' brilliant, arbitrary color, more intense than the scientific color of the Neo-Impressionists and the nondescriptive color of Gauguin and Van Gogh, and to the direct, vigorous brushwork with which Matisse and his friends had been experimenting the previous year at St.-Tropez and Collioure on the Riviera in the south of France. The Fauves accomplished the liberation of color toward which, in their different ways, Cézanne, Gauguin, Van Gogh, Seurat, and the Nabis had been experimenting. Using similar means, the Fauves were intent on different ends. They wished to use pure color squeezed directly from the tube, not to describe objects in nature, not simply to set up retinal vibrations, not to accentuate a romantic or mystical subject, but to build new pictorial values apart from all these. Thus, in a sense, they were using the color of Gauguin and Seurat, freely combined with their own linear rhythms, to reach effects similar to those which Cézanne constantly sought. It is no accident that of the artists of the previous generation, it was Cézanne whom Matisse revered the most.

Earlier in 1905, at the Salon des Indépendants, Matisse had already exhibited his large Neo-Impressionist composition *Luxe, calme et volupté* (color plate 48, page 116), a title taken from a couplet in Baudelaire's poem *L'Invitation au voyage*:

> *Là, tout n'est qu'ordre et beauté,*
> *Luxe, calme et volupté.*

"There, all is only order and beauty / Richness, calm, sensuality."

In this important work, which went far along the path to abstraction, he combined the mosaic landscape manner of Signac (who bought the painting) with figure organization that recalls Cézanne's many compositions of bathers (color plate 25, page 74), one of which Matisse owned. At the left of this St.-Tropez beach scene, Matisse depicted his wife, Amélie, beside a picnic spread. But this mundane activity is

146. HENRI MATISSE. *The Dessert (after Jan Davidsz. de Heem)*. 1893. Oil on canvas, 28⅞ x 39½" (73.3 x 100.3 cm). Musée Matisse, Nice-Cimiez, France © 1997 SUCCESSION H. MATISSE/ARTISTS RIGHTS SOCIETY (ARS), NEW YORK

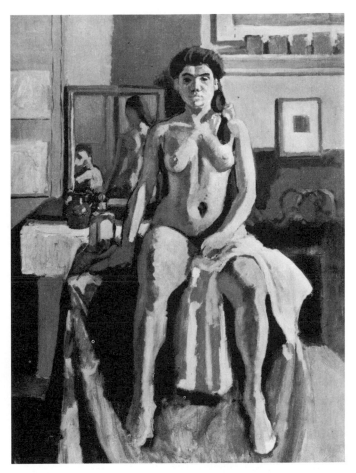

147. HENRI MATISSE. *Male Model (L'Homme Nu; "Le Serf"
Academie bleue; Bevilacqua).* Paris, 1900. Oil on canvas,
39⅛ x 28⅝" (99.3 x 72.7 cm). The Museum of Modern Art,
New York KAY SAGE TANGUY AND ABBY ALDRICH ROCKEFELLER FUNDS
© 1997 SUCCESSION H. MATISSE/ARTISTS RIGHTS SOCIETY (ARS), NEW YORK

148. HENRI MATISSE. *Carmelina.* c. 1903–4. Oil on canvas,
31½ x 25¼" (80 x 64.1 cm). Museum of Fine Arts, Boston
TOMPKINS COLLECTION
© 1997 SUCCESSION H. MATISSE/ARTISTS RIGHTS SOCIETY (ARS), NEW YORK

transported to a timeless, arcadian world populated by languid nudes relaxing along a beach that has been tinged with
dazzling red. With *Luxe,* therefore, Matisse offered a radical
reinterpretation of the grand pastoral tradition in landscape
painting, best exemplified in France by Claude Lorrain and
Poussin (see fig. 11). As in many of the paintings that post-
date this work, his idyllic world is exclusively female. Matisse
was wary of all theories in art, and though this bold painting
could hardly be considered an orthodox example of Neo-
Impressionism, his experimentation with the staccato strokes
of that style gave way to other modes.

At the Salon d'Automne of 1905, Matisse also exhibited
The Open Window and a portrait of Madame Matisse called
Woman with the Hat (color plates 49, 50, page 117). *The
Open Window* is perhaps the first fully developed example of a
theme favored by Matisse throughout the rest of his life. It is
simply a small fragment of the wall of a room, taken up principally with a large window whose casements are thrown wide
to the outside world—a balcony with flowerpots and vines,
and beyond that the sea, sky, and boats of the harbor at
Collioure. It was at this Mediterranean port, during the summer of 1905, that Matisse and Derain produced the first Fauve
paintings. In *The Open Window* the inside wall and the casements are composed of broad, vertical stripes of vivid green,

blue, purple, and orange, while the outside world is a gay pattern of brilliant small brushstrokes, ranging from stippled dots
of green to broader strokes of pink, white, and blue in sea and
sky. This diversity of paint handling, even in adjoining passages
within the same picture, was typical of Matisse's early Fauve
compositions. Between his painterly marks, Matisse left bare
patches of canvas, reinforcing the impact of brushstrokes that
have been freed from the traditional role of describing form in
order to suggest an intense, vibrating light. By this date, the
artist had already moved far beyond Bonnard or any of the
Neo-Impressionists toward abstraction.

In Neo-Impressionism, as in Impressionism, the generalized, all-over distribution of color patches and texture had
produced a sense of atmospheric depth inviting optical
entrance, at the same time that it also asserted the physical
presence and impenetrability of the painting surface. Matisse,
however, structured an architectural framework of facets and
planes that are even broader and flatter than those of Cézanne, suppressing all sense of atmosphere; internal illumination is replaced with a taut, resistant skin of pigment that
reflects light instead of generating it internally. Rather than
penetration, this tough, vibrant membrane of color and pattern solicits exploration, a journey over and across but rarely
beyond the picture plane. And even in the "Impressionist"

view through the window, the handling is so vigorously self-assertive that the scene appears to advance more than recede, as if to turn inside out the Renaissance conception of the painting as a simulacrum of a window open into the infinite depth of the real world. As presented by Matisse, the window and its sparkling view of a holiday marina become a picture within a picture, a theme the artist often pursued, transforming it into a metaphor of the modernist persuasion that painting must not duplicate the perceptual world but rather transcend and re-create it conceptually.

Matisse's *Woman with the Hat* caused even more furor than *The Open Window* because of the seemingly wild abandon with which the artist had applied paint over the surface, not only the background and hat but also in the face of the sitter, whose features are outlined in bold strokes of green and vermilion. Paradoxically, much of the shock value sprang from the actual verisimilitude of the painted image, its closeness to a recognizable subject an obvious measure of the degree of distortion imposed upon reality by the painter. At a later, more developed state of abstraction, modernist art was to become so independent of its sources in the material world as to seem less an attack on the familiar than an autonomous and thus a relatively less threatening, purely aesthetic object. In 1905 Matisse could liberate his means mainly by choosing motifs that inspired such freedom—the luminous, relaxed, holiday world of the French Riviera, a flag-decked street, or one of the period's fantastic hats.

Shortly after exhibiting *Woman with the Hat*, Matisse painted another portrait of Madame Matisse that in a sense was even more audacious, precisely because it was less sketch-like, more tightly drawn and structured. This is the work entitled *Portrait of Mme Matisse/The Green Line* (color plate 51, page 117), with a face dominated by a brilliant pea-green band of shadow dividing it from hairline to chin. At this point Matisse and his Fauve colleagues were building on the thesis put forward by Gauguin, the Symbolists, and the Nabis: that the artist is free to use color independently of natural appearance, building a structure of abstract color shapes and lines foreign to the figure, tree, or still life that remains the basis of the structure. Perhaps Matisse's version was more immediately shocking because his subject was so simple and familiar, unlike the exotic scenes of Gauguin or the mystical fantasies of Redon, in which such arbitrary colorism seemed more acceptable. With its heavy, emphatic strokes and striking use of complementary hues, the painting is actually closest to portraits by Van Gogh (see fig. 81).

The artist's experiments with arcadian figure compositions climaxed in the legendary *Bonheur de vivre* (color plate 52, page 118), a painting filled with diverse reminiscences of past art, from *The Feast of the Gods,* by Giovanni Bellini and Titian, to Persian painting, from prehistoric cave paintings to a composition by Ingres. In this large work (it is nearly eight feet wide), the artist has blended all these influences into a masterful arrangement of figures and trees in sinuous, undulating lines reminiscent of contemporary Art Nouveau design. *Le Bonheur de vivre* is filled with a mood of sensual languor; figures cavort with Dionysian abandon in a landscape that pulsates with rich, riotous color. Yet the figure groups are deployed as separate vignettes, isolated from one another spatially as well as by their differing colors and contradictory scales. As Matisse explained, *Le Bonheur de vivre* "was painted through the juxtaposition of things conceived independently but arranged together." He made several sketches for the work, basing his vision on an actual landscape at Collioure, which he painted in a lush sketch without figures that still contains some of the broken color patches of Neo-Impressionism (color plate 53, page 119). The circle of ecstatic dancers in the distance of *Le Bonheur de vivre,* apparently inspired by the sight of fishermen dancing in Collioure, became the central motif of Matisse's painting from 1909–10, *Dance (II)* (color plate 63, page 155). *Le Bonheur de vivre* is an all-important, breakthrough picture; it made a considerable impression on Picasso, whose *Les Demoiselles d'Avignon* (color plate 89, page 166), produced the following year, came about partly in response to it. Though radically different in spirit and style from Picasso's notorious picture, Matisse's painting was, like *Les Demoiselles,* intended as a major statement, a kind of manifesto of his current ideas about art. While Picasso's canvas was not exhibited publicly until 1937, *Le Bonheur de vivre* was bought immediately by the American collectors and writers Gertrude and Leo Stein, who introduced the two men to one another, and whose apartment was eventually filled with work by them and other avant-garde artists. In their collection, Picasso could study *Le Bonheur de vivre* at length.

In 1906 Matisse, Derain, and Vlaminck began to collect art objects from Africa, which they had first seen in ethnographic museums, and to adapt those forms into their art. Of all the non-Western artistic source material sought out by European modernists, none proved so radical or far-reaching as the art of Subsaharan Africa. Modern artists appropriated the forms of African art in the hope of investing their work with a kind of primal truth and expressive energy, as well as a touch of the exotic, what they saw as the "primitive," or, in Gauguin's word, the "savage." Unlike the myriad other influences absorbed from outside contemporary European culture—Asian, Islamic, Oceanic, medieval, folk, and children's art—these African works were not only sculptural, as opposed to the predominantly pictorial art of Europe, but they also embodied values and conventions outside Western tradition and experience. Whereas even the most idealized European painting remained ultimately bound up with perceptual realities and the ongoing history of art itself, African figures observed no classical canon of proportions and contained no history or stories that Europeans understood. With their relatively large, mask-like heads, distended torsos, prominent sexual features, and squat, abbreviated, or elongated limbs, they impressed Matisse and his comrades with the powerful plasticity of their forms (mostly unbound to the literal representation of nature), their expressive carving, and their iconic force.

When finally examined by such independent artists as Matisse and Picasso (first introduced to African art by Matisse), the figures seemed redolent of magic and mystery, the very qualities that European progressives, from Gauguin on, had been struggling to recover for new art. Obviously, such objects, once they reached European borders, had been stripped of their original contexts and functions, sometimes as a result of European colonization of the very people who made them. The response on the part of progressive Western artists, who had virtually no understanding of the original conditions that shaped these works, was largely ethnocentric. They assimilated African forms into a prevailing aesthetic determined by their search for alternatives to naturalism and for a more stylized, abstract conception of the figure.

An early example of African influence in Matisse's art occurred most remarkably in *Blue Nude: Memory of Biskra* (color plate 54, page 119), which, as the subtitle implies, came forth as a souvenir of the artist's recent visit to Biskra, a lush oasis in the North African desert. The subject of the painting—a reclining female nude—is a dynamic variation of a classic Venus pose, with one arm bent over the head and the legs flexed forward. Matisse had made the reclining nude central to *Le Bonheur de vivre*, and such was the interest it held for him that he then restudied it in a clay sculpture (see fig. 174). Now, working from memory and his own sculpture, and evidently encouraged by the example of African sculpture, he abstracted his image further. These influences produced the bulbous exaggeration of breasts and buttocks; the extreme contrapposto that makes torso and hips seem viewed from different angles, or assembled from different bodies; the scarified modeling, or vigorously applied contouring in a brilliant, synthetic blue, and the masklike character of the face. But however much these traits may evince the impact of African art, they have been translated from the plastic, freestanding, iconic sculpture of Africa into a pictorial expression by means of the Cézannism that Matisse had been cultivating all along. This can be seen in the dynamic character of the whole, wherein rhyming curves and countercurves, images and afterimages, and interchanges of color and texture make figure and ground merge into one another.

In its theme, *Blue Nude* belongs to the tradition in nineteenth-century painting of the odalisque, or member of a North African harem (color plate 1, page 33). This subject of the exotic, sexually available woman, exploited by Delacroix, Ingres, and countless lesser Salon artists, was destined for visual consumption by a predominantly male audience. But the level of abstraction Matisse imposes on his subject, as with Cézanne's bathers, moves the figure beyond the explicitly erotic. Responding to the charges of ugliness made against *Blue Nude*, Matisse said: "If I met such a woman in the street, I should run away in terror. Above all, I do not create a woman, *I make a picture.*" Inspired by the example of what he called the "invented planes and proportions" of African sculpture, Matisse did not restructure the human face and figure in the overt, aggressive manner of Picasso. Rather, he used these sources in his own subtle, reflective fashion, assimilating and synthesizing until they are scarcely discernible.

With *Le Luxe II*, a work of 1907–8 (color plate 55, page 120), Matisse signaled a move away from his Fauve production of the preceding years. It is a large painting, nearly seven feet tall, that Matisse elaborately prepared with full-scale charcoal and oil sketches. The oil sketch *Le Luxe I* is much looser in execution, and closer to Fauvism than are the flat, unmodulated zones of color in the final version. Though *Le Luxe II* explores a theme similar to that of *Luxe, calme et volupté* and *Le Bonheur de vivre* (color plate 48, page 116; color plate 52, page 118), the figures are now life-size and dominate the landscape. Coloristically, *Le Luxe II* is far more subdued, relying on areas of localized color bound by crisp lines. Despite the artist's abandonment of perspective, except for the arbitrary diminution of one figure, and modeling in light and shadow, the painting is not merely surface decoration. The figures, modeled only by the contour lines, have substance; they exist and move in space, with the illusion of depth, light, and air created solely by flat color shapes that are, at the same time, synonymous spatially with the picture plane. The actual subject of the painting remains elusive. The crouching woman, a beautiful, compact shape, seems to be tending to her companion in some way (perhaps drying her feet?), while another rushes toward the pair, proffering a bouquet of flowers. Matisse may have implied a mythological theme, such as Venus's birth from the sea, but, typically, he only hints at such narratives.

ANDRÉ DERAIN (1880–1954) met the older Matisse at Carrière's atelier in 1900, as already noted, and was encouraged by him to proceed with his career as a painter. He already knew Maurice de Vlaminck, whom Derain in turn had led from his various careers as violinist, novelist, and bicycle racer into the field of painting. Unlike Vlaminck, Derain was a serious student of the art of the museums who, despite his initial enthusiasm for the explosive color of Fauvism, was constantly haunted by a more ordered and traditional concept of painting.

Although Derain's Fauve paintings embodied every kind of painterly variation, from large-scale Neo-Impressionism to free brushwork, most characteristic, perhaps, are works like *London Bridge* (color plate 56, page 120). It was painted during a trip commissioned by the dealer Vollard, who wanted Derain to make paintings that would capture the special atmosphere of London: Claude Monet's many views of the city had just been successfully exhibited in Paris. To compare Derain's painting with the Impressionist's earlier work, *Bridge at Argenteuil* (color plate 10, page 36), is to understand the transformations art had undergone in roughly thirty years, as well as the fundamental role Impressionism had played in those transformations. While in London, Derain visited the museums, studying especially the paintings of Turner, Claude, and Rembrandt, as well as African sculpture. Unlike Monet's view of the Seine, his painting of the Thames is a brilliantly synthetic color arrangement of harmonies and dissonances. The background sky is rose-pink; the buildings silhouetted against it are complementary green and blue. By reiterating large color areas in the foreground and background and tilting the perspective, Derain delimits the depth of his image.

149. ANDRÉ DERAIN. *The Painter and His Family.* c. 1939.
Oil on canvas, 69½ x 48⅜" (176.5 x 122.9 cm).
Private collection © 1997 ADAGP, PARIS/ARTISTS RIGHTS SOCIETY (ARS), NEW YORK

In the summer of 1906, several months after the trip to London, Derain spent time in L'Estaque, the famous site of paintings by Cézanne (color plate 21, page 40). But his grand panorama of a bend in the road (color plate 57, page 120) takes its cue from Gauguin (color plates 29, 30, page 77) in its brilliant palette and the evocation of an idealized realm far from the urban bustle of London's waterways. With this work Derain travels to new extremes of intensity and antinaturalism in his color, a world in which the hues of a single tree can shift dramatically half a dozen times.

Derain was essentially an academic painter who happened to become involved in a revolutionary movement, participated in it effectively, but was never completely happy in the context. In 1906 he became a friend of Picasso and was drawn into the vortex of proto-Cubism. But throughout the years of World War I (1914–18), in which Derain fought, and later, when Picasso and Braque were expanding their discoveries in Synthetic Cubism and exploring alternative styles, Derain was moving back consistently from Cézanne to that artist's sources in Poussin and Chardin. His direction was away from the very experiments of the twentieth century he had helped to develop, back through the innovators of the

right: 150. MAURICE DE VLAMINCK. *Picnic in the Country.* 1905. Oil on canvas, 35 x 45⅝" (88.9 x 45.6 cm). Private collection, Paris
© 1997 ADAGP, PARIS/ARTISTS RIGHTS SOCIETY (ARS), NEW YORK

late nineteenth century, to the Renaissance. Derain's return to optical realism formed part of a larger "call to order" around World War I, when a number of French artists and writers sought to renew their art through the more classical forms of Western art. Picasso's classicizing work from the second half of the 1910s and early 1920s belongs to this phenomenon (see figs. 339, 341–345), as does, in the preceding chapter, the work of Maillol (see figs. 141, 142).

Perhaps most indicative of this gradual but thoroughgoing change in Derain's outlook is that in 1930 he sold several African sculptures from his collection to buy Renaissance and antique artworks. His new conservatism resulted in a very uneven output of landscapes, nudes, still lifes, and portraits and a serious decline after World War II in his reputation, one that is now being reexamined. *The Painter and His Family* (fig. 149) from about 1939, is an allegory of the painter's world composed in the manner of Dutch seventeenth-century interiors. Derain depicts himself at work, presumably painting the still life on the table. Surrounding him are a curious menagerie and his devoted family members—his wife, reading in the foreground; his sister-in-law, who enters through a door at the back of the room (perhaps in a quote from Velázquez's *Las Meninas*); and, just behind him, his niece, who plays the role of inspirational muse. The contrived content and the archaizing style of hard, simplified forms place the artist somewhere between contemporaries like Balthus, who painted Derain's portrait in 1936, and the earlier naïve paintings of Henri Rousseau (see figs. 77, 78). Given the portrait's traditional pictorial antecedents, as well as its somber tonality and docile imagery, it is hard to believe its maker was ever labeled a "wild beast."

MAURICE DE VLAMINCK (1876–1958). The career of Maurice de Vlaminck presents many parallels with Derain's, even though the artists were so different in personality and in their approach to the art of painting. Vlaminck was born in Paris of Flemish stock, a fact of considerable importance to him personally and one that may have contributed to the nature of his painting. Impulsive and exuberant, he was a self-

taught artist with anarchist political leanings who liked to boast of his contempt for the art of museums. From the time he met Derain and turned to painting, he was enraptured with color. Thus the Van Gogh exhibition at the Bernheim-Jeune Gallery in Paris in 1901 was a revelation to him. At the Salon des Indépendants in the same year, Derain introduced him to Matisse, but it was not until 1905, after exposure to the work of both artists, that Vlaminck's work reached its full, albeit limited, potential, despite his false claims to have been the leader of the Fauves. Van Gogh remained Vlaminck's great inspiration, and in his Fauve paintings Vlaminck characteristically used the short, choppy brushstrokes of the Dutchman to attain a comparable kind of coloristic dynamism, as in *Picnic in the Country* (fig. 150). Two figures are isolated within a coil of swirling color patches; they are foreigners from the world of nature, picnicking in a forest of paint. His small but dramatic *Portrait of Derain* (color plate 58, page 153) is one of several likenesses the Fauves made of one another. Vlaminck has here moved beyond the directional, multicolored daubs of *Picnic in the Country* to a boldly conceived image in which Derain's face is predominantly an intense, brilliant red with black contours, yellow highlights, and a few strokes of contrasting green shadow along the bridge of the nose, recalling Matisse's *Portrait of Mme Matisse/The Green Line* (color plate 51, page 117).

By 1908 Vlaminck too was beginning to be affected by the new view of Cézanne that resulted from the exhibitions after Cézanne's death in 1906. And although for a time he was in touch with the new explorations of Picasso and Braque leading toward Analytic Cubism, and even used various forms of simplification based on their ideas, in actuality he, like Derain, was gradually retreating into the world of representation by way of Cézanne. By 1915 Vlaminck had moved toward a kind of expressive realism that he continued to pursue for the rest of his life.

RAOUL DUFY (1877–1953) was shocked out of his reverence for the Impressionists and Van Gogh by his discovery of Matisse's 1905 painting *Luxe, calme et volupté*, when, he said, "Impressionist realism lost all its charm." In a sense, he remained faithful to this vision and to Fauve color throughout his life. His *Street Decked with Flags, Le Havre* (color plate 59, page 153) takes up a subject celebrated by the Impressionists, but the bold, close-up view of the flags imposes a highly abstracted geometric pattern on the scene.

Influenced by Georges Braque, his fellow painter from Le Havre, Dufy after 1908 experimented with a modified form of Cubism, but he was never really happy in this vein. Gradually he returned to his former loves—decorative color and elegant draftsmanship—and formulated a personal style based on his earlier Fauvism. His pleasurable subjects were the horse races and regattas of his native Le Havre and nearby Deauville, the nude model in the studio, and the view from a window to the sea beyond. He maintained a rainbow, calligraphic style until the end of his life, applying it to fabrics, theatrical sets, and book illustrations as well as paintings.

GEORGES ROUAULT (1871–1958). The one Fauve who

151. GEORGES ROUAULT.
Prostitute before a Mirror. 1906. Watercolor on cardboard, 27⅝ x 20⅞" (70.2 x 53 cm). Musée National d'Art Moderne, Centre National d'Art et de Culture Georges Pompidou, Paris

was almost exclusively concerned with a deliberately Expressionist subject matter is perhaps not to be considered a Fauve at all. This is Georges Rouault, who exhibited three works in the 1905 Salon d'Automne and thus is associated with the work of the group, although his paintings were not actually shown in the room with theirs. Throughout his long and productive career, Rouault remained deeply religious, deeply emotional, and profoundly moralistic. He came from a family of craftsmen, and he himself was first apprenticed to a stained-glass artisan. In the studio of Gustave Moreau he met Matisse and other future Fauves, and soon became Moreau's favorite pupil, for he followed most closely Moreau's own style and concepts.

By 1903 Rouault's art, like that of Matisse and others around him, was undergoing profound changes, reflecting a radical shift in his moral and religious outlook. Like his friend the Catholic writer and propagandist Léon Bloy, Rouault sought subjects to express his sense of indignation and disgust over the evils that, as it seemed to him, permeated bourgeois society. The prostitute became his symbol of this rotting society (fig. 151). He depicted her with a fierce loathing, rather than objectively, or with the cynical sympathy of Toulouse-Lautrec—who, it must be added, represented individuals rather than Rouault's general types. Rouault invited prostitutes to pose in his studio, painting them as degenerate creatures, with attributes such as stockings or corsets to indicate their profession. In many of these studies, done in watercolor, the woman is set within a confined space, to focus attention on the figure. Some of the poses and the predominantly blue hues of these watercolors stem in part from Rouault's admiration for Cézanne. The women's yellow-white bodies, touched with light-blue shadows, are modeled sculpturally with heavy, freely brushed outlines. Here the masklike grimace of the face, exuding an aura of vanity and decay, is reflected in the mirror, like a twisted paraphrase of the classical Venus, who contemplates her beauty in a looking glass. Rarely in the history of art has the nude been painted with such revulsion. It was perhaps not until the 1920s, with the work of the Berlin artist George Grosz, who bore witness through his art to the moral failings of postwar German society, that prostitutes were interpreted with comparable vehemence.

Rouault's moral indignation further manifested itself, like Daumier's, in vicious caricatures of judges and politicians. His counterpoint to the corrupt prostitute was the figure of the circus clown, sometimes the gay nomad beating his drum, but more often a tragic, lacerated victim. As early as 1904 he had begun to depict subjects taken directly from the Gospels—the Crucifixion, Jesus and his disciples, and other scenes from the life of Christ. He represented the figure of

152. GEORGES ROUAULT. *It Is Hard to Live . . .*, Plate 12 from Miserere. 1922. Etching, aquatint, drypoint, and roulette over heliogravure, printed in black, plate 18⅞ x 14³⁄₁₆" (47.9 x 36 cm). The Museum of Modern Art, New York

Christ as a tragic mask of the Man of Sorrows deriving directly from a crucified Christ by Grünewald or a tormented Christ by Bosch. Rouault's religious and moral sentiments are perhaps most movingly conveyed in a series of fifty-eight prints, titled Miserere, that was commissioned by his dealer Vollard (whose heirs the artist later had to sue to retrieve the contents of his studio). For years, Rouault devoted himself to the production of the etchings and aquatints of Miserere, which were printed between 1922 and 1927, but not published until 1948, when the artist was seventy-seven. The figure in *It Is Hard to Live . . .* (fig. 152) is reminiscent of Rouault's depictions of Christ's passion and, like other prints in the series, makes reference to the artist's suffering from the horrors of World War I, as well as to his faith in salvation. Technically, the prints are masterpieces of graphic compression. The black tones, worked over and over again, have the depth and richness of his most vivid oil colors.

The characteristics of Rouault's later style are seen in *The Old King* (color plate 60, page 153). The design has become geometrically abstract in feeling. Colors are intensified to achieve the glow of stained glass. A heavy black outline is used to define the rather Egyptian-style profile of the king's head and the square proportions of his torso. Paint is applied heavily with the underpainting glowing through, in the manner of Rembrandt. Rouault also captures some of the Dutch master's mood in this serene image of a world-weary ruler, who clutches a flower in his hand, one of the few traces of white in the entire painting. Although Rouault never entirely gave up the spirit of moral indignation expressed in the virulent satire that marked his early works, the sense of calm and the hope of salvation in the later paintings mark him as one of the few authentic religious painters of the modern world.

For Matisse, Fauvism was only a beginning from which he went on to a rich, productive career that spans the first half of the twentieth century. Derain and Vlaminck, however, did little subsequently that had the vitality of their Fauve works. It is interesting to speculate on why these young men should briefly have outdone themselves, but the single overriding explanation is probably the presence of Henri Matisse—older than the others, more mature and, ultimately, more gifted as an artist. But in addition to Matisse and Rouault, there was also Georges Braque, who, after discovering his first brief and relatively late inspiration in Fauvism, went on to restudy Cézanne, with consequences for twentieth-century art so significant that they must await a subsequent chapter on Cubism.

Developments in photography contemporary with the years of Fauvism include the first photographers to take up a color process, after one was finally made public in 1904 and commercially feasible in 1907 by **AUGUSTE** and **LOUIS LUMIÈRE** (1862–1954; 1864–1948) of Lyons, France (color plate 61, page 154). By coating one side of a glass plate with a mixture of tiny, transparent starch particles, each dyed red, green, or blue (the three primary photographic colors), and the other side with a thin panchromatic emulsion, the Lumières created a light-sensitive plate that, once exposed, developed, and projected on a white ground, reproduced a

153. Jacques-Henri Lartigue.
In My Room: Collection of My Racing Cars. 1905. Gelatin-silver print,
5⅜ x 7" (13.7 x 17.8 cm) © Association des Amis de J.-H. Lartigue, Paris

full-color image of the subject photographed. The Auto-chromes—as the Lumières called the slides made by their patented process—rendered images in muted tonalities with the look of a fine-grained texture, an effect that simply heightened the inherent charm of the bright subject matter. This was especially true for a Fauve generation still entranced with Neo-Impressionism. Their invention was immediately taken up by professional photographers, especially the so-called Pictorialists, such as Edward Steichen, as well as amateurs around the world.

Dufy's privileged, Belle Époque world of regattas and racecourses was also celebrated by the French contemporary photographer **Jacques-Henri Lartigue** (1894–1986), once the fast-action handheld camera had been introduced in 1888 and then progressively improved. Such developments encouraged experimentation and enabled this affluent child artist (he began making photographs at age seven) to capture not only his family and friends at their pleasures, but also the gradual advent of such twentieth-century phenomena as auto races and aviation. At his death in 1986 Lartigue, who was a painter professionally, left behind a huge number of photographs and journals that document a charmed life of holidays, swimming holes, and elegantly clad ladies and gentlemen. He was eleven when he aimed his camera at the toy cars in his bedroom (fig. 153), and his image adopts a child's angle on the world. The tiny cars, at eye level, take on strange dimensions, all of which is compounded by the mysteriously draped fireplace that looms above them. Owing to his view of photography as a pursuit carried on for private satisfaction

and delight, Lartigue did not become generally known until a show at New York's Museum of Modern Art in 1963, when his work took an immediate place in the history of art as a direct ancestor of the "straight" but unmistakably distinctive vision of such photographers as Brassaï and Henri Cartier-Bresson.

Matisse's Art after Fauvism

Fauvism was a short-lived but tremendously influential movement that had no definitive conclusion, though it had effectively drawn to a close by 1908. The direction of Matisse's art explored in *Le Luxe II* is carried still further in *Harmony in Red* (color plate 62, page 154), a large painting destined for the Moscow dining room of Sergei Shchukin, Matisse's important early patron who, along with Ivan Morozov, is the reason Russian museums are today key repositories of Matisse's greatest work. This astonishing painting was begun early in 1908 as *Harmony in Blue* and then repainted in the fall, in a radically different color scheme. Here Matisse returned to the formula of *Dinner Table,* which he had painted in 1896–97 (color plate 47, page 116). A comparison of the two canvases reveals dramatically the revolution that had occurred in this artist's works—and in fact in modern painting—during a ten-year period. Admittedly, the first *Dinner Table* was still an apprentice piece, a relatively conventional exploration of Impressionist light and color and contracted space, actually more traditional than paintings executed by the Impressionists twenty years earlier. Nevertheless, when submitted to the Salon de la Nationale it was severely criticized by the conservatives as being tainted with Impressionism. In *Harmony in Red* we have moved into a new world, less empirical and more abstract than anything ever envisioned by Gauguin, much less the Impressionists. The space of the interior is defined by a single unmodulated area of red, the flatness of which is reinforced by arabesques of plant forms that flow across the walls and table surface. These patterns were actually derived from a piece of decorative fabric that Matisse owned. Their meandering forms serve to confound any sense of volumetric space in the painting and to create pictorial ambiguities by playing off the repeated pattern of flower baskets against the "real" still lifes on the table. This ambiguity is extended to the view through the window of abstract tree and plant forms silhouetted against a green ground and blue sky. The red building in the extreme upper distance, which reiterates the color of the room, in some manner establishes the illusion of depth in the landscape, yet the entire scene, framed by what may be a window sill and cut off by the picture edge like other forms in the painting, could also itself be a painting on the wall. In essence Matisse has again—and in an even greater degree than in *Le Luxe II*—created a new, tangible world of pictorial space through color and line.

In two huge paintings of the first importance, *Dance (II)* and *Music,* both of 1909–10 (color plates 63, 64, page 155), and both commissioned by Shchukin, Matisse boldly outlined large-scale figures and isolated them against a ground of intense color. The inspiration for *Dance* has been variously traced to Greek vase painting or peasant dances. Specifically, the motif was first used by Matisse, as we have seen, in the background group of *Le Bonheur de vivre.* In *Dance,* the colors have been limited to an intense green for the earth, an equally intense blue for the sky, and brick-red for the figures. The figures are sealed into the foreground by the color areas of sky and ground, and by their proximity to the framing edge and their great size within the canvas, but they never-

theless dance ecstatically in an airy space created by these contrasting juxtaposed hues and by their own modeled contours and sweeping movements. The depth and intensity of the colors change in different lights, at times setting up visual vibrations that make the entire surface dance. *Music* is a perfect foil for the kinetic energy of *Dance,* in the static, frontalized poses of the figures arranged like notes on a musical staff, each isolated from the others to create a mood of trancelike withdrawal. Matisse's explanation of Fauvism as "the courage to return to a 'purity of means' " still holds true here. In both paintings the arcadian worlds of earlier painters (see fig. 11) have been transformed into an elemental realm beyond the specificities of time and place. When they were exhibited in the 1910 Salon d'Automne, these two extraordinary paintings provoked little but negative and hostile criticism, and Shchukin at first withdrew his commission, though he soon changed his mind.

In *The Red Studio* (color plate 65, page 156), Matisse returned to the principle of a single, unifying color that he had exploited in *Harmony in Red.* The studio interior is described by a uniform area of red, covering floor and walls. The space is given volumetric definition only by a single white line that indicates the intersection of the walls and floor, and by the paintings and furniture lined up along the rear wall. Furnishings—table, chair, cupboard, and sculpture stands—are dematerialized, ghostlike objects outlined in white lines. The tangible accents are the paintings of the artist, hanging on or stacked against the walls, and (in the foreground) ceramics, sculptures, vase, glass, and pencils. (*Le Luxe II* can be seen at the upper right.) By 1911, when this was painted, Picasso, Braque, and the other Cubists, as we shall see, had in their own ways been experimenting with the organization and contraction of pictorial space for some five years. Matisse was affected by their ideas, but sought his own solutions. Later works by Matisse are discussed in chapter 13.

8 Expressionism in Germany

In its most general sense, *Expressionism* is a term that has been applied by art historians to tendencies recurring in the arts since antiquity. The aim of all art is, of course, to communicate, to express ideas or sentiments that evoke responses in the viewer. But in the early years of the twentieth century, the term originated as a means of describing art that conveyed, through a wide range of styles and subject matter (or in the case of abstract art, no subject at all), the subjective emotions, the innermost feelings of the artist—what Vasily Kandinsky called "inner necessity." Herwarth Walden, critic, poet, musician, and the founder in 1910 of the German avant-garde periodical *Der Sturm* (the storm), drew the distinction between new, revolutionary tendencies after the turn of the century and Impressionism. Expressionist artists built on the discoveries of the Post-Impressionists, who rejected Impressionist devotion to optical veracity and turned inward to the world of the spirit. They employed many languages to give visible form to their feelings, but generally they relied on simple, powerful forms that were realized in a manner of direct, sometimes crude expression, designed to heighten the emotional response of the viewer. The essence of their art was the expression of inner meaning through outer form.

Important forerunners of Expressionism included Van Gogh, Munch, Klimt, Hodler, Ensor, Böcklin, Klinger, and Kubin. However, the young Expressionists in Germany also drew inspiration from their own native traditions of medieval sculpture and folk or children's art, as well as the art of other cultures, especially Africa and Oceania. German Expressionism involved the formation of individual groups of artists in several cities, primarily Dresden, Munich, and Berlin. We will see that separate developments also took place in Vienna. The period of German Expressionism began in 1905 with the establishment of the new artists' alliance known as Die Brücke (the bridge) in Dresden, and lasted until the end of World War I, when radical Dada artists in Germany rejected all forms of Expressionism in their turn. Many Expressionist artists welcomed World War I as a new beginning and the destruction of an old, moribund order. But as the horror in the trenches dragged on, the war took its toll on artists: some died in battle; others suffered psychological trauma and profound spiritual disillusionment.

PAULA MODERSOHN-BECKER (1876–1907) was born in Dresden and settled in the artists' colony at Worpswede near Bremen in 1897. Although she was not associated with any group outside of the provincial school of painters at Worpswede, Modersohn-Becker was in touch with new developments in art and literature through her friendship with the poet Rainer Maria Rilke (who had been Rodin's secretary), as well as through a number of visits to Paris. In these she discovered successively the French Barbizon painters, the Impressionists, and then, in 1905 and 1906, the works of Van Gogh, Gauguin, and Cézanne. Following this last trip, she embarked on a highly productive and innovative period, distilling and simplifying forms while heightening the expressiveness of color. In her extensive letters and diaries, the artist wrote of her desire for an art of direct emotion, of poetic expression, of simplicity and sensitivity to nature: "Personal feeling is the main thing. After I have clearly set it down in form and color, I must introduce those elements from nature that will make my picture look natural." She worried about the implications of marriage and motherhood for her professional life and, ironically, died prematurely following the birth of her only child, leaving us with only a suggestion of what she might have achieved. But in works such as *Self-Portrait with Camelia Branch* (fig. 154) there is a clear grasp of the broad simplification of color areas she had seen in Gauguin as well as in the Egyptian and early classical art she had studied at the Louvre.

EMIL NOLDE (1867–1956) was the son of a farmer from northwestern Germany, near the Danish border. The reactionary, rural values of this area had a profound effect on his art and his attitude toward nature. Strong emotional ties to the landscape and yearning for a regeneration of the German spirit and its art characterized the popular *völkisch* tradition, which was later incorporated into the nationalist and racist rhetoric of the Nazis. Early in his career, Nolde depicted the landscape and peasants of this region in paintings somewhat reminiscent of the French artist Millet. As an adult, he even took as his surname Nolde, the name of his native village (he was born Emil Hansen), to underscore a strong identifica-

154. PAULA MODERSOHN-BECKER. *Self-Portrait with Camelia Branch.* 1907. Oil on paperboard, 27½ x 11¾" (62 x 30 cm). Museum Folkwang, Essen, Germany

tion with the land. Unfortunately, Nolde ultimately failed to distinguish between such ties to his beloved ancestral home and the later Nazi exploitation of these ideas.

As a youth, Nolde studied woodcarving and worked for a time as a designer of furniture and decorative arts in Berlin. His first paintings, of mountains transformed into giants or hideous trolls, drew on qualities of traditional Germanic fantasy. The commercial success of some of these images enabled Nolde to return to school and to take up painting seriously in Munich, where he encountered the work of contemporary artists such as Adolph Menzel and Arnold Böcklin, and then in Dachau and Paris. While in Paris during the years 1899–1900, he, like so many art students, worked his way gradually from the study of Daumier and Delacroix to Manet and the Impressionists, and his color took on a new brilliance and violence as a result of his exposure to the latter. In 1906 he accepted an invitation to become a member of the Dresden group Die Brücke. Essentially a solitary, Nolde left Die Brücke after a year and devoted himself more and more to a personal form of Expressionist religious paintings and prints.

Among the first of Nolde's visionary religious paintings was *The Last Supper* of 1909 (color plate 66, page 157). One thinks back to Rembrandt's *Christ at Emmaus* or even Leonardo da Vinci's *Last Supper*, but Nolde's mood is different from the quiet restraint of these earlier works. Nolde's figures are crammed into a practically nonexistent space, the red of their robes and the yellow-green of their faces flaring like torches out of the surrounding shadow. The faces themselves are skull-like masks that derive from the carnival processions of Ensor. Here, however, they are given intense personalities—no longer masked and inscrutable fantasies but individualized human beings passionately involved in a situation of extreme drama. The compression of the group packed

within the frontal plane of the painting—again stemming from Ensor—heightens the sense of impending crisis.

Nolde believed in the ethnic superiority of Nordic people. In 1920 he became a member of the National Socialist party. His art was for a time tolerated by Joseph Goebbels, Hitler's minister of propaganda, but by 1934 it was officially condemned by the Nazis as stylistically too experimental—"primitive" and "un-German." In 1936, like other Expressionist artists in Germany, he was banned from working. Over one thousand of Nolde's works were among the sixteen thousand sculptures, paintings, prints, and drawings by avantgarde artists that were confiscated from German museums by Nazi officials. In 1937 many of these were included in a massive exhibition called "Entartete Kunst," or "Degenerate Art" (see fig. 318). Designed to demonstrate the kind of so-called decadent art that offended the Nazi government, which preferred watered-down classicist depictions of muscular Aryan workers, pretty nudes, or insipid genre scenes, the exhibition contained art in a tremendous range of styles and featured work by most of the artists discussed in this chapter. During four months in Munich, the exhibition was seen by two million visitors, a staggering attendance even by the standard of today's "blockbuster" exhibitions. Many of the works exhibited, including several by Nolde, were later destroyed by the Nazis, or lost during World War II.

Die Brücke

In 1905 Ernst Ludwig Kirchner, Erich Heckel, Karl Schmidt-Rottluff, and Fritz Bleyl formed an association they called Die Brücke—the bridge linking "all the revolutionary and fermenting elements." These young architecture students, all of whom wanted to be painters, were drawn together by their opposition to the art that surrounded them, especially academic art and fashionable Impressionism, rather than by any preconceived program. Imbued with the spirit of Arts and Crafts and Jugendstil, they rented an empty shop in a workers' district of Dresden, in eastern Germany, and began to paint, sculpt, and make woodcuts together. The influences on them were many and varied: the art of medieval Germany, of the French Fauves, of Edvard Munch, of non-Western sculpture. Although they worked in many media, probably their intensive study of the possibilities of woodcut did most to formulate their styles and to clarify their directions. For them, Van Gogh was the clearest example of an artist driven by an "inner force" and "inner necessity"; his paintings presented an ecstatic identification or empathy of the artist with the subject he was interpreting. The graphic works of Munch were widely known in Germany by 1905, and the artist himself was spending most of his time there. His obsession with symbolic subjects struck a sympathetic chord in young German artists, and from his mastery of the graphic techniques they could learn much. Among historic styles, the most exciting discovery was art from Africa and the South Pacific, of which notable collections existed in the Dresden Ethnographic Museum.

In 1906 Nolde and Max Pechstein joined Die Brücke. In the same year Heckel, then working for an architect, persuaded a manufacturer for whom he had executed a show-room to permit the Brücke artists to exhibit there. This was the historic first Brücke exhibition, which marked the emergence of twentieth-century German Expressionism. Little information about the exhibition has survived, since no catalogue was issued, and it attracted virtually no attention.

During the next few years the Brücke painters exhibited together and produced publications designed by members and manifestos in which Kirchner's ideas were most evident. The human figure was studied assiduously in the way Rodin had studied the nude: not posed formally but simply existing in the environment. Despite developing differences in style among the artists, a hard, Gothic angularity permeated many of their works.

The Brücke painters were conscious of the revolution that the Fauves were creating in Paris and were affected by their use of color. However, their own paintings maintained a Germanic sense of expressive subject matter, a characteristically jagged, Gothic structure and form. By 1911 most of the Brücke group were in Berlin, where a new style appeared in their works, reflecting the increasing consciousness of French Cubism as well as Fauvism, given a Germanic excitement and narrative impact. By 1913 Die Brücke was dissolved as an association, and the artists proceeded individually.

ERNST LUDWIG KIRCHNER (1880–1938) was the most creative member of Die Brücke. In addition to the extraordinary body of painting and prints he left, Kirchner, like Erich Heckel, made large, roughly hewn and painted wooden sculptures. These works in a primitivist mode were a result of the artists' admiration for African and Oceanic art. Kirchner's early ambition to become an artist was reinforced by his discovery of the sixteenth-century woodcuts of Albrecht Dürer and his Late Gothic predecessors. Yet his own first woodcuts, done before 1900, were probably most influenced by Félix Vallotton and Edvard Munch. Between 1901 and 1903 Kirchner studied architecture in Dresden, and then, until 1904, painting in Munich. Here he was attracted to Art Nouveau designs and repelled by the retrograde paintings he saw in the exhibition of the Munich Secession. Like so many of the younger German artists of the time, he was particularly drawn to German Gothic art. Of modern artists, the first revelation for him was the work of Seurat. Going beyond Seurat's researches, Kirchner undertook studies of nineteenth-century color theories that led him back to Johann Wolfgang von Goethe's essay, *History of the Theory of Colors*.

Kirchner's painting style about 1904 showed influences from the Neo-Impressionists but a larger, more dynamic brushstroke related to that of Van Gogh, whose work he saw, along with paintings by Gauguin and Cézanne, in the exhibition of the Munich Artists' Association held that year. On his return to Dresden and architecture school in 1904 Kirchner became acquainted with Heckel, Schmidt-Rottluff, and Bleyl, and with them, as already noted, he founded the Brücke group.

For subject matter Kirchner looked to contemporary life—landscape, portraits of friends, and nudes in natural settings—rejecting the artificial trappings of academic studios.

155. ERNST LUDWIG KIRCHNER. *Street, Berlin*. 1913.
Oil on canvas, 47½ x 35⅞" (120.6 x 91.1 cm).
The Museum of Modern Art, New York PURCHASE
PHOTOGRAPH © 1995 THE MUSEUM OF MODERN ART, NEW YORK

In both Dresden and Berlin, where he moved in 1911, he recorded the streets and inhabitants of the city and the bohemian life of its nightclubs, cabarets, and circuses. *Street, Dresden* (color plate 67, page 157) of 1908 is an assembly of curvilinear figures who undulate like wraiths, moving toward and away from the viewer without individual motive power, drifting in a world of dreams. Kirchner probably had Munch's street painting *Evening on Karl Johan* in mind when he made this work, and, as was frequently his habit, he reworked it at a much later date. In Berlin he painted a series of street scenes (fig. 155) in which the spaces are confined and precipitously tilted, and the figures are elongated into angular shards by long feathered strokes. Kirchner made rapid sketches of these street scenes, then worked up the image in more formal drawings in his studio before making the final paintings. He aimed to capture the animation of his first impressions, and the strong sense of movement may indicate his borrowing from Italian Futurist painting (see chapter 11), as well as the impact of Cubism. In the charged atmosphere of the street, and the distorted Gothic forms, Kirchner describes the condition of modern urban life, one of close physical proximity coupled with extreme psychological distance.

If Kirchner's *Market Place with Red Tower* (fig. 156) is compared with one of the French Cubist painter Robert De-

156. ERNST LUDWIG KIRCHNER. *Market Place with Red Tower.* 1915. Oil on canvas, 47½ x 35⅞" (120.7 x 91.1 cm). Museum Folkwang, Essen, Germany

157. ERICH HECKEL. *Two Men at a Table.* 1912. Oil on canvas, 38⅛ x 47¼" (96.8 x 120 cm). Kunsthalle, Hamburg
© 1997 VG BILD-KUNST, BONN/ARTISTS RIGHTS SOCIETY (ARS), NEW YORK

launay's Eiffel Tower studies, the difference between the German and the French vision becomes evident. Delaunay's *Eiffel Tower in Trees* of 1910 (see fig. 230) represents an expressive interpretation of Cubism in which the expression is achieved by the dynamism of abstract color shapes and lines. Kirchner's *Market Place* is expressive in a more explicit way. It reveals the artist's knowledge of Cubist works, but uses Cubist geometry with caution, combining it with defined perspective space distorted in the manner of Van Gogh for the similar end of creating a claustrophobic effect of compression—what Kirchner called the "melancholy of big city streets."

When World War I broke out Kirchner enlisted in the field artillery, but he suffered a mental and physical breakdown and was discharged in 1915. He recuperated in Switzerland, where he continued to live and work near the town of Davos until his death by suicide in 1938. During this late period in his career Kirchner continued to renew his style, painting many of his older themes along with serene Alpine landscapes and sympathetic portrayals of Swiss peasant life. The Swiss landscape can be seen through the window in Kirchner's moving *Self-Portrait with Cat* (color plate 68, page 157), in which the artist's weary countenance betrays his protracted struggle with illness and depression.

ERICH HECKEL (1883–1970) was a more restrained Expressionist whose early paintings at their best showed flashes of psychological insight and lyricism. After 1920 he turned more and more to the production of colorful but essentially Romantic-Realist landscapes. A painting such as *Two Men at a Table* (fig. 157) evokes a dramatic interplay in which not only the figures but the contracted, tilted space of the room is charged with emotion. This painting, dedicated "to Dostoyevsky," is almost a literal illustration from the Russian novelist's *Brothers Karamazov.* The painting of the tortured Christ, the suffering face of the man at the left, the menace of the other—all refer to Ivan's story, in the novel, of Christ and the Grand Inquisitor.

OTTO MÜLLER (1874–1930). Müller's colors are the most delicate and muted of all the Brücke painters, his paintings the most suggestive of an Oriental elegance in their organization. His nudes are attenuated, awkwardly graceful figures whose softly outlined, yellow-ocher bodies blend imperceptibly and harmoniously into the green and yellow foliage of their setting (fig. 158). He was impressed by ancient Egyptian fresco paintings and developed techniques to emulate their muted tonalities. The unidealized, candid depiction of nudes in open nature was among the Brücke artists' favorite subjects. They saw the nude as a welcome release from nineteenth-century prudery and a liberating plunge into primal experience. As Nolde proclaimed, "Primordial peoples live in their nature, are one with it and are a part of the entire universe." The relative gentleness of Müller's treatment of this theme found an echo in the contemporary photographs of the German-born photographer HEINRICH KÜHN (1866–1944), who, like early twentieth-century modernists in painting, sought to flatten space and create a more two-dimensional design—a Pictorial effect, as the photographers would have said—by viewing his subject or scene from above (fig. 159).

MAX PECHSTEIN (1881–1953) had a considerable head start in art by the time he joined Die Brücke in 1906. He had

158. OTTO MÜLLER. *Bathing Women*. 1912.
Crayon on paper, 17⅜ x 13⅜" (43.7 x 33.5 cm).
Bildarchiv Rheinisches Museum, Cologne
© 1997 VG BILD-KUNST, BONN/ARTISTS RIGHTS SOCIETY (ARS), NEW YORK

159. HEINRICH KÜHN. *The Artist's Umbrella*. 1910. Photogravure
on heavy woven paper, 9 x 11⅜" (23 x 28.9 cm).
The Metropolitan Museum of Art, New York
THE ALFRED STIEGLITZ COLLECTION, 1949

studied for several years at the Dresden Academy, and in general enjoyed an earlier success than the other Brücke painters. In 1905 he had seen a collection of wood carvings from the Palau Islands in Dresden's ethnographic museum, and these had a formative influence on his work. In 1914 he traveled to these islands in the Pacific to study the art at first hand. Pechstein was the most eclectic of the Brücke group, capable of notable individual paintings that shift from one style to another. The early *Indian and Woman* (color plate 69, page 158) shows him at his dramatic best in terms of the exotic subject, modeling of the figures, and Fauve-inspired color. Pechstein's drawing is sculptural and curvilinear in contrast to that of Müller or Heckel, and it was not tinged by the intense anxiety that informed so much of Kirchner's work.

KARL SCHMIDT-ROTTLUFF (1884–1976). In 1910 Karl Schmidt-Rottluff portrayed himself as the very image of the arrogant young Expressionist—in a green turtleneck sweater, complete with beard and monocle, against a roughly painted background of a yellow doorway flanked by purplish-brown curtains (fig. 160). Aside from Nolde, Schmidt-Rottluff was the boldest colorist of the group, given to vivid blues and crimsons, yellows and greens, juxtaposed in jarring but effective dissonance. Although never a fully abstract painter, he was probably the member of Die Brücke who moved furthest and most convincingly in the direction of abstract structure (color plate 70, page 158). Schmidt-Rottluff had been

traumatized by his service in Russia during World War I, and was later commissioned to redesign the imperial German eagle, casts of which were placed on buildings throughout Germany. Despite this service to his country, when the Nazis came to power he was dismissed from his position as professor of art in Berlin in 1933 and forbidden to work. His *Self-Portrait with Monocle* was seized from the museum that had acquired it in 1924 and included in the 1937 Degenerate Art show staged by the Nazis.

160. KARL SCHMIDT-ROTTLUFF. *Self-Portrait with Monocle*. 1910.
Oil on canvas, 33⅛ x 30" (84.1 x 76.2 cm). Staatliche
Museen zu Berlin, Preussischer Kulturbesitz, Nationalgalerie
© 1997 VG BILD-KUNST, BONN/ARTISTS RIGHTS SOCIETY (ARS), NEW YORK

right: 161. LOVIS CORINTH. *Death and the Artist,* from the Totentanz (Dance of Death) series. 1921. Soft-ground etching and drypoint on Japanese paper, plate 9½ x 7" (24 x 17.8 cm), sheet 11¾ x 9⅜" (30 x 24.7 cm). National Gallery of Art, Washington, D.C.
GIFT IN MEMORY OF SIGBERT H. MARCY AND IN HONOR OF THE FIFTIETH ANNIVERSARY OF THE NATIONAL GALLERY OF ART

Expressionist Prints

One of the principal contributions of German Expressionism was the revival of printmaking as a major form of art. During the nineteenth century a large number of the experimental painters and sculptors had made prints, and toward the end of the century, in the hands of artists like Toulouse-Lautrec, Gauguin, Redon, Klinger, and Ensor, printmaking assumed an importance as an independent art form beyond anything that had existed since the Renaissance. Several early twentieth-century artists outside Germany—Picasso, Munch, Rouault— were important printmakers as well as painters. In Germany, however, a country with a particularly rich tradition of print-making, this art form occupied a special place, and its revival contributed to the character of painting and sculpture. Expressionist artists sought a direct engagement with their materials and adapted their technique to the exigencies of the medium in order to achieve maximum expressive power. It was especially the simplified forms of the woodcut, a form of printmaking with a long and distinguished history in German art, that they felt conveyed most authentically the desired emotion. The artists of the Brücke brought about a veritable renaissance of this medium.

LOVIS CORINTH (1858–1925). Considerably older than the artists of Die Brücke, the painter and printmaker Lovis Corinth formed a link between German Romanticism and later German Expressionist art. In his lengthy academic training in Germany and France he learned that a firm grounding in draftsmanship was fundamental to his art. Like his favorite artist, Rembrandt, one of Corinth's preferred subjects was his own visage, which he explored in *Death and the Artist* (fig. 161). This print belongs to his series Dance of Death, his interpretation of a medieval theme that reminds us of our own inevitable mortality. Although Corinth made woodcuts occasionally, he favored lithography, etching, and especially drypoint, a technique that produces a rich, velvety line.

Of the Brücke group, Nolde was already an accomplished etcher by the time he made his first woodcuts. His *Prophet* of 1912 (fig. 162) owes something to Late Gothic German woodcuts as well as the prints of Munch. Nolde here uses intense black-and-white contrast and bold, jagged shapes, exploiting the natural grain of the woodblock and gouging it with angular cuts. His lithographs, which differ in expression from the rugged forms of this woodcut, include *Female Dancer* (color plate 71, page 158), one of the greatest of all German Expressionist prints. Nolde appreciated the artistic freedom afforded him by the lithographic medium, which he used in experimental ways, brushing ink directly on the stone printing matrix to create thin, variable washes of color. Nolde,

162. EMIL NOLDE. *Prophet.* 1912. Woodcut, 14¾ x 10½" (37.4 x 26.6 cm). Kunstmuseum, Düsseldorf im Ehrenhof, Grasphische Sammlung, 1947–54

who was interested in the body as an expressive vehicle, had made sketches in the theaters and cabarets of Berlin, as had Kirchner. But the sense of frenzied emotion and wild abandon in *Female Dancer* evokes associations with some primal, tribal ritual rather than with urban dance halls. Nolde had studied the work of non-European cultures in museums such as the Berlin Ethnographic Museum, concluding that Oceanic and African art possessed a vitality that much Western art lacked. He argued for their study as objects of aesthetic as well as scientific interest and made drawings of objects that were then incorporated into his still-life paintings. In 1913, shortly after making this lithograph, Nolde joined an official ethnographic expedition to New Guinea, then a German colony in the South Pacific, and later to the Far East. He made sketches of the landscapes and local inhabitants on his journeys, from which he returned highly critical of European colonial practices.

Kirchner, in both black-and-white and color woodcuts, developed an intricate, linear style that looked back to the woodcuts of Dürer and Martin Schongauer. His powerful portrait of the Belgian Art Nouveau architect Henry van de Velde (fig. 163), with its complex surface patterns made of characteristic V-shaped gouges, contrasts with the spare composition of Heckel's color woodcut *Standing Child* (color plate 72, page 159). Heckel's adolescent subject was a girl named Fränzi, who, with her sister Marcella, was a favorite model of the Brücke group. The artist reserved the color of the paper for the model's skin and employed three woodblocks—for black, green, and red inks—in the puzzle technique invented by Munch, for the brilliant, abstracted landscape behind her. It is a gripping image for its forthright design and the frank, precocious sexuality of the sitter.

Like Nolde, Pechstein chose dance as his subject for a color woodcut of 1910 (color plate 73, page 159), a work he may have been inspired to make after seeing a Somali dance group perform in Berlin that year. The dancers are portrayed against a colorful backdrop that resembles the kind of hangings with which the Brücke artists decorated their studios. Pechstein sought a deliberately crude execution here, with schematically hewn figures and a surface covered with irregular smudges of ink.

KÄTHE KOLLWITZ (1867–1945) devoted her life and her art, both printmaking and sculpture, to a form of protest or social criticism. She was the first of the German Social Realists who developed out of Expressionism during and after World War I. Essentially a Realist, and powerfully concerned with the problems and sufferings of the underprivileged, she stands somewhat aside from Die Brücke or other Expressionists, such as Kandinsky or Der Blaue Reiter. When she married a doctor, his patients, predominantly the industrial poor of Berlin, became her models.

As a woman, Kollwitz had been prevented from studying at the Academy, so she attended a Berlin art school for women. There she was introduced to the prints and writings of Max Klinger (see fig. 126) and decided to take up printmaking rather than painting. The graphic media appealed to

163. ERNST LUDWIG KIRCHNER. *Head of Henry van de Velde.* 1917. Woodcut, 19⅝ x 15¾" (50 x 40 cm). Staatliche Graphische Sammlung, Munich

Kollwitz for their ability to convey a message effectively and to reach a wide audience, because each image could be printed many times. Like Klinger, she produced elaborate print

164. KÄTHE KOLLWITZ. *Death Seizing a Woman* (1934) from the series Death. 1934–36. Lithograph, printed in black, composition: 20 x 14⅞₆" (50.8 x 36.7 cm). The Museum of Modern Art, New York

cycles on a central theme. Such is the case with *Death Seizing a Woman* (fig. 164), which belongs to her last great series of lithographs. Kollwitz had first explored the theme of a mother cradling her dead child much earlier and had no doubt witnessed such loss among her husband's patients. But the death of her youngest son (who, poignantly, had served as an infant model in the earlier works) in World War I transformed the subject into one of personal tragedy.

Although best known for her black-and-white prints and posters of political subjects, Kollwitz could be an exquisite colorist. Her lithograph of a nude woman (color plate 74, page 159), with its luminous atmosphere and quiet mood, so unlike her declarative political prints, demonstrates her extraordinary skill and sensitivity as a printmaker.

Der Blaue Reiter

The Brücke artists were the first manifestation of Expressionism in Germany but not necessarily the most significant. While they were active, first in Dresden and then in Berlin, a movement of more far-reaching implications was germinating in Munich around one of the great personalities of modern art, Vasily Kandinsky.

VASILY KANDINSKY (1866–1944) was born in Moscow and studied law and economics at the University of Moscow. Visits to Paris and an exhibition of French painting in Moscow aroused his interest to the point that, at the age of thirty, he refused a professorship of law in order to study painting. He then went to Munich, where he was soon caught up in the atmosphere of Art Nouveau and Jugendstil then permeating the city.

Munich, since 1890, had been one of the most active centers of experimental art in all Europe. Kandinsky soon was taking a leading part in the Munich art world, even while undergoing the more traditional discipline of study at the Academy and with older artists. In 1901 he formed a new artists' association, Phalanx, and opened his own art school. In the same year he was exhibiting in the Berlin Secession, and by 1904 had shown works in the Paris Salon d'Automne and Exposition Nationale des Beaux-Arts. That year the Phalanx had shown the Neo-Impressionists, as well as Cézanne, Gauguin, and Van Gogh. By 1909 Kandinsky was leading a revolt against the established Munich art movements that resulted in the formation of the Neue Künstler Vereinigung (NKV, New Artists Association). In addition to Kandinsky, the NKV included Alexej von Jawlensky, Gabriele Münter, Alfred Kubin, and, later, Franz Marc. Its second exhibition, in 1910, showed the works not only of Germans but also of leading Parisian experimental painters: Picasso, Braque, Rouault, Derain, and Vlaminck.

During this period, Kandinsky was exploring revolutionary ideas about nonobjective or abstract painting, that is, painting without literal subject matter, that does not take its form from the observed world. He traced the beginnings of this interest to a moment of epiphany he had experienced in 1908. Entering his studio one day, he could not make out any subject in his painting, only shapes and colors: he then

165. GABRIELE MÜNTER. *Peasant Woman of Murnau with Children.* c. 1909–10. Oil on glass, 8½ x 7¾" (21.6 x 19.5 cm). Städtische Galerie im Lehnbachhaus, Munich
© 1997 VG BILD-KUNST, BONN/ARTISTS RIGHTS SOCIETY (ARS), NEW YORK

realized it was turned on its side. In 1911 a split in the NKV resulted in the secession of Kandinsky, accompanied by Marc and Münter, and the formation of the association Der Blaue Reiter (The Blue Rider), a name taken from a book published by Kandinsky and Marc, which had in turn taken its name from a painting by Kandinsky. The historic exhibition held at the Thannhauser Gallery in Munich in December 1911 included works by Kandinsky, Marc, August Macke, Heinrich Campendonck, Münter, the composer Arnold Schönberg, the Frenchmen Henri Rousseau and Robert Delaunay, and others. Paul Klee, already associated with the group, showed with them in a graphics exhibition in 1912. This was a much larger show. The entries were expanded to include artists of Die Brücke and additional French and Russian artists: Roger de La Fresnaye, Kazimir Malevich, and Jean Arp, an Alsatian.

In his book *Concerning the Spiritual in Art*, published in 1911, Kandinsky formulated the ideas that had obsessed him since his student days in Russia. Always a serious student, he had devoted much time to the problem of the relations between art and music. He had first sensed the dematerialization of the object in the paintings of Monet, and this direction in art continued to intrigue him as, through exhibitions in Munich and his continual travels, he learned more about the revolutionary new discoveries of the Neo-Impressionists, the Symbolists, the Fauves, and the Cubists. Advances in the physical sciences had called into question the "reality" of the world of tangible objects, strengthening his conviction that art had to be concerned with the expression of the spiritual rather than the material. Despite his strong scientific and legal interests, Kandinsky was attracted to Theosophy, spiritism, and the occult. There was always a mystical core in his thinking—something he at times attributed to his Russian roots. This sense of an inner creative force, a product of the spirit rather than of external vision or manual skill, enabled him to arrive at an art entirely without representation other than colors and shapes. Deeply concerned with the expression of harmony in visual terms, he wrote, "The harmony of color and form must be based solely upon the principle of the proper contact with the human soul." And so, along with František Kupka, Delaunay, Piet Mondrian, and Malevich, Kandinsky was one of the first, if not—as traditionally thought—the very first, modern European artist to break through the representational barrier and carry painting into total abstraction. But while other pioneers of pictorial nonobjectivity worked in a Cubist-derived geometric mode, Kandinsky worked in a painterly, improvisatory, Expressionist, biomorphic manner that has had many illustrious heirs, from the Surrealist art of André Masson, Joan Miró, and Matta to the environmentally scaled compositions of Jackson Pollock and beyond, to many contemporary practitioners of gestural painting.

As he was moving toward abstraction, the early paintings of Kandinsky went through various stages of Impressionism and Art Nouveau decoration, but all were characterized by a feeling for color and many for a fairytale quality of narrative, reminiscent of his early interest in Russian folktales and mythology. Kandinsky pursued this line of investigation in the rural setting of the Bavarian village of Murnau, where he lived for a time with Münter. There, **GABRIELE MÜNTER** (1877–1962) took up *Hinterglasmalerei*, a local form of folk art in which a painting is done on the underside of a sheet of glass (fig. 165), and Kandinsky followed her example. The archaism of the style and its spiritual purity forecast the more radical simplifications soon to come. Like his contemporaries in Germany and France, Kandinsky was interested in native Russian art forms and in 1889 had traveled on an ethnographic expedition to study the people of Vologda, a remote Russian province north of Moscow. He was moved by the interiors of the peasants' houses, which were filled with decorative painting and furniture. "I learned not to look at a picture from outside," he said, "but to move within the picture, to live in the picture."

Blue Mountain, No. 84 (fig. 166) is a romantic work of stippled color dots organized within a few large, flat, outlined shapes of mountains and trees, with silhouetted riders on horseback forming a moving pattern on the frontal plane. Technical characteristics can be traced to the Neo-Impressionism of Seurat, while the decorative formula suggests Art Nouveau.

Because of his wish to associate his work with an image-free art form, Kandinsky began using titles derived from music, such as "Composition," "Improvisation," or "Impression." He made ten major paintings titled "Composition," which he considered to be his most complete artistic statements, expressive of what he called "inner necessity" or the artist's intuitive, emotional response to the world. A close examination of *Sketch for Composition II* (color plate 75, page 160) reveals that the artist is still employing a pictorial

166. VASILY KANDINSKY. *Blue Mountain, No. 84.* 1908–09. Oil on canvas, 41⅜ x 37⅞" (105.1 x 96.2 cm). Solomon R. Guggenheim Museum, New York
© 1997 ADAGP, PARIS/ARTISTS RIGHTS SOCIETY (ARS), NEW YORK

vocabulary filled with standing figures, riders on horseback, and onion-domed churches, but they are now highly abstracted forms in the midst of a tumultuous, upheaving landscape of mountains and trees that Kandinsky painted in the high-keyed color of the Fauves. Although Kandinsky said this painting had no theme, it is clear that the composition is divided into two sections, with a scene of deluge and disturbance at the left and a garden of paradise at the right, where lovers recline as they had in Matisse's *Le Bonheur de vivre* (color plate 52, page 118). Kandinsky balances these opposing forces to give his all-embracing view of the universe.

In general, Kandinsky's compositions revolve around themes of cosmic conflict and renewal, specifically the Deluge from the biblical book of Genesis and the Apocalypse from the book of Revelation. From such cataclysm would emerge, he believed, a rebirth, a new, spiritually cleansed world. In *Composition VII* (color plate 76, page 160), an enormous canvas from 1913, colors, shapes, and lines collide across the pictorial field in a furiously explosive composition. Yet even in the midst of this symphonic arrangement of abstract forms, the characteristic motifs Kandinsky had distilled over the years can still be deciphered, such as the glyph of a boat with three oars at the lower left, a sign of the biblical floods. He did not intend these hieroglyphic forms to be read literally, so he veiled them in washes of brilliant color. Though the artist carefully prepared this large work with many preliminary drawings and oil sketches, he preserved a sense of spontaneous, unpremeditated freedom in the final painting.

In 1914 the cataclysm of World War I forced Kandinsky to return to Russia, and, shortly thereafter, another phase of his long and productive career began (discussed in chapter 11). In looking at the work of the other members of Der Blaue Reiter up to 1914, we should recall that the individuals involved were not held together by common stylistic principles but rather constituted a loose association of young artists, enthusiastic about new experiments and united in their oppositions. Aside from personal friendships, it was Kandinsky who gave the group cohesion and direction. The yearbook *Der Blaue Reiter*, edited by Kandinsky and Franz Marc, appeared in 1912 and served as a forum for the opinions of the group. The new experiments of Picasso and Matisse in Paris were discussed at length, and the aims and conflicts of the new German art associations were described. In the creation of the new culture and new approach to painting, much importance was attached to the influence of so-called primitive and naïve art.

FRANZ MARC (1880–1916). Of the Blaue Reiter painters, Franz Marc was the closest in spirit to the traditions of German Romanticism and lyrical naturalism. In Paris in 1907 he sought personal solutions in the paintings of Van Gogh, whom he called the most authentic of painters. From an early date he turned to the subject of animals as a source of spiritual harmony and purity in nature. It became for him a symbol of that more primitive and arcadian life sought by so many of the Expressionist painters. Through his friend the painter August Macke, Marc developed, about 1910, enthusiasms for color whose richness and beauty were expressive also of the harmonies he was seeking. The great *Blue Horses* of 1911 is one of the masterpieces of Marc's mature style (color plate 77, page 161). The three brilliant blue beasts are fleshed out sculpturally from the equally vivid reds, greens, and yellows of the landscape. The artist has here used a close-up view, with the bodies of the horses filling most of the canvas. The horizon line is high, so that the curves of the red hills repeat the lines of the horses' curving flanks. Although the modeling of the animals gives them the effect of sculptured relief projecting from a uniform background, there is no real spatial differentiation between creatures and environment except that the sky is rendered more softly and less tangibly to create some illusion of distance. In fact, the artist uses the two whitish tree trunks and the green of foliage in front of and behind the horses to tie foreground and background together. At this time Marc's color, like that of Kandinsky, had a specifically symbolic rather than descriptive function. Marc saw blue as a masculine principle, robust and spiritual; yellow as a feminine principle, gentle, serene, and sensual; red as matter, brutal and heavy. In the mixing of these colors to create greens and violets, and in their proportions one to the other on the canvas, the colors formed abstract shapes and took on spiritual or material significance independent of the subject.

During the years 1911–12 Marc was absorbing the ideas and forms of the Cubists and finding them applicable to his own concepts of the mystery and poetry of color. His approach to art was religious in a manner that could be described as pantheistic, although it is suggestive that in his paintings only animals are assimilated harmoniously into nature—never humans. Marc, with Macke, visited Delaunay in Paris in 1912 and saw his particular experiments with color and form, the same year found Marc impressed by an exhibition of Italian Futurists he saw in Munich. Out of the influences of Kandinsky, of Delaunay's color abstraction, and of the Futurists' use of Cubist structures, Marc evolved his later style—which, tragically, he was permitted to explore for only two or three more years until he was killed in the war in 1916.

In *Stables* (color plate 78, page 161) the artist combined his earlier curvilinear patterns with a new rectangular geometry. The horses, massed in the frontal plane, are dismembered and recomposed as fractured shapes that are dispersed evenly across the surface of the canvas. The forms are composed parallel to the picture plane rather than tilted in depth. This treatment was in accordance with the experiments of Delaunay in his Simultaneous Disk series (color plate 101, page 203). At this stage, however, Marc's paintings were still far less abstract than Delaunay's, and their sense of nervous energy made them akin to the work of the Futurists. Intense but light-filled blues and reds, greens, violets, and yellows flicker over the structurally and spatially unified surface to create a dazzling illusion. Marc's use of color at this stage owed much to Macke, who frequently used the theme of people reflected in shop windows as a means of distorting space and multiplying images.

Marc, unlike Kandinsky, was predominantly a painter of imagery derived from the material world. But in a group of small compositions from 1914 and in sketches that he made in the notebook he carried, there was evidence of a move to abstraction. Some of his later works carry premonitions of the world conflict that ended his life. In a virtually abstract painting, *Fighting Forms* of 1914 (color plate 79, page 162), Marc returned to curvilinear pattern in a violent battle of black and red color shapes, of light and darkness. The forms are given such vitality that they take on the characteristics of forces in an ultimate encounter.

AUGUST MACKE (1887–1914). Of the major figures in Der Blaue Reiter, August Macke, despite his close association with and influence on Franz Marc, should probably not be considered an Expressionist at all. Macke, like Marc, was influenced by Kandinsky, Delaunay, and the Futurists, and perhaps more immediately by the color concepts of Gauguin and Matisse. Since he was killed in September 1914, one month after the beginning of World War I, his achievement must be gauged by the work of only four years.

After some Fauve- and Cubist-motivated exercises in semi-abstraction, Macke began to paint city scenes in high-keyed color, using diluted oil paint in effects close to that of watercolor. *Great Zoological Garden*, a triptych of 1912 (color plate 80, page 162), is a loving transformation of a familiar scene into a fairyland of translucent color. Pictorial space is delimited by foliage and buildings that derive from the later watercolors of Cézanne. The artist moves easily from passages as abstract as the background architecture to passages as literally representational as the animals and the foreground figures. The work has a unity of mood that is gay, light, and charming, disarming because of the naïve joy that permeates it.

Macke occasionally experimented with abstract organization, but his principal interest during the last two years of his life continued to be his cityscapes, which were decorative colored impressions of elegant ladies and gentlemen strolling in the park or studying the wares in brightly lit shop windows. In numerous versions of such themes, he shows his fascination with the mirrorlike effects of windows as a means of transforming the perspective space of the street into the fractured space of Cubism.

ALEXEJ VON JAWLENSKY (1864–1941) was well established in his career as an officer of the Russian Imperial Guard before he decided to become a painter. After studies in Moscow he took classes in the same studio in Munich as Kandinsky. Although not officially a member of Der Blaue Reiter, Jawlensky was sympathetic to its aims and continued for years to be close to Kandinsky. After the war he formed the group called Die Blauen Vier (the Blue Four), along with Kandinsky, Klee, and Lyonel Feininger.

By 1905 Jawlensky was painting in a Fauve palette, and his drawings of nudes of the next few years are suggestive of Matisse. About 1910 he settled on his primary theme, the portrait head, which he explored thenceforward with mystical intensity. *Mme Turandot* is an early example (fig. 167), painted in a manner that combines characteristics of Russian folk painting and Russo-Byzantine icons—qualities that were to become more and more dominant.

In 1914 Jawlensky embarked on a remarkable series of paintings of the human head that occupied him intermittently for over twenty years. Each of these "mystic heads" assumed a virtually identical format: a large head fills the frame, with features reduced to a grid of horizontal and vertical lines and planes of delicate color, more a schema for a face than an actual visage. In the twenties Jawlensky made a series, Constructivist Heads, (color plate 81, page 163), in which the eyes are closed, casting an introspective, meditative mood over these images, which, for the artist, were expressions of a universal spirituality. Like Kandinsky's compositions, they were variations on a theme, but in their restrictive, repetitive structure, they are closer in spirit to the neoplastic work of Piet Mondrian (color plate 200, page 388). In the mid-1930s the geometry gradually loosened, and Jawlensky dissolved the linear structure into dark, forcefully brushed abstractions that he called "Meditations."

PAUL KLEE (1879–1940) was one of the most varied, influential, and brilliant talents in the twentieth century. His stylistic development is difficult to trace even after 1914, since the artist, from this moment of maturity, continually reexamined his themes and forms to arrive at a reality beyond the visible world. He developed a visual language seemingly limitless in its invention and in the variety of its formal means. His complex language of personal signs and symbols evolved through private fantasy and a range of interests from plant and zoological life to astronomy and typography. Also interested in music, he wanted to create imagery infused with the rhythms and counterpoint of musical composition, saying that color could be played like a "chromatic keyboard."

Klee was born in Switzerland. The son of a musician, he was initially inclined toward music, but having decided on the career of painting, went to Munich to study in 1898. During the years 1903–6 he produced a number of etchings (fig. 168) that in their precise, hard technique suggest the graphic tradition of the German Renaissance, in their mannered linearity Art Nouveau, and in their mad fantasy a personal vision reflecting the influence of the Expressionist printmaker Alfred Kubin (see fig. 127). These were also among the first of Klee's works in which the title formed an integral part of his creative work. Klee brought tremendous verbal skill and wit to his paintings, drawings, and prints, using letters and words literally as formal devices in his compositions, which he then gave a literary dimension with poetic and often humorous titles.

Klee traveled extensively in Italy and France between 1901 and 1905 and probably saw works by Matisse. Between 1908 and 1910 he became aware of Cézanne, Van Gogh, and the beginnings of the modern movement in painting. In 1911 he had a one-person exhibition at the Thannhauser Gallery in Munich and in the same year met the Blaue Reiter painters Kandinsky, Marc, Macke, Jawlensky, and Münter. Over the next few years he participated in the Blaue Reiter exhibitions, wrote for *Der Sturm*, and, in Paris again in 1912, met Delaunay and saw further paintings by Picasso, Braque, and Henri Rousseau.

He took a trip with Macke to Tunis and other parts of North Africa in 1914. Like Delacroix and other Romantics before him, he was affected by the brilliance of the region's light and the color and clarity of the atmosphere. To catch the

167. ALEXEJ VON JAWLENSKY. *Mme Turandot*. 1912. Oil on canvas, 23 x 20" (58.4 x 50.8 cm).
Collection Andreas Jawlensky, Locarno, Switzerland

168. PAUL KLEE. *Two Gentlemen Bowing to One Another, Each Supposing the Other to Be in a Higher Position (Invention 6).* 1903. Etching on paper, sheet 4⅝ x 8⅛" (11.8 x 20.7cm). Solomon R. Guggenheim Museum, New York © 1997 VG BILD-KUNST, BONN/ARTISTS RIGHTS SOCIETY (ARS), NEW YORK

quality of the scene Klee, like Macke, turned to watercolor and a form of semiabstract color pattern based on a Cubist grid, a structure he frequently used as a linear scaffolding for his compositions (color plate 82, page 163). Although he made larger paintings, he tended to prefer small-scale works on paper like this one. Klee had a long and fertile career, the remainder of which will be traced in chapter 17.

LYONEL FEININGER (1871–1956). Although Lyonel Feininger was born an American of German-American parents, as a painter he belongs within the European orbit. The son of distinguished musicians, he was, like Klee, early destined for a musical career. But before he was ten he was drawing his impressions of buildings, boats, and elevated trains in New York City. He went to Germany in 1887 to study music, but soon turned to painting. In Berlin between 1893 and 1905 Feininger earned his living as an illustrator and caricaturist for German and American periodicals, developing a brittle, angular style of figure drawing related to aspects of Art Nouveau, but revealing a very distinctive personal sense of visual satire. The years 1906–8, in Paris, brought him in touch with the early pioneers of modern French painting. By 1912–13 he had arrived at his own version of Cubism, particularly the form of Cubism with which Marc was experimenting at the same moment. Feininger was invited to exhibit with Der Blaue Reiter in 1913. Thus, not surprisingly, he and Marc shared the sources of Orphism and Futurism, which particularly appealed to the romantic, expressive sensibilities of both. Whereas Marc translated his beloved animals into luminous Cubist planes, Feininger continued with his favorite themes of architecture, boats, and the sea. In *Harbor Mole* of 1913 (color plate 83, page 163), he recomposed the scene into a scintillating interplay of color facets, geometric in outline but given a sense of rapid change by the transparent, delicately graded color areas. In this work Feininger stated in a most accomplished manner the approach he was to con-

tinue, with variations, throughout the rest of his long life: that of strong, straight-line structure played against sensuous and softly luminous color. The interplay between taut linear structure and romantic color, with space constantly shifting between abstraction and representation, created a dynamic tension in his paintings.

In 1919 Feininger was invited to join the staff of the Bauhaus, the great experimental German school of architecture and design. He remained associated with the institution until it was closed by the Nazis in 1933, although he had stopped teaching full-time in the 1920s because the schedule interfered with his own painting. This continued steadily in the direction he had chosen, sometimes emerging in shattered architectural images, sometimes in serene and light-filled structures full of poetic suggestion. He returned to the United States in 1937 and settled in New York, a difficult move for an artist who had made his reputation in Europe. Once he was "spiritually acclimatized," as he said, to his new home, he moved in a new stylistic direction, one less restricted by the "straightjacket" of Cubism and characterized by a freer application of color.

In 1912 Herwarth Walden opened a gallery in Berlin for avant-garde art, the Sturm Galerie, where he exhibited Kandinsky and Der Blaue Reiter, Die Brücke and the Italian Futurists, and grouped as French Expressionists Braque, Derain, Vlaminck, Auguste Herbin, and others. Also shown were Ensor, Klee, and Delaunay. In 1913 came the climax of the Sturm Galerie's exhibitions, the First German Autumn Salon, including 360 works. Henri Rousseau's room had twenty paintings, and almost the entire international range of experimental painting and sculpture at that moment was shown. Although during and after the war the various activities of Der Sturm lost their impetus, Berlin, between 1910 and 1914, was a rallying point for most of the new European ideas and revolutionary movements, largely through the leadership of Walden.

Color plate 58. MAURICE DE VLAMINCK.
Portrait of Derain. 1906. Oil on cardboard,
10¾ x 8¾" (27.3 x 22.2 cm).
The Metropolitan Museum of Art, New York

Color plate 60. GEORGES ROUAULT. *The Old King.* 1916–36.
Oil on canvas, 30¼ x 21¼" (76.8 x 54 cm).
Carnegie Museum of Art, Pittsburgh PATRONS ART FUND

Color plate 59. RAOUL DUFY. *Street Decked with Flags, Le Havre.*
1906. Oil on canvas, 31⅞ x 25⅞" (81 x 65.7 cm).
Musée National d'Art Moderne, Paris

Color plate 61. LUMIÈRE BROTHERS.
Young Lady with an Umbrella. 1906–10.
Autochrome photograph. Fondation Nationale
de la Photographie, Lyons, France
©SOCIÉTÉ LUMIÈRE

below: Color plate 62. HENRI MATISSE.
Harmony in Red (The Dessert). 1908.
Oil on canvas, 70⅞" x 7'2⅝" (180 cm x 2.2 m).
The Hermitage Museum, St. Petersburg
© 1997 SUCCESSION H. MATISSE/ARTISTS RIGHTS SOCIETY
(ARS), NEW YORK

Color plate 63. HENRI MATISSE. *Dance (II)*. 1909–10. Oil on canvas, 8'5⅝" x 12'9½" (2.6 x 3.9 m).
The Hermitage Museum, St. Petersburg © 1997 SUCCESSION H. MATISSE/ARTISTS RIGHTS SOCIETY (ARS), NEW YORK

Color plate 64. HENRI MATISSE. *Music*. 1909–10. Oil on canvas, 8'5⅝" x 12'9¼" (2.6 x 3.9 m).
The Hermitage Museum, St. Petersburg © 1997 SUCCESSION H. MATISSE/ARTISTS RIGHTS SOCIETY (ARS), NEW YORK

above, left: Color plate 65. HENRI MATISSE.
The Red Studio. Issy-les-Moulineaux, 1911.
Oil on canvas,
71¼" x 7'2¼" (181 cm x 2.2 m).
The Museum of Modern Art, New York
MRS. SIMON GUGGENHEIM FUND © 1997
SUCCESSION H. MATISSE/ARTISTS RIGHTS SOCIETY (ARS), NEW YORK

left: Color plate 66. EMIL NOLDE.
The Last Supper. 1909. Oil on canvas,
33⅞ x 42⅛" (86 x 107 cm).
Statens Museum for Kunst, Copenhagen

above: Color plate 67.
ERNST LUDWIG KIRCHNER. *Street, Dresden*.
1908 (dated 1907 on painting). Oil on
canvas, 59¼" x 6'6⅞" (150.5 cm x 2 m).
The Museum of Modern Art, New York
PURCHASE

right: Color plate 68. ERNST LUDWIG
KIRCHNER. *Self-Portrait with Cat*. 1919–20.
Oil on canvas, 47¼ x 33½" (120 x 85 cm).
Busch-Reisinger Museum, Harvard University
Art Museums, Cambridge, Massachusetts
MUSEUM ASSOCIATION FUND

right: Color plate 70. KARL SCHMIDT-ROTTLUFF.
Rising Moon. 1912. Oil on canvas,
34½ x 37½" (87.6 x 95.3 cm).
The Saint Louis Art Museum
BEQUEST OF MORTON D. MAY
© 1997 VG BILD-KUNST, BONN/ARTISTS RIGHTS SOCIETY
(ARS), NEW YORK

Color plate 69. MAX PECHSTEIN.
Indian and Woman. 1910. Oil on canvas,
32¼ x 26¼" (81.9 x 66.7 cm).
The Saint Louis Art Museum
BEQUEST OF MORTON D. MAY
© 1997 VG BILD-KUNST, BONN/
ARTISTS RIGHTS SOCIETY (ARS), NEW YORK

Color plate 71. EMIL NOLDE. *Female Dancer.*
1913. Color lithograph, 20⅞ x 27⅛"
(53 x 69 cm). Nolde-Stiftung Seebüll,
Germany

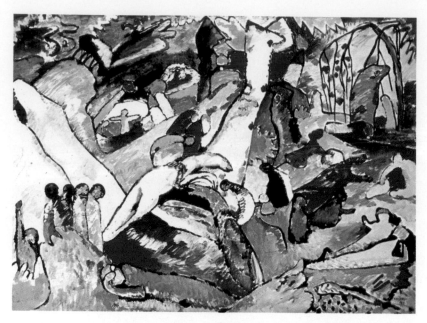

Color plate 75. VASILY KANDINSKY. *Sketch for Composition II.* 1909–10.
Oil on canvas, 38⅜ x 51¾" (97.5 x 131.5 cm).
Solomon R. Guggenheim Museum, New York
© 1997 ADAGP, PARIS/ARTISTS RIGHTS SOCIETY (ARS), NEW YORK

Color plate 76. VASILY KANDINSKY. *Composition VII.* 1913. Oil on canvas,
6'6¾" x 9'11⅛" (2 x 3 m). Tretyakov Gallery, Moscow
© 1997 ADAGP, PARIS/ARTISTS RIGHTS SOCIETY (ARS), NEW YORK

Color plate 77. FRANZ MARC. *The Large Blue Horses.* 1911. Oil on canvas, 40¾ x 70⅞" (103.5 x 180 cm).
Walker Art Center, Minneapolis
GIFT OF THE T. B. WALKER FOUNDATION. GILBERT M. WALKER FUND, 1942

Color plate 78. FRANZ MARC. *Stables.* 1913–14. Oil on canvas, 29⅛ x 62¼" (74 x 158.1 cm).
Solomon R. Guggenheim Museum, New York

right: Color plate 79. FRANZ MARC. *Fighting Forms.* 1914. Oil on canvas, 35⅞ x 51¾" (91.1 x 131.4 cm). Bayerische Staatsgemäldesammlungen, Munich

Color plate 80. AUGUST MACKE. *Great Zoological Garden.* 1912. Oil on canvas, 51⅛ x 7'6¾" (130 cm x 2.3 m). Museum am Ostwall, Dortmund, Germany

Color plate 81. ALEXEJ VON JAWLENSKY. *Love,* from the Constructivist Heads series. 1925. Oil on paper, 23¼ x 19⅝" (59 x 49.8 cm). Städtische Galerie im Lenbachhaus, Munich

© 1997 VG BILD-KUNST, BONN/ARTISTS RIGHTS SOCIETY (ARS), NEW YORK

Color plate 82. PAUL KLEE. *Hammamet with the Mosque.* 1914. Watercolor on paper, 8⅛ x 7½" (20.6 x 19.1 cm). The Metropolitan Museum of Art, New York

BERGGRUEN KLEE COLLECTION, 1984 © 1997 VG BILD-KUNST, BONN/ARTISTS RIGHTS SOCIETY (ARS), NEW YORK

Color plate 83. LYONEL FEININGER. *Harbor Mole.* 1913. Oil on canvas, 31¾ x 39¾" (80.6 x 101 cm). Private collection

© 1997 VG BILD-KUNST, BONN/ARTISTS RIGHTS SOCIETY (ARS), NEW YORK

Color plate 84. EGON SCHIELE. *The Self Seer II (Death and the Man)*. 1911. Oil on canvas, 31⅝ x 31½" (80.3 x 80 cm). Private collection

Color plate 85. OSKAR KOKOSCHKA. *The Tempest*. 1914. Oil on canvas, 71¼" x 7'2⅝" (181 cm x 2.2 m). Öffentliche Kunstsammlung Basel, Kunstmuseum
© 1997 PRO LITTERIS, ZÜRICH/ARTISTS RIGHTS SOCIETY (ARS), NEW YORK

Color plate 86. CONSTANTIN BRANCUSI.
Bird in Space. 1925. Marble, stone, and
wood, 71⅝ x 5⅜ x 6⅜" (181.9 x 13.7 x
16.2 cm), stone pedestal 17⅝ x 16⅜ x
16⅜" (44.8 x 41.6 x 41.6 cm), wood
pedestal 7¾ x 16 x 13⅝" (19.7 x 40.6 x
34 cm). National Gallery of Art,
Washington, D.C.

Color plate 87. PABLO PICASSO. *La Vie.* 1903.
Oil on canvas, 6'5⅜" x 50⅝" (2 m x 128.5 cm).
The Cleveland Museum of Art

Color plate 88.
PABLO PICASSO.
Family of Saltimbanques.
1905. Oil on canvas,
6'11¾" x 7'6⅜" (2.1 x
2.3 m). National Gallery
of Art, Washington, D.C.
CHESTER DALE COLLECTION
© 1997 ESTATE OF PABLO
PICASSO/ARTISTS RIGHTS SOCIETY (ARS),
NEW YORK

Color plate 89.
PABLO PICASSO.
*Les Demoiselles
d'Avignon.* June–July
1907. Oil on canvas,
8' x 7'8" (2.4 x 2.3 m).
The Museum of Modern
Art, New York
ACQUIRED THROUGH THE LILLIE P. BLISS
BEQUEST © 1997 ESTATE OF PABLO
PICASSO/ARTISTS RIGHTS SOCIETY
(ARS), NEW YORK

Color plate 90.
GEORGES BRAQUE. *Viaduct at L'Estaque.* 1907. Oil on canvas, 25⅝ x 31¾" (65.1 x 80.6 cm). The Minneapolis Institute of Arts
THE JOHN R. VAN DERLIP FUND, FIDUCIARY FUND AND GIFT OF FUNDS FROM MR. AND MRS. PATRICK BUTLER AND VARIOUS DONORS © 1997 ADAGP, PARIS/ARTISTS RIGHTS SOCIETY (ARS), NEW YORK

bottom, left: Color plate 91.
GEORGES BRAQUE. *Houses at L'Estaque.* August 1908. Oil on canvas, 28¾ x 23⅝" (73 x 60 cm). Kunstmuseum, Bern
HERMANN AND MARGRIT RUPF FOUNDATION © 1997 ADAGP, PARIS/ARTISTS RIGHTS SOCIETY (ARS), NEW YORK

bottom, right: Color plate 92.
GEORGES BRAQUE. *Violin and Palette.* 1909. Oil on canvas, 36⅛ x 16⅞" (91.7 x 42.9 cm). Solomon R. Guggenheim Museum, New York
© 1997 ADAGP, PARIS/ARTISTS RIGHTS SOCIETY (ARS), NEW YORK

Color plate 93. PABLO PICASSO. *Portrait of Daniel-Henry Kahnweiler.* 1910. Oil on canvas, 39½ x 28⅝" (100.6 x 72.8 cm).
The Art Institute of Chicago

Color plate 95. PABLO PICASSO. *Green Still Life.* Avignon, summer 1914. Oil on canvas, 23½ x 31¼" (59.7 x 79.4 cm).
The Museum of Modern Art, New York

Color plate 94. PABLO PICASSO.
Still Life with Chair Caning. 1912.
Oil and oilcloth on canvas, edged with rope,
10⅝ x 14⅝" (27 x 37 cm).
Musée Picasso, Paris

Expressionism in Austria

To understand the various directions in which Expressionism moved during and after World War I, we may look at two artists who were associated in some degree with Expressionism in Austria, where its practitioners were never formally organized into a particular group. Like Klee and Feininger, the Austrians Egon Schiele and Oskar Kokoschka each had his highly individualized interpretation of the style.

EGON SCHIELE (1890–1918) lived a short and tragic life, dying in the worldwide influenza epidemic of 1918. He was a precocious draftsman and, despite opposition from his uncle (who was his guardian after his father died insane, an event that haunted him), studied at the Vienna Academy of Art. His principal encouragement came from Gustav Klimt, at his height as a painter and leader of the avant-garde in Vienna when they met in 1907. The two artists remained close, but although Schiele was deeply influenced by Klimt, he did not fully share his inclination toward the decorative.

Schiele's many self-portraits rank among his supreme achievements and range from self-revelatory documents of personal anguish to records of a highly self-conscious and youthful bravado. *Schiele, Drawing a Nude Model before a Mirror* (fig. 169) demonstrates his extraordinary natural skill as a draftsman. The intense portrayal, with the artist's narrowed gaze and the elongated, angular figure of the model, differs from even the most direct and least embellished of Klimt's portraits. Here Schiele explores Matisse's familiar theme of the artist and model, but does so to create an atmosphere of psychological tension and explicit sexuality.

170. EGON SCHIELE. *The Painter Paris von Gütersloh.* 1918. Oil on canvas, 55½ x 43½" (141 x 110.5 cm). The Minneapolis Institute of Arts
GIFT OF THE P. D. MCMILLAN LAND COMPANY

Schiele's models included his sister Gerti, as well as friends, professional models, and prostitutes. But when, albeit innocently, in 1912 he sought out children for models in his small village, he was imprisoned for twenty-four days for "offenses against public morality."

A comparison of paintings by Klimt and Schiele is valuable in illustrating both the continuity and the change between nineteenth-century Symbolism and early twentieth-century Expressionism. Klimt's painting *Death and Life* (color plate 40, page 113) is heavy with symbolism presented in terms of the richest and most colorful patterns. Schiele's *The Self Seer II* (color plate 84, page 164), painted in 1911, is comparable in subject, but the approach could not be more different. The man (a self-portrait) is rigidly frontalized in the center of the painting and stares out at the spectator. His face is a horrible and bloody mask of fear. The figure of Death hovers behind, a ghostly presence folding the man in his arms in a grim embrace. The paint is built up in jagged brushstrokes on figures emerging from a background of harsh and dissonant tones of ocher, red, and green.

In Schiele's portraiture, other than the commissioned portraits he undertook occasionally to support himself, there persisted the same intensity of characterization. But in his later work the tense linear quality we have seen gave way to painterly surfaces built up with abundant, expressive brushwork. In one of his last works Schiele dramatically portrays his friend, the artist Paris von Gütersloh (fig. 170), with hands held aloft, fingers spread, as though the hands of the artist possessed some mysterious, supernatural power.

169. EGON SCHIELE. *Schiele, Drawing a Nude Model before a Mirror.* 1910. Pencil on paper, 21¾ x 13⅞" (55.2 x 35.2 cm). Graphische Sammlung Albertina, Vienna

OSKAR KOKOSCHKA (1886–1980), like Schiele, was a product of Vienna, but he soon left the city, which he found gloomy and oppressive, for Switzerland and Germany, embracing the larger world of modern art to become one of the international figures of twentieth-century Expressionism. Between 1905 and 1908 Kokoschka worked in the Viennese Art Nouveau style, showing the influence of Klimt and Aubrey Beardsley. Even before going to Berlin at the invitation of Herwarth Walden in 1910, he was instinctively an Expressionist. Particularly in his early "black portraits," he searched passionately to expose an inner sensibility—which may have belonged more to himself than to his sitters. Among his very first images is the 1909 portrait of his friend, the architect Adolf Loos (fig. 171), who early on recognized Kokoschka's talents and provided him with moral as well as financial support. The figure projects from its dark background, and the tension in the contemplative face is echoed in Loos's nervously clasped hands. The romantic basis of Kokoschka's early painting appears in *The Tempest* (color plate 85, page 164), a double portrait of himself with his lover, Alma Mahler, in which the two figures, composed with flickering, light-saturated brushstrokes, are swept through a dream landscape of cold blue mountains and valleys lit only by the gleam of a shadowed moon. The painting was a great success when Kokoschka exhibited it in the 1914 New Munich Secession. The year before, he wrote to Alma about the work, then in progress: "I was able to express the mood I wanted by reliving it. . . . Despite all the turmoil in the world, to know that one person can put eternal trust in another, that two people can be committed to themselves and other people by an act of faith."

Seriously wounded in World War I, Kokoschka produced little for several years, but his ideas and style were undergoing constant change. In 1924 he abruptly set out on some seven years of travel, during which he explored the problem of landscape, combining free, arbitrary, and brilliant Impressionist or Fauve color with a traditional perspective-space organization. Throughout his long life Kokoschka remained true to the spirit of Expressionism—to the power of emotion and the deeply felt sensitivity to the inner qualities of nature and the human soul. In 1933, in financial difficulties, the painter returned from his long travels. He went first to Vienna and then to Prague. During this period his works in public collections in Germany were confiscated by the Nazis as examples of "degenerate art." In 1938, as World War II approached, London became his home, and later Switzerland, although whenever possible he continued his restless traveling.

171. OSKAR KOKOSCHKA. *Portrait of Adolf Loos*. 1909. Oil on canvas, 29⅛ x 35⅞"
(74 x 91.1 cm). Staatliche Museen zu Berlin, Preussischer Kulturbesitz, Nationalgalerie

9 The Figurative Tradition in Early Twentieth-Century Sculpture

uguste Rodin lived until 1917, his reputation increasing until by the time of his death it had become worldwide. Aristide Maillol did not die until 1944 and continued to be a productive sculptor until his death. Antoine Bourdelle lived until 1929. The figurative tradition of these artists was, as we saw in chapter 6, reinforced by the growing recognition accorded the sculpture of older artists such as Edgar Degas and Paul Gauguin. While these men challenged some of the most time-honored sculptural traditions, they did so within the confines of the figurative mode. In his startling *Little Dancer Fourteen Years Old* (color plate 45, page 115), Degas's introduction of actual materials heightened the psychological relationship between object and viewer. Subsequently, in major works such as *The Burghers of Calais* (see fig. 137), Rodin explored new spatial concepts to break down traditional barriers between sculpture and its audience. Gauguin, in his painted wood reliefs or ceramics such as *Oviri* (see figs. 79, 130), moved away from the notion of sculpture as a public art form toward one of hermetic, Symbolist content that drew on non-European sources. These tendencies proved critical to virtually all of the artists under discussion in this chapter.

Despite revolutionary experiments being carried on by Cubist and abstract sculptors in the early part of the twentieth century, the figurative tradition was central to the development of modernism, and it continued to dominate the scene in Europe and the United States. Unlike early twentieth-century experimentalists in construction and assemblage, who regarded sculptural form as a means of shaping space (see figs. 215, 218), the artists discussed in this chapter held to the notion of sculpture as mass or volumes in space. They used traditional methods of creating those volumes—modeling clay or carving marble and wood. In their hands, the figure is a source of seemingly inexhaustible expression, ranging from the relatively conservative efforts of Georg Kolbe to the radical innovations of Constantin Brancusi. While Kolbe still belongs within the classical tradition, others among his contemporaries, namely Derain, Matisse, and Brancusi, deliberately turned to nonclassical sources,

172. ANDRÉ DERAIN. *Crouching Man.* 1907. Stone, 13 × 11" (33 × 27.9 cm). Galerie Louise Leiris, Paris
© 1997 ADAGP, PARIS/ARTISTS RIGHTS SOCIETY (ARS), NEW YORK

such as Oceanic or pre-Columbian art, for inspiration.

Among the twentieth-century painter-sculptors, Matisse depicted the human figure throughout his life, while Picasso began to desert naturalistic representation with a group of primitivist wood carvings, dated 1907, in which the influence of Iberian, archaic Greek, and African sculpture is even more explicit than in his paintings of that year. Also in 1907, **ANDRÉ DERAIN** (1880–1954) carved *Crouching Man* (fig. 172), transformed from a single block of stone—a startling example of proto-Cubist sculpture.

HENRI MATISSE (1869–1954). David Smith, one of the outstanding American sculptors of the twentieth century (see figs. 537, 538), frequently contended that modern sculpture

left: 173. HENRI MATISSE. *The Serf*. 1900–1904. Bronze, 36 x 13⅞ x 12¼" at base (91.4 x 35.3 x 31.1 cm). Hirshhorn Museum and Sculpture Garden, Smithsonian Institution, Washington, D.C. © 1997 SUCCESSION H. MATISSE/ARTISTS RIGHTS SOCIETY (ARS), NEW YORK

below: 174. HENRI MATISSE. *Reclining Nude, I (Nu couché, I; Aurore)*. Collioure, winter 1906–7. Bronze, 13½ x 19¾ x 11¼" (34.3 x 50.2 x 28.6 cm). The Museum of Modern Art, New York ACQUIRED THROUGH THE LILLIE P. BLISS BEQUEST PHOTOGRAPH © 1995 THE MUSEUM OF MODERN ART, NEW YORK © 1997 SUCCESSION H. MATISSE/ARTISTS RIGHTS SOCIETY (ARS), NEW YORK

was created by painters. Although this cannot be taken as literally true, given the achievements of Rodin, Brancusi, Raymond Duchamp-Villon, and Jacques Lipchitz, certainly major contributions to sculpture were made by such painters as Degas, Gauguin, Picasso, Amadeo Modigliani, and Matisse. Matisse said he "sculpted like a painter," and that he took up sculpture "to put order into my feelings and to find a style to suit me. When I found it in sculpture, it helped me in my painting." Throughout his career, Matisse explored sculpture as a medium parallel to his painted oeuvre, often setting up a dialogue between the two.

Matisse had studied for a time with Bourdelle (see fig. 143), and was strongly influenced by Rodin, then at the height of his reputation. Matisse's *The Serf* (fig. 173) was

begun in 1900, but not completed until 1904. Although sculpted after a well-known model, Bevilacqua, who had posed for Rodin and of whom Matisse also made a painting in a Cézannesque style (see fig. 147), it was adapted in attitude and concept (although on a reduced scale) from a Rodin sculpture called *Walking Man*. In *The Serf* Matisse carried the Expressionist modeling of the surface even further than Rodin, and halted the forward motion of the figure by adjusting the position of the legs into a solidly static pose and truncating the arms above the elbow.

Aside from the great Fauvist figure compositions seen in chapter 7, Matisse in the years between 1905 and 1910 was studying the human figure in every conceivable pose and in every medium—drawing, painting, linoleum cut, and lithography, as well as sculpture. The sculpture entitled *Reclining Nude, I* (fig. 174) is the most successful of the early attempts at a reclining figure integrated with surrounding space through its elaborate, twisting pose. The artist, working under the influence of African sculpture, used this pose to dramatic effect in *The Blue Nude* (color plate 54, page 119), painted at about the same time; it is interesting to compare his resolution of this spatial problem in the two media. Matisse was to experiment over the years with a number of sculpted variants of this reclining figure, generally more simplified and geometrized, and she frequently appeared in his later paintings.

Matisse also tried his hand at a very different kind of sculptural arrangement in the small *Two Women* (fig. 175), which consists of two figures embracing, one front view, and

opposite: 175. HENRI MATISSE. *Two Women*. 1907. Bronze, 18¼ x 10½ x 7⅞" (46.6 x 25.6 x 19.9 cm). Hirshhorn Museum and Sculpture Garden, Smithsonian Institution, Washington, D.C.
© 1997 SUCCESSION H. MATISSE/ARTISTS RIGHTS SOCIETY (ARS), NEW YORK

right: 176. HENRI MATISSE. *La Serpentine*. Issy-les-Moulineaux, autumn 1909. Bronze, 22¼ x 11 x 7½" (56.5 x 28.9 x 19 cm), including base. The Museum of Modern Art, New York
GIFT OF ABBY ALDRICH ROCKEFELLER
PHOTOGRAPH © 1995 THE MUSEUM OF MODERN ART, NEW YORK
© 1997 SUCCESSION H. MATISSE/ARTISTS RIGHTS SOCIETY (ARS), NEW YORK

the young girl drapes herself over the pillar. This is a work obviously related to the swirling figures of *The Dance II* (color plate 63, page 155), but Matisse actually derived his image from a photograph of a nude model in this curious pose.

Matisse's most ambitious excursion in sculpture was the creation of the four great Backs, executed several years apart, between 1909 and 1931 (fig. 177). While most of his sculptures are small-scale, these monumental reliefs are more than six feet high. They are a development of the theme stated in *Two Women*, now translated into a single figure in bas-relief, seen from the back. But Matisse has here resorted to an upright, vertical surface like that of a painting on which to sculpt his form. *Back I* is modeled in a relatively representational manner, freely expressing the modulations of a muscular back, and it reveals the feeling Matisse had for sculptural form rendered on a monumental scale. *Back II* is simplified in a manner reflecting the artist's interest in Cubism around 1913. *Back III* and *Back IV* are so reduced to their architectural components that they almost become abstract sculpture. The figure's long ponytail becomes a central spine that serves as a powerful axis through the center of the composition. Here, as in *The Blue Nude*, Matisse has synthesized African and Cézannesque elements, this time, however, in order to acknowledge, as well as resist, the formal discoveries made by the Cubists.

Matisse's other major effort in sculpture was a series of five heads of a woman, done between 1910 and 1916. *Jeannette I* and *Jeannette II* are direct portraits done from life in the freely expressive manner of late Rodin bronzes. But with

the other back view. Although it may have some connection with Early Renaissance exercises in anatomy in which artists depicted athletic nudes in front and back views, there is, in its blocky, rectangular structure, a closer connection with some frontalized examples of African sculpture. This influence is not surprising, given that Matisse began to collect African sculpture shortly before he made *Two Women*, which is based on a photograph, one clipped from an ethnographic magazine, depicting two Tuareg women from North Africa.

In *La Serpentine* (fig. 176), he used extreme attenuation and distortion to emphasize the boneless relaxation with which

177. HENRI MATISSE. From left to right: *Back I*. 1909. Bronze relief, 6'2⅜" x 3'8½" x 6½" (188.9 x 113 x 16.5 cm). *Back II*. 1913. Bronze, 6'2¼ "x 47" x 8" (188.6 x 119.4 x 20.3 cm). *Back III*. 1916–17. Bronze, 6'2" x 44¼" x 6" (188 x 112.4 x 15.2 cm). *Back IV*. c. 1930. Bronze, 6'2" x 45" x 7" (188 x 114.3 x 17.8 cm). Hirshhorn Museum and Sculpture Garden, Smithsonian Institution, Washington, D.C.
© 1997 SUCCESSION H. MATISSE/ARTISTS RIGHTS SOCIETY (ARS), NEW YORK

above, left: 178. HENRI MATISSE. *Jeannette V* (Jean Vaderin, 5th state). Issy-les-Moulineaux, 1916. Bronze, height 22⅞ x 8⅜ x 10⅝" (58.1 x 21.3 x 27.1 cm). The Museum of Modern Art, New York ACQUIRED THROUGH THE LILLIE P. BLISS BEQUEST
PHOTOGRAPH © 1995 THE MUSEUM OF MODERN ART, NEW YORK
© 1997 SUCCESSION H. MATISSE/ARTISTS RIGHTS SOCIETY (ARS), NEW YORK

above, right: 179. WILHELM LEHMBRUCK. *Standing Woman.* 1910. Bronze, (cast in New York, 1916–17, from original plaster), height 6'3⅛" (1.9 m), base diameter 20½" (52 cm). The Museum of Modern Art, New York ANONYMOUS GIFT

180. WILHELM LEHMBRUCK. *Seated Youth.* 1917. Composite tinted plaster, 40⅝ x 30 x 45½" (103.2 x 76.2 x 115.5 cm). National Gallery of Art, Washington, D.C.
ANDREW W. MELLON FUND

Jeannette III, Jeannette IV, and *Jeannette V* (fig. 178), he worked from his imagination, progressively transforming the human head first into an Expressionist study, exaggerating all the features, and then into a geometric organization of shapes. Picasso had produced his Cubist *Woman's Head* in 1909 (see fig. 207), and this work may have prompted Matisse's experiments. In turn, Picasso probably had Matisse's examples in mind when making his so-called Bosgeloup heads in the early 1930s. In Matisse's work, serial sculptures such as the *Backs* and the *Jeannettes* do not necessarily constitute a set of progressive steps toward the perfection of an idea. Rather, they are multiple but independent states, each version a definitive solution in and of itself.

WILHELM LEHMBRUCK (1881–1919). With the exception of such older masters as Maillol, the tradition of Realist or Realist-Expressionist sculpture flourished more energetically outside France than inside, after 1910. In Germany the major figure was Wilhelm Lehmbruck, who, after an academic training, turned for inspiration first to the Belgian sculptor of miners and industrial workers, Constantin Meunier (1831–1905), and then to Rodin. During the four years he spent in Paris, between 1910 and 1914, Lehmbruck became acquainted with Matisse, Modigliani, Brancusi, and Aleksandr Archipenko. A *Standing Woman* of 1910 (fig. 179), in which the drapery is knotted above the knees in a formula derived from classical Greek sculpture, illustrates the immediate impact of Maillol, as well as of antiquity. The details of the face are somewhat blurred, in a manner that Lehmbruck may have learned from the study of the fourth-century Greek sculpture of Praxiteles. This treatment gives the figure a peculiarly withdrawn quality, a sense of apartness and of inward contemplation, recalling not only the Praxitelean mode but also some of the mystical Madonnas of Gothic sculpture.

The emotional power of Lehmbruck's work comes not from his studies of the past but from his own sensitive and melancholy personality. His surviving works are few, for the entire oeuvre belongs to a ten-year period. In *Seated Youth* (fig. 180), his last monumental work, the artist utilized extreme elongation—possibly suggested by figures in Byzantine mosaics and Romanesque sculpture—to carry the sense of contemplation and withdrawal expressed in *Standing Woman* still further. His later work represents a departure from Maillol and from a nineteenth-century tradition that emphasized volume and mass. Lehmbruck's figure, with its long, angular limbs, is penetrated by space, and exploits for expressive ends the abstract organization of solid and voids. With its mood of dejection and loss, *Seated Youth* expresses the trauma and sadness Lehmbruck experienced in Germany during the First World War—suffering that finally, in 1919, led him to suicide.

left: 181. GEORG KOLBE. *Standing Nude.* 1926.
Bronze, height 50½" (128.3 cm).
Walker Art Center, Minneapolis
© 1997 VG BILD-KUNST, BONN/ARTISTS RIGHTS SOCIETY (ARS), NEW YORK

above: 182. ERNST BARLACH. *The Avenger.* 1914.
Bronze, 17¼ x 23 x 8¼" (43.8 x 58.4 x 21 cm).
Hirshhorn Museum and Sculpture Garden,
Smithsonian Institution, Washington, D.C.

The human figure was so thoroughly entrenched as the principal vehicle of expression for sculptors that it was even more difficult for them to depart from it than for painters to depart from landscape, figure, or still life. It was also difficult for sculptors to say anything new about the subject, or startlingly different from what had been said at some other point in history.

GEORG KOLBE (1877–1941). Other leading German sculptors of the early century included Georg Kolbe, Ernst Barlach, and Käthe Kollwitz. Kolbe began as a painter but developed an interest in sculpture during a two-year stay in Rome. His later encounter with Rodin's work in Paris had considerable impact on his work. For his figures or portrait heads, Kolbe combined the formal aspects of Maillol's figures with the light-reflecting surfaces of Rodin's. After some essays in highly simplified figuration, Kolbe settled into a lyrical formula of rhythmic nudes (fig. 181)—appealing, but essentially a Renaissance tradition with a Rodinesque broken surface. His depictions in the 1930s of heroic, idealized, and specifically Nordic figures now seem uncomfortably compatible with the Nazis' glorification of a "master race."

ERNST BARLACH (1870–1938). In addition to making sculpture carved from wood or cast in bronze, Ernst Barlach was an accomplished poet, playwright, and printmaker. Although, like Lehmbruck, he visited Paris, he was not as strongly influenced by French vanguard developments. His early work took on the curving forms of Jugendstil and

aspects of medieval German sculpture. In 1906 he traveled to Russia, where he was deeply impressed by the peasant population, and his later sculptures often depicted Russian beggars or laborers. Once he was targeted as a "degenerate artist" by Nazi authorities in the 1930s, Barlach's public works, including monumental sculptures for cathedrals, were dismantled or destroyed. He was capable of works of sweeping power and the integration in a single image of humor and pathos and primitive tragedy, as in *The Avenger* (fig. 182). His was a storytelling art, a kind of socially conscious Expressionism that used the outer forms of contemporary experiment for narrative purposes.

KÄTHE KOLLWITZ (1867–1945)—best known for her graphics (color plate 74, page 159; see fig. 164)—was capable of an equally intense expression in sculpture, with which she was increasingly involved throughout her last years. Given the intrinsically sculptural technique of carving and gouging a block of wood to make a woodcut, it is not surprising that a printmaker might be drawn to sculpture, though Kollwitz never actually made sculpture in wood. Like Lehmbruck, she learned of Rodin's work on a trip to Paris and admired the sculpture of Constantin Meunier, whose subjects of workers struck a sympathetic chord. But perhaps the most significant influence on her work in three dimensions was that of her friend Barlach. The highly emotional tenor of her work, whether made for a humanitarian cause, or in memory of her son killed in World War I, arose from

a profoundly felt grief that transmitted itself to her sculpture and prints. *Lamentation: In Memory of Ernst Barlach (Grief)* (fig. 183) is an example of Kollwitz's relief sculpture, a moving, close-up portrait of her own grieving.

CONSTANTIN BRANCUSI (1876–1957). In France during the early years of the twentieth century, the most individual figure—and one of the greatest sculptors of the time—was Constantin Brancusi. Born in Romania, the son of peasants, he left home at the age of eleven. Between 1894 and 1898 he was apprenticed to a cabinetmaker and studied in the provincial city of Craiova, and then at the Bucharest Academy of Fine Arts. In 1902 he went to Germany and Switzerland, arriving in Paris in 1904. After further studies at the École des Beaux-Arts under the sculptor Antonin Mercié, he began exhibiting, first at the Salon de la Nationale and then at the Salon d'Automne. Rodin, impressed by his contributions to the 1907 Salon d'Automne, invited him to become an assistant. Brancusi stayed only a short time. As he later declared, "Nothing grows under the shade of great trees."

The sculpture of Brancusi is in one sense isolated, in another universal. He worked with few themes, never really deserting the figure, but he touched, affected, and influenced

most of the major strains of sculpture after him. From the tradition of Rodin's late studies came the figure *Sleep* (fig. 184), in which the shadowed head sinks into the matrix of the marble. The theme of the Sleeping Muse was to become an obsession with Brancusi, and he played variations on it for some twenty years. In the next version (fig. 185), and later, the head was transformed into an egg shape, with the fea-

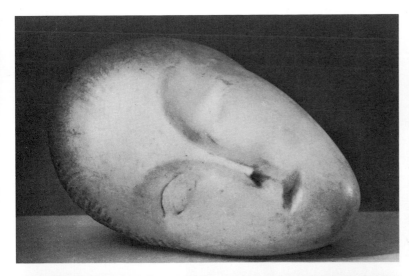

above: 186. CONSTANTIN BRANCUSI. *The Newborn.* 1915.
Marble, 5⅝ x 8½" (14.3 x 21.6 cm). Philadelphia Museum of Art

above: 187. CONSTANTIN BRANCUSI.
The Beginning of the World. c. 1920. Marble, metal, and
stone, 29⅝ x 11⅜" (75.2 x 28.9 cm).
Dallas Museum of Art

tures lightly but sharply cut from the mass. As became his custom with his basic themes, he presented this form in marble, bronze, and plaster, almost always with slight adjustments that turned each version into a unique work.

In a subsequent work, the theme was further simplified to a teardrop shape in which the features largely disappeared, with the exception of an indicated ear. To this 1911 piece he gave the name *Prometheus.* The form in turn led to *The Newborn* (fig. 186), in which the oval is cut obliquely to form the screaming mouth of the infant. He returned to the polished egg shape in the ultimate version, entitled *The Beginning of the World* (fig. 187), closely related to a contemporary work called *Sculpture for the Blind.* In this, his ultimate statement about creation, Brancusi eliminated all reference to anatomical detail. A similar work, as well as *The Newborn,* can be seen at the bottom of Brancusi's photograph of his studio taken around 1927 (fig. 188). The artist preferred that his work be discovered in the context of his own studio, amid tools, marble dust, and incomplete works.

He was constantly rearranging his work, grouping the sculptures into a carefully staged environment. Brancusi left the entire content of his studio to the French state; a reconstruction of it is in the Georges Pompidou Center in Paris. The artist took photographs of his own works (a privilege he rarely allowed others) in order to disseminate their images beyond Paris and to depict his sculptures in varying, often dramatic light. His many pictures of corners of his studio convey the complex interplay of his sculptural shapes and the cumulative effect of his search for ideal form. "Why write [about my sculptures]," he once said. "Why not just show the photographs?"

This tale of the egg was only one of a number of related themes that Brancusi continued to follow, with a hypnotic concentration on creation, birth, life, and death. In 1912 he made his first marble portrait of a young Hungarian woman named Margit Pogany, who posed several times in his studio. He portrayed her with enormous oval eyes and hands held up to one side of her face. This form was developed further

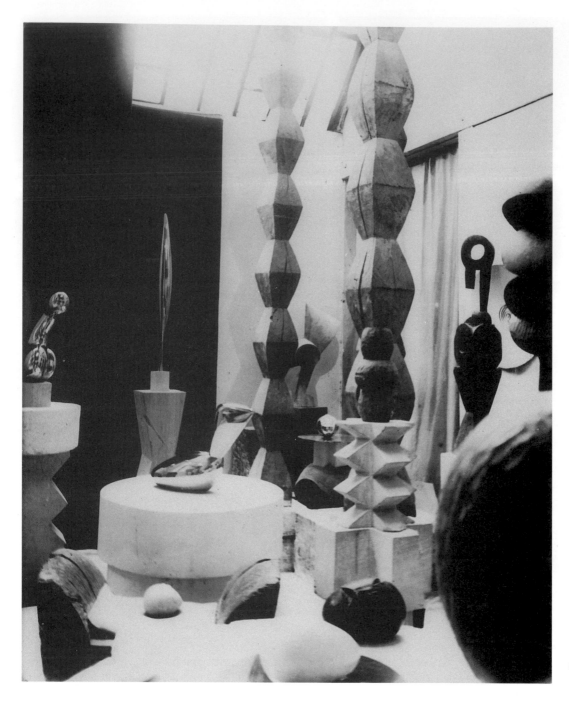

188. Constantin
Brancusi.
The Studio. c. 1927.
Photograph, mounted
on glass,
23¾ x 19¾"
(60.3 x 50.2 cm).
Pascal Sernet Fine Art,
London
© 1997 ADAGP, Paris/Artists Rights
Society (ARS), New York

in a number of drawings and in later variations in marble and polished bronze. In this example, *Madamoiselle Pogany III,* from 1931 (fig. 189), Brancusi refined and abstracted his original design. He omitted the mouth altogether and reduced the eyes to an elegant, arching line of a brow that merges with the nose. *Torso of a Young Man* (fig. 190) went as far as Brancusi ever did in the direction of geometric form. In this polished bronze from 1924, he abstracted the softened, swelling lines of anatomy introduced in his earlier wood versions of the sculpture into an object of machinelike precision. Brancusi was clearly playing on a theme of androgyny here, for while the *Torso* is decidedly phallic, it could also constitute an interpretation of female anatomy.

An artist who shared Brancusi's quest for the essence of things and discovered it in images of remarkable purity and cool elegance was the American photographer **Edward Wes-**

ton (1886–1958). Indeed, Weston's picture of a pair of gleaming nautilus shells, set against a velvet-black ground (fig. 191), seems to be nature's own organic response to the formal perfection of *Madamoiselle Pogany.* Here the photographer truly achieved his goal of a subject revealed in its "deepest moment of perception." Weston rigorously controlled form through his selection of motif, exposure time, and use of the ground-glass focusing screen of a large-format camera. In this way he could previsualize his prints and eliminate the random effects of light, atmosphere, and moment. He created a timeless image, leaving behind the Pictorialism of earlier aesthetic photography and entering the mainstream of modern art.

The subjects of Brancusi were so elemental and his themes so basic that, although he had few direct followers, little that happened subsequently in sculpture seems foreign to him. *The Kiss,* 1916 (fig. 192), depicts with the simplest of means

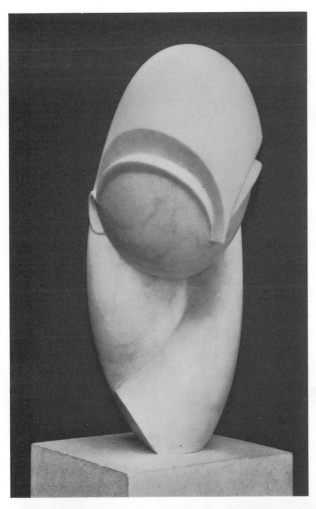

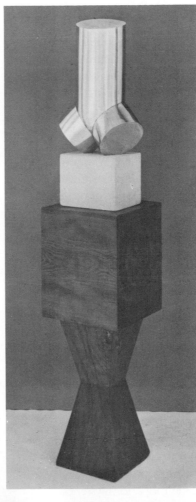

far left:
189. CONSTANTIN BRANCUSI.
Mademoiselle Pogany III. 1931.
Marble on limestone base,
17¾ x 7¼ x 11½"
(45.1 x 18.4 x 29.2 cm).
Philadelphia Museum of Art
LOUISE AND WALTER ARENSBERG COLLECTION
© 1997 ADAGP, PARIS/ARTISTS RIGHTS SOCIETY (ARS),
NEW YORK

left: 190. CONSTANTIN BRANCUSI.
Torso of a Young Man. 1924. Polished
bronze, 18 x 11 x 7" (45.7 x 28 x
17.8 cm), stone and wood base, height
40⅜" (102.6 cm).
Hirshhorn Museum and Sculpture
Garden, Smithsonian Institution,
Washington, D.C.
© 1997 ADAGP, PARIS/ARTISTS RIGHTS SOCIETY (ARS),
NEW YORK

below, left: 191. EDWARD WESTON.
Shells. 1927. Gelatin-silver print
© 1981 ARIZONA BOARD OF REGENTS,
CENTER FOR CREATIVE PHOTOGRAPHY, TUCSON

below: 192. CONSTANTIN BRANCUSI.
The Kiss. 1916. Limestone, 23 x 13 x
10" (58.4 x 33 x 25.4 cm).
Philadelphia Museum of Art
LOUISE AND WALTER ARENSBERG COLLECTION © 1997 ADAGP,
PARIS/ARTISTS RIGHTS SOCIETY (ARS), NEW YORK

an embracing couple, a subject Brancusi had first realized in stone in 1909. Although this had been the subject of one of Rodin's most famous marbles, in the squat, blockish forms of *The Kiss* Brancusi made his break with the Rodinesque tradition irrefutably clear. He was also aware of developments in Cubist sculpture, as well as works in a primitivist vein, such as the crudely carved wooden figures of the German Expressionists and especially Derain (fig. 172).

The artist was particularly obsessed with birds and the idea of conveying the essence of flight. For over a decade he progressively streamlined the form of his 1912 sculpture *Muiustra* (or "master bird," from Romanian folklore), until he achieved the astonishingly simple, tapering form of *Bird in Space* (color plate 86, page 165). The image ultimately became less the representation of a bird's shape than that of a bird's trajectory through the air. Brancusi designed his own bases and considered them an integral part of his sculpture. In *Bird in Space,* the highly polished marble bird (he also made bronze versions) fits into a stone cylinder that sits atop a cruciform stone base. This, in turn, rests on a large X-shaped wooden pedestal. He made several variations on these forms, such as that designed for *Torso of a Young Man.* These bases augment the sense of soaring verticality of the bird sculptures. In addition, they serve as transitions between the mundane physical world and the spiritual realm of the bird, for Brancusi sought a mystical fusion of the disembodied light-reflecting surfaces of polished marble or bronze and the solid, earthbound mass of wood. He said, "All my life I have sought the essence of flight. Don't look for mysteries. I give you pure joy. Look at the sculptures until you see them. Those nearest to God have seen them."

In his wood sculptures, although he occasionally strove for the same degree of finish, Brancusi usually preferred a primitive, roughed-out totem. In *King of Kings* (fig. 193), which he had previously titled *The Spirit of Buddha,* the great, regal shape comprises superimposed forms that are reworkings of the artist's wooden pedestals. Brancusi intended this sculpture for a temple of his design in India, commissioned by the Maharajah of Indore, who owned three of the artist's *Birds in Space.* The temple was never built. In fact, Brancusi's only outdoor work on a vast scale was a sculptural ensemble installed in the late 1930s in Tirjiu-Jiu, Romania, not far from his native village. This included the immense cast-iron *Endless Column* (fig. 194), which recalls ancient obelisks but also draws upon forms in Romanian folk art. The rhomboid shapes of *Endless Column* also began as socle designs that were, beginning in 1918, developed as freestanding sculptures. Wood versions of the sculpture can be seen in Brancusi's photograph of his studio (fig. 188). *Endless Column* was the most radically abstract of Brancusi's sculptures, and its reliance upon repeated modules became enormously important for Minimalist artists of the 1960s.

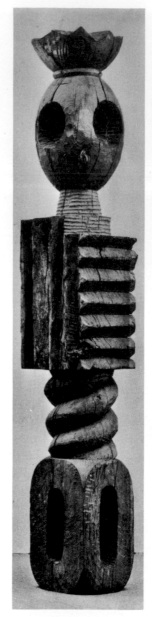

above: 193. CONSTANTIN BRANCUSI. *King of Kings.* c. 1930. Oak, height 9'10" (3 m). Solomon R. Guggenheim Museum, New York

© 1997 ARTISTS RIGHTS SOCIETY (ARS), NEW YORK/ADAGP, PARIS

right: 194. CONSTANTIN BRANCUSI. *Endless Column.* 1937–38. Cast iron, height 98' (29.9 m). Tirgu-Jiu, Romania

© 1997 ADAGP, PARIS/ARTISTS RIGHTS SOCIETY (ARS), NEW YORK

10 Cubism

The various modes of Cubism that Pablo Picasso developed jointly with Georges Braque in France between 1908 and 1914 offered a radically new way of looking at the world. Their shared vision had an inestimable impact on the vanguard art that followed it. Although Cubism was never itself an abstract style, the many varieties of nonobjective art it helped usher in throughout Europe are unthinkable without it. From Italian Futurism to Dutch Neo-Plasticism to Russian Constructivism, the repercussions of the Cubist experiment were thoroughly international in scope. But the legacy of Cubism was not exhausted in the first quarter of the twentieth century, when these movements took place. Its lingering influence can be felt in much art after World War II, in works as diverse as the paintings of Willem de Kooning (see fig. 518), the sculpture of David Smith (see fig. 539), the multimedia constructions of Robert Rauschenberg (see fig. 611), the photographs of David Hockney (see fig. 608), and the architecture of Frank Gehry (see fig. 914).

Cubism altered forever the Renaissance conception of painting as a window into a world where three-dimensional space is projected onto the flat picture plane by way of illusionistic drawing and one-point perspective. The Cubists concluded that reality has many definitions, and that therefore objects in space—and indeed, space itself—have no fixed or absolute form. Together, Picasso and Braque translated those multiple readings of the external world into new pictorial vocabularies of remarkable range and invention. The English critic Roger Fry, one of their early supporters, said of the Cubists, "They do not seek to imitate form, but to create form, not to imitate life but to find an equivalent for life." By 1911 the Cubist search for new forms was gaining momentum well beyond the studios of Picasso and Braque, eventually spawning a large school of Cubist painters in Paris. The work of these later practitioners of Cubism is discussed in the second half of this chapter.

The invention of Cubism was truly a collaborative affair, and the close, mutually beneficial relationship between Picasso and Braque was arguably the most significant of its kind in the history of art. According to Braque, it was "a union based on the independence of each." We have seen how much of the modern movement was the result of collective efforts between like-minded artists, as in Die Brücke or Der Blaue Reiter in Germany or the Fauves in France. But there was no real precedent for the intense level of professional exchange that took place between Picasso and Braque, especially from 1908 to 1912, when they were in close, often daily contact. "We were like two mountain climbers roped together," Braque said. Picasso called him "pard," in humorous reference to the Western novels they loved to read (especially those about Buffalo Bill), or "Wilbur," likening their enterprise to that of the American aviation pioneers, Wilbur and Orville Wright. Braque was a Frenchman, Picasso a Spaniard, and their temperaments, both personal and artistic, were strikingly dissimilar. Whereas Picasso was impulsive, prolific, and anarchic, Braque was slow, methodical, and meditative. He had little of Picasso's personal magnetism or colossal egotism, and the tendency toward lyrical painterliness in his work stood in strong contrast to Picasso's Expressionist sensibility. Their close working relationship was brought to an end by World War I. In August 1914 Braque was called up for active military duty; as a Spanish expatriate, Picasso was not called and remained behind, staying mostly in Paris and southern France during the war.

Already during his lifetime **PABLO RUIZ PICASSO** (1881–1973) achieved legendary status. The critic John Berger went so far as to say that by the 1940s, if Picasso wanted to possess anything, all he had to do was draw it. His career dominated three-quarters of the century, during which he produced a staggering output of art in virtually every conceivable medium, from small-scale prints, drawings, and ceramics to mural-size paintings and monumental public sculpture. The catalogue raisonné of his paintings and drawings alone runs to thirty-three volumes. With a combination of virtuoso technical facility and a virtually unparalleled ability to invent new forms, Picasso could easily compose in several styles simultaneously. As a result, the single-minded focus of his Cubist years, when he collaborated with a fellow artist more closely than at any other time in his career, presents something of an anomaly.

Nearly as well known as Picasso's art are the stormy details of his personal life, which was littered with emotionally destructive relationships, whether with lovers, children, or friends. More than is so for his contemporaries (including Braque), knowledge of Picasso's life is essential to a thorough understanding of his art, for autobiography infiltrates his imagery in overt and covert ways. As William Rubin, one of the leading Picasso specialists, has written, the biographical references in Picasso's art "often constitute a distinct and consistent substratum of meaning distinguishable from, but interwoven with, the manifest subject matter." No other twentieth-century artist discussed in this book has been studied in as much depth as Picasso. To justly serve the complex nature of his long and multifaceted career, his oeuvre, of which Cubism proper really only encompasses six years, is addressed in the course of several chapters in this text.

Picasso was born in the small, provincial town of Málaga, on the southern coast of Spain. Though an Andalusian by birth (and a resident of France for most of his life), he always identified himself as a Catalan, from the more industrialized, sophisticated city of Barcelona, where the family moved when he was thirteen. Picasso's father was a painter and art teacher who, so the fable goes, was so impressed by his adolescent son's talents that he handed over his brushes to him and renounced painting forever. Many such self-aggrandizing legends originated with Picasso himself or his friends and were perpetuated by his biographers. Nevertheless, the artist is recorded as saying, "In art one must kill one's father," and he eventually did revolt against all that his father tried to teach him.

Though Picasso was no child prodigy, by the age of fourteen he was making highly accomplished portraits of his family. In the fall of 1895, when his father took a post at Barcelona's School of Fine Arts, the young Picasso was allowed to enroll directly in the advanced courses. The following year, he produced his first major figure composition, *The First Communion*, which was included in an important exhibition in Barcelona. At the urging of his father, Picasso spent eight or nine months in Madrid in 1897, where he enrolled at but rarely attended the Royal Academy of San Fernando, the most prestigious school in the country. In Madrid, Picasso studied Spain's old masters at the great Prado Museum, admiring especially Diego Velázquez, whose work he copied. Several decades later, Picasso made his own highly personal interpretations of the old masters he saw in the museums (color plate 265, page 481).

Picasso suffered serious poverty in Madrid and chafed under the strictures of academic education, already setting his sights on Paris. First, however, he returned for a time to Barcelona, a city that provided a particularly stimulating environment for a young artist coming of age at the turn of the century. As the capital of Catalonia, a region fiercely proud of its own language and cultural traditions, Barcelona was undergoing a modern renaissance. Its citizens were vying to establish their city as the most prosperous, culturally enlightened urban center in Spain. At the same time, poverty, unemployment, and separatist sentiments gave rise to political unrest and strong anarchist tendencies among the Catalans. As in Madrid, the latest artistic trend in Barcelona was *modernista*, a provincial variation on Art Nouveau and Symbolism (see figs. 109, 110) that was practiced by young progressive artists and that left its mark on Picasso's early work.

Upon his return to Barcelona, Picasso declined to resume his studies at the Academy or to submit to his father's tutelage. Essential to his development was Els Quatre Gats (The Four Cats), a tavern that provided a fertile meeting ground for Catalan artists and writers. Picasso illustrated menus for Els Quatre Gats in the curving, stylized forms of *modernista* and made portraits of the habitués that were exhibited on the tavern walls. These incisive character studies amply illustrate that he had arrived at a personal mode well beyond his previous academic exercises. His watercolor from 1898–99, *The End of the Road* (fig. 195), contains aspects of the *modernista* style in its undulating silhouettes. On the path at the right, stooped figures shuffle toward a gateway over which an enormous angel of death presides. To their left, the wealthy travel by carriage to the same destination. "At the end of the road

195. PABLO PICASSO. *The End of the Road.* 1898–99. Oil washes and conté crayon on laid paper, 17⅞ x 11¾" (45.4 x 29.9 cm). Solomon R. Guggenheim Museum, New York

196. PABLO PICASSO. *Le Moulin de la Galette*. 1900. Oil on canvas, 35½ x 46" (90.2 x 116.8 cm). Solomon R. Guggenheim Museum, New York

THANNHAUSER COLLECTION, GIFT OF JUSTIN K. THANNHAUSER, 1978
© 1997 ESTATE OF PABLO PICASSO/ARTISTS RIGHTS SOCIETY (ARS), NEW YORK

death waits for everybody," the artist said, "even though the rich go in carriages and the poor on foot." Picasso's work can often be linked to events in his life, and his biographer John Richardson has observed that this image stems from the death in 1895 of Picasso's younger sister, Conchita. This early trauma was compounded by the family's inability to afford a suitable funeral. In more general terms, Picasso may have been responding to the poverty around him, including the ranks of mutilated veterans who had returned from Spain's colonial wars and were reduced to begging in the streets. But the pessimistic mood of *The End of the Road* was also in step with the prevailing *mal du siècle* that permeated international Symbolist art toward the end of the century. We have seen this melancholic soul sickness in the works of artists such as Aubrey Beardsley and Edvard Munch (see figs. 105, 120). Picasso's preoccupation with poverty and mortality was to surface more thoroughly in the work of his so-called Blue Period (color plate 87, page 165).

In 1900 one of Picasso's paintings was selected to hang in the Spanish Pavilion of the Universal Exposition in Paris. At age nineteen he traveled to the French capital for the first time with his friend and fellow painter Carles Casagemas. Picasso, who spoke little French, took up with a community of Spanish (mostly Catalan) artists in Paris and soon found benefactors in two dealers, a fellow Catalan, Pedro Mañach, and a Frenchwoman, Berthe Weill, also an early collector of works by Matisse. She sold the first major painting Picasso made in Paris, *Le Moulin de la Galette* (fig. 196), a vivid study of closely packed figures in the famous dance hall in Montmartre. Although he no doubt knew Renoir's Impressionist rendition of the same subject (color plate 11, page 37), Picasso rejected the sun-dappled congeniality of the Frenchman for a nocturnal and considerably more sinister view. This study of figures in artificial light relates to the dark, tenebrist scenes typical of much Spanish art at the time, but the seamy

ambience recalls the work of Degas or Toulouse-Lautrec, both shrewd chroniclers of Parisian café life.

Picasso's companion during this sojourn in Paris, Carles Casagemas, became completely despondent over a failed love affair after the artists arrived back in Spain. He eventually returned to Paris without Picasso and, in front of his friends, shot himself in a restaurant. Picasso experienced pangs of guilt for having abandoned Casagemas (though he soon had an affair with the very woman who had precipitated the suicide), and he later claimed that it was the death of his friend that prompted the gloomy paintings of his Blue Period. Between 1901 and 1904, in both Barcelona and Paris, Picasso used in many works a predominantly blue palette for the portrayal of figures and themes expressing human misery—frequently hunger and cold, the hardships he himself experienced as he was trying to establish himself. The thin, attenuated figures of Spain's sixteenth-century Mannerist painter, El Greco, found echoes in these works by Picasso, as did the whole history of Spanish religious painting, with its emphasis on mourning and physical torment.

One of the masterpieces of this period is the large allegorical composition Picasso made in Barcelona, *La Vie* (*Life*) (color plate 87, page 165), for which he prepared several sketches. For the gaunt couple at the left Picasso depicted Casagemas and his lover, though X rays reveal that the male figure was originally a self-portrait. They receive a stern gaze from the woman holding a baby at the right, who emerged from studies of inmates Picasso had made in a Parisian women's prison (where he could avail himself of free models), many of them prostitutes suffering from venereal disease. By heavily draping the woman, he gives her the timeless appearance of a madonna, thus embodying in a single figure his famously ambivalent and polarized view of women as either madonnas or whores. In their resemblance to a shamed Adam and Eve, the lovers (like their equivalent in the painting on the wall behind them) also yield a religious reading. From events in his own life Picasso built a powerful image around the universal themes of love, life, and death. The pervasive blue hue of *La Vie* turns flesh into a cold, stony substance while compounding the melancholy atmosphere generated by the figures. Many Symbolist painters, and Romantics before them, understood the richly expressive potential of the color blue, including Gauguin in his great canvas from 1897, *Where Do We Come From? What Are We? Where Are We Going?* (color plate 30, page 77), a painting that also confronts life's largest questions. Gauguin, whose work made a significant impression on the young Spaniard, died in the distant Marquesas Islands while Picasso was at work on *La Vie*.

In the spring of 1904 Picasso settled again in Paris; he was never to live in Barcelona again. He moved into a tenement on the Montmartre hill dubbed by his friends the Bateau Lavoir ("laundry boat," after the laundry barges docked in the Seine) that was ultimately made famous by his presence there. Until 1909 Picasso lived at the Bateau Lavoir in the midst of an ever-growing circle of friends: painters, poets, actors, and critics, including the devoted Max Jacob, who was to die in a concentration camp during World War II, and the poet Guillaume Apollinaire, who was to become the literary apostle of Cubism. In 1921 Picasso paid tribute to both men in his two versions of *The Three Musicians*, discussed in chapter 14 (color plate 153, page 321).

Among the first works Picasso made upon arriving in Paris in 1904 is a macabre pastel in which a woman with grotesquely elongated fingers and hunched shoulders kisses a raven. The subject evokes Edgar Allan Poe, a favorite writer among fin-de-siècle Parisian artists, particularly Odilon Redon, whose own brilliant pastels may have been a source for Picasso's (color plate 26, page 74). The model for *Woman with a Crow* (fig. 197) was an acquaintance of Picasso who had trained her pet bird to scavenge crumbs from the floors of the Lapin Agile, a working-class cabaret in Montmartre frequented by the artist and his entourage. Her skeletal form is silhouetted here against an intense blue, but the chilly hues and downcast mood of the Blue Period were giving way to the new themes and the warmer tonalities that characterize the next phase of Picasso's work, the Rose Period.

Between 1905 and 1906 Picasso was preoccupied with the subject of acrobatic performers who traveled from town to town, performing on makeshift stages. Called *saltimbanques*, these itinerant entertainers often dressed as Harlequin and Pierrot, the clownish characters from the Italian tradition of *commedia dell'arte* who compete for the amorous attentions of Columbine. These subjects were popular in France as well as Spain in nineteenth-century literature and painting, and Picasso certainly knew pictorial variations on the themes by artists such as Daumier and Cézanne. His own interest in circus subjects was stimulated by frequent visits to Paris's Cirque Médrano, located near his Montmartre studio.

Like his predecessor Daumier, Picasso chose not to depict the saltimbanque in the midst of boisterous acrobatics but

197. PABLO PICASSO. *Woman with a Crow.* 1904.
Charcoal, pastel, and watercolor on paper,
25½ x 19½" (64.8 x 49.5 cm).
The Toledo Museum of Art

during moments of rest or quiet contemplation. His masterpiece of this genre is *Family of Saltimbanques* (color plate 88, page 166); at seven and one-half feet tall, it was the largest painting he had yet undertaken and stands as a measure of his enormous ambition and talent in the early years of his career. He made several preparatory studies for the work, and X rays reveal that at least four variations on the final composition exist beneath the top paint layer. The painting began as a family scene, with women engaging in domestic chores and Harlequin (the figure seen here at the far left) watching a young female acrobat balance on a ball. But in the end Picasso settled upon an image virtually devoid of activity, in which the characters, despite their physical proximity, hardly take note of one another. Stripped of anecdotal details, the painting gains in poetry and mystery. The acrobats gather in a strangely barren landscape painted in warm brown tones that the artist loosely brushed over a layer of blue paint. An enigmatic stillness has replaced the heavy pathos of the Blue Period pictures, just as the predominant blues of the earlier work have shifted to a delicately muted palette of earth tones, rose, and orange. The identities of the individual saltimbanques have been ascribed to members of Picasso's circle, but the only identification experts agree upon is that of Picasso, who portrayed himself, as he often did, in the guise of Harlequin, easily identifiable by the diamond patterns of his motley costume. These homeless entertainers, who eked out a living through their creative talents, existing on the margins of society, have long been understood as surrogates for the artist.

Before completing this large canvas, Picasso exhibited over thirty works at a Paris gallery, including several saltimbanque subjects. Little is known about the sales, if any, that resulted from the show, but subsequently Picasso tended to refrain from exhibiting his work in Paris, except in small shows in private galleries (his favorite venue was his own studio). Although he was still seriously impoverished, his circumstances improved somewhat during these years as his circle of admirers—critics, dealers, and collectors—continued to grow.

One remarkable member of that circle was the American writer Gertrude Stein, who, with her brothers Leo and Michael, had lived in Paris since 1903 and was building one of the foremost collections of contemporary art in the city. She met Picasso in 1905 and introduced him the following year to Matisse, whose work the Steins were collecting. In their apartment Picasso had access to one of the great early Matisse collections, including the seminal *Bonheur de vivre* (color plate 52, page 118). He was keenly aware of Matisse as a potential artistic rival, particularly following the uproar over the 1905 Salon d'Automne, in which the Fauves made their public debut. Gertrude Stein said she sat for Picasso ninety times while he painted her portrait (fig. 198). Her ample frame, characteristically draped in a heavy, loose-fitting

199. PABLO PICASSO. *Two Nudes*. Paris, late 1906.
Oil on canvas, 59⅝ x 36⅝" (151.5 x 93 cm).
The Museum of Modern Art, New York
GIFT OF ALFRED H. BARR, JR.
© 1997 ARTISTS RIGHTS SOCIETY (ARS), NEW YORK/ADAGP, PARIS

dress, assumes an unconventional pose for a female sitter, one that clearly conveys the self-assured nature of a woman who freely proclaimed her own artistic genius. She fills her corner of space like some powerful, massive work of sculpture. Picasso abandoned the portrait over the summer while he vacationed in Spain and repainted Stein's face from memory upon his return. Its masklike character, with eyes askew, anticipates the distortions of coming years.

By the end of 1905 Picasso's figures had taken on an aura of beauty and serenity that suggests a specific influence from ancient Greek white-ground vase paintings. This brief classical phase was prompted by several factors, including his contented (at least temporarily) relationship with a young woman named Fernande Olivier, his studies of antiquities in the Louvre, and the 1905 retrospective of the work of Ingres, the French Neoclassical painter who died in 1867 (color plate 1, page 33). Picasso's subjects during this period were primarily nudes, either young, idealized women or adolescent boys, all portrayed in a restrained terra-cotta palette evoking the ancient Mediterranean. By the end of 1906, following a trip with Olivier to the remote Spanish village of Gósol, this classical ideal had evolved into the solid, thick-limbed anatomies of *Two Nudes* (fig. 199), a mysterious composition

198. PABLO PICASSO. *Gertrude Stein*. 1905–6. Oil on canvas, 39¼ x 32" (99.7 x 81.3 cm). The Metropolitan Museum of Art, New York BEQUEST OF GERTRUDE STEIN, 1946 © 1997 ESTATE OF PABLO PICASSO/ARTISTS RIGHTS SOCIETY (ARS), NEW YORK

in which we confront a woman and what is nearly but not quite her mirror reflection. The image is drained of the sentimentalism that inhabited the artist's Blue and Rose Period pictures. Significantly, Picasso was making sculpture around this time and earlier in 1906 had seen a show at the Louvre of Iberian sculpture from the sixth and fifth centuries B.C.E. These works were of particular interest to him since they had been excavated fifty miles from his hometown in Spain. In 1907 he actually purchased two ancient Iberian stone heads, which he returned in 1911, after discovering they had been stolen from the Louvre. In addition, the 1906 Salon d'Automne had featured a large retrospective of Gauguin's work, including the first important public presentation of his sculpture (see fig. 130). As an artist who plundered the art of non-Western and ancient cultures for his own work, Gauguin provided Picasso with a crucial model. Picasso's own preoccupation with sculptural form at this time, his encounter with the blocky contours of the ancient Iberian carvings, and the revelatory encounter with Gauguin's "primitivizing" examples left an indelible mark on paintings such as *Two Nudes*. It has even been proposed that the corpulence of Gertrude Stein influenced the hefty proportions of the models in this work. Certainly the nudes' simplified facial features, with their wide, almond-shaped eyes, recall the Stein portrait.

Picasso's large canvas, *Les Demoiselles d'Avignon* (color plate 89, page 166), has been called the single most important painting of the twentieth century. It is an astonishing image that virtually shattered every pictorial and iconographical convention that preceded it. Robert Rosenblum, one of the leading historians of Cubism, called this momentous work a "detonator of the modern movement, the cornerstone of twentieth-century art." In 1988 the Musée Picasso in Paris devoted a large exhibition exclusively to this picture. Nearly nine decades after its creation, during which scholars have generated hundreds of pages scrutinizing its history and pondering its meaning, the painting still has the power to assault the viewer with its aggressively confrontational imagery. Long regarded as the first Cubist painting, the *Demoiselles*, owing largely to the pioneering studies of Leo Steinberg, is now generally seen as a powerful example of Expressionist art—an "exorcism painting," Picasso said, in which he did not necessarily initiate Cubism, but rather obliterated the lessons of the past. In fact, this type of multifigure composition did not persist in high Cubism, which was the domain of the single figure, the still life, and to a lesser degree, the landscape. Nor did this violent, expressive treatment of the subject find a place in a style famous for emotional and intellectual control.

The five *demoiselles*, or young ladies, represent prostitutes from Avignon Street, in Barcelona's notorious red-light district, which Picasso knew well. The title, however, is not his; Picasso rarely named his works. That task he left to dealers and friends. The subject of prostitution, as we saw with Manet's scandalous *Olympia* (see fig. 43), had earned a prominent place in vanguard art of the nineteenth century. Cézanne responded to Manet's painting with his own idiosyncratic versions of the subject. Other French artists, especially Degas and Toulouse-Lautrec, had portrayed the interior life of brothels, but never with the vehemence or enormous scale seen here, where the theme of sexuality is raised to an explosive pitch. When confronted with the *Demoiselles*, it is difficult to recall that the gentle *Saltimbanques* painting (color plate 88, page 166) was only two years old. When compared to *Two Nudes* (fig. 199), the anatomies of the *Demoiselles* seem crudely flattened and simplified, reduced to a series of interlocking, angular shapes. Even the draperies—the white cloth that spills off the table in the foreground and the bluish curtains that meander around the figures—have hardened into threatening shards. The composition is riddled with deliberately disorienting and contradictory points of view. The viewer looks down upon the table at the bottom of the canvas, but encounters the nudes head-on. Eyes are presented full face, while noses are in complete profile. The seated figure at the lower right faces her cohorts but manages to turn her head 180 degrees to address the viewer. The two central figures offer themselves up for visual consumption by assuming classic Venus poses, while their companion to the left strides into the composition like a fierce Egyptian statue. All of the prostitutes stare grimly ahead, with emotions seemingly as hardened as the knife-edge forms themselves. The three women at the left wear the simplified "Iberian" features familiar from *Two Nudes*. Most shocking of all, however, are the harshly painted masks that substitute for faces on the two figures at the right. Here, Steinberg has written, sexuality is "divested of all accretions of culture—without appeal to privacy, tenderness, gallantry, or that appreciation of beauty which presupposes detachment and distance."

The *Demoiselles* evolved over several months in 1907, during which Picasso prepared his work in the manner of a traditional Salon artist, executing dozens of drawings, including those seen in figures 200 and 201. Through these preparatory works we are able to observe Picasso's progress as he honed his composition from a complex narrative to the startlingly iconic image we encounter today. The first of the drawings shown here, executed in the spring of 1907, reveals Picasso's early idea for the painting. At the far left is a figure the artist identified as a medical student. He holds a book in his right arm (in earlier studies it was a skull) and draws back a curtain with his left. In this curiously draped, stagelike setting, which Rosenblum called a "tumultuous theater of sexuality," the student encounters a provocative quintet of naked women, one of whom theatrically opens more curtains at the right. They surround a lone sailor, who is seated before a table holding a plate of spiky melons and a phallic *porrón*, a Spanish wine flask. A second still life on a round table projects precipitously into the room at the bottom of the sheet. Even on this small scale the women appear as aggressive, sexually potent beings, compared to their rather docile male clients. For Picasso, a frequenter of brothels in his youth, the threat of venereal disease was of dire concern in an era in which such afflictions could be fatal. The inclusion of a skull-toting medical student (who resembled Picasso in related sketches) probably reflects the artist's morbid conflation of sex and

200. PABLO PICASSO. *Medical Student, Sailor, and Five Nudes in a Bordello* (study for *Les Demoiselles d'Avignon*). March–April 1907. Graphite and pastel on paper, 18⅞ x 25" (47.7 x 63.5 cm). Öffentliche Kunstsammlung, Kunstmuseum, Basel
© 1997 ESTATE OF PABLO PICASSO/ARTISTS RIGHTS SOCIETY (ARS), NEW YORK

201. PABLO PICASSO. Compositional study with five figures for *Les Demoiselles d'Avignon*. June 1907. Watercolor, 6¾ x 8¾" (17.2 x 22.2 cm). Philadelphia Museum of Art
A. E. GALLATIN COLLECTION © 1997 ESTATE OF PABLO PICASSO/ARTISTS RIGHTS SOCIETY (ARS), NEW YORK

death and his characteristically misogynist view of women as the carriers of life-threatening disease.

In a tiny watercolor made about two months after the drawing (fig. 201), Picasso transformed the medical student into a woman in a similar pose and eliminated the sailor altogether. The prostitutes, for the most part, have shifted their attention from their clients to the viewer. It is this conception, deprived of the earlier narrative clues, that makes up the large painting. As Rubin has observed, Picasso did not need to spell out his allegory to get his meaning across; he could let these predacious seductresses stand alone as symbols of sex and death. In this watercolor, executed in delicate washes of pink with jagged, electric-blue contour lines, Picasso also

made drastic changes in his formal conception of the subject. The space is compressed, the figures are highly schematized, and rounded contours have given way to sharp, angular forms. Even the table is now a triangle, jutting into the room.

Equipped with a prodigious visual memory and keen awareness of past art, Picasso could have had several paintings in mind as he set to work on his large canvas in June. In 1905 he had seen Ingres's round painting *Turkish Bath* of 1863, a titillating Salon piece that displays intertwined female nudes in myriad erotic poses. While staying in Gósol in 1906 Picasso produced his own version of a harem subject. But in that painting, unlike the *Demoiselles*, the sitters (all likenesses of Olivier) are represented in conventional bather poses, washing themselves or brushing their hair. Matisse's spectacular painting *Bonheur de vivre* (color plate 52, page 118) had surely made a great impact on Picasso in the 1906 Salon des Indépendants, the juryless Salon that had been an important showcase for avant-garde art since 1884. The following year he had probably also seen Matisse's *Blue Nude* (color plate 54, page 119). It is possible that Picasso undertook the *Demoiselles* partly as a response to these radical compositions by his chief rival. Most important of all for the artist (and for the subsequent development of Cubism) was the powerful example of Cézanne, who died in 1906 and whose work was shown in a retrospective at the 1907 Salon d'Automne. It is important to recall that many of the French master's early depictions of bathers (see fig. 72) were fraught with an anxiety and violence that no doubt attracted Picasso, but a major instigating force for the *Demoiselles* were Cézanne's later bather paintings (color plate 25, page 74). Indeed, certain figures in Picasso's composition have been traced to precedents in Cézanne's works. Picasso, however, chose not to cast his models as inhabitants of a timeless pastoral, but as sex workers in a contemporary brothel.

The work of many other painters has been linked to Picasso's seminal painting, from El Greco to Gauguin, but more significant than these stimuli was the artist's visit in June 1907 to Paris's ethnographic museum, where he saw numerous examples of African and Oceanic sculpture and masks. According to Apollinaire (who no doubt embellished the artist's words), Picasso once said that African sculptures were "the most powerful and the most beautiful of all the products of the human imagination." He had already seen such works previously, during studio visits to artists who were collecting African art, including Matisse, Derain, and Vlaminck (for a discussion of the ways in which Western artists approached non-Western art, see chapter 7). But his encounter at the museum, which took place while he was at work on the *Demoiselles*, provoked a "shock," he said, and he went back to the studio and reworked his painting (though not specifically to resemble anything he had seen at the museum). It was at this time that he altered the faces of the two figures at the right from the "Iberian" countenances to violently distorted, depersonalized masks. Picasso not only introduced dramatically new, incongruous imagery into his painting, but he rendered it in a manner unlike anything else in the picture. When he added the women's masklike visages at the right he used dark hues and rough hatch marks that have been likened to scarification marks on African masks. By deploying such willfully divergent modes throughout his painting Picasso heightened its disquieting power. Rubin has even argued that the stylistic dissimilarity is central to an understanding of the work's sexual theme. As our eye scans from left to right in the painting, we may see what Rubin has called Picasso's "progressively darkening insights" into his own concept of woman's nature. African art provided him with a model for distorting the female form in order to express his own innermost anxieties.

When the French artist GEORGES BRAQUE (1882–1963) first saw *Les Demoiselles d'Avignon* in Picasso's studio in 1907, he compared the experience to swallowing kerosene and spitting fire. Most of Picasso's supporters, including Apollinaire, were horrified, or at least stymied, by the new painting. Derain ventured that one day Picasso would be found hanged behind his big picture. According to the dealer Daniel-Henry Kahnweiler, the *Demoiselles* "seemed to everyone something mad or monstrous." As more and more visitors saw it in the studio (it was not exhibited publicly until 1916), the painting sent shock waves throughout the Parisian artistic community. For Braque, the painting had direct consequences for his art, prompting him to apply his current stylistic experiments to the human figure.

Just one year younger than Picasso, Braque was the son and grandson of amateur painters. He grew up in Argenteuil, on the Seine River, and the port city of Le Havre in Normandy, both famous Impressionist sites. In Le Havre he eventually befriended the artists Raoul Dufy and Othon Friesz, who, like Braque, became associates of the Fauve painters. As an adolescent Braque took flute lessons and attended art classes in the evening at Le Havre's École des Beaux-Arts. Following his father's and grandfather's trade, he was apprenticed as a house painter and decorator in 1899 and at the end of 1900 he went to Paris to continue his apprenticeship. Some of the decorative and trompe l'oeil effects that later entered his paintings and collages were the result of this initial training. After a year of military service and brief academic training, Braque settled in Paris, but he returned frequently to Le Havre to paint landscapes, especially views of the harbor. Like his artist contemporaries in Paris, he studied the old masters in the Louvre and was drawn to Egyptian and archaic sculpture. He was attracted by Poussin among the old masters, and his early devotion to Corot continued. During the same period he was discovering the Post-Impressionists, including Van Gogh, Gauguin, and the Neo-Impressionist Signac, but it was ultimately the art of Cézanne that affected him most deeply and located him on the path to Cubism.

In the fall of 1905 Braque was impressed by the works of Matisse, Derain, and Vlaminck at the notorious Salon d'Automne where Fauvism made its public debut, and the following year he turned to a bright, Fauve-inspired palette. Unlike Picasso, Braque submitted work regularly to the Salons in the early years of his career. His paintings were included in the 1906 Salon des Indépendants, but he later

destroyed all the submitted works. In the company of Friesz and Derain, he spent the fall and winter of 1906 near Marseilles at L'Estaque, where Cézanne had painted (color plate 21, page 40). He then entered a group of his new Fauve paintings in the same Salon des Indépendants of March 1907 that included Matisse's *Blue Nude*. One of his works was purchased by Daniel-Henry Kahnweiler, who was to become the most important dealer of Cubist pictures. Though the record is not certain, Braque and Picasso probably met around this time. It was inevitable that their paths should cross, for they had several friends in common, including Derain, Matisse, and Apollinaire.

During his brief exploration of Fauvism, Braque developed his own distinctive palette, beautifully exemplified in the clear, opalescent hues of *Viaduct at L'Estaque* (color plate 90, page 167). Despite the bright coloration of this landscape, made on the site in the fall of 1907, its palette is tame compared with the radical color experiments of Matisse's Fauve work (color plate 51, page 117). Owing largely to lessons absorbed from Cézanne, Braque was not willing to forgo explicit pictorial structure in his landscapes. The composition of *Viaduct at L'Estaque* is rigorously constructed: the hills and houses, though abbreviated in form, retain a palpable sense of volume, created through Cézannist patches of color and blue outlines, rather than through traditional chiaroscuro. Braque built these forms up toward a high horizon line and framed them within arching trees, a favorite device of Cézanne. These tendencies in Braque's work are often referred to as "Cézannism."

At the end of 1907 Braque visited Picasso's studio with Apollinaire, possibly for the first time. By the end of the next year he and Picasso were regular visitors to one another's studios. On this visit Braque saw the completed *Demoiselles* and the beginnings of Picasso's most recent work, *Three Women* (fig. 203). Until this point Braque had rarely painted the human form, but early in 1908 he made a multifigure painting, since lost, which was clearly inspired by Picasso's *Three Women*. That spring he produced *Large Nude* (fig. 202), a painting that signals his imminent departure from Fauvism. Like Picasso, Braque had seen Matisse's *Blue Nude* the previous fall (color plate 54, page 119), and his massively muscular figure, set down with bold brushwork and emphatic black outlines, bears the memory of that work. But Braque repudiated the rich pinks and blues of *Viaduct at L'Estaque* for a subdued palette of ochers, browns, and gray-blue. While the model's simplified facial features and the angular drapery that surrounds her recall the *Demoiselles,* Braque was not motivated by the sexualized themes of Picasso's painting. With *Large Nude,* he said he wanted to create a "new sort of beauty . . . in terms of volume, of line, of mass, of weight."

A comparison between *Viaduct at L'Estaque* (color plate 90, page 167), and Braque's painting from August 1908, *Houses at L'Estaque* (color plate 91, page 167), makes clear the transformations that took place in his style after the fall of 1907. With the suppression of particularizing details, the houses and trees become simplified, geometric volumes that are experienced at close range and sealed off from surround-

202. GEORGES BRAQUE. *Large Nude.* 1908.
Oil on canvas, 55⅛ x 39" (140 x 99.1 cm).
Collection Alex Maguy, Paris

ing sky or land. Braque explained that in such pictures he wished to establish a background plane while "advancing the picture toward myself bit by bit." Rather than receding into depth, the forms seem to come forward, approximating an appearance of low-relief sculpture, or bas-relief. Braque's illusion of limited depth is not dependent upon traditional, single-point linear perspective. Cézanne, he later said, was the first to break with that kind of "erudite, mechanicized perspective." Instead, he achieves illusion by the apparent volume of the buildings and trees—their overlapping, tilted, and shifting shapes create the effect of a scene observed from various positions. Despite many inconsistencies (of the sort that abound in Cézanne's work), including conflicting orthogonals and vantage points, or roof edges that fail to line up or that disappear altogether, Braque managed a wholly convincing, albeit highly conceptualized, space. Color is limited to the fairly uniform ocher of the buildings, and the greens and blue-green of the trees. Whereas Cézanne built his entire organization of surface and depth from his color, Braque, in this work and increasingly in the paintings of the next few years, subordinates color in order to focus on pictorial structure.

With *Houses at L'Estaque* Braque established the essential syntax of early Cubism. He is generally credited with arriving at this point single-handedly, preliminary to the steady exchanges with Picasso that characterize the subsequent development of high Cubism. These works at L'Estaque confirmed Braque's thorough abandonment of Fauvism. "You can't remain forever in a state of paroxysm," he said. In fact,

Fauvism was a short-lived experiment for most of its artists, partly because painters such as Derain, Dufy, and Vlaminck fell increasingly under the spell of Cézanne's work. In the world of the Parisian avant-garde during this remarkably fertile period, there was a growing sense of polarization between the followers of Matisse and the Cézannist and Cubist faction. Gertrude Stein described the competing camps as "Picassoites" and "Matisseites."

In the fall of 1908 Braque took his new works from L'Estaque back to Paris, where he showed them to Picasso and submitted them to the progressive Salon d'Automne. The jury, which included Matisse, Rouault, and the Fauve painter Albert Marquet, rejected all his entries. According to Apollinaire, who told the story many ways over the years, it was upon this occasion that Matisse referred to Braque's "little cubes." By 1912 this had become the standard account for the birth of the term "Cubism." Following his rejection, Braque exhibited twenty-seven paintings in November 1908 at the Kahnweiler Gallery, with a catalogue by Apollinaire. Louis Vauxcelles, the critic who had coined the term "Fauve," reviewed the exhibition and, employing Matisse's term, observed that Braque "reduces everything, places and figures and houses, to geometrical schemes, to cubes."

The remarkable artistic dialogue between Braque and Picasso began in earnest toward the end of 1908, when they started visiting one another's studios regularly to see what had materialized during the day. "We discussed and tested each other's ideas as they came to us," Braque said, "and we compared our respective works." At this time Picasso completely revised *Three Women* (fig. 203), a painting he had begun the year before in response to Braque's Cubist paintings from

203. PABLO PICASSO. *Three Women.* 1907–8. Oil on canvas, 6'6¾ x 7'6½" (2 x 2.3 m). The Hermitage Museum,
St. Petersburg

L'Estaque. Those new experiments led Picasso toward a more thorough absorption of "Cézannism," as he developed his own variation on the bas-relief effects he saw in Braque's work. The three women here seem to be a reprise of the two central figures of the *Demoiselles*, although these mammoth beings seem even less like flesh and blood and appear as though chiseled from red sandstone. In earlier states of the painting, known from vintage photographs, the faces were boldly inscribed with features inspired by African masks, but Picasso altered those features in the final state. The women no longer stare boldly ahead, as in the *Demoiselles*, but seem to hover on the edge of deep slumber, suggesting, as Steinberg has surmised, the sexual awakening of some primitive life form. Picasso here employs Cézanne's *passage*, a technique whereby the edges of color planes slip away and merge with adjacent areas. For example, on the proper left leg of the central figure (meaning her left, not ours) is a gray triangle that dually functions as a facet of her thigh and part of her neighbor's breast. This *passage* mitigates any sense of clear demarcation between the figures and their environment, making, in Steinberg's words, "body mass and surround consubstantial." Such ambiguity of figure and ground, facilitated by an opening up of contours, is a fundamental characteristic of Cubism.

Like many of Matisse's and Picasso's greatest paintings, *Three Women* is in the collection of Russia's Hermitage Museum in St. Petersburg. The work was purchased by the brilliant Russian collector Sergei Shchukin, who was brought to Picasso's studio by Matisse in September 1908. He purchased *Three Women* from Leo and Gertrude Stein, who, because of strained relations with one another, began to sell their collection in 1913.

Although all of Picasso's paintings discussed to this point have depicted the figure, he frequently worked in the genre of still life throughout the development of Cubism, as did Braque. Materials drawn from the café and the studio, the milieux most familiar to these artists, became standard motifs in their work. Newspapers, bottles of wine, or food—the objects literally within their reach—were often elliptically coded with personal, sometimes humorous allusions to their private world.

Picasso's majestic painting *Bread and Fruit Dish on a Table* (fig. 204) is an especially intriguing example of still life, since the artist initiated it as a figurative composition and then gradually transformed the human figures into bread, fruit, and furniture. We know from small studies that Picasso was planning a painting of several figures seated and facing the viewer from behind the drop-leaf table seen here (just like one in Picasso's studio). William Rubin has proposed that Picasso used turpentine to rub out most of the figures' upper halves from his canvas, but retained their legs, thus explaining the strange, planklike forms under the table. Other shapes from the old composition were retained to take on new identities in the still life. For example, the loaves of bread at the right, mysteriously propped up against a backdrop, were once the arms of a figure. At the opposite end of the table sat a woman, whose breasts were retained as fruit in a bowl. Thus,

204. PABLO PICASSO. *Bread and Fruit Dish on a Table*. 1909. Oil on canvas, 64⅝ x 52¼" (164 x 132.5 cm). Öffentliche Kunstsammlung Basel, Kunstmuseum
DONATION DR. H.C. RAOUL LA ROCHE, 1952
© 1997 ESTATE OF PABLO PICASSO/ARTISTS RIGHTS SOCIETY (ARS), NEW YORK

Picasso makes a sly visual pun on a hackneyed symbol of fecundity and womanhood, while balancing this rounded, feminine form with the phallic loaves at the opposite end of the table. Rubin, who reads these gendered forms as stand-ins for Picasso and his companion, Fernande Olivier, has proposed that even through the apparently neutral medium of still life Picasso could "tap the sentiments and particulars of his most intimate life and thought." Just as in *Family of Saltimbanques* and *Demoiselles d'Avignon*, he concludes, the anecdotal elements fall away as the pictorial components are gradually concentrated in a more strictly frontal and iconic composition. Rubin has interpreted the stylistic duality at play in this composition—with its highly finished, almost naïve rendition at the left and the much looser handling at the right—as Picasso's preoccupation at the time with the work of Rousseau (on the left) and Cézanne (on the right).

Picasso spent the summer of 1909 in the Spanish hill town of Horta de Ebro (today Horta de San Juan), where his early Cubist style reached its apogee. Among the landscapes he made that summer is *Houses on the Hill, Horta de Ebro* (fig. 205) in which the stucco houses of the village, seen from above, are piled up into a pyramidal structure of such refinement and delicate coloration that the Expressionist intensity of *Demoiselles* seems a distant memory. Painted in facets of pale gold and gray, the highly geometricized houses are presented in multiple, often contradictory perspectives. Rather than creating forms that recede predictably into the distance,

Picasso often reversed the orthogonals to project toward the viewer. His juxtaposition of light and dark planes enhances the overall sculptural configuration but fails to convey any single or consistent light source. The result is a shimmering, prismatic world where space and mass are virtually synonymous. Even the sky has been articulated into a crystalline pattern of intersecting planes—realized through subtle shifts in value of light gray or green—that suggests a distant mountain range.

205. PABLO PICASSO. *Houses on the Hill, Horta de Ebro.* Horta de Ebro, summer 1909. Oil on canvas, 25⅝ x 31⅞" (65 x 81 cm). The Museum of Modern Art, New York
NELSON A. ROCKEFELLER BEQUEST. © 1997 ESTATE OF PABLO PICASSO/ARTISTS RIGHTS SOCIETY (ARS), NEW YORK

During this crucial summer Picasso also made several paintings and drawings of Olivier, including portrait studies (fig. 206) that articulate the sculptural mass of her head into a series of curving planes. After returning to Paris in September he worked in the studio of a friend, the Catalan sculptor Manuel Hugué, known as Manolo, and sculpted a head based on his Horta studies of Olivier. Several bronzes (fig. 207) were then cast at a later date from the artist's original plaster casts. Picasso, who is arguably one of the most inventive sculptors of all time, had experimented with sculpture intermittently for several years. With *Woman's Head (Fernande)* he recapitulates in three dimensions the pictorial experiments of the summer by carving Olivier's features into facets as geometric as one of his Horta hillsides. Despite its innovative treatment of plastic form as a ruptured, discontinuous surface, this sculpture still depends upon the conventional methods and materials of modeling clay into a solid mass and making plaster casts (two were made) from the clay model that are then used to cast the work in bronze. *Woman's Head* exerted considerable influence on other Cubist sculp-

206. PABLO PICASSO. *Woman's Head (Fernande).* Summer 1909. Ink and watercolor on paper, 13⅛ x 10⅛" (33.3 x 25.5 cm) The Art Institute of Chicago
ALFRED STIEGLITZ COLLECTION
© 1997 ESTATE OF PABLO PICASSO/ARTISTS RIGHTS SOCIETY (ARS), NEW YORK

207. PABLO PICASSO. *Woman's Head (Fernande).* Paris, fall 1909. Bronze, 16¼ x 9¾ x 10½" (41.3 x 24.7 x 26.6 cm). The Museum of Modern Art, New York
PURCHASE. PHOTOGRAPH © 1995 THE MUSEUM OF MODERN ART, NEW YORK
© 1997 ESTATE OF PABLO PICASSO/ARTISTS RIGHTS SOCIETY (ARS), NEW YORK

tors, such as Aleksandr Archipenko and Henri Laurens (see figs. 218, 222), but Picasso later said it was "pointless" to continue in this vein. In 1912 he embarked upon a much more radical solution for Cubist sculpture.

While Picasso was concentrating on his sculpture, Braque spent the winter of 1909 in Paris, composing several still lifes of musical instruments. Braque was an amateur musician, and the instruments he kept in his studio were common motifs in his paintings throughout his career. He preferred violins, mandolins, or clarinets to other objects, for they have, he said, "the advantage of being animated by one's touch." Despite the intensified fragmentation that Braque had by now adopted, the still-life subjects of *Violin and Palette* (color plate 92, page 167) are still easily recognizable. Within a long, narrow format he has placed, in descending order, his palette, a musical score propped up on a stand, and a violin. These objects inhabit a shallow, highly ambiguous space. Presumably the violin and music stand are placed on a table, with the palette hanging on a wall behind them, but their vertical disposition within the picture space makes their precise orientation unclear. Although certain forms, such as the scroll at the top of the violin neck, are rendered naturalistically, for the most part the objects are not modeled continuously in space but are broken up into tightly woven facets that open into the surrounding void. At the same time, the interstices between objects harden into painted shards, causing space to appear as concrete as the depicted objects. It was this "materialization of a new space" that Braque said was the essence of Cubism.

At the top of *Violin and Palette* Braque depicted his painting palette, emblem of his metier, hanging from a carefully drawn nail. The shadow cast by the nail reinforces the object's existence in three-dimensional space. By employing this curious detail of trompe l'oeil, Braque calls attention to the ways in which his new system departs from conventional means of depicting volumetric shapes on a flat surface. In so doing, he declares Cubism's defiance of the Renaissance conceptions of space that had been under assault since Manet, in which art functions as a mirror of the three-dimensional world, and offers in its place a conceptual reconfiguration of that world. This kind of artistic legerdemain, in which different systems of representation simultaneously exist within one painting, was soon to be most fully exploited in Cubist collage and papier collé.

Picasso also treated musical themes in his Cubist compositions. One of the best-known examples is his *Girl with a Mandolin (Fanny Tellier)* (fig. 208), made in Paris in the spring of 1910. Still working from nature, Picasso hired a professional model named Fanny Tellier to pose nude in his studio. When this portrait is compared with Braque's *Violin and Palette,* Picasso's style appears somewhat less radical than Braque's. The figure still retains a sense of the organic and, in some areas, is clearly detached from the background plane by way of sculptural modeling, seen especially in the rounded volumes of the model's right breast and arm. Picasso's chromatic range, however, is even more restricted than that found in Braque's still life. Employing soft gradations of gray and

208. Pablo Picasso. *Girl with a Mandolin (Fanny Tellier).* Paris, late spring 1910. Oil on canvas, 39½ x 29" (100.3 x 73.7 cm). The Museum of Modern Art, New York Nelson A. Rockefeller Bequest
Photograph © 1995 The Museum of Modern Art, New York
© 1997 Estate of Pablo Picasso/Artists Rights Society (ARS), New York

golden brown, Picasso endowed this lyrical portrait with a beautifully tranquil atmosphere. Robert Rosenblum has rightly compared *Girl with a Mandolin* with Camille Corot's meditative depictions of women in the studio with musical instruments. In the fall of 1910 both Picasso and Braque saw an exhibition of Corot's work in Paris and were much impressed.

By 1910 Picasso and Braque had entered the period of greatest intensity in their collaborative, yet highly competitive, relationship. Over the next two years they produced some of the most cerebral, complex paintings of their careers, working with tremendous concentration and in such proximity that some of their compositions are virtually indistinguishable from one another. For a time they even refrained from signing their own canvases on the front in order to downplay the individual nature of their contributions. "We were prepared to efface our personalities in order to find originality," Braque said. "Analytic Cubism," in which the object is analyzed, broken down, and dissected, is the term used to describe this high phase of their collaboration.

One of the masterpieces of Analytic Cubism is Picasso's *Portrait of Daniel-Henry Kahnweiler* (color plate 93, page 168), made a few months after *Girl with a Mandolin.* The painting belongs to a series of portraits of his primary art dealer that Picasso made in 1910. Kahnweiler was an astute

German dealer who had been buying works by Braque and Picasso since 1908. He was also a critic who wrote extensively about Cubist art, including the first serious theoretical text on the subject, which interprets the style in semiological terms as a language of signs. In this portrait, for which Kahnweiler apparently sat about twenty times, the modeled forms of *Girl with a Mandolin* have disappeared, and the figure, rather than disengaged from the background, seems merged with it. Here the third dimension is stated entirely in terms of flat, slightly angled planes organized within a linear grid that hovers near the surface of the painting. Planes shift in front of and behind their neighbors, causing space to fluctuate and solid form to dissolve into transparent facets of color. Although he delineated his figure not as an integrated volume but as a lattice of lines dispersed over the visual field, Picasso nevertheless managed a likeness of his subject. In small details—a wave of hair, the sitter's carefully clasped hands, a schematic still life at the lower left—the painter particularized his subject and helps us to reconstruct a figure seated in a chair. Though Picasso kept color to the bare minimum, his canvas emits a shimmering, mesmerizing light, which he achieved by applying paint in short daubs that contain generous admixtures of white pigment. The Italian critic Ardengo Soffici, whose early writings about Picasso and Braque helped bring the Italian Futurists in contact with Cubism, referred to the "prismatic magic" of such works.

The degree to which the pictorial vocabularies of Picasso and Braque converged during the Analytic phase of Cubism can be demonstrated through a comparison of two works from 1911. Picasso's *Accordionist* (fig. 209) dates from the astonishingly productive summer of that year, which he spent with Braque in Céret, a village in the French Pyrenees. At a nearly identical state of exploration is Braque's *The Portuguese (The Emigrant)* (fig. 210), begun in Céret after Picasso's departure and completed in Paris. Both artists reorder the elements of the physical world within the shallow depths of their respective compositions, using the human figure as a pretext for an elaborate scaffolding of shifting, interpenetrating planes. The presence of the figure is made evident principally by a series of descending diagonal lines that helps concentrate the geometric structure down the center of the picture. That structure opens up toward the edges of the canvas where the pictorial incident diminishes, occasionally revealing the painting's bare, unpainted ground. Both

209. PABLO PICASSO. *Accordionist.* summer 1911. Oil on canvas, 51¼ x 35¼" (130.2 x 89.5 cm). Solomon R. Guggenheim Museum, New York
© 1997 ESTATE OF PABLO PICASSO/ARTISTS RIGHTS SOCIETY (ARS), NEW YORK

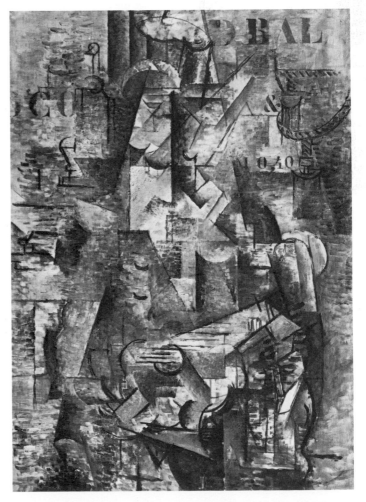

210. GEORGES BRAQUE. *The Portuguese (The Emigrant).* 1911. Oil on canvas, 46⅛ x 32" (117.2 x 81.3 cm). Öffentliche Kunstsammlung Basel, Kunstmuseum DONATION DR. H.C. RAOUL LA ROCHE, 1952
© 1997 ADAGP, PARIS/ARTISTS RIGHTS SOCIETY (ARS), NEW YORK

Picasso and Braque at this stage used a stippled, delicately modulated brushstroke so short as to be reminiscent of Seurat, though their color is restricted to muted gray and brown. Variation in values in these hues and in the direction of the strokes creates almost imperceptible change and movement between surface and depth.

However indecipherable their images, Braque and Picasso never relinquished the natural world altogether and they provided subtle clues that aid in the apprehension of their obscure subject matter. In the *Accordionist* (which represents a young girl, according to Picasso, not a man, as critics have often presumed) curvilinear elements near the bottom edge of the painting stand for the arms of a chair, while the small circles and stair-step patterns toward the center indicate the keys and bellows of an accordion. *The Portuguese (The Emigrant),* Braque said, shows "an emigrant on the bridge of a boat with a harbor in the background." This would explain, at the upper right, the transparent traces of a docking post and sections of nautical rope. In the lower portion are the strings and sound hole of the emigrant's guitar. Braque introduced a new element with the stenciled letters and numbers in the painting's upper zone. Like other forms in *The Portuguese,* the words are fragmentary. The letters D BAL at the upper right, for example, may derive from *Grand Bal,* probably a reference to a common dance-hall poster. While Braque had incorporated a word into a Cubist painting as early as 1909, his letters there were painted freehand as a descriptive local detail. For *The Portuguese* he borrowed a technique from commercial art and stenciled his letters, which here function not as illusionistic representations but as autonomous signs dissociated from any context that accounts for their presence. Inherently flat, the letters and numbers exist "outside of space," Braque said. And because they are congruent with the literal surface of the painting, they underscore the nature of the painted canvas as a material object, a physical fact, rather than a site for an illusionistic depiction of the real world. At the same time, they make the rest of the image read as an illusionistically shallow space. Braque's introduction of words in his painting, a practice soon adopted by Picasso, was one of many Cubist innovations that had far-reaching implications for modern art. Like so many of their inventions, the presence in the visual arts of letters, words, and even long texts is today commonplace (see figs. 614, 862; color plate 479, page 772).

In May 1912 the Cubists' search for alternative modes of representation led Picasso to the invention of collage, initiated by a small but revolutionary work, *Still Life with Chair Caning* (color plate 94, page 168). Within this single composition Picasso combined a complex array of pictorial vocabularies, each imaging reality in its own way. Painted on an oval support, a shape Picasso and Braque had already adopted for earlier Cubist paintings, this still life depicts objects scattered across a café table. The fragment of a newspaper, *Le Journal,* is indicated by the letters JOU that, given the artist's penchant for puns, could refer to *jouer,* meaning "to play" and implying that all this illusionism is a game. Over the letters Picasso depicted a pipe in a naturalistic style, while at the right he added the highly abstracted forms of a goblet, a knife, and a slice of lemon.

The most prominent element here, however, is a section of common oilcloth that has been mechanically printed with a design simulating chair caning. Picasso glued the cloth directly to his canvas. While this trompe l'oeil fabric offers an exacting degree of illusionism, it is as much a fiction as Picasso's painted forms, since it remains a facsimile of chair caning, not the real thing. After all, Picasso once said, art is "a lie that helps us understand the truth." But he did surround his painting with actual rope, in ironic imitation of a traditional gold frame. The rope could have been suggested by a wooden table molding (see the table in color plate 95, page 168, for example) or the kind of cloth edged with upholstery cord that Picasso had placed on a table in his studio. Combined with the oilcloth, a material from the actual world, the rope encourages a reading of the painted surface itself as a horizontal tabletop, a logical idea, given the presence of still-life objects. Such spatial ambiguity is one of the many ways Picasso challenged the most fundamental artistic conventions. His incorporation of unorthodox materials, ones not associated with "high" or "fine" art, offered a radically new connection to the external world and an alternative to the increasingly hermetic nature of his Analytic Cubist compositions. Around the same time he and Braque were reintroducing color to their Cubist paintings with bright, commercial enamel paint.

Following Picasso's invention of collage, Braque originated papiers collés (pasted papers). While this technique also involved gluing materials to a support, it is distinguished from collage in that only paper is used and, as Rubin has explained, "the sign language of the entire picture is in harmony with what is glued on." Collage, on the other hand, capitalized on disparate elements in jarring juxtaposition. It was this medium's potential for improbable, provocative combinations (see figs. 335, 336) that ultimately attracted the Dada and Surrealist artists.

Braque made his first papier collé, *Fruit Dish and Glass* (fig. 211), in September 1912, during an extended stay with Picasso in Sorgues, in the south of France. It has been proposed that Braque deliberately postponed his experiment until Picasso, who regarded any idea as fair game for stealing, had left Sorgues for Paris. Once he was satisfied with the results, Braque presented them to Picasso. "I have to admit that after having made the papier collé," he said, "I felt a great shock, and it was an even greater shock for Picasso when I showed it to him." Picasso soon began making his own experiments with Braque's "new procedure."

The precipitating event behind Braque's invention was his discovery in a storefront window of a roll of wallpaper printed with a *faux bois* (imitation wood-grain) motif. Braque had already used a house painter's comb to create *faux bois* patterns in his paintings, a technique he passed on to Picasso, and he immediately saw the paper's potential for his current work. Like the oilcloth in *Still Life with Chair Caning,* the *faux bois*

211. GEORGES BRAQUE. *Fruit Dish and Glass.* September 1912.
Charcoal and pasted paper on paper,
24⅜ x 17½" (62 x 44.5 cm).
Private collection

212. PABLO PICASSO. *Guitar, Sheet Music, and Wine Glass.* Fall
1912. Pasted papers, gouache, and charcoal on paper, 18⅞ x
14¾" (47.9 x 37.5 cm). McNay Art Museum, San Antonio, Texas
BEQUEST OF MARION KOOGLER MCNAY

offered a ready-made replacement for hand-wrought imagery. In *Fruit Dish and Glass* Braque combined the *faux-bois* paper with a charcoal drawing of a still life that, judging from the drawn words ale and bar, is situated in the familiar world of the café. With typical ambiguity, the cutouts of pasted paper play multiple roles, both literal and descriptive. Although their location in space is unclear, the *faux-bois* sections in the upper portion of the still life are signs for the wall of the café, while the rectangular piece below represents a wooden table. In addition, the *faux bois* enabled Braque to introduce color into his otherwise monochrome composition and to enlist its patterns as a kind of substitute chiaroscuro. At this point Braque was thinking of the papiers collés in relation to his paintings, for virtually the same composition exists as an oil.

Once Picasso took up the challenge of Braque's papier collé, the contrast in sensibility between the two became obvious. While Braque relied on his elegant Cubist draftsmanship to bind the imagery of his papiers collés together, Picasso immediately deployed more colorful, heterogenous materials with greater irony and more spatial acrobatics. In one of his earliest papiers collés, *Guitar, Sheet Music, and Wine Glass* of 1912 (fig. 212), a decorative wallpaper establishes the background for his still life (and implies that the depicted guitar is hanging on a wall). Picasso bettered Braque

by actually painting a simulated wood-grain pattern on paper, which he then cut out and inserted into his composition as part of a guitar. But the instrument is incomplete; we must construct its shape by decoding the signs the artist provides, such as a section of blue paper for the bridge and a white disk for the sound hole.

As in Picasso's collage *Still Life with Chair Caning*, different signifying systems or languages of representation are here at work, complete with charcoal drawing (the glass at the right) and a guitar-shaped cutout of painted *faux bois* paper. We have seen the way in which the Cubists represented newspapers in their painted compositions. The next step was to glue the actual newsprint to the surface of a papier collé. In this example, even the headline of the Paris newspaper *Le Journal*, LA BATAILLE S'EST ENGAGÉ (the battle is joined), has a dual message. It literally refers to the First Balkan War then being waged in Europe and thus adds a note of contemporaneity. Some specialists have argued that its deeper meaning is personal, that it is Picasso's friendly challenge to Braque on the battlefield of a new medium; others have read it as more directly political, an effort to confront the realities of war. By appropriating materials from the vast terrain of visual culture, even ones as ephemeral as the daily newspaper, and incorporating them unaltered into their works of art, Braque and

Picasso undermined definitions of artistic authenticity that made it the exclusive province of drawing and painting, media that capture the immediacy of the individual artist's touch. As always, this was accomplished with irony and humor—here again, JOU implies "game" or "play." It was a challenge that had profound consequences for twentieth-century art, one that in many ways still informs debates in contemporary art.

The inventions in 1912 of collage and papier collé, as well as Cubist sculpture (discussed below), essentially terminated the Analytic Cubist phase of Braque and Picasso's enterprise and initiated a second and more extensive period in their work. Called Synthetic Cubism, this new phase lasted into the 1920s and witnessed the adoption and variation of the Cubist style by many new practitioners, whose work is addressed later in this chapter. If earlier Cubist works analyzed form by breaking it down and reconstructing it as lines and transparent planes, the work after 1912 generally constructs an image from many diverse components. Picasso and Braque began to make paintings that were not primarily a distillation of observed experience but, rather, were built up by using all plastic means at their disposal, both traditional and experimental. In many ways, Synthetic Cubist pictures are less descriptive of external reality, for they are generally assembled with flat, abstracted forms that have little representational value until they are assigned one within the composition. For example, the bowl-shaped black form in figure 212 is ambiguous, but once inserted in the composition, it stands for part of a guitar.

The dynamic new syntax of Synthetic Cubism, directly influenced by collage, is exemplified by Picasso's *Card Player* (fig. 213), 1913–14. Here the artist has replaced the shimmering brown-and-gray scaffolding of *Accordionist* (fig. 209) with flat, clearly differentiated shapes in bright and varied color. We can make out a mustachioed card player seated at a wooden table, indicated by the *faux bois* pattern, on which are three playing cards. Below the table Picasso schematically inscribed the player's legs and on either side of the sitter he included an indication of wainscoting on the back wall. Here the cropped letters JOU, by now a signature Cubist conceit, have an obvious relevance. Picasso emulates in oil the pasted forms of collage and papiers collés by partially obscuring with a succeeding layer of paint each of the flat shapes that make up the figure.

After 1913 the principal trend in paintings by both Picasso and Braque was toward enrichment of their plastic means and enlargement of their increasingly individual Cubist vocabularies. In Picasso's *Green Still Life* (color plate 95, page 168), we can easily identify all the customary ingredients of a Cubist still life—compote dish, newspaper, glass, bottle, and fruit. Following the intellectual rigors of Analytic Cubism, the artist seems to delight here in a cheerfully decorative mode. Alfred Barr, an art historian and the first director of the Museum of Modern Art, called this decorative phase in Picasso's work Rococo Cubism. Here Picasso has disposed his still-life objects in an open, tangible space and, using commercial enamels, adds bright dots of color that resemble shiny

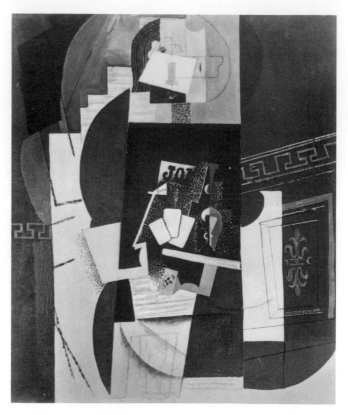

213. PABLO PICASSO. *Card Player*. Paris, winter 1913–14. Oil on canvas, 42½ x 35¼" (108 x 89.5 cm). The Museum of Modern Art, New York ACQUIRED THROUGH THE LILLIE P. BLISS BEQUEST. PHOTOGRAPH © 1995 THE MUSEUM OF MODERN ART, NEW YORK © 1997 ESTATE OF PABLO PICASSO/ARTISTS RIGHTS SOCIETY (ARS), NEW YORK

sequins. The entire ground of the painting is a single hue, a bright emerald green, which Picasso allows to infiltrate the objects themselves. In so doing, he reiterates the flatness of the pictorial space and deprives the objects of volume. Picasso painted *Green Still Life* in Avignon during the summer of 1914. Perhaps the optimistic mood of his work at this time reflects the calm domesticity he was enjoying there with Eva Gouel, his companion since 1911. But this sense of well-being was soon shattered. In August 1914 war was declared in Europe; by the end of 1915 Eva was dead of tuberculosis.

Braque, who was called up to serve in the French military (along with Guillaume Apollinaire and André Derain), said good-bye to Picasso at the Avignon railroad station, bringing to an abrupt end one of the greatest collaborations in the history of art. The war terminated the careers of many of Europe's most talented young artists and writers, including Apollinaire and Raymond Duchamp-Villon in France, Franz Marc and August Macke in Germany, and Umberto Boccioni in Italy. Picasso remained in Paris until 1917, when he went to Rome with the writer Jean Cocteau and worked on stage designs for the Ballets Russes. The postwar work of Braque and Picasso is discussed in chapter 14.

Cubist Sculpture

Even in the hands of great modernist innovators such as Jean Arp or Constantin Brancusi (see figs. 300, 185), sculpture was still essentially conceived as a solid form surrounded by void. These sculptors created mass either through subtractive methods—carving form out of stone or wood—or additive ones—building up form by modeling in wax or clay. By these definitions, Picasso's *Woman's Head (Fernande)* (fig. 207), regardless of its radical reconception of the human form, is still a traditional solid mass surrounded by space. Like collage, constructed sculpture is assembled from disparate, often unconventional materials. Unlike traditional sculpture, its forms are penetrated by void and create volume not by mass, but by containing space. With the introduction of constructed sculpture the Cubists broke with one of the most fundamental characteristics of sculptural form and provoked a rupture with past art equal to the one they fostered in painting.

Documents prove that constructed Cubist sculpture was invented by Braque and that, as Rubin has demonstrated, it predated the first papiers collés. Nevertheless, no such works from his hand exist. Because he did not preserve the works, it is assumed that he thought of them principally as aids to his papiers collés, not as an end in themselves. According to Kahnweiler, Braque made reliefs in wood, paper, and cardboard, but the only visual record of his sculpture is a 1914 photograph of a work taken in his Paris studio (fig. 214). This unusual composition represents a familiar subject, a café still life, assembled from paper (including newsprint) and cardboard on which Braque either drew or painted. He installed

his sculpture across a corner of the room, thus incorporating the real space of the surrounding studio into his work. This corner construction possibly influenced the corner-hung Counter-Reliefs of the Russian Constructivist artist Vladimir Tatlin (see fig. 249), who may have visited Picasso's or Braque's studio during a 1914 trip to Paris.

Though constructed sculpture was Braque's invention, it was Picasso, typically, who made the most thorough use of it. His first attempt in the new technique is a guitar made from sheet metal (fig. 215), which he first prepared in October 1912 as a maquette made simply of cardboard, string, and wire. Picasso's use of an industrial material such as sheet metal was highly unorthodox in 1912; in the years since, it has become a common sculptural medium. Clearly, the guitar was a significant impetus behind his papiers collés, for photographs taken in the artist's studio show it surrounded by works in this medium. In many ways, Picasso's guitar sculpture is the equivalent in three dimensions of the same motif in his papiers collé, *Guitar, Sheet Music, and Wine Glass* (fig. 212), made several weeks later.

Through a technique of open construction, the body of the cardboard guitar has largely been cut away. Volume is expressed as a series of flat and projecting planes, resulting in a quality of transparency heretofore alien to sculpture. Picasso made the guitar sound hole, a void in a real instrument, as a

214. GEORGES BRAQUE. 1914. The artist's Paris studio with a paper sculpture, now lost

215. PABLO PICASSO. Maquette for *Guitar*. Paris, October 1912. Construction of cardboard, string, and wire (restored), 25¾ x 13 x 7½" (65.1 x 33 x 19 cm). The Museum of Modern Art, New York

projecting cylinder, just as he designed the neck as a concavity. The morphology of the guitar was directly inspired by a Grebo mask from the Ivory Coast in Africa, in which the eyes are depicted as hollow projecting cylinders. Picasso owned two such examples, one of which he bought in August 1912, during a visit to Marseilles with Braque. *Guitar* was revolutionary in showing that an artist's conceptual inventions of form are signs of or references to reality, not mere imitations of it. They can be entirely arbitrary and may bear only a passing resemblance to what they actually denote. Thus the sign for the guitar's sound hole was stimulated by a completely unrelated source—the eyes of an African mask.

We know from a 1913 photograph that Picasso inserted his cardboard guitar into a relief still-life construction on the wall of his studio. In an even more fascinating sculptural ensemble, also recorded only in a contemporary photograph, he drew a guitarist on a canvas and gave the figure projecting cardboard arms that held an actual guitar. A real table with still-life props on it completed the composition. In this remarkable tableau Picasso closed the breach that separated painting and sculpture, uniting the pictorial realm with the space of the external world. Much of modern sculpture has addressed this very relationship, in which a sculpture is regarded not as a discrete work of art to be isolated on a pedestal, but rather as an object coexisting with the viewer in one unified space.

In two subsequent constructions, *Mandolin and Clarinet*, 1913, and *Still Life*, 1914 (figs. 216, 217), both assembled from crudely cut, partially painted pieces of wood, Picasso

furthered the explorations initiated by *Guitar. Mandolin and Clarinet*, composed mostly of scrap wood, demonstrates the artist's uncanny ability to envision the formal potential of found materials. Wood, which he and Braque had previously represented illusionistically, is now present as itself. In *Still Life* a curved section of wood stands for a table, on whose edge Picasso glued actual upholstery fringe. On the table he depicts a still life of food, a knife, and a glass. The black horizontal element at the upper left may be the bottom edge of a picture frame, referring us back to the painted dimension.

Unlike his constructed sculptures, which were essentially reliefs, Picasso's *Glass of Absinthe* (color plate 96, page 201) was conceived in the round and was originally modeled by conventional methods in wax. Absinthe is a highly addictive, stupor-inducing liquor, so lethal that by the end of 1914 it was outlawed in France. Upon completion of the wax model of *Glass of Absinthe*, Kahnweiler had the work cast in bronze in an edition of six, to each of which the artist added a "found object," a perforated silver spoon, as well as a bronze sugar cube. He hand-painted five variants and coated one with sand, a substance that first Braque and then he had added to their paint medium. The bright color patterns on this example are similar to those found in the artist's contemporary paintings (color plate 95, page 168). Picasso cut deep hollows in the glass to reveal the interior, where, as in the previous *Still Life*, the level of fluid reads as a horizontal plane. In this adaptation of collage methods to the medium of sculpture, Picasso adapts objects from the real world for expressive purposes in the realm of art. This lack of discrimination between

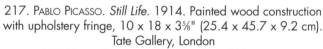

216. PABLO PICASSO. *Mandolin and Clarinet.* Fall 1913. Painted wood construction with cardboard, paper, and pencil marks, 22⅞ x 14⅛ x 9" (58 x 36 x 23 cm). Musée Picasso, Paris
© 1997 ESTATE OF PABLO PICASSO/ARTISTS RIGHTS SOCIETY (ARS), NEW YORK

217. PABLO PICASSO. *Still Life.* 1914. Painted wood construction with upholstery fringe, 10 x 18 x 3⅝" (25.4 x 45.7 x 9.2 cm). Tate Gallery, London
© 1997 ESTATE OF PABLO PICASSO/ARTISTS RIGHTS SOCIETY (ARS), NEW YORK

the "high" art media of painting and sculpture and the "low" ephemera of the material world remained a characteristic of Picasso's sculpture (see figs. 386, 562) and proved especially relevant for artists in the 1950s and 1960s (see figs. 610, 640).

ALEKSANDR ARCHIPENKO (1887–1964) has a strong claim to priority as a pioneer Cubist sculptor. Born in Kiev, Ukraine, he studied art in Kiev and Moscow until he went to Paris in 1908. Within two years he was exhibiting his highly stylized figurative sculptures, reminiscent of Gauguin's, with the Cubist painters at the Paris Salons. By 1913 he had established his own sculpture school in the French capital and his works were shown at Berlin's Der Sturm Gallery and in the New York Armory Show (see chapter 18). In 1913–14 Archipenko adapted the new technique of collage to sculpture, making brightly polychromed constructions from a variety of materials. He based some of the figures from this period on performers he saw at the Cirque Médrano, the circus that had also been one of Picasso's favorite haunts. In *Médrano II*, 1913 (color plate 97, page 201), Archipenko assembled a dancing figure from wood, metal, and glass, which he then painted and attached to a back panel, creating a stagelike setting. This work was singled out for praise by Apollinaire in his review of the 1914 Salon des Indépendants. In the manner of constructed sculpture, Archipenko articulated volume by means of flat colored planes. Thus, the dancer's entire torso is a single plank of wood to which her appendages are hinged. Her skirt is a conical section of tin joined to a curved pane of glass, on which the artist painted a delicate ruffle. Over the next few years, Archipenko developed this idea of a figure against a backdrop in his so-called sculpto-paintings, which are essentially paintings that incorporate relief elements in wood and sheet metal.

Walking Woman, 1918–19 (fig. 218) typifies Archipenko's later freestanding sculptures. In 1915 he began using the void as a positive element in figurative sculpture, effectively reversing the historic concept of sculpture as a solid surrounded by space. He believed that sculptural form did not necessarily begin where mass encounters space, but "where space is encircled by material." Here the figure is a cluster of open spaces shaped by concave and convex solids. Although Picasso's Cubist constructions were more advanced than those of Archipenko, the latter's Cubist figures had more immediate influence, first because they were more widely exhibited at an early date, and second because they applied Cubist principles to the long-familiar sculptural subject of the human figure, which revealed the implications of mass-space reversals more readily.

The great talents of RAYMOND DUCHAMP-VILLON (1876–1918) were cut short by his early death in the war in 1918. He was one of six siblings, four of whom became artists—the other three are Marcel Duchamp and Jacques Villon (whose work is discussed below) and Suzanne Duchamp. In 1900 Duchamp-Villon abandoned his studies in medicine and took up sculpture. At first influenced by Rodin, like nearly all sculptors of his generation, he moved rapidly into the orbit of the Cubists. In his most important

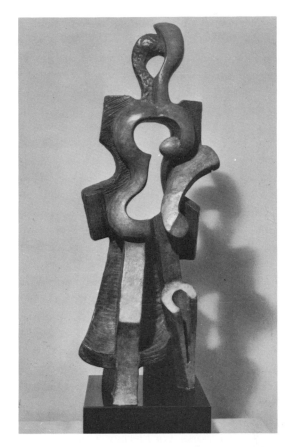

218. ALEKSANDR ARCHIPENKO. *Walking Woman.* 1918–19. Bronze, height 26⅜" (67 cm). Courtesy Frances Archipenko Gray

work, the bronze *Horse,* 1914 (fig. 219), the artist endows a preindustrial subject with the dynamism of a new age. Through several versions of this sculpture he developed the image from flowing, curvilinear, relatively representational forms to a highly abstracted representation of a rearing horse.

219. RAYMOND DUCHAMP-VILLON. *The Horse.* 1914. Bronze (cast c. 1930–31), 40 x 39½ x 22⅜" (101.6 x 100.1 x 56.7 cm). The Museum of Modern Art, New York VAN GOGH PURCHASE FUND

The diagonal planes and spiraling surfaces resemble pistons and turbines as they move, unfold, and integrate space into the mass of the sculpture. "The power of the machine imposes itself upon us to such an extent," Duchamp-Villon wrote, "that we can scarcely imagine living bodies without it." With its energetic sense of forms being propelled through space, *Horse* has much in common with the work of the Italian Futurist Umberto Boccioni, who had visited the studios of Archipenko and Duchamp-Villon in 1912 and whose sculptures (see fig. 240) were shown in Paris in 1913.

Cubism was only one chapter in the long career of JACQUES LIPCHITZ (1891–1973), though no other sculptor explored as extensively the possibilities of Cubist syntax in sculpture. Born in Lithuania in 1891 to a prosperous Jewish family, Lipchitz arrived in Paris in 1909 and there studied at the École des Beaux-Arts and the Académie Julian. In 1913 he met Picasso and, through the Mexican artist Diego Rivera, began his association with the Cubists. He also befriended a fellow Lithuanian, Chaim Soutine, as well as Amadeo Modigliani, Brancusi, Matisse, and, in 1916, Juan Gris. During this time he developed an abiding interest in the totemic forms of African sculpture. In 1913 he introduced geometric stylization into his figure sculptures and by 1915 was producing a wide variety of Cubist wood constructions as well as stone and bronze works. For Lipchitz, Cubism was a means of reexamining the essential nature of sculptural form and of asserting sculpture as a self-contained entity, rather than an imitation of nature.

In a series of vertical compositions in 1915–16, including *Man with a Guitar* (fig. 220), Lipchitz edged to the brink of abstraction. Conceived as a structure of rigid, intersecting planes, *Man with a Guitar* is an austere rendition of a familiar Cubist theme and was derived from the artist's earlier and more legible wood constructions of detachable figures. Late in life, Lipchitz, who often compared these abstracted sculptures to architecture, wrote about them in his autobiography, *My Life in Sculpture:* "I was definitely building up and composing the idea of a human figure from abstract sculptural elements of line, plane, and volume; of mass contrasted with void completely realized in three dimensions."

Lipchitz preferred modeling in clay to carving in stone, and most of his later sculptures, which gradually increased in scale, were developed as clay models and plaster casts that were then cast in bronze or carved in stone by his assistants. A bronze from 1928, *Reclining Nude with Guitar* (fig. 221), shows how the artist was still producing innovative forms within an essentially Cubist syntax. It evolved from his efforts to create a modern alternative to Rodin's *Thinker* (see fig. 133, upper center). In the blocky yet curving masses of the sculpture, Lipchitz struck a characteristic balance between representation and abstraction by effecting, in his words, "a total assimilation of the figure to the guitar-object." This analogy between the guitar and human anatomy was one already exploited by Picasso. From the 1930s to the end of his life, expressiveness became paramount in Lipchitz's sculpture and led to the free, rather baroque modeling of his later works.

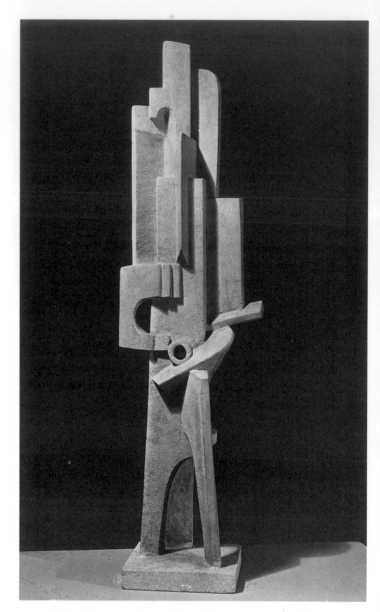

220. JACQUES LIPCHITZ. *Man with a Guitar.* 1915. Limestone, 38¼ x 10½ x 7¾" (97.2 x 26.7 x 19.5 cm). The Museum of Modern Art, New York

221. JACQUES LIPCHITZ. *Reclining Nude with Guitar.* 1928. Bronze, 16 x 29¾ x 13" (40.6 x 75.6 x 33 cm). Hirshhorn Museum and Sculpture Garden, Smithsonian Institution, Washington, D.C.

Color plate 96. PABLO PICASSO. *Glass of Absinth*. Paris, spring 1914.
Painted bronze with silver absinth spoon, 8½ x 6½ x 3⅜"
(21.6 x 16.4 x 8.5 cm); base diameter 2½" (6.4 cm).
The Museum of Modern Art, New York

Color plate 97. ALEKSANDR ARCHIPENKO. *Médrano II*. 1913.
Painted tin, wood, glass, and painted oilcloth,
49⅞ x 20¼ x 12½" (133.2 x 51.4 x 31.8 cm).
Solomon R. Guggenheim Museum, New York

Color plate 98. HENRI LAURENS. *Guitar and Clarinet.* 1920. Polychromed stone, 12⅝ x 14½ x 3½" (36.7 x 36.8 x 8.9 cm). Hirshhorn Museum and Sculpture Garden, Smithsonian Institution, Washington, D.C.
© 1997 ADAGP, PARIS/ARTISTS RIGHTS SOCIETY (ARS), NEW YORK

Color plate 99. JUAN GRIS. *Still Life and Townscape (Place Ravignan).* 1915. Oil on canvas, 45⅞ x 35⅛" (116.5 x 89.2 cm). Philadelphia Museum of Art
LOUISE AND WALTER ARENSBERG COLLECTION
© 1997 ADAGP, PARIS/ARTISTS RIGHTS SOCIETY (ARS), NEW YORK

Color plate 100.
JUAN GRIS. *Guitar with Sheet of Music.* 1926. Oil on canvas, 25⅝ x 31⅞"
(65.1 x 81 cm).
Private collection
© 1997 ADAGP, PARIS/ARTISTS RIGHTS SOCIETY (ARS), NEW YORK

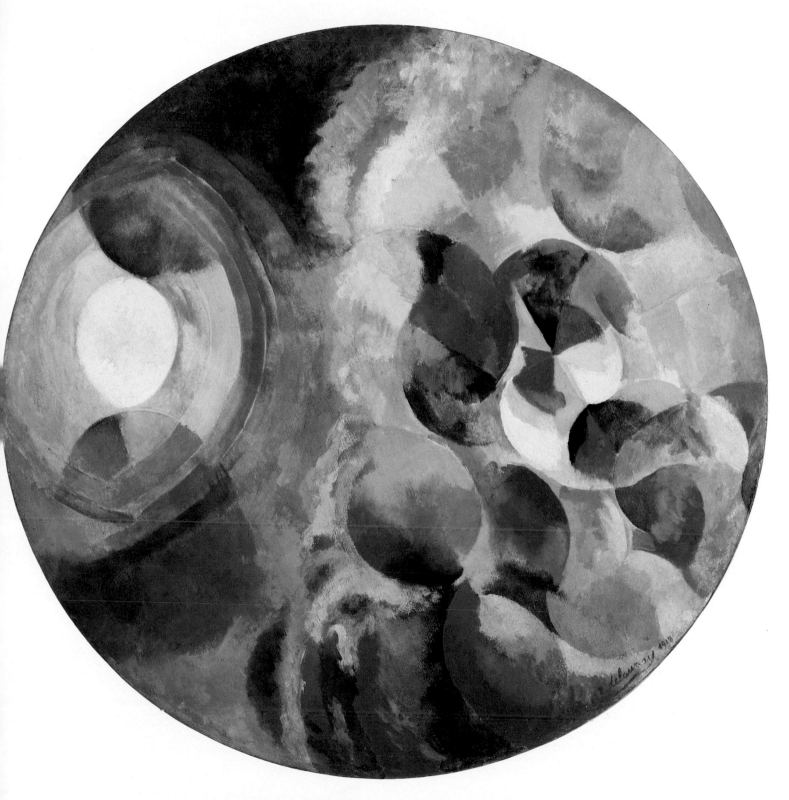

Color plate 101. ROBERT DELAUNAY. *Simultaneous Contrasts: Sun and Moon.* 1913 (dated 1912 on painting). Oil on canvas, diameter 53" (134.6 cm). The Museum of Modern Art, New York

Color plate 102. FERNAND LÉGER. *The City*. 1919.
Oil on canvas, 7'7" x 9'9" (2.3 x 3 m).
Philadelphia Museum of Art

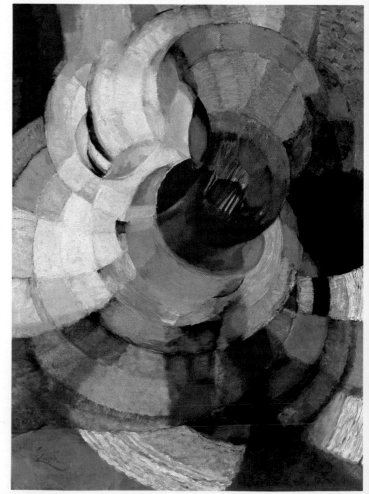

Color plate 103. FRANTIŠEK KUPKA.
Disks of Newton (Study for *Fugue in Two Colors*).
1911–12. Oil on canvas, 39⅜ x 29" (100 x 73.7 cm).
Philadelphia Museum of Art

left: Color plate 104. SONIA DELAUNAY. *Blanket*. 1911.
Appliquéd fabric, 42⅞ x 31⅞" (109 x 81 cm).
Musée National d'Art Moderne, Centre National d'Art
et de Culture Georges Pompidou, Paris

Color plate 105. MARCEL DUCHAMP.
Nude Descending a Staircase, No. 2. 1912. Oil on canvas,
58 x 35" (147.3 x 88.9 cm). Philadelphia Museum of Art
LOUISE AND WALTER ARENSBERG COLLECTION
© 1997 ADAGP, PARIS/ARTISTS RIGHTS SOCIETY (ARS), NEW YORK

Color plate 106. FRANCIS PICABIA. *Catch as Catch Can*. 1913.
Oil on canvas, 39⅝ x 32¼" (100.6 x 81.9 cm).
Philadelphia Museum of Art
LOUISE AND WALTER ARENSBERG COLLECTION
© 1997 ADAGP, PARIS/ARTISTS RIGHTS SOCIETY (ARS), NEW YORK

right: Color plate 108. GINO SEVERINI.
Dynamic Hieroglyphic of the Bal Tabarin. 1912.
Oil on canvas with sequins,
63⅝ x 61½" (161.6 x 156.2 cm).
The Museum of Modern Art, New York
ACQUIRED THROUGH THE LILLIE P. BLISS BEQUEST
© 1997 ADAGP, PARIS/ARTISTS RIGHTS SOCIETY (ARS), NEW YORK

above: Color plate 107.
GIACOMO BALLA. *Street Light (Lampada—Studio di luce).* Dated 1909 on canvas.
Oil on canvas,
68¾ x 45¼" (174.7 x 114.5 cm).
The Museum of Modern Art, New York
HILLMAN PERIODICALS FUND
© 1997 ESTATE OF GIACOMO BALLA/VAGA, NEW YORK, NY

right: Color plate 109. GINO SEVERINI. *Spherical Expansion of Light (Centrifugal).* 1913–14.
Oil on canvas, 24 x 20" (61 x 50.8 cm).
Collection Riccardo and Magda Jucker, Milan
© 1997 ADAGP, PARIS/ARTISTS RIGHTS SOCIETY (ARS), NEW YORK

below: Color plate 112. NATALIA SERGEEVNA GONCHAROVA. *Linen.* 1912. Oil on canvas, 35 x 27½" (89 x 70 cm). Tate Gallery, London

above: Color plate 110. CARLO CARRÀ. *Patriotic Celebration (Free-Word Painting).* 1914. Pasted paper and newsprint on cloth, mounted on wood, 15¼ x 12" (38.7 x 30.5 cm). Private collection, Italy

below: Color plate 111. UMBERTO BOCCIONI. *The City Rises.* 1910. Oil on canvas, 6'6½" x 9'10½" (2 x 3 m). The Museum of Modern Art, New York

Color plate 113. LYUBOV POPOVA. *Early Morning*. 1914. Oil on canvas, 27¾ x 35" (70.5 x 88.9 cm). The Museum of Modern Art, New York

Color plate 115. EL LISSITZKY. *Prounenraum* (Proun Room), created for Berlin Art Exhibition. 1923, reconstructed 1965. Wood, 9'10⅛" x 9'10⅛" x 8'6⅜" (3 x 3 x 2.6 m). Stedelijk Van Abbemuseum, Eindhoven, the Netherlands

Color plate 114. KAZIMIR MALEVICH. *Morning in the Village after a Snowstorm*. 1912. Oil on canvas, 31¾ x 31⅞" (80.6 x 80.9 cm). Solomon R. Guggenheim Museum, New York

222. Henri Laurens. *Bottle and Glass.* 1918. Painted wood and
iron, 24⅜ x 13⅜ x 8¼" (62 x 34 x 21 cm).
Musée National d'Art Moderne, Centre d'Art et de Culture
Georges Pompidou, Paris
© 1997 ADAGP, Paris/Artists Rights Society (ARS), New York

He moved to the United States to escape World War II
in Europe and produced many large-scale public sculptures
based on classical and biblical subjects.

Henri Laurens (1885–1954) was born, lived, and died
in Paris. He was apprenticed in a decorator's workshop and
for a time practiced architectural stone carving. By the time
of fully developed Analytic Cubism, he was living in Montmartre, immersed in the rich artistic milieu of the expanded
Cubist circle in Paris. First introduced to the group by Léger,
he was especially close to Gris and Braque, whom he met in
1911. After his initial exposure to Cubism, Laurens absorbed
the tenets of the style slowly. His earliest extant constructions
and papiers collés, closely related media in his oeuvre, date to
1915. Like Archipenko, whose Cubist work he knew well,
Laurens excelled at polychromy in sculpture, integrating
color into his constructions, low reliefs, and freestanding
stone blocks. Referring to the long tradition of color in sculpture from antiquity to the Renaissance, he emphasized its
function in reducing the variable effects of light on sculptural surfaces. As he explained, "When a statue is red, blue, or
yellow, it remains red, blue, or yellow. But a statue that has
not been colored is continually changing under the shifting
light and shadows. My own aim, in coloring a statue, is that
it should have its own light." His 1918 construction, *Bottle
and Glass* (fig. 222), is a very close transcription in three
dimensions of the cut and pasted papers of a collage made the

previous year. Laurens used concave, convex, and cutout
forms to signify volume and void. Thus, the wine bottle is
indicated as both a white cylinder and a cutout silhouette in
wood.

By 1919 Laurens had abandoned construction and embarked on a series of still-life sculptures in rough, porous
stone. Drawing on his background in architectural stone
carving and inspired by his love of French medieval sculpture,
he carved *Guitar and Clarinet* (color plate 98, page 202),
1920, in very low relief and then reinforced these flattened
volumes with delicate, matte color. In both style and iconography, the series has many affinities with Gris's paintings of
the same period, as well as Lipchitz's relief sculptures of 1918.
After 1920 Laurens relaxed the geometries of his Cubist style,
adopting more curvilinear modes devoted to the depiction of
the human figure.

The Development of Cubist Painting in Paris

In 1911, while Picasso and Braque were working together
so closely that it is difficult to distinguish between their
works, a number of other emergent Cubist painters in
France began to formulate attitudes destined to enlarge the
boundaries of the style. Of these, Albert Gleizes and Jean
Metzinger also became talented critics and expositors of
Cubism. Their book *On Cubism*, published in 1912, was
one of the first important theoretical works on the movement. In Paris Gleizes and Metzinger met regularly with
Robert Delaunay, Fernand Léger, and Henri Le Fauconnier at the home of the Socialist writer Alexandre
Mercereau, at the café Closerie des Lilas and, on Tuesday
evenings, at sessions organized by the journal *Vers et prose*.
At the café these painters met with older Symbolist writers
and younger, enthusiastic critics, such as Guillaume
Apollinaire and André Salmon. (They were not, however,
close to Cubism's inventors, Braque and Picasso.) In 1911
the group arranged to have their paintings hung together
in the Salon des Indépendants. The concentrated showing
of Cubist experiments created a sensation, garnering violent attacks from most critics, but also an enthusiastic
champion in Apollinaire. That year Archipenko and Roger
de La Fresnaye joined the Gleizes-Metzinger group, together with Francis Picabia and Apollinaire's friend, the
painter **Marie Laurencin** (1881–1956). Laurencin had
portrayed Apollinaire (seated in the center), herself (at the
left), Picasso, and his lover Fernande Olivier in her 1908
group portrait (fig. 223), whose naïve style is reminiscent
of Rousseau's.

The three brothers Jacques Villon, Marcel Duchamp, and
Raymond Duchamp-Villon, the Czech František Kupka, and
the Spaniard Juan Gris were all new additions to the Cubist
ranks. In the Salon des Indépendants of 1912 the "Salon
Cubists" were so well represented that protests against their
influence were lodged in the French parliament. Although
Picasso and Braque did not take part in these Salons, their
innovations were widely known. By then Cubist art had been

223. Marie Laurencin. *Group of Artists.* 1908. Oil on canvas, 25½ x 31⅞" (64.8 x 81 cm). The Baltimore Museum of Art
The Cone Collection © 1997 ADAGP, Paris/Artists Rights Society (ARS), New York

224. Juan Gris. *Homage to Pablo Picasso.* 1912. Oil on canvas, 36¾ x 29¼" (93.3 x 74.3 cm). The Art Institute of Chicago
Gift of Leigh B. Block © 1997 VEGAP, Madrid/Artists Rights Society (ARS), New York

shown in exhibitions in Germany, Russia, England, Spain, and the United States. The Italian Futurists, as we shall see, brought an Expressionist variant of Cubism back to Italy in 1911. In 1912 Kandinsky published works of Le Fauconnier and other Cubists in his yearbook *Der Blaue Reiter,* and Paul Klee visited the Paris studio of Delaunay. Marc and Macke soon followed.

As these artists began to push the original Cubist concepts to new limits, they demonstrated that Cubism was a highly flexible and adaptable idiom. Delaunay was developing his art of "simultaneous contrasts of color" based on the ideas of the color theorist Michel Eugéne Chevreul, who had so strongly influenced Seurat, Signac, and the Neo-Impressionists. In a large exhibition at the Parisian gallery La Boëtie, held in October 1912 and entitled (probably by Jacques Villon) "Section d'Or" (golden section), entries by Marcel Duchamp, Francis Picabia, and Kupka signaled innovative alternatives to the Cubism of Picasso and Braque.

During 1912 and 1913 Gleizes and Metzinger's *On Cubism* was published, as well as Apollinaire's book *Les Peintres cubistes* and writings by various critics tracing the origins of Cubism and attempting to define it. Even today, when we are accustomed to the meteoric rise and fall of art movements and styles, it is difficult to appreciate the speed with which Cubism became an international movement. Beginning in 1908 with the isolated experiments of Braque and Picasso, by 1912 it was branching off in new directions and its history was already being written.

Juan Gris (1887–1927) was born in Madrid but moved permanently to Paris in 1906. While Gris was not one of

Cubism's inventors, he was certainly one of its most brilliant exponents. Since 1908 he had lived next to his friend Picasso at the old tenement known as the Bateau Lavoir and had observed the genesis of Cubism, although he was then occupied in making satirical illustrations for French and Spanish journals. However, by 1912 he had produced a masterly and personal interpretation of Cubism with *Homage to Pablo Picasso* (fig. 224), which is akin to studies of heads made by Picasso in 1909 (see fig. 206), but which displays a mathematical control quite unrelated to anything that Picasso, Braque, or the other Cubists had attempted. Gris's sensibility differed significantly from Picasso's. He was not drawn to the primitive, and his work is governed by an overriding refinement and logic. In fact, he later made compositions based on the classical notion of ideal proportion called the golden section. With his portrait of Picasso, Gris turns the shifting planes and refracted light of Cubism into a regular grid that functions as an organizing armature within the painting.

Gris was closer to Braque and Picasso than were other members of the Cubist circle and he was aware of their latest working procedures. By 1912 he was making distinctive collages and papiers collés with varied textural effects, enhanced by his rich color sense. He tended to avoid Picasso's heterogenous media and subversive approach to representation, preferring to marshal his pasted papers, which he usually applied to canvas, into precise, harmonious arrangements. *The Table,* 1914 (fig. 225), depicting a typical Cubist still-life subject, incorporates Braque's *faux bois,* along with hand-applied charcoal and paper. Although the canvas is rectangular, the objects are contained within an oval format (the table

225. JUAN GRIS. *The Table.* 1914. Pasted and printed papers and charcoal on paper, mounted on canvas, 23½ x 17½" (59.7 x 44.5 cm). Philadelphia Museum of Art

A. E. GALLATIN COLLECTION © 1997 VEGAP, MADRID/ARTISTS RIGHTS SOCIETY (ARS), NEW YORK

itself is also rectangular) that seems to exist by virtue of a bright, elliptical spotlight. To achieve the layered complexity of this composition, Gris reordered the data of the visible world within at least two independent spatial systems. The foreshortened bottle of wine at the far right (which partially disappears under the *faux-bois* paper) and the open book at the bottom of the canvas are described as solid volumetric forms and exist in a relatively traditional space, one that positions the viewer above the table, looking down. The open book is drawn, but the text is an actual page from one of Gris's favorite detective novels. Toward the center of the work, a glass, bottle, and book, lightly inscribed in charcoal, are virtually transparent. They tilt at a precarious angle on what appears to be a second wooden table, introducing an altogether new space that is synonymous with the flat, upright surface of the painting. Perhaps the headline in the pasted newspaper, LE VRAI ET LE FAUX CHIC (real and false chic), refers not just to French fashion, but to the illusionistic trickery at play here.

Gris ceased making collages after 1914 but continued to expand his Cubist vocabulary in painting and drawing. In *Still Life and Townscape (Place Ravignan)* (color plate 99, page 202) he employed a common Cubist device by mixing alternative types of illusionism within the same picture. In the lower half, a room interior embodies all the elements of Synthetic Cubism, with large, intensely colored geometric planes interlocking and absorbing the familiar collage components: the fruit bowl containing an orange; the newspaper, *Le Journal;* the wine label, médoc. However, this foreground pattern of tilted color shapes leads the eye up and back to a window that opens out on a uniformly blue area of simple trees and buildings. Thus the space of the picture shifts abruptly from the ambiguity and multifaceted structure of Cubism to the cool clarity of a traditional Renaissance painting structure: the view through the window.

Gris was instrumental in bringing to Cubism a light and color that produced a decorative lyricism. At the end of his short life, however, he was beginning to explore an austere manner in which objects were simplified to elemental color shapes. In *Guitar with Sheet of Music* (color plate 100, page 202), a work of 1926, he reduced the familiar objects of Cubist still life to stark, simple shapes and eliminated from his colors the luminous iridescence of earlier works. Here paint is applied as a matte surface that reinforces the sense of architectural structure and at the same time asserts its own nature as paint. The bright primary colors—the blue of the sky, the yellow of the guitar's body, the red of the cloth—are framed by neutral tones. The table and window frame are composed of low-keyed brownish tones that merge imperceptibly, all within the deep shadow of the room. Within this subdued but beautifully integrated color construction the artist has created, through the linear geometry of his edges, a tour de force of shifting planes and ambiguous but structured space. The window frame merges at the lower left with the edge of the table. Meanwhile, the diagonal yellow line that bounds the red cloth on the left is drawn perfectly straight, yet is made to bend, visually, by the deeper value of the cloth as it (presumably) folds over the edge of the table. In one of the last works of his brief career, Gris integrates effects of architectural recession, sculptural projection, and flat painted surface into a tightly interlocked pictorial harmony.

The contributions of **ALBERT GLEIZES** (1881–1953) and Metzinger to the Cubist movement are usually overshadowed by the reputation of their theoretical writings, specifically the pioneering book *On Cubism,* in which much of the pictorial vocabulary of Cubism was first elucidated. Gleizes in particular was instrumental in expanding the range of the new art, in both type of subject and attitude toward that subject. His monumental 1912 painting *Harvest Threshing* (fig. 226) opens up to a vast panorama, the generally hermetic world of the Cubists. In its suggestions of the dignity of labor and the harmonious interrelationship of worker and nature, Gleizes explores a social dimension foreign to the unpeopled landscapes of Picasso and Braque. Although the palette remains within the subdued range—grays, greens, and browns, with flashes of red and yellow—the richness of Gleizes's color gives the work romantic overtones suggestive of Jean-François Millet (see fig. 18). In his later works, largely a response to the breakthrough Simultaneous Disk paintings of Robert Delaunay (color plate 101, page 203), Gleizes turned to a form of lyrical abstraction based on curvilinear shapes, frequently inspired by musical motifs.

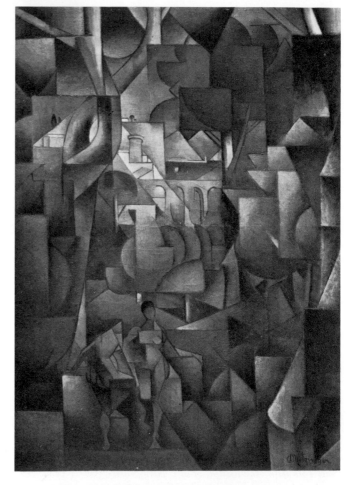

The early Cubist paintings of **JEAN METZINGER** (1883–1956) are luminous variations on Cézanne, whose posthumous 1907 retrospective in Paris left an indelible impression on countless artists. *The Bathers* (fig. 227), no doubt informed by Gris's immediate example, illustrates Metzinger's penchant for a carefully constructed rectangular design, in which the romantic luminosity of the paint plays against the strict geometry of the linear structure. For a time Metzinger toyed with effects derived from the Italian Futurists, later moving easily into the Synthetic Cubist realm of larger shapes and more vivid color. He received his first solo exhibition in 1919 at Léonce Rosenberg's Galerie de l'Effort Moderne in Paris, which was dedicated to a rigorous form of Cubism dubbed "crystal" Cubism, an apt designation for Metzinger's paintings at that time. Toward the end of World War I he began to assimilate photographic images of the figure or architectural scenes into a decorative Cubist frame. His later style, strongly influenced by Fernand Léger's work of the 1920s, was an idealized realism for which Cubist pattern was a decorative background—a formula that enmeshed many of the lesser Cubists.

FERNAND LÉGER (1881–1955). One almost inevitable development from Cubism was an art celebrating the ever-expanding machine world within the modern metropolis, for the geometric basis of much Cubist painting provided analogies to machine forms. In his 1924–25 essay, "The Machine Aesthetic: Geometric Order and Truth," Léger wrote: "Each artist possesses an offensive weapon that allows him to intimidate tradition. I have made use of the machine as others have used the nude body or the still life." Léger was not interested in simply portraying machines, for he opposed traditional realism as sentimental. Rather, he said he wanted to "create a beautiful object with mechanical elements."

Léger was from Normandy, in northwest France, where his family raised livestock. This agrarian background contributed to his later identification with the working classes and to his membership in the Communist Party in 1945. Once he settled in Paris in 1900, Léger shifted his studies from architecture to painting and passed from the influence of Cézanne to that of Picasso and Braque. One of his first major canvases, *Three Nudes in the Forest* (fig. 228), is an unearthly habitation of machine forms and wood-chopping robots. Here Léger seems to be creating a work of art out of Cézanne's cylinders and cones, but it is also clear that he

knew the work of artists such as Gleizes and Metzinger. The sobriety of the colors, coupled with the frenzied activity of the robotic figures, whose faceted forms can barely be distinguished from their forest environment, creates an atmosphere symbolic of a new, mechanized world.

In 1913 Léger began a series of boldly nonobjective paintings, called Contrastes de Formes (contrasts of forms), featuring planar and tubular shapes composed of loose patches of color within black, linear contours (fig. 229). The metallic-looking forms, which interlock in rhythmical arrangements like the gears of some great machine, give the illusion of projecting toward the viewer. At first these shapes were non-

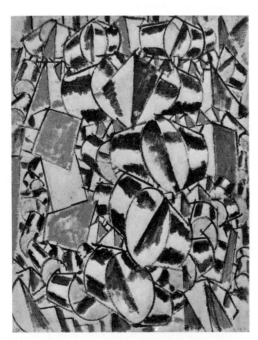

229. FERNAND LÉGER. *Contrast of Forms.* 1913.
Oil on canvas, 51 x 38" (129.5 x 96.5 cm).
Philadelphia Museum of Art

referential, but the artist later used the same morphology to paint figures and landscapes.

Léger's experience as a frontline soldier in World War I was a powerful, transformative event for his life and art. He was struck by the camaraderie of the soldiers in the trenches, and while impressed by the technology of modern warfare, he was stunned by its destructive power. His combat experience strengthened his identification with the common man and reinforced his resolve to abandon abstract painting, dedicating himself to the subjects of contemporary life.

With his monumental postwar canvas *The City* (color plate 102, page 204), 1919, Léger took another major step in the exploration of reality and the painted surface. In earlier canvases he had modeled his forms heavily, so that they seemed to project out from the surface of the canvas. Now he used the vocabulary of Synthetic Cubist collage to explore a variety of new forms. The rose-colored column in the foreground of *The City* is modeled volumetrically, for example, and some forms, set at an angle with dark edges, suggest tilted planes and the illusion of perspectival depth. On the other hand, forms such as the large stenciled letters are emphatically two-dimensional. Literal elements—machines, buildings, robot figures mounting a staircase, and signs—commingle with abstract shapes to produce fast-tempoed, kaleidoscopic glimpses of the urban industrial world. Léger's later work is discussed in chapter 14.

Orphism

Unlike Léger, **ROBERT DELAUNAY** (1885–1941) moved rapidly through his apprenticeship to Cubism and by 1912 had arrived at a formula of brilliantly colored abstractions with only the most tenuous roots in naturalistic observation. His restless and inquisitive mind had ranged over the entire terrain of modern art, from the color theories of Chevreul and the space concepts of Cézanne to Braque and Picasso. In 1911 and 1912 he exhibited with Der Blaue

Reiter group in Munich and at Der Sturm Gallery in Berlin in 1912. This early contact with the German avant-garde, and especially the paintings and writings of Kandinsky, helped to make Delaunay one of the first French artists to embrace abstraction. He also joined the Cubists at the highly controversial Salon des Indépendants of 1911, where Apollinaire declared his painting the most important work in the show. Delaunay avoided the figure and the still life, two of the mainstays of orthodox Cubism, and in his first Cubist paintings took as themes two great works of Parisian architecture, the Gothic church of Saint-Séverin and the Eiffel Tower. He explored his subjects in depth, and each exists in several variations. The series of paintings on the subject of the Eiffel Tower was begun in 1909. Like a work by Cézanne, the 1910 *Eiffel Tower in Trees* (fig. 230) is framed by a foreground tree, depicted in light hues of ocher and blue, and by a pattern of circular cloud shapes that accentuate the staccato tempo of the painting. The fragmentation of the tower and foliage introduces not only the shifting viewpoints of Cubism, but also rapid motion, so that the tower and its environment seem to vibrate. With his interest in simultaneous views and in the suggestion of motion, Delaunay may be reflecting ideas published by the Futurists in their 1910 manifesto.

Delaunay called his method for capturing light on canvas through color "simultaneity." Like František Kupka, he adapted the color theories of Chevreul to his art. In *Window on the City No. 3* (fig. 231), 1911–12, which belongs to a

231. ROBERT DELAUNAY. *Window on the City No. 3.* 1911–12. Oil on canvas, 44¾ x 51½" (113.7 x 130.8 cm). Solomon R. Guggenheim Museum, New York

series based on views of Paris, Delaunay applied his own variation of Chevreul's simultaneous contrast of colors for the first time. He used a checkerboard pattern of color dots, rooted in Neo-Impressionism and framed in a larger pattern of geometric shapes, to create a dynamic world of fragmented images. To all intents and purposes this is a work of abstraction, though its origin is still rooted in reality. In the spring of 1912 he began a new series, called Windows, in which pure planes of color, fractured by light, virtually eliminate any recognizable vestiges of architecture and the observed world. Throughout this period, however, he continued to paint clearly figurative works alongside his abstractions.

In his Disk paintings of 1913 (color plate 101, page 203) Delaunay abandoned even the pretense of subject, creating arrangements of vividly colored circles and shapes. While these paintings, which Delaunay grouped under the title *Circular Forms,* have no overt relationship to motifs in nature, the artist did, according to his wife, Sonia Delaunay, observe the sun and the moon for long periods, closing his eyes to retain the retinal image of the circular shapes. His researches into light, movement, and the juxtaposition of contrasting colors were given full expression in these works.

Born in the Czech Republic (then Bohemia, in the Austro-Hungarian Empire), FRANTIŠEK KUPKA (1871–1957) attended the art academies of Prague and Vienna. The many forces in those cities that shaped his later art include exposure to the spiritually oriented followers of Nazarene art, decorative Czech folk art and other forms of ornament, knowledge of color theory, an awareness of the art of the Vienna Secession, and a lifelong involvement with spiritualism and the occult. Settling in Paris in 1896, he discovered chronophotography and the budding art form of cinematography, and these contributed to his notion of an art, still based on the human figure, that suggested a temporal dimension through sequential movement. In this he shared concerns with the Futurists, though he developed his ideas before he saw their work in Paris in 1912.

230. ROBERT DELAUNAY. *Eiffel Tower in Trees.* 1910. Oil on canvas, 49⅞ x 36½" (126.7 x 92.7 cm). Solomon R. Guggenheim Museum, New York

Throughout his life Kupka was a practicing medium and maintained a profound interest in Theosophy, Eastern religions, and astrology. The visionary, abstract art he developed in Paris, beginning in 1910, was based on his mystical belief that forces of the cosmos manifest themselves as pure rhythmic colors and geometric forms. Such devotion to the metaphysical realm placed him at odds with most of the Cubist art he encountered in Paris. Though he arrived at a nonobjective art earlier than Delaunay, and though he was the first artist to exhibit abstract works in Paris, Kupka was then relatively unknown compared with the Frenchman, and his influence was not immediately felt. In 1911–12, with the painting he entitled *Disks of Newton* (color plate 103, page 204), Kupka created an abstract world of vibrating, rotating color circles. While his prismatic circles have much in common with Delaunay's slightly later Simultaneous Disks, it is not entirely clear what these two artists knew of one another's work.

In 1912 Apollinaire named the abstract experiments of Delaunay, Léger, Kupka, and others Orphism, a term that displeased the highly individualistic Kupka. Apollinaire defined the style as one based on the invention of new structures "which have not been borrowed from the visual sphere," though the artists he designated had all relied on the "visual sphere" to some degree. It was, however, a recognition that this art was in its way as divorced as music from the representation of the visual world or literal subject. Kandinsky, it will be recalled, equated his abstract paintings—done at about the same time—with pure musical sensation (color plate 76, page 160).

SONIA DELAUNAY (1885–1979), born Sonia Stern in a Ukrainian village in 1885, the same year as Robert Delaunay, moved to Paris in 1905. She married Robert in 1910 and, like him, was destined to become a pioneer of abstract painting. Her earliest mature works were Fauve-inspired figure paintings, and her first abstract work (color plate 104, page 205), dated 1911, was a blanket that she pieced together with scraps of material after the birth of her son. This remarkable object, which Delaunay said was based on examples made by Russian peasant women (and has much in common with the geometric pieced crazy quilts of nineteenth-century America), foreshadowed the abstract paintings she began to make by 1913. In those paintings, called Simultaneous Contrasts, Delaunay explored the dynamic interaction of brilliant color harmonies. While they had much in common with Robert's work, they gave evidence of a creative personality quite distinct from his. After World War I Sonia Delaunay devoted more time to textile design. When worn, her clothing literally set into motion the geometric patterns she had first explored in 1911. Her highly influential designs were sold throughout America and Europe.

Within a few years of one another, artists at various points on the globe achieved an art of nonrepresentation in which the illusion of nature was completely eliminated and whose end purpose was organization of the artist's means—color, line, space, and their interrelations and expressive potentials. Those artists included Kupka and the Delaunays in France; Kandinsky in Germany; Mikhail Larionov and, subsequently, Kazimir Malevich in Russia; Percy Wyndham Lewis in England; and Piet Mondrian and the artists of de Stijl in the Netherlands. The developments in Russia and the Netherlands, for which the invention and dissemination of Cubist styles were crucial, are examined in chapter 11. The quest for an abstract vocabulary continued during the 1920s, but World War I had disrupted creative effort everywhere. New forces, new attitudes, were beginning to make themselves felt.

Among the first artists to desert Cubism in favor of a new approach to subject and expressive content was **MARCEL DUCHAMP** (1887–1968), one of the most fascinating, enigmatic, and influential figures in the history of modern art. As previously noted, Duchamp was one of four siblings who contributed to the art of the twentieth century, though none was nearly so significant as Marcel. Until 1910 he worked in a relatively conventional manner based on Cézanne and the Impressionists. In 1911 he painted his two brothers in *Portrait of Chess Players* (fig. 232), obviously after coming under the influence of Picasso and Braque's Cubist explorations. The space is ambiguous and the color is low-keyed and virtually monochromatic, partly because Duchamp painted it by gaslight, not daylight. The linear rhythms are cursive rather than rectangular, while the figures are characterized by tense energy and individuality; their fragmentation gives them a fantastic rather than a physical presence. Even at the moment when he was learning about Cubism, Duchamp was already revolting against it, transforming it into something entirely different from the conceptions of its creators. The game of chess is a central theme in Duchamp's work. In his later career, when he had publicly stopped making art, chess assumed a role tantamount to artistic activity in his life.

232. MARCEL DUCHAMP. *Portrait of Chess Players*. 1911.
Oil on canvas, 39¾ x 39¾" (101 x 101 cm).
Philadelphia Museum of Art LOUISE AND WALTER ARENSBERG COLLECTION
© 1997 ADAGP, PARIS/ARTISTS RIGHTS SOCIETY (ARS), NEW YORK

During 1911 Duchamp also painted the first version of *Nude Descending a Staircase,* a work destined to become internationally notorious as a popular symbol of modern art. In the famous second version, 1912 (color plate 105, page 205), the androgynous, mechanized figure has been fragmented and multiplied to suggest a staccato motion. Duchamp was fascinated by the art of cinema and by the nineteenth century chronophotographs of Étienne-Jules Marey, which studied the body's locomotion through stop-action photographs, similar to those of Eadweard Muybridge and Thomas Eakins (see figs. 52, 64). Marey placed his subjects in black clothing with light metal strips down the arms and legs that would, when photographed, create a linear graph of their movement. Duchamp similarly indicated the path of his figure's movement with lines of animation and, at the elbow, small white dots. He borrowed the Cubists' reductive palette, painting the figure in what he called "severe wood colors." Cubists on the selection committee for the 1912 Salon des Indépendants objected to the work for its literary title and traditional subject. Duchamp refused to change the title and withdrew his painting from the show. Throughout his life he used and subverted established artistic modes; in *Nude* he freely adapted Cubist means to a peculiarly personal expressive effect. The rapidly descending nude is not a static form analyzed as a grid of lines and planes as is, for example, Picasso's *Portrait of Daniel-Henry Kahnweiler* (color plate 93, page 168). Instead, by depicting the figure in several successive moments at once, Duchamp implies movement through space. Although the Futurists were also interested in the pictorialization of movement, they regarded the nude as an outdated subject, ill-suited to their celebration of the new industrial age. Duchamp admired some of the works he saw in the first Futurist exhibition held in Paris in February 1912, but he had little sympathy for the group's rhetoric and unrelenting faith in technology. He said that while his painting had some "futuristic overtones," it was clearly more indebted to Cubism. *Nude Descending a Staircase* was shown in New York at the enormous 1913 Armory Show, the first international exhibition of modern art in the United States. The painting outraged American critics; one derided it as an "explosion in a shingle factory." Thanks to this exhibition, by the time Duchamp came to America in 1915 his reputation was already established.

Duchamp was an instinctive Dadaist, an iconoclast in art even before the creation of the Dada movement, so it is natural that he should have been immediately recognized by the first Dadaists and Surrealists as a great forerunner. From an attack on Cubism that involved a new approach to subject he passed, between 1912 and 1914, to an attack on the nature of subject painting, and finally to a personal reevaluation of the very nature of art. This phase of his career is properly examined in chapter 13.

Duchamp's elder brother Gaston, who took the name JACQUES VILLON (1875–1963), was at the other extreme from Marcel, a committed Cubist. In 1906 he moved to Puteaux, on the outskirts of Paris, where he had a house that

233. JACQUES VILLON. *Yvonne D. in Profile.* 1913. Drypoint on paper, first state, 21⅝ x 16⁵⁄₁₆" (54.9 x 41.4 cm). Boston Public Library
© 1997 ADAGP, PARIS/ARTISTS RIGHTS SOCIETY (ARS), NEW YORK

adjoined Kupka's. Puteaux, and particularly Villon's studio, became a remarkable nucleus of Cubist activity (the other being Montmartre, where Picasso and Braque worked) that eventually attracted Villon's brothers, Gleizes, Léger, Picabia, and Metzinger, among others. Villon established a personal, highly abstract, and poetic approach to Cubism, one he maintained throughout his long life. In addition to painting, he also made prints, a medium well suited to his particular Cubist idiom. The crystalline structure of jagged triangular shapes in the drypoint print depicting his sister, *Yvonne D. in Profile* (fig. 233), 1913, is indicated solely through parallel lines. They switch directions and densities, creating volumetric rhythms and a sense of shifting light over the surface.

FRANCIS PICABIA (1879–1953) began as a Cubist, though he later turned to Dada (see chapter 13). Until 1912 he was involved with the Cubist activities of Gleizes and Metzinger and exhibited with the Section d'Or, after which he turned to Orphism. In 1913 he spent five months in New York, where his works were included in the Armory Show and where he became a kind of self-appointed spokesman for the European avant-garde. There he met the American photographer Alfred Stieglitz, who gave him a show at his 291 Gallery. (Picabia subsequently contributed to Stieglitz's journal, *Camera Work.*) Upon his return to Paris he made a number of abstract paintings, including *Catch as Catch Can* (color plate 106, page 205). The image is supposedly a recollection of having watched a Chinese wrestler one evening in the company of Apollinaire, but the writhing, metallic planes of subdued color hardly coalesce into any identifiable imagery. Picabia abandoned Cubism in 1915, the year that, with Duchamp and Man Ray, he founded the American wing of Dada (see chapter 13).

11 Towards Abstraction

bstraction is the most dramatic and far-reaching development in the history of twentieth-century art. Ever since the latter part of the nineteenth century a number of artists were consciously or unconsciously moving toward a conception of painting as an entity unto itself rather than an imitation of, an illusion of, the physical world. In 1890 the French Symbolist painter Maurice Denis uttered a prophetic statement when he said that a painting, before it is anything else, is "essentially a flat surface covered with colors assembled in a particular order." We saw in chapter 10 that the originators of Cubism, Picasso and Braque, never exploited their style for the purposes of total abstraction. Although their Cubist pictures may represent highly abstracted interpretations of the material world, they were not in themselves abstract. We also saw, however, that other artists schooled in Cubism who were working in Paris, namely Delaunay, Kupka, and, for a time, Léger, made paintings that were not dependent upon recognizable forms. Slightly later in Zurich, the Dada artist Jean Arp was making collages with no discernible source in the real world (see fig. 298). Simultaneously, in other art centers in Europe and in the United States, a number of vanguard artists were abandoning the realm of appearances in pursuit of absolute, pure form. In Russia and the Netherlands, in particular, abstraction found even more fertile ground than in France; it was in these countries that early twentieth-century abstraction found its most expansive and most radical manifestations, with implications not merely for painting and sculpture but for architecture as well as graphic, industrial, and even fashion design. In the hands of these Utopian-minded pioneers, abstraction was not simply self-referential, art-for-art's-sake formalism. Rather it represented a powerful new way of perceiving and ultimately transforming the world. The artists fervently believed that art lost none of its expressive power or meaning when divorced from the tangible world. In fact, through the formal expression of pure sensation they hoped to discover a universal visual language able to transcend mundane experience and place the viewer in touch with an alternative, ultimately spiritual world.

Many terms have been employed to describe the nonimitative works of art produced throughout the twentieth century, including abstract, nonobjective, and nonrepresentational. All these terms refer to an art that depends solely on color, line, and shape for its imagery rather than motifs drawn from observable reality.

Futurism in Italy

In their paintings and sculptures the Italian Futurist artists sometimes verged on total abstraction (color plate 109, page 206), but for the most part they, like the Cubists, drew upon imagery derived from the physical world. Futurism was first of all a literary concept, born in the mind of the poet and propagandist Filippo Tommaso Marinetti in 1908 and announced in a series of manifestos in 1909, 1910, and subsequently. It began as a rebellion of young intellectuals against the cultural torpor into which Italy had sunk during the nineteenth century, and as so frequently happens in such movements, its manifestos initially focused on what they had to destroy before new ideas could flourish. The first manifesto, written by Marinetti, demanded the destruction of the libraries, museums, academies, and cities of the past, which he called mausoleums. It extolled the beauties of revolution, of war, of the speed and dynamism of machine technology: ". . . A roaring motorcar, which looks as though running on shrapnel, is more beautiful than the *Victory of Samothrace*." Much of the spirit of Futurism reflected the flamboyant personality of Marinetti himself; rooted in the anarchist and revolutionary fervor of the day, it attacked the ills of an aristocratic and bourgeois society and celebrated progress, energy, and change. In the field of politics it unfortunately was to become a pillar of Italian fascism.

In late 1909 or early 1910 the painters Umberto Boccioni, Carlo Carrà, and Luigi Russolo joined Marinetti's movement. Later it also included Gino Severini, who had been working in Paris since 1906, and Giacomo Balla. The group drew up a second manifesto in 1910. This again attacked the old institutions and promoted the artistic expression of motion, metamorphosis, and the simultaneity of vision itself. The painted moving object was to merge with its environment, so that no clear distinction could be drawn between the two: "Everything is in movement, everything rushes forward, everything is in constant swift change. A figure is never stable in front of us but is incessantly appearing and disappearing. Because images persist on the retina, things in movement multiply, change form, follow one another like vibrations within the space they traverse. Thus a horse in swift course does not have four legs: It has twenty, and their movements are triangular." The targets of the Futurists' critique included all forms of imitation, concepts of harmony and good taste, all art critics, all traditional subjects, tonal painting, and that perennial staple of art, the painted nude.

At this point, the paintings of the various Futurists had little in common. Many of their ideas still came from the unified color patterns of the Impressionists and even more explicitly from the Divisionist techniques of the Neo-Impressionists. But unlike Impressionism, Fauvism, and Cubism, all of which were generated by a steady interaction of theory and practice, Futurism emerged as a full-blown and coherent theory, the illustration of which the artists then set out to realize in paintings. The Futurists were passionately concerned with the problem of establishing empathy between the spectator and the painting, "putting the spectator in the center of the picture." In this they were close to the German Expressionists, who also sought a direct appeal to the emotions. Futurist art extolled metropolitan life and modern industry. This did not, however, result in a machine aesthetic in the manner of Léger, since the Italians were concerned with the unrestrained expression of individual ideals, with mystical revelation, and with the articulation of action. Despite their identity of purpose, Futurist art cannot be considered a unified style.

The Futurist exhibition in Milan in May 1911 was the first of the efforts by the new group to make its theories concrete. In the fall of that year, Carrà and Boccioni visited Paris. Severini took them to meet Picasso at his studio, where they no doubt saw the latest examples of Analytic Cubist painting. What they learned is evident in the repainted version of Boccioni's *The Farewells* (fig. 234), a 1911 work that is the first in a trilogy of paintings titled *States of Mind*, about arrivals and departures at a train station. Within a vibrating, curving pattern the artist introduces a Cubist structure of interwoven facets and lines designed to create a sense of great tension and velocity. Boccioni even added the shock of a literal, realistic collage-like element in the scrupulously rendered numbers on the cab of the dissolving engine. It will be recalled that in 1911 Braque and Picasso first introduced lettering and numbers into their Analytic Cubist works (see figs. 210, 212). Boccioni's powerful encounter with Cubism reinforced his own already developing inclinations, and he fully absorbed the French style into his own dynamic idiom.

An exhibition of the Futurists, held in February 1912 at the gallery of Bernheim-Jeune in Paris, was widely noticed and reviewed favorably by Apollinaire himself. It later circulated to London, Berlin, Brussels, the Hague, Amsterdam, and Munich. Within the year, from being an essentially provincial movement in Italy, Futurism suddenly became a significant part of international experimental art.

234. UMBERTO BOCCIONI. *States of Mind I: The Farewells.* 1911. Oil on canvas, 27¾ x 37⅞" (70.5 x 96.2 cm). The Museum of Modern Art, New York

GIACOMO BALLA (1871–1958), the oldest of the group and the teacher of Severini and Boccioni, early in the century painted realistic pictures with social implications. He then became a leading Italian exponent of Neo-Impressionism and in this context most strongly influenced the younger Futurists.

Balla's painting *Street Light* (color plate 107, page 206) is an example of pure Futurism. Using v-shaped brushstrokes of complementary colors radiating from the central source of the lamp, he created an optical illusion of light rays translated into dazzling colors so intense that they appear to vibrate. Balla, working in Rome rather than Milan, pursued his own distinctive experiments, particularly in rendering motion through simultaneous views of many aspects of objects. His *Dynamism of a Dog on a Leash* (fig. 235), with its multiplication of legs, feet, and leash, has become one of the familiar and delightful creations of Futurist simultaneity. The little dachshund scurries along on short legs accelerated and mul-

235. GIACOMO BALLA. *Dynamism of a Dog on a Leash (Leash in Motion).* 1912. Oil on canvas, 35 x 45½" (88.9 x 115.6 cm). Albright-Knox Art Gallery, Buffalo

tiplied to the point where they almost turn into wheels. This device for suggesting rapid motion or physical activity later became a cliché of the comic strips and animated cartoons. Balla eventually returned to more traditional figure painting.

Also at work within the ambience of Futurist dynamism was the Italian photographer and filmmaker **ANTONIO GIULIO BRAGAGLIA** (1889–1963), who, like Duchamp and Balla, had been stimulated by the stop-action photographs of Muybridge, Eakins, and Marey. Bragaglia, however, departed from those and shot time exposures of moving forms (fig. 236), creating fluid, blurred images of continuous action. These, he thought, constituted a more accurate, expressive record than a sequence of discrete, frozen moments. In 1913 Bragaglia published a number of his "photodynamic" works in a book entitled *Fotodinamismo futurista*.

GINO SEVERINI (1883–1966), living in Paris since 1906, was for several years more closely associated with the growth of Cubism than the other Futurists. Because of his presence there, he served as a conduit between his Italian colleagues and French artistic developments. His approach to Futurism is summarized in *Dynamic Hieroglyphic of the Bal Tabarin* (color plate 108, page 206), a gay and amusing distillation of Paris nightlife. The basis of the composition lies in Cubist faceting put into rapid motion within large, swinging curves. The brightly dressed chorus girls, the throaty chanteuse, the top-hatted, monocled patron, and the carnival atmosphere are all presented in a spirit of delight, reminding us that the Futurists' revolt was against the deadly dullness of nineteenth-century bourgeois morality. This work is a tour de force involving almost every device of Cubist painting and collage, not only the color shapes that are contained in the Cubist grid, but also elements of sculptural modeling that create effects of advancing volumes. Additionally present are the carefully lettered words VALSE, POLKA, BOWLING, while real sequins are added as collage on the women's costumes. The color has Impressionist freshness, its arbitrary distribution a Fauve boldness. Many areas and objects are mechanized and finely stippled in a Neo-Impressionist manner. Severini even included one or two passages of literal representation, such as the minuscule Arab horseman (upper center) and the tiny nude riding a large pair of scissors (upper left).

The sense of fragmented but still dominating reality that persisted in Severini's Cubist paintings between 1912 and 1914 found its most logical expression in a series of works on the subject of transportation, which began with studies of the Paris métro. With the coming of the war, the theme of the train flashing through a Cubist landscape intrigued Severini as he watched supply trains pass by his window daily, loaded with weapons or troops. *Red Cross Train Passing a Village* (fig. 237), one of several works from the summer of 1915 on this theme, is his response to Marinetti's appeal for a new pictorial expression for the subject of war "in all its marvelous mechanical forms." *Red Cross Train* is a stylization of motion, much more deliberate in its tempo than *Bal Tabarin*. The telescoped but clearly recognizable train, from which balloons of smoke billow, cuts across the middle distance. Large, hand-

236. ANTONIO GIULIO BRAGAGLIA. *The Cellist.* 1913.
Gelatin-silver print

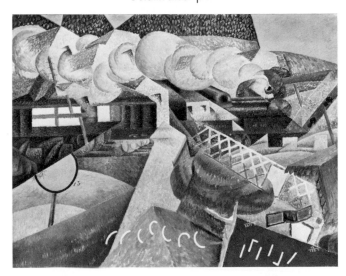

237. GINO SEVERINI. *Red Cross Train Passing a Village.* 1915.
Oil on canvas, 35¼ x 45¾" (89.5 x 116.2 cm).
Solomon R. Guggenheim Museum, New York
© 1997 ADAGP, PARIS/ARTISTS RIGHTS SOCIETY (ARS), NEW YORK

some planes of strong color, sometimes rendered with a Neo-Impressionist brushstroke, intersect the train and absorb it into the painting's total pattern. The effect is static rather than dynamic and is surprisingly abstract in feeling. During these years Severini moved toward a pure abstraction stemming partly from the influence of Robert Delaunay. In *Spherical Expansion of Light (Centrifugal)* (color plate 109, page 206) he organized what might be called a Futurist abstract pattern of triangles and curves, built up of Neo-Impressionist dots. For some of the works in this series Severini extended the exuberant color dots onto his wooden frame.

In 1916, in stark contrast to his work that preceded it, Severini made a highly naturalistic portrait of his wife nursing their son. He did not immediately follow this direction in his subsequent work, which generally consisted of still lifes in a Synthetic Cubist style. But after 1920 he turned to a highly distilled form of classicism, in which the Cubist fracturing of space gave way to traditional modeling in illusionistic depth.

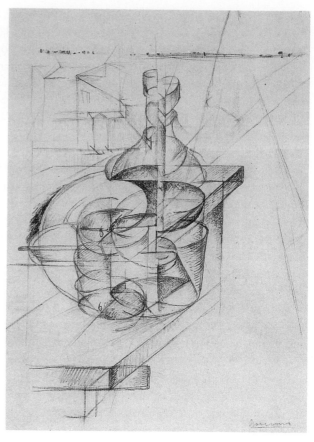

above: 238. UMBERTO BOCCIONI. *Development of a Bottle in Space.*
1912. Silvered bronze (cast 1931),
15 x 23¾ x 12⅞" (38.1 x 60.3 x 32.7 cm).
The Museum of Modern Art, New York ARISTIDE MAILLOL FUND

right: 239. UMBERTO BOCCIONI. *Table + Bottle + House.* 1912.
Pencil on paper, 13⅛ x 9⅜" (33.4 x 23.9 cm).
Civico Gabinetto dei Disegni, Castello Sforzesco, Milan

CARLO CARRÀ (1881–1966) was important in bridging the gap between Italian Futurism and a slightly later Italian movement called the Metaphysical School (see fig. 296). From 1912, however, Carrà moved steadily toward an orthodox version of Cubism that expressed little of the ideals of speed and dynamism extolled in the Futurist manifestos. He returned to painting the nude, a subject that led him toward a form of massive, sculptural modeling. It was only a short step from this work to the metaphysical painting of Giorgio de Chirico, which Carrà began to investigate around 1916. Meanwhile, in the propagandistic collage *Patriotic Celebration* (color plate 110, page 207), Carrà employed radiating colors, words and letters, Italian flags, and other lines and symbols to extol the king and army of Italy, and to visually simulate the noises of sirens and mobs. He used these "free words," as Marinetti did, to affect and stimulate the spirit and imagination directly through their visual associations, without the intervention of reason. A similar development was taking place among Russian Futurist poets and artists.

UMBERTO BOCCIONI (1882–1916) was perhaps the most talented of the Futurists. In his monumental work *The City Rises* (color plate 111, page 207) he sought his first "great synthesis of labor, light, and movement." Dominated by the large, surging figure of a horse before which human figures fall like ninepins, it constitutes one of the Futurists' first major statements: a visual essay on the qualities of violent action, speed, the disintegration of solid objects by light, and their reintegration into the totality of the picture by that very same light.

The greatest contribution of Boccioni during the last few

years of his short life was the creation of Futurist sculpture. During a visit to Paris in 1912, he went to the studios of Archipenko, Brancusi, and Duchamp-Villon and saw sculpture by Picasso. Immediately upon his return to Milan he wrote the *Technical Manifesto of Futurist Sculpture* (1912), in which he called for a complete renewal of this "mummified art." He began the manifesto with the customary attack on all academic tradition. The attack became specific and virulent on the subject of the nude, which still dominated the work not only of the traditionalists but even of the leading progressive sculptors, Rodin, Bourdelle, and Maillol. Only in the Impressionist sculpture of Medardo Rosso (see fig. 144), then the principal Italian sculptor of stature, did Boccioni find exciting innovations. Yet he recognized that, in his concern to capture the transitory moment, Rosso was bound to represent the subject in nature in ways that paralleled those of the Impressionist painters.

Taking off from Rosso, Boccioni sought a dynamic fusion between his sculptural forms and surrounding environment. He emphasized the need for an "absolute and complete abolition of definite lines and closed sculpture. We break open the figure and enclose it in [an] environment." He also asserted the sculptor's right to use any form of distortion or fragmentation of figure or object, and insisted on the use of every kind of material—"glass, wood, cardboard, iron, cement, horsehair, leather, cloth, mirrors, electric lights, etc., etc."

Boccioni's Futurist sculpture and his manifesto were the first of several related developments in three-dimensional art, among them Constructivist sculpture, Dada and Surrealist assemblage, kinetic sculpture, and even the Pop sculptural

240. UMBERTO BOCCIONI. *Unique Forms of Continuity in Space.* 1913. Bronze (cast 1931). 43⅞ x 34⅞ x 15¾" (111.2 x 88.5 x 40 cm). The Museum of Modern Art, New York
ACQUIRED THROUGH THE LILLIE P. BLISS BEQUEST.
PHOTOGRAPH © 1995 THE MUSEUM OF MODERN ART, NEW YORK

environments of the 1960s (color plate 302, page 528; see fig. 640). His *Development of a Bottle in Space* (fig. 238) enlarged the tradition of the analysis of sculptural space. The bottle is stripped open, unwound, and integrated with an environmental base that makes the homely object, only fifteen inches high, resemble a model for a vast monument. In a related drawing (fig. 239), possibly made in preparation for the sculpture, the forms of a bottle and glass are opened up, set in motion with rotational, curving lines, and penetrated by the flat planes of the table on which they sit.

Boccioni's most impressive sculpture, *Unique Forms of Continuity in Space* (fig. 240), was also his most traditional and the one most specifically related to his paintings. The figure, made up of fluttering, curving planes of bronze, moves essentially in two dimensions, like a translation of his painted figures into relief. It has something in common with the ancient Greek *Victory of Samothrace* so despised by Marinetti: the stances of both figures are similar—a body in dramatic mid-stride, draperies flowing out behind, and arms missing.

World War I killed several of the most talented experimental artists including Boccioni. Afterwards Futurism lost much of its impetus although many of the more propagandistic ideas and slogans were integrated into the rising tide of fascism and used for political and social ends.

Abstraction in Russia

In the attempt to evaluate the Russian achievement in the early part of this century, a number of points must be kept in mind. Russia, since the eighteenth century of Peter and Catherine the Great, had maintained a tradition of royal patronage of the arts and had close ties with the West. Russians who could afford it traveled frequently to France, Italy, and Germany, and through books and periodicals they were aware of new developments in European art. Russian literature and music attained great heights during the nineteenth century, while theater and ballet also made substantial strides and began to draw on the visual arts in interesting collaborations.

Art Nouveau and the ideas of French Symbolists and Post-Impressionist Nabis made themselves widely felt in the late 1880s through the movement known as the World of Art *(Mir Iskusstva);* Russian artists mingled these influences with Byzantine and Russian painting and decorative traditions.

In 1890, World of Art was joined by Sergei Diaghilev, destined to become perhaps the greatest of all impresarios of the ballet as well as an enthusiast for modern art in general. A few years later Diaghilev was launched on his career arranging exhibitions, concerts, theatrical and operatic performances, and ultimately the Ballets Russes, which opened in Paris in 1909 and went from success to success. From then on, Diaghilev drew on many of the greatest names in European painting to create his stage sets.

After *World of Art* periodical, first published in 1898 came other avant-garde journals. Reading these journals was yet another way that Russian artists could watch and absorb the progress of Fauvism, Cubism, Futurism, and their offshoots. Great collections of the new French art were formed by the enlightened collectors Sergei Shchukin and Ivan Morozov in Moscow. By 1914, Shchukin's collection contained over two hundred works by French Impressionist, Post-Impressionist, Fauve, and Cubist painters, including more than fifty by Picasso and Matisse. Morozov's collection included Cézannes, Renoirs, Gauguins, and many works by Matisse. Both Shchukin and Morozov were most generous in opening their collections to Russian artists, and the effect of the Western vanguard on these creative individuals was incalculable. Through such modern art collections and the exhibitions of the Jack of Diamonds group, an alternative exhibition society founded in 1910 in Moscow, Cubist experiments were known in St. Petersburg and Moscow almost as soon as they were inaugurated. By the time Marinetti, whose Futurist manifesto had long been available in Russia, visited Russia in 1914, the Futurist movement in that country was in full force.

Only within roughly the last two decades, and especially with *glasnost* under Mikhail Gorbachev and the dissolution of the Soviet Union, has scholarship on Russian avant-garde art really come into its own. Before that time, study of Russian modernism was restricted and sometimes suppressed; it was particularly difficult for Western scholars, who were often denied access to archival materials and works of art during the Cold War due to official policies of censorship and control of art that dated back to the government of the Soviet leader Joseph Stalin, who, in 1932, decreed that Socialist Realism (naturalistic art that celebrated the worker) was the only acceptable form of art. Moreover, the kinds of small, independent art groups that had until then flourished in Russia, particularly among the avant-garde, were proscribed. Paintings by many of the greatest artists were relegated to storerooms, hidden, or even destroyed; they could not be lent to outside museums. Such was the fate of the works of Kazimir Malevich, for example, an artist who chose not to leave the country after Stalin's rise to power unlike many of his contemporaries. Though known in the West through works that had been brought to New York, Amsterdam, and a few other cities, his paintings had not been seen in Russia for decades when they were at last exhibited there in 1988. Throughout this period, works by the Russian avant-garde were protected and studied by a small number of important Russian scholars, curators, and art collectors, sometimes against very difficult circumstances, but only relatively recently have they been able to collaborate freely with colleagues outside Russia to produce the kinds of exhibitions and publications necessary for a full understanding of this extraordinary art.

Rayonism, Cubo-Futurism, and Suprematism

Lifelong companions and professional collaborators, **MIKHAIL LARIONOV** (1881–1964) and **NATALYA SERGEYVNA GONCHAROVA** (1881–1962) are among the earliest proponents of the Russian avant-garde. Larionov was a founding member of the Jack of Diamonds group in 1910, but by the following year he and Goncharova had left it and formed a rival organization, the Donkey's Tail, claiming the need for a contemporary Russian art that drew less from Western Europe (since the Jack of Diamonds exhibitions included work by Western European artists) and more from indigenous artistic traditions. (Nevertheless, in 1912 they both participated in the second Blaue Reiter exhibition in Munich, as well as in a historic Post-Impressionist show in London.) They turned to Russian icons and *lubok* (folk) prints for inspiration and made works in what they termed a Neo-Primitive style. In 1912 Larionov created Rayonism based on his studies of optics and theories about how intersecting rays of light reflect off of the surface of objects (fig. 241). Rayonist works were first shown in December 1912 and then in a 1913 exhibition called "The Target." Although Rayonism was indebted to Cubism, and also related to Italian Futurism in its emphasis on dynamic, linear forms, Larionov and his circle emphasized its Russian origins. His paintings were among the first nonobjective works of art made in Russia. In them Larionov sought to merge his studies of nineteenth-century color theories with more recent scientific experiments (such as radiation). His 1913 manifesto titled *Rayonism*, which extolled this style as the "the true liberation of painting," was critical as the first published discourse in Russia on nonobjectivity in art.

Like Larionov, Goncharova was drawn to ancient Russian art forms. In addition to icons and *lubok* she studied tradi-

tional examples of embroidery. This nonhierarchical distinction between craft and "high" art is a key characteristic of the Russian avant-garde and is perhaps one of the reasons why so many women had positions at the artistic forefront equal to those of their male colleagues. After her Neo-Primitive phase, Goncharova produced paintings in a Futurist and Rayonist vein. Her colorful 1912 canvas *Linen* (color plate 112, page 207), reveals a knowledge of Cubist painters such as Gleizes

left: 241. MIKHAIL LARIONOV. *Blue Rayonism.* 1912. Oil on canvas, 25½ x 27½" (64.8 x 69.9 cm). Collection Boris Tcherinsky, Paris
© 1997 ADAGP, PARIS/ARTISTS RIGHTS SOCIETY (ARS), NEW YORK

below left: 242. NATALYA SERGEYVNA GONCHAROVA. *Icon Painting Motifs.* 1912. Watercolor on cardboard, 19½ x 13⅝" (49.5 x 34.6 cm). State Tretyakov Gallery, Moscow
GIFT OF GEORGE COSTAKIS
© 1997 ARTISTS RIGHTS SOCIETY (ARS), NEW YORK/ADAGP, PARIS

right: 243. LYUBOV POPOVA. Stage design for Vsevolod Meyerhold's *The Magnanimous Cuckold.* 1922. India ink, watercolor, paper collage, and varnish on paper, 19¾ x 27¼" (50.2 x 69.2 cm). State Tretyakov Gallery, Moscow

and Metzinger, but the Cyrillic letters add a distinctively Russian touch. She also continued to make paintings in a folk style, as can be seen in *Icon Painting Motifs* (fig. 242). The conceptual nature of such images explains, to some degree, how an ancient heritage of icons and flat-pattern design prepared Russians to accept total abstraction.

Larionov and Goncharova left Russia in 1915 to design for Diaghilev. They settled in Paris, became French citizens in 1938, and married in 1955. Neither artist produced more than a few significant Rayonist paintings, but their ideas were instrumental for an art that synthesized Cubism, Futurism, and Orphism and that contained "a sensation of the fourth dimension." This pseudo-scientific concept of a fourth spatial dimension, popularized by scientists, philosophers, and spiritualists gained currency among the avant-garde across Europe and in the United States in the early twentieth century.

One of the very strongest artists to emerge within the milieu of the prerevolutionary Russian avant-garde was the tragically short-lived **LYUBOV POPOVA** (1889–1924), whose sure hand and brilliant palette are evident even in her earliest work. The daughter of a wealthy bourgeois family, Popova was able to travel extensively throughout her formative years, visiting Paris in 1912 and studying under the Cubists Le Fauconnier and Metzinger. In 1913 she began working in the studio of Vladimir Tatlin, the Constructivist (figs. 249, 250). She also knew Goncharova and Larionov and like them was interested in Russian medieval art. When war broke out in 1914, Popova was in Italy, whence she returned to Russia and resumed working in Tatlin's studio. Unlike Tatlin, she was not concerned with the construction of objects in space, but rather found her expressive mode in painting. She developed a mature Cubo-Futurist style (the term is Malevich's), showing a complete assimilation of Western pictorial devices into her own dynamic idiom. Her still life *Objects from the Dyer's Shop* (color plate 113, page 208) contains a rich chromatic scheme that probably reflects her study of Russian folk

art. Here the Cubist-derived language of integrated, pictorialized form and space may have received its most authoritative expression outside the oeuvre of the two founding Cubists themselves. In late 1915 or early 1916, partly in response to Malevich's Suprematist canvases, Popova began to compose totally abstract paintings that she called Painterly Architectonics. In these powerful compositions she paid increasing attention to the facture, building up her paint surfaces with strong textures.

In the aftermath of the 1917 October Revolution, Russian artists were involved in serious debate about the appropriate nature of art under the new Communist regime. Constructivist artists abandoned traditional media such as painting and dedicated themselves to "production art." Their intention was to merge art with technology in products that ranged from utilitarian household objects to textile design, propagandistic posters, and stage sets for political rallies. Popova, who renounced easel painting in 1921, made designs for the theater, including a 1922 set for *The Magnanimous Cuckold* (fig. 243), a play staged by Vsevolod Meyerhold, one of the leading figures of avant-garde theater. The set was organized according to the principles that informed Popova's abstract paintings. Like a large Constructivist environment, it consisted of arrangements of geometric shapes within a structure of horizontal and vertical elements.

More than any other individual, even Delaunay or Kupka, it was **KAZIMIR MALEVICH** (1878–1935) who took Cubist geometry to its most radical conclusion. Malevich studied art in Moscow, where he visited the collections of Shchukin and Morozov. He was painting outdoors in an Impressionist style by 1903 and for a brief time experimented with the Neo-Impressionist technique of Seurat. In 1910 he exhibited with Larionov and Goncharova in the first Jack of Diamonds show in Moscow and subsequently joined their rival Donkey's Tail exhibition in which he exhibited Neo-Primitivist paintings of heavy-limbed peasants in bright Fauvist colors. By 1912 he was painting in a Cubist manner. In *Morning in the Village after a Snowstorm* (color plate 114, page 208) cylindrical figures of peasants move through a mechanized landscape, with houses and trees modeled in light, graded hues of red, white, and blue. The snow is organized into sharp-edged metallic-looking mounds. The resemblance to Léger's earlier machine Cubism is startling (see fig. 228), but it is questionable whether either man saw the other's early works for, unlike some of his colleagues, Malevich never traveled to Paris. For the next two years he explored different aspects and devices of Cubism and Futurism. He called his highly personalized amalgamation of the two styles Cubo-Futurism. In 1913 he designed Cubo-Futurist sets and costumes for an experimental performance billed as the "First Futurist Opera," called *Victory over the Sun*. This highly unorthodox opera, staged twice in St. Petersburg, anticipated the early Dada performances that took place in Zurich in 1916 (see fig. 297). The actors were mostly nonprofessionals who recited or sang their lines, accompanied by an out-of-tune piano. The nonnarrative texts of *Victory over the Sun* were called *zaum*, meaning "transrational" or "beyond-the-mind," and were intended to divest words of all conventional meaning. In 1913–14 Malevich created visual analogues to these semantic experiments in a number of paintings, labeling the style Transrational Realism. Through the juxtaposition of disparate elements in his compositions, he mounted a protest "against logic and philistine meaning and prejudice." In certain paintings of 1914, autonomous colored planes emerge from a matrix of Cubo-Futurist forms. In Malevich's abstract work of the following year, these planes came to function as entirely independent forms suspended on a white ground. "In the year 1913," the artist wrote, "in my desperate attempt to free art from the burden of the object, I took refuge in the square form and exhibited a picture which consisted of nothing more than a black square on a white field." Like Larionov (and a number of modern artists, for that matter), Malevich had a tendency to date his paintings retrospectively and assign them impossibly early dates. It was not until 1915 that he unveiled thirty-nine totally nonrepresentational paintings, whose style he called Suprematism, at a landmark exhibition in Petrograd (now St. Petersburg) called "0, 10 (Zero–Ten). The Last Futurist Exhibition" (fig. 244). Included in the exhibition was the painting *Black Square*, hung high across the corner of the room in the traditional place of a Russian icon. This emblem of Suprematism, the most reductive, uncompromisingly abstract painting of its time, represented an astonishing conceptual leap from Malevich's work of the previous year. The artist, in his volume of essays entitled *The World of Non-Objectivity*, defined Suprematism as "the supremacy of pure feeling in creative art." "To the Suprematist," he wrote, "the visual phenomena of the objective world are, in themselves, meaningless; the significant thing is feeling, as such, quite apart from the environment in which it is called forth."

As with Kandinsky and his first abstract paintings (color plate 76, page 160), the creation of this simple square on a plain ground was a moment of spiritual revelation to

244. KAZIMIR MALEVICH. Installation photograph of his paintings in "0, 10 (Zero–Ten). The Last Futurist Exhibition" in Petrograd (St. Petersburg), December 1915

Malevich. For the first time in the history of painting, he felt, it had been demonstrated that a painting could exist completely independent of any reflection or imitation of the external world—whether figure, landscape, or still life. Actually, of course, he had been preceded by Delaunay, Kupka, and Larionov in the creation of abstract paintings, and he was certainly aware of their efforts, as well as those of Kandinsky, who moved to largely nonobjective imagery in late 1913. Malevich, however, can claim to have carried abstraction to an ultimate geometric simplification—the black square. It is noteworthy that the two dominant wings of twentieth-century abstraction—the painterly Expressionism of Kandinsky and the hard-edge geometric purity of Malevich—should have been founded by two Russians. And each of these men had a conviction that his discoveries were spiritual visions rooted in the traditions of Old Russia.

In his attempts to define this new Suprematist vocabulary, Malevich tried many combinations of rectangle, circle, and cross, oriented vertically and horizontally. His passionate curiosity about the expressive qualities of geometric shapes next led him to arrange clusters of colored rectangles and other shapes on the diagonal in a state of dynamic tension with one another. This arrangement of forms implied continuous motion in a field perpetually charged with energy. Malevich established three stages of Suprematism: the black, the red or colored, and the white. In the final phase, realized in monochromatic paintings of 1917 and 1918 (fig. 245), the artist achieved the ultimate stage in the Suprematist ascent toward an ideal world and a complete renunciation of materiality, for white symbolized the "real concept of infinity." This example displays a tilted square of white within the canvas square of a somewhat different shade of white—a reduction of painting to the simplest relations of geometric shapes.

Malevich understood the historic importance of architecture as an abstract visual art and in the early 1920s, when he had temporarily abandoned painting, he experimented with drawings and models in which he studied the problems of form in three dimensions and his vision of Suprematist cities,

245. KAZIMIR MALEVICH. *Suprematist Composition: White Square on White.* 1918. Oil on canvas, 31¼ x 31¼" (79.4 x 79.4 cm). The Museum of Modern Art, New York

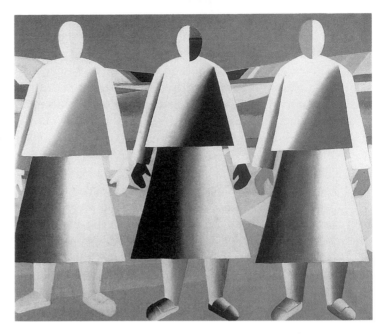

246. KAZIMIR MALEVICH. *Girls in the Field.* c. 1928. Oil on canvas, 41¾ x 49¼" (106 x 125 cm). State Russian Museum, St. Petersburg

planets, and satellites suspended in space. His abstract three-dimensional models, called Arkhitektons, were significant to the growth of Constructivism in Russia and, transmitted to Germany and Western Europe by his disciples, notably El Lissitzky, influenced the design teachings of the Bauhaus.

In the 1920s Malevich's idealist views were increasingly at odds with powerful conservative artistic forces in the Soviet Union who promoted Socialist Realism as the only genuine proletarian art. Eventually, this style was officially established as the only legitimate form of artistic expression. By the end of 1926 Malevich was dismissed from his position as director of GINKhUK, the Institute of Artistic Culture, and in 1930 he was even imprisoned for two months and interrogated about his artistic philosophy. In his late work Malevich returned to figurative style though in several works between 1928 (fig. 246) and 1932 he combined echoes of his early Cubo-Futurist work with Suprematist concepts. Perhaps the empty landscapes and faceless automaton figures of the last years express, as one discerning critic wrote in 1930, "the 'machine' into which man is being forced—both in painting and outside it."

Of the artists emerging from Russian Suprematism the most influential internationally was EL (ELEAZAR) LISSITZKY, (1890–1941) who studied architectural engineering in Germany. On his return to Russia at the outbreak of World War I, Lissitzky took up a passionate interest in the revival of Jewish culture, illustrating books written in Yiddish and organizing exhibitions of Jewish art. Like Marc Chagall, he was a major figure in the Jewish Renaissance in Russia around the time of the 1917 Revolution, which brought about the fall of the Czarist regime. In 1919 Chagall appointed Lissitzky to the faculty of an art school in Vitebsk that he headed. There Lissitzky, much to the disappointment of Chagall, became a disciple of the art of another faculty member, Malevich, developing his own form of abstraction, which he called Prouns

(fig. 247). The exact origin of this neologism is unclear, but it may be an acronym for "Project for the Affirmation of the

247. EL LISSITZKY. *Construction 99 (Proun 99).* 1924–25. Oil on wood, 50½ x 38¾" (128.3 x 98.4 cm). Yale University Art Gallery, New Haven GIFT OF THE SOCIÉTÉ ANONYME

New" in Russian. Lissitzky's Prouns are diverse compositions made up of two- and three-dimensional geometric shapes floating in space. The forms, sometimes depicted axonometrically, represent the artist's extension of Suprematist theories into the realm of architecture. Indeed, Lissitzky extended the Proun literally into the third dimension in 1923 with his *Prounenraum (Proun Room)* (color plate 115, page 208), consisting of painted walls and wood reliefs in a room that the viewer was to walk through in a counterclockwise direction. The artist wanted the walls to dissolve visually to allow the Proun elements to activate the space. The room was destroyed, but was reconstructed from Lissitzky's original documents for an exhibition in the Netherlands in 1965.

Lissitzky left Moscow in 1921 for Berlin, where he was associated with the Dutch abstractionist Theo van Doesburg and the Hungarian László Moholy-Nagy. He was one of the key artists who brought together Russian Suprematism and Constructivism, Dutch de Stijl, and the German Bauhaus and who, through Moholy-Nagy, later transmitted these ideas to a generation of students in the United States and elsewhere. In 1925 Lissitzky resettled permanently in Moscow, where in the 1930s he became an effective propagandist for the Stalinist regime. Once he abandoned abstraction, he made photographs as well as typographic, architectural and exhibition designs. Among his most memorable images is a photographic self-portrait from 1924, *The Constructor* (fig. 248). In a double exposure, the artist's face is superimposed on an image of his hand holding a compass over grid paper—a reminder of his role as an architect.

As we saw in our discussion of Der Blaue Reiter (see chapter 8), **VASILY KANDINSKY** (1866–1944) was a great figure in the transmission of Russian experiments in abstraction and construction to the West. Forced by the outbreak of war to leave Germany, he went back to Russia in 1914. In the first years after the Russian Revolution, the new Soviet government actively encouraged experimentalism and new forms in the arts to coincide with the new society communism was attempting to construct. In 1918 Kandinsky was invited by Tatlin to join the Department of Visual Arts (IZO) of Narkompros (NKP, the People's Commissariat for Enlightenment) in Moscow, and subsequently helped to reorganize Russian provincial museums. He remained in revolutionary Russia for seven years but eventually found his spiritual conception of art coming into conflict with the utilitarian doctrines of the ascendant Constructivists. In 1921 he left the Soviet Union for good, conveying many of his and his colleagues' innovations to the new Bauhaus School in Weimar, Germany (see chapters 16, 17).

Meanwhile, until 1920, Kandinsky continued to paint in the manner of free abstraction that he had first devised during the period 1910–14 (color plate 75 and color plate 76, page 160). That year he began to introduce, in certain paintings, regular shapes and straight or geometrically curving lines. During 1921 the geometric patterns began to dominate, and the artist moved into another major phase of his career. There can be no doubt that Kandinsky had been affected by the geometric abstractions of Malevich, Aleksandr Rodchenko, Tatlin, and the Constructivists. Despite the change from free forms to color shapes with smooth, hard edges, the tempo of the paintings remained rapid, and the action continued to be the conflict of abstract forms. *White Line, No. 232* (color plate 116, page 257) is a transitional work: the major color areas are still painted in a loose, atmospheric manner, but they are accented by sharp, straight lines and curved forms in strong colors.

Russian Constructivism

The word *Konstruktivizm* (Constructivism) was first used by a group of Russian artists in the title of a small 1922 exhibition of their work in Moscow. It is a term that has been applied broadly in a stylistic sense to describe a Cubist-based art that was developed in many countries. In general, that art is characterized by abstract, geometric forms and a technique in which various materials, often industrial in nature, are assembled rather than carved or modeled. But Constructivism originally referred to a movement of Russian artists after the revolution of 1917 who enlisted art in the service of the new Soviet system. These artists believed that a full integration of art and life would help foster the ideological aims of the new society and enhance the lives of its citizens. Such utopian ideals were common to many modernist movements, but only in Russia were the revolutionary political regime and the revolution in art so closely linked. The artists not only made constructed objects, but were major innovators in such areas as typographic design, textiles, furniture, and theatrical design (color plate 117 and color plate 118, page 258; fig. 255).

Constructivism was one of the significant new concepts to develop in twentieth-century sculpture. From the beginning of its history, sculpture had involved a process of creating

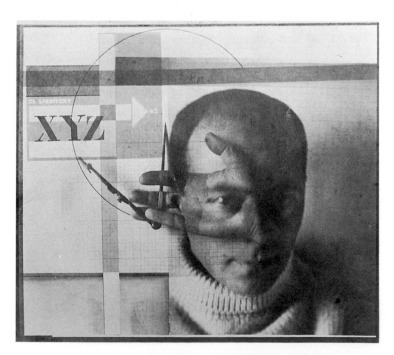

248. EL LISSITZKY. *The Constructor (Self-Portrait)*. 1924.
Gelatin-silver print, 4½ x 4⅞" (11.3 x 12.5 cm).
Stedelijk Van Abbemuseum, Eindhoven, the Netherlands

form by taking away from the amorphous mass of the raw material (the carving of wood or stone), or by building up the mass (modeling in clay or wax, that would later be cast in metal). These approaches presuppose that sculpture is mainly an art of mass rather than of space. Traditional techniques persisted well into the twentieth century, even in the work of so revolutionary a figure as Brancusi. The first Cubist sculpture of Picasso, the 1909 *Woman's Head* (see fig. 207), with its deep faceting of the surface, still respected the central mass.

True constructed sculpture, in which the form is assembled from elements of wood, metal, plastic, and other materials such as found objects, was a predictable consequence of the Cubists' experiments in painting. Even before Picasso, however, Braque made Cubist construction sculptures from pieces of paper and cardboard. Though the works have not survived, an early photograph shows a constructed still life mounted across a corner of the artist's studio (see fig. 214). It is tantalizing (but speculative) to think that Tatlin may have seen such a work during his trip to Paris, before embarking on his Counter-Reliefs. Picasso's 1912 *Guitar* and 1913 *Mandolin and Clarinet* (see figs. 215, 216) are groundbreaking Cubist constructions of cardboard, wire, string, and wood.

The subsequent development of constructed sculpture, particularly in its direction toward complete abstraction, took place outside France. Boccioni's Futurist sculpture manifesto of 1912 recommended the use of unorthodox materials, but his actual constructed sculpture remained tied to literal or Cubist subjects. Archipenko's constructed *Médrano* figures (color plate 97, page 201), executed between 1912 and 1914, were experiments in space-mass reversals, but the artist never deserted the subject—figure or still life—and soon reverted to a form of Cubist sculpture modeled in clay for casting in bronze.

In France and Italy the traditional techniques of sculpture—modeling, carving, bronze casting—were probably too powerfully imbedded to be overthrown, even by the leaders of the modern revolution. Sculptors who had attended art schools—Brancusi and Lipchitz for example—were trained in technical approaches unchanged since the eighteenth century. Modern sculpture there emerged from the Renaissance tradition more gradually than modern painting. Possibly as a consequence of this evolutionary rather than revolutionary process, even the experimentalists continued to utilize traditional techniques. The translation of Cubist collage into three-dimensional abstract construction was achieved in Russia first, then in Holland and Germany.

The founder of Russian Constructivism was **VLADIMIR TATLIN** (1895–1953). In 1914 he visited Berlin and Paris, where he saw Cubist paintings in Picasso's studio, as well as the constructions in which Picasso was investigating the implications of collage for sculpture. The result, on Tatlin's return to Russia, was a series of reliefs constructed from wood, metal, and cardboard, with surfaces coated with plaster, glazes, and broken glass. His exhibition of these works in his studio was among the first manifestations of Constructivism, just as the reliefs were among the first com-

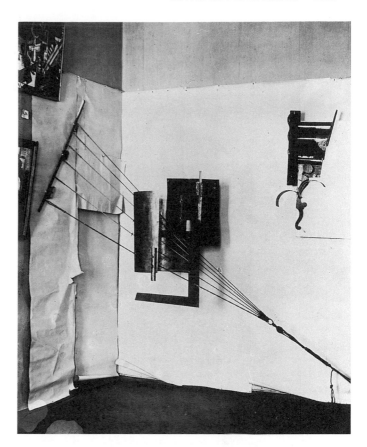

249. VLADIMIR TATLIN. *Counter-Relief*. 1915. Iron, copper, wood, and rope, 28 x 46½" (71 x 118 cm). Reconstruction. State Russian Museum, St. Petersburg
© 1997 ESTATE VLADIMIR TATLIN/VAGA, NEW YORK, NY

plete abstractions, constructed or modeled, in the history of sculpture. Tatlin's constructions, like Malevich's Suprematist paintings, were to exert a profound influence on the course of Constructivism in Russia.

Most of Tatlin's first abstract reliefs have disappeared, and the primary record of them is a number of drawings and photographs that document his preoccupation with articulating space. The so-called Counter-Reliefs, begun in 1914, were released from the wall and suspended by wires across the corner of a room (again, the location in Russian homes for icons), as far removed from the earthbound tradition of past sculpture as the technical resources of the artist permitted. One of these reliefs (fig. 249) has been assembled from Tatlin's original parts in St. Petersburg's State Russian Museum. Because the reliefs are made from ordinary materials, rather than traditional sculptural media such as bronze or marble, and because they are not isolated on a base, they tend to inhabit the space of the viewer more directly than conventional sculpture.

Tatlin developed a repertoire of forms in keeping with what he believed to be the properties of his chosen materials. According to the principles of what he called the "culture of materials" and "truth to materials," each substance, through its structural laws, dictates specific forms, such as the flat geometric plane of wood, the curved shell of glass, and the rolled cylinder or cone of metal. For a work of art to have significance, Tatlin came to believe that these principles must be

250. VLADIMIR TATLIN. Model for *Monument to the Third International.* 1919–20. Wood, iron, and glass, 20' (6.09 m). State Russian Museum, St. Petersburg
© 1997 ESTATE VLADIMIR TATLIN/VAGA, NEW YORK, NY

251. ALEKSANDR RODCHENKO. *Construction of Distance.* 1920. Wood. Private collection

considered in both the conception and the execution of the work, which would then embody the laws of life itself.

Like many avant-garde artists, Tatlin embraced the Russian Revolution. Thereafter, he cultivated his interest in engineering and architecture, an interest that saw its most ambitious result in his twenty-foot high model for a *Monument to the Third International* (fig. 250) which was exhibited in Petrograd (St. Petersburg) and Moscow in December 1920. Had the full-scale project been built, it would have been approximately thirteen hundred feet high, much taller than the Eiffel Tower and the biggest sculptural form ever conceived at that time. It was to have been a metal spiral frame tilted at an angle and encompassing a glass cylinder, cube, and cone. These glass units, housing conferences and meetings, were to revolve, making a complete revolution once a year, once a month, and once a day, respectively. The industrial materials of iron and glass and the dynamic, kinetic nature of the work symbolized the new machine age. The tower was to function as a propaganda center for the Communist Third International, an organization devoted to the support of world revolution, and its rotating, ascending spiral form was a symbol of the aspirations of communism and, more generally, of the new era. It anticipated, and in scale transcended,

all subsequent developments in constructed sculpture encompassing space, environment, and motion and has come to embody, probably more than any other invention, the ideals of Constructivism.

After the consolidation of the Soviet system in the 1920s, Tatlin readily adapted his nature-of-materials philosophy to the concept of production art, which held that in the classless society art should be rational, utilitarian, easily comprehensible, and socially useful, both aesthetically and practically. At its best this idealist doctrine inspired artists to envision a world in which even the most mundane objects would be beautifully designed. But as the Soviet Union grew intolerant of radical art, mandating the banalities of the Socialist Realist style, the ideals of production art turned to dogma. This doctrine worked tragically against such spiritually and aesthetically motivated artists as Malevich and Kandinsky, driving the latter out of Russia altogether. Tatlin, dedicated to Constructivist principles, went on to direct various important art schools and enthusiastically applied his immense talent to designing workers' clothing, furniture, and even a Leonardesque flying machine called the *Letatlin.*

By 1915–16, **ALEKSANDR RODCHENKO** (1891–1956) had become familiar with the work of Malevich and Tatlin and he soon began to make abstract paintings and to experiment with constructions. By 1920 he, like Tatlin, was turning more and more to the idea that the artist could serve the revolution through a practical application of art in engineering, architecture, theater, and industrial and graphic design. In his

252. ALEKSANDR RODCHENKO. *Hanging Construction*. 1920.
Wood. Location unknown

253. ALEKSANDR RODCHENKO. *Assembling for a Demonstration*.
1928. Gelatin-silver print, 19½ x 13⅞" (49.5 x 35.3 cm).
The Museum of Modern Art, New York
MR. AND MRS. JOHN SPENCER FUND.
COPY PRINT © 1995 THE MUSEUM OF MODERN ART, NEW YORK

Construction of Distance (fig. 251) he massed rectangular blocks in a horizontal-vertical grouping as abstract as a Neo-Plastic painting by Mondrian (color plate 201, page 389) and suggestive of the developing forms of the International Style in architecture (see chapter 16). Moreover, Rodchenko's example later proved highly relevant to Minimalist artists working in the United States in the 1960s (see fig. 699). His *Hanging Construction* (fig. 252) is a nest of concentric circles, which move slowly in currents of air. The shapes, cut from a single piece of plywood, could be collapsed after exhibition and easily stored. Rodchenko also made versions (none of which survives) based on a triangle, a square, a hexagon, and an ellipse. This creation of a three-dimensional object with planar elements reveals the Constructivists' interest in mathematics and geometry. It is also one of the first works of constructed sculpture to use actual movement, in a form suggestive of the fascination with space travel that underlay many of the ideas of the Constructivists. Apparently Rodchenko liked to shine lights on the constructions so that they would reflect off the silver paint on their surfaces, enhancing the sense of dematerialization in the work.

Rodchenko was ardently committed to the Soviet experiment. After 1921 he devoted himself to graphic, textile, and theater design. His advertising poster from 1924 (color plate 117, page 258) typifies the striking typographical innovations of the Russian avant-garde. Rodchenko also excelled at photography. Commenting on images like the one reproduced here, he wrote in 1928: "In photography there is the old

point of view, the angle of vision of a man who stands on the ground and looks straight ahead or, as I call it, makes 'belly-button' shots. . . . The most interesting angle shots today are those 'down from above' and 'up from below,' and their diagonals." By 1928 artists had long since discovered the value of the overhead perspective as a device for realizing a more abstract kind of image (see figs. 204, 434). In the vertiginous view of *Assembling for a Demonstration* (fig. 253) Rodchenko constructed a composition of sharp diagonals, light-dark contrasts, and asymmetrical patterns. Given the time and the place in which it was made, Rodchenko's photograph seems a metaphor for a new society where outdated perspectives have given way to dramatic new ones.

VARVARA FEDOROVNA STEPANOVA (1894–1958), like her husband, Rodchenko, gave up painting (at least until the 1930s) in order to devote herself to production art. This did not mean traditional decorative arts but rather functional materials manufactured in an equal partnership between artist and industrial worker. Within their Utopian framework, these new art forms were intended to aid in the creation of a new society. Tatlin's phrase "Art into Life" was the rallying cry of the Constructivists. Stepanova, who made designs for the state textile factory in Moscow, created striking fabrics in repetitive, geometric patterns that suited industrial printing methods. She designed clothing, as did Rodchenko and Tatlin, for the new man and woman (color plate 118, page 258), with an emphasis on comfort and ease of movement for the worker. The severe economic crisis that crippled Russia during the years of civil war following the revolution thwarted the implementation of many Constructivist goals. Not surprisingly, Stepanova's sophisticated designs, grounded as they were in a modernist sensibility, were received with greater enthusiasm when exhibited in Paris in 1925 than among the working rank and file in Moscow.

The artists who adopted the name Constructivist in 1921 worked in three-dimensional form, but the origin of the aesthetic in Tatlin's philosophy of materials was in the *faktura* (texture of paint) and the paint surface—its thickness, glossiness, technique of application. *Faktura* could be considered and treated as autonomous expression, as texture that generates specific forms. In this way the narrative function of figurative art was replaced by a self-contained system. As early as 1913 OLGA ROZANOVA (1886–1918) had asserted that the painter should "speak solely the language of pure plastic experience." In 1917, as if to illustrate the principle, she painted a remarkable picture (color plate 119, page 258), its composition simply a wide, lavishly brushed green stripe running up, or down, the center and cutting through a creamy white field of contrasted but equally strong scumbled texture. The result seems to reach across the decades to the 1950s and Barnett Newman's more monumental but scarcely more radical Zip paintings (color plate 251, page 458).

The Constructivist experiments of Tatlin, Rodchenko, and Stepanova came to an end in the early 1930s, as the Soviet government began to discourage abstract experiment in favor of practical enterprises useful to a struggling economy. Many of the Suprematists and Constructivists left Russia in the early 1920s. The most independent contributions of those who remained, including Tatlin and Rodchenko, were to be in graphic and theatrical design.

After the pioneering work in Russia, Constructivism developed elsewhere. The fact that artists like Kandinsky, Naum Gabo, and Anton Pevsner left Russia and carried their ideas to Western Europe was of primary importance in the creation of a new International Style in art and architecture.

The two leading Russian figures in the spread of Constructivism were the brothers ANTON PEVSNER (1886–1962) and NAUM GABO (1890–1977). Pevsner was first of all a painter whose history summarized that of many younger Russian artists. His exposure to nonacademic art first came about through his introduction to traditional Russian icons and folk art. He then discovered the Impressionists, Fauves, and Cubists in the Morozov and Shchukin collections. In Paris between 1911 and 1914, he knew and was influenced by Archipenko and Modigliani. Between 1915 and 1917 he lived in Norway with his brother Naum and on his return to Russia after the revolution taught at the Moscow Academy.

Naum Gabo (Naum Neemia Pevsner), who changed his name to avoid confusion with his elder brother, went to Munich in 1910 to study medicine but turned to mathematics and engineering. In Munich he became familiar with the scientific theories of Albert Einstein, among others, attended lectures by the art historian and critic Heinrich Wölfflin, and read Kandinsky's *Concerning the Spiritual in Art*. He left Germany when war broke out and settled for a time in Norway. There, in the winter of 1915–16, he began to make a series of heads and whole figures of pieces of cardboard or thin sheets of metal, figurative constructions transforming the masses of the head into lines or plane edges framing geometric voids. The interlocking plywood shapes that make up *Constructed Head No. 1* (fig. 254) establish the interpenetration of form and space without the creation of a surface or solid mass.

In 1917, after the Russian Revolution, Gabo returned to Russia with Pevsner. In Moscow he was drawn into the orbit of the avant-garde, meeting Kandinsky and Malevich and discovering Tatlin's constructions. He abandoned the figure and began to make abstract sculptures, including a motor-propelled kinetic object consisting of a single vibrating rod, as well as constructions of open geometric shapes in wood, metal, and transparent materials, as in *Column* (fig. 255). Originally conceived in 1920–21, these tower-shaped sculptures, like Tatlin's *Monument* (fig. 250), were part of Gabo's experiments for a visionary architecture. It is instructive to note that Ludwig Mies van der Rohe's model for a glass skyscraper (see fig. 417) was made in Germany at virtually the same time.

Between 1917 and 1920 the hopes and enthusiasms of the Russian experimental artists were at their peak. Most of the abstract artists were initially enthusiastic about the revolution, hoping that from it would come the liberation and triumph of progressive art. By about 1920, however, Tatlin and the group around him had become increasingly doctrinaire in their insistence that art should serve the revolution in specific, practical ways. Under government pressure, artists had to abandon or subordinate pure experiment in painting and sculpture and turn their energies to engineering, industrial and product design.

Gabo's reaction to these developments was recorded in the *Realistic Manifesto*. It was signed by Gabo and Pevsner and distributed in August 1920 at a Moscow exhibition of their work, but Gabo drafted it and was principally responsible for the ideas it contained. In many ways the manifesto was the culmination of ideas that had been fermenting in the charged atmosphere of Russian abstract art over the previous several years. At first a supporter of the revolutionary regime, Gabo found himself increasingly at odds with those members of the avant-garde who denounced art in favor of utilitarian objects to aid in the establishment of the socialist state. In the *Realistic Manifesto*, in which the word "realistic" refers to the creation of a new, Platonic reality more absolute than any imitation of nature, Gabo distinguished between his idealistic though "politically neutral" art and the production art of

254. NAUM GABO. *Constructed Head No. 1*. 1915. Reassembled 1985. Triple-layered plywood, height 21¼" (54 cm). Städelsches Kunstinstitut und Städtische Galerie, Frankfurt

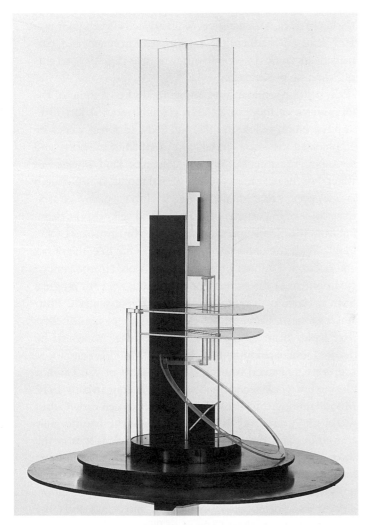

255. NAUM GABO. *Column*. c. 1923, reconstructed 1937. Wood, painted metal, and glass (later replaced with Perspex), 41½ x 29 x 29" (105.3 x 73.6 x 73.6 cm). Solomon R. Guggenheim Museum, New York

Tatlin, Rodchenko, and others. With revolutionary fervor, he proclaimed that the art of the future would surpass what he regarded as the limited experiments of the Cubists and Futurists. He called on artists to join forces with scientists and engineers to create a sculpture whose powerful kinetic rhythms embodied "the renascent spirit of our time." "The most important idea in the manifesto," Gabo later wrote, "was the assertion that art has its absolute, independent value and a function to perform in society whether capitalistic, socialistic, or communistic—art will always be alive as one of the indispensable expressions of human experience and as an important means of communication."

Until 1921 vanguard artists were allowed the freedom to pursue their new experiments. But as civil war abated, the Soviet state began to impose its doctrine of Socialist Realism. The result was the departure from Russia of Kandinsky, Gabo, Pevsner, Lissitzky, Chagall, and many other leading spirits of the new art throughout the 1920s who, as we shall see in chapter 17, continued to develop their ideas in the West.

De Stijl in the Netherlands

World War I marked the terminal point of the first waves of twentieth-century invention and interrupted or ended the careers of many artists. During the war, nations were isolated and the French ceased to dominate experimentation in art. A striking instance of the effect of this isolation is to be seen in the notable native development of art in the Netherlands, neutral during World War I and thus culturally as well as politically removed from both sides.

In general, Dutch artists and architects seem to have moved rather cautiously into the twentieth century. In architecture, Hendrik Berlage was the first important innovator and, as such, almost alone (color plate 123, page 260). International Art Nouveau had less impact in Holland than in Belgium, Austria, or France, except for a few designers and painters. So the bold, abstract work that was produced during World War I by the modernist group known as de Stijl (the Style) seems all the more remarkable. The genius of this movement was Piet Mondrian; other members included the Dutch painters Theo van Doesburg and Bart van der Leck, the Hungarian painter Vilmos Huszar, the Belgian sculptor Georges Vantongerloo, and the architects Gerrit Rietveld, Cornelis van Eesteren, J. J. P. Oud (see figs. 285, 286), and Robert van 't Hoff (see fig. 284).

Van Doesburg was the leading spirit in the formation and promotion of the group and the creation of its influential journal, also called *de Stijl*, devoted to the art and theory of the group and published from 1917 to 1928. De Stijl was dedicated to the "absolute devaluation of tradition . . . the exposure of the whole swindle of lyricism and sentiment." The artists involved emphasized "the need for abstraction and simplification": mathematical structure as opposed to Impressionism and all "Baroque" forms of art. They created art "for clarity, for certainty, and for order." Their works began to display these qualities, transmitted through the straight line, the rectangle, or the cube, and eventually through colors simplified to the primaries red, yellow, and blue and the neutrals black, white, and gray. For van Doesburg and Mondrian these simplifications had symbolic significance based on Eastern philosophy and the mystical teachings of Theosophy, a popular spiritual movement that was known among de Stijl artists partly through Kandinsky's book, *Concerning the Spiritual in Art.* Mondrian was particularly obsessed by the mystical implications of vertical-horizontal opposition and spent the rest of his life exploring them, producing in the process some of the most extraordinary works of art of the twentieth century. Despite van Doesburg's efforts to present the de Stijl group as a unified and coherent entity, it included many individual talents who did not strictly adhere to a single style. Nevertheless, the artists, designers, and architects of de Stijl shared ideas about the social role of art in modern society, the integration of all the arts through the collaboration of artists and designers, and an abiding faith in the potential of technology and design to realize new Utopian living environments based on abstract form. Together they developed an art based rigorously on theory, dedicated to formal purity, logic, balance, proportion, and rhythm.

The de Stijl artists were well aware of parallel developments in modern art in France, Germany, and Italy—Fauvism, Cubism, German Expressionism, and Italian Futurism. But they recognized as leaders only a few pioneers, such as Cézanne in painting and Frank Lloyd Wright and Berlage in architecture. They had little or no knowledge of the Russian experiments in abstraction until the end of the war, when international communications were reestablished. At that time, contact developed between the Dutch and Russian avant-garde, especially through El Lissitzky.

In addition to their work as painters, Mondrian and van Doesburg were influential theoreticians. In 1917 van Does-

burg published a book, *The New Movement in Painting,* and Mondrian published in *de Stijl* a series of articles, including "Neo-Plasticism in Painting," which in 1920, after his return to France, he expanded into his book *Le Néo-Plasticisme,* one of the key documents of abstract art. The term refers to the abstract style he had developed by that year.

By 1926 van Doesburg was seeking a new and individual variation on de Stijl, which he named Elementarism. In it he continued to use a composition consisting of rectangles, but tilted them at forty-five degrees to achieve what he felt to be a more immediately dynamic form of expression. At this point, differences between the two artists were so significant that Mondrian left the de Stijl group.

The transition from fairly conventional naturalistic paintings to a revolutionary modern style during and after World War I may be traced in the career of **PIET MONDRIAN** (1872–1944). Born Pieter Mondriaan, he was trained in the Amsterdam Rijksacademie and until 1904 worked primarily as a landscape painter. He then came under the influence of Toorop and for a time painted in a Symbolist manner. His early work evinces a tendency to work in series—which proved central to the development of his abstract work—and to focus on a single scene or object, whether a windmill, thicket of trees, a solitary tree, or an isolated chrysanthemum. Early landscapes adhered to a principle of frontality and, particularly in a series of scenes with windmills (fig. 256), to cut-off, close-up presentation.

By 1908 Mondrian was becoming aware of some of the innovations of modern art. His color blossomed in Fauve-inspired blues, yellows, pinks, and reds. In forest scenes he emphasized the linear undulation of saplings; in shore- and seascapes, the intense, flowing colors of sand dunes and water. For the next few years he painted motifs such as church facades presented frontally, in nearly abstract planes of arbitrary color or in patterns of loose red and yellow spots deriving from the Neo-Impressionists and the brushstrokes of Van Gogh. With any of his favorite subjects—the tree, the dunes and ocean, the church or windmill, all rooted in the familiar environment of the Netherlands—one can trace his progress from naturalism through Symbolism, Impressionism, Post-Impressionism, Fauvism, and Cubism to abstraction. In *Blue Tree* (color plate 120, page 259), an image to which he devoted several paintings and drawings, Mondrian employed expressive, animated brushwork reminiscent of Van Gogh's, causing the whole scene to pulsate with energy. In early 1912 Mondrian moved to Paris. Though he had seen early Cubist works by Picasso and Braque in Holland, he became fully conversant with the style in the French capital. It is in the wake of this experience that he emerges as a major figure in modern art. He himself came to regard the previous years as transitional. During his first years in Paris, Mondrian subordinated his colors to grays, blue-greens, and ochers under the influence of the Analytic Cubism of Picasso and Braque. But he rarely attempted the tilted planes or sculptural projection that gave the works of the French Cubists their defined, if limited, sense of three-dimensional spatial existence; his most

256. PIET MONDRIAN. *Mill by the Water.* c. 1905.
Oil on canvas mounted on cardboard,
11⅞ x 15" (30.2 x 38.1 cm).
The Museum of Modern Art, New York PURCHASE

257. PIET MONDRIAN. *Flowering Apple Tree.* 1912.
Oil on canvas, 31 x 42⅜" (78.5 x 107.5 cm).
Gemeente Museum, the Hague, the Netherlands

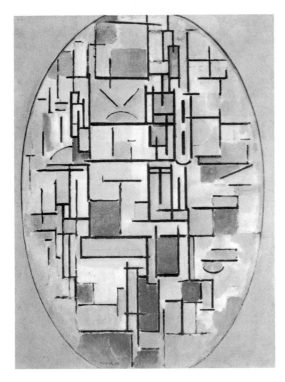

258. PIET MONDRIAN. *Color Planes in Oval.*
1913–14. Oil on canvas, 42⅜ x 31" (107.6 x
78.7 cm). The Museum of Modern Art, New York
PURCHASE.
© 1997 MONDRIAN HOLTZMAN TRUST

Cubist paintings still maintained an essential frontality. Mondrian was already moving beyond the tenets of Cubism to eliminate both subject and three-dimensional illusionistic depth.

As early as 1912 the tree had virtually disappeared into a linear grid that covered the surface of the canvas (fig. 257). Mondrian at this time favored centralized compositions, evident here in the central density of the pattern, which gradually loosens toward the edges. He also articulated this through an oval canvas (inspired by the Cubists), while the linear structure became more rectangular and abstract, as in *Color Planes in Oval* of 1913–14 (fig. 258). Despite its highly abstracted forms, this work belongs to a series of compositions Mondrian based on his drawings of Parisian building facades. By 1914 the artist had begun to experiment with a broader but still subtle range of colors, asserting their identity within a structure of horizontal and vertical lines.

In Paris Mondrian was profoundly affected by the example of Cubism, but he gradually began to feel that the style "did not accept the logical consequences of its own discoveries: it was not developing abstraction toward its ultimate goal, the expression of pure reality. I felt that this could only be established by pure plastics (plasticism)." In this statement, made in 1942, he emphasized the two words that summarize his lifelong quest—"plastic" and "reality". To him "plastic expression" meant simply the action of forms and colors. "Reality" or "the new reality" was the reality of plastic expression, or the reality of forms and colors in the painting. Thus, the new reality was the presence of the painting itself, as opposed to an illusionistic reality based on imitation of nature or romantic or expressive associations.

Gradually, as the artist tells it, Mondrian became aware that "(a) in plastic art reality can be expressed only through the equilibrium of *dynamic movements* of form and color; (b) pure means afford the most effective way of attaining this." These ideas led him to develop a set of organizational principles in his art. Chief among them were the balance of unequal opposites, achieved through the right angle, and the simplification of color to the primary hues plus black and white. It is important to recall that Mondrian did not arrive at his final position solely through theoretical speculation, but through a long and complex development in his painting.

In 1914 he returned to Holland, where he remained when war broke out. Between 1914 and 1916 he eliminated all vestiges of curved lines, so that the structure became predominantly vertical and horizontal. The paintings were still rooted in subject—a church facade, the ocean, and piers extending into the ocean—but these were now simplified to a pattern of short, straight lines, like plus and minus signs, through which the artist sought to suggest the underlying structure of nature. During 1917 and later he explored another variation (color plate 121, page 259)—rectangles of flat color of varying sizes, suspended in a sometimes loose, sometimes precise rectangular arrangement. The color rectangles sometimes touch, sometimes float independently, and sometimes overlap. They appear positively as forms in front of the light background. Their interaction creates a surprising illusion of depth and movement, even though they are kept rigidly parallel to the surface of the canvas.

Mondrian soon realized that these detached color planes created both a tangible sense of depth and a differentiation of foreground and background; this interfered with the pure reality he was seeking. This discovery led him during 1918 and 1919 to a series of works organized on a strict rectangular grid. In the so-called Checkerboards (fig. 259), rectangles of equal size and a few different colors are evenly distributed across the canvas. By controlling the strength and tone of his colors, Mondrian here neutralized any distinction between figure and ground, for the white and gray rectangles do not recede behind the colored ones or assume a subordinate role as support. Mondrian did not develop this compositional

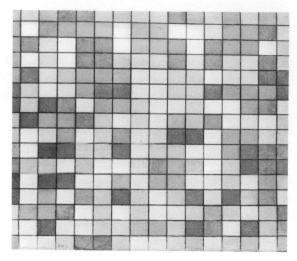

259. PIET MONDRIAN. *Composition with Grid 9;*
Checkerboard with Light Colors. 1919. Oil on canvas,
33⅞ x 41¾" (86 x 106 cm).
Gemeente Museum, the Hague, the Netherlands
© 1997 MONDRIAN HOLTZMAN TRUST

260. THEO VAN DOESBURG. *Card Players.*
1916–17. Tempera on canvas,
46½ x 58" (118.1 x 147.3 cm).
Private collection
© 1997 ARTISTS RIGHTS SOCIETY (ARS), NEW YORK/BEELDRECHT, AMSTERDAM

solution further, for he felt the modular grid was too prominent. Finally, therefore, he united the field of the canvas by thickening the dividing lines and running them through the rectangles to create a linear structure in tension with the color rectangles. In 1920, after returning to Paris, Mondrian came to the fulfillment of his Neo-Plastic artistic ideals. His paintings of this period express abstract, universal ideas: the dynamic balance of vertical and horizontal linear structure and simple, fundamental color. He continued to refine these ideas throughout the 1920s, 1930s, and early 1940s, until, as we shall see in chapter 17, he achieved works of monumental purity and simplicity (color plate 200, page 388; color plate 202, page 389). Mondrian's ultimate aim was to express a visual unity through an "equivalence of opposites"; this in turn expressed the higher mystical unity of the universe.

As already noted, THEO VAN DOESBURG (1883–1931) was the moving spirit in the formation and development of de Stijl. During his two years in the army (1914–16) he studied the new experimental painting and sculpture and was particularly impressed by Kandinsky's essay, *Concerning the Spiritual in Art.* In 1916 he experimented with free abstraction in the manner of Kandinsky, as well as with Cubism, but was still searching for his own path. This he found in his composition *Card Players* (fig. 260), based on Cézanne's painting (see fig. 73), but simplified to a complex of interacting shapes based on rectangles, the colors flat and reduced nearly to primaries. Fascinated by the mathematical implications of his new abstraction, van Doesburg explored its possibilities in linear structures, as in a later version of *Card Players* (fig. 261). Even Mondrian was affected by the fertility of van Doesburg's imagination. When artists work together as closely as the de Stijl painters did during 1917–19, it is extremely difficult and perhaps pointless to establish absolute priorities. Van Doesburg, Mondrian, and their colleague van der Leck for a time were all nourished by one another. However, each was beginning to go in his own direction in the 1920s, when

van Doesburg's attention turned toward architecture. He followed de Stijl principles until he published his *Fundamentals of the New Art* in 1924. He then began to abandon the rigid vertical-horizontal formula of Mondrian and de Stijl and to introduce diagonals. This heresy as well as fundamental differences over the nature of Neo-Plastic architecture led to Mondrian's resignation from de Stijl. Though by 1918 Mondrian himself made lozenge-shaped paintings with diagonal edges (see fig. 450), a development he discussed at length with van Doesburg, the lines contained within his compositions always remained strictly horizontal and vertical. In a 1926 manifesto in *de Stijl,* van Doesburg named his new departure Elementarism, and argued that the inclined plane reintroduced surprise, instability, and dynamism. In his murals at the Café l'Aubette, Strasbourg (fig. 262)—decorated in collaboration with Jean Arp and Sophie Taeuber-Arp (see fig. 299)—van Doesburg made his most monumental statement of Elementarist principles. He tilted his colored rectangles at forty-five degree angles and framed them in uniform strips of color. The tilted rectangles are in part cut off by the ceiling and lower wall panels. Across the center runs a long balcony with steps at one end that add a horizontal and diagonal to the design. Incomplete rectangles emerge as triangles or irregular geometric shapes. In proper de Stijl fashion van Doesburg designed every detail of the interior, down to the ash trays. Here he realized his ideals about an all-embracing, total work of abstract art, saying that his aim was "to place man within painting instead of in front of it and thereby enable him to participate in it."

In the richly diverse and international art world of the 1920s, van Doesburg provided a point of contact between artists and movements in several countries. He even maintained a lively interest in Dada. His Dada activities included his short-lived Dada magazine *Mecano,* his Dada poems written under the pseudonym of I. K. Bonset, and a Dada cultural tour of Holland with his friend Kurt Schwitters.

In sculpture the achievements of de Stijl were not comparable to those of the Russian Constructivists and were, in fact, concentrated principally in the works of the Belgian **GEORGES VANTONGERLOO** (1886–1965). Vantongerloo was not only a sculptor but also a painter, architect, and theoretician. His first abstract sculptures (fig. 263), executed during 1917 and 1918, were conceived in the traditional sense as masses carved out of the block, rather than as constructions built up of separate elements. They constitute notable transformations of de Stijl painting into three-dimensional design. Later, Vantongerloo turned to open construction, sometimes in an architectural form and sometimes in free linear patterns. In his subsequent painting and sculpture he frequently deserted the straight line in favor of the curved, but throughout his career Vantongerloo maintained an interest in a mathematical basis for his art, to the point of deriving compositions from algebraic equations.

Until his death in 1931, van Doesburg promoted de Stijl abroad, traveling across Europe and seeking new adherents to the cause as older members defected. His efforts contributed to the establishment of de Stijl as a movement of international significance. With its belief in the integration of the fine and applied arts, the de Stijl experiment paralleled that of the Bauhaus in many ways. Van Doesburg was in Weimar in the early 1920s, lecturing and promoting de Stijl ideas at the Bauhaus, and fomenting dissent among the school's younger members. Although he was not a member of the faculty, he probably contributed to an increased emphasis at the Bauhaus on rational, machine-based design. (On the Bauhaus see chapters 16 and 17.)

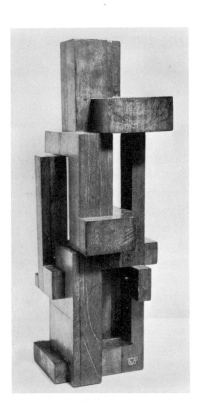

left: 261. THEO VAN DOESBURG. *Composition IX (Card Players)*. 1917. Oil on canvas, 45⅝ x 41¾" (115.9 x 106 cm). Gemeente Museum, the Hague, the Netherlands
© 1997 ARTISTS RIGHTS SOCIETY (ARS), NEW YORK/BEELDRECHT, AMSTERDAM

below: 262. THEO VAN DOESBURG, SOPHIE TAEUBER, and JEAN ARP. Interior, Café l'Aubette, Strasbourg. 1926–28, (destroyed 1940)

263. GEORGES VANTONGERLOO. *Construction of Volume Relations*. 1921. Mahogany, 16⅛ x 5⅝ x 5¾" (41 x 14.4 x 14.5 cm), including base. The Museum of Modern Art, New York
GIFT OF SILVIA PIZITZ. PHOTOGRAPH © 1995 THE MUSEUM OF MODERN ART, NEW YORK

12 Early Twentieth-Century Architecture

The United States

FRANK LLOYD WRIGHT (1867–1959), the most important American architect of the first half of the twentieth-century, studied engineering at the University of Wisconsin, where he read the work of the English critic John Ruskin and was particularly drawn to rational, structural interpretation in the writings of Eugène-Emmanuel Viollet-le-Duc. In 1887 he was employed by the Chicago architectural firm of Adler and Sullivan (color plate 38, page 113), with whom he worked until he established his own practice in 1893. There is little doubt that many of the houses built by the Sullivan firm during the years Wright worked there represented his ideas. Wright's basic philosophy of architecture was expressed primarily through the house form. The 1902–6 Larkin Building in Buffalo, New York (figs. 268, 269), was his only large-scale structure prior to Chicago's Midway Gardens (1914) and the Imperial Hotel in Tokyo (fig. 271). All three, incidentally, have been destroyed.

At the age of twenty-two Wright designed his own house in Oak Park, Illinois (1889), then a quiet community thirty minutes by train from downtown Chicago. His earliest houses, including his own, reflect influences from the shingle-style houses of H. H. Richardson and McKim, Mead & White (see figs. 92, 93) and developed the open, free-flowing floor plan of the English architects Philip Webb and Richard Norman Shaw. Wright used the characteristically American feature of the veranda, open or screened, wrapping around two sides of the house, to enhance the sense of outside space that penetrated to the main living rooms. Also, the cruciform plan, with space surrounding the central core of fireplace and utility areas (kitchen, landing, etc.), had an impact on Wright that affected his approach to house design as well as to more monumental design projects.

In the 1902–3 Ward Willits House in Highland Park, Illinois (fig. 264), Wright made one of his first individual and mature statements of the principles and ideas he had been formulating during his apprentice years. The house demonstrates his growing interest in a Japanese aesthetic. He was a serious collector of Japanese prints (about which he wrote a

book), and before his trip to Japan in 1905, he probably visited the Japanese pavilions at the 1893 World's Columbian Exhibition in Chicago. In the Willits House the Japanese influence is seen in the dominant wide, low-gabled roof and the vertical striping on the facade. The sources, however, are less significant than the welding together of all elements of the plan, interior and exterior, in a single integration of space and mass and surface. From the compact, central arrangement of fireplace and utility units, the space of the interior flows out in an indefinite expansion carried, without transition, to the exterior and beyond. The essence of the design in the Willits House and in the series of houses by Wright and his followers to which the name Prairie Style has been given, is a predominant horizontal accent of rooflines with deep, overhanging eaves echoing the flat prairie landscape of the Midwest. The earth tones of the prototypical Prairie house were intended to blend harmoniously with the surroundings, while the massive central chimney served both to break the horizontal, low-slung line of the roof and to emphasize the hearth as the spiritual and psychological center of the house.

The interior of the Wright Prairie house (figs. 267, 270) is characterized by low ceilings, frequently pitched at unorthodox angles, a cavelike sense of intimacy, and constantly changing vistas of one space flowing into another. The interior plastered walls of the Willits House were trimmed simply in

264. FRANK LLOYD WRIGHT. Ward Willits House, Highland Park, Illinois. 1902–1903

265. FRANK LLOYD WRIGHT. Robie House, Chicago. 1909

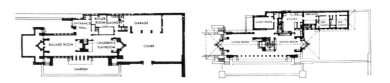

266. FRANK LLOYD WRIGHT. Plans of ground floor and first floor, Robie House

wood, imparting a sense of elegant proportion and geometric precision to the whole. Wright also custom-designed architectural ornaments for his houses, such as light fixtures, leaded glass panels in motifs abstracted from natural forms, and furniture, both built-in and freestanding (color plate 122, page 260). Though his emphasis on simple design and the honest expression of the nature of materials is dependent in part on Arts and Crafts ideals, Wright fervently supported the role of mechanized production in architectural design. He regarded the machine as a metaphor of the modern age but did not believe that buildings should resemble machines. Paramount in his house designs was the creation of a suitable habitat, in harmony with nature, for the prosperous, middle-class nuclear family. Wright understood the way in which his domestic dwellings embodied the collective values and identity of a community. These early residences were designed for Chicago's fast-growing suburbs, where structures could extend horizontally, as opposed to the city, where Victorian town houses were built on several stories to accommodate narrow urban plots. The sense of horizontal sprawl still present in suburban housing can be traced in part to Wright's innovations.

The masterpiece of Wright's Prairie Style is the 1909 Robie House in Chicago (figs. 265, 266, 267). The house is centered around the fireplace and arranged in plan as two sliding horizontal sections on one dominant axis. The horizontal roof cantilevers out on steel beams and is anchored at the center, with the chimneys and top-floor gables set at right angles to the principal axis. Windows are arranged in long, symmetrical rows and are deeply imbedded into the brick masses of the structure. The main, horizontally oriented lines

267. FRANK LLOYD WRIGHT. Dining Room, Robie House

of the house are reiterated and expanded in the terraces and walls that transform interior into exterior space and vice versa. The elements of this house, combining the outward-flowing space of the interior and the linear and planar design of exterior roofline and wall areas with a fortresslike mass of chimneys and corner piers, summarize other experiments that

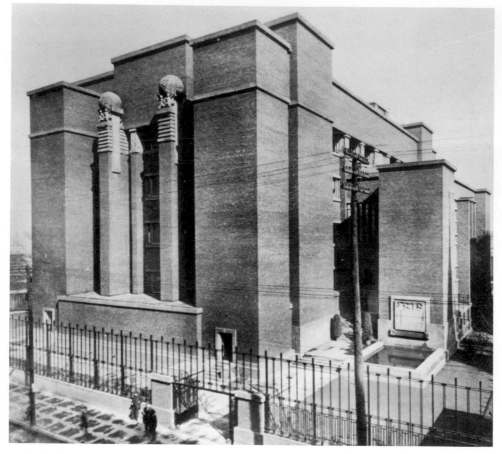

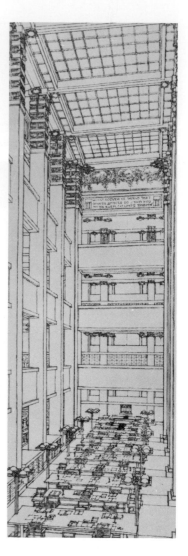

above: 268. FRANK LLOYD WRIGHT. Larkin Building, Buffalo, New York. 1902–6. Demolished 1950

right: 269. FRANK LLOYD WRIGHT. Interior, Larkin Building

Wright had carried on earlier in the Larkin Building (figs. 268, 269).

The Larkin Building represented radical differences from the Prairie house in that it was organized as rectangular masses in which Wright articulated mass and space into a single, close unity. The Larkin Building was, on the exterior, a rectangular, flat-roofed structure, whose immense corner piers protected and supported the window walls that reflected an open interior well surrounded by balconies. The open plan, filled with prototypes of today's "work stations," was Wright's solution for a commodious, light-filled office environment. The Larkin Building literally embodied the architect's belief in the moral value of labor, for its walls were inscribed with mottos extolling the virtues of honest work. This was one of those early modern industrial structures that embodied tremendous possibilities for the development of innovative kinds of internal, expressive space in the new tall buildings of America. The economics of industrial building soon destroyed these possibilities.

Wright was a pioneer of the international modern movement and historically, his experiments in architecture as organic space in the form of abstract design antedate those of most of the early twentieth-century European architectural pioneers. His designs had been published in Europe in 1910 and 1911, in two German editions by Ernst Wasmuth, and were studied by H. P. Berlage, Walter Gropius, and other architects. His works were known and admired by artists and architects of the Dutch de Stijl group, Robert van 't Hoff, J. J. P. Oud, Theo van Doesburg, Georges Vantongerloo, and Piet Mondrian. His design had common denominators not only with the classical formalism of these artists but also with the shifting planes and ambiguous space relationships of Cubism. On the other hand, Wright's work was influenced by what he saw during a long trip to Europe in 1909–10, particularly the architecture of the leading Vienna Secessionists, Joseph Maria Olbrich (figs. 279, 280) and Otto Wagner (fig. 277). At the very moment he was becoming a world figure, however, Wright was entering upon a period of neglect and even vilification in his own country. His highly publicized personal problems helped drive him from the successful midwestern practice he had built up in the early years of the century. These included his decision to leave his wife and six children for another woman, Mamah Cheney. In 1914 Cheney was murdered by a servant who also set fire to the home Wright had built for himself in Wisconsin, Taliesin (the name of a mythic Welsh poet meaning "shining brow"), which he subsequently rebuilt (fig. 270). Even more significant than these personal factors were cultural and social changes that, by 1915, had alienated the patronage for experimental architecture in the Midwest and increased the popu-

270. FRANK LLOYD WRIGHT. Interior, Taliesin,
Spring Green, Wisconsin. 1925

272. McKIM, MEAD & WHITE. Pennsylvania Station,
New York. 1906–10. Demolished 1966

271. FRANK LLOYD WRIGHT. Imperial Hotel, Tokyo.
c. 1912–23. Demolished 1968

273. McKIM, MEAD & WHITE. Interior train concourse,
Pennsylvania Station. Demolished 1966

larity of historical revivalist styles. Large-scale construction of mass housing soon led to vulgarization of these styles.

Wright himself, about 1915, began increasingly to explore the art and architecture of ancient cultures, including the Egyptian, Japanese, and Maya civilizations. The Imperial Hotel in Tokyo (fig. 271), a luxury hotel designed for Western visitors, occupied most of his time between 1912 and 1923 and represents his most ornately complicated decorative period, filled with suggestions of Far Eastern and Pre-Columbian influence. In addition, it embodied his most daring and intricate structural experiments to that date, experiments that enabled the building to survive the wildly destructive Japanese earthquakes of 1923 (only to be destroyed by the wrecking ball in 1968). For twenty years after the Imperial Hotel,

Wright's international reputation continued to grow. Though he frequently had difficulty earning a living, he was indomitable, and wrote, lectured, and taught, secure in the knowledge of his own genius and place in history, and invented brilliantly whenever he received commissions. Aside from Wright, a number of his followers, and a few isolated architects of talent, American architecture between 1915 and 1940 was largely in the hands of academicians and builders.

Nevertheless, two landmark structures were erected during this period on either coast. New York City's Pennsylvania Station (figs. 272, 273), built between 1906 and 1910 and demolished in 1966, represents one of the most tragic architectural losses of the twentieth century. "Until the first blow fell," said an editorial in *The New York Times*, "no one was

274. Bernard Maybeck. Palace of Fine Arts, Panama-Pacific International Exposition, San Francisco. 1915

convinced that Penn Station really would be demolished or that New York would permit this monumental act of vandalism." Penn Station was designed by the beaux-arts architectural firm of McKim, Mead & White, who designed classically inspired civic buildings throughout America in the early years of this century (see figs. 93, 98). Penn Station was one of the firm's most ambitious undertakings. The exterior (fig. 272), a massive Doric colonnade based on the ancient Roman Baths of Caracalla, presented a grand modern temple that underscored the power of the railroad as a symbol of progress. Inside, a visitor's first impression upon arrival was the dramatically vaulted spaces in the train concourse (fig. 273), its glass-and-steel construction recalling the crystal palaces of the previous century (see fig. 88). In 1966 the city responded to the destruction of Penn Station by establishing the New York City Landmarks Preservation Commission.

At the same time, on the other side of the continent, a remarkable act of preservation was taking place. The Palace of Fine Arts in San Francisco (fig. 274), designed by the architect **Bernard Maybeck** (1862–1957), was originally built in 1915 for the Panama-Pacific International Exposition. Celebrating the opening of the Panama Canal, the San Francisco Exposition was a world's fair that emulated the famous 1893 World's Columbian Exposition (see fig. 98). It was visited by nineteen million people during ten months in 1915, providing an economic boost to a city that had experienced the devastating earthquake and fire of 1906. Sited along a lagoon, Maybeck's popular ensemble included an open-air rotunda, a curving Corinthian peristyle, and a large gallery that housed the fair's paintings and sculptures. The unconventional Maybeck made classicism into his own personal idiom, violating canonical rules and covering his structures with richly imaginative classicizing ornament. He sought a mood of melancholy and past grandeur, for he envisioned the Palace as "an old Roman ruin, away from civilization . . . overgrown with bushes and trees." It was a mood

275. Auguste Perret. Apartment house, 25 Rue Franklin, Paris. 1902–3

somehow appropriate for a world in the midst of a bloody war. Though it was the only building not torn down after the fair, the Palace of Fine Arts was never intended as a permanent structure and it soon began to show signs of decay. Between 1962 and 1967 it was completely reconstructed out of concrete. Today it is still used as a city museum.

France

In continental Europe before 1914, the most significant architectural innovations were in Germany and Austria. These were fed by new developments, in painting and sculpture as well as in architecture, that originated in France, Holland, and Belgium. French architecture during this period was dominated by the Beaux-Arts tradition, except for the work of two architects of high ability, Auguste Perret and Tony Garnier. Both were pioneers in the use of reinforced concrete.

In his 1902–3 apartment building (fig. 275) **Auguste Perret** (1874–1954) covered a thin reinforced-concrete skeleton with glazed terra-cotta tiles decorated in an Art Nouveau foliate pattern. The structure is clearly revealed and allows for large window openings on the facade. The architect increased daylight illumination by folding the facade around a front wall and then arranging the principal rooms so that all had outside windows. The strength and lightness of the material also substantially increased openness and spatial flow.

Perret's masterpiece in ferroconcrete building is probably his Church of Notre Dame at Le Raincy, near Paris, built in

1922–23 (fig. 276). Here he used the simple form of the Early Christian basilica—a long rectangle with only a slightly curving apse, a broad, low-arched nave, and side aisles just indicated by comparably low transverse arches. Construction in reinforced concrete permitted the complete elimination of structural walls. The roof rests entirely on widely spaced slender columns, and the walls are simply constructed of stained glass (designed by the painter Maurice Denis) arranged on a pierced screen of precast-concrete elements. The church at Le Raincy remains a landmark of modern architecture not only in the effective use of ferroconcrete but also in the beauty and refinement of its design.

Austria and Germany

Experimental architecture in Spain, Britain, France, and the United States during the first part of the twentieth century has been traced in terms of the work of a few isolated architects of genius—Gaudí in Spain, Mackintosh in Britain, Perret and Garnier in France, and Sullivan and Wright in the United States. Of course there were also architects of high talent in other countries: Peter Behrens, Hans Poelzig, and others in Germany; Horta and van de Velde in Belgium; Wagner, Adolf Loos, Hoffmann, and Joseph Maria Olbrich in Austria; and Hendrik Petrus Berlage in the Netherlands. In Germany before 1930, largely as a result of enlightened governmental and industrial patronage, architectural experimentation, instead of depending only on brilliant individuals, was coordinated and directed toward the creation of a "school" of modern architecture. In this process, the contributions of certain patrons such as Archduke Ernst Ludwig of Hesse and the AEG (German General Electrical Company) were of the greatest importance.

In Vienna certain vanguard architects inveighed against the excesses of Art Nouveau in favor of clarity of proportion and simplicity of design, prompted in part by the ideals of the Arts and Crafts movement. OTTO WAGNER (1841–1918), the founder of Viennese modernism, was an academic architect during the early part of his life. His stations for the Vienna subway (1896–97) were simple and functional buildings dressed with Baroque details. But his 1894 book on modern architecture, Wagner had already demonstrated his ideas about a new architecture that used the latest materials and adapted itself to the requirements of modern life. His motto, "Necessity alone is the ruler of art," anticipated twentieth-century functionalism. In the hall of his 1905 Vienna Post Office Savings Bank (fig. 277) Wagner used unadorned metal and glass to create airy, light-filled, and unobstructed space. There is a certain stylistic similarity here to the auditorium that Victor Horta built in his Maison du Peuple in Brussels (see fig. 112), but also a considerable advance in the elimination of structural elements for the creation of a single unified and simple space.

left: 276. AUGUSTE PERRET. Church of Notre Dame, Le Raincy, France. 1922–23

277. OTTO WAGNER. Post Office Savings Bank, Vienna. 1905

ADOLF LOOS (1870–1933) was also active in Vienna early in the century. After studying architecture at Dresden, Loos worked in the United States for three years from 1893 (when he attended the World's Columbian Exposition in Chicago), taking various odd jobs while learning about the new concepts of American architecture, particularly the skyscraper designs of Sullivan and other pioneers of the Chicago School. Settling in Vienna in 1896, he followed principles established by Wagner and Sullivan favoring a pure and functional architecture and wrote extensively against the Art Nouveau decorative approach of the Vienna Secession. His 1910 Steiner House in Vienna (fig. 278) anticipated the unadorned cubic forms of the so-termed International Style of architecture that was to develop from the concepts of J. J. P. Oud, Walter Gropius, Ludwig Mies van der Rohe, and Le Corbusier in the 1920s and 1930s. Zoning rules allowed for only one story above street level, so Loos employed a barrel-vaulted roof on the front of the house that was so deep it allowed for two more levels facing the garden, which is the view seen here. The garden facade is symmetrical; simple, large-paned windows arranged in horizontal rows are sunk into the planar surfaces of the rectilinear facade. Reinforced concrete was here applied to a private house for almost the first time. Although the architecture and ideas of Loos never gained wide dissemination, the Steiner House is a key monument in the creation of the new style.

Among the Art Nouveau architects attacked by Loos were Olbrich and JOSEF HOFFMANN (1870–1956) who, with the painter Gustav Klimt, were the founders of the Vienna Secession movement that had broken with the powerful, academic art mainstream in Vienna. Hoffmann's masterpiece is the luxurious house he built for the Stoclet family in Brussels, the Palais Stoclet (see figs. 116, 117). Although this splendid mansion is characterized by severe rectangular planning and facades, and broad, clear, white areas framed in dark, linear strips (under the influence of Charles Rennie Mackintosh), its lavish interior design, an expression of the Secessionist belief in a total decorative environment, stands in stark contrast to the sobriety of Loos's Steiner House.

Hoffmann's importance as a pioneer is perhaps greater for his part in establishing the Wiener Werkstätte (Vienna Workshops) than for his achievement as a practicing architect. The workshops, which originated in 1903, continued the craft traditions of William Morris and the English Arts and Crafts movement, with the contradictory new feature that the machine was now accepted as a basic tool of the designer. For thirty years the workshops exercised a notable influence, teaching fine design in handicrafts and industrial objects. At about the same time, a cabinetmaker, Karl Schmidt, had started to employ architects and artists to design furniture for his shop in Dresden. Out of this grew the Deutsche Werkstätte (German Workshops), which similarly applied the principles of Morris to the larger field of industrial design. From these and other experiments, the Deutscher Werkbund (German Work or Craft Alliance) emerged in 1907. This was the immediate predecessor of the Bauhaus, one of the schools

278. ADOLF LOOS. Garden view, Steiner House, Vienna. 1910

279. JOSEPH MARIA OLBRICH.
Poster for the Second Exhibition of the Vienna Secession. 1898.
Color lithograph, 34 x 21" (86.3 x 53.3 cm).
Private collection

281. Peter Behrens. AEG Turbine Factory, Berlin. 1908–9

280. Joseph Maria Olbrich. Hochzeitsturm (Wedding Tower), Darmstadt, Germany. 1907

most influential in the development of modern architecture and industrial design.

Like Hoffmann, Joseph Maria Olbrich (1867–1908) was an original member of the Vienna Secession. In 1898 he designed the Secession Building, a structure of massive pylons covered with white stucco and surmounted by a dome of gilded laurel leaves. The building housed the second Secessionist exhibition and is featured on Olbrich's poster announcing the show (fig. 279). Relations between Austrian and German architecture were extremely close in the early years of the century. When Archduke Ernst Ludwig of Hesse wished to effect a revival of the arts by founding an artists' colony at Darmstadt in Germany, he employed Olbrich to design most of the buildings, including the Hochzeitsturm (Wedding Tower) (fig. 280). The Wedding Tower, which still dominates Darmstadt, was so named because it commemorated the archduke's second marriage; it was intended less as a functional structure than as a monument and a focal point for the entire project. Although it was inspired by the American concept of the skyscraper, its visual impact owes much to the towers of German medieval churches. Its distinctive "five-fingered gable" is symbolic of an outstretched hand. The manner in which rows of windows below the

gable are grouped within a common frame and wrapped around a corner of the building was an innovation of particular significance for later skyscraper design.

Peter Behrens (1868–1940), more than any other German architect of the early twentieth century, constituted a bridge between tradition and experiment. He began his career as a painter, producing Art Nouveau graphics, and then moved from an interest in crafts to the central problems of industrial design for machine production. Behrens turned to architecture as a result of his experience in the artists' collaborative at Darmstadt, where he designed his own house, the only one in the colony not designed by Olbrich. In 1903 he was appointed director of the Düsseldorf School of Art, and in 1907 the AEG company, one of the world's largest manufacturers of generators, motors, and lightbulbs, hired him as architect and coordinator of design of everything from products to publications. This appointment by a large industrial organization of an artist and architect to supervise and improve the quality of all its products was unusual and important in the history of architecture and design.

One of his first buildings for AEG, a landmark of modern architecture, is the Turbine Factory in Berlin (fig. 281). Although the building is given a somewhat traditional appearance of monumentality by the huge corner masonry piers and the overpowering visual mass of the roof (which again belies its actual structural lightness), it is essentially a glass-and-steel structure. Despite the use of certain traditional forms, this building is immensely important in its bold structural engineering, in its frank statement of construction, and in the social implications of a factory built to provide the maximum of air, space, and light. It is functional for both the processes of manufacture and the working conditions of the employees—concerns rarely considered in earlier factories.

The hiring of Behrens by AEG and similar appointments of consultants in design by other industries mark the emergence of some sense of social responsibility on the part of large-scale industry. Certain enlightened industrialists began to grasp the fact that well-designed products were not necessarily more expensive to produce than badly designed ones. Also, in the rapidly expanding industrial scene and the changing political environment of early twentieth-century Europe, the public image presented by industry was assuming increasing importance. The established industrial practice was one of short-sighted and often brutal exploitation of natural and human resources. In these early years of the century it was evident that industry had a powerful role to play in public affairs and even in promoting a national image. Such ideas may have influenced Behrens's transformation of his functional glass-and-steel Turbine Factory into a virtual monument to the achievements of modern German industry. This nationalistic attitude is, of course, disturbing in light of the industrial progress realized under the Third Reich, especially given the evidence of pro-Nazi activities in Behrens's last years. Nevertheless, Behrens is important not only as an architect but as a teacher of a generation that included Gropius, Le Corbusier, and Mies van der Rohe—all of whom worked with him early in their careers.

The Netherlands and Belgium

In the Netherlands, HENDRIK PETRUS BERLAGE (1856–1934), though he considered himself a traditionalist, was passionately devoted to stripping off the ornamental accessories of academic architecture and expressing honest structure and function. He characteristically used brick as a building material, the brick that, in the absence of stone and other materials, has created the architectural face of the Low Countries. His best-known building, the Amsterdam Stock Exchange (color plate 123, page 260), is principally of brick, accented with details of light stone. The brick is presented, inside and out, without disguise or embellishment, as is the steel framework that supports the glass ceiling. The general effect, with the massive corner tower and the low arcades of the interior, is obviously inspired by Romanesque architecture, in some degree seen through the eyes of the American architect H. H. Richardson, whose work he knew and admired (see fig. 97).

Berlage, in his writings, insisted on the primacy of interior space. The walls defining the spaces had to express both the nature of their materials and their strength and bearing function undisguised by ornament. Above all, through the use of systematic proportions, Berlage sought a total effect of unity and repose built from diversity, and thus ultimately an enduring style analogous to that created by the Greeks of the fifth century B.C.E. He conceived of an interrelationship of architecture, painting, and sculpture, but with architecture in the dominant role. The exchange building complex included other exchange halls (with decorations by artist Jan Toorop), a post office, a coffeehouse, a police station, and an administrative office. The building has fortunately been preserved despite early structural problems and eventual obsolescence as a stock exchange. Today the spaces house the Beurs van Berlage Foundation and the Dutch Philharmonic Orchestra.

Berlage's approach to architecture was also affected by Frank Lloyd Wright, whose work he knew through publications and then saw on a trip to the United States. He was enthusiastic about Wright and particularly about the Larkin Building (fig. 268), with its analogies to his own Amsterdam Stock Exchange. What appealed to Berlage and his followers about Wright was his rational approach—his efforts to control and utilize the machine and to explore new materials and techniques in the creation of a new society.

The complex relationship among English Pre-Raphaelite painting, William Morris and the English Arts and Crafts movement, French Impressionist, Neo-Impressionist, and Symbolist painting, Art Nouveau design, and the beginnings of modern rational architecture encountered earlier in the chapter on Art Nouveau (see fig. 103) is summarized in the fertile career of the Belgian HENRY CLEMENS VAN DE VELDE (1863–1957). Not only was van de Velde a painter, craftsman, industrial designer, and architect, he was also a critic. He had an extensive influence on German architecture and design and was a socialist who wanted to make his designs available to the working classes through mass production. Trained as a painter, first in Antwerp and then in Paris, he was in touch with the Impressionists and interested in Symbolist poetry. Back in Antwerp, painting in a manner influenced by Seurat, he exhibited with *Les Vingt*, the avant-garde Brussels group (see chapter 3). Through them he discovered Gauguin, Morris, and the English Arts and Crafts movement. As a result, he enthusiastically took up the graphic arts, particularly poster and book design and then, in 1894, turned to the design of furniture. All the time, he was writing energetically, preaching the elimination of traditional ornament, the assertion of the nature of materials, and the development of new, rational principles in architecture and design.

In 1899 van de Velde moved to Germany and, over the next few years, turned to architecture. In the 1913–14 Werkbund Theater in Cologne (figs. 282, 283) van de Velde made contributions to solving the problems of theater design, particularly in the stage area. The exterior revealed the emergence of a personal style in the way in which the molded forms of the facades defined the volume of the interior. With its strongly sculptural exterior, the Werkbund Theater has been seen as the definitive break with the linearity of Art Nouveau and as a predecessor of later Expressionist architecture. Following the end of World War I, the houses and other buildings on which he worked, notably the Kröller-Müller Museum (1936–38) at Otterlo the Netherlands, are characterized by austerity and refinement of details and proportions—evidence, perhaps, of the reciprocal influence of younger experimental architects who had emerged from his original educational systems.

Not the least of van de Velde's achievements by any means was the educational program he developed at the Weimar School, which he founded in 1906 under the patronage of the duke of Saxe-Weimar. This program put its emphasis on creativity, free experiment, and escape from dependence on past traditions. When van de Velde left Weimar in 1914 because of the war's outbreak he recommended the young architect Walter Gropius as his successor for the directorship. Gropius

the army hospital in Ferrara, they formed the association known as the *Scuola Metafisica*, or Metaphysical School. *The Great Metaphysician* (color plate 130, page 263) is a key work of the Ferrarese sojourn, the climax of the artist's visions of loneliness and nostalgia, his fear of the unknown, his premonitions of the future, and his depiction of a reality beyond the physical realm. *The Great Metaphysician* combines the architectural space of the empty city square and the developed mannequin figure. The deep square is sealed in with low classical buildings. The buildings to the right and left are abstract silhouettes extending dark, geometric shadows over the brown area of the square. Only the rear buildings and the foreground monument are brightly lit. This monument on a low base is a looming construction of elements from the studio and drafting table, crowned by a blank mannequin head. It is clear from these paintings that the Metaphysical School shared little of the Italian Futurists' faith in machine technology and its desire to capture the dynamism of modern life. Rather, de Chirico and his followers sought an elegiac and enigmatic expression, one that did not reject the art of the Italian past.

After 1920, just as de Chirico began to be recognized as a forerunner of new movements in many parts of Europe and the United States, he suddenly turned against the direction of his own painting and settled on an academic classicism that he continued to pursue. As a result, he quarreled violently with the Surrealists who had hailed his early work as the most crucial forerunner of their own, but who dismissed his later classically oriented style.

Though the theme of the horse had always been a staple in de Chirico's repertory of images, in the mid-1920s he began a long series of paintings featuring horses on the beach (fig. 295). These scenes usually evoked a classical past, with the horses surrounded by ancient Greek ruins. But here the subject is given a strange contemporary feeling, with bathers in modern attire. In his later career de Chirico did not actually repudiate his own early works. He took pride in his reputation, became jealously possessive of his achievements, and continued to make copies of his Metaphysical paintings, often on commission. However, as we have seen again and again, like others, he abandoned experimentation after World War I and attempted to rejoin the mainstream of Renaissance painting. In the case of de Chirico, this withdrawal into reaction made the artist something of a hero to certain Postmodernists of the 1980s. Since that time there has been a growing interest in the artist's late work.

CARLO CARRÀ (1881–1966), a Futurist and a Metaphysical painter, was sent to Ferrara for military service where he met de Chirico in 1917. Carrà had been replacing the coloristic fluidity of his Futurist paintings with a more disciplined style akin to Analytic Cubism. In the early war years he applied Marinetti's concept of free words to collages with propagandistic intent (color plate 110, page 207). The form proposed by Marinetti, and practiced for a time by Carrà as well as by Severini, specifically influenced poets and artists among the Dadaists and Surrealists.

295. GIORGIO DE CHIRICO. *Horses and Bathers (Cavalli e Bagnanti)*. 1936. Tempera on paper, 12 x 9¾" (30.5 x 22.9 cm). Private collection
© 1997 FOUNDATION GIORGIO DE CHIRICO/VAGA, NEW YORK, NY

The directions of Carrà's paintings and collages after 1912 indicate his search for a new content and for forms more plastic and less fragmented than those of Futurism. He painted pictures in 1917 that are almost pastiches of de Chirico's, but in *The Drunken Gentleman* (fig. 296), dated 1916 but probably painted in 1917, he developed an individual approach. The objects—sculptured head, bottle, glass, etc.—are modeled with clear simplicity in muted color gradations of gray and white and are given strength and solidity by the heavy impasto of the paint. Out of elementary still-life props, the artist created his own metaphysical reality. In 1918 Carrà published *Pittura Metafisica*, a book about the Metaphysical School. De Chirico, feeling justifiably that he was not given adequate credit, became embittered, and ended both their friendship and the Metaphysical School as a formal movement.

Zurich Dada: 1916–19

During World War I, Zurich, in neutral Switzerland, was the first important center in which an art, a literature, and even a music and a theater of the fantastic and the absurd arose. In 1915 a number of personalities, almost all in their twenties, exiles from the war that was sweeping over Europe, converged on this city. This international group included the German writers Hugo Ball and Richard Huelsenbeck, the Rumanian poet Tristan Tzara, the Romanian painter and sculptor Marcel Janco, the Alsatian painter, sculptor, and poet Jean (Hans) Arp, the Swiss painter and designer Sophie Taeuber, and the German painter and experimental filmmaker Hans Richter. Many other poets and artists were associated with Zurich Dada, but these were the leaders whose demonstrations, readings of poetry, noise concerts, art exhibitions, and writings attacked the traditions and preconcep-

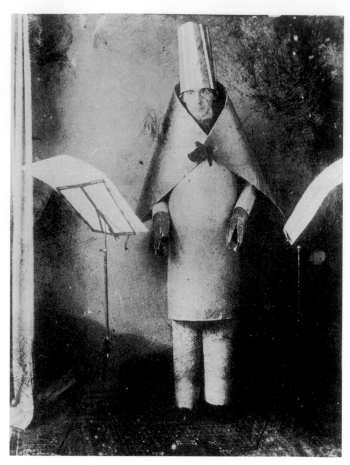

296. CARLO CARRÀ. *The Drunken Gentleman*. 1917 (dated 1916 on canvas). Oil on canvas, 23⅝ x 17½" (60 x 44.5 cm). Collection Carlo Frua de Angeli, Milan
© 1997 ESTATE OF CARLO CARRÀ/VAGA, NEW YORK, NY

297. Hugo Ball reciting the poem *Karawane* at the Cabaret Voltaire, Zurich. 1916. Photograph, 28⅛ x 15¾" (71.5 x 40 cm). Kunsthaus, Zurich

tions of Western art and literature. Thrown together in Zurich, these young men and women expressed their reactions to the spreading hysteria and madness of a world at war in forms that were intended as negative, anarchic, and destructive. From the very beginning, however, the Dadaists showed a seriousness of purpose and a search for new vision and content that went beyond any frivolous desire to outrage the bourgeoisie. This is not to deny that in the manifestations of Dada there was a central force of wildly imaginative humor, one of its lasting delights—whether manifested in free-word-association poetry readings drowned in the din of noise machines, in absurd theatrical or cabaret performances (fig. 297), in nonsense lectures, or in paintings produced by chance or intuition uncontrolled by reason. Nevertheless, it had a serious intent: the Zurich Dadaists were making a critical re-examination of the traditions, premises, rules, logical bases, even the concepts of order, coherence, and beauty that had guided the creation of the arts throughout history.

Hugo Ball, a philosopher and mystic as well as a poet, was the first actor in the Dada drama. With the nightclub entertainer Emmy Hennings, in February of 1916, he founded the Cabaret Voltaire, in Zurich, as a meeting place for these free spirits and a stage from which existing values could be attacked. Interestingly enough, across the street from the Cabaret Voltaire lived Lenin, who with other quiet, studious Russians was planning a world revolution. Ball and Hennings

were soon joined by Tzara, Janco, Arp, and Huelsenbeck.

The term "Dada" was coined in 1916 to describe the movement then emerging from the seeming chaos of the Cabaret Voltaire, but its origin is still doubtful. The popular version advanced by Huelsenbeck is that a French-German dictionary opened at random produced the word "dada," meaning a child's rocking horse or hobbyhorse. Richter remembers the *da, da, da da* ("yes, yes") in the Romanian conversation of Tzara and Janco. *Dada* in French also means a hobby, event, or obsession. Other possible sources are in dialects of Italian and Kru African. Whatever its origin, the name Dada is the central, mocking symbol of this attack on established movements, whether traditional or experimental, that characterized early twentieth-century art. The Dadaists used many of the formulas of Futurism in the propagation of their ideas—the free words of Marinetti, whether spoken or written; the noise-music effects of Luigi Russolo to drown out the poets; the numerous manifestos. But their intent was antithetical to that of the Futurists, who extolled the machine world and saw in mechanization, revolution, and war the rational and logical means, however brutal, to the solution of human problems.

Zurich Dada was primarily a literary manifestation, whose ideological roots were in the poetry of Arthur Rimbaud, in the theater of Alfred Jarry, and in the critical ideas of Max Jacob and Guillaume Apollinaire. In painting and sculpture, until Picabia arrived, the only real innovations were the free-form reliefs and collages "arranged according to the laws of chance" by Jean Arp. With few exceptions, the paintings and sculptures of other artists associated with Zurich Dada broke little new ground. In abstract and Expressionist film—principally through the experiments of Hans Richter and Viking Eggeling—and in photography and typographic design the Zurich Dadaists did make important innovations.

The Dadaists' theatrical activity, however, was an important precursor to the 1960s "happening" and other forms of performance art that developed after World War II, particularly that of the Fluxus group (see figs. 765, 615, 616). In fact, the effects of Dada are so pervasive in today's culture that one can find instructions for composing Dada poetry on the Internet. Arp described a typical evening at the Cabaret Voltaire thus: "Tzara is wiggling his behind like the belly of an Oriental dancer. Janco is playing an invisible violin and bowing and scraping. Madame Hennings, with a Madonna face, is doing the splits. Huelsenbeck is banging away non-stop on the great drum, with Ball accompanying him on the piano, pale as a chalky ghost. We were given the honorary title of Nihilists." The performers may have been wearing Marcel Janco's masks, made of painted paper and cardboard (color plate 131, page 264). Ball said these masks not only called for a suitable costume but compelled the wearer to act unpredictably, to dance with "precise, melodramatic gestures, bordering on madness." The Dadaists' theatrical antics obviously had precedents in the Russian Futurist performances such as the 1913 "Victory over the Sun," for which Malevich designed costumes.

Hugo Ball introduced abstract poetry at the Cabaret Voltaire with his poem *O Gadji Beri Bimba* in June 1916. He is seen in fig. 297 wrapped in cardboard costume and reciting his "sound poem," *Karawane*, from the two flanking music stands. Ball's thesis, that conventional language had no more place in poetry than the outworn human image in painting, produced a chant of more or less melodic syllables without meaning such as: "zimzim urallala zimzim zanzibar zimlalla zam. . . . " The frenzied reactions of the audience did not prevent the experiment from affecting the subsequent course of poetry. But one cannot evaluate Zurich Dada by a tally of its concrete achievements or stylistic influences.

The Zurich Dadaists were violently opposed to any organized program in the arts, or any movement that might express the common stylistic denominator of a coherent group. Nevertheless, three factors shaped their creative efforts. These were *bruitisme* (noise-music, from *le bruit*—"noise"—as in *le concert bruitiste*), simultaneity, and chance. *Bruitisme* came from the Futurists, and simultaneity from the Cubists via the Futurists. Chance, of course, exists to some degree in any act of artistic creation. In the past the artist normally attempted to control or to direct it, but it now became an overriding principle. All three, despite the artists' avowed negativism, soon became the basis for their revolutionary approach to the creative act, an approach still found in poetry, music, drama, and painting.

1916 witnessed the first organized Dada evening at a public hall in Zurich and the first issue of the magazine *Collection Dada*, which Tzara would eventually take over and move to Paris. The following year, the first public Dada exhibitions were held at the short-lived Galerie Dada. Such activities ushered in a new, more constructive phase for Zurich Dada, broadened from the spontaneous performances of the Cabaret Voltaire. But with the end of the war, Zurich Dada was drawing to an end. The initial enthusiasms were fading; its participants were scattering. Ball abandoned Dada, and Huelsenbeck, who had opposed attempts to turn Dada into a conventional, codified movement, left for Berlin. Tzara remained for a time in Zurich and oversaw the last Dada soirée in 1919, after which he moved to Paris. The painter Francis Picabia arrived in 1918, bringing contact with similar developments in New York and Barcelona. Picabia, together with Marcel Duchamp and Man Ray, had contributed to a Dada atmosphere in New York around Alfred Stieglitz's 291 Gallery and his periodical *291*. With the help of the collector Walter Arensberg, Picabia began to publish his own journal of protest against everything, which he called *391*. After a journey to Barcelona, where he found many like-minded expatriates, Picabia visited Zurich, drawn by the spreading reputation of the originators of Dada. Returning to Paris at the end of the war, he became a link, as did Tzara, between the postwar Dadaists of Germany and France. Dada, which was perhaps more a state of mind than an organized movement, left an enormous legacy to contemporary art, particularly the Neo-Dada art of the 1950s and early 1960s (see figs. 610, 614).

JEAN (HANS) ARP (1887–1966). The major visual artist to emerge from Zurich Dada was the Alsatian Jean (or Hans) Arp. Arp was born in Strasbourg, then a German city but subsequently recovered by France. He studied painting and poetry, and in Paris in 1904 he discovered modern painting, which he then pursued in studies at the Weimar School of Art and the Académie Julian in Paris. He also wrote poetry of great originality and distinction throughout his life. The disparity between his formal training and the paintings he was drawn to brought uncertainty, and he spent the years 1908–10 in reflection in various small villages in Switzerland. The Swiss landscape seems to have made a lasting impression on him, and the abstraction to which he eventually turned was based on nature and living organic shapes.

In Switzerland, Arp met Paul Klee, and after his return to Germany he was drawn into the orbit of Kandinsky and the Blaue Reiter painters (color plate 75, page 160; color plate 76, page 160). In 1912 he exhibited in Herwarth Walden's first Autumn Salon, and by 1914, back in Paris, he belonged to the circle of Picasso, Modigliani, Apollinaire, Max Jacob, and Delaunay.

Arp took an unusually long time finding his own direction. Since he destroyed most of his pre-1915 paintings, the path of his struggle is difficult, if not impossible, to trace. He experimented with geometric abstraction based on Cubism, and by 1915, in Zurich, he was producing drawings and collages whose shapes suggest leaves and insect or animal life but which were actually abstractions. With SOPHIE TAEUBER (1889–1943), whom he met in 1915 and married in 1922, he jointly made collages, tapestries, embroideries, and sculptures. Through his collaboration with her, Arp further clarified his ideas: "These pictures are Realities in themselves, without meaning or cerebral intention. We . . . allowed the elementary and spontaneous to react in full freedom. Since the disposition of planes, and the proportions and colors of these planes seemed to depend purely on chance, I declared that these works, like nature, were ordered according to the laws of chance, chance being for me merely a limited part of an unfathomable *raison d'être*, of an order inaccessible in its totality. . . . " Also emerging at this time was the artist's conviction of the metaphysical reality of objects and of life itself—some common denominator belonging to both the lowest and the highest forms of animals and plants. It may have been his passion to express his reality in the most concrete terms possible, as an organic abstraction (or, as he preferred to say, an organic *concretion*) that led him from painting to collage and then to relief and sculpture in the round.

In 1916–17 Arp produced some collages of torn, rectangular pieces of colored papers scattered in a vaguely rectangular arrangement on a paper ground (fig. 298). The story told of their origin is that Arp tore up a drawing that displeased him and dropped the pieces on the floor, then suddenly saw in the arrangement of the fallen scraps the solution to the problems with which he had been struggling. Arp continued to experiment with collages created in this manner, just as Tzara created poems from words cut out of newspapers, shaken and scattered on a table. Liberated from rational thought processes, the laws of chance, Arp felt, were more in tune with the workings of nature. By relinquishing a certain amount of control, he was distancing himself from the creative process. This kind of depersonalization, already being explored by Marcel Duchamp in Paris, had profound consequences for later art.

The lack of emphasis on the uniqueness of artistic creation allowed for fruitful collaboration between Taeuber and Arp, who dubbed their joint productions "Duo-Collages." Partly as a result of Taeuber's training in textiles, the couple did not restrict themselves to the fine art media traditionally reserved for painting and sculpture. In their desire to integrate art and life they shared an outlook with contemporaries in Russia and artists such as Sonia Delaunay (color plate 104, page 205). Taeuber developed a geometric vocabulary in her early compositions made in Zurich (color plate 132, page 264). These rigorous abstractions, organized around a rhythmical balance of horizontals and verticals, had a decisive influence on Arp's work. He said that when he met Taeuber in 1915, "she already knew how to give direct and palpable shape to her inner reality. . . . She constructed her painting like a work of masonry. The colors are luminous, going from rawest yellow to deep red." Between 1918 and 1920 Taeuber made four remarkable heads in polychromed wood, two of which are portraits of Arp (fig. 299). These heads, humor-

298. JEAN ARP. *Collage Arranged According to the Laws of Chance.* 1916–17. Torn and pasted paper, 19⅛ x 13⅝" (48.6 x 34.6 cm). The Museum of Modern Art, New York

ously reminiscent of hat stands, are among the few works of Dada sculpture made in Zurich. At the same time, Raoul Hausmann was creating similar mannequin-like constructions in Berlin (see fig. 312).

By 1915, Arp was devising a type of relief consisting of thin layers of wood shapes. These works, which he called "constructed paintings," represent a medium somewhere between painting and sculpture and give evidence of Arp's awareness of Cubist collage, which he would have seen in Paris in 1914–15. To make his reliefs, Arp prepared drawings and gave them to a carpenter who cut them out in wood. While not executed with the same aleatory methods employed for the collages, the relief drawings were the result of Arp's willingness to let his pencil be guided unconsciously, without a set goal in mind. The curving, vaguely organic forms that resulted, which evoked the body and its processes or some other highly abstracted form in nature, have been called "biomorphic," a term used to describe the abstract imagery of Arp's later work as well as that of many Surrealists who followed his lead. Arp developed a vocabulary of biomorphic shapes that had universal significance. An oval or egg shape, for example, was for him a "symbol of metamorphosis

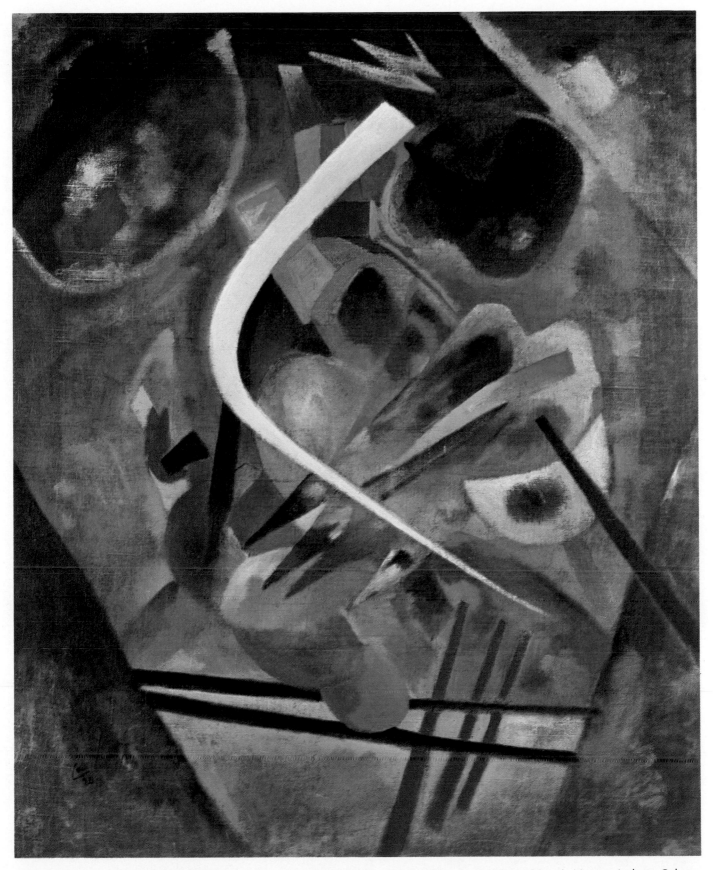

Color plate 116. VASILY KANDINSKY. *White Line, No. 232.* 1920. Oil on canvas, 38⅝ x 31½" (98.1 x 80 cm). Museum Ludwig, Cologne

Color plate 117.
ALEKSANDR RODCHENKO.
Untitled advertising poster.
1924. Gouache and pho-
tomontage on paper,
27½ x 33⅞"
(69.7 x 86.1 cm).
Rodchenko-Stepanova
Archive, Moscow

below, left: Color plate 118.
VARVARA FEDOROVNA
STEPANOVA. Design for
sportswear. 1923.
Gouache and ink on paper,
11⅞ x 8½" (30.2 x 21.7 cm).
Collection, Alexander
Lavrentiev

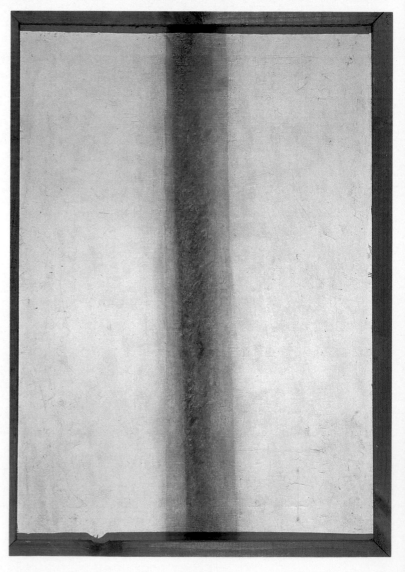

right: Color plate 119. OLGA ROZANOVA.
Untitled (Green Stripe). 1917–18. Oil on
canvas, 28 x 20⅞" (71 x 53 cm). Rostovo-
Yaroslavskij Arkhitekturno-Khudozhestrennyj
Muzej-Zapovednik, Rostovo-Yaroslavskij, Russia

Color plate 120. Piet Mondrian.
Blue Tree. c. 1908.
Oil on composition board,
22⅜ x 29½" (56.8 x 74.9 cm).
Dallas Museum of Art
Gift of the James H. and Lillian Clark Foundation
© 1997 Mondrian Holtzman Trust

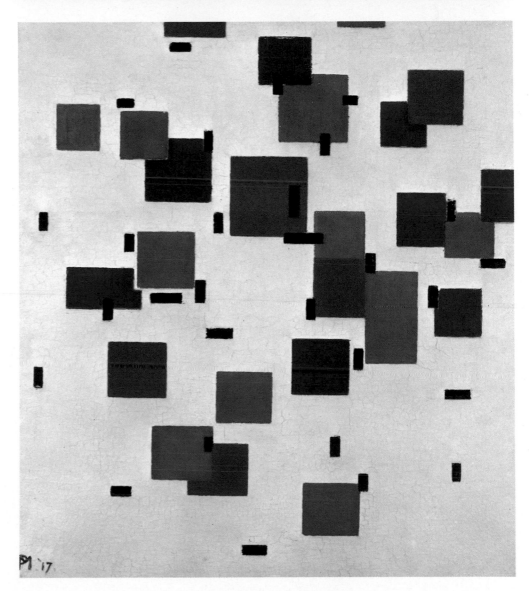

Color plate 121. Piet Mondrian.
Composition in Color A. 1917. Oil
on canvas, 19½ x 17⅜" (49.5 x
44.1 cm). Kröller-Müller Museum,
Otterlo, the Netherlands
© 1997 Mondrian Holtzman Trust

Color plate 122. FRANK LLOYD WRIGHT.
Susan Lawrence Dana House, Springfield, IL. 1902–04.
Interior perspective, dining room. Pencil and watercolor
on paper, 25 x 20⅜" (63.5 x 51.7 cm).
Avery Architectural and Fine Arts Library,
Columbia University, New York
FRANK LLOYD WRIGHT COLLECTION

Color plate 123. HENDRIK PETRUS BERLAGE.
Stock Exchange, Amsterdam. 1898–1903

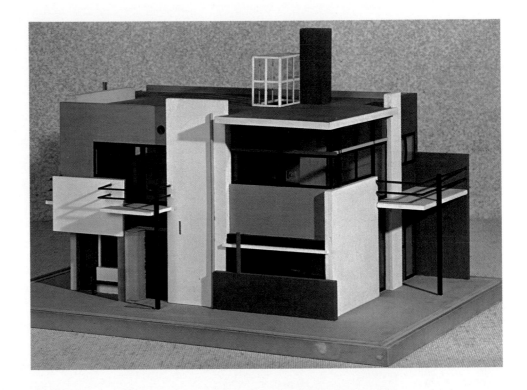

Color plate 124.
GERRIT RIETVELD. Model of the
Schröder House. 1923–24.
Glass and wood,
17⅜ x 28⅜ x 19¼" (44.1 x
72.1 x 48.9 cm).
Stedelijk Museum, Amsterdam
© 1997 ARTISTS RIGHTS SOCIETY (ARS), NEW
YORK/BEELDRECHT, AMSTERDAM

below: Color plate 125.
GERRIT RIETVELD. Living and
dining area, Schröder
House, with furniture by
Rietveld

right: Color plate 128.
MARC CHAGALL.
The Green Violinist.
1923–24. Oil on canvas,
6'5¾" x 42¾"
(1.9 m x 108.6 cm).
Solomon R. Guggenheim
Museum, New York
© 1997 ADAGP, PARIS/ARTISTS
SOCIETY (ARS), NEW YORK

Color plate 126.
MARC CHAGALL. *Paris
Through the
Window.* 1913. Oil
on canvas, 52⅜ x
54¾" (133 x 139.1 cm).
Solomon R.
Guggenheim Museum,
New York
© 1997 ADAGP, PARIS/ARTISTS
RIGHTS SOCIETY (ARS), NEW YORK

Color plate 127. MARC CHAGALL. *Birthday.* 1923. Oil on canvas,
31⅞ x 39⅜" (81 x 100 cm). Private collection
© 1997 ADAGP, PARIS/ARTISTS RIGHTS SOCIETY (ARS), NEW YORK

above, right: Color plate 129. GIORGIO DE
CHIRICO. *The Melancholy and Mystery of a
Street*. 1914. Oil on canvas, 34¼ x 28⅛"
(87 x 71.4 cm).
Private collection

right: Color plate 130.
GIORGIO DE CHIRICO.
The Great Metaphysician. 1917. Oil on
canvas, 41⅛ x 27½" (104.5 x 69.8 cm).
The Museum of Modern Art, New York

Color plate 131. MARCEL JANCO. *Mask.* 1919. Paper, card-board, string, gouache, and pastel, 17¾ x 8⅝ x 2" (45 x 22 x 5 cm). Musée National d'Art Moderne, Centre d'Art et de Culture Georges Pompidou, Paris

Color plate 132. SOPHIE TAEUBER. *Rythmes Libres.* 1919. Gouache and watercolor on vellum, 14¾ x 10⅞" (37.6 x 27.5 cm). Kunsthaus, Zurich

Color plate 133.
JEAN ARP. *Fleur Marteau.*
1916. Oil on wood,
24⅜ x 19⅝"
(61.9 x 49.8 cm).
Fondation Arp,
Clamart, France

Color plate 134. MARCEL
DUCHAMP. *The Passage
from Virgin to Bride.*
Munich, July–August
1912. Oil on canvas,
23⅜ x 21¼" (59.4 x
54 cm). The Museum of
Modern Art, New York

above: Color plate 135.
MARCEL DUCHAMP. *Tu m'*.
1918. Oil and graphite
on canvas, with brush,
safety pins, nut and bolt,
27½" x 10'2¾" (69.9 cm
x 3.2 m). Yale University
Art Gallery, New Haven
BEQUEST OF KATHERINE S. DREIER
© 1997 ADAGP, PARIS/ARTISTS RIGHTS
SOCIETY (ARS), NEW YORK

Color plate 136.
MARCEL DUCHAMP.
*The Bride Stripped Bare
by Her Bachelors, Even
(The Large Glass).*
1915–23. Oil, lead wire,
foil, dust, and varnish on
glass, 8'11" x 5'7" (2.7
x 1.7 m). Philadelphia
Museum of Art
BEQUEST OF KATHERINE S. DREIER
© 1997 ADAGP, PARIS/ARTISTS RIGHTS
SOCIETY (ARS), NEW YORK

O Tannenbaum im deutschen Raum, wie krumm sind deine Äste!

above: Color plate 137. MAN RAY. *Seguidilla.* 1919. Airbrushed gouache, pen and ink, pencil, and colored pencil on paper board, 22 x 27⅞" (55.9 x 70.8 cm). Hirshhorn Museum and Sculpture Garden

JOSEPH H. HIRSHHORN PURCHASE FUND AND MUSEUM PURCHASE © 1997 MAN RAY TRUST/ADAGP, PARIS/ARTISTS RIGHTS SOCIETY (ARS), NEW YORK

above, right: Color plate 138. JOHN HEARTFIELD. *Little German Christmas Tree.* 1934. Photomontage, 10¼ x 7⅞" (26 x 20 cm). The Metropolitan Museum of Art, New York

THE HORACE W. GOLDSMITH FOUNDATION GIFT
© 1997 VG BILD-KUNST, BONN/ARTISTS RIGHTS SOCIETY (ARS), NEW YORK

Color plate 139. KURT SCHWITTERS. *Picture with Light Center.* 1919. Collage of cut-and-pasted papers and oil on cardboard, 33¼ x 25⅞" (84.5 x 65.7 cm). The Museum of Modern Art, New York

PURCHASE
© 1997 VG BILD-KUNST, BONN/ARTISTS RIGHTS SOCIETY (ARS), NEW YORK

Color plate 140. MAX ERNST. *Celebes.* 1921. Oil on canvas,
51⅛ x 43¼" (129.9 x 109.9 cm). Tate Gallery, London
© 1997 ADAGP, PARIS/ARTISTS RIGHTS SOCIETY (ARS), NEW YORK

Color plate 141. GEORGE GROSZ. *Dedication to Oskar
Panizza.* 1917–18. Oil on canvas, 55⅛ x 43¼"
(140 x 109.9 cm). Staatsgalerie, Stuttgart
© ESTATE OF GEORGE GROSZ/VAGA, NEW YORK, NY

Color plate 142. MAX BECKMANN. *Self-Portrait in Tuxedo.* 1927.
Oil on canvas, 55½ x 37¾" (141 x 96 cm). Busch-Reisinger Museum,
Harvard University Art Museums, Cambridge, Massachusetts
© 1997 VG BILD-KUNST, BONN/ARTISTS RIGHTS SOCIETY (ARS), NEW YORK

Color plate 143. MAX BECKMANN. *Departure.* 1932–33. Oil on canvas, triptych, center panel 7'1¾" x 45⅜" (2.2 m x 115.2 cm); side panels each 7'1¾" x 39¼" (2.2 m x 99.7 cm). The Museum of Modern Art, New York GIVEN ANONYMOUSLY (BY EXCHANGE)

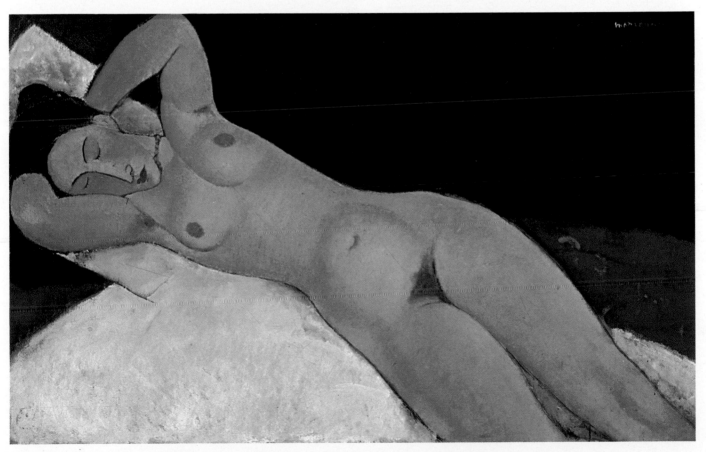

Color plate 144. AMEDEO MODIGLIANI. *Nude.* 1917. Oil on canvas, 28¾ x 45¾" (73 x 116.2 cm).
Solomon R. Guggenheim Museum, New York

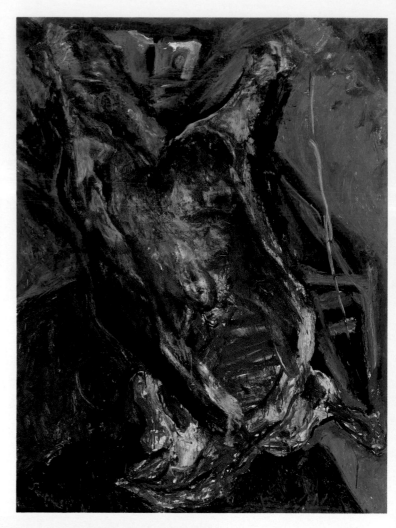

above, left: Color plate 145. CHAIM SOUTINE. *Woman in Red.* c. 1924–25. Oil on canvas, 36 x 25" (91.4 x 63.5 cm). Private collection

above, right: Color plate 146. CHAIM SOUTINE. *Side of Beef.* c. 1925. Oil on canvas, 55¼ x 42⅜" (140.3 x 107.6 cm). Albright-Knox Art Gallery, Buffalo, New York

Color plate 147. SUZANNE VALADON. *Blue Room.* 1923. Oil on canvas, 35⅜ x 45⅝" (90 x 115.9 cm). Musée National d'Art Moderne, Centre d'Art et de Culture Georges Pompidou, Paris

Color plate 148. HENRI MATISSE. *Piano Lesson.* Issy-les-Moulineaux, late summer, 1916. Oil on canvas, 8'1½" x 6'11¾" (2.45 x 2.13 m). The Museum of Modern Art, New York

Color plate 149. HENRI MATISSE. *Decorative Figure Against an Ornamental Background.* 1925–26. Oil on canvas, 51⅛ x 38½" (129.9 x 97.8 cm). Musée National d'Art Moderne, Centre d'Art et de Culture Georges Pompidou, Paris

left: Color plate 150. HENRI MATISSE. *Interior with a Phonograph.* 1924. Oil on canvas, 39¾ x 32" (101 x 81.3 cm). Private collection

above: Color plate 151. HENRI MATISSE. *Large Reclining Nude (The Pink Nude).* 1935. Oil on canvas, 26 x 36½" (66 x 92.7 cm). The Baltimore Museum of Art

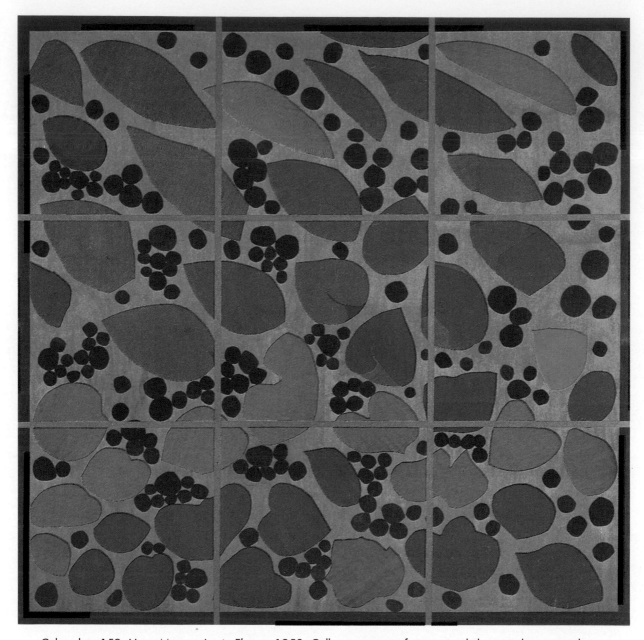

Color plate 152. HENRI MATISSE. *Ivy in Flower.* 1953. Collage maquette for a stained-glass window: gouache on cut and pasted paper, and pencil on colored paper, 9'3⅞" x 9'4⅝" (2.8 x 2.9 m). Dallas Museum of Art

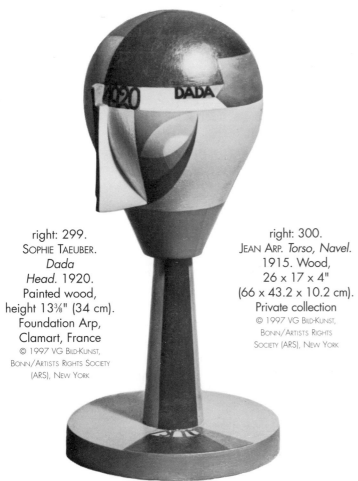

right: 299.
SOPHIE TAEUBER.
*Dada
Head.* 1920.
Painted wood,
height 13⅜" (34 cm).
Foundation Arp,
Clamart, France
© 1997 VG BILD-KUNST,
BONN/ARTISTS RIGHTS SOCIETY
(ARS), NEW YORK

right: 300.
JEAN ARP. *Torso, Navel.*
1915. Wood,
26 x 17 x 4"
(66 x 43.2 x 10.2 cm).
Private collection
© 1997 VG BILD-KUNST,
BONN/ARTISTS RIGHTS
SOCIETY (ARS), NEW YORK

and development of bodies." A viola shape suggested a female torso and then an accent was provided by a cut-out circle that became a navel (fig. 300). Arp's later painted reliefs suggested plants, exotic vegetables, crustaceans, or swarming amoebae, with strong implications of life, growth, and metamorphosis. He used the term "Formes terrestres" or "earthly forms" to describe these reliefs. A particular shape might suggest a specific object and thus give the relief its name, as in *Fleur Marteau (Hammer Flower)* (color plate 133, page 265). Although the origin of the shapes was initially intuitive (a line doodled on a piece of paper), the contour lines were as organic as the living organism that inspired them.

Although one of the founders of Zurich Dada who exerted tremendous influence on subsequent art, Arp did not perform in the wild theatrical presentations at the Cabaret Voltaire (unlike Taeuber who danced in the Dada performances). "He never needed any hullabaloo," said Huelsenbeck, "Arp's greatness lay in his ability to limit himself to art." He also contributed drawings and poems to the Dada publications between 1916 and 1919. In Arp's later, freestanding sculptures in marble or bronze, discussed in the chapter on Surrealism, the suggestion of head or torso became more frequent and explicit. When asked in 1956 about a 1953 piece entitled *Aquatic* (see fig. 357)—which, reclining, suggests some form of sea life, and standing on end, a particularly sensuous female torso—he commented, "In one aspect or another, my sculptures are always torsos."

In the same way that de Chirico is a forerunner of the wing of Surrealism that uses illusionistic techniques and recognizable images, Arp is a forerunner of the other wing that uses abstract biomorphic shapes and arbitrary, nondescriptive color to create a world of fantasy beyond the visible world. His work later influenced such artists as Joan Miró, André Masson, and Alexander Calder (see figs. 365, 366, 683).

New York Dada: 1915–20

New York Dada during World War I largely resulted from the accident that Marcel Duchamp and Francis Picabia both had come to New York and found a congenial environment at 291, the avant-garde gallery founded by Alfred Stieglitz, who had introduced the American public to such European masters as Rodin, Toulouse-Lautrec, Henri Rousseau, Matisse, and Picasso. (The "Stieglitz circle" of American artists is discussed in chapter 18.)

In 1915, assisted by Duchamp and Picabia, Stieglitz founded the periodical *291* to present the anti-art ideas of these artists. Thus, chronologically earlier than the Dada movement in Zurich, comparable ideas were fermenting independently in a small, cohesive group in New York. Aside from the two Europeans, the most important figure in the group was the young American artist Man Ray. In addition, the remarkable collectors Walter and Louise Arensberg, whose salons regularly attracted many of the leading artists and writers of the day, were important patrons of Marcel Duchamp, the artist of greatest stature and influence in the group.

MARCEL DUCHAMP (1887–1968). The enormous impact made on twentieth-century art by Marcel Duchamp is best summarized by the artist Richard Hamilton: "All the branches put out by Duchamp have borne fruit. So widespread have been the effects of his life that no individual may lay claim to be his heir, no one has his scope or his restraint." Duchamp, a handsome, charismatic man of astonishing intellect, devoted a lifetime to the creation of an art that was more cerebral than visual. By the beginning of World War I he had rejected the works of many of his contemporaries as "retinal" art, or art intended only to please the eye. Although a gifted painter, Duchamp ultimately abandoned conventional methods of making art in order, as he said, "to put art back at the service of the mind." Duchamp lived a simple but peripatetic existence, traveling back and forth between Europe and America for most of his adult life.

Duchamp's inquiry into the very nature of art was first expressed in such paintings as *Nude Descending a Staircase, No. 2* (color plate 105, page 205), which used Cubist faceting to give, he said, "a static representation of movement." Here, the Paris exhibition of the Futurists in February 1912 had helped the artist to clarify his attitudes, although his intention was at the opposite extreme of theirs. Their dynamism, their "machine aesthetic," was an optimistic, humorless exaltation of the new world of the machine, of speed, flight, and efficiency, with progress measured in these terms. Duchamp, though he used some of their devices, expressed disillusionment through satirical humor.

During the summer of 1912, which he spent in Munich, Duchamp turned to the painting of machines of his own creation. While still in France, he made *The King and Queen Traversed by Swift Nudes* in which the figures are not only mechanized but are machines in operation, pumping some form of sexual energy from one to another. Subsequently in Munich the artist pursued his fantasies of sexualized machines in a series of paintings and drawings, including *The Passage from Virgin to Bride* (color plate 134, page 265) and the initial drawing for the great painting on glass, *The Bride Stripped Bare by Her Bachelors, Even* or *The Large Glass* (color plate 136, page 266). The nature of these works is that of a machine organism, suggesting anatomical diagrams of the respiratory, circulatory, digestive, or reproductive systems of higher mammals. In *The Passage*, Duchamp abandons all vestiges of human anatomy. The organic becomes mechanized, and human flesh is supplanted by tubes, pistons, and cylinders. The term "mechanomorphic" was eventually coined to describe Duchamp's distinctive grafting of machine forms onto human activity. Thus, while he restored traditional symbols of inviolable purity and sanctified consummation, i.e., the virgin and the bride, he destroyed any sense of convention by presenting them as elaborate systems of anatomical plumbing. Doubting the traditional validity of painting and sculpture as legitimate modes of contemporary expression, and determined to undermine Cubism, Duchamp still created beautifully rendered, visually seductive works of art. His

recognition of this fact no doubt contributed to his decision to abandon painting at the age of twenty-five. "From Munich on," Duchamp said, "I had the idea of *The Large Glass*. I was finished with Cubism and with movement—at least movement mixed up with oil paint. The whole trend of painting was something I didn't care to continue."

During 1912 the so-called Armory Show was being organized in New York, and the American painters Walt Kuhn and Arthur B. Davies and the painter-critic Walter Pach were then in Paris selecting works by French artists. Four paintings by Duchamp were chosen, including *Nude Descending a Staircase* and *The King and Queen Traversed by Swift Nudes*. When the Armory Show opened in February 1913, Duchamp's paintings, and most particularly the *Nude*, became the *succès de scandale* of the exhibition. Despite the attacks in the press, all four of Duchamp's paintings were sold, and he suddenly found himself internationally notorious.

Duchamp was meanwhile proceeding with his experiments toward a form of art or nonart based on everyday subject matter—with a new significance determined by the artist and with internal relationships proceeding from a relativistic mathematics and physics of his own devising. Although he had almost ceased to paint, Duchamp worked intermittently toward a climactic object: *The Large Glass* (color plate 136, page 266), intended to sum up the ideas and forms he had explored in *The Passage from Virgin to Bride* and related paintings. For this project, he made, between 1913 and 1915, the drawings, designs, and mathematical calculations for *Bachelor Machine* and *Chocolate Grinder, No. 1* (fig. 301), later to become part of the male apparatus accompanying the bride in *The Large Glass*.

By far, Duchamp's most outrageous assault on artistic tra-

301. MARCEL DUCHAMP. *Chocolate Grinder, No. 1.* 1913. Oil on canvas, 24¾ x 25⅝" (62.9 x 65.1 cm).
Philadelphia Museum of Art LOUISE AND WALTER ARENSBERG COLLECTION.

302. Marcel Duchamp. *Bottle Rack* and *Fountain*. 1917. Installation photograph, Stockholm, 1963

dition was his invention in 1913 of the "readymade," defined by André Breton as "manufactured objects promoted to the dignity of art through the choice of the artist." Duchamp said his selection of common "found" objects, such as a bottle rack (fig. 302), was guided by complete visual indifference, or "anaesthesia," and the absence of good or bad taste. They demonstrated, in the most irritating fashion to the art world of Duchamp's day, that art could be made out of virtually anything and that it required little or no manipulation by the artist. Within Duchamp's vocabulary, his famously irreverent addition of a mustache and goatee to a reproduction of the *Mona Lisa* was a "rectified" readymade. An "assisted" readymade also required some intervention by the artist, as here, when Duchamp mounted an old bicycle wheel on an ordinary kitchen stool (fig. 303). Duchamp also introduced a kinetic dimension in his spinning bicycle wheel. He then experimented briefly with mechanical motion in his Rotative Plaques of the 1920s and his Roto-reliefs of the 1930s. Despite Russian Constructivist experiments with moving sculptures toward the end of World War I, it was not until the early 1930s, when Alexander Calder developed the "mobile," a term invented by Duchamp, that an artist fully explored the potential of motion in art (see fig. 455).

Because the readymade could be repeated indiscriminately,

303. Marcel Duchamp. *Bicycle Wheel.* New York, 1951 (third version, after lost original of 1913). Assemblage: metal wheel, 25½" (63.8 cm) diameter, mounted on painted wood stool, 23¾" (60.2 cm) high; overall 50½ x 25½ x 16⅝" (128.3 x 64.8 x 42.2 cm). The Museum of Modern Art, New York

Duchamp decided to make only a small number yearly, saying "for the spectator even more than for the artist, *art is a habit-forming drug* and I wanted to protect my readymade against such *contamination*." He stressed that it was in the very nature of the readymade to lack uniqueness, and since readymades are not originals in the conventional sense, a "replica will do just as well." To extend the perversity of this logic, Duchamp remarked: "Since the tubes of paint used by an artist are manufactured and readymade products we must conclude that all paintings in the world are assisted *readymades*." Duchamp limited the number of readymades to avoid any taint of predictable art activity that would result in a commodity to be bought and sold. This attitude struck a chord among a broad spectrum of artists in the late 1950s and 1960s. So-called Neo-Dada, Pop, Minimalist, and Conceptual artists, for example, in their own way undermined the cult of originality that surrounded objects crafted by the artist's hand. They incorporated found objects (see fig. 640), left the fabrication of their work to others (see fig. 699), or sometimes avoided the object altogether (see fig. 758).

For Duchamp, the conception, the "discovery," was what made a work of art, not the uniqueness of the object. One glimpses in the works discussed so far the complex process of Duchamp's thought—the delight in paradox, the play of visual against verbal, and the penchant for alliteration and double and triple meanings. In a deliberate act of provocation, Duchamp submitted a porcelain urinal, which he turned ninety degrees and entitled *Fountain* (fig. 302), to the 1917 exhibition of the New York Society of Independent Artists. The work was signed by R. Mutt, a pun on the plumbing fixture manufacturer J. L. Mott Iron Works. Needless to say, the association with the then popular Mutt and Jeff cartoons did not escape Duchamp. Although the exhibition was in principle open to any artist's submission without the intervention of a jury, the work was rejected. Duchamp resigned from the association, and *Fountain* became his most notorious readymade. Despite Duchamp's antiaesthetic attitude, readymades have taken on a perverse beauty of their own. "I threw the bottle rack and urinal in their faces as a challenge," he said, "and now they admire them for their aesthetic beauty."

In 1913–14 Duchamp carried out an experiment in chance that resulted in *3 Stoppages Étalon (3 Standard Stoppages)* (fig. 304) and was later applied to *The Large Glass*. In a spirit that mocked the notion of standard, scientifically perfect measurement, Duchamp dropped three strings, each one meter in length, from a height of one meter onto a painted canvas. The strings were affixed to the canvas with varnish in the shape they assumed to "imprison and preserve forms obtained through chance." These sections of canvas and screen were then cut from the stretcher and laid down on glass panels, and three templates were cut from wooden rulers in the profile of the shapes assumed by the strings. The idea of the experiment—not the action itself—was what intrigued Duchamp. *3 Stoppages Étalon* is thus a remarkable document in the history of chance as a controlling factor in the creation of a work of art.

In 1918, Duchamp made *Tu m'* (color plate 135, page 266), his last painting on canvas, for Katherine Dreier, a daring collector and leading spirit in American avant-garde art. The painting has an unusually long and horizontal format, for it was destined for a spot above a bookcase in Dreier's library. It includes a compendium of Duchampian images: cast shadows, drawn in pencil, of a corkscrew and two readymades, *Bicycle Wheel* and *Hat Rack*; a pyramid of color samples (through which an actual bolt is fastened); a trompe l'oeil tear in the canvas "fastened" by three real safety pins; an actual bottle brush; a sign painter's hand (rendered by a professional sign painter), as well as the outlines at the left and right of *3 Stoppages Étalon*. Together with Dreier and Man Ray, Duchamp eventually founded the Société Anonyme, an important organization that made publications, lectures, and exhibitions while building an important collection of modern art.

When the United States entered World War I, Duchamp relocated for several months to Argentina and then to

304. MARCEL DUCHAMP. *3 Stoppages Étalon*. 1913–14.
Assemblage: 3 threads glued to 3 painted canvas strips, each mounted on a glass panel; 3 wood slats, shaped along one edge to match the curves of the threads; the whole fitted into a wooden box, 3 painted canvas strips, each 5¼ x 47¼" (13.3 x 120 cm); each mounted on a glass panel, 7¼ x 49⅜ x ¼" (18.4 x 125.4 x .6 cm); three wood slats, 2½ x 43 x ⅛" (6.2 x 109.2 x .2 cm), 2½ x 47 x ⅛" (6.1 x 119.4 x .2 cm), 2½ x 43¼ x ⅛" (6.3 x 109.7 x .2 cm); overall, fitted into wood box, 11⅛ x 50⅞ x 9" (28.2 x 129.2 x 22.7 cm)].
The Museum of Modern Art, New York

305. MARCEL DUCHAMP. *Boîte en valise (Box in a Valise)*. 1941. Leather valise containing miniature replicas, photographs, and color reproductions of works by Duchamp. Philadelphia Museum of Art THE LOUISE AND WALTER ARENSBERG COLLECTION. © 1997 ADAGP, PARIS/ARTISTS RIGHTS SOCIETY (ARS), NEW YORK

Europe. In 1920 he returned to New York, bringing a vial he called *50 cc of Paris Air* as a gift for Walter Arensberg. During this period he invented a female alter ego, Rrose Sélavy, which when pronounced in French sounds like "Eros, c'est la vie" or "Eros, that's life." Duchamp inscribed works with this pseudonym and was photographed several times by Man Ray in the guise of his feminine persona. Such gestures were typical of his tendency to break down gender boundaries, and they demonstrate the degree to which Duchamp's activities and his personality were as significant, if not more so, than any objects he made.

In 1922, after yet another round of travel, Duchamp settled in New York and continued working on *The Large Glass* (color plate 136, page 266). He finally ceased work on it in 1923. This painting on glass, which was in gestation for several years, is the central work of Duchamp's career. The glass support dispensed with the need for a background since, by virtue of its transparency, it captured the "chance environment" of its surroundings. *The Large Glass* depicts an elaborate and unconsumated mating ritual between the bride in the upper half of the glass, whose machinelike form we recognize from *The Passage from Virgin to Bride* (color plate 134, page 265), and the uniformed bachelors in the lower half. These forms are rendered with a diagrammatic precision that underscores the pseudoscientific nature of their activities. Despite their efforts, the bachelors fail to project their "love gasoline" (the sperm gas or fluid constantly ground forth by the rollers of the chocolate machine) into the realm of the bride, so the whole construction becomes a paradigm of pointless erotic activity. Breton described it as "a mechanistic and cynical interpretation of the phenomenon of love." To annotate and supplement this cryptic work, Duchamp assem-

bled his torn scraps of notes, drawings, and computations in another work titled *The Green Box*. This catalogue of Duchampian ideas was later published by the artist in a facsimile edition. Duchamp allowed New York dust to fall on *The Large Glass* for over a year and then had it photographed by Man Ray calling the result *Dust Breeding;* then he cleaned everything but a section of the cones, to which he cemented the dust with a fixative. The final touch came when the *Glass* was broken while in transit and was thereby webbed with a network of cracks. Duchamp is reported to have commented with satisfaction, "Now it is complete."

Back in Paris in the mid-1930s, Duchamp devised a work of art that made all of his inventions easily portable. The *Boîte en valise (Box in a Valise)* (fig. 305) was a kind of leather briefcase filled with miniature replicas of his previous works, including all the aforementioned objects. It provided a survey of Duchamp's work, like a traveling retrospective. When, like so many other European artists, Duchamp sought final refuge in America from the war in Europe, he came equipped with his "portable museum." Friends such as the American artist Joseph Cornell, who also made art in box form (see fig. 260), helped him assemble the many parts of *Boîte*. Like *The Green Box,* it was made into a multiple edition. Duchamp intended that the viewer participate in setting up and handling the objects in the valise. In this way, the viewer completes the creative act set in motion by the artist.

Although Duchamp declared that he had ceased all formal artistic activity in order to devote himself to chess (at which he excelled), he worked in secret for twenty years on a major sculptural project, completed in 1966. *Étant Donnés* (fig. 306), one of the most disturbing and enigmatic works of the century, only came to public light after the artist's death when

306. MARCEL DUCHAMP. *Étant Donnés (Given: 1. The Waterfall 2. The Illuminating Gas)*. 1944–66. Mixed media, height 7' 11½" (2.4 m). Philadelphia Museum of Art

GIFT OF THE CASSANDRA FOUNDATION.

© 1997 ADAGP, PARIS/ARTISTS RIGHTS SOCIETY (ARS), NEW YORK

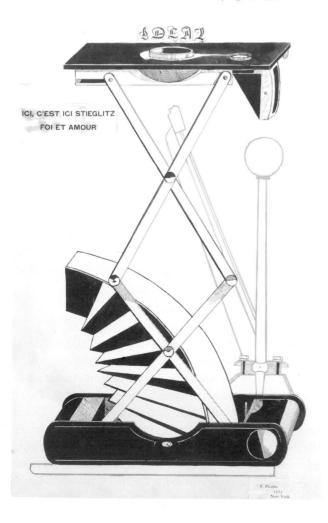

it was installed in the Philadelphia Museum of Art, which owns most of Duchamp's major works. *Étant Donnés*, a mixed-media assemblage built around the realistic figure of a nude woman sprawled on the ground, can only be viewed by one person at a time through a peephole in a large wooden door.

FRANCIS PICABIA (1879–1953). Picabia was born in Paris of wealthy Cuban and French parents. Between 1908 and 1911 he moved from Impressionism to Cubism. He joined the Section d'Or briefly and then experimented with Orphism and Futurism (color plate 106, page 205). In New York in 1915, he collaborated with Marcel Duchamp in establishing the American version of proto-Dada and, in the spirit of Duchamp, took up machine imagery as an emblematic and ironic mode of representation. "Almost immediately upon coming to America," Picabia said, "it flashed on me that the genius of the modern world is in machinery, and that through machinery art ought to find a most vivid expression." In this "mechanomorphic" style Picabia achieved some of his most distinctive work, particularly a series of Machine Portraits of himself and his key associates in New York. Thus, he saw Stieglitz (fig. 307) as a broken bellows camera, equipped with an automobile brake in working position and a gear shift in neutral, signifying the frustrations experienced by someone trying to present experimental art in philistine America. Both the Gothic letters and the title—*IDEAL*—confirm the conceptual or heraldic form of the portrait, while also establishing a witty contrast between the commonplace, mechanical imagery and the ancient, noble devices of traditional heraldry. Made for publication in Stieglitz's journal *291*, Picabia's 1915 portraits are modest in size, materials, and ambition, but in other works the artist developed his machine aesthetic—his "functionless" machines—into quite splendidly iconic paintings. They are still tongue-in-cheek, however, in their commentary on the seriocomic character of human sexual drives, and a work such as *Amorous Procession* (fig. 308) has many parallels, both thematic and formal, with Duchamp's *The Large Glass,* then under way in New York.

Returning to Europe in 1916, Picabia founded his journal *391* in Barcelona and published it intermittently in New York, Zurich, and Paris until 1924. After meeting the Zurich Dadaists in 1918, he was active in the Dada group in Paris. He reverted to representational art and, after the emergence of Surrealism, painted a series of Transparencies in which he superimposed thin layers of transparent imagery delineating classically beautiful male and female images, sometimes accompanied by exotic flora and fauna. Though the Transparencies were long dismissed as decadently devoid of either content or formal interest, they have assumed new importance among the acknowledged prototypes of 1980s Neo-Expressionism (see chapter 25).

With the end of World War II, Picabia resumed abstract

left: 307. FRANCIS PICABIA. *Ideal*. 1915. Pen and red and black ink on paper, 29⅞ x 20" (75.9 x 50.8 cm). The Metropolitan Museum of Art, New York

ALFRED STIEGLITZ COLLECTION, 1949. © 1997 ADAGP, PARIS/ARTISTS RIGHTS SOCIETY (ARS), NEW YORK

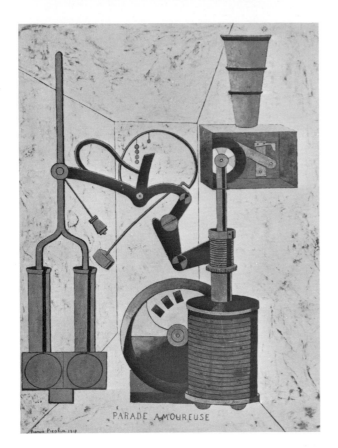

308. FRANCIS PICABIA. *Amorous Procession.* 1917.
Oil on cardboard, 38¼ x 29⅛" (97.2 x 74 cm).
Collection Morton G. Neumann Family, Chicago
© 1997 ADAGP, PARIS/ARTISTS RIGHTS SOCIETY (ARS), NEW YORK

309. MORTON SCHAMBERG. *God.* c. 1918. Miter box and
plumbing trap, height 10½" (26.7 cm).
Philadelphia Museum of Art LOUISE AND WALTER ARENSBERG COLLECTION

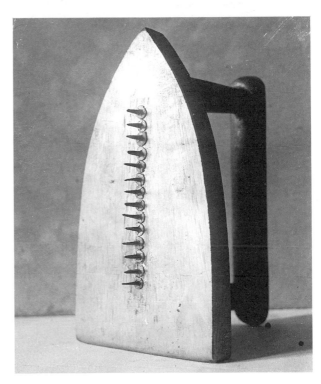

310. MAN RAY. *Gift.* Replica of lost original of 1921. Flatiron
with nails, height 6½ x 3⅝ x 3¾" (16.5 x 9.2 x 9.5 cm).
Collection Morton G. Neumann Family, Chicago
© 1997 MAN RAY TRUST/ADAGP, PARIS/ARTISTS RIGHTS SOCIETY (ARS), NEW YORK, NY

painting. His artistic qualities tend to be obscured by his personality. Exuberant and wealthy, with a concomitant magnificence of gesture, Picabia loved controversy, wished to be in the forefront of every battle, and had a cultivated sense of the ridiculous.

MORTON SCHAMBERG (1881–1918). Relatively unburdened by tradition, some American artists who met Duchamp and Picabia in New York had little difficulty entering into the Dada spirit. Indeed, one of the most daring of all Dada objects was made by Philadelphia-born Morton Schamberg, who mounted a plumbing connection on a miter box and sardonically titled it *God* (fig. 309). Though known primarily for this work, Schamberg was actually a painter and photographer who made a number of abstract, machine-inspired compositions in oil before his life was cut short by the massive influenza epidemic in 1918.

MAN RAY (1890–1976). No less ingenious in his ability to devise Dada objects was Man Ray (born Emmanuel Radnitsky), also from Philadelphia, who went on to pursue a lengthy, active career in the ambience of Surrealism. He gathered with Duchamp and Picabia at the Arensbergs' home and by 1916 had begun to make paintings inspired by a Dada machine aesthetic and three-dimensional constructions made with found objects. By 1919, Man Ray, who was always looking for ways to divest himself of the "paraphernalia of the traditional painter," was creating the first paintings made with an airbrush, which he called "Aerographs." The airbrush, normally reserved for commercial graphic work, made possible the soft tonalities in the dancing fans and cones of *Seguidilla* (color plate 137, page 267). The artist was delighted with his

new discovery. "It was wonderful," he said, "to paint a picture without touching the canvas."

In 1921, disappointed that Dada had failed to ignite a full-scale artistic revolution in New York, Man Ray moved to Paris. His exhibition in December of that year was a Dada event. The gallery was completely filled with balloons that viewers had to pop in order to discover the art. *Gift* (fig. 310), which exists today only in replicated form after a lost 1921 original, was made in the spirit of Duchamp's slightly altered or "assisted" readymades. With characteristic black humor, Man Ray subverted an iron's normal utilitarian function by attaching fourteen tacks to its surface, transforming this familiar object into something alien and threatening. The work was made for avant-garde composer Erik Satie, hence its title. Man Ray, who had taken up photography in 1915 through his association with Stieglitz, invented cameraless photographic images that he called "Rayographs." These were made by placing objects on or near sensitized paper that was then exposed directly to light. In the proper Dadaish manner, the technique was discovered accidentally in the darkroom. By controlling exposure and by moving or removing objects, the artist used this "automatic" process to create images of a strangely abstract or symbolic character (fig. 311).

Man Ray became an established figure in the Parisian avant-garde, gaining fame with his Dada films, with his experimental photographs, his photographs of artists and art, and his fashion photography (see fig. 399). By the mid-1920s, he was a central figure of the Surrealist circle, and his work of those years is taken up in chapter 15. Several of his assistants became important photographers in their own right: Berenice Abbott (see fig. 496), Bill Brandt (see fig. 408), and Lee Miller (see fig. 553).

German Dada: 1918–23

In 1917, with a devastating war, severe restrictions on daily life, and rampant inflation, the future in Germany seemed completely uncertain. This atmosphere was very conducive to the spread of Dada. Huelsenbeck, returning from Berlin to Zurich, joined a small group including the brothers Wieland and Helmut Herzfelde (who changed his name to John Heartfield as a pro-American gesture), Hannah Höch, the painters Raoul Hausmann, George Grosz, and, later, Johannes Baader. Huelsenbeck opened a Dada campaign early in 1918 with speeches and manifestos attacking all phases of the artistic status quo, including Expressionism, Cubism, and Futurism. A Club Dada was formed, to which Kurt Schwitters, later a major exponent of German Dada, was refused membership because of his association with Der Sturm Gallery, regarded by Huelsenbeck as a bastion of Expressionist art and one of Dada's chief targets. After the war came a period of political chaos, which did not end with the establishment of the Weimar Republic (1918–1933). Disillusionment among members of the Dada group set in as they realized that the so-called socialist government was actually in league with business interests and the imperial military. From their disgust with what they regarded as a bankrupt Western culture, they turned to art as a medium for social and political activity. Their weapons were mostly collage and photomontage. Berlin Dada, especially for Heartfield and Grosz, quickly took on a left-wing, pacifist, and Communist direction. The Herzfelde brothers' journal, *Neue Jugend,* and their publishing house, Malik-Verlag, utilized Dada techniques for political propaganda (color plate 138, page 267; fig. 319). George Grosz made many savage social and political drawings and paintings for the journal. One of the publications financed by the Herzfelde brothers was *Everyman His Own*

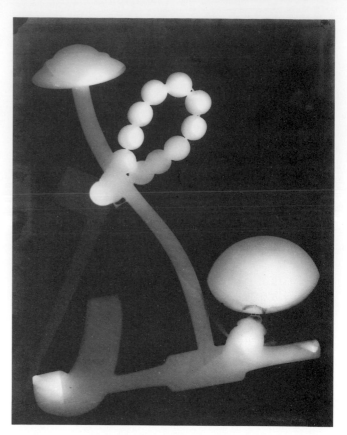

311. MAN RAY. *Untitled.* 1922. Gelatin-silver print (Rayograph), 9⅜ x 7" (23.8 x 17.78 cm). The Museum of Modern Art, New York GIFT OF JAMES THRALL SOBY. COPY PRINT © 1995 THE MUSEUM OF MODERN ART, NEW YORK. © 1997 MAN RAY TRUST/ADAGP, PARIS/ARTISTS RIGHTS SOCIETY (ARS) NEW YORK

Football. Although the work was quickly banned, its title became a rallying cry for German revolutionists. In 1919 the first issue of *Der Dada* appeared, followed in 1920 by the first international Dada-Messe, or Dada Fair, where the artists covered the walls of a Berlin gallery with photomontages, posters, and slogans like "Art is dead. Long live the new machine art of Tatlin." The rebellious members of Dada never espoused a clear program, and their goals were often ambiguous and sometimes contradictory. For, while they used art as a means of protest, they also questioned the very validity of artistic production.

The Zurich Dada experiments in noise-music and in abstract phonetic poetry were further explored in Berlin. Hausmann, the chief theoretician and writer of the group (nicknamed "the Dadasoph"), claimed the invention of a form of optophonetic poem, involving "respiratory and auditive combinations, firmly tied to a unit of duration," expressed in typography by "letters of varying sizes and thicknesses which thus took on the character of musical notations." In his *Spirit of Our Time (Mechanical Head)* (fig. 312), 1919, Hausmann created a kind of three-dimensional collage. To a wooden mannequin head he attached real objects, including a metal collapsing cup, a tape measure, labels, and a pocketbook. Through his use of common found objects, Hausmann partook of the iconoclastic spirit of Duchamp's readymades and implied that human beings had been reduced to mindless robots, devoid of individual will.

In the visual arts, a major invention or discovery was photomontage, created by cutting up and pasting together photographs of individuals and events, posters, book jackets, and a variety of typefaces in new and startling configurations—anything to achieve shock. The source material for this medium was made possible by the tremendous growth in the print media in Germany. Hausmann and Höch (who were lovers and occasional collaborators), along with Grosz, claimed to have originated Dada photomontage, although it was a technique that had existed for years in advertising and popular imagery. The technique derived from Cubist collage or papier collé, except that, in the Dada version, subject was paramount. Even this form has antecedents in Futurist collages (color plate 110, page 207). However, photomontage proved to be an ideal form for Dadaists and subsequently for Surrealists.

HANNAH HÖCH (1889–1978). Höch's large photomontage of 1919–20, with its typically sardonic title, *Cut with the Kitchen Knife Dada Through the Last Weimar Beer Belly Cultural Epoch of Germany* (fig. 313), was included in the Dada Messe, despite the efforts of Grosz and Heartfield to exclude her work. The dizzying profusion of imagery here demonstrates the ways in which photomontage relies on material appropriated from its normal context, such as magazine illustration, and introduces it into a new, disjunctive context, thereby investing it with new meaning. Höch here presents a satirical panorama of Weimar society. She includes

313. HANNAH HÖCH. *Cut with the Kitchen Knife Dada Through the Last Weimar Beer Belly Cultural Epoch of Germany*. 1919–20. Photomontage, 44⅞ x 35½" (114 x 90.2 cm). Staatliche Museen zu Berlin, Preussischer Kulturbesitz, Nationalgalerie
© 1997 VG BILD-KUNST, BONN/ARTISTS RIGHTS SOCIETY (ARS), NEW YORK

photographs of her Dada colleagues, Communist leaders, dancers, and sports figures, and Dada slogans in varying typefaces. The despised Weimar government leaders at the upper right are labeled "anti-Dada movement." At the very center of the composition is a photo of a popular dancer who seems to toss her own, out-of-scale head into the air. The head is a photo of expressionist printmaker Käthe Kollwitz (color plate 74, page 159; see fig. 164). Although the Dadists reviled the emotive art of the Expressionists, it seems likely that Höch had nothing but respect for this left-wing woman artist. Throughout the composition are photographs of gears and wheels, both a tribute to technology and a means of imparting a sense of dynamic, circular movement everywhere. One difference between Höch and her colleagues is the preponderance of female imagery in her work, indicative of her interest in the new roles of women in postwar Germany, which had just granted women the vote in 1918, two years before the United States.

JOHN HEARTFIELD (1891–1968). Heartfield made photomontages of a somewhat different variety. He composed images from the clippings he took from newspapers, retouching them in order to blend the parts into a facsimile of a single, integrated image. These images were photographed and made into photogravures for mass reproduction. Beginning in 1930, Heartfield contributed illustrations regularly to the left-wing magazine *AIZ* or *Workers' Illustrated Newspaper*.

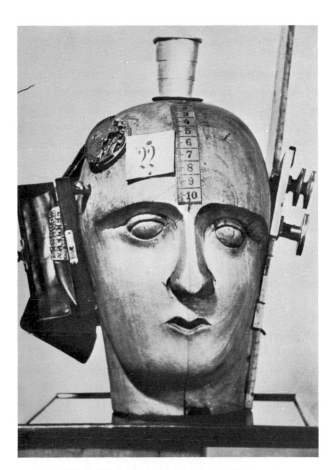

312. RAOUL HAUSMANN. *The Spirit of Our Time (Mechanical Head)*. 1919. Wood, leather, aluminum, brass, and cardboard, 12⅝ x 9" (32.1 x 22.9 cm). Musée National d'Art Moderne, Centre d'Art et de Culture Georges Pompidou, Paris
© 1997 ADAGP, PARIS/ARTISTS RIGHTS SOCIETY (ARS), NEW YORK

For one of his most jarring images of protest, made after Hitler became German chancellor (and had assumed dictatorial powers), he altered the words of a traditional German Christmas carol and twisted the form of a Christmas tree into a swastika tree (color plate 138, page 267). "O Tannenbaum im deutchen Raum, wie krumm/sind deine Äste!" (O Christmas tree in German soil, how crooked are your branches!) the heading reads. A text below states that in the future all trees must be cut in this form.

Somewhat apart from the Berlin Dadaists was the Hanoverian artist KURT SCHWITTERS (1887–1948) who completed his formal training at the Academy in Dresden and painted portraits for a living. He quarreled publicly with Huelsenbeck and was denied access to Club Dada because of his involvement with the apolitical and pro-art circle around Herwarth Walden's Der Sturm Gallery. He was eventually reconciled with other members of the group, however, and established his own Dada variant in Hanover under the designation *Merz*, a word in part derived from Com*merz*bank. "At the end of 1918," he wrote, "I realized that all values only exist in relationship to each other and that restriction to a single material is one-sided and small-minded. From this insight I formed Merz, above all the sum of individual art forms, Merz-painting, Merz-poetry." Schwitters was a talented poet, impressive in his readings, deadly serious about his efforts in painting, collage, and construction. These always involved a degree of deadpan humor delightful to those who knew him, but disturbing to unfamiliar audiences.

Schwitters's collages were made of rubbish picked up from the street—cigarette wrappers, tickets, newspapers, string, boards, wire screens, whatever caught his fancy. In these socalled *Merzbilder* or *Merzzeichnungen* (*Merz*-pictures or *Merz*-drawings) he was able to transform the detritus of his surroundings into strange and wonderful beauty. In *Picture with Light Center* (color plate 139, page 267) we see how Schwitters could extract elegance from these lowly found materials. He carefully structured his circular and diagonal elements within a Cubist-derived grid, which he then reinforced by applying paint over the collage, creating a glowing, inner light that radiates from the picture's center. For his 1920 assemblage, *Merzbild 25A (Das Sternenbild) (The Stars Picture)* (fig. 314), Schwitters did not restrict himself to the two-dimensional printed imagery we have seen in the photomontages of Heartfield and Höch. He incorporates rope, wire mesh, paint, and other materials which indicate the artist's strong concern for physicality in his surfaces. But his are not accidental juxtapositions or ones made purely for formal effect. The snatches of text from German newspapers in this *Merzbild* can be decoded as referring to recent political events in Germany.

Schwitters introduced himself to Raoul Hausmann in a Berlin café in 1918 by saying, "I am a painter and I nail my pictures together." This description applies to all his relief constructions, since he drew no hard-and-fast line between them and papiers collés, but most specifically to his series of great constructions that he called *Merz-Column* or *Merzbau* (fig.

314. KURT SCHWITTERS. *Merzbild 25A (Das Sternenbild*. 1920. Assemblage, 41 x 31⅛" (104.1 x 79.1 cm). Kunstsammlung Nordrhein-Westfalen, Düsseldorf
© 1997 VG BILD-KUNST, BONN/ARTISTS RIGHTS SOCIETY (ARS), NEW YORK

315), the culmination of Schwitters' attempts to create a *Gesamtkunstwerk*, or total work of art. He began the first one in his house in Hanover around 1920 as an abstract plaster sculpture with apertures dedicated to his Dadaist and Constructivist friends and containing objects commemorating them: Mondrian, Gabo, Arp, Lissitzky, Malevich, Richter, Mies van der Rohe, and van Doesburg. The *Merzbau* grew throughout the 1920s with successive accretions of every kind of material until it filled the room. Having then no place to go but up, he continued the environmental construction with implacable logic into the second storey. "As the structure grows bigger and bigger," Schwitters wrote, "valleys, hollows, caves appear, and these lead a life of their own within the overall structure. The juxtaposed surfaces give rise to forms twisting in every direction, spiraling upward." When he was driven from Germany by the Nazis and his original *Merzbau* was destroyed, Schwitters started another one in Norway. The Nazi invasion forced him to England, where he began again for the third time. After his death in 1948, the third *Merzbau* was rescued and preserved in the University of Newcastle. The example of Schwitters was crucial for later artists who sought to create sculpture on an environmental scale. They include figures as diverse as Louise Nevelson (color plate 261, page 463), Red Grooms (see fig. 616), and Louise Bourgeois (color plate 259, page 462).

Early in the 1920s, the de Stijl artist Theo van Doesburg had become friendly with Schwitters, and they collaborated on a number of publications, and even made a Dada tour of

315. KURT SCHWITTERS.
Hanover Merzbau. Destroyed.
Photo taken c. 1931
© 1997 VG BILD-KUNST, BONN/ARTISTS RIGHTS SOCIETY (ARS), NEW YORK

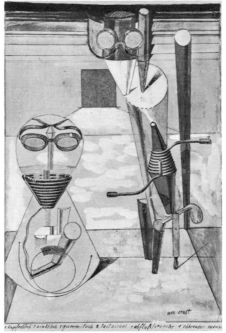

316. MAX ERNST. *1 Copper Plate 1 Zinc Plate 1 Rubber Cloth 2 Calipers 1 Drainpipe Telescope 1 Pipe Man.* 1920. Gouache, ink, and pencil on printed reproduction, 9½ x 6½" (24.1 x 16.7 cm). Whereabouts unknown © 1997 ADAGP, PARIS/ ARTISTS RIGHTS SOCIETY (ARS), NEW YORK

below: 317. MAX ERNST. *Here Everything Is Still Floating.* 1920. Pasted photoengravings and pencil on paper, 4⅛ x 4⅞" (10.5 x 12.4 cm). The Museum of Modern Art, New York
PURCHASE. PHOTOGRAPH © 1995 THE MUSEUM OF MODERN ART, NEW YORK. © 1997 ADAGP, PARIS/ARTISTS RIGHTS SOCIETY (ARS), NEW YORK

Holland. Schwitters more and more became attracted to the geometric abstractionists and Constructivists, for despite the fantasy and rubbish materials in his works, they demonstrated relationships and proportions that rival Mondrian's Neo-Plastic paintings in their subtlety and rigor.

When **MAX ERNST** (1891–1976) saw works by modern artists such as Cézanne and Picasso at the 1912 Sonderbund exhibition in Cologne, he decided to forgo his university studies and take up art. He went to Paris in 1913, and his works were included that year in the First German Autumn Salon at Der Sturm in Berlin, when the artist was only twenty-two years old. In 1919–20, Jean Arp established contact with Ernst in Cologne, and the two were instrumental in the formation of yet another wing of international Dada. The two had met in Cologne in 1914, but Ernst then served in the German army. At the end of the war, he discovered Zurich Dada and the paintings of de Chirico and Klee. Ernst's early paintings were rooted in a Late Gothic fantasy drawn from Dürer, Grünewald, and Bosch. The artist was also fascinated by German Romanticism in the macabre forms of Klinger and Böcklin. This "gothic quality" (in the widest sense) remained a most consistent characteristic of Ernst's fantasies. In 1919–20, a staggeringly productive period, he produced collages and photomontages that demonstrated a genius for suggesting the metamorphosis or double identity of objects, a topic later central to Surrealist iconography. In an ingenious work from 1920 (fig. 316), Ernst invented his own mechanistic forms as stand-ins for the human body. As he frequent-

ly did during this period, the artist took a page from a 1914 scientific text illustrating chemistry and biology equipment and, by overpainting certain areas and inserting his own additions, he transformed the goggles and other laboratory utensils into a pair of hilarious creatures before a landscape. The composition bears telling comparison with Picabia's mechanomorphic inventions (fig. 308).

During the winter of 1920–21, Ernst and Arp collaborated on collages entitled *Fatagagas,* short for *Fabrication des Tableaux garantis gasométriques.* Ernst usually provided the collage imagery while Arp and other occasional collaborators provided the name and accompanying text. In the *Fatagaga* titled *Here Everything Is Still Floating* (fig. 317), an anatomical drawing of a beetle becomes, upside-down, a steamboat floating through the depths of the sea. Some of the *Fatagagas* were sent to Tristan Tzara in Paris for illustration in his ill-fated publication, *Dadaglobe.* Ernst had been in close contact with the Parisian branch of Dada and many of the artists who would develop Surrealism. By the time he moved to Paris in 1922, he had already created the basis for much of the Surrealist vocabulary.

Despite the close friendship between Ernst and Arp, their approaches to painting, collage, and sculpture were different. Arp's was toward abstract, organic Surrealism in which figures or other objects may be suggested but are rarely explicit. Ernst, following the example of de Chirico and fortified by his own "gothic" imagination, became a principal founder of the wing of Surrealism that utilized Magic Realism—that is, precisely delineated, recognizable objects, distorted and transformed, but nevertheless presented with a ruthless realism that throws their newly acquired fantasy into shocking relief. *Celebes* (color plate 140, page 268) is a mechanized monster whose trunk-tail-pipe-line sports a cowskull-head above an immaculate white collar. A headless classical torso beckons to the beast with an elegantly gloved hand. The images are unrelated on a rational level; some are threatening (the elephant) while others are less explicable (the beckoning torso). The rotund form of the elephant was actually suggested to Ernst by a photograph of gigantic communal corn bin made of clay and used by people in southern Sudan. While not constructed as a collage, *Celebes,* with its many disparate motifs, is informed by a collage aesthetic as well as by the early paintings of de Chirico; it appeals to the level of perception below consciousness. Ernst, who worked in a remarkably broad range of media and styles, became a major figure of Surrealism (see figs. 358, 359, 360, 361).

Paris Dada: 1919–22

By the time Ernst left Cologne in 1921, Dada as an organized movement was dead, except in Paris. Duchamp had largely given up creative activity (unless his very existence might be so described) apart from occasional motion experiments, the completion of his *Large Glass,* and participation in commemorative exhibitions and publications. In Germany only Kurt Schwitters continued, a one-man Dada movement. Many of the original Dadaists, if they had not done so already, left Germany in the 1930s. But their efforts were not forgotten in the "Degenerate Art" exhibition organized by the Nazis in 1937 in Munich (fig. 318). On the "Dada wall," Grosz's statement, "Take Dada seriously! It's worth it!," is derisively inscribed next to works by Schwitters and enlarged details from works by Kandinsky (whom the organizers mistook for a Dada artist).

By 1918 writers in Paris, including Louis Aragon, André Breton, and Georges Ribemont-Dessaignes, were contributing to the Zurich periodical *Dada,* and shortly Tristan Tzara began writing for Breton and Aragon's journal *Littérature.* Picabia, after visiting the Dadaists in Zurich, published No. IX of *391* in Paris, and in 1919 Tzara himself also arrived there. Thus, by the end of the war, a Dada movement was growing in the French capital, but it was primarily a literary event. While in painting and sculpture Dada was largely an imported product, in poetry and theater it was in a tradition of the irrational and absurd that extended from Baudelaire,

318. Room 3 including Dada wall in Degenerate Art Exhibition, Munich, 1937

Rimbaud, and Mallarmé to Alfred Jarry, Guillaime Apollinaire, and Jean Cocteau. Paris Dada in the hands of the literary men consisted of frequent manifestos, demonstrations, periodicals, events, and happenings more violent and hysterical than ever before. The artists Arp, Ernst, and Picabia took a less active part in the demonstrations, although Picabia contributed his own manifestos. The original impetus and enthusiasm seemed to be lacking, however, and divisive factions arose, led principally by Tzara and by Breton. The early Dadaists, like the Bolsheviks of the Russian Revolution, sought to maintain a constant state of revolution, even anarchy, but new forces led by Breton sought a more constructive revolutionary scheme based on Dada. In its original form Dada expired in the wild confusion of the Congress of Paris called in 1922 by the Dadaists. The former Dadaists, including Picabia, who followed Breton, were joined by powerful new voices, among them Cocteau and Ezra Pound. By 1924 this group was consolidated under the name Surrealism.

The "New Objectivity" in Germany

In Germany the end of World War I brought a period of chaotic struggle between the extreme Left and the extreme Right. The Social Democrat government could control neither the still-powerful military, nor the skyrocketing inflation. Experimental artists and writers were generally leftwing, hopeful that from chaos would emerge a more equable society. Immediately following the Social Democrat revolution of 1918, some of the leaders of German Expressionism formed the November Group and were soon joined by Dadaists. In 1919 the Workers' Council for Art was organized. It sought more state support, more commissions from industry, and the reorganization of art schools. These artists included many of the original Expressionists, such architects as Walter Gropius and Erich Mendelsohn, and leading poets, musicians, and critics. Many of the aims of the November Group and the Workers' Council were incorporated into the program of the Weimar Bauhaus.

In their exhibitions, the November Group re-established contact with France and other countries, and antagonisms among the various new alignments were submerged for a time on behalf of a common front. The Expressionists were exhibited, as well as representatives of Cubism, abstraction, Constructivism, and Dada, while the new architects showed projects for city planning and large-scale housing. Societies similar to the November Group sprang up in many parts of Germany, and their manifestos read like a curious blend of Socialist idealism and revivalist religion.

The common front of Socialist idealism did not last long. The workers whom the artists extolled and to whom they appealed were even more suspicious of the new art forms than was the capitalist bourgeoisie. It was naturally from the latter that new patrons of art emerged. By 1924 German politics was shifting inevitably to the right, and the artists' Utopianism was in many cases turning into disillusionment and cynicism. There now emerged a form of Social Realism in painting to which the name the New Objectivity (*Die*

Neue Sachlichkeit) was given. Meanwhile, the Bauhaus, since 1919 at Weimar, had tried to apply the ideas of the November Group and the Workers' Council toward a new relationship between artist and society, and in 1925 it was forced by the rising tide of conservative opposition to move to Dessau. Its faculty members still clung to postwar Expressionist and Socialist ideas, although by 1925 their program mostly concerned the training of artists, craft workers, and designers for an industrial, capitalist society.

GEORGE GROSZ (1893–1959). The principal painters associated with the New Objectivity were George Grosz, Otto Dix, and Max Beckmann—three artists whose style touched briefly at certain points but who had essentially different motivations. Grosz studied at the Dresden Academy from 1909 to 1911 and at the Royal Arts and Crafts School in Berlin off and on from 1912 to 1916, partially supporting himself with drawings of the shady side of Berlin night life. These caustic works prepared him for his later violent statements of disgust with postwar Germany and humankind generally. In 1913 he visited Paris and, despite later disclaimers, was obviously affected by Cubism and its offshoots, particularly by Robert Delaunay and Italian Futurism. After two years in the army (1914–16), he resumed his caricatures while convalescing in Berlin. They reveal an embittered personality, now fortified by observations of autocracy, corruption, and the horrors of a world at war. Recalled to the army in 1917, Grosz ended his military career in an asylum on whose personnel and administration he made devastating comment.

After the war Grosz was drawn into Berlin Dada and its overriding left-wing direction. He made stage designs and collaborated on periodicals, but continued his own work of political or social satire. *Dedication to Oskar Panizza* of 1917–18 (color plate 141, page 268) is his most Futurist work. The title refers to the writer Panizza, whose work had been censored at the end of the nineteenth century. But the larger subject, according to Grosz, was "Mankind gone mad." He said the dehumanized figures represent "Alcohol, Syphilis, Pestilence." Showing a funeral turned into a riot, the painting is flooded with a blood or fire red. The buildings lean crazily; an insane mob is packed around the paraded coffin on which Death sits triumphantly, swigging at a bottle; the faces are horrible masks; humanity is swept into a hell of its own making, and the figure of Death rides above it all on a black coffin. Yet the artist controls the chaos with a geometry of the buildings and the planes into which he segments the crowd.

The drawing *Fit for Active Service* (fig. 319) was used as an illustration in 1919 for *Die Pleite (Bankruptcy)*, one of the many political publications with which Grosz was involved. The editors, Herzfelde, Grosz, and Heartfield, filled *Die Pleite* with scathing political satire, which occasionally got them thrown into prison. This work shows Grosz's sense of the macabre and his detestation of bureaucracy, with a fat complacent doctor pronouncing his "O.K." of a desiccated cadaver before arrogant Prussian-type officers. The spare economy of the draftsmanship in Grosz's illustrations is also

319. GEORGE GROSZ. *Fit for Active Service (The Faith Healers)*. 1916–17. Pen, brush, and India ink, sheet 20 x 14⅜" (50.8 x 36.5 cm). The Museum of Modern Art, New York A. CONGER GOODYEAR FUND. PHOTOGRAPH © 1995 THE MUSEUM OF MODERN ART, NEW YORK. © 1997 ESTATE OF GEORGE GROSZ/VAGA, NEW YORK, NY

320. GEORGE GROSZ. *Republican Automatons*. 1920. Watercolor, 23⅝ x 18⅝" (60 x 47.3 cm). The Museum of Modern Art, New York ADVISORY COMMITTEE FUND. PHOTOGRAPH © 1995 THE MUSEUM OF MODERN ART, NEW YORK. © 1997 ESTATE OF GEORGE GROSZ/VAGA, NEW YORK, NY

evident in a number of paintings done by the artist in the same period. *Republican Automatons* (fig. 320) applies the style and motifs of de Chirico and the Metaphysical School to political satire, as empty-headed, blank-faced, and mutilated automatons parade loyally through the streets of a mechanistic metropolis on their way to vote as they are told. In such works as this, Grosz comes closest to the spirit of the Dadaists and Surrealists. But he expressed his most passionate convictions in drawings and paintings that continue an Expressionist tradition of savagely denouncing a decaying Germany of brutal profiteers and obscene prostitutes, and of limitless gluttony and sensuality in the face of abject poverty, disease, and death. Normally Grosz worked in a style of spare and brittle drawing combined with a fluid watercolor. In the mid-1920s, however, he briefly used precise realism in portraiture, close to the New Objectivity of Otto Dix (fig. 322).

Grosz was frequently in trouble with the authorities, but it was Nazism that caused him to flee. In the United States during the 1930s, his personality was quite transformed. Although he occasionally caricatured American types, these were relatively mild, even affectionate. America for him was a dream come true, and he painted the skyscrapers of New York or the dunes of Provincetown in a sentimental haze. His pervading sensuality was expressed in warm portrayals of Rubenesque nudes.

The growth of Nazism rekindled his power of brutal commentary, and World War II caused Grosz to paint a series expressing bitter hatred and deep personal disillusionment. In the later works, sheer repulsion replaces the passionate convictions of his earlier statements. He did not return to Berlin until 1958, and he died there the following year.

OTTO DIX (1891–1969). The artist whose works most clearly define the nature of the New Objectivity was Otto Dix. Born of working-class parents, he was a proletarian by upbringing as well as by theoretical conviction. Dix's combat experience made him fiercely antimilitaristic. His war paintings, gruesome descriptions of indescribable horrors are rooted in the German Gothic tradition of Grünewald (see fig. 8). His painting *The Trench* (fig. 321) traveled throughout Germany as part of an exhibition mounted by a group called "No More War." It was stored for a time by Ernst Barlach (see fig. 182), but it was eventually destroyed, perhaps when the Nazis set fire to works of "degenerate" art in Berlin toward the end of the war. Like Grosz, Dix was exposed to influences from Cubism to Dada, but from the beginning he was concerned with uncompromising realism. This is a symptom of the postwar reaction against abstraction, a reaction most marked in Germany but also evident in most of Europe and in America. Even many of the pioneers of Fauvism and Cubism were involved in it. However, the superrealism of Dix was not simply a return to the past. In his portrait of the Laryngologist, *Dr. Mayer-Hermann* (fig. 322), the massive figure is seated frontally, framed by the vaguely menacing machines of his profession. Although the painting includes nothing bizarre or extraneous, the overpowering confrontation gives a sense of the unreal. For this type of "superreal-

321. OTTO DIX. *The Trench*. 1923. Destroyed 1943–45. Oil on canvas

ism" (as distinct from Surrealism) the term "Magic Realism" was coined: a mode of representation that takes on an aura of the fantastic because commonplace objects are presented with unexpectedly exaggerated and detailed forthrightness. Dix remained in Germany during the Nazi regime, although he was forbidden to exhibit or teach and was imprisoned for a short time. After the war he turned to a form of mystical, religious expression.

The portraiture of artists like Dix may very well have been influenced by the contemporary photographs of **AUGUST SANDER** (1876–1964), a German artist and former miner who became a natural and powerful exponent of the New Objectivity. Trained in fine art at the Dresden Academy, Sander set about to accomplish nothing less than a comprehensive photographic portrait of "Man in Twentieth-Century Germany." He was convinced that the camera, if honestly and straightforwardly employed, could probe beneath appearances and dissect the truth that lay within. Sander photographed German society in its "sociological arc" of occupations and classes, presenting cultural types in the environments that shaped them (fig. 323). In 1929 he published the first volume of his magnum opus—one of the

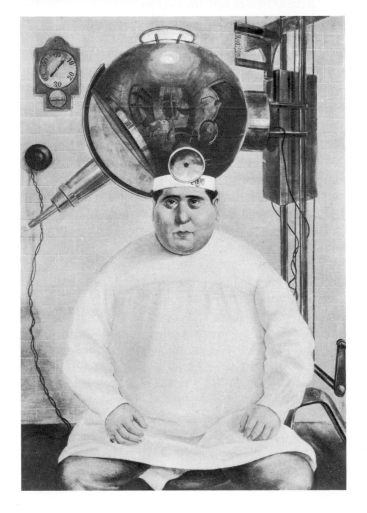

322. OTTO DIX. *Dr. Mayer-Hermann*. 1926.
Oil and tempera on wood, 58¾ x 39" (149.2 x 99.1 cm).
The Museum of Modern Art, New York

323. AUGUST SANDER. *Wandering People (Fahrendes Volk)*, from the series Citizens of the 20th Century (Menschen des 20 Jahrhunderts), from the portfolio, *Traveling People*. Düren (Cologne), Germany. 1930. Gelatin-silver print, 8¼ x 10" (21.0 x 25.5 cm). The Museum of Modern Art, New York
GIFT OF THE PHOTOGRAPHER.
© 1997 ARTISTS RIGHTS SOCIETY (ARS), NEW YORK/VG BILD-KUNST, BONN

most ambitious in the history of photography—under the title *Antlitz der Zeit (Face of Time)*, only to see the book suppressed and the plates (but not the negatives) destroyed in the early 1930s by the Nazis, who inevitably found the photographer's ideas contrary to their own pathological views of race and class. For his part, Sander wrote: "It is not my intention either to criticize or to describe these people, but to create a piece of history with my pictures."

The German photographer most immediately identified with the New Objectivity was **ALBERT RENGER-PATZSCH** (1897–1966). Like Paul Strand in the United States (see fig. 477), Renger-Patzsch avoided the double exposures and photographic manipulations of Man Ray (fig. 311) and Moholy-Nagy, as well as the artificiality and soulfulness of the Pictorialists, to practice "straight" photography closely and sharply focused on objects isolated, or abstracted, from the natural and man-made worlds (fig. 324). But for all its stark realism, such an approach yielded details so enlarged and crisply purified of their structural or functional contexts that the overall pattern they produce borders on pure design. Still, Renger-Patzsch insisted upon his commitment to factuality and his "aloofness to art for art's sake."

MAX BECKMANN (1884–1950). Beckmann was the principal artist associated with the New Objectivity but could only briefly be called a precise Realist in the sense of Dix. Born of wealthy parents in Leipzig, he was schooled in the Early Renaissance painters of Germany and the Netherlands, and the great seventeenth-century Dutch painters. After studies at the Weimar academy and a brief visit to Paris, Beckmann settled in 1903 in Berlin, then a center of German Impressionism and Art Nouveau. Influenced by Delacroix and by the German academic tradition, he painted large religious and classical murals and versions of contemporary disasters such as the sinking of the *Titanic* (1912), done in the mode of Géricault's *Raft of the Medusa*. By 1913 Beckmann was a well-known academician. His service in World War I

brought about a nervous breakdown and, as he later said, "great injury to his soul." The experience turned him toward a search for internal reality. Beckmann assumed many guises in over eighty-five self-portraits made throughout his life. The *Self-Portrait with Red Scarf* of 1917 (fig. 325) shows the artist in his Frankfurt studio, haunted and anxious.

324. ALBERT RENGER-PATZSCH. *Irons Used in Shoemaking, Fagus Works*. c. 1925. Gelatin-silver print. Galerie Wilde, Cologne

325. MAX BECKMANN. *Self-Portrait with Red Scarf*. 1917. Oil on canvas, 31½ x 23⅝" (80 x 60 cm). Staatsgalerie, Stuttgart

Beckmann's struggle is apparent in a big, unfinished *Resurrection,* on which he painted sporadically between 1916 and 1918. Here he attempted to join his more intense and immediate vision of the war years to his prewar academic formulas. This work, though unsuccessful, liberated him from his academic past. In two paintings of 1917, *The Descent from the Cross* (fig. 326) and *Christ and the Woman Taken in Adultery,* he found a personal expression, rooted in Grünewald, Bosch, and Brueghel, although the jagged shapes and delimited space also owed much to Cubism. Out of this mating of Late Gothic, Cubism, and German Expressionism emerged Beckmann's next style.

Although he was not politically oriented, Beckmann responded to the violence and cruelty of the last years of the war by painting dramas of torture and brutality—symptomatic of the lawlessness of the time and prophetic of the state-sponsored genocide of the 1930s. Rendered in pale, emotionally repulsive colors, the figures could be twisted and distorted within a compressed space, as in late medieval representations of the tortures of the damned (in Beckmann's work, the innocent), and the horror heightened by explicit and accurate details. Such works, which impart symbolic content through harsh examination of external appearance, were close to and even anticipated the New Objectivity of Grosz and Dix.

In *Self-Portrait in Tuxedo* from 1927 (color plate 142, page 268), Beckmann presents a view of himself quite different from the one ten years earlier. Now a mature figure, he appears serious and self-assured, debonair even. The composition is striking in its elegant simplicity, with deep blacks, for which Beckmann is understandably admired, set against the blue-gray wall and the artist's stark white shirt. In the later 1920s, as is clear from this image, Beckmann, moving from success to success, was regarded as one of Germany's leading artists. Nevertheless, when the Nazis came to power in 1933 he was stripped of his teaching position in Frankfurt, and 590 of his works were confiscated from museums throughout Germany. On the opening day of the 1937 "Degenerate Art" show in Munich, which included several works by Beckmann (including figs. 325 and 326), the artist and his wife fled to Amsterdam. In 1947, after years of hiding from the Nazis (he never returned to Germany), Beckmann accepted a teaching position at Washington University, St. Louis, Missouri, where he filled a position vacated by the Abstract Expressionist painter Philip Guston (color plate 243, page 455; color plate 244, page 455) and became a highly influential teacher. He remained in the United States for the rest of his life.

Throughout the 1930s and 1940s, Beckmann continued to develop his ideas of coloristic richness, monumentality, and complexity of subject. The enriched color came from visits to Paris and contacts with French artists, particularly Matisse and Picasso. However, his emphasis on literary subjects having heavy symbolic content reflected his Germanic artistic

326. MAX BECKMANN. *The Descent from the Cross*. 1917. Oil on canvas, 59½ x 50¾" (151.2 x 128.9 cm). The Museum of Modern Art, New York

roots. The first climax of his new, monumental-symbolic approach was the large 1932–33 triptych *Departure* (color plate 143, page 269). Beckmann made nine paintings in the triptych format, obviously making a connection between his work and the great ecclesiastical art of the past. Alfred Barr described *Departure* as "an allegory of the triumphal voyage of the modern spirit through and beyond the agony of the modern world." The right wing shows frustration, indecision, and self-torture; in the left wing, sadistic mutilation, torture of others. Beckmann said of this triptych in 1937: "On the right wing you can see yourself trying to find your way in the darkness, lighting the hall and staircase with a miserable lamp dragging along tied to you as part of yourself, the corpse of your memories, of your wrongs, of your failures, the murder everyone commits at some time of his life—you can never free yourself of your past, you have to carry the corpse while Life plays the drum." Also, despite his disavowal of political interests, the left-hand panel must refer to the rise of dictatorship that was driving liberal artists, writers, and thinkers underground.

The darkness and suffering in the wings are resolved in the brilliant sunlight colors of the central panel, where the king, the mother, and the child set forth, guided by the veiled boatman. Again, Beckmann: "The King and Queen have freed themselves of the tortures of life—they have overcome them. The Queen carries the greatest treasure—Freedom—as her child in her lap. Freedom is the one thing that matters—it is the departure, the new start."

It is important to emphasize the fact that Beckmann's allegories and symbols were not a literal iconography to be read by anyone given the key. The spectator had to participate actively; and the allusions could mean something different to each viewer. In addition, the allusions in Beckmann's work could be very complex. His last triptych, *The Argonauts* (fig. 327), was completed the day before he died. Beckmann originally intended the work as an allegory about the arts. Painting is represented in the left panel where an artist, presumably a surrogate self-portrait, works at a canvas. The women playing instruments in the right panel symbolize music, while the figures in the center depict poetry. Prompted by a dream to rename the painting *The Argonauts,* Beckmann gave his image a more mythological content. The model in the left-hand wing becomes Medea; the central figures represent Jason, the poet Orpheus, and the sea god Glaucus; and the women at the right form an ancient Greek chorus. But the triptych also has a biographical dimension, since during his years in Amsterdam Beckmann associated with a group of writers and artists called the Argonauts. Thus, Jason's heroic search for the golden fleece becomes an allegory about the voyage of Beckmann's own life.

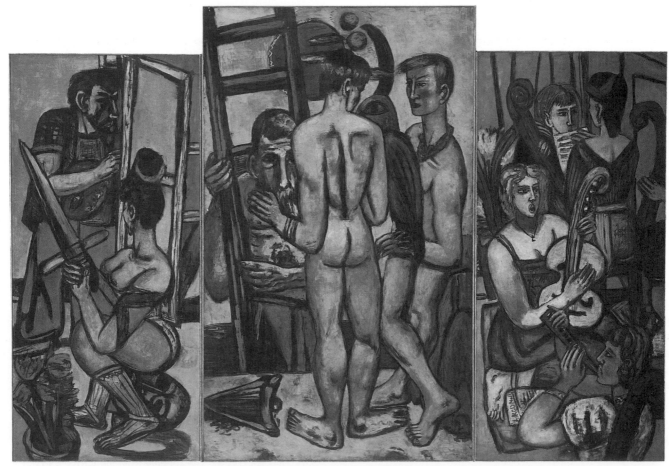

327. MAX BECKMANN. *The Argonauts.* 1949–50. Triptych, oil on canvas: left panel 6'1" x 33½" (1.8 m x 85.1 cm), center panel 6' 9" x 48" (2.1 m x 122 cm), right panel 6' 1" x 33½" (1.8 m x 85 cm).
National Gallery of Art, Washington, D.C. GIFT OF MRS. MAX BECKMANN. © 1997 VG BILD-KUNST, BONN/ARTISTS RIGHTS SOCIETY (ARS), NEW YORK

14 The School of Paris Between the Wars

André Derain was a major innovator in the early twentieth century who, by the outbreak of World War I, had turned to a classicizing style (see fig. 149). "After the war," he said, "I thought I would never be able to paint—that after all that—no one would be interested." In post-1918 Paris, in many quarters, years of tragic warfare had dampened enthusiasm for any experimentation, and cultural movements turned away from "scandalous" avant-gardism toward an art based on the values of the classical tradition. This general trend toward classicism in all the arts in the 1920s was characterized by poet and playwright Jean Cocteau as a "call to order." So profound was the longing for stability that even Dada, once it finally hit the French capital in the late teens, soon fizzled and gave way, as we shall see in the next chapter, to the less nihilistic, if thoroughly antirationalist, movement known as Surrealism. But far from transforming Paris into a citadel of reaction, this new common sensibility ushered in a mood of enlightened tolerance. Official taste grew less aggressively hostile to modernism; progressive painters and sculptors felt free to relax from formal experimentation and to settle for a period of consolidation and refinement of the extraordinary gains already made. Whole troupes of gifted artists in retreat from revolution in Russia, inflation in Germany, and provincialism in America were thus lured to Paris, assuring the continuing status of Paris as the glittering, cosmopolitan capital of world art.

The School of Paris embraced a wide variety of artists between the two world wars who intersected one another not only through their common base in France, but also in their independence from narrowly defined aesthetic categories. While distinctly modernist in their willingness to distort imagery for expressive purposes, they nonetheless regarded figurative imagery as fundamental to the meaning of their work. This diverse group could hardly have been more distinguished, including as it did the likes of Braque, Léger, Matisse, and Picasso. But perhaps even more representative of the polyglot School of Paris than those major figures is a subgroup known as *les maudits*—Modigliani from Italy, Soutine from Russia, and the Frenchman Maurice Utrillo. *Les mau-*

dits means "the cursed"—not only by poverty and alienation but also by the picturesque, ultimately disastrous bohemianism of their disorderly life-styles. Much of this gravitated around the sidewalk cafés of Montparnasse and Saint-Germain-des-Prés, the Left Bank quarters favored by the Parisian avant-garde since the mid-teens, when Picasso and his entourage abandoned Montmartre to the tourists.

AMEDEO MODIGLIANI (1884–1920) was born of well-to-do parents in Livorno, Italy. His training as an artist was often interrupted by illness, but he managed to get to Paris by 1906. Although he was essentially a painter, around 1910, influenced by his friend Brancusi, he began to experiment with sculpture. Modigliani's tragically short life has become the quintessential example of bohemian artistic existence in Paris, so much so that the myths and anecdotes surrounding his biography have helped to undermine a serious understanding of his work. For fourteen years he worked in Paris, ill from tuberculosis, drugs, and alcohol, but drawing and painting obsessively. He spent what little money he had quickly and tended to support himself by making drawings in cafés that he sold for a few francs to the sitter.

With very few exceptions the experiments Modigliani carried out in sculpture between approximately 1910 and 1914 (he rarely dated his works) are carved stone heads (fig. 328). He believed that modern sculpture should not be modeled in clay, but should result only from direct carving in stone. Beyond the powerful example of Brancusi, the sculptural sources that appealed to Modigliani were rooted in European medieval art and the art of a diverse range of cultures including Africa, Egypt, Archaic Greece, and India. Most of the heads have highly abstracted, geometricized features, with the characteristic long nose, almond eyes, and columnar neck. As we see here, Modigliani sometimes left the block of stone only roughed in at the back, making the sculptures strictly frontal in orientation. This tendency may have had something to do with his apparent intention to group the heads into some kind of decorative ensemble. It has been suggested that Modigliani abandoned the demanding métier of sculpture because of frail health and the expense of the materials. While both of these factors may have played a role

328. AMEDEO MODIGLIANI.
Head. c. 1911–13. Limestone,
height 25 x 6 x 8½"
(63.5 x 15.2 x 21 cm).
Solomon R. Guggenheim
Museum, New York

Modigliani also painted anonymous students or children—anyone who would pose for him without a fee. For his paintings of the nude, however, he frequently relied on professional models. Modigliani developed a fairly standard formula for his compositions of reclining nudes (color plate 144, page 269). The attenuated figure is normally arranged along a diagonal and set within a narrow space, her legs eclipsed by the edge of the canvas. The figure is outlined with a flowing but precise line, while the full volumes are modeled with almost imperceptible gradations of flesh tones. The artist was fond of contrasting draped white sheets against deep Venetian red, as seen in this 1917 example. Indeed, the influence here of Titian is unmistakable, particularly that of his painting *Venus of Urbino.* This sixteenth-century work is in the Uffizi, which Modigliani frequented during his student days in Florence. His friend and fellow Italian, Severini, detected a "Tuscan elegance" in the work of Modigliani, who kept reproductions of old-master paintings on the walls of his Paris studio. For his paintings of nudes, Modigliani frequently assumed a perspective above the figure, thus establishing a "gaze" that implies the subject's sexual availability to the artist and (presumably male) viewers. With their blatant sexuality and coy, flirtatious expressions, Modigliani's less successful nudes could come perilously close to a modernized variety of academic eroticism. In the artist's day, the works were deemed indecent enough to be confiscated from an exhibi-

in his decision, it seems likely that he wanted to be a painter first and foremost.

In subject, Modigliani's paintings rarely departed from the depiction of single portrait heads, torsos, or full figures against a neutral background. His career was so abbreviated that it is difficult to speak of a stylistic development, although one can observe a gradual transition from work inspired by the Symbolist painters, in which the figure is securely integrated in the space of an interior, toward a pattern of linear or sculptural detachment. Modigliani knew Picasso and occasionally visited his studio, but he never really succumbed to the lure of Cubism.

Although all of his portraits resemble one another through their elegantly elongated features or their reference to forms deriving from Archaic art, the portraits of his friends, among whom were notable artists and critics of the early twentieth century, are sensitively perceived records of specific personalities, eccentricities, and foibles.

Modigliani's closest friend in Paris was his fellow painter Chaim Soutine, who had none of the Italian's dashing good looks (fig. 329). His portrayal of this Lithuaninan peasant, with his disheveled hair and clothes, and narrow, unfocused eyes, conveys a psychological presence lacking in many of Modigliani's likenesses.

329. AMEDEO MODIGLIANI. *Chaim Soutine.* 1916.
Oil on canvas, 36⅛ x 23½" (91.8 x 59.7 cm).
National Gallery of Art, Washington, D.C. CHESTER DALE COLLECTION

tion by the police for their excessive exposure of pubic hair.

CHAIM SOUTINE (1894–1943) was the tenth of eleven children from a poor Lithuanian-Jewish family who somehow developed a passion for painting and drawing. He managed to attend some classes in Minsk and Vilna and, through the help of a patron, arrived in Paris before the war's outbreak. He studied at the École des Beaux-Arts, but it was through the artists whom he met at the old tenement known as La Ruche that he began to find his way. Modigliani in particular took the young artist under his wing. He showed him Italian paintings in the Louvre and helped improve Soutine's unrefined table manners and virtually non-existent French. "He gave me confidence in myself," Soutine said.

Soutine constantly destroyed or reworked his paintings, so that it is rather difficult to trace his development from an early to a mature style. An early formative influence on his work was Cézanne, an artist also greatly admired by Modigliani. Van Gogh also provided a crucial example, although the energy and vehemence of Soutine's brushstroke surpassed even that of the Dutchman. The essence of Soutine's Expressionism lies in the intuitive power, the seemingly uncontrolled but immensely descriptive brush gesture (color plate 145, page 270). In this quality he is closer than any artist of the early twentieth century to the Abstract Expressionists of the 1950s, especially de Kooning (see fig. 519), who greatly admired Soutine.

In 1919 Soutine was able to spend three years in Céret, a town in the French Pyrenees. The mountain landscape inspired him to a frenzy of free, Expressionist brush paintings that transcended in exuberance anything produced by the Fauves or the German Expressionists. In *Gnarled Trees* (fig. 330), Soutine conveys a sense of complete immersion in the landscape with this all-over composition, in which the buildings on the heaving hillside are barely decipherable through the trees and writhing coils of paint. Soutine painted furiously during this period, and the result was his first success. After his return to Paris in 1922, a large number of his canvases were bought by the American collector Albert C. Barnes, providing Soutine with a steady income.

In his portraits and figure paintings of the 1920s there is frequently an accent on a predominant color—red, blue, or white—around which the rest of the work is built. In *Woman in Red* (color plate 145, page 270) the sitter is posed diagonally across an armchair over which her voluminous red dress flows. The red of the dress permeates her face and hands and is picked up in her necklace and in the red-brown tonality of the ground. Only the deep blue-black of the hat stands out against it. Soutine endows the static figure with tremendous vitality and movement, as the undulating surface rhythms swerve back and forth across the canvas. The hands are distorted as though from arthritis, and the features of the face are twisted into a slight grin. As in many of Soutine's portraits, the first impression is of a cruel caricature. Then one becomes aware of the sitter's highly individual, and in this case ambiguous, personality. *Woman in Red* appears at once comic, slightly mad, and knowingly shrewd.

330. CHAIM SOUTINE. *Gnarled Trees.* c. 1920.
Oil on canvas, 38 x 25" (96.5 x 63.5 cm).
Private collection, New York
© 1997 ADAGP, PARIS/ARTISTS RIGHTS SOCIETY (ARS), NEW YORK

Soutine's passion for Rembrandt was commemorated in a number of extraordinary paintings that were free adaptations of the Dutch master's compositions. In *Side of Beef* (color plate 146, page 270) Soutine transcended Rembrandt's 1655 *Butchered Ox*, in the Louvre, with gruesome, bloody impact. Unlike Rembrandt, Soutine isolated his subject, stripped it of any anecdotal context, and pushed it up to surface. The story of the painting is well known—Soutine poured blood from the butcher's shop to refurbish the rapidly decaying carcass while his assistant fanned away the flies, and nauseated neighbors screamed for the police. Oblivious to the stench, Soutine worked obsessively before his subject, recording every detail in vigorously applied, liquid color. In this and in many other paintings, Soutine demonstrated how close are the extremes of ugliness and beauty.

The life and art of **MAURICE UTRILLO** (1883–1955) provide perhaps the greatest paradox among *les maudits*. He was the son of Suzanne Valadon, a famous model for Renoir, Degas, and Toulouse-Lautrec, who later became an accomplished painter herself (color plate 147, page 270). Utrillo began to imbibe as a youngster, was expelled from school, and by the age of eighteen was in a sanatorium for alcoholics.

During Utrillo's chaotic adolescence—probably through

his mother's influence—he became interested in painting. He was self-trained, like his mother, but his art is hardly reminiscent of hers. Although his painting was technically quite accomplished, and one can recognize stylistic influences from the Impressionists or their successors, his work is closer to the purposely naive or primitive type of painting then developing out of a recognition of Rousseau's paintings (see figs. 77, 78). (The word "primitive," in this context, is used to signify a nonacademic, often self-taught, folk art vision.)

The first paintings that survive are of the suburb of Montmagny where Utrillo grew up, and of Montmartre after he moved there. Painted approximately between 1903 and 1909, the brush texture is coarse and the color, sober. About 1910 Utrillo began to focus obsessively and almost exclusively on the familiar scenes of Montmartre painted street by street. On occasional trips away from Paris he painted churches and village scenes that were variations on his basic theme (fig. 331). Asnières, the village depicted here, was one of the sites along the Seine favored by the Impressionists. Utrillo admired the work of Sisley and Pissarro, but unlike them, he preferred the unremarkable village streets to the banks of the river. The composition of *Street in Asnières* typifies Utrillo's preferred format. Buildings along a narrow street, one devoid of pedestrians, disappear into a long perspective. To emulate the rough texture of the chalky, whitewashed buildings, he frequently added plaster to his zinc white oil paint. In fact, this period is often referred to as his "white" period. His favorite scenes recur over and over again, his memory refreshed from postcards. In the late teens, following the

white period, his color brightened and figures occurred more frequently. From 1930 until his death in 1955, Utrillo, now decorated with the Legion of Honor, lived at Le Vésinet outside Paris. During these twenty-five years of rectitude, carefully regimented by his wife, he did little but repeat his earlier paintings.

SUZANNE VALADON (1867–1939) was born in central France, the child of a peasant woman who moved with her infant daughter to the Montmartre district of Paris. As a teenager, Valadon joined the circus as a performer but soon left because of an injury. She became a popular model in a neighborhood filled with artists and was soon posing for Puvis de Chavannes, Renoir, and Toulouse-Lautrec. It was through an instinctive aptitude for drawing that this self-taught artist, a woman from a working-class background, achieved a remarkable degree of success in the male-dominated art world of turn-of-the-century Paris. By the 1890s, some of her drawings had even been purchased by Degas, who introduced her to the technique of etching. Valadon's repertoire of subjects was virtually the opposite of her son's. She made still lifes and the occasional landscape, but the bulk of her oeuvre was devoted to portraits and nudes. The subject of the richly decorative canvas *Blue Room* (color plate 147, page 270) seems to fall somewhere between those two genres. While she assumes a traditional odalisque pose, this ample, unidentified woman, unconventionally dressed in striped pants, has little in common with the bold courtesans of Manet (see fig. 43) or the alluring nudes of Modigliani (color plate 144, page 269). What is more, Valadon's model

331. MAURICE UTRILLO.
Street in Asnières.
1913–15.
Oil on canvas,
29 x 36¼"
(73.7 x 92.1 cm).
Private collection

332. HENRI MATISSE. *Variation on a Still Life by de Heem.* Issy-les-Moulineaux, late 1915. Oil on canvas, 71¼" x 7'3" (181 cm x 2.2 m). The Museum of Modern Art, New York
GIFT AND BEQUEST OF FLORENE M. SCHOENBORN
© 1997 SUCCESSION H. MATISSE/ARTISTS RIGHTS SOCIETY (ARS), NEW YORK

right: 333. HENRI MATISSE. *Music Lesson.* 1917. Oil on canvas, 8'1⅛" x 6'7" (2.4 x 2 m). The Barnes Foundation, Merion, Pennsylvania
© 1997 SUCCESSION H. MATISSE/ARTISTS RIGHTS SOCIETY (ARS), NEW YORK

follows intellectual pursuits (note the books on the bed), and she smokes a cigarette, hardly the habit of a "respectable" woman, without the slightest self-consciousness. While her frank presentation of a female subject recalls depictions of prostitutes by her friend Toulouse-Lautrec, Valadon's painting seems a deliberate riposte to the clichéd interpretations of the theme by many generations of male artists. *Blue Room* was shown in the 1923 Salon d'Automne and purchased by the French state three years later.

A painter thoroughly immersed in the art of Cézanne, **HENRI MATISSE** (1869–1954) inevitably found himself responding, albeit in full color, to the structured interpretation of Cézanne's style realized by Braque and Picasso. This was evident in the 1911 *Red Studio,* already seen as possibly the culmination of Matisse's prewar Fauvist period (color plate 65, page 157). Subsequently, the great colorist went on to produce in 1915 what may have been his most orthodox Cubist painting in *Variation on a Still Life by de Heem* (fig. 332), where he demonstrated and clarified for himself the development that had taken place since his first free copy executed in 1893–95 (see figs. 145, 146). Now the palette has become much lighter; the space is closed in and frontalized; the tablecloth and the architecture of the room are geometrized, as are the fruits and vessels on the table. The result is a personal interpretation of Synthetic Cubism. His studies of Cubism were extremely helpful, for they encouraged him to simplify his pictorial structures, to control his intuitively decorative tendency, and to use the rectangle as a counterpoint for the curvilinear arabesque that came most naturally to him.

Piano Lesson of 1916 (color plate 148, page 271), unquestionably, is one of the artist's most austere interpretations of a Cubist mode. This work is a large, abstract arrangement of geometric color planes—lovely grays and greens. A tilted foreground plane of rose pink constitutes the top of the piano, on which a gray metronome sits; its triangular shape echoed in the angled green plane that falls across the window. The planes are accented by decorative curving patterns based on the iron grille of the balcony and the music stand of the piano. The environment is animated by the head of the child sculpturally modeled in line evoking Matisse's 1908 sculpture *Decorative Figure,* which appears in so many of his paintings. Another of his works, the painting *Woman on a High Stool,* 1914, is schematically indicated in the upper right, but its nature, as a painting or a figure in the room, remains unclear, heightening the sense of spatial ambiguity. The stiff figure seems to preside over the piano lesson like a stern instructor. Here again, Matisse's interpretation of Cubism is intensely personal, with none of the shifting views and fractured forms that mark the main line of Picasso, Braque, and Juan Gris. Its frontality and its dominant uniform planes of geometric color areas show that Matisse could assimilate Cubism in order to achieve an abstract structure of space and color.

The following year Matisse undertook another version of this theme with *Music Lesson* (fig. 333), a painting identical in size to *Piano Lesson,* and depicting the same room in the artist's house at Issy-les-Moulineax, southwest of Paris. Matisse has transformed that room into a pleasant domestic setting for a portrait of his family. The three Matisse children are in the foreground while Madame Matisse sews in the gar-

334. HENRI MATISSE. *Large Reclining Nude (Grand Nu Couché)*. 1935. State I of 22 states; photographs sent by the artist to Etta Cone, in letters of September 19 and November 16, 1935. The Baltimore Museum of Art
CONE ARCHIVE © 1997 SUCCESSION H. MATISSE/ARTISTS RIGHTS SOCIETY (ARS), NEW YORK

den. Everyone is self-absorbed and takes little heed of the artist or the beautiful garden view. The artist implies his own presence through his open violin case on the piano, his painting on the wall in the upper right corner, and his sculpture, *Reclining Nude 1/Aurora* (see fig. 174), here enlarged and flesh-colored, which sits by the pond before a lush landscape. In the wake of *Piano Lesson,* Matisse replaced his previous Cubist austerity with a lush indulgence in color and decorative forms.

In late 1917 Matisse moved to Nice, in the south of France, where he then spent much time off and on for the rest of his life. There, in spacious, seaside studios, he painted languorous models, often in the guise of exotic "odalisques," under the brilliant light of the Riviera. These women reveal little in the way of individual psychologies; they are objects of fantasy and visual gratification in a space seemingly sealed off from the rest of the world. Throughout most of the 1920s, Matisse supplanted heroic abstraction with intimacy, charm, and a mood of sensual indulgence, all suffused with color as lavish as an Oriental carpet. Yet, even within this deceptively relaxed manner, the artist produced images and compositions of grand, even solemn magnificence. The *Decorative Figure Against an Ornamental Background* of 1925–26 (color plate 149, page 271) epitomizes this most Rococo phase of Matisse's work. Owing to the highly sculptural modeling of his model, Matisse manages to set the figure off against the patterned Persian carpet, which visually merges with a background wall that has been decorated with the repetitive floral motif of French baroque wallpaper and the elaborate volutes of a Venetian mirror. Such solidly modeled figures appear in his sculptures of this period and also in the lithographs he was then making.

During the 1920s Matisse also revived his interest in the window theme, first observed in the 1905 Fauve *The Open Window* (color plate 49, page 117) and often with a table, still life, or figure set before it. Through this subject, in fact, one may follow the progress that Matisse made from Fauve color to Cubist reductiveness and back to the coloristic, curvilinear style of the postwar years. In *Interior with a Phonograph* (color plate 150, page 271), a painting of 1924, Matisse gave full play to his brilliant palette, while at the same time constructing a richly ambiguous space, created through layers of rectangular openings. Just to the left of the composition's center, the artist once again portrayed himself in mirror reflection as in the 1903–4 *Carmelina* (see fig. 148), with the nude now replaced by the still life of fruit and flowers.

In 1908, Matisse had expressed his desire to create "an art of balance, of purity and serenity devoid of troubling or depressing subject matter, an art which would be . . . something like a good armchair in which to rest from physical fatigue." Indulgent as this may seem, Matisse was almost puritanical in his self-effacing and unsparing search for the ideal. Thus, in the wake of his 1920s luxuriance, the sixty-six-year-old artist set out once again to renew his art by reducing it to essentials. The evidence of this prolonged and systematic exercise has survived in a telling series of photographs, made during the course of the work—state by state—that Matisse carried out toward the ultimate realization of his famous *Large Reclining Nude* (formerly called *Pink Nude*) of 1935 (color plate 151, page 271; fig. 334). He progressively selected, simplified, and reduced—not only form but also color, until the latter consisted of little more than rose, blue, black, and white. The odalisque thus rests within a pure, rhythmic flow of uninterrupted curves. While this cursive sweep evokes the simplified volumes of the figure, it also fixes them into the gridded support of the chaise longue. Here is the perfect antithetical companion to the Fauve *Blue Nude* of 1907 (color plate 54, page 119), a sculptured image resolved by far less reductive means than those of the *Large Reclining Nude.*

Like the grand French tradition of the decorative that preceded him—from Le Brun at Versailles through Boucher and Fragonard in the eighteenth century to Delacroix, Renoir,

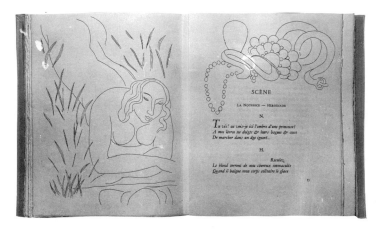

335. HENRI MATISSE. *Herodotus* (plate, page 54) from *Poésies* by Stéphane Mallarmé. Lausanne, Albert Skira. 1932. Etching, printed in black, plate 13⅛₆ x 9¹⁵⁄₁₆" (33.2 x 25.2 cm). The Museum of Modern Art, New York

and Monet in the nineteenth—Matisse had ample opportunity to be overtly decorative when, in 1932, he began designing and illustrating special editions of great texts, the first of which was *Poésies* by Stéphane Mallarmé (fig. 335), published by Albert Skira. In the graphics for these volumes, Matisse so extended the aesthetic economy already seen in the *Large Reclining Nude* that, even when restricted to line and black-and-white, he could endow the image with a living sense of color and volume.

But Matisse's most extraordinary decorative project during the 1930s was his design for a large triptych on the subject of the dance (fig. 336) that was commissioned by the American collector Albert C. Barnes, the same man who had bought the works of Soutine. The triptych would occupy three lunettes in the Barnes Foundation that housed the collection in Merion, Pennsylvania, which already owned more works by Matisse than any museum. Matisse began to paint the murals in Nice, only to discover well into the project that he was using incorrect measurements. So he began anew. (In 1992, in an amazing discovery that seems virtually impossible in the late twentieth century, this unfinished but splendid composition was discovered rolled up in Matisse's studio in Nice.) When Matisse returned to work on the second version, he discovered that his usual technique of brushing in every color himself was enormously time-consuming. In order to expedite his experiments with form and color

arrangements, he cut large sheets of painted paper and pinned them to the canvas, shifting them, drawing on them, and, ultimately, discarding them once he decided on the final composition. This method of cutting paper, as we shall see, became an artistic end in itself during the artist's later years. For both compositions, Matisse turned to his own previous treatment of the dance theme (color plate 63, page 155). In the final murals, the dancing figures are rendered in light gray, to harmonize with the limestone of the building, and are set against an abstract background consisting of broad, geometric bands of black, pink, and blue. The limbs of the figures are dramatically sliced by the curving edges of the canvas, which somehow enhance rather than inhibit their ecstatic movements. Matisse, of course, understood that his image had to be legible from a great distance, so the forms are boldly conceived, with no interior modeling to distract from their simplified contours. The result is one of the most successful mergings of architecture and painting in the twentieth century. Fortunately, Matisse felt the need to paint a third version of the murals, which now belongs, as does the first, to the Musée National d'Art Moderne in Paris.

When war struck Europe again, Matisse reacted as he had in 1916–18 and countered the inhumanity of the moment with an art radiant with affirmative, humanist faith. Even as age and physical disability put him in bed he proceeded to invent a new technique and through it to discover a transcen-

336. HENRI MATISSE. *Merion Dance Mural* 1932–33. Oil on canvas: left 11'1¾" x 14'5¾" (3.4 x 4.4 m), center 11'8⅛" x 16'6⅛" (3.6 x 5 m), right 11'1⅜" x 14'5" (3.4 x 4.4 m). The Barnes Foundation, Merion, Pennsylvania

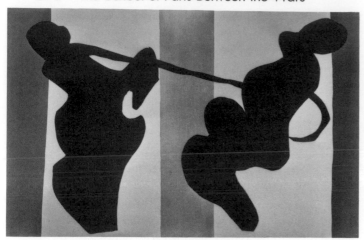

337. HENRI MATISSE. *The Cowboy*. Plate 14 from *Jazz*. Paris, Teriade. 1947. Pochoir, printed in color, composition 16⅝ x 25⅝" (42.2 x 65.1 cm). The Museum of Modern Art, New York

dent burst of expressive freedom. With Matisse, the hand could hardly have been more secure or the conception more youthful (color plate 152, page 272; fig. 337). The new Color Field painters emerging in New York toward the end of the 1950s were especially inspired by this late work, an art created by a man lying on his back, drawing with charcoal, fastened to the end of a long wand, on paper attached to the walls and ceiling above. In this way, Matisse produced vast compositions that he then executed by cutting out forms from paper he had hand-painted. Once assembled into figurative and abstract-decorative compositions that were very often mural in scale, the découpages, or paper cutout works, can be seen as the apotheosis of the artist's initial Cézanne-inspired Fauve aesthetic. They consist of planes of pure color flattened by their interlocking relationships and still conjuring dimensionality by the density of their color and the monumental simplicity of their arabesque shapes. Moreover, by "cutting into color," Matisse was manipulating his material in a way that translated

338. HENRI MATISSE. Chapel of the Rosary of the Dominican Nuns, Vence, France. Consecrated June 25, 1951

into pictorial terms his experience in handling clay for sculpture. With this, he at last attained the long-sought synthesis of sculpture and painting, of drawing and color. The irrepressible gaiety of the cutouts could be expressed to dazzling impact on a mural scale (color plate 152, page 272) or through the artist's marvelous illustrations for the book *Jazz* (fig. 337), which carried a text composed and hand-printed by Matisse. Following the experience of *Jazz*, Matisse began making cutouts in earnest, eventually to the exclusion of all other forms of artistic activity.

Beginning in 1948, Matisse used the cutout technique to design the stained glass and priests' chasubles for the Dominican chapel in the French Mediterranean village of Vence (fig. 338). Like the Barnes murals, the Vence chapel was a collaboration between art and architecture, but now Matisse extended his talents for decoration beyond painting to vestments, mural-sized drawings on ceramic tiles, and even a spire on the roof of the chapel.

PABLO PICASSO (1881–1973). In 1914, Braque, Léger, Apollinaire, and Derain were all at the front. Picasso recalls, "When the mobilization started I accompanied Braque and Derain to the station. I have never seen them again." While his remark is somewhat overstated, his close working relationship with Braque was over, and the death of Apollinaire in 1918 from influenza meant the loss of one of Cubism's most eloquent champions. As a foreigner, Picasso could remain in Paris, but he would work more now in isolation. After the period of discovery and experiment in Analytic Cubism, collage, and the beginning of Synthetic Cubism (1908–14), there was a pause during which the artist used Cubism for decorative ends (color plate 95, page 168; color plate 96, page 201). Then, suddenly, during a sojourn in Avignon in 1914 he began a series of realistic portrait drawings in a sensitive, scrupulous linear technique that recalls his longtime admiration for the drawings of Ingres. But works such as the portrait of Picasso's dealer Ambroise Vollard (fig. 339) were in the minority during 1915–16, when the artist was still experimenting with painting and constructions in a Synthetic Cubist vein.

The war years and the early 1920s saw a return to forms of Realism on the part of other French painters—Derain, Vlaminck, Dufy—originally associated with Fauvism or Cubism. In Picasso's case, however, the drawings and, subsequently, paintings in a realist mode constituted a reexamination of the nature of Cubism. He obviously felt confident enough of his control of the Cubist vocabulary to attack reality from two widely divergent vantage points. With astonishing agility, he pursued Cubist experiments side by side with realistic and classical drawings and paintings in the early 1920s.

In 1917 Picasso agreed, somewhat reluctantly because of his hatred for travel or disruption of his routine, to go to Rome as part of a group of artists including the poet and dramatist Jean Cocteau, the composer Erik Satie, and the choreographer Léonide Massine, to design curtains, sets, and costumes for a new ballet, *Parade*, being prepared for Diaghilev's Russian Ballet Company. The stimulus of Italy, where the artist stayed for two months visiting museums,

339. PABLO PICASSO. *Ambroise Vollard.* 1915.
Pencil on paper, 18⅜ x 12⁹⁄₁₆"
(46.7 x 31.9 cm).
The Metropolitan Museum of Art, New York
ELISHA WHITTELSEY COLLECTION, 1947.
© 1997 ESTATE OF PABLO PICASSO/ARTISTS RIGHTS SOCIETY (ARS), NEW YORK

340. PABLO PICASSO. *The American Manager.* 1917.
Reconstruction. Realized by Kermit Love for The Museum of
Modern Art, 1979. Paint on cardboard, fabric, paper, wood,
leather, and metal, 11'2¼" x 8' x 44½" (3.4 x 2.4 x 1.1 m).
The Museum of Modern Art, New York
© 1997 ESTATE OF PABLO PICASSO/ARTISTS RIGHTS SOCIETY (ARS), NEW YORK

churches, and ancient sites like Pompeii, contributed to the resurgence of classical style and subject matter in his art. He also met and married a young dancer in the corps de ballet, Olga Koklova, who became the subject of many portraits in the ensuing years. *Parade* brought the artist back into the world of the theater, which, along with the circus, had been an early love and had provided him with subject matter for his Blue and Rose Periods. Dating back to the late eighteenth century, a traditional "parade" was a sideshow performance enacted outside a theater to entice the spectators inside. The ballet was built around the theme of conflict: on the one hand, the delicate humanity and harmony of the music and the dancers; on the other, the commercialized forces of the mechanist-Cubist managers (fig. 340) and the deafening noise of the sound effects. Cocteau, who saw the play as a dialogue between illusion and reality, the commonplace and the fantastic, intended these real sounds—sirens, typewriters, or trains—as the aural equivalent of the literal elements in Cubist collages (color plate 94, page 168). The mixture of reality and fantasy in *Parade* led Apollinaire in his program notes to refer to its *sur-réalisme,* one of the first recorded uses of this term to describe an art form. *Parade* involved elements of shock that the Dadaists were then exploiting in Zurich. Picasso designed a backdrop for *Parade* that is reminiscent of his 1905 circus

paintings in its tender, romantic imagery and Rose Period coloring. Like Satie's soothing prelude, it reassured the audience, which was expecting Futurist cacophony. But they were shocked out of their complacency once the curtain rose. When *Parade* was first performed in Paris on May 18, 1917, the audience included sympathetic viewers like Gris and Severini. Most of the theatergoers, however, were outraged members of the bourgeoisie. The reviews were devastating, and *Parade* became a *succès de scandale,* marking the beginning of many fruitful collaborations between Picasso, as well as other artists in France, with the Ballets Russes.

During this period, Picasso's early love of the characters from the Italian *commedia dell'arte* and its descendant, the French circus (color plate 88, page 166), was renewed. Pierrots, Harlequins, and musicians again became a central theme. The world of the theater and music took on a fresh aspect that resulted in a number of brilliant portraits (fig. 341) and figure drawings done in delicately sparse outlines.

During the early postwar period Picasso experimented with different attitudes toward Cubism and representation. In the latter, the most purely classical style was manifested before 1923 in drawings that have the quality of white-ground Greek vase paintings of the fifth century B.C.E. Already by 1920 such drawings embodied a specifically Greek

341. PABLO PICASSO. *Portrait of Igor Stravinsky*. 1920. Pencil
and charcoal on paper, 24¼ x 18¾"
(61.6 x 47.6 cm). Musée Picasso, Paris
© 1997 ESTATE OF PABLO PICASSO/ARTISTS RIGHTS SOCIETY (ARS), NEW YORK

342. PABLO PICASSO. *Nessus and Dejanira*. 1920.
Pencil on paper, 8¼ x 10¼" (21 x 26 cm).
The Museum of Modern Art, New York
ACQUIRED THROUGH THE LILLIE P. BLISS BEQUEST. © 1997 ESTATE OF PABLO PICASSO/ARTISTS RIGHTS
SOCIETY (ARS), NEW YORK/VG BILD-KUNST, BONN

have been compared with Michelangelo's sybils in the Sistine
Ceiling, do not embody the traditional proportions of the
classical canon, they are endowed with a classical sense of
dignity and timelessness. Picasso prepared his compact, rigorously constructed image with several studies. The composi-

or Roman subject matter he continued to explore thereafter.
Nessus and Dejanira (fig. 342), for example, depicts a story
from Ovid's *Metamorphoses* in which the centaur, Nessus,
tries to rape Dejanira, the wife of Herakles. Picasso said that
his trips to the Mediterranean in the south of France inspired
such ancient mythological themes. This subject so intrigued
the artist that he made versions in pencil, silverpoint, watercolor, and etching. The representational line drawings also
revived his interest in the graphic arts, particularly etching and
lithography. He produced prints at the beginning of his
career and some during the earlier Cubist phase and the war
years. And from the early 1920s, printmaking became a major
phase of his production, to the point where Picasso must be
regarded as one of the great printmakers of the century.

During 1920, Picasso also began to produce paintings,
frequently on a large scale, in which his concept of the classical is in a different and monumental vein, which he pursued
for the next four years in painting and then continued in
sculpture in the early 1930s. The years of working with flat,
tilted Cubist planes seemed to build up a need to go to the
extreme of sculptural mass. The faces are simplified in the
heavy, coarse manner of Roman rather than Greek sculpture,
and when Picasso placed the figures before a landscape, he
simplified the background to frontalized accents that tend to
close rather than open the space. Thus, the effect of high
sculptural relief is maintained throughout. While the stone
giantesses of *Three Women at the Spring* (fig. 343), which

343. PABLO PICASSO. *Three Women at the Spring*. Fontainebleau,
summer 1921. Oil on canvas, 6'8¼" x 68½"
(2 m x 174 cm). The Museum of Modern Art, New York
GIFT OF MR. AND MRS. ALLAN D. EMIL. PHOTOGRAPH © 1995 THE MUSEUM OF MODERN ART,
NEW YORK. © 1997 ESTATE OF PABLO PICASSO/ARTISTS RIGHTS SOCIETY (ARS), NEW YORK

344. PABLO PICASSO. *Two Women Running on the Beach
(The Race)*. 1922. Gouache on plywood,
13⅜ x 16¾" (34 x 42.5 cm).
Musée Picasso, Paris
© 1997 ESTATE OF PABLO PICASSO/ARTISTS RIGHTS SOCIETY (ARS), NEW YORK

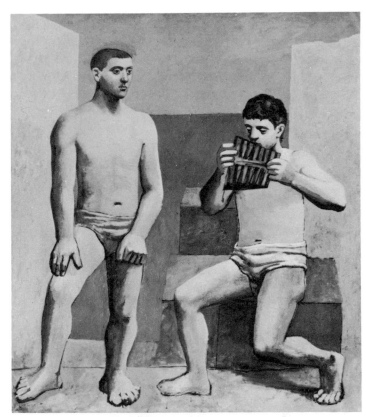

345. PABLO PICASSO. *The Pipes of Pan*. 1923. Oil on canvas,
6'6½" x 68½" (2 m x 170 cm). Musée Picasso, Paris
© 1997 ESTATE OF PABLO PICASSO/ARTISTS RIGHTS SOCIETY (ARS), NEW YORK

346. PABLO PICASSO. *By the Sea*. 1920. Oil and charcoal on wood,
31⅞ x 39⅜" (81 x 39.4 cm). Private collection
© 1997 ESTATE OF PABLO PICASSO/ARTISTS RIGHTS SOCIETY (ARS), NEW YORK

tional elements rotate around the central placement of a jug
and the women's enormous hands, which are extended in
slow, hieratic gestures. Occasionally, Picasso put his figures
into lumbering activity, as in *Two Women Running on the
Beach (The Race)* (fig. 344). The colors of this painting, in
accordance with its classical mood, are light and bright—
blues and reds, with raw pink flesh tones. Despite their massive limbs, these ecstatic women bound effortlessly across the
landscape. It has been suggested that the elated mood of
some of Picasso's classical paintings has to do with a happy
domestic life with Olga and their son Paulo, born in 1921. In
1924, Picasso enlarged this tiny composition to create a backdrop for Diaghilev's ballet *The Blue Train*. We should recall
that such massive figures were not new in Picasso's career, for
during his proto-Cubist period (1906–8), he had produced
a number of comparably substantial and primitivizing renditions of the nude (see fig. 199).

Picasso continued to paint for two or three more years in a
classical vein, but now with more specific reference to the spirit of Athens in the fifth century B.C.E. One of the outstanding
works of this phase is *The Pipes of Pan* (fig. 345). The two
youths are presented as massive yet graceful athletes, absorbed
in their own remote and magical world. The rectilinear forms
of the background deny any sense of spatial recession, turning
the whole into a kind of Mediterranean stage set.

In the early 1920s, Picasso also produced a few strange
and almost inexplicable figure pieces, of which the most curious is *By the Sea* (fig. 346). He habitually spent his summers
on the beach, which gave him ample opportunity to observe
the body in uninhibited motion. Though the subject and the
colors are classical, the artist has used fantastic anatomical distortions, particularly of the running and standing figures, that
predict the next, or Surrealist, stage of his work.

The incredible aspect of Picasso's art during this ten-year

period (1915–25) is that he was also producing major Cubist
paintings. He explained this stylistic dichotomy in a 1923
interview: "If the subjects I have wanted to express have suggested different ways of expression I have never hesitated to
adopt them." After the decorative enrichment of the Cubist
paintings around 1914–15 (color plate 95, page 168),
including Pointillist color and elaborate, applied textures, the
artist moved back, in his Cubist paintings of 1916–18, to a

347. PABLO PICASSO. *Harlequin with Violin (Si tu veux).* 1918.
Oil on canvas, 56 x 39½" (142.2 x 100.3 cm).
The Cleveland Museum of Art
PURCHASE, LEONARD C. HANNA, JR., BEQUEST.
© 1997 ESTATE OF PABLO PICASSO/ARTISTS RIGHTS SOCIETY (ARS), NEW YORK

The other direction in which Picasso took Cubism involved sacrifice of the purity of its Analytic phase, for the sake of new exploration of pictorial space. The result was, between 1924 and 1926, a series of the most colorful and spatially intricate still lifes in the history of Cubism. The *Mandolin and Guitar* (color plate 154, page 321) illustrates the artist's superb control over the entire repertory of Cubist and perspective space. The center of the painting is occupied by the characteristic Cubist still life—the mandolin and guitar twisting and turning on the table, encompassed in mandorlas of abstract color-shapes: vivid reds, blues, purples, yellows, and ochers against more subdued but still-rich earth colors. The tablecloth is varicolored and patterned, while the corner of the room with the open window is a complex of tilted wall planes and receding perspective lines creating a considerable illusion of depth, which, however, seems constantly to be turning back upon itself, shifting and changing into flat planes of color. The whole is infiltrated by an intense, Mediterranean light, made palpable by the deeply colored shadow that falls across the red tile floor at the right.

It is interesting to recall the extent to which the subject of the open window, often with a table, still life, or figure before it, had also been a favorite of Matisse (color plate 150, page

greater austerity of more simplified flat patterns, frequently based on the figure (fig. 347). Simultaneously, he made realistic versions of this subject, a favorite with him. From compositions involving figures, there gradually evolved the two versions of the *Three Musicians* (color plate 153, page 321), both created in 1921, the same year Picasso made *Three Women at the Spring*. The three figures in both versions are seated frontally in a row, against an enclosing back wall, and fixed by the restricted foreground, but the version in New York is simpler and more somber. They are a superb summation of the Synthetic Cubist style up to this point, but they also embody something new. The figures are suggestive not so much of gay musicians from the *commedia dell'arte*, as of some mysterious tribunal. It has been persuasively argued that the three musicians represent covert portraits of Picasso's friends—the writer Max Jacob (the monk at the right), who had recently withdrawn from the world and taken up residence in a monastery, and Apollinaire (Pierrot, the "poet" at the left), who had died in 1918. Picasso portrays himself, as he often did, in the guise of Harlequin, situated in the middle between his friends. The somber, somewhat ominous mood of the painting is thus explained as the work becomes a memorial to these lost friends.

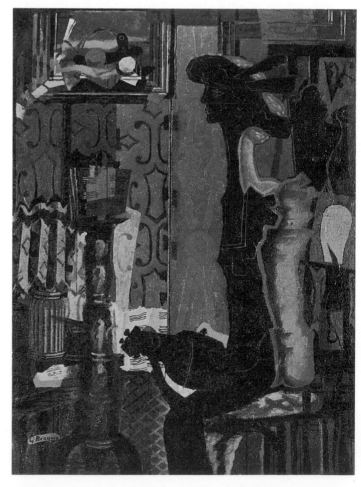

348. GEORGES BRAQUE. *Woman with a Mandolin.* 1937. Oil on canvas, 51¼ x 38¼" (130.2 x 97.2 cm). The Museum of Modern Art, New York MRS. SIMON GUGGENHEIM FUND. PHOTOGRAPH © 1995 THE MUSEUM OF MODERN ART, NEW YORK. © 1997 ADAGP, PARIS/ARTISTS RIGHTS SOCIETY (ARS), NEW YORK

271). Because of certain relationships between Matisse's work and Picasso's contemporary paintings of similar subjects, it is possible that Picasso may have been looking at his colleague's work. Yet his approach was entirely different, based as it was on Cubism.

Picasso, in some Cubist still lifes of the mid-1920s, introduced classical busts as motifs suggesting the artist's simultaneous exploration of classic idealism and a more expressionist vision. In *Studio with Plaster Head* (color plate 155, page 321), the Roman bust introduces a quality of mystery analogous to De Chirico's use of classical motifs (see fig. 294). While the trappings of classicism are here, the serenity of Picasso's earlier classical paintings has vanished. While the work belongs to a tradition of still lifes representing the attributes of the arts (including the prominent architect's square at the right), the juxtapositions become so jarring as to imbue the composition with tremendous psychological tension. At the back of the still life, fragments of his small son's toy theater imply that the entire still life is an ancient Roman stage set. The head is almost a caricature, which Picasso manages to twist in space from a strict profile view at the right to a fully frontal one in the deep blue-gray shadow at the left. The broken plaster arm at the left grips a baton so intensely it feels animated. Just before Picasso made this still life, he had completed his large, anguished painting, *Three Dancers* (color plate 182, page 332). This work takes us into another stage in Picasso's development (discussed in chapter 15). At this moment the Surrealists were emerging as an organized group, claiming Picasso as one of their own.

GEORGES BRAQUE (1882–1963) and Picasso parted company in 1914, never to regain the intimacy of the formative years of Cubism. After his war service and the slow recovery from a serious wound, Braque returned to a changed Paris in the fall of 1917. Picasso was in Italy, and other pioneers of Cubism or Fauvism were widely scattered. However, Juan Gris and Henri Laurens (see figs. 225, 222) were still in Paris, working in a manner that impressed Braque and inspired him in his efforts to find his own way once more. By 1918–19 a new and personal approach to Synthetic Cubism began to be apparent in a series of paintings, of which *Café-Bar* (color plate 156, page 322) is a document for Braque's style for the next twenty years. The painting is tall and relatively narrow, a shape which Braque had liked during his first essays in Cubism (color plate 92, page 167) and which he now began to utilize on a more monumental scale. The subject is a still life of guitar, pipe, journal, and miscellaneous objects piled vertically on a small, marble-topped pedestal table, or *guéridon*. This table was used many times in subsequent paintings. The suggested elements of the environment are held together by a pattern of horizontal green rectangles patterned with orange-yellow dots, placed like an architectural framework parallel to the picture plane. Compared with his prewar Cubist paintings, *Café-Bar* has more sense of illusionistic depth and actual environment. Also, although the colors are rich, and Braque's feeling for texture is more than ever apparent, the total effect recalls, more than Picasso's, the subdued, nearly monochrome tonal-

ities that both artists worked with during their early collaboration (see figs. 209, 210). This is perhaps the key to the difference between their work in the years after their collaboration. Whereas the approach of Picasso was experimental and varied, that of Braque was conservative and intensive, continuing the first lessons of Cubism. Even when he made his most radical departures into his own form of classical figure painting during the 1920s, the color remained predominantly the grays, greens, browns, and ochers of Analytic Cubism. The principal characteristic of Braque's style emerging about 1919 was that of a textural sensuousness in which the angular geometry of earlier Cubism began to diffuse into an overall fluid pattern of organic shapes.

This style manifested itself in the early 1920s in figure paintings in which the artist seems to have deserted Cubism as completely as Picasso did in his classical figure paintings. The difference again lay in the painterliness of Braque's oils and drawings, as compared with Picasso's Ingres-like contour drawings and the sculptural massiveness of his classical paintings. In Braque's paintings and figure drawings of the 1920s, the artist effected his personal escape from the rigidity of his earlier Cubism and prepared the way for his enriched approach to Cubist design of the later 1920s and 1930s.

The most dramatic variation on his steady, introspective progress occurred in the early 1930s when, under the influence of Greek vase painting, Braque created a series of line drawings and engravings with continuous contours. This flat, linear style also penetrated to his newly austere paintings of the period: scenes of artist and model or simply figures in the studio. These works exhibit some of the most varied effects of Braque's entire career. The dominant motif of *Woman with a Mandolin* (fig. 348) is a tall, dark silhouette of a woman seated before a music stand. Behind her, a profusion of patterns on the wall recalls the wallpaper patterns of Braque's earlier papiers collés. Because the woman's profile, very similar to ones in related compositions, is painted a deep black-brown, she becomes a shadow, less material than the other richly textured elements in the room. She plays the role of silent muse in this monumental painting that becomes Braque's meditation on art and music, one that was probably inspired by Corot's treatment of the same subject.

During the years of World War II Braque made several still life variations on the *vanitas* theme, in which a skull traditionally symbolizes the transitoriness of existence (color plate 157, page 322). It is a subject that Cézanne, Braque's great mentor, had frequently represented. Here Braque inserts the skull into the context of the studio. In fact, it seems to rest on his easel, implying that we are looking at a painted "illusion" rather than a "real" object. Braque has mixed sand with his oil medium, a practice he frequently applied to endow his surfaces with an almost sculptural quality. While Braque continued to explore the theme of the studio in his paintings after World War II, he also turned to landscape, specifically inspired by Van Gogh, a genre he had explored only intermittently since his early Cubist days.

Unlike Braque, RAOUL DUFY (1877–1953) joined Matisse in remaining true to the principles of Fauve color but, in the years following the "Wild Beast" episode (color plate 59, page 153), refined them into a sumptuous signature style. Even more than Matisse, he was a decorator with a connoisseur's taste for the hedonist joys of the good life. During the 1920s and 1930s, Dufy achieved such popularity, not only in paintings, watercolors, and drawings, but also in ceramics, textiles, illustrated books, and stage design, that his art, with its rainbow palette and stylish, insouciant drawing, seems to embody the hallmark look of the period. *Indian Model in the Studio at L'Impasse Guelma* (color plate 158, page 322) shows many of the characteristics of the artist's mature manner. The broad, clear area of light blue that covers the studio wall forms a background to the intricate pattern of rugs and paintings whose exotic focus is an elaborately draped Indian model. The use of descriptive line and oil color laid on in thin washes is such that the painting—and this is true of most of Dufy's works—seems like a delicately colored ink drawing.

Dufy combined influences from Matisse, Rococo, Persian and Indian painting, and occasionally modern primitives such as Henri Rousseau, transforming all these into an intimate world of his own. The artist could take a panoramic view and suggest, in minuscule touches of color and nervous arabesques of line, the movement and excitement of his world—crowds of holiday seekers, horse races, boats in sun-filled harbors, the circus, and the concert hall.

FERNAND LÉGER (1881–1955). In his paintings of the 1920s and 1930s Léger took part in the so-called "call to order," a tendency toward classicism that characterized much French art after World War I. This trend emerged in the years following the aesthetic upheaval of early modernist experiments and the military violence of the war, which Léger had experienced firsthand. One of his major figure compositions from the postwar period is *Three Women (Le Grand Déjeuner)* (color plate 159, page 323). The depersonalized figures, who stare fixedly and uniformly at the spectator, are machine-like volumes modeled out from the rigidly rectangular background. Léger labored over *Three Women,* the largest of three very similar versions of the same composition, and even repainted the figure at the right, who was originally the same marble-like gray color of her companions. He regarded the subject as timeless and eternal and said that his artistic sources were David, Ingres, Renoir, and Seurat. Although *Three Women* reflects Léger's engagement with classicism, it is also resolutely modern. In the abstract, rectilinear forms of the background one detects an awareness of the contemporary paintings of Mondrian, who was then living in Paris.

In the austere, Art Deco elegance of *The Baluster,* 1925 (fig. 349), Léger's art had much in common with the 1920s Purism of Amédée Ozenfant and Le Corbusier (figs. 351, 352). Léger remained very close to these artists after their initial meeting in 1920. Meanwhile, the inherent monumentality of its simple forms and iconic frontality also places *The Baluster* at one with the classicizing tendencies of Léger's art at this time.

During the last twenty years of his life, Léger revisited a

349. FERNAND LÉGER. *The Baluster.* 1925. Oil on canvas, 51 x 38¼" (129.5 x 97.2 cm). The Museum of Modern Art, New York PHOTOGRAPH © 1995 THE MUSEUM OF MODERN ART, NEW YORK. © 1997 ADAGP, PARIS/ARTISTS RIGHTS SOCIETY (ARS), NEW YORK

few basic themes through which he sought to sum up his experiences in the exploration of humanity and its place in the contemporary industrial world. It was a social as well as a visual investigation of his world as an artist, a final and culminating assessment of his plastic means for presenting it. In a painting from 1948–49, *Homage to Louis David* (fig. 350), Léger shows the happy middle-class family on a Sunday outing, the women in shorts and bathing suits, the men overdressed for the occasion, and the entire family stiffly posed for a fictive itinerant photographer. Like all Léger's late paintings, the picture has a Cubist structural frame, combined with a realistic if stylized presentation. In titling this, his idealized vision of the working class, Léger recalls David, whom he no doubt admired for his revolutionary politics and the unsentimental, austere nature of his art.

In *The Great Parade* of 1954 (color plate 160, page 323), Léger brought to culmination a series in which figure and environment are realistically drawn in heavy black outline on a white ground, over which float free shapes of transparent red, blue, orange, yellow, and green. This huge work also reflects a lifelong obsession with circus themes, as well as an interest in creating a modern mural art. Léger worked on the large canvas for two years. "In the first version," he recalled, "the color exactly fitted the forms. In the definitive version one can see what force, what vitality is achieved by using color on its own."

Despite his increased interest in illustrative subject and

350. FERNAND LÉGER. *Homage to Louis David*. 1948–49. Oil on
canvas, 60½" x 6'¾" (150 cm x 1.8 m). Musée National d'Art
Moderne, Centre d'Art et de Culture Georges Pompidou, Paris
© 1997 ADAGP, PARIS/ARTISTS RIGHTS SOCIETY (ARS), NEW YORK

351. AMÉDÉE OZENFANT. *Guitar and Bottles*. 1920.
Oil on canvas, 31⅞ x 39¼" (81 x 99.7 cm).
Solomon R. Guggenheim Museum, New York
PEGGY GUGGENHEIM COLLECTION, VENICE, 1976

social observation, these late works are remarkably consistent
with Léger's earliest Cubist paintings based on machine
forms. The artist was one of a few who never really deserted
Cubism but demonstrated, in every phase of his prolific output, its potential for continually fresh and varied expression.

Purism

The variant on Cubism called Purism was developed around
1918 by the architect and painter Charles-Édouard Jeanneret,
who in 1920 began to use the pseudonym LE CORBUSIER
(1887–1965), and the painter AMÉDÉE OZENFANT (1886–
1966). Unlike Picasso and Braque, who were not theoretically inclined, many later Cubist-inspired artists were prone
to theorizing and they published their ideas in journals and
manifestos. In their 1918 manifesto entitled *After Cubism*,
Ozenfant and Jeanneret attacked the then current state of
Cubism as having degenerated into a form of elaborate decoration. In their painting they sought a simple, architectural
structure and the elimination of decorative ornateness as well
as illustrative or fantastic subjects. To them the machine
became the perfect symbol for the kind of pure, functional
painting they hoped to achieve—just as, in his early, minimal
architecture, Le Corbusier thought of a house first as a
machine for living. Purist principles are illustrated in two still-
life paintings by Ozenfant (fig. 351) and Le Corbusier (fig.
352). Executed in 1920, the year Purism reached its maturity,
both feature frontally arranged objects, with colors subdued
and shapes modeled in an illusion of projecting volumes.
Symmetrical curves move across the rectangular grid with the
antiseptic purity of a well-tended, brand-new machine. Le

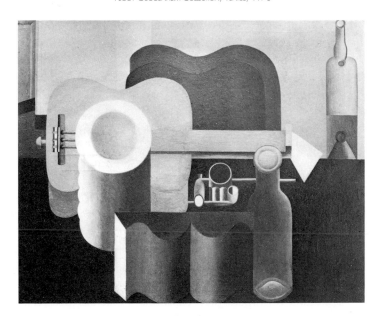

352. LE CORBUSIER (Charles-Édouard Jeanneret). *Still Life*. 1920.
Oil on canvas, 31⅞ x 39¼" (81 x 99.7 cm).
The Museum of Modern Art, New York
VAN GOGH PURCHASE FUND. PHOTOGRAPH © 1995 THE MUSEUM OF MODERN ART, NEW YORK.
© 1997 ADAGP, PARIS/FLC ARTISTS RIGHTS SOCIETY (ARS), NEW YORK

Corbusier continued to paint throughout his life, but his theories gained significant expression only in the great architecture he produced. Ozenfant had enunciated his ideas of
Purism in *L'Élan*, a magazine published from 1915 to 1917,
before he met Le Corbusier, and in *L'Esprit nouveau*, published with Le Corbusier from 1920 to 1925. Ozenfant, who
eventually settled in the United States, later turned to teaching and mural painting.

15 Surrealism

In 1917 Apollinaire referred to his own drama *Les Mamelles de Tirésias,* and also to the ballet *Parade* produced by Diaghilev, as Surrealist. The term was commonly used thereafter by the poets André Breton and Paul Éluard, as well as other contributors to the Paris journal *Littérature*. The concept of a literary and art movement formally designated as Surrealism, however, did not emerge until after the demise of Dada in Paris.

During the first years after World War I, French writers had been trying to formulate an aesthetic of the nonrational stemming from Arthur Rimbaud, the Comte de Lautréamont, Alfred Jarry, and Apollinaire. By 1922 Breton was growing disillusioned with Dada on the ground that it was becoming institutionalized and academic, and led the revolt that broke up the Dada Congress of Paris. Together with Philippe Soupault, he explored the possibilities of automatic writing in his 1922 Surrealist texts called *The Magnetic Fields*. Pure psychic automatism, one of the fundamental precepts of Surrealism, was defined by Breton as "dictation of thought, in the absence of any control exercised by reason, and beyond any aesthetic or moral preoccupation." While this method of composing without any preconceived subject or structure was designed for writing, its principles were also applied by the Surrealists to drawing.

After 1922 Breton assumed the principal editorship of *Littérature* and gradually augmented his original band of writers with artists whose work and attitudes were closest to his own: Picabia, Man Ray, and Max Ernst. This group met Breton regularly at his or Paul Éluard's home, or at some favorite café, where they discussed the significance of the marvelous, the irrational, and the accidental in painting and poetry. One of the exercises in which the artists engaged during their gatherings was the so-called "exquisite corpse" (fig. 353). The practice was based on an old parlor game in which one participant writes part of sentence on a sheet of paper, folds the sheet to conceal part of the phrase, and passes it to the next player, who adds a word or phrase based on the preceding contribution. Once the paper made it around the room in this manner, the provocative, often hilarious sentence was read aloud. One such game produced "The exquisite corpse will drink the young wine," hence the name. When this method was adapted for collective drawings, the surprising results coincided with the Surrealist love of the unexpected. The element of chance, of randomness and coincidence in the formation of a work of art had for years been explored by the Dadaists. Now it became the basis for intensive study for the Surrealists, whose experience of four years of war made them attach much importance to their isolation, their alienation from society and even from nature.

353. ANDRÉ BRETON, VALENTINE HUGO, GRETA KNUTSON, AND TRISTAN TZARA. *Exquisite Corpse.* c. 1930. Ink on paper, 9¼ x 12¼" (23.5 x 31.1 cm). Morton G. Neumann Family Collection

From the meetings between writers and painters emerged Breton's *Manifesto of Surrealism* in 1924, containing this definition:

SURREALISM, *noun, masc., pure psychic automatism by which it is intended to express, either verbally or in writing, the true function of thought. Thought dictated in the absence of all control exerted by reason, and outside all aesthetic or moral pre-*

occupations. ENCYCL. Philos. Surrealism is based on the belief in the superior reality of certain forms of association heretofore neglected, in the omnipotence of the dream, and in the disinterested play of thought. It leads to the permanent destruction of all other psychic mechanisms and to its substitution for them in the solution of the principal problems of life.

This definition emphasizes words rather than plastic images, literature rather than painting or sculpture. Breton, a serious student and disciple of Freud from whose teachings he derived the Surrealist position concerning the central significance of dreams and the subconscious, conceived of the Surreal condition as a moment of revelation in which are resolved the contradictions and oppositions of dreams and realities. In the second manifesto of Surrealism, issued in 1930, he said: "There is a certain point for the mind from which life and death, the real and the imaginary, the past and the future, the communicable and the incommunicable, the high and the low cease being perceived as contradictions."

Two nineteenth-century poets were major symbols and prophets for the Surrealists. Isidore Ducasse, known as COMTE DE LAUTRÉAMONT (1846–1870), and ARTHUR RIMBAUD (1854–1891) were outspoken critics of established nineteenth-century poets and of forms and concepts of nineteenth-century poetry. Unknown to each other, both lived almost in isolation, wandered from place to place, and died young; yet they wrote great poetry while still adolescents, tearing at the foundations of Romantic verse. They had irresistible appeal to the Surrealists. The idea of revolt permeates Lautréamont's poetry: revolt against tradition, against the family, against society, and against God. The ideas of Rimbaud at seventeen years of age seemed to define for the Surrealists the nature of the poet and poetry. Like the Christian mystic, the poet must attain a visionary state through rigid discipline—but here a discipline of alienation, monstrosities of love, suffering, and madness. Like the Surrealists later, Rimbaud was concerned with the implications of dreams, the subconscious, and chance.

Breton, an influential and charismatic figure, was the self-appointed but generally acknowledged leader of Surrealism. He was ruthless in the role as he invited artists to join the group only to disinvite them when they failed to toe the party line. As Breton formulated a dogma of Surrealist principles, schisms and heresies appeared, for few of Surrealism's exponents practiced the word with scrupulous literalness. In fact, the works of the major Surrealists have little stylistic similarity, aside from their departure from traditional content.

The first group exhibition of Surrealist artists was in 1925 at the Galerie Pierre. It included Arp, de Chirico, Ernst, Klee, Man Ray, Masson, Miró, and Picasso. In that year Yves Tanguy joined the group. A Surrealist gallery was opened in 1927 with an exhibition of these artists joined by Marcel Duchamp and Picabia. Except for René Magritte, who joined later that year, and Salvador Dalí, who did not visit Paris until 1929, this was the roster of the first Surrealist generation.

JEAN (HANS) ARP (1887–1966), one of the original Zurich Dadaists (color plate 161, page 324), was active in Paris Dada during its brief life, showing in the International Dada Exhibition of 1922, at the Galerie Montaigne, and contributing poems and drawings to Dada periodicals. When official Surrealism emerged in 1924, Arp was an active participant. At the same time, he remained in close contact with German and Dutch abstractionists and Constructivists. He also contributed to Theo van Doesburg's *De Stijl* and with El Lissitzky edited *The Isms of Art*. He and Sophie Taeuber eventually made their home in Meudon, outside of Paris.

During the 1920s Arp's favorite material was wood, which the artist made into painted reliefs. He also produced paintings, some on cardboard with cutout designs making a sort of reverse collage. Although Arp had experimented with geometric abstractions between 1915 and 1920 (see figs. 298, 300), frequently collaborating with Sophie Taeuber, he abandoned geometric shapes by 1920. His art would increasingly depend on the invention of biomorphic, abstract forms, based on his conviction that "art is a fruit that grows in man, like a fruit on a plant, or a child in its mother's womb." Yet Arp championed the geometric work of van Doesburg and Sophie Taeuber and collaborated with these two artists in decorating ten rooms of the Café l'Aubette in his native city, Strasbourg (fig. 354). His murals for the café (now destroyed) were the boldest, freest, and most simplified examples of his organic abstraction. They utilized his favorite motifs: the navel and mushroom-shaped heads, sometimes sporting a mustache and round-dot eyes. Despite the anthropomorphic connotations of the titles, such as *Rising Navel and Two Heads* and *Navel-Sun*, these murals were biomorphic abstractions. *Rising Navel and Two Heads* was simply three flat, horizontal, scalloped bands of color, with two color shapes suggesting flat mushrooms floating across the center band. *Navel-Sun* was a loosely circular white shape floating on a blue background (though most of the actual color scheme is lost). There are no parallels for these large, boldly abstracted shapes until the so-termed Color Field painters of the 1950s and 1960s (color plate 338, page 574).

354. JEAN ARP, SOPHIE TAEUBER, AND THEO VAN DOESBURG. *Rising Navel and Two Heads.* 1927–28. Mural: oil on plaster. Café l'Aubette, Strasbourg. Destroyed 1940

Of particular interest among Arp's reliefs and collages of the late 1920s and early 1930s were those entitled or subtitled *Objects Arranged According to the Laws of Chance, or Navels* (fig. 355). These continued the 1916–17 experiment in *Collage with Squares Arranged According to the Laws of Chance* (see fig. 298), although, now, most of the forms were organic rather than geometric. Arp, who made several versions of this wooden relief in 1930–31, often employed the oval shapes seen here, which were for him variations on a navel, his universal emblem of human life. He also produced string reliefs based on chance—loops of string dropped accidentally on a piece of paper and seemingly fixed there. Arp, it would appear, may have given the laws of chance some assistance, and for this he had a precedent in Duchamp's *3 Stoppages Étalon* (see fig. 304).

Arp's great step forward in sculpture occurred at the beginning of the 1930s when he began to work completely in the round, making clay and plaster models for later bronze or marble sculptures. Arp shared Brancusi's devotion to mass rather than space as the fundamental element of sculpture (see fig. 192), and he, in turn, influenced major sculptors of the next generation—among them Alexander Calder, Henry Moore, and Barbara Hepworth. He thus stands at the opposite pole from the Constructivists Tatlin or Gabo (see figs. 249, 255). In developing the ideas of his painted reliefs, he had introduced seemingly haphazard detached forms that hovered around the matrix like satellites. *Head with Three Annoying Objects* (fig. 356) was first executed in plaster, Arp's preferred medium for his first sculptures in the round. On a large biomorphic mass rest three objects, identified as a mustache, a mandolin, and a fly. The three strange forms were named in Arp's earliest title for the work, derived from one of his stories in which he describes himself waking up with three

356. JEAN ARP. *Head with Three Annoying Objects*. 1930 (cast 1950). Bronze, 14⅛ x 10¼ x 7½" (35.9 x 26 x 19.1 cm). Estate of Jean Arp © 1997 VG BILD-KUNST, BONN/ARTISTS RIGHTS SOCIETY (ARS), NEW YORK

"annoying objects" on his face. The objects were not secured to the "head," for Arp intended that they be moved around by the spectator, introducing yet another element of chance in his work. A number of artists in the early 1930s were exploring the notion of kineticism in painting or sculpture with interchangeable parts, including Alberto Giacometti (fig. 392) and Alexander Calder (see figs. 453, 454). Arp applied titles to his works, after he had made them, with the intention of inspiring associations between images and ideas or, as he said, "to ferret out the dream." Thus, the meaning of the work, like the arrangement of the forms themselves, was open-ended. *Head with Three Annoying Objects*, like many of Arp's plasters, was not cast in bronze until much later, in this case not until 1950. To his biomorphic sculptural forms Arp applied the name "human concretion," for even forms that are not derived from nature, he said, can still be "as concrete and sensual as a leaf or a stone." In 1930 van Doesburg had proposed the name "concrete art" as a more accurate description for abstract art. He contended that the term "abstract" implied a taking away from, a diminution of, natural forms and therefore a degree of denigration. What could be more real, he asked, more concrete than the fundamental forms and colors of nonrepresentational or nonobjective art? Although van Doesburg's term did not gain universal recognition, Arp used it faithfully, and as "human concretions," it has gained a specific, descriptive connotation for his sculptures in the round. "Concretion signifies the natural process of condensation," Arp said, "hardening, coagulating, thickening, growing together. Concretion designates the solidification of a mass. Concretion designates curdling, the curdling of the earth and the heavenly bodies. Concretion designates solidification, the mass of the stone, the plant, the animal, the man. Concretion is something that has grown."

This statement could be interpreted as a manifesto in opposition to the Constructivist assertion of the primacy of space in sculpture. More pertinent, however, is its emphasis

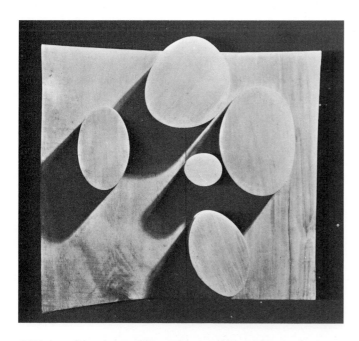

355. JEAN (HANS) ARP. *Objects Arranged According to the Laws of Chance, or Navels*. 1930. Varnished wood relief, 10⅜ x 11⅛ x 2¼" (26.3 x 28.3 x 5.4 cm). The Museum of Modern Art, New York
PURCHASE © 1997 VG BILD-KUNST, BONN/ARTISTS RIGHTS SOCIETY (ARS), NEW YORK

on sculpture as a process of growth, making tangible the life processes of the universe, from the microscopic to the macrocosmic. Thus a human concretion was not only an abstraction based on human forms but also a distillation in sculpture of life itself.

The art of Jean Arp took many different forms between 1930 and 1966, the year of the artist's death. Abstract (or concrete) forms suggesting sirens, snakes, clouds, leaves, owls, crystals, shells, starfish, seeds, fruit, and flowers emerged continually, suggesting such notions as growth, metamorphosis, dreams, and silence. *Aquatic* (fig. 357), a work of 1953 carved in marble, represents one of his more explicitly figurative sculptures. Here the suggestion of metamorphosis is particularly striking. The artist, who had long since dispensed with notions of fixed orientation in his work, tells us it may be displayed horizontally or vertically: vertically it is a female nude, horizontally, a finny monster.

While developing his freestanding sculpture, Arp continued to make reliefs and collages. To make his new collages, Arp tore up his own previously made work (color plate 161, page 324). In this example from 1937, he essentially recomposed the work by connecting the new forms with curvilinear lines drawn in pencil. This new liberating collage technique, a catharsis from the immaculateness of the sculptures, was described by Arp:

> I began to tear my papers instead of curving them neatly with scissors. I tore up drawings and carelessly smeared paste over and under them. If the ink dissolved and ran, I was delighted. . . . I had accepted the transience, the dribbling away, the brevity, the impermanence, the fading, the withering, the spookishness of our existence. . . . These torn pictures, these papiers déchirés brought me closer to a faith other than earthly. . . .

Many of these torn-paper collages found their way to the United States and offered a link between organic Surrealism and postwar American Abstract Expressionism.

MAX ERNST (1891–1976). The career of Max Ernst as a painter and sculptor, interrupted by four years in the German army began when he moved to Paris from Cologne in 1922. At the end of 1921, the Dadaist Ernst had resumed painting after having devoted himself to various collage techniques since 1919 (see figs. 316, 317). The manner of *Celebes,* the chief painting of his pre-Paris period (color plate 140, page 268), is combined with the technique of his Dada assemblages in *Two Children Are Threatened by a Nightingale* (color plate 162, page 324), a 1924 dream landscape in which two girls—one collapsed on the ground, the other running and brandishing a knife—are frightened by a tiny bird. The fantasy is given peculiar emphasis by the elements in actual relief—the house on the right and the open gate on the left. A figure on top of the house clutches a young girl and seems to reach for the actual wooden knob on the frame. Contrary to Ernst's usual sequence, the title of this enigmatic painting (inscribed in French on the frame) preceded the image. In his biographical summary, speaking in the third person, the artist noted: ". . . He never imposes a title on a painting. He waits until a title imposes itself. Here, however, the title existed *before* the picture was painted. A few days before, he had written a prose poem which began: à la tombée de la nuit, à la lisière de la ville, deux enfants sont menacés par un rossignol (as night falls, at the edge of town, two children are threatened by a nightingale) . . . He did not attempt to illustrate this poem, but that is the way it happened."

In 1925, Ernst became a full participant in the newly established Surrealist movement. That year he began to make drawings that he termed *frottage* (rubbing), in which he used the child's technique of placing a piece of paper on a textured surface and rubbing over it with a pencil. The resulting image was largely a consequence of the laws of chance, but the transposed textures were consciously reorganized in new contexts, and new and unforeseen associations were aroused. Not only did *frottage* provide the technical basis for a series of unorthodox drawings; it also intensified Ernst's perception of the textures in his environment—wood, cloth, leaves, plaster, and wallpaper.

Ernst applied this technique, combined with scraping or *grattage,* to his paintings of the late 1920s and the 1930s. The 1927 canvas, *The Horde* (fig. 358), is indicative of the increasingly ominous mood of Ernst's Surrealist paintings. The monstrous, treelike figures are among the many frightening premonitions of the conflagration that would overtake Europe in the next decade. In 1941, Ernst managed to escape the war in Europe and settle in New York City where his presence, along with other Surrealist refugees, would have tremendous repercussions for American art. His antipathy toward the rise of fascism was given fullest expression in *Europe after the Rain* (color plate 163, page 324), which has been aptly described as a requiem for the warravaged continent. In this large painting Ernst employed yet another technique, one invented by Oscar Dominguez (fig. 377), which Ernst called "decalcomania." He placed paper or glass on the wet painted surface and then pulled it away to achieve surprising textural effects.

Of all the menageries and hybrid creatures Ernst invented,

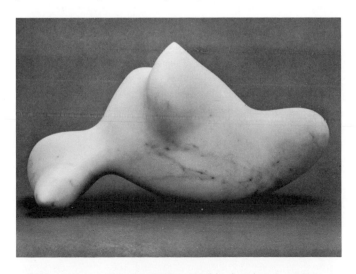

357. JEAN ARP. *Aquatic.* 1953. Marble, height 13 x 25 ⁵⁄₁₆ x 9 ³⁄₁₆" (33 x 64.1 x 23.5 cm). Walker Art Center, Minneapolis

left:
358. MAX ERNST. *The Horde.*
1927. Oil on canvas,
44⅞ x 57½" (114 x 146.1 cm).
Stedelijk Museum, Amsterdam
© 1997 ADAGP, PARIS/ARTISTS RIGHTS SOCIETY
(ARS), NEW YORK

below:
359. MAX ERNST. *Surrealism and Painting.* 1942. Oil on canvas,
6'4¾" x 7'7¾" (2 x 2.3 m).
The Menil Collection, Houston
© 1997 ADAGP, PARIS/ARTISTS RIGHTS SOCIETY
(ARS), NEW YORK

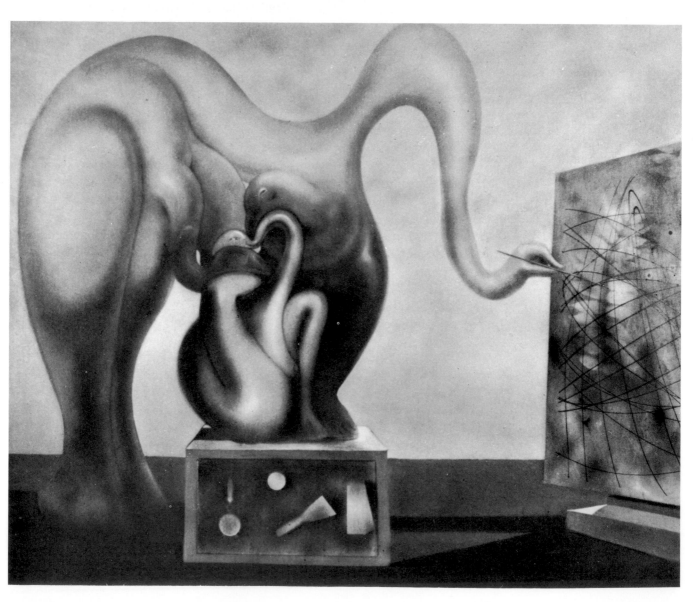

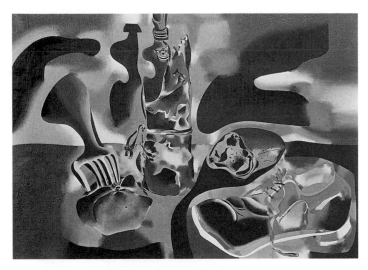

363. JOAN MIRÓ. *Still Life with Old Shoe.* Paris, January 24–May 29, 1937. Oil on canvas, 32 x 46" (81.3 x 116.8 cm). The Museum of Modern Art, New York

GIFT OF JAMES THRALL SOBY. PHOTOGRAPH © 1995 THE MUSEUM OF MODERN ART, NEW YORK. © 1997 ADAGP, PARIS/ARTISTS RIGHTS SOCIETY (ARS), NEW YORK

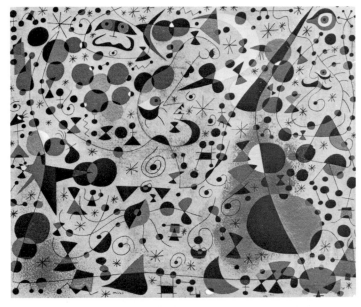

364. JOAN MIRÓ. *The Poetess.* December 31, 1940. Gouache and oil wash on paper, 15 x 18" (38.1 x 45.7 cm). Private collection, New York © 1997 ADAGP, PARIS/ARTISTS RIGHTS SOCIETY (ARS), NEW YORK

the Spanish Pavilion at the Paris International Exposition, in 1937 (fig. 381). The imagery of *Still Life with Old Shoe* is explicit and reflects Miró's revulsion against events in his beloved Spain, which he expressed through humble objects and shapes that convey decay, disease, and death.

With the outbreak of World War II, Miró had to flee from the home he had set up in Normandy and go back to Spain. The isolation of the war years and the need for contemplation and reevaluation led him to read mystical literature and to listen to the music of Mozart and Bach. "The night, music

right: 365. JOAN MIRÓ. *Lunar Bird.* 1966. Bronze, 7'8⅛" x 6'10¾" x 61½" (2.3 m x 2.1 m x 156.2 cm). The Hirshhorn Museum and Sculpture Garden, Smithsonian Institution, Washington, D.C.

© 1997 ADAGP, PARIS/ARTISTS RIGHTS SOCIETY (ARS), NEW YORK

and stars," he said, "began to play a major role in suggesting my paintings." Between January 1940 and September 1941 he worked on a series of twenty-three small gouaches entitled Constellations (fig. 364), which are among his most intricate and lyrical compositions. In the sparkling compositions the artist was concerned with ideas of flight and transformation as he contemplated the migration of birds, the seasonal renewal of butterfly hordes, the flow of constellations and galaxies. The artist dispersed his imagery evenly across the surface in an "allover" pattern. In 1945 the Constellations were shown at New York's Pierre Matisse Gallery, where they affected the emerging American Abstract Expressionist painters who were then seeking alternatives to Social Realism and Regionalism (color plate 235, page 451; fig. 535). As the leader of the organic-abstract wing of Surrealism, Miró had a major impact on younger American painters.

By the end of World War II, Miró was working on a large scale, with bold shapes, patterns, and colors. In his work of the 1960s Miró showed an interest in the American Abstract Expressionists, who earlier had been influenced by him. The mural-sized painting *Blue II* (color plate 168, page 326) of 1961 is an almost pure example of Color Field painting—a delicately inflected blue ground punctuated only by a vertical red slash and a trail of black oval shapes. Since the 1920s the color blue had held metaphorical significance for Miró. It is a favorite color of his native Catalonia, and a hue he particularly associated with the world of dreams. In addition to mural-size paintings, Miró created sculpture on a monumental scale in the 1960s. To make *Moonbird* (fig. 365), he enlarged a small bronze *Bird* from 1944. With its massive, crescent-shaped head, phallic protuberances, and powerful stance, the

366. ANDRÉ MASSON. *Battle of Fishes*. 1926. Sand, gesso, oil, pencil, and charcoal on canvas, 14¼ x 28¾" (36.2 x 73 cm). The Museum of Modern Art, New York PURCHASE. PHOTOGRAPH © 1995 THE MUSEUM OF MODERN ART, NEW YORK. © 1997 ADAGP, PARIS/ARTISTS RIGHTS SOCIETY (ARS), NEW YORK

creature has all the presence of an ancient cult statue.

One of the artists who would eventually emigrate from Paris to New York was **ANDRÉ MASSON** (1896–1987). Among the first Surrealists, Masson was the most passionate revolutionary, a man of violent convictions who had been deeply spiritually scarred by his experiences in World War I. Physically wounded almost to the point of death, he was hospitalized for a long time, and after partially recovering, he continued to rage against the insanity of humankind and society until he was confined for a while to a mental hospital. Masson was by nature an anarchist, his convictions fortified by his experience. He joined the Surrealist movement in 1924, though his anticonformist temperament caused him to break with the controlling Breton, who was constantly welcoming new converts and excommunicating others whom he felt had strayed from his Surrealist orthodoxy. Masson first left the group in 1929, rejoined it in the 1930s, and left again in 1943.

The paintings done by Masson in the early 1920s reveal the artist's debt to Cubism, particularly that of Juan Gris (see fig. 225). In 1925 Masson was regularly contributing automatic drawings to the Surrealist periodical, *La Révolution surréaliste*, founded by Breton in 1925. These works directly express his emotions and contain various images having to do with the sadism of human beings and the brutality of all living things. *Battle of Fishes* (fig. 366) is an example of Mason's sand painting, which he made by freely applying adhesive to the canvas, then throwing sand over the surface and brushing away the excess. The layers of sand would suggest forms to the artist, "although almost always irrational ones," he said. He then added lines and small amounts of color, sometimes directly from the paint tube, to form a pictorial structure around the sand. Here the imagery is aquatic, though the

artist has called the fish anthropomorphic. As in the *frottage* technique of Ernst or Arp's string collages, Masson here allows chance to help determine his composition, though his firm belief that art should be grounded in conscious, aesthetic decisions eventually led him away from this method.

In the late 1930s, after living principally in Spain for three years, Masson temporarily painted in a style closer to the more naturalistic wing of Surrealism. He painted monstrous, recognizable figures from mythological themes in a manner influenced by Picasso and possibly Dalí. To escape World War II in Europe he lived for a time in the United States, where, he said, "Things came into focus for me." He exhibited regularly in New York, exercising influence on some of the younger American painters, particularly Jackson Pollock. He would later say that Pollock carried automatism to an extreme that he himself "could not envision." Living in rural Connecticut, Masson reverted to a somewhat automatist, Expressionist approach in works such as *Pasiphaë* (color plate 169, page 326). The classical myth of the Minotaur provided one of Masson's recurrent themes in the 1930s and 1940s. (It was he who named the Surrealist review after this part-man, part-bull beast.) Because she displeased Poseidon, Pasiphaë was made to mate with a bull, giving birth to the Minotaur. Masson said he wanted to represent the violent union of woman and beast in such a way that it is impossible to tell where one begins and the other ends. This subject, which had become a major theme in Picasso's work, also appealed to the young Pollock, who made a painting of Pasiphaë the same year Masson did.

YVES TANGUY (1900–1955) had literary interests and close associations with the Surrealist writers Jacques Prévert and Marcel Duhamel in Paris after 1922. With Prévert he dis-

left: 367. YVES TANGUY. *The Furniture of Time.* 1939.
Oil on canvas, 46 x 35¼" (116.7 x 89.4 cm).
The Museum of Modern Art, New York
JAMES THRALL SOBY BEQUEST. PHOTOGRAPH © 1995 THE MUSEUM OF MODERN ART, NEW YORK.
© 1997 ADAGP, PARIS/ARTISTS RIGHTS SOCIETY (ARS), NEW YORK

painter Kay Sage, Tanguy continued to evoke his barren, destroyed universe. In one of his last paintings, *Multiplication of the Arcs* (color plate 171, page 327), the objects fill the ground area back to a newly established horizon line, to create an enigmatic, foreboding scene that the art historian James Thrall Soby called "a sort of boneyard of the world."

Surrealism in the 1930s

From its inception, Surrealism in painting tended in two directions. The first, represented by Miró, André Masson, and, later, Matta, is biomorphic or abstract Surrealism. In this tendency, automatism—"dictation of thought without control of the mind"—is predominant, and the results are generally close to abstraction, although some degree of imagery is normally present. Its origins were in the experiments in chance and automatism carried on by the Dadaists and the automatic writing of Surrealist poets.

The other direction in Surrealist painting is associated with Salvador Dalí, Yves Tanguy, and René Magritte. It presents, in meticulous detail, recognizable scenes and objects that are taken out of natural context, distorted and combined in fantastic ways as they might be in dreams. Its sources are in the art of Henri Rousseau, Chagall, Ensor, de Chirico, and nineteenth-century Romantics. These artists attempted to use images of the subconscious, defined by Freud as uncontrolled by conscious reason (although Dalí, in his "paranoiac-critical methods," claimed to have control over his subconscious). Freud's theories of the subconscious and of the significance of dreams were, of course, fundamental to all aspects of Surrealism. To the popular imagination it was the naturalistic Surrealism associated with Dalí, Tanguy, and others of that group that signified Surrealism, even though it has had less influence historically than the abstract biomorphism of Miró, Masson, and Matta.

Surrealism was a revolutionary movement not only in literature and art but also in politics. The Surrealist period in Europe was one of deepening political crisis, financial collapse and the rise of fascism, provoking moral anxiety within the avant-garde. Breton's periodical, *La Révolution surréaliste,* maintained a steady communist line during the 1920s. The Dadaists at the end of the war were anarchists, and many future Surrealists joined them. Feeling that government systems guided by tradition and reason had led mankind into the bloodiest holocaust in history, they insisted that government was undesirable, that the irrational was preferable to the rational in art and in all of life and civilization. The Russian Revolution and the spread of communism provided a channel for Surrealist protests during the 1920s. Louis Aragon and, later, Paul Éluard joined the Communist Party, while Breton, after exposure to the reactionary bias of Soviet communism or Stalinism in art and literature (discussed in chapter 11), took

covered the writings of Lautréamont and other prophets of Surrealism. In 1923, inspired by two de Chirico paintings in the window of the dealer Paul Guillaume, he decided to become a painter. Entirely self-taught, Tanguy painted in a naive manner until 1926, when he destroyed many of these canvases. After meeting André Breton, Tanguy began contributing to *La Révolution surréaliste* and was adopted into the Surrealist movement. Tanguy's subsequent progress was remarkable, in both technique and fantastic imagination. *Mama, Papa Is Wounded!* (color plate 170, page 327) was the masterpiece of this first stage and exhibited all the obsessions that haunted him for the rest of his career. The first of these was an infinite-perspective depth, rendered simply by graded color and a sharp horizon line—a space that combines vast emptiness and intimate enclosure, where ambiguous organic shapes, reminiscent of Arp's biomorphic sculptures, float in a barren dreamscape.

After making a visit to Africa in 1930 and 1931, where he saw brilliant light and color and desert rock formations, Tanguy began to define his forms precisely. It has been proposed that the vast spaces in his paintings during the 1930s, so strongly suggestive of a submarine world, were inspired by Tanguy's childhood memories of summers spent on the Brittany coast. *The Furniture of Time* (fig. 367) offers no points of reference for measuring the space in which the curious, bonelike objects drift aimlessly and uneasily. Tanguy endowed his forms with a tangible, literal presence by virtue of his highly illusionistic manner, yet those forms have no equivalent in nature. His unique contribution was this co-opting of realist techniques to conjure up abstract images.

After moving to the United States in 1939 and settling in Woodbury, Connecticut, with his wife, the American Surrealist

a Trotskyist position in the late 1930s. Picasso, who made Surrealist work in the 1930s, became a Communist in bitter protest against the fascism of Franco. Although schisms were occurring among the original Surrealists by 1930, new artists and poets were being recruited by Breton. Dalí joined in 1929, the sculptor Alberto Giacometti in 1931, and René Magritte in 1932. Other later recruits included Paul Delvaux, Henry Moore, Hans Bellmer, Oscar Domínguez, and Matta Echaurren. The list of writers was considerably longer. Surrealist groups and exhibitions were organized in England, Czechoslovakia, Belgium, Egypt, Denmark, Japan, Holland, Romania, and Hungary. Despite this substantial expansion, internecine warfare resulted in continual resignations, dismissals, and reconciliations. Also, artists such as Picasso were adopted by the Surrealists without joining officially.

During the 1930s the major publication of the Surrealists was the lavish journal *Minotaure*, founded in 1933 by Albert Skira and E. Tériade. The last issue appeared in 1939. Although emphasizing the role of the Surrealists, the editors drew into their orbit any of the established masters of modern art and letters who they felt had made contributions. These included artists as diverse as Matisse, Kandinsky, Laurens, and Derain. Picasso made the cover design for the first issue (fig. 368), a collage having in its center a classical drawing of the Minotaur holding a short sword. His association with the journal may have fostered his interest in the beast.

368. PABLO PICASSO. Maquette for the cover of *Minotaure*. Paris, May, 1933. Collage of pencil on paper, corrugated cardboard, silver foil, ribbon wallpaper painted with gold paint and gouache, paper doily, burnt linen, leaves, tacks, and charcoal on wood, 14⅜ x 11⅜" (36.5 x 28.7 cm) (irregular). The Museum of Modern Art, New York GIFT OF MR. AND MRS. ALEXANDRE P. ROSENBERG. PHOTOGRAPH © 1995 THE MUSEUM OF MODERN ART, NEW YORK. © 1997 ESTATE OF PABLO PICASSO/ARTISTS RIGHTS SOCIETY (ARS), NEW YORK

Surrealism made its debut in the United States in 1931 with an exhibition at the Wadsworth Atheneum in Hartford, Connecticut, organized by James Thrall Soby. It traveled to New York in 1932 to the Julien Levy Gallery. In the 1930s American interest in Surrealism tended to focus on the dream imagery of Salvador Dalí, which was merged, to varying degrees, with this country's indigenous realist traditions. This was the era of the Great Depression in America, when many artists reinforced the sense of political isolationism by looking to local cultural traditions as subjects for their art, suspiciously viewing Surrealism as a product of leftist European artists. But by 1942, the locus of Surrealism had effectively shifted to New York as artists fled the war in Europe. The city's Museum of Modern Art had already presented a landmark exhibition in 1936 called "Fantastic Art, Dada, and Surrealism," and a small number of important dealers such as Levy, Pierre Matisse, and Sidney Janis, provided support for the movement. The American Surrealist journals *View* and *VVV* were born in the early 1940s, serving as venues for exchange between European Surrealists and American followers of the movement. In 1942, an exhibition called "First Papers of Surrealism" (referring to the artist-immigrants' papers) was held in an old New York mansion with an installation by Marcel Duchamp, who crisscrossed the room with two miles of string. The result resembled a giant spider web through which one could barely distinguish the works of art.

Another famous installation of Surrealist art was designed in 1942 (fig. 369) by the architect **FREDERICK KIESLER** (1890–1965) for the gallery, Art of This Century, the brainchild of Peggy Guggenheim. Guggenheim had befriended many of the Surrealists while living in Europe before the war and was building what would become one of the world's greatest collections of modern art. She was one of several remarkable American women who played instrumental roles in the development of avant-garde art. For her Art of This Century gallery, Kiesler designed the Surrealist room with curved walls from which paintings were hung on cantilevered arms that moved. Sculptures such as Giacometti's *Woman with Her Throat Cut* (fig. 393) can be seen on the biomorphic chairs Kiesler designed. Not surprisingly, the installation caused a great stir in New York art circles. Guggenheim's unique venture lasted for five years (it closed when she returned to Europe), but not before she had mounted numerous groundbreaking exhibitions, fostered the careers of some of the most promising young Americans (most notably, Jackson Pollock), and provided a kind of modern-day salon (or "laboratory for new ideas" as she called it) to encourage exchange between those Americans and their European forebears.

Born in Chile as Roberto Sebastiano Antonio Matta Echaurren, **MATTA** (b. 1911) studied architecture in the Paris office of Le Corbusier until, in 1937, he showed some of his drawings to Breton, who immediately welcomed him into the Surrealist fold. When he migrated to the United States in 1939 (on the same boat as Tanguy), Matta was relatively little known despite his association with the Surrealists in Paris.

369. FREDERICK KIESLER. Design of the Surrealist gallery, Art of this Century, 1942. Installation photo by Berenice Abbott from the periodical *VVV*. Courtesy Frederick Kiesler Archives

His one-man show at the Julien Levy Gallery in 1940 was so completely outside the experience of most American experimental artists that its impact was momentous. Although Matta—the last painter, along with Arshile Gorky, claimed for Surrealism by André Breton—was much younger than most of his expatriate European colleagues and less well known, his paintings marked the step forward that American artists were seeking. Over the next few years, together with the American painter Robert Motherwell, Matta, who unlike many emigré artists from Europe spoke English fluently, helped to forge a link between European Surrealism and the American movement to be called Abstract Expressionism. Matta's painting at that point is exemplified in the 1942 *Disasters of Mysticism* (color plate 172, page 328). Although this has some roots in Masson and Tanguy, it is a powerful excursion into unchartered territory. In its ambiguous, automatist flow, from brilliant flame-light into deepest shadow, it suggests the ever-changing universe of outer space. Soby, who owned *Disasters of Mysticism*, detected in its dark, volcanic spaces a debt to sixteenth-century nocturnal mysticism, exemplified in painting by Mathias Grünewald's masterpiece, the *Isenheim Altarpiece* (see fig. 8).

WILFREDO LAM (1902–1982) brought a broad cultural background to his art and, once he arrived in Paris in 1938, to the Surrealist milieu. Born in Cuba to a Chinese father and a European-African mother, he studied art first in Havana and, as of 1923, in Madrid. In Spain he fought on the side of the defeated Republicans during the Civil War, a fact that endeared him to Picasso when he moved to Paris. Through Picasso, Lam was introduced to members of the Parisian Surrealist circle, including Breton. He explored imagery in his paintings heavily influenced by Picasso, Cubism, and a continuing interest in African art. Lam left Paris because of the war, eventually traveling with Breton to Martinique, where they were joined by Masson. The Frenchmen moved on to New York while Lam returned to Cuba in 1942. His rediscovery of the rich culture of his homeland helped to foster the emergence of his mature style. "I responded always to the presence of factors that emanated from our history and our geography, tropical flowers, and black culture." While in Cuba, Lam made the large, sumptuously colored painting *The Jungle* (color plate 173, page 328). Based in part on the Cuban religion of Santería, which blends mystical and ritualistic traditions from Africa with old-world Catholicism, *The Jungle* presents a dense tropical landscape populated by odd, long-limbed creatures whose disjointed heads resemble African masks. These human-animal hybrids, who assemble in a lush dreamscape, give evidence of the artist's Surrealist background. Lam returned to France in 1952, though he continued to travel widely and visit Cuba. His art was especially important to postwar European and American artists, especially the CoBrA painters of northern Europe (see chapter 20).

The artist who above all others symbolizes Surrealism in the public imagination is the Spaniard **SALVADOR DALÍ** (1904–1989). Not only his paintings, but his writings, his utterances, his actions, his appearance, his mustache, and his genius for publicity have made the word "Surrealism" a common noun in all languages, denoting an art that is irrational, erotic, mad—and fashionable. Dalí's life itself was so completely Surrealist that his integrity and his pictorial accomplishment have been questioned, most bitterly by other Surrealists. The primary evidence must be the paintings them-

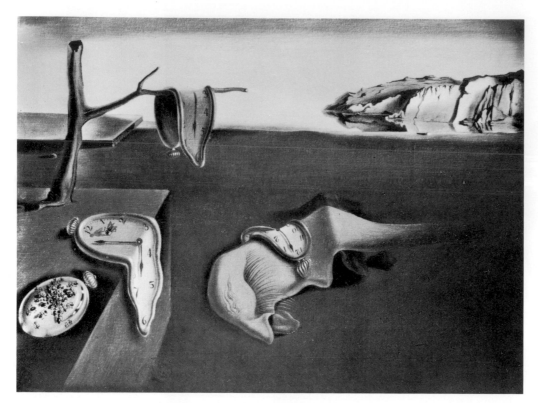

370. SALVADOR DALÍ. *The Persistence of Memory (Persistance de la mémoire).* 1931. Oil on canvas, 9½ x 13" (24.1 x 33 cm). The Museum of Modern Art, New York

GIVEN ANONYMOUSLY. PHOTOGRAPH © 1995 THE MUSEUM OF MODERN ART, NEW YORK. © 1997 DEMART PRO ARTE/ARTISTS RIGHTS SOCIETY (ARS), NEW YORK

selves, and no one can deny the immense talent, the power of imagination, or the intense conviction that they display.

Dalí was born at Figueras, near Barcelona. Like Picasso, Miró, and, before them, Gaudí, he is a product of the rich Catalan culture. A substantial part of the iconography of Dalí, real or imagined, is obviously taken from episodes in his childhood and adolescence. The landscape of these early years is constantly in his paintings, which are marked by the violence of his temperament—ecstatic, filled with fantasy, terror, and megalomania. Dalí encountered Italian and French art, Impressionism, Neo-Impressionism, and Futurism before he went to Madrid's Academy of Fine Arts in 1921. He followed the normal academic training, developing a highly significant passion for nineteenth-century academicians and artists such as Millet, Böcklin, and particularly the meticulous narrative painter Ernest Meissonier. His discoveries in modern painting during the early 1920s were Picasso, de Chirico, Carrà, and the Italian Metaphysical School, whose works were reproduced in the periodical *Valori Plastici.* More important for his development was the discovery of Freud, whose writings on dreams and the subconscious seemed to answer the torments and erotic fantasies he had suffered since childhood. Between 1925 and 1927 he explored several styles and often worked in them simultaneously: the Cubism of Picasso and Gris, the Purism of Le Corbusier and Ozenfant, the Neoclassicism of Picasso, and a precise realism derived from Vermeer.

In 1929 Dalí visited Paris for several weeks and, through his friend Miró, met the Surrealists. He moved to the French capital for a time the following year (though his main residence was Spain) to become an official member of the movement. Breton regarded the dynamic young Spaniard as a force for renewal in the group, which was plagued by endless ideological conflict. Dalí became deeply involved with Gala Éluard, the wife of the Surrealist writer Paul Éluard, and he married her in 1934. He had by now formulated the theoretical basis of his painting, described as "paranoiac-critical": the creation of a visionary reality from elements of visions, dreams, memories, and psychological or pathological distortions.

Dalí developed a precise miniature technique, accompanied by a particularly disagreeable, discordant, but luminous color. His precise trompe l'oeil technique was designed to make his dreamworld more tangibly real than observed nature (he referred to his paintings as "hand-painted dream photographs"). He frequently used familiar objects as a point of departure—watches, insects, pianos, telephones, old prints or photographs—many having fetishist significance. He declared his primary images to be blood, decay, and excrement. From the commonplace object Dalí set up a chain of metamorphoses gradually or suddenly dissolving and transforming the object into a nightmare image, given conviction by the exacting technique. He was determined to paint as a madman—not in an occasional state of receptive somnambulism, but in a continuous frenzy of induced paranoia. In collaboration with Luis Buñuel, Dalí turned to the cinema and produced two documents in the history of Surrealist film, *Un chien andalou (an Andalusian Dog)* (1929) and *L'Age d'or (the Golden Age)* (1930). The cinematic medium held infinite possibilities for the Surrealists, in the creation of dissolves, metamorphoses, and double and quadruple images, and Dalí made brilliant use of these.

The microscopic brand of realism that Dalí skillfully deployed to objectify the dream reached full maturity in a tiny painting from 1929, *The Accommodations of Desire* (color plate 174, page 329). The work is actually a collage. Dalí cut out two illustrations of lions' heads from a book and placed them on what appear to be stones. These strange forms, which cast dark shadows in the desolate landscape, serve as backdrops to other imagery, including an army of ants. These insects swarming over a hand provide one of the most disturbing images in *Un chien andalou,* which had been made earlier the same year, and appear again, this time over a pocket watch, in one of Dalí's later paintings (see fig. 370). Dalí's own painted incarnations of a lion's head appear at the upper left of *Accommodations* and with the group of figures at the top of the composition. In his mostly fictional autobiography of 1942, *The Secret Life of Salvador Dalí,* the artist said he undertook this painting in the midst of his courtship with Gala (who was then still married to Paul Éluard). The experience of intense sexual anxiety led him to paint this work, in which "desires were always represented by the terrorizing images of lions' heads."

In his book *La Femme visible* (1930), Dalí wrote: "I believe the moment is at hand when, by a paranoiac and active advance of the mind, it will be possible (simultaneously with automatism and other passive states) to systematize confusion and thus to help discredit completely the world of reality." By 1930 Dalí had left de Chirico's metaphysical landscapes for his own world of violence, blood, and decay. He sought to create in his art a specific documentation of Freudian theories applied to his own inner world. He started a painting with the first image that came into his mind and went on from one association to the next, multiplying images of persecution or megalomania like a true paranoiac. He defined his paranoiac-critical method as a "spontaneous method of irrational knowledge based upon the interpretative-critical association of delirious phenomena."

Dalí shared the Surrealist antagonism to formalist art, from Neo-Impressionism to Cubism and abstraction. *The Persistence of Memory* of 1931 (fig. 370) is a denial of every twentieth-century experiment in abstract organization. Its miniature technique (the painting is just over a foot wide) goes back to the Flemish art of the fifteenth century, and its sour greens and yellows recall nineteenth-century chromolithographs. The space is as infinite as Tanguy's, but rendered with hard objectivity. The picture's fame comes largely from the presentation of recognizable objects—watches—in an unusual context, with unnatural attributes, and on an unexpected scale. Throughout his career Dalí was obsessed with the morphology of hard and soft. Here, lying on the ground, is a large head in profile (which had appeared in several previous paintings) seemingly devoid of bone structure. Drooped over its surface, as on that of a tree and shelf nearby, is a soft pocket watch. Dalí described these forms as "nothing more than the soft, extravagant, solitary, paranoiccritical Camembert cheese of space and time." He painted the work one night after dinner, when, after all the guests were gone, he contemplated the leftover Camembert cheese melting on the table. When he then looked at the landscape in progress in his studio—with shoreline cliffs and a branchless olive tree—the image of soft watches came to him as a means of representing the condition of softness. This pictorial meta-

morphosis, in which matter is transformed from one state into another, is a fundamental aspect of Surrealism. Softness as a condition in sculpture would be fully explored in the 1960s by the American artist Claes Oldenburg (see fig. 619).

Dalí's painting during the 1930s vacillated between an outrageous fantasy and a strange atmosphere of quiet achieved less obviously. *Gala and the Angelus of Millet Immediately Preceding the Arrival of the Conic Anamorphoses* (color plate 175, page 329) exemplifies the first aspect. In the back of a brilliantly lighted room is Gala, grinning broadly, as though snapped by an amateur photographer; in the front sits an enigmatic male, actually a portrait of Vladimir Lenin. Over the open doorway is a print of Millet's *Angelus,* a painting that, in Dalí's obsessive, Freudian reading, had become a scene about predatory female sexuality. Around the open door a monstrous comic figure (Maxim Gorky) emerges from the shadow, wearing a lobster on his head. There is no rational explanation for the juxtaposition of familiar and phantasmagoric, but the nightmare is undeniable. Dalí painted *Soft Construction with Boiled Beans: Premonitions of Civil War* (fig. 371) in Paris in 1936, on the eve of the Spanish Civil War. It is a horrific scene of psychological torment and physical suffering. A gargantuan figure with a grimacing face is pulled apart, or rather pulls itself apart ("in a delirium of auto-strangulation" said the artist), while other body parts are strewn among the excretory beans along the ground. Despite the anguish expressed in his painting, Dalí reacted to the civil war in his native country—as he did to fascism in general—with characteristic self-interest. Indeed, he quarreled bitterly and broke with the

371. SALVADOR DALÍ. *Soft Construction with Boiled Beans: Premonitions of Civil War.* 1936. Oil on canvas, 39¼ x 39" (100 x 99 cm). Philadelphia Museum of Art

372. SALVADOR DALÍ. *Christ of St. John of the Cross*. 1951. 6'8⅝" x 45⅝" (2 m x 115.9 cm). Glasgow Art Gallery and Museum, Kelvingrove © 1997 DEMART PRO ARTE/ARTISTS RIGHTS SOCIETY (ARS), NEW YORK

Surrealists in 1934 over his refusal to condemn Hitler.

In 1940 Dalí, like so many of the Surrealists, moved to the United States where he painted society portraits and flourished in the role of Surrealist agent provocateur. His notoriety and fashionable acceptance reached their height with designs for the theater, jewelry, and objets d'art. In 1941 came a transition in his painting career. He announced his determination to "become Classic," to return to the High Renaissance of Raphael, though his paintings still contained a great deal of Surrealist imagery. He and Gala resettled permanently in Spain in 1948. After 1950 Dalí's principal works were devoted to Christian religious art exalting the mystery of Christ and the Mass. In the 1951 painting (fig. 372), the dramatically foreshortened figure of Christ floats above the familiar landscape of the bay of Port Lligat, Spain, where Dalí lived. Unlike so many of the early miniaturist paintings on panel, this work is executed on a monumental scale.

In contrast to Dalí, RENÉ MAGRITTE (1898–1967) has been called the invisible man among the Surrealists. George Melly, in the script for a BBC film on Magritte, wrote: "He is a secret agent; his object is to bring into disrepute the whole apparatus of bourgeois reality. Like all saboteurs, he avoids detection by dressing and behaving like everybody else." Thanks perhaps to the artist's anonymity, due in part to Breton's lukewarm support of his art, the works of Magritte have had gradual but overwhelming impact.

After years of sporadic study at the Brussels Academy of Fine Arts, Magritte, like Tanguy, was shocked into realizing his destiny when in 1923 he saw a reproduction of de Chirico's 1914 painting *The Song of Love*. In 1926 he emerged as an individual artist with his first Surrealist paintings inspired by the work of Ernst and de Chirico. Toward the end of that year, Magritte joined other Belgian writers, musicians, and artists in an informal group comparable to the Paris Surrealists. He moved to a Paris suburb in 1927, entering one of the most highly productive phases of his career, and participated for the next three years in Surrealist affairs. Wearying of the frenetic, polemical atmosphere of Paris after quarreling with the Surrealists, and finding himself without a steady Parisian dealer to promote his work, he returned to Brussels in 1930 and lived there quietly for the rest of his life. Because of this withdrawal from the art centers of the world, Magritte's paintings did not receive the attention they deserved. In Europe he was seen as a marginal figure among the Surrealists, and in the United States, in the heyday of painterly abstraction after World War II, his meticulous form of realism seemed irrelevant to many who were drawn instead to the work of Masson or Miró. But subsequent generations, such as that of the Pop artists, appreciated Magritte's genius for irony, uncanny invention, and deadpan realism. Important exhibitions were held in Europe and the United States after World War II, climaxing in the retrospective at New York's Museum of Modern Art in 1965. Another major survey exhibition in 1992 brought the work of this brilliant artist once again to the forefront, and his unique contribution to Surrealism as well as his tremendous impact on art and commercial design is now more fully recognized.

The perfect symbol (except that Magritte disliked attributions of specific symbolism) for his approach is the painting entitled *The Treachery (or Perfidy) of Images* (fig. 373). It portrays a briar pipe so meticulously that it might serve as a tobacconist's trademark. Beneath, rendered with comparable precision, is the legend *Ceci n'est pas une pipe* (This is not a pipe). This delightful work confounds pictorial reality and underscores Magritte's fascination with the relationship of language to the painted image. A similar idea is embodied in *The False Mirror* (color plate 176, page 330): the eye as a false mirror when it views the white clouds and blue sky of nature. This might be another statement of the artist's faith, for *The False Mirror* introduces the illusionistic theme of the landscape that is a painting, not nature. One of Magritte's best-known images, it was transformed by CBS into their famous corporate logo.

Color plate 153.
PABLO PICASSO. *Three Musicians.* Fontainebleau, summer 1921. Oil on canvas, 6'7" x 7'3¾" (2 x 2.2 m). The Museum of Modern Art, New York

Color plate 154. PABLO PICASSO. *Mandolin and Guitar.* 1924. Oil and sand on canvas, 56⅛ x 67¾" (142.6 cm x 2 m). Solomon R. Guggenheim Museum, New York

Color plate 155. PABLO PICASSO. *Studio with Plaster Head.* Juan-les-Pins, summer 1925. Oil on canvas, 38⅝ x 51⅝" (97.9 x 131.1 cm). The Museum of Modern Art, New York

Color plate 156. GEORGES BRAQUE.
Café-Bar. 1919. Oil on canvas, 63¼ x 31⅞"
(160.7 x 81 cm).
Öffentliche Kunstsammlung Basel, Kunstmuseum
© 1997 ADAGP, PARIS/ARTISTS RIGHTS SOCIETY (ARS), NEW YORK

Color plate 157. GEORGES BRAQUE. *Vase, Palette, and Skull.* 1939. Oil and sand on canvas, 36 x 36" (91.4 x 91.4 cm). Kreeger Museum, Washington, D.C.
© 1997 ADAGP, PARIS/ARTISTS RIGHTS SOCIETY (ARS), NEW YORK

right: Color plate 158. RAOUL DUFY.
*Indian Model in the Studio at L'Impasse
Guelma.* 1928. Oil on canvas, 31⅞ x 39⅜"
(81 x 100 cm). Private collection
© 1997 ADAGP, PARIS/ARTISTS RIGHTS SOCIETY (ARS), NEW YORK

Color plate 159. FERNAND LÉGER. *Three Women (Le Grand Déjeuner)*. 1921. Oil on canvas, 6'¼" x 8'3"
(1.8 x 2.5 m). The Museum of Modern Art, New York

Color plate 160. FERNAND LÉGER. *The Great Parade*. 1954. Oil on canvas, 9' x 13'1" (2.7 x 4 m).
Solomon R. Guggenheim Museum, New York

Color plate 161. JEAN ARP. *Composition*. 1937. Collage of torn paper, india ink wash, and pencil, 11¾ x 9" (29.8 x 22.9 cm). Philadelphia Museum of Art

Color plate 162. MAX ERNST. *Two Children Are Threatened by a Nightingale*. 1924. Oil on wood with wood construction, 27½ x 22½ x 4½" (69.8 x 57.1 x 11.4 cm). The Museum of Modern Art, New York

Color plate 163. MAX ERNST. *Europe after the Rain*. 1940–42. Oil on canvas, 21½ x 58⅛" (54.6 x 147.6 cm). Wadsworth Atheneum, Hartford, Connecticut ELLA GALLUP SUMNER AND MARY CATLIN SUMNER COLLECTION FUND

Color plate 164. JOAN MIRÓ. *The Harlequin's Carnival*. 1924–25. Oil on canvas, 26 x 36⅝" (66 x 93 cm). Albright-Knox Art Gallery, Buffalo, New York
ROOM OF CONTEMPORARY ART FUND
© 1997 ADAGP, PARIS/ARTISTS RIGHTS SOCIETY (ARS), NEW YORK

Color plate 165. JOAN MIRÓ. *Dog Barking at the Moon*. 1926. Oil on canvas, 28¾ x 36¼" (73 x 92 cm). Philadelphia Museum of Art
A. E. GALLATIN COLLECTION © 1997 ADAGP, PARIS/ARTISTS RIGHTS SOCIETY (ARS), NEW YORK

above: Color plate 166. JOAN MIRÓ. *Painting*. 1933. Oil on canvas, 68½" x 6'5¼" (174 cm x 2 m). The Museum of Modern Art, New York
LOULA D. LASKER BEQUEST (BY EXCHANGE)
© 1997 ADAGP, PARIS/ARTISTS RIGHTS SOCIETY (ARS), NEW YORK

Color plate 167. JOAN MIRÓ. *Object*. 1936. Assemblage: stuffed parrot on wood perch, stuffed silk stocking with velvet garter and doll's paper shoe suspended in hollow wood frame, derby hat, hanging cork ball, celluloid fish, and engraved map, 31⅞ x 11⅞ x 10¼" (81 x 30.1 x 26 cm). The Museum of Modern Art, New York
GIFT OF MR. AND MRS. PIERRE MATISSE
© 1997 ADAGP, PARIS/ARTISTS RIGHTS SOCIETY (ARS), NEW YORK

above: Color plate 168.
JOAN MIRÓ. *Blue II.* 1961.
Oil on canvas, 8'10¼" x
11'7¾" (2.7 x 1.4 m).
Musée National d'Art
Moderne, Centre d'Art et
de Culture Georges
Pompidou, Paris

Color plate 169. ANDRÉ
MASSON. *Pasiphaë.* 1943.
Oil and tempera on can-
vas, 39¾ x 50" (101 x
127 cm). Private collection

Color plate 170. YVES TANGUY.
Mama, Papa Is Wounded! (Maman, Papa est blessé).
1927. Oil on canvas, 36¼ x 28¾" (92.1 x 73 cm).
The Museum of Modern Art, New York
PURCHASE
© 1997 ESTATE OF YVES TANGUY/ARTISTS RIGHTS SOCIETY (ARS), NEW YORK

below: Color plate 171. YVES TANGUY.
Multiplication of the Arcs. 1954. Oil on canvas,
40 x 60" (101.6 x 152.4 cm). The Museum of
Modern Art, New York
MRS. SIMON GUGGENHEIM FUND
© 1997 ESTATE OF YVES TANGUY/ARTISTS RIGHTS SOCIETY (ARS), NEW YORK

above: Color plate 172. MATTA. *Disasters of Mysticism.* 1942. Oil on canvas, 38¼ x 51¾" (101.6 x 152.4 cm). Estate of James Thrall Soby, New Canaan, Connecticut
© 1997 ADAGP, PARIS/ARTISTS RIGHTS SOCIETY (ARS), NEW YORK

left: Color plate 173. WILFREDO LAM. *The Jungle.* 1943. Gouache on paper mounted on canvas, 7'10¼" x 7'6½" (2.4 x 2.3 m). The Museum of Modern Art, New York
INTER-AMERICAN FUND
© 1997 ADAGP, PARIS/ARTISTS RIGHTS SOCIETY (ARS), NEW YORK

above: Color plate 174. SALVADOR DALÍ. *Accommodations of Desire*. 1929. Oil and collage on panel, 8⅝ x 13¾" (22 x 34.9 cm). Private collection
© 1997 DEMART PRO ARTE/ARTISTS RIGHTS SOCIETY (ARS), NEW YORK

left: Color plate 175. SALVADOR DALÍ. *Gala and the Angelus of Millet Immediately Preceding the Arrival of the Conic Anamorphoses*. 1933. Oil on panel, 9⅜ x 7⅜" (23.8 x 18.7 cm). National Gallery of Canada, Ottawa
© 1997 DEMART PRO ARTE/ARTISTS RIGHTS SOCIETY (ARS), NEW YORK

right: Color plate 176
RENÉ MAGRITTE.
The False Mirror. 1928.
Oil on canvas, 21¼ x
31⅞" (54 x 81 cm). The
Museum of Modern Art,
New York
PURCHASE © 1997 C. HERSCOVICI,
BRUSSELS/ARTISTS RIGHTS SOCIETY (ARS),
NEW YORK

below: Color plate 177.
RENÉ MAGRITTE.
The Human Condition
1933. Oil on canvas,
39⅜ x 31⅞"
(100 x 81 cm).
National Gallery of Art,
Washington, D.C.
GIFT OF THE COLLECTORS COMMITTEE
© 1997 C. HERSCOVICI,
BRUSSELS/ARTISTS RIGHTS SOCIETY
(ARS), NEW YORK

Color plate 178. RENÉ MAGRITTE. *The Dominion of Light.* 1952. Oil on canvas, 39½ x 32" (100.3 x 81.3 cm). Collection Lois and Georges de Menil

Color plate 179. PAUL DELVAUX. *Entrance to the City.* 1940. Oil on canvas, 63 x 70⅞" (160 x 180 cm). Private collection

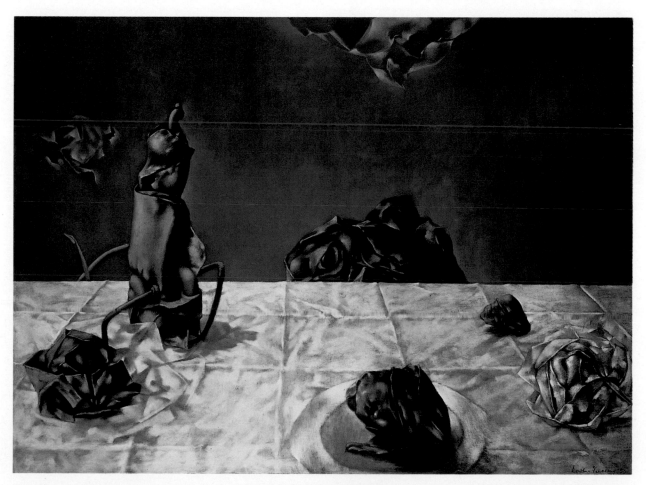

Color plate 180. DOROTHEA TANNING. *Some Roses and Their Phantoms.* 1952. Oil on canvas, 29⅞ x 40" (75.9 x 101.6 cm). Collection Sir Roland Penrose, London

Color plate 181. LEONORA CARRINGTON. *Self-Portrait (The White Horse Inn)*. 1936–37. Oil on canvas, 25½ x 32⅛" (65 x 81.5 cm). Private collection

below, left: Color plate 182. PABLO PICASSO. *Three Dancers (The Dance)*. 1925. Oil on canvas, 7'⅝" x 56¼" (2.1 m x 142.9 cm). Tate Gallery, London

below, right: Color plate 183. PABLO PICASSO. *Large Nude in Red Armchair*. May 5, 1929. Oil on canvas, 6'4¾" x 51¼" (1.9 m x 130.2 cm). Musée Picasso, Paris

above: Color plate 184. PABLO PICASSO. *Painter and Model.* 1928.
Oil on canvas, 51⅛ × 64¼" (130 × 163.2 cm).
The Museum of Modern Art, New York THE SIDNEY AND HARRIET JANIS COLLECTION
© 1997 ESTATE OF PABLO PICASSO/ARTISTS RIGHTS SOCIETY (ARS), NEW YORK

below: Color plate 185. PABLO PICASSO.
Crucifixion. 1930. Oil on plywood, 20⅛ × 26"
(51.1 × 66.1 cm). Musée Picasso, Paris
© 1997 ESTATE OF PABLO PICASSO/ARTISTS RIGHTS SOCIETY (ARS), NEW YORK

above: Color plate 186. PABLO PICASSO. *Seated Bather*. Paris, early 1930. Oil on canvas, 64¼ x 51" (163.2 x 129.5 cm). The Museum of Modern Art, New York
MRS. SIMON GUGGENHEIM FUND
© 1997 ESTATE OF PABLO PICASSO/ARTISTS RIGHTS SOCIETY (ARS), NEW YORK

above, right: Color plate 187. PABLO PICASSO. *Girl Before a Mirror*. Boisgeloup, 1932. Oil on canvas, 64 x 51¼" (162.3 x 130.2 cm). The Museum of Modern Art, New York
GIFT OF MRS. SIMON GUGGENHEIM
© 1997 ESTATE OF PABLO PICASSO/ARTISTS RIGHTS SOCIETY (ARS), NEW YORK

right: Color plate 188. PABLO PICASSO. *Still Life with Steer's Skull*. April 5, 1942. Oil on canvas, 51¼ x 38¼" (130.2 x 97.2 cm). Kunstsammlung Nordrhein-Westfalen, Düsseldorf
© 1997 ESTATE OF PABLO PICASSO/ARTISTS RIGHTS SOCIETY (ARS), NEW YORK

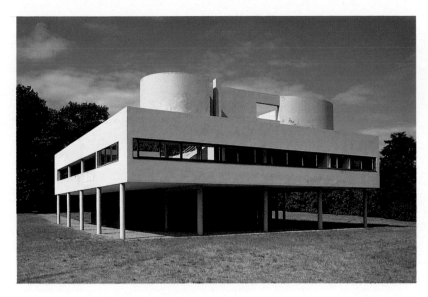

Color plate 189. LE CORBUSIER. Villa Savoye, Poissy, France. 1928–30

Color plate 190. WILLIAM VAN ALEN. Chrysler Building,
New York. 1928–30

Color plate 191. FRANK LLOYD WRIGHT. Edgar K. Kaufmann House
(Fallingwater), Bear Run, Pennsylvania. 1934–37

Color plate 192. JOSEF ALBERS. *City.* 1928. Sandblasted colored glass, 11 x 21⅝" (27.9 x 54.9 cm). Kunsthaus, Zürich

Color plate 193. JOSEF ALBERS. *Homage to the Square: Apparition.* 1959.
Oil on board, 47½ x 47½" (120.7 x 120.7 cm).
Solomon R. Guggenheim Museum, New York

373. RENÉ MAGRITTE. *The Treachery (or Perfidy) of Images.*
1928–29. Oil on canvas, 23¼ x 31½" (60 x 80 cm). Los Angeles
County Museum of Art

PURCHASED WITH FUNDS PROVIDED BY THE MR. AND MRS. WILLIAM PRESTON HARRISON COLLECTION.
© 1997 ADAGP, PARIS/ARTISTS RIGHTS SOCIETY (ARS), NEW YORK

374. RENÉ MAGRITTE. *Portrait.* 1935. Oil on canvas,
28⅞ x 19⅞" (73.3 x 50.2 cm).
The Museum of Modern Art, New York

GIFT OF KAY SAGE TANGUY. PHOTOGRAPH © 1995 THE MUSEUM OF MODERN ART,
NEW YORK. © 1997 ADAGP, PARIS/ARTISTS RIGHTS SOCIETY (ARS), NEW YORK

After 1926 Magritte changed his style of precise depiction
very little, except for temporary excursions into other sometimes alarmingly divergent manners, and he frequently
returned to earlier subjects. In 1933, he reconciled with the
Parisian Surrealists and throughout the decade made many of
the key works of his career, usually in compositions of the
utmost clarity and simplicity. In *The Human Condition* (color

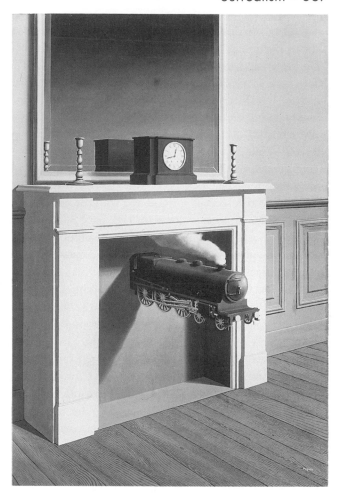

375. RENÉ MAGRITTE. *Time Transfixed.* 1938. Oil on canvas,
57½ x 38½" (146.1 x 97.8 cm). The Art Institute of Chicago

THE JOSEPH WINTERBOTHAM COLLECTION.
© 1997 ADAGP, PARIS/ARTISTS RIGHTS SOCIETY (ARS), NEW YORK

plate 177, page 330), we encounter a pleasant landscape
framed by a window. In front of this opening is a painting on
an easel that "completes" the very landscape it blocks from
view. The problem of real space versus spatial illusion is as old
as painting itself, but it is imaginatively treated in this picture
within a picture. Magritte's classic play on illusion implies that
the painting is less real than the landscape, when in fact both
are painted fictions. The tree depicted here, as the artist
explained, exists for the spectator "both inside the room in
the painting, and outside in the real landscape. Which is how
we see the world: we see it as being outside ourselves even
though it is only a mental representation of it that we experience inside ourselves."

Magritte's visual jokes can border on the gruesome. The
succinctly titled *Portrait* (fig. 374) is a still life of a succulent
slice of ham accompanied by wine bottle, glass, knife, and
fork, all drawn with the utmost precision. The table on which
they rest is a vertical rectangle of dark gray. Out of the center
of the slice of ham stares a human eye. Magritte re-created
this scene as a life-size, three-dimensional installation for a
1945 exhibition in Brussels. In another painting from the
1930s called *Time Transfixed* (fig. 375), the artist introduced
the unexpected into an otherwise unremarkable scene: a tiny
locomotive emerges from a fireplace, transforming the latter

into a kind of domestic railway tunnel. The trains in de Chirico's paintings (see fig. 294) may have contributed to Magritte's vision, as did the fireplace in a friend's house. The effect here of ominous mystery is not unlike that of the Italian's, but Magritte accomplished it with characteristic restraint and an almost geometric precision.

Much of Magritte's iconography was established early in the 1930s, but his variations have been infinitely ingenious. In certain pictures of the 1950s, he calcified all the parts—figures, interiors, landscapes, objects—into a single rock texture. A midnight street scene is surmounted with the blue sky and floating clouds of noon (color plate 178, page 331); jockeys race over cars and through rooms; or an elegant horsewoman passing through a forest is segmented by the trees. The basic forms and themes continued through the 1960s.

PAUL DELVAUX (1897–1994), like Magritte a longtime resident of Brussels, came to Surrealism slowly. Beginning in 1935, he painted women, usually nude but occasionally clothed in chaste, Victorian dress. Sometimes lovers appear, but normally the males are shabbily dressed scholars, strangely oblivious to the women. *Entrance to the City* (color plate 179, page 331) gives the basic formula: a spacious landscape; nudes wandering about, each lost in her own dreams; clothed male figures, with here a partly disrobed young man studying a large plan; here also a pair of embracing female lovers and a bowler-hatted gentleman (reminiscent of Magritte) reading his newspaper while walking. The chief source seems to be fifteenth-century painting, and even the feeling of withdrawal in the figures suggests the influence of Piero della Francesca (see fig. 4) translated into Delvaux's peculiar personal fantasy. Delvaux's dreamworld is filled with a nostalgic sadness that transforms even his erotic nudes into something elusive and unreal.

Another artist who, like Delvaux, belonged to a later, second wave of Surrealism is HANS BELLMER (1902–1975). In 1923–24 this Polish-born artist trained at the Berlin Technical School, studying drawing with George Grosz and establishing contact with other figures related to German Dada. In the 1930s, he began assembling strange constructions called "dolls" (fig. 376), adolescent female mannequins whose articulated and ball-jointed parts—heads, arms, trunks, legs, wigs, glass eyes—could be dismembered and reassembled in every manner of erotic or masochistic posture. The artist then photographed the fetishistic objects, whose implied sadism appealed to the Surrealists. This led to publication of Bellmer's photographs in a 1935 issue of *Minotaure*, a career for the artist in the Surrealist milieu of Paris, and a place for his camera-made images in the history of photography. Bellmer also vented his rather Gothic imagination in a number of splendid, if sometimes unpublishable, drawings. His particular brand of photography, whereby he invented actual objects or tableaux as subjects for his camera, would find its 1980s equivalent in artists such as Cindy Sherman (see fig. 864). OSCAR DOMINGUEZ (1906–1957), a Spaniard noted for his assemblages of found objects, joined the Surrealists in 1934. He contributed the technique known

376. HANS BELLMER. *Doll.* 1935. Wood, metal, and papier-mâché. Manoukian Collection, Paris

377. OSCAR DOMINGUEZ. Untitled. 1936. Gouache transfer (decalcomania) on paper, 14⅛ x 11½" (35.9 x 29.2 cm). The Museum of Modern Art, New York PURCHASE. PHOTOGRAPH © 1995 THE MUSEUM OF MODERN ART, NEW YORK. © 1997 ADAGP, PARIS/ARTISTS RIGHTS SOCIETY (ARS), NEW YORK

as "decalcomania": inks or watercolor paints transferred from one sheet of paper to another under pressure (fig. 377). The process appealed to the Surrealists—particularly Max Ernst—for its startling automatic effects (color plate 163, page 324).

As a movement, Surrealism was not particularly hospitable to women, except as a terrain onto which male artists projected their erotic desires. Nevertheless, it attracted a number

378. MERET OPPENHEIM. *Object (Le Déjeuner en fourrure) (Luncheon in Fur)*. 1936. Fur-covered cup, diameter 4⅜" (10.9 cm); saucer, diameter 9⅜" (23.7 cm); spoon, length 8" (20.2 cm). Overall height 2⅞" (7.3 cm). The Museum of Modern Art, New York

of gifted women artists in the 1930s. MERET OPPENHEIM (1913–1985), a native of Berlin, arrived in Paris as an art student in 1932 when she was not even twenty years old. She soon met Alberto Giacometti, who introduced her to members of the Surrealist circle, and began to take part in the group's exhibitions in 1935, and she is recorded in a number of photographs by Man Ray. Although she produced a large body of work—paintings, drawings, sculpture, and poetry—until her death in 1985, Oppenheim is known almost exclusively for the notorious *Object (Le Déjeuner en fourrure) (Luncheon in Fur)* (fig. 378), a cup, plate, and spoon covered with the fur of a Chinese gazelle. The idea for this "fur-lined tea cup," which has become the very archetype of the Surrealist object, germinated in a café conversation with Picasso about Oppenheim's designs for jewelry made of fur-lined metal tubing. When Picasso remarked that one could cover just about anything with fur, Oppenheim quipped, "Even this cup and saucer. . . ." In Oppenheim's hands, this emblem of domesticity and the niceties of social intercourse metamorphosed into a repellent, hairy object. Such inventions fulfilled brilliantly Breton's notion of the reconciliation of opposites by bringing together in a single form materials of the most disparate nature imaginable.

The career of American artist DOROTHEA TANNING (b. 1910) was somewhat overshadowed by that of her husband, Max Ernst, whom she married in 1946. Tanning had met Ernst in New York, where she moved after her brief art studies in Chicago. Like so many American artists, she was deeply moved by the experience of the important exhibition "Fantastic Art, Dada, and Surrealism," held at the Museum of Modern Art in 1936. She also came into contact with Surrealist refugee artists from Paris, including Ernst and Breton. By the early 1940s, Tanning had produced her first mature paintings, and in 1944 she was given a solo exhibition at the Julien Levy Gallery, the most important commercial showcase for Surrealist art in New York. In 1946, she moved

with Ernst to Sedona, Arizona, and later settled in France. Tanning's skills as a painter and her penchant for the bizarre are given rich expression in the 1950 still life *A Few Roses and Their Phantoms* (color plate 180, page 331). Across a table covered with strange, rose-hybrid formations and a tablecloth marked by grid-shaped folds, a monstrous creature raises its head. Tanning's meticulous style intensifies the hallucinatory quality of the scene. As she claimed, "It's necessary to paint the lie so great that it becomes the truth."

Slightly younger than Tanning is British-born LEONORA CARRINGTON (b. 1917), who has distinguished herself both as a writer and a painter. Carrington scandalized her wealthy Catholic parents by moving to Paris with Max Ernst at the age of twenty and becoming an associate of the Surrealists. During the war, when Ernst was interned by the French as an enemy alien (and imprisoned for a time in a cell with Hans Bellmer), Carrington suffered a mental collapse, about which she wrote eloquently after her recovery. After living in New York for two years, she settled in Mexico in 1942. The enigmatic imagery that surrounds the artist in her *Self-Portrait (The White Horse Inn)* (color plate 181, page 332) is closely tied to Celtic mythology and memories of her childhood, themes she also explored in her literary works. In two of her short stories the protagonists befriend, respectively, a hyena and a rocking horse, both with magical powers. Carrington, who apparently kept a rocking horse in her Paris apartment, created a lead character in one of the stories who is capable of transforming herself into a white horse. In the painting, Carrington, wearing her white jodhpurs, reaches out to a hyena heavy with milk, while a toy horse levitates inexplicably above her head. Given the subject of Carrington's own fable, the galloping white horse seen through the window becomes a kind of liberated surrogate self. In a larger context, this arresting painting abounds with classic Freudian dream imagery, in which horses are symbols of sexual desire. After she moved to Mexico, Carrington cast her mystical subjects of alchemy and the occult in imagery reminiscent of Hieronymous Bosch.

Pablo Picasso and Surrealism

In 1925, André Breton claimed PABLO PICASSO (1881–1973) as "one of ours," but sensibly exempted the great artist (at least for a time) from adopting a strict Surrealist regime. Picasso was never a true Surrealist. He did not share the group's obsession with the subconscious and the dreamworld, though his paintings could certainly plumb the depths of the psyche and express intense emotion. The powerful strain of eroticism and violence in Picasso's art also helped ally him to the Surrealist cause. He had considerable sympathy for the group, exhibiting with them and contributing drawings to the second number of *La Révolution surréaliste* (January 15, 1925). In the fourth number of the magazine (July 15, 1925) Breton, the editor, included an account of *Les Demoiselles d'Avignon* (color plate 89, page 166) and reproduced collages by Picasso and his new painting, *Three Dancers (The Dance)* (color plate 182, page 332). We have already

379. PABLO PICASSO. *Painter and Model Knitting*
(executed in 1927) from *Le Chef-d'oeuvre inconnu*,
by Honoré de Balzac. Paris, Ambroise Vollard, 1931. Etching,
printed in black, plate 7⁹⁄₁₆ x 10⅞" (19.2 x 27.6 cm).
The Museum of Modern Art, New York
THE LOUIS E. STERN COLLECTION.
© 1997 ESTATE OF PABLO PICASSO/ARTISTS RIGHTS SOCIETY (ARS), NEW YORK

seen in *Studio with Plaster Head* (color plate 155, page 321) how a new emotionalism was entering Picasso's work in the mid-1920s. Made like that still life in the summer of 1925, *Three Dancers* was an even more startling departure from the artist's classically lyrical drawings of dancers of the previous months. Before a French window, three figures perform an ecstatic dance. The dancer in the center seems frozen, pinned in space in a crucified position. To her right the figure is jagged and angular. The figure to the left is disjointed and bends back in a state of frenzied abandon, baring her teeth in a terrifying grimace. Given her relationship to the "crucified" dancer, this figure has been compared to a mourning woman at the foot of the Cross. The subject of the Crucifixion would preoccupy the artist in the early 1930s (color plate 185, page 333). Picasso used every device in the Synthetic Cubist vocabulary—simultaneous views, full face and profile, hidden or shadow profiles, faceting, interpenetration, positive and negative space—to create his image. The anguish expressed in the painting can be attributed to the increasing marital tension between the artist and his wife Olga Koklova (who was a dancer), as well as to the recent death of Picasso's close friend, the Catalan painter Raymon Pichot, whose profile can be seen here silhouetted against the window.

Violence and anxiety (a favorite emotional coupling of the Surrealists) also permeate *Large Nude in Red Armchair* (color plate 183, page 332), 1929. It is a timeless subject—a figure seated in a red chair draped in fabric and set against green flowered wallpaper (these saturated hues will persist in Picasso's work of the 1930s). But the woman's body is pulled apart as though made of rubber. With knobs for hands and feet and black sockets for eyes, she throws her head back not in sensuous abandon but in a toothy wail, resulting in what Breton called "convulsive" beauty. While he did it to rather different effect, the photographer André Kertész subjected

the female body to a similar kind of elasticity in the 1930s (fig. 400). In many ways, Picasso's paintings can be read as running commentaries on his relationships, and his paintings of the late 1920s, when his marriage was disintegrating, were filled with women as terrifying monsters. As the poet Paul Éluard observed, "He loves with intensity and kills what he loves." In 1932, when Picasso began to make portraits of a young woman named Marie-Thérèse Walker (color plate 187, page 334) the explicit eroticism in his paintings shifted to a gentler mood.

One of Picasso's favorite subjects was the artist in his studio. He returned to it periodically throughout his career, often exploring the peculiarly voyeuristic nature of the painter's relationship to the model. His interest may have been aroused by a commission from the dealer Ambroise Vollard to provide drawings for an edition of Honoré de Balzac's novel, *The Unknown Masterpiece*. The illustrated edition (fig. 379) was published in 1931. This story concerns a deranged painter who worked for ten years on the portrait of a woman and ended with a mass of incomprehensible scribbles. Picasso's interpretation may illustrate his mistrust in complete abstraction, for he had already warned against trying "to paint the invisible." In a highly geometric composition of 1928 (color plate 184, page 333) reality and illusion as presented in the drawing are reversed. The painter and model are highly abstracted, Surrealist ciphers, while the "painted" portrait on the artist's canvas is a more "realistic" profile.

Picasso rarely painted the religious subjects so dear to Spanish artists, so the small, brilliant *Crucifixion* (color plate 185, page 333) comes as a surprise in 1930. He had, however, been making drawings related to this Christian theme, and as we have seen, the subject played a role in *Three Dancers*. The artificial palette of bright reds, yellows, and greens seems unsuited to the somber subject, but Picasso's is hardly a conventional treatment. In a small canvas he managed to cram a crowd of disjointed figures in such a dizzying variety of scales and styles that is extremely difficult to read the picture. We can recognize certain familiar aspects of the story. For example, a tiny red figure at the top of the ladder nails Christ's hand to the Cross. The mourning Virgin is located just before the Cross, her face distorted in the same grimace we encountered in *Large Nude in Red Armchair*. In the foreground two soldiers throw dice for Christ's clothes, but the anatomies of the figures are drastically distorted by unexpected shifts in scale. *The Crucifixion*, in the end, is a highly personal version of this sacred subject, one that drew on many sources, including ancient legends and rituals. What is clear is its sense of tortured agony that looks back to the *Isenheim Altarpiece* (see fig. 8) and forward to *Guernica*. Picasso's drawings after the Isenheim were published in 1933 in the first issue of *Minotaure*, for which he designed the cover (fig. 368). Also included in that issue was a suite of drawings called An Anatomy, in which the female figure is constructed in terms of abstract shapes or household objects. A woman's head, for example, becomes a goblet, or her breasts, a pair of teacups. An Anatomy is generally regarded

380. PABLO PICASSO. *Straw Hat with Blue Leaves*. May 1, 1936. Oil on canvas, 24 x 19¾" (61 x 50.2 cm). Musée Picasso, Paris
© 1997 ESTATE OF PABLO PICASSO/ARTISTS RIGHTS SOCIETY (ARS), NEW YORK

as one of Picasso's most thoroughly Surrealist works.

In 1928 Picasso made sketches of the figure compartmentalized into strange bonelike forms. Variations on these highly sexualized "bone figures" continued into the 1930s. One startling example is *Seated Bather* (color plate 186, page 334), which is part skeleton, part petrified woman, and all monster, taking her ease in the Mediterranean sun. While she has the nonchalance of a bathing beauty, *Seated Bather* has the predatory countenance of a praying mantis, one of the Surrealists favorite insect images (because the female sometimes devours her mate after the sexual act). Careful inspection of the right-hand portion of *Crucifixion* (color plate 185, page 333) proves that this strange head first made an appearance in that work. *Seated Bather* is a tour de force of the use of negative space. Picasso was, after all, an artist deeply involved with sculptural form both in painting and in three-dimensional objects.

In 1932 Picasso painted some of his most beautifully lyrical paintings of women, a change of heart inspired by his love affair with Marie-Thérèse. Although she had been involved with Picasso since 1927, when she was only seventeen, Marie-Thérèse did not become the overt subject of Picasso's paintings until their liaison was more public. His magical portrait of her (color plate 187, page 334), which assimilates classical repose with Cubist space, is a key painting of the 1930s. Such moments of summation for Picasso alternate with cycles of fertile and varied experiment. *Girl Before a Mirror* brings together brilliant color patterns and sensual curvilinear rhythms. The maiden, rapt in contemplation of her mirror image, sees not merely a reversed reflection but some kind of mysterious alter ego. In the mirror her clear blond features become a disquieting series of dark, abstracted forms, as though the woman is peering into the depths of her own soul.

The lengths to which Picasso could go to effect a Surrealist dislocation of human anatomy can be found in a another depiction of a female subject, *Straw Hat with Blue Leaves* (fig. 380). Here the woman's head has become a bulbous-shaped form (characteristically analogous to breasts) on which the eyes float to the periphery. This bizarre form is topped with a cheerful straw hat decorated with bright blue leaves. But it is unclear whether the tall purple shape next to the head is a vase or the figure's neck. This strange image so intrigued American artist Jasper Johns that he incorporated it into a number of his compositions from the 1980s (color plate 300, page 528).

Picasso's major painting of the 1930s and one of the masterworks of the twentieth century is *Guernica* (fig. 381). Executed in 1937, the painting was inspired by the Spanish Civil War—specifically, the destruction of the Basque town of Guernica by German bombers in the service of Spanish Fascists. Picasso was profoundly disturbed by the conflict

381. PABLO PICASSO. *Guernica*. 1937. Oil on canvas, 11'6" x 25'8" (3.5 x 7.8 m). Museo Nacional Centro de Arte Reina Sofía, Madrid
© 1997 ESTATE OF PABLO PICASSO/ARTISTS RIGHTS SOCIETY (ARS), NEW YORK

382. PABLO PICASSO.
Minotauromachy.
March 23, 1935. Etching
and engraving on copper
plate, printed in black; plate
19½ x 27⅜"
(49.6 x 69.6 cm).
The Museum of Modern Art,
New York.

from the beginning, and in a sense had been unconsciously preparing for this statement since the end of the 1920s. Some of its forms first appear in the *Three Dancers* of 1925. The figure of the Minotaur, the bull-man monster of ancient Crete and a Surrealist subject as we have seen, first appeared in Picasso's work in a collage of 1928, and in 1933 he made a series of etchings, collectively known as the Vollard Suite, in which the Minotaur is a central character. During 1933 and 1934 Picasso also drew and painted bullfights of particular savagery. These explorations climaxed in the etching *Minotauromachy* of 1935 (fig. 382), one of the great prints of the twentieth century and a demonstration of this artist's consummate skill as a draftsman. It presents a number of figures reminiscent of Picasso's life and his Spanish past, including: recollection of the prints of Goya, the women in the window, the Christ-like figure on the ladder, the little girl holding flowers and a candle (a symbol of innocence), the screaming, disemboweled horse (that becomes the central player in *Guernica*) carrying a dead woman (dressed as a toreador), the Minotaur groping his way. The place of the Minotaur (or the bull) in Picasso's iconography is ambiguous. It may be a symbol of insensate, brute force, sexual potency and aggression, a symbol of Spain, or sometimes his role is that of the artist himself.

In January 1937, Picasso, who supported the Loyalists in the Spanish Civil War, created two etchings, each composed of a number of episodes, and accompanied by a poem, *Dream and Lie of Franco.* The dictator Franco is shown as a turnip-headed monster, and the bull the symbol of resurgent Spain. The grief-stricken woman in *Guernica* who shrieks over the body of her dead child first appeared in this etching. In May and June 1937 Picasso painted his great canvas *Guernica* for the Spanish Pavilion of the Paris World's Fair. It is a huge painting in black, white, and gray, a scene of terror and devastation. Although he used motifs such as the scream-

ing horse or agonized figures derived from his Surrealist distortions of the 1920s, the structure is based on the Cubist grid and returns to the more stringent palette of Analytical Cubism (color plate 93, page 168). With the exception of the fallen warrior at the lower left, the human protagonists are women, but the central victim is the speared horse from *Minotauromachy. Guernica* is Picasso's most powerfully Expressionist application of Cubism and one of the most searing indictments of war ever painted.

Picasso made dozens of studies for *Guernica,* of individual motifs as well the entire composition, which went through several changes. Its impact on the artist himself may be seen in innumerable works during the next decades. One image that continued to haunt him in the wake of *Guernica* is that of the weeping woman (fig. 383). In the etching shown here, the woman holds a handkerchief to her face with spiky fingers while tears fall from her displaced, comma-shaped eyes like long nails. The sense of palpable, uncontrollable sorrow expressed in this and related works was, on a political level, Picasso's continuing reaction to the disastrous events in his native Spain. On a personal level, it is a response to the distress he witnessed (and instigated) among the women in his life. By the end of 1936, despite having just fathered a child with Marie-Thérèse, Picasso had begun an affair with a dark-haired beauty named Dora Maar. Maar was an intense, sharply intelligent woman who spoke fluent Spanish. She was a trained photographer and documented the various stages of *Guernica* through photographs. The collective psychological trauma experienced by these women (as well as by the artist's estranged wife Olga) as they vied for position in the artist's life is given full expression in the harrowing, jagged forms of the weeping women series.

Picasso remained in France after the outbreak of World War II and subsequent German occupation of the country. Many of his paintings and sculptures from this period reveal

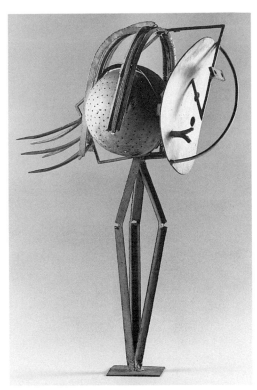

left: 383. PABLO PICASSO. *Weeping Woman with Handkerchief.* July 1937. Etching, aquatint, and dry-point on paper, 27¼ x 19½" (69.2 x 49.5 cm). Musée Picasso, Paris

© 1997 ESTATE OF PABLO PICASSO/ARTISTS RIGHTS SOCIETY (ARS), NEW YORK

below: 384. PABLO PICASSO. *Woman in the Garden.* 1929–30. Bronze after welded and painted iron original, height 6'10¾" (2.1 m). Musée Picasso, Paris

© 1997 ESTATE OF PABLO PICASSO/ARTISTS

right: 385. PABLO PICASSO. *Head of a Woman.* 1929–30. Painted iron, sheet metal, springs, and found colanders, 39⅜ x 14⁵⁄₁₆ x 23¼" (100 x 37 x 59 cm). Musée Picasso, Paris

© 1997 ESTATE OF PABLO PICASSO/ARTISTS RIGHTS SOCIETY (ARS), NEW YORK

his response to the war (and of Spain's final capitulation to Franco's fascism), through stark, spare imagery. Animal skulls appear frequently, first in 1939, after the artist learned of the death of his mother, and again in 1942 (color plate 188, page 334) following the death of his friend, the sculptor Julio González. Picasso often commemorated the death of a loved

one with a significant painting, and this dark, majestic still life is filled with ominous imagery. A steer's skull, painted in shades of gray, is set against the cross-shaped mullions of a French door, beyond which lies a black night. Picasso thus created a variation on traditional memento mori still lifes, which include a human skull to remind us of life's transitory nature.

In 1928 Picasso's Surrealist bone paintings revived his interest in sculpture, which, except for a few sporadic assemblages, he had abandoned since 1914. With the technical help of his old friend Julio González, a skilled metalworker and sculptor, he produced welded iron constructions that, together with similar constructions produced at the same time by González himself, marked the emergence of direct-metal sculpture as a major, modern medium. Picasso's *Woman in the Garden* (fig. 384), a large, open construction in which the figure consists of curving lines and organic-shaped planes, is one of the most intricate and monumental examples of direct-metal sculpture produced to that date. The woman's face is a triangle with strands of windswept hair and the by-now familiar gaping mouth. The bean-shaped form in the center stands for her stomach, while the disk below it is one of Picasso's familiar signs for a woman's sex. The original version of this work was made of welded and painted iron. Picasso commissioned González to make a bronze replica. Other metal constructions undertaken in collaboration with González around this time include the *Head of a Woman* (fig. 385), assembled from found objects, such as rods, springs, and, for the back of the head, common household colanders. Judging from the projecting features of the face, it is likely that Picasso had examples of African masks in mind when he made this work.

Carried away by enthusiasm, Picasso modeled in clay or plaster until by 1933 his studio at Boisgeloup was filled: massive heads of women, reflecting his contemporaneous curvilinear style in painting and looking back in some degree to his

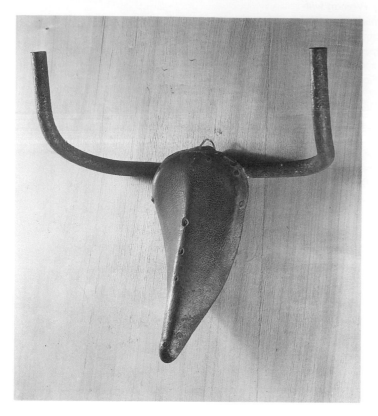

386. PABLO PICASSO. *Bull's Head.* 1943. Bicycle saddle and handle-bars, 13¼ x 17⅛ x 7½" (33.5 x 43.5 x 19 cm). Musée Picasso, Paris

Greco-Roman style; torsos transformed into anthropomorphic monsters; surprisingly representational animals—a cock and a heifer; figures assembled from found objects organized with humor and delight. Of the found-object sculptures perhaps the most renowned is the *Bull's Head* of 1943 (fig. 386), which consists of an old bicycle seat and handlebars—a wonderful example of the way Picasso remained alert to the formal and expressive potential of the most commonplace objects.

The full range of Picasso's experiments in sculpture becomes clear when we realize that *Bull's Head* was made shortly before *Man with a Sheep* (fig. 387). This work, which was created through the more conventional medium of plaster for bronze, is one of the artist's most moving conceptions. On the one hand the figure could refer to the Early Christian figure of the Good Shepherd, Christ the protector of the oppressed, a figure that may be traced back to classical pagan religions as well as to the Old Testament. On the other, it could recall ancient rituals of animal sacrifice. Picasso insisted, "There is no symbolism in it. It is just something beautiful." This sculpture stood prominently in the artist's Paris studio at the time of the Liberation of France by the Allied forces where, in the words of one Picasso scholar, it "came to epitomize an act of faith in humanity."

JULIO GONZÁLEZ (1876–1942), another son of Barcelona, had been trained in metalwork by his father, a goldsmith, but for many years he practiced as a painter. In Paris by 1900, he came to know Picasso, Brancusi, and Charles Despiau, but after the death of his brother Jean in 1908, he lived in isolation for several months. In 1910, González

began to make masks in a metal repoussé technique. In 1918, he learned the technique of oxyacetylene welding and by 1927 was producing his first sculptures in iron. In the late 1920s he made a series of masks (fig. 388) from sheets of flat metal. The sharp, geometric contours in the mask, as González's drawings make clear, are derived from strong angular shadows cast over faces in the hot sun. Their openwork construction had no real precedents in figure sculpture. Then in 1928, Picasso asked him for technical help in constructed sculpture, his new interest. It may be said that González began a new age of iron for sculpture, which he described in his most famous statement: "The age of iron began many centuries ago by producing very beautiful objects, unfortunately for a large part, arms. Today, it provides as well, bridges and railroads. It is time this metal ceased to be a murderer and the simple instrument of a super-mechanical science. Today the door is wide open for this material to be, at last, forged and hammered by the peaceful hands of an artist." By 1931 González was working in direct, welded iron, with a completely open, linear construction in

387. PABLO PICASSO. *Man with a Sheep.* 1943. Bronze, height 7'3" (2.2 m). Musée Picasso, Paris

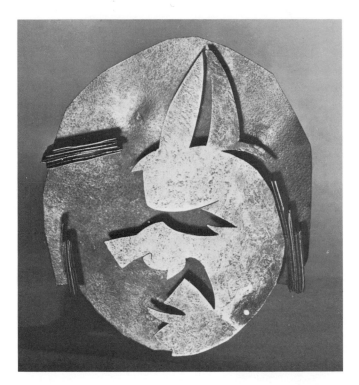

which the solids were merely contours defining the voids. The difference between these constructions and the earlier ones of Gabo and the Russian Constructivists lies not only in the technique—which was to have such far-reaching effects on younger sculptors—but also in the fact that, however abstract his conceptions, González was always involved with the figure. *Monsieur Cactus I (Cactus Man I)* (fig. 389), as the name implies, is a bristly and aggressive individual, suggesting a new authority in the artist's work, which death cut short. The original version of this work was made of welded iron; it was then cast in bronze in an edition of eight. Scholars have posited that the angry, aggressive forms of *Monsieur Cactus I* may embody the artist's reactions to the recent fascist defeat of resistance forces in his beloved Barcelona, signaling the end of civil war in Spain.

Side by side with these openwork, direct-metal constructions, González continued to produce naturalistic heads and figures, for instance, the large wrought-iron and welded *Montserrat* (fig. 390), a heroic figure of a Spanish peasant woman, symbol of the resistance of the Spanish people against fascism. *Montserrat* is the name of the mountain range near Barcelona, a symbol of Catalonia. This work was commissioned by the Spanish Republic for the same Spanish Pavilion that housed *Guernica* at the 1937 international exhibition in Paris.

The sculpture of Picasso and González after the 1920s has had a special pertinence to contemporary sculpture for its exploration of the techniques of direct metal and the use of the found object. For example, when American sculptor David Smith came upon González's work in the early 1930s, he was inspired to make welded metal sculpture (see fig. 537).

If González was the pioneer of the new iron age, **ALBERTO GIACOMETTI** (1901–1966) was the creator of a new image of humanity. After studies in Geneva and a long

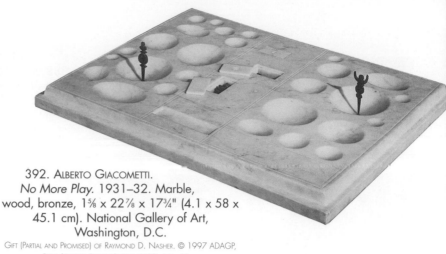

392. ALBERTO GIACOMETTI.
No More Play. 1931–32. Marble,
wood, bronze, 1⅝ x 22⅞ x 17¾" (4.1 x 58 x
45.1 cm). National Gallery of Art,
Washington, D.C.
GIFT (PARTIAL AND PROMISED) OF RAYMOND D. NASHER. © 1997 ADAGP,
PARIS/ARTISTS RIGHTS SOCIETY (ARS), NEW YORK

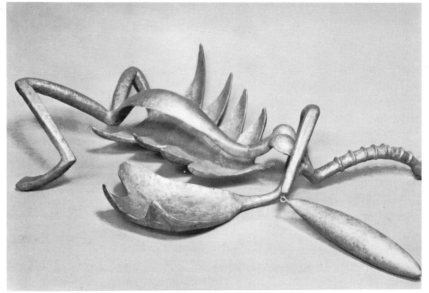

391. ALBERTO GIACOMETTI. *Femme cuiller (Spoon
Woman).* 1926–27. Bronze, height 57" (144.8 cm).
Solomon R. Guggenheim Museum, New York
© 1997 ADAGP, PARIS/ARTISTS RIGHTS SOCIETY (ARS), NEW YORK

393. ALBERTO GIACOMETTI. *Woman with Her Throat Cut (Femme égorgée).*
1932 (cast 1949). Bronze, 8 x 34½ x 25" (20.3 x 87.6 x 63.5 cm).
The Museum of Modern Art, New York
PURCHASE. © 1997 ARTISTS RIGHTS SOCIETY (ARS), NEW YORK/ADAGP, PARIS

stay with his father, a painter, in Italy, where he saturated himself in Italo-Byzantine art and became acquainted with the Futurists, he moved to Paris in 1922. Here he studied with Bourdelle for three years and then set up a studio with his brother Diego, an accomplished technician who continued to be his assistant and model to the end of his life.

The first independent sculptures reflected awareness of the Cubist sculptures of Lipchitz and Laurens and, more importantly, of African, Oceanic, and prehistoric art. Giacometti's *Femme cuiller* of 1926–27 (fig. 391) is a frontalized, Surrealist-primitive totem, with a spiritual if not a stylistic affinity to the work of Brancusi. Probably inspired by a Dan spoon from West Africa that the artist saw in the Paris ethnographic museum, *Femme cuiller* has the totemic presence of an ancient fertility figure.

At the end of the 1920s Giacometti was drawn into the orbit of the Paris Surrealists. For the next few years he made works that reflected their ideas, and until 1935 (when he was expelled by the Surrealists) he exhibited with Miró and Arp at the Galerie Pierre. The sculptures he produced during these years are among the masterpieces of Surrealist sculp-

ture. *No More Play* (fig. 392) is an example of Giacometti's tabletop compositions. Here the base becomes the actual sculpture on which movable parts are deployed like pieces in a board game (as the title implies). In two of the rounded depressions carved in the lunarlike surface are tiny male and female figurines in wood. Between them are three cavities, like coffins with movable lids. One of these contains an object resembling a skeleton. The viewer can interact with the work, moving aside the lids of the "tombs" to discover their contents and crossing over the usually inviolable space between object and viewer.

Woman with Her Throat Cut (fig. 393), a bronze construction of a dismembered female corpse, bears a family resemblance to Picasso's 1930 *Seated Bather* (color plate 186, page 334). The spiked, crustacean form of the sculpture infers the splayed and violated body of a woman. According to the artist, the work should be shown directly on the floor, without a base. From a high vantage point, looking down, the Surrealist theme of sexual violence is even more palpable. Although the work, like *No More Play*, can be moved by the viewer, the phallic shaped form should be placed in the

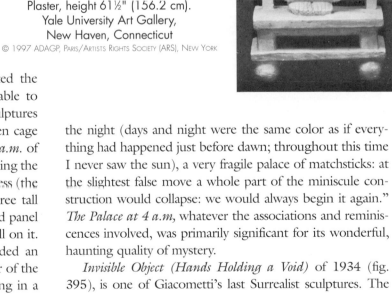

394. ALBERTO GIACOMETTI. *The Palace at 4 a.m.* 1932–33. Construction in wood, glass, wire, and string, 25 x 28¼ x 15¾" (63.5 x 71.8 x 40 cm). The Museum of Modern Art, New York PURCHASE. PHOTOGRAPH © 1995 THE MUSEUM OF MODERN ART, NEW YORK. © 1997 ADAGP, PARIS/ARTISTS RIGHTS SOCIETY (ARS), NEW YORK

right: 395. ALBERTO GIACOMETTI. *Invisible Object (Hands Holding a Void).* 1934. Plaster, height 61½" (156.2 cm). Yale University Art Gallery, New Haven, Connecticut © 1997 ADAGP, PARIS/ARTISTS RIGHTS SOCIETY (ARS), NEW YORK

leaflike "hand" of the figure, for Giacometti related the sculpture to the common nightmare of being unable to move one's hand. In a number of his Surrealist sculptures Giacometti experimented with the format of an open cage structure, a series that climaxed in *The Palace at 4 a.m.* of 1932–33 (fig. 394), a structure of wooden rods defining the outlines of a house. At the left a woman in a long dress (the artist's recollection of his mother) stands before three tall rectangular panels. She seems to look toward a raised panel on which is fixed a long, oval spoon shape with a ball on it. To the right, within a rectangular cage, is suspended an object resembling a spinal column, and in the center of the edifice hangs a narrow panel of glass. Above, floating in a rectangle that might be a window, is a sort of pterodactyl—"the skeleton birds that flutter with cries of joy at four o'clock." This strange edifice was the product of a period in the artist's life that haunted him and about which he has written movingly: ". . . when for six whole months hour after hour was passed in the company of a woman who, concentrating all life in herself, made every moment something marvelous for me. We used to construct a fantastic palace in

the night (days and night were the same color as if everything had happened just before dawn; throughout this time I never saw the sun), a very fragile palace of matchsticks: at the slightest false move a whole part of the miniscule construction would collapse: we would always begin it again." *The Palace at 4 a.m*, whatever the associations and reminiscences involved, was primarily significant for its wonderful, haunting quality of mystery.

Invisible Object (Hands Holding a Void) of 1934 (fig. 395), is one of Giacometti's last Surrealist sculptures. The elongated female personage half sits on a cage/chair structure that provides an environment and, when combined with the plank over her legs, pins her in space. With a hieratic gesture, she extends her hands to clasp an unseen object, as though partaking in some mysterious ritual. The work has been connected to sources in Oceanic and ancient Egyptian art. The form of a single, vertical, attenuated figure is significant for Giacometti's later work (see chapter 20).

Photography and Surrealism

Paris in the 1920s and 1930s was especially fertile ground for photography. From many countries, several of the medium's greatest practitioners came together with painters and sculptors in an era of remarkably fruitful exchange between the arts. The 1920s, in particular, was a highly experimental period for photography, one that witnessed, among other related media, the explosion of cinema. The Surrealists valued photography for its ability to capture the bizarre juxtapositions that occur naturally in daily experience. A sense of Surrealist dislocation could be made even more shocking when documented in a photograph than in, for example, a literally rendered painting by Dalí. The Surrealists regularly featured photography, including vernacular snapshots by amateurs, in their various journals. Their embrace of the medium, however, was not unequivocal. Breton viewed photography as a threat to what he regarded as the more important activity of painting, and mistrusted its emphasis on the external world rather than internal reality. Nevertheless, photography was even used by artists in the Surrealist movement who were not necessarily photographers. Max Ernst, for example, incorporated pho-

tographs into his collages; Bellmer's creations, as we have seen, were wholly dependent upon photography; and Magritte made photographs of his family and friends, a practice apart from his paintings. Others, who were not card-carrying Surrealists but who shared a kindred vision with the movement's official practitioners, were claimed by the Surrealists.

EUGÈNE ATGET (1857–1927). By no means a practicing Surrealist, Atget was one of the artists co-opted by the group through Man Ray, who sensed that his vision lay as much in fantasy as in documentation. After having failed at the military and at acting, Atget taught himself photography and began a business to provide photographs, on virtually any subject, for use by artists. By 1897, he had begun to specialize in images of Paris and conceived what a friend called "the ambition to create a collection of all that which both in Paris and its surroundings was artistic and picturesque. An immense subject." Over the next thirty years Atget labored to record the entire face of the French capital, especially those aspects of it most threatened by "progress." Although he was often commissioned by various official agencies to supply visual material on Paris, Atget proceeded with a single-minded devotion to transform facts into art, with the result that his unforgettable images go beyond the merely descriptive to evoke a dream-

396. EUGÈNE ATGET. *Saint-Cloud.* 1915–19. Albumen-silver print, 7 x 9⅜" (17.8 x 23.8 cm).
The Museum of Modern Art, New York

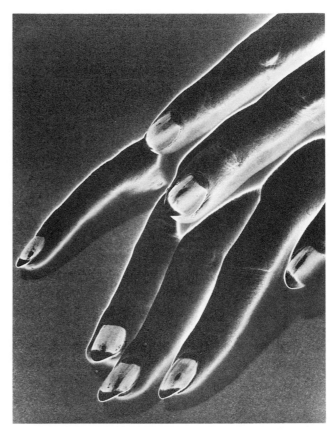

right: 397. EUGÈNE ATGET. *Magasin, avenue des Gobelins.*
1925. Albumen-silver print, 9⅜ x 7" (24 x 18 cm).
The Museum of Modern Art, New York

398. MAN RAY. *Fingers.* 1930. Solarized gelatin-silver print
from negative print, 11½ x 8¾" (29.2 x 22.2 cm). The
Museum of Modern Art, New York

like world that is also profoundly real. A mesmerized nostalgia for a lost and decaying classical past inhabits his photographs, as in the romantically melancholic image of the old royal gardens at Saint-Cloud (fig. 396).

Among the approximately ten thousand pictures by Atget that survive—images of streets, buildings, historic monuments, architectural details, parks, peddlers, vehicles, trees, flowers, rivers, ponds, the interiors of palaces, bourgeois apartments, and ragpickers' hovels—are a series of shop fronts (fig. 397) that, with their grinning dummies and superimposed reflections from across the street, would fascinate the Surrealists as "found" images of dislocated time and place. Man Ray was so taken with *Magasin* that he arranged for it to be reproduced in *La Révolution surréaliste* in 1926. The sense it evokes of a dreamworld, of a strange "reality," of threatened loss, could only have been enhanced by equipment and techniques that were already obsolescent when Atget adopted them and all but anachronistic by the time of his death. As John Szarkowski wrote in *Looking at Pictures:*

> Other photographers had been concerned with describing specific facts (documentation), or with exploiting their individual sensibilities (self-expression). Atget encompassed and transcended both approaches when he set himself the task of understanding and interpreting in visual terms a complex, ancient, and living tradition. The pictures that he made in the service of this concept are seductively and deceptively simple, wholly poised, reticent, dense with experience, mysterious, and true.

The photographer most consistently liberated by Surrealism was the one-time New York Dadaist **MAN RAY** (1890–1976) (see figs. 310, 311), who after 1921 lived in Paris and became a force within the circle around André Breton. In the course of his long and active career, Man Ray worked not only in photography but also in painting, sculpture, collage, constructed objects, and film. Despite his preference for the art of painting, it was through photography and film that Man Ray found the most successful expression of his aesthetic goals. From 1924 through the 1930s he contributed his photographs regularly to the Surrealist journals as well as to books by Surrealist writers.

In Paris Man Ray continued to experiment with unorthodox methods in the darkroom. In his Sabattier prints (so named for the early inventor of the technique but generally known as "solarizations"), a developed but unfixed print is exposed to light and then developed again, producing a reversal of tones along the edges of forms and transforming the mundane into something otherworldly (fig. 398). Regarding one of his most famous paintings, *Observatory Time—The Lovers* executed in 1930–32, Man Ray wrote:

> One of these enlargements of a pair of lips haunted me like a dream remembered: I decided to paint the subject on a scale of superhuman proportions. If there had been a color process enabling me to make a photograph of such dimensions and showing the lips floating over a landscape, I would certainly have preferred to do it that way.

399. MAN RAY. *Observatory Time—The Lovers*. 1936. Halftone reproduction. Published in *Harper's Bazaar,* November 1936
COURTESY *HARPER'S BAZAAR.* © 1997 MAN RAY TRUST/ADAGP, PARIS/ARTISTS RIGHTS SOCIETY (ARS), NEW YORK

To support himself while carrying out such ambitious painting projects, Man Ray made fashion photographs and photographic portraits of his celebrated friends in art, literature, and society. For *Harper's Bazaar* in 1936 he posed a model wearing a couturier beach robe against the backdrop of *Observatory Time—The Lovers* (fig. 399), thereby grafting on to this elegant scene of high fashion a hallucinatory image of eroticism and Freudian association so revered by the Surrealists.

ANDRÉ KERTÉSZ (1894–1985). A lyrical, life-affirming *joie de vivre*, tethered to a rigorous, original sense of form, characterizes the work of Kertész. Unlike his friend and fellow Hungarian Brassaï, Kertész was already a practicing photographer when he arrived in France from Budapest in 1925. It was there, however, that he became a virtuoso of the 35-mm camera, the famous Leica introduced in 1925, and emerged as a pioneer of modern photojournalism. Though the handheld camera had been used for years by amateurs, the desire for spontaneity among professionals meant that, by the 1930s, the Leica became the preferred camera of photojournalists. In his personal work, Kertész used the flexible new equipment not so much for analytical description, which the large-format camera did better, as for capturing the odd, fleeting moment or elliptical view when life is most, because unexpectedly, revealed. Photographed in the studio of a sculptor friend, Kertész's image of a woman in a pose like a human pinwheel (fig. 400) playfully mimics the truncated limbs of the nearby sculpture.

The kind of elastic distortion achieved by Picasso through sheer force of imagination opened a rich vein of possibilities for photography, equipped as this medium was with every

sort of optical device. Kertész created his funhouse effects (what he called "Distortions") by photographing nude models reflected in a special mirror (fig. 401). Despite the visual affinities with Surrealist-inspired imagery, Kertész never claimed any allegiance with the Surrealists and was not asked by Breton to enlist in their ranks. In 1936 he immigrated to America, where he free-lanced for the leading illustrated magazines, with the exception of *Life.* The editors of that publication, one crucial to the development of American photography in the 1930s (see fig. 494), told him his photographs "spoke too much."

BRASSAÏ (Gyula Halász) (1899–1984) took his pseudonym from his native city, Brasso, in the Transylvanian part of Romania. Following his days as a painting student in Budapest and Berlin, Brassaï arrived in the French capital in 1924 and promptly fell in love with its streets, bars, and brothels, artists, poets, and writers, even its graffiti. He sought out these subjects nightly and slept throughout the day. Once introduced to the small-format camera by his friend Kertész, Brassaï proceeded to record the whole of the Parisian human comedy, doing so as faithfully and objectively as possible (fig. 402). That desire for objectivity prompted Brassaï to turn down an invitation from Breton to join the Surrealists. He watched for the moment when character seemed most naked and most rooted in time and place. Brassaï published the results of his nocturnal Parisian forays in the successful book *Paris de nuit* (1933).

MAURICE TABARD (1897–1984) was born in France but came to the United States in 1914. After studying at the New York Institute of Photography and establishing himself as a portrait photographer in Baltimore, Tabard returned to Paris in 1928 to make fashion photographs. He soon came to know photographers Man Ray, Henri Cartier-Bresson, and Kertész, as well as the painter Magritte, and became a well-known figure in Parisian avant-garde circles. Like Man Ray, he made solarized photographs and continued his creative photography alongside his more commercial fashion work. The methods by which Tabard obtained his Surrealistic imagery differed significantly from those of artists like Kertész or Cartier-Bresson. While these artists found their images by chance or, in the case of Kertész's Distortions, a special mirror, Tabard engaged in the elaborate manipulation of his craft in the darkroom. His haunting photographs involve double exposures and negative printing, sometimes combining several techniques within the same image. In an untitled work (fig. 403), Tabard locked the figure into place with a ladder-shaped shadow pattern that slyly mimics the very shape of a roll of negative film. Other superimposed imagery, including what appears to be the shadow of a tennis racket at the lower left, contributes to an overall sense of disorienting complexity.

Born in Mexico City, MANUEL ALVAREZ BRAVO (b. 1902) was a self-taught photographer whose work contributed to the artistic Renaissance that flourished in Mexico during the 1930s. He befriended many of Mexico's leading avant-garde artists during this period, including the muralists

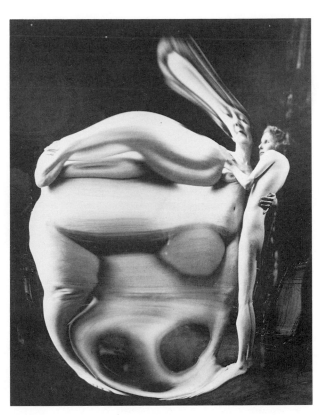

400. ANDRÉ KERTÉSZ. *Satiric Dancer, Paris.* 1926.
Gelatin-silver print.
Courtesy Susan Harder Gallery, New York

401. ANDRÉ KERTÉSZ. *Distortion No. 4.* 1933.
Gelatin-silver print.
Courtesy Susan Harder Gallery, New York

402. BRASSAÏ. *Dance Hall.* 1932. Gelatin-silver print, 11½ x
9¼" (29.2 x 23.5 cm). The Museum of Modern Art, New York

403. MAURICE TABARD. *Untitled.*
1929. Gelatin-silver print.
Collection Robert Shapazian, Fresno, California

404. Manuel Alvarez Bravo. *Laughing Mannequins.* c. 1932. Gelatin-silver print, 7⁵⁄₁₆ x 9⁹⁄₁₆" (18.6 x 24.3 cm). The Art Institute of Chicago Julien Levy Collection, Gift of Jean Levy and the Estate of Julien Levy

Diego Rivera, David Alfaro Siqueiros, and José Clemente Orozco, whose work, often recorded in Alvarez Bravo's photographs, is discussed in chapter 18. In addition, Mexico attracted many exceptional photographers from abroad. Among those foreigners was Tina Modotti (see fig. 505), who encouraged Alvarez Bravo's work and put him in contact with the American photographer Edward Weston (see fig. 191). Alvarez Bravo also met Paul Strand (see fig. 477) in Mexico in 1933 and befriended Cartier-Bresson (figs. 406, 407) during the Frenchman's visit to Mexico the following year. His work represents a distinctive blend of cultural influences, fusing imagery and traditions indigenous to Mexico with current ideas imported from Europe and America, including those of the Surrealists. His work was illustrated in the last issue of the Surrealist journal *Minotaure*. Like Cartier-Bresson, Alvarez Bravo's penchant for discovering the visual poetry inherent in the quotidian world could result in delightful and mysterious found images. He happened upon a group of mannequins, a favorite Surrealist stand-in for the human figure, which he captured in a photograph of an outdoor market from the early thirties (fig. 404). The smiling cardboard mannequins, themselves merely mounted photographs, repeat the same woman's face, which unlike their real counterparts below, meet the viewer's gaze.

LISETTE MODEL (1901–1983). Perhaps closer in spirit to the Austrian Expressionists than to the Surrealists is the work of Lisette Model, a photographer born in turn-of-the-century Vienna to an affluent family of Jewish and Catholic descent. Model wanted first and foremost to become a concert pianist. In Vienna, she studied music with the composer Arnold Schönberg and through his circle became aware of the activities of the Viennese avant-garde. In 1926, following the loss of the family's fortune during the war, she moved to France with her mother. It was in Paris, in 1933, that she decided to abandon music and become a professional photographer, learning the rudiments of the technique from her younger sister, her friend Florence Henri, and Rogi André, a photographer (and former wife of Kertész) who advised her to photograph only those things that passionately interested her. Model was, above all, passionately interested in people. She regarded the world as a vast cast of characters from which to construct her own narratives. With her twin-lens Rolleiflex in hand, she photographed, during a visit to Nice to see her mother, wealthy vacationers soaking up the Riviera sun along that city's Promenade des Anglais (fig. 405). Model wielded her camera like an invasive tool. Catching her wary subjects off-guard, she approached them closely, sometimes squatting to record their reactions at eye level. She then cropped her images in the darkroom, augmenting the confrontational nature of the close-up effect. In 1935 her photographs from Nice were published in the left-wing journal *Regards* as disparaging examples of a complacent middle class, "the most hideous specimens of the human animal." While it is not clear whether Model intended the photographs to be interpreted in this way, she did allow their publication again in 1941 in the weekly *PM,* one of the American publications for which she worked once she settled in New York in 1938. *PM* published the photos to demonstrate the French characteristics that, in its opinion, led to the country's capitulation in World War II. This photograph was illustrated under the heading "Cynicism." By the 1950s Model had become an influential teacher; her students included the American photographer Diane Arbus (see fig. 661).

HENRI CARTIER-BRESSON (b. 1908). Photography, with its power to record, heighten, or distort reality, a power vastly increased with the invention of the small, handheld 35mm Leica, proved a natural medium for artists moved by the Surrealist spirit. One of the most remarkable of these was France's Henri Cartier-Bresson, who, after studying painting with the Cubist André Lhote, took up photography and worked as a photojournalist covering such epochal events as the Spanish Civil War (1936–39). But his is a photojournalism broadly defined, for his magical images, especially those made before World War II, have little to do with a discernible narrative or straight reportage of visual fact. From a very young age Cartier-Bresson was powerfully influenced by the Surrealist theories of André Breton. He matured quickly as a photographer, and in the first half of the 1930s, in photographs made in France, Spain, Italy, and Mexico, he created

406. HENRI CARTIER-BRESSON. *Cordoba, Spain.* 1933. Gelatin-silver print © HENRI CARTIER-BRESSON

some of the most memorable images of the twentieth century.

Freed from the encumbrance of a tripod, the photographer could exploit new opportunities, pursuing the action as it unfolded. "I prowled the streets all day," Cartier-Bresson has said of his practice, "feeling very strung up and ready to pounce, determined to 'trap' life—to preserve life in the act of living. Above all, I craved to seize the whole essence, in the confines of a single photograph, of some situation that was in the process of unrolling itself before my eyes." Citing Man Ray, Atget, and Kertész as his chief influences, Cartier-Bresson allowed his viewfinder to "discover" a composition within the world moving about him. Once this had been seized upon, the photographer printed the whole uncropped negative, an image that captured the subject in "the decisive moment" (the title of his 1952 book). *Cordoba, Spain* (fig. 406) is a witty juxtaposition of art and life, as in the collages of his friend Max Ernst. A woman clasps a hand to her bosom, unconsciously repeating the gesture in the corset advertisement behind her. The woman squints into the photographer's lens while the eyes of the poster model are "blinded" by an advertisement that has been pasted over her face. In this deceptively simple image, the artist sets up a complex visual dialogue between art and life, illusion and reality, youth and old age. Also made in Spain, the artist's photographs of children playing among the ruins in Seville, Spain

405. LISETTE MODEL. *Promenade des Anglais.* 1934. Gelatin-silver print on paper, mounted on pasteboard, 13½ x 10⅞" (34.2 x 27.6 cm). The National Gallery of Canada, Ottawa

407. HENRI CARTIER-BRESSON. *Seville, Spain.* 1934.
Gelatin-silver print, 7⅝ x 11⅜" (19.4 x 29 cm)

(fig. 407), foretell with uncanny accuracy the devastation that would overrun the country three years later in the Spanish Civil War. Like actors within a bizarre stage set, the children move with balletic grace amidst what looks like postwar rubble. Little wonder why Cartier-Bresson's pictures, with their momentary equipoise among form, expression, and content, have served as models of achievement to photographers for more than half a century.

Although England's **BILL BRANDT** (1904–1983) was trained as a photographer in the late 1920s in Paris, where he worked in the studio of Man Ray, he made mostly documentary photographs in the 1930s, including a famous series on the coal miners in the North of England during the depression. After World War II, he rediscovered his early days in Surrealist Paris and gave up documentation in favor of re-exploring the "poetry" of optical distortion, or what he called "something beyond reality." Between 1945 and 1960, he photographed nudes with a special Kodak camera that had an extremely wide-angle lens and a tiny aperture. The distortions transformed the nudes into Surreal dream landscapes. He exploits the same distorted perspective in *Portrait of a Young Girl, Eaton Place, London* from 1955 (fig. 408). The dreamlike effect is enhanced by the artist's characteristically high-contrast printing style, where portions of the room disappear into blackness. Brandt also continued to make portraits after the war, recording the likenesses of many of the greatest luminaries in the world of literature and art, including leading Surrealists such as Magritte, Arp, Miró, and Picasso.

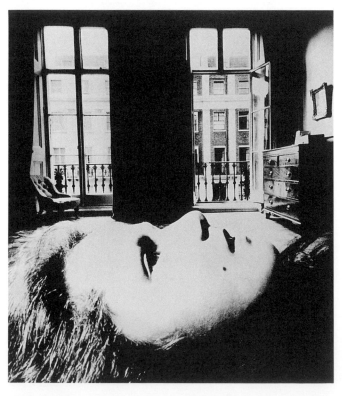

408. BILL BRANDT. *Portrait of a Young Girl, Eaton Place, London.*
1955. Gelatin-silver print, 17 x 14¾" (43.2 x 37.5 cm).
Courtesy Edwynn Houk Gallery, Chicago

16 Modern Architecture Between the Wars

One of the most remarkable cultural phenomena in Europe of the 1920s was the Bauhaus, the school established in 1919 by the architect Walter Gropius. In the disastrous wake of World War I, the Staatliches Bauhaus was formed in Weimar, Germany, from two schools of arts and crafts. As was the case with the English Arts and Crafts and the Deutscher Werkbund traditions, Gropius was convinced of the need for unity of architect, artist, and craftsman. The program of Gropius was a departure in its insistence not that the architect, the painter, and the sculptor should work with the craftsperson, but that they should first of all be craftspeople. The concepts of learning by doing, of developing an aesthetic on the basis of sound craft skills, and of breaking down barriers between "fine art" and craft were fundamental to the Bauhaus philosophy.

Despite Gropius's belief in architecture as the supreme art form, the Bauhaus originally provided no course in architecture, for Gropius felt that students should be skilled in the basic crafts before graduating to the design of buildings. The core of Bauhaus teaching was a division of courses into workshops, which taught those craft skills, and a mandatory foundation course designed to encourage the students' creative powers and liberate them from past experiences and prejudices. This course, initially developed by the Swiss painter Johannes Itten, introduced the student to materials and techniques through elementary but fundamental practical experiments. Under the influence of Itten, the foundation course also included the investigation of Eastern philosophies and mystical religions. Eventually, Gropius opposed Itten's mystical bent, insisting that craft be reconciled with industrial production for the modern machine age. As he wrote in 1923, the year he persuaded Itten to resign, "The Bauhaus believes the machine to be our modern medium of design and seeks to come to terms with it." This shift toward a machine aesthetic was no doubt due in part to the influence of Constructivism (see chapter 11) and the presence in Weimar by 1921 of the de Stijl artist Theo van Doesburg.

The first proclamation of the Bauhaus declared: "Architects, painters, and sculptors must recognize anew the composite character of a building as an entity. . . . Art is not a 'profession.' There is no essential difference between the artist and the craftsman. The artist is an exalted craftsman. . . . Together let us conceive and create the new building of the future, which will embrace architecture and sculpture and painting in one unity and which will rise one day toward heaven from the hands of a million workers like the crystal symbol of a new faith." This initial statement reflected a nostalgia for the guild systems and collective community spirit that built the great Gothic cathedrals, as well as the socialist thought then current in Germany and throughout much of Europe. Suspicion of this political attitude caused antagonism toward the school among the more conservative elements in Weimar, an antagonism that finally in 1925 drove the Bauhaus to its new home in Dessau.

Over the years the Bauhaus attracted one of the most remarkable art faculties in history. Vasily Kandinsky, Paul Klee, Lyonel Feininger, Georg Muche, and Oskar Schlemmer were among those who taught painting, graphic arts, and stage design. Pottery was taught by Gerhard Marcks, who was also a sculptor and graphic artist. When Johannes Itten left in 1923, the foundation course was given by László Moholy-Nagy, a Hungarian painter, photographer, theater and graphic designer, and, through his writings and teaching, the most influential figure after Gropius in developing and spreading the Bauhaus idea. (Work produced by many of these painters, photographers, and sculptors during their tenure at the Bauhaus and after is addressed in chapter 17.) In addition to the star-studded faculty, the Bauhaus frequently attracted distinguished foreign visitors, such as, in 1927, the Russian Suprematist painter Kazimir Malevich (see figs. 244, 245). When the Bauhaus moved to Dessau, several former students joined the faculty—the architects and designers Marcel Breuer and Herbert Bayer, and the painter and designer Josef Albers (color plate 193, page 336), who reorganized the foundation course. Of the artist-teachers, Kandinsky, Klee, and Feininger were to become recognized as major twentieth-century painters. Moholy-Nagy, through his books *The New Vision* and *Vision in Motion* and his directorship of the New Bauhaus, founded in Chicago in 1937

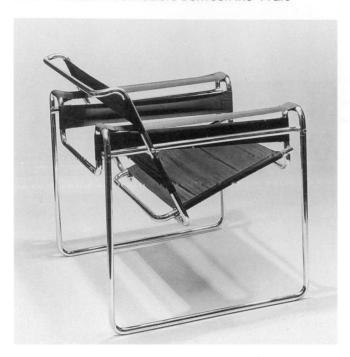

409. MARCEL BREUER. Armchair, Model B3. Dessau, Germany. Late 1927 or early 1928. Chrome-plated tubular steel with canvas slings, 28⅛ x 30¼ x 27¾" (71.4 x 76.8 x 70.5 cm). The Museum of Modern Art, New York GIFT OF HERBERT BAYER

(now the Institute of Design of the Illinois Institute of Technology), greatly influenced the teaching of design in the United States. Josef Albers would become one of the most important art teachers in the United States, first at the remarkable school of experimental education, Black Mountain College in North Carolina, and subsequently at Yale University. Marcel Breuer, principally active at the Bauhaus as a furniture designer, ultimately joined Gropius in 1937 on the faculty of Harvard University and practiced architecture with him. After Breuer left this partnership in 1941, his reputation steadily grew to a position of world renown (see figs. 719, 745).

The Bauhaus curriculum was divided into two broad areas: problems of craft and problems of form. Each course had a "form" teacher and a "craft" teacher. This division was necessary because a faculty could not be found during the first four years who were capable of integrating the theory and practice of painting, sculpture, architecture, design, and crafts—although Klee taught textile design and Marcks pottery. With the move to Dessau, however, and the addition of Bauhaus-trained staff members, the various parts of the program were integrated.

In Dessau the accent on craft declined while the emphasis on architecture and industrial design were substantially increased; and the architecture students were expected to complete their training in engineering schools. Gropius said: "We want to create a clear, organic architecture, whose inner logic will be radiant and naked, unencumbered by lying façades and trickeries; we want an architecture adapted to our world of machines, radios and fast motor cars, an architecture whose function is clearly recognizable in the relation of its form. . . . [W]ith the increasing strength of the new materi-

als—steel, concrete, glass—and with the new audacity of engineering, the ponderousness of the old methods of building is giving way to a new lightness and airiness."

The greatest practical achievements at the Bauhaus were probably in interior, product, and graphic design. For example, Marcel Breuer created many furniture designs at the Bauhaus that have become classics, including the first tubular-steel chair (fig. 409). He said that, unlike heavily upholstered furniture, his simple, machine-made chairs were "airy, penetrable," and easy to move. Though initially women were to be given equal status at the Bauhaus, Gropius grew alarmed at the number of women applicants and restricted them primarily to weaving, a skill deemed suitable for female students. Gunta Stölz and Anni Albers were major innovators in the area of textile design at the school's weaving workshop. In ceramic and metal design, a new vocabulary of simple, functional shapes was established. The courses in display and typographic design under Bayer, Moholy-Nagy, Tschichold, and others revolutionized the field of type. Bauhaus designs have passed so completely into the visual language of the twentieth century that it is now difficult to appreciate how revolutionary they were on first appearance. Certain designs, such as Breuer's tubular chair and his basic table and cabinet designs, Gropius's designs for standard unit furniture, and designs by other faculty members and students for stools, stacking chairs, dinnerware, lighting fixtures, textiles, and typography so appealed to popular tastes that they are still manufactured today.

Gropius resigned his position in 1928 and named as his successor Hannes Meyer, a Marxist who placed less emphasis on aesthetics and creativity than on rational, functional, and socially responsible design. Meyer was forced to leave the Bauhaus in 1930, and Mies van der Rohe (Gropius's first choice in 1928) assumed the directorship. Mies's work as an architect is discussed below. Inevitably, activities at the Bauhaus aroused the suspicions of the reactionary political forces that finally brought about its closing in 1933.

The Architecture of Gropius

After spending two years in the office of Peter Behrens (see fig. 281), WALTER GROPIUS (1883–1969) established his own practice in Berlin. In 1911 he joined forces with his partner Adolph Meyer (1881–1929) to build a factory for the Fagus Shoe Company at Alfeld-an-der-Leine (fig. 410). The Fagus building represents a sensational innovation in its utilization of complete glass sheathing even at the corners. In effect, Gropius here had invented the curtain wall that would play such a visible role in the form of subsequent large-scale twentieth-century architecture.

Gropius and Meyer were commissioned to build a model factory and office building in Cologne for the 1914 Werkbund Exhibition of arts and crafts and industrial objects (fig. 411). Gropius felt that factories should possess the monumentality of ancient Egyptian temples. For one façade of their "modern machine factory," the architects combined massive brickwork with a long horizontal expanse of open glass sheathing, the latter most effectively used to encase the exterior spiral staircases at the corners (clearly seen in the view reproduced here). The pavilions at either end have flat overhanging roofs derived from Frank Lloyd Wright (see fig. 265), whose work was known in Europe after 1910, and the entire building reveals the elegant and disciplined design that became a prototype for many subsequent modern buildings.

During the years that he was director of the Bauhaus,

above:
410. WALTER GROPIUS AND ADOLPH MEYER. Fagus Shoe Factory, Alfeld-an-der-Leine, Germany. 1911–25

below:
411. WALTER GROPIUS AND ADOLPH MEYER. Model Factory at the Werkbund Exhibition, Cologne. 1914

right:
412. WALTER GROPIUS AND ADOLPH MEYER. Design for the Chicago Tribune Tower. 1922

Gropius continued his own architectural practice in collaboration with Meyer until Meyer's death. One of the projects, unfulfilled, was the design for the Chicago Tribune Tower in 1922 (fig. 412). The highly publicized competition for this tower, with over two hundred and fifty entries from an international assortment of architects, provides a cross section of the eclectic architectural tendencies of the day, ranging from strictly historicist examples based on Renaissance towers to the modern styles emerging in Europe. The traditionalists won the battle with the highly effective neo-Gothic tower designed by the American architect RAYMOND HOOD (1881–1934) (probably in collaboration with John Mead Howells) (fig. 413). The design of Gropius and Meyer, in the spare rectangularity of its forms, its emphasis on skeletal structure, and its wide tripartite windows, was actually based on

413. RAYMOND HOOD. Chicago Tribune Tower. 1922–25

the original skyscraper designs of Sullivan and the Chicago School (see figs. 99, 100) and at the same time looked forward to the skyscraper of the mid-twentieth century.

When the Bauhaus was moved to Dessau in 1925, Gropius closed his Weimar office, ending his partnership with Meyer. His most important architectural achievement at the Bauhaus was the design for the new buildings at Dessau (fig. 414). These buildings, finished in 1926, incorporated a complex of classrooms, studios, workshops, library, and living quarters for faculty and students. The workshops consisted of a glass box rising four stories and presenting the curtain wall, the glass sheath or skin, freely suspended from the structural-steel elements. The form of the workshop wing suggests the uninterrupted spaces of its interior. On the other hand, in the dormitory wing, the balconies and smaller window units contrasting with clear expanses of wall surface imply the broken-up interiors of individual apartments.

The asymmetrical plan of the Bauhaus is roughly cruciform, with administrative offices concentrated in the broad, uninterrupted ferroconcrete span of the bridge linking workshops with the classrooms and library. In every way the architect sought the most efficient organization of interior space. At the same time he was sensitive to the abstract organization of the rectangular exterior—the relation of windows to walls, concrete to glass, verticals to horizontals, lights to darks. The Bauhaus combined functional organization and structure with a geometric, de Stijl-inspired design. Not only were the Bauhaus buildings revolutionary in their versatility and in the application of abstract principles of design on the basis of the interaction of verticals and horizontals, but they also embodied a new concept of architectural space. The flat roof of the Bauhaus and the long, uninterrupted planes of white walls and continuous window voids create a lightness that opens up the space of the structure. The interior was furnished with designs by Bauhaus students and faculty, including Breuer's tubular-steel furniture. The building was seri-

414. WALTER GROPIUS. Workshop Wing, Bauhaus, Dessau, Germany. 1925–26

ously damaged in World War II and underwent limited renovation in the 1960s. In the 1970s, it was finally restored to its original appearance. Since the reunification of Germany, it has become easier to visit this landmark building, which has become the focus of new studies and a site for historical exhibitions related to the Bauhaus.

Gropius left the Bauhaus in 1928 to devote his full time to architectural practice. Until he was forced by the Nazis' rise to power to leave Germany in 1934—first for England and then, in 1937, for the United States—much of his building was in low- or middle-cost housing. In his pioneering European works, Gropius helped provide the foundation of what would later be dubbed the International Style. (His profound influence on postwar American architecture is discussed in chapter 23.)

The International Style

After World War I, communication among architects was reestablished so rapidly and stylistic diffusion was so widespread that it became difficult to speak of national styles. Rather, centers of experimentation arose where architects and artists from all over now converged. Major forces in the formation of the style were de Stijl art and architecture in Holland, the new experiments in German architecture, and though he never considered himself a participant, Frank Lloyd Wright. The first manifestation of what came to be called the International Style took place in 1927 at the Deutsche Werkbund Weissenhofsiedlung Exhibition in Stuttgart, organized by Mies van der Rohe. The presentation included display housing designed by, among others, Mies, Gropius, Le Corbusier, and the de Stijl architect J. J. P. Oud (see figs. 285, 286). The actual term International Style was given prominence by an exhibition of advanced tendencies in architecture held at New York's Museum of Modern Art in 1932. The show, a collaborative effort between museum director Alfred H. Barr, Jr., architectural historian Henry-Russell Hitchcock, and architect Philip Johnson, attempted to define and codify the characteristics of the style, although the exhibition's strictly formalist approach paid virtually no heed to the underlying ideologies and individual formal vocabularies that gave rise to modern architecture in Europe. The first principle of the new architecture of structural steel and ferroconcrete was elimination of the bearing wall. The outside wall became a skin of glass, metal, or masonry constituting an enclosure rather than a support. Thus, one could speak of an architecture of volume rather than of mass. Window and door openings could be enlarged indefinitely and distributed freely to serve both function—activity, access, or light—and design, exterior or interior. The regular distribution of structural supports led to rectangular regularity of design and away from the balanced axial symmetry of classical architecture.

Other principles involved the general avoidance of applied decoration and the elimination of strong contrasts of color on both interiors and exteriors. The International Style resulted in new concepts of spatial organization, particularly that of a free flow of interior space, as opposed to the stringing together of static symmetrical boxes that up to then had been necessitated by interior bearing walls. Finally, the International Style lent itself to urban planning and low-cost mass housing —to any form of large-scale building involving inexpensive, standardized units of construction.

Unquestionably the experiments of the pioneers of modern architecture in the use of new materials and in the stripping away of accretions of Classical, Gothic, or Renaissance tradition resulted in various common denominators that may be classified as a common style. However, the individual stamp of the pioneers is recognizable even in their most comparable architecture and can hardly be reduced to a single style.

LE CORBUSIER (1887–1965). Among the generation of architectural pioneers who rose to prominence during the 1920s, Le Corbusier, the artistic pseudonym of the Swiss Charles-Édouard Jeanneret, was a searching and intense spirit, a passionate but frustrated painter, a brilliant critic, and an effective propagandist for his own architectural ideas. He studied in the tradition of the Vienna Workshops, learned the properties of ferroconcrete with Perret in Paris, and worked for a period with Behrens in Berlin (where he no doubt met Gropius and Mies van der Rohe). He moved to Paris from his native Switzerland in 1916. Although he condemned all forms of historical revivalism, he did not reject tradition, and his architecture evolved through an adherence to the basic principles of classicism. While he never became a painter of the first rank, his interest in and knowledge of Cubism and its offshoots affected his attitude toward architectural space and structure. Le Corbusier's principal exploration throughout much of his career was the reconciliation of human beings with nature and the modern machine. This was addressed largely through the problem of the house, to which he applied his famous phrase, "a machine for living." By exploiting the lightness and strength of ferroconcrete, his aims were to maximize the interpenetration of inner and outer space and create plans of the utmost freedom and flexibility. A drawing of 1914–15 states the problem and his solution (fig. 415). This is a perspective drawing for the skeleton of a house to be mass-produced of inexpensive, standardized

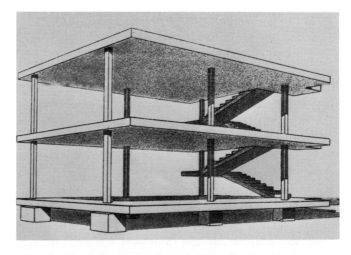

415. LE CORBUSIER. Perspective drawing for Domino Housing Project. 1914–15 © 1997 ADAGP, PARIS/ARTISTS RIGHTS SOCIETY (ARS), NEW YORK

materials. The structure consisted of six slender pillars standing on a broad, flat base and supporting two other floors or areas that may be interpreted as an upper floor and a flat roof. The stories are connected by a freestanding, minimal staircase. The ground floor is raised on six blocks, suggestive of his later use of stilts or piers.

This drawing is important for showing how early Le Corbusier established his philosophy of building. Outer walls, windows, or complete glass sheaths can simply be hung on this frame. Inner partitions can be distributed and shaped in any manner the architect desires. The entire structure can be repeated indefinitely either vertically or horizontally, with any number of variations. (Le Corbusier did not invent the system of ferroconcrete screen-wall construction; Behrens and Gropius had already constructed buildings involving the principle.) Le Corbusier's "Five Points of a New Architecture," published in 1926, were (1) the pilotis—supporting narrow pillars to be left free to rise through the open space of the house; (2) the free plan—composing interior space with non-bearing interior walls to create free flow of space and also interpenetration of inner and outer space; (3) the free façade—the wall as a nonsupporting skin or sheath; (4) the horizontal strip window running the breadth of a façade; (5) the roof garden—the flat roof as an additional living area. These points could provide an elementary outline of the International Style.

The masterpiece among Le Corbusier's early houses was the Villa Savoye at Poissy (color plate 189, page 335, and fig. 416), thirty miles from Paris. Along with Mies's German Pavilion (fig. 420), the Villa Savoye is generally regarded as a paradigm of the International Style. The three-bedroom house, beautifully sited in an open field, is almost a square in plan, with the upper living area supported on delicate piers or pilotis. The enclosed ground level has a curved-glass end wall containing garage and service functions, set under the hovering second storey. The Savoye family, arriving from Paris, would drive right under the house—the curve of the ground floor was determined by the radius of a car. Although today's suburban homes are loosely designed around the automobile, in 1929 this design concept was based on the notion of the car as the ultimate machine and the idea that the approach up to and through the house carried ceremonial significance.

In the main living area on the second level, the architect has brilliantly demonstrated his aim of integrating inside and outside space. The rooms open on a terrace, which is protected by half walls or windbreaks above horizontal openings that continue the long, horizontal line of the strip windows. The horizontal elements are tied together in sections by a central ramp that moves through each level and in- and out-of-doors. The complex of volumes and planes in the Villa Savoye relate to Le Corbusier's own Purist painting (see fig. 352). One historian has written of this building, "The visitor wandering through the interiors might glimpse cylindrical forms through layers of semi-reflecting glass and sense how Cubist ambiguities enlivened the play of surfaces: it was like entering the fantasy world behind the picture plane of a Purist still life."

Le Corbusier, like Wright, had few major commissions

416. LE CORBUSIER. Interior, Villa Savoye, Poissy, France. 1929–30

417. LUDWIG MIES VAN DER ROHE. Model for a glass skyscraper. 1922

during the 1920s, but he continually advanced his ideas and his reputation through his writings and through his visionary urban-planning projects. These large-scale housing projects, a response to the growing urban populations and housing shortages of postwar France, were never actually built. In 1922 he drew up a plan for a contemporary city (the "Ville Contemporaine") of three million inhabitants, involving rows of gleaming glass skyscrapers placed on stilts to allow for pedestrian passage. They were connected by vast highways and set in the midst of parks. In his 1925 "Plan Voisin" for Paris, Le Corbusier envisioned an enormous urban-renewal project that would have replaced the historic buildings north of the Seine with a complex of high-rise buildings. Like the Ville Contemporaine, this radiant modern city was the architect's drastic antidote to the traffic-congested streets of modern Paris and the soot-filled slums of the nineteenth-century city. It was based on the Utopian notion common among the modern pioneers that, armed with the right city planning and the appropriate faith in technology, architecture could revolutionize patterns of living and improve the lives of modern city dwellers on a physical, economic, and even spiritual level. But Le Corbusier's buildings were designed to house the industrial and intellectual elite, not the poor. In the face of today's massive urban crises, his desire to create cities where "the air is clean and pure" and "there is hardly any noise" seems naively idealistic, but his urban schemes were prophetic in the way they anticipated elements of today's cityscapes.

Le Corbusier's writings, also, have been tremendously influential in modern world architecture. His trenchant book *Vers une architecture* (1923) was immediately translated into English and other languages and has since become a standard treatise. In it he extolled the beauty of the ocean liner, the airplane, the automobile, the turbine engine, bridge construction, and dock machinery—all products of the engineer, whose designs had to reflect function and could not be embellished with nonessential decoration. Le Corbusier dramatized the problems of modern architecture through incisive comparisons and biting criticisms and, in effect, spread the word to a new generation. The pure, elegant geometry of Le Corbusier's International Style would give way in the years after World War II to a new, more organic vocabulary. (These later works are examined in chapter 23.)

LUDWIG MIES VAN DER ROHE (1886–1969). The spare, refined architecture of Mies van der Rohe, built on his edict that "Less is more," is synonymous with the modern movement and the International Style. He has arguably had a greater impact on the skylines of American cities than any other architect. His contribution lies in the ultimate refinement of the basic forms of the International Style, resulting in some of its most famous examples.

Some of the major influences on Mies were his father, a master mason from whom he initially gained his respect for craft skills; then Peter Behrens, in whose atelier he worked for three years; and Frank Lloyd Wright. From Wright, Mies gained his appreciation for the open, flowing plan and for the predominant horizontality of his earlier buildings. He was affected not only by Behrens's famous turbine factory, but also by Gropius's 1911 Fagus Factory (fig. 410), with its complete statement of the glass curtain wall. Gropius had been in Behrens's office between 1907 and 1910, and the association between Gropius and Mies that began there, despite a certain rivalry, continued.

Mies's style remained almost conventionally Neoclassical until after World War I. Then, in the midst of the financial and political turmoil of postwar Germany, he plunged into the varied and hectic experimentation that characterized the Berlin School. After 1919 most of the new ideas fermenting in the arts during the war began to converge on Berlin, which became one of the world capitals for art and architecture. These ideas included German Expressionism, Russian Suprematism and Constructivism, Dutch de Stijl, and international Dadaism. Contact was re-established between German artists and French Cubists and Italian Futurists. The Bauhaus school created by Gropius at Weimar in 1919 was in continuous and close contact with Berlin.

In 1921 and 1922 Mies completed two designs for skyscrapers (fig. 417), which, although never built, established the basis of his reputation. One was triangular in plan, the second a free-form plan of undulating curves. In these he proposed the boldest use yet envisaged of a reflective, all-glass sheathing suspended on a central core. As Mies wrote, "Only in the course of their construction do skyscrapers show their bold, structural character, and then the impression made by their soaring skeletal frames is overwhelming. On the other hand, when the facades are later covered with masonry, this impression is destroyed and the constructive character denied. . . . The structural principle of these buildings becomes clear when one uses glass to cover non-loadbearing walls. The use of glass forces us to new ways." No comparably daring design for a skyscraper was to be envisaged for thirty or forty years. Because there was no real indication of either the structural system or the disposition of interior space, these projects still belonged in the realm of visionary architecture, but they were prophetic projections of the skyscraper.

Mies's other unrealized projects of the early 1920s included two designs for country houses, both in 1923, the first in brick (figs. 418, 419) and the second in concrete. The brick

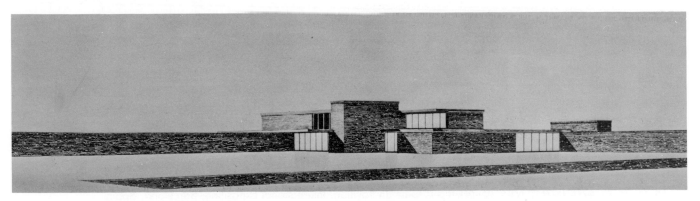

418. LUDWIG MIES VAN DER ROHE. Elevation for brick country house. 1923 © 1997 VG BILD-KUNST, BONN/ARTISTS RIGHTS SOCIETY (ARS), NEW YORK

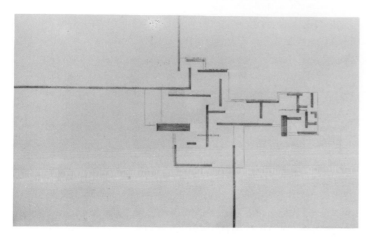

419. LUDWIG MIES VAN DER ROHE. Plan for brick country house.
1923 © 1997 VG BILD-KUNST, BONN/ARTISTS RIGHTS SOCIETY (ARS), NEW YORK

country-house design so extended the open plan made famous by Frank Lloyd Wright that the freestanding walls no longer enclose rooms: They create spaces that flow into one another. Mies fully integrated the interior and exterior spaces. As is clear from the plan, two of the walls extend from inside the house to the exterior, running right off the page and defying any sense of traditional enclosure. The plan of this house, drawn with the utmost economy and elegance, and the abstract organization of planar slabs in the elevation exemplify Mies's debt to the principles of de Stijl. In fact, the plan has often been compared to the composition of a 1918 painting by Mies's friend Theo van Doesburg.

One of the last works executed by Mies in Europe, before the rise of Nazism limited his activity and then forced his migration to the United States, was the German Pavilion for the Barcelona International Exposition in 1929 (fig. 420). Mies was in charge of Germany's entire contribution to the Exposition. The "Barcelona Pavilion," destroyed at the end of the exposition, has become one of the classics of Mies's

career and perhaps the pre-eminent example of the International Style. Here was the most complete statement to date of all the qualities of refinement, simplification, and elegance of scale and proportion that Mies, above all others, brought to modern architecture. In this building he contrasted the richness of highly polished marble wall slabs with the chrome-sheathed slender columns supporting the broad, overhanging flat roof. Thus, the walls are designed to define space rather than support the structure. In a realization of the open plan he had designed for the brick country house, the marble and glass interior walls stood free, serving simply to define space. But in contrast to the earlier work, the architect put limits on the space of pavilion and court by enclosing them in end walls. This definition of free-flowing interior space within a total rectangle was to become a signature style for Mies in his later career. The pavilion was furnished with chairs (known as the Barcelona chair), stools, and glass tables designed only by Mies. In the Barcelona Pavilion, Mies demonstrated that the International Style had come of age, had come to a maturity that permitted comparison with the great styles of the past. Fortunately, in 1986, to celebrate the centenary of the artist's birth, the pavilion was completely reconstructed in Barcelona according to the original plans.

Mies became director of the Bauhaus in 1930 but had little opportunity to advance its program. After moving from Dessau to Berlin in that year, the school suffered increasing pressure from the Nazis until it was finally closed in 1933. In 1937, with less and less opportunity to practice, Mies left for the United States, where in the last decades of his life he was able to fulfill, in a number of great projects, the promise apparent in the relatively few buildings he actually built in Europe (see chapter 23 for his works in the United States).

At its best, the uncompromising rationalism of Mies's architecture could produce compelling examples of pristine, streamlined form. In lesser hands, as is apparent in skylines

420. LUDWIG MIES VAN DER ROHE. German Pavilion, International Exposition, Barcelona, Spain.
1929. Reconstructed 1986

across the United States, Mies's minimalist forms could become redundant and impersonal, being almost brutal glass and steel monuments to consumer capitalism or drab, monotonous apartment dwellings. In the words of architecture critic Ada Louise Huxtable, "Mies's reductive theories, carried to their conceptual extreme, contained the stuff of both sublimity and failure, to which even he was not immune." By the early 1970s, the modern movement, and particularly the International Style as represented above all by Mies and Le Corbusier, would encounter a protracted backlash, opening the door to the era of Postmodernism (see chapter 26).

In the 1930s a number of events and individuals pointed the way to a new modern era in American design. Following the 1932 exhibition at The Museum of Modern Art in New York that gave the International Style its name, exhibitions of the Chicago School (1935), Le Corbusier (1935), and the Bauhaus (1938) all took place. Also in the 1930s, a number of new skyscrapers were built that broke the eclecticism of the skyscraper form and reintroduced aspects of the Chicago School or innovations of the Bauhaus and the International Style.

During the first half of the twentieth century, most experiments in modern architecture were carried out on individual houses. This is understandable, since the cost of building a skyscraper or an industrial complex is so exorbitant that it took a half-century before a greater number of patrons dared to gamble on modern buildings. The first European architects to come to America in the 1920s, William Lescaze, **RICHARD NEUTRA** (1892–1970), and Rudolf Schindler, devoted much of their careers to house architecture. The Austrians Schindler and Neutra worked for Wright and were partners for a time. They each built a house in California for Dr. Philip Lovell, combining aspects of Wright's house design with that of the International Style. The Neutra house (fig. 421)—placed spectacularly on a mountainside—was built of steel girders on a foundation of reinforced concrete. Through its open terraced construction, Neutra took every advantage of the amenities of landscape and climate and (along with Schindler) created a distinct style of southern California architecture.

Skyscraper Design in the United States

The historic and prophetic nature of the Chicago Tribune Tower competition was largely overlooked, even by the critics of Raymond Hood's neo-Gothic design, for whom the most effective solution was that of the second-prize winner, **ELIEL SAARINEN** (1873–1950) (fig. 422). The leading Finnish architect of the early twentieth century (before Alvar Aalto, whose work is discussed in chapter 23), Saarinen belongs chronologically with the first generation of architectural pioneers. His design for the Chicago Tribune Tower is nearly as rooted in the Middle Ages as that of Hood's, but it incorporates a greater degree of abstraction in the detailing. At a moment when American builders were turning away from outright revival styles but were not yet prepared to

421. RICHARD NEUTRA. Dr. Lovell's "Health" House, Los Angeles. 1927

422. ELIEL SAARINEN. Design for the Chicago Tribune Tower. 1922

left: 423. WILLIAM VAN ALEN. Chrysler Building, New York. 1928–30

right: 424. REINHARD AND HOFMEISTER, CORBETT, HARRISON, HARMON AND MACMURRAY, HOOD AND FOUILHOUX. Rockefeller Center, New York. 1931–39

accept radical solutions, Saarinen's qualified modernism had great appeal and influence. Indeed, in his next major building, Hood, who was the competition winner, was influenced by Saarinen's Chicago Tribune proposal. Saarinen moved permanently to the United States in 1923. His late works, dating after 1937, were done in collaboration with his son Eero, who became a leading architect in America by mid-century (see fig. 752).

One of the most elegant silhouettes of the New York skyline is the Chrysler Building (color plate 190, page 335, and fig. 423), designed in 1928 by the Beaux-Arts-trained architect **WILLIAM VAN ALEN** (1882–1954). Until 1931, when the Empire State Building was erected, the Chrysler Building was New York's tallest skyscraper. The structure, a masterpiece of Art Deco design, gradually tapers to a pinnacle made of stainless steel, which is mounted on a stepped, scalloped base. At various points the exterior is decorated with giant American eagles (color plate 190, page 335) and Chrysler's corporate logos, and the interior features lavish marble detailing. On the verge of the Depression, the Chrysler Building

paid tribute to American commercialism on a grand scale.

The most comprehensive complex of skyscrapers from this period is Rockefeller Center in New York, begun in 1931 and finished in 1939 (fig. 424). The center was proposed by John D. Rockefeller, Jr., to house the Metropolitan Opera Company within a large commercial complex. The original plan, completed in 1932, occupied three city blocks and consisted of fourteen buildings, theaters, and open public spaces, with the tall, slender RCA (now GE) Building in the center. After the war the complex expanded to include twenty-one office buildings. Although the original buildings have elements of Gothic detail, these have been simplified or altogether eliminated in the newer buildings of the 1950s and 1960s. Rockefeller Center is of significance not only for its contribution to harmonious, rational skyscraper design, but even more for its planning concept. Introduced here are large, open areas for pedestrians between the office buildings, many recreational facilities, an elaborate theater (Radio City Music Hall), radio and television studios, a second theater, shops, restaurants, and a skating rink. The rink is placed below street

432. LÁSZLÓ MOHOLY-NAGY. *Light-Space Modulator.* 1922–30. Kinetic sculpture of steel, plastic, wood, and other materials with electric motor, 59½ x 27½ x 27½" (151.1 x 69.9 x 69.9 cm). Busch-Reisinger Museum, Harvard University Art Museums, Cambridge, Massachusetts GIFT OF SIBYL MOHOLY-NAGY. © 1997 VG BILD-KUNST, BONN/ARTISTS RIGHTS SOCIETY (ARS), NEW YORK

433. LÁSZLÓ MOHOLY-NAGY. *Untitled.* c. 1940. Photogram, silver bromide print, 20 x 16" (50 x 40 cm). The Art Institute of Chicago GIFT OF GEORGE AND RUTH BANFORD. © 1997 VG BILD-KUNST, BONN/ARTISTS RIGHTS SOCIETY (ARS), NEW YORK

Gropius made him a professor in the Weimar Bauhaus, where he taught the important foundation course with Josef Albers. A student, the photographer Paul Citroën (fig. 435), described Moholy-Nagy's arrival: "Like a vigorous, eager dog, Moholy burst into the Bauhaus circle, ferreting out with unfailing scent the still unsolved, still tradition-bound problems in order to attack them." In sharp contrast to the intuitive, mystical teaching methods of his predecessor Johannes Itten, Moholy-Nagy, a committed exponent of the Constructivist alliance of art and technology, stressed objectivity and scientific investigation in the classroom. Until 1928 he was a principal theoretician in applying the Bauhaus concept of art to industry and architecture. With Gropius's resignation and the ensuing emphasis at the Bauhaus on practical training and industrial production over art and experimentation, Moholy-Nagy left the school, eventually to reside in Amsterdam and London, painting, writing, and making art in all media. In 1937 he was appointed director of the New Bauhaus in Chicago, which is now a part of the Illinois Institute of Technology. His two principal books, *The New Vision* and *Vision in Motion,* are major documents of the Bauhaus method. His stylistic autobiography, *Abstract of an Artist* (added to the English edition of *The New Vision*), is one of the clearest statements of the modern artist's search for a place in technology and industry.

By 1921 Moholy-Nagy's interests began to focus on elements that dominated his creative expression for the rest of his life—light, space, and motion. He explored transparent and malleable materials, the possibilities of abstract photography, and the cinema. Beginning in 1922, Moholy-Nagy pioneered in the creation of light-and-motion machines built from reflecting metals and transparent plastics. The first of these kinetic, motor-driven constructions, which he named "light-space modulators," was finally built in 1930 (fig. 432). When the machines are set in motion their reflective surfaces cast light on surrounding forms. During the 1920s Moholy-Nagy was one of the chief progenitors of mechanized kinetic sculpture, but a number of his contemporaries, including Gabo, Tatlin (see fig. 250), Rodchenko (see fig. 252), and somewhat later, Alexander Calder (figs. 453, 455) were also exploring movement in sculpture. Moholy-Nagy's legacy can be found in the work of several contemporary artists who have exploited light as a medium, including Robert Irwin (see fig. 688), Dan Flavin (see fig. 687), and, more recently, Christian Boltanski (see fig. 872).

Moholy-Nagy also made abstract paintings, clearly influenced by El Lissitzky and Malevich, three-dimensional constructions, and worked in graphic design. Always alert to ways of exploring the potential of light for plastic expression, Moholy-Nagy also became an adventurous photographer. Apparently unaware of Man Ray's technically identical Rayograms (see fig. 398), Moholy-Nagy developed his own cameraless images in 1922, just a few months after the American. Both artists' discoveries attest to the highly experimental nature of photography in the 1920s. Moholy-Nagy and his wife Lucia, who collaborated with him until 1929,

called their works "photograms" (fig. 433). However, unlike Man Ray, with his interest in the surreality of images discovered by automatic means, Moholy-Nagy remained true to his Constructivist aesthetic and used the objects placed on light-sensitive paper as "light modulators," materials for exercises in light. As in the example shown on the previous page, the overlapping, dematerialized shapes of the photogram form abstract compositions that are set in motion purely by the manipulation of light. Moholy-Nagy regarded the camera as an instrument for extending vision and discovering forms otherwise unavailable to the naked eye. In the "new vision" of the world presented in his 1925 Bauhaus book entitled *Malerei Fotografie Film (Painting Photography Film)* the artist included not only his own photographic works but also scientific, news, and aerial photographs, all of them presented as works of art. In addition to the abstract photograms, he also made photomontages as well as photographic images made with a camera. Like Rodchenko (see fig. 253), he was partial to sharply angled, vertiginous views that bring to mind the dynamic compositions of Constructivist painting. About his 1928 aerial view taken from Berlin's radio tower (fig. 434), itself a symbol of the new technology, Moholy-Nagy wrote: "The receding and advancing values of the black and white, grays and textures, are here reminiscent of the photogram." To him, the camera was a graphic tool equal to any as a means of rendering reality and disclosing its underlying purity of form. And so he wrote in a frequently quoted statement: ". . . the illiterate of the future will be ignorant of camera and pen alike."

Like the Dada artists, the Constructivists found photomontage a fertile process. Although they often produced effects similar to those of the Dadaists, the Constructivists' purposes in juxtaposing images—altogether various in context, subject, scale, proportion, and tonal value—were quite different, usually related to propagandistic or formal problems, rather than to a search for dissonance, mystery, or subversion (color plate 138, page 267; see fig. 313).

Cutting and pasting were precisely how the Bauhaus photographer **Paul Citroën** (1896–1983) composed the famous 1923 photomontage entitled *Metropolis* (fig. 435). Within the large thirty-by-forty-inch format, Citroën crushed together such a towering mass of urban imagery that Moholy-Nagy called it "a gigantic sea of masonry," stabilized, however, by Constructivism's controlling principle of abstract design.

Naum Gabo (1890–1977). Following his departure from Russia in 1922, where the abstract art he had helped to evolve (see fig. 255) proved incompatible with the utilitarian policies of the Soviet regime, Naum Gabo lived in Germany until 1932, perfecting his Constructivist sculpture and con-

435. Paul Citroën. *Metropolis.* 1923. Collage, printed matter, and postcards, 30 x 23" (76.2 x 58.4 cm). Printroom of the University of Leiden, the Netherlands

tributing importantly though indirectly to Moholy-Nagy's ideas on light, space, and movement at the Bauhaus. Though not a member of the Bauhaus faculty, Gabo lectured there in 1928, published an important article in *Bauhaus* magazine, and was in contact with several Bauhaus artists. In 1922 eight of his sculptures were included in a huge exhibition of Russian art that traveled to Berlin and Amsterdam, helping to bring his work and that of other Russian abstract artists before an international audience. Like Moholy-Nagy, Gabo was a major practitioner of Constructivism in many media. But both artists demonstrated how the formal aspects of the new style were assimilated in the West without the attendant political trappings. As we saw in chapter 11, while Gabo sympathized with the initial aims of the revolution in Russia, he did not harness his art to specific ideas of collectivism and utilitarianism. Nevertheless, he shared Moholy-Nagy's Utopian belief in the transformative powers of art and encouraged modern artists to look to technology and the machine for forms and materials to appropriately express the aims of the new social order.

In Germany, Gabo continued the research he had begun in Russia. He explored the possibilities of new media, particularly glass and recently developed plastic materials such as celluloid, to exploit a sense of planar transparency in his sculpture. For Diaghilev's 1927 ballet *La Chatte*, Gabo and his brother Antoine Pevsner designed a set filled with large, geometric sculptures in a transparent material that shone, in the words of one observer, with "quicksilver radiance." In 1931 he took part in the international competition for the Palace of Soviets, a never-realized building that was to be a proud symbol in Moscow of the new Soviet Union. Several architects from the West also took part, including Le Corbusier and Gropius. Gabo, who had been searching for a form of architectural expression through his constructed sculpture, proposed a daring, winged structure of reinforced concrete.

In 1932, after Nazi storm troopers came to his studio, Gabo left Germany for Paris, where he was active in the *Abstraction-Création* group, organized partly as an antidote to the influence of Surrealism. His next move was to England, where in the years from 1936 to 1946 he was active in the circle of abstract artists centering on Herbert Read, Ben Nicholson (fig. 459), Barbara Hepworth (fig. 463), and Henry Moore (fig. 461). Thereafter Gabo lived in the United States until his death in 1977.

Gabo's principal innovation of the 1940s was a construction in which webs of taut nylon string were attached to interlocking sheets of Perspex, a clear plastic first marketed in 1935 (fig. 436). Over many years, Gabo made several versions of *Linear Construction in Space, No. 1* on different scales. In these he attained a transparent delicacy and weightlessness unprecedented in sculpture. Like a drawing in space, the nylon filaments reflect light and gracefully articulate the void as positive form, transforming the space, according to Gabo, into a "malleable material element." Though probably not a direct influence on Gabo, Henry Moore had created abstract sculptures incorporating tautly stretched string as of

above:
436. NAUM GABO. *Linear Construction in Space, No. 1 (Variation).* Conceived 1942–43 (possibly executed 1948). Plastic and nylon thread, 24½ x 24½" (62.2 x 62.2 cm). The Phillips Collection, Washington, D.C.

left:
437. NAUM GABO. Construction for the Bijenkorf department store, Rotterdam, the Netherlands. 1956–57. Prestressed concrete, steel ribs, stainless steel, bronze wire, and marble, height 85' (25.9 m)

1937, and Barbara Hepworth followed with her own examples a few years later (see fig. 463).

Once he emigrated to the United States in 1946, Gabo was able to realize his ambition to create large-scale public sculpture. His major architectural-sculptural commission was a monument for the new Bijenkorf department store in Rotterdam (fig. 437). Designed by the Bauhaus architect

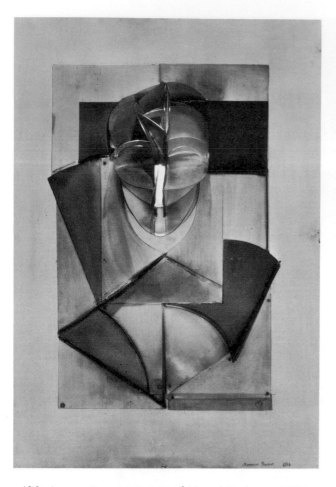

438. ANTOINE PEVSNER. *Portrait of Marcel Duchamp*. 1926.
Celluloid on copper, 37 x 25¾" (94 x 65.4 cm).
Yale University Art Gallery, New Haven SOCIÉTÉ ANONYME COLLECTION.
© 1997 ADAGP, PARIS/ARTISTS RIGHTS SOCIETY (ARS), NEW YORK

439. ANTOINE PEVSNER. *Projection into Space*. 1938–39.
Bronze and oxidized copper, 21½ x 15¾ x 19½"
(54.6 x 40 x 50.2 cm). Private collection, Switzerland
© 1997 ADAGP, PARIS/ARTISTS RIGHTS SOCIETY (ARS), NEW YORK

Marcel Breuer, the store was part of a massive postwar campaign to rebuild the city that had been devastated by German bombs in 1940. Gabo's solution for the busy urban site was a soaring, open structure consisting of curving steel shafts that frame an inner abstract construction made of bronze wire and steel. Gabo likened the sculpture, with its tremendous weight anchored in the ground, to the form of a tree.

ANTOINE PEVSNER (1886–1962), unlike his younger brother, Gabo, never taught at the Bauhaus, but worked within the Constructivist tradition in close relation to Gabo's. Pevsner left Russia in 1923 and settled permanently in Paris, where he exhibited with his brother in 1924. At this time he began to work seriously in abstract constructed sculpture. Commissioned by the Société Anonyme in 1926 (see chapter 13), the *Portrait of Marcel Duchamp* (fig. 438) is an open plastic construction with obvious connections to Gabo's *Constructed Head No. 1* of 1915 (see fig. 254). By the end of the 1920s Pevsner had nearly abandoned construction in transparent plastics in favor of abstract sculpture in bronze or copper. In 1932 he joined the *Abstraction-Création* group in Paris and, like Gabo, contributed to their journal.

In the 1938–39 *Projection into Space* (fig. 439), Pevsner realized his idea of "developable surface" or sculpture realized from a single, continuously curving plane. *Projection into Space*

is polished bronze, not cast but hammered out to machine precision. Despite the geometric precision of his work, Pevsner denied any mathematical basis in its organization. The structure of *Projection into Space* invites viewing from multiple vantage points, carrying the eye of the spectator around the perimeter, while the dynamic, spiraling planes give the illusion of movement in a static design. This quality was explored in such Futurist sculptures as Boccioni's *Development of a Bottle in Space* (see fig. 238), but the elimination of all representation in Pevsner's work heightens the sense of dynamism.

Pevsner carried out several large architectural commissions. The *Dynamic Projection in the 30th Degree* (fig. 440) at the University of Caracas, Venezuela, enlarged from a smaller bronze version, is more than eight feet high. This sculpture centers around a sweeping diagonal form that thrusts out from the center at a thirty-degree angle from the compressed waist of the vertical, hourglass construction. In this form, Pevsner combines solidity of shape with the freedom of a vast pennant seemingly held rigid by the force of a tornado wind. He massed his linear ridges so closely that the entire structure is composed of a tight mass of thin reeds that appear to spin in space.

JOSEF ALBERS (1888–1976). After Johannes Itten left the Bauhaus, the foundation course was given by Moholy-Nagy

440. ANTOINE PEVSNER. *Dynamic Projection in the 30th Degree.* 1950–51. Bronze, height over 8' (4.53 m). Aula Magna, University City, Caracas, Venezuela (Carlos Raúl Villanueva, architect)

and German artist Josef Albers, each teaching in accordance with his own ideas. After studying widely in Germany, Albers entered the Bauhaus as a student and, in 1923, was appointed to the faculty. While at the Bauhaus he met and married Anni Fleischmann, a gifted textile artist. In addition to his work in the foundation course, Albers taught furniture design and headed the glass workshop. He remained on the Bauhaus faculty until the Nazis closed the Bauhaus in 1933 and then emigrated to the United States. Thanks to the efforts of the American architect Philip Johnson, then a curator at The Museum of Modern Art in New York, Albers was offered a position at Black Mountain College, an experimental school in North Carolina. From the time of his arrival in 1933 he became a major influence in the training of American artists, architects, and designers. After leaving Black Mountain College in 1949, he continued to exert his influence on art students at Yale, where he taught from 1950–60.

Albers's early apprenticeship in a stained-glass workshop contributed to his lifelong interest in problems of light and color within geometric formats. In his glass paintings of the 1920s one can observe the transition from organic, free-form compositions of glass fragments to grid patterns, as in *City* (color plate 192, page 336), in which the relations of each color strip to all the others are meticulously calculated. To create these compositions, Albers invented a painstaking technique of sandblasting and painting thin layers of opaque glass, which he then baked in a kiln to achieve a hard, radiant surface. The title of this work highlights its resemblance to an

International Style skyline. In fact, Albers adapted this composition in 1963 for a fifty-four-foot-wide mural, which he called *Manhattan*, that was commissioned for New York's Pan Am Building. The precise geometry and industrial methods of Albers's work had special relevance for the 1960s when impersonal, hard-edge Minimalism and Op art (color plate 349, page 578; see fig. 679) were gaining momentum over the painterly emotionality of Abstract Expressionism.

With remarkable assiduity, Albers developed his increasingly reductive vocabulary throughout the 1940s, exploring issues of perception, illusionism, and the often ambiguous interaction of abstract pictorial elements. Beginning in 1950 he settled on the formula that he entitled Homage to the Square (color plate 193, page 336), where he relentlessly explored the relationships of color squares confined within squares. For the next twenty-five years, in a seemingly endless number of harmonious color combinations, Albers employed the strict formula of Homage to the Square in paintings and prints of many sizes. Within this format, the full, innermost square is, in the words of one observer, "like a seed; the heart of the matter, the core from which everything emanates." As can be seen in this example from 1959, the interior square is not centered, but rather positioned near the bottom of the canvas. According to a predetermined asymmetry, the size of the color bands at the bottom is doubled on each side of the square and tripled at the top. By restricting himself to this format, Albers sought to demonstrate the subtle perceptual ambiguities that occur when bands of pure color are juxta-

441. Oskar Schlemmer. *Study for The Triadic Ballet.*
c. 1921–23. Gouache, brush and ink, incised enamel,
and pasted photographs on paper, 22⅝ x 14⅝" (57.5 x
37.1 cm). The Museum of Modern Art, New York

Gift of Mr. and Mrs. Douglas Auchincloss. Photograph © 1995 The Museum of Modern Art, New York. © 1997 VG Bild-Kunst, Bonn/Artists Rights Society (ARS), New York

posed. Colors may advance or recede according to both their own intrinsic hue and in response to their neighbor, or they may even appear to mix optically with adjacent colors. And by the painted miter effect, seen here on the outermost corners, the artist creates a sense of illusionistic depth on a flat surface. As *Apparition,* one subtitle (color plate 193, page 336) suggests, Albers's goal was to expose the "discrepancy between physical fact and psychic effect."

German artist **Oskar Schlemmer** (1888–1943) taught design, sculpture, and mural painting at the Bauhaus from 1920 to 1929. During his years at the Bauhaus, Schlemmer's real passion was theater. In 1923 he was appointed director of theater activities at the school. He then moved with the Bauhaus from Weimar to its new Dessau location in 1925 and set up the experimental Theater Workshop. Schlemmer's art—whether painting, sculpture, or the costumes he designed for his theatrical productions—was centered on the human body. While he applied the forms of the machine to the figure, he emphasized the need to strike a balance between humanist interests and the growing veneration at

the Bauhaus for technology and the machine.

For his best-known performance, *The Triadic Ballet,* Schlemmer encased the dancers in colored, geometric shapes made from wood, metal, and cardboard. Designed to limit the range of the dancers' movements, the costumes transformed the performers into abstracted, kinetic sculptures. Indeed, for exhibition purposes, Schlemmer mounted his "spatial-plastic" costumes as sculptures. As can be seen in his collage *Study for The Triadic Ballet* (fig. 441), Schlemmer envisioned the stage as an abstract, gridded space through which the performers moved according to mathematically precise choreography. The deep perspective space depicted here is similar to that of Schlemmer's paintings from this period. *The Triadic Ballet* was first performed in 1922 in the artist's native Stuttgart and was subsequently presented in several European cities. Most of Schlemmer's sculptures were polychromed relief constructions, but in 1923 he made two freestanding sculptures, one of which was the original plaster for *Abstract Figure* (fig. 442). For this large, imposing torso, Schlemmer sought the same clarity of form and geometric precision he reserved for his theatrical designs. With its gleaming surfaces and streamlined forms, the sculpture shines like the chassis of a new automobile.

When Gropius resigned as director, his replacement, Hannes Meyer, increasingly politicized the curriculum at the school and stressed practical, industrial training over artistic production. This trend eventually led to Schlemmer's resignation. Not long after Schlemmer left the Bauhaus in 1929, the school came under attack from local Nazi authorities in Dessau, and the Nazis destroyed the murals and wall reliefs the artist had executed for the workshop building at Weimar. Despite repeated Nazi assaults on his reputation for creating "degenerate" and "Bolshevist" art, Schlemmer remained in Germany until his death in 1943.

A principal exponent of the "machine" doctrine was the sculptor **Rudolf Belling** (1886–1972), a native of Berlin and one of the founders in 1918 of Germany's November Group (see chapter 13). He was influenced by Archipenko, who lived in Berlin in the early 1920s and whose works were known there much earlier. Belling produced stylized figurative sculptures in a Cubist idiom as well as works in which the head or figure was translated into highly polished machine forms (fig. 443). Because of Germany's deteriorating political climate in the 1930s, Belling left the country and taught for a time in New York before emigrating to Turkey in 1937. His work tended toward the figurative but he also made small- and large-scale abstract works. Belling finally returned to Germany in 1966.

Willi Baumeister (1889–1955), like Schlemmer, was a pupil of Adolf Hoelzel at the Stuttgart Academy. He collaborated with both artists on a mural project in 1914 and frequently exhibited with Schlemmer in Germany. Throughout the interwar years, he was in contact with an international range of artists from Le Corbusier and Léger to Malevich and Moholy-Nagy. He exhibited in Paris in the 1920s, and became a member of the Parisian groups *Cercle et Carré* and

442. OSKAR SCHLEMMER. *Abstract Figure*. 1923. Bronze (cast 1962 from original plaster), 42⅛ x 26⅜" (107 x 67 cm). Collection Frau Tut Schlemmer, Stuttgart
© 1997 VG BILD-KUNST, BONN/ARTISTS RIGHTS SOCIETY (ARS), NEW YORK

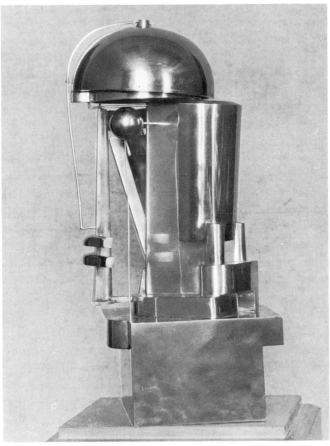

443. RUDOLF BELLING. *Sculpture*. 1923. Bronze, partially silvered, 18⅞ x 7¾ x 8½" (48 x 19.7 x 21.5 cm). The Museum of Modern Art, New York A. CONGER GOODYEAR FUND. PHOTOGRAPH © 1995 THE MUSEUM OF MODERN ART, NEW YORK. © 1997 VG BILD-KUNST, BONN/ARTISTS RIGHTS SOCIETY (ARS), NEW YORK

Abstraction-Création during 1930 and 1931. Although he never served on the Bauhaus faculty, he associated with several artists at the school and contributed to its magazine.

Baumeister's mature work was affected by Cubism and Purism as well as by Schlemmer's machine-based figurative style. Following the war, during which he served in the German army, Baumeister made a series of shallow relief constructions called *Mauerbilder* (wall pictures). In one example from 1923 (color plate 194, page 385), geometric forms surround a highly schematized figure. While some of these forms are painted directly on the wooden support, others are sections of wood that have been applied to the surface. A decade earlier, Archipenko, whose work was no doubt known to Baumeister, was experimenting with similar techniques in Paris (color plate 97, page 201). With its futuristic vision, Baumeister's composition is also reminiscent of Léger's great painting *The City* (color plate 102, page 204). By the early 1930s organic shapes appeared among the artist's machine forms, evidence of his interest in the Surrealism of Joan Miró and the paintings of Paul Klee. In works like *Stone Garden I* (fig. 444) Baumeister added a textural element by building up his surfaces sculpturally with sand and plaster. During the Nazi

444. WILLI BAUMEISTER. *Stone Garden I*. 1939. 39⅜ x 31⅞" (100 x 81 cm). Collection Felicitas-Karg Baumeister, Stuttgart
© 1997 VG BILD-KUNST, BONN/ARTISTS RIGHTS SOCIETY (ARS), NEW YORK

regime, when he was branded a "degenerate artist," he painted in secret, advancing his own ideas toward freer abstraction—ideograms with elusive suggestions of figures. These signs came from the artist's imagination, but were based on his studies of the art of ancient civilizations, such as Mesopotamia.

Introduced earlier in the context of Der Blaue Reiter (see chapter 8), **PAUL KLEE** (1879–1940) served in the German army in World War I immediately after the breakup of the group in 1914. In 1920 Gropius invited him to join the staff of the Bauhaus at Weimar, and he remained affiliated with the school from 1921 to 1931, an enormously productive period for Klee. Becoming a teacher prompted the artist to examine the tenets of his own painting and strengthened his resolve to discover and elaborate rational systems for the creation of pictorial form. He recorded his theories in his copious notes and in publications of great significance for modern art, including his 1925 Bauhaus book, *Pedagogical Sketchbook*.

Klee's art, as differentiated from that of Mondrian or even of Kandinsky in his later phase, was always rooted in nature and was only on occasion completely abstract—principally in his Bauhaus years when he was especially close to Kandinsky and to the traditions of Russian Suprematism and Constructivism and Dutch de Stijl. He drew upon his great storehouse of naturalistic observations as his raw material, but his paintings were never based on immediately observed nature except in his early works and sketchbook notations. Yet even his abstract paintings have a pulsating energy that appears organic rather than geometric, seemingly evolved through a natural process of dynamic growth and transformation.

In both his teaching and his art, Klee wanted to bring about a harmonious convergence of the architectonic and the poetic. With their heightened emphasis on geometric structure and abstract form, Klee's paintings and drawings (he saw no need to distinguish between the two) from the Bauhaus years reveal formal influences from his Constructivist colleagues. Yet Klee was above all an individualist, and however much he absorbed from the Bauhaus Constructivists, his poetic and intuitive inventions provided a counterweight to the more scientific and objective efforts at the school.

Like Mondrian and Kandinsky, Klee was concerned in his teachings and his painting with the geometric elements of the work of art—the point, the line, the plane, the solid. To him, however, these elements had a primary basis in nature and growth. It was the process of change from one to the other that fascinated him. To Klee, the painting continually grew and changed in time as well as in space. In the same way, color was neither a simple means of establishing harmonious relationships, nor a method of creating space in the picture. Color was energy. It was emotion that established the mood of the painting within which the line established the action.

Klee saw the creative act as a magical experience in which the artist was enabled in moments of illumination to combine an inner vision with an outer experience of the world, to render an image that was parallel to and capable of illuminating the essence of nature. His art grew out of the traditions of

Romanticism and specifically of Symbolism, but he departed from both traditions in that his inner truth, his inner vision, was revealed not only in the subject, the color, and the shapes as defined entities, but even more in the process of creation. To start the creative process, Klee, a consummate draftsman, would begin to draw like a child; he said children, like the insane and "primitive" peoples, had the "power to see." He let the pencil or brush lead him until the image began to emerge. As it did, of course, his conscious experience and skills came back into play in order to carry the first intuitive image to a satisfactory conclusion. At that conclusion, some other association, recollection, or fantasy would result in the poetic and often amusing titles that then became part of the total work. Because he placed such value on inner vision and

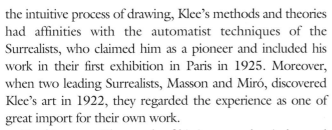

the intuitive process of drawing, Klee's methods and theories had affinities with the automatist techniques of the Surrealists, who claimed him as a pioneer and included his work in their first exhibition in Paris in 1925. Moreover, when two leading Surrealists, Masson and Miró, discovered Klee's art in 1922, they regarded the experience as one of great import for their own work.

Teaching gave Klee much of his iconography. As he used arrows to indicate lines of force for his students, these arrows began to creep into his work, where they are both formal elements and mysterious vectors of emotion (fig. 445). Color charts, perspective renderings, graphs, rapidly rendered heads, plant forms, linear patterns, checkerboards—elements of all descriptions became part of that reservoir of subconscious visual experience, which the artist then would transform into magical imagery. Usually working on a small scale in a deliberately naive rendition, Klee scattered or floated these elements in an ambiguous space composed of delicate and subtle harmonies of color and light that add mystery and strangeness to the concept. *Dance, Monster, to My Soft Song!* (fig. 446) shows a mad-eyed, purple-nosed head on an infinitesimal body floating above a tiny pianist, a childlike drawing integrated with the most sophisticated gradations of yellow ocher and umber.

During the 1920s Klee produced a series of black pictures in which he used the oil medium, sometimes combined with watercolor. Some of these were dark underwater scenes where fish swam through the depths of the ocean surrounded by exotic plants, abstract shapes, and sometimes strange little human figures. These works, including *Around the Fish* (fig. 445), are Surrealist in tone, embodying an arrangement of irrelevant objects, some mathematical, some organic. Here the precisely delineated fish on the oval purple platter is surrounded by objects, some machine forms, some organic, some emblematic. A schematic head on the upper left grows on a long stem from a container that might be a machine and is startled to be met head-on by a red arrow attached to the fish by a thin line. A full and a crescent moon, a red dot, a green cross, and an exclamation point are scattered throughout the black sky—or ocean depth—in which all these disparate signs and objects float. This mysterious, noctural still life typifies Klee's personal language of hieroglyphs and his use of natural forms merely as a point of departure into a fantastic realm. "The object grows beyond its appearance through our knowledge of its inner being," he wrote, "through the knowledge that the thing is more than its outward aspect suggests."

Made during Klee's Bauhaus years, *In the Current Six Thresholds* (color plate 195, page 385) is an arrangement of vertical and horizontal rectangles, in which gradations of deep color are contained within precisely ruled lines. Yet the total effect is organic rather than geometric, dynamic rather than static, with the bands continuing to divide as they move leftward. This and several similar abstract compositions of brilliantly colored strata were inspired by Klee's trip to Egypt in 1928–29 and are a response to the vast expanse and intense light of the desert landscape. Most of Klee's paintings were in combinations of ink and watercolor, media appropriate to the delicate quality of fluidity that he sought; but this particular work combines oil and tempera, media he used to obtain effects of particular richness and density.

Desiring more time for his art and frustrated by the demands of teaching and the growing politicization within the Dessau Bauhaus, Klee left the school in 1931 and took a less demanding position at the Dusseldorf Academy. During the early 1930s he applied his own variety of Neo-Impressionism in several compositions, culminating in a monumental painting from 1932, *Ad Parnassum* (fig. 447). Klee covered the canvas with allover patterns of thick, brightly colored daubs of paint. These are marshaled into distinctive forms by way of a few strong lines that conjure up a pyramid beneath a powerful sun. Like Kandinsky, Klee believed that pictorial composition was analogous to music and that sound can "form a synthesis with the world of appearances." Both he and his wife were musicians, and he applied the principals of musical composition, with all its discipline and mathematical precision, to his paintings and to his teaching. The title of this majestic work derives from a famous eighteenth-century treatise on musical counterpoint called, in Latin, *Gradus ad Parnassum (Stairway to Parnassus)*, Parnassus being a mountain sacred to the Greek god Apollo and the Muses. Klee orchestrated color and line in his painting according to his own system of pictorial polyphony, creating a luminous, harmonious vision aimed at elevating the viewer into a higher, Parnassian realm.

In 1933, the year Hitler assumed power in Germany, Klee was dismissed from his position at the Dusseldorf Academy and returned to his native Bern in Switzerland. Figures, faces (sometimes only great peering eyes), fantastic landscapes, architectural structures (sometimes menacing) continued to appear in his art during the 1930s. Perhaps the principal characteristic of the last works was Klee's use of bold and free black linear patterns against a colored field. One of his last works, *Death and Fire* (color plate 196, page 386), 1940, is executed with brutal simplicity, thinly painted on an irregular section of rough burlap. The rudimentary drawing, like the scrawl of a child, delineates a harrowing, spectral image, expressive perhaps of the burdens of artistic isolation, debilitating illness, and the imminent threats of war and totalitarianism. Unlike so many other members of the European avant-garde, Klee, who died within months of the war's out-

break, did not emigrate to the United States. However, his work became known to American audiences and especially to American artists through several exhibitions held throughout the 1930s and 1940s.

VASILY KANDINSKY (1866–1944) renewed his friendship with Klee when he returned to Germany from Russia in 1921, and in 1922 he joined the Weimar Bauhaus. One of the school's most distinguished faculty members, he taught a course called "Theory of Form" and headed the workshop of mural painting (regarded as superior to traditional easel painting at the school). Kandinsky's style underwent significant changes during his tenure at the Bauhaus. While he still adhered to the mystical Theosophical beliefs expressed in his seminal book from 1911, *Concerning the Spiritual in Art*, he had come to value form over color as a vehicle for expression, and his paintings evolved toward a more objective formal vocabulary. Even before coming to the Bauhaus, under the influence of Russian Suprematism and Constructivism, Kandinsky's painting had turned gradually from free expressionism to a form of geometric abstraction. This evolution is evident in *White Line, No. 232* (color plate 116, page 257), a painting Kandinsky made in 1920 while still in Russia, which contains some regular shapes, straight lines, and curving shapes with sharply defined edges.

By 1923, in *Composition VIII* (fig. 448), regular, hard-

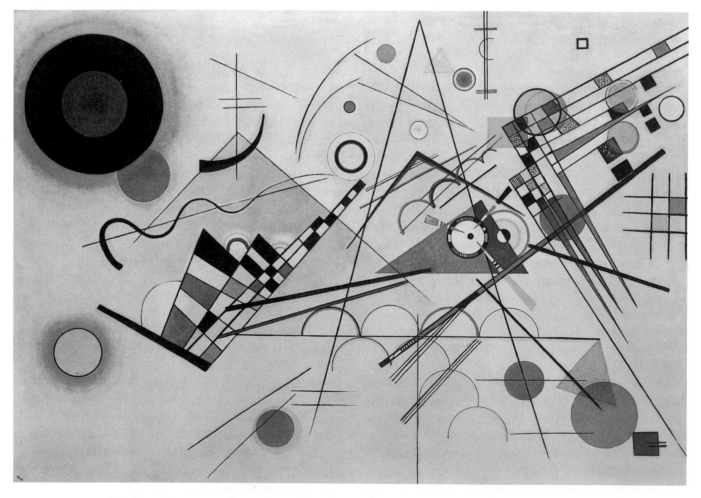

448. VASILY KANDINSKY. *Composition VIII*. 1923. Oil on canvas, 55⅛ x 79⅛" (140 x 201 cm). Solomon R. Guggenheim Museum, New York GIFT, SOLOMON R. GUGGENHEIM. © 1997 ADAGP, PARIS/ARTISTS RIGHTS SOCIETY (ARS), NEW YORK

edged shapes had taken over. As discussed in chapter 8, Kandinsky regarded the works he named Compositions as the fullest expression of his art. Ten years had elapsed between *Composition VII*, 1913 (color plate 76, page 160) and *Composition VIII*, and a comparison of the two paintings elucidates the changes that had taken place during this period. The deeply saturated colors and tumultuous collision of painterly forms in the 1913 picture are here replaced by clearly delineated shapes—circles, semicircles, open triangles, and straight lines—that float on a delicately modulated background of blue and creamy beige. The emotional climate of the later painting is far less heated than that of its predecessor, and its rationally ordered structure suggests that harmony has superseded the apocalyptic upheavals that inhabit the artist's prewar paintings.

Kandinsky fervently believed that abstract forms were invested with great significance and expressive power, and the spiritual basis of his abstract forms set him apart from Bauhaus teachers like Moholy-Nagy. "The contact of the acute angle of a triangle with a circle," he wrote, "is no less powerful in its effect than that of the finger of God with the finger of Adam in Michelangelo's [*Creation of Man*] painting." The circle, in particular, was filled with "inner potentialities" for the artist, and it took on a prominent role in his work of the 1920s. In *Several Circles, No. 323* (color plate 197, page 386), the transparent color circles float serenely across one another above an indeterminate, gray-black ground, like planets orbiting through space. It is hardly surprising that the artist revered the circle as a "link with the cosmic" and as a form that "points most clearly to the fourth dimension."

In 1926 Kandinsky published, as a Bauhaus book, *Point and Line to Plane*, his textbook for a course in composition. As compared with Klee's *Pedagogical Sketchbook* of the previous year, Kandinsky's book attempts a more absolute definition of the elements of a work of art and their relations to one another and to the whole. Here the artist affirmed the spiritual basis of his art, and his correspondence of the time reveals a combination of the pragmatic and the mystical.

Kandinsky continued his association with the Bauhaus until the school was closed in 1933. At the end of that year he moved permanently to Paris, where he was soon involved in the *Abstraction-Création* group of artists, and became friendly with Miró, Arp, and Pevsner. This new environment heightened Kandinsky's awareness of the Surrealist activities of the 1920s, and one senses certain qualities akin to abstract Surrealism in the many works created after his move to France.

In general Kandinsky moved toward freer, more biomorphic shapes and colors in the final years of his life. The lyrical, coloristic aspect of his painting began to resurface in Paris and to merge with the architectonic approach of his Bauhaus paintings. The edges of forms remained sharp, but their shapes seem to have emerged from a microscopic world. Such imagery probably stems from the artist's interest in biology and embryonic life forms as symbols of life and regeneration. In *Composition IX, No. 626*, 1936 (color plate 198, page 387), one of two final Compositions he made in

Paris, Kandinsky geometrically structured the canvas by painting identical triangles at either end and, in between, a parallelogram subdivided into four smaller identical parallelograms. On this rigidly defined ground the artist scattered a free assortment of dancing little shapes: circles, checkerboard squares, long, narrow rectangles, and amoeba-like figures. Kandinsky made his forms translucent—we can read ground colors beneath overlapping shapes—and his palette is a vivid array of candy-colored pastels. This device of playing small, free forms against a large geometric pattern intrigued the artist during these years. The contrast of freedom and control in Kandinsky's late work came from his lifelong concentration on the relations between intuitive expression and calculated abstract form.

Abstraction in Paris During the 1920s and 1930s

During the 1920s in France, abstract painting and sculpture were largely imported products, struggling for survival against the influences of the classical revival or "call to order" that swept the country between the wars; older modernists who worked in figurative modes—namely Matisse, Picasso, and the Cubist school painters; and the sensation created by the Surrealists. Of the Cubists, Robert Delaunay, one of the pioneers of abstraction (see fig. 230) during the 1920s, was principally interested in Cubist figuration, although after 1930 he returned to abstraction in painting and sculpture. Léger, too, painted a number of abstract architectural murals during the 1920s, although he was wary of pure abstraction as practiced by lesser talents than an artist like Mondrian. But the influence of Dutch de Stijl and Russian Constructivism and Suprematism was more apparent in Germany than in France during the 1920s, partly because of the Bauhaus and partly because France, less receptive to imported influences in modern art, did not easily relinquish its long classical tradition. When the de Stijl artists held an exhibition in 1923 at the gallery of Léonce Rosenberg, it aroused little general interest, although it had an immediate impact on Léger's work. Mondrian, then living in Paris, found no French buyers for his abstract work.

In 1925, the same year as the first Surrealist exhibition, a large exhibition of contemporary art was organized in Paris by a Polish painter, Victor–Yanaga Poznanski, at the hall of the Antique Dealers Syndicate. Among those included were Arp, Baumeister, Brancusi, the pioneer American abstractionist Patrick Henry Bruce, Robert and Sonia Delaunay, van Doesburg, Goncharova, Gris, Klee, Larionov, Léger, Miró, Moholy-Nagy, Mondrian, Ben Nicholson, Ozenfant, and Jacques Villon. The stated purpose of the exhibition was "not to show examples of every tendency in contemporary painting, but to take stock, as completely as circumstances permit, of what is going on in *non–imitative plastic art*, the possibility of which was first conceived of by the Cubist movement." There were obvious omissions, such as the Czech Kupka, one of the first abstract painters in Paris (color plate 103, page 204); still, it was a surprisingly comprehensive showing of international abstract tendencies. In the face of the rising tides of representation and Surrealism, however, it did not attract wide attention. In addition, Albert Gleizes, a primary force behind the show, had hoped to demonstrate that Cubism was not only the great progenitor of the various forms of abstraction but was still a vital force in vanguard art; instead, the late Cubists were overshadowed by the foreign innovators.

The next major development in the history of abstraction in Paris came in 1930, when Theo van Doesburg and the French artist Jean Hélion founded the group *Art Concret*, devoted to hard-edged abstraction, and the artist-critic Michel Seuphor and the Uruguayan painter Joaquín Torres-García founded a rival group and periodical entitled *Cercle et Carré* (*Circle and Square*). There were few French participants in the first exhibition of *Cercle et Carré*, held in April 1930 on the ground floor of the building in which Picasso lived. They included Arp (*persona grata* with both Surrealists and many abstractionists), Baumeister, Jean Gorin (a new French disciple of Mondrian), Kandinsky, Le Corbusier, Léger, Mondrian, Ozenfant, Pevsner, Schwitters, the American Futurist Joseph Stella, Sophie Taeuber-Arp, and Torres-García. Also shown, according to Seuphor, but not in the catalogue, were Moholy-Nagy, Hans Richter, and Raoul Hausmann.

The periodical and the exhibition, although short-lived, had a considerable impact. Their impetus and their mailing lists were taken over by Vantongerloo and Auguste Herbin, upon the formation in 1931 of a comparable group, *Abstraction-Création*, with a periodical in 1932 of the same name. *Abstraction-Création*, both in its exhibitions and its publications, was a force for abstract art until 1936, standing against the repressive regimes in Germany and Russia which were inhospitable towards the innovations of the avant-garde. The group embraced a broad spectrum of abstract styles but lacked the shared Utopian outlook that had unified members within the earliest groups devoted to abstraction. In the 1930s the rise of Nazism in Germany brought more practitioners of abstraction to Paris, where they joined artists from Russia, Holland, and other countries. By the mid-1930s, despite the general coolness of the French to geometric abstraction, *Abstraction-Création* on occasion had as many as four hundred members—from Suprematism, Constructivism, de Stijl, and their many offshoots.

AUGUSTE HERBIN (1882–1960) had pursued a long, slow path toward abstraction through Fauvism and Cubism until 1920. During the early 1920s he returned, like so many others, to representation, but by the end of that decade his work had again become abstract, remaining so until his death. Stimulated in part by the examples of Kupka (color plate 103, page 204) and Delaunay (color plate 101, page 203), and supported by his involvement with *Abstraction-Création*, Herbin devoted himself to an abstract syntax of curved lines and circular movements. Throughout the 1930s he painted brightly colored arabesques in a flat, hard-edge style (color plate 199, page 387).

PIET MONDRIAN (1872–1944), whose evolution to pure abstraction through Cubism was discussed in chapter 11, lived in Paris between 1919 and 1938 and was by far the city's most influential figure in abstract painting. Although he did not take the initiative in the organization of *Cercle et Carré* or *Abstraction-Création*, his presence and participation were of the greatest significance for the growth and spread of abstraction. As already noted, he had resigned from de Stijl in 1925 after disagreements with van Doesburg about the nature of de Stijl architecture (though the artists reconciled in 1929). While Mondrian envisioned, like his de Stijl colleagues, the integration of all the arts, he fervently believed, unlike some, in the superiority of painting. A prolific writer and theoretician, Mondrian fully formulated his ideas during the 1920s and, in the last decade of his life, during which he moved from Paris to New York (where he died in 1944), he produced some of the greatest work of his career.

In 1920, with his first full-fledged painting based on the principles of Neo-Plasticism (meaning roughly "new image" or "new form"), Mondrian had found his solution to the long unsolved problem—to express universals through a dynamic and asymmetrical equilibrium of vertical and horizontal structure, with primary hues of color disposed in rectangular areas. These elements gave visual expression to Mondrian's beliefs about the dynamic opposition and balance between the dual forces of matter and spirit, theories that grew out of his exposure to Theosophy and Hegelian dialectics. It was only through the pure "plastic" expression of what Mondrian called the "inward" that humans could approach "the divine, the universal."

In his painting, Mondrian was disturbed by the fact that up to this point, in most of his severely geometric paintings—including the floating color-plane compositions and the so-called grid or checkerboard paintings—the shapes of red, blue, or yellow seemed to function visually as foreground forms against a white and gray background (color plate 121, page 259) and thus to interfere with total unity. The solution, he found, lay in making the forms independent of color, with heavy lines (which he never thought of as lines in the sense of edges) moving through the rectangles of color. By this device he was able to gain a mastery over color and space that he would not exhaust for the next twenty years.

A mature example of Neo-Plasticism is seen in the 1921–25 *Tableau II* (color plate 200, page 388). Here is the familiar palette of red, blue, yellow, black, and two shades of gray. The total structure is emphatic, not simply containing the color rectangles but functioning as a counterpoint to them. Both red and gray areas are divided into larger and smaller rectangles with black lines. The black rectangles, since they are transpositions of lines to planes, act as further unifying elements between line and color. The edges of the painting are left open. Along the top and in the lower left corner, the verticals do not quite go to the edge, with the consequence that the grays at these points surround the end of the lines. Only in the lower right does the black come to the edge, and this is actually a black area through which the line moves slightly to the left of the edge. In all other parts a color, principally gray, but red and yellow at the upper left, forms

449. Piet Mondrian's studio at rue du Départ, Paris. 1926. © 1997 Mondrian Holtzman Trust

the outer boundary. The result that Mondrian sought—an absolute but dynamic balance of vertical and horizontal structure, using primary hues and black and white—is thus achieved. Everything in the painting holds its place. By some purely visual phenomenon, caused by the structure and the subtle disposition of color areas, the grays are as assertive as the reds or yellows; they advance as well as recede. The painting is not in any sense of the word flat. Everything is held firmly in place, but under great tension.

The open composition with emphasis on large white or light-gray areas predominated in Mondrian's production during the 1920s and, in fact, for the rest of his life. He usually worked on several pictures at once and often developed a single idea by working in series, devoting as many as eight paintings to variations on a single theme. He made several variations within a square format in the late 1920s, including *Composition with Red, Blue, and Yellow* (color plate 201, page 389), in which a large color square of red, upper right, is joined point to point with a small square of comparably intense blue by intersecting black lines. Surrounding them, equally defined by heavy black lines, are rectangles of off-white. In the lower right-corner is one small rectangle of yel-

low. The white areas, combined with the subordinate blue and yellow, effectively control and balance the great red square. Although Mondrian did not ascribe symbolic meaning to colors in the way that, for example, Kandinsky did, he told a collector that his paintings with a predominant red were "more real" than paintings with little or no color, which he considered "more spiritual."

Mondrian's studio in Paris (fig. 449) was a Spartan living space but a remarkable artistic environment that was famous throughout the European art world. For his ideal Neo-Plastic environment, the artist created geometric compositions on the studio walls with arrangements of colored squares. The easel seen at the left in this photograph was more for viewing than working (Mondrian preferred to paint on a horizontal surface). In 1930 Mondrian was visited by Hilla Rebay, an artist and zealous supporter of abstract art who was to help found the Museum of Non-Objective Painting in New York, eventually to become the Solomon R. Guggenheim Museum. "He lives like a monk," she observed, "everything is white and empty, but for red, blue and yellow painted squares that are spread all over the room of his white studio and bedroom. He also has a small record player with Negro

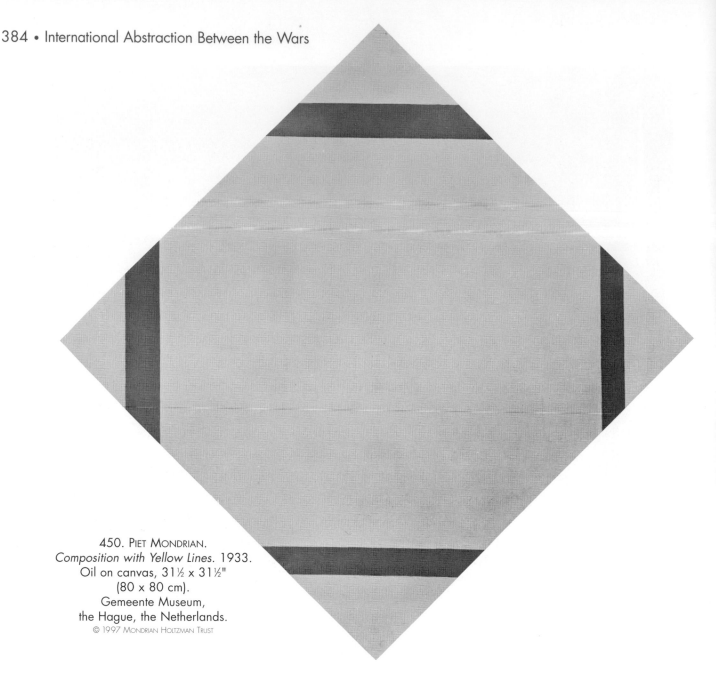

450. PIET MONDRIAN.
Composition with Yellow Lines. 1933.
Oil on canvas, 31½ x 31½"
(80 x 80 cm).
Gemeente Museum,
the Hague, the Netherlands.

music." Second only to painting was the artist's abiding passion for American jazz.

Composition with Yellow Lines (fig. 450) is among Mondrian's most beautiful so-called lozenge or diamond paintings. In this format, first used by Mondrian in 1918, the square painting is turned on edge but the horizontal and vertical axes are maintained internally. The shape inspired some of Mondrian's most austere designs, in which his desire to transgress the frame, to express a sense of incompleteness, was given tangible expression. *Composition with Yellow Lines* represents an ultimate simplification in its design of four yellow lines, delicately adjusted in width and cutting across the angles of the diamond. Mondrian's fascination with the incomplete within the complete, the tension of lines cut off by the edge of the canvas that seek to continue on to an invisible point of juncture outside the canvas, is nowhere better expressed than here. The viewer's desire to rejoin and complete the lines, to recover the square, is inescapable. The fascination of this magnificently serene painting also lies in another direction—the abandonment of black. Mondrian would fully develop this aspect in his very last works (color

plate 202, page 389). It is important to note that he also designed his own frames out of narrow strips of wood that he painted white or gray and set back from the canvas. He wished to emphasize the flat, nonillusionistic nature of his canvases, "to bring the painting forward from the frame . . . to a more real existence."

In 1932 Mondrian introduced a dramatic new element in one of his neo-plastic compositions when he divided a black horizontal into two thin lines. With this "double line" he set out to replace all vestiges of static form with a "dynamic equilibrium." This innovation brought a new, vibrant opticality to the work, with Mondrian multiplying the number of lines until, by the late 1930s, the paintings contained irregular grids of black lines with only one or a few planes of color. A midway point in this development is *Composition in White, Black, and Red,* 1936 (fig. 451), a structure of vertical and horizontal black lines on a white ground, with a small black rectangle in the upper left corner and a long red strip at the bottom. The lines, however, represent intricate proportions, both in the rectangles they define and in their own thickness and distribution. In the right lower corner, horizontal lines,

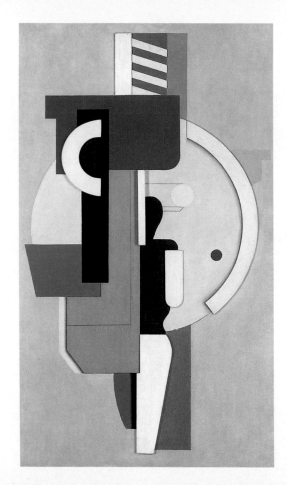

right: Color plate 194.
WILLI BAUMEISTER. *Wall Picture with Circle II.* 1923. Oil and wood on wood panel, 46½ x 27⅛" (118 x 69 cm). Hamburger Kunsthalle

below: Color plate 195.
PAUL KLEE. *In the Current Six Thresholds.* 1929. Oil and tempera on canvas, 17⅛ x 17⅛" (43.5 x 43.5 cm). Solomon R. Guggenheim Museum, New York

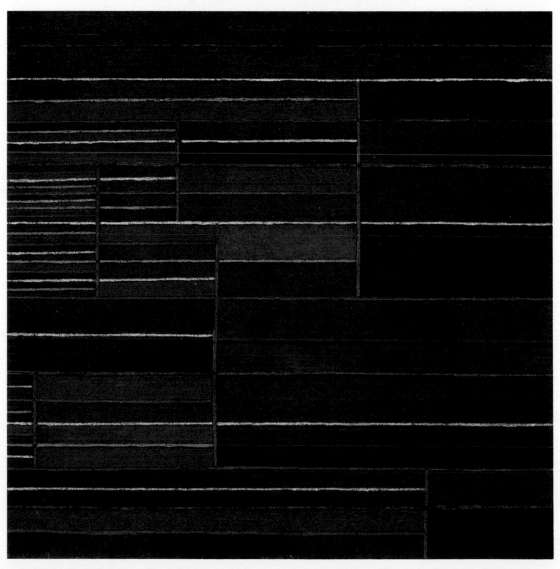

Color plate 196. PAUL KLEE. *Death and Fire.* 1940. Oil drawing in black paste on burlap, surrounded by colored paste ground, mounted on burlap, 18⅛ x 17¼" (46 x 43.8 cm). Kunstmuseum Bern

PAUL KLEE STIFTUNG
© 1997 VG BILD-KUNST, BONN/ARTISTS RIGHTS SOCIETY (ARS), NEW YORK

Color plate 197. VASILY KANDINSKY. *Several Circles, No. 323.* 1926. Oil on canvas, 55⅛ x 55⅛" (140 x 140 cm). Solomon R. Guggenheim Museum, New York

GIFT, SOLOMON R. GUGGENHEIM
© 1997 ADAGP, PARIS/ARTISTS RIGHTS SOCIETY (ARS), NEW YORK

Color plate 198. VASILY KANDINSKY. *Composition IX, No. 626.* 1936. Oil on canvas, 44⅝" x 6'4¾" (113.4 cm x 1.9 m). Musée National d'Art Moderne, Paris ATTRIBUT DE L'ETAT © 1997 ADAGP, PARIS/ARTISTS RIGHTS SOCIETY (ARS), NEW YORK

Color plate 199. AUGUSTE HERBIN. *Composition.* 1939. Oil on canvas, 18½ x 43¼" (47 x 109.9 cm). Private collection

Color plate 200. PIET MONDRIAN. *Tableau II*. 1921–25. Oil on canvas, 29½ x 25⅝" (74.9 x 65.1 cm).
Private collection, Zürich

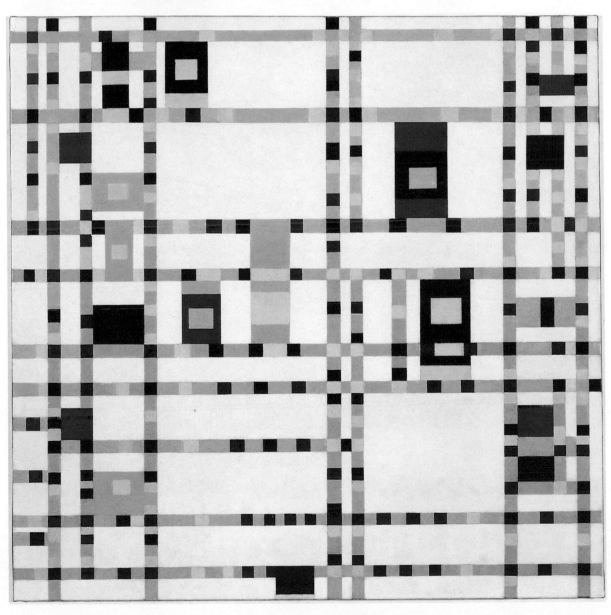

Color plate 203. STANLEY SPENCER. *The Resurrection of the Soldiers.* 1928–29. Oil on canvas, glued to the wall, 21' x 17'6"
(6.4 x 5.3 m). Sandham Memorial Chapel, Burghclere, England

Color plate 204.
JOHN SLOAN.
Hairdresser's Window.
1907.
Oil on canvas, 31⅞ x
26" (81 x 66 cm).
Wadsworth Atheneum,
Hartford, Connecticut
THE ELLA GALLUP SUMNER AND MARY
CATLIN SUMNER COLLECTION

Color plate 205.
MAURICE PRENDERGAST.
The Idlers. c. 1918–20.
Oil on canvas, 21 x 23"
(53.3 x 58.4 cm).
Maier Museum of Art,
Randolph-Macon
Woman's College,
Lynchburg, Virginia

right: Color plate 206.
GEORGE BELLOWS. *Cliff Dwellers*. 1913. Oil on canvas, 39½ x 41½" (100.3 x 105.4 cm). Los Angeles County Museum of Art
LOS ANGELES COUNTY FUNDS

below: Color plate 207.
ROMAINE BROOKS. *Self-Portrait*. 1923. Oil on canvas, 46¼ x 26⅞" (117.5 x 68.3 cm). National Museum of American Art, Washington, D.C.

below, right:
Color plate 208. MARSDEN HARTLEY. *Portrait of a German Officer*. 1914. 68¼ x 41⅜" (173.3 x 105.1 cm). The Metropolitan Museum of Art, New York
THE ALFRED STIEGLITZ COLLECTION, 1949

varied subtly in width, create a grid in which the white spaces seem to expand and contract in a visual ambiguity.

In 1938, with the approaching war, Mondrian left Paris for London, joining his friends Gabo, Ben Nicholson, and Barbara Hepworth. After two years he moved to New York, where he spent his final four years. Manhattan became the last great love of his life, perhaps because of the Neo-Plastic effect created by skyscrapers rising from the narrow canyons of the streets and the rigid grid of its plan. As early as 1917 Mondrian had said, "The truly modern artist sees the metropolis as abstract life given form: it is closer to him than nature and it will more easily stir aesthetic emotions in him." A major new stimulus to the artist was that of lights at night, when the skyscrapers were transformed into a brilliant pattern of light and shadow, blinking and changing. Mondrian also loved the tempo, the dynamism, of the city—the traffic, the dance halls, the jazz bands, the excitement of movement and change. He felt driven to translate these rhythms into his late paintings, where the earlier grid of black lines is replaced by a complex weave of colored lines.

The impact of the city and music is most evident in *Broadway Boogie-Woogie,* painted in 1942–43 (color plate 202, page 389). Here the artist returned to the square canvas but departed radically from the formula that had occupied him for more than twenty years. There is still the rectangular grid, but the black, linear structure balanced against color areas is gone. In fact, the process is reversed: the grid itself is the color, with the lines consisting of little blocks of red, yellow, blue, and gray. The ground is a single plane of off-white, and against it vibrate the varicolored lines as well as larger color rectangles.

451. PIET MONDRIAN. *Composition in White, Black, and Red.* 1936. Oil on canvas, 40¼ x 41" (102.2 x 104.1 cm). The Museum of Modern Art, New York

In New York, Mondrian devised a method of composing with commercially available colored tapes that he could easily shift around the canvas until he was satisfied enough to commit paint to canvas. This kind of experimental and intuitive approach had always governed Mondrian's working process. Although his art was guided by certain firmly held principles, it was far from formulaic. His paintings are filled with subtle textural effects and give evidence of repeated revisions as the artist painted layer upon layer, widening or narrowing his lines and experimenting with alternative colors. Even his whites are rarely the same hue. Although most art suffers in reproduction, the physical presence of Mondrian's paintings is especially difficult to capture even in the best photographs.

In evaluating the course of geometric abstraction from its beginnings, about 1911, to the present time, an understanding of the part played by Piet Mondrian is of paramount importance. He was the principal figure in the founding of geometric abstraction during World War I, for the ideas of de Stijl and their logical development were primarily his achievement. Mondrian's influence extended not only to abstract painting and sculpture but also to the forms of the International Style in architecture. Through the teachings of the Bauhaus in Germany and its offshoots in the United States, his theories were spread throughout the Western world. During the 1920s and 1930s in Paris it was probably the presence and inspiration of Mondrian more than any other single person that enabled abstraction to survive and gradually to gain strength—in the face of the revolution of Surrealism and the counterrevolution of representation, as well as economic depression, threats of dictatorship, and war.

In the United States Mondrian was a legendary figure inspiring not only the geometric abstractionists, who had been carrying on a minority battle against Social Realism and Regionalism, but also a number of younger artists who were to create a major revolution in American art. The emerging Abstract Expressionists of the early 1940s almost without exception had great respect for Mondrian, even though their painting took directions that would seem diametrically opposed to everything he believed in.

It was a formative encounter with Mondrian's art in 1930 that prompted American ALEXANDER CALDER (1898–1976) to become an abstract artist. Born in Philadelphia, Calder was the son and grandson of sculptors. After studying engineering, training that would have direct consequences for his art, he was gradually drawn into the field of art, studying painting at New York's Art Students League and working as an illustrator. Calder's early paintings of circus or sports scenes reflect the styles of his teachers, such as John Sloan (color plate 204, page 391). By the time he went to Paris for the first time in 1926, he had begun to make sculptures in wire and wood. In the French capital he eventually attracted the attention of avant-garde artists and writers (especially the Surrealists) with his *Circus,* a full-fledged, activated environment made up of tiny animals and performers that Calder assembled from wire and found materials and then set into motion. At the same time he made wooden sculptures and

452. ALEXANDER CALDER. *Romulus and Remus.* 1928. Wire and wood, 30½ × 124¼ × 26" (77.5 × 316.2 × 66 cm). Solomon R. Guggenheim Museum, New York GIFT OF THE ARTIST. © 1997 ADAGP, PARIS/ARTISTS RIGHTS SOCIETY (ARS), NEW YORK

portraits and caricatures constructed from wire. The wire sculptures are early demonstrations of Calder's marvelous technical ingenuity and playful humor. They could be quite large, such as the nine-foot *Romulus and Remus* of 1928 (fig. 452). Here Calder bent and twisted wire into a composition of such economy that the entire "torso" of the she-wolf is one continuous stretch of wire. The whole is like a three-dimensional line drawing (Calder made drawings related to the wire sculpture at the same time).

In 1930, during a subsequent sojourn in Paris, Calder visited Mondrian's rigorously composed studio (fig. 449), which deeply impressed him. "This one visit gave me a shock that started things," he said. "Though I had heard the word 'modern' before, I did not consciously know or feel the term 'abstract.' So now, at thirty-two, I wanted to paint and work in the abstract." He began to experiment with abstract painting and, more significantly, abstract wire constructions that

illustrated an immediate mastery of constructed space sculpture. These early abstract sculptures, which were exhibited in a 1931 solo exhibition in Paris, consisted of predominantly austere, geometric forms like open spheres; but they also contained a suggestion of subject—constellations and universes.

In 1931 Calder joined the *Abstraction-Création* group at Arp's suggestion and began to introduce motion in his constructions. At first he induced movement by hand cranks or small motors, but eventually the sculptures were driven merely by currents of air and Calder's carefully calibrated systems of weights and balances. There were precedents of kineticism in sculpture, as we have seen, in the work of Gabo, Rodchenko, and Moholy-Nagy, but no artist developed the concept as fully or ingeniously as Calder. His first group of hand and motor "mobiles" was exhibited in 1932 at the Galerie Vignon, where they were so christened by Marcel Duchamp. When Arp heard the name "mobile," he asked, "What were those things you

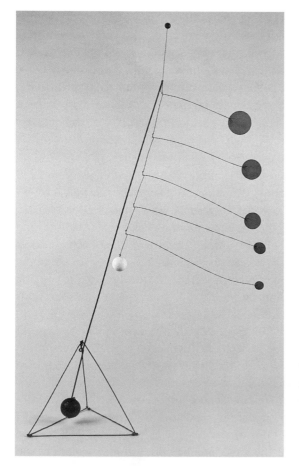

left: 453. ALEXANDER CALDER. *Object with Red Disks (Calderberry Bush).* 1932. Painted steel rod, wire, wood, and sheet aluminum, 7'4½" × 33" × 47½" (2.2 m × 83.8 cm × 120.7 cm). Whitney Museum of American Art, New York PURCHASE, WITH FUNDS FROM THE PERCY URIS PURCHASE FUND. © 1997 ADAGP, PARIS/ARTISTS RIGHTS SOCIETY (ARS), NEW YORK

454. ALEXANDER CALDER. *The White Frame.* 1934. Painted wood, sheet metal, wire, 7½ × 9' (2.3 × 2.7 m). Moderna Museet, Stockholm
© 1997 ADAGP, PARIS/ARTISTS RIGHTS SOCIETY (ARS), NEW YORK

455. ALEXANDER CALDER. *Lobster Trap and Fish Tail.* 1939. Hanging mobile: painted steel wire and sheet aluminum, approximately 8'6" high x 9'6" diameter (2.6 x 2.9 m). The Museum of Modern Art, New York

did last year—stabiles?" Thus was also born the word that technically might apply to any sculpture that does not move but that has become specifically associated with Calder's works. Calder's stabiles will be discussed further in chapter 22.

The characteristic works of the 1930s and 1940s—those on which the image of the artist has been created—are the wind-generated mobiles, either standing or hanging, made of plates of metal or other materials suspended on strings or wires, in a state of delicate balance. The earliest mobiles were relatively simple structures in which a variety of objects, cut-out metal or balls and other forms in wood, moved slowly in the breeze. A far greater variety of motion was possible than in the mechanically driven mobiles. For one thing, the element of chance played an important role. Motion varied from slow, stately rotation to a rapid staccato beat. In the more complex examples, shapes rotated, soared, changed tempo, and, in certain instances, emitted alarming sounds. In *Object with Red Disks (Calderberry Bush)* (fig. 453), a standing mobile from 1932 (the second title, by which the work is best known, was not Calder's), Calder counterbalanced the five flat red disks with wooden balls and perched the whole on a wire pyramidal base. When set in motion by the wind or a viewer's hand, this abstract construction moves through

space in preplanned yet not entirely predictable ways.

Among Calder's motorized mobiles were also reliefs, with the moving parts on a plane of wooden boards or within a rectangular frame. One of the earliest on a large scale is *The White Frame* of 1934 (fig. 454). In this, a few elements are set against a plain, flat background: a large, suspended disk at the right, a spiral wire at the left, and between them, suspended on wires, a white ring and two small balls, one red and one black. Put into motion, the large disk swings back and forth as a pendulum, the spiral rotates, and the balls drop unexpectedly and bounce from their wire springs. Whereas *The White Frame* (with the exception of the spiral) still reflects Mondrian, geometric abstraction, and Constructivism, Calder also used free, biomorphic forms reflecting the work of his friends Miró and Arp. From this point onward, Calder's production moved easily between geometric or Neo-Plastic forms and those associated with organic Surrealism.

By the end of the 1930s Calder's mobiles had become extremely sophisticated and could be made to loop and swirl up and down, as well as around or back and forth. One of the largest hanging mobiles of the 1930s is *Lobster Trap and Fish Tail* (fig. 455), a work that was commissioned for The Museum of Modern Art. Although the work is quite abstract, the subject association of the Surrealist-inspired, biomorphic forms is irresistible. The torpedo-shaped element at top could be a lobster cautiously approaching the trap, represented by a delicate wire cage balancing at one end of a bent rod. Dangling from the other end is a cluster of fan-shaped metal plates suggestive of a school of fish. The delicacy of the elements somewhat disguises the actual size, some nine and a half feet in diameter.

After World War II, as we shall discover in chapter 22, Calder would realize one of the most international of careers, expanding the abstract art he had already formulated in the 1930s to an architectural scale of unprecedented grandeur.

The Emergence of Abstraction in Great Britain

The most radical element in English art to emerge before World War I was the group who called themselves Vorticists (from "vortex"), founded in 1914 and led by the talented painter, writer, and polemicist PERCY WYNDHAM LEWIS (1886–1957). In the catalogue of the only exhibition of Vorticism, at the Doré Galleries in 1915, Lewis described the movement with all the zealous rhetoric typical of the early avant-garde: "By vorticism we mean (a) Activity as opposed to the tasteful Passivity of Picasso; (b) Significance as opposed to the dull and anecdotal character to which the Naturalist is condemned; (c) Essential Movement and Activity (such as the energy of a mind) as opposed to the imitative cinematography, the fuss and hysterics of the Futurists."

Of particular significance for Vorticism was the association of the American poet Ezra Pound, then living in England. He gave the movement its name and with Lewis founded the periodical *Blast,* whose subtitle read *Review of the Great English Vortex.* For the Vorticists, abstraction was the optimal artistic language for forging a link between art and life. As Lewis stated in *Blast,* they believed that artists must "enrich abstraction until it is almost plain life." Lewis was a literary man by character, and much of his contribution lay in the field of criticism rather than creation. He was influenced by Cubism from a very early date and, despite his polemics against it, by Futurism. In one of his few surviving pre-World War I paintings (fig. 456), Lewis composed a dynamic arrangement of rectilinear forms that arch through the composition as though propelled by an invisible energy force. This adaptation of a Cubist syntax to metallic, machine-like forms has much in common with Léger's paintings of exactly the same moment (see fig. 229). Lewis was primarily attempting, by every means in his power, to attack and break down the academic complacency that surrounded him in England. In the vortex he was searching for an art of "activity, significance, and essential movement, and a modernism that should be clean, hard, and plastic."

Vorticism was a short-lived phenomenon, and it produced few artists of real originality. Nevertheless, it was of great importance in marking England's involvement in the new experimental art of Europe. As with other artists associated with the movement, few of Lewis's early, Vorticist paintings

456. PERCY WYNDHAM LEWIS. *Composition.* 1913. Watercolor drawing on paper, 13½ x 10½" (34.3 x 26.7 cm). Tate Gallery, London

survive. After the war he abandoned abstraction in favor of a stylized figurative art. He made portraits of Edith Sitwell, Ezra Pound, and other literary figures, giving them, through a suggestion of Cubist structure, an appearance of modernism.

Also involved with Vorticism was the English photographer ALVIN LANGDON COBURN (1882–1966), already a committed modernist. In what Lewis and Pound called Vortographs, Coburn photographed through a kind of kaleidoscope, which actually consisted of three mirrors clamped together as a hollow triangle—to record the refracted image of crystal and wood pieces arranged on a glass tabletop (fig. 457). The results are thoroughly abstract photographs that have obvious parallels with Lewis' paintings.

STANLEY SPENCER (1891–1959). The two English painters of greatest stature between the wars, Stanley Spencer and Ben Nicholson, represent extremes of contrast. Stanley Spencer went through life seemingly oblivious to everything in avant-garde twentieth-century painting, taking inspiration, like the Pre-Raphaelites he admired, from the Italian painters of the quattrocento. In his earlier paintings with Christian themes, he developed an eccentric, personal style in his drawing reminiscent of early Italian or Flemish paintings, which enhanced his visionary effects. In the 1920s Spencer received a large mural commission for the Sandham Memorial Chapel in Burghclere, Hampshire. From 1927 to 1932 he labored over a cycle of paintings based on his war experiences, culminating in the huge central panel, *The Resurrection of Soldiers*

457. ALVIN LANGDON COBURN. *Vortograph #1*. 1917.
Silver print, 7⅞ x 5¾" (20 x 14.6 cm).
The Museum of Modern Art, New York

(color plate 203, page 390). This vast panorama depicts dead soldiers struggling from their graves amidst a forest of overturned crosses. Although Spencer maintained his religious convictions throughout his life, in his later works he also turned to obscure themes with erotic overtones, sometimes suggestive of paintings by Balthus (color plate 267, page 483). Even these usually have religious implications in the sense of a modern morality play.

The other outstanding English painter to emerge between the wars was **BEN NICHOLSON** (1894–1981), who was also a distinguished sculptor. Whereas Spencer might be considered narrowly English, Nicholson was an internationally minded artist. During the 1920s he practiced variants on the late Cubist styles of Braque and Picasso and also made his first abstract compositions. These early works reveal most of the characteristics of Nicholson's mature style. Their surfaces appear abraded, and color is subdued, matte, and delicately harmonious: a limited range of gray whites to grayed ochers, with a few accents of black and muted red, and rubbed pencil passages of dark gray. All the elements of still life—table, bottle, and bowls—are frontalized and flattened, and shapes are outlined with ruler and compass in fine and precise contour.

In 1932 Nicholson began spending time in Paris, where he became acquainted with Giacometti, Braque, Picasso, Kandinsky, and Brancusi, among others. Over the next few years, Nicholson championed the cause of English art abroad and became an important artistic conduit between the Con-

tinent and his native country, helping to attract foreign artists to London during the tense period leading up to the war. In 1933 he was invited to join the Parisian *Abstraction-Création* group and, significantly for his art, saw works by Calder and Mondrian for the first time. At the end of that year he made his first carved and painted relief, an abstract composition of irregular, free-floating circles over rectangular shapes that he carved into a shallow board. Nicholson had met Barbara Hepworth in 1931; they were married from 1938 to 1951. They shared a studio in London, and Nicholson's exposure to her sculptor's tools and methods, as well as his recent experience carving linocuts, led him to make reliefs and, eventually, sculpture. We have seen how any number of modern artists, from Arp (color plate 133, page 265) to Archipenko (color plate 97, page 201) to Baumeister (color plate 194, page 385), were interested in creating such hybrids between sculpture and painting.

A residual biomorphism, inspired partly by the work of Miró and Calder, is evident in Nicholson's early reliefs. Gradually, however, he eliminated such Surrealist overtones from his work, and the forms within his reliefs—circles, squares, and rectangles—became highly regularized in response to Mondrian's example. In the mid-1930s he avoided both color and drawn line in a series of White Reliefs (fig. 458). Here the total spatial effect is achieved through the multi-level interaction of overlapping rectangles, with an occasional circle in sunken relief. It has been suggested that Nicholson's exclusive use of white, a color connoting purity and infinity in abstract painting, soon followed his reading of an article written by van Doesburg and published by *Art Concrèt* that was devoted to the subject of white in painting.

Although they never realized anything as significant as the Bauhaus, English artists and architects throughout the 1930s were exploring potential alliances between their disciplines

458. BEN NICHOLSON. *White Relief*. 1936. Oil on carved board, 8 x 9¼" (20.3 x 23.5 cm). Private collection

based on the new Constructivist aesthetic. The white, recti-linear surfaces of Nicholson's paintings, stripped of all distracting detail, were regarded as ideal decor for the pared-down spaces of the new International Style architecture. Such ideas were specifically addressed in an important book Nicholson edited in 1937 with Gabo and the architect Leslie Martin (with collaboration from Barbara Hepworth), *Circle: International Survey of Constructive Art*—published in conjunction with a London exhibition, *Constructive Art*, and designed to counteract the sensation generated by a recent show of international Surrealism held in London. *Circle* was an important publication, since it introduced many of the Bauhaus ideas to Britain.

In the late 1930s Nicholson reintroduced color to his reliefs, a shift stimulated in part by the landscape at St. Ives, Cornwall, where he lived with Hepworth. The 1939 *Painted Relief* (fig. 459) is a meticulous adjustment of squares and rectangles in low but varying relief projections. The larger shapes are a uniformed, muted, sand color. Played against these are a square of dull ocher, a rectangle of red-brown and, top and bottom at the left, strips of pure white. The only lines are two circles drawn in graphite, placed slightly off-center in the major squares. The edges of the relief shapes also act as a precise, rectangular linear structure.

In the same period Nicholson continued to produce paintings involving the same principles of strict rectangularity, using the primary colors with black, white, and grays, sometimes varying the values and intensities, and combining light, delicate blues with intense yellows. Despite a commitment to abstraction until the end of his life, Nicholson also habitually drew from nature, resulting in images with an Ingres-like delicacy and precision.

Unlike Nicholson or, for that matter, any of the sculptors previously discussed in this chapter, the sculptor **HENRY MOORE** (1898–1986) was not an artist in the Constructivist mold. He did not assemble forms from the newest industrial materials but rather carved and modeled them in the traditional media of stone and plaster. His was a humanist art, grounded in the contours of nature and the human figure, however abstract the forms sometimes appear. Although he did not begin to assume a genuinely international stature before 1945, Moore was a mature artist by the early 1930s, in touch with the main lines of Surrealism and abstraction on the Continent, as well as all the new developments in sculpture, from Rodin through Brancusi to Picasso and Giacometti. By 1935 he had already made original statements and had arrived at many of the sculptural figurative concepts that he was to build on for the rest of his career.

The son of a coal miner, Moore studied at the Royal College of Art in London until 1925. The greatest immediate influence on him were the works of art he studied in English and European museums, particularly the Classical, pre-Classical, African, and pre-Columbian art he saw in the British Museum. He was also attracted to English medieval church sculpture and to artists of the Renaissance tradition— Masaccio, Michelangelo, Rodin, Maillol—who had a particular feeling for the monumental.

Between 1926 and 1930 the dominant influence on Moore was pre-Columbian art. The 1929 *Reclining Figure* (fig. 460) is one of the artist's first masterpieces in sculpture, a work that may owe its original inspiration to the Toltec sculpture of Chac-Mool, the Mexican Rain Spirit (with overtones of Maillol). The massive blockiness stems from a passionate devotion to the principle of truth to materials, that is,

459. BEN NICHOLSON. *Painted Relief.* 1939. Oil and pencil on composition boards mounted on painted plywood, 32⅞ x 45" (83.5 x 114.3 cm). The Museum of Modern Art, New York

460. HENRY MOORE.
Reclining Figure. 1929.
Hornton stone, height 22½"
(57.2 cm).
Leeds City Art Galleries,
England

461. HENRY MOORE.
Reclining Figure. 1939.
Elm wood,
37" x 6'7" x 30"
(94 cm x 2 m x 76.2 cm)
The Detroit Institute of Arts

FOUNDERS SOCIETY. PURCHASE WITH FUNDS FROM
THE DEXTER M. FERRY, JR., TRUSTEE CORPORATION

the idea that stone should look like stone, not flesh. In this and other reclining figures, torsos, and mother-and-child groups of the late 1920s, Moore staked out basic themes he then used throughout his life.

In the early 1930s the influence of Surrealist sculpture, particularly that of Picasso, became evident. Moore began to explore other materials, particularly bronze, and his figures took on a fluidity and sense of transparent surface appropriate to what was for him a new material. Similar effects were achieved in carved wood in which he followed the wood grain meticulously, as in *Reclining Figure* from 1939 (fig. 461), his largest carved sculpture to date. Less angular than the 1929 figure, this sculpture consists of a series of holes that pierce the undulating solids, transforming the body into a kind of landscape filled with caves and tunnels. A characteristic of the figures of the 1930s is this opening up of voids, frequently to the point where the solids function as space-enclosing frames.

In the mid 1930s Moore turned to abstract forms, and by opening up the masses and creating dispersed groups, he studied various kinds of space relationships. This began a continuing concern with a sculpture of tensions between void and solid, of forms enclosed within other forms. These exper-

iments were then translated back into the figures. The interest in spatial problems led Moore during the 1940s and 1950s to an ever greater use of bronze and other metals in which he could enlarge the voids of the figures. These developments in Moore's work during and after World War II are pursued in chapter 20.

BARBARA HEPWORTH (1903–1975), the other English sculptor of international stature before World War II, was an art student contemporaneously with Henry Moore. The two remained close through much of her career, with Moore's work exerting a direct influence on her early sculpture. In 1925, while Hepworth was in Italy with her first husband, the English sculptor John Skeaping, she learned marble-carving techniques from an Italian master craftsman. Throughout the 1920s she made highly accomplished figurative sculptures in wood and stone that reflected her admiration of Egyptian, Cycladic, and Archaic Greek art. After meeting Nicholson in 1931 Hepworth became a leader of the abstract movement in England. She and Nicholson were active in the Unit One group, which was attempting to establish a common front for modern artists and architects in England, and in 1933 they were invited to become members of the *Abstraction-Création* group in Paris. While in Paris, Hepworth visited the studios of Arp and Brancusi, encounters with significant implications for her later work. The relationship between Hepworth and Nicholson produced a rich cross-fertilization of ideas, media, and forms in the art of each artist. They were also close to the critic Herbert Read, then emerging as the leading advocate of modern art in England. When, during the 1930s, Gabo, Mondrian, Gropius, Moholy-Nagy, Breuer, and other European artists and architects moved to England, Hepworth and Nicholson were able to strengthen old friendships and form new ones among these pioneers of modernism.

Hepworth's sculpture began to shed its semblance to the real world after 1931, when she carved *Pierced Form*. In this alabaster sculpture, the stone is penetrated in the center by a large hole to make, in the artist's words, "an abstract form and space." This crucial rupture with the notion of sculpture as a solid, closed form was one of Hepworth's major contributions to modern sculpture, and the discovery would inform her subsequent work as well as that of Moore. Her exposure to the work of Arp (who had previously made his own penetrated sculptures) prompted Hepworth to make biomorphic, multi-part sculptures in the early 1930s, but by 1934 she adopted a more geometric syntax and eliminated, at least for a time, the remaining vestiges of naturalism from her work.

In one of her best-known works, *Two Segments and Sphere* (fig. 462), 1935–36, Hepworth heightens the tension between the geometric forms, with their smooth, perfected finish, through the precarious positioning of the sphere. The late 1930s produced some of Hepworth's most severely simplified sculptures, a number of them consisting of a single marble column, gently rounded or delicately indented to emphasize their organic, figurative source.

In 1939, stimulated by Moore's example, she also began to explore the use of strings stretched across voids, as in *Wave* (fig. 463). Hepworth's wood sculptures are marked by the same loving finish as her works in stone. The woods are beautiful and frequently exotic—mahogany, scented guarea, lagos ebony, or

462. BARBARA HEPWORTH. *Two Segments and Sphere*. 1935–36. Marble, 10½ x 10½ x 8½" (26.7 x 26.7 x 21.6 cm). Private collection

rosewood—worked to bring out their essential nature. Easier to work than stone, wood inspired her to an increasingly open type of composition, with voids penetrating the mass of the material. She also began at this time to use color, generally whites and blues in interior areas, to contrast with the natural wood of the exterior. In *Wave*, the blue interior refers to the ocean, for, like Nicholson, Hepworth was deeply affected by the Cornish landscape. "I used colour and strings in many of the carvings of this time. The colour in the concavities plunged me into the depth of water, caves, or shadows deeper than the carved concavities themselves. The strings were the tension I felt between myself and the sea, the wind or the hills."

For a period after World War II, as we shall see in chapter 20, the sense of a naturalistic subject reemerged in Hepworth's sculptures.

463. BARBARA HEPWORTH. *Wave*. 1943. Plane wood with blue interior and strings, 12 x 17½ x 8¼" (30.5 x 44.5 x 21 cm). Private collection, on loan to the Scottish National Gallery of Modern Art, Edinburgh

18 American Art Before World War II

In the years leading up to World War I, the American realist tradition was given new life within the ranks of the so-called "Ashcan School," a term invented by critics in the 1930s that loosely describes a group of artists in New York who favored, as the name implies, commonplace subjects, even ones that emphasized the seedy aspects of daily life. In an era when the United States was shifting from an agrarian to an industrially based economy, artists turned to the vitality of the city for their themes, sometimes documenting the lives of the country's urban inhabitants with a literalness that shocked viewers accustomed to the bland generalizations of academic art. Thus, the first modern American revolution in painting was not away from, but toward, realism.

The subsequent developments toward realism and new pictorial subject matter are explained in part by the fact that the academic spirit had become anathema to many young painters by the turn of the century, when the professional survival of an artist was largely contingent on membership in the National Academy of Design, the American equivalent of the French Academy of Arts. The National Academy of Design perpetuated the traditions of the French Academy, such as annual juried exhibitions. Although it merged with the more tolerant Society of American Artists in 1907, it remained steadfastly intolerant of new developments.

At the same time, important venues in New York, particularly Alfred Stieglitz's gallery known as 291 and, in 1913, the gigantic exhibition of modern art known as the Armory Show, introduced European modernists to American audiences and nurtured a number of American artists committed to experimental art. During the years of the Great Depression, the country's focus turned inward, giving rise to new varieties of naturalistic art based on intrinsically American themes. These were practiced by the so-called Regionalists, who recorded the rural life of the Midwest, and the more politically engaged Social Realists, who documented the social consequences of extreme economic change. Also a fertile period for American photography, the era before World War II witnessed the development of photojournalism, as well as social documentary and advertising photography.

The Eight

In response to the National Academy's rejection of their work for the 1907 spring exhibition, eight painters participated in a historic exhibition at the Macbeth Gallery in New York City in February 1908. The painter Robert Henri, who had moved to New York from Philadelphia in 1900, was the intellectual leader of this loosely constituted group called "The Eight." Since 1905, Henri had been a member of the Academy, but in 1907 he found himself so at odds with his colleagues there that he began to organize an alternative exhibition. Henri was joined by four Philadelphia artists, John Sloan, Everett Shinn, William Glackens, and George Luks, all illustrators who provided on-the-spot pictorial sketches for newspapers, a practice not yet replaced by the new art of photography. As a result of this vocation, they came to painting first as draftsmen whose subject matter portrayed the transient and everyday realities of American life. Three other artists also joined ranks—the landscape painters Ernest Lawson and Maurice Prendergast, and the Symbolist Arthur B. Davies—and the group was dubbed "The Eight" by a New York journalist. The exhibition of The Eight was a milestone in the history of modern American painting. Although the participating artists represented various points of view and contrasting styles, they were united by their mutual hostility to the entrenched art of the academicians with their rigid jury system, and by the conviction that artists had the right to paint subjects of their own choosing. The show received mixed reviews. Some criticized it for its "inappropriate" recording of the uglier aspects of the New York scene. Several of the exhibiting artists were to become leading members of the Ashcan School.

ROBERT HENRI (1865–1929) was a dedicated teacher and artistic leader who instinctively rebelled against the clichés of the American academic tradition. He had studied at the Pennsylvania Academy with Thomas Anshutz, a pupil of Eakins, and at the Académie Julian in Paris, where the spontaneous sketching techniques he had learned under Anschutz displeased his teachers. In 1900, Henri settled in New York, where he eventually joined the faculty at the New York School of Art, exhibited widely, and, by 1905, was highly enough regarded to be elected as a member of the National Academy of Design.

Although Henri painted many landscapes, he is primarily admired for the forthright expression and painterly freedom of his portraits, including *Laughing Child* (fig. 464), which was among the works he exhibited with The Eight. Not surprisingly, he made the portrait during a visit with some of his students to Haarlem, the Netherlands, home of the seventeenth-century painter Frans Hals. Like the Dutch master, Henri favored immediacy of expression over academic finish in his portraits. During the trip he painted several informal half-length portraits of children against dark backgrounds, skillfully capturing the fresh enthusiasm of his young subjects.

Following the exhibition of The Eight, Henri established his own independent art school, becoming one of the most influential teachers in the early years of the twentieth century. His instruction left its mark on a number of artists including Stanton Macdonald-Wright, Patrick Henry Bruce, and Stuart Davis who would become leading American abstractionists (although Henri himself cared little for abstract art), as well as leading realist painters George Bellows and Edward Hopper (color plate 206, page 392; color plate 221, page 413).

Henri encouraged his friend and protégé **JOHN SLOAN** (1871–1951) to portray what was most familiar to him—the streets and sidewalks of the American urban landscape. In fact, Sloan happened upon the scene depicted in *Hairdresser's Window* (color plate 204, page 391) on his way to visit Henri at his New York studio. He was so taken with the spectacle of a crowd gathering beneath the window of a hairdresser's shop that he returned home to paint it from memory and, in his words, "without disguise." In the signs adorning the façade, Sloan introduced amusing puns such as one announc-

ing the shop of "Madame Malcomb." Because Sloan and his fellow Ashcan School painters searched for artistic potential within the most commonplace subjects, including commercial advertisements, they can be regarded as early forerunners of 1960s Pop art. Andy Warhol, for example, admired Sloan's work (see color plate 309, page 531).

Sloan offered an alternative to the teeming life of city streets in *The Wake of the Ferry II*, 1907 (fig. 465), where the delicately brushed passages that describe a misty sky and blue-green water recall Henri but also Manet and Whistler. Here a solitary passenger looks out over New York Harbor from the stern of a commuter ferryboat. New York's waterways provided subjects rich with possibilities for other American painters, including Henri and Glackens, as well as for poets such as Walt Whitman, whose 1856 poem, "Crossing Brooklyn Ferry," contributed to Sloan's choice of subject. And in the year Sloan's picture appeared, American photographer Alfred Stieglitz captured a related subject with his camera (see fig. 470).

WILLIAM GLACKENS (1870–1938) had traveled to Paris with Henri in 1895 and, like his friend, sought out the work of Manet. Among the works he sent to the exhibition of The Eight was *Chez Mouquin* (fig. 466), in which an elegant couple—the wealthy collector James B. Moore and a companion—are shown at a popular New York restaurant that was frequented by The Eight and their circle. With its lively atmosphere and glittering mirrored reflections, *Chez Mouquin* embodies all the warmth, spontaneity, and psychological complexity of the Impressionists' depictions of the Parisian café world. In his paintings (as opposed to his illustrations), Glackens was generally drawn to convivial subjects rather than to the grim realities of New York streets.

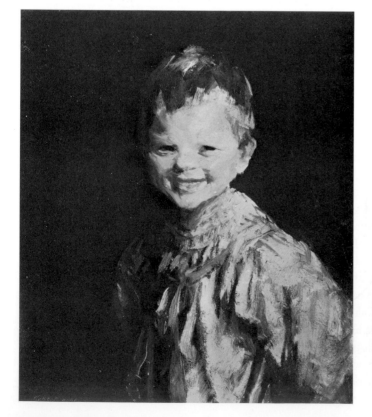

left: 464. ROBERT HENRI. *Laughing Child.* 1907. Oil on canvas, 24 x 20" (61 x 50.8 cm). Whitney Museum of American Art, New York
LAWRENCE H. BLOEDEL BEQUEST

below: 465. JOHN SLOAN. *The Wake of the Ferry II.* 1907. Oil on canvas, 26 x 32" (66 x 81.3 cm). The Phillips Collection, Washington, D.C.

466. WILLIAM GLACKENS. *Chez Mouquin.* 1905. Oil on canvas, 48¼ x 36¼" (122.6 x 92.1 cm). The Art Institute of Chicago
FRIENDS OF AMERICAN ART COLLECTION, 1925

Although the terms The Eight and Ashcan School are often used interchangeably, not all of the participants in the Macbeth Gallery show qualify as Ashcan artists. A case in point is the Boston artist **MAURICE PRENDERGAST** (1858–1924), who had trained in Paris in the 1890s and closely followed the developments of the French avant-garde. Like

Henri, Prendergast chronicled the lighter side of contemporary life—the pageantry of holiday crowds, city parks, or seaside resorts (color plate 205, page 391). He recorded these scenes in a distinctive, highly decorative style consisting of brightly colored patches of paint, reflecting his interest in Japanese prints and the French Nabis painters.

GEORGE BELLOWS (1882–1925), one of Henri's most accomplished students, developed a bold, energetic painting style that he deployed for subjects ranging from inner city street scenes; to tense, ringside views of boxing matches; to socialites at their leisure. His *Cliff Dwellers,* 1913 (color plate 206, page 392), is a classic Ashcan School depiction of daily life in Manhattan's Lower East Side. Forced from their apartments by stifling summer heat, the residents carry out their lives on stoops and sidewalks, seemingly trapped within the imposing walls of the tenement walk-ups. Bellows published a preparatory drawing for this work in *The Masses,* a radical socialist magazine for which Sloan also worked, assigning it a satirical caption, "Why don't they all go to the country for vacation?"

The social criticism inherent in Ashcan School art, such as the scenes of urban grime and disadvantage painted by Sloan and Bellows, was even more explicit in the photographic work of **JACOB AUGUST RIIS** (1849–1914), a Danish immigrant who took his camera into alleys and tenements on New York's Lower East Side to document the wretched conditions he found there (fig. 467). Originally a police reporter, Riis said of his camera, "I had use for it and beyond that I never went." To the dismay of New York society, he used his pictures to expose the poverty and starvation that he felt were direct results of the industrial revolution. His book *How the Other Half Lives: Studies Among the Tenements of New York,* which was published in 1890, became a classic reference for subsequent social documentary photographers (see figs. 659, 660).

467. JACOB A. RIIS. *"Five Cents a Spot" Lodging House, Bayard Street.* c. 1889. Gelatin-silver print. Museum of the City of New York
JACOB A. RIIS COLLECTION

468. LEWIS W. HINE. *Child in Carolina Cotton Mill.* 1908. Gelatin-silver print on masonite, 10½ x 13½" (26.7 x 34.3 cm). The Museum of Modern Art, New York PURCHASE.
COPY PRINT © 1995 THE MUSEUM OF MODERN ART, NEW YORK

Riis's counterpart in the industrial landscape was **LEWIS W. HINE** (1874–1940), who also used the camera as an instrument of social reform. After an initial series devoted to immigrants on Ellis Island, Hine spent the years 1908–16 as the staff photographer for the National Child Labor Committee visiting factories to photograph and expose child-labor abuses (fig. 468). Like Riis, Hine used the camera to produce "incontrovertible evidence" that helped win passage of laws protecting children against industrial exploitation.

Although American painter **ROMAINE BROOKS** (1874–1970) was a contemporary of the Ashcan artists, she lived in Europe for most of her adult life and, therefore, her work developed independently of trends in New York. Brooks survived a troubled childhood with a highly unstable mother to become a painter of considerable reputation in the 1920s, only to fall into almost total obscurity by the 1940s. As an adult she lived a peripatetic life in Europe, settling finally in Paris in 1905. Her paintings, mostly portraits, were shown at the Parisian gallery Durand-Ruel in 1910. In Paris, Brooks met American poet Natalie Clifford Barney, whose literary salon attracted such illustrious talents as Marcel Proust, Colette, and André Gide. Brooks conducted an open lesbian relationship with Barney and adopted a masculine manner of dress. In her arresting *Self-Portrait* (color plate 207, page 392), shown at the 1923 Salon des Indépendants, she projects herself as a stark silhouette with deeply shadowed eyes against an urban backdrop. With their delicate palette and fin-de-siècle mood, Brooks's paintings recall those of Whistler, though the somber and even grim aspect of her work is distinctly her own.

The Stieglitz Circle: 291

Meanwhile, in 1905, at the Little Galleries of the Photo-Secession at 291 Fifth Avenue in New York, the photographer Alfred Stieglitz, together with another photographer,

Edward Steichen, began to hold exhibitions of photography, and shortly thereafter of modern European paintings and sculpture, as well as African art. The artists whose work Stieglitz exhibited at the gallery known as 291 were more cosmopolitan than the Henri group, in closer contact with current events and artists in Europe, and less chauvinistic about Americanism in art. Also, Stieglitz and Steichen were in touch with avant-garde leaders in Paris, not only with artists but also with writers and collectors such as Gertrude Stein and her brothers Michael and Leo, the first collectors of Matisse and Picasso.

291 would serve as a rallying place for the American pioneers of international modernism like John Marin, Marsden Hartley, Max Weber, Arthur Dove, Georgia O'Keeffe, and Stanton Macdonald-Wright, all of whom are discussed below. *291* was also the title of a short-lived Dada magazine for, as we saw in chapter 13, 291 gallery was a focal point for the New York Dada movement. Between 1908 and 1917, works by many of the greatest European modernists were exhibited there, including Cézanne, Toulouse-Lautrec, Brancusi, Matisse, and Braque, as well as members of the American avant-garde, including John Marin, Charles Sheeler, Charles Demuth, Georgia O'Keeffe, and Elie Nadelman. At the same time, Stieglitz edited an important quarterly magazine on photography called *Camera Work*. After 291 closed in 1917, Stieglitz continued to sponsor exhibitions at various New York galleries. In 1925 he opened the Intimate Gallery, which was succeeded by An American Place in 1929. Stieglitz directed this gallery until his death in 1946, featuring exhibitions by America's leading modernists. The impact of Stieglitz and his exhibitions at 291 gallery was enormous. The miscellaneous group of Americans shown by him constituted the core of experimental art in the United States in the first half of the century and was the single most important factor, aside from the Armory Show, in the birth of the modern spirit in this country.

Well before he became the linchpin figure in the American avant-garde, **ALFRED STIEGLITZ** (1864–1946) lived in Berlin, where he began his long and distinguished career as a photographer and launched his crusade to establish photography as a fine art. He was an ardent advocate of Pictorialism in photography, a movement at one with the aestheticism of the late, Symbolist years in nineteenth-century art. The Pictorialists deplored the utilitarian banality of sharp-focus, documentary photography, as well as the academic pretension of Rejlander and Robinson (see figs. 23 and 26). Instead, they insisted upon the artistic possibilities of camera-made imagery, but held that these could be realized only when art and science had been combined to serve both truth and beauty. Many Pictorialists tried to emulate painting and prints through soft-focus or darkroom manipulation. Too often, however, these attempts to create a heightened sense of poetry resulted in murky, sentimentalized facsimiles of etchings or lithographs.

Because Stieglitz maintained a strict "truth to materials" position, creating highly expressive images without the aid of

469. ALFRED STIEGLITZ. *The Street, Fifth Avenue.* 1896. Gravure,
12 x 9" (30.5 x 22.9 cm). Albright-Knox Art Gallery, Buffalo
GENERAL PURCHASE FUNDS, 1911

470. ALFRED STIEGLITZ. *The Steerage.* 1907. Chloride print, 28 x 23⅜"
(11 x 9.2 cm). The Art Institute of Chicago
ALFRED STIEGLITZ COLLECTION, 1949

darkroom enhancement, he was an important forerunner of so-called "straight" photography. The straight photographer exploits the intrinsic properties of the camera to make photographs that look like photographs instead of imitations of paintings or fine-art prints. Stieglitz joined the Pictorialists in maintaining that their works should be presented and received with the same kind of regard accorded other artistic media. In pursuit of this goal, progressive photographers broke away from the older camera societies, with their commercial and technological preoccupations, and initiated an international secession movement that provided spaces for meetings and exhibitions, stimulated enlightened critical commentary, and published catalogues and periodicals devoted to photography. The most distinguished and far-reaching in its impact was New York's Photo-Secession group established by Stieglitz in 1902 to promote "the serious recognition of photography as an additional medium of pictorial expression."

Stieglitz achieved his pictorial goals through his choice of subject, light, and atmosphere in the world about him, allowing natural fog or snow, for instance, to produce the desired effects of atmosphere and mystery (fig. 469). As this image of a snow-covered New York street illustrates, Stieglitz, like the Ashcan painters, drew inspiration from the spectacle of the modern metropolis. But rather than the noisy clutter of humanity envisioned by Sloan or Bellows, Stieglitz was drawn

to a lone cloaked figure silhouetted beneath the bare branches of a towering tree.

In 1907, during an ocean liner voyage to Europe, Stieglitz discovered little of social or visual interest among his fellow first-class passengers and wandered over to steerage, where the cheapest accommodations were located. As he later explained, he was struck by the combination of forms he witnessed among the crowds on deck—"a round straw hat, the funnel leaning left, the stairway leaning right, the white drawbridge with its railings made of circular chains, white suspenders crossing on the back of a man." Stieglitz quickly fetched his camera and took what he came to regard as one of his greatest photographs, *The Steerage* (fig. 470). Avoiding even the slightest Pictorialist sentiment or anecdote, Stieglitz provided a straight document of the scene, which he said was not merely a crowd of immigrants but "a study in mathematical lines . . . in a pattern of light and shade." This tendency to seek abstract form in the world around him led to Stieglitz's most daring series: cloud formations, or what he called "Equivalents" (fig. 471). These recordings of turbulent skies express, Stieglitz said, his "most profound experience of life."

Stieglitz's longtime associate in New York's Photo-Secession group, 291 gallery, and *Camera Work* magazine was EDWARD STEICHEN (1879–1973), a photographer first introduced in chapter 6 as the creator of the hauntingly

above: 471. ALFRED STIEGLITZ. *Equivalent.* 1930. Chloride print, 4¹¹⁄₁₆ x 3⅝" (11.9 x 9.2 cm). The Art Institute of Chicago THE ALFRED STIEGLITZ COLLECTION

right: 472. EDWARD STEICHEN. *Gloria Swanson.* 1924. Gelatin-silver print, 16¹⁄₁₆ x 13½" (42.1 x 34.2 cm). The Museum of Modern Art, New York GIFT OF THE PHOTOGRAPHER. COPY PRINT © 1995 THE MUSEUM OF MODERN ART, NEW YORK

moonlit time exposure of Rodin's *Balzac* (see fig. 139). After World War I, during which he worked as an aerial photographer, the artist abandoned his soft-focus Pictorialism to practice and promote straight, unmanipulated photography. Steichen became chief photographer for *Vogue* and *Vanity Fair,* publications in which his celebrity portraits and fashion photos appeared regularly from 1923 to 1938, exerting tremendous influence on photographers on both sides of the Atlantic (fig. 472). In 1947, Steichen was appointed director of the Department of Photography at the Museum of

Modern Art, a position he held until his retirement in 1962.

In a 1910 group show at 291, Stieglitz included works by **MAX WEBER** (1881–1961), a painter who had recently returned from three years in Paris. For an American, Weber was unusually knowledgeable about recent developments in French art and, by the time of Stieglitz's show, had begun to develop his own highly individual Cubist style. The 1915 *Chinese Restaurant* (fig. 473) is Weber's response to a friend's suggestion that he make a subject of New York's Chinatown. Among the wealth of colorful patterns are over-

473. MAX WEBER. *Chinese Restaurant.* 1915. Oil on canvas, 40 x 48" (101.6 x 121.9 cm). Whitney Museum of American Art, New York PURCHASE

lapping fragments of curtains, tile floors, and figures—colliding visual recollections of a bustling, urban interior. Toward the end of World War I, the artist abandoned Cubism for a form of expressionist figuration related to that of Chagall and Soutine.

American painter **MARSDEN HARTLEY** (1877–1943) came to Steiglitz's attention with the Post-Impressionist landscapes he made in Maine in 1907–9. His works were shown for the first time at 291 in 1909. From 1912 to 1915 Hartley lived abroad, first in Paris then in Berlin. In Germany he was drawn to the mystically based art of Der Blaue Reiter, and the first works he made there are clearly inspired by Kandinsky. Hartley regarded Berlin, the capital of imperial Germany, as the most vital urban center in Europe. In 1914–15 he made a series of paintings in which the imagery is derived from German flags, military insignia, and all the regalia of the imperial court. The works were not a celebration of German nationalism but rather Hartley's response to the lively pageantry of the city. *Portrait of a German Officer* (color plate 208, page 392) also contains coded references to the artist's lover, a German officer who was killed on the Western Front. Like Weber, Hartley began with motifs from the visible world, broke them up, and reorganized them into a largely abstract composition. The exuberant energy, brilliant colors, and bold arrangement of forms are evidence of his interest in the Orphist paintings of Delaunay (color plate 101, page 203). In 1916, as war raged in Europe, Hartley's German works were shown at 291, where they were regarded by some reviewers as treacherously pro-German. These works would influence experimental American painters from Stuart Davis to Robert Indiana and Jasper Johns (see figs. 507, 637, 613). Hartley continued to travel widely after the war and in his later years concentrated primarily on Cézanne-inspired landscapes in Nova Scotia and in his native Maine.

JOHN MARIN (1870–1953). Of the other artists Stieglitz promoted—John Marin, Arthur Dove, and Georgia O'Keeffe—Marin is known for his superb skill as a watercolorist. Following his art studies in Philadelphia and New York, Marin spent most of the period from 1905 to 1910 in Europe. His introduction to Stieglitz in 1909 marked the beginning of a long and fruitful relationship. The late watercolors of Cézanne were an important source for Marin, and Cubist structure remained a controlling force in the cityscapes and landscapes he painted throughout his life. During the 1920s and 1930s he carried his interpretations of New York close to a form of expressionist abstraction, although his subject is almost always discernible. In the 1922 *Lower Manhattan* (color plate 209, page 409), an explosion of buildings erupts from a black circular form containing a sunburst center, a cutout form Marin attached to the paper with thread. He reinforced the transparent, angular strokes of the watercolor with charcoal, creating a work of graphic force that vibrates with the clamor and speed of the city. In summer, Marin painted the Maine coast. By 1949, when he composed the highly expressionist seascape, *The Fog Lifts* (color plate 210, page 409)—where the fog takes the abstract form

474. ARTHUR G. DOVE. *Nature Symbolized No. 2.* c. 1911.
Pastel on paper, 18 x 21⅝" (45.7 x 54.9 cm).
The Art Institute of Chicago ALFRED STIEGLITZ COLLECTION, 1949

of white rectangles—Marin had been voted the best painter in America in a poll of art-world professionals conducted by *Look* magazine.

ARTHUR G. DOVE (1880–1946) temporarily abandoned a career as an illustrator to go to Paris in 1907 in hopes of becoming a full-time painter. With astonishing speed, he seems to have absorbed much of the course of European modernism. The group of pastels he exhibited at 291 in 1912, including figure 474, were even closer to nonobjectivity than contemporaneous paintings by Kandinsky (color plate 76, page 160), whose work Dove knew, and were probably the earliest and most advanced statements in abstract art by an American artist. Throughout his life Dove was concerned more with the spiritual forces of nature than its external forms, and as his title suggests, the curving shapes of his pastels were inspired by the natural world. He later called his group of abstract pastels *The Ten Commandments*, as if together they constituted his spiritual and artistic credo.

Between 1924 and 1930, Dove produced a remarkable group of multimedia assemblages, including *Goin' Fishin'* (fig. 475), composed of fragmented fishing poles and denim shirtsleeves. The artist was aware of the evolution of Cubist collage, but his own efforts have more in common with nineteenth-century American folk art, trompe l'oeil painting, and Dada collage, though Dove never adopted the subversive, anti-art stance of the first Dadaists. His own collages range from landscapes made up literally of their natural ingredients—sand, shells, twigs, and leaves—to delicate and magical papiers collés. *Goin' Fishin'* can be seen in relation to the fascinating constructions of Joseph Cornell (see fig. 548), and even to the so-called "combine paintings" of Robert Rauschenberg (see fig. 611).

During the 1930s Dove continued to paint highly abstracted landscapes in which concentric bands of radiating color and active brushwork express both a quality of intense

475. ARTHUR G. DOVE. *Goin' Fishin'*. 1925.
Collage on wood panel, 19½ x 24" (49.5 x 61 cm).
The Phillips Collection, Washington, D.C.

light and the artist's sense of spiritual energy in nature. In the last years of his life he experimented with bold, geometric shapes, such as those in *That Red One* of 1944 (color plate 211, page 409), where the connection to the observable world is virtually severed.

The last exhibition at 291, held in 1917, was the first solo show of work by GEORGIA O'KEEFFE (1887–1986). O'Keeffe differed from most of the other American pioneers of modernism in that she never trained in Europe. Hartley said she was simply "modern by instinct," although she had been exposed, especially through exhibitions at 291, to the work of the major European modernists. The biomorphic, Kandinsky-like abstractions in charcoal that she made in 1915–16 so impressed Stieglitz that he soon gave her an exhibition. The two artists fell in love and married in 1924. Before he put away his camera for good in 1937, Stieglitz made more than 300 photographs of O'Keeffe, contributing to her almost cult status in twentieth-century American art.

Whether her subjects were New York skyscrapers, enormously enlarged details of flowers, bleached cow skulls, or adobe churches, they have become icons of American art. Her paintings involve such economy of detail and such skillful distillation of her subject, that, however naturalistically precise, they somehow become works of abstraction. Alternatively, her abstractions have such tangible presence that they suggest forms in nature. *Music—Pink and Blue, II* (color plate 212, page 410) is a breathtaking study in chromatic relationships and organic form. The title suggests the lingering influence of Kandinsky's equation of color with music and, ultimately, with emotion. Typically ambivalent, the forms resemble an enlarged close-up of the flowers O'Keeffe painted in the 1920s, although some have seen them as sexually charged. The artist rejected such readings of

her imagery, and partly in response to such interpretations, her work became more overtly representational in the 1920s, as in her tightly rendered views of New York skyscrapers made from the window of her Manhattan apartment (color plate 213, page 410). In 1929 O'Keeffe visited New Mexico for the first time and thereafter divided her time between New York and the Southwest, settling permanently in New Mexico in 1949. The desert motifs she painted there count among her best-known images (fig. 476).

Three important American photographers associated like O'Keeffe with the West are PAUL STRAND (1890–1976), Imogen Cunningham, and Ansel Adams. Photographer and filmmaker Paul Strand studied in his native New York City with Lewis Hine. He officially entered the Stieglitz circle in 1916 when his photographs were reproduced in *Camera Work* and he exhibited at 291 for the first time. By this time Strand had abandoned the Pictorialism of his early work and was making photographs directly inspired by Cubist painting. In 1917 he contributed an essay to the last issue of *Camera Work* in which he called for photography as an objective art "without tricks of process or manipulation." In 1930–32 Strand worked in New Mexico and photographed the same church that his friends O'Keeffe and Adams had already depicted (fig. 477). All three artists were drawn to

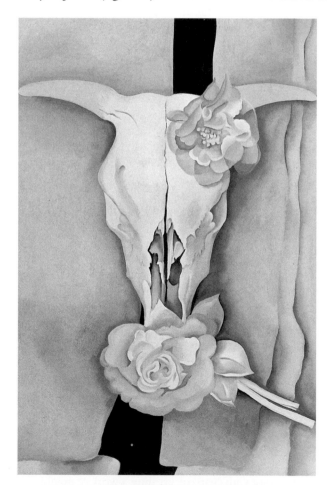

476. GEORGIA O'KEEFFE. *Cow's Skull with Calico Roses*. 1931.
Oil on canvas, 35⅞ x 24" (91.2 x 61 cm).
The Art Institute of Chicago

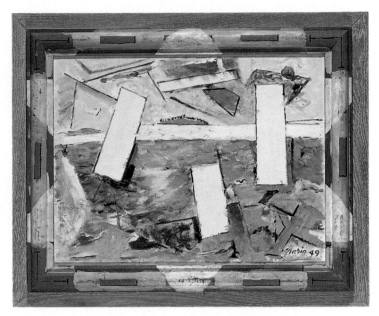

Color plate 209. JOHN MARIN. *Lower Manhattan (Composition Derived from Top of Woolworth Building)*. 1922. Watercolor and charcoal with paper cutout attached with thread on paper, 21⅝ x 26⅞" (54.9 x 68.3 cm). The Museum of Modern Art, New York
ACQUIRED THROUGH THE LILLIE P. BLISS BEQUEST

Color plate 210. JOHN MARIN. *The Fog Lifts*. 1949. Oil on canvas, 22 x 28" (55.9 x 71.1 cm). Wichita Art Museum, Kansas
THE ROLAND P. MURDOCK COLLECTION

Color plate 211. ARTHUR G. DOVE. *That Red One*. 1944. Oil on canvas, 27 x 36" (68.9 x 91.4 cm). William H. Lane Foundation, Leominster, Massachusetts

right: Color plate 213. GEORGIA O'KEEFFE. *Radiator Building—Night, New York*. 1927. Oil on canvas, 48 x 30" (121.9 x 76.2 cm). Carl van Vechten Gallery of Fine Arts, Fisk University, Nashville, Tennessee

ALFRED STIEGLITZ COLLECTION

© 1997 THE GEORGIA O'KEEFFE FOUNDATION/ARTISTS RIGHTS SOCIETY (ARS), NEW YORK

above: Color plate 212. GEORGIA O'KEEFFE. *Music—Pink and Blue, II*. 1919. Oil on canvas, 35½ x 29" (90.2 x 73.7 cm). Whitney Museum of American Art, New York

GIFT OF EMILY FISHER LANDAU IN HONOR OF TOM ARMSTRONG

© 1997 THE GEORGIA O'KEEFFE FOUNDATION/ARTISTS RIGHTS SOCIETY (ARS), NEW YORK

right: Color plate 214. STANTON MACDONALD-WRIGHT. *Abstraction on Spectrum (Organization, 5)*. c. 1914. Oil on canvas, 30⅛ x 24³⁄₁₆" (76.5 x 61.4 cm). Des Moines Art Center, Iowa

PURCHASED WITH FUNDS FROM THE COFFIN FINE ARTS TRUST, NATHAN EMORY COFFIN MEMORIAL COLLECTION

right: Color plate 215.
MORGAN RUSSELL.
*Synchromy in Orange:
To Form.* 1914.
Oil on canvas with painted
frame, 11'3" x 10'3"
(3.4 x 3.1 m).
Albright-Knox Art Gallery,
Buffalo, New York

GIFT OF SEYMOUR H. KNOX

below: Color plate 216.
CHARLES SHEELER.
Church Street El. 1920.
Oil on canvas,
16 x 19⅛"
(40.6 x 48.6 cm).
Cleveland Museum of Art

MR. AND MRS. WILLIAM H.
MARTLATT FUND

above: Color plate 217.
CHARLES DEMUTH. *I Saw the Figure 5 in Gold.* 1928.
Oil on composition board, 36 x 29¾" (91.4 x 75.6 cm).
The Metropolitan Museum of Art, New York
THE ALFRED STIEGLITZ COLLECTION, 1949

above, right: Color plate 218.
CHARLES DEMUTH. *Buildings, Lancaster.* 1930.
Oil on composition board, 24 x 20" (61 x 50.8 cm).
Whitney Museum of American Art, New York
ANONYMOUS GIFT

right: Color plate 219.
THOMAS HART BENTON.
City Building, from the mural series America Today. 1930.
Distemper and egg tempera on gessoed linen with oil glaze, 7'8" x 9'9" (2.3 x 3 m). Originally in the New School for Social Research, New York. Relocated in 1984. Collection the Equitable Life Assurance Society of the United States
© 1997 T.H. BENTON AND R.P. BENTON
TESTAMENTARY TRUSTS/LICENSED BY VAGA, NEW YORK, NY

right: Color plate 220.
GRANT WOOD. *Young Corn.*
1931. Oil on masonite
panel, 24 x 29⅞"
(61 x 75.9 cm).
Community School District,
Cedar Rapids, Iowa

Color plate 221. EDWARD HOPPER. *Early Sunday Morning.* 1930. Oil on canvas, 35 x 60" (88.9 x 152.4 cm).
Whitney Museum of American Art, New York

Color plate 222. HORACE PIPPIN. *Domino Players*. 1943. Oil on composition panel, 12¾ x 22" (32.4 x 55.9 cm).
The Phillips Collection, Washington, D.C.

Color plate 223. BEN SHAHN. *Liberation*. 1945. Tempera on cardboard mounted on composition board,
29¾ x 40" (75.6 x 101.6 cm). The Museum of Modern Art, New York JAMES THRALL SOBY BEQUEST © 1997 ESTATE OF BEN SHAHN/VAGA, NEW YORK, NY

Color plate 224. PETER BLUME.
The Eternal City. 1934–37
(dated 1937 on painting).
Oil on composition board,
34 x 47⅞" (86.4 x 121.6 cm).
The Museum of Modern Art,
New York

Color plate 225.
ROMARE BEARDEN. *The Dove.*
1964. Cut-and-pasted paper,
gouache, pencil, and colored
pencil on cardboard,
13⅜ x 18¾"
(34 x 47.6 cm).
The Museum of Modern Art,
New York

Color plate 226.
JACOB LAWRENCE.
No. 1, from The Migration of
the Negro Series. 1940–41.
Tempera on hardboard,
12 x 18" (30.5 x 45.7 cm).
The Phillips Collection,
Washington, D.C.

Color plate 227. DIEGO RIVERA. *Flower Day*. 1925. Oil on canvas, 58 x 47½" (147.4 x 120.6 cm). Los Angeles County Museum of Art LOS ANGELES COUNTY FUNDS

Color plate 228. FRIDA KAHLO. *Self-Portrait with Thorn Necklace*. 1940. Oil on canvas, 24 x 18¾" (61 x 47.6 cm). Art Collection, Harry Ransom Humanities Research Center, University of Texas, Austin

Color plate 229. STUART DAVIS. *Report from Rockport*. 1940. Oil on canvas, 24 x 30" (61 x 76.2 cm). The Metropolitan Museum of Art, New York MILTON AND EDITH A. LOWENTHAL COLLECTION. BEQUEST OF EDITH ABRAHAMSON LOWENTHAL

477. Paul Strand. *Ranchos de Taos Church, New Mexico.* 1931.
Gelatin-silver print © Aperture Foundation, Inc., Paul Strand Archive

478. Imogen Cunningham. *Two Callas.* 1929.
Gelatin-silver print © Imogen Cunningham Trust, 1970

the desert sun's ability to accentuate the abstract qualities and dignified monumentality of the adobe structure.

From her origins in soft-focus Pictorialism, **IMOGEN CUNNINGHAM** (1883–1976), like Strand, shifted to straight photography, producing a series of plant studies (fig. 478), in which blown-up details and immaculate lighting yield an image analogous to the near-abstract flower paintings of Georgia O'Keeffe. In 1932, along with Ansel Adams, Edward Weston, and other photographers in the San Francisco Bay area, Cunningham helped found Group f.64, a loosely knit organization devoted to "straight" photography as an art form. Their name refers to the smallest aperture on a large-format camera, producing sharp focus and great depth of field.

ANSEL ADAMS (1902–1984) discovered his passion for nature at the age of fourteen during a visit to California's Yosemite Valley, but it was his initial meeting with Paul Strand in 1930 that decided his career in photography. Three years later he met and engaged the support of Alfred Stieglitz. A masterful technician, Adams revealed himself to be an authentic disciple of Stieglitz when he wrote: "A great photograph is a full expression in the deepest sense, and is, thereby, a true expression of what one feels about life in its entirety. And the expression of what one feels should be set forth in terms of simple devotion to the medium—a statement of the utmost clarity and perfection possible under the conditions of creation and production." Adams's art is synonymous with spectacular unspoiled vistas of the West. In figure 479 he chose a close-up angle of the wintry landscape, filling the frame to reveal the textures of the earth. An experienced mountaineer and dedicated conservationist, Adams was an active member of the Sierra Club by the late 1920s. He helped establish the Department of Photography at New York's Museum of Modern Art in 1940 and the Center for Creative Photography at the University of Arizona, Tucson, in 1975.

479. Ansel Adams. *Frozen Lakes and Cliffs, The Sierra Nevada, Sequoia National Park, California.* 1932. Gelatin-silver print
Courtesy the Trustees of the Ansel Adams Publishing Rights Trust. All rights reserved

The Armory Show: 1913

In the early part of the twentieth century, an important catalyst affecting the subsequent history of American art was the International Exhibition of Modern Art held in New York at the 69th Regiment Armory on Lexington Avenue at 25th Street between February 17 and March 15, 1913. Organized by a group known as the Association of American Painters and Sculptors, the exhibition was generally intended to show a large number of European and American artists and to compete with the regular exhibitions of the National Academy of Design. Arthur B. Davies, a member of The Eight respected by both the Academy and the independents and highly knowledgeable in the international art field, was chosen as chairman and, with the painter Walt Kuhn, did much of the selection. They were assisted in several European cities by the painter and critic Walter Pach. Glackens headed up the selection of American works. The Armory Show proved to be a monumental though uneven exhibition focusing on the nineteenth- and early-twentieth-century painting and sculpture of Europe. The artists included ranged from Goya to Delacroix to Daumier, Courbet, Manet, the Impressionists, Van Gogh, Gauguin, Cézanne, Redon, the Nabis, and Seurat and his followers. Matisse and the Fauves were represented, as were Picasso, Braque, and the Cubists. German Expressionists were slighted, while the Orphic Cubists and the Italian Futurists withdrew. Stieglitz purchased the only Kandinsky in the show. Among sculptors, the Europeans Rodin, Maillol, Brancusi, Nadelman, and Lehmbruck, as well as the American Lachaise, were included, although sculpture generally received far less representation than painting. American painting of all varieties and quality dominated the show in sheer numbers, but failed in its impact relative to that of the exotic imports from Europe.

The Armory Show created a sensation, and while the lay press covered it extensively (at first with praise), the affair was savagely attacked by critics and American artists. Matisse was singled out for abuse, as were the Cubists, while Duchamp's *Nude Descending a Staircase* (color plate 105, page 205), it will be recalled, enjoyed a succès de scandale and was likened to a "pile of kindling wood."

As a result of the controversy, an estimated 75,000 people attended the exhibition. At Chicago's Art Institute, where the European section and approximately half the American section were shown, nearly 200,000 visitors crowded the galleries. Here the students of the Art Institute's school, egged on by their faculty, threatened to burn Matisse and Brancusi in effigy. In Boston, where some 250 examples of the European section were displayed at Copley Hall, the reaction was more tepid.

The Armory Show, an unprecedented achievement in this country, was a powerful impetus for the advancement of modernism in America. Almost immediately, there was evidence of change in American art and collecting. New galleries dealing in modern painting and sculpture began to appear. Although communications were soon to be interrupted by World War I, more American artists than ever were inspired to go to Europe for study. American museums began to buy and to show modern French masters, and a small but influential new class of collectors, which included Dr. Albert Barnes, Arthur Jerome Eddy, Walter Arensberg, Lillie P. Bliss, Katherine Dreier, and Duncan Phillips, among others, was born. From their holdings would come the nuclei of the great public collections of modern art that Americans know today.

Synchromism

In the year of the Armory Show, two Americans living in Paris, STANTON MACDONALD-WRIGHT (1890–1973) and MORGAN RUSSELL (1886–1953), launched their own movement, which they called Synchromism. The complicated theoretical basis of Synchromism began with the principles of French color theorists such as Chevreul, Rood, and Blanc. The color abstractions, or Synchromies, that resulted were close to the works of Delaunay, although the ever-competitive Americans vehemently differentiated their work from the Frenchman's. Their 1913 exhibitions in Munich and Paris were accompanied by brash manifestos, followed by a show the following year in New York. Macdonald-Wright, whose *Abstraction on Spectrum (Organization, 5)* (color plate 214, page 410) clearly owes much to Delaunay's earlier disk paintings (see color plate 101, page 203), wrote in 1916, "I strive to divest my art of all anecdote and illustration and to purify it to the point where the emotions of the spectator will be wholly aesthetic, as when listening to good music." Macdonald-Wright returned to the United States in 1916 and showed his work at 291 the following year. He moved to California in 1918, the year that more or less witnessed the end of Synchromism, and although he largely returned to figuration after 1920, his subsequent work was still informed by his early Synchromist experiments.

Russell, on the other hand, spent most of his life in France after 1909. His most important Synchromist work is the monumental *Synchromy in Orange: To Form* (color plate 215, page 411), which was included in the 1914 Salon des Indépendants in Paris. Although the densely packed, curving planes of brilliant color seem to construct a wholly abstract composition, the central planes actually contain the forms of a man, possibly Michelangelo's *Dying Slave* in the Louvre, after which Russell—always interested in sculpture—had made drawings. By virtue of juxtaposed advancing and receding colors, the painting has a distinctive sculptural appearance, an aspect that sets his work apart from that of Delaunay.

The Precisionists

A major manifestation of a new spirit in American art during the 1920s was the movement called "Precisionism," a term that was not coined until the 1940s. Never organized into a coherent group with a common platform, the artists involved have also been labeled "Cubo-Realists" and even "the Immaculates." All these terms refer to an art that is basically descriptive, but guided by geometric simplification that stems partly from Cubism. Precisionist paintings tend to be stripped of detail and are often based on sharp-focus photographs. The movement is in some ways the American equivalent of *Neue Sachlichkeit* or New Objectivity, the realist style practiced in Germany after the war (see chapter 13). The Germans, who had witnessed the horrors of war firsthand, tended to focus on politically charged images of the

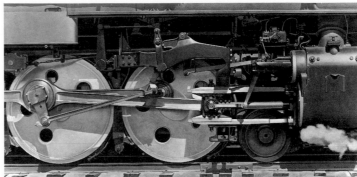

left: 480. CHARLES SHEELER. *Drive Wheels*. 1939. Gelatin-silver print, 6¹¹⁄₁₆ x 9¹¹⁄₁₆" (17 x 24.5 cm). Smith College Museum of Art, Northampton, Massachusetts GIFT OF DOROTHY C. MILLER '25 (MRS. HOLGER CAHILL), 1978

481. CHARLES SHEELER. *Rolling Power*. 1939. Oil on canvas, 15 x 30" (38.1 x 76.2 cm). Smith College Museum of Art, Northampton, Massachusetts PURCHASED, DRAYTON HILLYER FUND, 1940

human figure, while the Americans preferred the more neutral subjects of still life, architecture, and the machine. Given their hard-edge style and urban theme, Georgia O'Keeffe's New York views of the 1920s could be considered Precisionist (color plate 213, page 410), but Charles Demuth and, especially, Charles Sheeler, were the key formulators of the style.

CHARLES SHEELER (1883–1965), a Philadelphia-born and -trained artist, began working as a professional photographer around 1910, making records of new houses and buildings for architectural firms. In 1919 he moved to New York, where he had already befriended Stieglitz and Strand, and continued to participate in the salons organized by the collectors Walter and Louise Arensberg. There he met Duchamp, Man Ray, Picabia, and other artists associated with New York Dada. Sheeler photographed skyscrapers—double exposing, tilting his camera, and then transferring these special effects and the patterns they produced to his paintings. In 1920 he collaborated with Strand on a six-minute film entitled *Manahatta*, using a kaleidoscopic technique to celebrate the dizzying effect of life surrounded by a dense forest of skyscrapers. Sheeler's 1920 painting *Church Street El* (color plate 216, page 411), representing a view looking down from a tall building in Lower Manhattan, was based on a still from the film. In its severe planarity and elimination of details, *Church Street El* is closer to geometric abstraction than to the shifting viewpoints of Cubist painting. By cropping the photographic image, Sheeler could be highly selective in his record of the details provided by reality and thus create arbitrary patterns of light and shadow and flat color that transform themselves into abstract relationships. At the right in *Church Street El,* amidst the deep shadows and bright, geometric patches of sun, an elevated train moves along its tracks.

The relationship between Sheeler's camera work (he stopped making strictly commercial photographs in the early 1930s), and his canvases seems to have been virtually symbiotic, as can be seen in a comparison of the photograph entitled *Drive Wheels* and the painting known as *Rolling Power* (figs. 480, 481). At first glance, the work in oil comes across as an almost literal transcription of the photograph, which the

artist took in preparation for the painting. But Sheeler altered the composition and suppressed such details as the grease on the engine's piston box, all in keeping with Precisionism's love of immaculate surfaces and purified machine imagery. Sheeler made this painting for *Fortune* magazine as part of a commissioned series called "Power," consisting of six "portraits" of power-generating machines. Much of Sheeler's work in the 1930s and 1940s was a celebration of the modern industrial age, with majestic views of factories, power plants, and machines.

CHARLES DEMUTH (1883–1935), also a Pennsylvanian, was the other major figure of Precisionism. He completed his studies at the Pennsylvania Academy in 1910 and, during a trip to Paris in 1912, befriended Hartley, who became an important mentor. Demuth's first mature works were book illustrations and imaginative studies of dancers, acrobats, and flowers, free and organic in treatment, and seemingly at the other extreme from a Precisionist style. The work of Cézanne was an important guiding force during these years. Like Marin, Demuth was a consummate watercolorist, but where Marin used the medium for great drama and expressive purpose, Demuth applied his delicate washes of color with restraint. "John Marin and I drew our inspiration from the same sources," he once said. "He brought his up in buckets and spilled much along the way. I dipped mine out with a teaspoon but I never spilled a drop." In a number of landscapes executed in watercolor in Bermuda in 1917, using interpenetrating and shifting planes and suggestions of multiple views (fig. 482), Demuth began his experiments with abstract lines of force, comparable to those used by the Futurists (see fig. 241).

In the 1920s Demuth produced a series of emblematic portraits of American artists. Made up of images, words, and letters, these "posters," as Demuth called them, were related to some of the Dadaists' symbolic portraits, including Picabia's (see fig. 307). They led to what is probably Demuth's greatest painting, *I Saw the Figure 5 in Gold* (color

482. CHARLES DEMUTH. *Bermuda No. 2—The Schooner*. 1917.
Watercolor and pencil on paper, 10 x 13⅞" (25.4 x 35.2 cm).
The Metropolitan Museum of Art, New York
THE ALFRED STIEGLITZ COLLECTION, 1949

plate 217, page 412). The work is a tribute to his friend, the
poet William Carlos Williams; in fact, it is based on "The
Great Figure," a poem Williams wrote after seeing a roaring,
red fire engine emblazoned with a golden number 5:

> Among the rain
> and lights
> I saw the figure 5
> in gold
> on a red
> fire truck
> moving
> tense
> unheeded
> to gong clangs
> siren howls
> and wheels rumbling
> through the dark city.

Other work in the 1920s and 1930s included exquisite
watercolor still lifes as well as motifs drawn from the modern
industrial landscape. Demuth's last major works were oil paintings of vernacular architecture in his hometown, Lancaster,
Pennsylvania—its watertowers, factories, and grain elevators
(color plate 218, page 412). Like other artists in the postwar
era, Demuth was searching for intrinsically American subjects
and became interested in the advertising signs on buildings
and along highways, not as blemishes on the American landscape but as images with a certain abstract beauty of their own.
Demuth's work influenced that of somewhat younger artists,
such as Stuart Davis, who avowed his debt to the older artist,
as well as that of future generations, including Jasper Johns
and, quite specifically, Robert Indiana (see fig. 637).

American Scene Painting

The 1930s, a decade that opened after the Wall Street crash
of 1929 and closed with the onset of war in Europe, was a
period of economic depression and political liberalism in the
United States. Reflecting the spirit of isolationism that dominated so much of American thought and action after World
War I, art became socially conscious, nationalist, and regionalist. One of the seminal books of the decade was John
Steinbeck's *Grapes of Wrath* (1939), a Pulitzer Prize winning
novel celebrating the quiet heroism of Oklahoma Dust Bowl
victims during their migration to California. Among painters,
the Regionalists and Social Realists became the guardians of
conservatism, fighting a rearguard battle against modernism.
These artists concentrated on intrinsically "American"
themes in their painting, whether chauvinistic praise of the
virtues of an actually declining agrarian America, or bitter
attacks on the American political and economic system that
had produced the sufferings of the Great Depression.

A crucial event of this decade—indeed, a catalyst for the
development of American art transcending even the Armory
Show—was the establishment, by the United States government, of a Federal Art Project, a subdivision of the WPA
(Works Progress Administration). Just as the WPA provided
jobs for the unemployed, the Art Project enabled many of
the major American painters to survive during these difficult
times. A large percentage of the younger artists who created
a new art in America after World War II might never have
had a chance to develop their art had it not been for the
government-sponsored program of mural and easel painting.

The decade, however, witnessed little of an abstract tendency in American art. There were many artists of talent, and
many interesting directions in American art, some of them
remarkably advanced. Those who claimed the most attention
throughout the era were the painters of the so-called
American Scene, an umbrella term covering a wide range of
realist painting, from the more reactionary and nationalistic
Regionalists, through the generally left-wing Social Realists,
to what might be called the Magic Realists. Divergent as their
sociopolitical concerns may have been, the several factions
merged in their common preference for illustrational styles
and their contempt for "highbrow" European formalism.

The Regionalists

Dominant among the American Scene artists were the
Regionalists THOMAS HART BENTON (1889–1975), John
Steuart Curry, and Grant Wood. Of these artists, Missouri-
born Benton was the most ambitious, vocal, and ultimately
influential. Benton moved in modernist circles in New York
in the teens and twenties and even made abstract paintings
based on his experiments with Synchromism in Paris. By
1934, however, when his fame was such that his self-portrait
was reproduced on the cover of *Time* magazine, the artist had
vehemently rejected modernism. "I wallowed," he once said,
"in every cockeyed ism that came along, and it took me ten
years to get all that modernist dirt out of my system." Benton
was the darling of the right-wing critic Thomas Craven, who
promoted the Regionalists as the great exemplars of artistic
nationalism, and an object of scorn among leftist artists such
as Stuart Davis.

In 1930 Benton executed a cycle of murals on the theme

483. THOMAS HART
BENTON.
*The Ballad of the
Jealous Lover of Lone
Green Valley.* 1934.
Oil on masonite
panel, 42¼ x 53¼"
(107.3 x 135.3 cm).
The Spencer Museum
of Art, University of
Kansas, Lawrence

of modern technology for the New School for Social Research in New York. Called America Today, the series presents a cross section of working America through images based on the artist's travel sketches. Benton presented his vast subject by means of compositional montage, as in *City Building* (color plate 219, page 412), where heroic construction workers are seen against a backdrop of New York construction sites. Benton turned to the sixteenth-century art of Michelangelo and El Greco for the sinewy anatomies and mannered poses of his figures.

Benton earned the distinction of being the onetime teacher, at New York's Art Students League, of Jackson Pollock, who would transform the older master's rugged individualism, energy, and mural scale into the radical brand of modernism Benton so detested. A young Pollock (playing the harmonica) is depicted among the country musicians in Benton's 1934 painting *The Ballad of the Jealous Lover of Lone Green Valley* (fig. 483). As the old fiddler plays, a tale of jealousy and violence is enacted behind, and the landscape heaves up in abstract evocation of his music. Even in Benton's renditions of such traditional, rural themes, the lessons of Synchromism linger.

Although he studied in Paris for a time in 1923, Iowan **GRANT WOOD** (1892–1942) was never tempted by European modernist styles as was his colleague Benton. More formative was his trip in 1928 to Munich, where he discovered the art of the fifteenth-century Flemish and German primitives. The art from this historical period was also experiencing a resurgence of interest among some German painters, whose work had much in common with Wood's. Adapting this stylistic prototype to the Regionalist concerns he shared with Benton, Wood produced his most celebrated painting, *American Gothic* (fig. 484), which soon became a

national icon. Wood was initially drawn to the nineteenth-century "Carpenter Gothic" house that is depicted in the portrait's background. The models for the farmer couple (not husband and wife but two unrelated individuals) were actually Wood's sister and an Iowan dentist. The marriage of a

484. GRANT WOOD. *American Gothic.* 1930.
Oil on beaverboard, 29⅞ x 24⅞" (75.9 x 63.2 cm).
The Art Institute of Chicago

miniaturist, deliberately archaizing style with contemporary, homespun, Puritan content drew immediate attention to the painting, though many wondered if Wood intended to satirize his subject. Wood, who was certainly capable of pictorial satire, said the subjects were painted out of affection rather than ridicule.

Wood applied his exacting style to landscape as well (color plate 220, page 413), but his harmonious vision of cultivated nature, with its swelling hills, tidy crops, and delightfully puffy trees, was based on childhood recollections. Modern technology, not to mention the harsh realities of Depression-era drought and dust bowls, has no place here.

A more romantic approach to Regionalism was taken by the Ohio artist CHARLES BURCHFIELD (1893–1967). Following an early psychological crisis, Burchfield tended to project chronic fears and obsessions into his brooding paintings of rural and small-town America. Of *November Evening* (fig. 485), he said, "I have attempted to express the coming of winter over the Middle West as it must have felt to the pioneers—great black clouds sweep out of the west at twilight as if to overwhelm not only the pitiful human attempt at a town, but also the earth itself."

By contrast with the midwestern Regionalists, EDWARD HOPPER (1882–1967) was a mainly urban, New York artist. While he admired Burchfield's ability to discover a "hideous beauty" in the vulgar, eclectic jumble of the American landscape, he rejected any association with Benton and his fellow Regionalists, who, he said, caricatured America in their paintings. Hopper, a student of Henri, made several long visits to France between 1906 and 1910, at the exact moment of the Fauve explosion and the beginnings of Cubism. Seemingly unaffected by either of them, Hopper early on established his

basic themes of hotels, restaurants, or theaters, as well as their lonely inhabitants.

Much of the effect of Hopper's paintings is derived from his sensitivity to light—the light of early morning or of twilight, the dreary light filtered into a hotel room or office, or the garish light of a lunch counter isolated within the surrounding darkness. He could even endow a favorite landscape motif—the lighthouses of Cape Cod or Maine—with a quiet eeriness (fig. 486). One of his finest works, *Early Sunday Morning* (color plate 221, page 413) depicts a row of buildings on a deserted Seventh Avenue in New York, where a sharp, raking light casts deep shadows and a strange stillness over the whole. The flat façade and dramatic lighting have been linked to Hopper's interest in stage-set design. Despite his reputation as one of America's premier realist painters, Hopper brought a rigorous sense of abstract design to his compositions. In a late work from 1955, *Carolina Morning* (fig. 487), a woman in a red dress and hat stands in a shaft of intense light. Beyond her the seemingly endless landscape is described with great economy in almost geometric terms.

Two self-taught artists whose portrayals of American life, albeit from very different perspectives, found common ground with American Scene painting are Anna Mary Robertson Moses, known as GRANDMA MOSES (1860–1961) and Horace Pippin. The quintessential untutored artist whose work was steeped in the strong tradition of American folk art, Grandma Moses graduated to oil painting after making pictures in yarn. Her innate gift for creating charming, nostalgic landscapes eventually attracted collectors and critics, and in 1939 she was included in an exhibition called "Contemporary Unknown American Painters" at The Museum of Modern Art. Ironically, her work (fig. 488) began to attract

485. CHARLES BURCHFIELD. *November Evening.* 1934. Oil on canvas, 32⅛ x 52" (81.6 x 132.1 cm).
The Metropolitan Museum of Art, New York GEORGE A. HEARN FUND, 1934

486. EDWARD HOPPER. *Lighthouse Hill.* 1927.
Oil on canvas, 28¼ x 39½" (71.8 x 100.3 cm).
Dallas Museum of Art
GIFT OF MR. AND MRS. MAURICE PURNELL, 1958

487. EDWARD HOPPER. *Carolina Morning.* 1955. Oil on canvas,
30 x 40⅛" (76.2 x 102 cm).
Whitney Museum of American Art, New York

considerable attention in the 1940s and 1950s, just as the abstract art of the New York School came to dominate the international scene. This proves that there was a large sector of the American public that preferred this art to the more radical statements of the avant-garde. By her death at the age of 101, her name had long been a household word.

Like Grandma Moses, the autodidact **HORACE PIPPIN** (1888–1946) took up painting late in life. A black artist from West Chester, Pennsylvania, Pippin worked in isolation until his work was first exhibited in the late 1930s. He began to attract considerable attention at a time of growing interest in the work of self-taught artists. Pippin's range of subjects was

broad, from portraits, landscapes, and genre themes to a series on the life of abolitionist John Brown, as well as scenes based on his experiences as an infantryman in World War I, when his right arm, with which he later painted, was permanently injured. *Domino Players* (color plate 222, page 414), based on Pippin's childhood memories, shows the artist seated at a table between his female relatives. In such domestic scenes, Pippin offered a rare glimpse into an African-American household in the North around the turn of the century. Although he does not exclude the less picturesque aspects from his painting, he orchestrates the composition around stylized silhouettes and decorative color.

488. GRANDMA MOSES. *Checkered House.* 1943. Oil on canvas, 36 x 45"
(91.4 x 114.3 cm). IBM Corporation, Armonk, New York

489. RAPHAEL SOYER. *Office Girls.* 1936. Oil on canvas,
26 x 24" (66 x 61 cm).
Whitney Museum of American Art, New York PURCHASE

The Social Realists

Also active in American Scene were such confirmed urban
realists as **RAPHAEL SOYER** (1899–1987) and **ISABEL
BISHOP** (1902–1988), both of whom portrayed and digni-

490. ISABEL BISHOP. *Waiting.* 1938. Oil and tempera on gesso
panel, 29 x 22½" (73.7 x 57.2 cm).
Newark Museum, New Jersey
PURCHASE 1944, ARTHUR F. AGNER MEMORIAL COMMITTEE

fied Depression-wracked New York and its unemployed or
overworked and underpaid masses. Soyer captured the fragility and stoic forbearance of tired office workers preserving
their selfhood within the anonymous crowd pressing along
the city streets (fig. 489). Bishop, who had studied at the Art
Students League while still a teenager, worked for a time on
the Federal Arts Project, which employed a large number of
women artists. She made keenly observed sketches of secre-
taries and stenographers relaxing during their breaks in
Union Square, which she worked into paintings back in her
studio. Through a technically complex method of oil and
tempera, which she built up layer upon layer, she brought to
her humble subjects all the humanity of Rembrandt (fig.
490). Bishop's words of tribute to her friend, the painter
Reginald Marsh, could easily pertain to her own work. She
said he portrayed "little people, in unheroic situations" that
are "modeled in the grand manner."

Foremost among the Social Realist painters of the 1930s
was **BEN SHAHN** (1898–1969), whose work as an artist was
inextricably tied to his social activism. His paintings, murals,
and posters inveighed against fascism, social injustice, and the
hardships endured by the working-class poor. An accomplished photographer, Shahn often based his paintings on
his own photographs, although the matte, thinly brushed
tempera surfaces of his canvases have little to do with the
photographic realism of Sheeler, for example. Shahn gained
recognition in the early 1930s through his paintings on the
notorious case of Sacco and Vanzetti. These were bitter
denunciations of the American legal system, which had condemned the two Italian-American anarchists to death in 1927
(fig. 491). With the title *The Passion of Sacco and Vanzetti*,
Shahn obviously meant to draw a connection to Christian
martyrdom. *Liberation* (color plate 223, page 414) depicts
vacant-eyed children swinging before a stark backdrop of
wartime devastation, literally hanging on for dear life.

The Russian-born Social Realist **PETER BLUME** (1906–
1992) painted fantastic imagery in a clearly delineated style
that invites comparison with the work of Surrealist Salvador
Dalí (see fig. 371). But the term Magic Realism, referring to
an art that produces bizarre effects through the depiction of
reality, is perhaps a more suitable description of Blume's art.
Most American artists greeted Surrealism and all the concomitant theories of the unconscious with great circumspec-
tion during the 1930s. It was not until the following decade,
when European artists escaping fascism immigrated to the
United States, that Surrealism, and particularly the automatist, abstract style of Miró, made major inroads into American
art. Blume, who denounced Surrealism in print, looked
abroad and excoriated the evils of Italian fascism in his painting *The Eternal City* (color plate 224, page 415). He had
traveled to Italy in 1932 on a Guggenheim grant and said the
idea to paint this work came to him while standing in the
Roman forum as a strange light illuminated the scene. The
bilious green head of a jack-in-the-box Mussolini looms over
the ruins of classical civilization, while the Man of Sorrows
continues his eternal mourning on the right.

left: 491. BEN SHAHN. *The Passion of Sacco and Vanzetti.*
1931–32. Tempera on canvas, 7'1½" x 4' (2.1 x 1.2 m).
Whitney Museum of American Art, New York
GIFT OF EDITH AND MILTON LOWENTHAL.
© 1997 ESTATE OF BEN SHAHN/VAGA, NEW YORK, NY

left: 492. DOROTHEA
LANGE. *Migrant
Mother.* 1936.
Gelatin-silver print.
Library of Congress,
Washington, D.C.
FARM SECURITY ADMINISTRATION
COLLECTION

below: 493.
WALKER EVANS.
*Miner's Home, West
Virginia.* 1935.
Gelatin-silver print,
7⁹⁄₁₆ x 9½" (19.2 x
24.1 cm). National
Gallery of Art,
Washington, D.C.
GIFT OF MR. AND MRS.
HARRY H. LUNN, JR.

Recording American Life in the 1930s

In miraculous contrast to the stalled economy, as well as the failed banks and crops, a wealth of splendid documentary photographs were produced by a group of artists, many of whom were sponsored by the Federal Farm Security Administration (FSA). Established in 1935, the FSA aimed to educate the population at large about the drastic tolls the Great Depression had taken upon the country's farmers and migrant workers. The moving force behind the project was economist Roy Stryker, who hired photographers, including Ben Shahn, Dorothea Lange, and Walker Evans, to travel to the hard-pressed areas and document the lives of the residents.

DOROTHEA LANGE (1895–1965), following studies at Columbia University under Clarence H. White (see fig. 62), opened a photographic studio for portraiture in San Francisco. However, she found her most vital themes in the life of the streets, preparing her for the horrors of the Depression, which she encountered firsthand when she joined Stryker's FSA project. With empathy and respect toward her subjects, Lange photographed a tired, anxious, but stalwart mother with two of her ten children in a migrant worker camp (fig. 492). This unforgettable and often reproduced image has been called "The Madonna of the Depression." Considered the first "documentary" photographer, Lange redefined the term by saying, "a documentary photograph is not a factual photograph per se . . . it carries with it another thing, a quality [in the subject] that the artist responds to."

WALKER EVANS (1903–1975) studied in Paris, albeit literature rather than art, and like Lange, was hired by the FSA, making work, he later said, "against salon photography, against beauty photography, against art photography." In the summer of 1936 he and the writer James Agee lived in Alabama with sharecropper families. The result, a 1941 book called *Let Us Now Praise Famous Men,* containing Evans's photographs and Agee's writings, is one of the most moving documents in the

494. MARGARET BOURKE-WHITE. *Fort Peck Dam, Montana.*
Photograph for the first cover of *Life* magazine,
November 23, 1936. *Life* magazine © 1936, 1964 TIME, INC.

495. ALFRED EISENSTAEDT. *The Kiss (Times Square).* 1945
Life magazine
© TIME, INC.

history of photography. The straightest of straight photographers, Evans brought such passion to his work that in his images the simple, direct statement ends up being almost infinitely complex—as metaphor, irony, and compelling form. His portrait of a West Virginia miner's home tells volumes about the life of its inhabitants. The cheerful middle-class homemakers and the smiling Santa that cover the walls are sadly incongruent in this stark environment (fig. 493).

Another student of Clarence H. White and a hard-driving professional was **MARGARET BOURKE-WHITE** (1904–1971), who, as Alfred Eisenstaedt wrote, "would get up at daybreak to photograph a bread crumb." Really a photojournalist, Bourke-White was the first staff photographer recruited for Henry Luce's *Fortune* magazine, and when the same publisher launched *Life* in 1936, Bourke-White's photograph of Fort Peck Dam appeared on the cover of the first issue (fig. 494). As this image indicates, Bourke-White favored the monumental in her boldly simplified compositions of colossal, mainly industrial, structures. While emphasizing the positive in this example of her work—evidence of the country's accomplishments in a time of faltering confidence—the photographer focused her camera on human subjects as well—on the farmers of the drought-stricken Midwest during the Depression, in the war zones and prison camps of the early 1940s, and among black South Africans in the 1950s.

ALFRED EISENSTAEDT (1898–1995) was one of the first four staff photographers hired for *Life* magazine and another

pioneer in the field of pictorial journalism. His photograph of the hazing of a West Point cadet appeared on the cover of the second issue of the magazine. He began his career as an Associated Press photographer in Germany but immigrated to the United States in 1935 when prewar politics made it difficult to publish his frank pictures. Though he produced many pictorial narratives, he was a firm believer in the single image that presented the distillation, the essential statement, of an event. This photograph (fig. 495) of a sailor kissing a nurse in New York City's Times Square on Victory Day, 1945, sums up all the feelings of joy and relief at the end of World War II. In contrast to the images of the horrors of war that would flood the media from the 1960s on, this image continues to be reproduced on posters and cards and seems to stand for of the invincibility of the American spirit.

BERENICE ABBOTT (1898–1991) returned to New York in 1929 from Paris, where she had learned photography as an assistant to Man Ray. One of the first to appreciate the genius of Atget (see fig. 397), and the only photographer ever to make portrait studies of the aged French photographer, Abbott fell in love with Manhattan upon her rediscovery of the city, much as Atget had with Paris. First working alone and then for the WPA, she set herself the task of capturing the inner spirit and driving force of the metropolis, as well as its outward aspect. Though utterly straight in her methods, Abbott, owing to her years among the Parisian avant-garde, inevitably saw her subject in modernist, abstract terms (fig. 496).

496. BERENICE ABBOTT. *New York at Night.* 1933. Gelatin-silver print on masonite, 10½ x 12¾" (26.7 x 32.4 cm). The Museum of Modern Art, New York PURCHASE. COPY PRINT © 1995 THE MUSEUM OF MODERN ART, NEW YORK

497. WEEGEE. *The Critic.* Gelatin-silver print, 10⅝ x 13³⁄₁₆" (27 x 33.5 cm). The Art Institute of Chicago GIFT OF MRS. STUYVESANT PEABODY

WEEGEE (b. Arthur Fellig) (1899–1968), another advocate of straight photography, exploited the voyeuristic side of taking pictures. He worked primarily as a freelance newspaper photographer and published his sensational and satiric images of criminal arrests, murder victims, and social events in the tabloids of the day. Like Lisette Model (see fig. 405), with whom he worked at *PM* magazine, he captured events by surprising his subjects, often through the use of startling flash. *The Critic* (fig. 497) is an ironic image of two women going to the opera. They stare at the camera and are in turn stared at by the woman to the right who ogles their jewelry, flowers, and furs, all in marked contrast to her own disheveled appearance. In 1936, Weegee published *Naked City,* a book with many images of New York's underworld.

JAMES VAN DER ZEE (1886–1983). During the 1920s in New York, a renaissance of black culture was taking place in the neighborhood of Harlem (fig. 498). Writers, intellectuals, musicians, and artists looked to the roots of their own cultural traditions in this remarkably fertile period known as the Jazz Age, which was brought to a halt by the stock market crash

498. JAMES VAN DER ZEE. *Harlem Billiard Room.* n.d. Gelatin-silver print, 11 x 14" (27.9 x 35.6 cm). Collection The Studio Museum in Harlem Archives, the James Van Der Zee Collection and Donna Van Der Zee Collection

499. ROMARE BEARDEN. *Folk Musicians.* c. 1941–42.
Gouache and casein on composition board, 36⅜ x 46⅝"
(92.4 x 118.4 cm) COURTESY ESTATE OF ROMARE BEARDEN.
© 1997 ROMARE BEARDEN FOUNDATION/LICENSED BY VAGA, NEW YORK, NY

right: 500. JACOB LAWRENCE. *Daybreak—A Time to Rest.* 1967.
Egg tempera on hardboard, 30 x 24" (76.2 x 61 cm).
National Gallery of Art, Washington, D.C.
ANONYMOUS GIFT

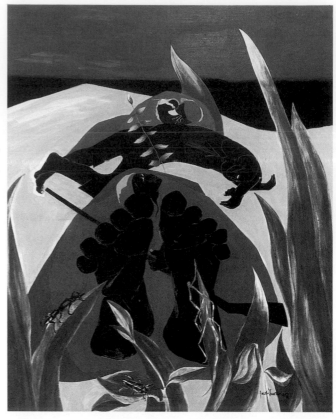

in 1929. While earning his living as a portrait photographer, Van Der Zee created a visual portrait of Harlem at the time, documenting places and events such as weddings, funerals, anniversaries, and graduations. With his background in art and music, Van Der Zee took occasion to manipulate his photographs, even to the point of painting his own backdrops and retouching his prints.

Artist **ROMARE BEARDEN** (1914–1988) was born in North Carolina but grew up in the thriving artistic milieu of the Harlem Renaissance. In a 1934 essay, "The Negro Artist and Modern Art," Bearden called on fellow African-American artists to create art based on their own unique cultural experiences. He became involved with the 306 Group, an informal association of African-American artists and writers who met in studios at 306 West 141st Street. In 1936 he studied for a time with the German artist George Grosz (see fig. 319) at the city's Art Students League, no doubt drawn to the artist's reputation as a political satirist. *Folk Musicians* (fig. 499) is characteristic of Bearden's early work. With its simple modeling and consciously "folk" manner, the painting coincides with the prevailing Social Realist trends. In the 1950s, following a trip to Paris, Bearden experimented for a time with an Abstract Expressionist style. Soon after adopting collage techniques in the 1960s he found a distinctive language of his own in the new medium. *The Dove* (color plate 225, page 415) is a compendium of images originating in Bearden's memories of life in the South and in Harlem. With such works the artist said he wanted to "establish a world through art in which the validity of my Negro experience could live and make its own logic."

Like Bearden, **JACOB LAWRENCE** (b. 1917) took the African-American experience as his subject, but with a preference for recounting black history through narrative series. In 1940–41 he made sixty panels on the theme of the migration by black men and women from the rural South to the industrial North during and after World War I (color plate 226, page 415). Lawrence thoroughly researched his subject and attached his own texts to each of the paintings, outlining the causes for the migration and the difficulties encountered by the workers in the new labor force as they headed for steel mills and railroads in cities like Chicago, Pittsburgh, and New York. Lawrence visually unified his panels by establishing a common color chord of red, green, yellow, and black and a style of boldly delineated silhouettes. The year after its creation, Lawrence's epic cycle went on a national museum tour organized by the Museum of Modern Art. By 1967, when he made a series on the life of black abolitionist Harriet Tubman (fig. 500), Lawrence had earned an international reputation and had begun a distinguished university career in teaching.

The Mexican Muralists

While many artists in the United States were looking to indigenous traditions and subject matter between the two world wars, a number of like-minded artists in Mexico turned to their own history and artistic heritage, contributing to a veritable renaissance of Mexican painting. The years 1910 to 1920 were ones of armed revolution in Mexico. This was the era of Emiliano Zapata and Pancho Villa, which began with the ousting of the longtime president Porfirio Díaz and ended with a new constitution that swept in major land and labor reforms. Beginning in the 1920s and continuing to mid-century, Mexican artists covered the walls of the country's schools, ministerial buildings, churches, and museums with vast murals, depicting subjects drawn from ancient and modern Mexico that celebrated a new Mexican cultural nationalism. The leading muralists were Diego Rivera, José Orozco, and David Siqueiros, each of whom also worked in the United States, where they struck a common chord with American artists. This was partially because they utilized the classical tradition of fresco painting and partially because theirs was an art of social protest with an obvious appeal to the left wing, a dominant force in American cultural life throughout the Depression decade. The muralists were enormously influential throughout Latin America and the United States.

Following his artistic education in Mexico City, **DIEGO RIVERA** (1886–1957) studied and lived in Europe between 1907 and 1921. There he met many of the leading vanguard artists, including Picasso and Gris, and was developing his own Cubist style by 1913. He went to Italy near the end of his European sojourn and was particularly impressed by Renaissance murals. Upon his return to his native country, Rivera began to receive important commissions for monumental frescoes from the Mexican government. In the murals, he attempted to create a national Mexican style reflecting both the history of Mexico and the socialist spirit of the Mexican revolution. Rivera turned away from the abstracting forms of Cubism to develop a modern neoclassical style for his mural art consisting of simple, monumental forms and bold areas of color. This style can also be seen in Rivera's occasional easel painting, such as *Flower Day,* 1925 (color plate 227, page 416). While the subject of this work relates to an enormous mural project Rivera undertook for the Ministry of Education in Mexico City, the massive figures and classically balanced composition derive from Italian art, as well as the Aztec and Mayan art Rivera consciously emulated.

In 1930 Rivera, a loyal Communist, came to the United States, where he carried out several major commissions that were ironically made possible by capitalist industry. Political differences between artist and patrons did lead to problems, however. Most notably, in 1931, Rivera's mural at Rockefeller Center, New York, was destroyed due to his Communist leanings. Foremost among the surviving work is a narrative fresco cycle for the Detroit Institute of Arts on the theme of the evolution of technology, culminating in automobile manufacture (fig. 501). Rivera toured the Ford motor plants for months, observing the workers and making preparatory sketches for his murals. The result is a tour de force of mural painting in which Rivera orchestrated man and machine into one great painted symphony. "The steel industry itself," he said, "has tremendous plastic beauty . . . it is as beautiful as the early Aztec or Mayan sculptures." In Rivera's vision, there is no sign of the unemployment or economic depression then crippling the country, nor the violent labor strikes at Ford that had just preceded Rivera's arrival in Detroit. At the same

501. DIEGO RIVERA. *Detroit Industry.* 1932–33. Fresco, north wall. The Detroit Institute of Arts GIFT OF EDSEL B. FORD

502. JOSÉ OROZCO. *The Epic of American Civilization: Modern Migration of the Spirit.* 1932–34. Mural, fourteenth panel. Baker Library, Dartmouth College, Hanover, New Hampshire TRUSTEES OF DARTMOUTH COLLEGE

503. DAVID SIQUEIROS. *Echo of a Scream.* 1937. Enamel on wood, 48 x 36" (121.9 x 91.4 cm). The Museum of Modern Art, New York GIFT OF EDWARD M. M. WARBURG. PHOTOGRAPH © 1995 THE MUSEUM OF MODERN ART, NEW YORK

time, Rivera reveals criticism of capitalism's potential effects. Even though he portrays the workers in a dignified manner, it is important to note that they are rendered in a dehumanized fashion.

JOSÉ CLEMENTE OROZCO (1883–1949), like Rivera and Siqueiros, received his training in Mexico City, at the Academy of San Carlos. Orozco lived in the United States between 1927 and 1934, when he also had important fresco commissions, notably from Dartmouth College in Hanover, New Hampshire. There, in the Baker Library, he created a panorama of the history of the Americas. Orozco was not a proselytizer in the same vein as Rivera, and he hardly shared his compatriot's faith in modern progress and technology. In his complex of murals for Dartmouth he showed the viewer the human propensity for greed, deception, or violence. The murals begin with the Aztec story of Quetzalcoatl, continue with the coming of the Spaniards and the Catholic Church, and conclude with the self-destruction of the machine age climaxed in a great Byzantine figure of a flayed Christ, who, upon arriving in the modern world before piles of guns and tanks, destroys his own cross (fig. 502). All of this is rendered in Orozco's fiery, expressionist style, with characteristically long, striated brushstrokes and brilliant color.

The third member of the Mexican triumvirate of muralists, DAVID ALFARO SIQUEIROS (1896–1974), executed fewer public murals throughout the 1930s than his countrymen, partly because his radical political activities kept him on the move. Siqueiros maintained a lifelong interest in adopting the means of modern technology to art. In 1932 he came to the United States, to experience a "technological civilization," he said, and taught at a Los Angeles art school where he execut-

ed murals using innovative materials and methods. Siqueiros's first major mural project came in 1939 for the Electricians' Syndicate Building in Mexico City.

The 1937 *Echo of a Scream* (fig. 503), painted the year Siqueiros joined the antifascist Republican Army in the Spanish Civil War, illustrates his vigorous pictorial attack. Beginning with an actual news photograph of a child in the aftermath of a bombing in the contemporaneous Sino-Japanese war, Siqueiros portrayed the "echo" of the child's cries as an enormous disembodied head above a field of ruins, creating a wrenching image of despair.

In 1936, Siqueiros established an experimental workshop in New York that attracted young artists from Latin America and the United States (including Jackson Pollock). The workshop was dedicated to modernizing the methods and materials of art production and to promoting the aims of the Communist Party of the United States. One participant described the use of new industrial paints in the workshop: "We sprayed [it] through stencils and friskets, embedded wood, metal, sand, and paper. We used it in thin glazes or built up into thick gobs. We poured it, dripped it, splattered it, hurled it at the picture surface. . . . What emerged was an endless variety of accidental effects." Using such experimental methods, Siqueiros hoped to establish bold new ways of creating modern propagandistic murals. While the experience of working with Siqueiros was formative, it was Orozco, of all the Mexican muralists, who left the strongest mark on Jackson Pollock. In general, the tradition of mural painting in the 1930s, of both the Mexicans and WPA artists like Benton, was crucial for the Abstract Expressionist artists coming of age during the ensuing decades. In their work, which is discussed in the next chapter, they transferred the large scale of public murals to their canvases.

For the career of Mexican painter **FRIDA KAHLO** (1907–1954), the defining moment came in 1925, when her spine, pelvis, and foot were crushed in a bus accident. During a year of painful convalescence, Kahlo, already a victim of childhood polio, took up painting. Through Tina Modotti, she met Diego Rivera, whom she married in 1929. She was with Rivera in Detroit when he painted the murals at the Institute of Arts, but Kahlo was never seduced by the monumental scale or public themes of mural art. She turned her attention inward and over the years created an intimate autobiography through an astonishing series of self-portraits. In an early example (fig. 504), she envisions herself in a long pink dress surrounded at the left by the cultural artifacts of Mexico and, at the right, the industrial trappings of the North. The hieratic and miniaturist technique she employs in this tiny painting on metal is in the folk tradition of Mexican votive paintings, which depict religious figures often surrounded by fantastic attributes.

When André Breton visited Kahlo and Rivera in Mexico in 1938, he claimed Kahlo as a Surrealist and arranged for her work to be shown, with considerable success, in New York and Paris. Soon thereafter she and Rivera divorced (they remarried a year later), and Kahlo made several self-portraits documenting her emotional grief. In *Self-Portrait with Thorn Necklace* (color plate 228, page 416) she assumes the role of martyr, impaled by her necklace from which a dead hummingbird (worn by the lovelorn) hangs.

Closely associated with the artist-reformers of Mexico was the Italian-born photographer **TINA MODOTTI** (1896–1942), who, with her longtime companion Edward Weston, arrived in Mexico in the early 1920s. She joined the Communist Party in 1927 and worked in Mexico until her expulsion in

below: 504. FRIDA KAHLO. *Self-Portrait on the Border Between Mexico and the United States.* 1932. Oil on sheet metal, 12¼ x 13¾" (31.1 x 34.9 cm). Private collection. Courtesy Christie's, New York

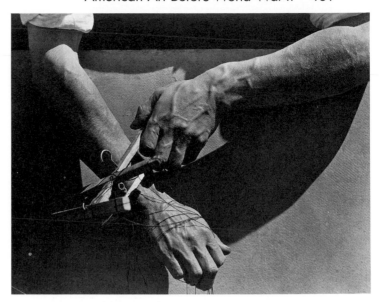

505. TINA MODOTTI. *Number 29: Hands of Marionette Player.* n.d. Gelatin-silver print, 7½ x 9½" (19.1 x 24.1 cm). The Museum of Modern Art, New York GIVEN ANONYMOUSLY. COPY PRINT © 1995 THE MUSEUM OF MODERN ART, NEW YORK

506. RUFINO TAMAYO. *Animals.* 1941. Oil on canvas, 30⅛ x 40" (76.5 x 101.6 cm). The Museum of Modern Art, New York INTER-AMERICAN FUND. PHOTOGRAPH © 1995 THE MUSEUM OF MODERN ART, NEW YORK

1930 for political activism. She was commissioned by the muralists to document their works in photographs that were circulated internationally, contributing to the celebrity of the movement. Her sharp-focus, unmanipulated images, such as her elegant close-ups of plants, bear the influence of Weston. But Modotti used her camera largely for socially relevant subjects, as when she photographed the hands and marionettes of a friend who used his craft for political ends (fig. 505). With a sure eye for the telling image, Modotti built a remarkable corpus of photographs until she gave up the profession to live a nomadic existence in pursuit of her political goals.

RUFINO TAMAYO (1899–1991), a native of Oaxaca, of Zapotec Indian descent, was another Mexican artist who came to prominence around the 1930s. He lived in New York in the late 1920s and again from 1936 to 1948, teaching and working for a time for the WPA, until the government banned foreign artists from the program. Although he occasionally received mural commissions, he differed from the three leading muralists from Mexico in that his aim was never to create a public art based on political ideologies: instead, he preferred themes of a more universal nature. His style was informed both by modernist modes, such as Cubism and Surrealism, and by Mexico's pre-Hispanic art. In the early 1940s he made a series of animal paintings (fig. 506), whose forms were inspired by popular Mexican art, but whose implied violence seems to register the traumas of war.

Toward Abstract Art

As we have seen in both the United States and Mexico, during the 1930s modernism, particularly Cubism and abstraction, seemed to have been driven underground, although the Federal Art Project in the United States did support many of the more experimental artists. Thus the seeds were sown that would germinate, after World War II, in the transformation of the United States from the position of a provincial follower to that of a full partner in the creation of new art and architecture.

A number of other events and developments of the 1920s and 1930s also helped to strengthen the presence of vanguard art in New York. As we saw in chapter 13, Katherine Dreier, the painter and collector advised by Marcel Duchamp, had organized the Société Anonyme in 1920 for the purpose of buying and exhibiting examples of the most advanced European and American art and holding lectures and symposia on related topics. In 1927 A. E. Gallatin, another painter-collector, put his personal collection of modern art on display at New York University under the name of The Gallery of Living Art. It remained on exhibition for fifteen years. (This collection, now at the Philadelphia Museum together with the even more important one of Walter C. Arensberg, includes examples by Brancusi, Braque, Cézanne, Kandinsky, Klee, Léger, Matisse, Miró, Mondrian, and many others.) The Gallery of Living Art was, in effect, the first museum with a permanent collection of exclusively modern art to be established in New York City. It was followed by the opening of The Museum of Modern Art in 1929, which followed its initial series of exhibitions by modern masters with the historic shows entitled "Cubism and Abstract Art" and "Fantastic Art, Dada, Surrealism," both in 1936. In 1935 the Whitney Museum held its first exhibition of American abstract art, and in 1936 the American Abstract Artists (AAA) group was organized. By 1939 Solomon R. Guggenheim had founded the Museum of Non-Objective Painting, which would eventually be housed in the great Frank Lloyd Wright building on Fifth Avenue (see fig. 709).

During the 1930s a few progressive European artists began to visit or immigrate to the United States. Léger made three visits before settling at the outbreak of World War II. Not only Duchamp, but Ozenfant, Moholy-Nagy, Josef Albers, and Hans Hofmann were all living in the United States before 1940. The last four were particularly influential as teachers. The group as a whole represented the first influx of European artists who spent the war years in the United States and who helped to transform the face of American art. Conversely, younger American painters continued to visit and study in Paris during the 1930s. Throughout the late 1930s and 1940s, the AAA held annual exhibitions dedicated to the promotion of every form of abstraction, but with particular emphasis on experiments related to the Constructivist and Neo-Plasticist abstraction being propagated in Paris by the *Cercle et Carré* and *Abstraction-Création* groups. These close contacts with the European modernists were of great importance for the subsequent course of modern American art.

The professional career of STUART DAVIS (1894–1964) encompassed six decades of modern art in the United States. Although Davis was deeply committed to socialist reforms, unlike his Mexican contemporaries or Americans like Ben Shahn, he did not exploit painting as a public platform for his political beliefs. Davis was committed to art with a broad, popular appeal and he negotiated the vast territory between pure abstraction and the conservative realism of the Regionalists. After leaving Robert Henri's School of Art, Davis exhibited five watercolors in the 1913 Armory Show and was converted to modernism as a result of seeing the European work on display. Davis's collages and paintings of the early 1920s are explorations of consumer-product packaging as subject matter (in this case cigarette wrappers, fig. 507), executed in a manner that demonstrates his assimilation of Synthetic Cubism as well as his recollection of nineteenth-century American trompe l'oeil painting. But his emphatically two-dimensional, deadpan treatment of imagery borrowed from popular culture, as well as his use of modern commercial lettering in his paintings, made Davis an important forerunner of 1960s Pop art (see color plate 309, page 531).

In his Egg Beater series of 1927–28, the first genuine abstractions to be initiated by an American artist in ten years, Davis attempted to rid himself of the last illusionistic vestiges of nature, painting an eggbeater, an electric fan, and a rubber glove again and again until they had ceased to exist in his eyes and mind except as color, line, and shape relations (fig. 508). Throughout the 1930s, Davis continued the interplay of clearly defined, if fragmented, objects with geometric abstract organization. His color became more brilliant, and he intensified the tempo, the sense of movement, the gaiety, and the rhythmic beat of his works through an increasing complication of smaller, more irregular, and more contrasted color shapes.

The culmination of these experiments may be seen in a 1940 work, *Report from Rockport* (color plate 229, page 416). Although more abstract than most paintings of the 1930s, *Report from Rockport,* inspired by the main square of Rockport, Massachusetts, includes recognizable bits such as buildings and gas pumps. There is a suggestion of depth, achieved by a schematic linear perspective, but the graphic shapes that twist and vibrate across the surface, with no clear representational role to play, reaffirm the picture plane.

In the 1952 painting *Rapt at Rappaport's* (color plate 230, page 449), the alliterative and slangy title invades the

left: 507. STUART DAVIS. *Lucky Strike*. 1921. Oil on canvas,
33¼ x 18" (84.5 x 45.7 cm).
The Museum of Modern Art, New York
GIFT OF THE AMERICAN TOBACCO COMPANY, INC.
PHOTOGRAPH © 1995 THE MUSEUM OF MODERN ART, NEW YORK.
© 1997 ESTATE OF STUART DAVIS/VAGA, NEW YORK, NY

above: 508. STUART DAVIS. *Egg Beater #1*. 1927. Oil on canvas,
27 x 38¼" (68.6 x 97.2 cm).
The Phillips Collection, Washington, D.C.
© 1997 ESTATE OF STUART DAVIS/VAGA, NEW YORK, NY

actual painting: the staccato letters flash across the bottom edge like a news bulletin. Large, flat color-shapes of red, blue, orange, and black are suspended against a green ground. This surface is animated with invented shapes characteristic of Davis, the torn S or 8 ribbon, the floating white circles or rectangles. In bright blue letters toward the center of the painting is the word "ANY," which Davis said "means any subject matter is equal to art, from the most insignificant to one of relative importance."

Until his death in 1964, Davis searched for ways to merge American subject matter with the forms of European Cubism. The result was a thoroughly new, uniquely American Cubist language. In 1943 he wrote in a leading art magazine, "Enormous changes are taking place which demand new forms, and it is up to artists living in America to find them." His enthusiasm for American innovation in painting did not extend, however, to the subjective art of the Abstract Expressionists in the 1950s. "I like popular art, Topical ideas, and not High Culture or Modernistic Formalism," he said. "I care nothing for Abstract Art as such, but only as it evidences a contemporary language of vision suited to modern life." With the emergence of Pop Art in the 1960s, along with more objective forms of abstraction, Davis's art took on new relevance in the history of modern American art.

GERALD MURPHY (1888–1964) was a wealthy business-man turned artist: on whom F. Scott Fitzgerald modeled the hero of *Tender Is the Night*. Murphy came into contact with the Parisian avant-garde by working on sets for the Ballets Russes. He scandalized the 1924 Salon des Indépendants by exhibiting an eighteen-foot-high painting of an ocean liner in an elegant Art Deco style. Murphy worked with painstaking slowness, producing only one painting a year, and no more. Of a total of probably fifteen, a mere seven are known to survive. Still, he had absorbed the principles of Synthetic Cubism and, like Stuart Davis, succeeded in turning them to his own purposes, which were to transform American popular imagery

509. GERALD MURPHY. *Razor*. 1924. Oil on canvas. 32⅝ x 36½"
(82 x 92 cm). Dallas Museum of Art, Foundation for the Arts
FOUNDATION FOR THE ARTS COLLECTION, GIFT OF THE ARTIST

into a high art of simple, emblematic force. His bold composition, *Razor,* 1924 (fig. 509), depicts a common matchbox above a heraldic arrangement of a razor and fountain pen. This subject has an autobiographical significance. Murphy developed a patent for a safety razor and was the son of the owner of Cross pens. Murphy belongs in a realm entirely of his own, but one that nevertheless looked clairvoyantly forward to the Pop art of the 1960s. He had no direct influence on the movement, however. He was not well known in the United States owing to his isolation abroad, and his rare canvases were not discovered until years after his death.

One of the artists who exhibited with the AAA was BURGOYNE DILLER (1906–1965), the first American to develop an abstract style based on Dutch de Stijl. Once enrolled at the Art Students League in 1929, Diller very rapidly assimilated influences from Cubism, German Expressionism, Kandinsky, and Suprematism, much of which he knew only through reproductions in such imported journals as *Cahiers d'art.* Finally, toward the end of 1933, he had his first direct experience of a painting by Mondrian, whose Neo-Plastic principles became the main source of Diller's interests. But far from servile in his dependence on Neo-Plasticism, Diller went his own way by integrating color and line, so that lines became overlapping color planes (color plate 231, page 449). Thus he created a complex sense of space that differs from that of Mondrian.

IRENE RICE PEREIRA (1907–1971) was another pioneer American abstractionist to appear in the 1930s. Taking her inspiration primarily from the Bauhaus, Pereira sought to fuse art and science, or technology, in a new kind of visionary image grounded in abstract geometry, but aspiring, like the Suprematist art of Malevich, toward an even more rarefied realm of experience. Boston-born and professionally active in New York, Pereira developed a seasoned style compounded of sharply rectilinear hard-edge and crystalline planes, often

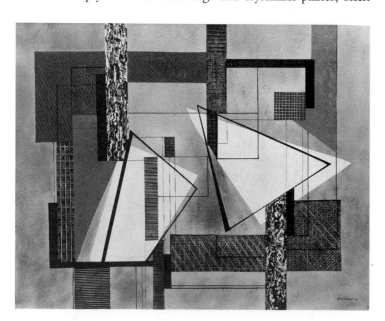

510. IRENE RICE PEREIRA. *Abstraction.* 1940. Oil on canvas, 30 x 38" (76.2 x 96.5 cm). Honolulu Academy of Arts
GIFT OF PHILIP E. SPALDING, 1949

rendered as if floated or suspended one in front of the other, with transparent tissues giving visual access to the textured, opaque surfaces lying below (fig. 510). The resulting effect is that of a complex, many-layered field of color and light.

Born and schooled in Hartford, Connecticut, MILTON AVERY (1885–1965) moved to New York in 1925. He characteristically worked in broad, simplified planes of thinly applied color deriving from Matisse but involving an altogether personal poetry. His subjects were his family and friends, often relaxing on holiday (color plate 232, page 450). In his later years he turned more to landscape, which, though always recognizable, became more and more abstract in organization. Avery was a major source for the American Color Field painters of the 1950s and 1960s, such as Mark Rothko, Adolph Gottlieb, and Helen Frankenthaler (see figs. 531, 535, 668). In a memorial service for his friend, Rothko said, "Avery had that inner power which can be achieved only by those gifted with magical means, by those born to sing."

AUGUSTUS VINCENT TACK (1870–1949) may have been the most maverick and removed of all the progressive American painters between the wars. He alone managed to integrate modernism's nonobjective and figural modes into organic, sublimely evocative, wall-size abstractions. His work formed a unique bridge between an older pantheistic tradition of landscape painting, last seen in the work of Ryder, and its new, brilliant flowering in the heroic abstractions of the postwar New York School, especially in the paintings of Clyfford Still (color plate 252, page 458). Duncan Phillips, Tack's faithful patron throughout the 1920s and 1930s, foresaw it all when he declared, in commenting on one of Tack's paintings: "We behold the majesty of omnipotent purpose emerging in awe-inspiring symmetry out of thundering chaos. . . . It is a symbol of a new world in the making, of turbulence stilled after tempest by a universal God."

Tack was trained in the 1890s by the American Impressionist John Twachtman, and his early work consisted largely of landscapes in the style of late-nineteenth-century American symbolism. In 1920, when he made his first paintings of the Rocky Mountains, Tack discovered in the majestic western landscape motifs capable of expressing his mystical feelings about nature. These he gradually transformed into the dramatic, nature-inspired abstractions of his mature work (color plate 233, page 450). Here is American Romanticism in the most advanced form it would take during the period, a view of the material world examined so intensely that it dissolved into pure emblem or icon, an aesthetic surrogate for the divine, ordering presence sensed within the disorder of raw nature. Tack's paintings exhibit uncanny affinities with the allover abstractions of the New York School of the 1950s.

American Sculpture Between the Wars

American sculpture during the first forty years of the twentieth century lagged behind painting in quantity, quality, and originality. Throughout this period the academicians predominated, and such modernist developments in sculpture as

511. GASTON LACHAISE. *Standing Woman (Elevation)*. 1912–27. Bronze, height 70" (177.8 cm). Albright-Knox Art Gallery, Buffalo

512. ELIE NADELMAN. *Man in the Open Air*. c. 1915. Bronze, 54½" (138.4 cm) high; at base 11¾ x 21½" (29.9 x 54.6 cm). The Museum of Modern Art, New York GIFT OF WILLIAM S. PALEY (BY EXCHANGE). PHOTOGRAPH © 1995 THE MUSEUM OF MODERN ART, NEW YORK

Cubism or Constructivism were hardly visible. Archipenko, who had lived in the United States since 1923 and who had taught at various universities, finally opened his own school in New York City in 1939. Alexander Calder, as we have seen, was the best-known American sculptor before World War II, but largely in a European context, since he had worked in Paris since 1926. His influence grew continuously in America in the following years. During the 1930s, in sculpture as in painting, despite the predominance of the academicians and the isolation of the progressive artists, a number of sculptors were emerging, including David Smith and Isamu Noguchi (see chapter 19), who helped to change the face of American sculpture at the end of World War II.

Among the first generation of modern American sculptors, two important practitioners were Europeans, trained in Europe, who immigrated to the United States early in their careers. These were GASTON LACHAISE (1882–1935) and Elie Nadelman. Lachaise was born in Paris and trained at the École des Beaux-Arts. By 1910 he had begun to experiment with the image that was to obsess him during the rest of his career, that of a female nude with enormous breasts and thighs and delicately tapering arms and legs (fig. 511). This

form derives from Maillol, but in Lachaise's versions it received a range of expression from pneumatic elegance to powerfully primitivizing works that recall the ancient sculpture of India, as well as prehistoric female fertility figures.

ELIE NADELMAN (1882–1946) received his European academic training at the Warsaw Art Academy. Around 1909 in Paris, where he made the acquaintance of artists including Picasso and Brancusi, Nadelman was reducing the sculpted human figure to almost abstract curvilinear patterns. After his move to the United States in 1914, he was given an exhibition by Stieglitz at 291. He progressively developed a style that might be described as a form of sophisticated primitivism (he was an avid collector of folk art), marked by simplified surfaces placed in the service of a witty and amusing commentary on contemporary life (fig. 512).

Son of a Chicago architect, sculptor JOHN STORRS (1885–1956) studied in Chicago, Boston, New York, and Paris and, beginning in 1915, was making geometricized figurative sculptures in stone. Their streamlined forms, partly inspired by the Native American art Storrs admired and collected, was at one with the prevailing Art Deco style in architecture and decorative arts. During the 1920s Storrs began to

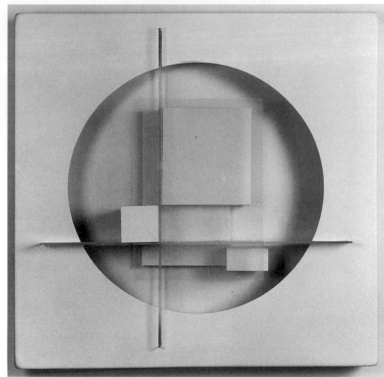

left: 513. JOHN STORRS. *Forms in Space #1*. c. 1927. Steel and copper, 20½ x 4 x 1⅝" (52.1 x 10.2 x 4.1 cm). The Metropolitan Museum of Art New York PURCHASE

above: 514. THEODORE ROSZAK. *Construction in White*. 1938. Painted wood and plastic, 49½ x 49½" (125.7 x 125.7 cm). Estate of Theodore Roszak COURTESY HIRSCHL & ADLER GALLERIES, NEW YORK

make sculpture directly inspired by modern architectural forms, initially those created by Frank Lloyd Wright. His later celebrations in sculpture of the modern skyscraper (fig. 513) parallel those of Stella, O'Keeffe, Sheeler, and Marin. Such works led to commissions for monumental sculpture to decorate actual buildings. Owing to the artist's long expatriation to France's Loire Valley, Storrs's achievement remained largely forgotten until the 1960s.

Although in the minority, sculpture in a Constructivist vein thrived in the hands of a few American artists in the 1930s, including Polish-born THEODORE ROSZAK (1907–1981). Having arrived in Chicago with his family as a young child, Roszak was trained as a conventional painter at that city's School of the Art Institute, but a change in his outlook occurred during a trip to Europe from 1929 to 1931. Upon returning to the United States, he settled in New York in 1931, working eventually for the WPA. He made de

Chiricoesque figure paintings with Cubist overtones in the early 1930s and, at the same time, began making Cubist sculptures and reliefs. By the middle of the decade, he was experimenting with abstract constructed sculpture as a result of his interest in machines and industrial design. *Construction in White* (fig. 514) is a whitewashed plywood box in which Roszak cut a circle and inserted intersecting sheets of plastic and wooden blocks. He said he wanted such works to stand as utopian symbols for perfection. Critical to Roszak were the teachings of the Bauhaus, particularly those of Moholy-Nagy, who had come to Chicago in 1937 to head the New Bauhaus. In the aftermath of World War II, Roszak became disenchanted with technology and began to make welded sculpture in steel. The rough surfaces and aggressive forms of these works reflected the violence and terror of war, situating his work in the arena of Abstract Expressionism, a movement discussed in the following chapter.

19 Abstract Expressionism and the New American Sculpture

ooking back on the 1940s, the American painter Barnett Newman recalled that artists of his generation "felt the moral crisis of a world in shambles, a world devastated by a great depression and a fierce world war, and it was impossible at that time to paint the kind of paintings that we were doing—flowers, reclining nudes, and people playing the cello. . . . This was our moral crisis in relation to what to paint." In 1942, immediately after the United States's entry into World War II, the dominant styles of painting were still Social Realism and Regionalism. The war contributed to a ferment that brought on the creation of the first major original direction in the history of American painting, Abstract Expressionism, or, as it is more generically called, the New York School. Attitudes conditioned by the war—a sense of alienation and a loss of faith in old systems and old forms of expression—led artists to explore a broad range of intellectual thought from existentialism to the theories of Sigmund Freud and Carl Jung.

The diverse group of artists involved were opposed to all forms of Social Realism and any art form that smacked of nationalism (although they too were conscious of shaking off the mantle of European influence). And much as they admired Mondrian, who was present in New York by 1940, they also rejected as trivial the pure geometric abstraction, promoted within the ranks of the American Abstract Artists (AAA), that his art had spawned in this country. On the other hand, the relationship of the new movement to the European Surrealists, some of whom also sought refuge from the war in New York, was critical. The Americans were particularly drawn to the organic, abstract, automatist Surrealism of Matta, Miró, and Masson. Matta, who was in New York from 1939 to 1948, helped introduce the idea of psychic automatism to the Abstract Expressionists. The Americans were less concerned with the new method as a means of tapping into the unconscious than as a liberating procedure that could lead to the exploration of new forms.

The art world of the 1940s was far from a coherent community. The experience of the Federal Arts Project of the WPA had provided a degree of camaraderie among artists, but largely the artists of the New York School gathered informally in their studios and in the cafeterias and taverns of Greenwich Village to discuss—and sometimes to battle over—the burning artistic issues of the day. Peggy Guggenheim's gallery, Art of This Century, featured European Surrealists but also became a venue for young artists like Clyfford Still, Robert Motherwell, and, especially, Jackson Pollock. A handful of other dealers and a few critics, notably formalist critic Clement Greenberg, as well as Harold Rosenberg and Thomas Hess, eventually championed the new movement. Recognition by major museums came more slowly. The Museum of Modern Art began to acquire works by a few of the artists in the mid-1940s and to include their work in group exhibitions devoted to contemporary American art. By 1958 a kind of international apotheosis of Abstract Expressionism took place with the museum's exhibition "The New American Painting," which was devoted to the movement and sent on tour to eight European countries.

Although Abstract Expressionism is as diverse as the artists involved, in a very broad sense two main tendencies may be noted. The first is that of the so-called gestural painters, concerned in different ways with the spontaneous and unique touch of the artist, his or her "handwriting," and the emphatic texture of the paint. It included such major artists as Pollock, Willem de Kooning, and Franz Kline. In 1952 Harold Rosenberg coined the phrase "action painting" to describe the process by which the spontaneous gesture was enacted on the canvas. "At a certain moment," he wrote, "the canvas began to appear to one American painter after another as an arena in which to act—rather than as a space in which to reproduce, re-design, analyze or 'express' an object, actual or imagined. What was to go on the canvas was not a picture but an event." What "action painting" failed to account for was the balance these artists struck between forethought and spontaneity, between control and the unexpected.

The other group consisted of the Color Field painters, concerned with an abstract statement in terms of a large, unified color shape or area. Here one can include Rothko, Newman, Still, and Gottlieb, as well as, to a degree, Motherwell and Reinhardt. In 1948 several of these painters founded an informal school called the "Subjects of the Artist." They were unit-

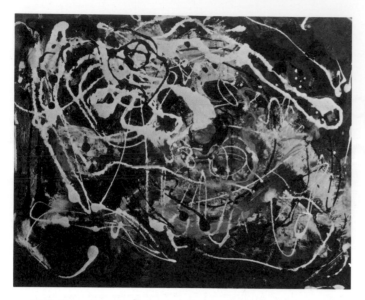

515. HANS HOFMANN. *Spring.* 1944–45 (dated 1940). Oil on wood panel, 11¼ x 14⅛" (28.5 x 35.7 cm).
The Museum of Modern Art, New York
GIFT OF MR. AND MRS. PETER A. RUBEL.
PHOTOGRAPH © 1995 THE MUSEUM OF MODERN ART, NEW YORK

ed by their belief that abstract art could express universal, timeless themes. As they had stated in a letter to the *New York Times,* "There is no such thing as a good painting about nothing." Even at its most abstract, art could convey a sense, in Rothko's works, of "tragedy, ecstasy and doom."

Gestural Painting

The career of **HANS HOFMANN** (1880–1966) encompassed two worlds and two generations. Born in Bavaria, Hofmann lived and studied in Paris between 1903 and 1914, experiencing the range of new movements from Neo-Impressionism to Fauvism and Cubism. He was particularly close to Robert Delaunay, whose ideas on color structure were a formative influence. In 1915 Hofmann opened his first school in Munich. In 1932 he moved to the United States to teach, first at the University of California, Berkeley, then at New York's Art Students League, and finally at his own Hans Hofmann School of Fine Arts in New York and in Provincetown, Massachusetts. Hofmann's greatest concern as a painter, teacher, and theoretician lay in his concepts of pictorial structure, which were based on architectonic principles rooted in Cubism. This did not prevent him from attempting the freest kinds of automatic painting. A number of such works executed in the mid-1940s (fig. 515), although on a small scale, anticipated Pollock's drip paintings.

During the last twenty-six years of his life, Hofmann made abstractions of amazing variety, ranging from a precise but painterly geometry to a lyrical expressionism. In his best-known works, Hofmann used thick, rectangular slabs of paint and aligned them with the picture's edge (color plate 234, page 451), demonstrating his ideas about color as a space-creating device. While the rectangles of color affirm the literal flatness of the picture, they also appear to advance and recede spatially, creating what Hofmann called a "push and pull"

effect. Like his fellow German emigré Josef Albers, Hofmann was one of the premier art educators and theoreticians in the United States, and his influence on a generation of younger artists such as Lee Krasner (fig. 523) was enormous.

Armenian-born **ARSHILE GORKY** (1905–1948) was a seminal figure in the early years of the New York School and a largely self-taught artist whose work constituted a critical link between European Surrealism and American Abstract Expressionism. Gorky arrived in the United States in 1920, a refugee from the Turkish campaign of genocide against the Armenians. His early experiments with both figuration and abstraction were deeply influenced by Cézanne and then Picasso and Miró. He achieved a distinctive figural mode in *The Artist and His Mother* (fig. 516), a haunting portrait based on an old family photograph in which the artist posed beside his mother, well before she died of starvation in his arms.

By the early 1940s Gorky was evolving his mature style, a highly original mode of expression that combined strange hybrid forms with rich, fluid color. His most ambitious painting, *The Liver Is the Cock's Comb,* 1944 (color plate 235, page 451), is a large composition that resembles both a wild and vast landscape and a microscopically detailed internal anatomy. Gorky combined veiled but recognizable shapes, such as claws or feathers, with overtly sexual forms, creating an erotically charged atmosphere filled with softly brushed, effulgent color. His biomorphic imagery owed much to Kandinsky and Miró, whose works Gorky knew well, and to the Surrealist

516. ARSHILE GORKY. *The Artist and His Mother.* c. 1929–36. Oil on canvas, 60 x 50" (152.4 x 127 cm).
National Gallery of Art, Washington, D.C.
AILSA MELLON BRUCE FUND.
© 1997 ESTATE OF ARSHILE GORKY/ARTISTS RIGHTS SOCIETY (ARS), NEW YORK

automatism of Masson and Matta, both in New York in the 1940s. The year he made this painting, Gorky met André Breton, the self-appointed leader of the Surrealists who took a strong interest in his work and arranged for it to be shown at a New York gallery in 1945.

Although its subject is deliberately ambiguous, the 1947 painting *The Betrothal II* (fig. 517) is one of three composi tions of the same title that Gorky built around the central image of a man on horseback. Over a delicate linear scaffolding the artist applied transparent washes of muted color, leaving visible the tracks of his brush. It was this kind of dynamic balance between representational and figurative motifs that appealed to Gorky's contemporaries like de Kooning. Following a series of traumatic events, Gorky hanged himself in his studio in 1948, just as his fellow Abstract Expressionists, whose work was indebted to his example, were gaining serious momentum in New York.

WILLEM DE KOONING (1904–1997) was a central figure of Abstract Expressionism, even though he was not one of the first to emerge in the public eye during the 1940s. Born in Rotterdam, the Netherlands, de Kooning underwent rigorous artistic training at the Rotterdam Academy. He came to the United States in 1926, where, throughout the 1930s, he slowly made the transition from house painter, commercial designer, and Sunday painter to full-time artist. Of the utmost importance was his early encounter with Stuart Davis, Gorky, and the influential Russian-born painter John Graham, "the three smartest guys on the scene," according to de Kooning. They were his frequent companions to The Metropolitan Museum of Art, where it was possible to see examples of

ancient and old-master art. Although he did not exhibit until 1948, de Kooning was an underground force among younger experimental painters by the early years of the decade.

One of the most remarkable aspects of de Kooning's talent was his ability to shift between representational and abstract modes, which he never held to be mutually exclusive. He continued to make paintings of figures into the 1970s, but even many of his most abstract compositions contain remnants of or allusions to the figure. "Even abstract shapes must have a likeness," he said. In the late 1940s he made a bold group of black-and-white abstractions in which he had fully assimilated the tenets of Cubism, which he made over into a dynamic, painterly idiom. Using commercial enamel paint, he made a largely black composition through which he spread a fluid network of white lines, occasionally allowing the medium to run its own course down the surface. Certain recognizable forms—a hat and glove at the upper right—can be detected. Elsewhere in the picture de Kooning subsumed figurative references within rhythmically flowing lines, creating vestigial forms that function as his characteristic shorthand for the human body. After *Painting* (fig. 518) was shown in de Kooning's first solo show in New York, it was acquired by The Museum of Modern Art, and de Kooning thus emerged as one of the pioneers of a new style.

While he was making his "black and white" compositions, de Kooning was also working on large paintings of women. In 1950 he began what became his most famous canvas, *Woman, I* (fig. 519). This monumental image of a seated woman in a sundress is de Kooning's overpowering, at times repellent, but hypnotic evocation of woman as sex symbol

519. WILLEM DE KOONING. *Woman, I.* 1950–52. Oil on canvas, 6'3⅞" x 58" (1.9 m x 147 cm). The Museum of Modern Art, New York PURCHASE. PHOTOGRAPH © 1995 THE MUSEUM OF MODERN ART, NEW YORK. © 1997 WILLEM DE KOONING REVOCABLE TRUST/ARTISTS RIGHTS SOCIETY (ARS), NEW YORK

and fertility goddess. Although the vigorous paint application appears entirely improvisatory, the artist labored over the painting for eighteen months, scraping the canvas down, revising it, and, along the way, making countless drawings—he was a consummate draftsman—of the subject. When his paintings of women were exhibited in 1953, de Kooning, who once said his work was "wrapped in the melodrama of vulgarity," was dismayed when critics failed to see the humor in them, reading them instead as misogynist. He said his images had to do with "the female painted through all the ages, all those idols," by which he could have meant a Greek Venus, a Renaissance nude, a Picasso portrait, or a curvaceous American movie star, with her big, ferocious grin. Over the years when he was questioned about his paintings of women, de Kooning emphasized the dual nature of their sexual identity, claiming they derived partly from the feminine within himself. By the middle of the 1950s the women paintings gave way to compositions sometimes called "abstract urban landscapes" that, with their slashing lines and colliding forms, reflect the lively, gritty atmosphere of New York City streets (color plate 236, page 452). Here de Kooning dragged charcoal through wet paint, churning up the surface to create a heated atmosphere that pulsates with an intense, metropolitan beat. *Gotham News* is a quintessential example of gestural Abstract Expressionism, with the hand of the artist, the emotion- packed "gesture," everywhere apparent.

In the 1960s de Kooning moved to a large, light-filled studio on Long Island where he continued his exploration of the woman theme, although, as he said, in a more "friendly and pastoral" vein. In *Two Figures in a Landscape* (fig. 520) the figures are barely decipherable amidst baroque flourishes of lush, liquid medium. De Kooning continued to paint brilliantly into the 1980s (he stopped in 1990 because of debilitating illness), and although the works are largely abstract, the suggestion of figures lurks in the curving folds of paint (color plate 237, page 452). Here, in his characteristic light, delicate palette of rose, yellow, green, blue, and white, de Kooning has decelerated his violent brushstroke into a grandly fluid, almost Rubenesque flow of graceful color shapes.

According to de Kooning, it was JACKSON POLLOCK (1912–1956), with the radical "drip" paintings he began to make in the late 1940s, who "broke the ice." Hailing from the West, becoming a huge force on the New York art scene, living hard, drinking hard, and then dying violently in a car accident at a young age, Pollock achieved a stature of mythic proportions in the 1950s, an international symbol of the new American painting after World War II. Pollock came to New York from Cody, Wyoming, by way of Arizona and California. Until 1932 he studied at the Art Students League with Thomas Hart Benton, who represented, Pollock said, "something against which to react very strongly, later on." Nevertheless, there is a relation between Pollock's abstract arabesques and Benton's rhythmical figurative patterns (see fig. 483). The landscapes of Ryder (see fig. 67) as well as the work of the Mexican muralists were important sources for Pollock's violently expressive early paintings, as were the automatic methods of Masson and Miró. In 1939 The Museum of Modern Art mounted a large Picasso exhibition, a catalytic event for artists like de Kooning and Pollock who were then

520. WILLEM DE KOONING. *Two Figures in a Landscape.* 1967. Oil on canvas, 5'10" x 6'8" (1.8 x 2 m). Stedelijk Museum, Amsterdam © 1997 WILLEM DE KOONING REVOCABLE TRUST/ARTISTS RIGHTS SOCIETY (ARS), NEW YORK

521. JACKSON POLLOCK. *Guardians of the Secret.* 1943. Oil on canvas, 4'1¾" x 6'3" (1.2 x 1.9 m). San Francisco Museum of Modern Art

struggling to come to terms with his work, learn from his example, and forge their own independent styles out of Cubism.

Pollock's paintings of the mid-1940s, usually involving some degree of actual or implied figuration, were coarse and heavy, suggestive of Picasso, but filled with a nervous, brutal energy all their own. In *Guardians of the Secret,* 1943 (fig. 521), schematic figures stand ceremonially at either end of a large rectangle, perhaps a table, an altar, or a funeral bier. A watchdog, with possible affinities to Tamayo's animals (see fig. 506), reclines below. Pollock, who covered his canvas with calligraphic, cryptic marks like some private hieroglyphic language, apparently began with legible forms that he gradually obscured. As he later told his wife, the painter Lee Krasner, "I chose to veil the imagery." That imagery has been analyzed within the broad spectrum of Pollock's visual interests, including ritual Navaho sand painting, African sculpture, prehistoric art, and Egyptian painting. Some have seen evidence of Jungian themes in the powerfully psychic content of the artist's early work. Pollock underwent psychoanalysis with Jungian analysts beginning in 1939. Carl Jung's theories of the collective unconscious as a repository for ancient myths and universal archetypes were a frequent topic of discussion in Abstract Expressionist circles. By 1947 the artist had begun to experiment with all-over painting, a labyrinthine network of lines, splatters, and paint drips from which emerged the great "drip" or "poured" paintings of the next few years. *Number 1, 1950 (Lavender Mist)* (color plate 238, page 453) is one of his most beautiful drip paintings, with its intricate web of oil colors mixed with black enamel and aluminum paint. One can trace the movements of the artist's arm, swift and assured, as he deployed sticks or dried-out brushes to drip paint onto the surface. His lines are divorced from any descriptive function and range from stringlike thinness to

coagulated puddles, all merging into a hazy, luminous whole that seems to hover above the picture plane rather than illusionistically behind it. These paintings, generally executed on a large canvas laid out on the floor (fig. 522), are popularly associated with so-called Action Painting. It was never the intention of the critic Harold Rosenberg, in coining this term, to imply that Action Painting was a kind of athletic exercise, but rather that the process of painting was as important as the completed picture. Despite the furious and seemingly haphazard nature of his methods, Pollock's painting was not a completely uncontrolled, intuitive act. There is no

522. Jackson Pollock at work in his Long Island studio, 1950
PHOTOGRAPH BY HANS NAMUTH

question that, in his paintings and those of other Abstract Expressionists, the elements of intuition and accident play a large and deliberate part, and this was indeed one of the principal contributions of Abstract Expressionism, which had found its own inspiration in Surrealism's psychic automatism. At the same time, however, Pollock's works are informed by skills honed by years of practice and reflection, just as the improvisatory talents of jazz musicians are paradoxically enhanced by regimented training.

Pollock's spun-out skeins of poured pigment contributed other elements that changed the course of modern painting. There was, first, the concept of the all-over painting where no part of the composition is given formal precedence over another; it is "non-hierarchical," with zones of pictorial interest evenly distributed over the surface. This "holism," together with the large scale of the works, introduced another concept—that of wall painting different from the tradition of easel painting. In 1946 Pollock, who had worked on mural projects for the WPA, said he wanted to paint pictures that "function between the easel and the mural," and as scholar Elizabeth Frank has written, "Over the next three years, Pollock made great art out of this 'halfway state,' preserving the tension between the easel picture, with its capacity to draw the viewer into a fictive world, and the mural, or wall picture, with its power to inhabit the viewer's own space." This was the final break from the Renaissance idea of painting detached from spectator, to be looked at as a self-contained unit. The painting became an environment that encompassed the spectator. The feeling of absorption or participation is heightened by the ambiguity of the picture space. The colors and lines, although never puncturing deep perspective holes in the surface, still create a sense of continuous movement, a billowing, a surging back and forth, within a limited depth. Pollock referred to being "in" his paintings when he worked, and in *No. 1, 1950 (Lavender Mist)* tracks of his literal presence are recorded in his own handprints, especially at the upper left edge. In 1950, partly to dispel his image in the popular press as a mad artist senselessly flinging paint at the canvas, Pollock allowed photographer Hans Namuth to film him at work at his Long Island studio. The result is an invaluable document of the artist's methods, which embraced both spontaneity and premeditation, control and exhilarating freedom. In many ways, Pollock departed from the tradition of Renaissance and modern painting before him, and although he had no direct stylistic followers, he significantly affected the course of experimental painting after him.

Pollock experienced a period of crisis and doubt in the wake of the success of his drip paintings, and in late 1950 he began to re-explore problems of the figure in several drawings and paintings, using black predominantly. Between 1953 and 1956 he returned to traditional brush painting, sometimes heavy in impasto, and involving images reminiscent of paintings of the early 1940s. In the magisterial *Portrait and a Dream*, 1953 (color plate 239, page 453), he constructed a visual dialogue between the two modes. At the left he dripped black paint in abstracted patterns that vaguely suggest human anatomies. At the right, in what many have read as a self-portrait, he resorted to overt figuration with brush-applied color. It is as if the figure lies latent within his webs of paint and, with a few adjustments of his arm, those webs could call up an image. "When you're painting out of your unconscious," Pollock said, "figures are bound to emerge." These later works suffered in critical opinion when compared to the great drip paintings—even some of Pollock's strongest supporters saw them as a retreat. The black-and-white canvases and the paintings that followed suggest a new phase of signal importance, unfortunately terminated by the artist's premature death in 1956.

Brooklyn-born LEE KRASNER (1908–1984) received academic training at several New York art schools before joining the Mural Division of the WPA in 1935. From 1937 to 1940 she was a student of Hans Hofmann, from whom she said she "learned the rudiments of Cubism." Subsequently, she joined the AAA, devoted to nonobjective art in the tradition of Mondrian, and exhibited Cubist abstractions in their group shows. Krasner became acquainted with Pollock in 1942, when John Graham included their work in a show of young artists he felt showed promise. Krasner and Pollock began to share a studio in 1942, married in 1945, and relocated to Long Island. Unlike Pollock, Krasner felt a general antipathy toward the Surrealists and their automatic methods, due in part, as art historian Barbara Rose has shown, to the fundamentally misogynist attitudes of the European Surrealist emigrés (compounded by the general marginalization of women within Abstract Expressionist circles). Throughout the 1940s and 1950s, working in the presence of Pollock's forceful personality, Krasner gradually moved away from Cubist-based forms to a concern for spontaneous gesture and large-scale all-over compositions, while remaining committed to a Mondrianesque sense of structure.

Between 1959 and 1962, following the traumas of Pollock's death as well as that of her mother, Krasner made a series of Umber paintings, so-called for their predominantly brownish hues, which the artist once referred to as "colorless." In *Polar Stampede* (color plate 240, page 454) she approached the canvas with a loaded brush, allowing the medium to splash and explode across the canvas until it resembled a vast glacial landscape that has been stirred up, as the title suggests, by the pounding hooves of wild animals. Krasner did not allow herself improvisational freedom to the degree practiced by Pollock, for even in such a boldly gestural abstraction, her strokes have a regular, rhythmical beat, and a sense of imposed structure.

Krasner had employed the collage medium throughout her career. In 1976 she cut up her own drawings and paintings, Cubist works made while she was a Hofmann student, and assembled the sliced images into new compositions (fig. 523). The sharp cutout forms of *Imperative* overlap and interpenetrate one another, ironically functioning like the Cubist shards in her original drawings. By mutilating her own past work, Krasner, always intensely self-critical, could reclaim it for another context and arrive at a powerful new synthesis.

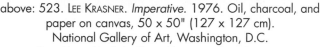

above: 523. Lee Krasner. *Imperative.* 1976. Oil, charcoal, and
paper on canvas, 50 x 50" (127 x 127 cm).
National Gallery of Art, Washington, D.C.
Gift of Mr. and Mrs. Eugene Thaw in Honor of the Fiftieth Anniversary.
© 1997 Pollock-Krasner Foundation/Artists Rights Society (ARS), New York

above, right: 524. Franz Kline. *Nijinsky (Petrushka).* c. 1948.
Oil on canvas, 33½ x 28" (85.1 x 71.1 cm).
Private collection

right: 525. Franz Kline. *Nijinsky.* 1950. Enamel on canvas,
46 x 35¼" (116.8 x 89.5 cm).
Collection Muriel Kallis Newman, Chicago

Franz Kline (1910–1962) was as fascinated with the
details and tempo of contemporary America as Stuart Davis,
but he was also deeply immersed in the tradition of Western
painting, from Rembrandt to Goya. Born in the coal mining
region of Eastern Pennsylvania, Kline studied art at academies
in Boston and London and, throughout the 1940s, painted
figures and urban scenes tinged with Social Realism. A passion for drawing manifested itself during the 1940s, particularly in his habit of making little black-and-white sketches,
fragments in which he studied single motifs, structure, or
space relations. One day in 1949, according to the painter
and critic Elaine de Kooning, Kline was looking at some of
these sketches enlarged through an opaque projector and
saw their implications as large-scale, free abstract images.
Although this moment of revelation has no doubt been
largely exaggerated, and although Kline's mastery of abstraction was gradual rather than instantaneous, he did formulate
his new Abstract Expressionist vocabulary with astonishing
speed. Even the first works in his new mode have nothing
tentative about them.

In 1950, when he made his first large-scale black-and-
white abstractions, Kline was well aware of the experiments of
the pioneer Abstract Expressionists, although he never shared
Pollock's interest in myth or Rothko's interest in the sublime,
and he tended to gamble less with spontaneous gesture than
his friend de Kooning. Instead he worked out his compositions in advance, with preliminary sketches, usually on pages
torn from the telephone book. *Nijinsky* (fig. 525) is an
abstract meditation on his own figurative representations of
the same subject (fig. 524), based on a photograph of the
great Russian dancer as Petrushka; at the same time, it was a
surrogate self-portrait.

above: 526. FRANZ KLINE. *Mahoning.* 1956. Oil on canvas,
6'8" x 8'4" (2 x 2.5 m).
Whitney Museum of American Art, New York
GIFT OF THE FRIENDS OF THE WHITNEY MUSEUM OF AMERICAN ART

right: 527. BRADLEY WALKER TOMLIN. *Number 8 (Blossoms).*
1952. Oil on canvas, 66 x 48" (167.6 x 121.9 cm).
The Phillips Collection, Washington, D.C.

In later works such as *Mahoning* (fig. 526), named for a town in Pennsylvania, curved forms give way to straight, girderlike strokes that hurtle across the full breadth of the eight-and-a-half-foot canvas. Using house-painter brushes on unstretched canvases tacked to his studio wall, Kline shaped rugged but controlled brushstrokes into powerful, architectural structures that have affinities with motifs in the industrial landscape he admired, such as trains, cranes, and bridges. In the paint texture as well as in the shaping of forms, there is an insistence on the equivalence of the whites that prevents the work from becoming simply a blown-up black drawing on a white ground, for, as Kline said, he also painted the white areas, sometimes on top of the black.

A number of artists in this period had pared their art down to the essentials of black and white. Both de Kooning and Pollock, as we have seen, renounced color for a time, as did Newman, Motherwell, Tomlin, and a number of other artists associated with the New York School. Kline reintroduced color to his compositions in the late 1950s but sadly died in the midst of these new experiments. Like de Kooning's, Kline's painted structures had an enormous impact on younger artists. Their art fostered a veritable school of gestural painters in the late 1950s, the so-called second generation of Abstract Expressionists, the best of whom forged their own individual styles from the pioneers' examples (see fig 667).

Until the mid-1940s BRADLEY WALKER TOMLIN (1899–1953) was one of the most sensitive and accomplished Americans working in an abstract Cubist style. Toward the end of the decade he completed a group of pic-

tographic compositions related to those of Adolph Gottlieb, including *All Souls Night* (color plate 241, page 454), where he superimposed signs resembling ancient hieroglyphs over loosely brushed and delicately colored backgrounds. In 1948 he began to experiment with a form of free calligraphy, principally in black and white, deriving from his interest in automatism, Zen Buddhism, and Japanese brush painting. From this moment, Tomlin returned during the last three or four years of his life to painterly abstractions in which delicate and apparently spontaneous dabs of paint create shimmering, all-over compositions (fig. 527). In these so-called petal paintings Tomlin discovered some of the lyricism and luminosity of Impressionist landscape, but he subjected his strokes to the pictorial discipline of a loosely implied grid.

MARK TOBEY (1890–1976) moved to Seattle from New York in 1922, working far from the artistic mainstream yet producing a unique body of work in step with the most advanced tendencies in New York. Tobey was drawn by a profound inner compulsion to a personal kind of abstract expression with strongly religious overtones. He was a convert to the Bahai faith, which, as he said, stresses "the unity of the world and the oneness of mankind." He began studying Chinese brush painting in 1923 and in the 1930s studied Zen Buddhism in China and Japan. In *Broadway* (fig. 528) Tobey applied the lessons of calligraphy to communicate his vivid memories of New York. His animated line, with its intricate, jazzy rhythms—he called it "white writing"—captures the lights, noise, and frenetic tempo of the city. Like de Kooning or Pollock, Tobey moved between figurative and

528. MARK TOBEY. *Broadway*. c. 1935. Tempera on Masonite, 26 x 19¼" (66 x 48.9 cm). The Metropolitan Museum of Art, New York ARTHUR H. HEARN FUND, 1942. © 1997 ARTISTS RIGHTS SOCIETY (ARS), NEW YORK/PRO LITTERAS, ZURICH

abstract modes throughout the 1940s. In his best-known pictures, such as *Universal Field* (color plate 242, page 454), the "white writing" becomes a kind of nonreferential calligraphy that is a small-scale, wrist-painted version of the broader, arm-gestured, all-over surfaces poured by Pollock. Tobey, who actually composed paintings from linear arabesques and fluid color before Pollock's all-over drip paintings, here creates a field charged with energy and light, as his drawn lines zip and dart through a shallow, electrified space.

PHILIP GUSTON (1913–1980) was expelled from Manual Arts High School in Los Angeles along with his classmate Jackson Pollock for lampooning the conservative English department in their own broadside. Pollock was readmitted but Guston set off on his own, working at odd jobs, attending art school briefly as well as meetings of the Marxist John Reed Club. He followed Pollock to New York and, in 1936, signed up for the mural section of the WPA. By the mid 1940s, drawing on his diverse studies of Cubism, the Italian Renaissance, and the paintings of de Chirico and Beckmann, Guston was painting mysterious scenes of figures in compressed spaces, works that earned him a national reputation.

Guston taught painting for a time in Iowa City and St. Louis and, in the late 1940s, began to experiment with loosely geometric abstractions. After many painful intervals of doubt, he made his first abstract gestural paintings in 1951. These contained densely woven networks of short strokes in vertical and horizontal configurations (color plate 243, page 455) that reveal an admiration for Mondrian. The heavy impasto, serene mood, and subtle, fluctuating light in his

muted reds, pinks, and grays caused some to see these works as lyrical landscapes, earning Guston the label "Abstract Impressionist." But Guston, who had negligible interest in the Impressionists, explained, "I think of painting more in terms of the drama of this process than I do of 'natural forces.'"

Throughout the 1950s Guston employed larger and more emphatic gestures in his abstractions, with a tendency to concentrate centralized color masses within a light and fluid environment. In 1962 the Guggenheim Museum mounted a major Guston retrospective, as did New York's Jewish Museum in 1966. But despite these external successes, Guston was experiencing a crisis in his art, with concerns about abstraction's limited potential for expressing the full gamut of human experience. This came at a time when the New York art world was undergoing a profound shift in mood, with young Pop artists and hard-edge abstractionists casting a critical eye over what they regarded as the emotional excesses of their artistic predecessors. In the late 1960s Guston returned to recognizable imagery, astonishing the art world with intensely personal, sometimes nightmarish images in a crude, cartoonish style. "I got sick and tired of all that purity," he said, "[I] wanted to tell stories." A gifted draftsman with a penchant for caricature as well as an abiding fondness for cartoons and, not insignificantly, for the writing of Franz Kafka and Samuel Beckett, Guston composed haunting scenes in a characteristic palette of pink, black, and red. At first he invented narratives featuring ominous hooded figures, their ambiguous identity ranging from marauding Ku Klux Klan members to the artist himself at his easel (color plate 244, page 455). In *Head and Bottle* (fig. 529), Guston

529. PHILIP GUSTON. *Head and Bottle*. 1975. Oil on canvas, 65½ x 68½" (166.4 x 174 cm). Collection Mrs. Sally Lilienthal, San Francisco COURTESY MCKEE GALLERY, NEW YORK

envisioned himself as a grotesque, disembodied head, flushed and stubble covered, but eye wide open. When they were shown in New York, Guston's figurative paintings were seen by many, including both former detractors and supporters, as a betrayal of abstraction and an acquiescence to current Pop trends. In many ways the full implications of Guston's powerfully moving late canvases were not realized until they provided the key inspiration for the New Image and Neoexpressionist painters of the early 1970s and 1980s.

ELAINE DE KOONING (1918–1989) is probably best known for the figurative paintings she made in the 1950s, but at her death a group of abstractions from the 1940s were discovered among her belongings that reveal her early experiments within an Abstract Expressionist mode. She spent the summer of 1948 at Black Mountain College in North Carolina with her husband, Willem de Kooning (they had married in 1943). At the invitation of Josef Albers, Willem had joined the distinguished faculty at the experimental school as a visiting artist. Elaine, who painted *Untitled, Number 15* (fig. 530) during that summer, always acknowledged the importance of her husband's paintings as well as those of their friend, Archile Gorky, to her own work: "Their reverence and knowledge of their materials, their constant attention to art of the past and to everything around them simultaneously, established for me the whole level of consciousness as the way an artist should be." In *Untitled, Number 15* she set suggestive, biomorphic forms amidst more geometric shapes, recalling her husband's work of the same moment (fig. 518). Following the summer at Black Mountain, de Kooning began to write on contemporary art for *ARTnews*. Her keen perceptions, as well as her vantage point as a practicing artist who knew her subject firsthand, endowed authority to her writing. In the 1950s and 1960s de Kooning displayed expressive tendencies in a group of male

530. ELAINE DE KOONING. *Untitled, Number 15*. 1948.
Enamel on paper, mounted on canvas,
32 x 44" (81.3 x 111.8 cm).
The Metropolitan Museum of Art, New York IRIS CANTOR GIFT

portraits (including several of John F. Kennedy). Her portrayal of art critic Harold Rosenberg (color plate 245, page 455), one of the champions of Abstract Expressionism, shows how she brought the gesturalism of the New York School to bear on figurative subjects.

GRACE HARTIGAN (b. 1922) was an even younger member of the New York School than Elaine de Kooning, but one who also witnessed the critical artistic developments in New York at close range. A so-called second-generation Abstract Expressionist, Hartigan was deeply affected as a young artist by her encounter with Pollock's drip paintings and Willem de Kooning's black-and-white paintings, both shown in New York in 1948. Like de Kooning, whose famous perambulations through downtown New York were fodder for his strident "urban landscapes" (color plate 236, page 452), Hartigan found inspiration in the most mundane aspects of city life. In *Giftwares*, 1955 (color plate 246, page 456), she depicted the tawdry offerings of a storefront window, transforming them into a glowing still life where the loosely brushed but legible forms are pulled up to the picture plane and dispersed evenly across the canvas. While Hartigan's rich palette is a kind of homage to Matisse, her homely subject matter prefigures work by artists of the 1960s such as Claes Oldenburg (see color plate 304, page 529). In the late 1950s Hartigan embarked on a series of distinctive gestural abstractions before resuming a highly abstracted brand of figuration the following decade.

Color Field Painting

Although the divisions between gestural and Color Field painting are somewhat artificial, there were both formal and conceptual differences between artists such as Kline and de Kooning on the one hand and Mark Rothko and Barnett Newman on the other. In 1943 Rothko, Newman, and Adolph Gottlieb stated their purpose in a letter to *The New York Times:* "We favor the simple expression of the complex thought. We are for the large shape because it has the impact of the unequivocal. We wish to reassert the picture plane. We are for flat forms because they destroy illusion and reveal truth." These artists fervently believed that abstract art was not "subjectless" and that, no matter how reductive, it could communicate the most profound subjects and could elicit a deep emotional response in the viewer. In their search for universal subject matter, for "symbols," as Rothko said, "of man's primitive fears and motivations," the artists looked to many sources, including Jungian theory, Surrealist practice, and the art of non-European societies.

MARK ROTHKO (1903–1970) immigrated to the United States with his Russian-Jewish family in 1913 and moved ten years later to New York, where he studied with Max Weber at the Art Students League. Throughout the 1930s he made figurative paintings on mostly urban themes and in 1935 formed an independent artists' group with Gottlieb called "The Ten." In 1940, in search of more profound and universal themes and impressed by his readings of Nietzsche and Jung, Rothko began to engage with ancient myths as a source

531. MARK ROTHKO. *No. 18.* 1948. Oil on canvas,
67¼ x 55⅞" (170.8 x 141.9 cm). Frances Lehman Loeb
Art Center, Poughkeepsie, New York
GIFT OF MRS. JOHN D. ROCKEFELLER III.
© 1997 KATE ROTHKO-PRIZEL AND CHRISTOPHER ROTHKO/ARTISTS RIGHTS SOCIETY (ARS), NEW YORK

of "eternal symbols." In their letter to *The New York Times,* Rothko, Newman, and Gottlieb proclaimed that "only that subject matter is valid which is tragic and timeless. That is why we profess spiritual kinship with primitive and archaic art." Rothko first made compositions based on classical myths and then, by the mid-1940s, painted biomorphic, Surrealist-inspired, hybrid creatures floating in primordial waters. These forms began to coalesce at the end of the decade into floating color shapes with loose, undefined edges within larger expanses of color (fig. 531). By 1949 Rothko had refined and simplified his shapes to the point where they consisted of color rectangles floating on a color ground (color plate 247, page 456). He applied thin washes of oil paint that contain considerable tonal variation and blurred the edges of the rectangles to create luminous color effects and a shifting, ambiguous space. Rothko explored this basic compositional type with infinite and subtle variation over the next twenty years. By the sheer sensuousness of their color areas and the sense of indefinite outward expansion without any central focus, the paintings are designed to absorb and engulf the spectator. Eventually, Rothko was painting on a huge scale, something that contributed to the effect of enclosing, encompassing color.

At the end of the 1950s Rothko began to move away from bright, sensuous color toward hues that were deep and somber, and, by implication, tragic (color plate 248, page 457). He received commissions for large public spaces, including a restaurant in the Seagram Building in New York, at Harvard University, and in a chapel in Houston, Texas (color plate 249, page 457). The latter series consists of fourteen large panels, almost uniform in their deep black and red-brown, virtually filling the entire space of available walls. The canvases are designed to be homogenous with the octagonal building (designed by Philip Johnson) and the changing light of the chapel interior. Thus, the forms have a less amorphous, harder-edge appearance than Rothko's earlier abstractions. They represent a total architectural-pictorial experience in a sense analogous to that attempted by the religious muralists of the Baroque seventeenth century. Rothko, who even adopted the format of the triptych, long associated with religious subjects, never intended his abstractions as mere formal experiments, but rather wanted to provoke emotional and even transcendental experiences for the viewer.

Like Rothko, **BARNETT NEWMAN** (1905–1970) studied at the Art Students League in the 1920s, but his artistic career did not begin in earnest until 1944 when he resumed painting after a long hiatus and after he had destroyed his early work. By the mid-1950s, he had received nowhere near the attention that Pollock, de Kooning, or Rothko had attracted, although it could be argued that, in the end, his work had the greatest impact on future generations. His mature work differed from that of other Abstract Expressionists (except Ad Reinhardt's) in its radical reductiveness and its denial of painterly surface. As the group's most thoughtful and most polemical theorist, Newman produced an important body of critical writing on the work of his contemporaries as well as subjects such as Pre-Columbian art.

In his early mythic paintings Newman, who had studied botany and ornithology, explored cosmic themes of birth and creation, of primal forms taking shape. Those works share many traits with Rothko's biomorphic paintings. By 1946, however, in canvases such as *Genesis—The Break* (color plate 250, page 457), the forms become more abstract and begin to shed their biological associations, although the round shape here is a recurring seed form. Thomas Hess, a leading champion of the Abstract Expressionists, said this painting (as the title implies) was about "the division between heaven and earth." In 1948 Newman made *Onement I,* which he regarded as the breakthrough picture that established his basic formula—a unified color field interrupted by a vertical line—a "Zip" as he called it—or, rather, a narrow, vertical contrasting color space that runs the length of the canvas. The nature of the Zip varied widely, from irregular hand-brushed bands to uninflected, straight edges made possible with the use of masking tape, but the impression is usually of an opening in the picture plane rather than simply a line on the surface. While Newman did not seek out the atmospheric effects that Rothko achieved in his mature works, he was capable of brushed surfaces of tremendous beauty and nuance.

Vir Heroicus Sublimis (color plate 251, page 458) is a mature, mural-size painting where the multiple zips differ in hue and value, sometimes in stark contrast to the brilliant monochrome field, sometimes barely distinguishable from it.

532. Barnett Newman in front of *The Stations of the Cross* at the Solomon R. Guggenheim Museum, New York, April 1966. Now at the National Gallery of Art, Washington, D.C.
ROBERT AND JANE MEYERHOFF COLLECTION. © 1997 BARNETT NEWMAN FOUNDATION/ARTISTS RIGHTS SOCIETY (ARS), NEW YORK

Newman wanted to maintain a human scale for his essentially life-affirming, humanist art. Indeed, his title for this work means "Heroic Sublime Man," and his ubiquitous Zip has been read as a sign for the upright human being. Between 1958 and 1964, Newman painted a series of fourteen canvases on the Passion of Christ called *The Stations of the Cross—Lema Sabachthani,* the first four of which are shown in figure 532 with Newman at the Guggenheim Museum. He orchestrated his Zip across the series, restricting himself to black and white on unprimed canvas, but varying his medium (Magna, oil, and acrylic) as well as the thickness and character of his Zips. Newman said the theme of the series was "the unanswerable question of human suffering," underscored by Christ's words in the Hebrew subtitle, meaning "Why did you forsake me?"

CLYFFORD STILL (1904–1980), who grew up in Washington state and Alberta, Canada, differed from the other Abstract Expressionists in that virtually all of his training and early artistic development took place outside of New York City. He had, however, exhibited at Peggy Guggenheim's Art of This Century beginning in 1945, thanks to Rothko, and at the Betty Parsons Gallery in 1947. In 1950, he moved to New York from San Francisco, where he had been teaching at the progressive California School of Fine Arts (now the San Francisco Art Institute). He remained in New York until 1961, when he took up residence in the Maryland countryside. Fiercely proud and independent, notoriously controlling about his work, and generally contemptuous of the "art world," Still preferred to exhibit infrequently and live at a considerable distance from the nexus of the art establishment in New York.

Since the late 1930s Still had painted in a freely abstract manner (which contained some vestigial figuration) involving large, flowing images executed in heavy, even coarse, paint textures that he realized with brushes and palette knives. He repeatedly disavowed any interest in Surrealism or the "myth-making" of his New York contemporaries and refused to give titles to his works (other than neutral letters or a date) to prevent any association with a specific subject. By 1947 he was working on a huge scale, sometimes eight by ten feet, with immense, crusty areas of color in a constant state of fluid though turgid movement (color plate 252, page 458). These abstractions have been described as awesome landscapes that combine the drama and vastness of the West—its mesas, canyons, and rivers—and the Romantic landscape imagery of Albert Pinkham Ryder (see fig. 67). Art historian Robert Rosenblum characterized the dramatic effect of Still's canvases as "abstract sublime." Although the artist resisted any direct association between his compositions and landscape, he once said his work reflected "man's struggle and fusion with nature." Over the next twenty years Still persisted in his basic image, a predominant color varying in value and shot through with brilliant or somber accents (fig. 533).

AD REINHARDT (1913–1967) was already painting geometric abstractions in the late 1930s. He was a superb cartoonist and parodist, a mercilessly incisive writer, and a serious student of Oriental art. As he wrote in a prominent art magazine in 1962, he was for "art-as-art." He regarded the talk of myth and tragedy among the Abstract Expressionists as histrionic nonsense and objected to any association of his art with the ideals of the group. After abstract experiments in the 1940s (fig. 534), Reinhardt began in the early 1950s to simplify his palette to a single color. The first groups of such paintings were all red or all blue (color plate 253, page 458), and the final group, on which he worked for the rest of his life, was all black. One first observes these later paintings simply as monochrome fields. With time, however (and especially with the black paintings), there begins to emerge a second,

Color plate 230. STUART DAVIS. *Rapt at Rappaport's.* 1952.
Oil on canvas,
52⅜ x 40⅜" (133 x 102.6 cm).
Hirshhorn Museum and Sculpture
Garden, Smithsonian Institution,
Washington, D.C.
© 1997 ESTATE OF STUART DAVIS/VAGA,
NEW YORK, NY

Color plate 231. BURGOYNE DILLER.
Second Themes. 1937–38. Oil on
canvas, 30⅛ x 30"
(76.5 x 76.2 cm).
The Metropolitan Museum of Art,
New York
GEORGE A. HEARN FUND, 1963
© 1997 ESTATE OF BURGOYNE DILLER/LICENSED BY VAGA,
NEW YORK, NY

above: Color plate 232.
MILTON AVERY. *Swimmers and Sunbathers.* 1945. Oil on canvas, 28 x 48⅛" (71.1 x 122.2 cm). The Metropolitan Museum of Art, New York

GIFT OF MR. AND MRS. ROY R. NEUBERGER, 1951
© 1997 MILTON AVERY TRUST/ARTISTS RIGHTS SOCIETY (ARS), NEW YORK

left: Color plate 233.
AUGUSTUS VINCENT TACK. *Night, Amargosa Desert.* 1935. Oil on canvas mounted on plywood panel, 7' x 48" (2.1 m x 121.9 cm). The Phillips Collection, Washington, D.C.

ACQUIRED 1937

right:
Color plate 234.
HANS HOFMANN.
The Gate. 1959–60.
Oil on canvas,
6'2⅝" x 48¼"
(1.9 m x 122.6 cm).
Solomon R.
Guggenheim
Museum,
New York

below:
Color plate 235.
ARSHILE GORKY. *The
Liver Is the Cock's
Comb*. 1944.
Oil on canvas,
6'1¼" x 8'2"
(1.9 x 2.5 m).
Albright-Knox Art
Gallery, Buffalo,
New York

above: Color plate 236.
WILLEM DE KOONING.
Gotham News. 1955. Oil,
enamel, charcoal, and news-
paper transfer on canvas,
69" x 6'7"
(175.3 cm x 2 m).
Albright-Knox Art Gallery,
Buffalo, New York
GIFT OF SEYMOUR H. KNOX
© 1997 WILLEM DE KOONING REVOCABLE
TRUST/ARTISTS RIGHTS SOCIETY (ARS),
NEW YORK

left: Color plate 237.
WILLEM DE KOONING.
Pirate (Untitled II). 1981.
Oil on canvas, 7'4" x
6'4¾" (2.2 x 1.9 m). The
Museum of Modern Art,
New York
SIDNEY AND HARRIET JANIS COLLECTION
FUND © 1997 WILLEM DE KOONING
REVOCABLE TRUST/ARTISTS RIGHTS SOCIETY
(ARS), NEW YORK

Color plate 238. JACKSON POLLOCK. *Number 1, 1950 (Lavender Mist)*. 1950. Oil, enamel, and aluminum on canvas, 7'4" x 9'11" (2.2 x 3 m). National Gallery of Art, Washington, D.C. AILSA MELLON BRUCE FUND

Color plate 239. JACKSON POLLOCK. *Portrait and a Dream*. 1953. Enamel on canvas, 58⅛" x 11'2½" (147.6 cm x 3.4 m). Dallas Museum of Art

Color plate 240. LEE KRASNER. *Polar Stampede*. 1960. Oil on cotton duck, 7'9⅝" x 13'3¾" (2.4 x 4 m). Collection Pollock Krasner Foundation. Courtesy Robert Miller Gallery, New York

Color plate 241. BRADLEY WALKER TOMLIN. *All Souls Night*. 1947. Oil on canvas, 42½ x 64" (106.7 x 162.6 cm). Courtesy Vivian Horan Fine Art, New York

Color plate 242. MARK TOBEY. *Universal Field*. 1949. Pastel and tempera on cardboard, 28 x 44" (73.9 x 116.2 cm). Whitney Museum of American Art, New York

right: Color plate 243. Philip Guston.
Zone. 1953–54. Oil on canvas,
46 x 48" (116.8 x 122.2 cm).
The Edward R. Broida Trust,
Los Angeles

below: Color plate 244. Philip Guston.
Studio. 1969. Oil on canvas, 48 x 42"
(121.9 x 106.7 cm). Private collection
Courtesy McKee Gallery, New York

Color plate 245. Elaine de Kooning. *Harold
Rosenberg #3.* 1956. Oil on canvas, 6'8" x 58⅞"
(2 m x 149.5 cm). National Portrait Gallery,
Smithsonian Institution, Washington, D.C.

Color plate 246. GRACE HARTIGAN. *Giftwares*. 1955. Oil on canvas, 63" x 6'9" (160 cm x 2.1 m). Neuberger Museum of Art, College at Purchase, State University of New York
GIFT OF ROY R. NEUBERGER

Color plate 247. MARK ROTHKO. *Untitled (Rothko number 5068.49)*. 1949. Oil on canvas, 6'9⅜" x 66⅜" (2.1 m x 168.6 cm). National Gallery of Art, Washington, D.C.
GIFT OF THE MARK ROTHKO FOUNDATION
© 1997 KATE ROTHKO, PRIZEL AND CHRISTOPHER ROTHKO/ARTISTS RIGHTS SOCIETY (ARS), NEW YORK

left: Color plate 248. MARK ROTHKO. *White and Greens in Blue.* 1957. Oil on canvas, 8'4" x 6'10" (2.5 x 2.1 m). Private collection

below: Color plate 249. MARK ROTHKO. North, Northeast, and East wall paintings in the Rothko Chapel, Houston, Texas. 1965–66 (opened in 1971). Oil on canvas.

Color plate 250. BARNETT NEWMAN. *Genesis—The Break.* 1946. Oil on canvas, 24 x 27" (61 x 68.6 cm). Collection DIA Center for the Arts, New York

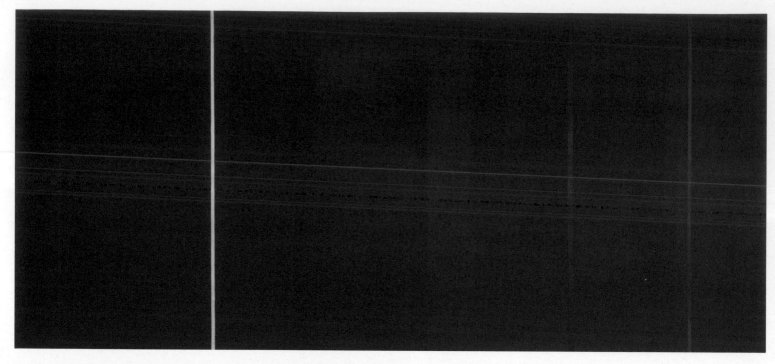

Color plate 251. BARNETT NEWMAN. *Vir Heroicus Sublimis.* 1950–51. Oil on canvas, 7'11⅜" x 17'9¼" (2.4 x 5.1 m).
The Museum of Modern Art, New York

GIFT OF MR. AND MRS. BEN HELLER © 1997 BARNETT NEWMAN FOUNDATION/ARTISTS RIGHTS SOCIETY (ARS), NEW YORK

right: Color plate 253.
AD REINHARDT. *Abstract Painting, Blue.* 1952. Oil on canvas, 6'3" x 28" (1.9 m x 71.1 cm). Carnegie Museum of Art, Pittsburgh

MUSEUM PURCHASE: GIFT OF WOMEN'S COMMITTEE © 1997 ESTATE OF AD REINHARDT/ARTISTS RIGHTS SOCIETY (ARS), NEW YORK

Color plate 252. CLYFFORD STILL. *Number 2.* 1949. Oil on canvas, 7'8" x 67" (2.3 m x 170.2 cm). Collection June Lang Davis, Medina, Washington

Color plate 254. ADOLPH GOTTLIEB. *Orb*. 1964. 7'6" x 60" (2.3 m x 152.4 cm). Dallas Museum of Art

above: Color plate 255. ROBERT MOTHERWELL. *Elegy to the Spanish Republic No. 34.* 1953–54. Oil on canvas, 6'8" x 8'4" (2 x 2.5 m). Albright-Knox Art Gallery, Buffalo, New York

GIFT OF SEYMOUR H. KNOX, 1957

© 1997 DEDALUS FOUNDATION/VAGA, NEW YORK, NY

below: Color plate 256. ROBERT MOTHERWELL. *Summer Open with Mediterranean Blue.* 1974. Acrylic on canvas, 48" x 9' (121.9 cm x 2.7 m). Collection Dedalus Foundation

© 1997 DEDALUS FOUNDATION/VAGA, NEW YORK, NY

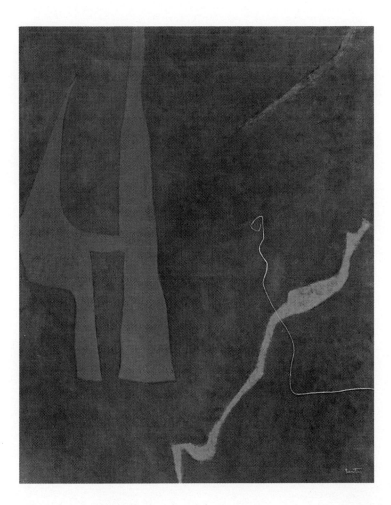

Color plate 257. WILLIAM BAZIOTES. *Dusk*. 1958. Oil on canvas, 60⅜ x 48¼" (153.4 x 122.6 cm). Solomon R. Guggenheim Museum, New York

below: Color plate 258. ISAMU NOGUCHI. *The Isamu Noguchi Garden Museum* Opened to the public in 1985. Long Island City, New York

Color plate 259. LOUISE BOURGEOIS. *Cell (You Better Grow Up)*. 1993. Steel, glass, marble, ceramic, and wood, 6'11" x 6'10" x 6'11½" (2.1 x 2.1 x 2.1 m). Collection the artist COURTESY ROBERT MILLER GALLERY, NEW YORK
© 1997 LOUISE BOURGEOIS/VAGA, NEW YORK, NY

Color plate 260. JOSEPH CORNELL.
Untitled (The Hotel Eden).
c. 1945. Mixed-media construction, 15⅛ x 15¾ x 4¾" (38.4 x
40 x 12.1 cm). National Gallery
of Canada, Ottawa

below: Color plate 261.
LOUISE NEVELSON.
Left: *An American Tribute to the
British People.* 1960–65. Wood
painted gold, 10'2" x 14'3" (3.1
x 4.3 m). Tate Gallery, London.
Right: *Sun Garden, No. 1.*
1964. Wood painted gold,
6' x 41" (1.8 m x 104.1 cm).
Private collection

Color plate 262. MARK DI SUVERO. *Aurora.* 1992–93. Steel, 16'4" x 19'6" x 26' (5 x 5.9 x 7.9 m). National Gallery of Art, Washington, D.C. GIFT OF THE MORRIS AND GWENDOLYN CAFRITZ FOUNDATION

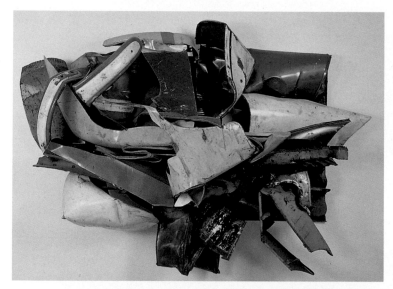

Color plate 263. JOHN CHAMBERLAIN. *Dolores James.* 1962. Welded and painted automobile parts, 6'4" x 8'1" x 39" (1.9 m x 2.5 m x 99.1 cm). Solomon R. Guggenheim Museum, New York

© 1997 JOHN CHAMBERLAIN/ARTISTS RIGHTS SOCIETY (ARS), NEW YORK

Color plate 264. ELIOT PORTER. *Dark Canyon, Glen Canyon.* 1965. Dye transfer print, 10³⁄₁₆ x 7⅞" (25.9 x 20 cm). The Art Institute of Chicago

ANONYMOUS GIFT IN HONOR OF HUGH EDWARDS

533. CLYFFORD STILL. *Untitled.* 1958. Oil on canvas, 9'5¾" x 13'3½" (2.9 x 4 m). Collection Lannan Foundation, Los Angeles

534. AD REINHARDT. *Number 18, 1948–49.* 1948–49. Oil on canvas, 40 x 60" (101.6 x 152.4 cm). Whitney Museum of American Art, New York

inner image—a smaller rectangle, square, or central cross consisting of a slightly different value or tone of the color, what art historian Yve-Alain Bois called "the fictive transformation of one color into two." These paintings are static, symmetrical, and, according to the artist, timeless. They reward slow looking and are powerfully hypnotic in their effect. Although Reinhardt applied his colors with brushes, he did so without leaving the slightest trace of his implement on the surface. His stated purpose was the elimination of all associations, all extraneous elements, all symbols and signs, and the refinement of paintings to a single dominant experience. During the 1960s and 1970s Reinhardt was much admired by Minimalist artists and became something of a cult figure among Conceptual artists, partly due to his writings and his extreme positions on the nature and function of the artist.

Born in New York City, **ADOLPH GOTTLIEB** (1903–1974) studied at the Art Students League with John Sloan and Robert Henri (see figs. 465, 464). While visiting Europe in 1921–22, he became aware of European experiments in

Cubism, abstraction, and expressionism. The friendships he formed with Rothko, Newman, and Milton Avery in the 1920s were crucial for his future development. During a stay in Arizona during 1937–38 Gottlieb made a number of curious paintings of objects picked up from the desert and arranged within rectangular compartments. These anticipated the Pictographs, irregular painted grids filled with two-dimensional ideograms, that occupied the artist between 1941 and 1953 (fig. 535). Gottlieb composed his paintings intuitively, drawing on automatist methods as well as on his interest in ancient myth and ritual. The Pictographs—the term itself implies prehistoric cave paintings—had European sources, from Mondrian, Miró, and Klee, to the Cubist grids of Uruguayan artist Torres-Garcia, but it was also African and Native American art, notably that of the Northwest Coast Kwakiutl Indians, that provided Gottlieb with a powerful lexicon of mysterious signs that he assembled to form a new, universal language. Their ominous mood may reflect the dark realities of war, as Gottlieb himself seemed to suggest: "Today when our aspirations have been reduced to a desperate attempt to escape from evil . . . our obsessive, subterranean, and pictographic images are the expression of the neurosis which is our reality."

Gottlieb's principal development of the late 1950s and 1960s was the series called Bursts (color plate 254, page 459),

below: 535. ADOLPH GOTTLIEB. *Voyager's Return.* 1946. 37⅞ x 29⅞" (96.2 x 75.9 cm). The Museum of Modern Art, New York

a sort of cosmic landscape consisting generally of an upper circular or ovoid element suggestive of a burning sun, below which is a broken, exploding element, open and dynamic in contrast with the closed form above. The Burst paintings combined aspects of gestural painting along with the expansive color of the Color Field painters. In general terms the Bursts seem to underscore life's fundamental dualities and conjure up landscapes at the dawn of civilization. More specifically, they carry the inescapable and recent memory of the atom bomb.

ROBERT MOTHERWELL (1915–1991), the youngest of the artists originally associated with Abstract Expressionism, received his early training in literature, art history, and philosophy at Stanford, Harvard, and Columbia. Motherwell's erudition made him an eloquent spokesman for Abstract Expressionism as well as a leading theorist of modern art. As a painter, he was largely self-trained with the exception of some formal study as a young man in California and later with the Surrealist Kurt Seligmann in New York. In 1941 Motherwell made the acquaintance of several European Surrealists in New York, including Ernst, Tanguy, Masson, and Matta. The aspect of Surrealism that most intrigued him was automatism, the concept of the intuitive, the irrational, and the accidental in the creation of a work of art. In many ways his early paintings represent an attempt to resolve the seeming contradictions between Mondrian (whom he had also met in New York) and the abstract Surrealists.

In 1943, at Matta's suggestion, the artist began to experiment with collage, introducing automatist techniques in roughly torn pieces of paper. One of the earliest examples is *Pancho Villa, Dead and Alive*, 1943 (fig. 536). Motherwell, who had traveled to Mexico with Matta, had been struck by a photograph of the revolutionary figure, dead and covered with blood. The stick figures representing Villa relate directly to a painting by Picasso from the 1920s, but other forms

536. ROBERT MOTHERWELL. *Pancho Villa, Dead and Alive.* 1943. Gouache and oil with cut-and-pasted papers on cardboard, 28 x 35⅞" (71.7 x 91.1 cm). The Museum of Modern Art, New York

in the collage were to become signature Motherwell images, such as the ovoid shapes held in tension between vertical, architectural elements. Motherwell's involvement with the collage medium was more extensive than that of any of his Abstract Expressionist colleagues. In the many examples he continued to make throughout his career, he used postcards, posters, wine labels, or musical scores that constitute a visual day-by-day autobiography.

In 1948 Motherwell made the first work of his great series of abstract paintings entitled Elegy to the Spanish Republic (color plate 255, page 460), inspired by his profound reaction to the defeat of the Spanish Republic by fascist forces. Until his death in 1991, he painted more than one hundred and fifty variants of the Elegy theme. These works stand in contrast to the brilliantly coloristic collages and paintings also created by the artist during the same period, for they were executed predominantly in stark black and white. "Black is death; white is life," the artist said. It is no coincidence that Picasso's monumental painting of protest against the Spanish Civil War, *Guernica* (see fig. 381), was black, white, and gray. The Elegies consisted for the most part of a few large, simple forms, vertical rectangles holding ovoid shapes in suspension. The scale increased as the series progressed, with some works on a mural scale. While the loosely brushed, organic shapes have been read literally by some, who have suggested that they might represent male genitalia referring to Spanish bull fights, Motherwell said he invented the forms as emblems for universal tragedy.

In 1967 the artist began a second great series known as the Opens or "wall and window" paintings (color plate 256, page 460). Motherwell once said he would rather look at "parks or town squares with walls" than at "raw nature." "There is something in me," he wrote, "that responds to the stark beauty of dividing a flat solid plane." The Opens range from relatively uniform, although never flat, areas of color to large-scale exercises in rhythmic, variegated brushwork. There is always movement and light within the ground, as well as the so-termed window motif common to these paintings. Although the Open series owed its inception partly to Motherwell's lifelong dialogue with the art of Matisse, and particularly to two of his starkly reductive compositions from 1914 involving views from windows, Motherwell's paintings also coincided with the significant emergence of Minimalist abstraction in New York.

A friend of Motherwell and Matta from the early 1940s, WILLIAM BAZIOTES (1912–1963) explored with them aspects of Surrealist automatism and was respected enough by the European Surrealists to be included in their group shows in New York. Automatism for Baziotes was not a matter of improvisational gesture, with all its implications of emotion and angst. Rather it guided a slow and meditative process during which allusive, biomorphic forms emerged as the painting proceeded. Devoted to fantastic, mythic subject matter, Baziotes was a cofounder of the Subjects of the Artist school in 1948. His painting was characterized by shifting, fluid, diaphanous color into which were worked biomorphic shapes

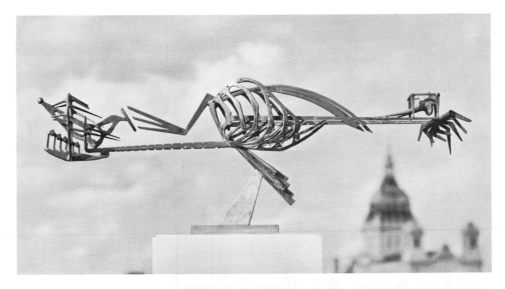

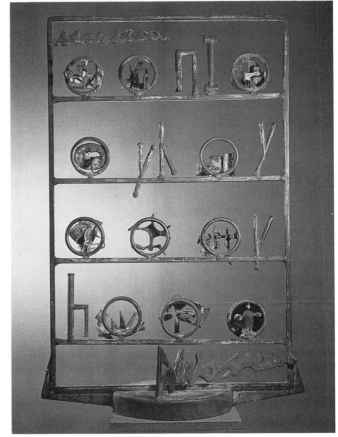

drawn with great sensitivity and suggestive of microcosmic or marine life. The strange forms in *Dusk* (color plate 257, page 461) may have something in common with prehistoric fossils, the kind the artist might have seen on one of his many visits to New York's American Museum of Natural History. Like a number of his fellow Abstract Expressionists, Baziotes was fascinated by paleontology and primitive life forms.

Postwar American Sculpture

Constructed sculpture—as contrasted with sculpture cast in bronze or modeled in stone—and particularly welded metal sculpture, constituted a major direction taken by American artists after World War II. DAVID SMITH (1906–1965), a pioneer sculptor of his generation working in welded steel, said of his medium: "The material called iron or steel I hold in high respect. . . . The metal itself possesses little art history. What associations it possesses are those of this century: power, structure, movement, progress, suspension, brutality." Smith was born in Decatur, Illinois, and between 1926 and 1932 he studied painting, initially with John Sloan, at the Art Students League in New York (he was never formally trained as a sculptor). In the early 1930s, intrigued by reproductions of Picasso's welded-steel sculptures of 1929–30 (see fig. 384) as well as the works of Julio González (see fig. 388) he saw in John Graham's collection, Smith began to experiment with constructed metal sculpture. He first learned to weld in an automobile plant in the summer of 1925, and in 1933 he established a studio in Brooklyn at Terminal Iron Works, a commercial welding shop. During World War II he worked in a locomotive factory, where he advanced his technical experience in handling metals, and the sheer scale of locomotives suggested to him possibilities for the monumental development of his metal sculpture. Moreover, these experiences established the kinship Smith felt with skilled industrial laborers.

During the 1930s and 1940s Smith endowed his sculpture with a Surrealist quality, derived from Picasso, González, and Giacometti. With the 1948 *Royal Bird* (fig. 537), he suggests the viciously aggressive skeleton of some great airborne creature of prey. Smith and his wife, the sculptor Dorothy Dehner,

had purchased a photograph of the fossils of a prehistoric flying creature, and both made use of the image in their work. His works of the early 1950s are like steel drawings in space, as in *The Letter* (fig. 538), whose metal glyphs recall Gottlieb's pictographic paintings (see fig. 535). Yet even when the sculptures were composed two-dimensionally, the natural setting, seen through the open spaces, introduced ever-changing suggestions of depth, color, and movement.

Toward the end of the war, Smith established a studio in Bolton Landing in upstate New York. The studio was a complete machine shop, and during the 1950s and 1960s Smith populated the surrounding fields with his sculptures, which were becoming more monumental in scale and conception.

left: 539. DAVID SMITH.
Sentinel I. 1956. Steel,
7'5⅝" x 16⅞" x 22⅝"
(2.3 m x 42.9 cm x
57.5 cm).
National Gallery of Art,
Washington, D.C.
GIFT OF THE COLLECTORS COMMITTEE.
© 1997 ESTATE OF DAVID
SMITH/VAGA, NEW YORK, NY

above: 540. DAVID SMITH.
left: *Cubi XVIII*. 1964. Stainless steel, height 9'8" (2.9 m).
Museum of Fine Arts, Boston ANONYMOUS DONATION
center: *Cubi XVII*. 1963. Stainless steel, height 9' (2.7 m).
Dallas Museum of Art
right: *Cubi XIX*. 1964. Stainless steel, height 9'5" (2.9 m).
The Tate Gallery, London © 1997 ESTATE OF DAVID SMITH/VAGA, NEW YORK, NY

During much of his career Smith worked on sculpture in series, with different sequences sometimes overlapping chronologically and individual sculptures within series varying widely in form. The Agricola constructions (1952–59) continued the open, linear approach (with obvious agrarian overtones), while the Sentinels (1956–61) were usually monolithic, figurative sculptures, at times employing elements from farm or industrial machinery. *Sentinel I* (fig. 539), for example, includes a tire rod and industrial steel step, while the "head" of the sculpture resembles street signs. In contrast with most sculptors who used found objects in the 1960s, Smith integrated such objects in the total structure so that their original function was subordinated to the totality of the new design. This was Constructivist sculpture in the original sense of Naum Gabo's *Realistic Manifesto,* in which voids do not merely surround mass but are articulated by form. Significant for future developments in sculpture is the fact that *Sentinel I* is not isolated on a pedestal but occupies our actual space.

In his last great series, begun in 1961 with the generic name of Cubi, Smith assembled monumental, geometric volumes in stainless steel (fig. 540). Always sensitive to the surfaces of his sculptures, Smith had often painted his sculptures in bright, flat color or with distinctive painterly textures. In the Cubi series he exercised particular care in the polishing of the surfaces to create effects of brilliant, all-over calligraphy out of the highly reflecting, light-saturated surfaces. Smith's relation to the subsequent development of Minimal art is critical (see fig. 696). Yet he differs from most sculptors from more recent decades who, in a deliberate insistence on anonymity, design their structures and then generally have them manufactured and painted by others—professional technicians. Smith, however, made his own works, only rarely employing shop assistants. Even the Cubi series, with its machinelike precision, was constructed and finished by the artist.

DOROTHY DEHNER (1901–1994) was overshadowed as an artist by her famous husband, David Smith, yet for nearly forty years she produced a distinguished body of sculpture whose origins should be considered within the Abstract Expressionist milieu. She met Smith while studying painting at the Art Students League in 1926 and slightly later became acquainted with Davis, Gorky, and Avery. Dehner worked in various figurative styles until the late 1940s, when she also began to exhibit her work publicly. Although many of her early abstract paintings and drawings demonstrate a sculptor's sensibility, Dehner did not start making sculpture until 1955, five years after she and Smith had separated. *Nubian Queen* (fig. 541), a bronze from 1960, is an open-frame, totemic construction containing mysterious, pictographic signs. Like her contemporaries in the New York School, Dehner sought

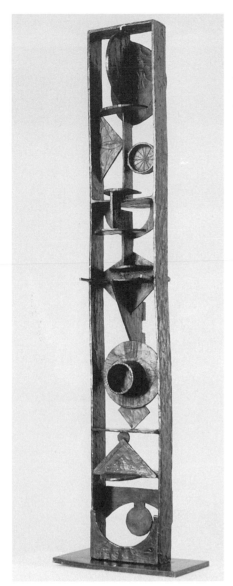

541. DOROTHY DEHNER. *Nubian Queen.* 1960. Bronze, 56 x 14 x 8" (142.2 x 35.6 x 20.3 cm). Courtesy Richard L. Eagan Fine Arts, New York

left: 542. SEYMOUR LIPTON. *Avenger.* 1957. Nickel-silver on Monel metal, length 23 x 18" (58.4 x 45.7 cm). Spencer Museum of Art, University of Kansas, Lawrence
GIFT OF SENIOR CLASS OF 1961

below: 543. IBRAM LASSAW. *Clouds of Magellan.* 1953. Welded bronze and steel, 52" x 6'10" x 18" (132 cm x 2.1 m x 45.7 cm). Collection Philip Johnson, New York

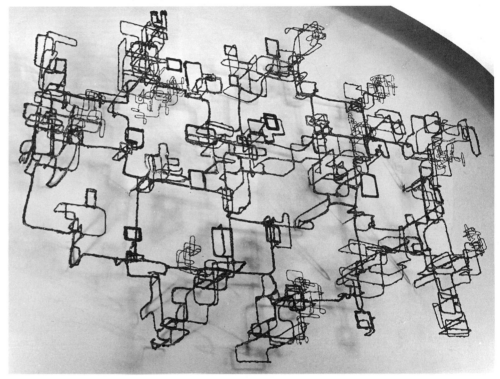

to develop a private language with universal implications, and like them, as her title implies, she liked to tap the art of ancient civilizations for forms and meaning. "I am an archaeologist of sorts," she said. "I tamper with the past."

Originally trained as a dentist, **SEYMOUR LIPTON** (1903–1981) made his first sculptures in the late 1920s and began showing his work and teaching sculpture in the 1940s. For his mature works, he hammered and welded sheets of Monel metal, nickel-silver, or bronze into large spatial volumes, suggestive at times of threatening plant forms or predatory creatures with sharp, protruding features (fig. 542). His works combine monumentality and repose with a sense of poetry and mystery. Spatially they offer an interplay of exterior and interior forms that emphasize the nature of the material.

As the works already seen would suggest, many American sculptors shared the dynamic, open-form aesthetics espoused by the Abstract Expressionists. **IBRAM LASSAW** (b. 1913) even managed to three-dimensionalize Pollock's all-over webs of flowing lines, doing so in reliefs and freestanding

constructions formed as intricate, welded cages (fig. 543). However, rather than Pollock's daring equilibrium of colliding opposites—freedom and control—Lassaw seemed more at one with the mystical harmony of Mark Tobey. And indeed he was a student of the German mystics and Zen philosophy.

ISAMU NOGUCHI (1904–1988) was born in Los Angeles of a Japanese father and an American mother, and then lived in Japan from the age of two to fourteen before returning to the United States. In 1927, following some academic training and his decision to become a sculptor, Noguchi went to Paris and served for a time as assistant to Brancusi, from whom he learned direct carving methods. He became acquainted with members of the American expatriate artistic community in Paris (assisting Calder occasionally with his Circus) as well as with the work of the Cubists, Surrealists, and Constructivists. On visits to China and Japan during the 1930s he studied brush painting and ceramic sculpture. He made his first abstract sculptures while in Paris, but his first thoroughly original works date to the 1940s, when he made

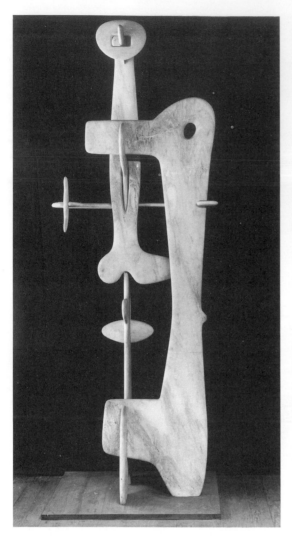

544. ISAMU NOGUCHI. *Kouros.* 1944–45. Pink Georgia marble on slate base, 10' x 34½" x 42" (3 m x 86.6 cm x 7 cm). The Metropolitan Museum of Art, New York FLETCHER FUND

545. ISAMU NOGUCHI. Stage set for *Herodiade* by Martha Graham. 1944. Set includes a clothesrack, chair, and mirror in painted plywood

Surrealistic carvings in marble or slate. In *Kouros* (fig. 544) he evoked the human form, not only by the Greek title *Kouros*, but also by the Miróesque biomorphism of the eyeholed and crossed, bonelike forms, all interlocked and arranged in the vertical orientation of a standing figure. The idea for sculptures of this type had originated with Noguchi's 1944 set design for *Herodiade,* a ballet by premier American choreographer Martha Graham (fig. 545). In the ballet, Salome dances before her mirror, in which she sees, according to Noguchi, "her bones, the potential skeleton of her body." As Noguchi's renown spread internationally in the 1950s, he began to receive commissions for public sculptures and gardens around the world. In Long Island City, where he had had a studio since 1961 (he later established a studio in Japan as well), Noguchi established a garden and museum devoted to his work (color plate 258, page 461). The works shown here give evidence of the many textures and forms the artist could mine from stone, from immaculately smooth and geometric forms to organic ones that incorporate accidental forms caused in the quarry process.

French-born **LOUISE BOURGEOIS** (b. 1911), best known

as a sculptor, is hard to classify because of her active work in a variety of media. She began as a painter-engraver in the milieu of Surrealism. Early in her artistic career, she made a disquieting group of paintings in which she, struggling at the time with the multiple roles of mother, wife, and artist, rendered women literally as houses or *Femmes-Maisons.* Bourgeois turned to carved sculpture in the late 1940s, creating roughly hewn, vertical forms in wood that she called "Personages." Their closest historical analogue is perhaps the Surrealist sculptures of Giacometti (see fig. 391). The Personages, Bourgeois said, are each endowed with individual personalities and are best experienced as a group. Indeed, the five vertical elements of *Quarantania I* (fig. 546) were originally separate sculptures that Bourgeois later decided to unite on a single base. Like all of Bourgeois's work, the Personages have powerful psychic associations rooted in her own biography. In the 1960s Bourgeois experimented with new materials such as latex, plaster, and marble to create globular, overtly sexual shapes that bubble forth as though in the process of being born from a raw matrix. Bourgeois could achieve a palpable sense of the organic within an essentially abstract composition. The highly

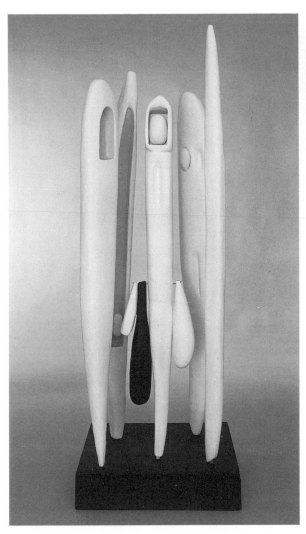

546. LOUISE BOURGEOIS. *Quarantania I*. 1947–53. Painted wood on wood base, 62⅜ x 11¾ x 12" (158.4 x 29.8 x 30.5 cm). The Museum of Modern Art, New York
GIFT OF RUTH STEPHAN FRANKLIN. © 1997 LOUISE BOURGEOIS/VAGA, NEW YORK, NY

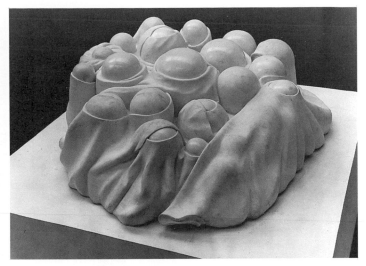

547. LOUISE BOURGEOIS. *Cumul I*. 1968. Marble, 22¼ x 50 x 48" (56.5 x 127 x 121.9 cm). Museé National d'Art Moderne, Centre d'Art et de Culture Georges Pompidou, Paris
© 1997 LOUISE BOURGEOIS/VAGA, NEW YORK, NY

polished cluster of forms in *Cumul I* (fig. 547), for example, seems to emerge from a thin membrane before our very eyes. While such works might appear to be related to Hans Bellmer's mammalian dolls (see fig. 376), Bourgeois never shared the Surrealists' interest in sexual transgression.

At the age of eighty, Bourgeois embarked on one of her most intriguing series, elaborate, multimedia environments she called Cells. Within the glass and chain-link enclosure of *Cell (You Better Grow Up)*, 1993 (color plate 259, page 462), are two mirrors. Between them is a roughly cut marble base on which sit exquisite carvings of an adult hand grasping those of a child. Other objects include furniture, perfume bottles, and a glass object that contains a tiny glass figurine. Bourgeois has said that the figurine and the child's hands are self-portraits and form part of a work about "the frightened world of a child who doesn't like being dependent and who suffers from it." Bourgeois's remarkable achievements did not receive their proper recognition until the 1980s, perhaps because her long involvement with performance and installation art seemed to strike a chord of common purpose with trends during that decade.

Also rooted in Surrealism is the work of **JOSEPH CORNELL** (1903–1972), a highly cultivated American who brought great distinction to the tradition of assemblage. In the early 1930s, Cornell became acquainted with the Julien Levy Gallery in New York, a center for the display of European Surrealism. His first experiments with collage were inspired by works of Max Ernst, and soon Levy was exhibiting his small constructions along with the works of the Surrealists. By the mid-1930s Cornell had settled on his formula of a simple box, glass-fronted usually, in which he arranged found objects such as cork balls, photographs, and maps. Cornell read widely, especially French nineteenth-century literature, but his interests were broad, from poetry and ballet to astronomy and other natural sciences. He haunted the penny arcades, libraries, dime stores, souvenir shops, art galleries, and movie houses of New York, searching out the ephemera that became the content of his enchanting assemblages. In his boxes Cornell, who lived reclusively and rarely spoke about his work, created a personal world filled with nostalgic associations—of home, family, childhood, and of all the literature he had read and the art he had seen. As one critic has noted, "He treated the ephemeral object as if it were the rarest heirloom of a legendary prince or princess."

Medici Slot Machine of 1942 (fig. 548) centers around reproductions of a Renaissance portrait, multiple images taken from the early cinema, and symbols suggesting relations between past and present. Everything is allusion or romantic reminiscence gathered together, as one idea or image suggested another, to create intimate and magical worlds. As in all his creations, Cornell carefully organized his imagery within a loosely geometric format.

To enhance their quality of nostalgia, Cornell often deliberately distressed his surfaces, sometimes placing his boxes in the elements to achieve a weathered, time-battered appearance. In *Untitled (The Hotel Eden)* (color plate 260, page 463), the ad for Hotel Eden is tattered and the paint on the

548. JOSEPH CORNELL. *Medici Slot Machine.* 1942. Construction, 13½ x 12 x 4¼" (34.3 x 30.5 x 10.8 cm). Private collection

right: 549. LOUISE NEVELSON. *Dawn's Wedding Chapel I.* 1959. Painted wood, 7'6" x 51" (2.3 m x 129.5 cm). Whitney Museum of American Art, New York

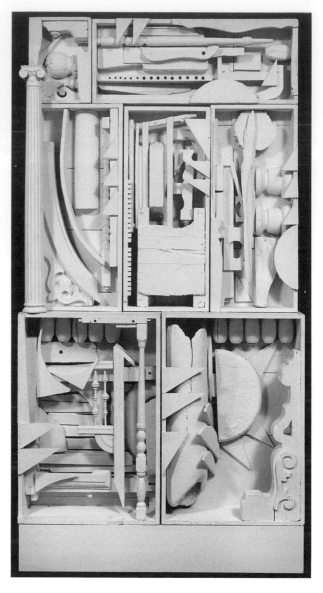

box is cracked and worn. Cornell usually built a number of boxes around specific themes, such as the Aviaries of the 1940s, which include images of exotic birds. As is typical of his work, this composition is rich with associations. The Hotel Eden ad suggests the garden of paradise while the spiral form at the upper left probably refers to a work by Cornell's friend Marcel Duchamp. Cornell was included in the historic 1936 exhibition at The Museum of Modern Art in New York, "Fantastic Art, Dada, and Surrealism," where his works were installed next to Miró's Surrealist *Object* from 1936 (color plate 167, page 325). Although the affinities between *Untitled (Hotel Eden)* and Miró's sculpture are obvious, Cornell objected to being classified with the Surrealists, saying that while he admired their art, he had no interest in the dream world and the subconscious and preferred to focus on "healthier possibilities." By the 1950s Cornell knew many of the Abstract Expressionist painters who were his contemporaries, but his quiet, small-scale works had less in common with the Sturm und Drang of their canvases than with the assemblage art of the next generation (see fig. 610).

LOUISE NEVELSON (1899–1988) also worked in a tradition of assemblage, but with results quite different from those

of Cornell. She studied voice, dance, and art in New York in the late 1920s. In 1931 she went to Europe, studying painting for a time with Hans Hofmann in Munich and playing small roles in movies in Berlin and Vienna. Nevelson always maintained a flair for the theatrical in her eccentric dress and flamboyant manner, and her mature work as a sculptor also manifested a distinctively theatrical sensibility. She liked to control the light and space within which the viewer experienced her work, which, like that of contemporary painters such as Pollock and Rothko, was environmental in its effect. In the early 1940s Nevelson began to make Surrealist-inspired assemblage sculptures from found objects and received her first solo exhibition in New York. Her characteristic works of the 1950s were large wooden walls fitted with individual boxes filled with scores of carefully arranged found objects—usually sawed-up fragments of furniture or woodwork rescued from old, destroyed houses. (It is not inconsequential that her Russian father was a builder who ran a lumber yard.) Nevelson then painted her sculpture a uniform matte black, white, or, in her later work, reflective gold. *Dawn's Wedding Chapel I* (fig. 549) formed part of a large installation at The Museum of Modern Art in 1959 called *Dawn's Wedding*

Feast. It included freestanding, totemic constructions as well as wall pieces, all suggesting a site for ancient ceremony and ritual. The shallow boxes were filled with assemblages of whitewashed balusters, finials, posts, moldings and other architectural remnants. Despite their composition from junk, they achieve a quality of decayed elegance, reminiscent of the graceful old houses from which the elements were mined.

In the 1960s Nevelson relied less on found materials and had boxes fabricated according to her specifications (color plate 261, page 463). Her work took on a more pristine character, with regularized, sometimes symmetrical forms, tendencies in sync with the Minimalist aesthetic of the decade. Despite the new immaculate precision, these works still have the altarlike quality of the wooden walls. In the later 1960s the prolific Nevelson experimented with new materials, including clear Plexiglas, and began to execute commissions for large-scale outdoor works in aluminum, steel, and Cor-Ten steel.

MARK DI SUVERO (b. 1933), much younger than the other artists discussed in this chapter, was still studying art at college in California when the Abstract Expressionists reached the pinnacles of fame in the mid-1950s. But only three years after arriving in New York in 1957, di Suvero was translating the bold gestures of Kline and de Kooning into a dynamic sculptural idiom. *Hankchampion,* 1960 (fig. 550), consists of massive, rough timbers that project in a precarious, even threatening manner. These early wood pieces sometimes incorporated chain, rope, and found objects, like barrels or tires, scavenged from the industrial landscape. Although di Suvero admires Rodin and Giacometti, the most meaningful and immediate artistic example for him is Smith, who changed sculpture, di Suvero said, from an "objet d'art type of thing to something that is really strong, American industrial art." Despite a drastic accident in 1960, when he was nearly crushed to death in an elevator, by the mid-1960s di Suvero was making works in steel so large they required a crane. Although he has made many small-scale works, he is best known for the monumental outdoor sculpture he has

made since that time, including *Aurora* (color plate 262, page 464), which is named after a poem about New York City by Federico García Lorca. These works, which enter the realm of architecture by virtue of their scale and I-beam construction, are as much at home in the urban environment as in a natural landscape. In the spirit of Constructivism, di Suvero's jutting, angled girders shape the void with energy-charged forms that have no sense of central mass. And in the spirit of Calder, some of his mammoth sculptures have moving parts that invite audience participation.

JOHN CHAMBERLAIN (b. 1927) also began to make sculpture in the early 1960s, works that seemed like three-dimensional expressions of Abstract Expressionism's explosive, painterly styles and that imply at times an almost violent approach to the medium. Chamberlain has used parts of junked automobiles, with their highly enameled, colored surfaces, to create all-over, abstract constructions of surprising beauty. Like di Suvero, he uses industrial tools, cutting and crushing shapes that he then pieces together, like an assemblage, through a trial-and-error process. The works can be either freestanding sculptures or, as in *Dolores James* (color plate 263, page 464), wall-mounted reliefs. Chamberlain's spontaneous, not entirely predictable methods parallel the Abstract Expressionists' balance of control and accident. "There's no formula," he said. For example, one night, when he was nearing completion of *Dolores James,* he returned late to his studio and heaved an eight-pound sledgehammer at the sculpture, achieving exactly the final touch he sought. Unlike the Abstract Expressionists, Chamberlain avoided ascribing profound meaning to his works. In this he is closer to many artists of his own generation who also made, or make, art from society's detritus.

Developments in American Photography

Depression, political upheaval, and war shaped the vision of photographers in the 1940s just as much as that of painters. Meanwhile, photography—by then ubiquitous not only in

550. MARK DI SUVERO. *Hankchampion.* 1960. Wood and chains, nine wooden pieces, overall 6'5¼" x 12'5" x 8'9" (2 x 3.8 x 2.7 m). Whitney Museum of American Art, New York

GIFT OF MR. AND MRS. ROBERT C. SCULL

551. ROBERT CAPA. *Normandy Invasion, June 6, 1944.*
Gelatin-silver print
© ROBERT CAPA

552. UNKNOWN PHOTOGRAPHER. *Mt. Vesuvius, 1944, After the
Eruption of 1944.* Gelatin-silver print. The Imperial War
Museum, London

news coverage and advertising but also in professional journals, books, and special exhibitions—had exerted an unavoid-
ably powerful effect on the vision of everyone, painters
perhaps most of all. Even some of the field pictures made by
the Hungarian-American photographer **ROBERT CAPA**
(1913–1954), during first the Spanish Civil War and then
World War II (fig. 551), seemed to parallel Abstract Expressionism's fluid organization and its conception of the picture
as an expression of personal, tragic experience universalized
through abstract gesture. Capa, who was killed by a land
mine while photographing the war in Indochina, was deeply
committed to the process of the photojournalistic essay,
which he felt could be done only by being as close to the subject matter as possible. It was as if photojournalism, scientific
aerial photography, and microphotography helped prepare
the human eye for the pictorial formulations of Abstract
Expressionism (fig. 552). Another photographer who witnessed the carnage of war firsthand was **LEE MILLER** (1907–
1977), an American who spent most of her adult life in
Europe, first in the Parisian circle of Man Ray between 1929
and 1932 and, after 1942, as an official U.S. forces war cor-
respondent. She was present in 1945 at the liberation of
Dachau and Buchenwald concentration camps, and the harrowing scenes she recorded there are among the most painfully revelatory images of the war (fig. 553). Some of them were
soon featured in *Vogue* magazine, where her work had been
published since the late 1920s, with the heading "Believe It."

The grandeur and metaphor of Abstract Expressionist
painting encouraged photographers to reinvestigate the
potential of both straight and manipulated procedures for creating abstract imagery. Among the photographers who used
straight photography to discover imagery capable of yielding
abstract forms potent with mystical feeling was **MINOR
WHITE** (1908–1976). In works like the one seen here (fig.
554), White's vision paralleled stylistic developments in the art
of Abstract Expressionism's gestural painters, as did his devotion to photography as a form of self-expression. "I photo-

graph not that which is," he said, "but that which I AM." As
a teacher and writer, and founder, as well as a longtime editor,
of *Aperture* magazine, White had a vast and enduring influence on a whole generation of younger photographers.
AARON SISKIND (1903–1991), too, lost interest in subject
matter as a source of emotive content and turned instead
toward flat, richly textured patterns discovered on old weathered walls, segments of advertising signs, graffiti, and peeling
posters (fig. 555). In his preoccupation with the expressive

553. LEE MILLER. *Dead SS Guard in the Canal at Dachau.* 1945.
Modern gelatin-silver print, 1988. Lee Miller Archives, England

554. MINOR WHITE. *Sun in Rock, Devil's Slide, 1947.* October 12, 1947. Gelatin-silver print, 7⅜ x 9⅝" (18.7 x 24.5 cm). The Museum of Modern Art, New York

GIFT OF SHIRLEY BURDEN. COURTESY THE MINOR WHITE ARCHIVE, PRINCETON UNIVERSITY. COPY PRINT © 1995 THE MUSEUM OF MODERN ART, NEW YORK

555. AARON SISKIND. *Chicago 1949.* Gelatin-silver print
© AARON SISKIND

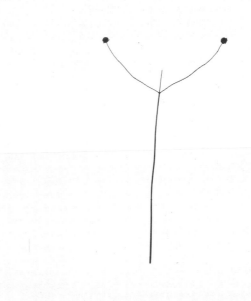

556. HARRY CALLAHAN. *Weed Against the Sky, Detroit.* 1948. Gelatin-silver print
© HARRY CALLAHAN. COURTESY PACE/MACGILL GALLERY, NEW YORK

qualities of two-dimensional design, Siskind found himself in sympathetic company among painters like Kline and Willem de Kooning, whose influence he acknowledged. Another photographer whose work might look at home with New York gestural paintings was ELIOT PORTER (1901–1990). Porter portrayed nature in the tradition of Ansel Adams, but he, unlike most noncommercial photographers of his generation, worked in color (color plate 264, page 464). To attain such quality, Porter, unlike most color photographers, made his own color-separation negatives and dye transfers. HARRY CALLAHAN (b. 1912), a friend and colleague of Siskind's, has generated photographs of such uncompromising spareness

and subtlety that they often border on abstraction. His principal influence came from Ansel Adams and the former Bauhaus faculty at the Institute of Design in Chicago, where he taught beginning in 1946. He later founded the photography department at the Rhode Island School of Design. Callahan has worked continuously with three deeply felt personal themes—his wife, Eleanor; the urban scene; and simple, unspoiled nature. His study of a spindly weed (fig. 556) is entirely straightforward, yet making sense of the form requires close scrutiny. In his later Cape Cod landscape, Callahan shifted from the isolated close-up of a motif in nature to seemingly limitless space (fig. 557). This distilled view of the world, with its emphatic horizontality and nearly monochrome tonality, recalls many a late Rothko. "I think that every artist continually wants to reach the edge of nothingness," Callahan said.

Obviously, not all photographers active in the 1940s and 1950s were combing the landscape for abstract forms of expression. New York photographer HELEN LEVITT (b. 1913) began recording ordinary street life with her hand-held Leica in the late 1930s. Crucial to her was the discovery of the work of Henri Cartier-Bresson, and her magically complex image of children playing with a broken mirror in the street (fig. 558) is more reminiscent of the Frenchman's "decisive moment" than the social indictments of documentary photographers such as Lewis Hine (see fig. 468). Levitt

above: 557. HARRY CALLAHAN. *Cape Cod.* 1972. Gelatin-silver print, 9⅝ x 9¾" (24.4 x 24.8 cm) © HARRY CALLAHAN. COURTESY PACE/MACGILL GALLERY, NEW YORK

right: 558. HELEN LEVITT. *New York.* c. 1940. Gelatin-silver print. Courtesy Fraenkel Gallery, San Francisco

below, right: 559. ROY DECARAVA. *Shirley Embracing Sam.* 1952. Gelatin-silver print 12¹⁵⁄₁₆" x 10¹⁄₁₆" (32.9 x 28.1 cm). Collection The DeCarava Archive

included this and other scenes from Spanish Harlem in a book called *A Way of Seeing*, 1965. It included an essay by James Agee, the writer who had already collaborated on a book with Levitt's friend, Walker Evans (see chapter 18).

Self-taught photographer **ROY DECARAVA** (b. 1919) grew up in the midst of the Harlem Renaissance. By 1952 his photographs had earned him a prestigious Guggenheim grant, the first ever awarded to an African American. "I want to photograph Harlem through the Negro people," he wrote in his proposal, "Morning, noon, night, at work, going to work, at play, in the streets, talking, kidding, laughing, in the home, in the playground, in the schools. . . . I want to show the strength, the wisdom, the dignity of the Negro people." One of the resulting photographs is a tender and moving portrait of an embracing couple (fig. 559), which seems to sum up in a single image all the humanity the artist experienced in an entire community. In 1955, DeCarava, like Levitt, joined forces with a noted writer, in this case the poet Langston Hughes, to publish his Harlem photographs as a book, aptly titled *The Sweet Flypaper of Life.*

20 Postwar European Art

If the American artists discussed in the previous chapter were coming to terms with the traumatic aftershocks of the war, their European counterparts had experienced those events at closer range, thus undergoing an even deeper sense of despair and disillusionment, factors that had profound implications for their art.

Painting and Sculpture in France

The School of Paris was enriched in the postwar years by the continuing activity of the older established artists such as Picasso and Matisse who, for the most part, lived and worked outside of the artistic mainstream in Paris. Returning from the United States were Léger, Chagall, Ernst, Masson, Picabia, Hélion, and others, including Duchamp, who, after 1945, commuted between Paris and New York. With Giacometti, Dubuffet, and Bacon, figuration received a new impetus in the years immediately following World War II. Nevertheless,

560. PABLO PICASSO. *The Charnel House.* Paris 1944–45; dated 1945. Oil and charcoal on canvas, 6'6⅝" x 8'2½" (2 x 2.5 m). The Museum of Modern Art, New York

various forms of free abstraction dominated the painting in Europe just as they had in the United States.

PABLO PICASSO (1881–1973). When World War II came to a close in 1945, Picasso was living in Paris, internationally recognized as the preeminent artist of the century and the subject of exhibitions on a regular basis around the globe. For the next eight or nine years he continued his involvement with a young painter named Françoise Gilot, who, after leaving him, published an account of their life together that enraged the artist. In the mid-1950s Picasso settled in the south of France with another woman, Jacqueline Roque, and followed his own pursuits, relatively unaffected by the radical new developments in contemporary art in Europe and the United States. By the end of the decade he was seventy-eight years old and had outlived Matisse and many of his oldest friends.

Picasso, who had joined the Communist Party in 1944, summed up his feelings about the war in *The Charnel House* (fig. 560), 1944–45, a painting that, like *Guernica* (see fig. 381), consists of a grisaille palette of black, white, and gray. A tabletop still life, rendered in a late Cubist style and left as an open line drawing, indicates a domestic setting. Beneath it, the twisting, anguished forms of a family massacred in their home could be the twentieth-century echo of Daumier's *Rue Transnonain* (see fig. 38).

A theme that had preoccupied Picasso throughout his career, but especially during his later years, was that of the artist as voyeur. Painted toward the end of his relationship with Gilot, *The Shadow* of 1953 (fig. 561) is one of two compositions Picasso made showing his own shadow cast before the image of a nude woman sleeping on a bed. While the rectangular shape that frames the artist suggests a doorway, it also mimics the shape of a canvas and implies a fictive, painted, and strangely distant presence of the woman. Besides painting, the ever-prolific Picasso also explored the media of ceramics, prints, and sculpture during his last years. The head of the playful *Baboon and Young* (fig. 562), a bronze from 1951, was cast using two toy automobiles that belonged to Picasso's son, Claude, and the whole figure was derived from this chance resemblance.

During these years, Picasso engaged in an extended dia-

above: 561. PABLO PICASSO.
The Shadow.
29 December 1953.
Oil and charcoal on canvas,
49½ x 38"
(125.7 x 96.5 cm).
Musée Picasso, Paris
© 1997 ESTATE OF PABLO PICASSO/ARTISTS
RIGHTS SOCIETY (ARS), NEW YORK

right: 562. PABLO PICASSO.
Baboon and Young. 1951 (cast
1955). Bronze, after found
objects, 21 x 13¼ x 20¾" (53.3
x 33.3 x 52.7 cm). The Museum
of Modern Art, New York
MRS. SIMON GUGGENHEIM FUND. PHOTOGRAPH
© 1995 THE MUSEUM OF MODERN ART,
NEW YORK. © 1997 ESTATE OF PABLO PICASSO/
ARTISTS RIGHTS SOCIETY (ARS), NEW YORK

logue with several great artists of the past, including Velázquez, Courbet, Manet, and, Delacroix, through his own variations on their best-known works. In the several versions he made of Delacroix's *Women of Algiers*, a painting he knew in the Louvre, Picasso may have been paying tribute not only to France's leading Romantic painter, but to Matisse as well. The latter, famous for his Orientalist themes and languid odalisques, had died a few months before Picasso embarked on this sumptuously colored and highly subjective interpretation of Delacroix's masterpiece (color plate 265, page 481). Picasso envisioned his own imminent death in the *Self-Portrait* (fig. 563), 1972, made less than a year before he died at age ninety-one. His stubble-covered, emaciated skull, with its bluish cast and hypnotically staring eyes, is a harrowingly probing self-image.

ALBERTO GIACOMETTI (1901–1966), gave Surrealism its most important sculptor (see Chapter 15). In 1934, however, he broke with the Surrealists, convinced that his abstracted constructions were carrying him too far from the world of actuality. He returned to a study of the figure, working from the model, with the intent of rendering the object and the space that contained it exactly as his eye saw them. In 1939 he decided to make a new start:

> I began to work from memory. . . . But wanting to create from memory what I had seen, to my terror the sculptures became smaller and smaller, they had a likeness only when they were small, yet their dimensions revolted me, and tirelessly I began again, only to end several months later at the same point. A large figure seemed to me false and a small one equally unbearable, and then often they became so tiny that with one touch of my knife they disappeared into dust. But head and figures seemed to me to have a bit of truth only when small. All this changed a little in 1945 through drawing. This led me to want to make larger figures, but then to my surprise, they achieved a likeness only when tall and slender. . . .

When Giacometti returned to Paris from Switzerland after the war, his long struggle, during which he finished few works, was reaching resolution. The 1947 *Head of a Man on a Rod* (fig. 564) presents a human head, tilted back with mouth agape and impaled on a metal rod. The violence and sense of anguish in the head, which seems to utter a silent scream, is related to the artist's earlier Surrealist works (see fig. 393), but the peculiar format may have been aided by his recollection of a painted human skull from New Ireland (in the western Pacific Ocean) that he had seen displayed on a rod in an ethnographic museum in Basel, Switzerland. Typical of the artist's postwar sculpture, the surface retains all the bumps and gouges formed by his hand in the original wet plaster, thus resembling both a horrible laceration of the flesh and a rugged, rocky landscape.

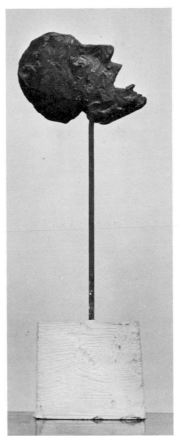

563. PABLO PICASSO. *Self-Portrait*. 30 June 1972. Pencil and colored pencil on paper, 25¾ x 20" (65.4 x 50.8 cm). Courtesy Fuji Television Gallery, Tokyo
© 1997 ESTATE OF PABLO PICASSO/ARTISTS RIGHTS SOCIETY (ARS), NEW YORK

564. ALBERTO GIACOMETTI. *Head of a Man on a Rod*. 1947. Bronze, height 21¾" (55.2 cm). Private collection
© 1997 ADAGP, PARIS/ARTISTS RIGHTS SOCIETY (ARS), NEW YORK

565. ALBERTO GIACOMETTI. *Chariot*. 1950. Bronze, 57 x 26 x 26⅛" (144.8 x 65.8 x 66.2 cm). The Museum of Modern Art, New York PURCHASE. PHOTOGRAPH © 1995 THE MUSEUM OF MODERN ART, NEW YORK. © 1997 ADAGP, PARIS/ARTISTS RIGHTS SOCIETY (ARS), NEW YORK

Giacometti's characteristic sculptural type in the postwar period was the tall, extremely attenuated figure that appeared singly or in groups, standing rigid or striding forward like an Egyptian deity. The figures ranged in scale from a few inches high to fully life-size, as in *Man Pointing* (color plate 266, page 482). Giacometti first conceived this work, distinctive for its active gesture, as half of a pair of male figures, so that the man seemed to be pointing the direction for his companion. The artist may have disliked the implied narrative and decided to isolate the pointing figure.

Early on he had developed (perhaps from his studies of ancient sculpture) a passion for color in sculpture, and one of the most striking aspects of the later works is the patinas he used on his bronzes. *Chariot* (fig. 565), which suggests an ancient Etruscan bronze, was originally designed as a commemorative public sculpture in Paris, but was ultimately rejected. Giacometti said the idea came from watching nurses wheel medicine carts about while he was hospitalized in 1938. He placed the figure alone, high on a chariot, he said, "in order to see it better and to situate it at a precise distance from the floor." Even when placed in groups, such as in environments evoking city squares, Giacometti's figures still maintain a sense of the individual's isolation (fig. 566). Giacometti insisted that his sculptures were not embodiments of modern

566. ALBERTO GIACOMETTI. *The Forest (Composition with Seven Figures and One Head)*. 1950. Painted bronze, 22 x 24 x 19¼" (55.9 x 61 x 48.9 cm). The Reader's Digest Collection
© 1997 ADAGP, PARIS/ARTISTS RIGHTS SOCIETY (ARS), NEW YORK

567. ALBERTO GIACOMETTI. *Bust of Diego*. 1955. Bronze, 22¼ x
12¼ x 5⅞" (56.5 x 31 x 15 cm). Foundation Alberto
Giacometti, Kunsthaus, Zürich, Switzerland

was born into a distinguished Polish family and raised in Paris, Geneva, and Berlin. His father was a painter and an art historian, his mother was also a painter, and the poet Rainer Maria Rilke was a close family friend. Balthus settled in the French capital in 1924. He copied the old masters in the Louvre but did not undergo any formal artistic training. Although he both exhibited in 1934 at the Galerie Pierre, a Parisian gallery associated with the Surrealists, and was intrigued by the sexual implications of dreams, Balthus was not seduced by the Surrealist love of the irrational. Since the 1930s he has painted figures in enigmatically narrative compositions, especially semi-aware pubescent girls, whom he places in situations of riveting mystery, ambiguous eroticism, and a light that transfixes time (color plate 267, page 483). The atmosphere of these lavishly brushed paintings seems all the more charged for being contained within an architectonic structure as classically calm and balanced as those of Piero della Francesca and Poussin. The result is an eerie sense of the anxiety and decadence that haunted the renascent Europe of the postwar years.

After a brief period of study at the Académie Julian in 1918, which he found uncongenial, **JEAN DUBUFFET** (1901–1985) continued to draw and paint privately, but much of his time was taken up with music, the study of languages, experimental theater, and puppetry. He had his first one-man exhi-

anguish and alienation and rejected the Existentialist readings of his postwar work by such illustrious figures as Jean-Paul Sartre and Samuel Beckett. Yet Sartre's description of one of Giacometti's sculptures seems very apt here: "Emptiness filters through everywhere, each creature secretes his own void."

Giacometti's favorite models continued to be his brother Diego, his mother, and his wife, Annette. Diego is the subject of dozens of sculptures and paintings, ranging from relatively representational works to corrugated, pitted masks. In the mid-1950s he made several busts of Diego (fig. 567) that present a slice of a head so thin that the frontal view is practically nonexistent. As the critic David Sylvester has written, "There is virtually no transition between front view and side view."

From the 1940s Giacometti also drew and painted with furious energy. He sought to pin down space, which could only be suggested in the sculptures. Like the sculptures, every painting was a result of constant working and reworking. In their subordination of color and emphasis on the action of line, the paintings have the appearance of drawings or even engravings. In *The Artist's Mother* (fig. 568), the figure is almost completely assimilated into the network of linear details of the conventional living room.

A notoriously reclusive painter closely allied to Giacometti since the early 1930s is Balthasar Klossowski, Count de Rola, known by his childhood nickname **BALTHUS** (b. 1908). He

568. ALBERTO GIACOMETTI. *The Artist's Mother*. 1950.
Oil on canvas, 35⅜ x 24" (89.9 x 61 cm).
The Museum of Modern Art, New York

Color plate 265. PABLO PICASSO. *Women of Algiers, after Delacroix.* 11 February 1955. Oil on canvas, 45 x 57½"
(114.3 x 146.1 cm). Collection Mrs. Victor Ganz, New York © 1997 ESTATE OF PABLO PICASSO/ARTISTS RIGHTS SOCIETY (ARS), NEW YORK

Color plate 266.
ALBERTO GIACOMETTI. *Man Pointing* (detail). 1947

Color plate 266.
ALBERTO GIACOMETTI. *Man Pointing*. 1947. Bronze, height 70½" (179.1 cm). The Museum of Modern Art, New York

Color plate 267. BALTHUS. *Les Beaux Jours (The Golden Days)*. 1944–46. Oil on canvas, 58¼" x 6'6¾" (148 cm x 2 m). Hirshhorn Museum and Sculpture Garden, Smithsonian Institution, Washington, D.C.

Color plate 268. JEAN DUBUFFET. *View of Paris: The Life of Pleasure*. 1944. Oil on canvas, 35 x 45¾" (88.9 x 116.2 cm). Private collection

Color plate 269. JEAN DUBUFFET. *Dhotel velu aux dents jaunes (Dhotel Hairy with Yellow Teeth)*. 1947. Mixed media on canvas, 45¾ x 35" (116.2 x 88.9 cm). Collection Morton G. Neumann Family, Chicago

above: Color plate 270. JEAN DUBUFFET. *Virtual Virtue.* 1963. Oil on canvas, 37¾ x 50⅞" (95.9 x 129.2 cm). Sammlung Ernst Beyeler, Basel
© 1997 ADAGP, PARIS/ARTISTS RIGHTS SOCIETY (ARS), NEW YORK

below: Color plate 271. JEAN FAUTRIER. *Nude.* 1960. Oil on canvas, 35 x 57½" (88.9 x 146.1 cm). Collection de Montaigu, Paris
© 1997 ARTISTS RIGHTS SOCIETY (ARS), NEW YORK/ADAGP, PARIS

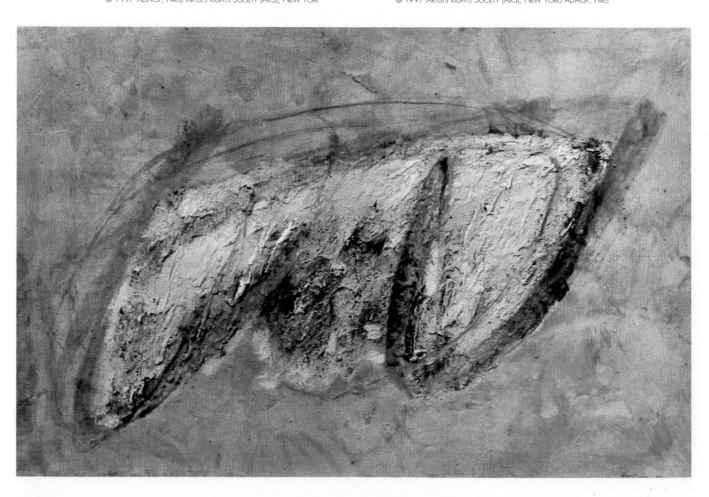

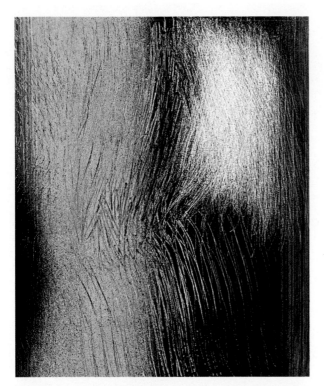

above: Color plate 272. HANS HARTUNG. *Untitled T. 1963-H26.* 1963. Oil on canvas, 39⅜ x 31⅞" (100 x 81 cm). Private collection

right: Color plate 273. PIERRE SOULAGES. *Painting.* 1953. Oil on canvas, 6'4¾" x 51⅛" (1.9 m x 130 cm). Solomon R. Guggenheim Museum, New York

below: Color plate 274. GEORGES MATHIEU. *Painting.* 1953. Oil on canvas, 6'6" x 9'10" (2 x 3 m). Solomon R. Guggenheim Museum, New York

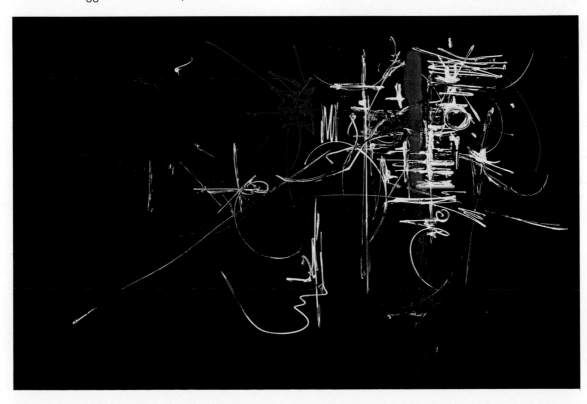

above: Color plate 276. GERMAINE RICHIER. *Praying Mantis.*
1949. Bronze, height 47¼" (120 cm).
Middelheim Sculpture Museum, Antwerp

© 1997 ADAGP, PARIS/ARTISTS RIGHTS SOCIETY (ARS), NEW YORK

Color plate 275. NICOLAS DE STAËL. *Untitled.* 1951. Oil
on canvas, 63 x 29" (160 x 73.7 cm). Private collection

© 1997 ADAGP, PARIS/ARTISTS RIGHTS SOCIETY (ARS), NEW YORK

Color plate 277. MAX BILL. *Nine Fields Divided by Means of Two
Colors.* 1968. Oil on canvas, 47 x 47" (119.4 x 119.4 cm).
Albright-Knox Art Gallery, Buffalo, New York

GIFT OF SEYMOUR H. KNOX

© 1997 BEELDRECHT, AMSTERDAM/ARTISTS RIGHTS SOCIETY (ARS), NEW YORK

above: Color plate 278.
GIORGIO MORANDI. *Still Life.*
1951. Oil on canvas, 8⅞ x
19⅝" (22.5 x 50 cm).
Kunstsammlung Nordrhein-
Westfalen, Düsseldorf

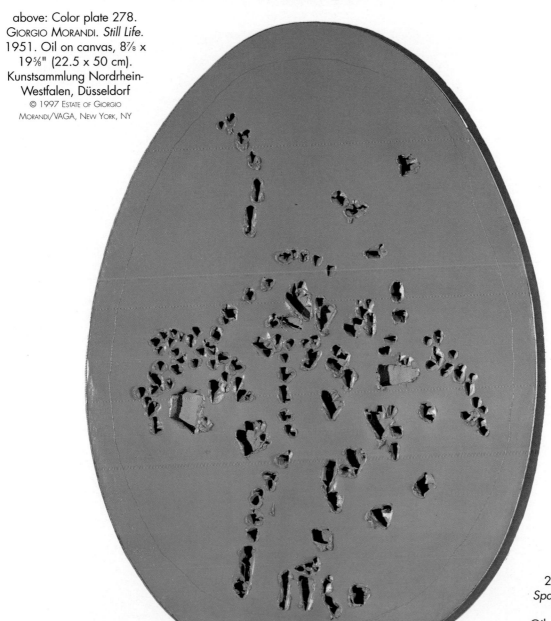

left: Color plate
279. LUCIO FONTANA.
*Spatial Concept: The End
of God.* 1963.
Oil on canvas, 70 x 48½"
(177.8 x 123.2 cm).
Gallerie dell'Ariete, Milan

Color plate 280. ALBERTO BURRI. *Composition.* 1953. Oil, gold paint, and glue on canvas and burlap, 34 x 39⅜" (86.4 x 100 cm). Solomon R. Guggenheim Museum, New York

Color plate 281. ANTONI TÀPIES. *Painting Collage.* 1964. Mixed media on canvas, 14 x 22" (35.6 x 55.9 cm). Galleria Toninelli, Rome

Color plate 282. EDUARDO CHILLIDA. (architect: Luis Peña Ganchegui; engineer: José María Elósegui). *Combs of the Wind.* 1977. Iron. Donostia Bay, San Sebastián, Spain

bition at the Galerie René Drouin in Paris in 1944, when he was forty-three years old. Of singular importance to the artist was discovering Dr. Hans Prinzhorn's book, *Bildnerei der Geisteskranken* (*Artistry of the Mentally Ill*). Here Dubuffet found a brutal power of expression that seemed to him much more valid than the art of the museums or even the most experimental new art. The art of the mentally ill and of children became the models on which he built his approach. He also admired Paul Klee, who had also used as sources works by people with no formal artistic training.

After 1944 Dubuffet became one of the most prolific painters in history and concurrently carried on an impressive program of writing on, cataloguing, and publishing his own work. The first of the new paintings, close in spirit to Klee, were panoramic views of Paris and Parisians, the buses, the metro, the shops, the back streets (color plate 268, page 483). The colors are bright and gay, the people drawn in a childlike manner, and the space composed as in primitive or archaic painting. "Remember," Dubuffet wrote, "that there is only one way to paint well, while there are a thousand ways to paint badly: they are what I'm curious about; it's from them that I expect something new, that I hope for revelations."

In the late 1940s Dubuffet made a series of caricatural portraits of the writers, intellectuals, and artists who were his friends. Typically subversive, his likenesses (for they do evoke a likeness despite their childlike rendering and wild deformation) defy all conventions of the portraiture genre. In *Dhotel velu aux dents jaunes* (*Dhotel Hairy with Yellow Teeth*) (color plate 269, page 483) he gave the subject a comically huge head inscribed with highly unflattering features and placed it atop a body the shape of a flattened watermelon. In his Corps de Dame series (fig. 569), the female body is presented frontally and two-dimensionally, as if passed over by a cement roller. Dubuffet rejected out of hand the Western tradition of the nude, and his paintings are an assault on "normal" standards of beauty. He observed that women's bodies had "long been associated (for Occidentals) with a very specious notion of beauty (inherited from Greeks and cultivated by the magazine covers); now it pleases me to protest against this aesthetic, which I find miserable and most depressing. Surely I aim for a beauty, but not that one."

Dubuffet had explored the variations of graffiti images and the textural effects of Paris walls with their generations of superimposed posters. The scratched, mutilated surface, built up in combinations of paint, paper, and sand, particularly appealed to him. Dubuffet's characteristic technique, what he called "*haute pâte*," was based on a thick ground made of sand, earth, fixatives, and other elements into which the pigment was mixed. Figures were incised into this ground, accidental effects were embraced, and the whole—scratched and scarred—worked in every way to give the picture a tangible, powerful materiality. Dubuffet also made pictures out of leaves, tinfoil, or butterfly wings, and his sculptures could be formed of slag, driftwood, sponges, or other found materials. In one of his most ingenious inventions, *L'Âme du Morvan* (*The Soul of Morvan*) (fig. 570), he managed to coax a strange little figure,

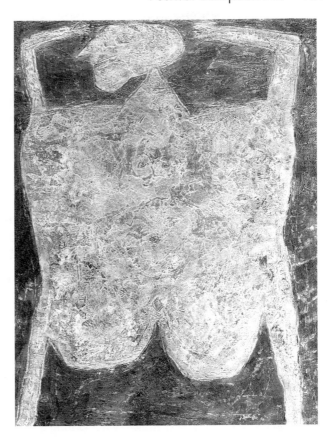

569. JEAN DUBUFFET. *Triumph and Glory* from the Corps de Dame series. 1950. Oil on canvas, 51 x 38½" (129.5 x 97.8 cm). Solomon R. Guggenheim Museum, New York
© 1997 ADAGP, PARIS/ARTISTS RIGHTS SOCIETY (ARS), NEW YORK

570. JEAN DUBUFFET. *L'Âme du Morvan* (*The Soul of Morvan*). May 1954. Grape wood and vines mounted on slag base, with tar, rope, and metal, 18⅜ x 15⅜ x 12¾" (46.7 x 39.1 x 32.4 cm). Hirshhorn Museum and Sculpture Garden, Smithsonian Institution, Washington, D.C.
GIFT OF MARY AND LEIGH B. BLOCK, BY EXCHANGE.
© 1997 ADAGP, PARIS/ARTISTS RIGHTS SOCIETY (ARS), NEW YORK

who stands before a bush on a base of slag, out of found grape vines (Morvan is a wine-producing district in France).

One of Dubuffet's main sources of inspiration was the great collection he began in 1945 of *Art brut* or "raw art." Numbering thousands of works in all media, the collection, today in the Collection de l'art brut in Lausanne, Switzerland, contained the art of the mentally disturbed, the "primitive," and the "naive"—"anyone," Dubuffet said, "who has never learned to draw." To him these works had an authenticity, an originality, a passion, even a frenzy, that were utterly lacking in the works of professional artists.

It was Dubuffet's habit to work on a single theme or a few related themes for an extended period. In the 1950s he turned to landscape imagery, but his landscapes are hardly conventional views of nature. Rather, they are representations of the soil, of the earth itself, like close-up views of geologic formations, barren of any vegetation. His works through the 1950s are predominantly in close-keyed browns, dull ochers, and blacks, verging toward monochrome and, in a series called Texturologies, complete abstraction.

In 1962 Dubuffet embarked on a vast cycle of paintings and, eventually, sculptures, to which he gave the invented name *"L'Hourloupe"* (he liked the sound of the word) and on which he worked for the next twelve years. During a phone conversation, Dubuffet doodled some curving, biomorphic cells in red ink that he filled in with blue parallel lines. These shapes became the formal vocabulary or the "sinuous graphisms" of the *hourloupe* language. From this modest beginning the artist developed a vast repertoire of all-over paintings, some on a scale transcending anything he had ever done previously. Semiautomatic and self-contained, the *hourloupe* forms flow and change like amoebae and are filled with suggestions of living organisms, even at times of human figures. In *Virtual Virtue* (color plate 270, page 484) dozens of heads and little figures are embedded in the undulating lines of an *hourloupe* matrix. With no sense of illusionistic space, they tumble and float freely in what becomes a teeming maze of *hourloupe* creatures.

In 1966 Dubuffet extended the *hourloupe* vocabulary into the realm of sculpture, eventually stripping away the classic *hourloupe* red and blue and composing simply in black and white. At first the sculptures were relatively small-scale, but Dubuffet adapted his technique to outdoor monumental sculpture, sometimes on an environmental scale (fig. 571). Beyond his marvelous corpus of painting and sculpture, Dubuffet has left a body of writings filled with humor and keen observations that apply not only to his work but to art in a larger, universal sense. "What I expect from any work of art," he wrote, "is that it surprises me, that it violates my customary valuations of things and offers me other, unexpected ones."

right: 571. JEAN DUBUFFET. *Group of Four Trees.* 1972.
Epoxy paint on polyurethane over metal structure,
38 x 40 x 34' (11.6 x 12.2 x 10.4 m).
The Chase Manhattan Bank, New York
© 1997 ADAGP, PARIS/ARTISTS RIGHTS SOCIETY (ARS), NEW YORK

Something of Dubuffet's serious whimsy was shared in the postwar years by the French photographer **ROBERT DOISNEAU** (1912–1994), who, like Brassaï, roamed the streets of Paris seeking and finding what he called "the unimaginable image" within "the marvels of daily life." Having discovered that nothing is more bizarre or amusing than the banal, Doisneau managed to brighten a depressed, existential age with his witty, but never snide, photographic commentary on the foibles, misadventures, and daily pursuits of the human—and particularly Parisian—race. In 1949 he made a series of photographs in the shop of an antique dealer in the city (fig. 572), capturing the varied responses of passersby to a provocative painting of a nude hanging in the window.

Before she became a photographer herself, the German-born French **GISÈLE FREUND** (b. 1912) wrote a doctoral thesis on nineteenth-century photography which was accepted at the Sorbonne. During World War II she lived for a time in South America, where she photographed Juan and Evita Perón. In Mexico City after the war she made an important photographic record of her friends Diego Rivera and Frida Kahlo. Upon moving back to France in 1952, Freund continued her program of recording the likenesses of the most renowned humanist writers, intellectuals, and artists. Her broad roster of subjects ranged from James Joyce to Simone de Beauvoir to the African-American writer Richard Wright (fig. 573), author of *Native Son,* who settled permanently in Paris after 1945.

L'Art Informel and Tâchisme in France

The emergence of postwar French abstraction coincided in time, if not in forms, with the origins of Abstract

572. ROBERT DOISNEAU.
From the series
The Sideways Glance.
1949. Gelatin-silver
print © ROBERT DOISNEAU

Expressionism in the United States. Although the artists on both sides of the Atlantic developed their highly individual styles independently of one another, there was a shared sense of the need to reject geometric abstraction and to create a spontaneous art that was guided by emotion and intuition rather than by rational systems of composition. *Art informel* (literally "formless art") is a term devised in 1950 by the French critic Michel Tapié to describe this new type of expression. He coined another term, *un art autre*—"another art," or "a different art," the essence of which is creation with neither desire for, nor preconceptions of, control, geometric or otherwise. *Tachisme* (from the French word *tache*, meaning the use of the "blot," the "stain," the "spot," or the "drip"), yet another term coined in the 1950s by another French critic, refers specifically to European versions of gestural painting.

JEAN FAUTRIER (1898–1964) was one of the leading pioneers of *Art informel*. After painting representationally through the 1920s and 1930s, he produced a moving series of small paintings on paper during World War II that he called Hostages. Incited by war atrocities that he had witnessed firsthand, the Hostage paintings were built-up masses of paint centralized on a neutral ground that resembled decayed human heads served on platters. These were followed by a series called Naked Torsos, whose shapes, albeit

573. GISÈLE FREUND. *Richard Wright.* 1959.
Gelatin-silver print © GISÈLE FREUND

574. BRAM VAN VELDE. *Painting.* 1955. Oil on canvas, 4'3" x 6'5" (1.3 x 2 m). Stedelijk Museum, Amsterdam

© 1997 ADAGP, PARIS/ ARTISTS RIGHTS SOCIETY (ARS), NEW YORK

abbreviated and abstracted, emerge from a heavy paste of clay, paint, glue, and other materials that are built up in a thick relief. *Nude* (color plate 271, page 484), 1960, was made the year Fautrier won the grand prize at the prestigious Venice Biennale. "No art form," the artist said, "can produce emotion if it does not mix in a part of reality."

Though Dutch by birth, **BRAM VAN VELDE** (1895–1981) spent most of his professional life in France, where after World War II he became especially close to Samuel Beckett, who wrote about his work. His paintings exhibit a remarkable consistency from the late 1940s until the end of his career. They tend to be composed around loosely delineated triangular, round, and ovoid forms that are situated in allover compositions and accentuated with patterns of dripping paint (fig. 574). Though he developed his art without knowledge of American action painting, van Velde was the European artist closest to aspects of his fellow Hollander, Willem de Kooning, who was one of the few Americans to exhibit any interest in van Velde's art when it was first shown in New York in 1948. Van Velde provided an unconscious link between the Americans and the younger painters of the CoBrA group (figs. 591, 592, 593, 594).

HANS HARTUNG (1904–1989) fled Nazi oppression in his native Germany in the early 1930s and settled in Paris, via Spain, in 1935. He experimented with free abstract drawings and watercolors as early as 1922, well before his exact contemporary, de Kooning, was formulating his Abstract Expressionist mode in America. Hartung's brittle, linear style owes something to the work of Kandinsky, whom he met in 1925. During the 1930s he explored abstract Surrealism as derived from Miró and Masson, but always with an emphasis on the spontaneous, gestural stroke. He also produced Surrealist col-

lages as well as constructed sculpture in the tradition of his father-in-law, Julio González. In the late 1940s Hartung developed his mature style of brisk graphic structures, usually networks of bold black slashes played against delicate, luminous washes of color. His improvisational marks, combined with his statement in 1951 that he "acted on the canvas" out of a desire to record the trace of his personal gesture, allied him in the minds of some with action painting. In the 1960s he tended to scratch into fresh paint with various instruments to achieve finely textured surfaces, in which light and shadow are played over one another in shimmering chiaroscuro effects (color plate 272, page 485).

PIERRE SOULAGES (b. 1919) first came to Paris in 1938 but did not settle there until 1946. Almost immediately he began painting abstract structures based on studies of tree forms. The architectural sense and the closely keyed color range that characterize his abstract style (color plate 273, page 485)—this sense of physical, massive structure in which the blacks stand forth like powerful presences—may have been inspired originally by the prehistoric dolmens of his native Auvergne, as it certainly was by the Romanesque sculpture of the area. Soulages has consistently maintained a subdued palette, feeling that "the more limited the means, the stronger the expression." His great sweeping lines or color areas are usually black or of a dark color. Heavy in impasto, they are applied with the palette knife or large, housepainter's brushes. Soulages thus achieved an art of power and elegance, notable not only in his forms but in the penetrating, encompassing light that unifies his paintings.

Berlin-born Alfred Otto Wolfgang Schulze, known by his pseudonym **WOLS** (1913–1951), was raised in Dresden, where he soon excelled at many things, including music and photography. Interested primarily in drawing and painting, however, the precocious Wols moved to Paris at the age of nineteen in 1932. There he briefly met members of the international Surrealist circle such as Arp and Giacometti. For part of the war he was, like other expatriate Germans, interned by the French. While detained he made bizarre watercolors and drawings of hybrid creatures that somewhat resemble Surrealist "exquisite corpse" drawings (see fig. 353). Following the war he developed an abstract style with some analogies to that of Fautrier. He built up his paint in a heavy central mass that was shot through with flashes of intense colors and automatic lines and dribbles (fig. 575). Wols's

abstractions create the impact of magnified biological specimens—he had been interested in biology and anthropology from an early age—or encrusted, half-healed wounds. Wols never fully recovered from his war-induced traumas. He abused alcohol and died at the age of 38.

GEORGES MATHIEU (b. 1921) has been described as the Salvador Dali of *Art informel*. After World War II he moved toward a calligraphic style that owes something to Hartung. His calligraphy (color plate 274, page 485), however, consists of sweeping patterns of lines squeezed directly from the tube in slashing, impulsive gestures. To him, speed of execution is essential for intuitive spontaneity. Typically, Mathieu turned to elaborate titles taken from battles or other events of French history, reflecting his insistence that he is a traditional history painter working in an abstract means. In love with spectacle, performance, and self-promotion, he has at times painted before an audience, dressed in armor and attacking the canvas as though it were the Saracen and he were Roland. Despite such exhibitionism, Mathieu became an artist of substantial abilities in the 1950s, one of the leading European exponents of brush, gesture, and action painting.

Born and educated in Montreal, **JEAN-PAUL RIOPELLE** (b. 1923) settled in Paris in 1947. He soon became part of the circle around the dealer Pierre Loeb (a leading supporter of Surrealism in the 1930s) that included Mathieu and Vieira da Silva. Between 1955 and 1979, he was the companion of the second-generation American Abstract Expressionist Joan Mitchell (see fig. 667). The individual mode he achieved in the 1950s (fig. 576) is essentially a kind of controlled chance, an intricate, allover mosaic of small, regular daubs of jewellike color, applied directly from tubes, shaped with palette knives, and held together by directional lines of force.

A Parisian artist difficult to classify but one of great sensitivity was Lisbon-born **MARIA ELENA VIEIRA DA SILVA** (1908–1992). By the 1940s she was preoccupied with the interaction of perspective and nonperspective space in paintings based on architectural views of the city or deep interior spaces. By 1950 these views became abstract, grid-based compositions. *The Invisible Stroller* (fig. 577) is related to

above: 575. WOLS. *Bird.* 1949.
Oil on canvas, 36¼ x 25⅜"
(92.1 x 64.5 cm).
The Menil Collection, Houston © 1997
ADAGP, PARIS/ARTISTS RIGHTS SOCIETY (ARS), NEW YORK

right: 576. JEAN-PAUL RIOPELLE. *Blue Knight.* 1953. Oil on canvas,
3'8⅞ x 6'4½" (1.1 x 1.9 m).
Solomon R. Guggenheim Museum,
New York
© 1997 ADAGP, PARIS/ARTISTS RIGHTS SOCIETY (ARS), NEW YORK

577. MARIA ELENA VIEIRA DA SILVA. *The Invisible Stroller.* 1951.
Oil on canvas, 52 x 66" (132.1 x 167.6 cm).
San Francisco Museum of Modern Art
GIFT OF MR. AND MRS. WELLINGTON S. HENDERSON.
© 1997 ADAGP, PARIS/ARTISTS RIGHTS SOCIETY (ARS), NEW YORK

Picasso's Synthetic Cubist paintings of the 1920s in its fluctu-ations of Renaissance perspective and the Cubist grid. From this point forward she gradually flattened the perspective while retaining a rigid, rectangular, architectural structure. The view of the city, seen from eye level, eventually became the view from an airplane—as one rises higher and higher, details begin to blur and colors melt into general tonalities.

NICOLAS DE STAËL (1914–1955) was an independently minded painter who did not wish to be associated with his contemporary abstractionists in France, seeing himself more as the heir to his friend Braque, as well as Matisse. De Staël

was born an aristocrat in prerevolutionary Russia, grew up in Belgium, where he attended the Royal Academy of Fine Arts, and moved to Paris in 1938. In the 1940s he was painting abstractly, but as his work developed, he became more and more concerned with the problem of nature versus abstrac-tion. De Staël seems to have been increasingly desirous of asserting the subject while maintaining abstract form. "A painting," he said, "should be both abstract and figurative: abstract to the extent that it is a flat surface, figurative to the extent that it is a representation of space." He began to use palette knives in 1949, building up his abstract composi-tions with thick, loosely rectangular patches of paint that he scraped into the desired form with a knife (color plate 275, page 486). In 1953 he traveled to Sicily and was struck by the intense quality of light. He made landscapes in brilliant hues (fig. 578)—in this example reds, oranges, and yellows—in which the forms are simplified to large planes of flat color accented by a few contrasting color spots. In the last year of his life he began painting even more literal scenes of boats and birds and oceans, in thin, pale colors. In 1955, de Staël, who had achieved a considerable reputation on both sides of the Atlantic, apparently reached an impasse in his work and jumped to his death from the terrace of his studio.

GERMAINE RICHIER (1904–1959) received her training from two sculptors who had studied with Rodin, Louis-Jacques Guiges and Bourdelle (see fig. 143). Her earliest sur-viving sculptures of the late 1930s and early 1940s are sensi-tive portrait heads in the Rodin tradition. But the macabre quality of her mature works, fully formulated in *Praying Mantis* (color plate 276, page 486), is uniquely her own. This strangely humanoid creature resembles a seated figure while maintaining all the predatory traits of a threatening insect. *The Hurricane* (fig. 579) is characteristic of Richier's race of brutish beings. This is a monster, powerful as a gorilla, with pitted, torn, and lacerated flesh, that in a curious way is both

578. NICOLAS DE STAËL.
Agrigente. 1954. Oil on can-vas, 34¾ x 50½" (88.3 x 128.3 cm). The Museum of Contemporary Art,
Los Angeles
THE RITA AND TAFT SCHREIBER COLLECTION.
© 1997 ADAGP, PARIS/ARTISTS RIGHTS SOCIETY (ARS), NEW YORK

left: 579. Germaine
Richier. *The Hurricane*.
1948–49. Bronze,
height 70⅛" (178.1 cm).
Private collection

left: 581. César.
The Yellow Buick.
1961.
Compressed
automobile,
59½ x 30¾ x
24⅞" (151.1 x
78.1 x 63.2 cm).
The Museum of
Modern Art,
New York

threatening and pitiful. Richier experimented with the bronze technique, using paper-thin plaster models, into the surface of which she pressed every sort of organic object. Sometimes she placed her figures in front of a frame, the background of which is either worked in relief or, in some larger works, painted by a painter friend such as da Silva. There are relations in her encrusted surfaces both to Giacometti's postwar sculpture, which was being formulated at exactly the same time (color plate 266, page 482), and to Dubuffet's contemporary figure paintings (fig. 569).

Whereas Richier stretched the possible limits of bronze casting, César Baldaccini, known as César (b. 1921), created comparable effects from the assemblage of old iron scraps and machine fragments. His work developed independently of trends in the United States but paralleled that of certain American sculptors, notably John Chamberlain (color plate 263, page 464). In his constructions, César tends toward abstraction, but he often returns to the figure. (Paradoxically the figure is also implicit in many of the abstract works.) His 1958 *Nude* (fig. 580) is a pair of legs with lower torso, eroded and made more horrible by the sense of life that remains. During the late 1950s and early 1960s (and with variants in new materials into the 1990s) the artist made what he called "compressions," assemblages of automobile bodies crushed

under great pressure then pressed into massive blocks of vari-colored materials, as they are in auto junkyards when the metal is processed for reuse (fig. 581). By this point in his career César was associated with New Realism in France, a movement discussed in the following chapter.

Concrete Art

In 1930, as we saw, Theo van Doesburg coined the word "concrete" as a substitute for "nonobjective" in its specific application to the geometric abstractions of de Stijl. This was followed in 1931 by the *Abstraction-Création* group in Paris, which, during the 1930s, sought to advance the principles of pure abstraction and of Mondrian's Neo-Plasticism. The concept of Concrete art was revived in 1947 in the Salon des Réalités Nouvelles, as the gallery of Denise René became an international center for the propagation of Concrete art. Among the leaders outside of France was Josef Albers, who went to the United States in 1933 and whose influence there on the spread of geometric abstraction (see chapter 17). He exhibited regularly with the American Abstract Artists group as well as with *Abstraction-Création* in Paris. Many aspects of postwar Constructivism, Color Field painting, Systemic painting, and Op art stem from the international tradition of de Stijl and Concrete art.

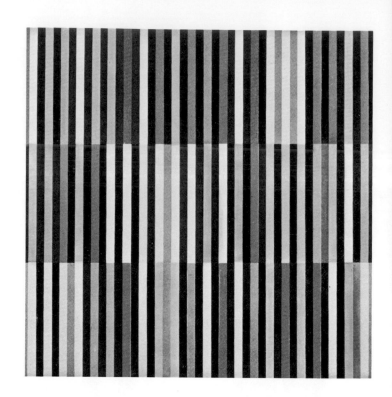

above: 582. MAX BILL. *Endless Ribbon from a Ring I.* 1947–49, executed 1960. Gilded copper on crystalline base, 14½ x 27 x 7½" (36.8 x 68.6 x 19.1 cm). Hirshhorn Museum and Sculpture Garden, Smithsonian Institution, Washington, D.C.
© 1997 BEELDRECHT, AMSTERDAM/ARTISTS RIGHTS SOCIETY (ARS), NEW YORK

right: 583. RICHARD PAUL LOHSE. *Pure Elements Concentrated in Rhythmic Groups.* 1949–56. Private collection

Important for the spread of Concrete art in Europe were developments in Switzerland, particularly in the work of **MAX BILL** (1908–1994) and Richard Paul Lohse. Though he was making nonobjective art by the late 1920s, Bill first began applying the term "concrete" to his work in 1936. He defined it as follows: "Concrete painting eliminates all naturalistic representation; it avails itself exclusively of the fundamental elements of painting, the color and form of the surface. Its essence is, then, the complete emancipation of every natural model; pure Creation." Bill studied at the Dessau Bauhaus between 1927 and 1929 when Josef Albers was teaching there. Subsequently, he developed into one of the most prolific and varied exponents of Bauhaus ideas as a painter, sculptor, architect, graphic and industrial designer, and writer. He was associated with *Abstraction-Création* in Paris and organized exhibitions of Concrete art in Switzerland. Bill insisted that painting and sculpture have always had mathematical bases, whether arrived at consciously or unconsciously. In the forms of Concrete art he found rationalism, clarity, and harmony—qualities symbolic of certain universal ideas.

In sculpture, the form that intrigued Bill the most and on which he played many variations is the endless helix or spiral (fig. 582). The metal ribbon form turns back upon itself with no beginning and no end and is, therefore, in a constant state of movement and transformation. In his paintings, Bill, like Albers, was systematically experimenting with color, particularly the interaction of colors upon one another (color plate 277, page 486). Much as his square and diamond-shaped paintings owe to the example of Mondrian, Bill's pristine forms and regular grids have more to do with geometry than they do with Mondrian's intuitive and painterly approach to picture making, situating his work in the context of so-called Hard-Edge abstract painting in the 1960s (see chapter 22).

Like Bill, **RICHARD PAUL LOHSE** (1902–1988) made mathematics the basis of his painting. He sought the simplest

possible unit, such as a small color-square. Using not more than five color hues but a great range of color values, Lohse built his units into larger and larger complexes through a mathematically precise distribution of hues and values in exact and intellectually predetermined relationships (fig. 583). This resulted in a composition or color structure that spreads laterally and uniformly, with no single focus, to the edges of the canvas, with the implication and the potential of spreading to infinity beyond the edges. Despite the rigidity of Lohse's system, the subtlety and richness of his colors and the shimmering quality deriving from their infinitely varied relationships give his works a lyrical, poetic feeling. His art is important as an immediate prototype for Op art (see fig. 679).

Postwar Art in Italy and Spain

In post-Fascist Italy, art underwent a period of intense debate in major urban centers, such as Milan, Bologna, and Rome, with artistic concerns rarely divorced from political ones. A strong realist movement existed, particularly among Communist artists who wielded painting as a political weapon. But alternatives arose among younger artists who proposed radical abstraction as an antidote to Italy's time-honored classical tradition. In the late 1950s various forms of *Art informel* spread throughout the country, supplanting more Cubist-based forms of abstraction. Despite the Italian proclivity for forming groups and issuing manifestos, *Art informel* was never a uniform movement in that country, and figuration, particularly in sculpture, was still a viable artistic force.

An artist who seemed to exist outside of the political and artistic fray was **GIORGIO MORANDI** (1890–1964), who emerged as the leading figurative painter in Italy in the years after the war. Despite his forays into landscape and, occasionally, figure painting, Morandi's primary subject throughout his long career was still life. At odds with much of the large-scale, highly emotive, and abstract art being produced after the war, this artist's quiet and modestly scaled still lifes drew as much upon Italian Renaissance traditions as contemporary developments. Old enough to have been exposed to Futurism (though his work was never truly Futurist), Morandi was included in the First Free Futurist Exhibition in Rome in 1914. In 1918–19, after discovering the works of Carrà and de Chirico, he made enigmatic still lifes that allied him with the Metaphysical School. Morandi lived quietly in his native Bologna, teaching, painting, and traveling only rarely. It was not until after World War II that he was acknowledged as Italy's leading living painter. The unremarkable elements of his postwar still lifes (color plate 278, page 487)—bottles, jars, and boxes—are typically massed together against a neutral ground to form a kind of miniature architecture, with an emphasis on abstract organization and subdued, closely valued tones. There is a meditative, serialized aspect to the artist's work, for he often cast similar arrangements in slightly altered light conditions. Morandi's still lifes, in which humble objects are reduced to their essence, continue the tradition of the great French eighteenth-century still-life painter Jean-Baptiste-Siméon Chardin and provided an important example to younger realist painters, notably the American Wayne Thiebaud (color plate 315, page 532).

The Italian contribution in postwar figuration came chiefly in sculpture, especially in the work of **MARINO MARINI** (1901–1980) and Giacomo Manzù. The sculpture of Marini, a native of Tuscany, was deeply rooted in history. For Italians, he said, "the art of the past is part and parcel of our daily life in the present." Besides ancient Roman and Etruscan art, he was interested in the sculpture of Archaic Greece and Tang dynasty China. He also traveled widely and was knowledgeable about modernist developments in Italy and elsewhere in Europe. Marini was a mature artist by the 1930s, already experimenting with his favorite themes, including the Horse and Rider. During the war, the equestrian subject, traditionally employed to commemorate noblemen or celebrate war heroes in city squares, became for Marini a symbol for suffering and disillusioned humanity. His horse (fig. 584), with elongated, widely stretched legs and upthrust head, screams in agony, in a manner that recalls Picasso's *Guernica* (see fig. 381). The rider, with outflung stumps for arms, falls back in a violent gesture of despair. "My equestrian figures," the artist said, "are symbols of the anguish that I feel when I survey contemporary events. Little by little, my horses become more restless, their riders less and less able to control them." By 1952, when he won the grand prize for sculpture at the Venice Biennale, Marini was recognized internationally as one of Europe's most prominent sculptors.

GIACOMO MANZÙ (1908–1991) was influenced by Medardo Rosso (see fig. 144), but his real love was the sculpture of the Italian Renaissance. In 1939, deeply affected by the pernicious spread of Fascism and the threat of war, he made a series of relief sculptures around the subject of the Crucifixion that was his not-so-veiled anti-Fascist statement.

The Cardinal series, begun in 1938 but represented here by a version from 1954 (fig. 585), takes a conventional subject and creates from it a mood of withdrawal and mystery. At the same time, Manzù utilizes the heavy robes for a simple and monumental sculptural volume. In the Renaissance tradition,

left: 584. MARINO MARINI. *Horse and Rider.* 1952–53. Bronze, height 6'10" (2.1 m). Hirshhorn Museum and Sculpture Garden, Smithsonian Institution, Washington, D.C.
© 1997 ESTATE OF MARINO MARINI/LICENSED BY VAGA, NEW YORK, NY

right: 585. GIACOMO MANZÙ. *Large Standing Cardinal.* 1954. Bronze, height 66½" (168.9 cm). Hirshhorn Museum and Sculpture Garden, Smithsonian Institution, Washington, D.C.

left: 586. GIACOMO MANZÙ. *Death by Violence II*. 1962. Bronze door panel, 39¾ x 28¼" (101 x 71.8 cm). St. Peter's Basilica, Vatican, Rome

below: 587. AFRO BASALDELLA. *For an Anniversary*. 1955. Oil on canvas, 4'11" x 6'6⅝" (1.5 x 2 m). Solomon R. Guggenheim Museum, New York
GIFT OF MR. AND MRS. JOSEPH PULITZER, JR., 1958

Manzù carried out a number of great religious commissions, notably the bronze door panels for St. Peter's Basilica in Rome and for Salzburg Cathedral. He received the commission for St. Peter's in 1952 and the doors were finally consecrated in 1964. Manzù solved the daunting task of designing doors for St. Peter's—he was, after all, entering the realm of Bramante, Michelangelo, and Bernini—with a series of low-relief sculptures on the general theme of death. Each door contains a large, vertical panel on a biblical theme above four smaller ones. In these lower panels Manzù chose to depict not the deaths of Christian martyrs but those of common people. In *Death by Violence II* (fig. 586), a man succumbs to a tortuous death observed by a compassionate female onlooker.

Italian *Art informel* found an advocate in **AFRO**, (1912–1976) as Afro Basaldella is generally known, though his work was never as radical as that of Lucio Fontana or Alberto Burri. His work was central to the Italian *Astratto-Concreto* (Abstract-Concrete) movement at mid-century, led by the influential Italian critic Lionello Venturi. During the 1950s, Afro created an atmospheric world of light and shadows, with subdued, harmonious colors and shapes that are in a constant state of metamorphosis (fig. 587). Of Italian postwar painters, Afro was closest to his American contemporaries, having lived in the United States in 1950 and admired the work of Gorky, Kline, and de Kooning. His later work of the 1960s took on a painterly appearance close to that of these gestural painters.

Among abstract artists in Italy in the years after the war, the protean **LUCIO FONTANA** (1899–1968) was the leading

exemplar. Born in Argentina and trained as a sculptor in Italy, he was producing abstract sculpture by the early 1930s (though he also continued to work figuratively into the 1950s). He joined the Paris-based *Abstraction-Création* group in 1935, the same year he signed the First Manifesto of Italian Abstract Artists. During World War II he returned to Argentina, where he published the White Manifesto, outlining his idea of a dynamic new art for a new age made with such modern materials as neon light and television—media whose potential would not be tapped again for art until the 1960s and 1970s. The White Manifesto, modeled in many ways on Futurist prototypes, anticipated the theory of what would be known as Spatialism (laid out in five more manifestos between 1947 and 1952), a concept that rejected the illusionistic or "virtual" space of traditional easel painting in favor of the free development of color and form in literal space. The Spatialist program was designed to emancipate art from past preconceptions, to create an art that would "transcend the area of the canvas, or the volume of the statue, to assume further dimensions and become . . . an integral part of architecture, transmitted into the surrounding space and using discoveries in science and technology."

Fontana also became a pioneer of environments in 1949 when he used free forms and black light to create vast Spatialist surrounds, experiments followed in the 1950s by designs made in collaboration with the architect Luciano Baldessari. In his desire to break down categories of painting, sculpture, and even architecture, Fontana described his works as *Concetti Spaziale* (CS), or "Spatial Concepts," thus inau-

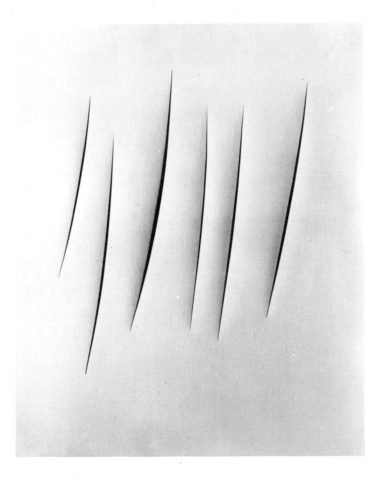

left: 588. LUCIO FONTANA. *Concetto Spaziale (Attese)*. c. 1960. Oil on canvas, 36 × 28¾" (91.4 × 73 cm). Folkwang Museum, Essen, Germany

below: 589. LUCIO FONTANA. *Spatial Conception Nature (5 Balls)*. 1959–60 (cast 1965). Bronze. Hirshhorn Museum and Sculpture Garden, Smithsonian Institution, Washington D.C.
GIFT OF JOSEPH H. HIRSHHORN, 1980

gurating Conceptualism, a style that would not actually take hold in Western art until the 1970s.

Beginning in 1949, Fontana perforated the canvas with *buchi* (holes) in abstract but not haphazard patterns, thereby breaking through the picture plane that had been the point of departure for painting throughout much of its history. In 1963 he made a series of egg-shaped canvases called Spatial Concept: The End of God (color plate 279, page 487) in which holes are torn in a brightly painted canvas to reveal another colored canvas below, all resembling a strange, extraterrestrial landscape. During the 1950s he experimented with matter, as did his fellow countrymen Afro and Alberto Burri, building up his penetrated surfaces with heavy paint, paste, canvas fragments, or pieces of glass. It was in 1958 that, feeling he was complicating and embellishing the works too much, Fontana slashed a spoiled canvas. He realized that with this simple act, he could achieve the integration of surface with depth that he had been seeking. These monochrome, lacerated canvases, called *Tagli* (slashes) (fig. 588) were subtitled *Attese*, which translates as "expectations," and are the works for which he is best known. Although Fontana's abstract art was tied to the general trends of *Art informel*, instead of making a positive gesture that adds medium to the painted surface, his incisions in a sense remove medium and transform the normally taut and inviolable surface of the canvas into a soft, malleable form that opens to a mysteriously dark interior. For Fontana, the gesture was less about creating a record of the individual artist's emotion—as with Mathieu or de Kooning—than it was an exploration of the

physical properties of a work of art and the emotions those properties evoke. Shortly after he made his first *Tagli*, Fontana applied his Spatial Concepts to bronze sculptures (as he already had to ceramics), creating roughly textured, rounded forms that are sliced and punctured (fig. 589).

ALBERTO BURRI (1915–1995) was a doctor in the Italian army in North Africa when he was captured by the British. Later he was turned over to the Americans, who interned him as a prisoner of war in Texas. During this confinement, Burri began drawing and painting. Abandoning medicine after his release, he continued his creative activities free of formal artistic training, and developed a style indebted at first to Mondrian and Miró. Burri eventually evolved a method with a strong emphasis on texture and materiality, incorporating thick paint, pastes, and tar into his paintings. He received his first exhibition in a Roman gallery in 1947 and in 1950 began to make *Sacchi* (color plate 280, page 488), paintings made of old sacks roughly sewn together and heavily splashed with paint. The results were often reminiscent of bloodstained bandages, torn and scarred—devastating commentaries on war's death and destruction. They have been related to Schwitters's collages (see fig. 314), but their impact is much more brutal. For a period in the mid-1950s Burri burned designs on thin wood panels like those used in orange crates. In them, as in the earlier works, there is a strange juxtaposition: on the one hand, the ephemeral material coupled with a sense of destruction and disintegration (the material sometimes disintegrates visibly when moved from place to place); on the other hand, the elegance and control with which the artist arranges his elements. Burri also worked in metal and plastics, melting and re-forming these media with a blowtorch to achieve results of vivid coloristic beauty combined with effects of disease and decay. His influence has been strong, particularly on the young Robert Rauschenberg (see fig. 610), who visited his studio in 1953.

Coming of age following the devastation of the Spanish Civil War and in the wake of the great twentieth-century Spanish painters—Picasso, Miró, Dalí—**ANTONI TÀPIES** (b. 1923) emerged as an important artist in Spain after World War II. By the early 1960s his reputation had spread well beyond Spain's borders and was bolstered by important exhibitions in Europe and the United States. Tàpies is a native of Barcelona, a gifted draftsman, and a man of broad erudition who drew from many sources for his art, including Catalan mysticism and Zen Buddhism. His early paintings, which incorporate collage and depict fantastic subjects, are highly reminiscent of Klee and Miró. Then in the mid-1950s, profound changes took place in his outlook and his art. "Suddenly, as if I had passed through the looking-glass," he wrote, "a whole new perspective opened up before me, as if to tell the innermost secret of things. A whole new geography lit my way from one surprise to the next." He began to make abstract compositions distinctive for their emphasis on thick, rough textures (color plate 281, page 488). He built up the surfaces of his paintings in a variety of ways, with combinations of paint, varnish, sand, and powdered marble, to create an effect of textured relief. His colors are generally subdued but rich, with surfaces worked in many ways—punctured, incised, or modeled in relief. These highly tactile surfaces became analogous to walls for the artist into which he crudely scratched lines or onto which he applied other materials. As such, they were metaphors for the struggles and sufferings of the artist and his fellow Spaniards during the war years, which he said, "seemed to inscribe themselves on the walls around me." (The fact that *tàpies* means walls in Spanish is not inconsequential.) Like many of his contemporaries, Tàpies has used "poor" materials—weathered wood, rope, burlap, cardboard, and found objects—to create crusty, battered surfaces expressive of his deeply spiritual, antimaterialistic values.

Originally trained as an architect, the Spanish artist **EDUARDO CHILLIDA** (b. 1924) turned to sculpture in his early twenties. Though he has carved work in wood and stone, Chillida has worked predominantly in iron, drawing upon the long blacksmithing tradition in his native Basque country. His earlier works are complex, interlocking configurations that can suggest the old tools from which they were forged (fig. 590). More recently he has forged massive, monumental bars in powerful architectural abstractions with extremely subtle relations among the massive, tilted rectangular shapes. His works on a monumental scale include a well-known sculpture for his hometown of San Sebastián called *Combs of the Wind* (color plate 282, page 488). At a place where Chillida spent much of his childhood, iron forms resembling massive pincers are dramatically sited on the beach and embedded into the living rock.

590. EDUARDO CHILLIDA. *Dream Anvil No. 10.* 1962. Iron on wooden base, 17⅛ x 20½ x 15⅛" (43.5 x 52.1 x 38.4 cm). EMANUEL HOFFMANN FOUNDATION, PERMANENT LOAN TO THE KUNSTMUSEUM, BASEL. © 1997 VEGAP, MADRID/ARTISTS RIGHTS SOCIETY (ARS), NEW YORK

CoBrA

In the postwar Netherlands—in the very home of Mondrian and de Stijl—a desire for expressionistic liberation from the constraints of anything rational, geometric, or Constructivist became particularly urgent. Inspired by the Danish painter Asger Jorn, three Dutch artists—Karel Appel, Cornelis Corneille, and George Constant—established the Experimental Group in 1948. They sought new forms of elemental expression, as much opposed to Mondrian and de Stijl as to the Academy. Through contacts with similar groups in Copenhagen and Brussels, this association evolved into the international group CoBrA (standing for Copenhagen, Brussels, and Amsterdam), led by Jorn and the Belgian poet Christian Dotrement. Most of the painters associated with the CoBrA group employed some sort of subject or figuration, often informed by their interests in the art of the untutored, what Dubuffet called *Art brut.* The most important unifying principle among these divergent artists was their doctrine of complete freedom of abstract expressive forms, their emphasis on brush gesture, and their rejection of rationalism and geometric abstraction. In the words of one artist, "The gallery had become a temple dedicated to the right angle and the straight line." In this, they were allied to Dubuffet and

Fautrier in France, for both of whom the artists of CoBrA had great respect, although the latter's means of expression were generally more violent, colorful, and dynamic. CoBrA, which lasted as a united movement from 1948 to 1951, was a critical artistic phenomenon for its internationalism. At the same time, it infused tremendous energy into abstract painting when an exhausted Europe was just emerging from the war.

KAREL APPEL (b. 1921) has consistently painted figures and portraits with heavy impasto and brilliant color. In 1956 he portrayed Willem Sandberg (fig. 591), who was then director of Amsterdam's Stedelijk Museum, which specializes in modern art. Sandberg was a heroic figure to young artists in Holland both for his support of experimental art—he championed CoBrA against intensely negative public sentiment—and for his efforts to save Dutch Jews from deportation by the Nazis during the war. Like Dubuffet, whose work he first saw in 1947 during a trip to Paris, Appel was drawn to the intuitive and the spontaneous, especially in the art of children, and he achieved a wonderful childlike quality in his own work (color plate 283, page 521). He was also inspired by Van Gogh, Matisse, Schwitters, and Picasso. "You have to

learn it all," he said, "then forget it and start again like a child. This is the inner evolution."

Appel established a certain control through his representation of an explicit face or figure. Another CoBrA painter, the Dane ASGER JORN (1914–1973), represents the most extreme reaction against Concrete art's principles of order and harmony. Jorn spent several formative years in Paris, where he received disciplined training at Léger's academy. Important for his future development was his exposure to the biomorphic Surrealism of Arp and Masson, memories of which remained a constant strain in his art. Like the American Abstract Expressionists, who learned from similar Surrealist examples, Jorn was interested in mythic subject matter, which he drew from specifically Nordic folkloric traditions.

Jorn produced paintings (fig. 592) that could be an illustration of Ruskin's famous epithet concerning Whistler—a pot of paint flung in the face of the public. Yet, Jorn's art still has its roots in figuration. Using largely undiluted primary colors with blacks, whites, and greens, Jorn smeared, slashed, and dripped the paint in a seemingly uncontrolled frenzy. Yet the control, established in large, swinging, linear movements,

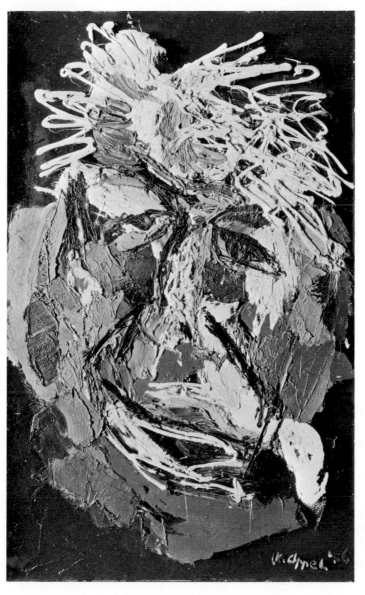

591. KAREL APPEL. *Portrait of Willem Sandberg.* 1956. Oil on canvas, 51¼ x 31⅞" (130.2 x 81 cm). Museum of Fine Arts, Boston
TOMPKINS COLLECTION.
© 1997 BEELDRECHT, AMSTERDAM/ARTISTS RIGHTS SOCIETY (ARS), NEW YORK

592. ASGER JORN. *Green Ballet*. 1960. Oil on canvas,
4'9" x 6'6¾" (1.4 x 2 m).
Solomon R. Guggenheim Museum, New York

and asymmetrically balanced color-shapes, remain sufficient to bring a degree of order out of chaos while not diminishing the explosive fury of the abstract expression (color plate 284, page 521).

The Belgian **PIERRE ALECHINSKY** (b. 1927) was a latecomer to the CoBrA group. When he first encountered their paintings at a 1949 exhibition in Brussels, he detected in them "a total opposition to the calculations of cold abstraction, the sordid or 'optimistic' speculations of socialist realism, and to all forms of division between free thought and the action of painting freely." By comparison with Appel or Jorn, particularly in his earlier works of the 1950s, Alechinsky produced an allover structure of tightly interwoven elements with a sense of order deriving from microcosmic organisms (fig. 593). His later works, on the other hand, have a curvilinear morphology that stems partly from Belgian Art Nouveau as well as from the artist's automatist practices.

An artist who continued the traditions of Austrian Art Nouveau and Expressionism was the Viennese Friedrich Stowasser, or **HUNDERTWASSER** (b. 1928). Impossible to classify through allegiance to any particular artistic group or cause, Hundertwasser has traveled extensively and has lived in France, Italy, and Austria. Though he arrived in Paris at the dawn of *Art informel*, he developed a style directly at odds with it, preferring instead to forge a visual language based on specific

forms, ones derived largely from the Austrians Gustav Klimt and Egon Schiele (see figs. 107, 170). His early work carries strong overtones of Paul Klee, and he shared that Swiss painter's interest in art produced by children. Perhaps the most distinctive feature of Hundertwasser's art is its jewellike color, as in the self-portrait (color plate 285, page 521), a brilliant pastiche of Art Nouveau decoration.

Among photographers in Central Europe, the Czech **JOSEF SUDEK** (1896–1976) was a leading practitioner. Called the "Poet of Prague," Sudek, who lost an arm while serving in World War I, established a successful photographic business in that city in the 1920s and at first made photographs in a Romantic, Pictorialist style. Once German troops marched into Prague in 1939, Sudek retreated into his studio, focusing his camera on whatever caught his eye through the window of his studio. He continued his habit of photographing still lifes on his studio windowsill or views into the garden or the city streets into the 1950s (fig. 594). "I believe that photography loves banal objects," Sudek said. "I am sure you know the fairy tales of Andersen: when the children go to bed, the objects come to life, toys, for example. I like to tell stories about the life of inanimate objects, to relate something mysterious: the seventh side of a dice."

Painting and Sculpture in England

Irish-born **FRANCIS BACON** (1909–1992) was a figurative painter of disquieting subject matter. After the war, Bacon achieved international stature on a par with that of Giacometti and Dubuffet. A wholly self-taught artist, Bacon left his native

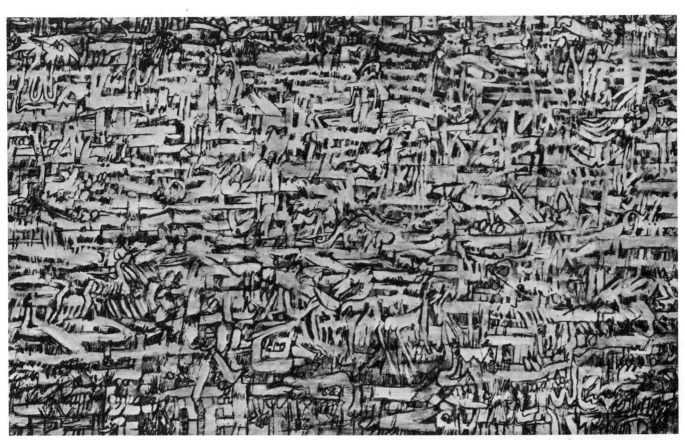

593. PIERRE ALECHINSKY. *Ant Hill*. 1954. Oil on canvas,
4'11½" x 7'9⅞" (1.5 x 2.4 m).
Solomon R. Guggenheim Museum, New York
© 1997 ADAGP, PARIS/ARTISTS RIGHTS SOCIETY (ARS), NEW YORK

Dublin as a teenager, traveling in Europe and settling eventually in London. He painted grim, violent, and grotesquely distorted imagery in the most seductive painterly style, producing what one writer described as "a terrible beauty." Bacon's devotion to the monstrous has been variously interpreted as a reaction to the plight of the world, or as a nihilist or existential attitude toward life's meaninglessness.

In *Painting* (color plate 286, page 522), 1946 (he destroyed most of his prewar work), Bacon created a horrific montage of images that, according to his explanation, emerged by accident. "I was attempting to make a bird alighting on a field," the artist said, "but suddenly the lines that I'd drawn suggested something totally different, and out of this suggestion arose this picture. I had no intention to do this picture; I never thought of it in that way. It was like one continuous accident mounting on top of another." From the deep shadows, cast by an umbrella, emerges the lower half of a head with a hideously grimacing mouth topped by a bloody moustache. The figure, whose torso virtually disappears into long strokes of paint, is enclosed within a round tubular structure (related to the artist's own earlier furniture designs); hunks of meat seem to be skewered on it. The sense of formal disjunction compounds the imagery's disturbing effect. At

594. JOSEF SUDEK. *The Window of My Studio*. 1954.
Gelatin-silver print © JOSEF SUDEK ESTATE

595. FRANCIS BACON. *Head VI*. 1949. Oil on canvas, 36⅝ x 30⅛" (93.2 x 76.5 cm). Arts Council Collection, London

tion to become a terrifying and universal depiction of ruthless, predatory power.

Bacon's sense of tradition extended from Giotto to Rembrandt and Van Gogh. He used paintings of the past as the basis of his works but transformed them through his own inward vision and torment. Beginning in 1949 he made a series of paintings after Velázquez's great *Portrait of Pope Innocent X* in the Doria Pamphili Gallery in Rome (which Bacon knew only through reproductions). In these works, the subject is usually shown seated in a large, unified surrounding space. The figure is blurred as though seen through a veil, while the perspective lines delineate a glass box within which the figure is trapped. *Head VI* (fig. 595) is the first of these papal images and the last of a series of six heads Bacon made in 1948–49. Typically, the pope's mouth is wide open, as if uttering a horrendous scream. The artist, who said he wanted to "make the best painting of the human cry," conflated the papal image with a famous still from *The Battleship Potemkin*, 1925, the Russian film by Sergei Eisenstein. The still is a closeup of a panic-stricken, screaming nanny who has been shot in the eye, her pince-nez shattered as she loses her grip on the baby stroller in her care. The exact meaning of the Pope series is deliberately obscured by the artist, but there is no question of the horrifying impact and beauty of the painting, as well as the grandness of the conception.

Bacon's subject was almost exclusively the human face or figure—especially the male nude, which he portrayed in states of extreme, sometimes convulsive effort. Like Max Beckmann (see color plate 143, page 269), he frequently made triptychs, a format he had employed for scenes related to human suffering. In *Triptych—May–June 1973* (fig. 596), Bacon created a harrowing yet touching posthumous portrait of his close friend, George Dyer, during a fatal illness. In three separate

the figure's feet is what appears to be a decorative Oriental carpet, as though this creature, who wears a yellow corsage, is sitting for a conventional formal portrait. Presiding over the entire scene is a gigantic beef carcass, strung up like a crucifixion beneath a grotesquely incongruous festoon of decorative swags. Some have detected in the figure a resemblance to despots of the day, such as Mussolini. Although *Painting* was made just after the war, it transcends any particular associa-

596. FRANCIS BACON. *Triptych—May–June 1973*. 1973. Oil on canvas, each panel 6'6" x 4'10" (1.9 x 1.5 m). Private collection, Switzerland COURTESY MARLBOROUGH GALLERY, LONDON

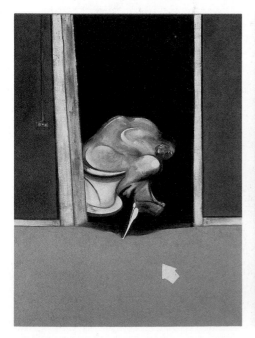

images, cinematic in their sense of movement and sequential action, Dyer's ghostly white figure is seen through darkened doorways. In the central canvas, where Dyer lurches toward the sink, the blackness spills out onto the floor as an ominous shadow. This is life at its most fundamental, when the human protagonist is stripped of all pretense and decorum.

Bacon's contemporary, the Englishman **GRAHAM SUTHERLAND** (1903–1980), also explored agonized expressions of human suffering. During the 1920s he made visionary, meticulously representational landscapes of rural England, mostly in etching, that reveal his passion for the works of the nineteenth-century English Symbolist painter, Samuel Palmer. In the mid-1930s his work became increasingly abstract and imaginative, partly as a result of his discovery of Surrealism. He made paintings based on forms found in nature—rocks, roots, or trees—that were divorced from their landscape context and, in the artist's words, "redefined the mind's eye in a new life and a new mould." He eventually transformed these found forms into monstrous figures that suggest some of Picasso's Expressionist-Surrealist work of the 1930s. The war years, when he worked as an official war artist, led to a certain reversion in the representation of devastated, bombed-out buildings and industrial scenes. By the end of World War II Sutherland had arrived at his characteristically jagged, thistlelike totems, frequently based quite literally on plant forms but transformed into menacing beasts or chimeras. A commission for the painting of a Crucifixion in St. Matthew's Church, Northampton, led him back to Grünewald's Isenheim Altarpiece (see fig. 8). In Grünewald's torn, lacerated figures he found a natural affinity (fig. 597). Along with his expressionist works, he carried on a successful career as a portraitist, painting many of the great figures of his time in literal but penetrating interpretations.

An important European heir to the tradition of truth to visual fact is the Berlin-born British artist **LUCIAN FREUD** (b. 1922), the grandson of the great pioneer psychoanalyst. Freud, whose Jewish family left Berlin for England when Hitler came to power in 1933, showed an aptitude for drawing at a young age and was producing remarkably accomplished work by the time he reached his early twenties. So resolute is this artist in his uncompromising devotion to what he sees that the resulting image, realized with an obsessive command of line and texture, can seem the product of a clinically cold and analytical mind (fig. 598). Still, in their very "nakedness," models whom the artist has revealed in excruciatingly close-up detail confront the viewer with an unmistakable presence and identity. Not since Germany's *Neue Sachlichkeit* painting of the 1920s has the human figure been examined with such an unflinching eye and such tight, linear precision. In the late 1950s Freud's paint handling began to loosen, although the empiricism of his gaze was undiminished. He has largely painted nudes—friends and family, never professional models—who assume strange and revealing poses, whose bodies are far from idealized, and whose flaccid flesh is tinged by blood that courses beneath the surface. Freud has always made close-up portraits—one of the

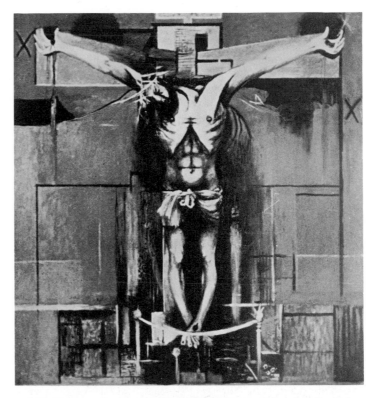

597. GRAHAM SUTHERLAND. *Crucifixion*. 1946. Oil on hardboard, 8' x 7'6" (2.4 x 2.3 m). Church of St. Matthew, Northampton, England © 1997 ARTISTS RIGHTS SOCIETY (ARS), NEW YORK/PRO LITTERAS, ZURICH

best is a 1952 portrait of Bacon—and he has sometimes turned his unsparing gaze on himself. In his self-portrait from 1985 (color plate 287, page 523), skin becomes a harsh terrain of myriad hues, with deep pink crevices and thick white highlights. When experiencing a painting by Freud "in the flesh," one is aware of pigment yet never loses the sense of depicted flesh as a living, mutable substance that painfully documents "the thousand natural shocks that flesh is heir to."

The major figures in sculpture in postwar England were **HENRY MOORE** (1898–1986) and Barbara Hepworth, two

598. LUCIAN FREUD. *Girl with White Dog*. 1951–52. Oil on canvas, 30 x 40" (76.2 x 101.6 cm). Tate Gallery, London

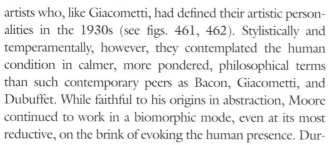

left: 599. HENRY MOORE. *Tube Shelter Perspective.* 1941. Chalk, pen, and watercolor, 8½ x 6½" (21.6 x 16.5 cm). Collection Mrs. Henry Moore, England

above: 600. HENRY MOORE. *Interior-Exterior Reclining Figure (Working Model for Reclining Figure: Internal and External Forms).* 1951. Bronze, height 14" (35.6 cm). Hirshhorn Museum and Sculpture Garden, Smithsonian Institution, Washington, D.C.

artists who, like Giacometti, had defined their artistic personalities in the 1930s (see figs. 461, 462). Stylistically and temperamentally, however, they contemplated the human condition in calmer, more pondered, philosophical terms than such contemporary peers as Bacon, Giacometti, and Dubuffet. While faithful to his origins in abstraction, Moore continued to work in a biomorphic mode, even at its most reductive, on the brink of evoking the human presence. Dur-

ing World War II he made thousands of drawings of the underground air-raid shelters in London (fig. 599). As huge numbers of people sought refuge from German bombs in the tunnels of the London underground, Moore was able to study reclining figures and figural groups that gave him many ideas for sculpture. During and after the war he continued to develop his theme of recumbent figures composed as an intricate arrangement of interpenetrating solids and voids (fig.

601. HENRY MOORE. *King and Queen.* 1952–53. Bronze, height 63½" (161.3 cm). Hirshhorn Museum and Sculpture Garden, Smithsonian Institution, Washington, D.C.

600). The early 1950s also brought forth new experiments in more naturalistic, albeit attenuated figures, such as *King and Queen* (fig. 601). First inspired by an ancient Egyptian sculpture of a royal couple, Moore's king and queen have masks for faces and flattened-out, leaflike bodies, yet maintain a sense of regal dignity. They have nothing to do with "present-day Kings and Queens," Moore said, but with the "archaic or primitive idea of a King."

After mid-century Moore received many international commissions for architectural sculpture. His works can be found on university and museum grounds and other public sites across America and Europe. Generally, these late works are dependent to some degree on Moore's earlier concepts of reclining or standing figures. They are executed for the most part in bronze, which is given a range of effect—from the most jagged and rocky to the most finished and biomorphic. In 1972 Moore made a large bronze titled *Sheep Piece,* here shown in the form of a bronze working model (color plate 288, page 523), which he installed in the meadows behind his studio in the English countryside. The sheep on his property (which he loved to sketch) liked to rub themselves against the sculpture or rest in its shadows. Because of the interaction of the two bronze elements in *Sheep Piece* and their soft, organic contours, the work has been related to mating animals, but a sexual dimension is implied, not explicit.

During the 1950s **BARBARA HEPWORTH** (1903–1975) discovered a promising new direction through a series of so-called Groups in white marble (fig. 602), small in scale and magical in their impact. She was inspired to make them after watching people stroll through Piazza San Marco during a 1950 trip to Venice, where her work was featured at the Biennale. The Groups were related to her early compositions and to her continuing interest in ancient, specifically Cycladic art. In modern terms they recall both Giacometti's groups of walking figures and the biomorphs that inhabit Tanguy's Surrealist landscapes (see fig. 367). Much of Hepworth's sculpture of the 1950s was a direct response to the landscape—the rocks, caves, cliffs, and sea—surrounding her home in Cornwall, along with the monumental Neolithic stone menhirs that populate the landscape. She preferred carving in wood or stone to modeling in clay and continued to work with wood and stone on both small and large scales after the war. After 1960 Hepworth received a number of commissions that permitted her to compose on a monumental scale, often in bronze. She was at her most abstract with works such as *Squares with Two Circles (Monolith)* (fig. 603). Its pierced, rectangular forms relate to the geometry of Ben Nicholson's reliefs (see fig. 458), but the large circular openings recruit the surrounding environment as an active participant in our perception of the work.

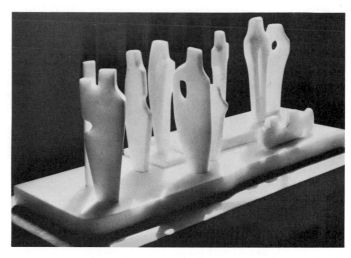

602. BARBARA HEPWORTH. *Group III (Evocation).* 1952. Marble, height 9" (23 cm). Private collection

603. BARBARA HEPWORTH. *Squares with Two Circles (Monolith).* 1963 (cast 1964). Bronze, 10'4" x 5'5" x 2'6" (3.2 x 1.7 x .8 m). Patsy R. and Raymond D. Nasher Collection, Dallas

21 Pop Art and Europe's New Realism

The shift in values that took place in the late 1950s was in part a reaction to the barrage of new products and accompanying mass-media explosion produced by the postwar consumer society. The Pop generation responded enthusiastically to the very media images that the Abstract Expressionists chose to disregard. As Andy Warhol said, "The Pop artists did images that anybody walking down Broadway could recognize in a split second—comics, picnic tables, men's trousers, celebrities, shower curtains, refrigerators, Coke bottles—all the great modern things that the Abstract Expressionists tried so hard not to notice." Many artists of the younger generation viewed the legacy of Abstract Expressionism as an oppressive mantle that had to be lifted. They were no longer concerned with the heroic "act" of painting, in "hot" gestures and emotions. Rather, they found subjects within the immediate environment of their popular culture and sought to incorporate them into their art through "cool," depersonalized means. Pop art, Happenings, environments, assemblage, and *Nouveau Réalisme* began to emerge simultaneously in several American and European cities.

In 1961, The Museum of Modern Art mounted "The Art of Assemblage," a show organized by William C. Seitz that surveyed modern art involving the accumulation of objects, from two-dimensional Cubist *papiers collés* and photographic montages, through every sort of Dada and Surrealist object, to junk assemblage sculpture, to complete room environments. Seitz described such works as follows: "(1) They are predominantly *assembled* rather than painted, drawn, modeled, or carved; (2) Entirely or in part, their constituent elements are pre-formed natural or manufactured materials, objects, or fragments not intended as art materials." In 1962, the Sidney Janis Gallery's "New Realists" exhibition opened. The show included a number of the recognized British and American Pop artists, representatives of the French *Nouveau Réalisme,* and artists from Italy and Sweden working in related directions. Although the American Pop artists had been emerging for several years, Janis's show was an official recognition of their arrival. The European participants seemed closer to the tradition of Dada and Surrealism;

the British and Americans more involved in contemporary popular culture and representational, commercial images. But Pop was not embraced by everyone. Proponents of abstract painting decried the arrival of commercialism in art as rampant vulgarity, a youthful attack on high culture. And, like many labels assigned by critics and historians, the Pop designation is rejected by most of the artists it has been used to describe, for the term groups together a very diverse population of highly individual artists.

Pop Art in Great Britain

Although Pop art has often been regarded as an American phenomenon, it was first introduced in England in the mid-1950s. The term was first used in print in 1958 by the English critic Lawrence Alloway but for a somewhat different context than its subsequent use. Late in 1952, calling themselves the Independent Group, Alloway, the architects Alison and Peter Smithson, the artist Richard Hamilton, the sculptor Eduardo Paolozzi, the architectural historian Reyner Banham, and others met in London at the Institute of Contemporary Arts. The discussions focused around popular (thus, Pop) culture and its implications—such entities as Western movies, science fiction, billboards, and machines. In short, they concentrated on aspects of contemporary mass culture and were centered on its current manifestations in the United States. As the Smithsons wrote in 1956, "Mass production is establishing our whole pattern of life—principles, morals, aims, aspirations, and standard of living. We must somehow get the measure of this intervention if we are to match its powerful and exciting impulses with our own."

RICHARD HAMILTON (b. 1922) defined Pop art in 1957 as "Popular (designed for a mass audience), Transient (short-term solution), Expendable (easily forgotten), Low cost, Mass produced, Young (aimed at youth), Witty, Sexy, Gimmicky, Glamorous and Big Business." A pioneer Pop work is his small collage made in 1956 and entitled *Just what is it that makes today's homes so different, so appealing?* (fig. 604). A poster of this work was used in a multimedia exhibition held at London's Whitechapel Art Gallery in 1956, enti-

604. RICHARD HAMILTON. *Just what is it that makes today's homes so different, so appealing?* 1956. Collage on paper, 10¼ x 9¾" (26 x 24.8 cm). Kunsthalle Tübingen, Sammlung Zundel

right: 605. EDUARDO PAOLOZZI. *Medea.* 1964. Welded aluminum, 6'9" x 6' (2.1 x 1.8 m). Rijksmuseum Kröller-Müller, Otterlo

tled "This is Tomorrow." Hamilton is a disciple of and respected authority on Marcel Duchamp, whose influence on Pop art cannot be overestimated. Another outstanding influence was Kurt Schwitters. Hamilton's collage shows a "modern" apartment inhabited by a pinup girl and her muscle-man mate, whose lollipop barbell prophesies the Pop movement on its label. Like Adam and Eve in a consumers' paradise, the couple have furnished their apartment with products of mass culture: television, tape recorder, an enlarged cover from a comic book, a Ford emblem, and an advertisement for a vacuum cleaner. Through the window can be seen a movie marquee featuring Al Jolson in *The Jazz Singer*. The images, all culled from contemporary magazines, provide a kind of inventory of visual culture.

An important point about Hamilton's, and subsequent Pop artists', approach to popular culture is that their purpose was not entirely satirical or antagonistic. They were not expressionists like George Grosz and the Social Realists of the 1930s who attacked the ugliness and inequities of urban civilization. In simple terms, the Pop artists looked at the world in which they lived and examined the objects and images that surrounded them with intensity and penetration, frequently making the viewer conscious of that omnipresence for the first time. This is not to say that the artists, and especially Hamilton, were unaware of the ways in which consumer mass culture is communicated to the public—its clichés, its manipulations: On the contrary, his work is often rich with irony and humor. As critic David Sylvester once

observed, "Hamilton never makes it clear how far he is being satirical, how far he is rather in love with the subject, and this equivocation is an aspect of the message."

Hamilton has been deeply involved with photography since the 1960s. His painting, *I'm dreaming of a white Christmas* (color plate 289, page 524) and a subsequent screenprint, *I'm dreaming of a black Christmas,* were based on a film still of Bing Crosby in the 1954 movie *White Christmas*. By effecting in paint the kind of chromatic reversal found in a photographic negative he creates not only a magical, looking-glass world but pointedly transforms the protagonist of *White Christmas* into a black man.

The sculptor **EDOUARDO PAOLOZZI** (b. 1924), an original member of the Independent Group, produced collages in the early 1950s made from comic strips, postcards, and magazine clippings that are today regarded as important forerunners of Pop art. Originally influenced by Giacometti, Dubuffet, and Surrealism, Paolozzi became interested in the relations of technology to art and emerged as a serious spokesman for Pop culture. In the mid-1950s he began to make bronze sculptures whose rough surfaces were cast from found objects, resulting in creatures that resemble towering, battered robots. Searching for forms and techniques consistent with his growing interest in machine technology, Paolozzi produced welded sculptures in the 1960s that vary from austere, simplified forms in polished aluminum to elaborate polychromed constructions (fig. 605). While the snakelike tubular forms of *Medea* may have been inspired by the ancient Hellenistic sculpture *Laocoön,* Paolozzi's humanoid machine seems to have materialized from the science fiction screen.

Of the other English Pop artists, PETER BLAKE (b. 1932) took subjects of popular idolatry, such as Elvis Presley, the Beatles, or pinup girls, and presented them in an intentionally naïve style. His early painting *On the Balcony* (fig. 606) was inspired by a Social Realist painting he saw in New York's Museum of Modern Art of working-class people holding masterpieces of modern art they could never afford to own. In his own variation on the theme, Blake's young people hold paintings, including, at the left, Manet's painting titled *On the Balcony.* But they are also surrounded by images from the popular media and advertising, as well as by paintings by Blake and his contemporaries.

An ambiguous but crucial figure in the development of English Pop is the American R. B. KITAJ (b. 1932), who studied at Oxford and at London's Royal College of Art under the GI Bill and has since lived mainly in England. Kitaj painted and continues to paint everyday scenes or modern historical events and personalities in broad, flat, color areas combined with a strong linear emphasis and a sense of fragmentation. Although it is difficult to define his work narrowly within the tradition of Pop art, despite its sometimes comic-strip look, his images influenced some of the English

Pop artists. Unique to Kitaj, however, is an emphatic disinterest in popular culture, counterbalanced by a serious, wide-ranging commitment to political and literary themes. Often mixed into his paintings is an evocative combination of images derived from sources too complex to yield a clear or specific message. The characteristic work seen here (fig. 607) shows a Parisian café scene in which Kitaj's hero, the German Jewish critic Walter Benjamin, can be identified, juxtaposed to a man wielding a red pick, as if to announce the calamities of 1940 that would drive Benjamin to suicide as the German army began occupying the French capital. Kitaj's eloquent words about Benjamin seem to pertain equally to his own work, "His wonderful and difficult montage, pressing together quickening tableaux from texts and from a disjunct world, were called citations by a disciple of his who also conceded that the picture-puzzle distinguished everything he wrote."

DAVID HOCKNEY (b. 1937), a student of Kitaj's at the Royal College, was so gifted and productive that he had already gained a national reputation by the time of his graduation. Emerging within the same period as the Beatles, Hockney—with his peroxide hair, granny glasses, gold lamé jacket, and easy charm—became something of a media event in his own right, a genial exponent of the go-go, hedonistic style of the 1960s. But Hockney drew neither imagery nor

left: 606. PETER BLAKE. *On the Balcony.* 1955–57. Oil on canvas, 47¾ x 35¾" (121.3 x 90.8 cm). Tate Gallery, London

607. R. B. KITAJ. *The Autumn of Central Paris (After Walter Benjamin).* 1972–73. Oil on canvas, 60 x 60" (152.4 x 152.5 cm). Private collection

608.
DAVID HOCKNEY.
The Brooklyn Bridge, Nov. 28th 1982. 1982.
Photographic collage,
42¹⁵⁄₁₆ x 22¹³⁄₁₆"
(109.1 x 57.9 cm).
Private collection

style from comics and ads; instead, he used his own life and the lives of his friends and lovers—their faces and figures, houses, and interiors. These he rendered in a manner sometimes inspired by Picasso and Matisse, but often, in the delicate *faux naïf* manner of children, or, again, in the style of an admired older contemporary such as Francis Bacon (color plate 286, page 522).

Hockney has long maintained a residence in southern California. There the swimming pool became a central image in his art, complete with its cool, synthetic hues. *A Bigger Splash* (color plate 290, page 524) is a painting that combines Hockney's signature equilibrium of elegant Matissean flat-pattern design and luxurious mood with a painterly virtuosity for the splash itself. One critic described this sense of suspended animation as the "epitome of expectant stillness," which is due partly to the artist's habit of working from photographs. Infatuated with the Polaroid and, later, with 35mm cameras, Hockney has used photography in more recent years to reexplore the riches of the space-time equation investigated by Braque and Picasso in Analytic Cubism (fig. 608). In his large collage of the Brooklyn Bridge, dozens of smaller sequential photographs compose a single motif, simulating the scanning sensation of actual vision.

Such composite treatments of subject matter have infiltrated Hockney's paintings as well, including the 1984 *A Visit with Christopher and Don, Santa Monica Canyon* (color plate 291, page 524). This panoramic composition re-creates a visit to the house of the artist's friends, the painter Don Bachardy and the writer Christopher Isherwood. The viewer's eye is carried through the fractured spaces of the house, where we see the residents at work at either end and where we periodically encounter the same fractured view of the Santa Monica Canyon. Like most of Hockney's work since 1980, the painting combines flat unmodulated zones with more painterly ones, all enriched by a palette of sumptuous color. In recent years this multi-faceted artist has designed stage sets and even directed performances for the opera.

Neo-Dada and Pop Art in the United States

Pop art, especially during the 1960s, had a natural appeal to American artists, who were living in the midst of an even more blatant industrial and commercial environment than that found in Great Britain. Once they realized the tremendous possibilities of their everyday surroundings in the creation of new subject matter, the result was generally a bolder and more aggressive art than that of their European counterparts, much of whose work derived from American motion pictures, popular idols, comic strips, or signboards.

Toward the end of the 1950s, as we have already noted, there were many indications that American painting was moving away from the heroic rhetoric and grand painterly gestures of Abstract Expressionism. (For example, de Kooning had parodied the gleaming smiles and exaggerated cleavage of American movie starlets in his paintings of women, one of which even included a cutout mouth from a popular magazine advertisement.) American art had a long tradition of interest in the commonplace that extended from the trompe l'oeil paintings of the nineteenth century (see fig. 56) through the Precisionist painters of the early twentieth century (see fig. 481). Marcel Duchamp's anti-art program led younger painters back to Dada and most specifically to Kurt Schwitters, who was a crucial model for several young artists. Two leading American artists, Robert Rauschenberg and Jasper Johns, were the most obvious heirs to Duchamp and Schwitters. The art of these important forerunners to Pop art has been referred to as Neo-Dada.

ROBERT RAUSCHENBERG (b. 1925), one of the most important artists in the establishment of American Pop's vocabulary, came of age during the Depression in Port Arthur, Texas. "Having grown up in a very plain environment," he once said, "if I was going to survive, I had to appreciate the most common aspects of life." After serving in the U.S. Navy in World War II and with the help of the GI Bill, Rauschenberg studied at the Kansas City Art Institute and at the Académie Julian in Paris. In 1948, feeling the need for a more disciplined environment, he went for the first of several times to Black Mountain College, an experimental school in North Carolina, to study with Josef Albers. Rauschenberg learned not so much a style as an attitude from the disciplined methods of this former Bauhaus professor, for the young artist's rough, accretive assemblages could hardly be further from the pristine geometry of Albers's abstrac-

609. ROBERT RAUSCHENBERG. *Untitled* (Glossy five-panel black painting). c. 1951 (one panel now lost or destroyed). Installation view, Stable Gallery, New York, 1954. Oil and newspaper on canvas, 7'3" x 14'3" (2.2 x 4.3 m)
PHOTOGRAPH BY ROBERT RAUSCHENBERG. © 1997 ROBERT RAUSCHENBERG/VAGA, NEW YORK, NY

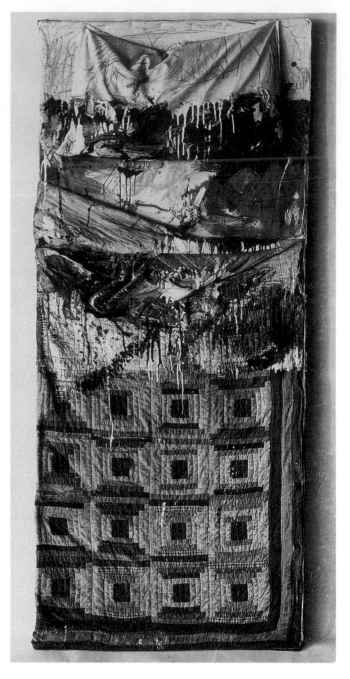

tions. Even more important in his development was the presence at Black Mountain College in the early 1950s of John Cage, a composer, writer, and devotee of Duchamp, whom Rauschenberg had met in New York, and the choreographer-dancer Merce Cunningham. Cage was an enormously influential figure in the development of Pop art, as well as offshoots such as Happenings, environments, and innumerable experiments in music, theater, dance, and the remarkably rich combinative art forms that merged from all of these media. Rauschenberg would later join Cage and Cunningham in several collaborative performances, designing sets, costumes, and lighting.

Rauschenberg has literally used the world as his palette, working in nearly every corner of the globe and accepting virtually any material as fodder for his art. He has always used photographic processes in one way or another and, in 1950, collaborating with his former wife Susan Weil, achieved an otherworldly beauty simply by placing figures on paper coated with cyanotype and exposing the paper to light (color plate 292, page 524). He has made radically reductive abstractions, including the all-white and all-black paintings he began while at Black Mountain College in the early 1950s (fig. 609), as well as paintings and constructions out of almost anything except conventional materials. An artist who values inclusivity, multiplicity, and constant experimentation, Rauschenberg scavenges his environment for raw materials for his art, excluding little and welcoming detritus of every sort. He once remarked, "Painting relates to both art and life. Neither can be made. (I try to act in that gap between the two.)"

After settling in New York, Rauschenberg became more

and more involved with collage and assemblage in the 1950s. By 1954 he had begun to incorporate objects such as photographs, prints, or newspaper clippings into the structure of the canvas. In 1955 he made one of his most notorious "combine" paintings, as he called these works, *Bed* (fig. 610). This work included a pillow and quilt over which paint was splashed in an Abstract Expressionist manner, a style Rauschenberg has never completely abandoned. His most spectacular combine painting, the 1959 *Monogram* (fig. 611), with a stuffed Angora goat encompassed in an automobile tire, includes a painted and collage-covered base—or rather a painting extended horizontally on the floor—in which free brush painting acts as a unifying element. Whereas *Bed* still bears a resemblance to the object referred to in its title, *Monogram* is composed of disparate elements that do not, as is true of Rauschenberg's work as a whole, coalesce into any single, unified meaning or narrative. The combine paintings clearly had their origin in the collages and constructions of

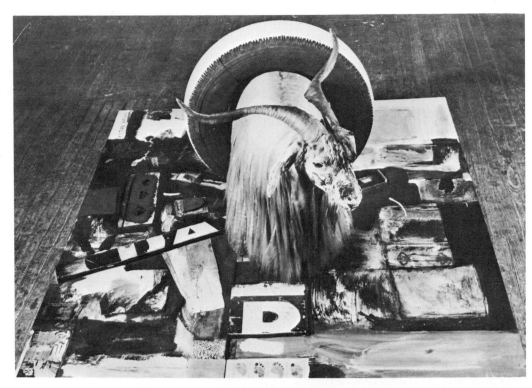

left: 610.
ROBERT RAUSCHENBERG. *Bed.*
1955. Combine painting:
oil and pencil on pillow,
quilt, and sheet on wood
supports, 6'3¼" x 31½" x 8"
(1.9 m x 80 cm x 20.3 cm).
The Museum of Modern
Art, New York

right: 611.
ROBERT RAUSCHENBERG.
Monogram. 1959.
Combine painting: oil and
collage with objects, 42 x
63½ x 64½" (106.7 x
161.3 x 163.8 cm).
Moderna Museet,
Stockholm

Schwitters and other Dadaists (see fig. 314). Rauschenberg's
work, however, is different not only in its great spatial expansion, but in its desire to take not so much a Dada, anti-art
stance as one that expands our very definitions of art.

In 1962, at virtually the same time as Andy Warhol,
Rauschenberg began to make paintings using a photo
silkscreen process. Collecting photographs from magazines
or newspapers, he had the images commercially transferred
onto silkscreens, first in black and white and, by 1963, in
color, and then set to work creating a kaleidoscope of images
(color plate 293, page 525). In the example shown here, the
subjects range from traffic signs, to the Statue of Liberty, to
a photograph of the Sistine Ceiling. They collide and overlap
but are characteristically arranged in a grid configuration and
are bound together by a matrix of the artist's gestural brushstrokes. In 1964, after Rauschenberg had won the prestigious grand prize at the Venice Biennale, and as a guarantee
against repeating himself, he called a friend in New York and
asked him to destroy all the silkscreens he had used to make
the silkscreen paintings.

Throughout the 1960s Rauschenberg, like many of the
original Pop artists, was involved in a wide variety of activities, including dance and performance. His natural inclination toward collaboration drew him to theater and dance.
Pelican, a dance he actually choreographed, was performed
in 1963 as part of a Pop festival in Washington, D.C. Because
the event took place in a roller-skating rink, Rauschenberg
planned a dance of three—including himself, another man
on roller skates, and Carolyn Brown, a professional from
Merce Cunningham's company who danced on point. The
men trailed large parachutes that floated like sails behind
them during the performance, which Rauschenberg dedicated to his heroes, the Wright brothers (fig. 612).

612. ROBERT RAUSCHENBERG performing in *Pelican,* 1963,
Washington, D.C. PHOTOGRAPH BY PETER MOORE

In more recent years this indefatigable and prolific artist
has extended his quest to make art from anything and from
anywhere. He has tended to work in series since the 1970s,
when he made the Cardboard series, numerous sculptures,
prints, and photographs, as well as dozens of works that
involved fabric. For the Hoarfrost series he employed a solvent process to transfer photographic images onto delicate
cloths that are draped unstretched on the wall (color plate
294, page 525). Rauschenberg said his aim was to "dematerialize the surface" with these works, whose diffuse images
tend to undulate as people pass by.

In 1984, Rauschenberg launched a seven-year project
called ROCI (Rauschenberg Overseas Culture Interchange),
during which he traveled to foreign and often politically sensitive countries to work for the cause of world peace.
Collaborating with local artisans and drawing upon indige-

nous traditions and motifs, Rauschenberg created a body of work in each of the ten countries he visited, including the former Soviet Union, Cuba, China, Tibet, Chile, and Mexico. He completed the series with ROCI USA, for which images were deposited through complex techniques onto a brightly polished surface of stainless steel (color plate 295, page 526). The artist then attached a crumpled section of the stainless steel that projects awkwardly from the main support. Typically, chance and accident are welcome, for the highly reflective surface records the activities of the viewers in the gallery.

JASPER JOHNS (b. 1930), a native of South Carolina, appeared on the New York art scene in the mid-1950s at the same moment as Rauschenberg (whom he had befriended in 1954), with paintings that were entirely different but equally revolutionary. These were, at first, paintings of targets and American flags executed in encaustic (pigment mixed with heated wax), with the paint built up in sensuous, translucent layers (fig. 613). In addition to targets and flags, he took the most familiar items—numbers, letters, words, and maps of the United States—and then painted them with such precision and neutrality that he made them objects in themselves rather than illusionistic depictions of familiar objects.

Johns, who destroyed all of his previous work, said the subject of the flag was suggested by a dream in which he was painting an American flag. Because a flag, like a target or numbers, is inherently flat, Johns could dispense with the illusionistic devices that suggest spatial depth and the relationship between figure and ground. At the same time, the ready-made image meant that he did not have to invent a composition in the traditional sense. "Using the design of the American flag took care of a great deal for me because I didn't have to design it," he said. "So I went on to similar things like targets—things the mind already knows. That gave me room to work on other levels." This element of impersonality, this apparent withdrawal of the artist's individual

invention, became central to many of the painters and sculptors of the 1960s. Johns's flags, despite their resemblance to actual flags, are clearly paintings, but with his matter-of-fact rendering he introduces the kind of perceptual and conceptual ambiguity with which Magritte had toyed in his painting *The Treachery (or Perfidy) of Images* (see fig. 373).

One of the artist's most enigmatic works is *Target with Plaster Casts* (color plate 296, page 526), another encaustic painting of a flat, ready-made design, but this time the canvas is surmounted by nine wooden boxes with hinged lids, which can either remain closed or be opened to reveal plaster-cast body parts. A strange disjunction exists between the neutral image of a target and the much more emotion-laden body parts, each a different color, just as it does between the implied violence of dismembered anatomical parts and their curious presentation as mere objects on a shelf.

In one of his best-known sculptures, *Painted Bronze,* Johns immortalized the commonplace by taking two Ballantine Ale cans and casting them in bronze (fig. 614). Yet upon close examination it becomes clear that Johns has painted the labels freehand, so what at first appears mass-produced, like a Duchamp ready-made, is paradoxically a unique, handwrought form. Moreover, although the cans look identical, one is in fact "open" and cast as a hollow form; the other is closed and solid.

From the end of the 1950s, Johns's paintings were marked by the intensified use of expressionist brushstrokes and by the interpolation of actual objects among the flat signs. In *Field Painting* (color plate 297, page 526), for example, the Ballantine Ale can reappears, this time as itself. The can is magnetically attached along with other objects—ones we might find around the artist's studio—to three-dimensional letters that are hinged to the center of the painting. These metal letters spell out "RED," "YELLOW," and "BLUE," although, like their flat counterparts painted on the canvas, they do not necessarily signify the color we read. Johns frequently presents alternative forms of representation in a single painting, for in addition to these signs, Johns includes color in a welter of Abstract Expressionist paint strokes. One of the foremost printmakers of the postwar era, Johns invoked the names of the primary colors again in a color lithograph titled *Souvenir* (color plate 298, page 527), where they encircle his portrait. His drawings and prints usually relate closely to his paintings and sculptures. In *Souvenir,* the various images—the back of a canvas, a flashlight, and a rearview bicycle mirror—are present as actual objects in a painting of the same name. As they often do in Johns's works, the themes here revolve around issues of perception, of looking, and of being looked at.

In the 1970s Johns made paintings that suggest a continuing dialogue between figuration and abstraction, as well as a penchant for complex, hermetic systems of representation. In 1972, in the large, four-panel painting *Untitled* (color plate 299, page 527), he introduced a pattern of parallel lines in the far left panel, so-called crosshatches, that have played a role in his work ever since. Johns has said that the crosshatch motif

613. JASPER JOHNS. *Flag.* 1954–55 (dated 1954 on reverse). Encaustic oil and collage on fabric mounted on plywood, 42¼ x 60⅝" (107.3 x 153.8 cm). The Museum of Modern Art, New York GIFT OF PHILIP JOHNSON IN HONOR OF ALFRED H. BARR, JR. PHOTOGRAPH © 1995 THE MUSEUM OF MODERN ART, NEW YORK. © 1997 JASPER JOHNS/VAGA, NEW YORK, NY

614. JASPER JOHNS. *Painted Bronze.* 1960. Painted bronze,
5½ x 8 x 4¾" (14 x 20.3 x 12.1 cm).
Museum Ludwig, Cologne
© 1997 JASPER JOHNS/VAGA, NEW YORK, NY

was not one he invented but one he spotted on a passing car, just as he said the flagstone pattern in the two center panels was something he glimpsed on a Harlem wall while riding to the airport in the late 1960s. These flagstone-pattern panels—one in oil, one in encaustic—re-create two slightly different sections of the same panel from an earlier painting called *Harlem Light.* If one panel is superimposed on the other so they match up, the panels form an "imagined square," as Johns described it in his typically enigmatic "Sketchbook Notes." At the far right, in startling contrast to these flat, abstract forms, are three-dimensional casts made from a woman's and a man's body that Johns painted and attached with wing-nuts to strips of wood.

In his more recent work, such as the watercolor *The Bath* (color plate 300, page 528), Johns has introduced coded autobiographical references among images derived from the works of artists he admires. Illusionistically "tacked" to the painting surface is a rendering after Picasso's Surrealist painting *Straw Hat with Blue Leaves* (see fig. 380), but Johns has sliced the painting in such a way that we are forced to reconstruct it. Between these two sections, the artist depicted his own bathtub faucet and, at the left, the floorboards of a former studio. The background of this drawing is derived from a detail from Matthias Grünewald's great *Isenheim Altarpiece* (see fig. 8). Johns incorporated imagery from this painting into a large number of his works in the 1980s. This particular motif derives from Grünewald's portrayal of a diseased demon from the panel dedicated to the Temptation of St. Anthony. Typically, however, Johns obscured the image by rotating the figure and rendering it in monochrome. Another detail from the *Isenheim Altarpiece* can barely be deciphered in the blue background of a 1993 encaustic painting, *Mirror's Edge 2* (color plate 301, page 528), while the demon resurfaces on the section of curling paper that Johns attached to the surface with trompe l'oeil masking tape. These forms

appear among motifs appropriated from Johns's own work. Borrowed imagery may, as in the case of two lithographs by Barnett Newman, be cast here in a fresh and ever-ambiguous context. Just as we peer closely at the imagery to bring it into focus, we ponder its meaning, searching for clues to explain the elaborate puzzle of images that simultaneously enrich and contradict one another and, like layers of experience, accumulate meaning the more we look.

Happenings and Environments

The concept of mixing media and, more importantly, integrating the arts with life itself was a fundamental aspect of the Pop revolution. A number of artists in several countries were searching for ways to extend art into a theatrical situation, but the leading spokesman for this attitude remains ALLAN KAPROW (b. 1927), who defined a Happening as

> an assemblage of events performed or perceived in more than one time and place. Its material environments may be constructed, taken over directly from what is available, or altered slightly: just as its activities may be invented or commonplace. A Happening, unlike a stage play, may occur at a supermarket, driving along a highway, under a pile of rags, and in a friend's kitchen, either at once or sequentially. If sequentially, time may extend to more than a year. The Happening is performed according to plan but without rehearsal, audience, or repetition. It is art but seems closer to life.

Kaprow's first public Happening, entitled *18 Happenings in 6 Parts,* was held at New York's Reuben Gallery in 1959. The Reuben Gallery, along with the Hansa and Judson galleries, was an important experimental center where many of the leading younger artists of the 1960s first appeared. Among those who showed at the Reuben Gallery, aside from Kaprow, were Jim Dine, Claes Oldenburg, George Segal, and Lucas Samaras.

Happenings may have had their ultimate origins in the Dada manifestations held during World War I. They might even be traced to the improvisations of the *commedia dell' arte* of the eighteenth century. The Japanese have made some claim to priority in Happenings, since the so-called Gutai Group of Osaka staged Happening-like performances as early as 1955, and from 1957 onward such performances were held in Tokyo as well as Osaka. And in Düsseldorf, the Zero Group was founded in 1957 in the interest of narrowing the gap between art and life, while similar developments were taking place in other art centers on the Continent. But the role of Abstract Expressionism was significant as well. Kaprow regarded Pollock's large drip paintings as environmental works of art that implied an extension beyond the edge of the canvas and into the viewer's physical space. Pollock, said Kaprow, "left us at the point where we must become preoccupied with and even dazzled by the space and objects of our everyday life." Pollock's contemporary, the composer John Cage, taught Kaprow at the New School for Social Research in New York City in the late 1950s. Cage helped to bridge art and life by experimenting with musical forms that involved unplanned audience participation and by welcoming chance noises as part of his music. His famous composition "4'33"" consisted of the pianist David Tudor lifting the lid of the piano and sitting for four minutes and thirty-three seconds without playing. The random sounds that occurred during the performance constituted the music.

Happenings, a term that not all of those involved embraced, could vary extremely in each instance and with each artist. They were not simply spontaneous, improvisational events but could be structured, scripted, and rehearsed performances. Sometimes they were performed for an audience, which participated to varying degrees according to the artist, sometimes for a camera, sometimes for no one at all. Because the instigators were for the most part visual artists, visual concerns tended to override dramatic or more conventionally theatrical ones, such as plot or character development. In the moment shown here from Kaprow's 1964 Happening *Household* (fig. 615), the participants are licking strawberry jam from the hood of a car that is soon to be set on fire.

Whatever its history and its exact characteristics, in a few years the idea of the Happening and the word itself became part of American folk culture. Many of the terms, images, and ideas of Pop art found instant popular acceptance, if not always complete comprehension, with the mass media—television, films, newspapers, and magazines. Even television and newspaper advertisements, fashion, and product design were affected, resulting in a fascinating cycle in which forms and images taken from popular culture and translated into works of art were then retranslated into other objects of popular culture.

Two months after Kaprow's first Happening at the Reuben Gallery, the artist **RED GROOMS** (b. 1937) staged—and performed in—*The Burning Building,* his most famous "play," as he called his version of Happenings, at another New York venue (fig. 616). Grooms, a native of Nashville, Tennessee,

settled in New York in 1957, the year he met Kaprow. In *The Burning Building* he played the Pasty Man who, in search of love's secret, pursues the Girl in the White Box into a firemen's den. In the still shown here, he has been socked in the head by a fireman, who carries him behind the curtain.

From his early work as a painter and performance artist, Grooms began to make multimedia works on an environmental scale, peopling entire rooms with cutout figures and objects painted in brilliant and clashing colors. He has continually enlarged his scope to embrace constructions such as the famous *Ruckus Manhattan,* a gargantuan environment he first built in the lobby of a New York office building (he completed several sections at the Marlborough Gallery). With a team of helpers, Grooms created a walk-in map of lower Manhattan in a spirit of wild burlesque, manic, big-city energy, and broad, slapstick humor (color plate 302, page 528).

GEORGE SEGAL (b. 1924) began as a painter of expressive, figurative canvases, but feeling hampered by the spatial limitations of painting, he began to make sculpture in 1958. For a number of years he was closely associated with Allan Kaprow (who staged his first Happening at Segal's New Jersey farm), a connection that stimulated his interest in sculpture as a total environment. By 1961, Segal was casting figures from live models, both nude and fully clothed. He wrapped his subjects in bandages, that he then covered in plaster, allowed to dry, and then cut open and resealed, a process that records quite literally the details of anatomy and expression, but retains the rough textures of the original bandages. The figure—or, as is often the case, the group of figures—is then set in an actual environment—an elevator, a lunch counter, a movie ticket booth, or the interior of a bus—with the props of the environment retrieved from junkyards (color plate 303, page 529). Although Segal painted the figures in many of his later sculptures, he left them white

615. ALLAN KAPROW. Photograph from *Household,* a Happening commissioned by Cornell University, 1964

in the early works, producing the effect of ghostly wraiths of human beings existing in a tangible world. The qualities of stillness and mystery in Segal's sculpture are comparable to the desolate psychological mood of Edward Hopper's paintings, while his many treatments of the female nude, including *Girl Putting on Scarab Necklace* (fig. 617), are reminiscent of Degas's paintings of bathers engaged in private moments. In the mid-1970s Segal altered his technique and began to use the body casts as molds from which either plaster or bronze casts could be made. As a result, he was able to make work suitable for outdoor public sculpture.

CLAES OLDENBURG (b. 1929), a native of Sweden who was brought to the United States as an infant, has created a brilliant body of work based on lowly common objects ranging from foodstuffs to clothespins to matchsticks. Through his exquisitely rendered drawings or through the sculpture he has executed in diverse media and scales, Oldenburg transforms the commonplace into something extraordinary. "Art," he has said, "should be literally made of the ordinary world; its space should be our space; its time our time; its objects our ordinary objects; the reality of art will replace reality."

Oldenburg's most important early work consisted of two installations made in New York, *The Street*, 1960, and *The Store*, 1961. The former was created of little more than common urban detritus—newspaper, burlap, and cardboard—that the artist tore into fragmented figures or other shapes and covered with graffiti-like drawing. Leaning against the walls or hanging from the ceiling, these ragged-edged objects collectively formed, according to the artist, a "metaphoric mural," that mimicked the "damaged life forces of the city street." Given Oldenburg's willingness to make art from the most common materials, it is no surprise that a crucial early model for him was Jean Dubuffet's *Art brut*. In its fullest manifestation, *The Store* (color plate 304, page 529) was literally presented in a storefront in Manhattan's Lower East Side, where the artist both made his wares and sold them to the public, thus circumventing the usual venue of a commercial gallery. He filled *The Store* floor to ceiling with sculptures inspired by the tawdry merchandise he saw regularly in downtown cafeterias and shop windows. They were made of plaster-soaked muslin placed over wire frames, which was then painted and attractively priced for such amounts as $198.99. Items from *The Store* depicted everything from women's lingerie, to fragments of advertisements, to food such as ice-cream sandwiches or hamburgers—all roughly modeled and garishly painted in parody of cheap urban wares. Within the installations of *The Street* and *The Store*, Oldenburg staged several Happenings between 1960 and 1962. Situating objects within an environment and sometimes using that environment as a context for Performances—a term he prefers to Happenings—has been characteristic of Oldenburg's art for decades.

The stuffed-fabric props that Oldenburg made for some of these performances (with his first wife doing much of the sewing), led to his first independent soft sculptures. Among his earliest examples were stuffed and painted canvas objects such as the nine-and-a-half-foot-long *Floor Cake* (fig. 618). Soon he was using canvas as well as shiny colored vinyl to translate hard, rigid objects, such as bathroom fixtures, into soft and collapsing versions (fig. 619). These familiar, prosaic toilets, typewriters, or car parts collapse in violation of their essential properties into pathetic, flaccid forms that have all the vulnerability of sagging human flesh. Oldenburg, who has described his work as "the detached examination of human beings through form," essentially models his soft sculptures through

direct manipulation, but also allows gravity to act upon their pliant forms. Thus, the works exist in a state of constant flux.

One of Oldenburg's images, the *Geometric Mouse*, is so synonymous with his name that it has come to represent a kind of surrogate self-portrait. Just as soft sculptures paradoxically metamorphosed hard forms into soft, the *Geometric Mouse*, originally derived from the shape of an old movie camera, is a hard (and highly abstracted) version of an organic subject. Oldenburg has executed the *Geometric Mouse* on several scales ranging from small tabletop versions to an eighteen-foot-high outdoor sculpture. *Geometric Mouse, Scale A* (fig. 620), made of steel and aluminum, consists of a few interlocked metal planes but still bears an eerie resemblance to its namesake.

In the mid-1960s, Oldenburg began to make drawings of common objects to be built to the scale of monuments: a pair of giant moving scissors to replace the Washington Monument, a colossal peeled banana for Times Square, or an upside-down Good Humor bar to straddle Park Avenue (with a bite removed to allow for the passage of traffic) (fig. 621). These so-called non-feasible proposals challenge traditional notions of public sculpture as heroic and commemorative in nature and replace them with irreverent and often hilarious solutions. In more recent years, Oldenburg has realized his fantastic proposals with "feasible" monuments and large-scale projects that have been built around the globe (see fig. 799 and color plate 392, page 674). Most of these have been conceived with his wife and collaborator Coosje van Bruggen.

JIM DINE (b. 1935), a gifted draftsman and printmaker

618. CLAES OLDENBURG. *Floor Cake (Giant Piece of Cake)*. 1962. Installation view, Sidney Janis Gallery, New York, 1962. Synthetic polymer paint and latex on canvas filled with foam rubber and cardboard boxes, 58⅜" x 9'6¼" x 58⅜" (1.5 x 2.9 x 1.5 m). The Museum of Modern Art, New York

619. CLAES OLDENBURG. *Soft Toilet*. 1966. Vinyl, Plexiglas, and kapok on painted wood base, 57¹⁄₁₆ x 27⅝ x 28¹⁄₁₆" (144.9 x 70.2 x 71.3 cm). Whitney Museum of American Art, New York

620. CLAES OLDENBURG. *Geometric Mouse, Scale A.* 1969.
Steel and aluminum, height 12' (3.7 m). Walker Art Center, Minneapolis
GIFT OF MILES Q. FITERMAN

621. CLAES OLDENBURG. *Proposed Colossal
Monument for Park Avenue, New York:
Good Humor bar.* 1965. Crayon and
watercolor on paper, 23½ x 17½"
(59.7 x 44.5 cm).
Collection of Carroll Janis, New York

originally from Cincinnati, was also one of the original organizers of Happenings, but he gave up this aspect of his work, he said, because it detracted from his painting. Like Rauschenberg, Johns, Kaprow, and others, Dine's art has roots in Abstract Expressionism, although he seems to treat it with a certain irony, combining gestural painting with actual objects, particularly tools (fig. 622). His objects recall both Duchamp's ready-mades and contemporaneous works by Johns (see fig. 614). In Dine's paintings the tools are not only physically present but frequently are also reiterated in painted images or shadows and are then referred to again through lettered titles. For Dine, the son and grandson of hardware store merchants, the tools as well as the painted image of the palette are self-referential signs, just as his many images of neckties and bathrobes are surrogate self-portraits (color plate 305, page 529).

LUCAS SAMARAS (b. 1936), a native of Greece who came to America in 1948, participated in early Happenings at the Reuben Gallery, where he had his first one-man show in 1960. Since then Samaras has worked in a broad range of media, including boxes and other receptacles. In his containers, plaques, books, and table settings, prosaic elements are

622. JIM DINE. *Hatchet with Two Palettes, State #2.* 1963. Oil
on canvas with wood and metal, 6' x 4'6" (1.8 x 1.4 m).
Collection Robert E. Abrams, New York

left: 623. LUCAS SAMARAS. *Box No. 3*. 1962–63. Mixed media. Collection Howard Libman Foundation

624. LUCAS SAMARAS. *Untitled*. 1965. Yarn and pins on wood chairs, 35½ x 35¾ x 19¼" (90.2 x 91 x 49 cm). Walker Art Center, Minneapolis ART ACQUISITION FUND, 1966

given qualities of menace by the inclusion of knives, razor blades, and thousands of ordinary but outward-pointing pins (fig. 623). The objects are meticulously made, with a perverse beauty embodied in the threat that is implied. They are not easily classified as Neo-Dada or Pop art; rather, they are highly expressionist works that carry with them a Surrealist brand of disturbance. Samaras sees in everyday objects a combination of menace and erotic love, and it is this sexual horror, yet attraction, that manifests itself in his work. Two chairs, one covered with pins and leaning back, the other painted and leaning forward, create the effect of a grim sexual dance of the inanimate and commonplace (fig. 624). To touch one of his objects is to be mutilated. Yet the impulse to touch it, the tactile attraction, is powerful.

In the mid-1960s Samaras began to construct full-scale rooms where every surface has been covered with mirrors (fig. 625). Upon entering these reflective environments the viewer is dazzled and disoriented as his or her image is infinitely multiplied in every direction. Such optical experiments have led Samaras to a series of what he calls "photo-transformations" in which his own Polaroid image, placed within his familiar environment, is translated into monstrous forms (color plate 306, page 530).

Like Samaras, **RICHARD ARTSCHWAGER** (b. 1923) resists easy classification, for the art of both men bridges several trends in the 1960s, from Pop to Minimalism to Conceptualism. In 1953, after spending three years in the army, earning a degree in physical science at Cornell University, and studying painting for a year in New York, Artschwager began to make commercial furniture. This technical experience, combined with a growing commitment to art over science, one kindled by an admiration for the work of Johns, Rauschenberg, and Oldenburg, led Artschwager to the fabrication of hybrid objects that occupy a realm somewhere between furniture and sculpture. *Chair* (fig. 626) is fabricated

625. LUCAS SAMARAS. *Room No. 2*. 1966. Wood and mirrors, 8 x 10 x 8' (2.4 x 3 x 2.4 m). Albright-Knox Gallery, Buffalo

above: Color plate 283.
KAREL APPEL. *Angry Landscape.* 1967.
Oil on canvas, 51¼" x 6'4¾"
(130.2 cm x 1.9 m). Private collection

below, left: Color plate 284.
ASGER JORN. *The Enigma of Frozen Water.*
1970. Oil on canvas, 63¾ x 51⅛"
(162 x 130 cm).
Stedelijk Museum, Amsterdam

below, right: Color plate 285.
HUNDERTWASSER. *House which was born in Stockholm, died in Paris, and myself mourning it.* 1965. Mixed media,
32 x 23⅔" (81.3 x 60.1 cm).
Private collection

Color plate 286. FRANCIS BACON. *Painting*. 1946. Oil on canvas, 6'6" x 52" (2 m x 132.1 cm).
The Museum of Modern Art, New York PURCHASE

Color plate 287. LUCIAN FREUD.
Reflection (Self-Portrait). 1983–85.
Oil on canvas, 22⅛ x 20⅛"
(56.2 x 51.2 cm).
Private collection

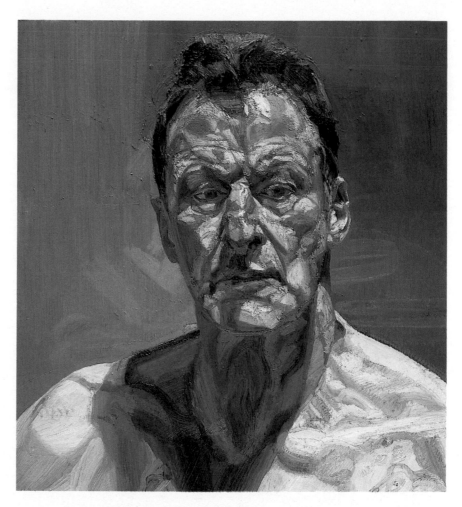

Color plate 288. HENRY MOORE.
Working model for *Sheep Piece*. 1971.
Bronze, length 56" (142 cm).
The Henry Moore Foundation

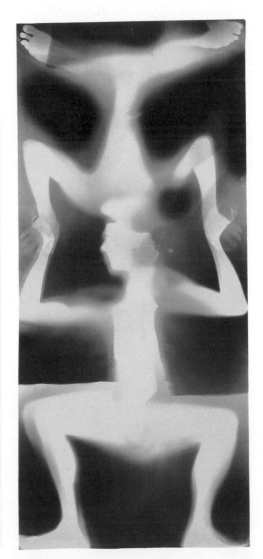

right: Color plate 291. DAVID HOCKNEY. *A Visit with Christopher and Don, Santa Monica Canyon 1984. 1984.* Oil on canvas, 6' x 20' (1.8 x 6.1 m). Courtesy the artist

below: Color plate 292. ROBERT RAUSCHENBERG and SUSAN WEIL. *Untitled (Double Rauschenberg).* c. 1950. Monoprint: exposed blueprint paper, 8'9" x 36" (2.7 m x 91.4 cm). Collection Cy Twombly, Rome

Color plate 289. RICHARD HAMILTON. *I'm dreaming of a white Christmas.* 1967–68. Oil on canvas, 42 x 63" (106.7 x 160 cm). Öffentliche Kunstsammlung Basel, Kunstmuseum ON PERMANENT LOAN FROM THE LUDWIG COLLECTION, AACHEN

Color plate 290. DAVID HOCKNEY. *A Bigger Splash.* 1967. Acrylic on canvas, 8'⅛" x 8'⅛" (2.4 x 2.4 m). Tate Gallery, London

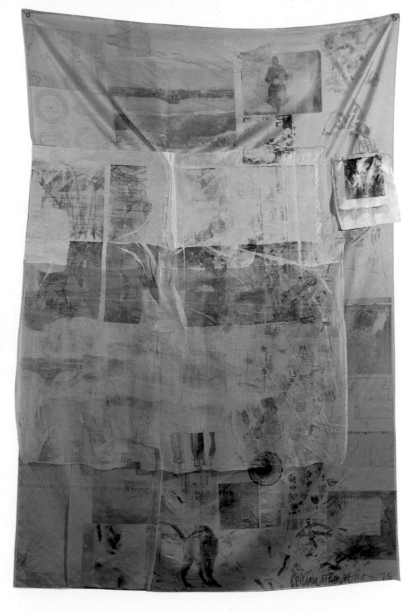

above: Color plate 293. ROBERT RAUSCHENBERG.
Estate. 1963. Oil and silkscreen ink on canvas, 8' x 70"
(2.4 m x 177.8 cm). Philadelphia Museum of Art

right: Color plate 294. ROBERT RAUSCHENBERG.
Blue Urchin from the Hoarfrost series. 1974. Collage and
solvent transfer on cloth, 6'4" x 49" (1.9 m x 124.5 cm).
Collection Paul Hottlet, Antwerp

Color plate 295. ROBERT RAUSCHENBERG. *Trowser/ROCI USA (Wax Fire Works)*. 1990. Acrylic, enamel, and fire wax on mirrored aluminum and stainless steel, 8'10" x 15'4" x 12¼" (2.7 m x 4.7 m x 31.1 cm). Private collection

© 1997 ROBERT RAUSCHENBERG/VAGA, NEW YORK, NY

Color plate 296. JASPER JOHNS. *Target with Plaster Casts*. 1955. Encaustic and collage on canvas with wood construction and plaster casts, 51 x 44 x 3½" (129.5 x 111.8 x 8.9 cm). Collection David Geffen

© 1997 JASPER JOHNS/VAGA, NEW YORK, NY

Color plate 297. JASPER JOHNS. *Field Painting*. 1964. Oil and canvas on wood with objects, 6' x 36¾" (1.8 m x 93.3 cm). Collection the artist

© 1997 JASPER JOHNS/VAGA, NEW YORK, NY

Color plate 298. JASPER JOHNS. *Souvenir*. 1970. Color lithograph on paper, 30⅝ x 22⅜" (77.8 x 56.8 cm). Philadelphia Museum of Art

GIFT OF THE ARTIST © 1997 JASPER JOHNS/VAGA, NEW YORK, NY

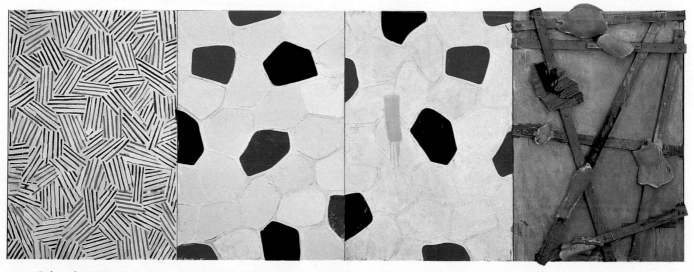

Color plate 299. JASPER JOHNS. *Untitled*. 1972. Oil, encaustic, and collage on canvas with objects, 6' x 16' (1.8 x 4.9 m). Museum Ludwig, Cologne

© 1997 JASPER JOHNS/VAGA, NEW YORK, NY

right: Color plate 300.
JASPER JOHNS. *The Bath.* 1988.
Watercolor and pencil on paper,
29⅛ x 37¾"
(74 x 95.9 cm).
Collection the artist
© 1997 JASPER JOHNS/VAGA, New York, NY

below, left: Color plate 301.
JASPER JOHNS. *Mirror's Edge 2.*
1992. Encaustic on canvas,
66 x 44⅛" (167.6 x 112.1 cm).
Collection Robert and Jane
Meyerhoff, Phoenix, Maryland
© 1997 JASPER JOHNS/VAGA, New York, NY

below, right: Color plate 302.
RED GROOMS AND THE RUCKUS
CONSTRUCTION COMPANY. Detail from
Ruckus Manhattan (World Trade
Center, West Side Highway).
1975–76. Mixed media.
Installation view Grand Central
Station, New York. 1993.
Courtesy MTA and Marlborough
Gallery, New York
© 1997 RED GROOMS/ARTISTS RIGHTS SOCIETY (ARS),
New York

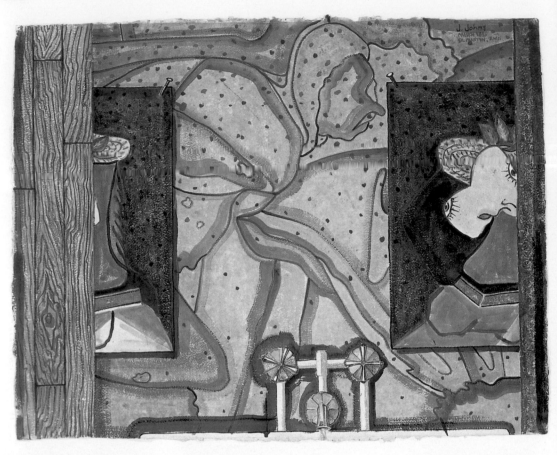

above: Color plate 303. GEORGE SEGAL.
The Diner. 1964–66. Plaster, wood, chrome,
masonite, and formica, 8'6" x 9' x 7'3"
(2.6 x 2.7 x 2.2 m).
Walker Art Center, Minneapolis

left: Color plate 304. CLAES OLDENBURG
seated in *The Store.* 107 East 2nd Street,
New York, December 1961.
Courtesy the artist

right: Color plate 305. JIM DINE. *Double
Isometric Self-Portrait (Scrape).* 1964.
Oil with metal rings and hanging chains on
canvas, 56⅞" x 7'½" (144.5 cm x 2.14 m).
Whitney Museum of American Art, New York

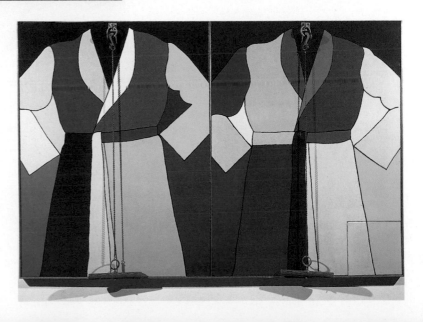

Color plate 306. LUCAS SAMARAS. *Photo Transformation.*
1973–74. SX-70 Polaroid, 3 x 3" (7.6 x 7.6 cm).
Private collection

Color plate 307. ROY LICHTENSTEIN. *Artist's Studio: The Dance.* 1974.
Oil and magna on canvas, 8' x 10'8" (2.4 x 3.3 m).
Collection Mr. and Mrs. S.I. Newhouse, Jr., New York

Color plate 308. ROY LICHTENSTEIN. *Bedroom at Arles.* 1992. Oil and magna on canvas, 10'6" x 13'9½" (3.2 x 4.2 m).
Robert and Jane Meyerhoff Collection, Phoenix, Maryland

Color plate 309.
ANDY WARHOL. *210 Coca-Cola Bottles.* 1962.
Silkscreen ink on synthetic polymer paint on canvas,
6'10½" x 8'9" (2.1 x 2.7 m).
Collection Martin and Janet Blinder
© 1997 ANDY WARHOL FOUNDATION FOR THE VISUAL ARTS/ARTISTS RIGHTS SOCIETY (ARS),
NEW YORK

Color plate 310. ANDY WARHOL. *Marilyn Monroe.* 1962.
Silkscreen ink on synthetic oil, acrylic, and silkscreen enamel on
canvas, 20 x 16" (50.8 x 40.6 cm).
Courtesy Leo Castelli, New York
© 1997 ANDY WARHOL FOUNDATION FOR THE VISUAL ARTS/
ARTISTS RIGHTS SOCIETY (ARS), NEW YORK

Color plate 311. ANDY WARHOL. *Camouflage Self-Portrait.* 1986. Silkscreen ink
on synthetic polymer paint on canvas, 6'10" x 6'10" (2.1 x 2.1 m).
The Metropolitan Museum of Art, New York
PURCHASE, MRS. VERA LIST GIFT © 1997 ANDY WARHOL FOUNDATION FOR THE VISUAL ARTS/
ARTISTS RIGHTS SOCIETY (ARS), NEW YORK

above: Color plate 312. JAMES ROSENQUIST.
Portion of *F-111*. 1965. Oil on canvas with aluminum,
overall 10 x 86' (3 x 28.2 m)
Private collection
© 1997 JAMES ROSENQUIST/VAGA, NEW YORK, NY

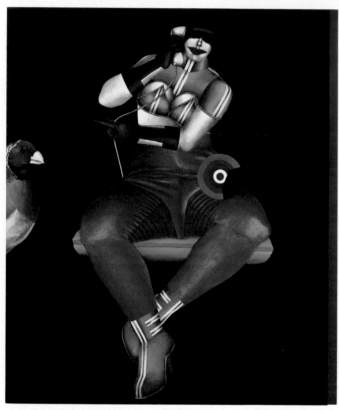

above: Color plate 313. RICHARD LINDNER. *Hello*. 1966.
Oil on canvas, 70 x 60" (177.8 x 152.4 cm).
Collection Robert E. Abrams

above, right: Color plate 314.
MARISOL. *The Family*. 1962. Painted wood and other materials
in three sections, overall 6'10⅝" x 65½" x 15½"
(2.2 m x 166.3 cm x 39.3 cm).
The Museum of Modern Art, New York
ADVISORY COMMITTEE FUND
© 1997 MARISOL ESCOBAR/LICENSED BY VAGA, NEW YORK, NY

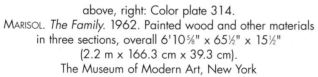

Color plate 315.
WAYNE THIEBAUD. *Pie Counter*. 1963. Oil on canvas,
30 x 36" (76.2 x 91.4 cm).
Whitney Museum of American Art,
New York

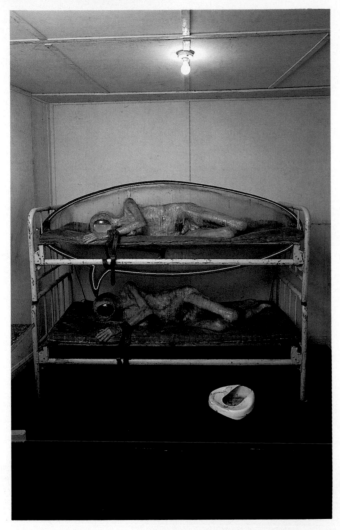

Color plate 317. JESS. *Will Wonders Never Cease: Translation #21*. 1969. Oil on canvas mounted on wood, 21 x 28⅛" (53.3 x 71.4 cm). Hirshhorn Museum and Sculpture Garden, Smithsonian Institution, Washington, D.C.

below: Color plate 318. EDWARD RUSCHA. *Standard Station, Amarillo, Texas*. 1963. Oil on canvas, 65" x 10' (165.1 cm x 3 m). Hood Museum of Art, Dartmouth College, Hanover, New Hampshire

above: Color plate 316. EDWARD KIENHOLZ. *The State Hospital*. 1966. Mixed-media tableau, 8 x 12 x 10' (2.4 x 3.7 x 3 m). Moderna Museet, Stockholm

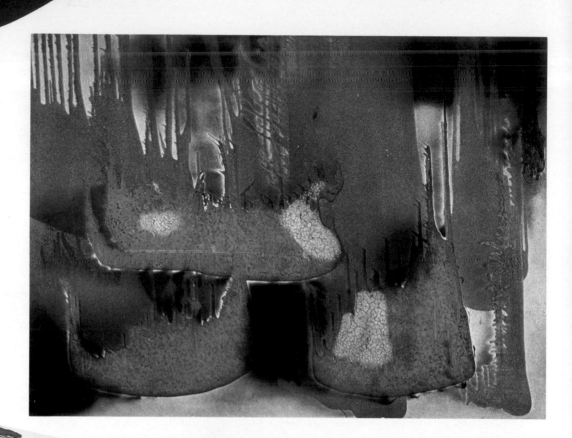

Color plate 319. LUIS JIMÉNEZ. *Man on Fire*.
1969. Fiberglass, 7'5" x 60" x 19"
(2.3 m x 152.4 cm x 48.3 cm).
National Museum of American Art,
Washington, D.C.

above: Color plate 321. YVES
KLEIN. *Fire Painting*.
1961–62. Flame burned into
asbestos with pigment,
43 x 30"
(109.2 x 76.2 cm).
Private collection

left: Color plate 322.
NIKI DE SAINT-PHALLE.
Black Venus. 1965–67.
Painted polyester,
9'2¼" x 35" x 24"
(2.8 m x 89 cm x 61 cm).
Whitney Museum of American
Art, New York

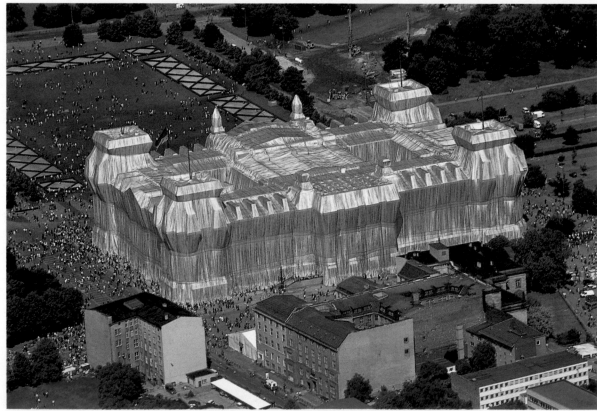

above: Color plate 323.
CHRISTO AND JEANNE-CLAUDE.
Running Fence, Sonoma and Marin Counties, California. 1972–76 (now removed). 2,050 steel poles set 62' apart; 65,000 yards of white woven synthetic fabric, height 18' (5.5 m), length 24½ miles (39.4 km)
© 1976 CHRISTO

left: Color plate 324.
CHRISTO AND JEANNE-CLAUDE.
Wrapped Reichstag, Berlin. 1971–95 (now removed)
© 1995 CHRISTO

626. RICHARD ARTSCHWAGER. *Chair*. 1966. Formica and wood,
59 x 18 x 30" (149.9 x 45.7 x 76.2 cm).
Private collection, London COURTESY MARY BOONE GALLERY, NEW YORK.
© 1997 RICHARD ARTSCHWAGER/ARTISTS RIGHTS SOCIETY (ARS), NEW YORK

in the artist's favorite medium, Formica (here in an elaborately marbleized pattern he used in the mid-1960s). Though superficially like a chair, one would be hard-pressed to use it as such. "I'm making objects for non-use," the artist once said, recall-

ing Duchamp's use-deprived ready-mades that transformed everyday objects into works of art. With its rigid, unforgiving surfaces, the geometricized *Chair* seems to share qualities with the abstract sculptures Minimalist artists were making at the same time (see fig. 696), yet its resemblance to a functional object runs entirely counter to their aims.

Moreover, Artschwager was also making representational paintings based on black-and-white photographs. For these he has used acrylic paint on Celotex, a commercial composite paper that is used less for fine art than for inexpensive ceilings. It has a fibrous texture or tooth and is embossed with distinct patterns, providing a prefabricated "gesture" for the painter. Building façades as well as architectural interiors figure prominently among the subjects of Artschwager's paintings. A large diptych from 1972, *Destruction III* (fig. 627), belongs to a series depicting the demolition of the grand Traymore Hotel in Atlantic City. Like his sculptures, Artschwager's painting presents a distant facsimile of its subject, removed from reality more than once by way of the intermediary photographic source that is then translated into the grainy textures of the Celotex support. This tension between illusion and reality has been a central theme of Artschwager's art to the present day.

Although the same age as Artschwager, **LARRY RIVERS** (b. 1923) had very different beginnings as an artist in New York. He came of age artistically during the height of Abstract Expressionism, but has always remained committed to figuration in one form or another. Originally trained as a musician, Rivers studied painting at Hans Hofmann's school in the late

627.
RICHARD ARTSCHWAGER.
Destruction III. 1972.
Acrylic on Celotex with
metal frames, two
panels, 6'2" x 7'4"
(1.9 x 2.2 m). Private
collection
COURTESY MARY BOONE GALLERY,
NEW YORK. © 1997 RICHARD
ARTSCHWAGER/ARTISTS RIGHTS SOCIETY
(ARS), NEW YORK

628. LARRY RIVERS. *Double Portrait of Berdie.* 1955. Oil on canvas, 5'10¾" x 6'10½" (1.8 cm x 2.1 m). Whitney Museum of American Art, New York
ANONYMOUS GIFT. © LARRY RIVERS/VAGA, NEW YORK, NY

629. LARRY RIVERS. *Dutch Masters and Cigars II.* 1963. Oil and collage on canvas, 8' x 5'7⅜" (2.4 x 1.7 m). Collection Robert E. Abrams © LARRY RIVERS/VAGA, NEW YORK, NY

1940s and made paintings in the manner of Bonnard and Matisse. By the mid-1950s, he had developed a style of portraiture remarkable for its literal and explicit character, including the nude studies of his mother-in-law (fig. 628). After 1960, Rivers began to incorporate stenciled lettering and imagery from commercial sources into his paintings, both of which signaled a departure from the naturalistic conventions that guided his early work. The Dutch Master cigar box, reproducing a painting by Rembrandt, became the subject of several works (fig. 629). In the 1963 version shown here, he appropriated the printed image but altered it by means of his characteristically schematic rendering and loose patches of paint. Rivers cannot exactly be described as a Pop artist, for he shares the concern for the everyday image but differs in the degree of painterly handling.

ROY LICHTENSTEIN (1923–1997) may have defined the basic premises of Pop art (if anyone could) more precisely than any other American painter. Beginning in the late 1950s, he took as his primary subject the most banal comic strips or advertisements and enlarged them faithfully in his paintings, using limited, flat colors (based on those used in commercial printing), and hard, precise drawing. The resulting imagery documents, while it gently parodies, familiar images of modern America. When drawing from comic strips, Lichtenstein preferred those representing violent action and sentimental romance and even incorporated the Benday dots used in photomechanical reproduction. *Whaam!* (fig. 630) belongs to a group of paintings Lichtenstein made in the early 1960s. By virtue of its monumental scale (nearly fourteen feet across) and dramatic subject (one based on heroic images in the comics of

630. ROY LICHTENSTEIN. *Whaam!* 1963. Oil and Magna on two canvas panels, 5'8" x 13'4" (1.7 x 4.1 m).
Tate Gallery, London PURCHASE

World War II battles), *Whaam!* becomes a kind of history painting for the Pop generation. The artist's depictions of giant, dripping brushstrokes, meticulously constructed and far from spontaneous, are a hilarious Pop riposte to the heroic individual gestures of the Abstract Expressionists (fig. 631).

Lichtenstein frequently turned his attention to the art of the past and made free adaptations of reproductions of paintings by Picasso, Mondrian, and other modern artists. In *Artist's Studio: The Dance* (color plate 307, page 530), he borrowed an image from Matisse's *Dance (II)* (color plate 63, page 155) as a background for his own studio still life. In

1909, Matisse had depicted his own painting, *Dance I*, in a similar fashion behind a tabletop still life. Although some of the imagery here may have originated with Matisse, the final effect is unmistakably Lichtenstein's.

Throughout his career, Lichtenstein excelled at sculpture, including freestanding bronze versions of his brushstrokes. *Expressionist Head* (fig. 632) recalls the angular anatomies and bold contours of German Expressionist prints (see fig. 152), but Lichtenstein, himself an accomplished printmaker, regularized the forms within heavy black lines, using bright, primary colors and replacing the characteristic Benday dots with emphatic hatch marks. Like most of Lichtenstein's sculpture, *Expressionist Head* is essentially a two-dimensional conception, like a freestanding painting in relief, rather than a volumetric structure in the round. In the early 1990s,

below, left: 631. ROY LICHTENSTEIN. *Big Painting No. 6*. 1965.
Oil and Magna on canvas, 7'8" x 10'9" (2.3 x 3.3 m).
Kunstsammlung Nordrhein-Westfalen, Düsseldorf

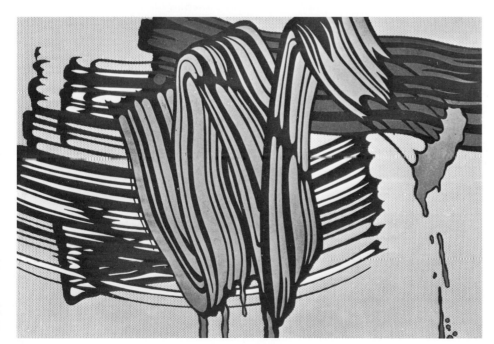

above: 632. ROY LICHTENSTEIN. *Expressionist Head*. 1980. Painted and patinated bronze with painted wooden base, number four of an edition of six, 55 x 41 x 18" (139.7 x 104.1 x 45.7 cm). Private collection

Lichtenstein made a series of large paintings of domestic interiors, some of them based on ads in the Yellow Pages. *Bedroom at Arles* (color plate 308, page 530) clearly derives from Van Gogh's famous painting *The Bedroom*, 1888, at the Art Institute of Chicago, which Lichtenstein knew through reproductions. Not only has he dramatically enlarged the scale of Van Gogh's canvas, but he has transformed the Dutchman's rustic bedroom into a modern bourgeois interior. For example, Van Gogh's cane-seated chairs have become bright yellow Barcelona chairs similar to those designed by architect Mies van der Rohe.

The paintings of Lichtenstein, as well as those of Wesselmann, Rosenquist, and Warhol, share an attachment to the everyday, commonplace, or vulgar image of modern industrial America. Often appropriating their subjects wholesale from their sources, they also treat their images in an impersonal, neutral manner. They do not comment on the scene or attack it like Social Realists, nor do they exalt it like advertisers. They seem to be saying simply that this is the world we live in, this is the urban landscape, these are the symbols, the interiors, the still lifes that make up our own lives. As opposed to assemblage artists who create their works from the refuse of modern industrial society, the Pop artists deal principally with the new, the "store-bought," the idealized vulgarity of advertising, of the supermarket, and of television commercials.

The artist who more than any other stands for Pop in the public imagination, through his paintings, objects, underground movies, and personal life, is **ANDY WARHOL** (1928–1987). Like several others associated with Pop, Warhol was first a successful commercial artist. Initially, he made, like Lichtenstein, paintings based on popular comic strips, but shortly after began to concentrate on the subjects for which he is best known, ones derived from advertising and commercial products—Coca-Cola bottles, Campbell's soup cans, and Brillo cartons (color plate 309, page 531). His most characteristic manner was repetition within a grid—endless rows of Coca-Cola bottles, literally presented, and arranged as they might be on supermarket shelves or an assembly line (Warhol appropriately dubbed his studio "The Factory"). "You can be watching TV and see Coca-Cola," Warhol wrote in his 1975 autobiography, "and you can know that the President drinks Coke, Liz Taylor drinks Coke, and just think, you can drink Coke, too. A Coke is a Coke and no amount of money can get you a better Coke than the one the bum on the corner is drinking." Warhol's signature image is the Campbell's soup can, a product he said he ate for lunch every day for twenty years, "over and over again." When he exhibited thirty-two paintings of soup cans in 1962 at a Los Angeles gallery, the number was determined by the variety of flavors then offered by Campbell's, and the can-

633. ANDY WARHOL. Installation view of *Campbell's Soup Cans*, Ferus Gallery, Los Angeles, 1962

634.
ANDY WARHOL. *Electric Chair*. 1965. Acrylic and silkscreened enamel on canvas, 22 x 28" (55.9 x 71.1 cm). Private collection © 1997 ANDY WARHOL FOUNDATION FOR THE VISUAL ARTS/ARTISTS RIGHTS SOCIETY (ARS), NEW YORK

vases were monotonously lined up around the gallery on a white shelf like so many grocery goods (fig. 633).

From these consumer products he turned to the examination of contemporary American folk heroes and glamorous movie stars, including Elvis Presley, Elizabeth Taylor, and Marilyn Monroe (color plate 310, page 531). He based these early paintings on appropriated images from the media; later he photographed his subjects himself. Using a mechanical photo-silkscreen process from 1962 on, Warhol further emphasized his desire to eliminate the personal signature of the artist and to depict the life and the images of his time without comment. As in his portrait of Monroe, Warhol allowed the layers of silkscreen colors to register imperfectly, thereby underscoring the mechanical nature of the process.

For a period beginning in 1962, Warhol utilized scenes of destruction or disaster taken from press photographs and presented them with the same degree of impersonality. In doing so he suggested that familiarity breeds indifference, even to such disturbing aspects of contemporary life. In his gruesome pictures of automobile wrecks, for example, the nature of the subject makes the attainment of a neutral attitude more difficult. *Electric Chair* (fig. 634), taken from an old photograph, shows the grim, barren, death chamber with the empty seat in the center and the sign SILENCE on the wall. Presented by the artist either singly or in monotonous repetition, the scene becomes a chilling image, as much through the abstract austerity of its organization as through its associations.

In the later 1960s Warhol turned more and more to the making of films, where his principle of monotonous repetition becomes hypnotic in its effect, and to the promotion of the rock band The Velvet Underground. At the same time, he increasingly made himself and his entourage the subject of his art. Equipped with his famous silver-sprayed wig and an

attitude of studied passivity, Warhol achieved cult status in New York, especially after he nearly died in 1968 from gunshot wounds inflicted by a woman declaring herself the sole member of S.C.U.M (the Society for Cutting Up Men).

Throughout the 1970s, Warhol worked primarily as a society portraitist, making garishly colored silkscreen paintings of friends, art world figures, and celebrities. Among his most haunting images are the camouflage self-portraits he made in the 1980s; the one illustrated here was executed the year before his death (color plate 311, page 531). Disguising his own well-known visage in a camouflage pattern was a highly appropriate gesture for this artist, who fabricated his famous persona for public consumption but who remained privately a shy, devout Catholic and devoted son. "If you want to know all about Andy Warhol," he once said, a little disingenuously, "just look at the surface: of my paintings and films and me, and there I am. There's nothing behind it."

For a period, JAMES ROSENQUIST (b. 1933) was a billboard painter, working on the scaffolding high over New York's Times Square. The experience of painting commercial images on an enormous scale was critical for the paintings he began to produce in the early 1960s. His best-known and most ambitious painting, *F-111*, 1965, is ten feet high and eighty-six feet long, capable of being organized into a complete room to surround the spectator (color plate 312, page 532). Named after an American fighter-bomber whose image traverses the entire eighty-six feet, *F-111* also comprises a series of fragmented images of destruction, combined with prosaic details including a light bulb, a plate of spaghetti, and, most jarring of all, a little girl who grins beneath a bullet-shaped hair dryer. Surrounded by this environment of canvas, visitors experienced the visual jolt of the painting's garish, Technicolor-like palette, augmented by reflective aluminum panels at either

635. TOM WESSELMANN. *Interior No. 2.* 1964. Mixed media, including working fan, clock, and fluorescent light, 60 x 48 x 5" (152.4 x 121.9 x 12.7 cm). Private collection
© 1997 TOM WESSELMANN/VAGA, NEW YORK, NY

636. TOM WESSELMANN. *Great American Nude #57.* 1964. Acrylic and collage on board, 48 x 65" (121.9 x 165.1 cm). Whitney Museum of American Art, New York
© 1997 TOM WESSELMANN/VAGA, NEW YORK, NY

end, and its barrage of imagery on a colossal scale. Between 1965, when it was first unveiled at the Leo Castelli Gallery in New York, and 1968, *F-111* traveled to several countries, gaining notoriety not only as an aggressive indictment of war, but as one of the seminal works of the Pop era.

TOM WESSELMANN (b. 1931) parodied American advertising in his 1960s assemblages, which combined real elements—clocks, television sets, air conditioners—with photomontage effects of window views, plus sound effects. In works such as *Interior No. 2* (fig. 635), the literal presence of actual objects dissolves the barriers between depicted forms and reality itself. Wesselmann's most obsessive subject has been the female nude, variations on which comprise a series he called Great American Nude (fig. 636). For decades, Wesselmann has painted this faceless American sex symbol, whose pristine and pneumatic form mimics the popular conventions of pinup magazines.

ROBERT INDIANA (Robert Clarke of Indiana) (b. 1938) could be considered the Pop heir to early-twentieth-century artists such as Charles Demuth and Stuart Davis. He painted a figure five in homage to the former (fig. 637) and, like Davis, he has painted word-images, in his case attaining a stark simplicity that suggests the flashing words of neon signs—EAT, LOVE, DIE. These word-paintings are composed of stenciled letters and precise, hard-edge color-shapes that relate Indiana to much abstract art of the 1960s. Unlike most of Pop art, which manifests little or no overt comment on society, Indiana's word-images are often bitter indictments, in code, of modern life, and sometimes even a devastating indictment of brutality, as in his Southern States series. However, it was with his classic rendering of "LOVE" that

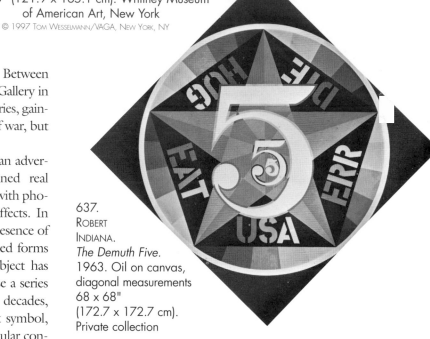

637. ROBERT INDIANA. *The Demuth Five.* 1963. Oil on canvas, diagonal measurements 68 x 68" (172.7 x 172.7 cm). Private collection

Indiana assured a place for himself in the history of the times (fig. 638). He continued to produce, among other new images, many LOVE variations on an ever-increasing scale, as in the aluminum sculpture seen here.

RICHARD LINDNER (1901–1978), the son of German Jews, immigrated first to France (where he was temporarily interned in a concentration camp after the outbreak of World War II), and later, in 1941, to New York. There, he first gained a reputation as an illustrator and teacher during the 1950s. Because the paintings he began to make in the early 1960s, with their figurative subject matter, slick finish, and urban imagery, bore a superficial resemblance to Pop art, his name has been associated with that movement. But his work is rooted in European literary and artistic traditions and is more appropriately related to the machine Cubism of

638. ROBERT INDIANA. *LOVE*. 1972. Polychrome aluminum, 6 x 6 x 3' (1.8 x 1.8 x .9 m). Multiple, Formerly Galerie Denise René, Paris

right: 639. WAYNE THIEBAUD. *Down Eighteenth Street (Corner Apartments)*. 1980. Oil and charcoal on canvas, 48⅜ x 35⅛" (122.9 x 89.2 cm). Hirshhorn Museum and Sculpture Garden, Smithsonian Institution, Washington, D.C.

Léger as well as to the work of earlier German artists such as Schlemmer. He created a bizarre and highly personal dreamworld of metallic, tightly corseted mannequins who at times seem to be caricatures of erotic themes, with bright, flat, even harsh colors (color plate 313, page 532). These fetishistic images, which are filled with personal and literary associations (his mother had run a custom-fitting corset business in Germany), recapitulate the artist's past.

Since the early 1960s, **MARISOL** (Marisol Escobar) (b. 1930), a Paris-born Venezuelan artist active in New York, has created life-size assemblages made of wood, plaster, paint, and found objects, which she sometimes arranges in environments. These are primarily portraits of famous personalities such as Lyndon Johnson, Andy Warhol, and John Wayne. Marisol based her more anonymous sitters in *The Family* on a photograph of a poor American family from the South (color plate 314, page 532). Their features are painted on planks of wood, but the artist added three-dimensional limbs and actual shoes for some of the figures to create her peculiar hybrid of painting and sculpture. Reminiscent of Walker Evans (see fig. 493), *The Family* contains a pathos alien to most Pop art.

Pop had also begun to emerge from the distinctive culture of California by the early 1960s, although the four California artists discussed here—Wayne Thiebaud, Edward Kienholz, Jess, and Ed Ruscha—made work that is only tangentially related to Pop. One of the most distinguished artists to have emerged in California in the 1960s is the painter **WAYNE**

THIEBAUD (b. 1920). He sees his work as part of a long realist tradition in painting, one that includes artists such as Chardin, Eakins, and Morandi, and has never been comfortable with attempts to tie his work to the Pop movement. But when his distinctive still lifes of mass-produced American foodstuffs were first exhibited in New York in 1962, they were regarded as a major manifestation of West Coast Pop and were seen by many as ironic commentaries on the banality of American consumerism. Thiebaud has specialized in still lifes of carefully arranged, meticulously depicted, lusciously brushed and colored, singularly perfect but unappetizing bakery goods (color plate 315, page 532), the kind one would find illuminated under harsh fluorescent lights on a cafeteria counter or under a bakery window. He insists that these works reflect his nostalgia and affection for the subject, rather than any indictment of American culture. While the repetitive nature of the motif is ostensibly monotonous, the paint handling has little in common with the flat, deadpan surfaces favored by Pop artists such as Warhol and Lichtenstein. Thiebaud uses the oil medium to re-create the very substance it depicts, whether cake frosting or mustard or pie meringue, building up his colorful paint layers into a thick, delectable impasto. Thiebaud, who has had a distinguished career as professor of art at the University of California, Davis, has also produced landscapes since the 1960s, often eliciting marvelous chromatic and spatial effects from the vertiginous streets of San Francisco (fig. 639).

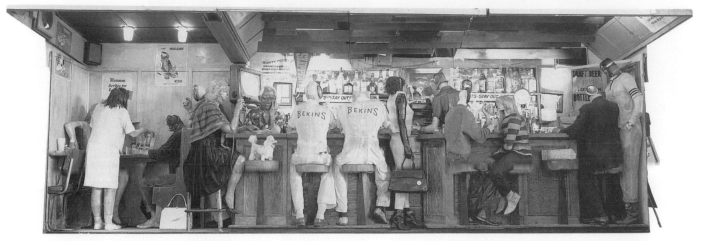

640. EDWARD KIENHOLZ. *The Beanery.* 1965. Mixed-media tableau, 7 x 6 x 22' (2.1 x 1.8 x 6.7 m). Stedelijk Museum, Amsterdam

In strong contrast to Thiebaud's traditional oil paintings is the subversive art of Los Angeles artist **EDWARD KIENHOLZ** (1927–1994). Like George Segal, he created elaborate tableaux, but where Segal's work is quietly melancholic, Kienholz's constructions embodied a mordant, even gruesome view of American life. For his famous bar, *The Beanery* (fig. 640), he produced a life-size environment, one where the viewer can actually enter and mingle with the customers. The patrons are created mostly from life casts of the artist's friends. Kienholz filled the tableau, as he called this and similar works, with such grotesque invention and such harrowing accuracy of selective detail that a sleazy neighbor-

641. JESS. *The Fifth Never of Old Lear.* 1974. Engraving on paper, 33 x 28" (83.8 x 71.1 cm).
Collection Mr. and Mrs. Harry W. and Margaret Anderson

hood bar-and-grill becomes a scene of nightmarish proportion. The seedy customers, who smoke, converse, or stare off into space, wear clocks for faces, all frozen at ten minutes after ten. "A bar is a sad place," Kienholz once said, "a place full of strangers who are killing time, postponing the idea that they're going to die." *The State Hospital* (color plate 316, page 533) is a construction of a cell with a mental patient and his self-image, both modeled with revolting realism and placed on filthy mattresses beneath a single glaring light bulb. The man on the upper mattress is encircled by a neon cartoon speech bubble, as though unable to escape reality even in his thoughts. The two effigies of the same creature—one with goldfish swimming in his glass-bowl head—make one of the artist's most horrifying "concept tableaux." Though he could not have known Duchamp's late work, *Étant Donnés* (see fig. 306), Kienholz adopted a very similar format for *The State Hospital,* which can be viewed only through a barred window in the cell, thus implicating the viewer as a voyeur in this painfully grim scene.

The West Coast artist Burgess Collins, known as **JESS** (b. 1923), came to public attention in 1963 when his comic-strip collage was included in "Pop Art U.S.A.", an exhibition at the Oakland Art Museum. But the reclusive Jess felt no kinship with the Pop artists who were his contemporaries, and most of his mature work is so saturated with romanticism, nostalgia, and fantasy that it seems a far cry from the cool detachment of Pop. After earning a degree in chemistry at the California Institute of Technology, Jess began to study painting in San Francisco under a number of prominent Bay area artists, including Clyfford Still, who, he said, taught him the "poetics of materials." The abstract paintings he made in the first half of the 1950s bear the imprint of this leading Abstract Expressionist. Jess's longtime companion and frequent collaborator was the Beat poet Robert Duncan, who died in 1988.

In 1959, Jess began a series of paintings he called Translations, for which he took preexisting images, usually old photographs, postcards, or magazine illustrations, and "translated [them] to a higher level emotionally and sometimes spiritually." The Translations, which he continued to make until 1976, are painstakingly crafted, built up into

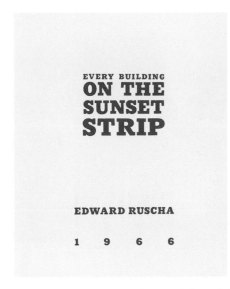

642. EDWARD RUSCHA. Title page (above left) and opening pages (above right) of *Every Building on the Sunset Strip*. 1966. Artist's book

thick, bumpy layers of oil paint where the impasto resembles a kind of painted relief (color plate 317, page 533). They usually feature an accompanying text. The words at the top of the canvas here come from a text by Gertrude Stein in which a boy learns from his father about the cruelty of collecting butterflies as specimens. The hauntingly tender image, painted in pastel colors in a deliberately anti-naturalistic, paint-by-numbers style, is based on an engraving from an 1887 children's book titled *The First Butterfly in the Net*. Since the early 1950s, Jess has also made elaborate works, which he calls "Paste-Ups" (fig. 641), that are inspired by Dada and Surrealist collage and photomontage, especially the work of Max Ernst. Composed of images clipped from magazines and other sources, the Paste-Ups contain layers of meaning as dense and complex as the tightly woven imagery.

The work of ED RUSCHA (b. 1937), who settled in Los Angeles in 1956, is not easily classified. His hard-edge technique and vernacular subject matter allied him with California Pop art in the 1960s, while his fascination with words and typography, exemplified by his printed books and word paintings, corresponded to certain trends in Conceptual art that surfaced in the following decade. In 1963 Ruscha, who had

already worked as a graphic designer of books, published *Twenty-six Gasoline Stations,* a book containing unremarkable photographs of the filling stations on Route 66, a road he had frequently traveled, between Oklahoma City, his birthplace, and Los Angeles. He based his large painting *Standard Station, Amarillo, Texas* (color plate 318, page 533) on one of those images, though he transformed the mundane motif into a dramatic composition, slicing the canvas diagonally, turning the sky black, and transforming the station into a sleek, gleaming structure that recalls Precisionist paintings of the 1930s.

Another of Ruscha's books, *Every Building on the Sunset Strip* (fig. 642), documents exactly what its title indicates in a series of deadpan photographs. "I want absolutely neutral material," he told one interviewer. "My pictures are not that interesting, nor the subject matter. They are simply a collection of 'facts'; my book is more like a collection of ready-mades." In the 1980s, Ruscha began to make dark, tenebrous paintings, in which mysterious objects float, like the words of his earlier paintings, in a nebulous space (fig. 643).

LUIS JIMÉNEZ (b. 1940), a Mexican-American artist from El Paso, Texas, is not associated specifically with California,

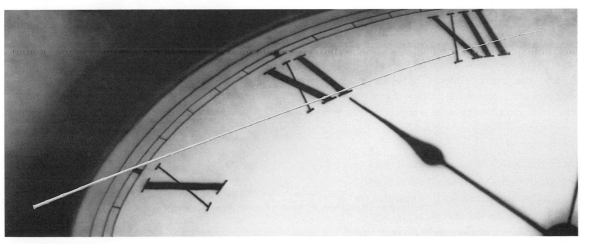

643. EDWARD RUSCHA. *Five Past Eleven*. 1989. Acrylic on canvas, 4'11" x 12'1⅝" (1.5 x 3.7 m). Hirshhorn Museum and Sculpture Garden, Smithsonian Institution, Washington D.C.

but rather the southwestern part of the United States where Mexican and American cultures meet and sometimes clash. Jiménez draws from his Mexican heritage, looking especially to the artistic examples of the Mexican muralists (see chapter 18), as well as the myths, technologies, and mass media of *"el Norte."* He shares certain concerns with Pop artists for, as he has said, "I am making what I would consider people's art and that means that the images are coming from popular culture and so is the material." Jiménez, who learned sign painting techniques in his father's El Paso shop, uses fiberglass for his sculpture, a material that, he has pointed out, is used to make canoes, cars, and hot tubs. He then applies a jet aircraft acrylic urethane finish with an airbrush, giving his surfaces a hard plastic, autobody glow that is repellent and seductive at the same time. It is a finish that, the artist says, "the critics love to hate." His dramatic *Man on Fire* (color plate 319, page 534), a subject previously treated by Orozco, is based on the heroic figure of Cuauhtémoc, the last Aztec emperor, whom the Spaniards tortured by fire.

Europe's New Realism

The movement known as *Nouveau Réalisme* developed in the 1950s when the prevailing trend in Europe was geometric abstraction and *Art informel,* the form of gestural abstraction discussed in chapter 20 that was the European counterpart to Abstract Expressionism. New Realism was officially founded in 1960 by the French critic Pierre Restany in the Parisian apartment of artist Yves Klein. A manifesto was issued and exhibitions of the group were held in Milan in 1960 and at Restany's Gallery J in Paris the following year. The latter was labeled "40 Degrees Above Dada," indicating a kinship, at least according to Restany, with the earlier movement. Though certain aims of the group were comparable to those of Pop art, such as their revolt against the hegemony of abstract painting, the results proved as different as the artists involved. In general, the Europeans were less interested in literal transcriptions of popular culture and were less indebted to the processes and imagery of commercial media. The original members included Klein, Martial Raysse, Arman, Jean Tinguely, Daniel Spoerri, Raymond Hains, Jacques de la Villeglé, and François Dufrêne. All of these, with the exception of Klein, were included in the 1961 exhibition "The Art of Assemblage," at The Museum of Modern Art, and in the "New Realists" show the following year at Sidney Janis. In the catalogue for the latter, Pierre Restany is quoted as saying

> In Europe, as well as in the United States, we are finding new directions in nature, for contemporary nature is mechanical, industrial and flooded with advertisements. . . . The reality of everyday life has now become the factory and the city. Born under the twin signs of standardization and efficiency, extroversion is the rule of the new world. . . .

A seminal figure of Pierre Restany's *Nouveau Réalisme* was **YVES KLEIN** (1928–1962), one of the brilliant talents of the postwar period, whose early death arrested a career full of promise. The son of painters, Klein was a restless and imaginative innovator with a genuine, original fantasy that lent

644. YVES KLEIN. *Leap into the Void.* Fontenay-aux-Roses, October 23, 1960. Photograph by Harry Shunk

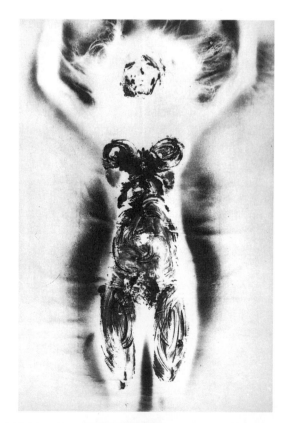

645. YVES KLEIN. *Shroud Anthropometry 20, "Vampire."*
c. 1960. Pigment on canvas, 43 x 30" (109.2 x 76.2 cm).
Private collection
© 1997 ADAGP, Paris/Artists Rights Society (ARS), New York

authenticity to his most scandal-provoking displays. He was a pioneer of later trends such as Conceptual art, Body art, Minimalist art, and Performance art. Klein and his followers were concerned with the dramatization of ideas beyond the creation of individual works of art, and the mystical basis of his art set his work apart from his American Pop contemporaries. He was a member of an obscure spiritual sect, the Rosicrucian Society, and held a black belt in judo from Japan. He even opened his own judo school in Paris in 1955, the year of his first public exhibition in Paris. In his blue monochrome abstractions (color plate 320, page 534), paintings that caused an uproar when first shown in Milan in 1957, he covered the canvas with a powdery, eye-dazzling, ultramarine pigment that he patented as IKB ("International Klein Blue"). For Klein, this blue embodied unity, serenity, and, as he said, the supreme "representation of the immaterial, the sovereign liberation of the spirit." Though not exactly identical, these flat, uninflected paintings carried no indication of the artist's hand on their surface and were a radical alternative to gestural abstraction.

In 1958 Klein attracted audiences to an exhibition of nothingness—*Le Vide* (The Void)—nothing but bare walls in a Parisian gallery. In 1960 he had himself photographed leaping off a ledge in a Paris suburb as a "practical demonstration of levitation" (fig. 644). The photograph was doctored to remove the judo experts holding a tarpaulin below, but Klein, a serious athlete, was obsessed with the notion of flight or self-levitation. In this altered image he presents less a prankish demonstration of levitation than an artistic leap of the imagination. "Today the painter of space ought actually to go into space to paint," he wrote, "but he ought to go there without tricks or fraud, not any longer by plane, parachute or rocket: he must go there by himself, by means of individual, autonomous force, in a word, he should be capable of levitating." The artist's blue "Anthropometries" (a term invented by Restany) of 1960 were realized with nude models, his "living brushes," who were covered with blue paint and instructed by Klein to roll on canvases to create imprints of their bodies (fig. 645). In the Cosmogonies he experimented with the effects of rain on canvas covered with wet blue paint. To make his fire paintings, he used actual flames to burn patterns into a flame-retardant surface and, in later works, added red, blue, and yellow paint (color plate 321, page 534). With his passion for a spiritual content comparable to that preached by Kandinsky, Mondrian, and Malevich, Klein was moving against the trend of his time with its insistence on impersonality, on the painting or sculpture as an object, and on the object as an end in itself.

Other artists associated with New Realism had interests ranging from assemblage to kinetic sculpture, to forms closer to those of American Pop. For Swiss artist **JEAN TINGUELY** (1925–1994), a friend and collaborator of Klein's, mechanized motion was the primary medium. "The machine allows me," he said, "above anything, to reach poetry." After the formulation of metamechanic reliefs and what he called "metamatics," which permitted the use of chance and sound in

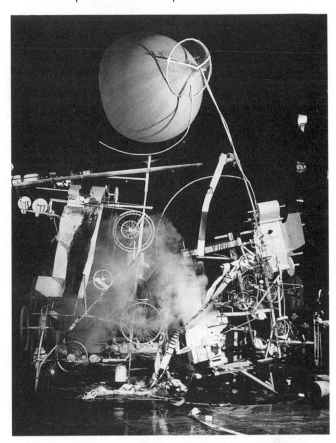

646. JEAN TINGUELY. *Homage to New York*. 1960. Mixed media. Self-destructing installation in the garden of The Museum of Modern Art, New York
© 1997 ARTISTS RIGHTS SOCIETY (ARS), NEW YORK/ADAGP, PARIS

motion machines, he introduced painting machines in 1955, another attempt at debunking the assumptions surrounding so-called Action Painting or French *"Tâchisme."* By 1959 he had perfected the metamatic painting machines, the most impressive of which was the *metamatic-automobile-odorante-et-sonore* for the first biennial of Paris. It produced some forty thousand paintings in an Abstract Expressionist style on a roll of paper that was then cut by the machine into individual sheets. In 1960 Tinguely came to New York, where he and his companion, the artist Niki de Saint-Phalle, collaborated with like-minded American artists including Johns, Rauschenberg, and John Cage. His 1960 *Homage to New York* (fig. 646), created in the garden of New York's Museum of Modern Art from refuse and motors gathered around the city, was a machine designed to destroy itself. In the era's spirit of Happenings, an audience watched for about half an hour as the machine smoked and sputtered, finally self-destructing with the aid of the New York Fire Department. A remnant is in the museum's collection.

Parisian-born **NIKI DE SAINT-PHALLE** (b. 1930) lived in New York from 1933 to 1951 and exhibited with the New Realists from 1961 to 1963. In the early 1960s, she began to make "Shot-reliefs," firing a pistol at bags of paint that had been placed on top of assemblage reliefs. As the punctured bags leaked their contents, "drip" paintings were created by chance methods. These works were both a parodic comment on Action Painting and a ritualistic kind of Performance art.

647. NIKI DE SAINT-PHALLE. *SHE—A Cathedral.* 1966. Mixed-media sculptural environment, 20 x 82 x 30' (6.1 x 25 x 9.1 m). Moderna Museet, Stockholm (fragmentary remains) © 1997 NIKI DE SAINT-PHALLE/ARTISTS RIGHTS SOCIETY (ARS), NEW YORK

Saint-Phalle is best known for her *"Nana"* or woman figures that came to represent the archetype of an all-powerful woman. These are rotund, highly animated figures made of papier mâché or plaster and painted with bright colors (color plate 322, page 534). *SHE* (*"hon"* in Swedish) (fig. 647), an eighty-two-foot, six-ton reclining *Nana,* was made in collaboration with Tinguely and a Swedish sculptor named Per-Olof Ultvedt for the Moderna Museet in Stockholm, where it was exhibited. Visitors entered the sculpture on a ramp between the figure's legs. Inside this giant womb they encountered a vast network of interconnected compartments with several installations, including a bottle-crunching machine, a planetarium, and a cinema showing Greta Garbo movies. The work was destroyed after three months, although the head remains in the permanent collection of the Moderna Museet.

ARMAN (Armand Fernández) (b. 1928) was a friend and in some degree a competitor of Klein's. In 1960, he filled a Parisian gallery with refuse, calling it *Le Plein* (Fullness), thus countering Klein's exhibition of nothingness or *Le Vide*. He either fractured objects such as violins into many slivers or made assemblages, which he called "accumulations," that consisted of reiterated objects—such as old sabres, pencils, teapots, or eyeglasses (fig. 648). He placed these objects under vitrines or imbedded them in transparent, quick-

setting polyester, as seen here. Later he created clear plastic figures to act as containers for the objects. Thus the transparent nude torso in figure 649 contains a cluster of inverted paint tubes that pour brilliantly colored streamers of paint down into the belly and groin.

Arman called one group of his assemblages *poubelles* (trash cans), a term that comments on the prodigal waste inherent in contemporary consumerism while also alluding to the potential for visual poetry in life's recycled detritus, as we have seen in the art of Schwitters and Rauschenberg. In 1982, Arman realized a work on a grand scale with *Long-Term Parking* (fig. 650), a sixty-foot concrete tower with some sixty complete automobiles embedded in it. The park-like setting outside Paris affirms the artist's contention that, far from expressing the anti-art bias present in Duchamp's original use of found objects, *Long-Term Parking* reflects a desire to restructure used or damaged materials into new, aesthetic forms as a metaphor for the hope that modern life may yet prove salvageable.

The work of **MARTIAL RAYSSE** (b. 1936) perhaps has the most in common with British and American Pop artists, though his beach scenes are more reminiscent of his native Côte d'Azur than Coney Island (fig. 651). In 1962, for an exhibition at Amsterdam's Stedelijk Museum, the artist created an installation called "Raysse Beach." Using life-size,

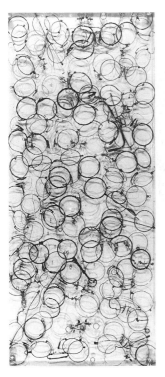

above, left: 648. ARMAN. *Argusmyope.* 1963.
Accumulation of old spectacles in polyester, 28⅛ x 13¾"
(71.4 x 34.9 cm).
Private collection, Brussels
© 1997 Arman/Artists Rights Society (ARS), New York

above, right: 649. ARMAN. *La Couleur de mon amour.* 1966.
Polyester with embedded objects, 35 x 12" (88.9 x 30.5 cm).
Collection Philippe Durand-Ruel, Paris
© 1997 Arman/Artists Rights Society (ARS), New York

650. ARMAN. *Long-Term Parking.* 1982. Sixty automobiles
embedded in cement, 60 x 20 x 20' (18.3 x 6.1 x 6.1 m).
Centre d'Art de Montcel, Jouy-en-Josas, France
© 1997 Arman/Artists Rights Society (ARS), New York

photographic cutouts of bathing beauties set in an actual
environment of sand and beach balls, he achieved a synthetic
re-creation of an expensive watering place, a kind of artificial
paradise for the Pop generation. Central to "Raysse Beach,"
as in the work of his American and British contemporaries,
are the stereotypical images of women drawn from advertising. After participating in Janis's New Realists exhibition in
1962, Raysse spent time in both New York and Los Angeles
and associated with American artists such as Oldenburg and
Rauschenberg whom he had already met in Paris.

Bulgarian-born **CHRISTO** (Christo Javacheff) (b. 1935)
escaped across the Iron Curtain to Western Europe, settling
in Paris in 1958. There he came into contact with the New
Realists and began to wrap found objects in cloth. In the
context of New Realism, Christo's work had the most in
common with that of Arman, although Arman chose to
expose his found objects while Christo elected to obscure
them within his packages, placing the emphasis less on the
contained than on the container. These wrapped forms could
range from more recognizable shapes like a chair, a woman,

right: 651. MARTIAL RAYSSE. *Tableau dans le style français.*
1965. Assemblage on canvas, 7'1⅞" x 4'6⅜"
(2.2 x 1.4 m). Collection Runquist
© 1997 Martial Raysse/Artists Rights Society (ARS), New York

653. CHRISTO AND JEANNE-CLAUDE. *Wrapped Kunsthalle, Bern.*
1968. Synthetic fabric, 27,000 square feet (2,508 square
meters) and rope, 10,000 feet (3,048 meters)

public and private interests. *Running Fence* (color plate 323, page 536), a 1972–76 work executed in the California coastal counties of Marin and Sonoma, entailed raising $3.2 million exclusively through the sale of Christo's preparatory drawings, as well as collages, project-related photographs, books, and films. A significant part of the process also involved persuading the ranchers who owned the land and participating in numerous public hearings and sessions in California courts to overcome local bureaucracies and calm the concerns of environmentalists. Christo and Jeanne-Claude managed sixty-five skilled workers who anchored twenty-four miles of infrastructure, and then three hundred and fifty students who strung 2,050 panels of white nylon fabric, measuring eighteen-by-sixty-eight feet, along steel cables and poles. As with the earlier wrapped objects, *Running Fence* made the familiar visible in a new way. When finally in place, it undulated like a luminous ribbon of light through fields, hills, and valleys (with passages left for cattle, cars, and wildlife), finally tapering off into the Pacific Ocean. *Running Fence* stood for two weeks and then was dismantled, leaving no ecological damage to either flora or fauna.

In 1995, the Christos realized a project that had been in gestation since 1971. They wrapped the Reichstag in Berlin, the traditional seat of the German parliament and a building with a long and violent history (color plate 324, page 536). Although the project came to fruition well after the Berlin Wall fell in 1989, the Reichstag appealed to Christo, a native of a Soviet-bloc country, as a symbol of the divisions of East and West. The tenacious Christos took on a formidable bureaucracy and marshaled a massive team, including ninety professional climbers, to pull 119,603 square yards of silver fabric around the building, resulting in an extraordinary apparition in the urban landscape. As with all of the Christos' projects, the wrapping of the Reichstag, in itself a work of

or a Volkswagen, to unidentified objects, in essence a new kind of abstract, highly enigmatic sculpture (fig. 652). Like Oldenburg's proposed colossal monuments, Christo's most ambitious projects were at first carried out only in drawings and collages. But in 1968, working side by side with his wife, **JEANNE-CLAUDE** (b. 1935), he actually wrapped his first building, the Kunsthalle in Bern, Switzerland (fig. 653).

The Christos are best known for their vast environmental projects that have involved wrapping not only whole buildings but even significant portions of open nature. Christo and Jeanne-Claude have had to excel at public relations as much as at plastic form, since to carry out their epic proposals they must engage the sympathetic attention of countless

Performance art, was heavily documented each step of the way through myriad photographs.

The Italian artist **Mimmo Rotella** (b. 1918) was an early practitioner of the New Realist technique known as *décollage*, which he began to use in 1953 (fig. 654). The antithesis of collage, *décollage* is the art of tearing up posters—promotions for films or commercial products—to create new compositions, not in pristine Pop art guise, but torn and tattered with one superimposed image intruding on another. A familiar feature of any urban landscape is the layers of half-torn posters—modern urban palimpsests—that decorate billboards and walls. Rotella stripped these posters, usually promoting Italian films, from city walls and mounted them on canvas only to tear them off layer after layer. "What a thrill, what fantasy, what strange things happen," he said, "clashing and accumulating between the first and last layer."

The idiosyncratic art of the Italian **Piero Manzoni** (1933–1963) ushered in much of the conceptually based work of the 1960s and 1970s that would follow his death. A significant event in Manzoni's career, which was even more short-lived than Klein's, was the 1957 exhibition in Milan of monochrome paintings. The young Manzoni was an agent provocateur in the Dada, or more appropriately, the Italian Futurist tradition, as when he signed cans purportedly containing his own excrement. He also made "living sculpture"

655. Piero Manzoni. *Achrome.* 1959. Kaolin on pleated canvas, 55⅛ x 47¼" (140 x 120 cm). Musée National d'Art Moderne, Centre d'Art et de Culture Georges Pompidou, Paris
© 1997 Piero Manzoni/VAGA, New York, NY

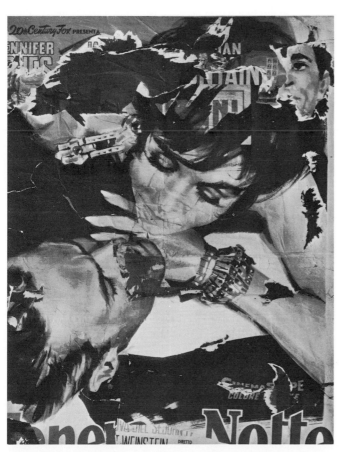

654. Mimmo Rotella. *Untitled.* 1963. Décollage, 47¼ x 39⅜" (120 x 100 cm). Location unknown
© 1997 Mimmo Rotella/VAGA, New York, NY

by signing the bodies of friends or nude models and providing certificates of authenticity, resulting in a kind of "living ready-made." Manzoni was involved with the Italian group *Arte nucleare* (Nuclear Art), which denounced the idea of personal style and called for an art of explosive force in the postnuclear age. Klein also exhibited with this group in Milan in 1957. That year, after seeing Klein's show, Manzoni ceased to make tar-based paintings and began his series of Achromes, monochrome works with neutral surfaces that were emphatically devoid of any imagery. The Achromes consisted of a whole range of media, including plaster or cotton balls, or pebbles mounted on canvas. *Achrome* of 1959 (fig. 655) is reminiscent of the work of an older Italian artist, Fontana (see fig. 589), for Manzoni simply manipulated the canvas by folding it into pleats.

One of the "living sculptures" Manzoni signed in 1962 was the Belgian **Marcel Broodthaers** (1924–1976), a practicing poet who began to make art objects in 1964. Like many of his contemporaries discussed in this chapter, Broodthaers staged Happenings and drew the materials for his reliefs and assemblages from the artifacts of everyday life. He also took part in exhibitions that included work by New Realist or Pop artists. Yet his was a highly individual and hermetic form of expression, one shaped less by the Pop model than by the Surrealist climate that dominated postwar Brussels, thanks largely to the example of his fellow countryman, Magritte. A case in point is *La Tour visuelle* (color plate 325,

656. MARCEL BROODTHAERS. *Ground plan of Musée d'art moderne* at Le Coq, on the North Sea coast of Belgium. View of the ground plan of the *Section Documentaire*, August 1969

page 569), a mysterious construction of small glasses covered with magazine reproductions of an eye which recalls Surrealist objects of the 1930s. Broodthaers, who detested commercialism and what he saw as fashion trends in art, accused the New Realists of acquiescing to the "industrial accumulations that our era produces." Much of his activity centered on Dada-inspired inquiries into the nature of art and subversive alternatives to traditional art institutions. In 1968 he designated his own fictitious *Musée d'art moderne* whose various "sections" were manifested in differing venues, from his house to temporary exhibitions. In 1970 he created his virtual museum on a Belgian beach by tracing its grand plan in the sand. Wearing "museum" hats, he and a colleague set up signs warning visitors not to touch the objects, only to have it all washed away by the tide (fig. 656).

Chronologically, the Colombian painter **FERNANDO BOTERO** (b. 1932) belongs with the Pop generation. He spent most of the 1960s in New York and later settled in Paris (in 1973). But Botero's distinctive figurative style of small-headed, corpulent men and women evolved from long-standing traditions in European painting (fig. 657). In Spain, where he attended Madrid's San Fernando Academy, Botero found his mentors in Goya and Velázquez, gaining from the one an ironic view of the human comedy and from the other a respect for solid, meticulously rendered form. But the love of lucid space and monumental quietude comes from Piero

657. FERNANDO BOTERO. *Portrait of a Family*. 1974. Oil on canvas, 7'9½" x 6'5" (2.4 x 2 m). Private collection

della Francesca, whose wall paintings and fresco technique Botero studied in and around Florence. "The deformation you see," the artist said, "is the result of my involvement with painting. The monumental and, in my eyes, sensually provocative volumes stem from this. . . . My concern is for formal fullness, abundance." He transformed the Latin American bourgeoisie into a fantasy realm of pneumatically obese figures and elaborate South American costumes, all in a feigned naïve or peasant manner and held together by a subversive, devilish wit.

The Snapshot Aesthetic in American Photography

As the vernacular sources and the mixed media of Pop artists would suggest, photography had become an even more dominant presence in Western culture by the early 1960s. A number of artists discussed in this chapter have made broad use of photography in their art, including Hamilton, Hockney, Rauschenberg, Warhol, Samaras, Ruscha, and Christo. At the same time, a growing number of young professional photographers were attracted to the common artifacts of everyday life and conveyed them in seemingly casual compositions devoid of pictorial convention and overt commentary. Their goal was not an exquisitely crafted and designed print but rather an authentic, directly communicated image. The familiarity and growing facility of light-sensitized imagery, due in part to the growing practice of photojournalism, tended to deheroicize the mystical pretensions and technical perfection sought by, for instance, Strand, Weston, and White. In place of such elevated straight photography came the "snapshot aesthetic," its subject matter derived from the most banal urban and suburban landscapes and its procedures as deceptively haphazard as such fare would require.

These new attitudes were most fully realized, appropriately enough, in the work of an artist from abroad, the Swiss American **ROBERT FRANK** (b. 1924). Aided by a Guggenheim grant, Frank in 1955 set out on a journey of discovery throughout the United States, using his 35mm Leica camera to take a fresh, detached, but ultimately critical, if not absolutely damning, look at postwar American society consumed by violence, conformity, and racial and social division (fig. 658). "Criticism," said Frank, "can come out of love." Frank published eighty-three of his photographs, edited from the original 20,000, in book form, first in France and then in the United States as *The Americans* (1959), with an introduction by Jack Kerouac. When judged by conventional standards of the medium, Frank's photographs look harsh, informal, blurred, and grainy, more like glimpses than concentrated views of a subject. Their individual strengths build within the cumulative effect of the book as a whole, the sequential nature of which led the artist to filmmaking. In the 1960s, *The Americans* became a celebrated icon of American

658. ROBERT FRANK. *Political Rally, Chicago, 1956.* From *The Americans*, 1958. Gelatin-silver print © ROBERT FRANK

photography and a model for young photographers who shared his outlook. Frank retained a sequential format in much of his later still photography, including the triptych *Moving Out* from 1984 (color plate 326, page 569). From the streets of America he turned inward to haunting meditations on his own life, as shown here with depictions of three abandoned rooms from different moments in his past. The text the artist attached to these images includes the words, "Just accept lost feelings—Shadows in empty room—Silence on TV."

Native-born **GARY WINOGRAND** (1928–1984) said he took photographs in order to see what the things that interested him would look like as photographs, clearly implying that his was no a priori vision but rather one discovered in the course of making pictures. This partly explains the look of his photographs, which record scenes so bluntly real, so familiar, that they are usually absorbed into the scanning process of normal vision. Winogrand was working as a photojournalist and commercial photographer when he encountered the work of Walker Evans and Frank whose approach to the medium helped shape his mature work. In 1969, he abandoned commercial photography to focus on his personal work and received his second Guggenheim grant, one to explore the "effect of media on events." A number of the prints made under the grant eventually formed a catalogue and an exhibition called "Public Relations" (1977). Given the years Winogrand worked on the project (1969–73), it is no surprise that the Vietnam War played a central role in this

659. GARY WINOGRAND. *Peace Demonstration, Central Park, New York*. 1969. Gelatin-silver print
COURTESY FRAENKEL GALLERY, SAN FRANCISCO

group of photographs (fig. 659). Winogrand forgot little of the aesthetic principles learned while a painting student at Columbia University, for however informal *Peace Demonstration* may appear, as in the tipping of the camera, the resulting image is a marvel of conflicting patterns somehow contained with great rigor.

Like Frank, LEE FRIEDLANDER (b. 1934) found telling imagery among the vast array of visual accident available in America's streets, whether it was a jazz band's procession in a New Orleans neighborhood, or abruptly cropped views of anonymous pedestrians in New York City. Friedlander, who said he wanted to record "the American social landscape and its conditions," exploited the layered reflections of shop windows, often including his own image in such works in a kind of deliberate appropriation of a standard faux pas of amateur photography (fig. 660). Even at their most banal, his pho-

tographs, such as this shot of a tawdry restaurant window featuring a portrait of President and Jacqueline Kennedy the year before the assassination, could possess a strange poignancy.

The work of Winogrand and Friedlander was featured in a 1967 exhibition at The Museum of Modern Art called "New Documents." According to John Szarkowski, the show's curator, such artists represented a new generation of photographers who were directing "the documentary approach toward more personal ends. Their aim has been not to reform life, but to know it." Also included in the show was the work of New Yorker DIANE ARBUS (1923–1971) whose startling photographs issued from her enormous curiosity about human nature in all its manifestations. Her primary mentor was Lisette Model (see fig. 405), but Arbus probed the psyche of her subjects even more deeply than her teacher. In the 1950s Arbus made grainy, 35mm pho-

660. LEE FRIEDLANDER. *Washington, D.C.* 1962. Gelatin-silver print
COURTESY FRAENKEL GALLERY, SAN FRANCISCO

top: 661. DIANE ARBUS. *Untitled #6.* 1970–71. Gelatin-silver print, 10 x 10" (25.4 x 25.4 cm)
COURTESY ROBERT MILLER GALLERY, NEW YORK

above: 662. BRUCE DAVIDSON. *Coney Island* from the series Brooklyn Gang. 1959. Gelatin-silver print

tographs of anonymous subjects she encountered in Coney Island or on the streets of New York City. From 1960 until her death she made photographs for magazines such as *Esquire* and *Harper's Bazaar*. Her best-known photographs are the portraits she made in the 1960s, often of traditionally forbidden subjects, of people existing on the fringes of society, including transvestites, nudists, circus performers, or residents in a home for the mentally retarded. Arbus's ability to gain the trust of her subjects resulted in sometimes painfully honest images (fig. 661). Her photographs are simultaneously disquieting and wondrous, for she captured a world at which we are not allowed to stare. As her daughter, Doon Arbus has written, "She was determined to reveal what others had been taught to turn their back on."

Frank's book *The Americans* provided a critical example for New York photographer **BRUCE DAVIDSON** (b. 1933), who was working for *Life* magazine by the late 1950s. Davidson favored gritty subjects drawn from New York's subways or Coney Island and sometimes worked in series such as one following the teenage members of a Brooklyn street gang (fig. 662). The Brooklyn Gang photo-essay was published in *Esquire* magazine in 1960 with an accompanying text by Norman Mailer extolling the rebelliousness of youth.

663. Photograph by Richard Avedon. *Zazi, street performer, Piazza Navona, Rome, July 27, 1946*

664. EDDIE ADAMS. *Vietnamese General Executing Vietcong*. 1968. Gelatin-silver print

Davidson's grainy snapshot presents the subject, teenagers primping before vending machines, without sentiment or criticism. "I force-developed the film," he said, "to increase the harsh contrast that reflected the tension in their lives."

RICHARD AVEDON (b. 1923) also spent time as a street photographer both in his native city of New York and also in Italy (fig. 663). But he is known for his elaborately staged fashion photographs from the 1950s that appeared in *Harper's Bazaar*, a magazine that attracted some of the best talents in photography, including, for example, Model, Brandt, Brassaï and Frank. Since the early 1950s Avedon has photographed many of the most celebrated personalities of his era, from Marilyn Monroe to Giacometti to the Beatles. In 1969 he switched from a Rolleiflex to a much less maneuverable eight-by-ten-inch view camera, sometimes printing in a square format. At the age of nineteen Avedon served in the merchant marines, where he had to take thousands of head shots for ID photos, a fact that may have determined a preference in later years for photographing his subjects straight on, at excruciatingly close range, and against a stark white background.

The 1940s and 1950s were the heyday of American photojournalism, with picture magazines such as *Life, Harper's,* and *Vogue,* among others, supporting the work of many of photography's leading practitioners. Some of the most memorable photographs of the 1960s came from professional photojournalists, including this gripping image from the war in Vietnam that was captured by the Associated Press photographer **EDDIE ADAMS** (b. 1933) (fig. 664). So powerful was the response to this photograph of an execution of a Vietcong as it was circulated across American newspapers that it served to escalate opposition to the war and to raise questions about the role of photographers who witness violent events and may even encourage them by their presence.

The serial arrangements, narrative content, and even the narcissism found in Warhol's art appeared in the photography of **DUANE MICHALS** (b. 1932), who acquired such traits from the same source as Warhol—Madison Avenue, where Michals has worked as a commercial photographer. Convinced that a single image offers little more than the record of a split second cut from the ongoing flow of time, Michals constructs "photostories," or playlets, to expose or symbolize the reality behind our perceptions of it (fig. 665). "I use photography," the artist says, "to help me explain my experience to myself. . . . I believe in the imagination. What I cannot see is infinitely more important than what I can see."

665. DUANE MICHALS. *Paradise Regained.* 1968. Series of gelatin-silver prints
© DUANE MICHALS, COURTESY SIDNEY JANIS GALLERY, NEW YORK

22 Sixties Abstraction

As postwar Western Europe and the United States increasingly relaxed from the economic stress and political turbulence of the thirties and forties, the rare synthesis of mystical aspiration and physical gesture realized by the Abstract Expressionist generation began to break apart, releasing its constituent elements to follow their own separate, if not altogether unrelated, paths. While some of these, like the Pop and New Realist trends seen in the last chapter, led to Neo-Dada attempts to make art ever more inclusive of the most ordinary reality, others manifested themselves in systematic campaigns of excluding from art all but its most essential properties. By the end of the 1960s, vanguard artists and critics would contend that art could be distilled into idea alone.

Something of the dual response to fifties art—to its creators' "elitist" sense of themselves as existential heroes and of their works as sacred objects—has already been seen in the career of Yves Klein, who, on the one hand, "organized" an exhibition that consisted of nothingness *(le vide)* and, on the other hand, made sculpture in the Duchampian "ready-made-aided" fashion, by mounting blue-sprayed sponges on pedestals. And when Klein made paintings by exposing a colored support to such external, non-art elements as fire and rainwater or by emptying the canvas of all but its irreducible condition of flat, monochrome, rectangular shape, he briefly merged extremes of both materiality and immateriality that would characterize sixties art (color plates 320, 321, page 534). Either way, the tendency involved simplifications so progressive that by the end of the decade the abstraction they produced would be known by such terms as Minimalism, ABC art, or Primary Structures. While all of this may have come about as a necessary correction of fifties heroics, the new abstractionists would evince heroics of their own, not only in the sheer size of their works but also in their ambition to provide an alternative to the dominant critical discourse of Clement Greenberg and his followers, who had defined modernism as the consistent and exclusive engagement of painting and sculpture with the qualities essential to them—namely, their flat, optical surfaces.

Pop art, thanks to its ability to entertain a newly affluent, status-conscious mass audience, may have earned its title by becoming the most popular development ever to have occurred in the higher reaches of culture. During the sixties, however, before the underlying complexity of Pop had been fully appreciated, it was abstraction that dominated within the world of art itself, mainly because its essential sobriety and internal processes of self-purification engaged the attention of the most serious critics and thinkers.

Post-Painterly Color Field Abstraction

A number of exhibitions held in the 1960s drew attention to certain changes that were occurring in American painting. Between 1959 and 1960, Clement Greenberg was instrumental in mounting a series of one-man exhibitions at the New York gallery French and Company for such artists as Barnett Newman, David Smith, Morris Louis, Kenneth Noland, and Jules Olitski. Over the course of the next few years, several museums followed suit with exhibitions designed to draw attention to new directions in American art. Among these shows were "American Abstract Expressionists and Imagists" at New York's Guggenheim Museum in 1961; "Toward a New Abstraction" at New York's Jewish Museum in 1963; "Post-Painterly Abstraction," organized in 1964 by Clement Greenberg for the Los Angeles County Museum of Art; and "Systemic Painting," curated in 1966 by Lawrence Alloway at the Guggenheim Museum.

These exhibitions illustrated that many young artists were attempting to break away from what they felt to be the tyranny of Abstract Expressionism, particularly in its emphasis on the individual brush gesture. Some of these artists turned to representational art forms ranging from minutely observed figuration to assemblages to Pop art, while others sought new means of retinal stimulation through Op art, or light and motion. Most suggestive was the apparent development of abstract painting to which a number of names have been applied, including Abstract Imagism, Post-Painterly Abstraction, and, most enduring of all, Color Field painting, sometimes simplified to Field painting. Still other terms have been applied to various aspects of this direction during the 1960s—such as Systemic, Hard-Edge, and Minimalist painting. These labels, all variously referring to abstract painting, do not necessarily encompass all the same artists or all the directions involved—which range from forms of allover painting to almost blank canvases. The emphasis throughout, however, is on pure, abstract *painting*, in distinction to figuration, optical illusion, object making, fantasy, light, motion, or any of the other tendencies away from the act of painting itself.

Among all the painters working against the current of gestural Abstract Expressionism in the late 1950s there was a general move (as Greenberg has pointed out) to openness of design and image. Many of the artists turned to the technique of staining raw canvas and moved toward a clarity and freshness that differentiated their works from those of the Abstract Expressionists, which were characterized by compression and brushwork. Aside from the Hard-Edge painters, others such as Helen Frankenthaler and Morris Louis were working in new directions, although with an increasing subordination of brush gesture and paint texture. The directions and qualities suggested in the work of these artists—some already well established, some just beginning to appear on the scene— were to be the dominant directions and qualities in abstract painting during the 1960s.

As noted above, probably no single, encompassing term can be found to describe this painting any more satisfactorily than Abstract Expressionism in the 1950s. For purposes of discussion we will use the broad term "Color Field painting" in our analysis of a number of individual artists who will illustrate the range of expression involved in the art of the 1960s.

Greenberg's inclusion of **SAM FRANCIS** (1923–1994) among his "Post-Painterly" abstractionists may seem somewhat surprising, since Francis during the 1950s was associated with American Abstract Expressionism and *Art informel* in Paris, where he lived until 1961. As Greenberg explained in the introduction to the exhibition catalogue of "Post-Painterly Abstraction," Abstract Expressionism was characterized by its "painterly" quality, a term inspired by the use of the German word *malerisch* by the Swiss art historian Heinrich Wölfflin to describe Baroque art. Just as Wölfflin contrasted the painterly quality of the Baroque with the crisp linearity of High Renaissance art, so Greenberg identified Post-Painterly Abstraction with more sharply defined compositions and less emphasis on evidence of the artist's gesture. According to Greenberg: "By contrast with the interweaving of light and dark gradations in the typical Abstract Expressionist picture, all the artists in this show move towards a physical openness of design, or towards linear clarity, or towards both."

Despite Francis's association with Abstract Expressionism or painterly art, however, his inclusion among the Post-Painterly abstractionists is not illogical, since the direction of his painting had been toward that openness and clarity of which Greenberg spoke. In fact, despite his continuing use of

666. SAM FRANCIS. *Shining Back*. 1958. Oil on canvas, 6'7⅜" x 53⅛" (2 m x 135 cm). Solomon R. Guggenheim Museum, New York

the brush gesture, the drip, and the spatter, his similarities to the Color Field painters were greater than his differences, and he even anticipated some of these artists in his use of extreme openness of organization. In contrast to his work of the early 1950s, one notes that *Shining Back* (fig. 666), a work of 1958, is far more open in its structure, although the drips and spatters are quite apparent. In later works, despite lingering vestiges of spatters, the essential organization is that of a few free but controlled color-shapes—red, yellow, and blue—defining the limits of a dominant white space. Francis's paintings of the early 1970s (color plate 327, page 570) increasingly emphasized the edge to the point where his paint spatters at times surround a clear or almost clear center area of canvas. There continues to be a great range in his painting within the general formula of lyrical abstraction. At times he uses a precise linear structure as a control for his free patterns of stains and spatters. Although Francis has been associated with the Color Field painters, it is important to note that he antedated the classification and should be thought of as a highly talented pure painter, who developed his style without concern for fashions or categories.

An artist also sometimes included within the category of

667. JOAN MITCHELL. *August, Rue Daguerre*. 1957. Oil on canvas, 6'10" x 69" (2.1 m x 175.3 cm). The Phillips Collection, Washington, D.C.

Color Field painters is **JOAN MITCHELL** (1926–1992), one of the younger artists associated with Abstract Expressionism during the 1950s, along with Francis, Norman Bluhm, Ray Parker, and others. From the 1950s forward, Mitchell demonstrated a particular interest in evoking her relationship to landscape, initially depicting both natural and urban environments. *August, Rue Daguerre* (fig. 667) reflects the artist's experience of Paris at a specific time and place. As the painting's title suggests, memory played an important role in Mitchell's work. Mitchell moved permanently to France in 1959, living first in Paris and then in Vétheuil. In France, Mitchell continued to pursue her own directions in painting, undeterred by the popularity of Pop and Op art. In the mid-1960s, Mitchell expunged references to urban settings from her work and from that time onward turned predominantly to nature for inspiration in her painting (color plate 328, page 570).

Another transitional figure between Abstract Expressionism and Color Field painting is **HELEN FRANKENTHALER** (b. 1928). Frankenthaler became the first American painter after Jackson Pollock to see the implications of the color staining of raw canvas to create an integration of color and ground in which foreground and background cease to exist. *Mountains and Sea* (fig. 668), Frankenthaler's first "stained" painting, marked a turning point in her career. Having just returned from a vacation in Nova Scotia, the artist found herself experimenting with composition and consistency of paint. She recounts:

Before, I had always painted on sized and primed canvas—but my paint was becoming thinner and more fluid and cried out to be soaked, not resting. In *Mountains and Sea*, I put in the charcoal line gestures first, because I wanted to *draw* in with color and shape the totally abstract memory of the landscape. I spilled on the drawing in paint from the coffee cans. The charcoal lines were original guideposts that eventually became unnecessary.

During a visit to her studio in 1953, Morris Louis was so affected by *Mountains and Sea* that he and Kenneth Noland began to stain canvases themselves, establishing their own staining techniques.

Frankenthaler employs an open composition, frequently building around a free-abstract central image and also stressing the picture edge (color plate 329, page 570). The paint is applied in uniformly thin washes. There is no sense of paint texture—a general characteristic of Color Field painting—although there is some gradation of tone around the edges of color-shapes, giving them a sense of detachment from the canvas. The irregular central motifs float within a rectangle, which, in turn, is surrounded by irregular light and dark frames. These frames create the feeling that the center of the painting is opening up in a limited but defined depth.

In 1960 Frankenthaler made her first prints. Since then, she has worked with a variety of printmaking techniques in addition to painting, using each of these media to explore pictorial space through the interaction of color and line on a particular surface. One of her most successful prints is *Essence Mulberry*, executed in 1977 (color plate 330, page 571). Inspired by an exhibit of medieval prints at the Metropolitan Museum of Art, Frankenthaler combined the shades of mulberry, blue, yellow, and brown. The effect of the blended colors is as delicate and luminescent as that of her paintings.

MORRIS LOUIS (1912–1962) was one of the most talented new American painters to emerge in the 1950s. Living in Washington, D.C., somewhat apart from the New York scene and working almost in isolation, he and a group of artists that included Kenneth Noland were central to the development of Color Field painting. The basic point about Louis's work and that of other Color Field painters, in contrast to most of the other new approaches of the 1960s is that they continued a tradition of pure painting exemplified by Pollock, Newman, Still, Motherwell, and Reinhardt. All of these artists were concerned with the classic problems of pictorial space and the statement of the picture plane. Louis characteristically applied extremely liquid paint to an unstretched canvas, allowing it to flow over the inclined surface in effects sometimes suggestive of translucent color veils. The importance of Frankenthaler's example in Louis's development of this technique has been noted. However, even more so than Frankenthaler, Louis eliminated the brush gesture, although the flat, thin pigment is at times modulated in billowing tonal waves (color plate 331, page 571). His "veil" paintings consist of bands of brilliant, curving color-shapes submerged in translucent washes through which they emerge principally at the edges. Although subdued, the

668. HELEN FRANKENTHALER. *Mountains and Sea.* 1952. Oil on unprimed canvas, 7'2⅞" x 9'9¼" (2.2 x 3 m). Collection the artist, on extended loan to the National Gallery of Art, Washington, D.C.

resulting color is immensely rich. In another formula the artist used long, parallel strips of pure color arranged side by side in rainbow effects. Although the separate colors here are clearly distinguished, the edges are soft and slightly interpenetrating (color plate 332, page 571).

JULES OLITSKI (b. 1922) might be seen as a man of romantic sensibility expressing himself through the pure sensuousness of his large, assertive color areas. In earlier works he, like so many of his contemporaries, explored circle forms, but his were generally irregular, off-center, and interrupted by the frame of the picture to create vaguely organic effects reminiscent of Jean Arp. Moving away from these forms, Olitski began, in about 1963, to saturate his canvas with liquid paint, over which he rolled additional and varying colors. In later works, he sprayed on successive layers of pigment in an essentially commercial process that still admitted a considerable degree of accident in surface drips and spatters (color plate 333, page 572). The dazzling, varied areas of paint are defined by edges or corners and perhaps internal spots of roughly modeled paint that control the seemingly limitless surfaces. Sometimes apparently crude and coloristically disturbing, Olitski's flamboyant works are nevertheless arresting.

Much has been made of the Color Field painters' emphasis on the canvas edge in contrast to the traditional pattern of centralization. This move to the framing edge of the canvas in painting has perhaps some analogies to the accent on horizontality in contemporary abstract sculpture, which similarly may be considered a reaction against the long tradition of centralized, or vertical, focus in sculpture.

In the 1960s, the young LARRY POONS (b. 1937) created an intriguing form of Systemic painting with optical-illusionistic implications (color plate 334, page 572). His small, brightly colored ellipses vibrate against their highly saturated background. Poons's painting, created through a process of staining, was simultaneously identified as Color Field painting by Greenbergian formalists and as Optical art by other critics when William Seitz included the artist's work in the 1965 show "The Responsive Eye" at The Museum of Modern Art in New York. Poons subsequently placed more emphasis on the painterly surface of his work. His heavily textured canvases frequently have long, vertically dragged brushstrokes, densely arranged in rich color patterns. Like Olitski and the other Color Field painters, Poons created large-scale works that would envelop the visual field of the viewer, provoking a purely optical response to the image—the Greenbergian ideal for modernist painting.

As the trend toward progressively more radical abstraction intensified throughout the sixties, urged on by doctrinaire critical support, commercial hype, and media exposure, only exceptionally strong artists could maintain their aesthetic independence from the dominant mode. Several of the most interesting figures to emerge during this period enjoyed the advantage—for purposes of their autonomy—of regular residence remote from the competitive New York art scene.

In the course of his career, which began in the 1940s, the work of California artist RICHARD DIEBENKORN (1922–1993) moved through three distinct phases. During the first he worked in an Abstract Expressionist vein, guided by the

669. RICHARD
DIEBENKORN.
*Man and Woman in
Large Room.* 1957.
Oil on canvas,
71 x 62½"
(180.3 x 158.8 cm).
Hirshhorn Museum and
Sculpture Garden,
Smithsonian Institution,
Washington, D.C.

examples of Clyfford Still and Mark Rothko, who both taught at the California School of Fine Arts (now the San Francisco Art Institute). While the outward forms of his style would change, the artist early on discovered his most fundamental concern: to abstract from his perception of things seen in order to realize painterly, atmospheric distillations of the western American landscape. From the outset, he learned to translate sensuous, visual experience into broad fields of color structured in relation to the flatness and rectilinearity of the picture plane by a framework of firm but painterly geometry (color plate 335, page 572). Even as an Abstract Expressionist, Diebenkorn was, like most of his generation, a cool formalist, and in the context of his art he eschewed pri-

vate, autobiographical associations in favor of meanings more generally evocative of the interaction between inner and outer worlds. This can be seen to particular advantage in *Man and Woman in Large Room* (fig. 669), executed after the artist shifted from abstraction to figuration, a change that occurred in 1954 while he was in close contact with two other California painters, David Park and Elmer Bischoff. The picture functions almost as a visual metaphor because the figures, despite the expressionist potential of their juxtaposition and handling, participate in the overall structure of the total image while also enriching its sober, contemplative mood with an undercurrent of emotional pressure. In the early 1960s Diebenkorn moved on from this second, or figurative,

670. CY TWOMBLY.
Untitled. 1969.
Crayon and oil on
canvas, 6'6" x
8'7" (2 x 2.6 m).
The Whitney
Museum of
American Art,
New York
GIFT OF MR. AND MRS. RUDOLF
B. SCHULHOF

671. Cy Twombly. *Untitled*. 1978 (Roma). Wood, cloth, wire, nails, mat oil paint, 17" x 7'3⅝" x 7¾" (43.2 cm x 2.2 m x 19.7 cm). Private collection. On loan to the Kunsthaus, Zürich, Switzerland

phase and, in 1967, following his relocation to Santa Monica, began his third mode, devoted to a series of majestic non-figurative abstractions known as the Ocean Park paintings (color plate 336, page 573). In variation after variation, and with Matisse as a primary source, in addition to Mondrian, Monet, and of course Abstract Expressionism, Diebenkorn purified and monumentalized his personal vocabulary of mist- and light-filled color planes emanating from but rigorously contained within a softly drawn architectural scaffolding. Pentimenti, odd and oblique angles in the structural borders, and the expansiveness and close harmony of the luminous colors work without benefit of human images to generate a sense of tension within the pervasive calm, a sense of presence within figuratively empty fields.

After World War II and following a semester of study at Black Mountain College, the slightly younger **Cy Twombly** (b. 1928) established a more or less permanent base in Rome, where he too succeeded in calmly pivoting his art on the convergence point of freedom and control, lucidity and opacity. A poet as well as a painter and a sculptor, Twombly found his characteristic image in a slate-gray ground covered with white graffiti, drawings intermingled with words and numbers, like chalk lessons half-erased on the blackboard at the end of a busy school day (fig. 670). Sometimes snatches of personal verse, often brief quotations from a classical source, always legible at first sight, though never quite complete or entirely coherent on close examination, Twombly's scribbles and scrawls activate the surface with gestures as decisive as Pollock's, but, unlike those of the Abstract Expressionists, they remain indeterminate. Additionally, Twombly's images suggest a hidden narrative, evoked in bold but enigmatic terms, such as the tragic love affair of Hero and Leander, depicted in a three-panel series testifying to their passion and to the fury and progressive calming of the ocean that ended their romance (color plate 337, page 573). Like many of his paintings, Twombly's sculptures also find frequent inspiration

in classical sources, as in his rendition of an elongated, delicately balanced chariot (fig. 671).

Hard-Edge Painting

The term "Hard-Edge" was first used by the California critic Jules Langsner in 1959, and then given its current definition by Lawrence Alloway in 1959–60. According to Alloway, Hard-Edge was defined in opposition to geometric art, in the following way: "The 'cone, cylinder, and sphere' of Cézanne fame have persisted in much 20th-century painting. Even where these forms are not purely represented, abstract artists have tended toward a compilation of separable elements. Form has been treated as discrete entities," whereas "forms are few in hard-edge and the surface immaculate. . . . The whole picture becomes the unit; forms extend the length of the painting or are restricted to two or three tones. The result of this sparseness is that the spatial effect of figures on a field is avoided." The important distinction drawn here between Hard-Edge and the older geometric tradition is the search for a total unity in which there is generally no foreground or background, no "figures on a field." During the 1950s Ellsworth Kelly, Ad Reinhardt, Leon Polk Smith, Alexander Liberman, Sidney Wolfson, and Agnes Martin (most of them exhibiting at the Betty Parsons Gallery in New York) were the principal pioneers. Barnett Newman (also showing at Parsons) was a force in related but not identical space and color explorations.

Ellsworth Kelly (b. 1923), who matured artistically in Paris following World War II, has come to be considered a leader of the Hard-Edge faction within Color Field painting, although he himself has expressed some discomfort with this label. He explained to Henry Geldzahler in 1963, "I'm interested in the mass and color, the black and white—the edges happen because the forms get as quiet as they can be." Despite the abstract appearance of much of Kelly's painting, the artist drew extensively on his observation of the natural forms around him, as in *Window, Museum of Modern Art,*

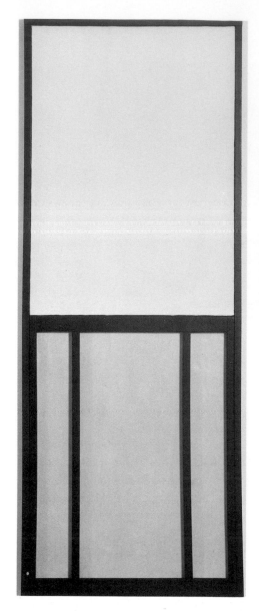

left: 672. ELLSWORTH KELLY. *Window, Museum of Modern Art, Paris.* 1949. Oil on wood and canvas, two joined panels, 50½ x 19½ x ¾" (128.7 x 49.5 x 1.9 cm). Private collection

above: 673. ELLSWORTH KELLY. *Red Curve II.* 1972. Oil on canvas, 45" x 14' (114.3 cm x 4.3 m). Stedelijk Museum, Amsterdam

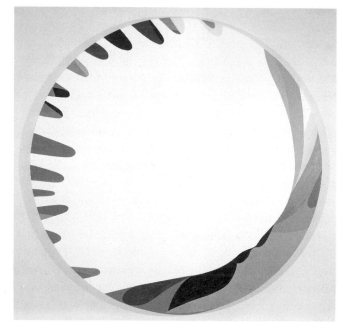

674. JACK YOUNGERMAN. *Roundabout.* 1970. Acrylic on canvas, diameter 8' (2.4 m). Albright-Knox Art Gallery, Buffalo
GIFT OF SEYMOUR H. KNOX. © 1997 JACK YOUNGERMAN/VAGA, NEW YORK, NY

Paris of 1949 (fig. 672). In 1954 the artist returned to the States, settling in New York among a group of artists, including Jasper Johns, Robert Rauschenberg, Jack Youngerman, and Agnes Martin, who, like him, resisted the dominance of Abstract Expressionism. In New York, Kelly experimented with collage and continued to develop his interest in shape and color and the relationship of figure to ground. The influence of Arp was evident at times. His paintings of the early 1960s frequently juxtaposed fields of equally vibrant color, squeezing expansive shapes within the confines of a rectangular canvas (color plate 338, page 574). But just as the tension in these canvases suggests, shapes frequently exploded from the rectangular frame, becoming artworks themselves that undermined the distinction between painting and sculpture (fig. 673). Kelly has continued to work in a similar vein up to the present day. He does not differentiate between his paintings and his sculptures, and some of his works are difficult to classify. Thus *Untitled (Mandorla)* of 1988 (color plate 339, page 574), a bronze distillation of a form common in medieval painting, gains much of its effect from its relief projection. Kelly also creates freestanding sculptures, works that use forms similar to those of his paintings, although projected on an environmental scale and constructed industrially of Cor-ten steel.

Like Kelly, **JACK YOUNGERMAN** (b. 1926) developed his art in Paris during the postwar years, courtesy of the GI Bill, before returning to join the New York School in the 1950s. He then became known for bringing a special Matisse-like rhythm and grace to the Constructivist tradition by rendering leaf, flower, or butterfly forms with the brilliantly colored flatness and clarity of Hard-Edge painting (fig. 674). In such work, he looked back not only to Arp and Matisse, but also to the rhapsodic, nature-focused art of painters who worked in the United States, such as Dove, O'Keeffe, and Gorky. For a while, he designed his undulant silhouettes so that figure and ground, somewhat in the manner of contemporary Op art, appear to reverse, with negative and positive switching back and forth.

Although **KENNETH NOLAND** (b. 1924), as noted, was close to Louis during the 1950s and, like him, sought through Color Field painting an essential departure from the brush-gesture mode of De Kooning or Kline, his personal solutions were quite different from those of his fellow Washingtonian. Using the same thin pigment to stain unsized canvas, Noland made his first completely individual statement when, as he said, he discovered the center of the canvas. From this point, between the mid-1950s and 1962, his principal image was the circle or a series of concentric circles exactly centered on a square canvas (color plate 340, page 574). Since this relation of circle to square was necessarily ambiguous, in about 1963 Noland began to experiment with different forms, first ovoid shapes placed above center, and then meticulously symmetrical chevrons starting in the upper corners and coming to a point just above the exact center of the bottom edge (fig. 675). The chevrons, by their placement as well as their composition, gave a new significance to the shape of the canvas and created a total, unified harmony in which color and structure, canvas plane and edges, are integrated.

In 1964 Noland was a featured painter at the United States pavilion of the Venice Biennale, and in 1965 he was given a one-man exhibition at the Jewish Museum. It was about this time that Noland, working within personally defined limits of color and shape relationships, systematically expanded his vocabulary. The symmetrical chevrons were followed by asymmetrical examples. Long, narrow paintings with chevrons only slightly bent led to a series in which he explored systems of horizontal strata, sometimes with color variations on identical strata, sometimes with graded horizontals, as in *Graded Exposure* (color plate 341, page 575). In the next few years the artist moved from the solution of *Graded Exposure* to a formula of vertical or horizontal canvases in which his use of the grid allowed a dominant emphasis to fall on the framing edge. In his subsequent work, Noland has continued to explore the relationship of form and color to the edge of the canvas itself, manipulating the shape of the canvas as well as the painted surface.

Owing to both his own qualities and the range of this new abstraction, **AL HELD** (b. 1928) relates to Color Field as well as to Hard-Edge painting. He remains one of the strongest of the sixties abstractionists both in his forms and in his use of color, but, unlike most of the other Color Field painters, who eliminated brushstrokes and paint texture, Held built up his paint to create a texture that added to the total sense of weight and rugged power. He worked over his paint surfaces, sometimes layering them to a thickness of an inch, although since 1963 he has sanded down the surface to a machine precision. For a period in the 1960s, Held based paintings on letters of the alphabet (fig. 676). In these, again,

above: 675. KENNETH NOLAND. *Golden Day*. 1964. Acrylic on canvas, 6 x 6' (1.8 x 1.8 m). Private collection

right: 676. AL HELD. *Ivan the Terrible*. 1961. Acrylic on canvas, 12' x 9'6" (3.7 x 2.9 m). André Emmerich Gallery, New York

677. AL HELD. *Flemish IX.* 1974. Acrylic on canvas, 6' x 60"
(1.8 m x 152.4 cm). Sable-Castelli Gallery, Toronto
© 1997 AL HELD/VAGA, NEW YORK, NY

the open portions of the letters were subordinated, both to hold the edges of the canvas and to establish a sense of great scale.

If the flat, unmodulated, or thinly stained paint surface must be considered a criterion of Color Field painting, Held should not really be associated with this direction. Unlike the Color Field artists, Held focused on the powerful presence of colored forms on the canvas, rather than on the less tangible, more self-consciously optical effect of stained color itself.

In the later 1960s, Held overtly rebelled against modernist derisions of illusionism and forcibly resisted the pervasive reductive trend in contemporary painting. He introduced complex, apparently three-dimensional shapes into his paintings. The insistent exploration of a kind of rigid geometric abstraction led Held through the refinement of his means to a black-and-white structure. In a group of paintings of 1974 and early 1975 he presented an architecture of box-like structures outlined in white against a uniform black ground (fig. 677). The title "Flemish" in the series suggests that Held, like many of his contemporaries in the 1970s, had been looking at old master paintings, in this instance conceivably the works of Jan van Eyck or Rogier van der Weyden.

Along with the black paintings with white linear structures, he exhibited their counterparts, consisting of black lines on a flat white ground. In earlier examples the perspective structures were heavily outlined. In a number of the white paintings, however, some lines became much more delicate, mixed with weightier ones. Transparent, curvilinear elements became increasingly dominant as part of the total spatial interplay.

When Held reintroduced color, he did so with a vengeance, adopting a high-keyed intensity combined with an expanded scale to give his newer paintings an almost overwhelming impact (color plate 342, page 575). The color combined with his handling of line produces the paradoxical effect of clarity and ambiguity. Throughout these powerful and challenging pictures, images constantly fail to conform to what experience and tradition have prepared us to anticipate in logical-seeming spatial structures. And so, as within the perpetual movement of Pollock's calligraphic complexity, the eye gives up trying to sort out the irresolutions and accepts the special unity or indivisibility of the total configuration.

Optical Painting (Op Art)

More or less related tendencies in the painting of the 1960s may be grouped under the heading of Optical (or Retinal) painting. "Optical" in this context should not be confused with the quality of opticality attributed by Greenbergian formalists to modernist painting. For modernists, opticality conveyed the sense of a flat, non-illusionistic image that seemed to deny the necessity of physical support, as opposed to the optical stimulation of Op art, which implied the presence of an illusion generated by the stimulation of the retina. What is called Op art overlaps at one end with light sculpture or construction (in its concern with illusion, perception, and the physical and psychological impact of color) and with the effects of light experiments on the spectator. At the other end, it impinges on some, though not by any means all, aspects of Color Field painting in its use of brilliant, unmodulated color in retinally stimulating combinations, especially as seen in the art of Larry Poons. Op art came to the forefront of the New York art world in 1965 with William Seitz's exhibition, "The Responsive Eye," at The Museum of Modern Art. Although a relatively wide range of artists were included, among them certain Color Field painters and Hard-Edge abstractionists whom we have already encountered, such as Larry Poons, Morris Louis, Kenneth Noland, and Ellsworth Kelly, it was immediately clear that Op art represented something new. Op art actively engaged the physiology and psychology of seeing with eye-teasing arrangements of color and pattern that seemed to pulsate. Such works generated strong associations with science and technology. Interestingly, although Op art enjoyed much popular interest in the United States, it generated more serious critical acclaim in Europe, with light sculptor Julio Le Parc claiming first prize at the Venice Biennale in 1966.

Optical illusion is not new in the history of art; nor is the overlap it implies between artistic rendition and scientific theories of vision. In the course of the last five hundred years, for example, one thinks back to the discovery, or rediscovery, of linear and atmospheric perspective in the fifteenth century, to the interest of many Impressionist and Post-Impressionist painters, like Georges Seurat, in nineteenth-century theories of color and perception, and to the experimentation of several Bauhaus artists, like László Moholy-Nagy and Josef Albers, with similar questions. In our examination of the Optical and kinetic art of the 1960s we survey some of the approaches taken by artists during this period to forms of art involving optical illusion or some other specific aspect of perception. Of course, as we have already seen, such concerns were not limited to Op and kinetic artists alone.

The Hungarian-French painter VICTOR VASARELY (1908–1997) was the most influential master in the realm of Op art. Although his earlier paintings belonged in the general tradition of Concrete art, Vasarely in the 1940s devoted himself to Optical art and theories of perception. Vasarely's art theories, first presented in his 1955 *Yellow Manifesto,* involved the replacement of traditional easel painting by what he called "kinetic plastics." To him, "painting and sculpture become anachronistic terms: it is more exact to speak of a bi-, tri-, and multidimensional plastic art. We no longer have distinct manifestations of a creative sensibility, but the development of a single plastic sensibility in different spaces." Vasarely sought the abandonment of painting as an individual gesture, the signature of the isolated artist. Art to him in a modern, technical society had to have a social context. He saw the work of art as the artist's original idea rather than as an object consisting of paint on canvas. This idea, realized in terms of flat, geometric-abstract shapes mathematically organized, with standardized colors, flatly applied, could then be projected, reproduced, or multiplied into different forms—murals, books, tapestries, glass, mosaic, slides, films, or television. For the traditional concept of the work of art as a unique object produced by an isolated artist, Vasarely substituted the concept of social art, produced by the artist in full command of modern industrial communication techniques for a mass audience.

Vasarely was a pioneer in the development of almost every optical device for the creation of a new art of visual illusion. His *Photographismes* are black-and-white line drawings or paintings. Some of these were made specifically for reproduction, and in them Vasarely frequently covered the drawing with a transparent plastic sheet. The plastic sheet has the same design as the drawing but in a reverse, negative-positive relationship. When the two drawings—on paper and on plastic—are synchronized, the result is simply a denser version. As the plastic is drawn up and down over the paper, the design changes tangibly before our eyes. Here, literal movement creates the illusion. This and many further devices developed by Vasarely and other Op artists produced refinements of processes long familiar in games of illusion or halls of mirrors.

In his Deep Kinetic Works, Vasarely translated the principle of the plastic drawings into large-scale glass constructions, such as the 1953 *Sorata-T* (fig. 678), a standing triptych, six and one-half by fifteen feet in dimension. The three transparent glass screens may be placed at various angles to create different combinations of the linear patterns. Such art lends itself to monumental statements that Vasarely was able to realize in murals, ceramic walls, and large-scale glass constructions. What he called his "Refractions" involve glass or mirror effects with constantly changing images.

Since Vasarely was aware of the range of optical effects

possible in black and white, a large proportion of his work is limited to these colors, referred to by the artist simply as "B N" (*blanche noir*: white black). It was, nevertheless, in color that the full range of possibilities for Optical painting could be realized, and Vasarely was well aware of this. In the 1960s his color burst out with a variety and brilliance unparalleled in his career. Using small, standard color-shapes—squares, triangles, diamonds, rectangles, circles, sometimes frontalized, sometimes tilted, in flat, brilliant colors against equally strong contrasting color grounds—he set up retinal vibrations that dazzle the eye and bewilder perception (color plate 343, page 576).

The British artist BRIDGET RILEY (b. 1931) and the American artist RICHARD ANUSZKIEWICZ (b. 1930) have been responsible for producing highly acclaimed examples of Op art. Working primarily in black and white and different values of gray, with repeated, serial units frontalized and tilted at various angles, Riley produces extremely effective illusions, seemingly making the picture plane weave and billow before our eyes. Her use of variations in tone (which Vasarely

678. VICTOR VASARELY. *Sorata-T.* 1953.
Triptych, engraved glass slabs, 6'6" x 15'
(2 x 4.6 m). Private collection

679. BRIDGET RILEY. *Drift 2*. 1966. Emulsion on canvas,
7'7½" x 7'5½" (2.3 x 2.3 m).
Albright-Knox Art Gallery, Buffalo

680. RICHARD ANUSZKIEWICZ. *Inflexion*. 1967. Liquitex on canvas, 60 x 60" (152.4 x 152.4 cm). Collection the artist

rejected) accentuates the illusion (fig. 679). By contrast, Anuszkiewicz, a student of Josef Albers, often uses color and straight lines to achieve pulsating visual effects in his painting—frequently defying our ability to distinguish background and foreground in his works or to identify whether a given plane is moving forward or receding. In *Inflexion* (fig. 680) he skillfully reversed the direction of the radiating lines on each side of the square, thus making what at a quick glance looks like a traditional perspective box suddenly begin to turn itself inside out.

Israeli-born **YAACOV AGAM** (Jacob Gipstein) (b. 1928) is identified with a form of Optical painting, or rather painted relief, in which the illusion is created by the movement of the spectator. *Double Metamorphosis II* (figs. 681, 682), one of the largest of these, consists of a raised aluminum surface which is carefully painted in order to change, or metamorphose, into another image. The "double" metamorphosis refers, first, to the process of change and, second, to the existence of three different combinations for the viewer. Viewed from the left, a grid of colorful shapes is seen (fig. 681); viewed from the front, new shapes and colors with less "solidity" are visible (fig. 682); viewed from the right, a pattern of black-and-white lines and circles appears. The device used here is similar to that employed by some Renaissance painters intrigued by the illusionistic possibilities of linear perspective. (Hans Holbein the Younger in his 1533 painting *The Ambassadors* introduced in the foreground an exceedingly foreshortened skull—a memento mori. If one stands at the side of the picture and looks across it, the skull resumes its normal shape.) The interesting fact about Agam's relief paintings, beyond their technical virtuosity, is that each successive view resolves itself effectively.

Op art has also inspired a movement toward allover painting, in which a network or a mosaic of color strokes or dots covers the canvas and seems to expand beyond its limits. Larry Poons, during the 1960s, created a seemingly haphazard but actually meticulously programmed mosaic of small oval shapes that vibrate intensely over a ground of strong color (color plate 334, page 572). The organization of Poons's color-shapes, worked out mathematically on graph paper, was another indication of the tendency toward systems or to what, in the 1960s, was called Systemic painting.

Motion and Light

Two directions explored sporadically since early in the twentieth century gained new impetus in the 1960s. These are motion and light used literally, rather than as painted or sculptured illusions. Duchamp's and Gabo's pioneering experiments in motion and Moholy-Nagy's in motion and light have already been noted. Before World War II (and since, as we shall see), Calder was the one artist who made a major art form of motion. Except for further explorations carried on by Moholy-Nagy and his students during the 1930s and 1940s, the use of light as a medium was limited to variations on "color organs," in which programmed devices of one kind or another produced shifting patterns of colored lights. These originated in experiments carried on in 1922 at the Bauhaus by Ludwig Hirschfeld-Mack (1893–1965) and by the American Thomas Wilfred (1889–1968), a brilliant, creative talent and inventor of the color organ. The last years of the 1960s saw a tremendous revival in the use of light as an art form and in so-called mixed media, where the senses are assaulted by live action, sound, light, and motion pictures at the same time.

Color Plate 325. MARCEL BROODTHAERS.
La tour visuelle. 1966. Glass, wood, and magazine
reproductions, 33½ x 31½" (85 x 80 cm).
The Scottish National Gallery of Modern Art,
Edinburgh

Color plate 326. ROBERT FRANK. *Moving Out.*
1984. Silver-gelatin developed-out prints
with acrylic paint,
9¹⁵⁄₁₆ x 47¹³⁄₁₆" (50.6 x 121.4 cm).
National Gallery of Art,
Washington, D.C.

ROBERT FRANK COLLECTION, ANONYMOUS GIFT

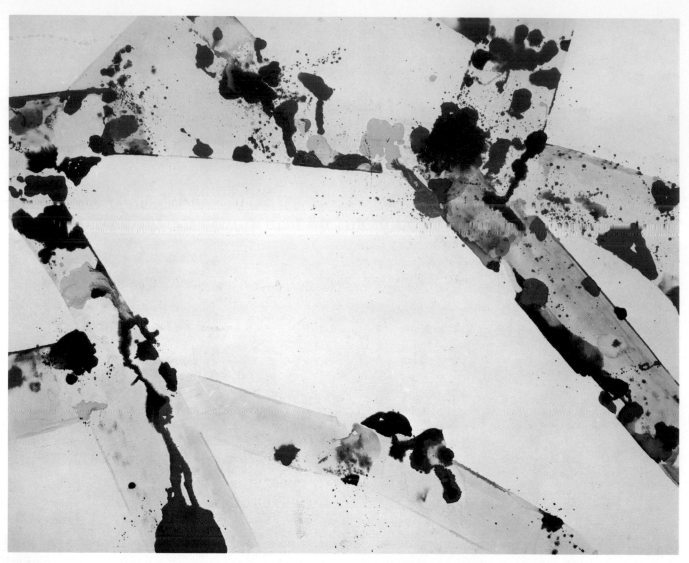

Color plate 327. SAM FRANCIS. *Untitled, No. 11.* 1973. Acrylic on canvas, 8 x 10' (2.4 x 3 m).
André Emmerich Gallery, New York

above: Color plate 328. JOAN MITCHELL. *Land.* 1989. Oil on canvas, 9'2¼" x
13'1½" (2.8 x 4 m). National Gallery of Art, Washington, D.C.

GIFT OF LILA ACHESON WALLACE

right: Color plate 329. HELEN FRANKENTHALER. *Interior Landscape.* 1964.
Acrylic on canvas, 8'8¾" x 7'8¾" (2.7 x 2.4 m).
San Francisco Museum of Modern Art GIFT OF THE WOMEN'S BOARD

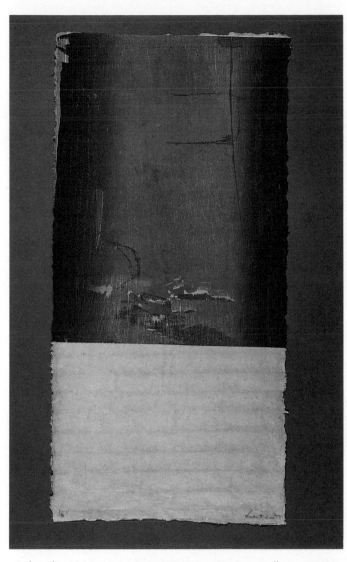

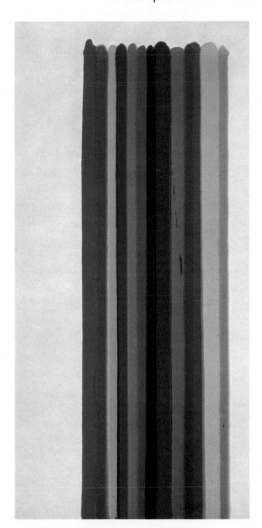

Color plate 330. HELEN FRANKENTHALER. *Essence Mulberry*. 1977. Woodcut, edition of 46, 39½ x 18½" (100.3 x 47 cm). Printed and published by Tyler Graphics Ltd.

© 1977 HELEN FRANKENTHALER/TYLER GRAPHICS LTD.

Color plate 332. MORRIS LOUIS. *Moving In*. 1961. Acrylic on canvas, 7'3½" x 41½" (2.2 m x 105.4 cm). André Emmerich Gallery, New York

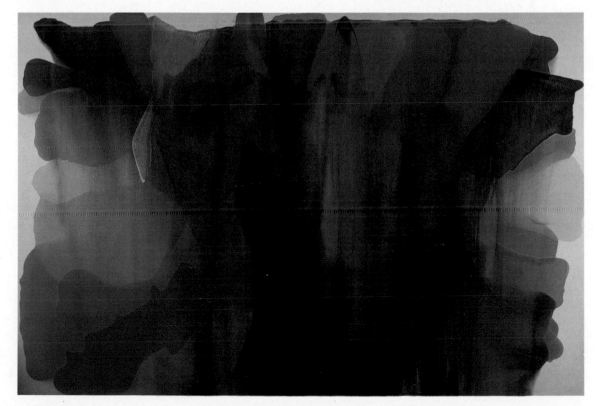

Color plate 331. MORRIS LOUIS. *Kaf*. 1959–60. Acrylic on canvas, 8'4" x 12' (2.5 x 3.7 m). Collection Kimiko and John G. Powers, New York

Color plate 333. JULES OLITSKI. *High a Yellow*. 1967. Acrylic on canvas, 7'8½" x 12'6" (2.3 x 3.8 m).
Whitney Museum of American Art, New York © 1997 JULES OLITSKI/VAGA, NEW YORK, NY

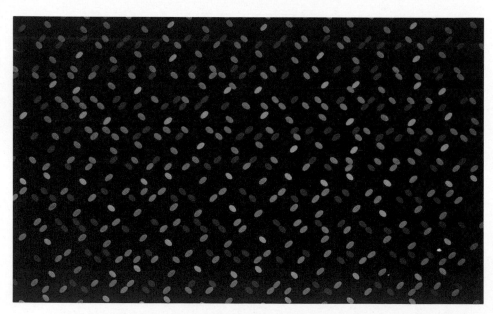

Color plate 334. LARRY POONS. *Nixes Mate*. 1964. Acrylic on canvas, 70" x 9'4"
(177.8 cm x 2.8 m). Formerly collection Robert C. Scull, New York
© 1997 LARRY POONS/VAGA, NEW YORK, NY

Color plate 335. RICHARD DIEBENKORN. *Berkeley No. 22*. 1954. Oil on canvas, 59 x 57"
(149.8 x 144.9 cm). Hirshhorn Museum and
Sculpture Garden, Smithsonian Institution,
Washington, D.C.
REGENTS' COLLECTIONS ACQUISITION PROGRAM

Color plate 336.
RICHARD DIEBENKORN.
Ocean Park No. 54.
1972. Oil on canvas,
8'4" x 6'9" (2.5 x
2.1 m). San Francisco
Museum of Modern Art
GIFT OF THE FRIENDS OF GERALD
NORDLAND

Color plate 337. CY TWOMBLY. *Hero and Leander.* Triptych (Rome), 1981–84. Oil and chalk on canvas, panel one 66⅛" x 6'8¾"
(168 cm x 2.1 m); panels two and three each 68½" x 6'11⅞" (174 cm x 2.1 m).
Collection Galerie Karsten Greve, Cologne, Germany

above: Color plate 338. ELLSWORTH KELLY. *Orange and Green*. 1966. Liquitex (matte varnish) on canvas, 7'4" x 65" (2.2 m x 165.1 cm).
Collection Robert and Jane Meyerhoff, Phoenix, Maryland

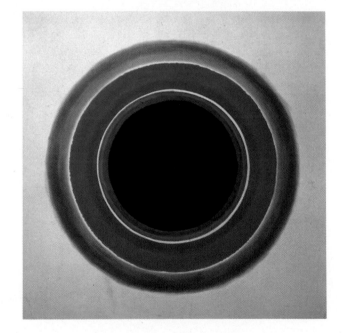

above, right: Color plate 339. ELLSWORTH KELLY. *Untitled (Mandorla)*. 1988. Bronze, 8'5" x 53½" x 21" (2.6 m x 135.9 cm x 53.3 cm). Private collection

right: Color plate 340. KENNETH NOLAND. *A Warm Sound in a Gray Field*. 1961. 6'10½" x 6'9" (2.1 x 2.06 m).
Private collection, New York
© 1997 KENNETH NOLAND/VAGA, NEW YORK, NY

top: Color plate 341. KENNETH NOLAND. *Graded Exposure*. 1967. Acrylic on canvas, 7'4¾" x 19'1" (2.3 x 5.8 m). Collection Mrs. Samuel G. Rautbord, Chicago

above: Color plate 342. AL HELD. *Mantegna's Edge*. 1983. Mural, acrylic on canvas, length 55' (16.8 m). Southland Center, Dallas

left: Color plate 343. VICTOR VASARELY. *Vega Per.* 1969. Oil on canvas, 64 x 64" (162.6 x 162.6 cm). Honolulu Academy of Arts

below, left: Color plate 344. ALEXANDER CALDER. *La Grande Vitesse.* 1969. Painted steel plate, height 55' (16.8 m). Calder Plaza, Vandenberg Center, Grand Rapids, Michigan

above, right: Color plate 345. CHRYSSA. *Ampersand III.* 1966. Neon lights in Plexiglas, height 30¼" (76.8 cm). Collection Robert E. Abrams Family

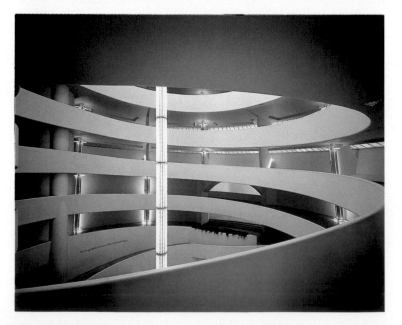

left: Color plate 346. DAN FLAVIN. *Untitled (to Tracy to cele-brate the love of a lifetime),* 1992, expanded version of *Untitled (to Ward Jackson, an old friend and colleague who, during the Fall of 1957 when I finally returned to New York from Washington and joined him to work together in this museum, kindly communicated),* 1971. A system of two alternating modular units in fluorescent light, sixteen white bulbs, each 24" (61 cm) long, four each of pink, yellow, green, and blue, each 8' (2.4 m) long. Solomon R. Guggenheim Museum, New York

Color plate 347. ROBERT IRWIN. *9 Spaces, 9 Trees.* July 7, 1983. Cast concrete planters; Kolorgaard ⅝"
(1.59 cm) aperture, plastic-coated blue fencing; Visuvius plum trees; and Sedum Oreganium ground cover.
Public Safety Building Plaza, Seattle, Washington

SPONSORED BY THE CITY OF SEATTLE, THE SEATTLE ARTS COMMISSION, AND THE NATIONAL ENDOWMENT FOR THE ARTS

Color plate 348. ANTHONY CARO. *Midday.* 1960. Painted steel, 7'7¾" x 35⅞" x 12'1¾"
(2.3 m x 91.1 cm x 3.7 m). The Museum of Modern Art, New York MR. AND MRS. ARTHUR WEISENBERG FUND

Color plate 349. FRANK STELLA. *Jasper's Dilemma*. 1962–63. Alkyd on canvas, 6'5" x 12'10" (2 x 3.9 m).
Collection Alan Power, London © 1997 FRANK STELLA/ARTISTS RIGHTS SOCIETY (ARS), NEW YORK

Color plate 350. FRANK STELLA. *Agbatana III*. 1968. Fluorescent acrylic on canvas, 9'11⅞" x 14'11⅞" (3 x 4.6 m).
Allen Memorial Gallery, Oberlin College, Oberlin, Ohio

RUTH C. ROUSCH FUND FOR CONTEMPORARY ART AND NATIONAL FOUNDATION OF THE ARTS AND HUMANITIES GRANT © 1997 FRANK STELLA/ARTISTS RIGHTS SOCIETY (ARS), NEW YORK

Color plate 351. FRANK STELLA. *The Pequod Meets the Jeroboam: Her Story, Moby Dick Deckle Edges.* 1993. Lithograph, etching, aquatint, relief, and mezzotint, 70 x 65¼" (177.8 x 165.7 cm). Printed and published by Tyler Graphics Ltd.

Color plate 352.
DONALD JUDD.
Untitled (Progression).
Edition of seven, 1965.
Red lacquer on
galvanized iron,
5 x 69 x 8½"
(12.7 x 175.3 x 21.6 cm).
The Saint Louis Art
Museum, Missouri
GIFT OF MR. AND MRS. JOSEPH A. HELMAN
© 1997 ESTATE OF DONALD JUDD/VAGA,
NEW YORK, NY

Color plate 353.
DONALD JUDD.
Untitled. 1977.
Concrete, outer ring
diameter 49'3" (15 m), height
35" (88.9 cm); inner ring
diameter 44'3" (13.5 m),
height 35"–6'10"
(88.9 cm–2.1 m).
Collection the City of Münster,
Germany
© 1997 ESTATE OF DONALD JUDD/VAGA,
NEW YORK, NY

Color plate 354.
SOL LEWITT.
*Wall Drawing No. 652,
on Three Walls,
Continuous Forms with
Color in Washes
Superimposed.* 1990.
Color-ink wash on wall,
approx. 30 x 60'
(9.1 x 18.3 m).
Shown here as installed
temporarily at the Addison
Gallery, Andover,
Massachusetts, 1993.
Now in the Indianapolis
Museum of Art
GIFT OF THE DUDLEY SUTPHIN FAMILY
© 1997 SOL LEWITT/ARTISTS RIGHTS
SOCIETY (ARS), NEW YORK

Color plate 355. CARL ANDRE. *37 Pieces of Work*. Fall 1969. Aluminum, copper, steel, lead, magnesium, and zinc, 1,296 units (216 of each metal), each unit 12 x 12 x ¾" (30.5 x 30.5 x 1.9 cm), overall 36 x 36' (11 x 11 m). Installation view in the rotunda of the Solomon R. Guggenheim Museum, New York, 1970. Now in the Crex Collection, Hallen für neue Kunst, Schaffhausen, Switzerland © 1997 CARL ANDRE/VAGA, NEW YORK, NY

Color plate 356. RICHARD SERRA. *Belts*. 1966–67. Vulcanized rubber, blue neon, 18'1¾" x 6'2⅞" x 17⅜" (5.5 m x 1.9 m x 44 cm). Solomon R. Guggenheim Foundation, New York THE PANZA COLLECTION © 1997 RICHARD SERRA/ARTISTS RIGHTS SOCIETY (ARS), NEW YORK

Color plate 357. AGNES MARTIN. *Night Sea.* 1963. Oil and gold leaf on canvas, 6 x 6' (1.8 x 1.8 m). Private collection

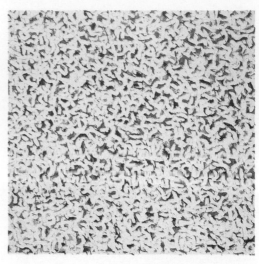

Color plate 358. ROBERT RYMAN. *Untitled.* 1962. Oil on linen canvas, 69½ x 69½" (176.5 x 176.5 cm). Collection the artist

Color plate 359. BRICE MARDEN. *The Dylan Painting.* 1966. Oil and wax on canvas, 60" x 10' (152.4 cm x 3 m). Collection Helen Portugal, New York

© 1997 BRICE MARDEN/ARTISTS RIGHTS SOCIETY (ARS), NEW YORK

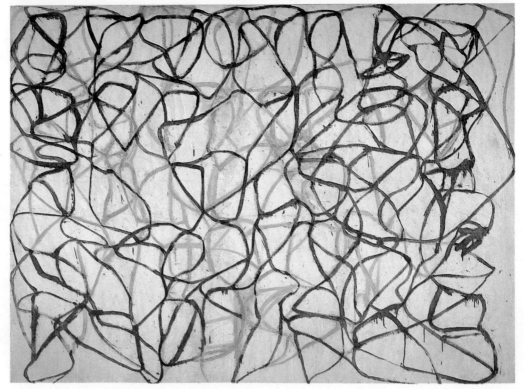

Color plate 360. BRICE MARDEN. *Cold Mountain 5 (Open).* 1988–91. Oil on linen, 9 x 12' (2.7 x 3.7 m). Collection Robert and Jane Meyerhoff, Phoenix, Maryland

© 1997 BRICE MARDEN/ARTISTS RIGHTS SOCIETY (ARS), NEW YORK

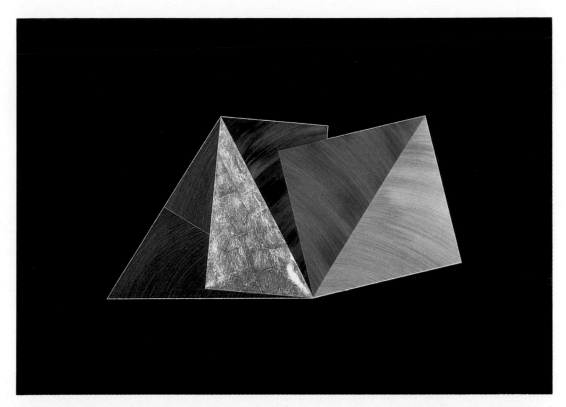

Color plate 361. DOROTHEA ROCKBURNE. *The Glory of Doubt, Pascal.* 1988–89. Watercolor and gold leaf on prepared acetate and watercolor on museum board, 44¹³⁄₁₆" x 6'1¾" (114 cm x 1.9 m). Courtesy André Emmerich Gallery, New York © 1997 DOROTHEA ROCKBURNE/ARTISTS RIGHTS SOCIETY (ARS), NEW YORK

Color plate 362. JO BAER. *Untitled (Wraparound Triptych—Blue, Green, Lavender).* 1969–74. Oil on canvas, each panel 48 x 52" (122 x 132 cm). Courtesy the artist

above: Color plate 363. FRANK LLOYD WRIGHT. Taliesin West, Scottsdale, Arizona. 1937–38

below: Color plate 364. LE CORBUSIER. Interior, Notre-Dame-du-Haut, Ronchamp, France. 1950–54

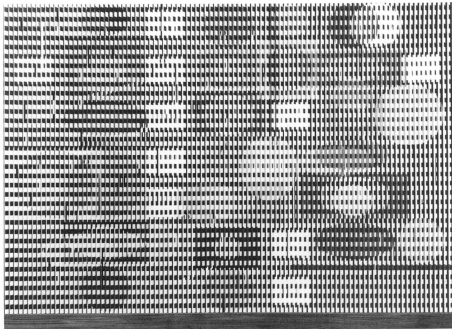

left: 681. YAACOV AGAM. *Double Metamorphosis II.*
1964. Oil on corrugated aluminum, in eleven
parts, 8'10" x 13'2¼" (2.7 x 4 m).
The Museum of Modern Art, New York
GIFT OF MR. AND MRS. GEORGE M. JAFFIN. PHOTOGRAPH © 1995
THE MUSEUM OF MODERN ART, NEW YORK. © 1997 ADAGP, PARIS/ARTISTS
RIGHTS SOCIETY (ARS), NEW YORK

above: 682. YAACOV
AGAM. *Double
Metamorphosis II.*
Frontal view
© 1997 ADAGP, PARIS/ARTISTS RIGHTS
SOCIETY (ARS), NEW YORK

Gabo revived his interest in mechanical motion in the early 1940s, and about 1950 Nicolas Schöffer began using electric motors to activate his constructions. During the 1950s and 1960s the interest in light and motion suddenly snowballed through a series of exhibitions and the organization of a number of experimental groups. Although kinetic and light art has become a worldwide movement, the first great impetus came from Europe, particularly from France, Germany, and Italy. Beginning in 1955 a center for its presentation was the Denise René Gallery in Paris, formerly the stronghold for the promotion of Concrete art. In that year a large exhibition held there included kinetic works by Duchamp, Calder, Vasarely, Agam, Pol Bury, Jean Tinguely, and Jesus Rafaël Soto. The first phase of the new movement climaxed in the great exhibition held sequentially at the Stedelijk Museum in Amsterdam, the Louisiana Museum in Denmark, and the Moderna Museet in Stockholm. This was a vast and somewhat chaotic assembly representative of every aspect of the history of light and motion, actual or illusionistic, back to the origins of the automobile and to Eadweard Muybridge's nineteenth-century photographic studies of human and animal figures in motion. Although the exhibition did not draw any conclusions, it did illustrate dramatically that the previous few years had seen a tremendous acceleration of interest in these problems. The popular curiosity about this art is evinced by the fact that the exhibitions were seen by well over one hundred thousand persons.

During the 1960s exhibitions of light and motion proliferated throughout the world, and the number of artists involved in different aspects increased spectacularly. A few outstanding examples will illustrate the range of possibilities achieved.

Although not usually labeled as a 1960s kinetic artist, **ALEXANDER CALDER** (1898–1976), whose prewar career we reviewed in Chapter 17, merits further discussion here for his important example and continued influence in the area of kinetic sculpture. From the late 1940s he had developed a growing interest in monumental forms, and in the transformation of the mobile to a great architectural-sculptural wind machine whose powerful but precisely balanced metal rods, tipped with large, flat, organically shaped disks, encompass and define large areas of architectural space. Since the mobile, powered by currents of air, could function better outdoors than indoors, Calder began early to explore the possibilities for outdoor mobiles. Critical to the development of these works was his reconceptualization of the mobile as something that could function as an autonomous, standing structure, interrelating stable and mobile forms in an organic whole. Calder, in the late 1950s and 1960s, created many large, standing mobile units that rotate in limited but impressive movement over a generally pyramidal base, essentially neutral in form.

As *The Spiral* (fig. 683) suggests, the development of Calder's moving sculptures bore an important relationship to the development of nonmoving elements in his structures. One of Calder's most impressive achievements of the 1960s occurred in his great "stabiles"—his large-scale, nonmoving, metal-plate constructions, usually painted black. However,

above: 683. ALEXANDER CALDER. *The Spiral*. 1958.
Standing mobile, sheet metal and metal rods, height 30'
(9.1 m). UNESCO, Paris

right: 684. GEORGE RICKEY. *Four Lines Up*. 1967.
Stainless steel, height 16' (4.9 m).
Collection Mr. and Mrs. Robert H. Levi, Lutherville, Maryland

one of the most impressive of these stabiles is the glowing red *La Grande Vitesse* (color plate 344, page 576) in Grand Rapids, Michigan, a triumphantly expansive work with which Calder filled an otherwise deadly public space with Miró-like charm and radiant splendor. While suggesting a great exotic flower in its prime or a Martha Graham dancer in full arabesque, the assemblage of red biomorphic planes boldly displays the structure of its rivets and struts and testifies to the dynamic potential contained within it.

Among the mobilists, one of the most interesting to emerge in the United States since World War II is **GEORGE RICKEY** (b. 1907). Rickey composes long, tapering strips or leaf clusters of stainless steel in a state of balance so delicate that the slightest breeze or touch of the hand sets them into a slow, stately motion or a quivering vibration (fig. 684). During the 1970s Rickey enlarged his vocabulary with the introduction of large-scale rectangular, circular, and triangular forms in aluminum, so precisely balanced that they maintain the possibility of imperceptible and increasingly intricate patterns in movement. His is an art of motion entirely different from but, in its own way, as original as that of Calder.

JESUS RAFAËL SOTO (b. 1923) is not literally a motion artist except insofar as his constructions, consisting of exquisitely arranged metal rods, are sensitive to the point that even a change in atmosphere will start an oscillation (fig. 685). A Venezuelan resident of Paris, Soto began as an illusionistic painter and then moved to a form of linear construction during the 1960s. He has developed a personal statement of great elegance, sometimes translated into a form of large-scale architecture involving spectator participation. In exhibitions at the Solomon R. Guggenheim Museum in New York and the Hirshhorn Museum and Sculpture Garden in Washington, D.C., he created out of his typical plastic strings entire room environments the spectator could enter.

The arts of light and motion encouraged a number of new artists' organizations and manifestos during the late 1950s and 1960s. In 1955, in connection with the exhibition "Movement" at the Denise René Gallery in Paris, Victor Vasarely, the pioneer of French Op art, issued his *Yellow Manifesto,* discussed earlier, outlining his theories of perception and color. Bruno Munari (b. 1907), who was producing kinetic works in Italy as early as 1933, was also an impor-

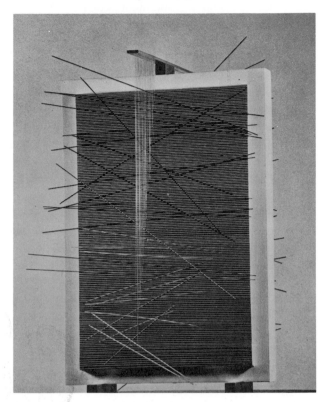

below: 686. JULIO LE PARC. *Continuel-Lumière avec Formes
en Contorsion*. 1966. Motorized aluminum and wood,
6'8" x 48½" x 8" (2 m x 123.2 cm x 20.3 cm).
Howard Wise Gallery, New York

tant theoretician. Very active in Spain, Equipo 57 (founded in 1957) represented a group of artists who worked as an anonymous team in the exploration of motion and vision. This anonymity, deriving from concepts of the social implications of art and also, perhaps, from the examples of scientific or industrial research teams, was evident at the outset in the Group T in Milan, founded in 1959; Group N in Padua, Italy, founded in 1960; and Zero Group in Düsseldorf, Germany, founded in 1957 by the kinetic artist Otto Piene. In most cases the theoretical passion for anonymity soon dimmed, and the artists began to emerge as separate individuals.

Of the light experimentalists, the Argentine **JULIO LE PARC** (b. 1928) has been one of the most imaginative, varied, and visually—even aesthetically—successful in his employment of every conceivable device of light, movement, and illusion (fig. 686). The awarding to him of the painting prize at the 1966 Venice Biennale was not only a recognition of his talents but also an official recognition of the new media. Since 1958 Le Parc has lived in Paris where, in 1960, he was a founder of the Groupe de Recherche d'Art Visuel (GRAV). With its home base at the Denise René Gallery, this group carried on research in light, perception, movement, and illusion. Le Parc became a major bridge between various overlapping but normally separate tendencies in the art of the

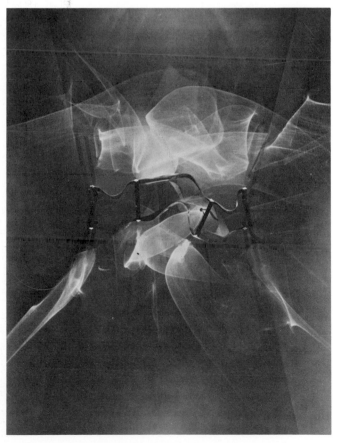

1960s, not only light and movement, but also different forms of optical, illusionistic painting, and programmed art.

The Greek artist CHRYSSA (b. 1933) bridged the gap between Pop art and light sculpture. She first explored emblematic or serial relief forms composed of identical, rhythmically arranged elements of projecting circles or rectangles. Then in the early 1960s she made lead reliefs derived from newspaper printing forms. From this Chryssa passed to a kind of found-object sculpture in which she used fragments of neon signs. The love of industrial or commercial lettering, whether from newspapers or signs, became a persistent aspect of her works. In this she was recording the American scene in a manner analogous to the Pop artists or to the earlier tradition of Stuart Davis and Charles Demuth. Soon her fascination with the possibilities of light, mainly neon tube light, completely took over her construction of elaborate light machines so that, in some curious manner, they resemble contemporary American industrial objects (color plate 345, page 576).

Perhaps the most significant group effort in the United States was EAT—Experiments in Art and Technology—which developed from a series of performances involving Robert Rauschenberg, the engineer Billy Klüver, and the composer John Cage. Presented in 1966, "9 Evenings: Theater and Engineering" involved dance, electronic music, and video projection. These presentations may be traced back to experiments carried on by Cage since the 1940s as well as a collaborative exhibition held at the Denise René Gallery in Paris in 1955. Experiments by the Groupe de Recherche d'Art Visuel, sponsored by the René Gallery during the 1950s and

1960s, also anticipated the EAT group in the utilization of sophisticated technology and effects of light and movement. EAT perhaps extended the collaborative effort furthest in its attraction of large-scale financial support, particularly for its Pepsi-Cola Pavilion at the Osaka World's Fair of 1970.

A number of artists, including DAN FLAVIN (1933–1996), LARRY BELL (b. 1939), and ROBERT IRWIN (b. 1928), combined an interest in technology and light with the principles of Minimalism, to be discussed at greater length in the final section of this chapter. Dan Flavin exploited the fluorescent fixtures with which he worked for their luminosity, but also took advantage of their status as objects themselves (fig. 687). In his works the glass tubes of fluorescent light functioned as sculpture both when illuminated and when unlit, creating different effects. In turn, he recognized that light itself could transform an environment, as his 1992 installation at the newly renovated Guggenheim demonstrated (color plate 346, page 576). In a spirit similar to that of Dan Flavin, Larry Bell produced large vacuum-coated glass boxes that not only intrigue viewers with their shimmering iridescence, but force their audience to confront them as objects, rather than as pure visual effect. The art of Robert Irwin transforms the environment of the viewer even more radically. Working with tightly stretched semitransparent textile scrims, which are lighted from behind, he creates an eerie, isolated environment with an effect that can only be described as hypnotic (fig. 688). One is drawn toward what appears to be an impenetrable void that, upon contemplation, increasingly surrounds the spectator. Irwin's interest in the void may have its roots in the empty gallery of Yves Klein,

687. DAN FLAVIN. *Pink and Gold.* 1968. Light columns. Installation view. Formerly Dwan Gallery, New York

688. ROBERT IRWIN. *Untitled.* 1965–67. Acrylic automobile lacquers on prepared, shaped aluminum with metal tubes.
Walker Art Center, Minneapolis GIFT OF THE ARTIST. © 1997 ROBERT IRWIN/ARTISTS RIGHTS SOCIETY (ARS), NEW YORK

but Irwin greatly developed the earlier concept. Irwin has also worked outdoors, creating public sculptures inspired by the environment in which they are installed and intended to invite the participation of their audiences (color plate 347, page 577).

Modernism vs. Minimalism

It will already have become apparent to the reader of this chapter that several trends characterize painting and sculpture of the 1960s, with a number of artists blurring the distinction between these two media. What has been discussed less explicitly is exactly what was at stake ideologically in the various approaches to art making during this decade, particularly in the realm of sculpture. We have alluded to the influence of the modernist, formalist critic Clement Greenberg with regard to postwar, particularly American, painting. In the early 1960s, with the publication of a collection of his essays, *Art and Culture,* and the continued success of the Abstract Expressionist artists whose work he had championed since the 1940s, his influence was at a height. However, at precisely the same moment that his critical viewpoint prevailed, a number of artists began to question the criteria upon which his valuations of excellence rested. According to Greenberg, the best modern art should continue the historical trajectory of painting since the time of Manet, which he understood to involve a progressive evolution toward flatness as artists became increasing effective at exploiting those qualities peculiar to the medium of paint. Somewhat ironically, for Greenberg even sculpture was submitted to the same criterion of displaying

opticality rather than illusionistic volume. The most serious, protracted assault on these ideas was mounted by those artists who have now come to be known as Minimalists. But in order to understand the nature of this attack and its implications, it is first necessary to consider the work of a paradigmatic 1960s modernist sculptor, Anthony Caro.

As a young sculptor, **ANTHONY CARO** (b. 1924) was particularly influenced by the example of Henry Moore, for whom he worked as an assistant between 1951 and 1953. Subsequently, Caro began to experiment with new materials, particularly metals, and with the process of welding. He was encouraged to pursue this direction by a lengthy trip in the fall of 1959 to the United States, where he met with Clement Greenberg, David Smith, and Kenneth Noland. Taking their advice to heart, Caro returned to England, where he worked actively in a modernist vein, producing sculptures that stressed less their presence as physical objects and more their dematerialized optical appearance. Greenberg's 1961 essay on the sculpture of David Smith clarifies the goal toward which Caro's work strived: "To render substance entirely optical, and form, whether pictorial, sculptural or architectural, as an integral part of ambient space—this brings anti-illusionism full circle. Instead of the illusion of things, we are now offered the illusion of modalities: namely, that matter is incorporeal, weightless and exists only optically like a mirage." In similar fashion, Caro's works often seem to dematerialize in front of the eye, registering their surfaces rather than any sense of bulk. The bright planes of *Midday* (color plate 348, page 577) appear almost to float.

Elements of *Riviera* (fig. 689) seem to hover effortlessly in the air.

The Reaction Against Modernism: Minimalism

If Pop, Op, and kinetic art can be considered departures from the priorities of modernism, as was theorized in the writings of Clement Greenberg and his followers, Minimalism offered a particularly poignant alternative to Greenbergian ideals. The term "Minimalism" was coined in 1965 to characterize an art of extreme visual reduction, but many Minimalist artists resisted its application to their work. Although the pronounced simplicity of these works may seem to be in keeping with Greenbergian principles of truth to the medium and the evolution of flatness—that is, the elimination of the illusion of depth on a two-dimensional surface within the history of modern art—Minimalism actually served to undermine the principles of modernism articulated by Greenberg. Minimalist artists discussed their ideas in print, providing an immediately available intellectual justification for their dissatisfaction with modernist criticism. With Greenberg and his followers advocating that painting embrace those qualities unique to the medium—flatness, pictorial surface, and the effect of pure opticality—and that sculpture aspire to similar goals—an emphasis on surface and optical effect—Minimalist artists began agitating against the supremacy of painting and, above all, began to stress the primacy of the object itself. Greenberg's protégé Michael Fried perceived the seriousness of the conflict when he observed in 1967 that "There is . . . a sharp contrast between the literalist [i.e., Minimalist] espousal of objecthood—almost, it seems, as an art in its own right—and modernist painting's self-imposed imperative that it defeat or suspend its own objecthood through the medium of shape." Minimalist artists denied the modernist tenet that art must function autonomously, and instead considered the importance of a work's environment. They often took into account theories of perceptual psychology and emphasized the importance of the audience's interaction with their pieces, arguing that art need not be absorbed from a single viewpoint in a purely optical fashion. Minimalist art considered not only the eyes of the spectator, but also the body.

Perhaps one of the most significant artists of the 1960s, FRANK STELLA (b. 1936), or rather his art, straddled the line between the modernism advocated by Greenberg and Minimalism, at least as far as adherents of each of these positions were concerned. One of the youngest and most talented of the artists associated with the new American painting, Stella first gained wide recognition in 1960 with a number of works exhibited by New York's Museum of Modern Art during one of its periodic group shows of American artists, on this occasion entitled "Sixteen Americans." The "black" paintings shown there were principally large, vertical rectangles, with an absolutely symmetrical pattern of light lines forming regular, spaced rectangles moving inward from the canvas edge to the cruciform center. Significantly, these lines were not formed by adding white pigment to the canvas—rather, they marked the areas where Stella had *not* laid paint down. Stella's paintings, in their balanced symmetry and repetition of identical motifs, were related to experiments by Minimal or Primary Structure sculptors (color plate 349, page 578). They had, however, a compelling power of their own, which led modernist critics to praise their optical qualities. Over the next few years, using copper or aluminum paint, the artist explored different shapes for the canvas, which were suggested by variations on his rectilinear pattern. In these, Stella used deep framing edges, which gave a particular sense of object solidity to the painting. In 1964 the artist initiated a series of "notched V" compositions, whose shapes resulted from the joining of large chevrons (fig. 690). After some explorations of more coloristic rectangular stripe patterns with at times optical effects, Stella, about 1967, turned to brilliantly chromatic shapes, interrelating protractor-drawn semicircles with rectangular or diamond effects.

689. ANTHONY CARO. *Riviera*. 1971–74. Rusted steel, 10'7" x 27' x 10' (3.2 x 8.2 x 3 m). Collection Mr. and Mrs. Bagley Wright, Seattle

690. FRANK STELLA. *Quathlamba*. 1964. Metallic powder in polymer emulsion on canvas, 6'5" x 13'7" (2 x 4.1 m). Private collection

These "protractor" works (color plate 350, page 578) sometimes suggest abstract triptychs, with their circular tops recalling later Renaissance altarpieces. The apparent stress that these works placed on their qualities as objects was extremely appealing to Minimalists. At the same time, two friends and former schoolmates of Stella's, the Greenbergian critic Michael Fried and the Minimalist artist Carl Andre, to be discussed shortly, both struggled for his allegiances. Fried repeatedly claimed that the artist represented modernist principles; Andre worked closely with him and, when Stella's Black paintings were included in "Sixteen Americans," he wrote a statement about them at his request. Although Stella refused to connect himself firmly with one group or the other, his work continued to undermine any strict division between painting and sculpture.

In the 1970s Stella moved increasingly toward a form of three-dimensional painted relief, bold in color and dynamic in structure (fig. 691). It became increasingly difficult to tell if the works were paintings or sculptures, for by now the artist was using lacquer and oil colors on aluminum bases.

Since then, Stella has remained committed to his abstractions, even as figurative work has increasingly received significant critical attention. He offered a passionate defense of abstract art in 1983 when he gave a series of six lectures at Harvard University as the Charles Eliot Norton Professor of Poetry. However, his most recent works seem to challenge the distinction between abstraction and representation as Stella strives to produce reliefs that convey narrative significance in a nonfigurative fashion and yet incorporate forms that are suggestive of more naturalistic representation (color plate 351, page 579). While one cannot strictly classify Stella as an antimodernist, his works have clearly challenged the narrow critical categories defined by Greenbergian modernism both in terms of form and content. As the art historians Charles Harrison and Paul Wood have recently pointed out, Stella's overt reference to Herman Melville's *Moby Dick* in a number of works offering visual interpretations of poignant themes and moments in the novel apparently ignores the Greenbergian demand that art be autonomous from literary content. Nor does Stella's recent work fit tidily into the Greenbergian categories of painting and sculpture. If Stella's art no longer retains obvious visual affinities with 1960s

691. FRANK STELLA. Exhibition, 1971. Lawrence Rubin Gallery, New York

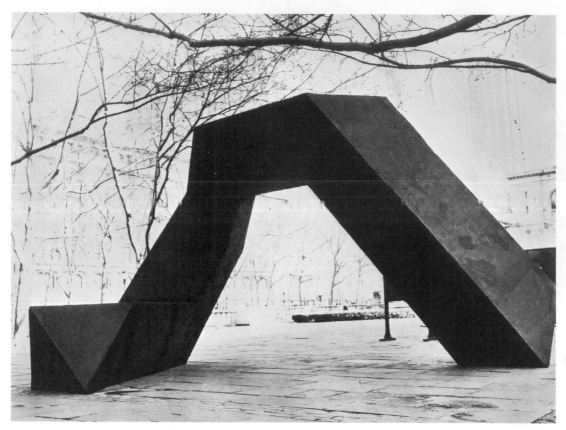

Minimalism, it nevertheless continues to reflect the legacy of the shift in critical discourse that Minimalism marked.

Despite the fact that **TONY SMITH** (1912–1980) matured artistically during the forties and fifties, his most important impact was felt during the 1960s. The sculptor came to be celebrated by other Minimalists for pronouncements about the nature of art in general and about his art in particular that helped to define the new approach of the group as a whole. Perhaps most influential was Smith's description of a nighttime drive down an unfinished segment of the New Jersey Turnpike in the 1950s. According to Smith:

> The drive was a revealing experience. . . . [I]t did something for me that art had never done. At first I didn't know what it was, but its effect was to liberate me from many of the views I had about art. . . . I thought to myself, it ought to be clear that's the end of art. Most painting looks pretty pictorial after that. There is no way you can frame it, you just have to experience it.

The notion that one had to experience art, not merely imbibe its significance by standing in front of it, manifested itself in Smith's large, abstract pieces. Smith's *Cigarette*, 1961–66, requires that the viewer walk around it and through it to fully absorb it (fig. 692). Despite its association with an everyday, disposable object, Smith's sculpture is by no means pictorial. Nor can it be easily cast aside or discarded. Smith's work subverts traditional categories of sculpture and experience. Smith's *Die* of 1962 (fig. 693) has perhaps more obvious affiliations with the common object from which its name derives. However, its scale has been radically increased and the cube bears none of a die's traditional markings. *Die* is not an object that can be cast thoughtlessly by a human hand. Instead, the steel sculpture cannot even be adequately

perceived, at least from a single, frontal viewpoint. Again, the spectator must move around the object, unable to grasp it in its entirety, for the closer one gets, the larger the sculpture grows, reminding its audience of the limits of perception itself.

DONALD JUDD (1928–1994) served as one of Minimalism's most important sculptors and theorists. With his art

above: 694. DONALD JUDD. *Untitled*. 1965. Galvanized iron, seven boxes, each 9 x 40 x 31" (22.9 x 101.6 x 78.7 cm) [9" (22.9 cm) between each box]. Hirshhorn Museum and Sculpture Garden, Smithsonian Institution, Washington, D.C.

above, right: 695. RONALD BLADEN. *The X*. 1967. Painted wood model to be made in metal, 22'8" x 24'6" x 12'6" (6.9 x 7.5 x 3.8 m). Fischbach Gallery, New York

criticism and, in particular, his 1965 essay "Specific Objects," Judd helped to define the convictions behind the Minimalist questioning of the traditional categories of painting and sculpture. As he wrote in 1965, "A work can be as powerful as it can be thought to be. Actual space is intrinsically more powerful and specific than paint on a flat surface. . . . A work need only be interesting." For Judd, the "specific object" could not be classified as either a painting or a sculpture, or even precisely described prior to its making, except in principle. The specific object was less about creating particular structures and more about an attitude toward art making.

Through his sculpture, Judd carried the objective attitude to a point of extreme precision. He repeated identical units, often quadrangular, at regular intervals (fig. 694; color plate 352, page 580). These are made of galvanized iron or aluminum, occasionally with Plexiglas. Although Judd sometimes painted the aluminum in strong colors, he at first used the galvanized iron in its original unpainted form, something that seemed to emphasize its neutrality. Progressively, however, Judd used color more frequently in his work, which he

began to have professionally manufactured in the mid-1960s.

Judd vehemently insisted that Minimal sculpture—most specifically his own—constituted a direction essentially different from earlier Constructivism or rectilinear abstract painting. The difference, as he saw it, lay in his search for an absolute unity or wholeness through repetition of identical units in absolute symmetry. Even Mondrian "composed" a picture by asymmetrical balance of differing color areas.

Judd's works raise fundamental questions concerning the nature and even the validity of the work of art, the nature of the aesthetic experience, the nature of space, and the nature of sculptural form. Progressively Judd's works, like those of most Minimalist sculptors, expanded in scale and impressiveness. In 1974 Judd introduced spatial dividers into his sculptures. Three years later he began to construct large-scale works in cement to be installed in the landscape, creating structures of order and harmony that resonated with their environment (color plate 353, page 580).

RONALD BLADEN (1918–1988), a Canadian, used monumental, architectural forms, frequently painted black, that loom up like great barriers in the space they occupy (fig. 695). Like Tony Smith, with whose work his has a certain affinity, Bladen made his mock-ups in painted wood, since the cost of executing these vast structures in metal would be exorbitant. Very often Minimalist artists envisioning large-scale structures must await patrons for the final realization of their projects.

ROBERT MORRIS (b. 1931), who during the 1960s was associated with Minimal sculpture, has since that time become a leader in a wide variety of sculptural, environmen-

696. ROBERT MORRIS. Exhibition, Green Gallery, New York, 1964. Left to right: *Untitled (Cloud)*, 1964, painted plywood; *Untitled (Boiler)*, 1964, painted plywood; *Untitled (Floor Beam)*, 1964, painted plywood; *Untitled (Table)*, 1964, painted plywood

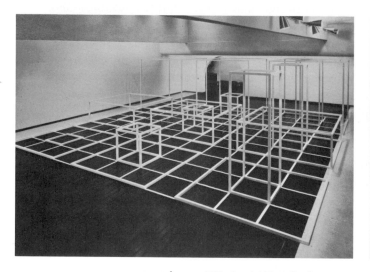

above: 697. SOL LEWITT. *Sculpture Series "A."* 1967. Installation view, Dwan Gallery, Los Angeles

above, right: 698. SOL LEWITT. *All Combinations of Arcs from Four Corners, Arcs from Four Sides, Straight Lines, Not-Straight Lines, and Broken Lines* (detail). 1976. White crayon on black wall. Installation view, John Weber Gallery, New York

tal, Conceptual, and Post-Minimalist forms. A student of Tony Smith, Morris, like Donald Judd, proved to be an important polemicist of Minimal art. His "Notes on Sculpture," a four-part collection of essays, provided an important statement about the heritage to which the Minimalists felt themselves heir—a tradition marked by the Constructivists, the work of David Smith, and the paintings of Mondrian— but perhaps even more important, also about the gestalt principles he believed to be at stake in Minimal sculpture. Gestalt theory focuses on human perception, describing our ability to understand certain visual relationships as shapes or units. Through his familiarity with these theories, Morris posited that one's body has a fundamental link to one's perception and experience of sculpture. According to Morris, "One knows immediately what is smaller and what is larger than himself. It is obvious, yet important, to take note of the fact that things smaller than ourselves are seen differently than things larger." Morris's sensitivity to the impact of scale and the corresponding implications of publicness or privateness led to his pioneering work in the organization of entire rooms into a unity of sculptural mass and space.

The monumental size of much Minimal sculpture led inevitably during the 1960s and 1970s to the concept of sculpture designed for a specific space or place. This in turn resulted in ideas such as the use of the gallery space itself as an element in an architectural-sculptural organization. An early experiment in this direction was Morris's 1964 exhibition at the Green Gallery in New York, where large, geometric sculptural modules were integrated within the room, whose space became an element of the total sculpture (fig. 696). The idea, of course, had been anticipated by such varied sculptural environments as Schwitters's *Merzbau* of 1925 and Yves Klein's *Le Vide* exhibition of the late 1950s in which the spectators provided the sculptural accents for the empty, white-walled gallery. The implications of sculpture in place have been explored and expanded enormously during the 1970s and extended to environments based on painting, Conceptualism, and Earthworks.

SOL LEWITT (b. 1928) is identified with a kind of serial Minimalism consisting of open, identical cubes integrated to form proportionately larger units. These cubes increase in scale until they dominate the architectural space that contains them (fig. 697). In such works, the physical essence is only the outline of the cubes, while the cubes themselves are empty space. Like Judd and Morris, LeWitt made important theoretical contributions to artistic practice during the later 1960s. His "Paragraphs on Conceptual Art," published in 1967, reiterated the discomfort of Minimalist artists with Greenbergian standards of quality and set the stage for the recognition of yet another digression from the modernism endorsed by Greenberg—Conceptual art, to be discussed in Chapter 24. LeWitt's 1967 essay argued that the most important aspect of a work of art was the idea behind it rather than its form. "The idea becomes a machine that makes the art," LeWitt stated. LeWitt's 1968 *Box in the Hole* put into practice his thesis. Created in the Netherlands, this work consisted of a metal cube that the artist buried, covered over, and preserved in a series of recorded photographs.

During the early 1970s LeWitt exploited gallery space in another way—by drawing directly on the walls. These drawings—frequently destroyed at the close of each exhibition—were generally accumulations of rectangles, drawn with a ruler and pencil, toned to various degrees of gray, and accompanied by written specifications. These specifications ensured that a given work could be executed by assistants in the artist's stead. Following LeWitt's conception, the internal lines of each rectangle, creating the tones, might be diagonal as well as vertical or horizontal, and the result was frequently

a geometric abstraction of considerable beauty (fig. 698). In recent years, LeWitt has continued his practice of creating wall installations, and the institution in which LeWitt realizes his work often preserves it (color plate 354, page 580). However, even if the walls are repainted, the piece is not destroyed; it continues to exist as a well-specified idea and can be reinstalled by following the artist's instructions.

Influenced early by the ideas of his friend Frank Stella and the sculptor Brancusi, the early work of **CARL ANDRE** (b. 1935) consisted of vertical wooden sculptures, given form by the use of a saw. By the early 1960s, Andre had moved away from carving to the construction of sculptures using identical units. The arrangement of the wooden units in *Pyramid* (fig. 699), first created in 1959 and later reconstructed in 1970, suggests a carved form, although the undulations of the surface were not created by the use of a saw. Other pieces from the early sixties, like *Well* (fig. 699), are more literal in their arrangement. After 1965 Andre began to make horizontally

699. CARL ANDRE. Background, *Pyramid*, 1959/70, wood; middle ground, *Well*, 1964/70, wood; foreground, *Lever*, 1966, firebrick. Installation view. Exhibition at the Solomon R. Guggenheim Museum, New York, 1970. Photograph courtesy Paula Cooper Gallery, New York

oriented sculptures, known as floorpieces. *Lever* (fig. 699), of 1966, initiated this new phase in his work. These pieces were made out of rugged, industrial materials not traditionally used in fine art. Combined timbers extended horizontally, bricks, styrofoam units, or identical metal squares assembled on the floor were sculptures intended to be walked on. One of Andre's most intricate floorpieces, *37 Pieces of Work*, was executed in 1969 on the occasion of his one-man show at the Guggenheim (color plate 355, page 581). The work consisted of a combination of tiles made from six metals: aluminum, copper, steel, magnesium, lead, and zinc. The tiles were a foot square and three-eighths of an inch thick. Each metal piece was placed first into a six-by-six-foot square and then used to create accompanying squares by being alternated with one other metal until every combination had been achieved. The title for the piece comes from the thirty-six-square pattern the metals formed plus the square that encompassed the whole. The work was particularly appropriate for the Guggenheim, where viewers could gaze down on it as they ascended or descended the museum's spiral ramp.

Significantly, since the mid-sixties Andre's sculpture has not been constructed in the studio, but rather in the exhibition space. The idea behind the pieces need only be realized where they are intended for display.

In contrast, **RICHARD SERRA** (b. 1939), after experimenting with different materials such as sheets of vulcanized rubber, created sculptures consisting of enormously heavy sheets of steel or lead that balanced against or on top of one another. Despite the various forms that Serra's work takes, it consistently interrogates and exploits the natural form of the material itself. Rather than manipulate his media, Serra looks to their own physical properties for inspiration. Serra's *Belts*, for example, take their form from the simple act of hanging the rubber loops on the wall (color plate 356, page 581). The incorporation of a neon tube into the work heightens one's awareness of the intricacies of the forms the draped rubber has created. Serra's *One Ton Prop (House of Cards)* (fig. 700) balances four five-hundred-pound lead sheets against one another. The seemingly casual arrangement of these slabs paradoxically accentuates their weight and communicates a sense of dangerous but exciting precariousness—heightened by the title's reference to the ephemeral, practically weightless house of cards it emulates. Like Andre, Serra has also been interested in a form of "scatter sculpture," in which series of torn lead sheets are scattered on the floor or molten lead is splashed along the base of a wall. Throughout the 1970s and 1980s Serra progressively enlarged the scale of his sculptures, took them out-of-doors, and combined his sheets of steel with a landscape environment, thus effecting the inevitable transition from sculpture placed in an interior environment to sculpture that becomes part of a landscape (see fig. 800).

Minimalism proper may seem antithetical to painting, due to the repeated emphasis of its adherents on the importance of the departure from traditional categories. However, a number of accomplished painters did emerge whose approach to their art links them with the sculptural

700. RICHARD SERRA. *One Ton Prop (House of Cards)*. 1969.
Lead antimony, four plates, 48 x 48 x 1" (122 x 122 x 2.54 cm).
The Museum of Modern Art, New York

701. AGNES MARTIN. *Flowers in the Wind*. 1963.
Oil on canvas, 6'3" x 6'3" (1.9 x 1.9 m).
Thomas Ammann Fine Arts, Zürich

Minimalists. These painters refuse to take the medium for granted and carefully deliberate how to engage it. Their work is distinguished not only by great attention to the painted surface, but also by recognition of the painting as a discrete object in and of itself.

Among the Minimalists, the senior painter remains **AGNES MARTIN** (b. 1912), who refined her art over many years, during which time she progressed from rather traditional still lifes to Gorky-like biomorphic abstractions, before arriving in the early 1960s at her mature, mandarin distillations of pure form. This occurred in New York City, where the artist had moved from New Mexico, taking with her a haunting memory of the desert's powdery air and light. This she evoked by the improbable process of honing and purifying her means until they consisted of nothing but large, square canvases gridded all over with lines so delicately defined and subtly spaced as to suggest not austere, cerebral geometry but trembling, spiritual vibrations, what Lawrence Alloway characterized as "a veil, a shadow, a bloom" (fig. 701; color plate 357, page 582). Declaring their physical realities and limitations yet mysteriously intangible, intellectually derived but romantic in feeling and effect, the paintings of Agnes Martin are the product of a mind and a sensibility steeped not only in the meditative, holistic forms of Reinhardt, Rothko, and Newman, but also in the paintings of Paul Klee and the landscapes, as well as the poetry, of Old China. While using Hard-Edge structure in a visually self-dissolving or -contradicting manner, Martin has no interest whatever in the retinal games played by the Op masters. Nor has she ever aspired to the heroics of the Abstract Expressionists. When Martin felt that she had lost her clarity of vision in 1967, she ceased painting and left New York, returning to New Mexico in 1968. After a period of solitude and contemplation, Martin resumed painting in the early 1970s. Her images retained their subtle, geometric character, but her colors became even softer and more luminous, making visible the artist's sense of life's essence as a timeless, shadowy emanation. "When I cover the square with rectangles," she explained in 1967, "it lightens the weight of the square, destroys its power."

Since the beginning of his career as an artist, **ROBERT RYMAN** (b. 1930) has been interested in "how paint worked." His fascination with paint extended from the way in which it was applied to the way in which it interacted with its support to the way in which various types of paint worked together. In his earliest work, Ryman explored a range of colors, gradually giving way to his exclusive use of white paint. *Untitled* of 1962 (color plate 358, page 582) represents one of the last instances of color in his work. Here multiple white strokes are layered over a background of blue and red, literally suppressing the bright hues while simultaneously reacting to them. By the mid-1960s the artist had dedicated himself to the use of white, believing that this color more than any other could highlight his manipulation of paint itself. By way of elucidating his great "silent" all-white paintings (fig. 702), Ryman said in a frequently quoted statement, "It's not a question of what to paint but how to paint." With these words he declared his commitment to pure painting. "To make" has often figured in Ryman's conversation, for he considers "making" a matter of knowing the language of his materials—canvas, steel, cardboard, paper, wood, fiberglass, Mylar interacting with oil, tempera, acrylic, epoxy, enamel—and of exploiting its syntax so that his paintings come alive with their own story. In *Classico III*, Ryman applied white polymer paint in an even film to twelve rectangles of hand-made Classico paper precisely assembled to form a larger rectangle gridded by shadows between the smaller units. With these positioned slightly off-center on a white ground, three types of rectangles in different scales and relationships echo and interact with one another, thus reelaborating the literal while seducing the eye by testing its visual acuity. Thus, in Ryman's art, monochromatic reductiveness has resulted in a subtle recomplication, paradoxically arising from the artist's sensibility to means and his consequent unwillingness to take the painting's support for granted. In *Paramount* (fig. 703) the artist has left the margins, the edges, of the canvas unpainted, drawing attention simultaneously to the paint and to the linen surface upon which it rests. The artist's attention to detail is such that even the metal fasteners that join the painting to the wall are considered part of the work. He carefully considers the relationship of the painting to the wall, considering the height at which it is hung and the distance at which it projects forward. By such rigorous reconsideration and elevation of material detail, by attention to issues of optical perception, as well as by the beauty of his painting touch, Ryman invests his pictures with a lyrical presence, all while pursuing a stern Minimalist program of precision and purification.

702. ROBERT RYMAN. *Classico III*. 1968. Polymer on paper, 7'9" x 7'5" (2.4 x 2.3 m). Stedelijk Museum, Amsterdam

703. ROBERT RYMAN. *Paramount*. 1981. Oil on linen with metal fasteners, 7'4" x 7' (2.2 x 2.1 m). Private collection. Courtesy Thomas Ammann Fine Arts, Zürich

In his reaction against the perceived excess of Abstract Expressionism, **ROBERT MANGOLD** (b. 1937), a student of the Hard-Edge painter Al Held, stressed the factuality, rather than the ineffability, of art by creating surfaces so perdurably hard, so industrially finished, and so eccentrically shaped that the object quality of the work could not be denied. The object quality of the work was indeed emphasized in a number of works in which the artist joined canvases, drawing attention to the importance of edges and creating a play between the real line formed by the junction of the two (or more) parts and the drawn line laid down by the artist. In paintings of the late sixties and early seventies, aerated, decorative color accentuated the neutrality of monotone grounds, while an "error" in the overcalculated, mechanical drawing subverted the dominance of the literal overall shape of the canvas from within (fig. 704). Destabilized by elusive color

and the imperfection of internal pattern, stringent, self-referring formalism gave way to spreading openness and unpredictability reminiscent of the outside world of nature and humanity. And so Mangold too struck a balance between impersonality and individualism, thereby bringing a welcome warmth to the pervasive cool of Minimalist aesthetics.

The precocious **BRICE MARDEN** (b. 1938) had hardly graduated from the Yale School of Art, where he too worked with Al Held, when he created his signature arrangement of rectangular panels combined in often symbolic order and painted with dense monochrome fields of beeswax mixed with oil (color plate 359, page 582). During his career, Marden has become a virtuoso in his ability to balance color, surface, and shape throughout a continuously developed and extended series of variations, progressing always toward ever-more quintessential possibilities within a set of purposeful restrictions. In

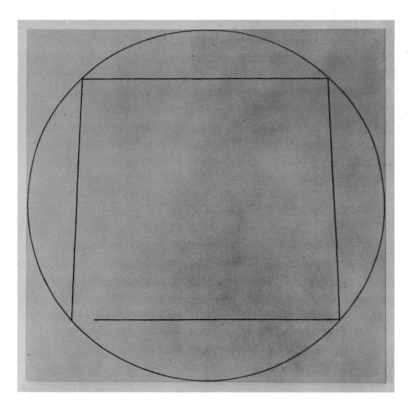

704. ROBERT MANGOLD. *Untitled*. 1973. Acrylic and pencil on Masonite,
24 x 24" (61 x 61 cm). John Weber Gallery, New York

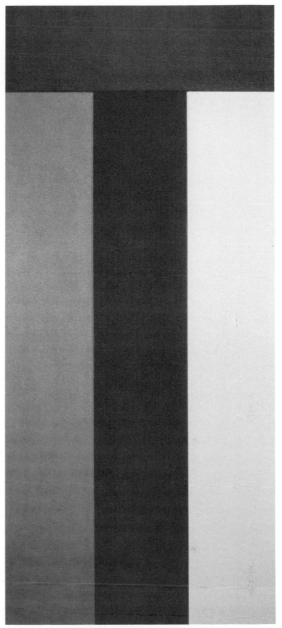

705. BRICE MARDEN. *Elements III*. 1983–84.
Oil on canvas, 7' x 36" (2.1 m x 91.4 cm).
Collection Douglas Cramer, Los Angeles

the late 1970s Marden began to introduce a combination of vertical and horizontal planes within his paintings (fig. 705). Significantly, despite the abstract quality of his images, Marden continues to negotiate an underlying relationship between his painting and nature. His series of Elements paintings, undertaken in 1981, incorporate red, yellow, blue, and green, colors that medieval alchemists used to symbolize, respectively, fire, air, water, and earth. More recently, Marden has introduced a curving, calligraphic line into his paintings. His study of Chinese calligraphy and corresponding interest in Asian culture inform his Cold Mountain paintings, named in honor of the ninth-century Chinese poet whose name comes from his sacred dwelling spot (color plate 360, page 582). The layering and interplay of lines and pigment comment on the process of painting itself. The rhythmic, allover disposition of the calligraphic marks produces a sense of balance, harmony, and precision much in keeping with his earlier painting. Here again, then, we find the Minimalist paradox of extreme simplicity capable of rewarding sustained contemplation with revelations of unsuspected spiritual complexity or sheer hedonist delectation.

Drawing directly on the wall or attaching thereto a flattened, geometric shape created merely by folding brown wrapping paper, **DOROTHEA ROCKBURNE** (b. 1921) would seem to have so reduced her means as to take Minimalism over the line into *Arte povera*, an Italian movement characterized by its use of humble materials (fig. 706). However, the direction proved quite different, fixed by the artist's preoccupation with the mathematics of set theory as an intellec-

tually pure strategy for creating an intricate interplay of simple geometric forms and real space. So dependent was this art on process itself that Rockburne created it directly on the gallery wall, thereby condemning her pieces to a poignant, even romantic life of fragility and impermanence. Subsequently, she found it possible to work in more durable materials and installations, as well as with richer elements, such as color and texture (color plate 361, page 583). Still, Rockburne remains loyal to her fundamental principle of "making parts that form units that go together to make larger units." And the logical yet unexpected way the scientifically derived but imaginative folding produces flat, prismatic shapes—built up, subdivided, creased, and integrated through drawing into the adjacent space—ends by seducing the eye while engaging the mind. It also reveals why the chal-

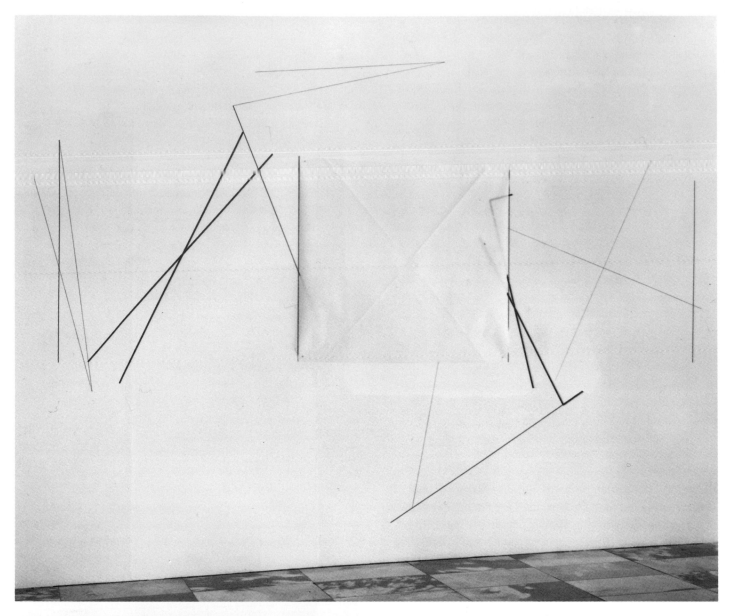

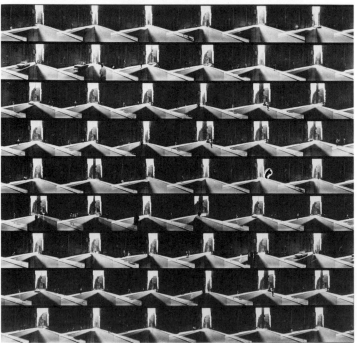

above: 706. DOROTHEA ROCKBURNE. *Indication Drawing Series: Neighborhood* from the wall series "Drawing Which Makes Itself." 1973. Wall drawing: pencil, and colored pencil with vellum, 13'4" x 8'4" (4.1 x 2.5 m). The Museum of Modern Art, New York GIFT OF J. FREDERICK BYERS III. © 1997 DOROTHEA ROCKBURNE/ARTISTS RIGHTS SOCIETY (ARS), NEW YORK

left: 707. RAY K. METZKER. *Untitled (Wanamaker's)*. 1966–67. Multiple printing, 32¼ x 34¾" (82 x 89.3 cm). The Museum of Modern Art, New York PURCHASE. COPY PRINT © 1995 THE MUSEUM OF MODERN ART, NEW YORK

lenge of working within Minimalism's puritanically self-denying regimen appealed to some of the most gifted artists to emerge in the 1960s, offering them variables as elementary as the themes of Beethoven and no less subject to an infinity of permutations and combinations.

During the 1960s, **JO BAER** (b. 1929) eliminated the brushwork and painterly texture of her earlier work to create pieces whose smooth surfaces showed little evidence of the artist's hand. By reducing the visual interest of the surface itself, Baer believed that the spectator could focus on the image as a whole. Like Robert Morris, Baer incorporated the principles of gestalt psychology into her work, using her images to test the limits of human perception, and thus drawing attention to the object nature of the canvas itself. In 1969 Baer began her Wraparound series, literally wrapping a painted black band outlined in color around the edge of the stretcher (color plate 362, page 583). Looking at the work from a certain angle, the viewer can perceive the whole band, but upon closer inspection it becomes evident that the flat solid that one perceived is actually painted on two adjacent surfaces. The canvas is no longer purely surface, but a three-dimensional entity, relying upon all of its surfaces to produce its particular effect. The colored bands in *Untitled (Wraparound Triptych—Blue, Green, Lavender)* play an extremely important role, for it is the borders of an image that first signal the human brain to the presence of a shape.

If Minimalism can be associated with a particular style, then its visual achievements, if not its theoretical principles, can be associated with photography as well.

The serialism, the interest in process as a key source of content, and the search for the complex within the simple, unified image—all fundamental to the Minimalist aesthetic—can be found in the fieldlike collages created by **RAY K. METZKER** (b. 1931), a former student of the influential Harry Callahan at the Institute of Design in Chicago (fig. 707). But like most photographers in the straight tradition, Metzker begins with a subject in the phenomenal world. Having photographed this not as a single, fixed image but rather as a series of related moments, he then combines, repeats, juxtaposes, and superimposes the shots until he has achieved a composite, gridlike organization reminiscent of Minimalist painting, a single visual entity in which the whole is different and more rewarding than its parts. Metzker says of his work, "Where photography has been primarily a process of selection and extraction, I wish to investigate the possibilities of synthesis."

BERND BECHER (b. 1931) and **HILLA BECHER** (b. 1934), German photographers married in 1961, together developed the documentary style for which they have become known. The Bechers turn their attention to the industrial transformation of the Western world, taking stark, closely cropped photographs of the architectural structures that index this change. Their photographs are frequently arranged into series of visually similar buildings and hung together in the format of a grid (fig. 708). Each photograph functions as an independent unit, but the virtually uniform lighting and the similarity in form and viewpoint produce an effect of serial repetition that could be compared to the sculptures of Judd or Andre. Unlike many Minimalist artists, however, the Bechers are not concerned with artistic effect alone, but use their photographs to raise political and social issues related to industrialization.

708. BERND AND HILLA BECHER. *Water Towers.* 1980. Black-and-white photographs, 61¼ x 49¼" (155.6 x 125.1 cm). Courtesy Sonnabend Gallery, New York

As the 1960s faded into the 1970s the innovations of Minimalism continued to be felt. Artists like Judd, Morris, and LeWitt helped lead a successful revolt against the dominance of modernism, as codified by Greenberg, demonstrating the legitimacy of art forms that were not part of this tradition. Conceptual art, Performance art, Process art, Earth art, and site-specific work are among the many new approaches to art making spawned, at least in part, by Minimalism. Furthermore, the interest these artists demonstrated in theoretical concerns from outside the realm of art and their ability to combine philosophical discourse with the creation of art helped to transform the relationship between the practice of art making and critical theory, paving the way for Postmodernism. In subsequent decades, both artists and critics have brought extra-artistic philosophical and literary theories to bear on the creation and discussion of meaningful art. The implications of this development will be discussed more fully in subsequent chapters.

23 The Second Wave of International Style Architecture

The achievements of modern architecture in the first third of the century were a series of relatively isolated experiments: Prairie Style, Secession, Deutscher Werkbund, de Stijl, Expressionism, Bauhaus, and International Style. Of all these initiatives of modernism, International Style was by far the most enduring, and, almost inevitably, was eventually the most diluted.

With the exception of new concepts still developing in the field of engineering, by 1930 many of the ideas of twentieth-century architecture had been stated in one form or another. The early 1930s saw a degree of retrogression of styles, particularly in such communist and fascist countries as Russia and Germany. In those countries and, to a lesser degree, in Italy, a ponderous Neoclassicist style took the place of innovation. Such a style was also used for Italian government buildings (although free experiment, particularly in engineering, was not entirely discouraged). Even in France, despite the achievements of Le Corbusier, official buildings, such as the 1937 Museum of Modern Art in Paris, frequently reverted to a form of Neoclassicism. (In contrast, during this period and since, Finland and the Scandinavian countries—Sweden, Norway, and Denmark—made notable contributions to modern design in architecture, furniture, crafts, and industrial products.)

In the United States, modernist continental European architecture was not well received, certainly not in American schools of architecture. In the 1930s most were still teaching in the Beaux-Arts tradition. Things began to change in about 1931, however, after an exhibition organized by the architectural historian Henry-Russell Hitchcock and the architect Philip Johnson was held in the newly opened Museum of Modern Art in New York. (In 1932, in collaboration with Hitchcock, Johnson wrote the catalog *The International Style*.) "The International Style: Architecture Since 1922" included examples of projects from fifteen countries. The term "International Style" was itself notable, for what had begun a decade earlier as the intimation of a new spirit of restless originality had crystallized into a style that the show's curators had been able to identify in buildings from many countries around the world. By the mid-1930s, with the wider dissemination of the principles of the new architecture, even more countries had at least one or two buildings that were clearly influenced by the designs of such early International Style pioneers as Le Corbusier and Mies van der Rohe. The spread of the modern movement in the early 1930s was furthered by private patronage and competitions, as well as by municipal and commercial bodies.

Not coincidentally, as in all the other arts and sciences in the mid-1930s, the United States benefited from the immigration of many scientists, artists, and architects escaping from Nazi repression of the avant-garde. Walter Gropius (see fig. 410) quit the Bauhaus in 1928 and moved to England in 1934, where he stayed for three years and collaborated with British International Style architect Maxwell Fry in the design of Impington Village College near Cambridge. In 1937 he left for the United States, to become professor and then chairman of the Department of Architecture at Harvard University. Marcel Breuer (see fig. 409) also arrived in America in 1937, joining Gropius at Harvard. They remained in partnership until 1941, when Breuer established his own practice.

Also in 1937, László Moholy-Nagy (see fig. 432) had been invited to Chicago to form a New Bauhaus. In the same year, Josef Albers (color plate 192, page 336) arrived in America to teach design and painting, first at Black Mountain College in North Carolina, and then at Yale University. In 1938, Mies van der Rohe (see fig. 417), who had arrived in the United States the year before, was invited to the Illinois Institute of Technology (then the Armour Institute) in Chicago. José Luis Sert migrated to the United States from Spain in 1939 and in 1958 became dean of the Harvard Graduate School of Design in Cambridge, Massachusetts. So it was largely as a result of Gropius's and Mies's efforts, as well as those of a few other distinguished émigrés, that architectural schools were transformed over the next twenty years and American architecture entered genuinely into the age of modern architectural experiment. In the same way, Moholy-Nagy, Albers, and others likewise had a powerful effect on industrial, product, and graphic design in the United States.

Later Styles of the Early Modernist Architects

Throughout the early 1930s, FRANK LLOYD WRIGHT (1867–1959) continued to critique modernism and the International Style through his writings and new building designs. While he was aware of Art Deco revivalism, which had a tremendous influence in the United States, as well as the extreme austerity of the International Style in Europe, his sympathies lay elsewhere. Despite his focus on the circular theme in the 1936–39 Johnson Building in Racine, Wisconsin (see figs. 429, 430), Wright did not desert the rectangle or the triangle. Taliesin West is the winter headquarters that he began to build for himself and the Taliesin Fellowship in 1937 near Phoenix, Arizona, which, characteristically, was still being built and rebuilt at the time of Wright's death in 1959. It is in plan and elevation a series of interlocked triangles. The architect's most dramatic assimilation of buildings into a natural environment, it becomes an outgrowth of the desert and mountain landscape (color plate 363, page 584).

Wright also used primary forms for his great unbuilt project, Broadacre City, and for his Usonian houses—innovative low-cost homes for middle-income families. He completed a number of buildings based on the circle, the climax of which is the Solomon R. Guggenheim Museum (fig. 709). First conceived in 1943, it was not actually completed until 1959, after Wright's death. The Guggenheim Museum, sited on Fifth Avenue across from Central Park, is the only building this leading American architect was ever commissioned to do in New York City, and it was many years before he could obtain the necessary building permits for so revolutionary a structure. As designed by Wright, it consisted of two parts: the main exhibition hall and a small administration wing (known as the monitor building), both circular in shape. A smooth, unbroken white band that stretches across the museum's Fifth Avenue façade connects the two sections, giving the appearance of seamless unity. The gallery proper is a continuous spiral ramp around an open central well. The building radiates outward toward the top, the ramps broadening as the building rises, in order to provide ample light and space. A skylight dome on graceful ribs provides natural, general lighting, and continuous strip-lighting around the ramps provides additional illumination. Permanent fins divide the exhibition areas of the ramps into equal bays, where the art is shown. Wright believed that architecture should essentially involve movement, not just be a fixed enclosure of space, and the Guggenheim's spiral form was the ultimate expression of his effort to get beyond the box. Explaining his concept to a skeptical public, Wright noted that his design provided "a greater repose, the atmosphere of the quiet unbroken wave: no meeting of the eye with abrupt changes of form."

As a museum, the Guggenheim offers easy and efficient circulation in one continuous spiral, in contrast to the traditional museum with galleries consisting of interconnected, rectangular rooms. Some of Wright's detractors have criticized the design of the Guggenheim's slanting walls and

709. FRANK LLOYD WRIGHT. Solomon R. Guggenheim Museum, New York. 1957–59. Addition 1989–1992

ramps for competing too strongly for the viewer's attention with the artwork. On the other hand, supporters argue that this museum succeeds in what few others even attempt: to be more than a passive site for curatorial activities and to actively engage the viewer's experience of the art, which can be seen across the rotunda as well as at close range.

In 1982 the architectural firm of Gwathmey Siegel accepted a commission to renovate the museum in order to accommodate the need for expanded gallery, storage, and administrative space. Gwathmey Siegel was charged with bringing the aging building into line with state-of-the-art museum technology. Further, it was to create an addition that would include gallery spaces capable of accommodating large-scale contemporary art. Work on the museum involved two distinct challenges: a restoration of the building to some semblance of Wright's intentions and an expansion plan that involved alterations of the original structure.

The restoration provoked almost as much controversy and publicity as had Wright's original design, and some critics have argued that a major icon of American architecture has been tragically compromised. The new addition (fig. 709) consists of a thin rectangular slab whose beige limestone façade is ornamented with a gridded pattern. Four new floors of galleries, three of which have double-height ceilings, make it possible to show large scale works for the first time. In its present renovated state, which was opened to the public in 1992, the great spiral—a monumental sculpture in its own right—was opened all the way to the re-exposed dome, where new ultraviolet filtering glass was installed. Formerly unused roof space was converted into a sculpture terrace providing sweeping views of Central Park, invoking the intrinsic connection between Wright's architecture and nature.

In stark contrast to Wright, who always maintained his role as "chief architect" supported by his Taliesin Fellowship, WALTER GROPIUS (1883–1969) soon put into effect his principles of collaboration, a "team approach" to design, and

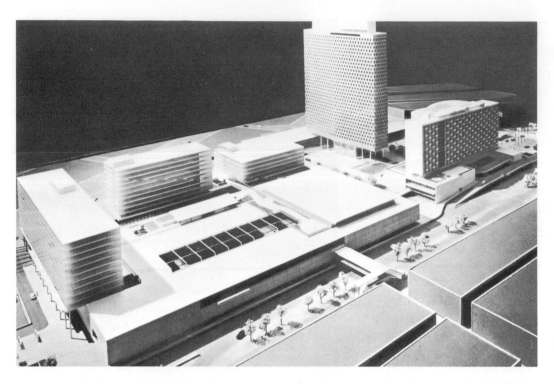

left:
710. WALTER GROPIUS.
Project for City Center for
Boston Back Bay. 1953

below:
711. MIES VAN DER ROHE.
Block plan for Illinois
Institute of Technology,
Chicago. 1940
© 1997 ARTISTS RIGHTS SOCIETY (ARS),
NEW YORK/VG BILD-KUNST, BONN

his ideas concerning the social responsibilities of the architect (see chapter 16). For Gropius, who had founded the Bauhaus in 1919 on these ideas, collaboration meant that architects, landscape architects, and city planners would work together to create a cohesive vision of the modern world. This view increasingly gained acceptance in the profession and would eventually triumph over Wright's approach. In 1945, with some of his students, Gropius formed The Architects' Collaborative (TAC) to design buildings in the United States and elsewhere. At Harvard he continued his design of school structures with the Harvard Graduate Center of 1949–50. Here he had the problem of placing a modern building among traditional structures dating from the eighteenth to the twentieth century. He made no compromises with tradition but used stone and other materials to enrich and give warmth and variety to his modern structure. He also commissioned artists such as Miró, Arp, and Albers to design murals, thus introducing to Harvard the Bauhaus concept of the integration of the arts. In his later years with TAC, Gropius moved again into full-scale architectural production. Perhaps the most ambitious of the later designs was the 1953 Boston Back Bay Center (fig. 710), unhappily never constructed. This was a large-scale project involving office buildings, a convention hall, a shopping center, an underground parking lot, and a motel. It involved two slab structures set in a T form, raised on stilts above intricate, landscaped pedestrian walks and driveways. The façades were to have been richly textured, and the entire complex represented a superior example of recent American experiments in the creation of total industrial, mercantile, or cultural centers—social and environmental experiments with which Gropius himself had been concerned in Germany during the 1920s.

Having completed relatively few buildings before leaving

712. MIES VAN DER ROHE. New National Gallery, Berlin, 1968.
Sculpture by Alexander Calder

713. LE CORBUSIER. Unité d'Habitation, Marseilles, France. 1947–52

Germany in 1937, **MIES VAN DER ROHE** (1886–1969) was finally able to put his ideas into practice in the United States, a country that since the Depression had witnessed virtually no new major public constructions. Big institutions were eager to put up new buildings and to develop a new architecture. In the skyscrapers he created, as well as in the new campus for the Illinois Institute of Technology (IIT), Mies's solutions had an incalculable influence on younger architects everywhere. The plan of the Institute (fig. 711), as drawn by Mies in 1940, has an abstract quality characterized by its rectangular complexes. Located in a deteriorating area on the south side of Chicago, the school complex, with its conception of architecture geared to industrialization and standardization, became a model of efficient organization and urban planning. In fact, it inspired large-scale slum clearance in the area, a process that was intended to, but ultimately did not, eliminate overcrowding and physical deterioration of older buildings there. Of steel-and-glass construction, the IIT buildings are organized on a modular principle. Each unit in itself and in relation to others exemplifies Mies's sensitivity to every aspect of proportion and detail of construction.

After having envisioned (but not built) two highly advanced skyscraper designs in 1919 and 1919–21, Mies was finally able to carry out some of his ideas in his 1951 apartments on Lake Shore Drive in Chicago (fig. 732) and in the 1956–57 Seagram Building, on which Philip Johnson collaborated (color plate 365, page 633). Both the Seagram Building, Mies's first New York building, and his 1964 Federal Center in Chicago represent the climax of a development in the United States that began with Sullivan and the Chicago School, was diverted during the age of eclecticism, and gradually came back to its original principles. The common element of this development is the use of steel I-beams, which were fundamental to Chicago School construction and which

Mies used almost decoratively on the exteriors of some of his skyscrapers. In his later works, such as the 1963–68 National Gallery in Berlin (fig. 712), Mies not only refined his modular system to a point of utmost simplicity, but, using a suspended frame structure, was able to realize his ideals of a complete and uninterrupted interior space divisible or subdivisible at will. For Mies, space was neutral and floating; the structure was a grid. Many imitators have tried to copy his forms through their continuous repetition of the steel-and-glass motif, but usually without his feeling for space, scale, and proportion. His perception of universal, open, and flexible space is as fundamental to modern architecture as is the free plan of Le Corbusier.

The characteristics of the later style of **LE CORBUSIER** (1887–1965) were the exploration of free, organic forms and the statement of the materials of construction. The enormous 1947–52 building in Marseilles known as the Unité d'Habitation (fig. 713), an apartment complex designed primarily for blue-collar workers, both carries out the architect's town-planning ideas and asserts its rough-surfaced concrete structure as if a massive sculpture. The building is composed of 337 small duplex apartments, with a two-storey living room, developed in his earlier housing schemes. It includes shops, restaurants, and recreation areas, to constitute a self-contained community (Community facilities are provided on the roof). Unités were also built in Nantes, France, and Berlin. In them, Le Corbusier abandoned the concept of concrete as a precisely surfaced machine-age material and presented it in its rough state, as it came from the molds. In doing so, he inaugurated a new style—almost a new age—in modern architecture, to which the name Brutalism was later given. Possibly related to the French word *brut* (uncut, rough, raw), the term has taken many forms, but involves fundamentally the idea of "truth to materials" and the blunt statement of

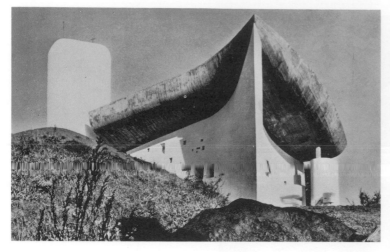

above: 714. LE CORBUSIER. Notre-Dame-du-Haut, Ronchamp, France.
1950–54

right: 715. LE CORBUSIER. Assembly Building, Chandigarh, India.
1959–62

their nature and essence. Frank Lloyd Wright had preached this doctrine from the beginning of his career, and Mies in his own way was as ruthlessly dedicated to the statement of glass and steel as Le Corbusier was to the statement of concrete. Brutalism, however, most characteristically manifests itself in reinforced concrete not only because of its texture but even more because of its innate sculptural properties.

Le Corbusier's seminal pilgrimage chapel of Notre-Dame-du-Haut at Ronchamp in France (color plate 364, page 584 and fig. 714), built 1950–54, is a brilliant example of his new sculptural style. Completed during the height of a new formalism in the United States (for example, the cultural center complex of buildings at Lincoln Center in New York City, discussed below), the Ronchamp chapel represented a radical departure from Le Corbusier's previous projects. Sensitively

and elegantly attuned to the hilltop site on which it was built, the structure is molded of white concrete topped by the dark, floating mass of the roof and accented by towers (inspired by Roman emperor Hadrian's villa at Tivoli, near Rome), which together serve as a geometric counterpoint to the main mass. The interior is lit by windows of varying sizes and shapes that open up from small apertures to create focused tunnels of light, evoking a rarefied spirituality. Its external, curvilinear forms and the mystery of its interior recall prehistoric grave forms. Since Ronchamp, form, function, symbolism, and plasticity all have taken on new meaning. This structure succeeded in relaxing architects' formerly traditional attitudes toward form and space; the "box" was finally exploded.

In the 1950s Le Corbusier was finally given the opportunity to put into effect his lifelong ideas on total city planning

716. ALVAR AALTO. Villa Mairea,
Noormarkku, Finland. 1939

in the design and construction at Chandigarh, a new capital for the Punjab in India, on which he worked into the 1960s. The city of Chandigarh is situated on a plain at the edge of hills, and the government buildings, comprising the Secretariat, the Assembly, and the High Court, rise to the north in the foothills, with the three buildings arranged in an arc around a central area incorporating terraces, gardens, and pools. The Court Building was conceived as a giant umbrella beneath which space was made available for court functions and for the general public. Access roads to the city are largely concealed, while changing vistas are achieved through ingeniously placed man-made mounds. The Assembly Building (fig. 715) is the most impressive of the three, with its ceremonial entrance portico and upward-curving roof set on pylons, the whole reflected in the pool and approached by a causeway. Le Corbusier's achievement is embodied not only in the individual buildings, but in their subtle visual relationships to one another. The total complex constitutes a fitting climax to Le Corbusier's career, and it fulfilled the request of the client, Prime Minister Jawaharlal Nehru, that the city be "symbolic of the freedom of India, unfettered by the traditions of the past."

Finland

ALVAR AALTO (1898–1976) created architecture that is characterized by a warmth and humanity in his use of materials, by buildings that are individualized, and by a harmony with human scale. In producing his buildings, Aalto never abandoned the timeless quality of the traditional architecture of his Finnish homeland. His structures seem to grow out of the site and to solve particular design problems, whether those of factory, house, school, or town plan. Aalto was also a leading furniture designer known for his bentwood chairs and tables.

He made his debut in 1937 with the interior of the Savoy Restaurant in Helsinki, where he introduced his perforated plywood screen and free-form furnishings, quite different in style from the understated luxury of the Villa Mairea (1939) in Noormarkku, Finland (fig. 716), designed for patrons Harry and Maire Gullichsen. The villa anticipates Aalto's own words of the following year, when he wrote that the function of architecture is to "bring the material world into harmony with human life." Villa Mairea is highly compatible with natural materials and clearly harmonizes with the fir forest that surrounds it. Aalto employed polished teak and other woods and rough masonry throughout in order to soften the modernist rigor of his design. Warm-toned wood is used inside in the paneled flooring and slatted ceilings, the columns, and in his own birchwood furniture.

Aalto's was the first major response against the machine aesthetic that was creative rather than reactionary. In non-residential buildings, traditional rectangular windows replaced the "orthodox" glass curtain wall. He expanded the vocabulary of avant-garde architecture to the point where the original sources of the new style all but vanished, and he established a distinctly personal alternative. Irregular, even picturesque, shapes and masses dominate his work, but the very informality of his buildings defies comparison with Mies or Le Corbusier. There are affinities in Aalto's work with the organic, site-specific forms of Wright, the exuberant expressionism of Erich Mendelsohn, and the cool rationality of Gropius. But his free-form brick and glass-brick walls and undulating wood vaults made fresh use of natural materials. Even the grandest structures have a human scale. His versatility helped establish a distinctive brand of modernism that still flourishes in Finland and Scandinavia.

As the leading architect of far northern Europe, Aalto had opportunities to experiment with public buildings, churches, and town planning, and he achieved one of his most personal expressions in the Town Hall at Säynätsalo (fig. 717). This is a center for a small Finnish town, and includes a council room, library, offices, shops, and homes. The brick structure with broken, tilted rooflines merges with

717. ALVAR AALTO. Town Hall, Säynätsalo, Finland. c. 1950–53

the natural environment. Heavy timbers, beautifully detailed, are used for ceilings to contrast with the brick walls inside. Somewhat romantic in concept, this center illustrates the combination of effective planning and respect for the natural environment characteristic of Aalto and the best northern European architecture.

Great Britain

After the pioneering work of the late nineteenth century by William Morris, Philip Webb, Charles Voysey, and Charles Rennie Mackintosh (see figs. 91, 94, 95), there was, as has been noted, little experimental architecture in Great Britain during the first decades of the twentieth century. The arrival in Britain in the 1930s of the German architects Gropius, Breuer, Mendelsohn, and the Russian Serge Chermayeff stimulated new thinking. But World War II interrupted any new work and, after the war, budgetary limitations prevented large-scale architectural development. Although few great individual structures have been built in England in the twentieth century, progress has been made in two areas: town and country planning, and the design of schools and colleges.

The work of **JAMES STIRLING** (1926–1992) one of the most imaginative and influential British architects of the postwar period is associated with the tradition of Le Corbusier. Stirling had a masterly ability to combine all sorts of materials—glass, masonry, and metal—in exciting new contexts. Like Louis Kahn's work (figs. 746, 747), Stirling's early buildings are not so much functional, in the original International Style sense, as they are expressive and personal. With James Gowan he designed the low-cost housing complex of

718. JAMES STIRLING. History Faculty, Cambridge University, England. 1968

Ham Common in England in 1958, when he was studying the later ideas of Le Corbusier and exploring the possibilities of ready-made products in new architectural contexts. In his best-known building, the 1968 Cambridge University History Faculty (fig. 718), Stirling used a cohesive approach to achieve a dense and complex spatial integration. Although Stirling was an important teacher and thinker in his field, he was often pessimistic about the state of modern architecture. This is ironic because his own accomplishments continually demonstrate modern architecture's possibilities, especially, as we shall see, in his Postmodern work (see fig. 903).

France

With the exception of works by the titanic figure of Le Corbusier and earlier ones by Auguste Perret (see fig. 276), French architecture of the twentieth century has had comparatively few moments of inspiration. This is in part the result of the continuing academic system of training for architects and the persisting influence of a bureaucratic old guard. Le Corbusier (born Swiss) produced a number of masterpieces in France, but he ultimately achieved greater standing elsewhere.

The most ambitious postwar structure erected in Paris was that of the Y-shaped building for the United Nations Educational Scientific and Cultural Organization (UNESCO) designed by the international team of architects that included the Hungarian **MARCEL BREUER** (1902–1981) and the French Bernard Zehrfuss with the Italian architect-engineer Pier Luigi Nervi. The total complex is excellently adapted to a difficult site and is adorned by a large number of sculptures, paintings, and murals by leading artists such as Miró, Calder, Picasso, Noguchi, and Henry Moore. Breuer developed the ideas and plan of the UNESCO building into his work on a research center for IBM-France at La Gaude Var (fig. 719). This is a huge structure of a comparable, double-Y plan, with windows set within deep, heavy, concrete frames, the whole on concrete supports. The effect is of openness and airiness combined with massive, rugged detail.

As a result of an important postwar international competition, the most sensational and hotly debated building to be designed in the 1960s (erected in Paris during the early 1970s) was the high-tech extravaganza officially known as Le Centre National d'Art et de Culture Georges Pompidou designed by **RENZO PIANO** (b. 1937) and **RICHARD ROGERS** (b. 1933). (It is referred to as the Pompidou Center after the then-president, and also as the Beaubourg after the market district on which it was built.) One of the few constructed buildings conceived by England's Archigram Group (an association that derived from the British avant-garde), the Pompidou Center (fig. 720) is an icon to Parisian popular culture, housing France's national museum of modern art, a public library, audiovisual center, rooftop restaurant, and, added recently, an Internet café. In its time, the design achieved a new standard in open-space flexibility. The structural and service functions were channeled along the perime-

ter walls, leaving the interior open to be freely subdivided, and suitable for a multitude of changeable institutional and public uses. The Pompidou Center takes Bauhaus pragmatism and technology as its point of departure, only to transform them into a surreal dream of pipes and trusses, marine funnels and vents, most of which festoon the exoskeleton with all the logic of a Rube Goldberg "machine." With its

exposed fretwork, its clearly expressed structure and services, and its great diagonal escalators, as well as its bright colors and open planning, the building, as one commentator remarked, is like "a child's Tinkertoy fantasy realized on an elephantine scale." Despite a gargantuan appetite for fuel and difficult-to-manage, barnlike galleries, the Pompidou Center is a great popular success, largely because of a five-storey outside escalator, whose transparent plastic tubes offer riders a moving, spectacular view of Paris and the immense, open plaza below. The latter has become a forum—like the parvis of a medieval cathedral—for the whole human carnival, from artists displaying their works to jugglers and, of course, ordinary tourists and even Parisians going about their daily business.

Germany and Italy

As we have seen, Germany until 1930 was probably the world leader in developing new architecture. During the Nazi regime, German architecture declined from its position of world leadership to academic mediocrity. After the war, morale was low for those architects and artists who had remained in Germany through the war years. While America was enjoying prosperity, Germany underwent a slow, traumatic period of rebuilding, and it took that country a full decade to recover the principles forcibly surrendered to the Nazis.

In Italy, modern architecture made a somewhat belated arrival, perhaps because the sense of the past was so strong. The visionary Futurist Antonio Sant' Elia was killed in World War I at the age of twenty-eight, before he had a chance to execute his concepts for cities of the future (see chapter 12). His ideas faded along with Futurism, and his influence was felt principally, through publications, on pioneers outside of Italy, importantly Le Corbusier and the architects of de Stijl and the Bauhaus.

Although progressive architecture was not formally suppressed by Italian Fascism as it was in Germany under the Nazis, the climate did not favor growth. The official governmental style looked back to showy, nationalistic monumentalism, of which the nineteenth-century monument to Victor Emmanuel II in Rome is the most conspicuous example. In 1933 a group of younger architects, *gruppo 7*, banded togeth-

er and called for a functional and rational architecture that would value form over surface in modern architecture. A few outstanding, though isolated buildings resulted from this and other aspects of Italian rationalism between the wars, notably the aesthetically pure 1932–36 Casa del Popolo (House of the People) at Como (fig. 721), designed by GIUSEPPE TERRAGNI (1904–42). Originally designed as a building wrapped around three sides of square court, the structure is a simple half-cube. The court, which became a large meeting hall, stands in radical contrast to the Renaissance cathedral opposite it. After World War II, Italy expanded its experiment beyond progressive architecture, moving into a position of European and even world leadership in many aspects of industrial, product, and fashion design.

The most influential of Italy's prewar design giants was the engineer-architect PIER LUIGI NERVI (1891–1979). Nervi had a resounding ability to translate engineering structure into architectural forms of beauty. Chronologically he belongs with the pioneers of modern architecture, and his fundamental theses were stated in a number of important buildings executed during the 1930s. Working with reinforced concrete, his main contribution was the creation of new shapes and spatial dimensions with this material. As in the work of Mies van der Rohe, Nervi's buildings show a subtle fusion of structure and space. But while Mies searched for free internal space, Nervi's aesthetic was dependent on an energetic display of the structural parts of a building. One of the most important commissions of his career was the design of an aircraft hangar in 1935, the first of a number of variants he built for the military from 1936 to 1941. These hangars, which involve vast, uninterrupted concrete roof spans, led him to study different ways to create such spans. He learned how to lighten and strengthen his materials and to integrate aesthetic and structural variations in his designs.

In the 1950s Nervi developed the technique of hydraulic prestressing of ferroconcrete, which made it possible to lighten the support of buildings. Two of the dramatic applications of his experiments with ferroconcrete are the small and large sports palaces built for the 1960 Olympic Games in Rome (figs. 722, 723). The smaller one, seen here, meant to hold four- to five thousand spectators, was built in 1956–57, in collaboration with the architect ANNIBALE VITELLOZZI. The building is essentially a circular roof structure in which the enclosed shell roof, 197 feet in diameter, rests lightly on thirty-six Y-shaped concrete supports. The edge of the roof is scalloped, both to accentuate the points of support and to increase the feeling of lightness. A continuous window band between this scalloped edge and the outside wall provides uniform daylight and adds to the floating quality of the ceiling. On the interior, the ceiling, a honeycomb of radiating, interlaced curves, floats without apparent support. The shell domes of the small and large sports palaces became prototypes for other such structures.

During the 1950s and 1960s, Nervi collaborated with architects all over the world, including Italy's own GIO PONTI (1891–1979). The distinguished Ponti was the chief architect

721. GIUSEPPE TERRAGNI. Casa del Popolo (House of the People), Como, Italy. 1932–36

left: 722. PIER LUIGI NERVI AND ANNIBALE VITELLOZZI. Sports Palace, Rome. 1956–57

below, left: 723. PIER LUIGI NERVI AND ANNIBALE VITELLOZZI. Interior, Sports Palace. 1956–57

below: 724. GIO PONTI, PIER LUIGI NERVI, and others. Pirelli Tower, Milan, Italy. 1956–59

of the Pirelli Tower in Milan (fig. 724), the result of a close collaboration between architects and engineers. Ponti developed slowly but consistently from Neoclassical beginnings toward an elaborate modernism. The Pirelli Tower, for which Nervi was the structural engineer, remains Ponti's masterpiece, one of the most lucid and individual interpretations of the skyscraper yet achieved. In contrast to most American skyscrapers built before the 1950s, it is a separate building surrounded by lower structures, designed to be seen from all sides. The architect was concerned with making it a total, finite, harmonious unit. A boat shape, rather than the customary rectangular slab, was used. The thirty-two-storey building is carried on tapering piers, resulting in the enlargement of the spaces as the piers become lighter toward the top. The entire building, with the diagonal areas at the ends enclosed and framing the side walls of glass, has a sculptural quality, a sense of solidity and mass unusual in contemporary skyscraper design.

Latin America, Australia, Japan

A spectacular example of the spread of the International Style occurred in Latin America after 1940. Le Corbusier's participation in the design of the Ministry of Education and Health in Rio de Janeiro between 1937 and 1943 provided the spark that ignited younger Latin American architects, among them **OSCAR NIEMEYER** (b. 1907) in Brazil. Le Corbusier's influence is apparent in most of the subsequent development of Brazilian architecture, but Niemeyer and his contemporaries added expressionist and Baroque elements—such as curving shapes and lavish use of color reflecting the Latin American tradition—to plans and façades. In his Pampulha buildings of the early 1940s, Niemeyer hints at an integration of painting and architecture and effectively introduces a wilder interplay of forms and levels into his buildings.

Niemeyer's greatest work, and one of the most dramatic commissions in the history of architecture, was the design of

725. OSCAR NIEMEYER. Palace of the Dawn, Brasília, Brazil. 1959

ning, such a project must be seen as important, despite its deficiencies and failures.

In Australia, too, there has been a comparable development of new architectural forms, from the slab skyscraper to the integrated educational institution and advanced experiments in housing and urban planning. The most spectacular modern structure is the sail-form, freestanding Sydney Opera House of 1972 on Bennelong Point (fig. 726), jutting out into Sydney Harbor. The architectural firm of Hall, Todd, and Littleton succeeded the Danish architect **JØRN UTZON** (b. 1918), who made the original design. This cultural center includes an opera hall, a theater, an exhibition area, a cinema, and a chamber music hall.

Postwar architects in Japan exploited possibilities that had been previously regarded as utopian. An independent style in modern Japanese architecture has been apparent only since the 1960s. Before then, most of the new building was in derivative commercial modern style, particularly in Tokyo, where there were few buildings of distinction or originality. Modern architecture in Japan, despite a long history of sporadic progressive examples extending back to Frank Lloyd Wright's Imperial Hotel (see fig. 295), was actually only at its inception after World War II. While assimilating the powerful influence of Le Corbusier, the tradition of Japanese architecture (which for centuries embodied so many elements that today are called modern), is helping to turn contemporary influences into something authentically Japanese.

One of the most significant architects of Japan is **KENZO TANGE** (b. 1913), whose work in the 1950s and 1960s combined rationalist and traditional directions in Japanese aesthetics. Like most of the younger Japanese architects, Tange

the entire city at Brasília, the new federal capital of Brazil (fig. 725). The general plan of the city was laid out by Lúcio Costa, who won the competition sponsored by Brazil's president in 1956. Niemeyer, as chief architect, was entrusted with the principal public buildings. Costa's lucid plan was essentially two axes in the form of a cross. Niemeyer designed large-scale civic buildings for the top of the cross, with residential sites along the two arms of the horizontal axis, incorporating places for work and recreation. The idea of building an entire utopian city far removed from any of the principal centers of commerce and industry probably doomed Brasília. At a time in the history of architecture when very little actual attention was being paid to concepts of total urban plan-

726. JØRN UTZON AND HALL, TODD, AND LITTLETON. Sydney Opera House, Bennelong Point, Sydney Harbor, Australia. 1972

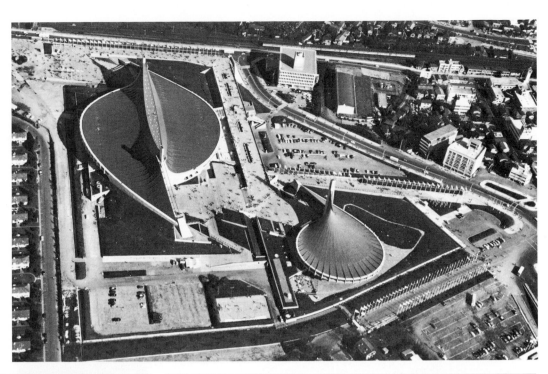

right: 727. KENZO TANGE. Aerial view, The National Gymnasiums, Tokyo. 1964

below, left: 728. KENZO TANGE. Yamanashi Press Center, Kofu, Japan. 1967

below, right: 729. TAKEO SATO. City Hall, Iwakuni, Japan. 1959

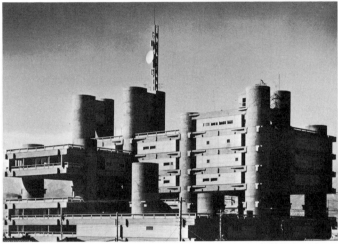

is a disciple of Le Corbusier and particularly of Le Corbusier's later, Brutalist style. Tokyo City Hall, a work of 1952–57, is a rather thin version of Le Corbusier's Unité d'Habitation (fig. 713), but with the Kasagawa Prefecture Office at Takamatsu (1955–58), Tange used a combination of massive, direct concrete treatment and horizontally accentuated shapes reminiscent of traditional Japanese architecture. Perhaps his most striking achievement is the pair of gymnasiums Tange designed for the Olympic Games held in Tokyo in 1964 (fig. 727). Both steel-sheeted roofs, somewhat resembling seashells, are suspended from immense cables, and their structure is at once daring and graceful. The larger roof hangs from two concrete masts, while the smaller is ingeniously suspended from a cable spiraling down from a single mast. The Brutalism of the buildings themselves, with their rough surface texture and frank structure, gives the complex—set as it is on a huge platform of undressed stone blocks—a sense of enduring drama.

Tange's Yamanashi Press Center in Kofu (fig. 728) is a fantastically monumental structure in which immense pylons provide for elevator and other services and act visually as tower supports for the horizontal office areas. The total impact is like

that of a vast Romanesque fortified castle. The interior, however, is open and flexible, with spaces defined by movable shoji screens that allow occupants to easily adapt their space to changing needs. Reminiscent of traditional timber post-and-beam constructions, the press building is one of the earliest examples of a reaction against the abstract, homogenous trends of modern architecture. Its very "uncompletedness" is an important aspect of its design; it is open-ended and not confined by more traditional architectural approaches that rely on finite elevations and concise imagery.

Another Japanese architect who followed the late Le Corbusier line is TAKEO SATO (1899–1972), whose City Hall in Iwakuni (fig. 729) merges the design of a traditional Japanese pagoda with modern building materials including concrete and glass. Similarly SACHIO OHTANI's (b. 1924) International Conference Building in Kyoto (fig. 730) is a massive structure that houses meeting and exhibition rooms, restaurants, administrative offices, shops, and recreation areas. Employing old trapezoidal and triangular design elements, it represents one of the most spectacular attempts yet made to combine on a vast scale ancient Japanese motifs and concepts with modern architectural forms.

730. SACHIO OHTANI. International Conference Building, Kyoto, Japan. 1955

731. WALLACE K. HARRISON. United Nations Headquarters, New York.
1947–50

The United States

Architects in the second half of the twentieth century have been able to develop new concepts of space, either employing new structural systems or using older methods, such as ferroconcrete, in new ways. Besides the skyscraper and other urban office buildings, some of the areas in which architects have worked on an unprecedented scale are industrial plants and research centers; university campuses; religious structures; cultural centers, including museums, theaters, and concert halls; and airports. Thus, in the following pages, architecture (located predominantly in the United States but in a few cases, internationally) is discussed by category, with some of the major creative and avant-garde talents placed under these headings.

Skyscrapers

In 1947, an international group of architects (including Le Corbusier and Niemeyer) met in New York under the general direction of **WALLACE K. HARRISON** (1895–1981) to design the permanent headquarters of the newly formed United Nations organization (fig. 731). This building, situated, at the edge of New York's East River and built between 1947 and 1950, was the first monument of the postwar renaissance in American architecture. Symbolically the most important building to be constructed in the aftermath of World War II, the headquarters complex represents the epit-

732. MIES VAN DER ROHE. Lake Shore Drive Apartments, Chicago. 1951

733. SKIDMORE, OWINGS & MERRILL. Lever House, New York. 1951–52

ome of modern functionalism and midcentury optimism. Aesthetically, it attempted to replace diverse national traditions with a design appropriate to the universal concept of the new world organization. The thirty-nine-storey Secretariat, a glass-and-aluminum curtain-wall building, introduced to the United States a concept of the skyscraper as a tall, rectangular slab; in this case the sides are sheathed in glass and the ends are clad in marble.

Le Corbusier was principally responsible for the main forms of the Secretariat, but its simplified, purified structure embodies the Bauhaus tradition of Gropius and **MIES VAN DER ROHE** (1886–1969), who designed the Lake Shore Drive Apartments in Chicago in 1951 (fig. 732). The skyscrapers of Louis Sullivan and the Chicago School fascinated Mies and drew him to study the skyscraper form. The basic problems—of steel structure, its expression, and its sheathing—were of longstanding interest. (He had already explored these problems in his earlier unbuilt skyscraper designs in 1919 and 1919–21.) The twin Lake Shore apartment buildings consist of two interrelated twenty-six-storey vertical blocks set at right angles to each other. The steel structural members accent the verticality. As is customary with Mies, the structures are built on a module and have floor-to-ceiling windows.

The inspiration of Mies and the United Nations building led immediately to the first large office building of the 1950s, Lever House (fig. 733), completed in 1952 by the architectural firm of **SKIDMORE, OWINGS & MERRILL,** with Gordon Bunshaft as chief architect. Lever House, which was intended as a monument-symbol for an international corporation, is a tall rectangular slab, occupying only a small part of a New York city block. It is raised on pylons that support the first enclosed level—the principal public areas—thus giving the total structure a vertical L-shape. The building is one of the first to have an all-glass-clad façade and windows of alternating horizontal ribbons of opaque green and light green tinted glass.

Across Park Avenue from Lever House is a masterpiece of skyscraper architecture, Mies van der Rohe's 1958 Seagram Building (color plate 365, page 633). In it Mies and Philip Johnson were given an unparalleled opportunity to create a monument to modern industry, comparable to the Gothic cathedrals that had been monuments to medieval religious belief. The building is a statement of the slab principle, isolated with absolute symmetry within a broad plaza lightened by balanced details of fountains. The larger skeleton structure is clearly apparent within the more delicate framing of the window elements that sheathe the building. The materials, bronze and amber glass, lend the exterior an opaque solidity that does not interfere with the total impression of light that fills the interior. The building is perhaps the consummate expression of Mies's emphasis on "the organizing principle of order as a means of achieving the successful relationship of the parts to each other and to the whole." In addition to its dramatic utilization of International Style principles of design, the Seagram Building is noted also for its detachment from

above: 734. PHILIP JOHNSON AND JOHN BURGEE. Pennzoil Place, Houston. 1972–76

right: 735. SKIDMORE, OWINGS & MERRILL. Sears Roebuck Tower, Chicago. 1966–69

the surrounding city streets. A deeply set plaza pushes the building away from the busy traffic, while the abrupt termination of the glass curtain—one floor above the plaza—appears to raise the building on stilts. No other building of modern times has had such influence on subsequent skyscraper design.

Through the late 1960s and 1970s, skyscrapers were constructed in increasing numbers and heights throughout the world. Even so, distinguished examples of the form were the exceptions. The Pennzoil Place building of 1976 in Houston (fig. 734) by **PHILIP JOHNSON** (b. 1906) and **JOHN BURGEE** (b. 1933) is an example of a glass-sheathed skyscraper that is both elegant and dramatic; it is a building admired by many.

Until 1974, the world's tallest skyscraper (1,454 feet) was the Sears Tower in Chicago (fig. 735), designed by Skidmore, Owings & Merrill. Conceived as a bundle of nine "tubes" lashed together for mutual bracing, the Sears Tower would seem to return to the original pragmatism of Bauhaus aesthetics, in that each of the independent but clustered tubes has its own height and setback, dictated by the internal needs of corporate activity. The blank anodized aluminum skin is highly functional—it resists weathering and soot better than other materials, while also symbolizing the sobriety of the retailing enterprise originally headquartered within the structure.

Domestic Architecture

After World War II, as roads and highways made outlying areas more accessible than ever, and as a war-based economy was replaced by peacetime values and normalization of life, the market for single-family houses exploded. The variations on the theme of the detached house in the International Style were endless and interchangeable from country to country. Much of the early American architecture of Gropius and Breuer consisted of homes following Bauhaus principles. For Gropius, standardization satisfied the needs of the masses for economical housing. Breuer's own second house at New Canaan, Connecticut, built in 1951, illustrates his ability to use materials that blend with the local environment. In this case Breuer selected principally concrete and fieldstone, with weathered wood beams used for the ceilings (fig. 736).

Mies van der Rohe built only one house in the United States—the Farnsworth House in Plano, Illinois—but Philip Johnson's own home in New Canaan, Connecticut, Glass House of 1949 (fig. 737), is a classic extension of Miesian principles of uninterrupted interior space. Few "boxes" have reached the level of rationalism and sophistication of the main house, which is a transparent glass "box." With its supreme structural clarity, the main house provides protection yet destroys the boundaries between nature and interior space, while the guesthouse, at the other extreme, is an enclosed structure.

736. MARCEL BREUER. House II, New Canaan, Connecticut. 1951

RICHARD NEUTRA (1892–1970), an émigré from Austria, developed his practice on the basis of the American preoccupation with building houses. A winter home in Palm Springs, California, for Edgar Kaufmann—the same client who in the 1930s had commissioned Fallingwater from Frank Lloyd Wright (color plate 191, page 335; see fig. 428) —was designed for a desert site by Neutra, who had been an employee of Wright in the 1920s. Set against a spectacular backdrop, the Kaufmann Desert House (1947) (fig. 738) is pristine and cool in its elegant rectilinearity. Glass is used generously, and the plan is arranged so that the space flows from interior to exterior with unfettered freedom. Walls can be pulled back to allow the interior to become part of the wide outdoor courtyard spaces. Neutra combined glass, stucco, and natural rock in this house, which is shaped in a cross of two intersecting axes. In his work and writing, Neutra advocated the link between architecture and the general health of the human nervous system. His physiological concerns addressed the beneficial impact of well-designed architecture and ambient environment. For Neutra, as for Wright, landscape was more important than historical precedents; Neutra's buildings, however, are not necessarily inextricable from their sites.

Of great significance to working architects were some

737. PHILIP JOHNSON. Glass House, New Canaan, Connecticut. 1949

738. Richard Neutra. Kaufmann House, Palm Springs, California. 1947

739. Frederick Kiesler. Plan of Endless House. 1958–59

unbuilt visionary house designs belonging to a new phase of imaginary architecture. Outstanding is the Endless House (fig. 739) designed by **Frederick Kiesler** (1890–1965), whom Philip Johnson called "the greatest non-building architect of our time." As early as 1925, Kiesler had conceived of a city in space built on a bridge structure, and an Endless Theater. From these, in 1934, he developed a "space house" and, after years of experiment, the so-termed Endless House, proposed many times in various versions but never built. The idea for an Endless Structure was to short-circuit any traditional divisions between floor, wall, and ceiling and to offer the inhabitant an interior that could be modified at will. In doing this he abandoned the rectangle and turned to

egg shapes freely modeled from plastic materials and proffering a continuously flowing interior space.

Such experiments represented an attempt not only to find new forms based on natural, organic principles but also to utilize new technical and industrial developments while potentially cutting building costs. Among the projects is the 1961 house and studio in Scottsdale, Arizona, built by **Paolo Soleri** (b. 1919). Literally a cave, it is the forerunner of Soleri's Dome House in Cave Creek, Arizona (1950). Soleri, born in Turin, Italy, came to the United States in 1947 and spent a year and a half in fellowship with Frank Lloyd Wright at Taliesin West in Arizona and Taliesin East in Wisconsin. His compact prototype city, Arcosanti (fig. 740), located in the

left: 740. PAOLO SOLERI. Arcosanti, Scottsdale, Arizona. Begun 1970

bottom, left: 741. CHARLES AND RAY EAMES. Eames Residence, Santa Monica, California. 1949

high desert of Arizona in a 4,060-acre preserve and now a National Landmark, has been under construction since 1970. For some thirty years, it has been an experiment, or "urban laboratory," to demonstrate his theory of arcology. Derived from architecture-ecology, an arcology is an integration of architecture, ecology, and urban planning. It involves a whole new concept of urban environment that would eliminate the automobile from within the city (walking would be the main way of getting around), develop renewable energy systems (solar and wind), utilize recycling, and allow access to and interaction with the surrounding natural environment. Using rounded concrete structures, this miniaturization of the city allows for the radical conservation of land, energy, and resources. The arcology concept proposes a highly integrated and compact three-dimensional urban form that is the opposite of the wasteful consumption of land, energy, time, and human resources of urban sprawl. (Only about two percent as much land as a typical city of similar population is needed.) Therefore, it presents a new solution to the ecological, economic, spatial, and energy problems of cities.

Less utopian than arcology is prefabrication—the mass manufacturing and prefabrication of individual parts. Although relatively few, there are instances of successful, large-scale prefabrication in the twentieth century, such as Nervi's 1948–49 Turin exhibition hall. Some manufacturers in the United States offer partially prefabricated small houses. The husband and wife architect-designers **CHARLES** (1907–1978) and **RAY EAMES** (1916–1988) (fig. 741), in a number of California houses, including their own in Santa Monica (1949), set out to prove that modern architecture could be both accessible and economical. Using prefabricated materials purchased through catalogues, the Eameses constructed residences consisting of light steel-skeleton cores covered with fitted plastic, stucco, or glass panels, as well as stock doors, windows, and accessories. These mass-produced materials were chosen for both their economic appeal and industrial aesthetic and represent a utilitarian domestic architectural style.

One of the greatest examples of prefabricated housing

742. MOSHE SAFDIE. Habitat '67, Montreal, Canada. 1967

design is by Israeli architect **MOSHE SAFDIE** (b. 1938). His megastructure, Habitat '67, commissioned by the 1967 World's Fair in Montreal, Canada, was conceived to provide fresh air, sunlight, privacy, and suburban amenities to residents of an urban location (fig. 742). It was designed as a permanent settlement of 158 dwellings made up of fifteen types of independent, interlocking prefabricated boxes. The boxes are staggered in order to provide open-deck space for each unit and to ensure versatile visual and spatial combinations. It is possible to add more prefabricated, mass-produced units to such a structure, rather than constructing them on site.

Another signal of the advent of an antimodernist aesthetic can be seen in the work of the self-styled New York School (not to be confused with the New York School of Painting that began with the Pollock–de Kooning–Rothko generation in the mid-1940s). Architects identified with the New York School include Peter Eisenman, John Hejduk, Michael Graves, Charles Gwathmey, and Richard Meier. The most heavily commissioned of these is **RICHARD MEIER** (b. 1934), whose glistening, white Cubist temples—supersophisticated exploitations of the modernist vocabulary—have proved irresistible to wealthy private clients, corporations, and cultural institutions. In Douglas House in Harbor Springs, Michigan (fig. 743), one sees Purist or de Stijl classicism so wittily elaborated as to become a new Mannerism, reversing simple closed forms and ribbon windows into intricate, openwork abstractions with double- and triple-height interiors and vast expanses of glass.

Cultural Centers, Theaters, and Museums

The most elaborate center to date in the United States is Lincoln Center for the Performing Arts in New York City (fig. 744), which opened in 1962 and represented a spectacular return to architectural formalism. The classical origins of its layout are not disguised. The focus of the principal structures on three sides of a monumental plaza is a fountain, designed by Philip Johnson. In addition to **MAX ABRAMOVITZ**'s (b. 1908) Philharmonic Hall (1962) (renamed Avery Fisher Hall), the Center comprises **PHILIP JOHNSON**'s (b. 1906) New York State Theater (1964), **EERO SAARINEN**'s (1910–1961) Vivian Beaumont Theater (1965), Skidmore, Owings & Merrill's Library-Museum of the Performing Arts (1965), and **WALLACE K. HARRISON**'s (1895–1981) Metropolitan Opera House (1966). A large subterranean garage lies beneath this group. Adjunct structures include a midtown branch of Fordham University and a home for the Juilliard School, completed in 1969 by Pietro Belluschi with Catalano and Westermann.

The buildings created by this notable assemblage of talent exemplify a new monumental classicism characteristic of public and official architecture in the 1960s. The project is impressive, despite the criticisms leveled against individual structures. The chief complaints, aside from problems of acoustics and sightlines, are focused on Lincoln Center's barren monumentality: The colossal scale is unrelieved by ornamental detail that would provide a human reference.

743. RICHARD MEIER. Douglas House, Harbor Springs, Michigan. 1971–73

744. Lincoln Center, New York. From left: New York State Theater, by Philip Johnson, 1964; Metropolitan Opera House, by Wallace K. Harrison, 1966; Philharmonic (now Avery Fisher) Hall, by Max Abramovitz, 1962

745. MARCEL BREUER. The Whitney Museum of American Art, New York. 1966

We saw earlier in this chapter, in the discussion about the renovation of Frank Lloyd Wright's Guggenheim Museum (fig. 709), that an art museum is actually a complicated design problem, involving basic questions of efficient circulation, adequate light—natural, artificial, or both—sufficient work and storage space, and, in most cases, rooms for other events, such as concerts, theatrical performances, and receptions. The Whitney Museum of American Art (fig. 745) designed by Marcel Breuer is another important museum design. Breuer's building, illustrative of his later Brutalism, is a development of the forms he used earlier for St. John's Abbey Church in Northfield, Minnesota (1953–61) and for the lecture hall of New York University's uptown campus in the Bronx (1956–61). The Whitney is a stark and impressive building, in which heavy, dark granite and concrete are used inside and out. The main galleries are huge, uninterrupted halls capable of being divided in almost any way by movable yet solid partitions and representing the utmost flexibility in installation space and artificial lighting. Natural daylight has been ignored except for that emanating from a few trapezoidal-shaped windows that function as relief accents.

The architect **LOUIS I. KAHN** (1901–1974) was responsible for the design of two important art buildings. Like Wright, Kahn initiated a new American architecture, but while Wright's work developed in the late nineteenth and early twentieth centuries, Kahn matured amidst the uncer-

tainty and loss of idealism of the 1930s and 1940s. Born in Estonia, Kahn emigrated with his family to Philadelphia in 1905. He received architectural training in the Beaux Arts tradition at the University of Pennsylvania. There followed a year in Europe, where he was impressed both by the monuments of classical antiquity and by his discovery of Le Corbusier. His later work would reflect a fusion of modernism's passion for technology and abstract form along with a profound awareness of history and its role in architecture. For a number of years, Kahn taught at Yale University, where he designed the Yale Art Gallery (1951–53), the university's first modern building and one of the first contemporary

746. LOUIS I. KAHN. Kimbell Art Museum, Fort Worth, Texas. 1972

747. LOUIS I. KAHN. Presidential Plaza of the National Assembly, Sher-e-Bangla Nagar, Bangladesh. Begun 1964

approaches to the design challenges of the art museum.

Kahn was not only a designer concerned with mechanical and visual relationships of extraordinary subtlety and aesthetic quality, but he was also a planner, both practical and poetical, whether for an individual building or an entire city. Perhaps the most visionary and completely realized of all of Kahn's buildings is his Kimbell Art Museum in Fort Worth, Texas, completed in 1972 (color plate 366, page 633; fig. 746)—one of the most dramatic as well as functionally effective art museums in the world. The museum design is based upon a parallel series of self-supporting cycloidal vaults that eliminate the need for interior supports and facilitate an unobstructed and very flexible use of the spacious interior. Indoor and outdoor spaces are integrated through the attachment of a sculpture garden that creates a fluid interchange between the museum and its environment, thereby altering the traditional notion of a museum as a place that cloisters works of art (fig. 746). The architect is also noted for the close attention paid to the issues of lighting and displaying works of art. By combining artificial and natural light, and breaking up the cavernous heights of gallery space with low-hanging lights and shorter partitioning walls, the Kimbell becomes both a practical space for the display and conservation of art, as well as an aesthetically pleasing environment that is at once monumental and intimate to the gallery visitor.

Like Le Corbusier's before him at Chandigarh, Kahn's experience with the Asian subcontinent unleashed a flood of innovation and invention. In 1962, East Pakistan (Bangladesh), determined to have its own capital, invited Kahn to design the entire city of Dacca, including the Assembly (fig. 747) and Supreme Court buildings, hostels, a school, a stadium, and a market. To protect the buildings from the harsh elements, Kahn enclosed them inside another structural layer. As he said of an earlier project, "I thought of wrapping ruins around the buildings." In this case, the "ruins" are large, free-standing concrete curtains pierced with geometric cutout shapes. For example, the curtain for the East Hostels for the members of Parliament (color plate 367, page 634), with its large semicircular cutout shapes, serves not to provide views to the exterior, but to shield the interiors from intense sun glare. The result is both primal and sophisticated. Like Le Corbusier's monuments for Chandigarh, Kahn's work for Dacca achieves an ancient aura without being culturally specific; the iconography of the cutout shapes seems almost indifferent to the history and artistic heritage of the patron.

The challenge of successfully designing a visitor center–museum for a highly specific and localized American community shows Richard Meier's versatility and his evolution as an architect. The differences between his rationalist Douglas House (fig. 743) and the three-storey Atheneum at Historic New Harmony in Indiana (color plate 368, page 634) are revealing. Now a living museum, New Harmony was a vigorous nineteenth-century American religious and utopian community established by the Harmony Society in 1814. The 1975–79 Atheneum is a gleaming-white metal structure set on a green lawn, a visual dialogue of solid and void,

projecting terraces, recessive spaces, and folded stairs that exclaim its modernity. The commission required a 180-seat auditorium, four exhibition galleries, observation terraces, and visitor and administrative facilities. Meier's work fulfills the program and stands in great contrast to the log cabins and brick buildings of the historic settlement. Meier does not belong to the tradition of expressionist architects interested in empathy and feeling. His is an architecture interested in visual effects, looking at and framing objects in space. Much like Le Corbusier's dictum that a house is a machine for living, the Atheneum is conceived as a device, a sophisticated instrument for controlling the experience of observing its collection. There is something detached in the way the building relates to the site of one of the most intriguing social experiments in American history, but its very detachment offers a completely unbiased vantage point.

IEOH MING PEI's (b. 1917) East Wing extension to the National Gallery of Art in Washington, D.C. (fig. 748), inaugurated in 1978, ten years after its initial conception, is an extraordinary venue for a broad range of museum activities. It not only is a site for temporary exhibitions and the museum's collection of twentieth-century art, but it also houses museum offices, the library, and the Center for Advanced Studies in the Visual Arts (CASVA). The awkward trapezoidal site posed several challenges for the architect, as did the need to integrate the new structure with the extant Beaux-Arts style West Building (designed by John Russell Pope). Pei resolved these problems by dividing the plot into triangular forms which became the building blocks of his design—reiterated throughout the building in floor tiles, skylights, stairs, and tables—and then connecting the entire construction to the West Building via an underground passage. The entrance to the East Building opens to an expansive, light-filled atrium with skylighted ceilings that provide the backdrop for several works of art commissioned especially for the building, including a monumental mobile by Alexander Calder (color plate 369, page 634). In contrast to the atrium's open, soaring space, the intimate interior is composed of individual galleries made from smaller rooms and lowered walls that provide a quiet space for the contemplation of the exhibited art. Viewed from outside, the East Building—dramatically set against the view of the Capitol Building—is easily recognized by its H-shape design which is formed by the stocky towers that extend from each of three vertices (fig. 748). Associated in his early years with both Gropius and Breuer, Pei is the consummate modernist, combining and refining the best principles of modern architecture in a continually inventive stream of eloquent commercial and public buildings, including the glass pyramid for the renovated Musée du Louvre in Paris in 1989 (see fig. 912).

Urban Planning and Airports

City planning has been a dream of architects since antiquity, and has only occasionally been realized, as in the Hellenistic cities built after the conquests by Alexander the Great, Roman forums, Renaissance and Baroque piazzas, the

above: 748. I. M. PEI. National Gallery of Art, East Wing, Washington D.C. 1978

Imperial Forbidden City of Peking, and Baron Haussmann's rebuilding of Paris in the nineteenth century. Le Corbusier, as indicated, was one of the visionaries of European planning in his designs for a new Paris, though it was only at Chandigarh (fig. 715) that he was able to realize some of his ideas. Brasília is perhaps the most complete realization of a new city plan in the twentieth century (fig. 725).

Most American efforts at urban planning or slum reconstruction have been brave but incomplete attempts, continually frustrated by antiquated codes and political or economic opposition. Among the most successful efforts at rebuilding the center of an American city are the urban renewal projects in Philadelphia (figs. 749, 750), which were begun in the 1940s. These have involved building new highways and coordinating the design of public and commercial buildings, as well as slum clearance and, most important in a city like Philadelphia, the preservation of historic monuments. The achievements in Philadelphia have been remarkable. Other comparable rehabilitation programs are being carried out, but the dream of the ideal city is still far from realized. Great American metropolises like New York, Miami, Phoenix, Atlanta, and Los Angeles continue to fight day-to-day battles concerning human welfare, traffic congestion, and water and air pollution.

Many of the best opportunities for a different kind of

urban planning were offered by the airports that proliferated in the 1950s and 1960s. Unfortunately, very few of these opportunities have been successfully realized. With some exceptions, the architecture proved routine and the solutions to problems such as circulation proved inadequate. New York's John F. Kennedy International Airport is a vast mix of miscellaneous architectural styles illustrating the clichés of modern architecture. A few individual buildings rise above the norm. One is EERO SAARINEN's (1910–1961) TWA terminal at JFK (fig. 751), with its striking airplane-wing profile and interior spaces, clearly suggestive of the ideal of flight. In 1961–62 Saarinen had the opportunity to design a complete terminal at the Dulles Airport near Washington, D.C. (fig. 752). The main building—with its upturned floating roof and the adjoining related traffic tower—incorporates an exciting and unified design concept. Saarinen here also resolved practical problems, such as transporting passengers directly to the plane through mobile lounges. Due to increased air traffic, Dulles was expanded in the mid-1990s to more than double its original size, reminding us that architectural form often derives from changing functions. In the 1960s, Saarinen had predicted such an expansion, and hence left behind explicit architectural plans for use at a future date. The final design, completed by the firm Skidmore, Owings & Merrill, remained loyal to Saarinen's

749. Independence Mall, Philadelphia, before reconstruction. 1950

750. Independence Mall, Philadelphia, after reconstruction. 1967

original plans; the building's design was extended by 320 feet on either end, thereby altering the airport's proportions but leaving the basic form undisturbed.

Architecture and Engineering

To many students of modern architecture and particularly of urban design, the solutions for the future seemed to lie less in the hands of the architects than in those of the engineers. Architects themselves were closely following new engineering experiments, particularly such new principles of construction as those advanced by **BUCKMINSTER FULLER** (1895–1983). Fuller was the universal man of modern engineering. As early as 1929 he designed a Dymaxion House (fig. 753), literally a machine for living which realized Le Corbusier's earlier con-

right: 751. EERO SAARINEN. Interior, TWA Terminal, John F. Kennedy International Airport, New York. 1962

below: 752. EERO SAARINEN. Dulles Airport, Chantilly, Virginia. 1961–62

753. BUCKMINSTER FULLER. Dymaxion House. 1929

754. BUCKMINSTER FULLER. American Pavilion, Expo '67, Montreal. 1967

cept. In the early 1930s, he built a practical three-wheeled automobile, which has been recognized as one of the few rational steps toward solving the problem of city traffic congestion, but which has never been put into production. These and many other inventions led ultimately to his geodesic dome structures, based on tetrahedrons, octahedrons, or icosahedrons. These domes, which can be created in almost any material and built to any dimensions, have been used for greenhouses, covers for industrial shops, and mobile, easily assembled living units for the American army. Although Fuller's genius had long been widely recognized, it is only in the second half of this century that he received an opportunity to demonstrate the tremendous flexibility, low cost, and ease of construction of his domes. There were many innovative structures at Expo '67, including Safdie's Habitat (fig. 742), but Fuller's geodesic dome (fig. 754) dominated the whole exposition. The structure was a prototype of what he called an "environmental valve" which encloses sufficient space for whole communities to live within a physical micro-

cosm. The American Pavilion at Expo '67 was a triumphant vindication of this engineer-architect whose ideas of construction and design promised to make most modern architecture obsolete.

This chapter has dealt with the years of massive building and rebuilding after World War II, when the International Style became so dominant that, in its second wave, it reigned virtually supreme. By the mid-1960s, critics of the International Style were expressing their disenchantment with its dogmatic loyalty to functionalism and its technology-driven aesthetic. Opposition to International Style/Bauhaus style in the 1970s took form in what began to be called Postmodernism. As we will see in chapter 26, Postmodernism's efforts to encompass the pluralism of a more completely global world, while simultaneously acknowledging the history and lessons of early styles and movements, have resulted in hybrid styles that fly in the face of the reductive, rational purity of the International Style.

24 The Pluralistic Seventies

Presented with a world torn by mounting political and social conflict, the younger artists who emerged during the late 1960s and early 1970s found little to satisfy their expressive needs in the extreme purity and logic of Minimal art. In the 1970s the New York critic John Perreault voiced the growing dissatisfaction with the Minimalist aesthetic: "Presently we need more than silent cubes, blank canvases, and gleaming white walls. We are sick to death of cold plazas, and monotonous 'curtain wall' skyscrapers . . . [as well as] interiors that are more like empty meat lockers than rooms to live in." With this, art entered what has come to be known as its Post-Minimal phase, a time, like so many others throughout history, when classical balance—in the case of Minimalism, the most literal, concrete kind of form and the most rarefied intellectual or even spiritual content—yielded to its expressive opposite. In part, it was the Minimalists' emphasis on the object, a commodity that could still be bought and sold, that certain artists rejected. To avoid the stigma of commercialism and to recover something of the moral distance traditionally maintained by the avant-garde in its relations with society at large, certain artists ceased to make objects altogether, except as containers of information, metaphors, symbols, and meaningful images. This ushered in Conceptualism, which considered a work finished as soon as the artist had conceived the idea for it and had expressed this, not in material, objective form, but rather in language, documentation, and proposals. Along with Conceptualism came Process art, which undermined Minimalist forms by subjecting them to such eroding forces of nature as atmospheric conditions, as well as to the physical force of gravity. Related to Process art were "scatter works," consisting of raw materials dispersed over the gallery floor, a manifestation that opposed formalism with formlessness; and Earthworks, for which artists abandoned the studio and gallery world altogether and operated in open nature, to create art by suggesting a formal dialectic with existing sites. But while some artists turned outward to nature, others turned inward, to their own bodies, which, on a more intimate scale, served as the site for formal procedures comparable to those worked upon the environment. The artist in his or her own person became central to Performance art, wherein ideas about the body and its dominant place in ideology were communicated in works of theater, consisting of paintings, songs, recitations, and dance, and often accompanied by instrumental or electronic music, light displays, and video.

Simultaneously, as some artists abandoned studio and gallery, others returned to those precincts and revived easel painting in a style so sharply focused that it was dubbed Photorealism. This mimesis derived not from the direct observation of the phenomenal world but rather from a purity of conception derived from an embrace of both painting and photography. Integral to the Photorealists was the primary source of the photograph, the very kind of documentation most favored by the iconoclastic Conceptualists.

From a modernist point of view, it appeared as if someone had opened Pandora's box in the 1970s and released all the demons that modernity had exorcized, not only illustration but also pattern, decoration, expressionism, and even narrative. Without a dominant mode to follow, younger artists would choose to work in more inclusive modes forbidden by a puritanical formalism. Rather than relying exclusively on the example provided by the plastic arts, artists during the 1970s turned increasingly to theoretical models as a means to develop their art. Particularly powerful were theories of language and meaning such as those provided by the writings of the analytic philosopher Ludwig Wittgenstein. During this period, the Marxist social and political critiques of such philosophers as Theodor Adorno, Herbert Marcuse, and Walter Benjamin—the Frankfurt school—began to have increasing resonance with artists and critics. Their thinking could easily be applied to the very system of commodification the art market traditionally supported. So open, however, did this "pluralistic" era become that even the formalists had their place.

The pluralism of artistic forms and motivations during the 1970s coincided with an accelerating revisionism in art history, a trend that reflected a desire on the part of many scholars to recover the other histories and arts of the nineteenth and twentieth centuries that modern accounts originally marginalized or repressed.

Conceptual Art

The gradual and progressive "dematerialization of the art object," as the critic Lucy Lippard called the anti-formalist movement, has its roots in the later 1960s, but became increasingly dominant in the early 1970s. In 1970 The Museum of Modern Art in New York mounted an exhibition called simply "Information." As the title suggests, the show acknowledged that ideas had once again come to the fore in art and, for many younger as well as older artists, taken precedence over concrete form. But already in 1967, the Minimalist sculptor Sol LeWitt had defined his own grid and cube works (see figs. 697, 698) as Conceptualist, while reinforcing his position with a theoretical exegesis that had vast influence on artists of kindred spirit. "In Conceptual Art," LeWitt wrote in 1967, "the idea or concept is the most important aspect of the work . . . all planning and decision[s] are made beforehand and the execution is a perfunctory affair. The idea becomes the machine that makes the art. . . ." But even though Minimalism took a preconceived, intellectual approach, creating art from such "ready-mades" as mathematical systems, geometric form, raw industrial materials, and factory production, and pushed reductive formalism just short of total self-elimination, Conceptual artists proposed that the next logical move lay in taking art beyond the object into the realm of language, knowledge, science, and worldly data.

As if to offer a textbook demonstration of how this could be accomplished, the American artist JOSEPH KOSUTH (b. 1945), a co-curator of the "Information" exhibition, made *One and Three Chairs* (fig. 755), which consists of a real chair accompanied by a full-scale photograph of it and a dictionary definition of "chair," together providing a progression from the real to the ideal and thereby encompassing all the essential properties of "chairness." The artist was concerned with

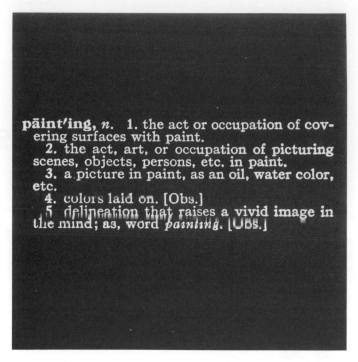

756. JOSEPH KOSUTH. *Art As Idea As Idea.* 1966. Mounted Photostat, 48 x 48" (121.9 x 121.9 cm).
Collection Roy and Dorothy Lichtenstein
© 1997 JOSEPH KOSUTH/ARTISTS RIGHTS SOCIETY (ARS), NEW YORK

exposing the mechanics of meaning, which in regard to objects involves linking a visual image with a mental concept. It is not inappropriate here to allude to the general language of semiotics, the science of signs, and say that the signifier (a specific representation of a concept or thing) and the signified (the actual object, chair) combine to produce the sign "chair." But by exhibiting, along with the signifier (a common folding chair), a surrogate for the visual representation (the photograph) and a surrogate for the mental concept (the dictionary definition), Kosuth transformed his analysis into art and thus created a new, higher, more transcendent sign, or metasign, which in turn invites further analysis. He would soon omit the first two steps of his analytical process and go straight for the metasign, in a sequence of Photostat blowups of dictionary definitions of art-related words (fig. 756). Once mounted and exhibited, the blowups not only achieve the character of visual art but also an objectlike aesthetic quality comparable to that of Minimalism's crisp, elegant beauty, as well as something of its serial repetitiveness.

Revisiting the early-twentieth-century idea later articulated by Donald Judd ("if someone says it's art, it's art"), Conceptualism soon spawned a vast and unruly variety of Post-Minimal works ranging from Performance art to Process and Land art, all united, however, by a common and unprecedented emphasis upon ideas and their expression through some medium other than a unique object—a permanent, portable, and therefore marketable commodity. In illustration of the principle that the "ideal Conceptual work," as Mel Bochner characterized it, could be described and experienced in its description and that it be infinitely repeatable, thus devoid of "aura" and uniqueness, the French artist DANIEL BUREN (b. 1938) reduced his painting to a uniform neutral and internal system of commercially printed vertical

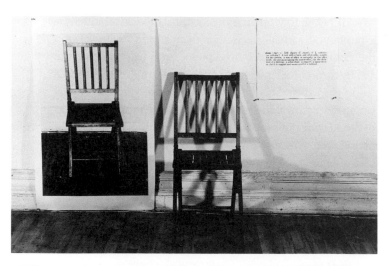

755. JOSEPH KOSUTH. *One and Three Chairs.* 1965. Wooden folding chair, photographic copy of a chair, and photographic enlargement of a dictionary definition of a chair; chair, 32⅜ x 14⅞ x 20⅞" (82 x 37.8 x 53 cm); photo panel, 36 x 24⅛" (91.5 x 61.1 cm); text panel, 24⅛ x 24½" (61.3 x 62.2 cm).
The Museum of Modern Art, New York
LARRY ALDRICH FOUNDATION FUND.
© 1997 JOSEPH KOSUTH/ARTISTS RIGHTS SOCIETY (ARS), NEW YORK

757. DANIEL BUREN. *Untitled.* 1973. Installation incorporating green and white stripes, Bleecker Street, NY (now destroyed)
© 1997 ADAGP, PARIS/ARTISTS RIGHTS SOCIETY (ARS), NEW YORK

stripes, a formal practice that readily lent itself to infinite replication and exhibition in any environment (fig. 757); the initial concept would never have to change. Variety came not in the form itself but rather in the context; however, the very sameness of the stripes robbed them of their intended neutrality or anonymity and conferred an individual stamp as irrefutable as a signature.

Of the many forms of "documentation" the Conceptualists found for their ideas, none served their purposes more perfectly than verbal language itself. "Without language, there is no art," declared **LAWRENCE WEINER** (b. 1940), who maintained that he cared little whether his "statements," a series of tersely phrased proposals, such as *A 36 x 36" Removal to the Lathing or Support of Plaster or Wallboard from a Wall* (fig. 758), were ever executed, by himself or anyone else. Fundamentally, he left the decision to implement the idea to the "receiver" of the work. "Once you know about a work of mine," he wrote, "you own it. There's no way I can climb into somebody's head and remove it."

Meanwhile, **DOUGLAS HUEBLER** (b. 1924) made the Conceptualists' attitude toward form stunningly clear in a famous pronouncement published in 1968: "The world is full of objects, more or less interesting; I do not wish to add any more. I prefer, simply, to state the existence of things in terms of time or of space." In *Location Piece #14* (fig. 759) Huebler proposed that photographs be taken from a plane window over each of the states between New York and Los Angeles. The piece would then be constituted of the twenty-four photographs, a map of the world, and the artist's state-

758. LAWRENCE WEINER. *A 36 x 36" Removal to the Lathing or Support of Plaster or Wallboard from a Wall.* 1967
Private collection © 1997 LAWRENCE WEINER/ARTISTS RIGHTS SOCIETY (ARS), NEW YORK

Location Piece #14
Global
Proposal*

    During a given 24 hour period 24 photographs will be made of an imagined point in space that is directly over each of 24 geographic locations that exist as a series of points 15 longitudinal degrees apart along the 45° Parallel North of the Equator.
    The first photograph will be made at 12:00 Noon at 0° Longitude near Contras, France. The next, and each succeeding photograph, will be made at 12:00 Noon as the series continues on to 15° Longitude East of Greenwich (near Senj, Yugoslavia) on to 30°, 45°, 60°, etc., until completed at 15° Longitude West of Greenwich. "Time" is defined in relationship to the rotation of the Earth around its axis and as that rotation takes 24 hours to be completed each "change" of time occurs at each 15° of longitude (Meridian); the *same* virtual space will exist at "Noon" over each location described by the series set for this piece. The 24 photographs will document the same natural phenomenon but the points from which they will be made graphically described 8,800 miles of linear distance and "fix" 24 hours of sequential time at one instant in real time.
    The 24 photographs, a map of the world and this statement will join altogether to constitute the form of this piece.
    *The owner of this work will assume the responsibility for fulfilling every aspect of its physical execution.

July, 1969   Douglas Huebler

759. DOUGLAS HUEBLER. *Location Piece #14.* 1969. Model photograph and text
© 1997 DOUGLAS HUEBLER/ARTISTS RIGHTS SOCIETY (ARS), NEW YORK

ART HISTORY

A young artist had just finished art school. He asked his instructor what he should do next. "Go to New York," the instructor replied, "and take slides of your work around to all the galleries and ask them if they will exhibit your work." Which the artist did.

He went to gallery after gallery with his slides. Each director picked up his slides one by one, held each up to the light the better to see it, and squinted his eyes as he looked. "You're too provincial an artist," they all said. "You are not in the mainstream." "We're looking for Art History."

He tried. He moved to New York. He painted tirelessly, seldom sleeping. He went to museum and gallery openings, studio parties, and artists' bars. He talked to every person having anything to do with art; travelled and thought and read constantly about art. He collapsed.

He took his slides around to galleries a second time. "Ah," the gallery directors said this time, "finally, you are historical!"

Moral: Historical mispronounced sounds like hysterical.

760. JOHN BALDESSARI. "Art History" from the book *Ingres and Other Parables*. 1972. Photograph and typed text, 8½ x 10⅞" (21.6 x 27.6 cm) © 1997 COLLECTION ANGELO BALDASSARRE, ITALY

ment, which concluded: "The owner of the work will assume the responsibility for fulfilling every aspect of its physical execution." But even if the "owner" assumed that responsibility, the photographs would have been indistinguishable from one another, leaving the viewer to make sense of them and bringing to bear issues of individual identity and personal response.

Yet more ephemeral, but touched with the charm of whimsy, were such proposals as the following by **ROBERT BARRY** (b. 1936), offered as serious artistic commentary:

> All the things I know
> But of which I am not
> At the moment thinking—
> 1:36 PM, June 1969.

While the Conceptualists may have denied the myriad sensory delights offered by traditional painting and sculpture, they discovered new possibilities within the relatively restricted field of language and linguistically analogous systems. Books, newspapers, magazines, catalogs, advertising, postal and telegraphic messages, charts, and maps were all seized upon and exploited as resources for information and opinions about art and almost anything else of world interest in the early 1970s. Also susceptible to Conceptualist exploitation was photography, inherent in any modern print medium, and now also readily available in the kinetic form of video, only made available to the public in the mid-1960s. The non-unique visual image quickly seemed almost as ubiquitous in Conceptual art as words.

JOHN BALDESSARI (b. 1931) composed brief but trenchant tales about art itself, usually accompanied by a reproduction of some key monument from the art-historical canon (fig. 760). In later works such as the photomontage *Heel* (color plate 370, page 635), however, the narrative element became extremely elliptical. The viewer may notice that almost all of the individuals shown in the side photographs

have injured feet, suggesting that the title may perhaps refer to the idea of an Achilles' heel, the legendary Greek hero's one vulnerable point. And we may also note that the individuals on the periphery could easily gather in the crowd depicted in the central image. But in order to grasp the connection that the artist apparently had in mind, it helps to know that he had been reading Elias Canetti's book *Crowds and Power*, on the relationship between crowds and individuals. In superimposing a unifying red line on the people who are drifting away from the main group, Baldessari was identifying what might be seen as the mob's Achilles' heel: Its vulnerable point is that its members may wander off, one by one, dissipating its power to act.

Books, like verbal communication, offer a sequential, cumulative experience, as opposed to the plastic art of traditional painting which, being static, has the potential for providing a total, all-at-once experience at the very instant of perception. Freed from the tedium of material restraints, the Conceptualists could now move even beyond the third dimension offered by sculpture to explore the fourth dimension of time. **ON KAWARA** (b. 1933), a Japanese artist resident in New York, made the passage of time itself the all-important subject by each day starting a small black painting that simply set forth the current date in white block letters (fig. 761). With every panel an equal component in the series, the work can reveal its full meaning only as a total conception.

The German artist **HANNE DARBOVEN** (b. 1941) recorded the passage of time and her experience of it by filling an enormous number of pages with a kind of abstract calligraphy and mysterious permutating numeration, derived in part from the days, weeks, and months of the calendar. As the artist's digits add up, multiply, and interweave, they eventually cover whole walls of gallery space (fig. 762), finally becoming a complete environment given over to a trancelike involvement with time's steady, inexorable advance.

Performance Art and Video

Some artists found the material limitations of the written word confining and preferred instead the temporality of Performance art. In the 1970s so many artists embraced Performance that it has been called the art form most characteristic of the period. To a generation more eager than ever to disavow the past, Performance meant venturing into an arena, specifically theater, where artists, owing partly to their lack of experience in the field, felt encouraged to proceed as if unfettered by rules or traditions. Not only did Performance liberate artists from the art object, but it also freed them to adopt whatever subject matter, medium, or material seemed promising for their purposes. Performance was not simply visual communication: It often incorporated words and called upon concepts of ritual and myth that had long been important to twentieth-century artists. Moreover, it enabled them to offer their work at any time, for any duration, at any kind of site, and in direct contact with their audience. This gave artists instant access to the receivers of their work—without the intervention of critics, curators, and dealers—and thus permitted them a new level of control over its display and destination. For all these reasons, Performance appeared to offer the maximum possibility for converting art from an object of consumption to a vehicle for ideas and action, a new form of visual communication.

No artist active in the 1970s realized the heroic potential of Performance more movingly than the charismatic and controversial German artist JOSEPH BEUYS (1921–1986). Shot down from his plane during World War II and given up for dead in the blizzard-swept Crimea, Beuys returned to peacetime existence determined to rehumanize both art and life by drastically narrowing the gap between the two. To achieve this, he employed personally relevant methods or materials in order to render form an agent of meaning. He began by piling unsymmetrical clumps of animal grease in empty rooms and then wrapping himself in fat and felt, an act that ritualized the materials and techniques the nomadic Crimean Tatars had used to heal the artist's injured body. Viewing his works "as stimulants for the transformation of the idea of sculpture or of art in general," Beuys intended them to provoke thoughts about what art can be and how the concept of art making can be "extended to the invisible materials used by everyone." He wrote about "Thinking Forms," concerned with "how we mold our thoughts," about "Spoken Forms," addressed to the question of "how we shape our thoughts into words," and finally about "Social Sculpture," meaning "how we mold and shape the world in which we live: *Sculpture as an evolutionary process; everyone is an artist.*" He continued, "That is why the nature of my sculpture is not fixed and finished. Processes continue in most of them: chemical reactions, fermentations, color changes, decay, drying up. Everything is in a *state of change.*"

In a Düsseldorf gallery in 1965 the artist created *How to Explain Pictures to a Dead Hare* (fig. 763), for which he sat in a bare room surrounded by his familiar media of felt, fat,

761. ON KAWARA. The Today Series of Date Paintings. 1966. Installation view PHOTOGRAPH COURTESY SPERONE WESTWATER GALLERY, NEW YORK

762. HANNE DARBOVEN. *24 Songs: B Form.* 1974. Ink on paper. Installation view, Sonnabend Gallery, New York

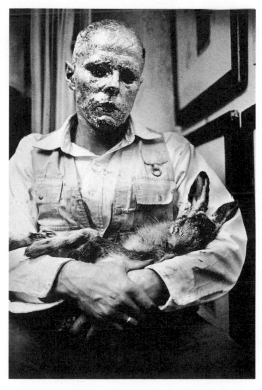

763. JOSEPH BEUYS. *How to Explain Pictures to a Dead Hare.* 1965. Performance at the Galerie Schmela, Düsseldorf, Germany

PHOTOGRAPH © 1986, WALTER VOGEL. © 1997 VG BILD-KUNST, BONN/ARTISTS RIGHT SOCIETY (ARS), NEW YORK

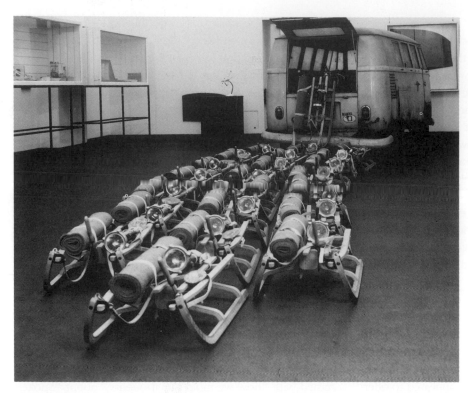

764. JOSEPH BEUYS. *The Pack*. 1969. Installation with Volkswagen bus and 20 sledges, each carrying felt, fat, and a flashlight. Staatliche Kunstsammlungen Kassel, Neue Galerie, Germany
© 1997 VG BILD-KUNST, BONN/ARTISTS RIGHT SOCIETY (ARS), NEW YORK

765. FLUXUS. Scene from *Festum Fluxurum Fluxus*. 1963
PHOTOGRAPH © MANFRED LEVE

wire, and wood, his face covered with gold leaf. A dead hare lay cradled in Beuys's arms, and he murmured urgently to it. To help explain this piece, Beuys said that in his oeuvre "the figures of the horse, the stag, the swan, and the hare constantly come and go: figures which pass freely from one level of existence to another, which represent the incarnation of the soul or the earthly form of spiritual beings with access to other regions." In this seemingly morbid performance, Beuys pointed up the complex and ambivalent feelings aroused in us by works of art that try to deal directly with such intractably unaesthetic subjects as death. The natural human reaction to the harmless creature held by the artist overturned any notion of "aesthetic distance." And the gold mask Beuys wore made

him seem not like an artist but rather a shaman or healer who, through magical incantation, could achieve a certain oneness with the spirits of animals.

Sculptural works such as *The Pack* (fig. 764) drew on Beuys's experience in a different way, endowing inanimate objects with associations both positive and negative and thereby transforming them. The sleds in this work, for example, may be part of a rescue operation, like the one that saved the artist's life during World War II, since each one carries such emergency gear as a flashlight, in addition to the felt and the animal fat that Beuys specifically associated with his own rescue. But at the same time *The Pack* bears some resemblance to the equipment of a military assault force or com-

Color plate 365.
MIES VAN DER ROHE AND
PHILIP JOHNSON. Seagram
Building, New York. 1958

Color plate 366.
LOUIS I. KAHN. Interior, west
gallery and bookstore,
Kimbell Art Museum, Fort
Worth, Texas. 1972

Color plate 367. LOUIS I. KAHN. East Hostels, Bangladesh Assembly, Dacca. 1962–74

Color plate 368. RICHARD MEIER. The Atheneum, west entrance, New Harmony, Indiana. 1975–79

Color plate 369. I. M. PEI. East Building, National Gallery of Art, Washington, D.C. 1968–78. At center, Alexander Calder. *Untitled*. 1976. Aluminum and steel, 29' x 10" x 76' (8.8 m x 25.4 cm x 23.2 m)

Color plate 370. JOHN BALDESSARI. *Heel*. 1986. Black-and-white photographs with oil tint, oil stick, and acrylic, mounted on board, 8'10½" x 7'3" (2.7 x 2.2 m). Los Angeles County Museum of Art GIFT OF THE ARTIST

Color plate 371. NAM JUNE PAIK. *TV Garden*. 1974–78. Video monitors (number varies) and plants, dimensions variable. Installation view, Whitney Museum of American Art, New York, 1982
PHOTOGRAPH © ESTATE OF PETER MOORE

Color plates 372a, 372b. VITO ACCONCI. *Instant House*. 1980. Flags, wood, springs, ropes, and pulleys, open 8' x 21' x 21' (2.4 x 6.4 x 6.4 m), closed 8' x 60" x 60" (2.4 m x 152.4 cm x 152.4 cm). Museum of Contemporary Art, San Diego
MUSEUM PURCHASE

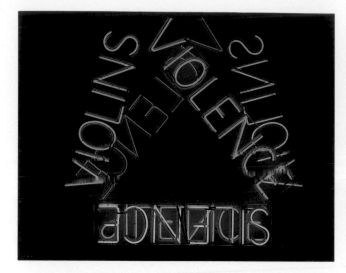

Color plate 373. BRUCE NAUMAN. *Violins Violence Silence.*
1981–82. Neon tubing with clear glass tubing suspension
frame, 62³⁄₁₆ x 65⅜ x 6" (158 x 166.1 x 15.2 cm).
Oliver-Hoffman Family Collection, Chicago.
Courtesy Leo Castelli Gallery, New York
© 1997 BRUCE NAUMAN/ARTISTS RIGHTS SOCIETY (ARS), NEW YORK

right: Color plate 375. GILBERT AND GEORGE.
Are You Angry or Are You Boring? 1977.
Color photograph, 7'11" x 6'7" (2.4 x 2 m).
Stedelijk van Abbemuseum, Eindhoven, the Netherlands

Color plate 374.
CHRIS BURDEN. *All the
Submarines of the
United States of
America.* 1987.
Installation with 625
miniature cardboard
submarines,
13'2" x 18' x 12'
(4 x 5.5 x 3.7 m),
length of each
submarine 8"
(20.3 cm).
Dallas Museum of Art
MATCHING GRANTS FROM
THE NATIONAL ENDOWMENT FOR THE
ARTS, THE 500 INC., THE JO LESCH
FUND, AND OTHER DONORS

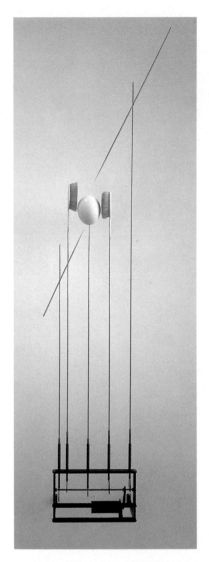

above: Color plate 376. REBECCA HORN. *Ostrich Egg Which Has Been Struck by Lightening.* 1995. Kunstmuseum, Bonn, Germany

above, right: Color plate 377. EVA HESSE. *Contingent.* 1969. Cheesecloth, latex, and fiberglass in eight panels. Installation, 12'7⁄8" x 9'4⅜" x 38½" (3.7 m x 2.9 m x 98 cm).
National Gallery of Australia, Canberra

right: Color plate 378. LYNDA BENGLIS. *Bounce.* 1969. Poured colored latex, size variable. Private collection

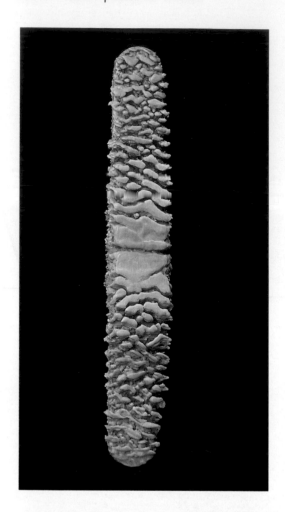

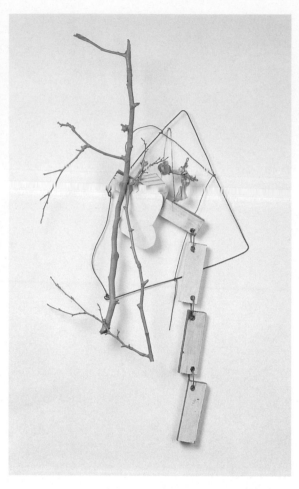

far left: Color plate 379.
Lynda Benglis.
Excess. 1971. Purified
beeswax, dammar resin,
pigments on Masonite and
pinewood, 36 x 5 x 4"
(91.4 x 12.7 x 10.2 cm).
Walker Art Center,
Minneapolis
Art Center Acquisition Fund, 1972
© 1997 Lynda Benglis/VAGA, New
York, NY

left: Color plate 380.
Richard Tuttle. *Monkey's
Recovery for a Darkened
Room (Bluebird)*. 1983.
Wood, wire, acrylic, paint,
mat board, string, and
cloth, 36 x 22 x 6½"
(91.4 x 55.9 x 16.5 cm).
Private collection

Color plate 381.
Sam Gilliam. *Carousel
Form II*. 1969. Acrylic on
canvas, 10' x 6'3" (3 x
1.9 m). Installation view,
Corcoran Gallery of Art,
Washington, D.C.
Collection the Artist

Color plates 382a, 382b. DIETER ROTH. Two screen prints from Six Piccadillies. 1969–70. Portfolio of six screen prints on board, each 19¾ x 27½" (50.2 x 69.9 cm). Courtesy Nolan/Eckman Gallery, New York

Color plate 384. NANCY HOLT. *Stone Enclosure: Rock Rings.* 1977–78. Brown Mountain stone, diameter of outer ring 40' (12.2 m), diameter of inner ring 20' (6.1 m), height of ring walls 10' (3 m). Western Washington University, Bellingham
FUNDING FROM THE VIRGINIA WRIGHT FUND, THE NATIONAL ENDOWMENT FOR THE ARTS, WASHINGTON STATE ARTS COMMISSION, WESTERN WASHINGTON UNIVERSITY ART FUND, AND THE ARTIST'S CONTRIBUTION

right: Color plate 383. ANA MENDIETA. *The Vivification of the Flesh* from the Amategram series. 1981–82. Gouache and acrylic on amate (bark) paper, 24¾ x 17" (62.9 x 43.2 cm). The Metropolitan Museum of Art, New York
PURCHASE, MRS. FERNAND LEVAL AND ROBERT MILLER GALLERY, INC., GIFT, 1983

Color plate 385. RICHARD LONG. *Ocean Stone Circle.* 1990. Stones, diameter 13' (3.96 m).
Courtesy Angles Gallery, Santa Monica, California

Color plate 386. MICHAEL SINGER. First Gate Ritual series.
1979. Bamboo and phragmites.
DeWeese Park, Dayton, Ohio
CITY BEAUTIFUL COUNCIL

Color plate 387. JOHN PFAHL. *Fat Man Atomic
Bomb/Great Gallery Pictographs.* 1984–85.
Two Cibachrome prints mounted on aluminum, 44 x 32"
(111.8 x 81.3 cm). Courtesy Janet Borden, Inc.

mando unit and in this way might assume a decidedly more aggressive than pacifist character.

In 1962 Beuys joined **FLUXUS,** a loosely knit, nonconformist international group noted for its Happenings, actions, publications, concerts, and mailing activities (fig. 765). By the end of 1965, however, he had severed relations with Fluxus, not finding the work sufficiently effective: "They held a mirror up to people without indicating how to change anything."

A different kind of Performance artist also once associated with Fluxus in Germany is **NAM JUNE PAIK** (b. 1932), who switched from straight electronic-music composition to the visual arts when he discovered the expressive possibilities of video. Paik proclaimed that "as collage technique replaced oil paint, the cathode ray tube will replace the canvas." Witty, charming, and therefore especially successful in collaboration, Paik achieved a certain notoriety for several of the pieces he created for the classical cellist Charlotte Moorman. One of these had the instrumentalist play the cello while wearing a bra made of two miniature TV sets, which took on a humanity by their association with one of the most intimate of personal garments (fig. 766). When Paik and Moorman performed *Opera Sextronique* in New York in 1967 the cellist removed all her clothing and promptly was arrested for, as the guilty verdict read, "an art which openly outrage[d] public decency." In contrast to his more provocative Performance works, some of Paik's video installation pieces can be quite lyrical, even contemplative. In *TV Garden* (color plate 371, page 635), the viewer is invited to stroll through a pastoral setting of potted plants, a garden decorated with television monitors. The glowing video images appear amid the foliage like blossoming flowers.

One form of Performance, Body art, often induced a

766. NAM JUNE PAIK. *TV Bra for Living Sculpture* (worn by Charlotte Moorman). 1969. Television sets and cello
PHOTOGRAPH © 1969 ESTATE OF PETER MOORE

forced intimacy between the performer and the audience, with results that could be amusing, shocking, or discomfiting. In *Seedbed* (fig. 767) **VITO ACCONCI** (b. 1940) spent hours during each performance day engaged in autoerotic activity beneath a gallerywide ramp, while visitors overhead heard via

767. VITO ACCONCI. *Seedbed.* 1972. Performance at the Sonnabend Gallery, New York

768. BRUCE NAUMAN. *Self-Portrait as a Fountain*. 1966–70.
Color photograph, 19¾ x 23¾" (50.2 x 60.3 cm). Edition of
eight. Courtesy Leo Castelli Gallery, New York
© 1997 BRUCE NAUMAN/ARTISTS RIGHT SOCIETY (ARS), NEW YORK

a loudspeaker the auditory results of his fantasizing. Here was
the artist reintroducing into his work the element of personal risk that came from rendering himself vulnerable to the
audience—by exposing himself engaged in an intensely personal activity and grappling with the potential audience alienation such drastic strategies could produce.

Subsequently, Acconci made sculptural objects that can
"perform" when activated by the viewer. The concerns
reflected in these later works have become more political than
psychological. His *Instant House* (color plate 372a, 372b,
page 635) lies flat on the floor in its collapsed position. When
the viewer sits on its swing, however, pulleys raise the house's
walls, making a toy building that looks like a child's version
of a military guardhouse. With the walls up, the viewer seated inside is surrounded by U.S. flags applied to the inner surfaces, while viewers outside see the red Soviet flag, with its
hammer and sickle. In Acconci's ironic reduction of Cold
War confrontation to the less intimidating proportions of a
playground game, the underlying result was to point to the
terrifyingly real implications of Cold War politics.

An artist who had the wit to make himself into a living pun
on Duchamp's *Fountain* is **BRUCE NAUMAN** (b. 1941) (fig.
768). Nauman's production over the years, however, has
been astonishingly diverse, and much more far-reaching in its
comments on other art and in its influence on younger artists
than this simple example would suggest. To choose just one
other facet of his art, Nauman has made works in neon. Some
of them blink on and off with disturbing imagery, such as
before-and-after images of a hanged man, and nothing could
more pointedly contrast with the austere, Minimal fluorescent light-fixture works of Dan Flavin (see fig. 687). Other of
Nauman's neon signs, such as *Violins Violence Silence* (color

plate 373, page 636), depict only words and as such are related to Conceptual art. Asking the viewer to read the neon lettering from either side, these signs juxtapose different words
and flip them around, backward and forward, until their
usual meanings are rendered confused and problematic; the
letters become unfamiliar and operate more like collections of
abstract shapes and bright colors. Unexpected relationships
emerge, such as the odd similarity between "violence" and
"violins."

Some Body artists risked going beyond contemporary
art's desired cutting edge into the realm of violent exhibitionism. **CHRIS BURDEN** (b. 1946) achieved international
fame in 1971 with a performance in Los Angeles that consisted of having a friend shoot him in the arm. As was critical
to much of Conceptual art, the performance was rendered
permanent in documentation. The straightforward factuality
of the photograph, capturing a crucial moment in another of
Burden's performances (fig. 769), makes the artist's ability to
endure seem all the more harrowing.

Burden distanced himself, however, from any threat of
actual danger in installations that, instead of enacting violence, sought to conceptualize it. He made *All the
Submarines of the United States of America* (color plate 374,
page 636) during the centennial of the 1887 launching of the

769. CHRIS BURDEN. *Doorway to Heaven*. November 15,
1973.
"At 6 p.m. I stood in the doorway of my studio facing the
Venice boardwalk. A few spectators watched as I pushed two
live electric wires into my chest. The wires crossed and
exploded, burning me, but saving me from electrocution."
PHOTOGRAPH BY CHARLES HILL, COURTESY OF RONALD FELDMAN FINE ARTS, INC., NEW YORK

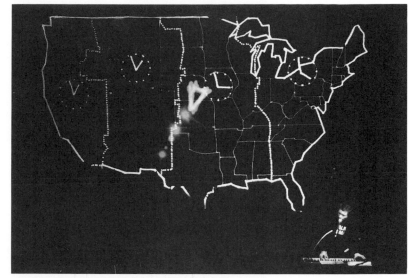

left: 770. GILBERT AND GEORGE. *The Singing Sculpture ("Underneath the Arches")*. 1969. Performance at the Sonnabend Gallery, New York, 1971

above: 771. LAURIE ANDERSON. *United States Part II*. Segment *"Let X = X."* October 1980. Performance at the Orpheum Theater, New York, presented by the Kitchen
PHOTOGRAPH © 1980 PAUL COURT

U.S. Navy's first submersible, the *SS1*. The work is made up of some 625 miniature models, representing each of the actual submarines launched by the navy during the ensuing hundred years, including its nuclear-missile–carrying Polaris submarines. The armada of tiny cardboard boats floats harmlessly in the gallery space, like a school of fish, but at the same time it enables the viewer to grasp a formidable reality—a fleet of warships capable of mass destruction.

A different kind of idea appeared when a pair of London-based artists known simply as **GILBERT** (b. 1943) **AND GEORGE** (b. 1942) transformed themselves into "living sculpture" and brought to art and the world of the late 1960s a much-needed grace note of stylish good humor. Indeed, they would probably have been quite successful performing in an old-fashioned English music hall, which, like all the rest of their material, Gilbert and George simultaneously parodied. For their best-known performance, *The Singing Sculpture* of 1971 (fig. 770), they covered their faces and hands with metallic paint, adopted the most outrageously proper English clothing and hairstyles, placed themselves on a tabletop, and proceeded to move and mouth as if they were wound-up marionettes rendering the recorded words and music of the prewar song that gave the piece its subtitle.

In their more recent work, Gilbert and George continue to feature themselves in their art, but now present only the documentary record of their presence in the form of grandly scaled wall compositions made of photographs and photograms, hand-dyed, mounted, and framed. They designate everything they do or make sculpture and explicitly favor homoerotic content, often expressed emblematically through voluptuous flower and vegetable imagery, while presenting it within a religio-aesthetic structure of stained-glass-colored mullions (color plate 375, page 637), and iconic cruciform patterns.

A Performance artist who actually does appear in music halls, as well as on big-time commercial records, while successfully preserving her place in the world of contemporary art is the multitalented **LAURIE ANDERSON** (b. 1947). As a second-generation Conceptualist and a child of the media age, Anderson can take for granted the intellectual rigor of her predecessors and, with greater command of far more spectacular means, transform the older artists' erudite, but rather amateurish, demonstrations into virtuoso performances. For *United States* (fig. 771), a four-part epic composed and first presented over the years 1978–82, Anderson marshaled the full array of her accomplishments—drawing, sculpture, singing, composing, violin playing, electronic effects—to deal in half-hour segments with the themes of transportation, politics, sociopsychology, and money. Juxtaposing images and text, sound and technological inventions, she swept her audience through a series of ironical "talking songs." The journey turned into a joy ride with such original devices as custom-made instruments, a slide show that magically came and went with the stroke of a neon violin bow, and red lips that suddenly floated in the dark. With humor, language, careful timing, style, and undeniable stage presence, Anderson has brought Conceptual art to a wider audience.

The German Performance artist **REBECCA HORN** (b.

772. REBECCA HORN. *The Feathered Prison Fan,* as it appears in the film *Der Eintänzer.* 1978.
White ostrich feathers, metal, motor, and wood, 39⅜ x 32⅝ x 12⅝" (100 x 82.9 x 32.1 cm).
Private collection, courtesy Thomas Ammann Fine Arts, Zürich
© 1997 VG BILD-KUNST, BONN/ARTISTS RIGHT SOCIETY (ARS), NEW YORK

1944) has been active in creating sculpture, installations, and films. In much of her work, Horn explores the physical limits of the human body, which she often extends by using fantastic theatrical props, such as a mask of feathers, or gloves with yard-long fingers, or bizarre winglike devices operated by the performer. The most ambitious of these props are conceived as machines, such as the mechanical sculpture *The Feathered Prison Fan* (fig. 772), which appeared in Horn's first feature-length film, *Der Eintänzer* (1978). The film is set in a ballet studio. In one scene, the fan's two huge, rotating sets of ostrich plumes, one on either side, envelop a ballerina.

Yet when the legs of the standing figure are exposed, the framework of feathers looks like a set of wings mounted on her seemingly headless torso, perhaps recalling the renowned Classical sculpture *Winged Victory,* or suggesting the feathered wings attached by the mythological Daedalus to his son Icarus—that is to say, evoking figures capable of flight. And indeed elsewhere, as in *Ostrich Egg Which Has Been Struck by Lightning* (color plate 376, page 637), the form of the egg calls to mind the most positive associations, symbolizing the artistic perfection sought by ballerinas as well as the formal absolutes of sculptors like Brancusi. The egg is placed

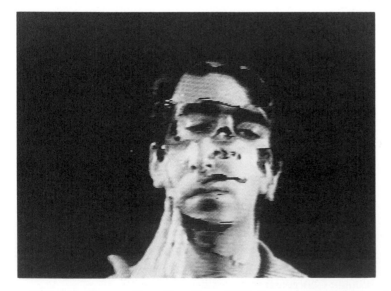

773. PETER CAMPUS. *Three Transitions.* 1973. Two-inch videotape in color with sound, 5 minutes. Courtesy Paula Cooper Gallery, New York

between two soft brushes that move up and down but never come into contact with it; the spearlike forms threaten to pierce the egg and just barely miss. It is sensual and erotic and, at the same time, tantalizes viewers, whose expectations are titillated and then denied.

During the seventies, Performance art extended itself into the new realm of video art, as video cameras and recorders became widely available and were used by artists such as Paik. **Peter Campus**'s (b. 1937) early video work included closed-circuit installations. Viewers approaching them, not realizing that they were on camera, would unexpectedly encounter projections of their own image. Observers found themselves suddenly turned into performers. In such works, Campus explored how we come to see ourselves in unanticipated ways, not only physically but psychologically, and took as one of his main themes the ever-shifting nature of the self-image. In the early 1970s he made a series of short videotapes, such as *Three Transitions* (fig. 773), that pursue related ideas of transformation, usually focusing on himself as performer. Among other strange metamorphoses acted out on the tape, Campus applies makeup to his face, and the cosmetics appear to dissolve his features away. In another sequence he sets fire to a mirror reflecting his face; the mirror burns up, leaving only a dark background. In such vignettes, the human face, the basis of our sense of identity, becomes little more than a changing mask. But though it is malleable enough, *Three Transitions* suggests, we can never see behind it.

Process Art

Like Performance, Process art countered the timelessness and structural stability of Minimal art with impermanence and variability. But while Performance artists operated in real time by making their bodies their material and personal action their means, artists interested in Process took perishability as an all-important criterion in their choice of materials and allowed the deteriorating effects of time to become their principal means. The Process artist's action concludes once he or she has selected the substance of the piece—ice, grass, soil, felt, snow, sawdust, even cornflakes—and has "sited" it, usually in a random way, by such means as scattering, piling, draping, or smearing. The rest is left to natural forces—time in tandem with gravity, temperature, atmosphere, and so forth—which suggests that now, in an art where creation and placement are an integral part of the same process, means have, in the truest sense of the word, become ends. Thus, as literal as Process art may be, it also constitutes a powerful metaphor for the life process itself.

In addition to Joseph Beuys, another artist fascinated with felt was the prolific and protean **Robert Morris** (b. 1931), who in the late 1960s subverted his own Minimalist sculptures by reinterpreting their Minimalist aesthetic in a heavy, charcoal-gray fabric that immediately collapses into shapelessness, however precisely or geometrically it may be cut (fig. 774). Soon Morris would be working with such insubstantial and transient materials as steam (fig. 775). He also countered the absolute, concrete clarity and order of his Minimal works

774. Robert Morris. *Untitled*. 1967–68. Felt, size variable. Installation view, 1968, Leo Castelli Gallery, New York. Private collection, New York
© 1997 Robert Morris/Artists Rights Society (ARS), New York

775. Robert Morris. *Untitled*. 1967–73. Four steam outlets placed at corners of square, 25 x 25' (7.6 x 7.6 m). Western Washington University, Bellingham
© 1997 Robert Morris/Artists Right Society (ARS), New York

with "scatter pieces," actually installations of what appear to be scraps left over from the industrial manufacture of metal as well as felt squares, cubes, cylinders, spheres, and grids. The order within this apparent disorder resulted from a "continuity of details," like that discernible in Pollock's allover gestur-

776. Photograph of Eva Hesse's Bowery studio in New York, 1966. Left to right: *Untitled*, 1965; *Ennead*, 1966; *Ingeminate*, 1965; *Several*, 1965; *Vertiginous Detour*, 1966; *Total Zero*, 1966; *Untitled*, 1966; *Long Life*, 1965; *Untitled*, 1966; *Untitled*, 1965; *Untitled or Not Yet*, 1966. Courtesy the Estate of Eva Hesse and Robert Miller Gallery, New York PHOTOGRAPH BY GRETCHEN LAMBERT

al painting—an art unified by the generalized, holistic continuum of its endlessly repeated drips and dribbles.

German-born EVA HESSE (1936–1970), who died at the tragically young age of thirty-four, saw Process, especially as evidenced in such unconventional, malleable materials as latex, rubber tubing, and fiberglass, as a way of subjecting Minimalist codes—serial order, modular repetition, anonymity—to a broader, less exclusive range of human values than those inherent in cerebral, male-prescribed grandeur and monumentality. Taking memory, sexuality, self-awareness, intuition, and humor as her inspiration, she allowed forms to emerge from the interaction of the processes inherent in her materials and such natural forces as gravitational pull (fig. 776). Thus, her pieces stretch from ceiling to floor, are suspended from pole to pole, sag and nod toward the floor, or hang against the wall. The works seem like dream objects, materializations of things remembered from the remote past, some even evoking the cobwebs that festoon old possessions long shut away in attics and basements. The pendulous, organic shapes of her sculptures provoke associations with gestation, growth, and sexuality, the kind of emotionally loaded themes that orthodox Minimalists cast aside. The late piece called *Contingent* (color plate 377, page 637), consisting of eight free-floating sections of cheesecloth that have

been covered with latex and fiberglass, takes the characteristic Minimalist concern with repeatable, serial objects and "humanizes" it. The luminous, translucent sheets function both as paintings and sculpture. Each of the hangings is allowed to become a distinct entity with its own texture and color, and each is suspended at a slightly different height; by these means, each one is given its own way of addressing the viewer who stands before it.

In the late 1960s LYNDA BENGLIS (b. 1941) became fascinated by the interrelationships of painting and sculpture and used Process as a route not only to new form but also to new content. She began to explore these issues by pouring liquid substances onto the floor, just as a sculptor might pour molten bronze (color plate 378, page 637). But instead of shaping it with a mold, she allowed the material to seek its own form. The artist simply mixed and puddled into the flow a startling array of fluorescent oranges, chartreuses, Day-Glo pinks, greens, and blues.

A process of a different kind, this one more organic, gave rise to the relief *Excess* (color plate 379, page 638). Here the use of beeswax suggests an association between the making of the work and bees' natural process of building up a honeycomb. The relief, too, was built up over time, growing bit by bit as the artist applied the pigmented wax medium. And the resulting form also displays unmistakable links with organic life: Its encrustations recall nothing so much as the furrowed underside of some immense caterpillar. In subsequent works, the artist has pursued pictorial effects in wall reliefs with strong plastic, sculptural qualities. Often they have fan or bow-knotted shapes and are finished with some sumptuously decorative materials, such as gold leaf or a patina of rich verdigris.

Although he came to prominence among ambitious Minimalist sculptors, many of whom have created powerful works in steel and stone, RICHARD TUTTLE (b. 1941) has dared to give viewers what Bruce Nauman, an admirer of his work, has called "a less important thing to look at." Working throughout his career with delicate and ephemeral materials, such as paper, string, and thin pieces of wire, Tuttle has been singled out for his work's unassertiveness: Dyed pieces of cloth in geometric shapes are simply attached to the wall of a gallery (fig. 777). When introduced into a specific venue such works do not aggressively transform their surroundings, as do the more imposing sculptural works of some of Tuttle's peers. Instead, with their subtle nuances of texture and color, his modest constructions invite an altogether different response from the viewer. They awaken a feeling for the sheer vulnerability of the domestic items around us, a sense of needing protection that can extend to ourselves as well. One critic has called Tuttle's work a "meditation on the extreme fragility of existence." Tuttle gains our sympathy for these frail, decorative pieces, like the one shown in color plate 380, page 638, by endowing them with so much visual life: a finely gauged interplay of primary colors, the skillful intermixing of geometric with organic forms, and an elegantly balanced yet lively composition.

The Washington-based artist SAM GILLIAM (b. 1933) fused painting and sculpture by flinging highly liquefied color onto stretched canvas, somewhat in the fashion of Jackson Pollock, though less in the spirit of controlled accident. He then suspended the support, minus its stretchers, from the ceiling (color plate 381, page 638). Thus draped and swagged, the material ceased to do what would traditionally be expected of painted canvas and became a free-form, plastic evocation.

Associated for a time with the Fluxus group, German-born DIETER ROTH (b. 1930) has lived in Switzerland, Iceland, Spain, and England. Roth became known for using food and other organic materials in unusual ways. In 1970 he had an exhibition of forty pieces of luggage, each filled with a different variety of cheese; during the run of the show, the cheeses rotted and the suitcases leaked, attracting hordes of flies. As Roth has said of another work of his made with foodstuffs, "Sour milk is like landscape, ever changing. Works of art should be like that—they should change like man himself, grow old and die."

777. RICHARD TUTTLE. Installation view of his exhibition at the Whitney Museum of American Art, New York, September 12–November 16, 1976. Left to right: *First Green Octagonal*. 1965. Dyed cloth, 54 x 22" (137.2 x 55.9 cm). Collection Jock Truman, New York. *Untitled*. 1967. Dyed cloth, approximately 38" (96.5 cm). Collection Mr. and Mrs. Ronald K. Greenberg. *Grey Extended Seven*. 1967. Dyed cloth, 39 x 59" (99.1 x 149.9 cm). Whitney Museum of American Art, New York
GIFT OF THE SIMON FOUNDATION AND THE NATIONAL ENDOWMENT FOR THE ARTS

above: 778. JANNIS KOUNELLIS. *Cotoniera.* 1967. Steel and cotton. Courtesy Sonnabend Gallery, New York

above, right: 779. JANNIS KOUNELLIS. *Untitled.* 1986. Forty-two miniature electric trains on circles of iron fixed to wooden columns, dimensions variable. Shown at the Museum of Contemporary Art of Chicago. Collection the artist

above: 780. HANS HAACKE. *Condensation Cube.* 1963–65. Acrylic plastic, water, and climatological conditions of the environment, 11¾ x 11¾ x 11⅝" (29.8 x 29.8 x 29.5 cm). Collection the artist

right: 781. HANS HAACKE. *Shapolsky et al., Manhattan Real Estate Holdings, a Real-Time Social System, as of May 1, 1971* (detail). 1971. 142 photographs, 2 maps, 6 charts. Edition of two. Collection the artist

214 E 3 St.
Block 385 Lot 11
5 story walk-up old law tenement

Owned by Harpmel Realty Inc., 608 E 11 St., NYC
Contracts signed by Harry J. Shapolsky, President('63)
                    Martin Shapolsky, President('64)
Principal Harry J. Shapolsky(according to Real Estate
Directory of Manhattan)

Acquired 8-21-1963 from John the Baptist Foundation,
c/o The Bank of New York, 48 Wall St., NYC,
for $237 600.- (also 7 other bldgs.)

$150 000.- mortgage at 6% interest, 8-19-1963, due
8-19-1968, held by The Ministers and Missionaries
Benefit Board of the American Baptist Convention,
475 Riverside Drive, NYC (also on 7 other bldgs.)

Assessed land value $25 000.-, total $75 000.- (including 212 and 216 E 3 St.) (1971)

While living in London, Roth became a friend of the British Pop artist Richard Hamilton. This association informs *Six Piccadillies* (color plate 382, page 639), a portfolio of six prints based on a postcard view of Piccadilly Circus. In this instance, Roth explored a different kind of process—the technical act of making a reproductive print, which involves separating the colors of the original image into the four basic colors used for commercial printing, creating plates to hold the inks, and reworking the image through a series of press proofs. By these means, Roth produced a series of variations on the postcard image, letting it "decompose" itself through the printed medium.

In one of his best-known works, JANNIS KOUNELLIS (b. 1936)—a Greek artist active in Italy and deeply influenced by the Process-related works of the Italians Alberto Burri and Lucio Fontana (color plate 280, page 488; see fig. 588)—stabled horses in a Roman art gallery as a way of dramatizing the contrast, and necessary relationship, between the organic world of nature and the human-created, artificial world of art. The same ideas inform *Cotoniera* (fig. 778), a piece in which the soft, white, perishable fluffiness of cotton was combined with the dark, indestructible rigidity of steel to create an ongoing dialectic of nature and industry. In an untitled work (fig. 779), the world of human creation is all that is made manifest, represented by no more than toy trains endlessly orbiting the pillars of a vacant building.

The anti-form movement found its most successful exponent in the German-American artist HANS HAACKE (b. 1936), who literally undermined the integrity of Minimalist forms by subjecting them to the eroding forces of nature, among them atmospheric conditions (fig. 780).

He went on to use Process as a means of making art play a much broader cultural role than that permitted by modernism's devotion to the art object. Sensitive to language, Haacke prefers the term "system" to "process" and characterizes his works as "real-time systems." In the 1970s he examined sociopolitical systems and their connections with the art world. In New York City, upon being invited to exhibit at the Solomon R. Guggenheim Museum, Haacke created *Shapolsky et al., Manhattan Real Estate Holdings, a Real-Time Social System, as of May 1, 1971* (fig. 781). In this Real Estate series, the artist captioned photographs of tenements, among other kinds of buildings, with business information about ownership, acquisition, and property values. Right away, the Real Estate pieces were seen as being so inflammatory that the Guggenheim canceled the show. While the cancellation may have preserved the political "neutrality" essential to a tax-free educational institution, it also guaranteed the very kind of public interest that could serve Haacke's overriding purpose—to jolt complacent viewers into helping to correct social injustice.

Following a long tradition set in Western civilization from Plato through Cézanne to the Minimalists, Canadian-born JACKIE WINSOR (b. 1941) visualizes perfection in what the 1960s had learned to call Primary forms—simple squares, cubes, cylinders, spheres, and grids. Although her works do not, once finished, evince the process of their own making, that process is quite significant. It involves prolonged, ritualistically repetitive activity and materials chosen for their power to endow ideal geometry with a mysteriously contradictory sense of latent, primitive energy. In a 1971–72 work entitled *Bound Grid* (fig. 782) that kind of energy was evident in the

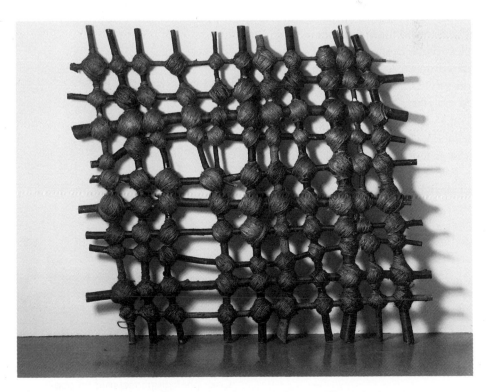

782. JACKIE WINSOR. *Bound Grid.* 1971–72. Wood and hemp, 7' x 7' x 8" (2.1 m x 2.1 m x 20.3 cm). Fonds National d'Art Contemporain, Paris

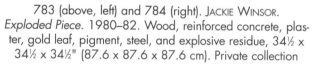

783 (above, left) and 784 (right). JACKIE WINSOR. *Exploded Piece.* 1980–82. Wood, reinforced concrete, plaster, gold leaf, pigment, steel, and explosive residue, 34½ x 34½ x 34½" (87.6 x 87.6 x 87.6 cm). Private collection

tension between a boldly simple grid form, made of crossed logs, and the slow, complex technique employed to lash the beams together. This entailed unraveling massive old ropes, returning them to their primary state as twine, and wrapping the crinkled, hairy strands round and round, for several days a week over months on end.

In 1980–82 Winsor set about activating energy in another, more startlingly dramatic manner. First she built a multi-layered interior of plaster, gold leaf, and fluorescent pigment contained within a cube made of hand-buffed, black concrete reinforced with welded steel. After bringing the piece to the requisite degree of perfection, Winsor added a further element —dynamite—and exploded it (fig. 783). Later she gathered up the fragments, reinforced the interior, and restructured the outer layer (fig. 784). In its final state *Exploded Piece* seems quiescent and contained, even though it bears the scars of the various stages—both the carefully measured and the immeasurably volatile—through which it passed in the course of its creation, destruction, and reconstruction. What physically happened to the form and its material constitutes the content of *Exploded Piece*.

The work of the Cuban-born artist **ANA MENDIETA** (1948–1985) centered on the drama of the body, treating

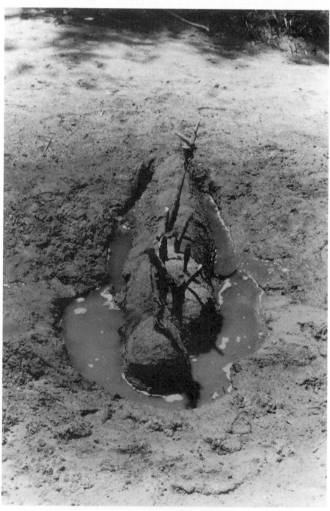

785. ANA MENDIETA. *Untitled,* from the Fetish series. 1977. Color photograph mounted on paperboard, 20 x 13¼" (50.8 x 33.7 cm). Whitney Museum of American Art, New York PURCHASE, WITH FUNDS FROM THE PHOTOGRAPHY COMMITTEE

female physiology as an emblem of nature's life cycle—its endless round of birth and decay and rebirth. Her sculptural Process pieces evoke the aura of ancient fertility rituals. These works were often developed from the direct imprint of the body, as when Mendieta outlined her own silhouette on the ground in gunpowder and then sparked it off, burning her form into the soil. The resulting image was a way, the artist said, of "joining myself with nature." That impulse is also evident in the pictograph-like figure outlines that Mendieta drew on amate (bark) paper, such as *The Vivification of the Flesh* (color plate 383, page 639). Here she inscribed a symbol of the body onto bark taken from a tree rooted in the earth. And in a work for the Fetish series (fig. 785), Mendieta created a mummy-shaped mound of mud resembling a barely buried corpse and surrounded it with a shallow ditch, as if carrying out an archaic burial ceremony and returning the deceased to the earth. At the same time, the irrigation ditch and the small branches stuck into the figure's chest suggest that the body has been in a sense "planted" in the ground, to germinate and be resurrected in the spring.

Earth and Site Works

At the same time that certain Process artists were integrating aspects of nature into their work, other Conceptualists acted on the idea of taking art out of both gallery and society and fixing it within far-off, uninhabited nature as huge, immobile, often permanent Land- or Earthworks. Insofar as pieces of this environmental character and scale were often not available to the general public, it was largely through documentation that they became known—which made such informational artifacts as photographs, maps, and drawings all the more important. Ironically, the documents often assumed a somewhat surprising fine-art pictorial quality, especially when presented in a conventional gallery setting. Then again, while Land artists may have escaped the ubiquitous marketing system of traditional art objects, they became heavily dependent on engineers, construction crews, earth-moving equipment, and even aerial-survey planes, the field equivalent of the factory-bound industrial procedures used by Minimalists, with all the high finance that that entailed. Still, the possibility of taking art into the wilderness held great meaning at the time, for just as Performance appeared to reintroduce an element of sacred ritual and mystery into a highly secularized modern society, Earth art seemed to formalize the revived interest in salvaging not only the environment but also what remained of such archaeological wonders as Stonehenge, Angkor Wat, and pre-Columbian burial mounds. The back-to-the-soil impulse may have first appeared in the post-studio works of Carl Andre, Robert Morris, and Richard Serra who began to democratize sculpture by adopting the most commonplace materials—firebricks, logs, metal squares, styrofoam, rusted nails—and by merely scattering them over the floor or assembling them on it. In Earth and Site works, the variables of selection and process inherent in the site took precedence over materials; they also shifted the perspective from that imposed by standing, vertical postures, with their anthropo-

morphic echoes of the human figure, to the bird's-eye overview allowed by arrangements that stayed flat to the ground.

When the American artist **WALTER DE MARIA** (b. 1935) initially felt the telluric pull, he responded by transporting earth directly into a Munich gallery (fig. 786). Once installed, *Munich Earth Room* consisted of 1,766 cubic feet of rich, aromatic topsoil spread some two feet deep throughout the gallery space. At the same time that this moist, brown-black rug kept patrons at a definite physical remove, it also provided a light-dark, textural contrast with the gleaming white walls of the gallery and filled the air with a fresh, country fragrance, both purely sensuous or aesthetic experiences. Eventually De Maria would expand the boundaries of his site to embrace vast tracts of fallow land and the entire sky above.

In *Lightning Field* (fig. 787) De Maria combined the pictorialness and ephemeral character of European Land art with the sublimity of scale and conception typical of American Earthworks. Here the natural force incited by the work is lightning, drawn by four hundred stainless-steel rods standing over twenty feet tall and arranged as a one-mile-by-one-kilometer grid set in a flat, New Mexican basin ringed by distant mountains. Chosen not only for its magnificent, almost limitless vistas and exceptionally sparse human population, but also for its frequent incidence of atmospheric electricity, the site offered the artist a prime opportunity to create a work that would involve both earth and sky, yet intrude upon neither, by articulating their trackless expanse with deliberately induced discharges of lightning. Few have ever been eyewitnesses to *Lightning Field* in full performance, but the photo-

786. WALTER DE MARIA. *Munich Earth Room.* 1968. Earth, 1,766 cubic feet (50 cubic meters). Installation view, Heiner Friedrich Gmbh., Munich, Germany

787. WALTER DE MARIA. *Lightning Field.* 1970–77. A permanent Earth sculpture: 400 stainless steel poles, with solid stainless-steel pointed tips, arranged in a rectangular grid array (16 poles wide by 25 poles long) spaced 220' (67.1 m) apart; pole tips form an even plane, average pole height 20'7" (6.3 m); work must be seen over a 24-hour period. Near Quemado, New Mexico

graphic documentation leaves little doubt about the sublime, albeit unpredictable, unrepeatable, and fugitive effects that can be produced by a work designed to celebrate the power and visual splendor of an awesome natural phenomenon.

In 1968, the same year that De Maria created his *Munich Earth Room,* ROBERT SMITHSON (1938–1973) relocated shards of sandstone from his native New Jersey to a New York City gallery and there piled them in a mirror-lined corner. In these gallery installations, Smithson utilized strategically positioned mirrors to endow the amorphous mass of organic shards with a new form (fig. 788). In general, Smithson saw the landscape as the illusion of having the same boundaries as the gallery, and the nonsite works were for him a synthesis of the organic, formless material from the landscape and such rigid, manufactured forms as mirrors. The juxtaposition of these materials represented the dialectic between entropy and order. And just as the transferal from nature to gallery would seem to have arrested the natural process of continuous erosion and excerpted a tiny portion from an immense, universal whole, the mirrored gallery setting expands the portion illusionistically, while also affirming its actual, and therefore infinite, potential for change in character and context. Fully aware of the multiple and contrary effects of his piece—of the dialectic it set up between site and nonsite—Smithson commented on the work's power as a metaphor for flux: "One's mind and the earth are in a constant state of erosion . . . ideas decompose into stones of unknowing."

Smithson moved from nonsite back to site and discovered a major inspiration in Utah's Great Salt Lake, which the artist saw as "an impassive faint violet sheet held captive in a stony matrix, upon which the sun poured down its crushing light." As this lyric phrase would suggest, the artist was a gifted and even prolific writer, whose essays about his Great Salt Lake creation have made *Spiral Jetty* the most famous and roman-

tic of all the Earthworks (fig. 789). He deposited 6,000 tons of earth into the lake, forming an enormous raised spiral. With its graceful curl and extraordinary coloration—pink, blue, and brown-black—the piece rewards the viewer with endless aesthetic delight, but the form, for all its purity, arose from Smithson's deep pondering of the site, combined with his fascination with entropy—the gradual degradation of matter and energy in the universe—and the possibilities of reclamation. At this particular point on the shore of the Great Salt Lake, Smithson found not only industrial ruin, in the form of wreckage left behind by oil prospectors, but also a landscape wasted and corroded by its own inner dynamism. As Smithson wrote, the gyre does not expand into a widening circle but winds inward; it is "matter collapsing into the lake mirrored in the shape of a spiral." Prophetic words, for in the ensuing years *Spiral Jetty* has disappeared—and reappeared—periodically amidst the changing water levels of the Great Salt Lake. The films and photographs that document the piece now provide the only reliable access to it. Tragically, Smithson's life was lost in a plane crash during an aerial inspection of a site in Texas.

One of the first to make the momentous move from gallery to wilderness was the California-born artist MICHAEL HEIZER (b. 1944), who, with the backing of art dealer Virginia Dwan and the aid of bulldozers, excavated a Nevada site to create the Earthwork *Double Negative* (fig. 790). Heizer is a Westerner and is sensitive to the immensity of the American landscape. In the Nevada desert he found what he called "that kind of unraped, peaceful religious space artists have always tried to put in their work." For *Double Negative* Heizer and his construction team sliced into the surface of Mormon Mesa and made two cuts to a depth of fifty feet, the cuts facing one another across a deep indentation to create a site fifteen hundred feet long and about fifty feet wide. But at the heart of this work resides a void, with the result that while providing an experience of great vastness, *Double Negative* does not so much displace space as enclose it. Here the viewer is inside and surrounded by the work, instead of outside and in confrontation with it.

For *Canceled Crop* (fig. 791), a work created in 1969 at

left: 788. ROBERT SMITHSON. *Chalk-Mirror Displacement.* 1969. Sixteen double-sided mirrors, each 10 x 60" (25.4 x 152.4 cm), and chalk: overall 10' x 10' x 10' (3 x 3 x 3 m). The Art Institute of Chicago

right: 789. ROBERT SMITHSON. *Spiral Jetty.* 1969–70. Black rock, salt crystal, and earth, diameter 160' (48.8 m), coil length 1,500' (457.2 m), width 15' (4.6 m). Great Salt Lake, Utah

above: 790. MICHAEL HEIZER. *Double Negative.* 1969–70. 240,000-ton displacement in rhyolite and sandstone, 1,500 x 50 x 30' (457.2 x 15.2 x 9.1 m). Mormon Mesa, Overton, Nevada. The Museum of Contemporary Art, Los Angeles

Finsterwolde, the Netherlands, **DENNIS OPPENHEIM** (b. 1938) plowed an "X" with 825-foot arms into a 422-by-709-foot wheat field. As if to comment on the binding artist-gallery-art cycle, Oppenheim said of his Dutch piece: "Planting and cultivating my own material is like mining one's own pigment. . . . I can direct the later stages of devel-

right: 791. DENNIS OPPENHEIM. *Canceled Crop.* 1969. Grain field, 1,000 x 500' (304.8 x 152.4 m). Finsterwolde, the Netherlands

right:
792. NANCY HOLT. *Stone Enclosure: Rock Rings* (detail of plate 384). 1977–78. Hand-quarried schist, outer ring diameter 40' (12.2 m), inner ring diameter 20' (6.1 m), ring walls height 10' (3.1 m). Western Washington University, Bellingham

below:
793. RICHARD LONG. *A Line in Scotland.* 1981. Framed work consisting of photography and text, 34½ x 49" (87.6 x 124.5 cm). Private collection, London

opment at will. In this case the material is planted and cultivated for the sole purpose of withholding it from a product-oriented system." In a work like this, Oppenheim emphasized the overridingly significant elements of time and experience in Conceptual art.

NANCY HOLT's (b. 1938) Land art involves architectural sculptures that are site specific—that is, absolutely integral with their surroundings. Holt's involvement with photography and camera optics led her to make monumental forms that are literally seeing devices, fixed points for tracking the positions of the earth, the sun, and the stars. One of her most important works to date is *Stone Enclosure: Rock Rings* (color plate 384, page 639), designed for and built as a permanent outdoor installation on the campus of Western Washington University at Bellingham, Washington. It consists of two con-

centric rings formed of stone walls two feet thick and ten feet high. The inner wall defines a tubular space at the center and the outer one an annular space in the corridor running between the two rings; the smaller ring measures twenty feet in diameter and the larger one forty feet. Piercing the walls are eight-foot arches and twelve circular holes three feet, four inches in diameter (fig. 792). Together these apertures give the spectator both physical and visual access into and even through a structure whose circular presence and carefully calculated perspectives evoke Stonehenge, a prehistoric monument apparently constructed, at least in part, as a device for marking the solar year.

In England, with its long history of nature worship in painting, poetry, and landscape gardening, a number of artists contemporary with Heizer and Smithson also looked to landscape as a means of creating art outside the market system. But these artists were confronted with very different issues—limited funds and a more intimate landscape that was densely populated, heavily industrialized, rigorously organized, and carefully protected. In response to these challenges, they chose by taste and necessity to treat nature with a featherlight touch and, in contrast to the more interventionist Americans, to work on a decidedly antiheroic scale.

RICHARD LONG (b. 1945) "intervenes" in the countryside mainly by walking through it, and indeed he has made walking his own highly economical means of transforming land into art. Along the way he expresses his ideas about time, movement, and place by making marks on the earth, by plucking blossoms from a field of daisies, or by rearranging stones, sticks, seaweed, or other natural phenomena (fig. 793). With these he effects simple, basic shapes: straight lines, circles, spirals, zigzags, crosses, and squares that he documents with photographs. A token of human intelligence is thus left on the site—like the stone markers put in place by prehistoric peoples—and is then abandoned to the weather. "A walk is just one more layer," Long asserts, "a mark laid upon the thousands of other layers of human and geographic history on the surface of the land." Then, in a reversal of this practice, these landmarks can also be gathered up and displayed in a museum setting (color plate 385, page 640), as were Smithson's nonsites. When the stones are arranged indoors, a token of the natural world is introduced into the human-made environment of buildings and communities. The walk through nature, which Long has now extended to a global enterprise, is completed as the walker returns home from the gallery.

In contrast to Nancy Holt, who constructs permanent sites for viewing transient phenomena, MARY MISS (b. 1944) builds deliberately fragile architectural sculptures as a means of stressing the ephemerality of experience. Common to both artists, however, is a preoccupation with time as it affects the perception of space, as well as a determination to create a viable public art by making the viewer more than a neutral receptor. In the publicly funded *Field Rotation* (fig. 794), sited at Governors State University in Park Forest South, Illinois, Miss took her primary inspiration from the terrain, an immense, flat field with a gently curved mound at the center. To explore this space and unleash its potential for yielding both a personal and shared expression of cultural experience, Miss used lines of posts to pattern the field as a kind of giant pinwheel. Its spokes or arms radiate outward from, while also converging toward, a "garden" sunk within the central hub or mound. In it, there is a pit shaped like a fortress and built up inside as a reticulated, cross-timbered lookout rising above a "secret" well filled with water. At the same time that the posts and their fanlike movement articulate the vast openness of the American landscape, the sunken garden which these paths lead both toward and away from provides sanctuary and retreat from the surrounding barrenness.

ALICE AYCOCK (b. 1946) too is less interested in building durable monuments than in siting and structuring her sculp-

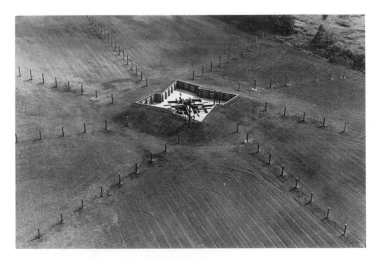

794. MARY MISS. *Field Rotation.* 1981. Steel, wood, and gravel, central structure 56' square (17.1 m square), depth 7' (2.1 m), sited on 4½ acres (1.8 hectares). Governors State University, Park Forest South, Illinois

795. ALICE AYCOCK. *A Simple Network of Underground Wells and Tunnels.* 1975. Concrete-block wells and tunnels underground, demarcated by a wall, 28 x 50' (8.5 x 15.2 m), area 20 x 40' (6.1 x 12.2 m). Marriewold West, Far Hills, New Jersey

tures to induce us to move and thus intensify our experience of the environment, the work, and ourselves. But in a manner all her own, Aycock is concerned with the psychological implications of the architectural sites she creates for the sake of reembracing nature. On a site she chose in Far Hills, New Jersey, Aycock built *A Simple Network of Underground Wells and Tunnels* (fig. 795), a structure, as the title would suggest, hardly visible at eye level, but filled with implications for the spectator courageous enough to enter it. To realize the work, Aycock began by marking off a twenty-eight-by-fifty-foot site with straight rows of cement blocks. Within this precinct she then excavated a twenty-by-forty-foot area and installed two sets of three seven-foot-deep wells connected by tunnels. After capping three of the shafts, the artist lowered ladders into two of the uncapped ones, thereby inviting the observer to descend and explore an underground labyrinth of dark, dank passages. Aycock uses Minimalism's cool, structuralist

796. MICHAEL SINGER. *Ritual Series/Syntax 1985* (to the memory of C'heng Man-ching). 1985. Granite and fieldstones, 12'1" x 8'6" x 56" (3.7 x 2.6 x 142 cm).
Collection the artist

vocabulary but in distinctively novel, even surreal ways, the better to evoke such structural precedents as caves, catacombs, dungeons, or beehive tombs and to make them the expressive agents of her own work.

Like many sculptors during the early 1970s, **MICHAEL SINGER** (b. 1945) first worked in natural environments with the materials and shifting conditions they offered (color plate 386, page 640). This took him from sites such as the beaver bogs near Marlboro, Vermont, to the saltwater marshes of Long Island, New York. Instead of imposing a preset idea on these sites in the manner of the Earth artists, however, he allowed nature and its laws to act on him and his work. In the mid-1970s, Singer returned to the studio and began making indoor pieces, confident that he could now maintain an aes-

797. JAMES TURRELL. Aerial view of *Roden Crater Project,* Sedona, Arizona, 1982. Cinder cone volcano, approx. height 540' (164.6 m), diameter 800' (243.8 m). Courtesy the artist

thetic distance from such dominant trends as Minimalism and Conceptualism. By the 1980s the resulting art had grown in complexity and poetic content, yet it preserved the essential qualities that Singer achieved almost from the start. Formally, they contained rough stones and beams of diverse shapes assembled on the ground or on doors in self-contained, though visibly accessible, structures held in serene balance—a balance so delicate and precarious that it seems redolent of the mystery emanating from altars and ritual gates (fig. 796).

The photographer **JOHN PFAHL** (b. 1939) has also been much involved with interventions at specific sites. Among his works along these lines is an extended series from 1974–78 called Altered Landscapes. Pfahl photographed natural landscapes that had been subtly changed by the addition of foreign objects, such as pieces of string, foil, or blue tape. In adding such modest items, he took care to ensure that the scene was never physically compromised.

Concern for the natural environment has prompted him to call attention to the threats posed to the earth by human indiscretions. In *Fat Man Atomic Bomb/Great Gallery Pictographs* (color plate 387, page 640) from the series Missile/Glyphs, Pfahl documents a marking of the land as harmless as his own—the petroglyphs, or pictorial inscriptions, carved or scored on rock by indigenous peoples. But he contrasts this with the horrific nuclear threat posed by modern missiles: A bulbous shape recalling "Fat Man," the military's familiar wartime name for the Nagasaki atomic bomb, has been added to the depiction of the site and looms ominously over a group of glyph figures. The small area of rock actually shown in the photograph seems to become a populated landscape, with a threatening mushroomlike form above. It is a restrained image that nonetheless carries tragic overtones.

JAMES TURRELL (b. 1943) was associated at the beginning of his career with the Californian Robert Irwin (see fig. 688), founder of what is usually called light and space art. But Turrell's installations, to a greater degree than Irwin's, explore the mysteries of how we actually perceive light, the basis of all visual experience. Specifically, Turrell manipulates our perception of light in order to reveal that our understanding of the space before us, and the way we see things within it, is fundamentally a kind of optical illusion. For example, in the early light sculpture *Afrum-Proto* (color plate 388, page 673), Turrell projected an intense beam of halogen light to form what appeared to be a three-dimensional, material object floating in the empty air. Turrell's most ambitious effort has been the site work called the *Roden Crater Project* (fig. 797), begun in 1974 and still in progress. By moving hundreds of thousands of cubic yards of earth and slightly reshaping the bowl of this extinct volcano, he is enhancing its ability to create optical effects, transforming it into a vast "viewing space." Far from city lights, visitors look up from the bottom of the crater; they witness with remarkable clarity the subtle yet extraordinary phenomena produced by changing atmospheric conditions in the clear Arizona sky. At times, the firmament over their heads can appear to solidify into an opaque disk that seems to rest on the rim of the crater. Under other conditions, the sky opens up as an illuminated dome of infinite depth, becoming what Turrell calls "celestial vaulting."

Associated with *Arte povera* in Italy—a group that attempted to illustrate the intersection of everyday life with the practice of art—the Italian sculptor and painter **MARIO MERZ** (b. 1925) has always been fascinated by the way that the world of nature and the world of modern civilization interact. Exploring this theme, he began making his "igloos" in 1968. Like much of his other work, the igloos fuse natural materials, such as mud and twigs, with industrial products, such as metal tubing and glass, to create rudimentary structures that look as if they were the shelters of some unknown nomadic people. The igloos' effect when shown in a museum is that their inhabitants have temporarily camped indoors. In the case of *Giap Igloo* (color plate 389, page 673), Merz added modern neon signs to the outside, which spell out (in Italian) a saying by the North Vietnamese general named in the work's title: "If the enemy masses his forces, he loses ground; if he scatters, he loses strength." With its military proverb and conspicuous association with Vietnam, this particular igloo, made in 1968 at the height of the war, suggests an improvised guerrilla fortification of sandbags.

Unlike most sculptors interested in the ramifications of the site, the Polish artist **MAGDALENA ABAKANOWICZ** (b. 1930) has focused on the human form. Her serial groups of figures owe some of their effectiveness to their complex surface textures, which give them each a certain organic, living quality, comparable to that of skin, as well as an individual identity. In the mid-1970s Abakanowicz planned a series of seated figures cast from plaster molds, one mold for the front and one for the back, but wound up making only frontal figures. A few years later she decided to use the back parts of the molds, too, which had been set aside. The result was the sculptural ensemble called *Backs* (color plate 390, page 673). Here, vulnerable-looking hunched-over forms, headless and with only remnants of arms and legs, gather in a crowd. The group produces an unsettling effect when installed at any site, regardless of whether it is a landscape or a museum. Outdoors, the *Backs* may recall knots of dispirited refugees resting by the side of the road, not an uncommon sight in Central Europe during the artist's lifetime. When shown in the confines of a gallery, the *Backs* become patient captives put on display, turned away from the museum visitors who have come to look.

A sense of vulnerability is also evident in *Zadra* (fig. 798), from the artist's cycle War Games. Working with huge tree trunks that foresters had cut down and left behind, Abakanowicz attacked the wood with ax and chainsaw, hacking it into forms associable with military aggression, such as cannon or projectile shapes. Yet she treated others of the dismembered trees as if they were war's helpless victims and compassionately bandaged them with burlap. (The bandage idea may owe something to the red-marked sackcloth collages made after World War II by Alberto Burri [color plate 280, page 488], who had been a doctor.) In this way, concerns related to the body were made manifest through natural materials from a particular site.

Monuments and Public Sculpture

The environmental impulse struck a deep, resonant chord in the consciousness of artists and patrons alike, and it continued to reverberate even as the trend developed away from interventionist procedures toward sculptural or architectural forms made independently of, but in relation to, the chosen site. Inevitably, public sculpture has produced conflicts, especially since much of it has been created under public commission or within the public domain, a situation that seems automatically to require that art strike some tenable balance between its own need for internal coherence and the needs of the immediate surroundings. At issue has been the degree of accommodation necessary between the essentially private, enigmatic character of modern art and the expectation that

798. MAGDALENA ABAKANOWICZ. *Zadra*, from the cycle War Games. 1987. Wood, steel, iron, and burlap, 4'3" x 3'3" x 27'4¾" (129.5 cm x 100.3 x 8.4 m). Hess Collection, Napa, California

© 1997 MAGDALENA ABAKANOWICZ/VAGA, NEW YORK, NY

799. CLAES OLDENBURG and COOSJE VAN BRUGGEN. *Batcolumn.*
1977. Welded steel bars painted gray, height 100' (30.5 m).
United States Social Security Administration, Chicago

art placed in the public arena be accessible—that is, expressive and meaningful—to the public at large. A dramatic precedent for daring, monumental sculpture in an outdoor public setting had been provided by the aged Picasso's nearly sixty-six-foot-high Chicago Civic Center sculpture of 1966 (color plate 391, page 673). That huge work's conspicuous location and popular acceptance did much to certify the validity of urban site works, as well as provide encouragement to young artists. But if site sculpture represented a return to a variant of the very "aesthetic object" that Conceptualism had set out to eliminate from art, the actual objects now produced by environmentally minded artists nonetheless remained relatively free of a modern purist egocentricity.

Monumental sculptures spread across the American urban landscape, largely owing to the United States General Services Administration's Art-in-Architecture program, which required one-half of one percent of the cost of constructing a new federal building to be allocated for the installation of artworks designed to enhance the new structure and its site. One of the most successful of the federally sponsored public monuments is **CLAES OLDENBURG** (b. 1929) and **COOSJE VAN BRUGGEN's** (b. 1942) *Batcolumn* (fig. 799). This large-scale project, as the collaborators prefer to call their monumental outdoor work, is a hundred-foot-tall open-work, diamond-grid steel shaft erected in 1977 directly in

front of the new Social Security Administration center in a former slum near the west end of Chicago's Loop. By recasting such a banal form on a monumental scale, the artists brought to fruition something of Oldenburg's earlier visionary "proposals" for revitalizing the urban world with familiar objects fantastically enlarged—immense baked potatoes, lipsticks, bananas, and Good Humor bars (see fig. 621). More recently the pair has realized *Spoonbridge and Cherry* (color plate 392, page 674) for the sculpture garden of the Walker Art Center in Minneapolis. In this work, an enormous aluminum spoon gracefully spans a pool of water. A red cherry, serving as a counterpoint to the spoon's curve, spurts water from its stem in summer. Oldenburg and Van Bruggen settled on the baseball-bat shape for the city of Chicago, an image mimicking the many heroic columns in the city's older Beaux Arts architecture, the soaring towers of the skyscraper architecture Chicago invented, the countless smokestacks celebrated by Carl Sandburg, and the game played and cheered with such passion on Wrigley Field. *Batcolumn* also makes a wry comment on art, specifically recalling Brancusi's *Endless Column* and Gabo's 1956–57 construction for the Bijenkorf department store in Rotterdam (see fig. 437).

The GSA program also commissioned the controversial *Tilted Arc* (fig. 800) by **RICHARD SERRA** (b. 1939), a 120-foot-long, 12-foot-tall, 72-ton slab of unadorned curved and tilted steel for the federal building complex (Federal Plaza) in New York City's Foley Square. Installed in 1981, the work was removed in 1989 in response to the demand of local civil servants, who found it objectionable. It was called a "hideous hulk of rusty scrap metal" and an "iron curtain" barrier not only to passage across the plaza but also to other activities that once took place there, such as jazz concerts, rallies, and simple lunchtime socializing. The removal occurred in the face of heavy opposition mounted by the professional art community, which supported the artist in his contention that to relocate a piece he deemed utterly site specific would be to destroy it, and that such action would constitute a breach of the moral, as well as legal, agreement the government made to maintain the work permanently as the artist conceived it.

During the heated public debate, little attention was given to the more positive, even lyrical, qualities of *Tilted Arc* as a physical object. For example, it could be argued that this surprisingly graceful sculptural "wall," with its elegantly leaning curvature, actually shielded the viewer from the barren vistas of Foley Square, at the same time helping to reshape what had surely been among the bleakest public plazas in New York City. And the worst insult hurled at the sculpture— "rusty scrap metal"—in fact pointed to one of its most distinctive characteristics, a feature related to Process art: An up-close examination showed that the steel's oxidizing surface was becoming more richly textured and complex as it aged and weathered. By the time it was dismantled, *Tilted Arc* had begun to look like an homage to the expansive, dappled surfaces of late Monet—a kind of industrial *Water Lilies.*

In contrast to the dispute over Serra's *Tilted Arc,* few

monuments created for urban sites have achieved such enthusiastic public acceptance as **MAYA YING LIN**'s (b. 1960) Vietnam Veterans Memorial (fig. 801), commissioned in 1981. Since its completion the memorial has been so highly acclaimed that it is difficult to remember how controversial the proposal originally was. In a competition to create a monument that would pay tribute to the 58,000 Americans killed in the Vietnam War and that would harmonize with its surroundings on the Mall in Washington, Lin daringly proposed two simple walls of polished black granite that would meet at an angle, an idea that at first seemed too abstract for some of

the interested parties. Several veterans' groups wanted to add a realistic sculpture of soldiers at the apex of the structure, which would have drastically compromised the original concept; the figure sculpture was eventually added, but at a distance from the granite wall that allowed both artworks to retain their individual integrity. In its finished state, the memorial's reflective black surface shows visitors their own faces as they search among the names of the casualties inscribed on the wall, an emotionally compelling effect that many have compared to an encounter between the living and the dead. In addition, visitors have the option of

801. MAYA YING LIN. Vietnam Veterans Memorial. 1982. Black granite, length 500' (152.4 m). The Mall, Washington, D.C.

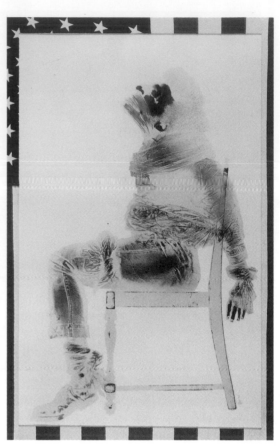

802. BEVERLY PEPPER. Four ductile iron sculptures installed at Central Park Plaza, New York, 1983. Left to right: *Volatile Presence,* height 27' (8.2 m); *Valley Marker,* height 25'8" (7.8 m); *Interrupted Messenger,* height 29' (8.8 m); *Measured Presence,* height 24' (7.3 m).
Courtesy André Emmerich Gallery, New York

803. DAVID HAMMONS. *Injustice Case.* 1970. Mixed-media print, 63 x 40½" (160 x 102.9 cm). Los Angeles County Museum of Art
MUSEUM PURCHASE FUND

making pencil rubbings of the names of their loved ones.

Since the 1960s, the American sculptor **BEVERLY PEPPER** (b. 1924) has worked in welded steel. Her monumental constructions have an imposing, if enigmatic, presence (fig. 802): They stand like totems left by some Neolithic group. Certain of the forms contribute to this effect, too, since they often appear to be based on ancient tools and spears. But other steel works seem more closely related to the modern industrial world that produced them, such as the drill and corkscrew shapes and the screwdriver form that are also visible in the examples illustrated here. This paradoxical interplay between the modern and the prehistoric gives Pepper's work much of its rich associative power. Considering her concern with elemental form, it is not entirely surprising that she would also create an outdoor environment in recent years with her *Cromlech Glen* (color plate 393, page 674). And yet, although she has somewhat reshaped the contours of the land in this case, her most notable intervention has been no more dramatic than arranging sandstone steps over small, sodded hills. Her modest outdoor setting in a public sculpture park is more respectful of the earth than is much Land art, and more immediately welcoming than many of her metal sculptures.

In celebrating the natural environment, **ALAN SONFIST** (b. 1946) has sought to re-create a sense of an earlier, unspoiled time. Sonfist was born in the South Bronx, in New York City, but happened to live near one of the area's few remaining wooded stretches. The compelling experience of seeing the vestiges of nature amid the gritty streets is reflected in his 1978 outdoor mural, *An American Forest,* at Tremont Avenue in the Bronx. The painting's marked contrast with the deteriorating buildings around it created a sharp sense of how different the American Eden had been before the land was settled and the cities built. In another, more Conceptual work, Sonfist put up memorial cardboard plaques at sites around the city where particular indigenous flora and fauna had long ago been displaced by sidewalks and office buildings. Some of his other work has a more international scope but still generally refers to an earlier, unspoiled state of nature. *Circles of Time* (color plate 394, page 675), created for the city of Florence, is a circular garden that allows the viewer to trace what Sonfist calls "the history of vegetation" in that part of Italy. Like the rings of a tree trunk, each successive concentric circle of this work, from the center to the outer rim, represents a later era. In the center is the virgin forest, a time before human intervention. Farther out lies a band representing the herb gardens of the ancient Etruscans. Next are bronze casts of endangered or extinct trees, then a ring of laurels, referring to the influence of Greece on Italy. A band of stones then demonstrates a historical Tuscan style of street pavement, and the present-day agriculture of modern Tuscany is memorialized at the edges.

DAVID HAMMONS (b. 1943) has used his art to present social and political issues in blunt terms: Among his first works were prints with intense and controversial content. His *Injustice Case* (fig. 803) shows, in negative, a bound and gagged captive tied to a chair. The figure could be any political prisoner in the world were the composition not framed by an American flag. The positive associations usually evoked by the flag are challenged, replaced by an implicit assertion that America does not offer "liberty and justice for all." It seemed natural that Hammons would later turn to public sculpture, gaining a broader audience for his continuing critique of American icons. In *Higher Goals* (fig. 804) Hammons applied bottle caps by the hundreds to telephone poles. The patterns of shiny, repeated caps give this unlikely urban monument a sense of the energy of street life. But the work also implies concern for the well-being of the city's youth. At the tops of the telephone poles are another emblem of street life: basketball hoops, placed far higher than anyone could possibly reach. "It's an anti-basketball sculpture," Hammons comments. "Basketball has become a problem in the black community because the kids aren't getting an education. That's why it's called *Higher Goals*. It means that you should have higher goals in life than basketball."

Figurative Art

Figurative artists, too, in their turn, succeeded in overturning the dominance of Minimalist abstraction, and in the 1970s they created new kinds of illusionism. But as they did, they revealed their roots in the rejected Minimalist aesthetic by translating its stillness and silence into the frozen, stop-action image of the camera—sources for which artists found in the work of Edward Hopper. And this was true even in the work of many painters who adamantly excluded photography per se from their creative process. Given its markedly static character, almost all of what also became known as the New Realism could be seen as a variety of still life, whatever its subject matter. Inevitably, therefore, genuine still-life motifs now made a stunning reappearance in painting, long after their early career in modernism, when artists favored inanimate, anonymous subjects as more conducive to cool, formalist manipulation or distortion than, for instance, the human image, with all its complicated moral and psychological associations.

In the United States, a pragmatic preoccupation with material reality, with the "old" realism, so to speak, had been too deeply ingrained to die out, even as Abstract Expressionism and its Minimal aftermath made American painting and sculpture more rigorously abstract than anything ever seen in the history of art. Throughout this very period, in fact, the most widely recognized of all living American painters was probably **ANDREW WYETH** (b. 1917), whose impeccably observed *Christina's World* (fig. 805) long remained, during an era dominated by abstraction, the most popular painting in the collection of The Museum of Modern Art in New York. Another American artist who never lost the realist faith was **ALICE NEEL** (1900–1984). After many years of subsisting on the fringes of mainstream art, Neel came into late, but considerable, fame as a portraitist of such searching, psychologically penetrating power that she seemed to be "stealing" the very souls of her sitters (fig. 806). In the same

804. DAVID HAMMONS. *Higher Goals.* 1982. Poles, basketball hoops, and bottle caps, height 40' (12.2 m). Shown installed in Brooklyn, New York, 1986

805. ANDREW WYETH. *Christina's World.* 1948. Tempera on gessoed panel, 32¼ x 47¾" (81.9 x 121.3 cm). The Museum of Modern Art, New York PURCHASE. PHOTOGRAPH © 1995 THE MUSEUM OF MODERN ART, NEW YORK

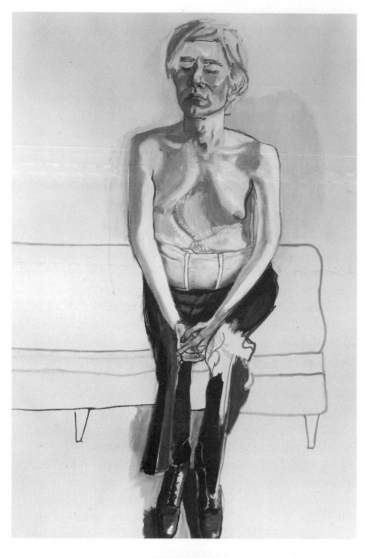

806. ALICE NEEL. *Andy Warhol.* 1970. Oil on canvas, 60 x 40" (152.4 x 101.6 cm). Whitney Museum of American Art, New York GIFT OF TIMOTHY COLLINS

generation as Neel was **FAIRFIELD PORTER** (1907–1975), who painted radiant, grandly composed intimist pictures (color plate 395, page 675; fig. 807). Albeit thoroughly American, his works seem aesthetically and spiritually at one with the calm, coloristic art of Vuillard and Bonnard. Commenting on his luminous environments, Porter said: "I was never one to paint *space.* I paint air." Although committed to figuration in his own painting, Porter regularly wrote criticism that early and unflaggingly supported the American Abstract Expressionists. This posed no conflict for Porter, since, as he stated, "the important thing for critics to remember is the 'subject matter' in abstract painting and the abstraction in representational work." Happy to work in the traditional genres of still life, landscape, and interior, as well as portraiture, Porter was convinced that the uncommon could be found in the commonplace, that "the extraordinary is everywhere."

A cultured realist who works with almost monastic dedication and attention to craft is the Spaniard **ANTONIO LÓPEZ GARCÍA** (b. 1936). His *Washbasin and Mirror* (fig. 808) gives meticulous attention to the most mundane objects, inviting comparison with certain of the Photorealists about to be discussed. At the same time, the geometric lucidity of this work—with the lowly bathroom tiles providing a strict compositional grid—carries on a tradition of firmly constructed shelf- and tabletop still lifes that can be traced back to Cézanne.

Although he began as a late Abstract Expressionist, **CHUCK CLOSE** (b. 1940) sought to reconcile the conceptual ramifications of the artist's mark with the likeness of the sitter. The camera-made image imposed the discipline of a fixed model that told precisely what the painting should look like even before it had been started (fig. 809). Furthermore, Close banished all color and worked exclusively from black-and-white images. He gave up thick, luxurious paint, and permitted himself only a few tablespoons of pigment for a huge, mural-size canvas. This was possible once he threw out his bristle brushes and adopted the airbrush, the better to eliminate all response to medium and surface; thus, his work achieves what appears to be the smooth, impersonal surface of a photographic print or slide. At the same time that Close used photographs as his point of departure, in lieu of the drawings, studies, or time-consuming observation employed by traditional realists, he committed himself to the long, arduous labor of transferring the image and all its information onto the canvas by means of a grid, a uniformly squared pattern comparable to the screen that makes tonal variations possible in photomechanical printing. The grid has served

807. FAIRFIELD PORTER. *The Mirror.* 1966. Oil on canvas, 6'½" x 5' (1.8 x 1.53 m). The Nelson-Atkins Museum of Art, Kansas City, Missouri
GIFT OF THE ENID AND CROSBY KEMPER FOUNDATION

808. Antonio López García. *Washbasin and Mirror.* 1967. Oil on wood, 38⅝ x 32⅞" (98.1 x 83.5 cm). Private collection, courtesy Marlborough Gallery, New York

painters for literally centuries as a device for transferring compositions from one surface to another, usually while also changing their scale, but only with Photorealism did the squares become so small that they could present the most minute facts, as well as overall form, so insistently as to become a dominant characteristic of the art.

But as though to counterbalance the mechanical impersonality of this process, Close decided to paint only the faces of himself and his friends, their heads full front and close up, and usually on a large scale. Trapped in a shallow, almost airless space and pitilessly exposed in all their physical imperfections, the subjects appear as much heroicized as exploited, as iconic as early American portraits, and as defiant as a police mug shot.

Eventually Close progressed toward color, at first by reproducing, one on top of the other, the cyan, magenta, yellow, and black separations of an image used in photomechanical color printing. The more Close experiments,

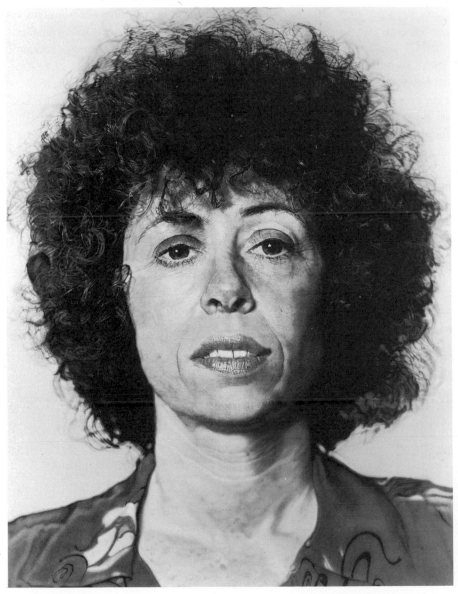

809. Chuck Close. *Linda.* 1975–76. Acrylic on linen, 9 x 7' (2.7 x 2.1 m). Akron Art Museum, Ohio

Museum Acquisition Fund, Anonymous Contribution and Anonymous Contribution in honor of Ruth C. Roush

810. CHUCK CLOSE. *Robert/Square Fingerprint II*. 1978.
Fingerprint impressions transferred to paper by pencil and
stamp-pad, 29½ x 22½" (74.9 x 57.2 cm).
Private collection

devices heightened the desired tension between the mechanical and manual aspects of the technique, as well as the fluid and fixed elements of the image, they so tipped the balance between image and system that the former seems all but swallowed by the now dominant and voracious grid. In 1988 Close suffered anterior spinal artery syndrome, which has partially paralyzed his body and confined him to a wheelchair. Despite this disability, the artist has continued to develop his art, producing new variations on the gridlike formats he uses to paint his large-scale portraits.

If there is an art with the quintessential Photorealist look, it may well be that of RICHARD ESTES (b. 1936). Estes does not work from a single photo but rather from two or more, combining them in various ways until his own painting has the feel he wants (color plate 397, page 676). Although his pictures have all the flat, planar qualities of broad sheets of reflective glass, those very surfaces, albeit transparent, reveal less what is depicted in depth than what is forward of the painting surface, achieving the effect of an inside-out world reflected from the viewer's own space (fig. 811). Reinforcing the sense of flatness is the uniformly sharp focus with which the artist, thanks to his photographic models, has been able to resolve the painting's two-dimensional surface, incorporating a vast quantity of visual information, however far or near. The richness of such an image reveals the importance of the camera to this new variety of realism, for never would a painter, however skilled, be able to render all that Estes has in a single work, or with such clarity and precision, merely by studying the subject directly. In the time required to complete the picture, everything would change—the light, the reflections, the weather, the season, the presence or absence of people and traffic, and certainly the commercial displays. The methods of traditional realism could not possibly freeze a modern urban jumble to one split second of time.

however, the more he seems to move away from realism as such toward greater involvement with process, which has inevitably carried his art back toward the realm of abstraction where it first began. Such a development is especially evident in the Fingerprint paintings, for which the artist filled each unit of the grid with an impression of his own fingerprint inked on a stamp pad (fig. 810). Other works have each cell filled with ovals or lozenge forms (color plate 396, page 675), breaking up the figure into geometric atoms in a manner surprisingly related to portraits by Gustav Klimt. While these

A mediating figure in the Photorealist style is the artist

811. RICHARD ESTES. *Double Self-Portrait.* 1976. Oil on canvas, 24 x 36" (60.8 x 91.5 cm). The Museum of Modern Art, New York

812. AUDREY FLACK. *Wheel of Fortune*. 1977–78.
Oil and acrylic on canvas, 8 x 8' (2.4 x 2.4 m).
Collection Louis, Susan, and Ari Meisel

right: 813. MALCOLM MORLEY. *Ship's Dinner Party*. 1966.
Magna color on canvas, 6'11¾" x 5'3¾" (2.1 x 1.62 m).
Museum van Hedendaagse Kunst, Utrecht, The Netherlands

AUDREY FLACK (b. 1931), who, while credited with painting the first genuine Photorealist picture, also revived certain thematic concerns guaranteeing an emotional and psychological content remote from the cool detachment characteristic of Photorealist art. Departing from her early career as an Abstract Expressionist, Flack turned to illusionism in a series creating her own contemporary versions of the seventeenth-century *vanitas* picture (fig. 812). These are still-life compositions positively heaped with the jewels, cosmetics, and mirrors associated with feminine vanity, mixed, as in old master works, with skulls, calendars, and burning candles, those ancient memento mori designed to remind human beings of their mortality and thus the futility of all greed and narcissism. Flack also composes in a traditional way, assembling the picture's motifs and arranging them to suit her purposes. Only then does she take up the camera, make a slide, project it onto the canvas, and paint over the image with an airbrush. Jammed with visual information and cropped like Baroque still lifes, Flack's pictures possess a lushness of color, a spatial complexity, and a vastness of scale that speak clearly for the time and place in which they were painted.

A British-born artist resident in New York since 1958, MALCOLM MORLEY (b. 1931) had an early engagement with Photorealism, but given his lack of interest in simulating the peculiar, glossy look of photographic images, he should be regarded as only tangentially related to the style. This, however, does not in any way diminish his importance as an artist,

for Morley's unrelenting search for meaning, both through painting and in life, has led him in a variety of fascinating but label-defying directions. One of these explorations found the artist using photographic material for his essential study or "drawing" for a painting. His choice, however, was not a projected image, but rather a color illustration from a magazine or a travel brochure (fig. 813). After dividing up this image into numbered squares, Morley would enlarge and transfer them, one by one, to the canvas, often painting parts of the image upside down as an added means of maintaining the superiority of the abstract process. In this instance, process offered the dual advantage of delivering the artist from involvement with the standard tricks of trompe-l'oeil painting, while ensuring that the abstractly rendered image would evince something of the reality reflected in its source. But to acknowledge the artificiality of that source, Morley often even retained the white border that normally surrounds a color reproduction or photographic print. Although he has denied overt political content, his choice of subject matter—Americans at play at a time of grave social unrest—is not without irony.

During the 1960s, the painter SYLVIA PLIMACK MANGOLD (b. 1938) concentrated on rendering convincing images of ordinary floors, showing the lines of the floorboards receding into illusionistic space. The wit of these images arose from their being domestic, indoor versions of Renaissance *veduta*, or views of a city square, with schematic lines of perspective fan-

814. SYLVIA PLIMACK MANGOLD. *One Exact One Diminishing on a Random Wood Floor.* 1976. Pencil, ink, and acrylic, 29⅞ x 39⅞" (75.9 x 101.3 cm). Collection Sidney and George Perutz

815. VIJA CELMINS. *Untitled (Big Sea No. 2).* 1969. Graphite on acrylic ground on paper, 34 x 45" (86.4 x 114.3 cm). AT&T Art Collection

ning out from the central vanishing point. Mangold began including mirrors in these works in 1972. *Opposite Corners* (color plate 398, page 676) shows one corner but also depicts a leaning mirror reflecting the opposite corner of the same room. At first the mirror's frame appears to be a doorway leading into the space of some adjoining chamber. But after a moment, it becomes clear that this receding "space" is in fact an illusion, as of course is the rest of the picture.

A few years after this group of works, Mangold began adding rulers to her images of floors (fig. 814). Some of these later works show one ruler within the image proper—lying in perspective on a receding floor—and also a second ruler, apparently out in the viewer's own space and abutting the

edge of the painting, as if fastened to the side of the canvas. Nothing could more economically demonstrate, and yet at the same time question, the laws of perspective that governed Western art from the Renaissance until modern times.

Born in Latvia, **VIJA CELMINS** (b. 1939) was taken to Germany as a child and then relocated to the United States in 1948. Her meticulously rendered, photograph-based drawings, paintings, and prints are also related to Photorealism. In a series of dispersed, allover compositions, she has worked from photographs of irregular natural surfaces—the desert, waves on the sea, and the craters of the moon. Here, the surface depicted, often with richly textured graphite, becomes nearly identical with the picture plane

816. VIJA CELMINS. *Concentric Bearings, D, 1984.* Published 1985. Mezzotint, aquatint, drypoint, and photogravure on Rives BFK paper, sheet 18 x 22⁷⁄₁₆" (45.7 x 57 cm). National Gallery of Art, Washington, D.C. GIFT OF GEMINI G.E.L. AND THE ARTIST

itself, undermining any sense of objective "distance" from the thing being shown. The result is often to turn the source photograph into a near-abstraction. The wave crests of *Untitled (Big Sea No. 2)* (fig. 815), for example, become so elaborately nuanced, through the thousands of individual marks made by the artist, that the camera's flat-footed realism is replaced by an infinitely complex pattern of surface incident. The surface becomes a vivid subject in its own right.

Celmins has also created portraits of individual objects, including World War II aircraft, based on found photographs. And she has combined quite distinct kinds of imagery in composite prints such as *Concentric Bearings, D* (fig. 816). The print is based on three separate, seemingly unrelated photographs: a warplane; a field of stars, from a NASA observatory photo of space; and a construction by Marcel Duchamp called a rotorelief. In the manner of some Conceptual art, the title, with its allusions to navigation and to planetary rotation, hints at elusive connections among the images.

Nowhere has the ambiguity and tension between the realness of artificiality and the artificiality of reality yielded more bizarre effects than in the work of **DUANE HANSON** (1925–1996), whose lifesize, freestanding sculptures (color plate 399, page 677) would seem, at first glance, to close the gap between art and life with a rude, decisive bang. To achieve such verisimilitude—in which artifice truly assumes the artlessness of literal reality—Hanson used direct casting from live models the way Superrealist painters use photography, as a means of shortcutting the preparatory stages so that the artist can devote the maximum time and energy to virtually duplicating the look of the model. So successful is the resulting counterfeit that the delight it can produce in viewers may suddenly turn to horror when what appears to be breathing flesh and blood fails to breathe, remaining as stiff and static as death. Having cast the figure in fiberglass-reinforced polyester resin, Hanson adds to the textured reality of pores, wrinkles, and bulges the coloration of skin, veins, and even boils and bruises. Such waxworks literalism stands in fascinating contrast to the figural sculptures of George Segal (see color plate 303, page 529), who also casts from live models, but avoids confusion between reality and make-believe by using the white mold as the final form. Furthermore, Segal restricts his use of actual objects to environmental props, while Hanson adds them to the figure as clothing and accessories.

The illusionist revival that produced Photorealism in painting and Superrealism in sculpture also regenerated naturalistic painting based on the traditional method of direct and prolonged examination of the model or motif. But even there, much new art evinced a distinctly photographic look, which hardly seems surprising, given the dominance of camera-made imagery in an age increasingly saturated by the media. One of the preeminent masters of directly perceived, sharp-focus realism is **PHILIP PEARLSTEIN** (b. 1924), a painter who belongs to the Pop, Minimalist, and Conceptual generation and even shares its preoccupation with banal subject matter and two-dimensionality, as well as cool, hard-edge handling. Pearlstein, however, reached his artistic maturity about 1970, and did so with an art devoted to the nude figure represented as a solid form in simulated depth (fig. 817). Pearlstein set about integrating modernism's flatness and cropping, characteristics initially fostered by the influence of Japanese prints. Thus, at the

817. PHILIP PEARLSTEIN. *Two Female Models with Regency Sofa.* 1974. Oil on canvas, 60 x 72" (152.4 x 182.9 cm). Courtesy Allan Frumkin Gallery, New York

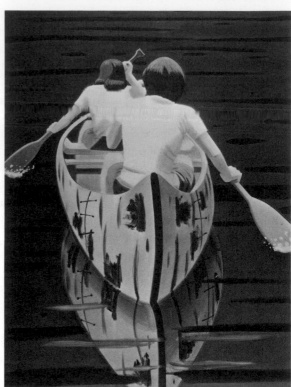

same time that he reproduces his figures in all their sexual explicitness, he also robs them of their potential for expressionist content by a process of objectification, by rendering only one part of the anatomy at a time. This results in the loss of bodily context and, often, extremities, a further depersonalization that becomes especially acute when it involves the head. The painter calls his pictorial manipulation of the figure "a sort of still-action choreography." Pearlstein declared:

> I have made a contribution to humanism in 20th-century painting—I rescued the human figure from its tormented, agonized condition given it by the expressionistic artists, and the cubist dissectors and distorters of the figure, and at the other extreme I have rescued it from the pornographers, and their easy exploitation of the figure for its sexual implications. I have presented the figure for itself, allowed it its own dignity as a form among other forms in nature.

If some painters associated with realism were moving toward a more direct account of visual appearances, paradoxically some photographers were going in the opposite direction, treating their subjects in a manner increasingly formal and abstract. JOEL MEYEROWITZ (b. 1938), who earlier in his career had earned recognition as a street photographer, now constructed images that in their immaculately refined sensibility seemed intent on portraying another world. Although *Porch, Provincetown* (color plate 400, page 677) of course

remains rooted in an actual time and place, it also brings to the fore highly abstracted elements, such as the conspicuously displayed trapezoid of the roof and the cleanly abutting planes of the sea and sky. The subtly graduated blush of twilight on the central column is so finely gauged as to make that cylindrical form look like a geometric sculpture, while the tinted atmosphere recalls painters of the sublime, from the American Luminists to Mark Rothko. Finally, the porch is shot from an angle that makes it appear to be suspended, without ground support, over the sea, as if in a Surrealist painting by René Magritte.

A reference to Surrealism can also be seen in the work of the Detroit-born photographer JERRY UELSMANN (b. 1934). Uelsmann combines two or more images into a composite work. By bringing together two ordinary representations in an unexpected way, the resulting image can have the disorienting, uncanny effect that the French nineteenth-century poet Lautréamont, admired by the Surrealists, compared to "the chance encounter on an operating table of a sewing machine and an umbrella." Uelsmann's *Untitled (Cloud Room)* (fig. 818) creates just this effect of "defamiliarizing" the ordinary. The banal Victorian interior, taken on its own, would be quite unremarkable, and the cloudy sky, too, is part of the everyday world around us. But when storm clouds gather indoors, their unexpected presence creates a strangely disquieting scene, indebted to the works of painters like Magritte and Delvaux. Uelsmann's technical mastery in printing the composite photographs enhances the effect; he joins the disparate images together so seamlessly as almost to convince us for a moment that we are witnessing something impossible but true.

The painter **ALEX KATZ** (b. 1927) made an early choice of figuration and stuck with it throughout abstraction's interregnum, turning some of abstraction's own sources of inspiration, such as Henri Matisse and Milton Avery, to his own purposes. Foremost among the formal abstract influences are broad fields of flat, clean color, elegantly brushed surfaces, radically simplified drawing, and epic scale (fig. 819). Selecting his portrait subjects from among family, friends, and art-world personalities, Katz monumentalizes them not only by the sheer size of the painted image but also by the technique of representing the sitters only from the shoulders up, as if they were antique emperors or empresses presented to the public in the form of portrait busts (color plate 401, page 677). But since the rendering has been done with a reductiveness reminiscent of movie billboards, a more apt association might be with the royalty of Hollywood rather than with that of monarchical Europe. Katz has declared: "I like to make an image that is so simple you can't avoid it, and so complicated you can't figure it out." To illustrate this quality of ambiguity, Katz here refers to the painting seen in color plate 401: "The original background was green grass," he said. "But when I painted green, it just didn't have enough light in it. It didn't show the light that I was seeing. So I turned it to yellow, switched to violet, and then to orange. Finally, that was a plausible light, the light I was really seeing, instead of just a bunch of pretty colors. It was a glowing summer light."

The painter **JANET FISH** (b. 1938) alludes to the seventeenth-century tradition of Dutch still life in her technical mastery, put to the purpose of revealing the beauty present in commonplace objects. Fish exploits art-historical precedents as a means of expanding the possibilities of her modernist experience, which began in Abstract Expressionism. That experience also includes a taste for Pop subject matter, usually glassware on the order of fruit jars, gin bottles, dishes, tumblers, and goblets. These she assembles in sufficient numbers to suggest the serial repetitiveness so meaningful to sensibilities nurtured during the 1960s. Yet she groups her objects with a sense of freedom and invention, as well as an allover distribution of accents, more akin to gestural painting than to Andy Warhol's serried rows of Coca-Cola bottles with their implied commentary on mass-produced overabundance (see color plate 309, page 531). Far from ambivalent toward the material world, Fish seems to be in love with it (color plate 402, page 678), or at least with those of its artifacts usable as vehicles for her interest in creating a dazzling array of chromatic and luminary effects. The qualities of her glasswork are particularly startling—reflections from sparkling, prismatic forms and jeweled overlays of transparent glass and liquid stand out against the plain background of a shallow, modernist space.

820. WILLIAM BAILEY. *Monte Migiana Still Life.* 1979. Oil on canvas, 54³⁄₁₆ x 60³⁄₁₆" (137.6 x 152.9 cm). Pennsylvania Academy of the Fine Arts, Philadelphia
PURCHASED WITH FUNDS FROM THE NATIONAL ENDOWMENT FOR THE ARTS (CONTEMPORARY ARTS PURCHASE) AND BERNICE MCILHENNY WINTERSTEEN, THE WOMEN'S COMMITTEE OF THE PAFA, MARION B. STROUD, MRS. H. GATES LLOYD, AND THEODORE T. NEWBOLD

In contrast to the scintillating light and color of Janet Fish, **WILLIAM BAILEY** (b. 1930) paints a tabletop still-life world of opaque crockery softly bathed in a low, suffused, even Romantically contemplative light (fig. 820). The sheer serenity of the effect must emanate as much from the artist's practice of re-creating his motifs from "memory" as from his preference for harmoniously balanced compositions of smooth, sensuous shapes and volumes juxtaposed in a limited space against the starkest of flat grounds. Acknowledging influences from the quietest still-life art of Giorgio Morandi (see color plate 278, page 487), the color theories of his Yale teacher Josef Albers (see color plate 193, page 336), and the rigorous geometries of Piet Mondrian (see fig. 451), Bailey transcends them all to realize an intensely personal vision made up of objective imagery refined to its essence within an abstract framework of ideal, classical simplicity.

An artist who shares this interest in the meticulously composed still life is the photographer **JAN GROOVER** (b. 1943) (color plate 403, page 678). Groover began as a painter of Minimalist abstractions in the early 1970s, but soon changed media because, she said, "With photography I didn't have to make things up; everything was already there." In her tabletop still lifes, which she began making in 1977, what Groover discovers as "already there" in ordinary objects are the clean, abstract shapes prized by geometric artists, shapes that are waiting to be revealed within the mundane things we handle every day. To bring these qualities out, she isolates the individual fruits, making each a separate sphere of delicate color. In addition, she selects culinary wares that are already something like miniature, polished Minimalist sculptures, such as the pastry funnel, which is revealed in her photographs as one

821. ALFRED LESLIE. The Killing Cycle (#5): *Loading Pier.*
1974–75. Oil on canvas, 9 x 6' (2.7 x 1.8 m).
Collection the Orchard Corporation

of those primal forms (the cone, the cylinder, and the sphere) that Cézanne called the basic elements of all art. In these deftly arranged, witty still lifes, small objects from the kitchen counter seem to take on epic proportions, emphasized by the large format in which Groover prints the photographs.

ALFRED LESLIE (b. 1927) achieved considerable success in the 1950s as a second-generation New York abstractionist. He reversed his direction during the following decade, however, and opted for an aggressive realism. At the same time, he sought a renewal of the traditional realist genres or subject types—specifically history painting—with all their narrative, symbolic, and didactic content. An exceptionally striking work is the fifth in a series of seven enormous paintings entitled The Killing Cycle (fig. 821), a decade-long project undertaken as a memorial to Frank O'Hara, a poet-critic friendly to Leslie's generation of American artists. O'Hara had been run down by a beach taxi while standing on the sandy shore of Fire Island late one night in the summer of 1966. To give the picture genuine specificity of time and place, Leslie spent months researching the circumstances of the accident; then, to endow it with universal significance, he based the composition and its powerful chiaroscuro effects on Caravaggio's great *Entombment of Christ* (see fig. 9) in the

Vatican collections. Thus detailed and enlarged, the picture transcends its initial inspiration to become an indictment of the mindless waste in modern life. Simultaneously, the image of the moribund O'Hara reverentially borne to his final rest by anonymous strangers assumes the quality of a Christian appeal for the brotherhood of all humankind.

Pattern and Decoration

While some saw Minimalism's aloof stillness, silence, and simplicity as potent with rarefied meaning, others could comprehend only a void, vast as the monumental scale preferred throughout mid- to late-twentieth-century art, and begging to be filled with a richly varied, even noisy abundance of pattern, decoration, *and* content. Here indeed was rebellion, for in the formalist's vocabulary of terms, the words "pattern" and "decoration" were fraught, still more than "illusionism," with pejorative connotations. And the taboo prevailed even though pattern—a systematic repetition of a motif or motifs used to cover a surface in a decoratively uniform manner—has an effect fully as flattening as any of the devices employed by the Minimalists. However, Pattern and Decoration, often abbreviated P&D, suffered the burden of connoting work done by anonymous artisans, not individuals of known genius, usually for the purpose of endowing mundane or utilitarian objects with an element of sensuous pleasure and delight, qualities thought to be secondary to the more intellectual concerns of high art. Thus, despite Duchamp, whose ready-mades proved that the context can make the art, skeptical purists wondered how objects formed from designs, materials, and techniques associated with crafts, folk art, and "women's work" could be transformed into vehicles of significant content.

The answer lay in the sources of inspiration for the P&D movement, which began to stir in the early 1970s and coalesced as a major development in the second half of the decade. First of all there was Matisse, a proud "decorator" in the grand French tradition, and his late, brilliantly colored and patterned paper cutouts. Then there were the arabesques and fretwork of Islam's intensely metaphysical art, the pantheistic interlaces of the medieval Celts, and the symbol-filled geometries of Native American art, all recently brought into stunning prominence by important exhibitions or installations. Finally, but perhaps most important of all, came the so-called craft works—carpets, quilts, embroidery, needlepoint, mosaic, wallpaper—not given much attention since Art Nouveau or the art of the Bauhaus. Apart from its irrefutable beauty, such patterned, decorative, and pragmatic folk art gained esteem for its content. Pattern painting, albeit based on the same grid as that underlying the formalist structures of Minimal art, appeared to resist self-containment in favor of an almost limitless embrace of surface. Because of its intrinsically more democratic character, Pattern and Decoration enjoyed broader and more enthusiastic support than did contemporaneous developments in Photorealism.

Initially, one of the strongest and most seasoned talents to transform traditionally undervalued decorative materials into ambitious, original art was **MIRIAM SCHAPIRO** (b. 1923), energized by feminism as well as by a love of ornamental work, however "low" or "high" its origins. Like so many artists of her generation, Schapiro had begun painting in an Abstract Expressionist style, but later turned to hard-edged abstract illusionism, using computers in the design process. In the late 1960s she joined the women's movement, which led her to become, with artist Judy Chicago, codirector of the Feminist Art Program, which in 1970 sponsored Womanhouse at the California Institute of the Arts in Los Angeles. For Womanhouse, students rehabilitated an old abandoned dwelling into an exclusively female environment, alive with artworks and performances based on women's dreams and fantasies. Since then, Schapiro has drawn her imagery and materials from the history of women's "covert" art, using buttons, threads, rickrack, sequins, yarn, silk, taffeta, cotton burlap, and wool, all sewn, embroidered, pieced, and appliquéd into useful but decorative objects. Reassembling them into dynamic, baroque, almost exploding compositions, the artist creates what she calls "femmages," a contraction of "female" and "image" that connotes something of the collage, découpage, and photomontage techniques used to make these grandly scaled projects. And the scale is important for Schapiro, since it reflects not only her modernist sensibility but also her desire to monumentalize "women's work."

In work after work, Schapiro articulated her new style, a synthesis of abstraction, feminist images (such as eggs, hearts, and kimonos), architectural framework, and decorative patterns, in the form of floral bouquets, lacy borders, and floating flowers. In one series she chose the fan shape—itself a decorative form—for wall-size canvases collaged with a wealth of patterned materials, ranging from cheap cotton to upholstery to luxurious Oriental fabrics, all combined to simulate the spreading folds of a real fan. *Black Bolero* (color plate 404, page 679) is a potpourri of art and craft, structure and decoration, abstraction and illusion, public politics and personal iconography, thought and sentiment.

Schapiro's codirector in the Feminist Art Program was **JUDY CHICAGO** (b. 1939). After producing Minimalist sculpture, in the late 1960s Chicago began to focus on feminist issues, often as a critique of the male-dominated art world of the time. Her best-known work, made in collaboration with four hundred other people, is *The Dinner Party* (color plate 405, page 679). This elaborate celebration of women's history was executed, with intentional exaggeration, in such stereotypically feminine and decorative artistic mediums as needlework and china painting. The entire ensemble is organized as a triangle, a primordial symbol of womanhood as well as equality. Within its enclosed floor area are inscribed the names of 999 women in history and legend, from the ancient world up to such modern figures as Isadora Duncan, Georgia O'Keeffe, and Frida Kahlo. Each side of the triangular table has thirteen place settings; in part this makes a wry comment on the Last Supper—an exclusively male dinner party of thirteen—and in part it acknowledges the number of witches in a coven. The thirty-nine settings pay tribute to thirty-nine notable women, and each name is placed on a

runner embroidered in a style appropriate to its figure's historical era. In *The Dinner Party*'s most provocative gesture, the plates on the table display painted and sculpted motifs based on female genitalia. Yet this, too, like the work as a whole, is also a comment on the gender history of art, since it might ostensibly allude to the famously anatomical flowers painted by O'Keeffe (see color plate 212, page 410).

In 1975 **ROBERT ZAKANITCH** (b. 1935) met Miriam Schapiro during a term of guest teaching at the University of California in San Diego, and early the following year in New York the two painters jointly organized a group called the Pattern and Decoration Artists. The group held its first meeting in a SoHo loft space, signaling not only to the art world at large but also to the participating artists that a greater variety of painting was being done than Conceptualism had allowed anyone to believe. This led to the "Ten Approaches to the Decorative" show in September 1976, sponsored by the Pattern and Decoration Artists themselves, and to a kind of collector interest that turned the recently opened Holly Solomon Gallery into a home for such P&D converts as Robert Kushner, Kim MacConnel, Valerie Jaudon, Ned Smyth, and Robert Zakanitch. When Zakanitch took up decorative imagery, he had been working as a Color Field abstractionist faithful to the Minimalist grid as the structural system of his painting. Once employing decoration, he retained both his color sophistication and his respect for structure, but translated the latter into a free, often organic lattice pattern and the former into floral motifs rendered with a lush, painterly, lavender-and-plum palette (fig. 822). Taken together, the painterliness, the opulent hues, the blooms, and the latticework invite obvious comparison with wallpaper designs, which in fact hold a preeminent place among the artist's acknowledged sources.

In 1974 **ROBERT KUSHNER** (b. 1949) went on an extended journey to Iran and Afghanistan. The experience was some-

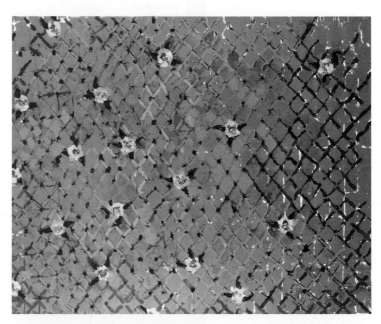

822. ROBERT ZAKANITCH. *Double Shirt*. 1981. Acrylic on canvas, 6 x 5' (1.8 x 1.5 m). Private collection, Sweden

thing of an epiphany for Kushner, who said: "On this trip, seeing those incredible works of genius, really master works which exist in almost any city, I really became aware of how intelligent and uplifting decoration can be." Although many of the masterworks happen to have been architectural, Kushner chose to express his love of arabesque fluency and grace not in mosaic or stone, but rather in loose, free-falling fabrics, decorated by their own "found" printed or woven patterns and Kushner's overpainting of leaf, vine, seed pod, floral, or, eventually, beast and human imagery. He disclosed a certain preference for a horizontal alignment of vertical panels, but instead of arranging his fabrics in a generally evenhanded sameness of length, he assembled them in whatever manner would produce a wearable costume. These works were then hung, unstretched, on the wall (color plate 406, page 680). When donned, these richly patterned and colored pieces, often in original atonal harmonies, yielded an effect of barbaric splendor unknown since the imperial era in Oriental civilization or perhaps the heyday of the Ballets Russes.

In his wall pieces (fig. 823), also structured from fabrics and airily billowing against the plane, Kushner made an unabashed avowal of his devotion to Matisse (see color plate 149, page 271). Like the Frenchman, Kushner counterbalanced his passion for Oriental embellishment with a Western humanist interest in the classical nude, flattened and all but lost in a crazy quilt of competing but subtly coordinated patterns, only to be slowly revealed, sometimes in mirror-image reversals, by the patterns themselves, as well as by the artist's own magisterial drawing.

VALERIE JAUDON (b. 1945), a Mississippi-born artist with a cosmopolitan background, began in Color Field painting and, by concentrating on the decorative aspect of that style, continued to be a steadfast abstractionist throughout her work in Pattern and Decoration. And to the degree that she is a purely nonobjective artist, Jaudon has remained closer to her Islamic sources, mingled with the interlaces of ancient Celtic Europe, than any of the practitioners seen thus far in the P&D movement. However, her patterns are not copied but freshly reinvented, ribbonlike paths traced over and under one another along arcs and angles reminiscent of Frank Stella (color plate 350, page 578). With their symmetrical order and complex weave of clean lines, which invite the eye but immediately lose it in the indivisibility of the total configuration, Jaudon's earlier maze patterns also became something of a hard-edged counterpart of the allover, holistic, or nonrelational compositions realized by Jackson Pollock. In subsequent work, Jaudon decelerated the *perpetuum mobile* of her interlaces by stabilizing them within a solidly grounded architectural image, capped by round domes, pointed steeples, and ogival arches (color plate 407, page 680). Adding to the sense of gravity and depth are the impastoed surfaces and the brilliant, strongly contrasted colors that distinguish one band or strap from another as well as from the background. Illusionism is denied, however, due to the emphatic linearity of the interlocking geometries and their resolutely flattened "folds" and "knots." Jaudon has said of her intricate art: "As far back as I can remember, everyone called my work decorative, and they were trying to put me down by saying it. Attitudes are changing now, but this is just the beginning. If we can only get over the strict modernist doctrines about purity of form, line, and color, then everything will open up."

Feminism also encouraged JOYCE KOZLOFF (b. 1942) in

823. ROBERT KUSHNER. *Sail Away*. 1983. Mixed fabric and acrylic, 7'3"x 17'2" (2.2 x 5.2 m). Courtesy Holly Solomon Gallery, New York

Color plate 388. JAMES TURRELL. *Afrum-Proto.* 1967. Quartz-halogen projection. Courtesy the artist

Color plate 389. MARIO MERZ. *Giap Igloo—If the Enemy Masses His Forces, He Loses Ground; If He Scatters, He Loses Strength.* 1968. Metal tubes, wire mesh, wax, plaster, and neon tubes, height 47¼" (120 cm), diameter 6'6¾" (2 m). Musée National d'Art Moderne, Centre d'Art et de Culture Georges Pompidou, Paris

Color plate 390. MAGDALENA ABAKANOWICZ. *Backs.* 1976–80. Burlap and resin, 80 pieces, each approx. 25¾ x 21¾ x 23¾" (65.4 x 55.2 x 60.3 cm). Installed near Calgary, Canada, 1982. Collection Museum of Modern Art, Pusan, South Korea

right: Color plate 391. PABLO PICASSO. *Chicago Monument.* 1966. Welded steel, height 65'9" (20 m). Civic Center, Chicago

Color plate 392. CLAES OLDENBURG and COOSJE VAN BRUGGEN. *Spoonbridge and Cherry*. 1985–88. Aluminum, stainless steel, and paint, 29'6" x 51'6" x 13'6" (9 x 15.7 x 4.1 m). Walker Art Center, Minneapolis

GIFT OF FREDERICK R. WEISMAN IN HONOR OF HIS PARENTS, WILLIAM AND MARY WEISMAN, 1988

Color plate 393. BEVERLY PEPPER. *Cromlech Glen*. 1985–90. Earth, sod, trees, and sandstone, 25 x 130 x 90' (7.6 x 39.6 x 27.4 m). Laumeier Sculpture Park, Saint Louis, Missouri

Color plate 394. ALAN SONFIST. *Circles of Time.* 1986–89. Bronze, rock, plants, and trees on three acres (1.2 hectares). Created in Florence, Italy. Courtesy the artist

Color plate 395. FAIRFIELD PORTER. *Under the Elms.* 1971–72. Oil on canvas, 62¹⁵⁄₁₆ x 46¼" (159.9 x 117.5 cm). Pennsylvania Academy of the Fine Arts, Philadelphia

GIFT OF MRS. FAIRFIELD PORTER

Color plate 396. CHUCK CLOSE. *Self-Portrait.* 1991. Oil on canvas, 8'4" x 7' (2.5 x 2.1 m). Collection Paine-Webber Group, Inc., New York. Courtesy PaceWildenstein, New York

below: Color plate 400. JOEL MEYEROWITZ. *Porch, Provincetown.* 1977. Color photograph, 10 x 8" (25.4 x 20.3 cm). Courtesy the artist

above: Color plate 399. DUANE HANSON. *Tourists.* 1970. Polychromed fiber-glass and polyester, 64 x 65 x 47" (162.6 x 165.1 x 119.4 cm). National Galleries of Scotland, Edinburgh

Color plate 401. ALEX KATZ. *The Red Band.* 1978. Oil on canvas, 6 x 12' (1.8 x 3.7 m). Private collection

above: Color plate 402.
JANET FISH. *Stack of Plates*. 1980. Oil on canvas, 48 x 70" (121.9 x 177.8 cm). Private collection

left: Color plate 403. JAN GROOVER. *Untitled*. 1979. Chromogenic color print, 18¹¹⁄₁₆ x 14¾" (47.5 x 37.5 cm). Courtesy Robert Miller Gallery, New York

Color plate 404. MIRIAM SCHAPIRO. *Black Bolero*. 1980. Fabric, glitter, synthetic polymer paint on canvas, 6 x 12' (1.8 x 3.7 m). Art Gallery of New South Wales, Australia

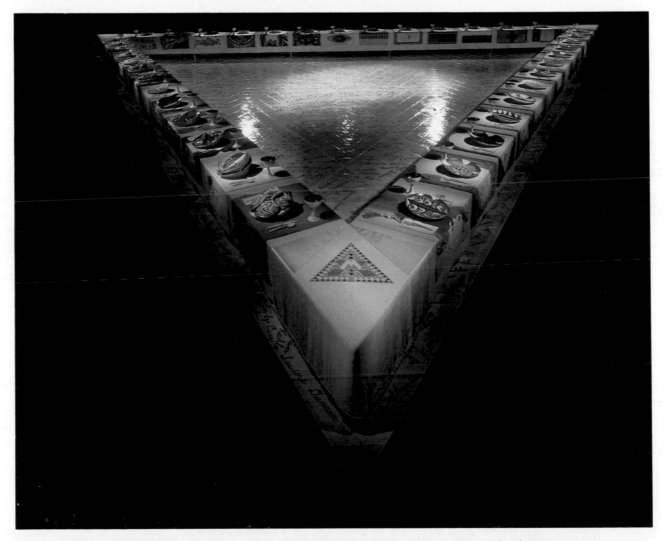

Color plate 405. JUDY CHICAGO. *The Dinner Party*. 1974–79. White tile floor inscribed in gold with 999 women's names; triangular table with painted porcelain, sculpted porcelain plates, and needlework, each side 48' (14.6 m).
Courtesy the artist

right: Color plate 406. ROBERT
KUSHNER. *Blue Flounce* from Persian
Line. 1974. Acrylic on cotton with
miscellaneous fabrics and fringe,
7'2" x 6'3" (2.2 x 1.9 m).
Courtesy Holly Solomon Gallery,
New York

below: Color plate 407. VALERIE
JAUDON. *Tallahatchee.* 1984. Oil
and gold leaf on canvas, 6'8" x
8' (2 x 2.4 m). Courtesy Sidney
Janis Gallery, New York

Color plate 408. JOYCE KOZLOFF. *Mural for Harvard Square Subway Station: New England Decorative Arts.* 1984. Tiles, overall 8 x 83' (2.4 x 25.2 m).
Cambridge, Massachusetts

Color plate 409.
SUSAN ROTHENBERG. *Bucket of Water.*
1983–84. Oil on canvas,
7' x 10'7" (2.1 x 3.2 m).
Private collection

right: Color plate 410.
ROBERT MOSKOWITZ. *Red Mill.* 1981.
Pastel on paper, 9'3" x 48¾"
(2.8 m x 123.8 cm). Museum of
Contemporary Art, San Diego

Color plate 411. DONALD SULTAN. *FOUR LEMONS, February 1, 1985*. 1985. Oil, spackle, and tar on vinyl tile, 8'1" x 8'1" (2.5 x 2.5 m). Collection the artist

Color plate 412. NEIL JENNEY. *Meltdown Morning*. 1975. Oil on panel, 25⅜" x 9'4½" (64.4 cm x 2.9 m). Philadelphia Museum of Art
PURCHASED: THE SAMUEL S. WHITE, 3RD, AND VERA WHITE COLLECTION (BY EXCHANGE) AND FUNDS CONTRIBUTED BY THE DANIEL W. DIETRICH FOUNDATION IN HONOR OF MRS. H. GATES LLOYD

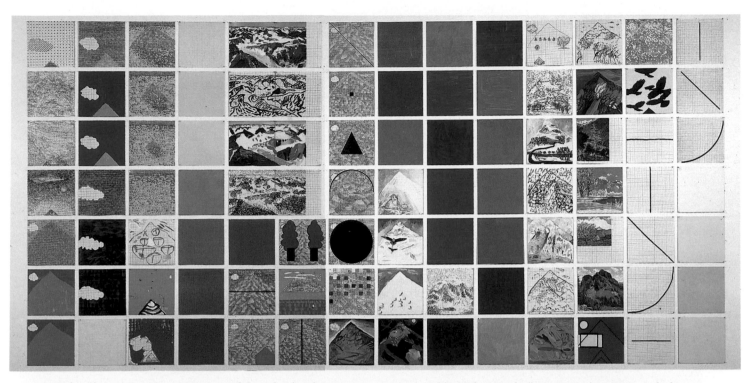

Color plate 413. JENNIFER BARTLETT. *Rhapsody* (detail). 1975–76. Enamel and baked enamel silkscreen grid on 987 steel plates, each 12 x 12" (30.5 x 30.5 cm), overall 7'6" x 153'9" (2.3 x 46.9 m). Collection Sidney Singer, New York

Color plate 414. JENNIFER BARTLETT. *4 P.M., Big Bunny.* 1992–93. Oil and silkscreen on canvas, 60 x 42" (152.4 x 106.7 cm). Courtesy Paula Cooper Gallery, New York

Color plate 415. JIM NUTT. *Slip.* 1990. Acrylic on canvas, enamel on wood frame, 29 x 28" (73.7 x 71.1 cm). Private collection, courtesy Phyllis Kind Gallery, New York and Chicago

above: Color plate 416. EDWARD PASCHKE. *Duro-Verde.*
1978. 48" x 8' (121.9 cm x 2.4 m). Virginia Museum
of Fine Arts, Richmond

below: Color plate 417. PAT STEIR. *Red Tree, Blue Sky, Blue
Water.* 1984. Three panels, each 60 x 60" (152.4 x 152.4 cm).
Private collection.
Courtesy Robert Miller Gallery, New York

above: Color plate 418. JOEL SHAPIRO. *Untitled*. 1987. Charcoal and chalk on paper, 53⅛ x 42⁹⁄₁₆" (134.9 x 108.1 cm). Whitney Museum of American Art, New York

PURCHASE, WITH FUNDS FROM MRS. NICHOLAS MILLHOUSE AND THE DRAWING COMMITTEE

left: Color plate 419. WILLIAM WEGMAN. *Blue Period*. 1981. Color Polaroid photograph. Private collection

Color plate 420. GEORG BASELITZ. *Torso 65*. 1990. Oil on canvas, 9'10⅛" x 8'2⅜" (3 x 2.5 m). Courtesy PaceWildenstein Gallery, New York

Color plate 421. MARKUS LÜPERTZ. *Singer with Black Mask*. 1987. Oil on canvas, 6'6¾" x 63" (2 m x 160 cm). Courtesy Galerie Michael Werner, Cologne and New York

Color plate 422. A. R. PENCK. *The Red Airplane*. 1985. Oil on canvas, 47" x 6'10½" (119.4 cm x 2.1 m). Private collection

above: Color plate 423. JÖRG IMMENDORFF. *Café Deutschland I.* 1977–78. Acrylic on canvas, 9'2" x 10'10" (2.8 x 3.3 m). Ludwig-Forum für Internationale Kunst, Aachen, Germany

left: Color plate 424. SIGMAR POLKE. *Hochstand.* 1984. Acrylic, lacquer, and cotton, 10'8" x 7'4" (3.3 x 2.2 m). Private collection, New York

above: Color plate 425. ANSELM KIEFER. *Germany's Spiritual Heroes*. 1973.
Oil and charcoal on canvas, 10'1" x 22'4½" (3 x 6.8 m).
The Eli Broad Family Foundation

Color plate 426. RAINER
FETTING. *Vincent*. 1984. Oil
and wood on canvas,
8'3⅝" x 6'2¾" (2.6 x
1.9 m). Collection the artist

824. JOYCE KOZLOFF. *Department of Cultural Affairs* installation. 1981. Three pilasters: clay, grout, and plywood, height 9'
(2.7 m). Collection Caroline Schneeback. *Tut's Wallpaper.* 1979. Two silkscreens on silk, 9' x 3'9" (2.7 x 1.14 m).
Right silkscreen: Collection the artist. Left silkscreen: Roth Collection

her decision to work in a decorative style. In her eagerness to dissolve arbitrary distinctions—between high and low art, between sophisticated and primitive cultures—Kozloff has copied her sources more faithfully than most of her peers in the P&D movement, depending on her "recontextualization" of the motifs for the transformation necessary to create new art. At first, she sought her patterns in the churrigueresque architecture of Mexico, but soon turned directly to its origins in Morocco and other parts of Islam. She then relocated them to independent picture or wall panels, composed of the most intricate, run-on, and totally abstract patterns, and executed them in gouache on paper or canvas or silkscreen on fabric (fig. 824). The architectural sources of her inspiration eventually drew the artist into large-scale environmental works, not for exterior display but rather for interiors, as large, complex installations that distributed patterns made of tile, fabric, and glass, among other materials, over every surface—ceilings and floors as well as walls.

Subsequently, Kozloff began to work on a monumental scale designing comprehensive tile decorations for public transport stations in San Francisco, Buffalo, Wilmington, and Cambridge, Massachusetts (color plate 408, page 681). The imagery ranges from purely abstract quilt and wall-stencil patterns to representations of clipper ships, milltowns, farm animals, tombstones, and the flowers and trees of open nature, all stylized, in the naïf manner of the old Yankee limners and decorators, to fit within the star shape of prefabricated tiles.

New Image Art

In December 1978 the Whitney Museum of American Art in New York mounted an exhibition entitled "New Image Painting." Though less than a critical triumph, the show provided a label for a number of strong emerging artists whose works had little in common other than recognizable but distinctly idiosyncratic imagery presented, for the most part, in untraditional, nonillusionistic contexts. Moreover, while honoring Minimalism's formal simplicity, love of system, and emotional restraint, they also tended to betray their roots in Conceptual art by continuing to be involved with elements of Process and Performance, narrative, Dada-like wit, and sociopsychological issues, sometimes even in verbal modes of expression. And while all the latter may, quite correctly, imply a Post-Minimalist sensibility, the New Imagists have almost invariably produced works that tend to prompt a more immediate appreciation of form, with content more evocative than obvious or specific in meaning. The great progenitor of the New Imagists was Philip Guston (see fig. 529), who by 1970 had abandoned his famous Abstract Expressionist style for a rather raucous form of figuration.

By the time the New Imagists matured in the 1970s—painters like Nicholas Africano, Jennifer Bartlett, Jonathan Borofsky, Susan Rothenberg, Donald Sultan, Robert Moskowitz, and Joe Zucker, or the sculptors Joel Shapiro and Barry Flanagan—the artistic legacy they fell heir to was a heavy one indeed, consisting as it did not only of Abstract

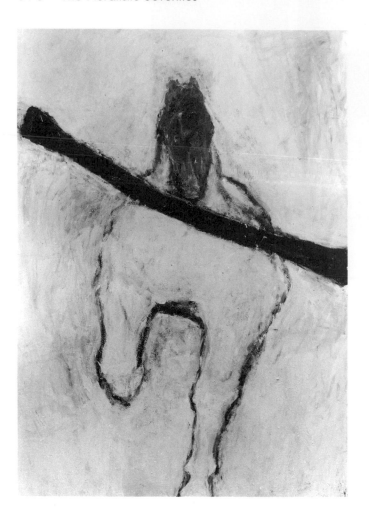

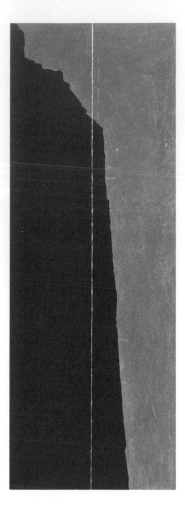

left: 825.
SUSAN ROTHENBERG.
Pontiac. 1979. Acrylic
and Flashe on canvas,
7'4" x 5'1"
(2.2 x 1.55 m).
Private collection

826. ROBERT MOSKOWITZ.
Seventh Sister. 1981.
9' x 3'3" (2.7 x 0.99 m).
Courtesy Blum Helman
Gallery, New York

Expressionism, but also the whole incredibly dense and varied sequence of reactions and counteractions that had come tumbling after. To assimilate such cultural wealth and yet move beyond it posed an awesome challenge, but the artists eager to accept it proved to be a hardy, clever, immensely well educated lot—many of them graduates of Yale University's School of Art in the time of such directors as Josef Albers and the Abstract Expressionist painter Jack Tworkov—and could never have approached the problems of *how* or *why* in the comparatively free, unself-conscious way of the first-generation New York School. But however complex and contradictory the situation, the younger artists who entered it arrived with ambition. Fortified with all the competitiveness that came with an art market exploding from the effects of media hype, these artists satisfied the rising demand for imagery and collectible objects after the latter's near demise under the impact of Minimalism and Conceptualism.

The striking originality of New Image art lay not in style, which varied widely from practitioner to practitioner, but rather in the particular techniques and ideas mixed and matched—appropriated—from the recent past, to the end that something familiar or banal would become unexpectedly fresh and strange, possibly even magical, thereby engendering a whole new set of possibilities.

In conjuring the kind of eroded or fragmentary figuration last seen in the art of Giacometti or, more robustly, De Kooning, **SUSAN ROTHENBERG** (b. 1945) further defied

Minimalism's immaculate, abstract surface by filling it with a dense, painterly, silver-gray facture reminiscent of Philip Guston's probing Abstract Expressionist brushwork and sooty palette. The motif with which she realized her first major success was the horse (fig. 825), a massive form steeped in myth yet intimately involved with human history, obedient and faithful but also potent and wild, and often a surrogate in Romantic painting for human feeling. At a time in the Post-Minimal period when the more advanced among younger artists were not quite ready to deal with the human face and figure, "the horse was a way of not doing people," Rothenberg has said, "yet it was a symbol of people, a self-portrait, really." Fundamentally, however, her concerns, like Giacometti's, were formal rather than thematic. In the case of Rothenberg, these centered, as in Minimal painting, on how to acknowledge and activate the flatness and objectivity of the painting surface, while also representing a solid recognizable presence.

In her more recent work, Rothenberg has become still more private and poetically spectral in her figuration (color plate 409, page 681), qualities that she manages to universalize by radiating her painterly surfaces with a new luminosity and a filtering of color.

ROBERT MOSKOWITZ (b. 1935) has said that "by adding an image to the painting, I was trying to focus on a more central form, something that would pull you in to such an extent that it would almost turn back into an abstraction." His method is to isolate an image of universal, even iconic familiarity—natural or cultural shrines like Rodin's *Thinker,* New York's Flatiron Building, or Yosemite National Park's Seventh Sister (fig. 826)—and then strip the form bare of all but its most basic, purified shape, making it virtually interchangeable with an abstract pictorial structure. "In all good work there is a kind of ambiguity, and I am trying to get the image just over that line," he said. By allowing it to travel no farther, however, the artist makes certain that, with the help of a clarifying title, the viewer feels sufficiently challenged to puzzle out and then rediscover the marvel of the monument long since rendered banal by overexposure (color plate 410, page 681). Aiding the cause of regenerated grandeur and refreshed vision are the epic scale on which Moskowitz casts his imagery and his tendency to cut images free of the ground, so that they seem truly out of this world, soaring and mythic. This effect is achieved within the enchanted realm of the artist's pictorial surface, a gorgeous display of painterly craft, alternately burnished, flickering, or luminous, glorifying the flatness of the plane and seducing the eye into exploring the image.

In his break with Minimalism, **DONALD SULTAN** (b. 1951), like most of his New Image peers, did not so much want to reject the past as to make his art contain more of it. Sultan wanted figuration with the monumental simplicity and directness of abstract art, an approach that would enable him to find an image that would clearly metamorphose out of the materials and mechanisms of its own creation. While teaching himself to draw—a skill seldom imparted in the figure-averse academies of the 1960s and early 1970s—and working as a handyman at the Denise René Gallery in New York, Sultan found himself retiling his employer's floor with vinyl. He immediately felt an affinity for the material's weight and thickness, as well as for the "found" colors and mottled patterns the tiles came in. Further, there was the unexpectedly sensuous way in which the cold, brittle substance became soft and pliant when warmed by cutting with a blowtorch, a process that, along with the medium and its special colors, would suggest a variety of new subject matter. Sultan also fell in love with the tar—called butyl butter—he used for fixing the tiles to their Masonite support, and began drawing and painting with it on the vinyl surface. This led to covering the plane with a thick layer of tar, which the artist would then paint on or carve into, in the latter instance achieving his image by revealing the vinyl color that lay below (fig. 827). Process had indeed now served as a means of putting more and more elements back into the art of painting. And it again served when Sultan hit upon what, so far, has been his most memorable motif, a colossal yellow lemon (color plate 411, page 682)—a luscious, monumental orb with nippled protrusions—whose color seems all the more dazzling for being embedded in tar, traces of which make the fruit's flat silhouette appear to swell into full, voluptuous volume.

A key figure among the New Imagists was **NEIL JENNEY** (b. 1945), who not so much abandoned the strategies he first learned in Minimal, Pop, and Process art as integrated them with beautifully illusionistic subject matter to achieve paintings of galvanic visual impact and chilling moral, even

827. DONALD SULTAN. *PLANT May 29, 1985.* 1985. Latex and tar on vinyl tile over Masonite, 8'¾" x 8'¾" (2.5 x 2.5 m). Hirshhorn Museum and Sculpture Garden, Smithsonian Institution, Washington, D.C.

admonitory force. In *Saw and Sawed,* for instance (fig. 828), we discover a version of the liquid Abstract Expressionist stroke and the cartoon-like drawing favored by the Pop artists, along with the visual-cum-verbal punning so beloved by both Pop and Conceptual artists, all set forth within a monochrome Color Field organization and a heavy, title-

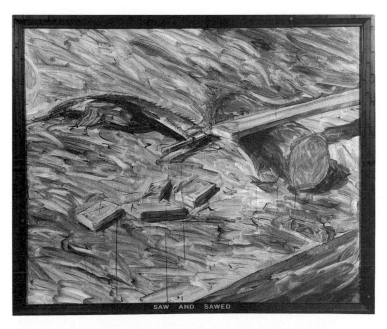

828. NEIL JENNEY. *Saw and Sawed.* 1969. Acrylic on canvas, 58½ x 70⅜" (148.6 x 178.8 cm). Whitney Museum of American Art, New York

829. JONATHAN BOROFSKY. Installation at the Whitney Museum of American Art, New York, December 21, 1984–March 10, 1985

bearing frame that together betray a Minimalist consciousness of the painting as a literal object. In tension with this literalness is the witty ambiguity of the meaning implied by the juxtaposition of words and images. Articulate and politically minded, Jenney has spoken of art as "a social science," and as a language through which to convey "allegorical truths." Gradually, as the visual and verbal jokes interact to engage the mind, and the painterly fluidity seduces the eye, attentive viewers find themselves drawn, through a chain of unexpected relationships, into a consideration of the consequences of human deeds, consequences of a serious environmental import. In subsequent paintings, Jenney reversed the relationship between his imagery and its enclosing frames until the latter—massive, dark, and funereal—allows a mere slit of an opening through which to glimpse a tantalizingly lyrical and illuminated nature (color plate 412, page 682). Along with the monumentalization of the coffin-like frame has come a vast enlargement of the label. And when this spells out a title like *Meltdown Morning*, the image so segmentally revealed assumes the character of longing for a paradise already lost beyond recovery, as well as mourning for the old world as it may appear—like a rare fragment presented in a museum showcase—following the nuclear winter.

By his own admission, **JONATHAN BOROFSKY** (b. 1942) has placed politics firmly at the center of his riotously pluralistic art, saying: "It's all about the politics of the inner self, how your mind works, as well as the politics of the exterior world." To express what has become an untidy multiplicity of concerns, Borofsky is at his most characteristic not in individual works but rather in whole gallerywide installations (fig. 829), where he can spread over ceilings and windows, around corners, and all about the floor the entire contents of his

mind, his studio, and even the history of his art, which began at the age of eight with a still life of fruit on a table. The first time he mounted such a display, at Paula Cooper's SoHo gallery in New York in 1975, the show, Borofsky said, "seemed to give people a feeling of being inside my mind." Mostly that mind seems filled with dreams or with the free associations so cherished by the Surrealists, all of which Borofsky has attempted to illustrate with an appropriate image or object. A representative installation would contain a giddy mélange of drawing, painting, sculpture, audio work, and written words, all noisily and messily driving home certain persistent themes. The drawings typically begin with automatist doodles in which the artist discovers and then develops images, such as a dog with pointed ears, blown up by an opaque projector to spill over and be painted onto walls and ceilings, as if one of the beasts guarding the gates of Dante's Hell were hovering above and threatening the whole affair. Among the other presences activating the installations are the giant "Hammering Men" (visible, in two versions, at the left in fig. 829), the motorized up-and-down motion of their arms inspired by the artist's father, on whose lap young Borofsky sat while listening to "giant stories." Another character took the artist into the realm of sexual politics, where he conjured images of an androgynous clown, half Emmett Kelly and half ballerina, dancing to the tune of "My Way." As a kind of interstitial tissue serving to link these disparate images, Borofsky littered the floor with crumpled fliers, which were in fact copies of an actual letter recounting the difficulties of ordinary life, a "found" work picked up by the artist from a sidewalk in California.

Another way that the artist has found to impose continuity on his work is through his obsessive counting, a rational

830. Jonathan Borofsky.
Self-Portrait. 1980.
Photograph. Edition of six.
Courtesy Paula Cooper
Gallery, New York

below: 831. Jennifer Bartlett.
Rhapsody (detail). 1975–76.
Enamel and baked enamel
silkscreen grid on steel plates,
987 plates, each 12 x 12"
(30.5 x 30.5 cm), overall
7'6" x 153'9"
(2.3 x 46.9 m).
Installation view, Paula
Cooper Gallery, New York.
Collection Sidney Singer,
New York

process of ordering, unlike the instinctual, random one seen elsewhere in his work, that he began in the late 1960s and has now taken beyond the three-million mark. Along the way he applies numbers to each of his pieces and sometimes even to himself (fig. 830), thus joining their separateness into one continuous, ongoing sequence. The record of this monotonous activity is displayed, in graph paper filled with numerals, stacked waist-high, and enshrined in Plexiglas.

Another New Image artist with a voracious appetite for exhibition space and the creative energy to fill it is **Jennifer Bartlett** (b. 1941), whose vast 987-unit painting entitled *Rhapsody* (color plate 413, page 683, and fig. 831) became, upon its presentation in 1976, one of the most sensational works of the decade. She simplified her painting process, eliminating wooden stretchers, canvas, and the paraphernalia of the oil paint, until she depended upon little more than the basic module of a one-foot-square steel plate, a flat, uniform surface commercially prepared with a coat of baked-on white enamel and an overprinted silkscreen grid of light gray lines. When finally assembled on a large loft-size wall, hundreds of the enameled steel plates yielded multipart compositions filled with color-dot paintings in an eye-dazzling display of brilliantly decorative abstract patterns.

For *Rhapsody* Bartlett expanded her repertoire of colors, not only by adding to it, but also by mixing and superimposing, and in a further attempt to include "everything," she decided to have figurative as well as nonfigurative images. For the former she chose the most "essential," emblematic forms of a house, a tree, a mountain, and the ocean, and for the latter a square, a circle, and a triangle—Cézanne's "cylinder, sphere, and cone" two-dimensionalized. She would also admit sections devoted purely to color, others to lines (horizontal, vertical, diagonal, curved), and some to different techniques of drawing (freehand, dotted, ruled). The painting finally climaxed in a 126-plate ocean sequence incorporating 54 different shades of blue.

832. JENNIFER BARTLETT. *Yellow and Black Boats.* 1985. Three canvases, each 10' x 6'6"
(3 x 2 m), overall 10' x 19'6" (3 x 2.9 m). Yellow Boat: Wood and enamel paint,
14 x 60½ x 31" (35.6 x 153.7 x 78.7 cm). Black Boat: Wood and flat oil-base paint,
6'6" x 5'8" x 3'2" (2 x 1.73 x 0.97 m) (height of mast included). Saatchi Collection, London

Bartlett went on to expand her basic themes in new directions, accommodating sculpture (fig. 832) and undertaking large public and private commissions. In her recent works (color plate 414, page 683), even the long-familiar grid contains new figurative puzzles; Bartlett has now incorporated into it rather enigmatic silkscreened images, in this case a rabbit, as well as the face of a clock, adding the narrative implications of time's passage.

A group of young figurative artists with a fondness for fantasy and grotesque exaggeration became known as the Chicago Imagists, since many of them studied at the school of the Art Institute of Chicago. JIM NUTT (b. 1938) began exhibiting in 1966 with a smaller group of these irreverent artists, which called itself the Hairy Who. At that time, Nutt's scribbly images of women were filled with adolescent aggressiveness. A more controlled work from a few years later, *It's a Long Way Down* (fig. 833), still portrays a woman in anxious, if cartoonish, terms. Nutt says that the painting has to do with the movies, and the gigantic female here, so hugely out of proportion with the tiny inset figures at the upper left and the fleeing form at the lower right, does seem to come from 1950s science-fiction epics like *Attack of the Fifty-Foot Woman.* The headless torso (or is it a mannequin?) at the bottom and the severed head plummeting through the skylight complete this scene of comic mayhem. In the more recent *Slip* (color plate 415, page 683), it is the portrait painting, rather than the movies, that has become the object of parody. Here Nutt's execution is quite ostentatiously derived from high modernist art. Indeed *Slip* is almost an homage to Picasso's portraits of women from the 1930s, which similarly reduce the face to

a collection of planes converging on a prominent nose.

Another artist initially associated with the Chicago Imagists is EDWARD PASCHKE (b. 1939). Paschke's figures of the 1970s may owe something to the particularity of Photorealism in their smoothly modulated contours, as with the exposed skin of the woman in *Duro-Verde* (color plate 416, page 684). But even though that work may depict a recognizable event—presumably a high-society party or masked ball—the outrageous colors that Paschke employs nullify any sense of illusionistic fidelity. The artificial, electric green is akin to garish neon lighting, and it recalls as well the distortions of a poorly tuned color TV set. Also puzzling is Paschke's deliberate overelaboration of the headgear and masks, perhaps intended to imply that this scene is more than it appears to be. Oddest of all are the dramatic poses of the figures: The frowning, lance-wielding male on the left and the pensive couple on the right suggest nothing so much as a Postmodernist version of the Expulsion from Paradise.

Perhaps more than any other contemporary artist, PAT STEIR (b. 1940) is conscious of the reality that art always comes from older art, giving culture a vernacular of current forms, which each painter or sculptor reinvests with subtly or sharply different meanings. A clear demonstration of the principle can be found in *Cellar Door* (fig. 834). The large, unmodulated black form at the left is a quotation of Malevich's *Black Square* (see fig. 244) of 1914–15, a key point of origin for subsequent geometric abstraction. As if to acknowledge Malevich's original challenge to traditional kinds of painting, Steir added a generic landscape to the right half of *Cellar Door* and then crossed it out, just as Malevich himself had once pasted a small reproduction of the Mona Lisa into

left: 833. JIM NUTT. *It's a Long Way Down.* 1971.
Acrylic on wood, 33⅞ x 24¾" (86 x 62.9 cm).
National Museum of American Art, Smithsonian
Institution, Washington, D.C.
GIFT OF S. W. AND B. M. KOFFLER FOUNDATION

below: 834. PAT STEIR. *Cellar Door.* 1972. Oil, crayon,
pencil, and ink on canvas, 6 x 9' (1.8 x 2.7 m).
Collection Mr. and Mrs. Robert Kaye, New York

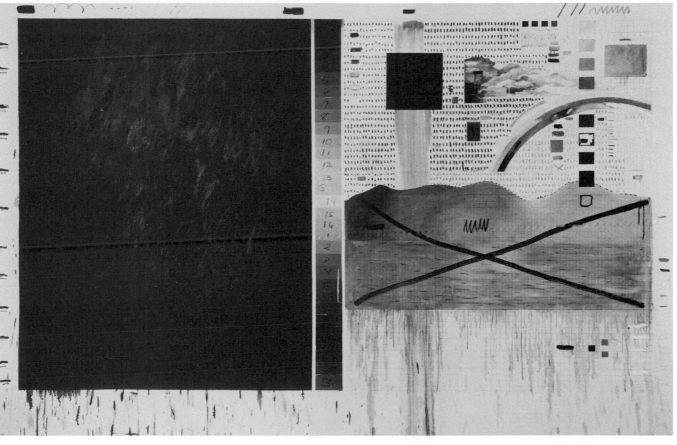

one of his own works and marked an "X" across the face. Yet Steir also maintains a certain ironic distance from Malevich, not least by the title she gives her painting. The black square in *Cellar Door* is not a mysterious geometric icon like the one that Malevich made, but rather something from ordinary experience—just the doorway to an unlighted basement.

Steir's *Red Tree, Blue Sky, Blue Water* (color plate 417, page 684) is a triptych. In this work, a "quotation" from Van Gogh (see fig. 80) on the left progresses across the three panels in a cinematic sequence, from distant shot to medium one and finally close-up. As they move in graduated stages from resolution to dissolution, the panels also illustrate Steir's conviction that realism and abstraction are the same thing. Adding to the complexity is the echo of Van Gogh's own quoted source: a Hiroshige woodcut. Furthermore, the bold red-on-blue coloration and the swirling, almost Art Nouveau configuration of the image in the right panel—a blowup of a blossom—would seem to betray a side glance at Mondrian's 1908 *Blue Tree* (color plate 120, page 259). But however absorbed she may be in the history of painting, Steir is also absorbed in the *act* of painting, which she evinces in an extraordinary lavishness of stroke, medium, and color.

In the context of the long figurative tradition of sculpture, **JOEL SHAPIRO** (b. 1941) conceivably might have had less difficulty than painters when he determined to warm the chill perfection of Minimalism's cubic forms with some sense of an activated human presence. Even so, Shapiro found it necessary to proceed only gradually, moving toward the figure through the more abstract objectivity of its habitations in an early 1970s series devoted to the house (fig. 835). Right away, however, Minimalism lost its monumentality as Shapiro miniaturized the image and pruned the form of all but its most primary or archetypal masses.

As for the humanity excluded from these modest abodes, Shapiro has so formed and inflected his figures that, on a larger scale, they share both the houses' clean lines and conflicted condition (fig. 836). Built of square-cut posts and cast in iron or beautiful reddish gold bronze—wood grain, knotholes, and all—these stick figures display the economy of a Sol LeWitt sculpture, only to break free of the box and pivot, writhe, dance, collapse, and crawl, as if to reenact in antiheroic, Constructivist terms the whole gamut of tormented pose and gesture invented a century earlier by Auguste Rodin (see fig. 136). Some of the assembled beams are even headless and otherwise dismembered in the manner of the French master's emotionally fraught personages. In the quasi-figural sculpture of Joel Shapiro, fragmentation becomes a symbol of spiritual or emotional incompleteness.

Some of Shapiro's drawings are even more evocative (color plate 418, page 685). In them, the ethereal smudging of the charcoal endows the geometric shapes with an aura of mystery, an effect enhanced by the use of deep colors. Beam-and-plank assemblages occupy a vast, atmospheric space that owes something to Romantic notions of the sublime.

The British sculptor **BARRY FLANAGAN** (b. 1941) arrived at his more recent Imagist art by way of painting, dance, and

835. JOEL SHAPIRO. *Untitled.* 1973–74.
Cast iron, 3 x 27¼ x 2⅝" (7.6 x 69.2 x 6.7 cm); clipboard base, 14¾ x 29¾ x 5" (37.5 x 75.6 x 12.7 cm).
Collection Paula Cooper Gallery, New York

836. JOEL SHAPIRO. *Untitled.* 1983–84. Bronze,
6'8¾" x 6' 8¾" x 4'4" (2.1 x 2.1 x 1.32 m). Number one of edition of three. Saint Louis Art Museum, Missouri
GIFT OF MR. AND MRS. BARNEY A. EBSWORTH

837. Barry Flanagan. *Hare on Bell with Granite Piers.* 1983. Bronze and granite, 7'11½" x 8'6" x 6'3" (2.4 x 2.6 x 1.9 m).
Number one of edition of five. Collection the Equitable Life Assurance Society of the United States, New York

various Conceptual installations produced with organic mate-rials and processes. And throughout a highly varied oeuvre the consistent factors have been those of poetic association, inven-tiveness, and good humor. Inevitably, therefore, Flanagan favors manipulatable substances, such as hessian, rope, felt, steel, stone, ceramic, or bronze, likely to "unveil" an image in the course of handling and shaping. In the early 1980s, while squeezing and rolling clay, Flanagan found a hare "unveiling" itself in the material, and the image proved to have all the fecundity in art that the creature displays in life (fig. 837). Leaping, strutting, dancing, balancing, or boxing, the slender, free-formed, but graceful bronze hare seems an emblem of Flanagan's own spontaneity, openness, and good humor.

WILLIAM WEGMAN (b. 1942) did not have to search hard for what may have been his most successful image; it simply came bounding into camera range in the form of a Weimaraner puppy, who grew up so stagestruck and histri-onically gifted that nothing could keep him either from the lights or from performing once they were turned on. Wegman chose to photograph the puppy in order to playful-ly insinuate the notion of charming canine turned surrogate

person. Naming his dog Man Ray, the artist punned him-self—Dada fashion—into reversing roles with the dog and thus becoming his—Man's—best friend. For some ten years, until the canine celebrity died, Wegman found the photo-genic Man Ray to be a gallant and sensitive actor in an on-going comedy of still photographs, made first with a Polaroid and then with a large-format camera (color plate 419, page 685). One by one they show Man Ray costumed and fully rehearsed to play through the whole range of ordinary and supposedly rational situations in contemporary society, all rendered farcical and pretentious—often hilariously so—by ironic contrast with the manifest dignity and intelligence of the noble Weimaraner.

If the Conceptual mock-seriousness and genial, impro-vised air of Wegman and his Weimaraner could be said to rep-resent the sunny side of the seventies, the radical chic of pho-tographer **ROBERT MAPPLETHORPE** (1946–1989) would have to signify the underside of the same period, a certain aspect of which he felt free to explore with concomitant tech-nical skill and attention to craft. Mapplethorpe's style was pure classicism—formal, serene, and timeless. And it prevailed with unwavering equanimity even though the subject was usually the mammon of high fashion, money, power, or the fast-track netherworld of drugs and homosexuality. Mapplethorpe approached it all with unflinching candor, his pictures of flowers as coolly exquisite yet overtly sexual as the black male anatomies that were his admitted personal obsession (fig. 838). Bathed in silken light, these elegant, heroic images make human flesh look as if it had been cast in bronze and then burnished to a dark, lustrous finish. Purely aesthetic effects—the results of precise analysis, framing, and lighting—appear all the more controlled for having been imposed upon content to which many, or most, viewers are certain to bring an array of individual, and heavily loaded, responses.

Despite the evidence of aesthetic distance and control, however, after Mapplethorpe's death a few figures prominent in public life accused his work of obscenity. In this atmosphere, a 1990 touring retrospective of Mapplethorpe's photographs, supported in part by the National Endowment for the Arts, encountered a political firestorm, as Senator Jesse Helms and others asserted that it was inappropriate for the government to help fund art that some citizens might find offensive. Officials of the Cincinnati Contemporary Arts Center were put on trial for obscenity simply for showing the exhibition. They were acquitted, but the controversy had a lasting impact on the fortunes of the NEA, and on the debate about public funding for the arts, over the ensuing decade.

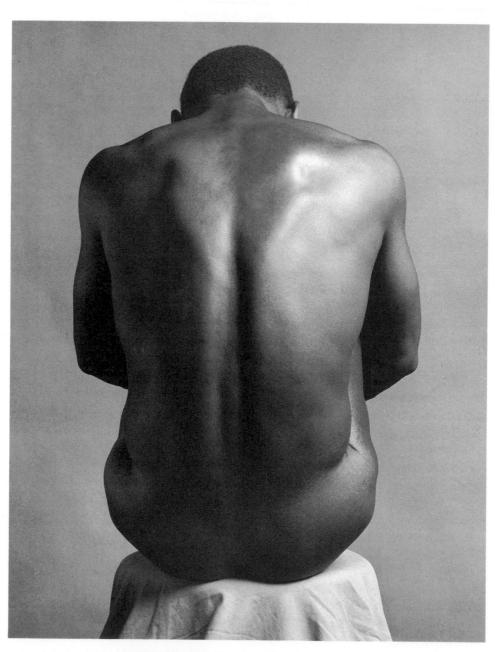

838. ROBERT MAPPLETHORPE. *Ajitto (Back)*. 1981.
Gelatin-silver print, 16 x 20" (40.6 x 50.8 cm) © THE ROBERT MAPPLETHORPE FOUNDATION, NEW YORK

25 The Retrospective Eighties

With the development of a myriad of new art approaches during the late 1960s and 1970s—including Conceptual art, Performance art, Land art, and Process art—and the powerful social and political critiques of feminists and civil rights activists, it became increasingly obvious that traditional histories and interpretations of "modernism" were simply inadequate. In the 1980s, a number of artists, and many more critics and theorists, declared a break with the modern era, identifying themselves as "Postmodernists." Postmodernism implied a dissatisfaction with the narrow confines modernity seemed to have imposed on art, apparently promoting the accomplishments of white, male artists of European descent and refusing to admit social and political concerns as viable concerns of "high" art. Postmodernism, by contrast, encouraged overtly polemical practices and the ironic distancing of conventions of the past. Hand in hand with the development of the critical stance of Postmodernism came the theoretical tools of Poststructuralism and deconstruction, which had gained increasing prominence since the early 1970s among artists, critics, and art historians. The term "Poststructuralism" indicates that this movement followed the development of Structuralism, a school of thought based on the premise that underlying structures in language and society could be identified and studied as vital aspects of culture. Poststructuralism refers to the work of a group of predominantly French philosophers including Jacques Derrida, Michel Foucault, and Roland Barthes who made it clear that structures of language and political institutions were simply conventions rather than natural facts. Understood as constructions rather than essential principles, structures of language and of institutions could then be analyzed and "deconstructed." The practice of deconstruction involves the minute examination of the assumptions underlying particular texts, which may be literary, but may also, by extension, refer to the larger discourses and social and political networks in which we live our lives. The insights of Poststructuralism combined with the technique of deconstruction provided critics and artists with the theoretical tools to analyze the underlying assumptions of

modernism itself, thereby supporting the critical break of Postmodernists with modernism.

Somewhat ironically, at the same time that Postmodern critiques of art making were being formulated, the audience and the market for new art were greater than ever. In this atmosphere, many younger artists felt that the way to go forward was, paradoxically, to go back—back to the great historic styles of the past that were venerated in textbooks. They therefore sought to address directly through their own work the artistic styles and canonical masterpieces of those earlier artists who to them seemed most significant. Some took for their models the German Expressionists of the years after World War I; others were more interested in the Abstract Expressionists who came to prominence in the 1940s. Still others looked to the Pop artists of the 1960s and their "appropriation" (or borrowing) of popular imagery from comic strips, advertisements, and the mass media in general. Some artists recalled difficult-to-categorize works, like the highly charged installations created by the Fluxus group or by Joseph Beuys. Others looked to those early abstract painters who had worked in a strict geometric mode, such as Malevich and Mondrian. And many artists found themselves drawn not simply to one, but simultaneously to several, of these very different exemplars from the past. Each of the principal "retrospective" approaches to art making that resulted will be discussed in turn in this chapter: the Neo-Expressionists, the cartoon and media appropriators, the new makers of installations, and the painters of a "neo-" abstraction.

Neo-Expressionism

With the beginning of the 1980s, the reaction against Minimalism assumed a new intensity, kindled mostly by young painters and so powered by a youthful love of bold gesture, heroic scale, mythic content, and rebellious figuration that critics dubbed it Neo-Expressionism. But the development could just as well have been called Neo-Surrealism, for while a number of the new artists did indeed defy Minimalist restraint and revive agitated, feeling-laden brushwork, almost all generated content by exploiting imagery so taboo, primal, or vulgar that descriptive realism seemed to

839. PABLO PICASSO. *Man, Guitar, and Bird-Woman.* 1970.
Oil on canvas, 63¾ x 51⅛" (161.9 x 129.9 cm).
Estate of the artist
© 1997 ESTATE OF PABLO PICASSO/ARTISTS RIGHTS SOCIETY (ARS), NEW YORK

pass into the dislocated, floating realm of dream, reverie, or nightmare. Yet even as the accessibility promised by identifiable subject matter was not quite as immediate as some might have hoped, collectors and museums, the media, and a good segment of critical opinion rejoiced that here, at long last, were big, vigorously worked, color-filled, meaningful canvases. In an art world hungry as much for excitement as for imagery, this was an event, made all the more newsworthy by the fact it had been spearheaded in Europe, where German and Italian artists were working with a self-confident strength and independence not seen since the original Expressionists, on the one hand, or the Futurists and Metaphysical artists, on the other. Now, for the first time in the postwar era, Continental artists could reclaim a full share of international attention and world leadership in new art, hand in hand with younger American painters.

The so-called Neo-Expressionists evinced a daring embrace of every possibility—metaphor, allegory, narrative, surfaces energized by and packed with photographic processes, broken crockery, or even oil paint—that made the New Image art of the 1970s seem yoked to modernist dogmas of emotional and formal reserve. Paradoxically, however, while previous avant-garde movements had tended to renounce the past, the Neo-Expressionists sought to liberate themselves from modernist restrictions by rediscovering those very elements within tradition most condemned by progressive, mainstream trends. For example the Americans, particularly, looked to the later, expressionist Picasso—the Picasso certain

modernists had thought unimportant—instead of the cerebral, discriminating artist of the heroic Analytic Cubist era. They flocked to the Guggenheim Museum's 1983 exhibition devoted to the Spanish master's final decade (1963–73), drinking in not only the liberated drawing and color, but also the unrestrained, autobiographical sexuality expressed by a still-vital octogenarian (fig. 839). Not coincidentally, sexuality loomed large—often blatantly—in Neo-Expressionist art as well.

Beyond these broad, and highly provisional, generalizations, the Neo-Expressionists—a title almost none of them finds appropriate—have little in common. Instead of forming a cohesive movement, they emerged, more or less independently, from their respective cultural and stylistic outposts.

Among the new German painters, the senior artist is GEORG BASELITZ (b. 1938), whose paintings are also perhaps the most distinctive, mainly for the topsy-turvy orientation of their imagery—usually upside down—and also for their full, ripe color and powerful, liberated brushwork (color plate 420, page 686). In a kind of keepsake gesture, Baselitz (born Georg Kern) renamed himself for his natal village, Deutschbaselitz in Saxony, after he moved out of what was then Russian-occupied East Germany and into West Berlin. There he soon became convinced that the prevailing mode of abstract painting did not fit his purposes. Instead, he decided to buck the trend by launching a one-man campaign to revive the kind of narrative and symbolic, emotive and rhapsodic, art in which Germans had traditionally achieved greatness (see

840. GEORG BASELITZ. *Ein Versperrter.* 1965. Oil on canvas,
63¾ x 51⅛" (161.9 x 129.9 cm).
Kunsthalle, Bielefeld, Germany

841. GEORG BASELITZ. *The Brücke Choir (Der Brückenchor)*. 1983. Oil on canvas, 9'2½" x 14'11" (2.8 x 4.5 m). Private collection

figs. 8 and 162). During the years 1964–66, when international art was dominated by American Color Field painting and Pop, Baselitz developed an iconography of heroic men—huge figures towering over a devastated landscape (fig. 840). Then, in a resolution of the problem of how to reintroduce recognizable imagery in a novel, personally determined, and evocative manner, Baselitz settled upon the form of a figure or figures left more or less intact but abstracted from reality by the upside-down posture and the disintegrating effects of brilliantly synthetic color and aggressive painterliness (fig. 841). Baselitz also began to paint not heroes but his own friends in the midst of such mundane activities as talking, eating, or drinking. But Baselitz's treatment of his subjects was far from ordinary, heroicizing them in scale or manipulating their bodies through distortions of anatomy and bizarre proportions.

Baselitz also makes sculpture, hacking it out of wood to register the medium's inherent qualities (fig. 842). While this may be a tribute to the late works of Michelangelo, it also calls to mind the German Expressionists, who excelled in woodcut yet sometimes suffered from having their work declared "degenerate" by the Nazis (see figs. 162, 163).

In his rebellion against all authoritarian "norms," most particularly those inherent in the monolithic consistency and coolness of 1960s American abstraction, MARKUS LÜPERTZ (b. 1941) made intensity the driving force behind his search for artistic independence, variability the character of his style, imaginative reference to earlier artists his process, and politics the content of his art (color plate 421, page 686). Although he rejected art sources and materials as consistently as he accepted them, Lüpertz saw Picabia as a precursor of his own stylistic restlessness (color plate 106, page 205; figs. 307, 308) and Picasso as a model of the obsessive quoter of established art. No less important, however, was Pollock, who for Lüpertz seemed a "Dionysiac" painter attuned to discovering the mysteries of his own personality within the unifying process of his art. Lüpertz uses abstracting means to invent

objects that, while simulating the look of real things, remain totally fictitious, or what might be called "pseudo-objects." In the early 1970s, the artist executed a "motif" series in which he introduced such recognizable but taboo emblems as German military helmets, uniforms, and insignia, and combined them with palettes and other attributes of the painting profession, thereby signifying the need to confront history and the imperative that art be grounded in concrete reality (fig. 843).

Having crossed from East- to West Germany, A. R. PENCK (b. 1939) struggled in his art not only with the cleft in

842. GEORG BASELITZ. *Untitled.* 1982–84. Limewood and oil, 8'3" x 28" x 18" (2.5 m x 71.1 cm x 45.7 cm). Scottish National Gallery of Modern Art, Edinburgh

PURCHASED WITH ASSISTANCE FROM THE NATIONAL ART COLLECTIONS FUND, WILLIAM LENG BEQUEST

843. Markus Lüpertz. *Helm 1.* 1970. Oil on canvas,
7'8½" x 6'3" (2.3 x 1.9 m).
Courtesy Mary Boone Gallery, New York

844. A. R. Penck. *The Crossing (Der Übergang).* 1963.
Oil on canvas, 37 x 47¼" (94 x 120 cm).
Ludwig-Forum für Internationale Kunst, Aachen, Germany

845. Sigmar Polke. *Bunnies.* 1966. Acrylic on linen,
59 x 39½" (150 x 100 cm).
Hirshhorn Museum and Sculpture Garden,
Smithsonian Institution, Washington, D.C.
Joseph H. Hirshhorn Bequest and Purchase Funds

German society but also with the need to come to terms with himself as an exile. The form this autodidactic artist found most effective for his purposes consisted of a stick figure—Everyman reduced to a cipher—set in multiples or singly in an allover tapestry or batik-like pattern against a field of solid color, with figures and ground so counterbalanced as to lock together in a dynamic positive-negative relationship (color plate 422, page 686). Mixing stick figures with a whole vocabulary of hieroglyphs, cybernetic symbols, graffiti signs, and anthropological or folkloric images, Penck expresses such personal and public concerns as the urgent need for individual self-assertion in a collective society. As a "crossover" himself, Penck has often painted poignant versions of the lone stick man walking a tightrope or a burning footbridge suspended above an abyss separating two islands of barrenness (fig. 844).

Almost in counterpoint to Penck's radically depersonalized imagery, **Jörg Immendorff** (b. 1945) paints in a detailed, if highly conceptualized, realist style. But like Penck, Immendorff is no less anguished over the theme of the individual entrapped by the contradictions of modern German life. In the Café Deutschland series, the central image is the artist himself, often seated at center between a pair of allegorical columns symbolizing the two halves of a polarized world (color plate 423, page 687). Together, the relatively public, accessible style, the autobiographical subject, and the lurid disco scene, complete with heavy, forward-pitching perspective, express, as Donald Kuspit has written, the ambivalence of "an egalitarian Maoist oriented towards the lost cause of a democratic China, and a successful bourgeois artist

846. ANSELM KIEFER. *Departure from Egypt*. 1984. Straw, lacquer, and oil on canvas, lead rod attached, 12'5" x 18'5" (3.8 x 5.6 m). Museum of Contemporary Art, Los Angeles

PURCHASED WITH FUNDS FROM DOUGLAS S. CRAMER, BEA GERSH, LENORE S. GREENBERG, FRED NICHOLAS, ROBERT A. ROWAN, AND PIPPA SCOTT

appropriated by the market." Nowhere is the surreal character of this dilemma more pronounced than in Immendorff's totemically figured chair-sculptures, carved from linden wood and then polychromed.

SIGMAR POLKE (b. 1941) could be said to epitomize the German artists of his generation, in that he appears to have been fully as interested in making a direct statement in response to American aesthetic hegemony, as in conceptualizing a new kind of art—an art sensitive to the unique complexities of life in contemporary Germany. In the 1960s, Polke adopted the benday process of mechanical reproduction not long after it had been integrated into painting by Roy Lichtenstein (see figs. 630, 631). But whereas the American artist aestheticized the dot system, not only by treating it as a kind of developed Pointillism but also, later, by using it to depict established icons of high art, the German simply hand-copied—"naïvely"—a blown-up media image, complete with off-register imperfections and cheap tabloid subject matter (fig. 845). Polke often works on an allover grid or otherwise mechanically patterned ground. Rauschenberg, of course, had incorporated ordinary textiles into his combine paintings, as part of the Synthetic Cubist technique of collage, but also as a Neo-Dada device for generating the shock effect once possible in the unorthodox juxtaposition of materials (see fig. 610). Polke, however, would develop the painterly "naturally" out of the mechanical and thus integrate the two techniques holistically. Then, in a further assimilation using photography, the artist developed his imagery by drawing from the underlying pattern of regularly repeating abstract motifs in photographs (color plate 424, page 687). Delineated like contour drawings and layered in a manner inspired by Picabia's Transparencies (see chapter 13), the images seem paradoxically to dematerialize at the same time that they emerge from the support upon which the artist has painted them, creating an illusionistically deep pictorial space even as it seems independent of all perspective systems. Moreover, the figuration phases in and out of focus, as if it were the product of photographic or darkroom manipula-

tions, or perhaps even of an accident with one of the toxic substances Polke is known to have used in creating his curiously synthetic and characteristic colors.

In the apocalyptic paintings of ANSELM KIEFER (b. 1945), called by art critic Robert Hughes "the best painter of his generation on either side of the Atlantic," German and Jewish history are frequently evoked symbolically through the motif of scorched earth. In keeping with the narrative style of Neo-Expressionism, his works are about fascism, yet his larger aims are based on achieving a kind of cultural redemption through understanding. Kiefer's preoccupation with the sorrows of his national heritage is evident in *Germany's Spiritual Heroes* (color plate 425, page 688). Inscribed on the painting's surface are the names of artists, including Arnold Böcklin and Joseph Beuys, as well as poets, the ruler Frederick the Great, and one composer: Richard Wagner. The work draws on the imagery of Wagner's tragic opera *Twilight of the Gods* (1874) to place the disastrous events of World War II in a cultural context. The great hall that we see in this painting recalls the mythological Valhalla, the dwelling place of gods and heroes, which is consumed by flames at the end of Wagner's opera. Its immolation is treated here as a prophecy of twentieth-century German history, in particular the mass destruction during the war. Wagner's example is especially pertinent in this regard given his status as the Nazis' favorite composer.

In pursuing such themes, Kiefer exercises the freedom to exploit and mix whatever media—straw and sand, overprocessed photographs, lacquer, and fire—might serve the expressive purpose of simulating ancient fields charred and furrowed from centuries of battle and seeming to stretch into infinity. Kiefer's elegiac themes continue in the later *Departure from Egypt* (fig. 846), one of his most grandiose landscapes, rolling away majestically, yet held to the plane by the hook of Aaron's rod. Even though appearing as if it had been swept by a firestorm, or mulched in blood and dung, tar and salt, the field rewards close inspection with an almost tactile sense of touch and texture.

847. ANSELM KIEFER. *Breaking of the Vessels.* 1990.
Lead, iron, glass, copper wire, charcoal, and Aqua-tec,
16' x 6' x 48½" (4.9 m x 1.8 m x 123.2 cm), weight 7½
tons. The Saint Louis Art Museum, Missouri
MUSEUM PURCHASE. FUNDS GIVEN BY MORE THAN TWENTY-FIVE DONORS THROUGHOUT THE ST. LOUIS AREA

Kiefer has also worked in sculpture. In recent years, he has made immense "books" out of lead and gathered them on shelves; the resulting assemblages can weigh several tons (fig. 847). Like Kiefer's paintings, his books contain the past, but as much through their material presence as through imagery or language. Indeed, these huge volumes have a massive solidity as sheer physical objects that, more than anything else, conveys a palpable sense of the weight of history upon us. The artist's use of lead also carries associations with time, both medieval and modern. Lead is the material that alchemists sought to convert into gold in the Middle Ages. It is used, nowadays, for shielding against the radioactivity of atomic devices—a modern transmutation of matter that in some sense has fulfilled the alchemist's dream, even as it has made great devastation possible. Through these material means, Kiefer evokes the nuclear threat of the post-World War II era, placing it within a broadly conceived history.

Younger than any of the German artists just discussed, the Berlin painter **RAINER FETTING** (b. 1949) typifies his generation in that he seems less driven by the divisiveness of postwar politics than by their concrete implications, especially as mapped onto the city of Berlin, which, divided for almost all of the eighties, was a focus of East-West tensions. In Fetting's work feeling translates into erotic imagery of almost hallucinating intensity and in brushwork of daredevil freedom, as if to suggest that hot-wired spontaneity is the only

way to express the life force within oneself and one's world when both are continuously threatened with annihilation. His *Vincent* (color plate 426, page 688) pays homage to a revered progenitor of such an emotionally driven, painterly art. And here, also, is the Expressionism of Kirchner—sexuality and fluid painterliness alike—pushed to new excesses of risk.

In a period of young artists' early and easy success, the Chicago-born **LEON GOLUB** (b. 1922) stood in stark contrast, a moralist in the tradition of Daumier and Beckmann who for some forty years stuck by his guns, all the while "famous for being ignored," until at last, in the early 1980s, the world's unhappy concerns had so evolved as to coincide with his own. These are the politics of power and the insidious, all-corrupting effects—violence and victimization—of power without accountability. The then–sixty-year-old Golub had spent the long years of his exile—out of fashion and out of the country—steadily gathering his artistic and intellectual forces, so that by the time the first pictures in the Mercenaries series came before an astounded public in 1982, they presented, as did the later White Squad pictures, a hardwon but fixating, frozen unity of raw form and brutal content (fig. 848).

By the time of the Mercenaries, Golub had begun to focus on the margins of society—the putative jungles of South America or Africa—where the emissaries of political power tend to be not anonymous armies or aloof public figures but "mercs," individual agents for hire. To bring home the evil of torture and terrorism, Golub hit upon the original and sophisticated idea of making his political criminals all too human, fascinating and specific in their cowboy swagger and guilty satisfaction, while at the same time reducing the intimidated victim to a dehumanized, undifferentiated state almost beneath empathy. The better to authenticate his images of unfettered police authority, Golub pieced them together from photos of political atrocities, complete with the ungainliness of men exuberantly, sensually caught up in their daily work. Furthermore, he made the petty tyrants ten feet tall, pressed them forward with a Pompeiian oxide-red ground, itself a primary emblem of imperial, political power, and cropped the antiheroes' legs off to spill the scene forward, as if into the viewer's own space. Far too committed and subtle to preach or sloganeer, Golub simply uses his long-seasoned pictorial means—a congruence of image, form, process, and psychology—to induce feelings of complicity in a system that manipulates some individuals to destroy human life and others to ignore the fact that it is happening.

Since the beginning of her career in Chicago in the late 1940s, **NANCY SPERO** (b. 1926) has pursued political and social themes. In the sixties her Vietnam drawings were among the most powerful statements against the war. *Search and Destroy* (fig. 849) from 1967 shows one of the helicopters that saw much combat in Southeast Asia, but Spero reimagines the aircraft in a fantastical, Goyaesque form, its nose like the lance of a huge swordfish, or perhaps a hypodermic needle, driven into the chest of a falling victim.

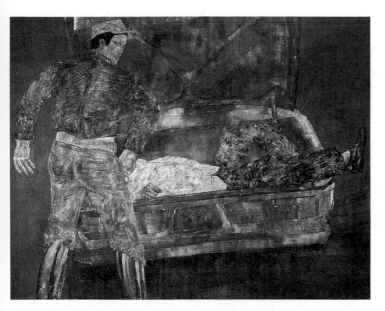

848. LEON GOLUB. *White Squad IV, El Salvador.* 1983.
Acrylic on canvas, 10' x 12'6" (3 x 3.8 m).
Saatchi Collection, London

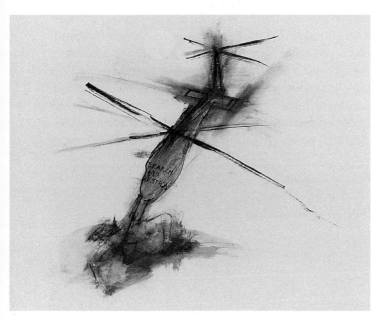

849. NANCY SPERO. *Search and Destroy.* 1967. Gouache
and ink on paper, 24 x 36" (61 x 91.4 cm).
Courtesy the artist

In subsequent work, Spero has expanded her drawings into elaborate sequences. These she has presented in the form of friezes or unrolled scrolls: panoramas of individual panels of print and collage, stretching up to 180 feet. These ribbons of images represent women from all eras and cultures. In *Goddess II* (color plate 427, page 721), ancient Greek-vase dancers and cavorting goddesses appear again and again, amid earlier and later sources. The diversity of the figures' printed textures and colors from one appearance to the next makes the frieze richly inclusive in its unfolding graphic development. Spero has said that her scrolls are "almost like a map. A map of the range of human experience

from birth to aging, war, and rape, to a celebratory dance of life—but depicted through images of women."

After making small drawings, artist's books, and videos in the seventies, the New York artist IDA APPLEBROOG (b. 1929) undertook a number of large, multipanel paintings in the eighties, among them *Noble Fields* (color plate 428, page 721). It is tempting to overinterpret these ambitious but ambiguous works. One critic, for example, saw in *Noble Fields* "a bald Medusa" who "sits for her portrait in an evening gown and arm cast, while little Oedy [Oedipus], nearby, eats a watermelon"—a description that some may find as puzzling as the painting. The mention of Oedipus does turn out to be illuminating, however, since the stresses and hostilities within a troubled family are among Applebroog's notable subjects. Nonetheless, she speaks of them with a vocabulary of private symbols that the viewer cannot hope to decipher fully. Fixed against empty backgrounds that offer no clue to their meaning, her mysterious gatherings of figures are painted in thin, somewhat unearthly colors that give them the air of characters half-remembered from a distant, traumatic past. At the same time, the multitude of seemingly unrelated images makes it impossible to offer a single, unified "reading" of the painting. In *Noble Fields,* our attention is pulled back and forth by the strip of images symmetrically disposed across the horizontal panel at the top; by the injured, supported male, the line of high-kicking dancers, and the dining female in the right panel; by the delicately painted monster in the left panel, who is also injured, though wearing a frilly dress; and by the child gorging himself on massive watermelons in the stark, lunar landscape at the center. These competing areas of interest are as difficult to reconcile with one another as are the members of a fractious family.

JOAN SNYDER (b. 1940) has long been known as a maker of virtuosic abstractions, full of painterly incident and visceral brushwork. *Love Your Bones* (color plate 429, page 721) was among the expressionistic "stroke" paintings that first brought her wide recognition in the early 1970s. In the 1980s, however, Snyder's political concerns came increasingly to the fore, in representational works as anguished as those created by the New Objectivity *(Neue Sachlichkeit)* painters after World War I. Unlike her earlier practice, with *Women in Camps* (fig. 850) she was now led to use photographic sources, creating a collage of discomforting vignettes of interned women, among them Holocaust victims and Palestinian refugees. She surrounded these figures with a heavily impastoed gray and black landscape, reflecting the kind of painterly involvement with materials present in her early work but now used with a new, political intent. It is in fact the very contrast between the small, trapped figures and the overwhelming, ominously painted areas around them that conveys the work's compassionate drama. Inscribed at the margins are the lines: "Women with babies, bags/Grandmas wearing babushkas and scarves and yellow stars/The Moon shone in Germany/The Moon shines in Palestine/And men are still seeking Final Solutions."

850. JOAN SNYDER. *Women in Camps.* 1988. Oil, acrylic, and ink on linen mounted on board with photographs, 22 x 43" (55.9 x 109.2 cm). Collection Mr. and Mrs. Richard Albright

Politically and socially committed, and uncensored in her approach, the British-born artist SUE COE (b. 1951) left English reticence far behind when she moved to New York in 1972, and there began aiming her agitprop collage-paintings at such targets as the CIA, male chauvinism, nuclear brinkmanship, apartheid in South Africa, and, ultimately, the policies of President Reagan (fig. 851). Coe takes her inspiration from the social and political inequities and righteous indignation expressed by such artists as Brueghel, Goya, Daumier, and Orozco. She is a true expressionist, using art to indict an unjust world with the same passion that inspired such earlier twentieth-century heroes of moral protest as Käthe Kollwitz, George Grosz, Otto Dix, and Max Beckmann. Formally, Coe has employed a whole battery of provocative techniques—a hellishly dark palette of graphite and gore; razor-sharp drawing; jammed, irrational spaces; collaged tabloid headlines—all of which she wields with a seemingly unshakable faith in the rightness of her perceptions. But, however extreme or fanatical these works may seem to viewers of a less political persuasion, one cannot help be moved by the poetic justice at work in a midnight world of evildoers metamorphosed into snakes and their hapless victims into angels, or in a painting overprinted with a banner like: "If animals believed in God, then the devil would look like a human being." As one critic wrote, "Sue Coe paints horror beautifully, ugliness elegantly, and monstrosity with precise sanity."

The San Francisco sculptor ROBERT ARNESON (1930–1992) delighted in poking fun at the pretensions of the art world. From his early Pop works of the sixties, such as his sculpted toilet bowls, to the caricature portrait busts for which he was best known, Arneson made light of artists' airs and affectations. He worked mainly in ceramic, a "craft" material not then considered suitable for serious sculpture, and in part this was another jab at "high art." In a gesture as self-revealing as it was satirical, he called an unflattering 1974 self-portrait *Search for Significant Subject Matter.* His caustic wit did not confine itself to the art world; in the eighties it was increasingly directed against prominent political figures. Here Arneson's satire grew harsher and took on greater moral

851. SUE COE. *Malcolm X and the Slaughter House.* 1985. Oil on paper, collage, 59⅞ x 49¼" (152.1 x 125.1 cm). Collection Don Hanson

left: 852. ROBERT ARNESON.
General Nuke.
Ceramic, bronze, and granite,
6'2½" x 32½" x 37"
(1.9 m x 82.6 cm x 94 cm).
Hirshhorn Museum and
Sculpture Garden,
Smithsonian Institution,
Washington, D.C.
© 1997 ESTATE OF ROBERT ARNESON/VAGA, NEW YORK, NY

right: 853. FRANCESCO CLEMENTE.
Francesco Clemente Pinxit (detail).
1981. Natural pigment on paper,
24 miniatures,
each approx. 8¾ x 6"
(22.2 x 15.2 cm).
Virginia Museum of
Fine Arts, Richmond
GIFT OF SYDNEY AND FRANCES LEWIS

urgency. Indeed, his savagely sarcastic portraits of politicians and military leaders, for whom he felt deep suspicion, can seem almost brutal in their vehemence. In *General Nuke* (fig. 852), the helmeted head of a belligerent commander is equipped with a ballistic missile for a nose. Its mouth drips blood, and the head rests on a pedestal that turns out, on closer inspection, to be composed of tiny corpses—the mass victims of the kind of nuclear war still widely feared in the early eighties, when Ronald Reagan called Russia an "evil empire." In this and a number of related works, Arneson's anti-war satire is worthy of the German New Objectivity painters, among them George Grosz.

Among the "three Cs" of Italy's *transavanguardia* a group completed by Sandro Chia and Enzo Cucchi—who first came to international notice at the 1980 Venice Biennale, the Neapolitan-born **FRANCESCO CLEMENTE** (b. 1952) is the most prolific. Clemente maintains studio-residences on three continents, in Rome, Madras, and New York, and every year migrates with his family to each for an extended period of work. While he absorbs and makes use of the imagery, ideas, and techniques native to the environment of each city, the artist also allows his disparate influences to overlap and cross-fertilize. This becomes especially apparent in works like the twenty-four miniatures of *Francesco Clemente Pinxit,* executed in India and in the classical Hindu manner, using natural pigments on paper (fig. 853).

In New York, Clemente taught himself oil painting. There, like Barnett Newman, Clemente created a series of pictorial works entitled The Fourteen Stations, interpreting the theme not with the traditional iconography of Christ's journey to the Cross, but rather through a dreamlike intermingling of personal psychology, cultural history, and religious symbolism. Throughout the Stations the artist's own naked body is ubiquitous. In Station No. IV, the figure has become what appears to be a Roman-collared priest, screaming like Bacon's Bishops (see color plate 286, page 522) and

lifting his white surplice to reveal a dark, egg-shaped void just above an intact scrotum (fig. 854).

In Rome, appropriately enough, Clemente has created frescoes. For one of them, Clemente, like a latter-day Duchamp, made use of an abandoned bicycle wheel and then partially frescoed it with the image of a jet plane. The bicycle wheel has subsequently appeared in other works, none of them more poetic, mysterious, or wonderfully lucid than a mural-sized, untitled canvas painted in Rome in 1984 (color plate 430, page 722). Here, the artist depicted a limpid underwater surface covered with rocks and pebbles, the arrangement as beautiful as any to be found in a clear stream —or an abstract painting. Concocted of pumice and pig-

854. FRANCESCO CLEMENTE. *The Fourteen Stations, No. IV.*
1981–82. Oil and encaustic on canvas, 6'6" x 7'4½" (2 x 2.2 m).
Private collection

855. SANDRO CHIA. *Three Boys on a Raft.* 1983.
Oil on canvas, 8'1" x 9'3" (2.5 x 2.8 m).
Collection PaineWebber Group, Inc.
© 1997 SANDRO CHIA/VAGA, NEW YORK, NY

856. ENZO CUCCHI. *Entry into Port of a Ship with a Red Rose.*
1985–86. Fresco, 9'2" x 13'1" (2.8 x 4 m).
Philadelphia Museum of Art
GIFT OF MR. AND MRS. DAVID N. PINCUS

below: 857. ENZO CUCCHI. *Vitebsk-Harar.* 1984.
Oil and polyurethane on canvas,
11'9¾" x 15' x 6'6¾" (3.6 x 4.6 x 2 m).
Courtesy Sperone-Westwater Gallery, New York

ment, the evanescent color evokes the classical fresco art of
Rome. Another "underwater" mural-size painting is dominated by a single huge figure (color plate 431, page 722).
Suspended, with eyes closed, is a being with combined male
and female reproductive roles. With Clemente's absorption in
the generative workings of the body, at first one might think
this large-headed figure a kind of highly developed fetus,
floating in amniotic fluid. The title *Semen,* however, suggests
a connection with swimming sperm. Together these clues
create an ambiguous, male/female double sense of potential
life awaiting fruition.

In his reaction to modernist conventions, **SANDRO CHIA**
(b. 1946), a Florentine now resident in New York, has relied
on Italian exuberance and a personal knowledge of the ersatz
Neoclassicism practiced during the Fascist 1920s and 1930s
by such artists as Giorgio de Chirico (see fig. 295) and Ottone
Rosai, themselves reactionaries against Futurism, Italy's first
modernist style. Adopting the older painters' conventions
—their absurdly muscle-bound nudes and self-consciously
mythic situations—but inflating the figures to balloon proportions while reducing the overblown narratives to the scale
of fairy tales, Chia parodies his sources at the same time that
he thumbs his nose at the authentic neoclassical rigor that
survived in Minimalism (fig. 855). Just as appealing as the
comic amplitude of these harmless giants is the Mediterranean warmth of Chia's color and the nervous energy of his
painterly surfaces (color plate 432, page 722)—qualities of
scale, palette, and touch that the artist has also managed to
translate into his heroic bronzes.

Unlike the world travelers Clemente and Chia, **ENZO
CUCCHI** (b. 1950) remains solidly rooted in the soil of his
native region near Ancona, a seaport on Italy's Adriatic

coast. There, for generations, the Cucchi family has worked
the land, a land whose circular compounds of farm buildings
and catastrophic landslides have often figured in the artist's
paintings. Because of his doomsday vision and his love of
richly textured surfaces, built up with what often seems a
waxy mixture of rusted graphite and coal dust, Cucchi has
sometimes been compared with Germany's Anselm Kiefer.
What sets him unmistakably apart, however, is the Italian
heritage he brings to his dramas; he grew up amid paintings
of saints and martyrs. In his *Entry into Port of a Ship with a
Red Rose* (fig. 856), austere yet emotively rendered crosses
glide across this large fresco in graceful clusters, as purpose-
fully as a fleet of sailboats. Such powerful compositional
effects, achieved with the sparest of means, have led certain
critics to align Cucchi's work with the early Renaissance fres-

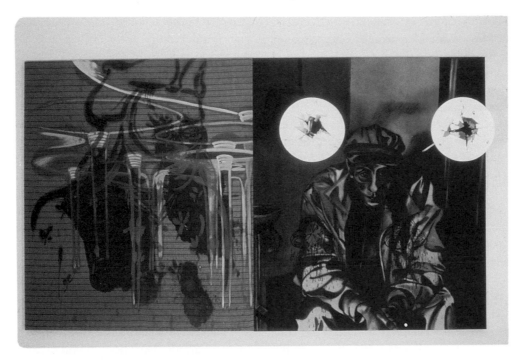

858. DAVID SALLE. *Miner.* 1985. Oil, acrylic, and table/fabric on canvas, 8' x 13'6¼" (2.4 x 4.1 m).
Collection Philip Johnson © 1997 DAVID SALLE/VAGA, NEW YORK, NY

coes of Giotto. Occasionally, however, a deadly calm settles over one of his apocalyptic landscapes (fig. 857), leaving a field empty of all but a congealed flood of pigment, in which float a ghostly presence, ranks of skulls, or the horrific head of a dead animal. Nonetheless, even these remains are granted an odd, elegiac poetry by the presence of an abandoned piano.

The artist most often credited with reigniting the eighties art scene was the Brooklyn-born, Texas-educated JULIAN SCHNABEL (b. 1951). For viewers starved by the lean visual diet of Minimalism and Conceptualism, Schnabel's epically scaled, thematic pictures—reviving the whole panoply of religious and cultural archetypes once the glory of traditional high art—had an electrifying effect.

In a move both innovative and violently expressive, Schnabel incorporated three dimensions into a flat surface by embedding broken crockery in plaster spread over a reinforced wood support (color plate 433, page 723). The idea for this startling process came to him during a visit to Barcelona, where he saw Antoni Gaudí's tiled mosaic benches in Güell Park (see color plate 42, page 114). However, rather than smooth and inlaid, Schnabel's shards project from the surface in an irregular, disjunctive arrangement that visually functions, edge to edge, as a field of enlarged Pointillist dots. At the same time that the china pieces serve to define shapes and even model form, they also break up the image and absorb it into the overall surface flicker like a primitive, encrusted version of the Art Nouveau figure-in-pattern effects of Gustav Klimt (see color plate 40, page 113), Gaudí's Austrian contemporary. By painting over, around, or even under the protruding ceramic, Schnabel has been able to create every kind of pictorial drama, from evocations of folkloric wit and charm to suggestions of mythic ritual. In Schnabel's

work the emotive effect arises less from the image, which is not immediately decipherable, than from an audacious stylistic performance, or process, that simultaneously structures and shatters both figure and surface.

Among the American Neo-Expressionists, one of the most controversial yet undeniably important artists is DAVID SALLE (b. 1952) (pronounced Sally). Salle maddens his critics but delights his supporters by forging an unpromising mix of secondhand, disassociated elements into a new, original conception. In the painting called *Tennyson* (color plate 434, page 723), the artist has disclosed his Conceptual background by lettering out the title in capitals, across a field dominated on the left by a found wooden relief of an ear, and on the right by a nude woman with her back to the viewer. The name of a proper Victorian poet, a piece of erotica, and a severed ear—at first, the different images may seem to cancel one another out and erase all meaning, until one recalls that in 1958 Jasper Johns reproduced Tennyson's name across the bottom of an abstract canvas painted in sober gray on gray, as if in tribute to the artistic integrity and lyrical, high-minded spirit of England's "good grey poet." The connection to Johns is reasserted by the disembodied ear, which could also evoke the poet's "ear for words," or Van Gogh's self-imposed mutilation, but also, and more tellingly, Jasper Johns's *Target with Plaster Casts.* Johns, we should recall, originally affixed a three-dimensional ear, among other anatomical parts, to *Target with Plaster Casts* (see color plate 296, page 526). Such correspondences generate others, and all combine as if in commentary on the naked woman, the sexism such imagery represents, and the role of the nude in the history of art.

For a large diptych entitled *Miner* (fig. 858), Salle painted the frontal image of a laborer, sitting with stooped shoulders and a tired, rather baleful expression on his face. In front of him at chest level float two immense diamond rings, conceivably the product of his labor and apparently objects forever beyond his reach. On either side of the subject's head are a pair of appliqués, cheap café tables with their tops smashed. Positioned and projected as they are, they could be the relics of a pub brawl, representations of the miner's headlights for working at some dark, stygian depth, or the glittering stone "headlights" set in the rings. Evidence of violence, remoteness, and despair set off vibrations unifying the disjunctive images, spaces, colors, textures, and styles in a mood of sadness and loss.

Few of the Neo-Expressionist painters address their art totally to the figure, in all its naked physical and psychological complexity. Foremost among such painters is **ERIC FISCHL** (b. 1948), whose immense fleshscapes have an unusually high-voltage effect on viewers (color plate 435, page 723). Even beyond the portrayal of nudity or the suggestions of alcoholism, voyeurism, onanism, homosexuality, bestiality, and incest, Fischl's paintings of nudes are concerned with the prurient fascination that most have with the nakedness of others. This can be seen in painting after painting set in the gardens, on the beaches, in the well-appointed rooms, and on the yachts of affluent America, all populated by flabby men, women, and children scattered about or clustered in scenes of boredom and isolation. Awkward and half-embarrassed, self-absorbed and yet furtively, even intently looking at one another's exposure, these figures communicate the individual's psychic nakedness as well.

The paintings in Fischl's India series (fig. 859) have been described as the work of a "postmodern tourist." Produced after a trip to northern India in 1988, the series implies a kind of critique of the "Orientalism" and "exoticism" often fea-

tured in European images of the Far East and Africa from the nineteenth century, just as Fischl's nudes are in part a critique of female nudes by Manet and others. Separating Fischl from his colonialist predecessors, however, is the fact that he lives in a decidedly postcolonial era, when Western imperialism has largely been replaced by Western tourism. But even if the beggars and the group of monkeys in *On the Stairs of the Temple* suggest little beyond a tourist's snapshot, the treatment of the woman at the center prompts a more complicated response. Clothed in pink and bathed in splendid light, she might at first seem to evidence at least some attempt by the artist to fathom another culture, with its different understanding of spiritual and physical beauty. And yet there is a real ambivalence communicated by the draped figure. The viewer's expectation that she be freely available as an object of visual delectation is at least partially confirmed by the subject's incarnation of the stereotype of veiled Other of the East.

The art of **ROBERT LONGO** (b. 1953) has been described as "apocalyptic Pop," huge "salon machines" made of over-lifesize imagery taken from or inspired by film, television, comic books, and advertisements, a secondhand world of kitsch and high culture mixed and matched to provoke voyeuristic responses to such basic themes as love, death, and violence. No less dramatic is the serendipitous variety of media that Longo and his Renaissance-like atelier of collaborators (usually named) have combined: sculpture (wood carving, stone intaglio, bronze casting), painting (acrylic, spray paint, and gold leaf), drawing, silkscreen, and photography.

Longo gained early recognition with a number of large black-and-white charcoal drawings, such as the 1979–82

859. ERIC FISCHL. *On the Stairs of the Temple.* 1989. Oil on linen, 9'7" x 11'8" (2.9 x 3.6 m). Collection Boris Leavitt

860. ROBERT LONGO. *Untitled* (Men in the Cities series).
1981. Charcoal and graphite on paper,
8' x 60" (2.4 x 152 cm).
Courtesy Metro Pictures, New York

861. TOM OTTERNESS. Frieze installation at Brooke Alexander, New
York. 1983. Cast plaster

series called Men in the Cities. For these impressive works the artist photographed single figures clad in sober business clothes and seized in the contorted throes of what could be either agony or ecstasy, dying or dancing (fig. 860). But while any such condition should create a sense of heat, the figures are frozen—and perhaps all the more haunting for the tension this ambiguity produces. Indeed, the technique by which they were made was a distancing one. It involves projecting the image to a large scale, combining features—head, legs, hands—from several such photographs to make more an archetype than a portrait, and then rendering the composite in charcoal and graphite on paper, an activity Longo carried out in concert with an assistant.

A Postmodernist par excellence, Kansas-born **TOM OTTERNESS** (b. 1952) looks back not only to sources in modernism but also to the far-off dawn of civilization, a primordial, if not-so-innocent, time when, in Otterness's conception, humanity had the undifferentiated, universal look of "Pillsbury dough boys," as well as an obsessive interest in

cavorting with such Primary forms as cones, cylinders, and spheres, when not engaged in every manner of primal sex. To contain the energies of his somewhat perverse little brutes, the artist tends to corral them into architectural friezes, an archaic form that assumes the character of an archaeological fragment (fig. 861). Offered segmentally (and sold by the foot), the friezes begin to look like three-dimensional comic strips and the actors like those in, for instance, "Alley Oop." The sex and symbolic geometry, the pagan Ur-folk rehearsing the primitive hilarities of contemporary Pop culture, the classical forms issued like yard goods—all are part of Otterness's desire to make his art a universally accessible language through which to comment on the human condition. Seeing parody and irreverence as more persuasive than heavy rhetoric, the artist has had his rotund ciphers romp through mythopoetic sagas of labor, mass uprisings, Edenic pleasures, and group bestiality. One by one, they lampoon such grave issues as aggression, sexism, and alienation in the technological age.

Appropriation

What in the eighties became widely known as "appropriation"—the act of borrowing directly from the work of others—takes many forms. Many of the Neo-Expressionists discussed above openly appropriated the painterly styles of the German Expressionists or the American Abstract Expressionists. In this section, we will examine some similar "high-art" borrowings carried out by artists other than the Neo-Expressionists. Often, however, appropriation has taken on a new meaning, referring specifically to the borrowing of visual styles or individual images not from earlier artists but from the mass media. It is this second kind of appropriation that this section will also explore.

By the eighties it had been clear for decades that commercial mass media—from newspapers and magazines to the movies and, most of all, television—had come to dominate every aspect of modern life, including its visual culture. Indeed, the Pop artists had embraced the vernacular of the media, making printed material like advertisements and comic strips the point of departure for their own imagery. In the eighties, however, artists' attitudes were quite different from those of their Pop forebears. The Pop artists brought

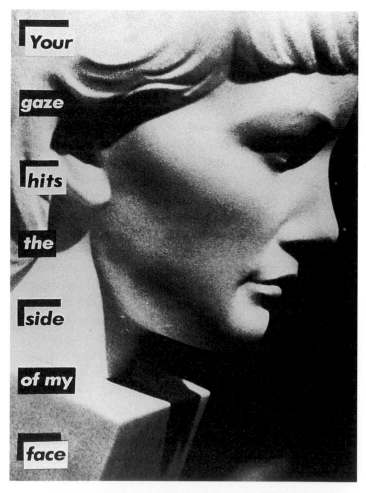

862. BARBARA KRUGER. *Untitled (Your Gaze Hits the Side of My Face).* 1981. Photograph, 60 x 40"
(152.4 x 101.6 cm). Collection Vijak Mahdavi and
Bernardo Nadal-Ginard. Courtesy Mary Boone

the icons of advertising and everyday consumer society under tight scrutiny, yet they had also in a sense glorified them—as with the household products turned into monumental sculptures by Oldenburg, the ads and comics elegantly "abstracted" by Lichtenstein, or the almost hagiographic paintings that Warhol made from publicity photos of Marilyn Monroe. But many eighties artists took a more critical and even hostile view of media culture. They often plucked their imagery directly from the lower pop-culture strata, from such sources as sex manuals, pulp romances, and the most tawdry television soap operas, transposing it, for maximum contrast, to the "elevated" sphere of high art, history, and politics.

Appropriation turned out to be an effective strategy for dealing with the power of the mass media and even stealing its thunder, so to speak. By extracting images from their familiar original contexts and mixing them with other images from different sources, appropriation inevitably strips iconography of its original import. But at the same time, by putting borrowed images into a new context—that is, by "recontextualizing" them—it also endows those images with a new and often unsettling impact that encourages viewers to see the original sources in a new light. This startling effect—making us see familiar images afresh, as if for the first time—is the source of appropriation's power as a critique. In the critical language of the day, this effect made appropriation art "deconstructive"—setting in motion an analytical process that takes apart and exposes the image-maker's designs on us. In the end, art of this kind gives us a more sophisticated awareness of just how easily we can be manipulated by visual images.

A notable example of the deconstruction process at work is found in the art of **BARBARA KRUGER** (b. 1945) (fig. 862). Her picture-and-text combinations reveal the most manipulative aspects of present-day media culture. Kruger, who in the mid-sixties had worked as a graphic designer at *Mademoiselle,* appropriates the "look" of a glossy magazine layout. Fashion magazines are an especially clear example of contrived visual meaning, since they freely crop and group their pictures and impose captions on them, in order to force the images into the particular "story" that the photo editor wants to tell. By borrowing these techniques, but by bringing them to the surface where we cannot miss them, Kruger makes it blatantly obvious that our response, not only to this picture—taken from a fifties photo annual—but to any media image, is largely dictated by the editorializing of its presenter. Our understanding of what we see is not a simple, direct response to the optical facts; rather, it is something that has been "constructed" for us.

But work like Kruger's suggests that artists can also turn the tables. If the media can manipulate images in order to convey new meanings, so can artists—by taking such images back, through appropriation, and giving them yet another new spin. In so doing, artists can open the viewer's eyes to the complex, ambiguous process by which the meanings of virtually all pictures are constructed. In fact, one can uncover a good deal about how a culture works and what it believes by "de-constructing" the way it uses visual images. This

above: 863. SHERRIE LEVINE. *After Piet Mondrian*. 1983.
Watercolor, 14 x 11" (35.6 x 27.8 cm).
Private collection

right: 864. CINDY SHERMAN. *Untitled Film Still*. 1979.
Photograph, 10 x 8" (25.4 x 20.3 cm).
Courtesy Metro Pictures, New York

process of deconstruction reveals how certain images function to enforce social myths rather than any underlying truth. By confronting these issues, the technique of appropriation brought art into the media-governed, image-saturated eighties—and, simultaneously, brought the eighties into art.

The quintessential Postmodernist among American artists may be **SHERRIE LEVINE** (b. 1947). Not only has she appropriated her art quite literally from art history, but she has done it with such audacity that it subverts the artist's own deconstructivist case against the modernist myth of originality. In 1981 Levine first startled the art world with an exhibition of her own unmanipulated photographs of well-known, textbook examples of photographs by such master cameramen as Edward Weston, Walker Evans, and Andreas Feininger. Her replication of these iconic images pointedly undermined the ideals of modernism—originality, authenticity, and even a masculine aesthetic. After completing this series, Levine came to realize that her subversion of these traditional standards of artistic value could be accomplished not only through photographic reproduction, but also through the insertion of her own, female touch. Not surprisingly, therefore, Levine was soon hand-painting watercolors and gouaches "after" art-book reproductions of paintings by Léger, Mondrian, Lissitzky, and Stuart Davis—complete with printing flaws (fig. 863).

Levine pursued similar goals in a subsequent series consisting of what were less pure appropriations than "generic paintings," as the artist calls them. A Hard-Edge series consists of small casein and wax panels offering two-color variations on a single theme of one narrow and three wide stripes (color plate 436, page 724). Here the model was a composite of all the great modernist stripe painters, again a pantheon of male artists, ranging from Barnett Newman through Ellsworth Kelly, Brice Marden, and Frank Stella to Daniel Buren. While the new works present the overall look of Minimal painting, they copy no one masterpiece or specific style. Instead, the artist takes on the male-dominated aesthetic of high modernism in a "generic" fashion. Despite Levine's evident iconoclasm, she is not without models, perhaps most significantly Marcel Duchamp, with whose ready-mades the artist has recently acknowledged an affinity.

Beginning in 1978, **CINDY SHERMAN** (b. 1954) allied the arts of Performance and photography; she cast herself as the star of an ongoing, inexhaustibly inventive series of still photographs sometimes characterized as "one-frame movie-making" (fig. 864). Complete with props, lighting, makeup, wigs, costume, and script, most of the pictures exploit, with the eventual goal of subverting, stereotypical roles played by women in films, television shows, and commercials of the 1950s and 1960s. Her characters have ranged from a bored suburban housewife to a Playboy centerfold.

In a 1980–82 series Sherman began working in color and

in near-lifesize dimensions. As the pictures grew larger, the situations became more concentrated, focusing on specific emotional moments rather than implying the fuller scenarios seen earlier. Setting, background (supplied by a projector system), clothes, wig, makeup, and lighting combined with the performance itself to create a unified mood: anxiety, ennui, flirtatiousness, or confidence. Often there is a disturbing detail, such as a piece of torn newspaper clutched by a plaid-skirted teenager sprawled on a linoleum kitchen floor (color plate 437, page 724). Like a Hitchcock heroine, the subject seems vulnerable to something outside the photographic frame.

More recently Sherman has appropriated images from old master art, as in her posed imitation of Carravaggio's *Bacchus* (color plate 438, page 724). Most of the works in this series, which the artist calls History Portraits, are based on reproductions from books about French or Italian art. However, Sherman's versions alter and distort the compositions on which they are based, and some of them seem to be composites or inventions, apparently unrelated to any particular source. Her project is actually based on the notion that "woman" does not exist apart from the way she is constructed through masquerade. However near or distant the quoted source, Cindy Sherman and her fellow

"image-scavengers" cause us to question the extent of our reliance on images and our relationship to the authority they wield.

KOMAR AND MELAMID Vitaly Komar (b. 1943) and Aleksandr Melamid (b. 1945) both grew up in the former Soviet Union. They began to collaborate in 1965 and were among the Conceptual artists active in Moscow in the seventies. The two left the U.S.S.R. for Israel in 1977 and settled in New York the following year.

In their works of the early eighties, they satirically appropriated the old-fashioned, academic Socialist Realism that was the officially approved style of the Stalinist period in which they spent their youth, much as the American Pop artists appropriated the style of comic strips and product advertisements. The difference is that Komar and Melamid satirized the failings of communism. Their *Double Self-Portrait as Young Pioneers* (fig. 865), for example, presents a heroic, romantically backlit portrait bust of Stalin on a high pedestal; the elevated head is near the top of the painting. In order to supply the needed measure of glory demanded in official depictions of the leader, the bust is given a trumpet fanfare by one of the artists and a salute by the other. Both young pioneers of what was supposed to be the new Russian Utopia are shown as uniformed little boys in short pants (as in fact they were when Stalin ruled)—but with the ironic addition of the spectacled, bearded, and mustached faces that the two men have as adults. By adding this incongruous element to the appropriated state style, the artists deflate the posture of hero worship that they were taught to revere in childhood.

The identical twin brothers **MIKE AND DOUG STARN** (b. 1961) work in collaboration. Although they have also painted, their principal medium is an elaborate variety of photocollage. In their work, they appropriate a photographic image and subject it to all manner of technical alterations. For instance, they might enlarge an image enormously, or make a toned print of it in the nineteenth-century manner. But they will also deliberately scratch the negative, or otherwise distress the image, crease or tear the print, stain it with glue, or cut it up. Then they will complete the work by mounting and framing the separate pieces, or tape them directly to the wall. Often the resulting large-scale photocollages are as imposing as murals.

In *Double Mona Lisa with Self-Portrait* (color plate 439, page 725) the Starns borrow an image often appropriated by such artists as Duchamp and Warhol. Instead of using a photograph that documents the painting alone, however, as those artists did, the Starns on a trip to Paris took a picture of the *Mona Lisa* surrounded by its large, cumbersome vitrine at the Louvre, with the rest of the gallery and its visitors, including the two artists themselves, reflected in the

865. KOMAR AND MELAMID. *Double Self-Portrait as Young Pioneers* from the Nostalgic Socialist Realism series. 1982–83. Oil on canvas, 6' x 50" (1.8 m x 127 cm). Courtesy Feldman Fine Art, Inc.

866. ALISON SAAR. *Sapphire*. 1986. Wood and mixed media, 27½ x 31½ x 13" (69.9 x 80 x 33 cm). Gherardi Collection

glass. Museumgoers can see that painting only at one remove, sealed behind a glaring barrier, and this has the ironic effect of distancing the work from the very people who have come to experience it firsthand. In place of an original that is thus remote and out of reach, the photocollage substitutes its own complex aesthetic structure for that of the *Mona Lisa*. But the cut-up and painstakingly reconstituted version that the Starns create, with its patchwork grid of Scotch-taped panels, openly announces itself as a homemade, "constructed" copy—totally at odds with the mystique of the precious, irreplaceable masterpiece. As if to further comment on issues of originality and reproduction, the Starns double Leonardo's famous image, making a twin of it, like the two artists in the glass.

The expressionistic figurative sculptures of **ALISON SAAR** (b. 1956) have been called "multicultural totems." They are often concerned with African-American experience and the central role of women within it. Like other artists of the eighties, Saar is more interested in the aesthetic form of the human body (as displayed, for example, by the sculptors of classical antiquity), than in its history. She is especially concerned with the changing conception of gender roles. Reflecting this interest, some of the sculptures incorporate small assemblages; the sense of a surrounding world filled with found objects helps to place the sculptures' concern with gender in a larger cultural context.

The rough-hewn effigy in *Sapphire* (fig. 866) opens up her hinged breasts to reveal a shrinelike inner chamber filled with relics. The hoard of talismanic items includes seashells, stones, leaves, a comb, and a figurine of an archer, all of them visible in the glow of a candle-shaped light bulb. These items, each bringing with it a different, valued association from the outside, are all cherished and preserved within the "site" of the maternal body. The sculpture, obviously, has less to do with

reproduction per se than with a larger sense of how the traditions of African-American culture have been nurtured and protected by generations of women. At the same time, *Sapphire* makes reference to the history of Western European art, in particular to the medieval reliquary—a repository housing sacred items—which often took on the shape of a revered individual.

Collage and assemblage, art forms developed early in the century, have always been based on the incorporation of individual objects. For her assemblages and installations, the Los Angeles artist **BETYE SAAR** (b. 1926), the mother of Alison Saar, appropriates diverse found materials. As collage makers from Schwitters to Rauschenberg have done, she uses objects already rich with meaning. In Saar's work they can range from artifacts of occult religion to ordinary printed images to computer circuit boards. When gathered into her smaller assemblages these items give an appearance not entirely unlike that of a Joseph Cornell box—but one made with resonances of outsider art as well as Cubism, folk art, and Dada. And unlike Cornell's boxes, the works seem ritualistic in intent, as if designed to be used in calling up the spirits of ancestors or carrying out ceremonial rites. "There is an apparent thread in my work," Saar has said. "This thread is curiosity about the mystical." The artist has always been interested in the mysterious power that well-loved objects have over us, calling one group of works her Personal Icons. The ordinary objects she uses, like beads, mirrors, and photographs, are made to assume talismanic significance by being put together—"recontextualized"—in a suggestive new arrangement. The assemblage *Sanctuary's Edge* (color plate 440, page 725), for instance, with its modern-day "relics," resembles the kind of altar used for personal devotions in the Middle Ages.

Graffiti and Cartoon Artists

As Neo-Expressionism opened art to the many possibilities once banned by Minimalist purism, and appropriation art drew on the commercial media, a number of even younger artists found themselves drawn into the energy and "authenticity" of one of the most "impure" of all graphic modes. This was graffiti, those "logos" (initials or nicknames) and scrawled "tags" (signatures) their teenage perpetrators called "writing" and had felt-tipped and spray-painted (or "bombed") all over New York City's subway trains regularly since about 1960. For most passengers, this takeover of public walls by an indigenous, compulsive graphism simply constituted vandalism, but as a look back at the paintings of Jackson Pollock, Jean Dubuffet, and Cy Twombly indicates, graffiti-like marks have long been used by major artists (see figs. 521, 569, 670). Not until the 1980s, however, did the underground world of self-taught graffiti writers, mainly from the Bronx and Brooklyn, come into direct, interactive contact with the aboveground, Manhattan-based realm of artists and art-school students. In 1983, the upwardly mobile graffitists received something like a high-art imprimatur when Rotterdam's Boymans–van Beuningen

Museum put on the first museum exhibition of graffiti art, and subsequently the art dealer Sidney Janis offered his "Post-Graffiti" show. In the latter event, one of the world's premier dealers in modern art displayed the scribbled canvases of graffiti writers in the very galleries where for years New Yorkers had been viewing such blue-chip masters as Mondrian and Marisol.

Now, years later, after the deaths of several prominent figures, it seems clear that the graffitists and cartoonists who profited the most from the once random but subsequently deliberate encounter of uptown and downtown art were those who brought to the experience the support of an art-school education. This, it would seem, equipped them with a basic pictorial language that needed only to be charged by the graffitists' urge to communicate, as well as by the example of the subway artists' powerful calligraphic thrust, in order to expand their own Pop-cartoon-TV vocabulary into something like a new, broadly expressive language.

The dean of the graffitists was KEITH HARING (1958–1990), who through his School of Visual Arts preparation and love of working in public spaces served as the all-important link between the subway world of self-taught graffiti "writers" and the younger "mainstream" artists responsive to an omnipresent, abrasive form of popular expression. Simultaneously as he made his presence felt in all the graffiti-and-cartoon forums, Haring was also traveling throughout New York City's subway system and leaving his artistic calling card on virtually every train platform. The ubiquitous format for his impromptu exhibitions was the black panel from which an advertising poster had been removed, providing a surface somewhat like that of a schoolroom blackboard (fig. 867). After 1980, Haring made thousands of chalk drawings "in transit," always quick, simple, strong, and direct, for the activity carried with it the risk of arrest for defacement of public property. He developed his well-known vocabulary of cartoon figures—radiant child, barking dog, flying saucer, praying man—intermingled with such universally readable signs as the cross, the halo, the pyramid, the heart, and the dollar, all rendered with such cheerfulness and generosity that their statements could be neither dismissed nor mistaken (fig. 868). While in his commercial work Haring seemed to prefer drawing with black sumi ink or Magic Marker on paper, oaktag, fiberglass, or vinyl tarpaulin, he remained ready to unleash his prodigious capacity for graphic invention on almost any kind of surface, including those provided by pottery vases or plaster casts of such cliché icons as Botticelli's Venus. On these, as well as on vast outdoor walls, he sometimes collaborated with genuine graffiti writers, most of whom otherwise preferred the moving, subterranean environment of the MTA trains.

A key figure was the Brooklyn-born Haitian-Hispanic artist JEAN-MICHEL BASQUIAT (1960–1988), a high-school dropout at seventeen and, even though self-educated, probably the most meteoric star among all the talents associated with graffiti. Like the other successful graffitists, Basquiat never actually "bombed" trains. Rather, while more or less

living on the streets in Lower Manhattan, he formed a partnership with a friend named Al Diaz and began Magic Marking poetic messages, illustrated with odd symbols, all over the city, signing their work SAMO© (standing for "same old shit"). Basquiat's imprimatur reveals something of the frustration he felt as a black artist trying to assert his art within the context of a predominantly white gallery system. In the summer of 1980, SAMO© exhibited at the "Times Square Show" and attracted favorable critical attention. Shortly thereafter, Basquiat struck out on his own, and soon his drawing revealed something more than graphic street smarts—actually what appeared to be a devotion to Picasso's *Guernica*. Basquiat's color-infused art is characteristically peopled by schematic figures with large, flattened, African-mask faces set against fields vigorously activated by such graffiti elements as words and phrases, arrows and grids, crowns, rockets, and skyscrapers (fig. 869). As with Polke and Salle, unity resides not in any logical relationship among the various images, signs, and symbols but rather, if at all, in the generalized vibrations set off by so many disjunctive elements, as well as in the pervasive toughness of it all. Quite apart from his sure sense of drawing, color, and composition, Basquiat may have found his greatest gift in an innate talent, which allowed him to balance the contradictory forces of "primitivism" and sophistication, immediacy and control, wit and aggressiveness. His death at a very early age makes it difficult

left: 867. KEITH HARING. *New York City Subway Drawing.* 1983. White chalk on black advertising space Courtesy Estate of Keith Haring

right: 868. KEITH HARING. *One-Man Show.* Installation view. 1982. Tony Shafrazi Gallery, New York

below: 869. JEAN-MICHEL BASQUIAT. *Grillo.* 1984. Oil on wood with nails, 8' x 17'6½" x 18" (2.4 m x 5.3 m x 45.7 cm). Stefan T. Edlis Collection, Chicago
© 1997 ADAGP, PARIS/ARTISTS RIGHTS SOCIETY (ARS), NEW YORK

to imagine what he might have accomplished as a fully mature artist.

While **KENNY SCHARF** (b. 1958) also worked briefly with graffiti artists, he takes his primary inspiration from the space-age fantasy world of children's animated TV cartoons, such as *The Jetsons,* a series generated virtually next door to where Scharf grew up in Los Angeles. At first, the artist developed his signature style three-dimensionally, as he "customized" all the electronic gear in the building used for the "Times Square Show," and then a series of Closets, complete rooms, like the one seen here, created for the 1985 Biennial Exhibition at New York's Whitney Museum (fig. 870). To customize radios, TVs, telephones, furniture, and whole environments, Scharf applies to these objects and spaces all manner of "salvage"—kitsch items on the order of plastic dinosaurs, toy robots, fake furs, ropes of tinsel, old hood ornaments, a deer head—and then proceeds to use spray paint to unify the mishmash forms and spaces.

Meanwhile, Scharf was also painting on canvas. In one vast work, entitled *When the Worlds Collide* and measuring some ten by seventeen feet (color plate 441, page 725), he fused, "in a kind of fun-house big bang," what critic Gerald Marzorati

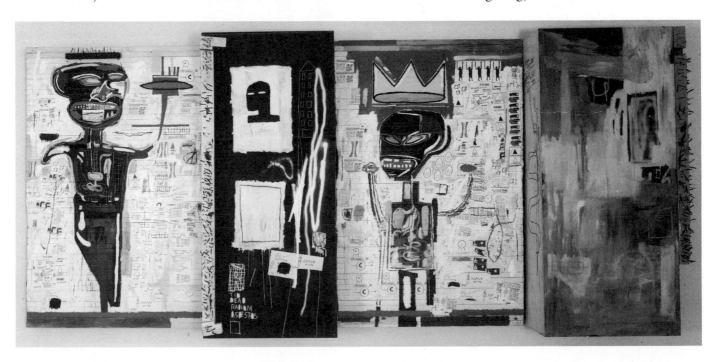

870. KENNY SCHARF. Installation for Whitney Biennial, 1985. Mixed media.
Whitney Museum of American Art, New York

called "the Saturday-morning innocence, the phone-doodle psychedelia, the magpie delirium, the electric chromatic dazzle. . . ." Here, against a spaced-out background divided into maplike fields of empyrean blue, soft-focus jellybeans, star-studded black holes, and loopy stretch springs, a huge, shiny, sharp-focus red mountain metamorphoses into a jolly monster with a waterfall tumbling over one shoulder and a moonscape visible in the depths of his broad grin. All about this goofy, improbable creature float whirling cannonballs, cotton-candy clouds, disembodied eyes and mouths, wriggling amoeboids, and an orange-colored, saber-toothed, cross-eyed pixie. In *When the Worlds Collide,* Tanguy-like illusionistic Surrealism or the doomsday world of Hieronymus Bosch has been reconceived by a California cartoonist, an artist certain that "you definitely cannot have too much fun" and equipped to go about it with the compositional and trompe-l'oeil ambitions of a Tiepolo. Scharf applies the style of hyperrealism to the subject matter of visual media in order to telescope in on the prominence of fantasy in American culture.

To the extent to which they share similar means and ends, the work of Scharf finds basic affinities with that of **RODNEY ALAN GREENBLAT** (b. 1960). Another child of the TV age, Greenblat has confessed to an early addiction not only to the

871. RODNEY ALAN GREENBLAT. *The Secret.* 1985. Pastel on paper, 57¼ x 43½" (145.4 x 110.5 cm).
Courtesy Gracie Mansion Gallery, New York

Saturday morning Hanna-Barbera cartoons but also to weekly evening sitcoms. Like Kenny Scharf, Greenblat grew up in Southern California and then entered the School of Visual Arts in New York. But unlike Scharf, Greenblat immediately began exhibiting in the galleries just emerging in the East Village. Perhaps owing to his later arrival, Greenblat was not directly involved in street or subway art; thus, while he shares a graffitist's compulsive approach to surfaces, filling them with a dense pattern of signs, symbols, calligraphic scrawls, and images, Greenblat applies them not to appropriated walls or "salvage," but rather to carpentered objects of his own crafting. He calls the process "Rodneyizing," a near, perhaps more polite, cousin of Scharf's "customizing," and a New Wave heir to Claes Oldenburg's and Red Grooms's ways of bringing the Pop spirit into art.

In his playful "cartooniverse" of Merrie Melodies sculpture, Loony Tunes furniture, and Walt Disney World castles and arks, Greenblat also allows time and space for paintings which narrate the sitcom adventures of Milo and Robot. In *The Secret* (fig. 871), a figure resembling Glinda the Good Witch seems to be advising the button-eyed protagonists to flee the menacing elements in the background—here not a Kansas tornado but something like Washington State's volcanic Mount St. Helens—and find refuge in a many-towered, Oz-like citadel pointed out on the far horizon.

Installations

Installation works of many different kinds were created throughout the sixties and seventies, whether by artists who also worked in Performance, such as Nam June Paik, Joseph Beuys, or Rebecca Horn, or by those with links to Conceptual art, such as Jonathan Borofsky and Christopher Burden. A special potential of environmental works, however, came to the fore in the eighties. Since we can walk around inside them, such works easily become surrogate scale models for us of the world at large, and installations made in this "retrospective" decade did increasingly seem like comments about the state of things in the Postmodern world. Some of these installations are specifically sites of memory, as in Christian Boltanski's work—places to reflect on the troubled history of the twentieth century and the role played by art within it. For many working in this mode, such as Ilya Kabakov, the lofty aspirations of early modernist art have collapsed into heaps of junk. At the same time, even those artists, like Bill Viola or Jenny Holzer, who have turned to new electronic technologies, have done so with a wary, skeptical irony, undercutting any great faith in the Brave New World of the future.

ILYA KABAKOV (b. 1933) grew up in the former Soviet Union during the harsh Stalinist era. Trained as an artist from an early age, for many years he earned his living as an illustrator of stories, most of them for children, while privately he made elaborate albums combining poetic, sometimes bizarre tales and drawings. Since the mid-eighties, Kabakov has created full-scale environmental pieces. They, too, contain elements of narrative, often revealing the difficulties and priva-

tions of Soviet daily life, but now conveying information through fantastical gatherings of objects, or what has been called Kabakov's "material poetry." After the collapse of the communist state, these works took on the character of archeological sites, where visitors could explore the cultural history of a vanished system.

In *The Man Who Flew into Space from His Apartment* (color plate 442, page 725), the hardships of ordinary people are ironically juxtaposed with the grandiose projects of the state—in this case, apparently, the highly touted Soviet space program, which launched the first human being into orbit. About this fantasy work Kabakov has said: "The person who lived here flew into space from his room, first having blown up the ceiling and the attic above it. He always, as far as he remembered, felt that he was not quite an inhabitant of this earth, and constantly felt the desire to leave it, to escape beyond its boundaries. And as an adult he conceived of his departure into space." The room we see, with a gaping hole in its roof, contains the debris left behind by the astronaut's improvised takeoff: broken dishes, the remains of some simple furniture, and stray items of worn clothing. Glued to the walls are various political posters as well as the traveler's blueprints, flight plans, and calculations; from the ceiling hangs the catapult that flung him away from it all, up into outer space. On a small model of a city, which has a light bulb for a sun, a strip of curved metal rising from an apartment building shows the planned trajectory of "the one who flew away."

A French artist who, like many of the Neo-Expressionists, deals with European memories of World War II is **CHRISTIAN BOLTANSKI** (b. 1944). Boltanski was born in Paris on Liberation Day (September 6, 1944), as the Allied armies entered the city, and his middle name is Liberté. He has often created works for such places as prisons, hospitals, and schools—institutional sites much concerned with the consequences, and the lessons, of the past. In a series of elegies that evoke the Holocaust, Boltanski hung up photographs of children and surrounded them with groups of small electric bulbs that cast a glow like candlelight in a place of worship (color plate 443, page 726). Some of his other works evoke feelings that are considerably less overwhelming, and more mixed. *The Shadows* (fig. 872) combines the macabre and the playful, suspending simple figurines in front of lights so that they cast large, ominous shadows on the walls. Little paper dolls hung by the neck and skeletons dangling in midair start to look like the indeterminate shapes a small child might be afraid of in the dark. At the same time, they give a knowing twist to the tradition of the Balinese puppet-shadow theater.

The American video artist **BILL VIOLA** (b. 1951) has also examined the theme of memory. Viola creates not only tapes but also installations that employ objects, video images, and recorded sound. He has long been interested in how the eye and brain process visual information, and he seeks to reveal their workings in his art. In *The Theater of Memory* (color plate 444, page 726), an uprooted tree lies diagonally across the floor of a darkened room. In its branches are dozens of

872. CHRISTIAN BOLTANSKI. *Shadows.* 1984. Cardboard cutouts, projector, and fan, dimensions variable.
Collection FRAC de Bourgogne, France © 1997 ADAGP, PARIS/ARTISTS RIGHTS SOCIETY (ARS), NEW YORK

lanterns. A video image is projected against one wall. At intervals, white noise, mostly static, erupts from the speakers, separated by long periods of silence. Of this installation Viola has said, "I remember reading about the brain and the central nervous system, trying to understand what causes the triggering of nerve firings that re-create patterns of past sensations, finally evoking a memory. I came across the fact that all of the neurons in the brain are physically disconnected from each other, beginning and ending in a tiny gap of empty space. The flickering pattern evoked by the tiny sparks of thought bridging these gaps becomes the actual form and substance of our ideas." His installation is a kind of working model of the brain's physiology, as if the viewer were standing inside the "theater" of a fantastically huge cranium. The lanterns are like the "tiny sparks of thought" that ignite along the branching "tree" of our intricate neural network; the projected video image evokes the "show" that memory replays before our mind's eye. It is a decidedly mechanistic, even clinical view of how human consciousness operates, but a certain romantic sense of mystery is nonetheless supplied by the dramatically lit tree, the archaic lanterns, and the darkened theatrical space.

London-born, American-trained **JUDY PFAFF** (b. 1946) is best known for her gallery-wide installations (fig. 873), works related not only to those of Jonathan Borofsky and the "scatterwork" Conceptualists, but also to the abstract illusionism of her teacher at Yale, Al Held. And walking into a Pfaff installation, viewers often sense that they have entered a giant allover Abstract Expressionist painting—one realized, in this case, as a lush, fluid underwater environment providing total immersion among all manner of strange, tropical flora and fauna, brilliantly colored and suspended in free-floating relationships. To produce her spectacular effects, Pfaff begins by assembling a huge collection of non-art materials—plaster, barbed wire, neon, contact papers, woods, meshes—from junkyards and discount outlets, hardware, lumber, and paint stores. Having transported these "mixed media" to the gallery, along with whatever few artifacts she may have prefabricated in her studio, Pfaff then sets to work in an improvisatory manner on site, first by setting up visual oppositions among painted and sculptural elements. With precariously counterbalanced rivalries as her guiding principle, she works through the space, adding and adjusting, establishing and upsetting formal relationships, all in a continuous flow of inspiration, editing, and fine-tuning.

Pfaff has also become an accomplished maker of portable, durable collages and constructed reliefs. These works often counter our expectations by changing the anticipated sizes of things. In *Supermercado* (color plate 445, page 726), which is more than eight feet high, the carefully counterbalanced, brightly painted geometric solids seem to make up an enormously enlarged still life of fruits, vegetables, and bowls (hence the "supermarket" of the title), with a brick wall behind it. Here modest still-life elements aspire to the monumental. Indeed, this gathering of subtly differentiated spheres, disks, and circles might as easily suggest a model of the solar system.

Color plate 427. NANCY SPERO. *Goddess II* (detail). 1985. Handprinting and collage on paper, 7'2" x 9'2" (2.2 x 2.8 m). Collection the artist

Color plate 428. IDA APPLEBROOG. *Noble Fields.* 1987. Oil on canvas, five panels, overall 7'2" x 11' (2.2 x 3.4 m). Solomon R. Guggenheim Museum, New York

GIFT, STEWART AND JUDY COLTON, 1987

Color plate 429. JOAN SNYDER. *Love Your Bones.* 1970–71, Oil, acrylic, and spray enamel on canvas, 6 x 12' (1.8 x 3.7 m). Neuberger Museum of Art, State University of New York, College at Purchase

GIFT OF ROY R. NEUBERGER

above: Color plate 430.
FRANCESCO CLEMENTE. *Untitled*. 1984.
Oil on canvas, 9' x 15'8" (2.7 x 4.8 m).
Thomas Ammann Fine Arts, Zürich

left: Color plate 431. FRANCESCO CLEMENTE.
Semen. 1983. Oil on linen, 7'9" x 13'
(2.4 x 4 m). Private collection.
Courtesy Sperone Westwater, New York

Color plate 432. SANDRO CHIA.
The Idleness of Sisyphus. 1981. Oil on canvas, in two
parts: top 6'9" x 12'8¼" (2.1 x 3.9 m), bottom
41" x 12'1¼" (104.5 cm x 3.9 m), overall
10'2" x 12'8¼" (3.1 x 3.9 m).
The Museum of Modern Art, New York

ACQUIRED THROUGH THE CARTER BURDEN, BARBARA JAKOBSON, AND SAIDIE MAY FUNDS
AND PURCHASE © 1997 SANDRO CHIA/VAGA, NEW YORK, NY

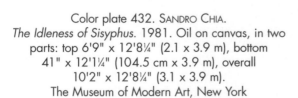

Color plate 433. JULIAN SCHNABEL. *The Sea*. 1981. Oil on wood, with Mexican pots and plaster, 9 x 13' (2.7 x 4 m). Saatchi Collection, London

above: Color plate 434. DAVID SALLE. *Tennyson*. 1983. Oil and acrylic on canvas, 6'6" x 9'9" (2 x 3 m). Private collection, New York

© 1997 DAVID SALLE/VAGA, NEW YORK, NY

Color plate 435. ERIC FISCHL. *The Old Man's Boat and the Old Man's Dog*. 1982. Oil on canvas, 7 x 7' (2.1 x 2.1 m). Collection Mr. and Mrs. Robert Lehrman, Washington, D.C.

Color plate 436. SHERRIE LEVINE. *Broad Stripe No. 12.* 1985.
Casein and wax on mahogany, 24 x 20" (61 x 50.8 cm).
Collection Lois and Richard Plehn, New York

above: Color plate 437. CINDY SHERMAN.
Untitled. 1981. Color photograph,
24 x 48" (61 x 121.9 cm).
Courtesy Metro Pictures, New York

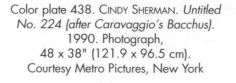

Color plate 438. CINDY SHERMAN. *Untitled
No. 224 (after Caravaggio's Bacchus).*
1990. Photograph,
48 x 38" (121.9 x 96.5 cm).
Courtesy Metro Pictures, New York

right: Color plate 439. MIKE AND DOUG STARN. *Double Mona Lisa with Self-Portrait.* 1985–88. Toned silver print and Ortho film, Scotch tape, and wood, 8'9" x 13'3½" (2.7 x 4.1 m). Museum of Fine Arts, Boston

GIFT OF CHARLENE ENGELHARD AND VIJAK MAHDAVI AND BERNARDO NADAL-GINARD © 1997 DOUG AND MIKE STARN/ARTISTS RIGHTS SOCIETY (ARS), NEW YORK

left: Color plate 442. ILYA KABAKOV. *The Man Who Flew into Space from His Apartment.* 1981–88. From the installation *10 Characters.* Six poster panels with collage, furniture, clothing, catapult, household objects, wooden plank, scroll-type painting, two pages of Soviet paper, diorama. Room dimensions 8' x 7'11" x 12'3" (2.4 x 2.4 x 3.7 m)

above: Color plate 440. BETYE SAAR. *Sanctuary's Edge.* 1988. Freestanding assemblage, 12¾ x 11⅞ x 3½" (32.4 x 30.2 x 8.9 cm). Collection Peter and Eileen Norton, Santa Monica, California

right: Color plate 441. KENNY SCHARF. *When the Worlds Collide.* 1984. Oil, acrylic, and enamel spray paint on canvas, 10'2" x 17'5¼" (3.1 x 5.3 m). Whitney Museum of American Art, New York

PURCHASE, WITH FUNDS FROM EDWARD R. DOWNE, JR., AND ERIC FISCHL © 1997 KENNY SCHARF/ARTISTS RIGHTS SOCIETY (ARS), NEW YORK

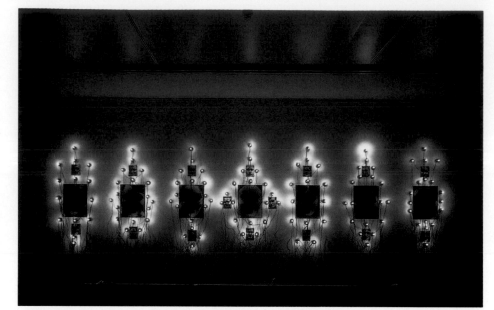

Color plate 443. CHRISTIAN BOLTANSKI. *Lessons of Darkness.* 1987. 23 black-and-white photographs and 117 lights, overall approx. 6'6¾" x 16'4¾" (2 x 5 m). Kunstmuseum, Bern, Switzerland

Color plate 444. BILL VIOLA. *The Theater of Memory.* 1985. Video/sound installation. Newport Harbor Art Museum, Newport Beach, California

Color plate 445. JUDY PFAFF. *Supermercado.* 1986. Painted wood and metal, 25 units, overall 8'4½" x 13'7¾" x 50" (2.6 m x 4.2 m x 127 cm). Whitney Museum of American Art, New York

PURCHASE, WITH FUNDS FROM THE LOUIS AND BESSIE ADLER FOUNDATION, INC., SEYMOUR M. KLEIN, PRESIDENT, AND THE SONDRA AND CHARLES GILMAN, JR., FOUNDATION, INC.

above: Color plate 446. SANDY SKOGLUND. *Radioactive Cats*. 1980. Cibachrome print, 30 x 40" (76.2 x 101.6 cm). Courtesy Janet Borden, Inc.

below: Color plate 447. JENNY HOLZER. *Untitled* (selection from Truism: Inflammatory Essays, The Living Series, The Survival Series, Under a Rock, Laments, and Mother and Child Text). LED electronic display signboard installation, 11" x 162' x 44" (27.9 cm x 49.4 m x 111.8 cm). Installation at the Solomon R. Guggenheim Museum, New York, 1989–90

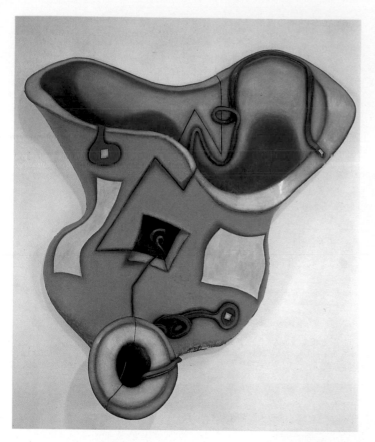

Color plate 449. HOWARD HODGKIN. *In the Bay of Naples.*
1980–82. Oil on wood, 54 x 60" (137.2 x 152.4 cm).
Private collection

above: Color plate 448. ELIZABETH MURRAY.
Careless Love. 1995–96.
Oil on shaped canvas with wood,
8'10½" x 8'3½" x 17"
(2.7 m x 2.5 m x 43.2 cm).
National Gallery of Art, Washington, D.C.
GIFT OF THE AARON I. FLEISCHMAN FOUNDATION

Color plate 450. GERHARD RICHTER. *Vase.*
1984. Oil on canvas, 7'4½" x 6'6¾"
(2.2 x 2 m). Courtesy Museum of
Fine Arts, Boston
JULIANA CHENEY EDWARDS COLLECTION

Color plate 451. TERRY WINTERS. *Good Government*. 1984. Oil on linen, 8'5¼" x 11'5¼" (2.6 x 3.5 m). Whitney Museum of American Art, New York
PURCHASE, WITH FUNDS FROM THE MNUCHIN FOUNDATION AND THE PAINTING AND SCULPTURE COMMITTEE

Color plate 452. SEAN SCULLY. *To Want*. 1985. Oil on canvas, 8'1⅛" x 9'5⅜" x 9⅜" (2.4 m x 2.9 m x 23.8 cm). Walker Art Center, Minneapolis
JUSTIN SMITH PURCHASE FUND

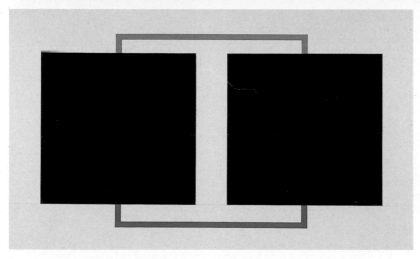

Color plate 453. PETER HALLEY. *Two Cells with Circulating Conduit*. 1985. Acrylic, Day-Glo acrylic, and Roll-a-Tex on canvas, 64" x 8'8" (162.6 cm x 2.6 m). Collection Cooper Fund, Inc. Courtesy the artist

Color plate 454.
NANCY GRAVES.
Wheelabout. 1985.
Bronze and steel with
polyurethane paint,
7'8½" x 70 x 32½"
(2.3 m x 177.8 cm x
82.6 cm).
Modern Art Museum of
Fort Worth, Texas
GIFT OF MR. AND MRS. SID R. BASS
© 1997 NANCY GRAVES/VAGA,
NEW YORK, NY

Color plate 455. DONALD LIPSKI. *The Starry Night*. Razor blades on wall. Installation at the
Capp Street Project, San Francisco, 1994

Color plate 456. MARTIN PURYEAR. *Old Mole*. 1985. Red cedar, 61 x 61 x 32" (154.9 x 154.9 x 81.3 cm). Philadelphia Museum of Art

PURCHASED: THE SAMUEL S. WHITE, 3RD, AND VERA WHITE COLLECTION (BY EXCHANGE) AND GIFT OF MR. AND MRS. C. G. CHAPLIN (BY EXCHANGE), FUNDS CONTRIBUTED BY MARION STOUD SWINGLE, AND FUNDS CONTRIBUTED BY FRIENDS AND FAMILY IN MEMORY OF MRS. H. GATES LLOYD

Color plate 457. CHRISTOPHER WILMARTH. *Susan Walked In*. 1972. Etched glass, steel, and steel cable, 60 x 60 x 30" (152.4 x 152.4 x 76.2 cm). Collection Susan Wilmarth, New York. Courtesy Sidney Janis Gallery, New York

Color plate 458. SIAH ARMAJANI. Irene Hixon Whitney pedestrian bridge. Bridge between Loring Park and Walker Art Center Sculpture Garden, Minneapolis. 1988. Length 375' (114.3 m)

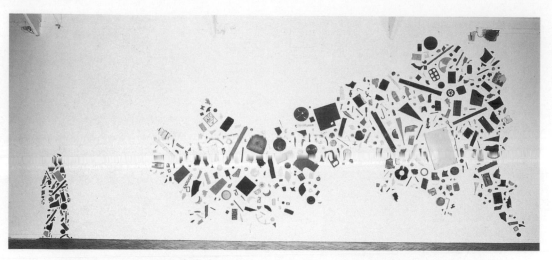

Color plate 459.
TONY CRAGG. *Britain Seen
from the North*. 1981.
Plastic and mixed media,
"figure" to left
66⅞ x 23" (170 x
58.4 cm), "Britain" to
right 12'1¼" x 22'10¾"
(3.7 x 7 m).
Tate Gallery, London

Color plate 460.
ANISH KAPOOR. *Part of the Red*.
1981. Bonded earth, resin, and
pigment, dimensions variable.
Kröller-Müller Museum, Otterlo,
The Netherlands

Color plate 461. ROBERT VENTURI
AND DENISE SCOTT BROWN.
National Gallery of Art,
London, with Sainsbury Wing
visible left of the main gallery
across Trafalgar Square.
1987–91

Color plate 462. CHARLES MOORE with U.I.G. and Perez Associates, Inc. Piazza D'Italia, New Orleans. 1975–80

Color plate 463. ROBERT A. M. STERN. Pool House, Llewelyn, New Jersey. 1981–82

below: Color plate 464. ROBERT A. M. STERN. Feature Animation Building, Walt Disney Company, Burbank, California. 1995

Color plate 465. JAMES STIRLING. Neue Staatsgalerie, Stuttgart, Germany. 1977–84

Color plate 466. ARATA ISOZAKI. Art Tower, Mito, Japan. 1990

Color plate 467. CESAR PELLI. World Financial Center Towers, New York City. View of Winter Garden flanked by Towers, with the North Tower of the World Trade Center in the background. 1979–88

Color plate 468. HELMUT JAHN. United Airlines Terminal of Tomorrow, O'Hare International Airport, Chicago. 1987–88

right: Color plate 470.
PEI COBB FREED & PARTNERS. Tower of Faces, United States Holocaust Memorial Museum, Washington, D.C. 1993–95

below: Color plate 469. TADEO ANDO. Chapel-on-the-Water, Hokkaido, Japan. 1988

Color plate 471. I.M. Pei. Rock and Roll Hall of Fame and Museum, Cleveland, Ohio. 1995

Color plate 472. Renzo Piano. Waiting room, Kansai International Airport, Osaka, Japan. 1994

above: Color plate 473. Frank O. Gehry. Chiat/Day Building and Claes Oldenburg and Coosje van Bruggen, *Binoculars*. Steel frame, concrete, and cement plaster painted with elastomeric paint, 45 x 44 x 18' (13.7 x 13.4 x 5.5 m). Venice, California. 1991

left: Color plate 474. Elizabeth Plater-Zyberk and Andres Duany. Seaside, Florida. Aerial plan view. 1980s.

right: 873. JUDY PFAFF. *Deepwater.* 1980. Installation view,
Holly Solomon Gallery, New York

An alumnus of CoLab, **JOHN AHEARN** (b. 1951) is an upstate New York artist who, with his associate Rigoberto Torres, has chosen to live in the South Bronx and there portray its citizens in an ongoing series of relief sculptures. Like George Segal, Duane Hanson, and John De Andrea before him, Ahearn makes life casts that comment on human relationships and society at large. What yields the Neo-Expressionist difference is the content of indomitable subjectivity radiating from Ahearn's faces, their intense gazes conveying ferocious dignity. The artist has said: "The basic foundation for the work is art that has a popular basis, not just in appeal, but in its origin and meaning. . . . It can mean something here, and also in a museum." Certainly the reliefs communicate a great deal in gallery settings, but by "here," Ahearn means the exterior walls above the very streets in which his subjects live and work (fig. 874). So popular are these open-air installations that the artist's neighbors compete to sit for twenty minutes with their heads covered by a fast-setting dental plastic. From the molds this creates, Ahearn then casts the form in fiberglass. But what brings the works to life is his immense gift for color, which allows Ahearn to evoke the ambient light and heat of skin itself.

874. JOHN AHEARN. *Homage to the People of the Bronx: Double Dutch at Kelly Street I: Freida, Jevette, Towana, Stacey.* 1981–82.
Oil on fiberglass, figures lifesize. Kelly Street, South Bronx, New York

Although generally considered a photographer, **SANDY SKOGLUND** (b. 1946) can easily be discussed in the context of artists who create installations. Skoglund stages elaborate, fanciful tableaux and then takes pictures of them. The installations themselves have occasionally been exhibited, but it is the photographs that have been widely shown. For *Radioactive Cats* (color plate 446, page 727), Skoglund made two dozen lively sculptures of felines, painted them acid green, and surrounded them with a drab, monochromatic apartment setting. Their science-fiction coloring and massed numbers give the foraging house cats the appearance of little domestic monsters, perhaps ready to pounce on their human cohabitants. "My work is based on this Frankensteinian model," Skoglund says, ". . . the theme of other animals or components—the world we have made—turning on us." And indeed, by calling these comically alarming situations in her own work "very Hitchcockian," she brings to mind Alfred Hitchcock's *The Birds,* the British director's famous suspense film about harmless animals suddenly turning deadly. More striking than Skoglund's pop-culture borrowings, however, is her surprisingly refined visual sensibility, which in this instance creates an elegantly decorative ensemble out of the most unlikely colors and shapes.

JENNY HOLZER (b. 1950) creates "language" works, related to the Conceptual art of the seventies, but based on the pithy, often political, sometimes cryptic sayings she composes. Early on, she began calling these sayings Truisms; a typical example is the one-liner, "Abuse of power comes as no surprise." Some of the sayings sound like replies to the home-spun wisdom of Benjamin Franklin in *Poor Richard's Almanac.* And although they can be as banal as fortune cookies, many have the punch of advertising slogans. At first, Holzer printed her aphorisms on posters pasted to the walls of public places, such as phone booths and bus shelters, but later she began running them on electric signboards in more prominent locations. In 1982 she obtained the use of the Spectacolor Board in New York's Times Square (fig. 875), and flashed pointed remarks like "Private property created crime" and "Protect me from what I want" to throngs of surprised pedestrians. Her subsequent installation pieces in museums and galleries have generally been more meditative and subjective, notably *Laments* for the 1990 Venice Biennale, but they have also at times been more spectacular. At the Guggenheim Museum in 1989, a ribbon of her color signboards ran along the edges of Frank Lloyd Wright's great spiral ramp (color plate 447, page 727). The lights spinning around the helical curve of the architecture surrounded the visitor with such urgent phrases as "You are a victim of the rules you live by."

Abstract Art

Perhaps the most surprising aspect of eighties culture was the strength and variety of art produced by painters and sculptors working with traditional media and a modernist vocabulary of form. There is no accepted umbrella term for this work. Neo-Geo was coined to describe the geometric abstractions of such painters as Peter Halley and Ross Bleckner, but it has little relevance to the other artists about to be discussed. The

875. JENNY HOLZER. Selection from Truisms. 1982. Spectacolor Board, 20 x 40' (6.1 x 12.2 m). Installation, Times Square, New York. Sponsored by Public Art Fund, Inc., New York

876. ROBERT MORRIS. *Untitled.*
1984. Painted cast Hydrocal,
pastel on paper,
7'6½" x 7'11" x 11"
(2.3 m x 2.4 m x 27.9 cm).
Collection Suzanne and
Howard Feldman, New York

difficulty of categorizing their art is an indication of its rich diversity. It has no single look or set of principles. What it does share is the following: a conviction that abstract painting and sculpture have not outlived their relevance, a desire to recover and expand abstraction's capacity to articulate human experience, and an interest in exploring the boundary between abstraction and representation.

In 1983 **ROBERT MORRIS** (b. 1931), the Minimalist, Conceptualist, and perennial seismographer of every aesthetic tremor, completed the first of his Firestorm paintings (fig. 876). With their massive sculptural frames embedded with visual references to human mortality, and their central panels pastel-painted like abstract, Turneresque visions of the Apocalypse, these powerful and shockingly unexpected works bring home, to resounding effect, a theme the artist maintains has always been at the core of his work, whatever its numerous and extreme stylistic shifts: the journey to death that is life. As if to unify his new expressionistic pictorial work with his old anti-form Conceptual mode, Morris, like Neil Jenney, sometimes provided an inscription to accompany the picture. For the work seen here, it refers to the tragic fire-bombing of Dresden at the end of World War II: "In Dresden, it was said afterwards that temperatures in the Altstadt [the "Old City"] reached 3,000 degrees. They spoke of 250,000 dead. Wild animals from the destroyed zoo were seen walking among those leaving the ruined city." Although few seemed to notice at the time, the grandly stoic, monumental geometry of Morris's Minimal work could be seen as

symbolic cenotaphs, coffins, or memorials, and his scatter pieces as matter undergoing its inevitable decay (see figs. 696, 774). Driven by the fear of nuclear holocaust, Morris abandoned his once arcane ways and joined the younger abstractionists in their determination to endow art with physical urgency.

"All my work," the American painter **ELIZABETH MURRAY** (b. 1940) asserts, "is involved with conflict—trying to make something disparate whole." A mid-seventies example of how she reconciled warring alternatives is *Beginner* (fig. 877), a large painting in which a swelling, Arp-like form, heavily impastoed in royal purple, balances against three sides of a rectangular field loosely brushed in gray on cocoa-brown. Tying the various large and simple elements together—colors, shapes, textures, and their respective moods—is the mediating pink of the small, relatively complicated "umbilical cord," coiled up like a pretzel in the dark fetal biomorph and projecting one straight end over into the lighter adjacent ground.

In 1979–80 Murray realized a major breakthrough in her art. Now she broke up not the image but rather the canvas, "shattering" it into separate shards or fragments (fig. 878). Further compounding the conflict was the wall, revealed through the cracks as one more participant in the battle for the viewer's eye. To make such disintegration whole again, Murray dipped into her academic background, reintroduced figuration, and imposed upon the entire archipelago of island canvases an image simple enough—brushes, palettes, cups, the

877. ELIZABETH MURRAY. *Beginner*. 1976. 9'5" x 9'6" (2.9 x 2.9 m). Saatchi Collection, London

painter's own hand—to be detected as a unifying factor. But this could be experienced only by observers sufficiently attentive to "read" across the relief map of fractures and fields, visually and conceptually piecing the whole together like a puzzle.

Subsequently, Murray has created large relief paintings with more unified imagery, even though fissures through the surface are still evident (color plate 448, page 728). These

878. ELIZABETH MURRAY. *Art Part*. 1981. Twenty-two canvases, overall 9'7" x 10'4" (2.9 x 3.1 m). Private collection

reliefs are generally based on such domestic items as old shoes with writhing shoelaces, or tables and cups, all given a certain awkwardness that can make them seem vulnerable and even endearing. The objects become stand-ins for their human users, and are thus laden with psychological overtones.

The English artist HOWARD HODGKIN (b. 1932) has developed over the course of several decades what might be called a Proustian subject. He paints heightened moments involving specific people, often the artist himself, in particular settings, usually enclosed. The people are remembered and made articulate through semiabstract pictorial equivalents. Hodgkin has said: "I am a representational painter, but not a painter of appearances. I paint representational pictures of emotional situations."

An important midcareer painting is the 1975 *Grantchester Road* (fig. 879), a characteristic work not only in its modest scale but also in its Cubist, stage-box space, flattened by the painted-in frame as well as by the Nabi-like polka-dot pattern and black Matissean planes echoed or rhymed throughout the picture. But the emphasis on flatness simply makes the depicted recession, slight though it may be in converging orthogonals and peephole views into blue infinity, all the more dramatic. For all its tautness and control, *Grantchester Road* displays a sense of pageantry: in the steep, balconied vastness of the domestic interior; in the opulence of its orange, mauve, and green palette; in its parentheses of sensuously undulant planes; and in the pulsating energy of the Neo-Impressionist spotting. The figure partially visible behind a dark post in the right foreground is a portrait of the artist, a voyeur semiconcealed within the lives and the spaces of his subjects.

In the 1980s Hodgkin came into full command of his pictorial language, a vocabulary of swaths, swirls, blobs, arrows, stripes, waves, and zigzags, free-floating in buoyant yet stable balance. His most astonishing breakthrough may have occurred in the Neapolitan series (color plate 449, page 728), pictures evoking memories of a brief, perhaps romantic, sojourn in Naples. In the work seen here, the suggestion of the Mediterranean Sea and sky is indicated by jeweled color not held to a strict horizon line. But the reality of that voluptuous world receives full expression in one broad aquamarine plane, a suave, buttery stroke slowly rising along a wavelike curve of pure rapture. Hodgkin here applied his increasingly passionate colors in soft-edged, lavalike flows, creamy glissades, vaporous plumes, and bubbling clusters of shaggy spots. Dissolved and emblematic as his representations became, Hodgkin in this painting acknowledged a pair of English presences in the mossy green cluster suspended from the upper edge of the frame, their heads emerging from the cellular mass like large smoky orbs. While the friends look upon the azure beauty of Naples's fabled bay, through a picture window like that of an Italian Renaissance painting, stars and fireworks stream away in wide, staccatoed ribbons of red and glittering white lights.

Like his compatriot A. R. Penck, Dresden-born and -educated GERHARD RICHTER (b. 1932) found himself so in

left: 879. HOWARD HODGKIN. *Grantchester Road.* 1975. Oil on wood, 49 x 57" (124.5 x 144.8 cm). Private collection

880. GERHARD RICHTER. *Scheune/Barn No. 549/1.* 1983. Oil on canvas, 27½ x 39½" (69.9 x 100.3 cm). Art Gallery of Ontario, Toronto

conflict with the social/aesthetic conditions of East Germany that he crossed over into the West in 1961. In 1962–63 Pop art first appeared in Germany, where its assault on puritanical formalism received vigorous reinforcement from Joseph Beuys and the Fluxus performance group. Catalyzed by the iconoclasm of Fluxus, Richter declared, "I'll paint a photo!" Since 1962 the artist has created scores of "photopaintings," all of them more distinctly "photographic" than Rauschenberg's transfers, and yet far more like real painting than Warhol's appropriations of media imagery. In his later photopaintings (fig. 880), Richter has felt freer to depart from the original, often a snapshot made by the artist himself, as long as he preserves something of a genuine photographic look, complete with such flaws as out-of-focus blurring, graininess, or exaggerated contrasts, all of which seem to be points of mediation between painting and photography. The calm, bucolic scenes of Richter's later work show the Central European landscape immersed in a soft, romantic atmosphere, a kind of filter that reveals as much about how painting and photography alike distort our vision of reality as it does about the world depicted.

In the late 1960s, Richter ran counter to the photopaintings with a return to pure abstraction, first in a series of heavily impastoed monochrome works called Gray Pictures (Graue Bilder). By the beginning of the 1980s, after investigating many different styles, he came to concentrate on full-color, vigorously worked expressionist abstractions, painted as a series and complementary to the cooler, more objective "classicism" of the photo landscapes. Now Richter paints his abstractions directly, without the intervention of photography, and evokes depth by investing the paintings with added layers of involvement and contradictory feeling (color plate 450, page 728). Although chromatically and gesturally similar to abstractions painted by De Kooning in the 1950s, Richter's nonrepresentational works issue from a different critical process and present a planar space layered in strata of relatively distinct elements, a structural conception already

seen as characteristic of much Postmodern art. In executing these works, Richter draws upon a vast repertoire of painterly techniques—troweling, scraping, flinging, scumbling, or merely brushing—that recall the artistic procedures of Hans Hofmann (see color plate 234, page 451).

It has been said that the American artist **TERRY WINTERS** (b. 1949) "dreams science into art," a tendency that particularly extends to the science of biology. This is an interest not without precedents in modern art. After examining illustrations in scientific journals, Vasily Kandinsky near the end of his career portrayed microorganisms in a way that turned them into abstract shapes floating in space. Winters has also undertaken to paint shapes based on such organisms (along with botanical forms like buds, seeds, tubers, and spores, among others), an interest in both artists' work that can be related to biomorphic Surrealism. But unlike the Russian, who generally focused in on one or two individual microscopic creatures, Winters suggests the dynamic interaction among the elements of a diverse community or colony as they grow, develop, and compete.

Winters spent three months working on the large painting now called *Good Government* (color plate 451, page 729), which he started without any particular title in mind. At the beginning, he struggled to bring coherence to its anarchic array of very different forms—from expansive blastulas to hard crystals to floating fragments of chromosomes. But at last, he says, he "thought it looked like one of those maps you saw in grammar school and it said 'good government' and everything was working together." Like a system of checks and balances, these abstracted organic forms make up a complex compositional scheme, while the carefully modulated textures of this painting keep its contending shapes in balanced, "organic" relationship with one another.

Born in Dublin and educated in London, **SEAN SCULLY**

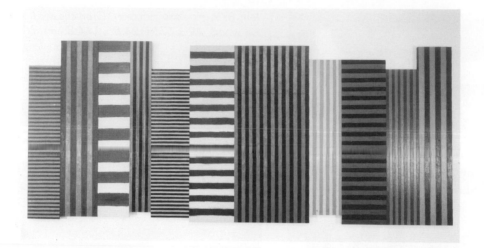

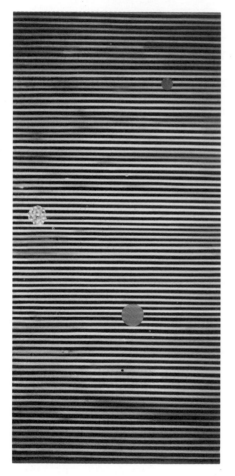

above: 881.
SEAN SCULLY. *Backs and Fronts*. 1981. Oil on canvas, 8 x 20' (2.4 x 6.1 m). Collection the artist

left: 882.
ROSS BLECKNER. *Seeing and Believing*. 1981. Oil and wax on canvas, 9'6" x 54" (2.9 m x 137 cm). Courtesy Mary Boone Gallery

right: 883.
ROSS BLECKNER. *Two Nights Not Nights*. 1988. Oil on canvas, 9' x 6' (2.7 x 1.8 m). Courtesy Mary Boone Gallery

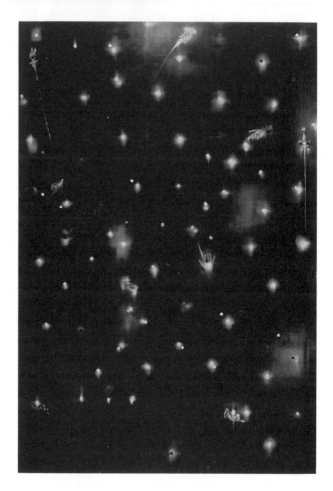

(b. 1945) moved to New York in 1973. Initially, the artist used stripes in large, densely woven "super-grids," painted in a style inspired by Mondrian, the English Vorticist David Bomberg, Agnes Martin, Brice Marden, and Sol LeWitt. Scully has worked ever since to make his stripes the vehicles of an emotional or spiritual content much greater than anything obtainable from pure retinal stimulation.

Beginning with his diptychs, Scully has progressively elaborated his simple vocabulary of striped panels into a language of such breadth and subtlety that each painting now emerges like "a physical event, a physical personality, just like people." His colors range from brilliant, clashing primaries to tawny mustards and creamy ochers, all warmed by long, sensuous

strokes freely wavering over traces of contrasting oilstick color, the latter set down freehand (color plate 452, page 729). Soon the diptychs became immense eight-by-ten-foot polyptychs, assembled from panels so various in their dimensions and depths, in their colors, widths, and directions of their banding, that they seem like interactive social beings abstracted into iconic versions of processional figures in a classical relief (figs. 881, 3). In this larger scheme, the colors and horizontal orientation assume the scale and atmosphere of landscape.

While the New York artist **ROSS BLECKNER** (b. 1949) can be compared to Scully in his use of vivid stripes to make large-scale, abstract paintings, he is also closely related to artists such as Sherrie Levine, who made "generic" versions of past styles in order to deconstruct modernist myths (fig. 863). Bleckner's *Seeing and Believing* (fig. 882) does not simply allude to sixties Op art; it simulates the hard-edged style in a way that reflects a Poststructuralist understanding of the history of twentieth-century abstraction. This had been regarded as a progressive unfolding of styles, each built upon the other, each "heroic and original" and unrepeatable. By throwing the "correct" sequence awry with the appropriation of an "incorrect" style, Bleckner showed that the past offers a variety of possibilities and has no fixed, necessary relationship to contemporary artistic practice.

As if to reinforce the point, in 1983 Bleckner started to exhibit somber paintings, *Two Nights Not Nights* (fig. 883), for example, in which evocations of spectral landscapes are combined with symbols—hands, torches, flowers, and urns—culled from the Baroque iconography of death and salvation. Such tenebrist paintings were among the earliest memorials to the victims of AIDS, which was first identified in 1982. Painted with wax and ground pigment rather than prepared paint, they are brilliant demonstrations of a return to old master techniques Bleckner shared with another CalArts-trained artist, Eric Fischl.

Bleckner's dazzling, striped paintings from the early eighties are the founding works of the mode of abstraction often called Neo-Geo. But it was a slightly younger artist from New York, **PETER HALLEY** (b. 1953), who became the undisputed leader of Neo-Geo. In a series of paintings based on Albers' *Homage to the Square*, Halley treated the square, in his own words, "as a sign for certain kinds of regimented or confining structures in the social landscape." His emphasis on the restrictive nature of post-industrial society reflects the influence of Michel Foucault's essay on confinement, while his equation of a seemingly abstract sign with "the social landscape" derives from Jean Baudrillard's notion of "hyper-reality." According to Baudrillard, two linked systems, international capitalism and mass media, force everyone to live in a glittering "empire of signs" where "hyper-real" images "simulate" and supplant ordinary reality. Paintings like *Two Cells with Circulating Conduit* (color plate 453, page 729) "simulate" this "empire" by presenting the high-tech equivalent of its "plumbing"—squares, to be understood as both battery and prison cells, connected by lines or "conduits" for the "circulation" of information and power. Halley, whose writings and interviews are often brought together with his paintings to form a single, pictorial-theoretical "package," contends his work collapses "the idea of the abstract and the representational" and gives abstraction a content. An important element in Halley's "package" is his use of Day-Glo paint, a "technological invention" with which he "simulates" traditional artist's color and evokes the spurious beauty of "the information age."

After graduating from the Yale School of Art, the sculptor and painter **NANCY GRAVES** (1940–1995) lived for a time in Italy. At a natural history museum in Florence, she happened upon a group of wax anatomical effigies made in the seventeenth century. In the effigies Graves saw the possibility for a new kind of sculpture. Like a taxidermist, but with a different purpose, she began making small stuffed animals and assembling them with the aid of found objects. In this menagerie she discovered her now-famous breakthrough image—the Bactrian, or double-humped, camel—a unique form that, because specific, left the artist "free to investigate the boundaries of art making" (fig. 884). Graves made several series of

884. NANCY GRAVES. *Camels VI, VII, and VIII.* 1968–69. Wood, steel, burlap, polyurethane animal skin, wax, and oil paint. *Camel VI*, 7'6" x 12' x 48" (2.3 m x 3.7 m x 121.9 cm); *Camel VII*, 8' x 9' x 48" (2.4 m x 2.7 m x 121.9 cm); *Camel VIII*, 7'6" x 10' x 48" (2.3 m x 3 m x 121.9 cm). National Gallery of Canada, Ottawa

left: 885. DONALD LIPSKI. *Building Steam #317*. 1985. Telephone, intercom, and crystal ball, 9½ x 6 x 5" (24.1 x 15.2 x 12.7 cm). Courtesy Germans Van Eck Gallery, New York

below: 886. MARTIN PURYEAR. *Night and Day*. 1984. Painted wood and wire, 6'11½" x 10' x 5" (2.1 m x 3 m x 12.7 cm). Patsy R. and Raymond D. Nasher Collection, Dallas

full-scale camels, at first fabricating them of table legs, market baskets, polyurethane, plaster, and painted skins. Gradually, however, she progressed toward engineered, portable forms built of steel beams and covered with hide, finally producing the only camels she has allowed to be used for museum exhibition. Although uncannily naturalistic, the Bactrians were not exercises in taxidermy but original creations made without reference to models and brought to life by the artist's own handwork and her meticulous regard for detail, texture, and proportion.

While investigating the form of the camel from the inside out, Graves discovered all manner of parts—fossils, skeletons, bones—that proved interesting in their own right as organic yet abstract forms, eminently suitable to aesthetic reordering. A commission for a bronze version of one of the resulting bone sculptures brought her together with the Tallix Foundry in Beacon, New York. Encountering a new technology launched the artist into a new phase of aesthetic experiment: open-form, polychrome, freestanding constructions combining natural and industrial shapes, often with a filigreed lightness (color plate 454, page 730). Graves would arrive at the foundry with shopping bags filled with fern fiddleheads, squid and crayfish, palmetto and monstera leaves, lotus pods and warty gourds, Chinese scissors, pleated lampshades, Styrofoam packing pellets, and potato chips. At Tallix

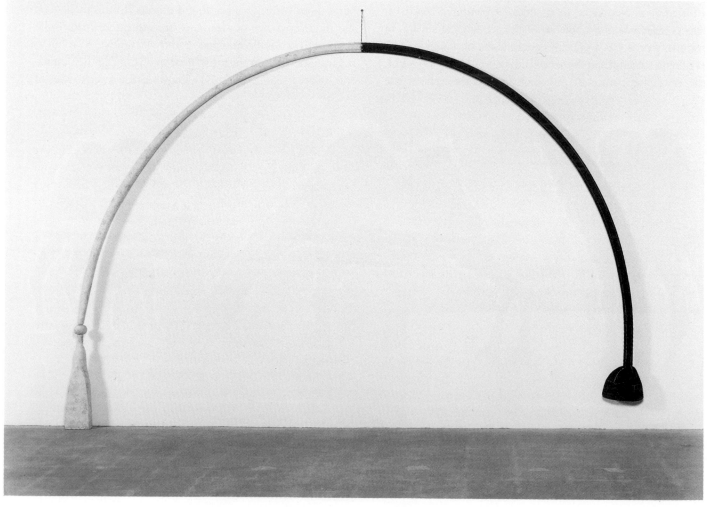

she expanded the repertoire of forms to include pieces of broken equipment, old drainage pipes and plumbing fixtures, or even spillages of molten metal. After a construction was assembled, Graves applied pigments either by patination or by hand painting, with brushwork as freewheeling and extemporaneous as the structure it embellishes.

For his sculpture, **DONALD LIPSKI** (b. 1947) takes parts and makes wholes, but he imagines the whole only after finding—or scavenging—the parts and discovering a kind of Surreal logic in some relationship among them (fig. 885). The artist's strategy is to raid junkyards, dustbins, dumpsters, and dime stores and then combine his gleanings in assemblages that astonish and delight with the perfect union of their incompatible parts. Thanks to his ingeniously associative mind, for example, Lipski can easily grasp how the shape and clarity of a crystal ball renders it ideal for nesting in the cradle of an old telephone. Wrenched from their usual context and reordered in brilliantly dysfunctional ways, commonplace items become strange and wonderful objects of contemplation. While Lipski's work may recall Marcel Duchamp and his readymades, its lack of anti-art or obvious political content and its lyrical obsessiveness give it a more certain kinship with the boxes of Joseph Cornell.

Like his sculptures, Lipski's installation *The Starry Night* (color plate 455, page 730) also presents ordinary things in unexpected ways. Here, nearly 25,000 double-edge razor blades were stuck directly into the Sheetrock walls of a gallery. Their swarming, swirling patterns across a large area resemble the cosmic vortex of the nocturnal sky in Van Gogh's famous painting. At the same time, however, and more disturbingly, the thousands of dangerously sharp blades protruding toward the viewer remind us of the artist's self-mutilation, and perhaps something of his tortured mental state at that time. With its complex reference combining visionary pleasure and psychological pain, Lipski's *Starry Night* communicates a surprisingly affecting sense of the dangers in an intensely romantic art.

While abjuring the reformative spirit of Minimalism, **MARTIN PURYEAR** (b. 1941) has accepted the mode's simplicity and gravity but gone on to invest his art with a human scale and animistic quality generated by craftsmanly process (fig. 886). Time spent with the Peace Corps in Sierra Leone reinforced Puryear's desire for "independence from technology" and gave him an appreciation of the "magic"—the special meaning or content—that seems to emanate from hand-honed objects. What resulted is an art in which modernist cerebration and tribal atavism, the traditions of Western high art and those of preindustrial craft, do not so much fuse as counterbalance one another in a dialectical standoff. The particular magic of *Old Mole* (color plate 456, page 731), for example, is to keep several suggestive tensions unresolved: a form that at first looks solid reveals itself as a weave of relatively weightless red cedar. And what appears to be an abstract shape quickly becomes in addition a kind of totem, a "mole" —a form that sits on the floor, looking alert and intelligent. Ultimately, the product of a craft technique is endowed with

887. CHRISTOPHER WILMARTH. *Insert Myself Within Your Story*, from the Breath series. 1979–80. Glass and steel, 12 x 11 x 9" (30.5 x 27.9 x 22.9 cm). Private collection, New York

a sculptural presence, comparable to the transformation that occurs in the work of Brancusi.

Although a child of Minimalism and a great admirer of both Tony Smith and Mark di Suvero, **CHRISTOPHER WILMARTH** (1943–1987) discovered early on that he was interested in engaging processes more intuitive and expressionist than the mathematical or industrial ones favored by his immediate predecessors. In this way he hoped to create a strongly formalist abstract art imbued with something of the humanity sensed in the works of Matisse and Brancusi. Wilmarth adopted the glass and steel, or bronze, frequently found in modernist sculpture, but cast them on a more intimate scale and layered, bent, and modified the materials until the resulting constructions look as if their eloquent interplays of light and shadow had evolved organically or lyrically, like poems (color plate 457, page 731). Throughout the 1970s the artist experimented with many variations on the thematic dualities of his materials—opacity contrasted with translucency, mass with evident ethereality—but in 1979 he encountered what was probably his greatest inspiration: the verse of Stéphane Mallarmé, as translated by the American poet Frederick Morgan. For the erotic poem entitled "Insert myself within your story" (fig. 887), Wilmarth, working with the craftsman Marvin Lipofsky, blew an ovoid shape in glass—rather like a free-form organic Brancusi egg or head—and then acid-etched the substance to make it seem alive with delicately suffused light and color. With the vitreous mass mounted on a steel support folded like a book and cut into by the outer "cover," the sculpture seems the aesthetic equivalent of a spirit wresting to free itself from brute matter.

Influenced both by the space-age environment of Cape Canaveral and by the classical preoccupation with drawing from the nude, **BRYAN HUNT** (b. 1947) brings to sculpture a duality of experience that makes it almost imperative that he mediate the constraints imposed by Minimalism. Looking to the past for the expressive power of traditional means—drawing, wood carving, modeling, bronze casting—he applies these age-old techniques in an attempt to humanize and to

right: 888. BRYAN HUNT. *Small Twist*. 1978. Bronze/black patina, 68 x 18 x 22" (172.7 x 45.7 x 55.9 cm). Courtesy Blum Helman Gallery, New York
© 1997 BRYAN HUNT/ARTISTS RIGHTS SOCIETY (ARS), NEW YORK

below: 889. SCOTT BURTON. *Pair of Rock Chairs*. 1980–81. Gneiss, a: 49¼ x 43½ x 40" (125.1 x 110.5 x 101.6 cm); b: 44 x 66 x 42½" (111.8 x 167.6 x 108 cm). The Museum of Modern Art, New York
ACQUIRED THROUGH THE PHILIP JOHNSON, MR. AND MRS. JOSEPH PULITZER, JR., AND ROBERT ROSENBLUM FUNDS

ennoble the ambiguous universality of abstract shapes chosen from concrete reality. So far one of his best-known images is that of a freestanding bronze waterfall, its tall, slender, serpentine shape a synthesis of nature, a sensuous female figure, and pure design (fig. 888). Adhering to the sculptural principle of truth to materials, Hunt chose to make his waterfalls of plaster and bronze because both materials pour. In recent work Hunt has mixed forms and media—now expanded to include welded metal—in open-form constructions that seem to translate waterfalls into classical, white-clad caryatids and dark metal armatures into drawings in space. Even though these complex works will eventually be cast in monolithic bronze, they are also to be patinated in the colors of their original plaster, wood, and steel materials, making them seem a compendium of the processes traditional to sculpture prior to the advent of the depersonalizing methods of Minimalism.

In his austere sculptures, which border on Primary forms but also possess the added dimension of functionality—tables, chairs, chaises longues—Scott Burton (1939–1989) returned Minimalism to its sources in the pragmatic design ideals of early modernism, such as de Stijl and the Bauhaus. At the same time, however, Burton was able to link furniture design, in a most unexpected way, with the Earthworks of the seventies.

The American artist achieved this unusual synthesis most fully in one of his favorite pieces, *Pair of Rock Chairs* (fig. 889). Starting with boulders that were, in the Duchampian sense, his ready-mades, Burton explored the contrast between carved and uncarved, or smooth and rough, areas of stone in a manner that recalled the "unfinished" works of such illustrious masters as Michelangelo, Rodin, and Noguchi. But beyond that, in *Rock Chairs* he so handled the two chunks of crude lava that from the back they are not only still rough, but still actually look like boulders—organic, even impressive parts of the natural environment. From the front, on the other hand, they display the smooth-cut shapes and proportions of functional, even inviting chairs. By deftly manipulating the carved/uncarved contrast, Burton overthrew any rigid opposition between art and nature. His minimal intervention into the raw stone created a most arresting hybrid—part natural object, part utilitarian object, part aesthetic object—with few direct precedents in the long history of sculpture. Although exquisitely refined works of sculpture, the *Rock Chairs* remain, at their most elementary level, rocks.

An artist often encountered in connection with Scott Burton in a variety of spheres—public, private, or gallery—and who shows an interest in blending the fine and applied arts is Siah Armajani (b. 1939), a self-described "architect/sculptor" born in Teheran but long since resettled in Minnesota. The dialectic between the extreme polarities of his native and adopted cultures runs through his art, filling it with no end of fascinating incongruities. To begin with, Armajani makes sculptures using the vocabulary of architecture, but the pieces have little or none of the functionality

890. Siah Armajani. *Closet Under Stairs.* 1985.
Painted wood, stain, and rope, 9'1½" x 40" x 7'1½"
(2.8 m x 101.6 cm x 2.2 m).
Hirshhorn Museum and Sculpture Garden,
Smithsonian Institution, Washington, D.C.
Acquired through the Joseph H. Hirshhorn Purchase Fund 1986

seen in the work of Burton. As *Closet Under Stairs* demonstrates (fig. 890), the space defined by the work cannot be penetrated even though the door is open, nor can the structure's second level be physically attained, since the stairs are tilted and the elevator "rostrum" has the scale of a toy. Consequently, the construction, like a painting or a relief sculpture, offers visual but not literal entrance and conceptual or metaphorical rather than kinetic experience.

Armajani has also shown an interest in the architectural engineering of bridges. Among the works he has created is the Irene Hixon Whitney pedestrian bridge, which links the sculpture garden at the Walker Art Center with Loring Park in the downtown part of Minneapolis (color plate 458, page 731). Stretching over sixteen lanes of traffic, the bridge reconnects two parts of the city that were divided twenty years earlier by the construction of a superhighway. The bridge's design itself suggests a kind of symbolic unity: that is, the two arches of the bridge, one convex and the other concave, unite to form a single sine curve.

Among the contemporary English sculptors of abstract tendency, the one with the widest international reputation is Tony Cragg (b. 1949), who has created scatter works composed of small plastic discards evenly distributed, like airy mosaics, across the floor or wall in abstract patterns of color and shape (color plate 459, page 732). Scavenging and assembling bright fragments from the beaches and dustbins

891. TONY CRAGG. *Suburbs.* 1990. Wood and rubber,
8'1½" x 13' x 13' (2.5 x 4 x 4 m).
Courtesy Marian Goodman Gallery, New York

892. RICHARD DEACON. *Tall Tree in the Ear.* 1983–84.
Galvanized steel, laminated wood, and blue canvas,
12'3⅝" x 8'2½" x 11'11" (3.7 x 2.5 x 3.6 m).
Saatchi Collection, London

of England, Cragg makes newly relevant the once-useful but now broken, dysfunctional, environmentally disfiguring products of industry. In addition to these assemblages, Cragg has fabricated freestanding sculptures. The large, bulbous forms of his *Suburbs* (fig. 891), carved in wood, have been compared by one critic to oversized door handles and rubber stamps. Yet they also have something of the figurative about

them, not least because these vertical entities are "posed" on carved bases, like those used by Brancusi, and have the rich surface texture of organic materials. The contours of the form at the right, moreover, are not unlike those of a Brancusi head. And even the "reclining" form at the front looks perhaps less like a door handle than like a figure sculpture that has fallen from its pedestal.

Born in Bombay, **ANISH KAPOOR** (b. 1954) was raised in London. He became known in the early eighties for sculptures of simple but refined geometric shapes coated with a thick sprinkling of powdered pigment (color plate 460, page 732). Although the ancient Greeks and Romans painted their sculptures, the practice has not been much observed in modern times, except for occasional modern works in steel, such as examples by Alexander Calder and David Smith. In works like theirs, however, the smooth coating of liquid paint dries and solidly adheres to the surface, wrapping itself around the structure like the skin of an animal. But Kapoor, by sprinkling his works with loose powder, seems to achieve the opposite effect, liberating color from any allegiance to the physical object it happens to rest on. The way that the granular pigment piles up in a vague "halo," or border area of loose grains around the form, indicates that color has no grip. Indeed the friable surface is made to seem vulnerable to the slightest breeze, which would blow away its layer of dust like a shifting sand dune. The work thus undermines the idea of "local" color, of a hue that is intrinsic, as green is to grass or blue to the sea and sky. Even when using blue—borrowing an intense shade of it from Yves Klein, for whom it signified infinity—Kapoor creates not something associated with the sky, but rather an abstract physical object that simply happens to be made of color.

In 1978–79 **RICHARD DEACON** (b. 1949) read Rainer Maria Rilke's *Sonnets to Orpheus* and conceived a profound admiration for the manner in which "the objects that appear in Rilke's poetry, whilst taking on connotations, retain the quality of actuality." And so at the same time that the many implications of ears, horns, and mouths found in Deacon's sculptures derive ultimately, if obliquely, from Rilke's central metaphor of Orpheus's head, the forms remain too ambiguous to suggest clear, precise images. This dual sense of familiarity and strangeness can be seen in *Tall Tree in the Ear* (fig. 892), where a large blue contour "drawing" in space seems to describe a very large dry bean, held upright by the silvery armature of what could be a pear. Together, the two- and three-dimensional forms conjure the shape and volume of the ear cited in the title, while their size evokes the immensity of the sound issuing from a tree, possibly one with branches full of noisy starlings or with leaves rustling in a strong wind. Mystery and poetry inform this art from beginning to end, for although calling himself a "fabricator," Deacon employs simple methods, such as riveting; uses unaesthetic materials, like steel and linoleum; and prepares no preliminary drawings or models. Instead, he permits initial idea, material, and process to interact to produce sculptures that announce their presence even as they imply absence.

26 Postmodernism in Architecture

The term *Postmodernism*, a term so often invoked throughout the last several chapters, originated not in the realms of painting and sculpture, but in the world of architecture. During the roaring years of building and rebuilding after World War II, the International Style became so dominant that nothing short of disaster seemed powerful enough to unseat its almost total hold on architects and clients. But in 1972, when the city of St. Louis dynamited Pruitt-Igoe (fig. 893), a public-housing project built according to the best theories of modern architecture, the rule of Bauhaus took a great fall. All measures to save this monument to rational, Utopian, Bauhaus planning had failed. Pruitt-Igoe had become an irredeemable social and economic horror scarcely twenty years after it had won the American Institute of Architects' award for its designer, Minoru Yamasaki (1912–1986), and despite

its use of new structural materials and techniques, its "industrial-society-universal" style, and the local community's intention for the project to effect social change. Complete with "streets in the air"— safe from automobile traffic—and access to "sky, space, and greenery," which Le Corbusier had deemed the "three essential joys of urbanism," it had been hoped that Pruitt-Igoe by example would inspire a sense of virtue in its inhabitants. By the end of the 1960s, however, the multimillion-dollar, fourteen-storey blocks had become so vandalized, crime-ridden, squalid, and dysfunctional that their manifold problems could only be resolved in demolition.

Modernist theory's aloof indifference to the small-scale, personal requirements—of privacy, individuality, context, and sense of place—was blamed for failing to make Pruitt-Igoe Housing a viable home for the persons whose lives it was meant to improve. The calamity in St. Louis, however,

893. MINORU YAMASAKI. Pruitt-Igoe public-housing project, St. Louis, at the moment of its demolition on July 15, 1972. Built 1952–53

was merely the "smoking gun" for critics who, since the mid-sixties, had attempted to indict Bauhaus modernism for alleged bankruptcy of both form and principle. These critics thought that the followers of Mies, Gropius, and Le Corbusier had created a movement far too narrowly ideological and impersonal—and certainly too self-referential in its inflexible insistence on formalism dictated purely by function and technology. Such a stance could not serve a society in which diversity would only increase

Postmodernist architects, in contrast, have chosen to accept some modernist theories and to reject others. Opposing the consistency of Bauhaus aesthetics, Postmodernism has a voracious appetite for an illogical and eclectic mix of history and historical imagery, for a wide variety of vernacular expression, for decoration and ornamentaion, and for metaphor, symbolism, and playfulness. At the same time, most architects working in the Postmodern mode accept industrial society and the use of new materials and construction methods, as well as the ideal of the betterment of society.

Rather than adhere to rigid ideological dogma, Postmodernists seek to build in relation to everything: the site and its established environment, the client's specific needs, including those of wit and adventure in living, historical precedent relevant to current circumstance, and communicable symbols for the whole enterprise. The best Postmodern architects demonstrate that the tools, methods, materials, and traditions that architects have always worked with are still valid yet debatable and open to realignments, rearrangements, and re-examinations in order to solve architectural problems—especially as new technologies and materials become available and are integrated. Thus, the work of Postmodernists seems almost unclassifiable except as a movement and style all its own, not merely "post-" something else. Instead of Le Corbusier and Mies van der Rohe, the Postmodernists are influenced by Antoni Gaudí (see figs. 109, 110) and Sir Edwin Lutyens, both moderns who escaped the monolithic imperatives of modernist dogma.

Among the first to appreciate Gaudí and Luytens anew and to voice and practice the revisionism that is Postmodernism was the Philadelphia architect and theorist **ROBERT VENTURI** (b. 1925). Often called the father of Postmodernism, Venturi did not train at Harvard University, whose School of Design, as noted earlier, had been transformed by Walter Gropius into a powerful Bauhaus-like force for modernism. Instead he went to Princeton University, where he studied under the distinguished Beaux Arts professor Jean Labatu. With such a grounding, Venturi gained historical points of reference for most examples of antimodern architecture. Moreover, it prepared Venturi to see in Rome, while there in 1954–56 as a Fellow at the American Academy in Rome, not purely a collection of great monuments but, instead, an urban environment characterized by human scale, sociable piazzas, and an intricate weave of the grand and the common. Supplemented by a period in the office of Louis I. Kahn, another admirer of the Roman model and a leading architect then moving from the International Style toward

more rugged, expressionist, or symbolic buildings (see figs. 746, 747), Venturi's Roman experience served as the basis of the new philosophy that would culminate in his seminal, and now famous, book *Complexity and Contradiction in Architecture* (1966). Here, with scholarship and wit, Venturi worked back through architectural history to Italian Baroque and Mannerist prototypes and argued that the great architecture of the past was not classically simple, but was often ambiguous and complex. He held, moreover, that modern architects' insistence upon a single style of unrelenting reductivism had proved utterly out of synchrony with the irony and diversity of modern times. To Mies van der Rohe's celebrated dictum, "Less is more," Venturi retorted "Less is a bore."

Complexity and Contradiction presented a broadside critique of mainstream modernism, at least as it had evolved in massive, stripped-down, steel-and-glass boxes that rose with assembly-line regularity throughout the postwar world. Even now, many critics consider Venturi's book the first and most significant written statement against the International Style. "Architects can no longer afford to be intimidated by the puritanically moral language of orthodox modern architecture," wrote Venturi.

> I like elements which are hybrid rather than "clear," distorted rather than "straightforward," ambiguous rather than "articulated," . . . boring as well as "interesting," conventional rather than "designed," accommodating rather than excluding . . . inconsistent and equivocal rather than direct and clear. I am for messy vitality over obvious unity. I include the non sequitur and proclaim the duality. I am for richness of meaning rather than clarity of meaning.

Venturi believed with total conviction that the modern movement had grown stale, that the successors to the great innovative modern architects had turned the latter's splendid revolution into a new Establishment, just as authoritarian and unresponsive to contemporary needs as the old one had been in its own time and way. Confronted with the pervasive gigantism, unaccommodating aloofness, sameness, and opportunism—giving less and charging more in the name of Utopian ideals—of the buildings designed by Bauhaus-inspired architects, Venturi decided that "Main Street is almost all right." In other words, rather than disdaining the ordinary landscape of average towns, we should embrace it, accommodate it, and improve upon it on its own terms.

With his wife, **DENISE SCOTT BROWN** (b. 1931), an architect and specialist in popular culture and urban planning and a partner in the firm of Venturi, Rauch and Scott Brown, Venturi wrote an even more provoking book in 1972, *Learning from Las Vegas*. In it they declared that a careful study of contemporary "vernacular architecture," like that along the automobile-dominated commercial "strip" roadways of Las Vegas, may be "as important to architects and urbanists today as were the studies of medieval Europe and ancient Rome and Greece to earlier generations." Venturi and Scott Brown had sought to learn from such a landscape because for them it constituted "a new type of urban form,

894. VENTURI, RAUCH, AND SCOTT BROWN. Best Products Company, Oxford Valley, Pennsylvania. 1977

radically different from that which we have known, one which we have been ill-equipped to deal with and which, from ignorance, we define today as urban sprawl. Our aim [was] . . . to understand this new form and to begin to evolve techniques for its handling." Since urban sprawl has come to stay, Venturi and Scott Brown are persuaded that architects should learn to love it, or at least discover how "to do the strip and urban sprawl well," from supermarket parking lots and hamburger stands to gas stations and gambling casinos. In their opinion, "the seemingly chaotic juxtaposition of honky-tonk elements expresses an intriguing kind of vitality and validity."

While such Pop taste and maverick views made Venturi the leading theoretician among younger architects, they also earned him the near-universal disdain of the modernist school and scorn of the kind heard from the critic Ada Louise Huxtable, who dubbed the quietly civilized Venturi the "guru of chaos." The Venturis' bold ideas also proved too much for insecure arbiters of major architectural commissions. Thus, until 1986, when Venturi, Rauch and Scott Brown won the coveted contract for the design of the extension to London's National Gallery on Trafalgar Square (color plate 461, page 732), the firm had to content itself with relatively small projects. Still, the partners made the most of every opportunity, even while insisting that they deliberately designed "dumb" buildings or "decorated sheds." The decorated shed concept may have received its most salient expression in the highway store designed for the venturesome Best Products Company (fig. 894), its long, low façade rising above the inevitable parking lot like a false front on Main Street in a Western frontier town, except for displaying huge, gaily colored flowers reminiscent of those made famous in the sixties by Andy Warhol. So far, however, the prototypical Venturi building is the one the architect built in 1962 for his mother in Chestnut Hill, Pennsylvania (fig. 895), a work whose apparent ordinariness is just enough "off" to clue the alert viewer that something more extraordinary has gone into

the design. Here we see the work of a mind of exceptional humor, sophistication, and irony.

In Chestnut Hill House, Venturi accepted the conventions of the "ugly and ordinary" American suburban crackerbox—its stucco veneer, wood frame, pitched roof, front porch, central chimney, and so forth—and parodied them so completely as to make the commonplace seem remarkable, enriching it with wit, light-hearted satire, and connotation. The process of transformation began with the scale of the façade, which the architect expanded until it rose several feet above the roof, thereby creating another false front. He then divided it at the center to reveal the conceit and widened the cleft at the bottom to create a recessed porch. Visually joining the two halves together, yet simultaneously emphasizing their division, are the arch-shaped strip of molding and the narrow stretched lintel above the entrance, both elements paper-thin like everything else about the façade (or like the crackerbox tradition itself, or even most modern curtain-wall construction). Meanwhile, the split in the front wall echoes the much older and more substantial stone architecture of the

895. ROBERT VENTURI. Chestnut Hill House, Philadelphia. 1962

Baroque era, which in turn joins with the expanded height of the façade to endow the humble crackerbox with a certain illusionistic grandeur. Such complexity and contradiction extend deep into the structure, to the chimney wall, for instance, which spreads wide like a great central mass, only to project above it the actually rather tiny chimney. Unity too has been honored with all the intricate ambiguities, as in the same size and number (five) of the asymmetrically disposed window panels on each side of the binary façade.

Hidden behind the deceptively simple exterior of Chestnut Hill House is a complicated interior, its rooms as irregularly shaped and small in scale as the façade is grand, in the Pop manner of Claes Oldenburg's giant hamburgers and baseball bats (see fig. 799). Yet the plan is tight and rational, not whimsical. For example, the crooked stairway may be broad at the base but it is narrow at the top, reflecting, in Venturi's view, the difference in scale between the "public" spaces downstairs and the "intimate" ones upstairs. By its very irregularity, the interior accommodates the books, family mementos, and overstuffed furniture long owned by the client as well as it expresses her patterns of living. Unlike Corbusier's "machines for living" or Frank Lloyd Wright's Usonian houses, Chestnut Hill House does not suggest or impose a living pattern for its inhabitants. And herein lies the essence of the Venturi concept: Chestnut Hill House makes no attempt to simulate a ship or an airplane, or to become some organic device to commune with nature; instead, Venturi's design proclaims its "houseness" with historically relevant allusions that are not overt symbols. "We don't think people want 'total design' as it is given to them by most modern architects," Scott Brown once said. "They want shelter with symbolism applied to it."

When Venturi and Scott Brown won the commission to design a new wing for the National Gallery in London, it was the most prominent commission their firm had ever received. They were to create a permanent exhibition space for the gallery's Early Renaissance collection, a large lecture hall, a computer information room, a video theater, and a gift shop and a restaurant. Located diagonally across from Trafalgar Square, their design for the Sainsbury Wing (color plate 461, page 732) integrates both modern and classicist impulses. The main entrance is skewed to face Trafalgar Square, and the Classical details of the façade echo those of the adjacent main building. Although not easy to see from the square, these Classical details have been subjected to subtle Postmodern idiosyncrasies and operations: The windows are blind and almost disappear into the stone façade, and among the regularly spaced pillars, there is a single round column that imitates the monument to Lord Nelson in Trafalgar Square. The western-facing façade on Whitcomb Street is plain brick. The huge glass wall of the grand staircase at the east side of the wing offers a view of the exterior of the old gallery. The top floor's neutral-toned galleries permanently housing the Renaissance collection, in their expertly scaled spaces, are respectful of both the old galleries and the artwork. Natural and artificial lighting are adjusted every two hours by computer, based on the amount of natural light filtering in through the glass "attics."

Venturi and Scott Brown's Sainsbury Wing demonstrates that the Postmodern style is flexible and capable of adapting to its site, its purpose, and other surrounding buildings.

Another student of Louis I. Kahn at the University of Pennsylvania and a close ally, if not an actual partner, of the Venturis is CHARLES MOORE (1925–1993), who looked not so much to Las Vegas for inspiration as to Disneyland, praising it as an outstanding public space as well as an embodiment of the American Dream. Concerned with the commercial or industrial sameness of the postwar world, and with it the lost "sense of place" that leaves dozens of American cities looking like Los Angeles, Moore became even more overtly historicist than Venturi and was fully committed to "the making of places." In this, he seemed more often than not to favor the "presence of the past" found in a Roman-Mediterranean kind of environment, and one of the most remarkable examples of how imaginatively he could develop the theme is the Piazza d'Italia in New Orleans (color plate 462, page 733; figs. 896, 897). Here Moore designed a public space to provide a small Italian-American community with an architectural focus of ethnic identity, a place exuberant enough to serve as the setting of the annual St. Joseph's festival, with its vibrant street life animated by temporary concessions. Piazza d'Italia is an early example of the playful Postmodern use of classical forms. An entirely decorative and symbolic construction, it is architecture that abandons the ambition for permanence. Working in close collaboration with architects well versed in the local culture, Moore took into account not only the taste and life patterns of the area's inhabitants, but also the mixed physical context into which the new piazza would have to fit. Thus, while the design incorporated the black-and-white lines of an adjacent modernist skyscraper in a graduated series of concentric rings, the circular form itself—essentially Italian Baroque in character—reaches into the community outside even as it draws that world in. Moore found the prototypes for his circular piazza in, among other places, Paris' Place des Victoires, the Maritime Theater at Hadrian's Villa near Tivoli, and, most of all, the famed Trevi Fountain in Rome, an open-air scenographic extravaganza combining a Classical façade, allegorical earth sculpture, and water.

At staggered intervals on either side of a cascading fountain at the Piazza d'Italia are various column screens representing the five Classical orders, the outermost and richest one, Corinthian, crowned by polished stainless-steel capitals supporting an entablature emblazoned with dedicatory Latin inscriptions. Over the waterfall stands a triumphal arch, polychromed in Pompeiian hues and outlined in neon tubing. With its eclectic mixture of historical reference and pure fantasy, such as water coursing down pilasters to suggest fluting, water spouts serving as Corinthian leaves, and stainless-steel Ionic order volutes sitting on neon necking, Piazza d'Italia combines archaeology, modernism, commerce, and theater to provide something for everyone—and above all a sense of place unlikely ever to be confused with Cleveland, Los Angeles, Sicily, or even the French Quarter in New Orleans itself.

About the same time as Venturi and Scott Brown began to

right: 896. CHARLES MOORE WITH U.I.G. AND PEREZ ASSOCIATES, INC. Piazza d'Italia, New Orleans. 1975–80

below, left: 897. CHARLES MOORE WITH U.I.G. AND PEREZ ASSOCIATES, INC. View from above, Piazza d'Italia

below, right: 898. HANS HOLLEIN. Austrian Travel Bureau, Vienna. 1978

fill commissions, Austria's **HANS HOLLEIN** (b. 1934) was in the American West, learning from Las Vegas and doing so in rebellion against the "Prussian dogma of modernism" at the Illinois Institute of Technology (where he had enrolled in 1958). Seeing all those open spaces and the expressive freedom they seemed to generate, Hollein had an epiphany. Since then, he learned to design with a pluralistic abandon worthy of his gaudy American models, but also with an exquisitely crafted, decorative elegance reflecting a subtle appreciation of Vienna's own Secessionist tradition (see figs. 277, 279, 280). Even more than the Americans, European Postmodernists such as Hollein have found few opportunities to work on a large scale. The Austrian architect has had to contain his drive

to invent and surprise within small jewel-box spaces—often old and refurbished interiors like that of the Austrian Travel Bureau in Vienna (fig. 898) of 1978. For this project Hollein adopted topographical themes and scenographic methods similar to those of Charles Moore in Piazza d'Italia—and in the same metaphorical spirit to make a pre-existing hall suggest all the many places to which the Travel Bureau could send its clients—in other words, evoking the concept of travel. Thus, the assemblage of allusive forms and images found here includes: a grove of brass palm trees clearly based on those in John Nash's kitchen for the early nineteenth-century Brighton Pavilion (see fig. 86); a broken "Grecian" column and pyramids; a model of the Wright brothers' biplane; plus rivers, mountains, and a ship's railing. Lined up along the central axis, like a miniature Las Vegas strip, are such temptations as an Oriental pavilion, a chessboard seating area, and a sales booth for theater tickets furnished with a cash window in the form of a Rolls-Royce radiator grille. Meanwhile, the hall itself

899. HANS HOLLEIN. Städtisches Museum Abteiberg. Mönchengladbach, Germany. 1972–82

has been styled to resemble a reconstruction of Otto Wagner's 1904–06 Post Office Savings Banks (see fig. 277), an ironic reference, since the whole purpose of the Travel Bureau is to encourage spending money, not saving it.

Hollein's first commission to design a museum came from the North German town of Mönchengladbach (fig. 899). Taking ten years to complete, from 1972 to 1982, Hollein's Städtisches Museum Abteiberg is a series of buildings that critic Charles Jencks called "a modern Acropolis." The museum is located on a hill next to a Romanesque church and, as Jencks suggested, it provides a modern-day equivalent to the church's spiritual role. Declaring that he "was never afraid to use materials in new contexts—plaster or aluminum or marble, and all this together," Hollein designed his museum as a tiny town within a town, an agglomeration of distinct but compatible structures. With its undulating red-brick terraces snuggled against the slope, its relaxed and vaguely mock-ancient but nonabrasive Disneyland replicas, and its odd and delightfully particular cutout corner and voluptuous semicircular marble stairs, the Mönchengladbach museum struck Jencks as "a beautifully detailed urban landscape dedicated to a form of collective belief we find valuable. Art may not be a very convincing substitute for religion, but it is certainly a suitable pretext for heroic architecture."

Another American architect profoundly influenced by Robert Venturi, as well as by Charles Moore (under whom

he studied at Yale) is **ROBERT A. M. STERN** (b. 1939). Like his mentors, Stern carries on an active practice in Postmodern architecture and also teaches, writes, and edits. Although fully as witty and sophisticated as Venturi and Moore, Stern is also more literal and fastidious in his use of historical precedents, and he is considerably more given to patrician sumptuousness. Indeed, he has developed into something of a one-person equivalent of McKim, Mead & White in that he seems never quite so truly himself as when designing the sort of grand Shingle-Style country house or holiday retreat that the turn-of-the-century New York firm once produced with understated aristocratic bravura. Thus, while reveling in the decorative revival quite as joyously as Venturi and Moore, until his commission to design a building for the Walt Disney Company, Stern has done so less in the Pop spirit that animates Moore's Piazza d'Italia and Venturi's Best Products Company's façade than in the fanciful manner of Hans Hollein's appropriations of Nash's Brighton Pavilion. This can be seen to splendid effect in the 1980–81 pool house (color plate 463, page 733) that is part of an estate in northern New Jersey, where stainless-steel "palm columns" evoke the famous palm supports in Nash's Brighton kitchen. And if the glistening material of these uprights produces a "wet" effect appropriate to the setting, so do the polychrome wall tiles derived from a kind of decor favored by the Secessionist Viennese (see fig. 107). Further enhancing the bathhouse

900. Philip Johnson and John Burgee.
AT&T Headquarters Building, New York. 1978–83

theme are the Art Deco figural quoins and the thick Tuscan columns, the latter reminiscent of Roman *terme*.

Stern's 1995 Feature Animation Building (color plate 464, page 733) in Burbank, California, designed for the Walt Disney Company (which also has commissioned projects from Robert Venturi and Frank O. Gehry), offers a host of references both to the architecture of Los Angeles and to the fantasy architecture found in Disney animation. It is a highly complex building that in many ways resembles a converted warehouse, which is what Disney chief executive Michael Eisner had in mind when commissioning the project. Stern wanted the building's lobby—a tall oval space with one curved and tilted glass wall—to evoke the eccentric fast-food joints and doughnut shops of the forties, fifties, and sixties that are still to be found in and around Los Angeles. The building's profile—the entrance is tucked under a conical spire—is a reference to the cone-shaped hat worn by Mickey Mouse in a segment of the film *Fantasia*. Along the south side of the building is a narrow gallery with a high roof that slopes upward toward the main façade, ending in a form like the prow of a ship that Stern called "the Mohawk" because it resembles a punk haircut; it also alludes to the Mad Hatter in Disney's animated version of *Alice in Wonderland*. Visible from a nearby freeway, giant stainless-steel letters above the front entrance spell the word ANIMATION. The work space is accommodated in three vast, connected barrel vaults that form the main bulk of the building. Under the vaults, painted black with stainless-steel studs that twinkle like stars, are girders and air shafts, which have been left exposed and painted white so that their industrial forms stand out against the dark ceiling.

In looking for the early signs of the Postmodern aesthetic, one finds an apparent scorn for a refined, finished quality. This shift can be seen in the career of **PHILIP JOHNSON**, who was an associate partner with Mies van der Rohe on the Seagram Building (color plate 365, page 633). By the mid-1950s, Johnson was turning against the order and rationalism of Miesian design, though his efforts to overturn the Modernist movement were not successful until, in association with John Burgee, he designed the New York headquarters building for American Telephone and Telegraph (figs. 900, 901, 902). Heightening the shock value of this famously revisionist skyscraper is the fact that Johnson, together with architectural historian Henry-Russell Hitchcock, in 1932 had written a book that introduced the International Style to the English-speaking world and thus made it a true architectural *lingua franca*. By the late 1970s, Johnson was the virtual dean of American architects, and his immense prestige was very persuasive to the stolid AT&T officers when they committed more than $200 million to a Postmodern design of

above: 901. Philip Johnson and John Burgee. Lobby, AT&T Headquarters Building, New York

902. PHILIP JOHNSON AND JOHN BURGEE.
AT&T Headquarters Building

almost perverse irony and elegance.

It was nicknamed the "Chippendale building" for the broken pediment at its top; its resemblance to a highboy chest in the Chippendale style was irresistible. With hindsight, it is not so surprising that Johnson's witty mind, always churning through a vast and erudite range of aesthetic references, perceived in the skyscraper an analogy to the highboy, a colossal stack of "drawers" set in a chest supported by tall legs and

topped by a scroll bonnet. Rising 647 feet without setback, it is divided into base, shaft, and granite-clad cornice. The shaft provides "drawers," or storeys; owing to their double height, there are only thirty-six "drawers" in what is a sixty-storey tower. With the AT&T building, Johnson reintroduced "history" into the vocabulary of modern high-rise architecture, a history rendered compatible with the commercial world of the 1960s and 1970s by reference not to palace or ecclesiastical structures, but rather to furniture. And herein may lie the essence of Postmodernism: the longing for symbolism and grandeur of the past disembodied from the faith that inspired the original forms. The consequence of such design is ingenious historical puns and self-mockery.

The AT&T building originally had a forest of "legs" at the street level, which sheltered a capacious, open loggia some five or six storeys high; it was dedicated as a public space until 1990, when the building was sold to the Sony Corporation. Within three years, Sony converted the loggia space to shops and renamed it Sony Plaza. In the original building, the historicism of the design produced a dizzying mix of themes. The façade (fig. 900), a central oculus-over-arch flanked by trabeated loggias, recalled the front face of the Pazzi Chapel, a Florentine Renaissance building known to every student of architecture and art history. Johnson and Burgee's interior, however, was a groin-vaulted Romanesque space (fig. 901). The broken pediment that capped the building still endow it with its renowned and idiosyncratic profile (fig. 902).

Already one of Britain's most important architects in the 1960s, when he worked in a somewhat Brutalist vein (see fig. 718), Sir **JAMES STIRLING** (1926–1994) seems to have been liberated by Postmodernism into full possession of his own expansive genius, best exemplified in the exciting addition to the Neue Staatsgalerie, or museum, in Stuttgart, built 1977–84 (color plate 465, page 734; figs. 903, 904). Here, in association with Michael Wilford (b. 1938), Stirling took advantage of the freedom allowed by revisionist aesthetics to solve the manifold problems presented by a difficult hillside site, hemmed in below by an eight-lane highway, above by a terraced street, and on one side by the original Renaissance

903. JAMES STIRLING. Neue Staatsgalerie, Stuttgart, Germany. 1977–84

904. JAMES STIRLING. Model, Neue Staatsgalerie

905. ARATA ISOZAKI. Gallery building, Art Tower. View of north end. Mito, Japan. 1990

Revival museum building. With so many practical considerations to satisfy, Stirling decided to work in "a collage of old and new elements . . . to evoke an association with a museum." And so while the three main exhibition wings repeat the U-shaped plan of the nineteenth-century structure next door, the external impression is that of a powerful Egyptian massing, ramped from level to level like the funerary temple of the Old Kingdom Queen Hatshepsut at Deir el-Bahri, but faced with sandstone and travertine strips more suggestive of medieval Italy. Visually, the ramps and the masonry unify a jumble of disparate forms clustered within the courtyard framed by the embracing U. Among these are an open rotunda surrounding a sculpture court, glazed and undulating walls of an entrance shaped rather like a grand piano, and a blue-columned, red-linteled, and glass-roofed high-tech taxi stand. Despite so many historical references and such diverse configurations, a sense of unity and distinctive place are real and even moving, as the ramps convey the pedestrian along a zigzag path leading from the lower highway into and steeply halfway round the inner circumference of the rotunda, and finally through a tunnel in the cleft walls to the upper terraced street behind the museum. Adding to this heady mix of sophisticated, urbanistic pluralism is the keenly British sensi-

bility of Stirling himself, who brought to the design the grace and allure of the eighteenth-century Royal Crescent at Bath.

Another figure linked to Postmodern classicism is the Japanese architect **ARATA ISOZAKI** (b. 1931). He began his career working for Kenzo Tange (see figs. 727, 728) but soon dispensed with any Corbusian influences he might have absorbed. A self-confessed "mannerist," Isozaki draws on a huge variety of sources—Eastern and Western, traditional and modernist—honestly quoting from and transposing them.

Situated in the old part of Mito, Japan, Isozaki's 1990 Art Tower is the cultural core of that city and an attraction for visitors from Tokyo. The building accommodates four major functions: a gallery of contemporary art, a concert hall, a theater, and the Mito centennial tower. The various parts are connected by a large square that occupies more than half the site. There is also a two-storey conference hall and a museum shop, café, and restaurant. The museum space has three glass roofs (fig. 905). A pyramid is set symmetrically at the center. The façades are set with porcelain tiles to give the plaza the feel of a European townscape. At the north end, a huge stone placed on the center axis of the square is suspended by thin cables. At the northwestern corner, sandwiched between the concert hall and the theater, is an entrance hall that serves as a connecting chamber for the building's functions. This vertical space links the plaza to the street. A 100-meter-high tower to the east of the plaza commemorates the centenary of Mito (color plate 466, page 734). This Brancusian "endless" column is made of titanium-paneled tetrahedrons; stacked together, their edges create a DNA-like double helix.

Raised in Argentina, internationally influential architect CESAR PELLI (b. 1926) came to the United States in 1952 to pursue a master's degree in architecture at the University of Illinois at Champaign-Urbana. Says Pelli, who was dean of the Yale University School of Architecture from 1977 to 1984, "A building is responsible to the place it occupies and takes in a city . . . the obligation is to the city. The city is more important than the building; the building is more important than the architect." His most visible buildings reference their locales in style and with motifs, and Pelli is able to invest commercial buildings with a public and sometimes even spiritual presence by means of forms and open spaces, especially when he can design public-space interiors.

The group of four towers called the World Financial Center, which range from thirty-three to fifty storeys tall, was completed in 1988 as a focus for New York City's Battery Park City, begun in 1979 and built on Hudson River landfill (across a main highway from the World Trade Center and, on the seaward side, by a small marina) (color plate 467, page 734). The towers, referentially Art Deco in style, have nearly flush façades of granite and glass, which also relate them to the World Trade Center skyscrapers. Each of the four Financial Center towers has a differently shaped roof: a stepped pyramid, a solid pyramid, a dome, and a cut pyramid. The cathedral-like lobbies of the individual towers are sumptuously decorated with marble, brass, and shiny black columns. The grandest public space is the large, glass-house-like Winter Garden, furnished, as it were, with giant palm trees and the setting for a continuous stream of concerts, performances, and exhibitions.

In the work of HELMUT JAHN (b. 1940), Postmodernism's serious playfulness gets articulated in large-scale projects. Jahn emigrated from Germany to Chicago in 1966, a staunch admirer of Mies van der Rohe. In 1967 he joined a leading Chicago architectural firm, C. F. Murphy, and by 1981 was principal architect in the renamed firm Murphy/Jahn. His $500-million United Airlines Terminal of Tomorrow at Chicago's O'Hare International Airport (color plate 468, page 735) is an example of Jahn's success in creating enormous public places that are interesting to experience, technically brilliant, yet not overwhelmingly futuristic. Awarded the A.I.A. Design Award of 1988, the building is an architectural echo of the past, reminiscent of the iron-and-glass structures of Victorian railway stations. One speeds through physically on people movers while, at the same time, being mentally transported through time to another era. As in Joseph Paxton's famous Crystal Palace of 1851 (see fig. 88), we see prefabricated metal and glass construction elevated to the scale of civic architecture. The corridors and departure/arrival gates are metal-framed, glass-paneled, and topped with soaring barrel vaults drenched with natural light. Yet Jahn's high-tech update on nineteenth-century engineering is a far cry from explicitly historicist buildings.

In any age, designing a house of worship is an especially demanding architectural challenge. In our time, Japanese architect TADAO ANDO (b. 1941) has designed some of the most effective places of worship in the modern idiom, using precision-cast reinforced concrete—partly to make his buildings earthquake safe—and siting the buildings and their precincts with extraordinary sensitivity to the surrounding topography. One such project is his 1988 Chapel-on-the-Water (color plate 469, page 735). Located on a small plain in the mountains on the northern island of Hokkaido, this church is part of a year-round resort. The structure consists—in plan—of two overlapping squares. The larger, partly projecting out into an artificial pond, houses the chapel, and the smaller contains the entry and changing and waiting rooms. A freestanding L-shaped wall wraps around the back of the building and one side of the pond. The chapel is approached from the back, and entry involves a circuitous route: a counterclockwise ascent to the top of the smaller volume through a glass-enclosed space open to the sky, with views of the pond and mountains. In this space are four large concrete crosses. The wall behind the altar is constructed entirely of glass, providing a dramatic panorama of the pond with a large cross set into the water. The glass wall can slide like a giant shoji screen, opening the interior of the chapel to nature. A 6,000-seat, semicircular theater, northwest of the chapel, is designed to accommodate open-air concerts and other events. Set on a fan-shaped artificial pond, the amphitheater is intersected by a long, bridgelike stage and a freestanding colonnade. A site wall points toward the church complex and turns at a sharp angle into the axis of the stage, tying the two projects together.

The work of the Iraqi architect ZAHA HADID (b. 1950) points in a very different direction from that of Stirling, Isozaki, Pelli, and Ando. Her theory and her buildings exploit the possibilities of deconstructionist theory, which challenges the standard functions, uses, and perceptions of a building, much in the way it has in the fields of literary criticism and art history. Deconstructionism and deconstructivist architects are a Postmodern phenomenon, although usually not regarded as mainstream Postmodernist. Born in Baghdad, Hadid studied mathematics in Beirut, was trained in architecture at the Architectural Association in London (1972–77), and established her own practice in 1979. She is one of the first experimental female architects to make a substantial reputation and impact in the second half of the twentieth century. Her Vitra Fire Station in Weil am Rhein (fig. 906), from 1993, is a prime example of deconstructivist architecture. The work of painter Kazimir Malevich (see figs. 244, 245) has served as a basis for her design work, and Hadid says that her inspiration derives from the principles of Suprematism—a concept developed by Malevich and expressed by the reduction of the picture to Euclidean geometric arrangements of forms in pure colors, designed to rep-

906. ZAHA HADID. Vitra Fire Station, Weil am Rhein, Germany. 1993

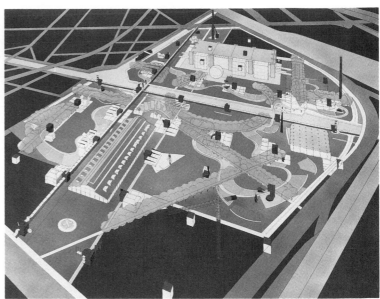

907. BERNARD TSCHUMI. Plan, Parc de la Villette, Paris. 1982–91

resent the supremacy of pure emotion. Thus many of her projects exist as unbuilt, Utopian investigations.

The Vitra Fire Station shows a high sense of individual personality in its vigorous use of space. The architectural concept has been developed into a layered series of walls that alternate between void and volume, giving the impression of speed and escape. Constructed of exposed reinforced concrete —a very suitable medium for Hadid to realize her sculptural expressiveness and ambitiously long spans and cantilevers— the whole building is movement suspended in time and space, expressing the fire fighter's tension between being on the alert and the possibility of exploding into action at any moment. Special attention was given to the sharpness of all edges, and any attachments, such as roof edges or claddings, were avoided lest they detract from the simplicity of the prismatic form and abstract quality of the architectural concept. This same absence of detail informed the frameless glazing, the large sliding panes enclosing the garage, and the treatment of the interior spaces. The surrounding landscape seems to stream into the building, testimony to Hadid's reputation as an assertive sculptural architect and to her unique ability to transcend building mass with pure anti-gravitational energy. Yet the question inevitably arises whether such bold, sculptural anti-functionalism is suitable to the actual purpose of the building, or for that matter to anything else.

Also deconstructivist and neither a park nor a building in the traditional sense, Parc de la Villette (1982–91)—an open space in a working-class district of Paris that more closely resembles a large architectural and environmental sculpture— is the first built work of deconstructonist theorist and teacher **BERNARD TSCHUMI** (b. 1944). His design for the master plan and structural elements of Parc de la Villette (which means "little city") (fig. 907) was selected from over 470 entries in an international competition. The 125-acre park is essentially completed. It consists of 70 acres of landscape, including walkways, gardens, and a canal, and thirty bright-

colored, geometrical "folies," small structures containing cultural and recreational facilities such as cinema, video workshop, information and day-care centers, and health club.

The organizing principle of the park's design is the superimposition of three independent ordering systems, each of which uses a separate vocabulary of either lines, points, or surfaces, so that there is no single or true master plan. In combination, the logic of each system loses its coherence, and accident and chance become determining factors in the resulting design. The most prominently ordered system, the 400-square-foot grid of points, is marked by the evenly spaced locations of the folie structures. Each of the folies is based on the form of the cube, which is then "cut away," or reconfigured, tranforming pure geometry into dissonance.

Tschumi, who is dean of the Graduate School of Architecture, Planning, and Preservation at Columbia University, uses the tools of architecture to question the validity of its own rules. Employing multiple organizing principles, he sets them up to collide with one another, each one serving to cancel, interrupt, and undermine the others—asserting that there is no pure space, that activity necessarily and unapologetically violates architectural purity. The three ordering systems of the park allow for almost infinite variations on this theme.

During much of the twentieth century, the machine has served as one of the most ubiquitous cultural metaphors, especially in regard to architectural form and theory. Designers have tried not only to emulate the machine's aesthetic qualities but to envision an architecture whose relationship to its function corresponds as directly as that of a machine to its purpose. The spare beauty of Le Corbusier's Villa Savoye (see fig. 416), for example, is due to his resoluteness in applying the machine metaphor. Tschumi, on the other hand, proposes that architecture is continually transformed by the quotidian events that take place in and around it, events too varied and complex to be described by any one architectural pro-

908. PETER EISENMAN. Wexner Center for the Visual and Performing Arts, Ohio State University, Columbus. 1983–1989

gram, such as the International Style, which was thought to be universal. Rather than imposing a structure on the disorderliness of urban life, Tschumi's Parc de la Villette is emblematic of the metropolis and its inherent intricacies. It becomes a stage set for an infinite number of human activities, both planned and spontaneous, authorized and illicit, which momentarily form part of the metropolis. Parc de la Villette thus affirms the random disorderliness and vibrancy of the city. With its keen awareness of the importance of events that surround it, Tschumi's architecture aspires to an inclusivity that reflects the haphazard complexity of the modern metropolis.

The mannerizing strain in the architecture of the 1980s and 1990s achieves extreme form in the work of the American deconstructivist theoretician and architect **PETER EISENMAN** (b. 1932). As a deconstructionst, Eisenman puts buildings together out of parts and elements that are not inherently unified or related, so that the realized structure may not serve the usual function or expectations for a building. Eisenman's first building to gain wide attention is the Wexner Center for the Visual and Performing Arts at Ohio State University, 1983–89 (fig. 908). The center strikingly exemplifies Eisenman's theory that architecture can be nonfunctional. The mannerisms of the Wexner Center come as much from the skewed and obstructed use of space as from any of its detailing. Indeed, the project is an intricate complex of construction and landscape, an "archaeological earthwork" as

Eisenman termed it, that radically breaks with the predictable order of the traditional bucolic American academic campus. Conceptually and physically, the design connects the academic community with the Columbus civic community through a right-angle interlock of the campus's and city's grid systems within the building itself. Symbolic of this is what the architect calls a "scaffolding" of square white steel pipes whose formation is a "metaphoric microcosm of the urban grid." The 140,000-square-foot building houses the experimental arts of computers, lasers, performance, and video and also exhibition galleries, café, bookstore, and administrative spaces. The most memorable image resulting from Eisenman's preoccupation with geometry and iconography is the rigorous coupling of architecture and landscape in a manner that revives, or at least harks back to, the formal geometric integration of garden and building of seventeenth-century European palaces.

The façade of Eisenman's Nunotani Building in Tokyo (fig. 909) illustrates perfectly a highly intentional parsing of walls, windows, storeys, and profiles. In its slipping and sliced exterior walls, its shaved-off window courses, and what can almost be seen as the first freeze-frame moments of its own demolition, this urban building offers itself as a monument to the notion of impermanence, a parody of tradition and technology.

Another architect and theorist working in the deconstructionist tradition of Postmodernism, **REM KOOLHAAS** (b. 1944), collaborated with Zaha Hadid in 1978 on the design

909. PETER EISENMAN. Nunotani Building, Tokyo, Japan. 1990–92

910. REM KOOLHAAS. The Netherlands Dance Theater, The Hague,
The Netherlands. 1987

911. NORMAN FOSTER. Telecommunications Tower,
Barcelona, Spain. 1992

of the Dutch Parliament Building, The Hague, in his native Holland. Koolhaas's studies for unbuilt, as well as built, projects are highly regarded by other architects. The Netherlands Dance Theater (fig. 910) is an example of Koolhaas's Postmodern vocabulary of style and technology, from the huge, semi-abstract mural topping the main structure to the mix of reinforced concrete, glass, metal, and masonry for the walls of its variously formed parts.

Less cerebral and more celebratorily high-tech is the Telecommunications Tower in Barcelona, Spain (fig. 911), the work of English architect Sir **NORMAN FOSTER** (b. 1935). The design of the tower follows the philosophy of Foster and Partners—that is, achieving maximum effect with minimal structural means, as well as achieving clarity and precision in the fitting together of components. The tower's precast con-

crete shaft is supported by three vertical steel trusses, and the whole structure is tethered to guy wires anchored in the mountainside. Modular floors and platforms are suspended from this structural skeleton in such a way as to accommodate future changes. Yet for all its functionality, the design of the Tower back-references an era of spaceship futurism with an innocent, almost playful World of Tomorrow effect.

Historicism was focused in the opposite direction—back in time—when Chinese-American architect **I. M. PEI** (b. 1917) executed the commission for modernizing certain parts of the Louvre Museum in Paris. I. M. Pei is popularly known for the controversy surrounding his Grand Louvre Pyramid (1988), constructed in the courtyard of the Louvre (fig. 912). The Pyramid deliberately turns the tradition and concept of pyramid inside out. A pyramid is supposed to be

912. I. M. Pei. Grand Louvre Pyramid. Paris. 1988

solid, dark, and solitary. Here it is clear glass, almost immaterial, a vast skylight hovering over streams of museum visitors as they are channeled into the Louvre galleries through the below-ground entrance corridors. Besides its associations of timelessness and its brilliant ingenuity in lighting an underground space, the ensemble is a superb example of how new buildings in old settings do not always have to accommodate themselves to the style of their "found" surroundings.

The expansion of existing museums and the proliferation of new museums in the last quarter century have generated some of the twentieth century's most exciting architecture. Still, creating a museum to document one of the most atro-

cious crimes in history was an exceptional architectural challenge. The United States Holocaust Memorial Museum was chartered by a unanimous act of Congress in 1980 and opened in 1993. Architect James Ingo Freed, of Pei Cobb Freed & Partners, fashioned a relationship between the architecture of the building and the exhibitions within, creating in effect a work of art. He visited a number of Holocaust sites, including camps and ghettos, to examine structures and materials. While architectural allusions to the Holocaust are not specific, subtle metaphors and symbolic reminiscences of history evoke thoughts and reflections. In Freed's words, "There are no literal references to particular places or occurrences from the historic event. Instead, the architectural form is open-ended." The building thus serves as a "resonator of memory."

At first glance, the exterior seems benign, which is part of the building's metaphoric message (fig. 913). Yet throughout, Freed has incorporated physical elements of concealment, deception, disengagement, and duality. The curved portico of the 14th Street entrance—with its squared arches, window grating, and cubed lights—is a mere front, a fake screen that actually opens to the sky, deliberately hiding the disturbing architecture of skewed lines and hard surfaces of the real entrance that lies behind it. This strategy of contrasting and juxtaposing appearance and reality is repeated throughout. Along the north brick walls, a different perspective reveals a roofline profile of camp guard towers, a procession of sentry boxes. Above the western entrance, a limestone mantle holds a solitary window containing sixteen solid "panes," framed by clear glass, reversing the normal order and obscuring the ability to look in or out. Inside, the physical layout ushers visitors through the exhibits in a sequence of experiences. The central, skylit, brick-and-steel Hall of Witness sets the stage of deadly efficiency that is echoed throughout this museum. The expansive use of exposed steel trusswork is an ever-present reminder of the death chambers.

Vertically piercing the entire structure is the Tower of Faces (color plate 470, page 735), a chimney-like shaft lined with photographs of the people of a village that was essentially annihilated in the Holocaust. One passes through the Tower of Faces on a bridge, surrounded above and below by the faces of men, women, and children for whom the chimney was the ultimate symbol of fear. From the main floor and on the way out, visitors climb an angled staircase to the second floor, which is highlighted by the hexagonal, light-filled Hall of Remembrance.

A very different monument and museum is a Pei project of 1995, the Rock and Roll Hall of Fame and Museum in Cleveland, Ohio (color plate 471, page 736). It is an elegant monument to the world's most subversive art form. All of Pei's work, from the East Building of the National Gallery of Art in Washington, D.C. (see fig. 748) to his Louvre Grand Pyramid in Paris, reflect an obsession with Euclidean geometry, a system that analyzes the physical world in terms of the cube, the sphere, and the pyramid. Pei's fascination with the pyramid, which he explored in the Louvre project, carried over into the Rock and Roll Museum. A 117-foot-high glass-and-steel tent leans at a 45-degree angle against a tower with cantilevered wings. The building is an architectural interpretation of the explosiveness of rock music. Pei himself has said

913. PEI COBB FREED & PARTNERS. United States Holocaust Memorial Museum, Washington, D.C. 1993

914. FRANK GEHRY. State of California Aerospace Museum, Santa Monica. 1982–86

that he took the explosion metaphor from rock's rebelliousness and energy.

The building has three major elements: the glass tent, the tower containing exhibition spaces, and a plaza with administrative offices, archives, storage, and more exhibition spaces. One of the wings houses a forty-four-foot-diameter cylindrical space for viewing films supported by a single column that plunges into the lake. The other wing comprises a 125-seat, cube-shaped auditorium that juts above the water without any visible means of support.

Ironically, it was Italian architect **RENZO PIANO** (b. 1937) who won the commission to design an all-new major airport, Kansai International Airport in Osaka, Japan, which opened in 1994. Piano came forward as a pioneer of Postmodernism when he collaborated in the 1970s with English architect Richard Rogers (1933–1996) in the design of the Centre Georges Pompidou, the controversial "inside-out" home of modern and contemporary art in Paris that showed his fascination and respect for the infrastructure of buildings (see fig. 720). Over the years, Piano's work has become more refined and restrained without a loss of any interest in technology and

materials. The soaring waiting halls at the Kansai International Airport (color plate 472, page 736) are examples of Piano's easy relationship with industrial-strength materials and how to use them elegantly on a very grand scale.

Canada-born American **FRANK O. GEHRY** (b. 1929) is another major figure whose approaches identify him with Postmodern design and whose work, invariably witty, is increasingly confident and masterly. His California Aerospace Museum (fig. 914), completed in 1984, is a showcase for jet-age technology. Through its good-natured combination of disparate forms, materials, and scale, the structure clearly evokes the spirit of flight and also acknowledges the real American city. Impaled on the cruciform strut above the forty-foot "hangar" door is the most sensational aspect of the building's design: a Lockheed F-104 Starfighter plane. The gigantic, thematically appropriate aircraft does not upstage the museum—no small feat considering the drama of the spectacle. It is simply a bold ornament applied to a bold building.

Of a similarly allusive style is Gehry's good-humored Chiat/Day Building of 1991 on Main Street in Venice, California (color plate 473, page 736). The headquarters for an advertising firm that has developed some of America's cleverest and most successful ad campaigns, as well as pioneering off-site creative work, the Chiat/Day Building is a collaboration between Gehry and sculptors Claes Oldenburg and Coosje van Bruggen (see fig. 799). Oldenburg and van Bruggen's mega-sized *Binoculars* forms the portal to an interior that is essentially open yet accommodating to impromptu meetings and plug-in work sites for its drop-in, drop-off employees.

Modernism's failure to generate feasible urban development has led to the rise of "edge cities" and "edge towns," now part of the landscape of post-industrial America. Most edge communities began not far from the downtown of a large city as tract-home suburbs with a shopping mall; then came office parks and other strip malls.

Miami-based architects and planners **ELIZABETH PLATER-ZYBERK** (b. 1950) and **ANDRES DUANY** (b. 1949) are acknowledged leaders of the New Urbanists, contemporary architects who seek to reform America's communities. Their remedy is as much moral as it is aesthetic. They believe that traditional town planning—a grid of streets lined with trees and front porches, studded with shops and parks—can heal the nation's fragmented sense of community. In their 1991 book, *Towns and Town-Making Principles: Andres Duany and Elizabeth Plater-Zyberk,* the Miami architect-planners champion a concern for the small-scale design decisions that make streets, blocks, and neighborhoods aesthetically pleasing. Since then, their firm, DPZ, has become much more heavily involved in urban planning.

Seaside, DPZ's eighty-acre resort town on the Florida Panhandle, took ten years to build its hundred different houses and buildings, including a fire station, restaurant pavilion, and town hall (color page 474, page 736). It is one of the new "old" towns, part of a movement that began in the early 1980s, first described as neo-traditional planning, and now known as New Urbanism. The tenets of the movement include creating pedestrian-oriented places organized around public open space—the antithesis of the postwar auto-oriented suburb. Residents aren't more than a few minutes' walk to any part of the development. Ultimately, New Urbanists aim to design cohesive, environmentally sensitive places that look very much like traditional towns or big-city neighborhoods.

DPZ's experience in laying out Seaside, a prime example of a New Urbanist development, offers an idea of the sorts of opportunities that New Urbanism is providing architects. Each element of a development is seen as a limb of a whole, and architects are now able to participate in town planning at large. They are designing not only the houses, but the streets, the blocks, the park, the whole community. A tendency of New Urbanism that disturbs its critics could derail the movement, however. Many of the projects do not provide for existing buildings in their "new neighborhoods." Old buildings found on the sites of the new "community of neighborhoods" are tagged to be removed or razed. The objection to this is not that the new neighborhoods—with their planned roads, parks, and houses—are not attractive, but that they do not have, or might not have, a true history or an urban vibrancy, that their lack of roots is being disguised by a thin simulation of nostalgia and history.

27 Epilogue

How can we define the 1990s? How give shape to an era still in progress? Without historical perspective, what can we make of the events and the artwork of this decade? While this chapter does not claim to be a definitive account of the 1990s, it does attempt to characterize what has transpired thus far. It takes as its focus art that has been circulated in the United States, although artists of different nationalities are considered. Because there has been no dominant style, medium, or movement, attention is given to the issues that have been most prevalent and to the strategies artists have adopted to address them. While it may be difficult to characterize the artists of this decade, a number of themes recur in their work. Among these are an interest in identity politics and gender; in exploring the provinces of public and private space, particularly in relation to exhibition procedures and the body; as well as in investigating the possibilities of a hybrid art, particularly the multimedia installation.

In the 1980s, a handful of artists—especially those whose work seemed to embody Poststructuralist ideas about media and representation—dominated both the market and the imagination of the critics. In the 1990s, the field has been more disparate. If anything, the decade has been characterized by conflict and decentralization. The 1990s began in conflict—conflict within the art world over identity politics, conflict between the art world and Congress over the funding of controversial art by the National Endowment for the Arts (NEA), and, of course, military conflict in the Persian Gulf. In addition, the robust art market of the previous decade gave way to much weaker support for the visual arts. As auction house sales have plummeted, so have sales in the galleries.

During the eighties, galleries in the SoHo (*south of Houston* Street) district of Manhattan seemed to multiply. In the 1990s, many art dealers closed their doors. A handful of SoHo's leading galleries abandoned the old neighborhood and moved to the budding art district of Chelsea, where rents were cheaper and where the Dia Center for the Arts had opened a mammoth exhibition space. Among those that moved were the Paula Cooper Gallery, the first gallery to open in SoHo in 1968, and Metro Pictures, which represented many of the most prominent artists of the 1980s. The shift in location was significant, signaling that SoHo was no longer the center of the international art world. But neither was Chelsea, and so the art world was left looking and feeling a good deal less cohesive.

The market was not the only factor in the decentralization of the New York art world. In the late 1980s and early 1990s, a segment of the art world that perceived itself to be overlooked confronted the dominant institutions and individuals. Debates regarding identity, marginality, and the role of race, class, and gender questioned the homogeneous view that had been presented by many museums and art journals. In 1990, three New York City museums—the New Museum of Contemporary Art, the Studio Museum in Harlem, and the Museum of Contemporary Hispanic Art—collaborated on an exhibition that challenged the prevailing view of the 1980s. "The Decade Show" featured artists of diverse ethnicity and focused on the significance of identity as an issue for artists outside of the so-called mainstream. The show strove to be inclusive and multivocal, representing African-American, Latin American, and Asian-American artists as well as a range of feminist perspectives.

By 1993, there was no longer such a clear distinction between the mainstream and the margins. In that year the Whitney Museum of American Art in New York made identity politics the focus of its Biennial Exhibition. The art world had embraced diversity, although defining one's identity became problematic. While artists and theorists continued to advocate identification with a community—be it based on race, ethnicity, gender, or sexual orientation—they also recognized that identity was hybrid. Identification with a race or nation was becoming more and more difficult as people were further displaced from their homelands and more and more countries outside of the United States imported Western culture through developing technology. Furthermore, the collapse of the Soviet Union and the fall of the Berlin Wall, in 1989, made the division between East and West less clear.

While artists of color have struggled to control representation of race, the recognition that race is a historical construction, that one's racial identity is determined by historical circumstances, has complicated that struggle. Adrian Piper, a light-skinned African-American artist, has been addressing her mixed-race identity since the late 1960s. During the "Decade Show," her video installation, *Cornered*, confronted passersby on the streets of SoHo from its position in the window of the New Museum. In the video, Piper explains that the definition of a "black" person in the United States is any individual with African ancestry, thereby suggesting that many people who have been "passing" as "white" may legally be defined as "black." Carrie Mae Weems, another African-American artist, approaches her racial identification in terms of her relationship to Africa. Her photographs documenting her visit to the Sea Islands, off the southeastern coast of the United States, suggest both her connection to an African past and her displacement from Africa's present-day realities as a citizen of the United States and a participant in its culture.

Rather than create alternative images to the traditional stereotypes of race, many artists focus on the impossibility of creating representations that correspond to actual experience. For example, Glenn Ligon uses stenciled texts that smudge and blur to make the point that African-American experience cannot be described by a language developed by and for a white society. Similarly, Lorna Simpson juxtaposes photographs with evocative words and phrases to suggest the shortcomings of both text and image in representing the lives of black women.

Just as there is no single direction for artists addressing race and national identity, feminism in the 1990s has been without a common focus. While the strict opposition between the feminist art movements of the 1970s and 1980s is often overstated, in the 1990s it has become difficult to discern any specific characteristics of a feminist art. That said, there are several artists whose work makes reference to issues and strategies of previous decades, especially to work produced in the 1970s. Annette Messager fragments representations of the female body in an effort to show the impossibility of presenting a universalized feminine experience. Ann Hamilton, whose work is not limited to a feminist interpretation, refers to women's labor in the fabrication of her large-scale installations.

One explanation for the decline in work that is expressly feminist is the shift in attention to the more general question of gender. In 1990, Judith Butler's influential book *Gender Trouble: Feminism and the Subversion of Identity*, in which she argued that gender is not only socially constructed, but performative, or based on a succession of repeated actions, made a strong impact on artists and critics alike. While the performance of gender in photography has had a long history, work produced in the 1990s is distinguished from earlier examples by an interest in the body as a site for the construction of multiple identities and as a physical mechanism. In addition, artists have benefited from advanced technology.

The Japanese artist Yasumasa Morimura photographs himself in drag, usually within the framework of a Western image tradition. In doing so, he equates the fictions of gender and race and challenges the distinction between East and West. Bedecked in ornate costumes and surrounded by elaborate theatrical settings, Morimura has represented himself as canonical figures in Western art history and as famous actresses from American, European, and Japanese films.

Matthew Barney challenges the equation of sexual anatomy with gender identification in his videos, installations, and photographs. Barney creates characters whose sexual ambiguity is suggested by a lack of genitals, by the presence of both male and female genitals, and/or by the nature of their hybrid appearance, a composite of human and beast. These creatures are often engaged in athletic activity, pointing to the physical construction of the body through training and also to the body's physical limits. Barney also uses viscous substances in his work, applying them both to the characters and to objects, alluding to sexual activity and the interior of the body.

Throughout the nineties, the body itself has been an important subject for artists in all media. Like Barney, Kiki Smith has explored the body's fluids and organs. In 1990 at the "Projects Room" of The Museum of Modern Art in New York, she installed bottles containing various substances, marked with the names of different body fluids, such as blood, saliva, and vomit. In her figurative work, she has exposed veins and arteries and represented liquids oozing from breasts and male genitalia.

The widespread interest in the body's fluids is no coincidence in an age plagued by a disease that travels through those fluids. AIDS has been both a subject for artwork and the inspiration for various artistic actions. In 1990, the organization Visual AIDS formed to bring attention to the AIDS crisis through strong visual symbols. Every year on World AIDS Day, which is December 1, museums and galleries across the country shroud works of art as a symbolic reminder of the creative lives lost to AIDS. This "Day Without Art" has recently been expanded to include "Night Without Light," for which cities across the country synchronize clocks to darken downtown skylines for fifteen minutes. Perhaps the most visible and widely known Visual AIDS project is the Red Ribbon Project, which came to public attention when movie stars donned the ribbons on the internationally televised Academy Awards broadcast.

Individual artists' responses to AIDS have varied from elegy to anger. David Wojnarowicz addressed his doubly ignored status as a homosexual and a person living with AIDS/HIV in his frank confessional writings and in his paintings. To his cartoon-inspired montage paintings he added sexual and AIDS-related imagery, for which he was singled out by Congress in the ongoing controversy over the NEA. In addition, his catalogue essay for "Witnesses," an exhibition addressing AIDS at Artists Space in New York, drew widespread attention and was one of the factors that caused the NEA to retract funding for that show. In the essay, Wojnarowicz attacked by name several public figures who had opposed homosexuality and/or funding for AIDS research and education.

No other event has shaken the American art world like the

challenge to the National Endowment for the Arts. Beginning in 1989 with attacks on the work of Andres Serrano and Robert Mapplethorpe, members of Congress have singled out artists whose work they perceive to threaten the traditional values of mainstream American culture. For example, Serrano's *Piss Christ,* a photograph of a plastic crucifix submerged in urine, was seen as an attack on religious values, while Mapplethorpe's explicitly homoerotic work was described as a threat to the American family.

The general response of the art world has been to defend itself on the basis of the First Amendment, arguing that any attempt to control work funded by the NEA would be an act of censorship. For its part, Congress has argued its responsibility to spend tax dollars wisely. In short, there has been a struggle between a right-wing element of Congress with an agenda to promote patriotism, religion, and traditional family life and a left-leaning art world that has given artists of color, feminists, and gay and lesbian activists a platform to speak. Ironically, when the Endowment was formed in 1965, it was seen as a symbol of freedom of expression, celebrated for giving artists the freedom to create. As it stands now, funding for individual artists has been completely eliminated and the overall budget greatly reduced. How these changes in funding will affect artists and arts institutions in the future remains to be seen.

The thirty-seven artists described here represent the multiplicity of ideas, practices, and issues that have circulated since the decade began. Artists of several generations are considered. The artists who have matured in the 1990s perhaps tell us the most about what is new about the world, offering fresh insight; those who have been working longer, such as Adrian Piper and Faith Ringgold, suggest the longevity of some current concerns. In addition, several artists who have been active for some time have recently become celebrated because, in the current climate, their work seems more relevant. This is true for Annette Messager, who has been active in France since the late 1960s, but who only recently has received widespread attention in the United States. Her popularity is perhaps due to her interest in exploring representations of the body, which is a current area of engagement in both the United States and France.

An additional difficulty in categorizing contemporary artists is that not all of them have chosen to work with traditional means. Artists like Stan Douglas and Tony Oursler have focused on media and technology in video installations, while challenges to the traditions of painting continue to be addressed by artists like Jonathan Lasker and Mark Tansey. In contrast to earlier decades, in which artistic styles and the formal strategies that constituted them were much easier to discern, the nineties thus far defy simple characterization. An examination of its artists within a different set of parameters clarifies the extent to which most of them are operating with a very different approach to art.

Color plate 475. LOUISE LAWLER. *Collage/Schwitters/Sotheby's.* 1991. Set of two Cibachromes with text on mats, each 19½ x 15" (49.5 x 38.1 cm). Courtesy the artist and Metro Pictures, New York

1. LOUISE LAWLER (b. 1947)
Collage/Schwitters/Sotheby's, 1991

LOUISE LAWLER's contribution to a 1983 group show was a matchbook imprinted, "Every time I hear the word culture I take out my checkbook—Jack Palance." With this vulgar object, Lawler challenged the art gallery's claim to be a place offering aesthetic objects in an atmosphere far removed from the promotional activities of business. While she has gone on to photography and installations, Lawler's basic concern remains not the making of art, but rather the unstable nature of the category of things designated as "art." To her, as to such Post-Minimal and "post-studio" artists as Daniel Buren and Hans Haacke, "art" is a construction that depends upon a grid of assumptions and practices for its privileged existence as a vehicle of "priceless," high culture. What distinguishes her work, and makes it important to the nineties, is its focus upon the presentational procedures of the art world.

In one installation, Lawler juxtaposed her own arrangement of art by the gallery's star artists (including Cindy Sherman and Robert Longo) with photographs she had taken of comparable arrangements of art in a museum, a corporate office, a private collection, and a coffee-table art-book. The sequence of arrangements, each modifying the look and meaning of the "art," demonstrated that no method of display or context is neutral. All of them embed objects within codes that fix significance and furnish viewers with guidelines for their responses. As Therese Lichtenstein explains, Lawler's work reveals the mechanisms by which "'natural' perception and 'cultural' cognition, seeing and reading" are bound together. The photographs in *Collage/Schwitters/Sotheby's* (color plate 475) were taken from the lowered viewpoint of someone bending to look at the labels and dots put on the collages to indicate their status at an upcoming auction. By cutting across the frames of Kurt Schwitters's art, the photographs also make a witty appropriation of the collage technique for which this artist is famed. His work appears as a set of fragments within a representation that inverts normal hierarchies of vision and comments on art's fate in late-twentieth-century commodity culture.

right: Color plate 476. JIMMIE DURHAM. *Red Turtle*. 1991.
Turtle shell, painted wood, paper,
61½ x 67 ½" (156.2 x 171.5 cm).
Collection Dr. and Mrs. Robert Abel, Jr., Delaware.
Courtesy Nicole Klagsbrun Gallery, New York
PHOTO COURTESY FRED SCRUTON, NEW YORK

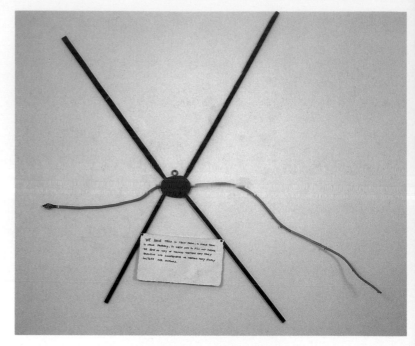

below, right: Color plate 477. JIMMIE DURHAM.
Original Re-Runs. 1993–94 installation at the Institute of
Contemporary Arts, London.
Courtesy Nicole Klagsbrun Gallery, New York
PHOTO COURTESY JIM DOW, NEW YORK

2. JIMMIE DURHAM (b. 1940)
Red Turtle, 1991
Original Re-Runs, 1993–94

JIMMIE DURHAM, a poet and artist active in the
American Indian Movement of the 1970s, disavows
any interest in trying to make a "new folk art." The
sculptures he assembles from discarded objects and
fragments of organic matter (color plate 476) are
"fake Indian artifacts" that bear ironic witness to the
shattered history of his people. Such work would
seem to have little in common with Lawler's cool,
"post-studio" photographs and installations, yet its
conceptual underpinnings are similar, and it too
acquires significance through a dismantling of norms
of display and consumption.

As objects that can be considered either art or
artifacts, Durham's "fakes" raise questions about
the supposedly objective principles used to classify
cultural production. In practice, the key marker of
difference has been racial identity. No matter how
refined, anything made by an American Indian was
defined as an artifact and relegated to an ethnographic or natural history museum. In Durham's installation *Original Re-Runs* (color plate 477), "stupid"
artifacts appeared in an ersatz ethnographic display
organized by signs of a colonial order sited, as Laura
Mulvey observes, "between religion and greed." But
binary opposition of victim and oppressor is not the
key to *Original Re-Runs* or any single piece by
Durham. He brings the battered signs and symbols
of Indian culture into quintessential Postmodern
configurations, which subvert past patterns of dualis-
tic thought and create a space where the meaning of
Indian culture can be renegotiated.

Color plate 478. ADRIAN PIPER. *Pretend #3*. 1990. Four enlarged photos, one photo of a pencil drawing on graph paper, silkscreened texts. Clockwise from top left: 47½ x 63¾" (120.7 x 161.9 cm); 11½ x 28¼" (29.2 x 71.8 cm); 66¾ x 28¼" (169.5 x 71.8 cm); 17½ x 36¼" (44.5 x 92.1 cm); 35 x 30" (89 x 76.2 cm). Collection Peter Norton. Courtesy the artist

below, right:
915. ADRIAN PIPER. *Political Self-Portrait #2 [Race]*. 1978. Photostat, 24 x 16" (61 x 40.6 cm). Collection Richard Sandor. Courtesy the artist

3. ADRIAN PIPER (b. 1948)
Political Self-Portrait #2 [Race], 1978
Pretend #3, 1990

In sharp contrast to Durham and many of the other minority artists discussed in this book, **ADRIAN PIPER** makes no reference to "native" artistic traditions in her work. Her art grows directly out of her training as a philosopher (Piper has a Ph.D. from Harvard University) and her experience of Performance and Conceptual art during the 1960s and 1970s. Her videotapes, photographic montages, and performances confront audiences with their complicity in the cruel, often hidden, racism of American society. For example, the words "Pretend not to know what you know" form a caption for the scenes of racial violence in a photographic montage of 1990 (color plate 478). The piece also contains, at the upper right corner, a picture of three monkeys that seem to "hear, see, and speak no evil."

During the course of her career, Piper has questioned the place of racial descent in the shaping of individual and communal identity. In an early self-portrait (fig. 915), she appears half black, half white, against a field of text that narrates her experiences as a light-skinned person, unable to claim total affiliation with either color group and faced with fear and mistrust on all sides. For her, racism is an impenetrable wall. "I tried," she writes in a later essay, "but couldn't crack the fears, fantasies, and stereotypes projected unto me." Art, she adds, is a "high, sunny window that holds out to me the promise of release into the night."

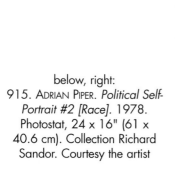

4. GLENN LIGON (b. 1960)
Untitled (I feel most colored when I am thrown against a sharp white background), 1990

In a project of 1992, the African-American painter **GLENN LIGON** used a corner of a room to present a quotation from James Baldwin that read, in part, "No true account really of black life can be held, can be contained, in the American vocabulary." To Ligon, who grew up in the Bronx, then studied at Wesleyan University, Middletown, Connecticut, "coming to voice" has entailed the making of paintings that combine abstraction with "found" language or quotations from literary sources ranging from Baldwin to Jean Genet and Mary Shelley. *Untitled (I feel most colored when I stand against a sharp white background)* (color plates 479a, 479b) is part of a series from 1988–90 in which all the paintings approximate the dimensions (6'8" x 2'6") of doors Ligon

found discarded in a studio at P.S. 1. The evocation of an architectural element adjusted to the norms of the human body adds to the disquieting impact of the account of racism inscribed on the surfaces of these black-and-white paintings. "I do not always feel colored," appears repeated over and over on one; "I remember the very day that I became colored" is the refrain on another. Ligon hand-stenciled the words onto the monochromatic backgrounds of each picture, working top to bottom, and letter to letter. It is a process that eventually reduces the message to indecipherable black smudges. The breakdown in communication can be understood as an allusion to Ligon's own estrangement from the "vocabulary" of mainstream white culture.

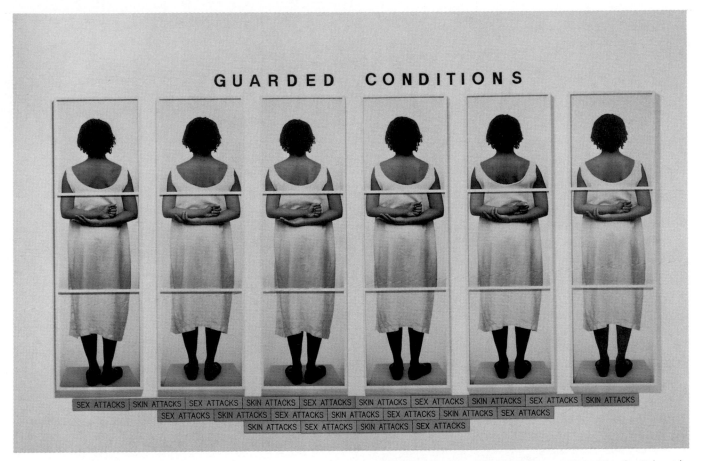

GUARDED CONDITIONS

916. LORNA SIMPSON. *Guarded Conditions*. 1989. 18 color Polaroid prints, 21 plastic plaques, plastic letters, overall dimensions 7'7" x 12'11" (2.3 x 3.9 m). Collection Museum of Contemporary Art, San Diego MUSEUM PURCHASE, CONTEMPORARY COLLECTORS FUND

5. LORNA SIMPSON (b. 1960)
Guarded Conditions, 1989
The Car, 1995

LORNA SIMPSON, a photographer from Brooklyn, New York, brings a feminist perspective to black identity politics. Neither the text nor the images of *Guarded Conditions* (fig. 916) succeed in portraying the woman who, back turned to the viewer, is identified simply by gender, color, and vulnerability to "sex attacks/skin attacks." The gap between the "real" woman and the figure in the picture is important inasmuch as it reveals the arbitrary power of the code of representation routinely applied to black women. Just as it dictates her appearance, so it assigns her a subordinate rank in society.

While Simpson continues to combine text and photographs in her recent work, its focus has shifted from racial stereotypes toward the isolation of modern urban life. The street depicted in *The Car* (color plate 480) appears uninhabited, yet according to the narrative accompanying the montage, there is someone in the car waiting for a sexual tryst. Much as in *Guarded Condition*, the individual is situated in a setting that imposes anonymity. Whatever he or she attains in the way of meaningful, subjective experience has to be, as Eleanor Heartney puts it, "teased out of an opaque and unresponsive world."

Color plate 480. LORNA SIMPSON. *The Car*. 1995. Serigraph on 12 felt panels with felt text panel, each panel 34 x 26" (86.4 x 66 cm), overall 8'6" x 8'8" (2.6 x 2.6 m). Courtesy Sean Kelly Gallery, New York

Color plate 481. CARRIE MAE WEEMS. *House/Field/Yard/Kitchen*. Detail from *From Here I Saw What Happened and I Cried*. 1995. From an installation of 34 monochrome c-prints with sand-blasted text on glass, each 26¾ x 22¾" (67.9 x 57.8 cm). Courtesy Pilkington Olsoff Fine Arts, New York

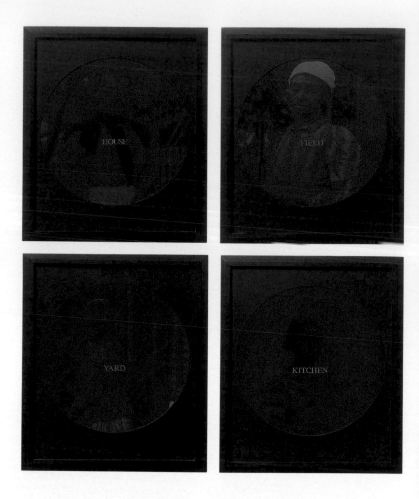

6. CARRIE MAE WEEMS (b. 1958)
Untitled (Blessing and Healing Oil), 1992
House/Field/Yard/Kitchen, 1995

CARRIE MAE WEEMS's approach to photography is informed by her study of folklore at the University of California, Berkeley. An early show of pictures of her family included an audiotape in which she narrates a history of her ancestors' migration from Mississippi to Oregon. To produce her best-known work, she traveled to the Sea Islands off the coast of South Carolina and Georgia, where blacks, living in seclusion, have preserved many aspects of their West African heritage. Not only did Weems photograph the domestic architecture, she also collected popular customs and sayings and reflected upon her own search for communal history. The outcome was an installation in which documentary photographs and texts (fig. 917) appeared alongside "souvenir plates" imprinted with Weems's sometimes ironic, always moving commentary on the experience of "looking for Africa." While she discovered a rich, vital heritage, she also had to come to terms with the reality of the African diaspora and her distance from the past. The souvenir plates, like Durham's "fake artifacts," signal a hard-won acceptance of the impossibility of a return to a bygone, "true" identity. One plate read, "WENT LOOKING FOR AFRICA and found uncombed heads, acrylic nails & Afrocentric attitudes Africans find laughable."

The work Weems made to complement an exhibition of photographs of slaves has been described as a "meditation on the violence that is exerted through point of view." The words "HOUSE," "FIELD," "YARD," "KITCHEN," that she superimposed upon enlarged reproductions of four pictures of women refer, for example, to the work and place they were forced to accept (color plate 481). The Los Angeles Getty Museum's decision to invite Weems to respond to its exhibition is indicative of the dramatic changes in exhibition policy sweeping through U.S. institutions during the nineties.

917. CARRIE MAE WEEMS. *Untitled (Blessing and Healing Oil)*, from the Sea Islands series. 1992. Two gelatin-silver prints, each 20 x 20" (50.8 x 50.8 cm); overall size 20 x 40" (50.8 x 101.6 cm). Courtesy Pilkington Olsoff Fine Arts, New York

Color plate 482. FAITH RINGGOLD. *Bitternest Part II: Harlem Renaissance Party.* 1988. Acrylic on canvas with tie-dyed and pieced fabric border, 8 x 7' (2.4 x 2.1 m). Collection National Museum of Art, Smithsonian Institution, Washington, D.C. © 1988 FAITH RINGGOLD. COURTESY ACA GALLERIES, NEW YORK/MUNICH

7. FAITH RINGGOLD (b. 1930)
Bitternest Part II:
Harlem Renaissance Party, 1988

FAITH RINGGOLD could have been considered in an earlier chapter of this book, but her placement in the Epilogue is appropriate because she played a key role in shaping the identity politics of the 1990s. Politically active in the civil rights movement, Ringgold went on to help organize the first protests against racism and sexism in the American art world. She gained notoriety as one of the Judson Three arrested for "desecrating" the U.S. flag at an exhibition about white oppression of minorities and was a founding member of the Women Students for Black Art Liberation. Painting has been important to Ringgold throughout her career, but "story quilts," based in part on an African-American tradition passed through her family, have become her most popular work. With titles such as *Who's Afraid of Aunt Jemima?* and *The Wake and Resurrection of the*

Bicentennial Negro, these quilts combine spirited social commentary with a lyrical combination of painting and sewing. *Bitternest Part II: Harlem Renaissance Party* (color plate 482) developed from a performance of the same name in which Ringgold cast herself as CeCe, a young woman living in Harlem during the 1920s. To make up for her inability to speak or hear, CeCe gives parties, like the one shown on the quilt, in which she dons elaborate costumes in order to communicate with her guests, Langston Hughes, Zora Neale Hurston, and other leaders of the Harlem Renaissance. Ringgold's quilts are pioneering examples of the hybrid art, merging performance with "high" and "low" media, that has become central to a number of younger artists, in particular Mike Kelley and Ann Hamilton.

Color plates 483a and
(below) 483b.
ANN HAMILTON. 'a round.'
1993. Installation and detail
at the POWER PLANT,
Toronto, Canada. Wrestling
dummies, canvas floor,
circular hand knitting.
Courtesy Sean Kelly Gallery,
New York

8. ANN HAMILTON (b. 1956)
'a round,' 1993

Though **ANN HAMILTON** studied sculpture at the Yale
School of Art, her undergraduate training in textile design
remained essential to her development. Over the past
decade, she has orchestrated a series of temporary installations in which a site is transformed through the addition of
a layer of material in combination with such ritualized
activity as weaving or sewing. Hundreds, if not thousands,
of objects may be used, and their modification almost
always requires intensive, small-scale labor. *Indigo Blue* featured 48,000 men's shirts; in *privations and excesses*,
750,000 pennies were laid in a bed of honey. The work,
carried out with the help of volunteers, blurs many of the
distinctions organizing the experience of modernist art. It
is collective and individual, focused on process as much as
on form.

Joan Simon observes that an installation by Hamilton
"invites viewers to be witnesses at temporal crossroads." At
MIT's List Visual Arts Center, this involved a stack of
obsolete patent books that carried a potent message about
the ephemeral nature of the technology to which the institute is devoted. *'a round'* (color plate 483a) was more subtle in its evocation of the history of its site, a former storage
area for the docks of Toronto, Canada. Wrestling dummies
piled high along the walls recalled the goods once stored
there, yet also suggested the goods' absorption into human
life. Memory of a system of exchange linking objects to
organic life was made yet more concrete by the presence of
a woman knitting with a seemingly endless length of yarn
(color plate 483b). While *'a round'* and similar installations
should not be restricted to a discussion of gender issues,
they have an important feminist subtext insofar as they elevate women's crafts and associate culture with metaphors
of gestation and birth.

Color plate 484. NICOLE EISENMAN. *Shipwreck*. Installation view at Jack Tilton Gallery, New York, 1996. Mixed media, dimensions variable. Courtesy Jack Tilton Gallery, New York

9. NICOLE EISENMAN (b. 1965)
Shipwreck, 1996

If artists of the 1990s agree about any one thing, it is that gender is produced through the performance of coded activity. The assumption goes hand in hand with the tacit understanding that the performance need not follow a rigid stereotype like the "mannish" lesbian of Romaine Brooks's generation. In the paintings and drawings of **NICOLE EISENMAN**, a graduate of the Rhode Island School of Design and "the ultimate bad girl," explicit references to same-sex desire appear alongside figures that subvert virtually all norms of male and female behavior. Eisenman has been compared to the New Image artist Jonathan Borofsky because of the ease with which she moves between private and public experience and her humorous, pluralistic approach to art making. *Shipwreck* (color plate 484), which makes a raucous sexual

allusion to the "little woman in the boat," spills off the wall onto the floor and includes a mural, an easel painting, a floor painting of drowning passengers, numerous drawings, three-dimensional wreckage, and flotsam and jetsam too numerous to list. Eisenman herself is to be seen at the center in the guise of a pirate in black eye patch, a figure now associated with both homosexuality and male impersonation by women. The role is apt for an uncloseted lesbian artist, who builds her work through unrestrained borrowing from movies, television, comics, and high art from Caravaggio forward. But challenging the art-historical canon is not Eisenman's main concern. Her work is based on the assumption that contemporary art, like the contemporary self, comes into being through knowing play with codes of meaning.

Color plate 485. YASUMASA MORIMURA. *Self-Portrait (Actress)/White Marilyn.* 1996. Ilfochrome/acrylic sheet, 37¼ x 47¼" (94.6 x 120 cm). Courtesy Luhring Augustine Gallery, New York

10. YASUMASA MORIMURA (b. 1951)
Self-Portrait (Actress)/White Marilyn, 1996

In 1985 the Japanese artist **YASUMASA MORIMURA** made a self-portrait in which he impersonated Van Gogh, complete with pipe and bandaged ear. Since then he has used costumes, makeup, and camera and computer technology to simulate dozens of masterworks from the Eastern and Western traditions. Invariably he is the photographer, set designer, costume manager, makeup artist, and actor or actress, as need be. In *Nine Faces,* he impersonated all the doctors in Rembrandt's *Anatomy Lesson of Dr. Tulp;* in another photograph from the Daughter of Art History series he impersonated both Manet's Olympia and her black maid. His most recent project, The Sickness Unto Beauty— Self-Portrait as Actress, involved the replication of key scenes from the *Seven Year Itch* (color plate 485) and other film classics of Hollywood, Europe, and Japan.

Debate over the significance of Morimura's appropriations has focused on the notion of a "third space" where issues of gender and nationality can be explored. His adaptation of female roles, for instance, has been interpreted as a response to the Western tendency to view Asia "as feminine, as Butterfly." To other critics, Morimura's art represents a welcome unsettling of restrictive codes of behavior, a procedure that has numerous parallels in contemporary art. But Norman Bryson contends that the "third space" should be viewed as a space created and dominated by the international media. According to him, the central concern of Morimura's art is capitalism's role in the construction of "postnational, post-gendered, de-essentialized identity."

918. NAN GOLDIN. *Ric and Randy Embracing, Salzburg.* 1994. Cibachrome print, 30 x 40" (76.2 x 101.6 cm). Courtesy Matthew Marks Gallery, New York

Color plate 486. NAN GOLDIN. *C Putting on Her Make-up at the Second Tip, Bangkok.* 1992. Cibachrome print, 30 x 40" (76.2 x 101.6 cm). Courtesy Matthew Marks Gallery, New York

11. NAN GOLDIN (b. 1953)
Ric and Randy Embracing, Salzburg, 1994
C Putting on Her Make-Up at the Second Tip, Bangkok, 1992

Many of NAN GOLDIN's photographs look like snapshots taken on the spur of the moment because someone is doing something memorable. *Ric and Randy Embracing, Salzburg* (fig. 918) is an instance of the intimacy she has achieved in her work, which now includes several books of photographs—*The Ballad of Sexual Dependency, The Other Side,* and, in collaboration with Nobuyoshi Araki, *Tokyo Love.* Goldin studied at the School of the Museum of Fine Arts, Boston, during the 1970s, but her interest in the drama of nightlife drew her away from the mainstream into the world of drag queens. After moving to New York City, she immersed herself in the bohemian scene of the East Village and began to photograph her friends. Goldin is frank about the autobiographical impetus behind her work: "This is my party. This is my family, my history." Her effort to document the private lives of a shifting set of lovers, friends, and acquaintances has been compared both to Robert Frank's *Americans* and Marcel Proust's *Remembrance of Things Past.* An important element in the evocative power of Goldin's photographs is their rich color (color plate 486). In addition, Goldin often presents her work as slide shows accompanied by music that helps arouse intense emotion in the audience. Her break with standard exhibition practice in favor of multimedia installation calls to mind Carrie Mae Weems. She also resembles Weems in her attention to the narrative dimension of photography. "Each of my pictures is a story," she says, "a story without end."

12. JEFF KOONS (b. 1955)
New Hoover Quadraflex, New Hoover Convertible, New Hoover Dimension 900, New Hoover Dimension 1000, 1981–86
Puppy, 1992

After studying at the Maryland Institute, JEFF KOONS went to New York City, where he worked, first, in the membership department of The Museum of Modern Art, and then in a Wall Street brokerage house. These experiences seem to have given Koons an acute awareness of art's affinity with the consumer goods that symbolize wealth, happiness, and control in late-twentieth-century society. His presentation of vacuum cleaners in a sealed, lighted vitrine has been viewed as a Pop version of a Minimalist sculpture or Duchamp ready-made (fig. 919). But the tensions implicit in Koons's work between the commodity, the aesthetic of display, and the fiction of art's "eternal" value prevent it from functioning as mere parody. The same can be said for his stainless-steel replications of cheap, inflatable children's toys or the colossal, flower puppy he installed at an eighteenth-century palace in Germany (color plate 487). Whether casting a lifeboat in bronze or creating a wooden sculpture of a post-card featuring a couple with their German shepherds, Koons distorts the scale, medium, and function of objects that are emblems of security for the middle class. Art critics' response to Koons has been complicated by his contradictory attitude toward consumer culture. While he appears to be trying to subvert commodity fetishism, his public statements have all struck quite a different note. According to Koons, his work "tries to function in the form of religion—not as religion

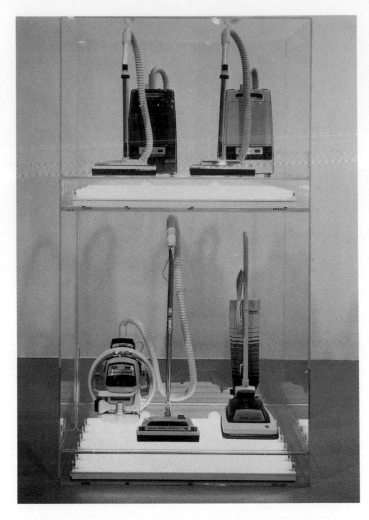

itself, but within what religion tries to do for individuals, so they can flourish in life." Perhaps the best description of Koons is the one proposed by critic Roberta Smith— "avant-gardist folk artist."

above, right: 919. JEFF KOONS. *New Hoover Quadraflex, New Hoover Convertible, New Hoover Dimension 900, New Hoover Dimension 1000.* 1981–86. Hoover vacuum cleaners, acrylic glass, and fluorescent light, 8'3" x 53½" x 28" (2.5 m x 136 cm x 71 cm). Private collection

right: Color plate 487. JEFF KOONS. *Puppy.* Installed at Schloss Arolsen, Germany, 1992. Flowering plants, steel, wood, and earth, 37'8" x 15'9" x 19'8" (11.5 x 4.8 x 6 m). Courtesy the artist

13. MIKE KELLEY (b. 1954)
Fruit of Thy Loins, 1990

Some of the most extreme "gender bending" of recent times has taken place in the performances, videos, and installations of **MIKE KELLEY**, an artist based in Los Angeles. An ongoing project started in 1987, "Half A Man," makes use of thrift-shop finds—felt banners, refinished furniture, Afghans, stuffed animals, hand-sewn and crocheted dolls—that are strongly associated with female handiwork and domesticity. But impersonation of the woman is not Kelley's goal. The "smart-aleck" adolescent is his usual role, and his concerns have to do with the treatment of childhood, on the one hand, and modernist conceptions of art and the artist, on the other. By using materials and themes traditionally relegated to the woman's sphere of activity, Kelley questions the stereotype of "hard" masculinity, which has one of its supreme symbols in the imposing modernist sculpture fabricated from industrial materials. There is, however, nothing "soft" about Kelley's use of old toys. "The stuffed animal," he observes, "is a pseudo-child, a cutified sexless being which represents the adult's perfect model of a child—a neutered pet." His installations strip away this web of loving yet manipulative pretense. The toys are arranged in ways that suggest irrepressible sexual urges or call attention to their neutered state. Several, like *Fruit of Thy Loins* (color plate 488), a stuffed rabbit giving birth, have been supplied with sexual functions. Inherently anarchic, Kelley's art has become a central source for current interest in the social coding of the body and the hitherto marginalized subject of youth culture.

Color plate 488. MIKE KELLEY. *Fruit of Thy Loins.* 1990. Stuffed animals, 39 x 21 x 12" (99.1 x 53.3 x 30.5 cm). Courtesy the artist and Metro Pictures, New York

920. ANNETTE MESSAGER.
My Vows (Mes Voeux).
1988–91. Photographs,
colored graphite on paper,
string, glass, black tape, and
pushpins over black paper or
black synthetic polymer
paint, height variable.
Museum installation
11'8¼" x 6'6¾" (3.6 x 2 m).
The Museum of Modern Art,
New York
GIFT OF THE NORTON FAMILY FOUNDATION.
© 1997 ARTISTS RIGHTS SOCIETY (ARS), NEW
YORK/ADAGP, PARIS

14. ANNETTE MESSAGER (b. 1943)
My Vows (Mes Voeux), 1988–91

The work of **ANNETTE MESSAGER**, the distinguished French artist, has generated new excitement during the past few years because of her pioneering representation of bodily experience. For Messager, as for many other women of her generation, it was crucial to find ways to overturn the authority of the female nude, an icon of Western culture and an impediment to all efforts to express "female" experience. Messager's response to the problem grew out of her contact with a concept of "feminine writing" proposed by two highly influential French theorists, Hélène Cixous and Luce Irigaray. They argued that women should write with an energy that breaks down restrictive, "masculine" structures and substitutes a multiplicity of points of view for closure. "Because we [women] are always open," writes Irigaray, "we will never travel all the way round our periphery: we have so many dimensions." At first glance, Messager's assemblage *My Vows* (fig. 920) appears abstract. Closer examination reveals that it is made from hundreds of small black-and-white photographs, hung from strings of varying length, each depicting a component of the female body: eye, ear, mouth, hand, nipple, pubic area, and so on. Though viewers can recognize these things, they cannot possibly deduce a single figure from them. The photographs are too numerous and many layered, too multifarious. They depict a female body that is capable of assuming an infinite number of shapes and identities.

Color plate 489. Matthew Barney. Production still from *Cremaster 4*. 1994. Courtesy Barbara Gladstone Gallery, New York

Color plate 490. Matthew Barney. *Cremaster 1: Goodyear Chorus*. 1995. C-print in self-lubricating plastic frame, 53¾ x 43½" (136.5 x 110.5 cm), edition of six. Courtesy Barbara Gladstone Gallery, New York

15. Matthew Barney (b. 1967)
Production still, *Cremaster 4*, 1994
Cremaster 1: Goodyear Chorus, 1995

Testing the limits of the body is the principal concern of the performances, videos, and installations of **Matthew Barney**, a young artist with a B.A. from Yale University, already acclaimed as the "mythographer of our closing millennium." He made his New York debut in 1991 with *Mile High Threshold: Flight with the Anal Sadistic Warrior*, a performance/installation piece in which he, nude except for climbing gear connected to his rectum, clambered across the ceiling of the gallery, leaving holes and stains to mark his trek. Other works, such as *Field Dressing (orifill)*, have juxtaposed the male sports hero with the beautiful bride, both roles played by Barney with much-admired skill. Since 1994, he has concentrated on the production of a series of lengthy, complex videos (color plate 489) that weave Anglo-Scot myth into a quasi-scientific sports story about the interplay of gendered identity and genital development. (Cremaster, the name of the series, refers to a muscle of the testis.) The videos are set on the Isle of Man, renowned for hybrid creatures and activities: i.e., a goat with three horns, an annual race for motorcycles with sidecars. The hero, played by Barney, is also a hybrid, at once gorgeous and perplexing (color plate 490). He has hooves and vestigial horns like a goat; combs his bright red, obviously dyed hair with meticulous care; and dresses and tap-dances like the star of a Hollywood musical extravaganza. When on view in a gallery, the Cremaster videos are accompanied by stills and props from the production, arranged in a system that can be compared to Weems's and Durham's adaptations of the display practices of ethnographic museums.

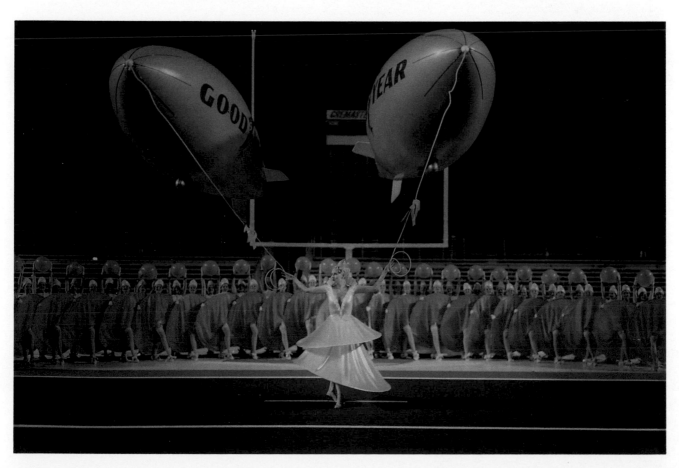

16. KIKI SMITH (b. 1954)
Untitled, 1990
Mother, 1992–93

In an interview, **KIKI SMITH** once asked, "What body parts could you lose and still think of yourself as yourself?" The question is basic to the sculptures and installations that brought her into international prominence during the early 1990s. An installation of 1990 at The Museum of Modern Art in New York confronted visitors with a dozen mirrored-glass jars labeled with the names of bodily fluids, from "saliva" to "blood." A year later, Smith exhibited two life-size wax figures, one male, the other female, hanging limply from metal supports (fig. 921). The sagging breasts of the woman seem to ooze milk; a rivulet of semen appears to run down the leg of her companion. Placed side by side, the statues awake deep-seated fears about the difficulty of holding the body together and presenting it to society as a proper, controlled self. The time Smith spent training in the Emergency Medical Service is evident in her depiction of the flesh—the soft, vulnerable insides of the body where psychological as well as physical trauma get lodged. Smith's *Virgin Mary* has a body of subcutaneous tissue; a length of papier-mâché intestines spills from the gut of another statue of the Madonna. Without skin, often fragmented, Smith's work exposes "the visceral truth of our emotional roots." The shattered glass of *Mother* (color plate 491) points to a separation of child and mother that leaves each mutilated and still far from independent.

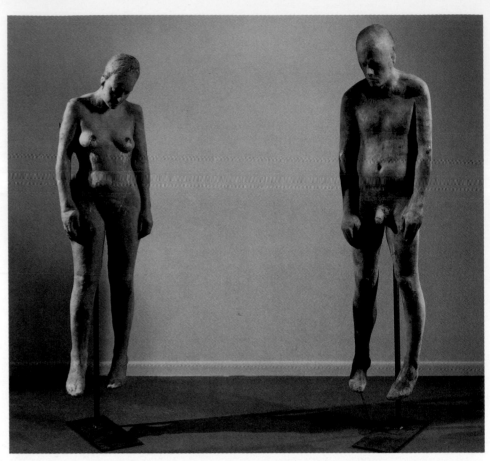

921. KIKI SMITH. *Untitled*. 1990. Beeswax and microcrystalline wax figures on metal stand. Installed height of female figure: 6'1½" (1.9 m). Installed height of male figure: 6'4¹⁵⁄₁₆" (2 m). Female figure: 64½ x 17⅜ x 15¼" (163.8 x 44.1 x 38.7 cm). Male figure: 69⁷⁄₁₆ x 20½ x 17" (176.4 x 52.1 x 43.2 cm). Stand for female figure: 61⅞ x 13⅞ x 19¹¹⁄₁₆" (157.2 x 35.2 x 50 cm). Stand for male figure: 61¾ x 14⁹⁄₁₆ x 19¹¹⁄₁₆" (156.8 x 37 x 50 cm).
Collection Whitney Museum of American Art, New York
PURCHASE WITH FUNDS FROM THE PAINTING AND SCULPTURE COMMITTEE

Color plate 491. KIKI SMITH. *Mother*. 1992–93. Cast glass and steel, dimensions variable.
High Museum of Art, Atlanta
PURCHASE FROM THE ELSON COLLECTION OF CONTEMPORARY GLASS

922. DAVID WOJNAROWICZ. *Untitled*. 1991 (printed 1993). Gelatin-silver print, 28½ x 28½" (72.4 x 72.4 cm), edition of 10. Courtesy Pilkington Olsoff Fine Arts, New York, and the Estate of David Wojnarowicz

17. DAVID WOJNAROWICZ (1954–92)
Untitled, 1991

The AIDS crisis has been a major factor in the growth of interest in the structure and meaning of the body, and references to the disease have become common. But the art world's response to the disease in the late 1980s and early 1990s took place amidst an impassioned debate that eventually included the NEA, a host of Washington politicians, a Roman Catholic cardinal, and several right-wing organizations. **DAVID WOJNAROWICZ**, a homosexual artist diagnosed as HIV-positive in 1987, was at the forefront of the scene.

He presented some of the earliest art that went beyond mourning and commemoration to make angry accusations against government and medical authorities for their neglect of AIDS victims. A 1989 show of photomontages and collages treated homophobia as the underlying cause of the slow pace with which help was being organized at either the local or national level. Later that same year, Wojnarowicz wrote an essay naming names, including John Cardinal O'Connor and Senator Jesse Helms, that made headlines and led to withdrawal of government support for an exhibition devoted to AIDS.

Wojnarowicz defies categorization; at one time or another, he was a performance artist, filmmaker, painter, photographer, and writer. What he called his "post-diagnostic" work became an extraordinary record of the experience of dying from a socially unacceptable disease. It consists of photomontages, journals, and an autobiography in comic-book style with drawings by James Romberger. *Untitled* (fig. 922) shows Wojnarowicz's face half-buried in earth. An appropriate caption for it comes from his book, *Memories That Smell Like Gasoline*: "I am disappearing. I am disappearing but not fast enough."

Color plate 492. FELIX GONZALEZ-TORRES. *Untitled (Portrait of Ross in L.A.)*. 1991. Multicolored candies, individually wrapped in cellophane, endless supply. Ideal weight, 175 lbs., dimensions vary with installation. Collection Howard and Donna Stone. Courtesy Andrea Rosen Gallery

18. FELIX GONZALEZ-TORRES
(1957–1996)
Untitled (Portrait of Ross in L.A.), 1991
Untitled (North), 1993

Though Cuban-born **FELIX GONZALEZ-TORRES** was an activist on gay and AIDS rights, his art differs from that of Wojnarowicz in spirit as well as form. There seems to be almost no anger, and the political commentary is oblique rather than explicit. In 1991, after his lover, Ross, died of AIDS, Gonzalez-Torres turned a photograph of his empty double bed into a poster, displayed, without any explanatory text, on twenty-four billboards throughout New York City. Like much of his art, the billboard raises questions about the public implications of private acts and leaves them deliberately unanswered. The billboard can but need not necessarily be seen as a defiant response to a 1986 Supreme Court ruling that moved sexual acts out of the zone of privacy, in which the citizen is completely free. In another memorial (color plate 492), audiences encountered an enormous "spill" of candy, each piece wrapped in gleaming paper, that equaled Ross's body weight. If anyone hesitated to take a candy, guards at the museum encouraged them to indulge, since a supply to make up the losses was on hand. The simplicity of the candy spill is deceptive, for it deftly engaged viewers in a process of sensation and thought that might concern any or all of the following: the difficulty of sorting out private and public experience; the ephemeral nature of pleasure; the tension between abundance and deprivation in our lives; the terrible losses caused by AIDS; Minimalist and Conceptual art. In 1992, Gonzalez-Torres began to use strings of low-watt incandescent bulbs to make sculpture (fig. 923). While each piece is unique, none has a fixed, permanent arrangement. "Play with it," Gonzalez-Torres told collectors. "Have fun. Give yourself that freedom."

923. FELIX GONZALEZ-TORRES. *Untitled (North)*. 1993. 15-watt light bulbs, porcelain light sockets, extension cords. Overall dimensions vary with installation. 12 parts: each 22'6" (6.9 m) long with 20' (6.1 m) extra cord. View of installation at the Milwaukee Art Museum. Courtesy Andrea Rosen Gallery, New York

Color plate 493. ANDRES SERRANO. *Piss Christ.* 1987. Cibachrome, silicone, Plexiglas, wood frame, 65 x 45⅛ x ¾" (165 x 115 x 1.9 cm) framed; 60 x 40" (152.4 x 101.6 cm) unframed. Edition of four. Courtesy Paula Cooper Gallery, New York

19. ANDRES SERRANO (b. 1950)
Piss Christ, 1987
The Morgue
(Knifed to Death I and II), 1992

If any single image could be said to be an icon of the 1989 battle over government support of the arts, it is **ANDRES SERRANO**'s *Piss Christ*, a large photograph of a wood-and-plastic crucifix submerged in urine (color plate 493). To the American Family Association and other champions of public morality, it was clearly blasphemous, while to those fighting for the artist's right to freedom of expression, *Piss Christ* was a "darkly beautiful" image that explores Catholicism's obsession with the body of Christ. The conflict might never have arisen had Serrano chosen a different title. *Piss Christ*, as Lucy Lippard points out, transforms the photograph "into a sign of rebellion or an object of disgust by changing the context in which it is seen." Though Serrano's approach to Catholic iconography may seem to be a perverse intrusion into Postmodern culture, it grows out of a Spanish aesthetic, formed during the Baroque era, that became important to the Surrealist artist Salvador Dalí and the filmmaker Luis Buñuel. Like them, Serrano produces images in which polar opposites—the sacred and the profane, violence and beauty—converge. He has photographed the homeless, the Ku Klux Klan, and cadavers in the New York City morgue. The depiction of the signs of death and forensic science in the diptych *The Morgue* (fig. 924) is made all the more haunting by Serrano's allusion to the meeting of human and divine hands in Michelangelo's *Creation of Man*.

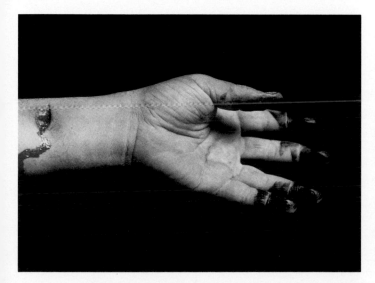

924. ANDRES SERRANO. *The Morgue (Knifed to Death I and II).* 1992. Diptych. Cibachrome, silicone, and Plexiglas, each panel 60 x 49½" (152.4 x 125.7 cm), edition of three. Courtesy Paula Cooper Gallery, New York

Color plate 494. TIM ROLLINS and K.O.S. *The Temptation of Saint Anthony—The Trinity.* 1989–90. Blood on paper and linen, 68" x 16' (172.7 cm x 4.9 m). Courtesy Mary Boone Gallery, New York

20. TIM ROLLINS (b. 1955) and K.O.S.
The Temptation of Saint Anthony —The Trinity, 1989–90

In 1980, **TIM ROLLINS** helped found Group Material, an organization for socially committed artists that had a part in shaping the careers of Andres Serrano and Felix Gonzalez-Torres. But it was his collaboration with high-school students from the South Bronx that catapulted Rollins into international fame during the late 1980s. Influenced by Paulo Freire's *Pedagogy for Liberation,* he started a class in which he and his mostly black and Hispanic students worked to develop a common language for the analysis of society and personal empowerment. The first project, which used George Orwell's *Animal Farm* as inspiration for caricatures of contemporary political figures, proved so successful it has been the model for Rollins's subsequent collaborations with groups of students, known after 1984 as K.O.S., Kids of

Survival. Each project begins with a classic text that is read in light of current issues, then "taken apart" yet again when its pages are incorporated into the art the students make under Rollins's guidance. While their interpretations of texts are often filtered through comic books, they also employ modernist abstraction. *The Temptation of Saint Anthony—The Trinity* (color plate 494) consists of three large paintings, each covered with pages from a novel by Gustave Flaubert that recounts Saint Anthony's struggles in a world where Satan is omnipresent and sexual desire produces monstrous afflictions of the flesh. The strongest link between the story and the triptych is the warm red color. Made from blood, it carries powerful associations with AIDS, linking the late-twentieth-century scourge to Saint Anthony's suffering.

925. SALLY MANN. *The Two Virginias, #4.* 1991. Gelatin-silver print. Courtesy Edwynn Houk Gallery, New York
© SALLY MANN

21. SALLY MANN (b. 1951)
The Two Virginias #4, 1991

The world of **SALLY MANN** seems light-years away from the social and political conflicts that came in the wake of gay liberation, AIDS, and minority rights advocacy. She lives in the rural South and, from 1984 to 1995, took photographs of her three children as they went about daily activities. She did this with the goal of portraying their "very complicated and rich lives." Yet publication of a selection of these photographs in the book *Immediate Family* caused a public uproar and led to allegations of child pornography and even abuse. The controversy focused on the implications of several photographs in which the children appear nude. Self-appointed spokespersons for a "code of moral decency" found the pictures suspect and called into question virtually all representations of undressed children. Significantly, there has been no similar dispute in Europe, where attitudes toward the body are more tolerant and admiration for

Mann's photography is widespread. It combines the snapshot intimacy of Nan Goldin's work with the tradition of narrative exemplified by the work of Julia Margaret Cameron. *The Two Virginias #4* (fig. 925), like many of Mann's pictures, unites two points of focus: the gleaming white hair and gnarled hands of the grandmother in the foreground and, in the background, the luminous little girl. The shift in perspective is noteworthy. It allows one image to capture both the girl's inner, private experience and her relationship to a family of towering adult figures. The play of similarities and differences between grandmother and child also suggests the passage of time and the fragility and beauty of childhood. "Tender, vertiginous, and scary" are the words Luc Sante uses to describe Mann's photographs of family life. Her most recent work concerns her marriage and the idyllic landscape where she grew up.

Color plate 495.
ROBERT GOBER.
Site-specific installation
at the Dia Center for
the Arts, New York,
September 24, 1992–
June 20, 1993.
Courtesy Dia Center for
the Arts, New York

below, right: 926.
ROBERT GOBER. *X Crib*.
1987. Enamel paint
and wood, 44 x 50½
x 33¼" (111.8 x
128.3 x 84.5 cm).
Private collection.
Courtesy Paula Cooper
Gallery, New York

22. ROBERT GOBER (b. 1954)
X Crib, 1987
Site-Specific Installation, Dia Center for the
Arts, New York, 1992–93

New York–based artist **ROBERT GOBER** is best known for
sculptures and installations that bring private, domestic experience into disquieting relationship with the sociopolitical
coding of identity. He started his career in 1978 with small-
scale replicas of nineteenth-century New England houses in
which he alluded to slavery and challenged the stereotype of
the heroic modern sculptor who makes grand, public work.
In this last regard, Gober resembles Mike Kelley. But over
the years, Gober has distinguished himself from Kelley and
many other contemporary artists by identifying with the
craftsperson, often the household carpenter, and making
meticulously crafted objects. These are usually either replicas
or slightly distorted versions of ordinary domestic fittings
like sinks, urinals, and beds. While Gober makes reference
to Duchamp, his work is also informed by feminist practice,
especially Cindy Sherman's use of simulated settings to
expose the way gender gets instilled inside the individual and
is made to seem "natural." *X Crib* (fig. 926), for instance,
was originally displayed along with other pieces of deformed
nursery furniture, offering ironic commentary on the apparatus that at once protects infants and gives them their first
lessons in the social coding of the body. To date, Gober's
most complex statement on the order that conditions
human experience has been an installation (color plate 495)
in which elements of a prison and a bathroom were integrated into a painted simulacrum of a woodland paradise. It
recalled the myth of Arcadia, yet it also suggested that this
ideal had been twisted into a device for validating a repressive puritanical system that uses the term "unnatural" to
condemn homosexuality and free self-expression.

23. ANNETTE LEMIEUX
(b. 1957)
Search, 1994

ANNETTE LEMIEUX has been called an "artist without a style" because she rejected the Neo-Expressionism of the early 1980s so totally that she erases all signs of her own identity and experience from her work. To her, art is a means for carrying out what the French philosopher Michel Foucault terms the "real political task": criticizing social institutions "in such a manner that the political violence which has always exercised itself obscurely through them will be unmasked." Visitors at her installation *Search* (color plates 496a, 496b) found themselves in a darkened space, crisscrossed with beams of light issuing from a moving phalanx of sinister, miniature vehicles, made by mounting surplus army helmets on wheels equipped with small motors. The installation was not meant to illustrate a particular event from history or a specific issue in contemporary society, such as police brutality. It dealt instead with the dehumanizing effects of the state's organization and deployment of power. The grotesque toy vehicles, surrogates for mechanized intelligence, patrolled the room without a will of their own or capacity to respond to their audience. Other installations have focused on war, starvation, and history's hold upon the present. A former studio assistant of David Salle, Lemieux follows his precedent by incorporating appropriated objects and images into her work.

Color plates 496a and 496b. ANNETTE LEMIEUX. *Search*. 1994 installation and detail. 18 helmets with headlamps, motors, and wheels, 24' x 11'6" (7.3 x 3.5 m). Courtesy McKee Gallery, New York

927. KATHARINA FRITSCH. *Rat-King (Rattenkönig)*. 1993 installation at Dia Center for the Arts, New York. Polyester resin, approximately 9' (2.7 m) high by 42½' (13 m) in diameter. Courtesy Dia Center for the Arts, New York

24. KATHARINA FRITSCH (b. 1956)
Rat-King (Rattenkönig), 1993

Like Lemieux and, to a greater extent, Robert Gober and Mike Kelley, the German artist **KATHARINA FRITSCH** makes installations that straddle the boundary between the adult social order and the world set apart as belonging to the child or the "childish." But in her case, the imagery generally comes from fairy tales, folklore, and collective phobias. An installation at the Carnegie Museum, Pittsburgh, Pennsylvania, centered on a ghostlike form and a pool of blood, while another consisted of a bright green, life-size replica of a stuffed elephant in a natural history museum. Inversions of scale, either from the miniature to the monumental or the converse, are a recurrent feature of Fritsch's work. For her, the object that is an exception to the rules and designated as a freak is an invaluable point of entry into the workings of collective identity. The sixteen larger-than-human figures of *Rat-King* (fig. 927) are arranged in a perfect circle. While each "king" appears to be poised for movement out into the world, he is bound to the pack by a knot binding all their tails together. Charles Wright views the installation as a macabre portrayal of the "familial hell" from which there is no exit. Fritsch based her portrayal of the animals on Edgar Allan Poe's notion of "indefinite definiteness." The clarity and precision with which the forms are rendered serve to make the fantastic rats appear "natural." When making the piece in New York, Fritsch found inspiration in a passage of Giorgio de Chirico's writing that goes, in part, "you will find again and again in New York . . . forgotten memories which return there as they return in the hours of semi-sleep."

Color plate 497. MONA HATOUM. *Light Sentence*. 1992. Wire-mesh lockers and motorized light bulb,
6'6" x 6'1" x 16'1" (2 x 1.9 x 4.9 m); installation space variable, minimum 15'11" x 25'11" (4.9 x 7.9 m); maximum
19'2" x 28'10" (5.8 x 8.8 m). Collection National Gallery of Canada, Ottawa

25. MONA HATOUM (b. 1952)
Light Sentence, 1992

The art of **MONA HATOUM**, a Palestinian now residing in London, is apt to make audiences uneasy because it reveals the uncertainties and dangers lurking in commonplace reality. This, for Hatoum, rarely involves consumer commodities or media imagery. Her work concerns the boundaries and fittings of the space in which individual identity is structured. Her contribution to a show devoted to the experience of colonialism was a gate charged with potentially fatal energy. The political implications of *Light Sentence* (color plate 497) are no less disturbing. It consists of thirty-six metal cages, of a type normally used to store personal items in locker rooms, arranged to form a prison-like complex. It is illuminated by a single light bulb, attached to a motorized cord, which moves slowly up and down, creating dramatic patterns of light and shadow. While the installation arouses deep feelings of dread, Hatoum thinks of it as also prompting an opposite reaction, "of poetry and beauty." "I want," she says, "to expose the contradictions which are always part of life."

Her investigation of the body envelope, the space that the individual regards as his or her own, acquired a feminist dimension in a recent installation, *Recollection*. Using thread and hundreds of balls woven from strands of her own hair, she demarcated a grid on the ceiling and floor of a room in Boston's Institute of Contemporary Art. Hatoum's adaptation of Minimalist forms like the grid and cube has parallels in the work of Eva Hesse, Roni Horn, and Rachel Whiteread who have all adapted Minimalist structures to articulate feminist or anti-authoritarian perspectives.

Color plate 498. GUILLERMO KUITCA. *Untitled*. 1995. Chalk and acrylic on canvas, 71" x 7'8" (180 cm x 2.3 m). Private collection. Courtesy Sperone Westwater, New York

26. GUILLERMO KUITCA (b. 1961)
Untitled, 1995

The Argentine painter **GUILLERMO KUITCA** approaches questions of identity and place from the perspective of someone always "other" and "elsewhere." While his detachment is no doubt bound up with his mixed ethnic heritage (his grandparents were Ukrainian Jews), it also reflects his experiences working in the theater and his immersion in the surrealistic writings of Jorge Luis Borges. The combination led Kuitca to conceive of reality as heterogeneous rather than single, and as requiring a distanced, ironic point of view if its arbitrary workings are to be apprehended. During the 1980s, he based a series of paintings of his parents' small apartment on the axonometric diagram that in architectural practice serves to clarify the layout of a building's interior spaces. Kuitca, however, turned the system against itself,

using it to show that he spent his childhood in a maze with secret recesses and cryptic symbols. Kuitca's work has become less autobiographical over the years and is now focused on charting "the volatile places where the exceedingly private and intensely public nature of human existence clash and converge." *Untitled* (color plate 498) is from a series, Puro Teatro, based on the seating plan of theaters. The brilliant blue color invokes psychological excitement of the sort associated with drama, but the absence of figures and the vertical tier of seats, each row with an implied price tag, introduce contradictory elements. They mark the place with signs of an anonymous system of class and suggest as well that seemingly "private" excitement cannot be separated from participation in mass spectacle.

Color plate 499. RIRKRIT TIRAVANIJA. *Untitled (for M.B.).* 1995. Plaster and enamel paint, dimensions variable.
Courtesy Gavin Brown's Enterprise, New York

27. RIRKRIT TIRAVANIJA (b. 1961)
Untitled (for M.B.), 1995

RIRKRIT TIRAVANIJA is a leading proponent of a "hybrid" art that breaks down the boundaries among different media and cultural traditions. His own background is diverse. He was born in Buenos Aires, grew up in Bangkok, migrated to Canada in his teens, and studied at four art schools, finishing up in New York at the Whitney Independent Study Program. In addition to this fund of personal experience as an "outsider," his art draws upon current theory of cross-cultural exchange and various types of Installation and Performance art, from Allan Kaprow's Happenings and Fluxus to Judy Pfaff and Mike Kelley. "Parallel spaces" is the way Tiravanija describes his installations, which usually entail the use of found materials to improvise a living arrangement within a gallery or museum setting. Thus far, the "parallel spaces" have included a café, a recreational lounge with gaming tables and a refrigerator, several dining rooms, a playhouse for children, and a reconstruction of his entire New York apartment.

The point for Tiravanija is only in part to offer a demonstration of the way the individual settles into a foreign place and marks out a space with signs of identity. The installations are always temporary and serve as settings for social activity. Color plate 499 reproduces the table set with woks filled with Thai food that Tiravanija cooked and offered free to all guests at his installation *Untitled (for M.B.).* In sharp contrast to Kuitca, Tiravanija treats the intersection between private and public experience as the site where communal ties can be built and celebrated.

Color plate 500. GABRIEL OROZCO. *The DS*. 1993. Sutured car, leather, coated fabric, 55" x 15'9" x 45" (140 cm x 4.8 m x 114 cm). Courtesy Marian Goodman Gallery, New York

28. GABRIEL OROZCO (b. 1962)
The DS, 1993

The "hybrid" art of the Mexican artist **GABRIEL OROZCO** has taken many forms during the last decade but almost invariably centers on a gesture that redefines a space and leaves a mark of its passage. The gestures are often modest: small balls of wet sand left on a rocky outcropping of Mexico's Pacific coast or, in another piece reminiscent of the work of Richard Long, photographs of the tracks left by his bicycle wheels as they passed through puddles on a New York street. But Orozco has also made monumental sculpture. In 1992, he rolled a ball of black Plasticine, equal in weight to his own body, through the streets of New York, allowing it to shift shape and pick up pieces of debris. When exhibited at the New Museum, the Plasticine ball (entitled *The Yielding Stone*) became a self-portrait that alluded to Orozco's experience of adaptation and change. Similarly, *The DS* (color plate 500) mixes personal history with ideas

of movement; only in this case, the history reaches back to Orozco's adolescence and his feelings when he drove a Citroën DS, the "dream machine" that symbolized French elegance and power during the 1950s and early 1960s. Orozco did not replicate the vintage sedan but, instead, purchased one, then gave it a new, hybrid identity by cutting it in three, removing the center piece, suturing the other two portions together, and restoring the paint and upholstery. Visitors were encouraged to sit in *The DS* at the gallery in Paris where it was shown. Most found the experience disorienting because of the contradictions they encountered: The car seems brand-new, yet is a phantom from the past; it looks fast, yet cannot move; and it has balance, but no center. To Francesco Bonami, *The DS* is above all an "outsider's" comment on the mixture of myth, nationalism, and technology that defined the modernist era.

928. RACHEL WHITEREAD. *House*. 1993, destroyed 1994. Concrete.
Commissioned by Artangel
PHOTO JOHN DAVIES, CARDIFF, WALES, UNITED KINGDOM

29. RACHEL WHITEREAD (b. 1963)
House, 1993, destroyed 1994

The English sculptor **RACHEL WHITEREAD** shares Orozco's interest in making objects that explore boundary conditions. As a student, she started to make casts that inverted normal sculptural procedure. Instead of reproducing the object, she reproduced its "negative space": the concave space inside a spoon, for instance, or the rectangle of space under a footed bathtub. By the late 1980s, she had started to apply her technique to domestic interiors and was producing "negative" casts of numerous pieces of furniture and eventually of a closet and an entire room. The pale color of the work and its materials, usually plaster, wax, or polyester resins, give them a general resemblance to George Segal's ghostly environments. But there are no human figures in Whiteread's sculptures, only impressions of the domestic framework of people's lives. The duality of presence and absence conveys a feeling close to what Freud referred to as the "uncanny," the haunting sense that the familiar and unfamiliar are confused. Whiteread's concern for the changes in the structure of neighborhoods led her to undertake the casting of the interior of an East London terrace house scheduled for imminent demolition. In the judgment of most art critics, *House* (fig. 928) was a masterpiece. It defied the laws of space and time by allowing viewers to stand outside a building on a public thoroughfare yet see the private spaces of family life. And, without being didactic, it challenged the policy of substituting high-rise towers for two-story dwellings, the traditional form of housing in Great Britain. Nonetheless, *House* was viewed as an eyesore and was torn down in 1994 amidst a media uproar on both sides of the Atlantic. Her most recent project is a monument to commemorate Viennese Jews killed in the Holocaust.

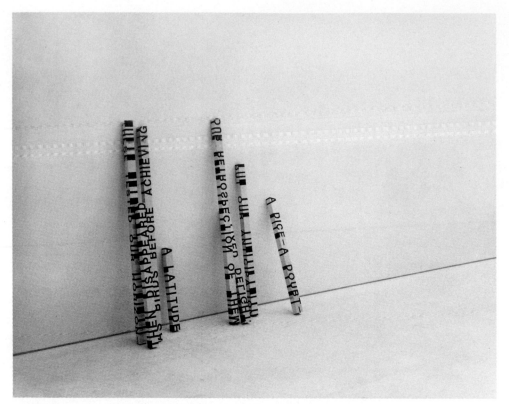

929. RONI HORN. *When Dickinson Shut Her Eyes—For Felix; No. 886.* 1993. Plastic and aluminum, dimensions variable. Courtesy Mary Boone Gallery, New York

Color plate 501. RONI HORN. *Cobbled Lead(s).* 1982–83. Installation for the courtyard of the Glyptothek, Munich, Germany. A 225-square-foot area of stone was removed from the cobblestone groundwork of the courtyard, and substituted by lead analogues, cast after the stones they replace. 600 lead pavers (20 short tons, 22 metric tons), triangular area 21 x 23 x 36' (6.4 x 7 x 11 m). Courtesy the artist and Matthew Marks Gallery, New York

30. RONI HORN (b. 1955)
Cobbled Lead(s), 1982–83
When Dickinson Shut Her Eyes— For Felix; No. 886, 1993

Though **RONI HORN** gained international recognition soon after she graduated from the Yale School of Art in 1978, her work is best discussed in the context of contemporary explorations of the habits of mind defining an individual's "inner geography." To produce *Cobbled Lead(s)* (color plate 501), Horn substituted minutely detailed lead analogues for the cobblestones of the courtyard of the Glyptothek, a museum in Munich. The shift in material from stone to metal was not dramatic, and the paving of a courtyard is, at any rate, an item most people take in through the peripheral vision reserved for the familiar, basic elements of their world. But it is precisely this type of ordinary vision that *Cobbled Lead(s)*, like many of Horn's other installations, seeks to isolate and restructure. Through adroit use of Minimalist and Conceptual strategies, her work focuses attention on the assumptions that transform the vast, indomitable universe into a safe, predictable habitat. The texts accompanying it are apt to have a sardonic wit and to involve a sharply focused moment of matter-of-fact perception. In a 1996 installation, for instance, Horn set a parallelogram of aluminum bars on the floor; black letters in the bars read "49 Miles," the distance between the earth's surface and the outer limit of its twilight atmosphere. Horn also employed aluminum bars in a piece (fig. 929) paying homage to Emily Dickinson, a poet who helped inspire her investigation of the metaphysical implications of seemingly small events. The words from a Dickinson poem are inscribed on the lengths of aluminum, but can be read only if viewers are willing to change their angle of vision.

Color plate 502. JONATHAN LASKER. *To Believe in Food.* 1991. Oil on linen, 8'4" x 6'3" (2.5 x 1.9 m). Private collection. Courtesy Sperone Westwater, New York

31. JONATHAN LASKER (b. 1948)
To Believe in Food, 1991

The paintings of **JONATHAN LASKER** reflect the profound changes Poststructural theory brought to the New York art world during the 1980s. For him, as for Ross Bleckner and Peter Halley, abstraction is not a universal language, but simply a code of representation that was granted inordinate privilege by twentieth-century culture. His work is distinctive because it bypasses the geometric structures anchoring 1980s Neo-Geo to focus on the vocabulary of 1950s and 1960s Field and Gesture painting. The harsh, bright yellow of *To Believe in Food* (color plate 502) serves as a foil for two types of mark making: fine, dark lines; and bold strokes of pale blue, rose, and brown paint. The isolation of the elements, together with the acid colors, makes Lasker's paint-

ing contradictory. It employs a mode of abstraction associated with spontaneous expression of powerful emotion, yet is suffused with strain and unease. The effect can be traced back to Lasker's studio procedure. Using ballpoint or felt-tip pens, he makes small-scale automatist drawings, then "copies" them onto paintings without any change, except scale and medium. Because of the self-quotation, Lasker's work short-circuits the concept of "authentic" gesture basic to modern abstraction and raises questions about our access to the unconscious. Neither clarified nor explained by the enlarged drawings, the unconscious becomes a baffling force that makes Lasker's ironies about abstraction all the more corrosive.

Color plate 503.
MARK TANSEY. *The Enunciation*. 1992. Oil on canvas, 7' x 64" (2.1 m x 163 cm). Museum of Fine Arts, Boston

ERNEST WADSWORTH LONGFELLOW FUND, THE LORNA AND ROBERT ROSENBERG FUND, AND ADDITIONAL FUNDS PROVIDED BY BRUCE A. BEAL, ENID L. BEAL, ROBERT L. BEAL, CATHERINE AND PAUL BUTTENWEISER, GRISWOLD DRAZ, JOYCE AND EDWARD LINDE, VIJAK MAHDAVI, BERNARDO NADAL-GINARD, MARTIN PERETZ, MRS. MYRON C. ROBERTS, AND AN ANONYMOUS DONOR

32. MARK TANSEY (b. 1949)
The Enunciation, 1992

While **MARK TANSEY** makes use of tactics of appropriation, his paintings never replicate any style or masterwork of the past. "Metarealist" is the term sometimes applied to them. They transpose issues and episodes from the history of art into scenes that are imaginary yet seem plausible, and raise questions about the scope and purpose of pictorial representation. Tansey explains his working procedure as "reverse collage" because he produces a unified picture from disparate, found motifs, most culled from magazines of the 1940s and 1950s. In *Leonardo's Wheel*, one of the Renaissance artist's famed hydraulic inventions appears atop Niagara Falls, seemingly in command of the water's torrential energy. *Enunciation* (color plate 503) depicts Marcel

Duchamp, seated alone in a train compartment, exchanging glances with his alter ego, Rrose Sélavy, shown diagonally opposite him at the window of a passing train. Her punning name (it can be read as *rose c'est la vie*, or "rose that's life") is elaborated in the picture's unusual, "quasi-feminine" tones, achieved through a color known as rose violet. Yet *Enunciation* should not be regarded as an unmasking of the myth of a modern hero. It endeavors, instead, to perform the task Tansey describes as "opening content," in particular a cluster of issues concerning pictorial art, sexual identity, and modern technology that Duchamp first "enunciated." Tansey's approach is grounded in assiduous study of deconstructionist and Poststructuralist theory.

Color plate 504. RAYMOND PETTIBON. 1995. Installation at David Zwirner Gallery, New York, September 14–October 14, 1995. Courtesy David Zwirner Inc., New York

33. RAYMOND PETTIBON (b. 1957)
Installation at David Zwirner Gallery, New York, 1995
No Title (The translator's have), 1996

Californian **RAYMOND PETTIBON** occupies a middle ground between Mark Tansey and Jonathan Lasker. Like Tansey, he makes representational art of a deconstructionist bent by "collaging" motifs from a wide variety of sources in high and low culture. But his central concern is the problem of subjectivity articulated by Lasker, and he too employs drawing so as to foreground the ambiguity between spontaneous self-expression and controlled artifice. Pettibon, however, rarely exhibits paintings, preferring to show pen-and-ink drawings, more than a hundred at a time, pushpinned to the wall in casual arrangements (color plates 504, 505). His infraction of the rules governing the display of drawings goes hand in hand with his assault upon the paradoxes in art's mediation of the private and public dimensions of the artist's personality. Though the drawings have a distinctive look, the naïveté of their execution defies criteria of originality and quality. They are, Pettibon says, "the kind of thing that wouldn't even be necessarily signed by an artist." Closer examination of the drawings discloses additional subversions of the ideal of the magisterial, self-expressive artist. A dizzying array of appropriations, from Gumby to Ad Reinhardt, runs through them, and in addition, they are interspersed with a handwritten text, purportedly a journal, laced with innumerable quotations, self-conscious asides to the "dear reader," and interjections like "vavoom" (color plate 505). The images and text complement one another yet constantly forestall the resolution of meaning and call attention to the tension between "real life" and life on paper. Since no Pettibon installation ever tells a complete story, his art is often associated with the tradition of the comic strip.

Color plate 505. RAYMOND PETTIBON. *No Title (The translator's have)*. 1996. Pen and ink on paper, 29¾ x 22" (75.6 x 55.9 cm). Courtesy Regen Projects, Los Angeles, and David Zwirner Inc., New York

Color plate 506. JEFF WALL. *A Sudden Gust of Wind (After Hokusai)*. 1993. Fluorescent light, color transparency, and display case, 7'6³⁄₁₆" x 12'4⁷⁄₁₆" (2.3 x 3.8 m). Tate Gallery, London
PHOTO TATE GALLERY, LONDON/ART RESOURCE, NEW YORK

34. JEFF WALL (b. 1946)
A Sudden Gust of Wind (After Hokusai), 1993

As a student in the art-history program of the Courtauld Institute, JEFF WALL developed a keen interest in the way nineteenth-century French artists transformed history painting into a vehicle for the depiction of contemporary life. His own project over the past two decades has been the fashioning of a comparable large-scale representation of the history of his own times. Choice of medium was an initial difficulty because Wall was searching for a means of making film and photography reflect the conditions of life under late capitalism. By his own account, the breakthrough came when he saw an illuminated sign and recognized it as "the perfect synthetic technology," related as much to "propaganda" as it is to photography, cinema, and painting. Wall began to enlarge color transparencies and mount them in light boxes,

producing mural-sized scenes, sometimes called "super-photographs," that have helped redefine the art of photography. Perhaps their closest parallel lies in Mark Tansey's paintings insofar as they use "reverse collage" to produce the illusion of a seamless reality. *A Sudden Gust of Wind (After Hokusai)* (color plate 506) consists of fifty pieces of film, shot over a five-month period, which were spliced together through digital technology. What is more, though shot on location near Vancouver, Canada, the scene features professional actors who were performing a script, in part inspired by a Japanese print. Such staging, together with the backlighting, infuse Wall's pictures with a dreamlike reality he associates with the media imagery infiltrating human experience in the late twentieth century.

Color plate 507. STAN DOUGLAS. *Evening.* 1994. Installation view of a continuous three-channel video projection with sound at the Renaissance Society, University of Chicago. Collection of Dakis Joannou, Athens, and Museum of Contemporary Art, Chicago. Courtesy David Zwirner, New York, and the Renaissance Society, University of Chicago

35. STAN DOUGLAS (b. 1960)
Evening, 1994

The African-Canadian artist **STAN DOUGLAS** speaks of his cross-media work as "archaeology" because of its preoccupation "with failed utopias and obsolete technologies." In a 1993 installation, he showed what appeared to be a vintage silent film, with subtitles and piano accompaniment, that wove the story of a missing Japanese worker into a synoptic history of corporate exploitation and the failure of a commune. In the film, the Japanese man is never found; similarly, Douglas's other work avoids narrative closure. The approach is indebted to Walter Benjamin's study of "forgotten" artifacts, but as Douglas explains, its central source is Samuel Beckett, who used "doubt" to write "from the margins." A video installation Douglas made in Paris, *Horschamps*, offers alternative versions of a jazz performance.

One version, done in the style of 1960s French television, tracks the musicians as they play their instruments, the other concentrates on their stances and moments of rest, arguably also key components of the performance. *Evening* (color plate 507) concerns the changes that swept through American news programs in the wake of the sociopolitical unrest of the sixties. The monitors show newscasts, dated January 1, 1969, and January 1, 1970, supposedly from ABC, NBC, and CBS, which Douglas made by combining staged sections and archival footage. All the earlier programs are more factual and delivered with a serious demeanor, while the later ones betray the symptoms of "infotainment." The anchormen are friendly, and disturbing content, like the killing of two Black Panthers, is relegated to the reports' margins.

Color plate 508. TONY OURSLER. *Untitled*. 1996. Installation view at Metro Pictures, New York, 1996. Video projection of eyes on 13 painted fiberglass globes with soundtracks, dimensions variable. Courtesy the artist and Metro Pictures, New York

36. TONY OURSLER (b. 1957)
Untitled, 1996

After studying at the California Institute for the Arts, TONY OURSLER returned to New York and used video to make pioneering examples of the "hybrid" narratives that would become a characteristic feature of 1990s art. They incorporated serious social commentary into phantasmagoric, popular fables and broke down the boundaries between video and the other arts. Oursler was not only the scriptwriter and cameraman, he also made the miniature sets and puppets with which the stories were performed. While he continues to combine the roles of video artist, dramatist, sculptor, and puppeteer, his current work consists of installations in which video equipment projects images onto figures and objects fashioned from common, found materials. Without the humor of the unusual juxtapositions, the scenes might well be unbearable because they take psychological disintegration as their major theme. According to Oursler, prolonged exposure to television induces "one of the predominant signs of schizophrenia: the inability to identify the perimeter of the body or to perceive the point at which the body ends and the rest of the world begins." This condition is depicted in an installation that seems at first glance to be a room-sized model of the universe (color plate 508). The globes, however, are giant eyes (an illusion produced through video projection). Some of the monstrous "world-eyes" watch television, others a video game or pornographic film. Duchamp's ironic portrayals of technology's impact on human identity are important antecedents for Oursler's work.

Color plate 509. GARY HILL. *And Sat Down Beside Her.* 1990. Detail from mixed-media installation consisting of three works: 1) Single-channel video and hanging TV tube with table, chair, lens, book, and speaker; 2) single-channel video and glass tube enclosing 1" (2.54 cm) TV tube, text applied on floor and speaker; 3) two-channel video and two 1" (2.54 cm) TV tubes with four lenses. Courtesy Donald Young Gallery, Seattle
PHOTO JEAN-MARC MEUNIER

37. GARY HILL (b. 1951)
And Sat Down Beside Her, 1990
I Believe It Is an Image in Light of the Other, 1991–92

In the video art of Seattle-based **GARY HILL**, critique of "techknowledge" leads to a complex exploration of the terms and conditions of subjective experience. Hill structures his "meta-narratives" through use of the "prepared video" introduced by Nam June Paik during the late 1960s. *And Sat Down Beside Her* (color plate 509) is a detail from an installation where tiny video cameras, stripped from their casings, were suspended on wires. The metaphor of the web-spinning, multi-eyed, as well as multi-legged spider is developed through the insertion of lenses that distort the projection of images and the specially built furniture. The chair and table are warped as if seen through a wide-angle lens. The fusion of projected imagery and tangible objects offers a troubling reminder of the media's power over the human perception of reality. What is yet more provocative is the installation's disclosure of the relationship between imagery, perception, and the individual's sense of the boundaries and shape of his or her body. For Hill, this provides the basis for philosophically oriented explorations of the elusive, even paradoxical nature of self-consciousness. Written texts almost invariably bleed into the visual imagery of his videos, and in addition, they have scripts Hill writes and performs himself (fig. 930). Continuously overlapping, the language and pictures portray Hill as a divided being, at once present and intelligible to himself, yet estranged from all knowledge. His work makes manifest "the impenetrable solitude which lies at the heart of consciousness."

930. GARY HILL. *I Believe It Is an Image in Light of the Other.* 1991–92. Detail from mixed-media installation, seven-channel video, modified TV tubes for projection, books, and speaker. Courtesy Donald Young Gallery, Seattle
PHOTO MARK B. MCLOUGHLIN

Bibliography

Since this is a general survey text shaped by a wide range of critical discourse, it is not practical to list all sources or background material. Books and museum or gallery catalogues have been pruned to a manageable list, and the listing consists almost entirely of books. In general, periodical articles have been omitted, except for the last chapters, because fewer monographic works exist on these contemporary artists.

The bibliography is organized as follows:

I. General

A. Surveys, Theory and Methodology
B. Dictionaries and Encyclopedias
C. Architecture, Engineering, and Design
D. Photography, Prints, and Drawings
E. Painting and Sculpture

II. Further Readings, arranged by chapter

Books and exhibition catalogues without discernible author(s) are listed alphabetically under the artist's last name in the appropriate stylistic section if a single artist's name is included in the title. In cases where more than one artist is listed in the title, the entry is alphabetized according to the first letters of the title.

ABBREVIATIONS
Exh. cat.   Exhibition Catalogue
LACMA   Los Angeles County Museum of Art
MMA   The Metropolitan Museum of Art, New York
MOMA   The Museum of Modern Art, New York
SRGM   Solomon R. Guggenheim Museum, New York
WMAA   Whitney Museum of American Art, New York

Works cited here are in English, except for selected important sources, catalogues, and documents.

I. GENERAL

A. SURVEYS, THEORY, METHODOLOGY

Arnheim, Rudolf. *Art and Visual Perception: A Psychology of the Creative Eye.* Enl. and rev. ed. Berkeley and Los Angeles: University of California Press, 1974.

Ashton, Dore. *The Unknown Shore: A View of Contemporary Art.* Boston: Little, Brown, 1962.

Atkins, Robert. *Artspeak: A Guide to Contemporary Ideas, Movements, and Buzzwords.* New York: Abbeville, 1990.

Baigell, Matthew. *A Concise History of American Painting and Sculpture.* New York: Harper & Row, 1984.

Barasch, Moshe. *Modern Theories of Art, 1: From Winckelmann to Baudelaire.* New York: New York University Press, 1990.

Barr, Jr., Alfred H. *Cubism and Abstract Art: Painting, Sculpture, Constructions, Photography, and Industrial Arts, Theatre, Films, Posters, Typography.* Cambridge, Mass.: Belknap Press of Harvard University Press, 1986.

———. *Defining Modern Art: Selected Writings of Alfred H. Barr.* Ed. I. Sandler and A. Newman. New York: Harry N. Abrams, 1986.

Battcock, Gregory. *Why Art: Casual Notes on the Aesthetics of the Immediate Past.* New York: Dutton, 1977.

Baudelaire, Charles. *Painter of Modern Life and Other Writings on Art.* Trans. and ed. J. Mayne. London: Phaidon, 1964.

Bearden, Romare. *A History of African American Artists: From 1792 to the Present.* New York: Pantheon Books, 1993.

Berger, John. *About Looking.* New York: Pantheon Books, 1980.

Besset, Maurice. *Art of the Twentieth Century.* New York: Universe Books, 1976.

Blanshard, Frances. *Retreat from Likeness in the Theory of Painting.* 2nd ed. New York: Columbia University Press, 1949.

Boime, Albert. *The Academy and French Painting in the Nineteenth Century.* London and New York: Phaidon, 1971.

Bois, Yve-Alain. *Painting as Model.* Cambridge, Mass.: MIT Press, 1991.

Bolton, Richard, ed. *Culture Wars: Documents From Recent Controversies in the Arts.* New York: New Press, 1992.

Bowness, Alan. *Modern European Art.* London: Thames & Hudson, 1972.

Britt, David, ed. *Modern Art: Impressionism to Post Modernism.* London: Thames & Hudson, 1989.

Broude, Norma, and Mary Garrard, eds. *The Expanding Discourse: Feminism and Art History.* New York: Harper & Row, 1992.

———, eds. *Feminism and Art History: Questioning the Litany.* New York: Harper & Row, 1982.

Brown, Milton W. *American Art: Painting, Sculpture, Architecture, Decorative Arts, Photography.* New York: Harry N. Abrams, 1979.

Brunette, Peter, and David Wills, eds. *Deconstruction and the Visual Arts: Art, Media, Architecture.* New York: Cambridge University Press, 1995.

Bryson, Norman. *Calligram: Essays in New Art History from France.* New York: Cambridge University Press, 1988.

———, ed. *Vision and Painting: The Logic of the Gaze.* New Haven, Conn.: Yale University Press, 1983.

———, et al., eds. *Visual Theory: Painting and Interpretation.* New York: Cambridge University Press, 1991.

Bürger, Peter. *Theory of the Avant-Garde.* Trans. M. Shaw. Minneapolis: University of Minnesota Press, 1984.

Cahn, Walter. *Masterpieces: Chapters on the History of an Idea.* Princeton, N.J.: Princeton University Press, 1979.

Canaday, John. *Mainstreams of Modern Art.* 2nd ed. New York: Holt, Rinehart and Winston, 1981.

Caws, Mary Ann. *The Eye in the Text: Essays on Perception, Mannerist to Modern.* Princeton, N.J.: Princeton University Press, 1981.

Chadwick, Whitney. *Women, Art and Society.* London: Thames & Hudson, 1990.

Clark, Kenneth. *Ruskin Today.* London: J. Murray, 1964.

Collingwood, R. G. *Essays in the Philosophy of Art.* Bloomington, Ind.: Indiana University Press, 1964.

Compton, Susan, ed. *British Art in the 20th Century: The Modern Movement.* Exh. cat. Munich: Prestel, 1986.

Cone, Michèle. *The Roots and Routes of Art in the 20th Century.* New York: Horizon Press, 1975.

Crane, Diane. *The Transformation of the Avant-Garde: The New York Art World, 1940–1985.* Chicago: University of Chicago Press, 1987.

Crary, Jonathan. *Techniques of the Observer: On Vision and Modernity in the 19th Century.* Cambridge, Mass.: MIT Press, 1990.

Craven, Wayne. *American Art: History and Culture.* Madison, Wis.: Brown and Benchmark, 1994.

Crimp, Douglas. *On the Museum's Ruins.* Cambridge, Mass.: MIT Press, 1993.

Crow, Thomas. *Modern Art in the Common Culture.* New Haven, Conn.: Yale University Press, 1996.

Danto, Arthur. *The Transfiguration of the Commonplace: A Philosophy of Art.* Cambridge, Mass.: Harvard University Press, 1981.

DeGeorge, Richard T., and Fernande M. DeGeorge, eds. *The Structuralists: From Marx to Lévi-Strauss.* Garden City, N.Y.: Anchor Books, 1972.

Driskell, David. *Two Centuries of Black American Art.* Exh. cat. New York: Knopf, 1976.

Dufrenne, Mikel, ed. *Main Trends in Aesthetics and the Sciences of Art.* New York: Holmes and Meier, 1979.

Eisenman, Stephen F., et al. *Nineteenth Century Art: A Critical History.* London: Thames & Hudson, 1994.

Ferguson, Russell, ed. *Discourses: Conversations in Postmodern Art and Culture.* Documentary Sources in Contemporary Art. Cambridge, Mass.: MIT Press, 1990.

Ferrier, Jean Louis, ed. *Art of Our Century: The Chronicle of Western Art, 1900 to the Present.* New York: Prentice-Hall, 1989.

Fineberg, Jonathan. *Art Since 1940: Strategies of Being.* Englewood Cliffs, N.J.: Prentice-Hall, 1995.

Fletcher, Valerie. *Dreams and Nightmares: Utopian Visions in Modern Art.* Exh. cat. Washington, D.C.: Hirshhorn Museum and Sculpture Garden, 1983.

Focillon, Henry. *The Life of Forms in Art.* New York: Zone Books, 1958.

Foster, Hal. *Return of the Real: The Avant-Garde at the End of the Century.* Cambridge, Mass.: MIT Press, 1996.

———, ed. *The Anti-Aesthetic: Essays on Postmodern Culture.* Seattle: Bay Press, 1983.

Frascina, Francis, ed. *Pollock and After: The Critical Debate.* New York: Harper & Row, 1985.

———, and Jonathan Harris, eds. *Art in Modern Culture: An Anthology of Critical Texts*. New York: Harper & Row, 1992.

———, and Charles Harrison, eds. *Modern Art and Modernism: A Critical Anthology*. New York: Harper & Row, 1982.

Fry, Roger. *Last Lectures*. Cambridge, England: The University Press, 1939.

———. *Vision and Design*. London: Chatto and Windus, 1957.

Gablik, Suzy. *Progress in Art*. London: Thames & Hudson, 1976.

Gage, John. *Colour and Culture: Practice and Meaning from Antiquity to Abstraction*. London: Thames & Hudson, 1993.

Goddard, Donald. *American Painting*. New York: MacMillan, 1990.

Goldwater, Robert. *Primitivism in Modern Art*. Enl. ed. Cambridge, Mass.: Harvard University Press, 1986.

Gombrich, Ernst H. *The Sense of Order: A Study in the Psychology of Decorative Art*. Ithaca, N.Y.: Cornell University Press, 1979.

Gordon, Donald. *Expressionism: Art and Idea*. New Haven, Conn.: Yale University Press, 1987.

Greenberg, Clement. *Clement Greenberg: The Collected Essays and Criticism*. 4 vols. Chicago: University of Chicago Press, 1986–93.

Guilbaut, Serge, ed. *Reconstructing Modernism: Art in New York, Paris, and Montreal, 1945–1964*. Cambridge, Mass.: MIT Press, 1990.

Hamilton, George H. *19th and 20th Century Art: Painting, Sculpture, Architecture*. New York: Harry N. Abrams, 1970.

Harris, Anne Sutherland, and Linda Nochlin. *Woman Artists: 1550–1950*. Exh. cat. Los Angeles: LACMA, 1976.

Harrison, Charles, and Paul Wood, eds. *Art in Theory, 1900–1990: An Anthology of Changing Ideas*. Cambridge, Mass.: Blackwell, 1992.

Hauser, Arnold. *The Social History of Art*. Trans. S. Godman. 4 vols. New York: Vintage Books, 1957.

Heller, Nancy G. *Women Artists: An Illustrated History*. New York: Abbeville, 1987.

Henderson, Linda Dalrymple. *The Fourth Dimension and Non-Euclidian Geometry in Modern Art*. Princeton, N.J.: Princeton University Press, 1983.

Herbert, Robert L., ed. *The Art Criticism of John Ruskin*. Garden City, N.Y.: Anchor Books, 1964.

———, ed. *Modern Artists on Art*. Englewood Cliffs, N.J.: Prentice-Hall, 1964.

Hertz, Richard. *Theories of Contemporary Art*. 2nd ed. Englewood Cliffs, N.J.: Prentice-Hall, 1993.

———, and Norman Klein, eds. *Twentieth-Century Art Theory: Urbanism, Politics and Mass Culture*. Englewood Cliffs, N.J.: Prentice-Hall, 1990.

Hildebrand, Adolf von. *The Problem of Form in Painting and Sculpture*. Ann Arbor, Mich.: University Microfilms, 1979.

Hoffman, Katherine. *Explorations: The Visual Arts Since 1945*. New York: HarperCollins, 1991.

Hofmann, Werner. *Turning Points in Twentieth-Century Art: 1890–1917*. Trans. C. Kessler. New York: G. Braziller, 1969.

Hunter, Sam, and John Jacobus. *Modern Art: Painting, Sculpture, Architecture*. 3rd ed. New York: Harry N. Abrams, 1992.

Jencks, Charles. *What is Post-Modernism?* 3rd ed. rev. London: Academy Editions, 1989.

Joachimides, Christos. *American Art in the 20th Century: Painting and Sculpture 1913–1933*. Exh. cat. London and Munich: Royal Academy of Arts and Prestel, 1993.

———, et al. *German Art in the 20th Century: Painting and Sculpture 1905–1985*. Exh. cat. London: Royal Academy of Arts and Weidenfeld & Nicholson, 1985.

Johnson, Ellen. *Modern Art and the Object*. London: Thames & Hudson, 1976.

Kaplan, Patricia, and Susan Manso, eds. *Major European Art Movements 1900–1945: A Critical Anthology*. New York: Dutton, 1977.

Kemp, Martin. *The Science of Art: Optical Themes in Western Art from Brunelleschi to Seurat*. New Haven, Conn.: Yale University Press, 1990.

Kepes, Gyorgy. *The New Landscape in Art and Science*. Chicago: P. Theobald, 1956.

———. *Sign, Image, Symbol*. New York: G. Braziller, 1966.

Kern, Stephen. *The Culture of Time and Space: 1880–1918*. Cambridge, Mass.: Harvard University Press, 1980.

Kleinbauer, W. Eugene. *Modern Perspectives in Western Art History: An Anthology of Twentieth-Century Writings on the Visual Arts*. Reprint of the 1971 ed. Toronto: University of Toronto Press, 1989.

Kozloff, Max. *Renderings: Critical Essays on a Century of Modern Art*. London: Studio Vista, 1968.

Kramer, Hilton. *The Age of the Avant-Garde: An Art Chronicle of 1956–1972*. New York: Farrar, Straus and Giroux, 1973.

Krauss, Rosalind. *The Optical Unconscious*. Cambridge, Mass.: MIT Press, 1993.

———. *The Originality of the Avant-Garde and Other Modernist Myths*. Cambridge, Mass.: MIT Press, 1986.

Kuh, Katharine. *The Artist's Voice*. New York: Harper & Row, 1962.

Kultermann, Udo. *The History of Art History*. New York: Abaris Books, 1993.

Kuspit, Donald B. *The Critic as Artist: The Intentionality of Art*. Ann Arbor, Mich.: UMI Research Press, 1984.

———. *The Cult of the Avant-Garde Artist*. New York: Cambridge University Press, 1993.

Langer, Cassandra. *Feminist Art Criticism: An Annotated Bibliography*. New York: G. K. Hall, 1993.

Levin, Kim. *Beyond Modernism: Essays on Art from the 70's and 80's*. New York: Harper & Row, 1988.

Lewis, Samella S. *African American Art and Artists*. Rev. ed. Berkeley, Calif.: University of California Press, 1994.

Liberman, Alexander. *The Artist in His Studio*. New York: Viking Press, 1960.

Lippard, Lucy. *From the Center: Feminist Essays on Women's Art*. New York: Dutton, 1976.

Lucie-Smith, Edward. *Art Today: From Abstract Expressionism to Superrealism*. 3rd ed. Oxford: Phaidon, 1989.

———. *Latin American Art of the 20th Century*. London: Thames & Hudson, 1993.

Lynton, Norbert. *The Story of Modern Art*. 2nd ed. Oxford: Phaidon, 1989.

McCoubrey, John. *American Art, 1700–1960: Sources and Documents*. Englewood Cliffs, N.J.: Prentice-Hall, 1965.

McShine, Kynaston. *An International Survey of Recent Painting and Sculpture*. Exh. cat. New York: MOMA, 1984.

Millon, Henry A., and Linda Nochlin, eds. *Art and Architecture in the Service of Politics*. Cambridge, Mass.: MIT Press, 1978.

Mitchell, W. J. T. *Picture Theory: Essays on Verbal and Visual Representation*. Chicago: University of Chicago Press, 1994.

———. *The Reconfigured Eye: Visual Truth in the Post-Photographic Era*. Cambridge, Mass.: MIT Press, 1992.

Nash, Steven A. *Arnason and Politics: A Commemorative Exhibition*. San Francisco: Fine Arts Museums, 1993.

Nelson, Robert, and Richard Shiff, eds. *Critical Terms for Art History*. Chicago: University of Chicago Press, 1993.

Nochlin, Linda. *The Body in Pieces: The Fragment as a Metaphor of Modernity*. London: Thames & Hudson, 1995.

———. *The Politics of Vision: Essays on Nineteenth-Century Art and Society*. London: Thames & Hudson, 1989.

———. *Women, Art, and Power: and Other Essays*. London: Thames & Hudson, 1989.

Norris, Christopher, and Andrew Benjamin. *What is Deconstruction?* New York: St. Martin's Press, 1988.

Osborne, Harold, ed. *Oxford Companion to Twentieth-Century Art*. New York: Oxford University Press, 1981.

Papadakes, Andreas, et al., eds. *Deconstruction: The Omnibus Volume*. New York: Rizzoli, 1989.

Parker, Rozsika, and Griselda Pollock. *Old Mistresses: Women, Art, and Ideology*. New York: Pantheon, 1981.

Pevsner, Nikolas. *Academies of Art: Past and Present*. Reprint of the 1940 ed. New York: Da Capo Press, 1973.

Pincus-Witten, Robert. *Postminimalism into Maximalism: American Art, 1966–1986*. Ann Arbor, Mich.: UMI Research Press, 1987.

Podro, Michael. *The Critical Historians of Art*. New Haven, Conn.: Yale University Press, 1982.

Poggioli, Renato. *The Theory of the Avant-Garde*. Trans. G. Fitzgerald. Cambridge, Mass.: Belknap Press of Harvard University Press, 1968.

Pollock, Griselda. *Vision and Difference: Femininity, Feminism, and the Histories of Art*. New York: Routledge, 1988.

Poppei, Frank. *Art of the Electronic Age*. London: Thames & Hudson, 1997.

Powell, Richard. *Black Art and Culture in the 20th Century*. London: Thames & Hudson, 1997.

———. *The Blues Aesthetic: Black Culture and Modernism*. Washington, D.C.: Washington Project for the Arts, 1989.

Preziosi, Donald. *Rethinking Art History: Meditations on a Coy Science*. New Haven, Conn.: Yale University Press, 1989.

Prown, Jules, and Barbara Rose. *American Painting: From the Colonial Period to the Present*. New ed. New York: Rizzoli, 1977

———, et al. *Discovered Lands, Invented Pasts: Transforming Visions of the American West*. Exh. cat. New Haven, Conn.: Yale University Press, 1992.

Rasmussen, Waldo, ed. *Latin American Artists of the Twentieth Century*. Exh. cat. New York: MOMA, 1993.

Read, Herbert. *Art Now: An Introduction to the Theory of Modern Painting and Sculpture*. New York: Pitman Publishing Corp., 1948.

Rees, A. L., and Frances Borzello. *The New Art History*. London: Camden Press, 1986.

Risatti, Howard, ed. *Postmodern Perspectives*. Englewood Cliffs, N.J.: Prentice-Hall, 1990.

Rodman, Selden. *Conversations with Artists*. New York: Capricorn Books, 1957.

Rose, Barbara. *American Art Since 1900*. Rev. ed. London: Thames & Hudson, 1975.

———. *Readings in American Art Since 1900: A Documentary Survey.* New York: Praeger, 1968.

Rosen, Charles, and Henri Zerner. *Romanticism and Realism: The Mythology of Nineteenth-Century Art.* New York: Viking, 1984.

Rosen, Randy, and Catherine C. Brawer. *Making Their Mark: Women Artists Move Into the Mainstream, 1970–85.* New York: Abbeville, 1989.

Rosenberg, Harold. *The Tradition of the New.* New York: Horizon Press, 1959.

Rubin, William, ed. *"Primitivism" in 20th-Century Art: Affinity of the Tribal and the Modern.* 2 vols. Exh. cat. New York: MOMA, 1984.

Schapiro, Meyer. *Theory and Philosophy of Art: Style, Artist, and Society.* New York: G. Braziller, 1994.

Selz, Peter. *Art in Our Times: A Pictorial History 1890–1980.* New York: Harry N. Abrams, 1981.

Shapiro, Theda. *Painters and Politics: The European Avant-Garde and Society: 1900–1925.* New York: Elsevier, 1976.

Smagula, Howard. *Currents: Contemporary Directions in the Visual Arts.* 2nd ed. Englewood Cliffs, N.J.: Prentice-Hall, 1989.

Stangos, Nikos, ed. *Concepts of Modern Art: From Fauvism to Postmodernism.* 3rd ed. World of Art. London: Thames & Hudson, 1994.

Stich, Sidra. *Made in USA: An Americanization in Modern Art, the '50s and '60s.* Berkeley, Calif.: University of California Press, 1987.

Stiles, Kristine, and Peter Selz. *Theories and Documents of Contemporary Art: A Sourcebook of Artist's Writings.* Berkeley and Los Angeles: University of California Press, 1996.

Tagg, John. *Grounds of Dispute: Art History, Cultural Politics and the Discursive Field.* Minneapolis: University of Minnesota Press, 1992.

Taylor, Joshua, ed. *Nineteenth-Century Theories of Art.* Berkeley, Calif.: University of California Press, 1991.

Tomkins, Calvin. *Post to Neo: The Art World of the 1980's.* New York: Holt, 1990.

Tuchman, Maurice. *The Spiritual in Art: Abstract Painting, 1890–1985.* New York: Abbeville, 1986.

Valéry, Paul. *Selected Writings.* New York: New Directions, 1964.

Varnedoe, Kirk, and Adam Gopnik. *High and Low: Modern Art and Popular Culture.* Exh. cat. New York: MOMA, 1990.

———, eds. *Modern Art and Popular Culture: Readings in High and Low.* New York: Harry N. Abrams, 1990. With Essays by J. E. Bowlt et al.

Wallis, Brian, ed. *Art After Modernism: Rethinking Representation.* Boston: Godine, 1984.

———, ed. *Blasted Allegories: An Anthology of Writings by Contemporary Artists.* Cambridge, Mass.: MIT Press, 1987.

Weiss, Jeffrey S. *The Popular Culture of Modern Art: Picasso, Duchamp, and Avant-Gardism.* New Haven, Conn.: Yale University Press, 1994.

Wheeler, Daniel. *Art Since Mid-Century: 1945 to the Present.* London: Thames & Hudson, 1991.

Whiting, Cécile. *Antifascism in American Art.* New Haven, Conn.: Yale University Press, 1989.

Wilkins, David G., Bernard Schultz, and Katheryn M. Linduff. *Art Past/Art Present.* 2nd ed. New York: Harry N. Abrams, 1994.

Wilmerding, John. *American Art.* Pelican History of Art. Harmondsworth: Penguin, 1976.

Witzling, Mara, ed. *Voicing Our Visions: Writings by Women Artists.* New York: Universe Books, 1991.

Wollheim, Richard. *Art and Its Objects.* New York: Cambridge University Press, 1980.

———. *Painting as an Art.* Princeton, N.J.: Princeton University Press, 1987.

Wood, Paul, et al. *Modernism in Dispute: Art Since the Forties.* New Haven, Conn.: Yale University Press, 1993.

Word as Image: American Art, 1960–1990. Exh. cat. Milwaukee, Wis.: Milwaukee Art Museum, 1990.

Worringer, Wilhelm. *Abstraction and Empathy: A Contribution to the Psychology of Style.* Trans. M. Bullock. New York: International University Press, 1980.

B. DICTIONARIES AND ENCYCLOPEDIAS

Baigell, Matthew. *Dictionary of American Art.* New York: Harper & Row, 1979.

Benezit, Emmanuel. *Dictionnaire critique et documentaire des peintres, sculpteurs, dessinateurs et graveurs.* 10 vols. Paris: Roger et F. Chernoviz, 1976.

The Britannica Encyclopedia of American Art. Chicago: Encyclopedia Britannica Education Corp., 1973.

Cummings, Paul. *Dictionary of Contemporary American Artists.* 6th ed. New York: St. Martin's Press, 1994.

Dunford, Penny. *Biographical Dictionary of Women Artists in Europe and America Since 1850.* Philadelphia: University of Pennsylvania Press, 1990.

Emanuel, Muriel, et al., eds. *Contemporary Artists.* 2nd ed. New York: St. Martin's Press, 1983.

Encyclopedia of World Art. 17 vols. New York: McGraw-Hill, 1959–68[?].

Heller, Jules, and Nancy Heller. *North American Women Artists of the Twentieth Century: A Biographical Dictionary.* New York: Garland, 1995.

Hughe, Rene, ed. *Larousse Encyclopedia of Modern Art from 1800 to the Present Day.* New York: Prometheus Press, 1965.

Jones, Lois Swan. *Art Information: Research Methods and Resources.* 3rd ed. Dubuque, Iowa: Kendall/Hunt, 1990.

Julier, Guy. *Dictionary of 20th-Century Design and Designers.* London: Thames & Hudson, 1993.

Lake, Carlton, and Robert Maillard, eds. *Dictionary of Modern Painting.* 3rd ed. New York: Tudor, 1964.

Lucie-Smith, Edward. *Lives of the Great Twentieth-Century Artists.* London: Thames & Hudson, forthcoming.

Maillard, Robert, ed. *New Dictionary of Modern Sculpture.* New York: Tudor, 1970.

Marks, Claude, ed. *World Artists 1980–1990.* New York: H. W. Wilson, 1991.

Norman, Geraldine. *Nineteenth-Century Painters and Painting: A Dictionary.* London: Thames & Hudson, 1977.

Osborne, Harold. *The Oxford Companion to Twentieth-Century Art.* Oxford: Oxford University Press, 1981.

Pevsner, Nikolaus, John Fleming, and Hugh Honour. *A Dictionary of Architecture.* Rev. enl. ed. Woodstock, N.Y.: Overlook Press, 1976.

Phaidon Dictionary of Twentieth-Century Art. Oxford: Phaidon, 1973.

Seuphor, Michel. *Dictionary of Abstract Painting, Preceded by a History of Abstract Painting.* New York: Paris Book Center, 1958.

Spalding, Frances. *20th Century Painters and Sculptors.* Woodbridge, Suffolk: Antique Collector's Club, 1990.

Stangos, Nikos. *Dictionary of Art and Artists.* Rev. ed. London: Thames & Hudson, 1994

Walker, John. *Glossary of Art, Architecture and Design Since 1945.* 3rd ed. Boston: G. K. Hall, 1992.

C. ARCHITECTURE, ENGINEERING, AND DESIGN

Arnheim, Rudolph. *The Dynamics of Architectural Form.* Berkeley and Los Angeles: University of California Press, 1977.

Banham, R. *The New Brutalism: Ethic or Aesthetic?.* London: Architectural Press, 1966.

———. *Theory and Design in the First Machine Age.* 2nd ed. New York: Praeger, 1960.

Behrendt, Walter. *Modern Building: Its Nature, Problems and Forms.* London: M. Kopkins, 1936.

Benevelo, Leonardo. *History of Modern Architecture.* 2 vols. Trans. H. J. Landry. Cambridge, Mass.: MIT Press, 1971.

Besset, Maurice. *New French Architecture.* New York: Praeger, 1967.

Blake, Peter. *The Master Builders.* New York: Knopf, 1960.

———. *No Place Like Utopia: Modern Architecture and the Company We Kept.* New York: Knopf, 1993.

Bouillon, Jean Paul. *Art Deco, 1903–1940.* New York: Rizzoli, 1989.

Brooks, H. Allen. *The Prairie School: Frank Lloyd Wright and His Mid-West Contemporaries.* Toronto: University of Toronto Press, 1972.

Collins, Michael, and Andreas Papadakis. *Post-Modern Design.* New York: Rizzoli, 1989.

Conrads, Ulrich, and Hans Sperlich. *The Architecture of Fantasy.* New York: Praeger, 1962.

Crook, J. Mordant. *The Dilemma of Style: Architectural Ideas from the Picturesque to the Post Modern.* Chicago: University of Chicago Press, 1987.

Curtis, William. *Modern Architecture Since 1900.* 2nd ed. Englewood Cliffs, N.J.: Prentice-Hall, 1987.

Dal Co, Francesco. *Figures of Architecture and Thought: German Architectural Culture, 1890–1920.* New York: Rizzoli, 1986.

———. *Transformations in Modern Architecture.* Greenwich, Conn.: New York Graphic Society, 1980.

Dormer, Peter. *Design Since 1945.* World of Art. London: Thames & Hudson, 1993.

Drexler, Arthur. *Transformations in Modern Architecture.* Exh. cat. New York: MOMA, 1979.

Eldredge, H. W., ed. *Taming Megalopolis.* 2 vols. New York: Praeger, 1967.

Ferebee, Ann. *A History of Design from the Victorian Era to the Present.* New York: Van Nostrand Reinhold, 1969.

Fitch, James. *American Building: The Historical Forces That Shaped It.* 2 vols. 2nd ed. Boston: Houghton Mifflin, 1966–72.

Frampton, Kenneth. *Modern Architecture: A Critical History.* London: Thames & Hudson, 1980.

———, and Yukio Futagawa. *Modern Architecture 1851–1945.* New York: Rizzoli, 1983.

Giedion, Sigfried. *Space, Time, and Architecture: The Growth of a New Tradition.* 4th ed. Cambridge, Mass.: Harvard University Press, 1963.

Heyer, Paul. *Architects on Architecture.* New York: Van Nostrand Reinhold, 1993.

Hitchcock, Henry-Russell. *Architecture: Nineteenth and Twentieth Centuries.* 4th ed. Pelican History of Art. Harmondsworth: Penguin, 1977.

Jencks, Charles. *Modern Movements in Architecture.* 2nd. ed. New York: Penguin, 1985.

Jensen, Rolf. *High Density Living.* London: L. Hill, 1966.

Joedicke, Jurgen. *Architecture Since 1945: Sources and Directions.* Trans. J. C. Palmes. New York: Praeger, 1969.

Johnson, Philip. *Deconstructionist Architecture.* Exh. cat. New York: MOMA, 1988.

Kaplan, Wendy, ed. *Designing Modernity: The Arts of Reform and Persuasion 1885–1945.* London: Thames & Hudson, 1995.

Kostof, Spiro. *History of Architecture: Settings and Rituals.* 2nd ed. New York: Oxford University Press, 1995.

Kultermann, Udo. *Architecture in the Twentieth Century.* New York: Van Nostrand Reinhold, 1993.

Lampugnani, Vittorio. *Dictionary of 20th-Century Architecture.* London: Thames & Hudson, 1988

Leach, Neil. *Rethinking Architecture: A Reader in Cultural Theory.* New York: Routledge, 1996.

Le Corbusier (Charles-Édouard Jeanneret). *Towards a New Architecture.* Trans. F. Etchells. London: Architectural Press, 1952.

Loyer, François. *Architecture of the Industrial Age.* New York: Skira, 1983.

Manhart, Marcia, and Tom Manhart, eds. *The Eloquent Object: The Evolution of American Art in Craft Media Since 1945.* Tulsa, Okla.: Philbrook Museum of Art, 1987.

Meeks, Carol. *The Railroad Station: An Architectural History.* New Haven, Conn.: Yale University Press, 1995.

Middleton, Robin and David Watkin. *Neoclassical and 19th Century Architecture.* 2 vols. History of World Architecture. New York: Electa/ Rizzoli, 1987.

Mignot, Claude. *Architecture of the Nineteenth Century in Europe.* New York: Rizzoli, 1984.

Mumford, Lewis. *The Culture of Cities.* New York: Harcourt, Brace and Co., c. 1938.

———. *Technics and Civilization.* New York: Rourledge, 1934.

Ockman, Joan, ed. *Architecture Culture, 1943–1968: A Documentary Anthology.* New York: Rizzoli, 1993.

Pehnt, Wolfgang. *Expressionist Architecture.* London: Thames & Hudson, 1973.

Peter, John. *Masters of Modern Architecture.* New York: G. Braziller, 1958.

Pevsner, Nikolaus. *Pioneers of Modern Design: From William Morris to Walter Gropius.* Harmondsworth: Penguin, 1975.

———. *The Sources of Modern Architecture and Design.* London: Thames & Hudson, 1977.

Pierson, William, and William H. Jordy. *American Buildings and Their Architects.* 4 vols. New York: Oxford University Press, 1986.

Pommer, Richard, ed. "Revising Modernist History: The Architecture of the 1920s and 1930s." *Art Journal* (Summer 1983) special issue.

Read, Herbert. *Art and Industry.*

Bloomington, Ind.: Indiana University Press, 1961.

Riesebero, Bill. *Modern Architecture and Design: An Alternative History.* Cambridge, Mass.: MIT Press, 1983.

Roth, Leland. *A Concise History of American Architecture.* New York: Harper & Row, 1980.

———. *Understanding Architecture: Its Elements, History and Meaning.* New York: Harper & Row, 1993.

Rudofsky, Bernard. *Architecture Without Architects: An Introduction to Non-Pedigreed Architecture.* Exh. cat. New York: MOMA, 1964.

Schuyler, Montgomery. *American Architecture and Other Writings.* Cambridge, Mass.: Belknap Press of Harvard University Press, 1964.

Scully, Vincent. *American Architecture and Urbanism.* Rev. ed. New York: H. Holy, 1988.

———. *Modern Architecture: The Architecture of Democracy.* Rev. ed. New York: G. Braziller, 1974.

Sharp, Dennis. *Modern Architecture and Expressionism.* New York: G. Braziller, 1966.

———. *The Rationalists: Theory and Design in the Modern Movement.* London: Architectural Press, 1979.

———, ed. *A Visual History of Twentieth-Century Architecture.* Greenwich, Conn.: New York Graphic Society, 1972.

Silver, Nathan. *Lost New York.* New York: Wing Books, 1967.

Smith, G. E. Kidder. *The New Architecture of Europe.* Cleveland and New York: World Pub. Co., 1961.

Tafuri, Manfredo. *Contemporary Architecture.* New York: Harry N. Abrams, 1977.

———. *Modern Architecture.* 2 vols. History of World Architecture. New York: Electa/Rizzoli, 1986.

Thackara, John, ed. *Design After Modernism: Beyond the Object.* London: Thames & Hudson, 1988.

Trachtenberg, Marvin, and Isabelle Hyman. *Architecture: From Pre-history to Post-Modernism.* New York: Harry N. Abrams, 1986.

Troy, Nancy. *Modernism and the Decorative Arts in France: Art Nouveau to Le Corbusier.* New Haven, Conn.: Yale University Press, 1991.

Tunnard, Christopher. *City of Man.* 2nd ed. New York: Scribner, 1970.

———, and Boris Pushkarev. *Man-Made America: Chaos or Control?* New Haven, Conn.: Yale University Press, 1963.

Vale, Brenda and Robert. *Green Architecture: Design for a Sustainable Future.* London: Thames & Hudson, 1996.

Vidler, Anthony. *The Architectural Uncanny: Essays in the Modern Unhomely.* Cambridge, Mass.: MIT Press, 1992.

Zevi, Bruno. *Towards an Organic Architecture.* London: Faber and Faber, 1950.

D. PHOTOGRAPHY, PRINTMAKING, AND DRAWING

Adams, Clinton. *American Lithographs, 1900–1960: The Artists and Their Prints.* Albuquerque, N.Mex.: University of New Mexico Press, 1983.

Adams, Robert. *Beauty in Photography: Essays in Defense of Traditional Values.* Millerton, N.Y.: Aperture, 1981.

Ades, Dawn. *Photomontage.* London: Thames & Hudson, 1986.

Barthes, Roland. *Camera Lucida, Reflections on Photography.* Trans. R. Howard. New York: Hill and Wang, 1981.

Benjamin, Walter. "The Work of Art in the Age of Mechanical Reproduction." In *Illuminations,* ed. H. Arendt. New York: Schocken Books, 1969.

Bolton, Richard, ed. *The Contest of Meaning: Critical Histories of Photography.* Cambridge, Mass.: MIT Press, 1989.

Braive, Michel. *The Photograph: A Social History.* New York: McGraw-Hill, 1966.

Bretell, Richard, et al. *Paper and Light: The Calotype in Great Britain and France 1839–1870.* Boston: D. R. Godine, 1984.

Buckland, Gail. *Reality Recorded: Early Documentary Photography.* Greenwich, Conn.: New York Graphic Society, 1974.

Buerger, Janet E. *The Era of the French Calotype.* Exh. cat. Rochester, N.Y.: International Museum of Photography at George Eastman House, 1982.

Bunnell, Peter, ed. *A Photographic Vision: Pictorial Photography, 1889–1923.* Salt Lake City: Peregrine Smith, 1980.

Burgin, Victor. *Thinking Photography.* London: MacMillan, 1990.

Castleman, Riva. *Modern Art in Prints.* New York: MOMA, 1973.

———. *Modern Artists as Illustrators.* Exh. cat. New York: MOMA, 1981.

———. *Prints from Blocks: From Gauguin to Now.* Exh. cat. New York: MOMA, 1983.

———. *Prints of the Twentieth Century.* London: Thames & Hudson, 1988.

Coe, Brian. *Colour Photography: The First Hundred Years, 1840–1940.* London: Ash & Grant, 1978.

———, and Paul Gates. *The Snapshot Photograph.* London: Ash & Grant, 1977.

Coke, Van Deren. *The Painter and the Photograph: From Delacroix to Warhol.* Rev. ed. Albuquerque, N.Mex.: University of New Mexico Press, 1972.

———, ed. *One Hundred Years of Photographic History: Essays in Honor of Beaumont Newhall.* Albuquerque, N.Mex.: University of New Mexico Press, 1975.

Daval, Jean-Luc. *Photography: History of an Art.* New York: Rizzoli, 1982.

Doty, Robert, ed. *Photography in America.* Exh. cat. New York: WMAA, 1974.

Evans, Martin Marix. *Contemporary Photographers.* 3rd ed. New York: St. James Press, 1995.

Fabian, Rainer, and Hans-Christian Adam. *Masters of Early Travel Photography.* London: Thames & Hudson, 1983.

Freund, Gisele. *Photography & Society.* Boston: D. R. Godine, 1980.

Friedman, Joseph Solomon. *The History of Colour Photography.* 2nd ed. London and New York: Focal, 1968.

Galassi, Peter. *Before Photography: Painting and the Invention of Photography.* Exh. cat. New York: MOMA, 1981.

Generations. Exh. cat. New York: SRGM, 1976.

Gernsheim, Helmut, and Alison Gernsheim. *The History of Photography: From the Camera Obscura to the Beginning of the Modern Era.* New York: McGraw-Hill, 1969.

Gidal, Tim N. *Modern Photojournalism: Origin and Evolution: 1910–1933.* New York: Collier, 1972.

Gilmour, Pat. *Modern Prints.* New York: Studio Vista/Dutton, 1970.

Goldberg, Vicki, ed. *Photography in Print: Writings from 1816 to the Present.* New York: Simon and Schuster, 1981.

Gover, C. Jane. *The Positive Image: Women Photographers in Turn of the Century*

America. Albany, N.Y.: SUNY Press, 1988.

Green, Jonathan, ed. *American Photography: A Critical History, 1945 to the Present*. New York: Harry N. Abrams, 1984.

———. *Camera Work: A Critical Anthology*. Millerton, N.Y.: Aperture, 1973.

———, ed. *The Snapshot*. Millerton, N.Y.: Aperture, 1974.

Greenough, Sarah, et al. *On the Art of Fixing a Shadow: One Hundred and Fifty Years of Photography*. Exh. cat. Chicago: Art Institute of Chicago, 1989.

Grundberg, Andy. *Photography and Art: Interactions Since 1946*. New York: Abbeville, 1987.

Hall-Duncan, Nancy. *The History of Fashion Photography*. New York: International Museum of Photography, 1977.

Hambourg, Maria Morris, et al. *The Waking Dream: Photography's First Century*. Exh. cat. New York: MMA, 1993.

Harker, Margaret. *The Linked Ring: The Secession Movement in Photography in Britain, 1892–1910*. London: Heinemann, 1979.

Hartmann, Sadakichi. *The Valiant Knights of Daguerre: Selected Critical Essays on Photography and Profiles of Photographic Pioneers*. Ed. H. W. Lawton and G. Knox. Berkeley, Calif.: University of California Press, 1978.

Hobgen, Carol, and Rowan Watson, eds. *From Manet to Hockney: Modern Artists' Illustrated Books*. Exh. cat. London: Victoria and Albert Museum, 1985.

Homer, William I. *Alfred Stieglitz and the American Avant-Garde*. Boston: New York Graphic Society, 1977.

———. *Alfred Stieglitz and the Photo-Secession*. Boston: New York Graphic Society, 1983.

Ives, Colta. *The Painterly Print*. New York: MMA, 1980.

Jammes, André, and Eugenia Janis. *The Art of French Calotype*. Princeton, N.J.: Princeton University Press, 1983.

Jeffrey, Ian. *Photography: A Concise History*. London: Thames & Hudson, 1981.

Johnson, William. *Nineteenth-Century Photography: An Annotated Bibliography, 1839–1879*. Boston: G. K. Hall, 1990.

Kozloff, Max. *Photography and Fascination: Essays*. Danbury, N.H.: Addison House, 1979.

———. *The Privileged Eye: Essays on Photography*. Albuquerque, N.Mex.: University of New Mexico Press, 1987.

Lavin, Maud, et al. *Montage and Modern Life, 1919–1942*. Exh. cat. Cambridge, Mass. and Boston: MIT Press and Institute of Contemporary Art, c. 1992.

Lewinski, Jorge. *The Camera at War*. London: W. H. Allen, 1978.

Lyons, Nathan, ed. *Photographers on Photography: A Critical Anthology*. Englewood Cliffs, N.J.: Prentice-Hall, 1956.

Maddow, Ben. *Faces: A Narrative History of the Portrait in Photography*. Boston: New York Graphic Society, 1977.

Mayor, A. Hyatt. *Prints and People: A Social History of Printed Pictures*. New York: MMA, 1971.

Mellor, David, ed. *Germany: The New Photography, 1927–33*. Exh. cat. London: Arts Council of Great Britain, 1978.

Naef, Weston. *Fifty Pioneers of Modern Photography: The Collection of Alfred Stieglitz*. Exh. cat. New York: MMA, 1978.

———, and James N. Wood. *Era of Exploration: The Rise of Landscape Photography in the American West: 1860–1885*. Exh. cat. Buffalo, N.Y.: Albright-Knox Gallery of Art, 1975.

Newhall, Beaumont. *The History of Photography from 1839 to the Present*. Rev. ed. Greenwich, Conn.: New York Graphic Society, 1982.

———, ed. *Photography: Essays and Images*. New York: MOMA, 1980.

Petruck, Peninah, ed. *The Camera Viewed: Writings on Twentieth-Century Photography*. 2 vols. New York: Dutton, 1979.

Phillips, Christopher, ed. *Photography in the Modern Era: European Documents and Critical Writings, 1913–1940*. Exh. cat. New York: MMA, 1989.

Pinney, Roy. *Advertising Photography*. New York: Hastings House, 1962.

Pollack, Peter. *Picture History of Photography*. Rev. ed. New York: Harry N. Abrams, 1970.

Rose, Bernice. *Allegories of Modernism: Contemporary Drawing*. Exh. cat. New York: MOMA, 1992.

Rosenblum, Naomi. *A World History of Photography*. Rev. ed. New York: Abbeville, 1989.

Rotzler, Willy. *Photography as Artistic Experiment: From Fox Talbot to Moholy-Nagy*. Garden City, N.J.: Amphoto, 1976.

Scharf, Aaron. *Art and Photography*. Harmondsworth and Baltimore: Penguin, 1974.

———. *Pioneers of Photography*. New York: Harry N. Abrams, 1976.

Solomon-Godeau, Abigail. *Photography at the Dock*. Minneapolis: University of Minnesota Press, 1991.

Sontag, Susan. *On Photography*. New York: Farrar, Straus and Giroux, 1973.

Stieglitz, Alfred, ed. *Camera Work: A Photographic Quarterly*. Reprint. New York: A. Stieglitz, 1969.

Stott, William. *Documentary Expression and Thirties America*. Reprint. Chicago: University of Chicago Press, 1986.

Stryker, Roy, and Nancy Wood. *In This Proud Land: America 1935–1943 as Seen in FSA Photographs*. Greenwich, Conn.: New York Graphic Society, 1973.

Szarkowski, John. *Looking at Photographs*. New York: MOMA and New York Graphic Society, 1973.

———. *The Photographer's Eye*. New York: MOMA and Doubleday, 1977.

———. *Photography Until Now*. Exh. cat. New York: MOMA, 1990.

Tagg, John. *The Burden of Representation: Essays on Photographies and Histories*. Amherst, Mass.: University of Massachusetts Press, 1988.

Tallman, Susan. *The Contemporary Print*. London: Thames & Hudson, 1996.

Trachtenberg, Allan, ed. *Classic Essays on Photography*. New Haven, Conn.: Leeter's Island Books, 1980.

Tsujimoto, Karen. *Images of America: Precisionist Painting and Modern Photography*. Exh. cat. San Francisco: San Francisco Museum of Modern Art, 1982.

Upton, Barbara, and John Upton. *Photography: Adapted from the Life Library of Photography*. 3rd ed. Boston: Little, Brown, 1980.

Vaizey, Marina. *The Artist as Photographer*. London: Sidgwick & Jackson, 1982.

Waldman, Diane. *Twentieth-Century American Drawing: Three Avant-Garde Generations*. Exh. cat. New York: SRGM, 1976.

Watrous, James. *A Century of American Printmaking, 1880–1980*. Madison, Wis.: University of Wisconsin Press, 1984.

Weaver, Mike, ed. *The Art of Photography,*

1839–1989. Exh. cat. New Haven, Conn.: Yale University Press, 1989.

Welling, William. *Photography in America: The Formative Years, 1839–1900*. New York: Crowell, 1978.

Witkin, Lee, and Barbara London. *The Photograph Collector's Guide*. Boston: New York Graphic Society, 1979.

Young, Mahonri Sharp. *Early American Moderns: Painters of the Stieglitz Group*. New York: Watson-Guptill, 1974.

E. PAINTING AND SCULPTURE

Alloway, Lawrence. *Topics in American Art Since 1945*. New York: Norton, 1975.

Andersen, Wayne. *American Sculpture in Process: 1930–1970*. Boston: New York Graphic Society, 1975.

Arnason, H. H. *Modern Sculpture from the Joseph H. Hirshhorn Collection*. Exh. cat. New York: SRGM, 1962.

Art of Our Time: The Saatchi Collection. London and New York: Lund Humphries and Rizzoli, 1984.

Ashton, Dore. *American Art Since 1945*. London: Thames & Hudson, 1982.

Auping, Michael, ed. *Abstraction, Geometry, Painting*. Exh. cat. New York: Harry N. Abrams, 1989.

Barr, Jr., Alfred H., ed. *Painting and Sculpture in the Museum of Modern Art: 1929–1967*. New York: MOMA, 1977.

Beardsley, John, and Jane Livingston. *Hispanic Art in the United States: Thirty Contemporary Painters*. Exh. cat. Houston and New York: Museum of Fine Arts and Abbeville, 1987.

Braun, Emily, ed. *Italian Art in the 20th Century: Painting and Sculpture, 1900–1988*. Exh. cat. Munich: Prestel, 1989.

Denis, Maurice. *Théories, 1890–1910: du symbolisme et de Gauguin vers un nouvel ordre classique*. 4th ed. Paris: Rouart and Watelin, 1920.

Dorival, Bernard. *The School of Paris in the Musée d'Art Moderne*. Trans. C. Brookfield and E. Hart. New York: Harry N. Abrams, 1962.

Eitner, Lorenz. *An Outline of Nineteenth Century European Painting: From David to Cézanne*. 2 vols. New York: Harper & Row, 1987.

Elsen, Albert. *Modern European Sculpture, 1918–1945: Unknown Beings and Other Realities*. New York: G. Braziller, 1979.

Farr, Dennis. *English Art: 1870–1940*. Oxford: Clarendon Press, 1978.

Finch, C. *Image as Language*. Harmondsworth: Finch, 1969.

Fink, Lois Marie, and Joshua C. Taylor. *Academy: The Academic Tradition in American Art*. Exh. cat. Chicago: University of Chicago Press, 1978.

From Hodler to Klee: Swiss Art of the Twentieth Century, Paintings and Sculptures. Exh. cat. London: Tate Gallery, 1959.

Haftmann, Werner. *Painting in the Twentieth Century*. 2 vols. Rev. ed. New York: Praeger, 1979.

Hamilton, George Heard. *19th and 20th Century Art*. New York: Harry N. Abrams, 1970.

———. *Painting and Sculpture in Europe, 1880–1940*. 2nd ed. Pelican History of Art. Harmondsworth: Penguin, 1981.

Hammacher, A. M. *Modern Sculpture: Tradition and Innovation*. Enl. ed. New York: Harry N. Abrams, 1988.

Hard Edge. [Albers, Arp. Bærtling, Herbin, Liberman, Lohse, Mortensen, Taeüber-Arp, Varsarely] Exh. cat. Paris: Galerie Denise

René, 1964.

Hughes, Robert. *Shock of the New*. 2nd ed. London: Thames & Hudson, 1991.

Hunter, Sam and John Jacobus. *Modern Art: Painting, Sculpture, Architecture*. 2nd ed. Englewood Cliffs, N.J.: Prentice-Hall, 1985.

Ito, Norihiro, et al., eds. *Art in Japan Today II, 1970–1983*. Tokyo: Japan Foundation, 1984.

Janson, H. W. *Nineteenth-Century Sculpture*. London: Thames & Hudson, 1985.

Joachimides, Christos, et al., eds. *German Art in the Twentieth Century: Painting and Sculpture, 1905–1985*. Exh. cat. London: Royal Academy of Arts, 1985.

Johnson, Ellen, ed. *American Artists on Art from 1940 to 1980*. New York: Harper & Row, 1982.

Krens, Thomas. *Refigured Painting: The German Image, 1960–1988*. Exh. cat. New York: SRGM, 1988.

Lane, John, and Susan Larsen, eds. *Abstract Painting and Sculpture in America: 1927–1944*. Exh. cat. Pittsburgh: Museum of Art, Carnegie Institute, 1983.

Lehmann-Haupt, Hellmut. *Art Under a Dictatorship*. New York: Oxford University Press, 1954.

Le Normand-Romain, Antoinette, et al. *Sculpture: The Adventure of Modern Sculpture in the Nineteenth and Twentieth Centuries*. New York: Rizzoli, 1986.

Lieberman, William S., ed. *Modern Masters: Manet to Matisse*. Exh. cat. New York: MOMA, 1975.

Lucie-Smith, Edward. *Art Now: From Abstract Expressionism to Superrealism*. New York: Morrow, 1977.

———. *Movements in Art Since 1945*. Rev. ed. London: Thames & Hudson, 1984.

Lynton, Norbert. *The Story of Modern Art*. Oxford: Phaidon, 1980.

Miller, Dorothy C., and William S. Lieberman. *The New Japanese Painting and Sculpture*. Exh. cat. New York: MOMA, 1966.

Nairne, Sandy, and Nicholas Serota. *British Sculpture in the Twentieth Century*. Exh. cat. London: Whitechapel Art Gallery, 1981.

Novak, Barbara. *American Painting in the Nineteenth Century: Realism, Idealism, and the American Experience*. 2nd ed. New York: Harper & Row, 1979.

———. *Nature and Culture: American Landscape and Painting, 1825–1875*. Rev. ed. New York: Oxford University Press, 1995.

Penny, Nicholas. *The Materials of Sculpture*. New Haven, Conn.: Yale University Press, 1993.

Pohribny, Arsén. *Abstract Painting*. Oxford and New York: Phaidon and Dutton, 1979.

Rosenblum, Robert. *Modern Painting and the Northern Romantic Tradition: Friedrich to Rothko*. London: Thames & Hudson, 1975.

———, and H. W. Janson. *Nineteenth-Century Art*. Englewood Cliffs, N.J.: Prentice-Hall, 1984.

Russell, John. *The Meanings of Modern Art*. Rev. ed. London: Thames & Hudson, 1991.

Schapiro, Meyer. *Modern Art: 19th and 20th Centuries: Selected Papers*. New York: G. Braziller, 1978.

Selz, Jean. *Modern Sculpture: Origins and Evolution*. Trans. Annette Michelson. New York: G. Braziller, 1963.

Selz, Peter. *Art in Our Times: A Pictorial History, 1890–1980*. New York: Harry N. Abrams, 1981.

Shone, Richard. *The Century of Change: British Painting Since 1900*. Oxford:

Phaidon, 1977.

Soby, James T., and Alfred H. Barr, Jr. *Twentieth-Century Italian Art*. Exh. cat. New York: MOMA, 1949.

Spalding, Frances. *British Art Since 1900*. London: Thames & Hudson, 1986.

Steinberg, Leo. *Other Criteria: Confrontations with Twentieth-Century Art*. New York: Oxford University Press, 1972.

Tapié, Michel and Tôre Haga. *Avant-Garde Art in Japan*. New York: Harry N. Abrams, 1962.

Tomkins, Calvin. *The Bride and the Bachelors: The Heretical Courtship in Modern Art*. New York: Viking Press, 1965.

Trier, Eduard. *Form and Space: Sculpture of the Twentieth Century*. Rev. ed. New York: Praeger, 1968.

Tuchman, Maurice. *The Spiritual in Art: Abstract Painting 1890–1985*. Los Angeles: LACMA, 1986.

Tucker, William. *The Language of Sculpture*. Reprint of the 1974 ed. London: Thames & Hudson, 1985.

Varnedoe, Kirk. *Northern Light: Realism and Symbolism in Scandinavian Painting 1880–1910*. Exh. cat. New York: Brooklyn Museum, 1982.

Waldman, Diane. *Collage, Assemblage, and the Found Object*. New York: Harry N. Abrams, 1992.

———. *Transformations in Sculpture: Four Decades of American and European Art*. New York: Solomon R. Guggenheim Foundation, 1985.

II. FURTHER READINGS, ARRANGED BY CHAPTER

1. THE PREHISTORY OF MODERN PAINTING

Adams, Steven. *The Barbizon School and the Origins of Impressionism*. London: Phaidon Press, 1994.

Ashton, Dore. *Rosa Bonheur: A Life and a Legend*. New York: Viking, 1981.

Baudelaire, Charles. *Eugène Delacroix: His Life and Work*. Trans. J. Bernstein. New York. Lear Publishers, 1947.

Bowness, Alan. *Gustave Courbet: 1819–1877*. Exh. cat. London: Arts Council of Great Britain, 1978.

Brookner, Anita. *Jacques-Louis David*. London: Oxford University Press, 1980.

Bryson, Norman. *Tradition and Desire: From David to Delacroix*. Cambridge and New York: Cambridge University Press, 1984.

Chu, Petra T. D., ed. *Courbet in Perspective*. Englewood Cliffs, N.J.: Prentice-Hall, 1977.

Crow, Thomas. *Emulation: Making Artists for Revolutionary France*. New Haven, Conn.: Yale University Press, 1995.

[Delacroix] *Eugène Delacroix (1798–1863): Paintings, Drawings, and Prints from North American Collections*. Exh. cat. New York: MMA, 1991.

[Delacroix] *The Journal of Eugène Delacroix*. New York: Garland Pub., 1972. Introduction by R. Motherwell.

Faunce, Sarah, ed. *Courbet Reconsidered*. Exh. cat. New Haven, Conn.: Yale University Press, 1988.

Fermigier, André. *Jean-François Millet*. Exh. cat. London: Hayward Gallery, 1976.

Fried, Michael. *Courbet's Realism*. Chicago: University of Chicago Press, 1990.

Gowing, Lawrence. *Turner: Imagination and Reality*. Exh. cat. New York: MOMA, 1966.

Harding, James. *Artistes Pompiers: French*

Academic Art in the 19th Century. New York: Rizzoli, 1979.

Hargrove, June, ed. *The French Academy: Classicism and Its Antagonists*. Newark, N.J.: University of Delaware Press, 1990.

Johnson, Dorothy. *Jacques-Louis David: Art in Metamorphosis*. Princeton, N.J.: Princeton University Press, 1993.

Johnson, Lee. *The Paintings of Eugène Delacroix: A Critical Catalogue, 1816–1831*. 2 vols. Oxford: Clarendon Press, 1981.

Landow, George P. *William Holman Hunt and Typological Symbolism*. New Haven, Conn.: Paul Mellon Centre for Studies in British Art, 1979.

Licht, Fred. *Goya: The Origins of the Modern Temper in Art*. New York: Universe Books, 1979.

———, ed. *Goya in Perspective*. Englewood Cliffs, N.J.: Prentice-Hall, 1973.

Murphy, Alexandra R. *Jean-François Millet*. Exh. cat. Boston: Museum of Fine Arts, 1984.

Nanteuil, Luc de. *David*. The Library of Great Painters. New York: Harry N. Abrams, 1985.

Rosenblum, Robert. *Jean-Auguste-Dominique Ingres*. New York: Harry N. Abrams, 1990.

Schnapper, Antoine. *David*. New York: Alpine Fine Arts Collection, 1980.

Stanton, Theodore, ed. *Reminiscences of Rosa Bonheur*. New York: Hacker Art Books, 1976.

Vigne, George. *Ingres*. Trans. J. Goodman. New York: Abbeville, c. 1995.

2. REALISM, IMPRESSIONISM AND EARLY PHOTOGRAPHY

Adler, Kathleen. *Manet*. Oxford: Phaidon, 1986.

———, and Tamar Garb. *Berthe Morisot*. Ithaca, N.Y.: Cornell University Press, 1987.

Barger, Susan M., and William B. White. *The Daguerreotype: Nineteenth-Century Technology and Modern Science*. Washington, D.C.: Smithsonian Institution Press, 1991.

Belloli, Andrea. *A Day in the Country: Impressionism and the French Landscape*. Exh. cat. Los Angeles: LACMA, 1984.

Boggs, Jean Sutherland. *Degas*. Exh. cat. Chicago: Art Institute of Chicago, 1996.

Boime, Albert. *The Art of the Macchia and the Risorgimento*. Chicago: University of Chicago Press, 1993.

Breeskin, Adelyn Dohme. *Mary Cassatt: A Catalogue Raisonné of the Oils, Pastels, Watercolors, and Drawings*. 2nd rev. ed. Washington, D.C.: Smithsonian Institution Press, 1979.

Brettell, Richard, and Joachim Pissarro. *The Impressionist and the City: Pissarro's Series Paintings*. New Haven, Conn.: Yale University Press, 1992.

Broude, Norma. *Edgar Degas*. New York: Rizzoli, 1993.

———. *Impressionism: A Feminist Reading*. New York: Rizzoli, 1991.

Buckland, Gail. *Fox Talbot and the Invention of Photography*. Boston: D. R. Godine, 1980.

Callen, Anthea. *The Spectacular Body: Science, Method, and Meaning in the Work of Degas*. New Haven, Conn.: Yale University Press, 1995.

Casteras, Susan, et al. *John Ruskin and the Victorian Eye*. New York: Harry N. Abrams, 1993.

Cikovsky, Jr., Nicolai. *Winslow Homer*. New York: Harry N. Abrams, 1990.

Clark, T. J. *The Absolute Bourgeois: Artists and Politics in France, 1848–1851*. London:

Thames & Hudson, 1982.

———. *The Painting of Modern Life: Paris in the Art of Manet and His Followers*. London: Thames & Hudson, 1985.

Clarke, Michael. *Corot and the Art of Landscape*. London: British Museum Press, 1991.

Cumming, Elizabeth, and Wendy Kaplan. *Arts and Crafts Movement*. World of Art. London: Thames & Hudson, 1991.

Degas Pastels. London: Thames & Hudson, 1992. Texts by J. S. Boggs and A. Maheux.

Denvir, Bernard. *The Impressionists: A Documentary Study*. London: Thames & Hudson, 1986.

———. *The Thames & Hudson Encyclopedia of Impressionism*. World of Art. London: Thames & Hudson, 1990.

Dorment, Richard. *James McNeill Whistler*. Exh. cat. New York: Harry N. Abrams, 1995.

Eisenman, Stephen F. *Gauguin's Skirt*. London: Thames & Hudson, 1997.

Fried, Michael. *Manet's Modernism, or, the Face of Painting in the 1860s*. Chicago: University of Chicago Press, 1996.

Gernsheim, Helmut. *Julia Margaret Cameron: Pioneer of Photography*. 2nd ed. London and New York: Aperture, 1975.

———, and Alison Gernsheim. *L. J. M. Daguerre: The History of the Diorama and the Daguerreotype*. Reprint. New York: Dover Publications, 1968.

Getlein, Frank. *Mary Cassatt: Paintings and Prints*. New York: Abbeville, 1980.

Goodrich, Lloyd. *Thomas Eakins*. 2 vols. Cambridge, Mass.: Harvard University Press, 1982.

Gordon, Robert, and Andrew Forge. *Monet*. New York: Harry N. Abrams, 1983.

Gosling, Nigel. *Nadar*. New York: Knopf, 1976.

Haas, Robert B. *Muybridge's Man in Motion*. Berkeley and Los Angeles: University of California Press, 1976.

Hambourg, Maria Morris. *Nadar*. Exh. cat. New York: Harry N. Abrams, 1995.

Hamilton, George. *Manet and His Critics*. Reprint of the 1954 ed. New Haven, Conn.: Yale University Press, 1986.

Hanson, Anne Coffin. *Manet and the Modern Tradition*. 2nd ed. New Haven, Conn.: Yale University Press, 1979.

Hendricks, George. *Edward Muybridge—The Father of the Motion Picture*. New York: Grossmann Publishers, 1975.

Herbert, Richard. *Impressionism: Art, Leisure and Parisian Society*. New Haven, Conn.: Yale University Press, 1988.

Higonnet, Anne. *Berthe Morisot*. New York: Harper & Row, 1990.

———. *Berthe Morisot's Images of Women*. Cambridge, Mass.: Harvard University Press, 1992.

Hills, Patricia. *John Singer Sargent*. Exh. cat. New York: WMAA, 1986. Texts by L. Ayres, A. Boime, W. H. Gerdts, S. Olson, G. A. Reynolds.

Hilton, Timothy. *The Pre-Raphaelites*. World of Art. Reprint of the 1970 ed. London: Thames & Hudson, 1985.

[Homer] *Winslow Homer and the New England Coast*. Exh. cat. New York: WMAA, 1985.

House, John. *Monet: Nature into Art*. New Haven, Conn.: Yale University Press, 1986.

Jenkyns, Richard. *Dignity and Decadence: Victorian Art and the Classical Inheritance*. Cambridge, Mass.: Harvard University Press, 1992.

Johns, Elizabeth. *Thomas Eakins and the Heroism of Modern Life*. Princeton, N.J.: Princeton University Press, 1984.

Kendall, Richard, and Griselda Pollock, eds. *Dealing with Degas: Representations of Women and the Politics of Vision*. New York: Universe Books, 1992.

Krell, Alan. *Manet and the Painters of Modern Life*. London: Thames & Hudson, 1996.

Lipton, Eunice. *Looking into Degas*. Berkeley, Calif.: University of California Press, 1986.

Lowry, Bates, and Isabel Barrett Lowry. *The Silver Canvas: Daguerreotype Masterpieces from the J. Paul Getty Museum*. London: Thames & Hudson, 1998.

[Manet] *The Complete Paintings of Manet*. New York: Harry N. Abrams, 1967. Introduction by P. Pool; Notes and Catalogue by S. Orienti.

[Manet] *Manet: 1832–1883*. Exh. cat. New York: MMA, 1983.

Mathews, Nancy M. *Mary Cassatt*. Library of American Art. New York: Harry N. Abrams, 1987.

Miller, David, ed. *American Iconology: New Approaches to Nineteenth-Century Art and Literature*. New Haven, Conn.: Yale University Press, 1993.

Moffett, Charles S. *The New Painting: Impressionism, 1874–1886*. Exh. cat. San Francisco: Fine Arts Museums, 1986.

Mosby, Dewey, and Darrel Sewell. *Henry Ossawa Tanner*. Exh. cat. Philadelphia and New York: Philadelphia Museum of Art and Rizzoli, 1991.

Needham, Gerald. *19th-Century Realist Art*. New York: Harper & Row, 1988.

Nochlin, Linda. *Impressionism and Post-Impressionism, 1874–1904: Sources and Documents*. Englewood Cliffs, N.J.: Prentice-Hall, 1976.

———. *Realism*. Harmondsworth.: Penguin, 1972.

———. *Realism and Tradition in Art, 1848–1900: Sources and Documents*. Englewood Cliffs, N.J.: Prentice-Hall, 1966.

Pollock, Griselda. *Mary Cassatt*. New York: Harper & Row, 1980.

The Pre-Raphaelites. Exh. cat. London: Tate Gallery, 1984.

Ratcliff, Carter. *John Singer Sargent*. New York: Abbeville, 1982.

Reff, Theodore. *Manet and Modern Paris*. Exh. cat. Washington, D.C.: National Gallery of Art, 1982.

Renoir. Exh. cat. London: Hayward Gallery, 1985. Texts by J. House, A. Distel, L. Gowing.

Renoir, Jean. *Renoir, My Father*. Trans. R. and D. Weaver. Boston: Little, Brown, 1962.

Rewald, John. *The History of Impressionism*. 4th rev. ed. New York: MOMA, 1973.

———, and Frances Weitzenhoffer, eds. *Aspects of Monet: A Symposium on the Artist's Life and Times*. New York: Harry N. Abrams, 1984.

Reynolds, Graham. *Victorian Painting*. Rev. ed. New York: Harper & Row, 1987.

Rosenblum, Robert. *Paintings in the Musée d'Orsay*. New York: Stewart, Tabori and Chang, 1989.

Rouart, Denis. *Renoir*. New York: Skira/Rizzoli, 1985.

Schaaf, Larry J. *Out of the Shadows: Herschel, Talbot and the Invention of Photography*. New Haven, Conn.: Yale University Press, 1992.

Sewell, Darrel. *Thomas Eakins: Artist of Philadelphia*. Exh. cat. Philadelphia: Philadelphia Museum of Art, 1982.

Spate, Virginia. *Claude Monet: Life and Work*. London: Thames & Hudson, 1992.

Stuckey, Charles F. *Claude Monet, 1840–1926*. Exh. cat. London: Thames & Hudson,

1995.

———, and William P. Scott. *Berthe Morisot, Impressionist*. New York: Hudson Hills Press, 1987.

Sutton, Denys. *Edgar Degas: Life and Work*. New York: Rizzoli, 1986.

Tinterow, Gary, and Henri Loyrette. *Origins of Impressionism*. Exh. cat. New York: MMA, 1994.

Tucker, Paul. *Monet at Argenteuil*. New Haven, Conn.: Yale University Press, 1981.

———. *Monet in the '90s: The Series Paintings*. Exh. cat. New Haven, Conn.: Yale University Press, 1989.

Wadley, Nicholas. *Renoir: A Retrospective*. New York: Hugh Lauter Levin Assoc., 1987.

Weisberg, Gabriel. *Beyond Impressionism: The Naturalist Impulse*. New York: Harry N. Abrams, 1992.

———. *The Realist Tradition: French Painting and Drawing, 1830–1900*. Exh. cat. Cleveland: Cleveland Museum of Art, 1981.

White, Barbara Ehrlich. *Renoir: His Life, Art, and Letters*. New York: Harry N. Abrams, 1984.

Wilmerding, John. *Winslow Homer*. New York: Praeger, 1972.

Wood, Christopher. *The Pre-Raphaelites*. New York: Viking, 1981.

3. POST-IMPRESSIONISM

Adriani, Gotz. *Cézanne: A Biography*. New York: Harry N. Abrams, 1990.

———. *Toulouse-Lautrec: Complete Graphic Works*. London: Thames & Hudson, 1988.

Andersen, Wayne. *Gauguin's Paradise Lost*. New York: Viking Press, 1971.

Balakian, Anna. *The Symbolist Movement: A Critical Appraisal*. New York: Random House, 1967.

Bonnard: The Late Paintings. Exh. cat. Washington, D.C.: Phillips Collection, 1984. Texts by J. Russell, S. A. Nash, J. Clair, A. Terrasse, H. Hahnloser-Ingold, J.-F. Chevrier.

Bonnard and His Environment. New York: MOMA, 1964. Texts by J. T. Soby, J. Elliott, M. Wheeler.

Bouret, Jean. *Henri Rousseau*. Greenwich, Conn.: New York Graphic Society, 1961.

Brettell, Richard, et al. *The Art of Paul Gauguin*. Exh. cat. Washington, D.C.: National Gallery of Art, 1988.

Broude, Norma. *Georges Seurat*. New York: Rizzoli, 1992.

Cachin, Françoise. *Cézanne*. Exh. cat. Philadelphia: Philadelphia Museum of Art, 1996.

[Cézanne] *The Complete Paintings of Cézanne*. Harmondsworth: Penguin, 1985. Introduction by I. Dunlop; Text by S. Orienti.

Dabrowski, Magdalena. *The Symbolist Aesthetic*. Exh. cat. New York: MOMA, 1980.

d'Argencourt, Louise, et al. *Puvis de Chavannes: 1824–1898*. Exh. cat. Ottawa: National Gallery of Canada, 1977.

Delevoy, Robert. *Symbolists and Symbolism*. New York: Rizzoli, 1978.

Denvir, Bernard. *Post-Impressionism*. World of Art. London: Thames & Hudson, 1992.

Druick, Douglas W., et al. *Odilon Redon: Prince of Dreams, 1840–1916*. Exh. cat. New York: Harry N. Abrams, 1994.

Frèches-Thory, Clair. *The Nabis: Bonnard, Vuillard and Their Circle*. New York: Harry N. Abrams, 1991.

Fry, Roger. *Seurat*. 3rd ed. London: Phaidon, 1965.

Gauguin, Paul. *The Intimate Journals of Paul*

Gauguin. Boston: KPI, 1985.

———. *Noa Noa: The Tahitian Journals.* Trans. O. F. Theis. New York: Dover Publications, 1985.

Gaunt, William. *Renoir.* Rev. enl. ed. Oxford: Phaidon, 1982. Text by K. Adler.

Gerhardus, Maly, and Dietfried Gerhardus. *Symbolism and Art Nouveau.* Trans. A. Bailey. Oxford: Phaidon, 1979.

Goldwater, Robert. *Paul Gauguin.* New York: Harry N. Abrams, 1984.

———. *Symbolism.* New York: Harper & Row, 1979.

Gowing, Lawrence. *Cézanne, the Early Years, 1859–1872.* New York: Harry N. Abrams, 1988.

Herbert, Robert L. *Georges Seurat, 1859–1891.* Exh. cat. New York: MMA, 1991.

Hirsh, S., ed. "Symbolist Art and Literature." *Art Journal* (Summer 1985) special issue.

Homer, William I. *Seurat and the Science of Painting.* Reprint. New York: Hacker Art Books, 1985.

Hulsker, Jan. *The Complete Van Gogh: Paintings, Drawings, Sketches.* New York: Harry N. Abrams, 1980.

Hyman, Timothy. *Bonnard.* London: Thames & Hudson, 1998.

Jullian, Philippe. *Dreamers of Decadence: Symbolist Painters of the 1890s.* Trans. R. Baldick. New York: Praeger, 1971.

Lemoyne de Forges, Marie-Therese, ed. *Signac.* Exh. cat. Paris: Musée National du Louvre, 1963.

Lewis, Mary Tompkins. *Cézanne's Early Imagery.* Berkeley, Calif.: University of California Press, 1989.

Lucie-Smith, Edward. *Symbolist Art.* London: Thames & Hudson, 1972.

Murray, Gale B. *Toulouse-Lautrec: A Retrospective.* Exh. cat. New York: Hugh Lauter Levin Assoc., 1992.

Rewald, John. *Paul Cézanne: The Watercolors, A Catalogue Raisonné.* Boston: Little, Brown, 1983.

———. *Post-Impressionism: From Van Gogh to Gauguin.* 3rd ed. rev. New York: MOMA, 1978.

———. *Seurat: A Biography.* New York: Harry N. Abrams, 1990.

Roskill, Mark. *Van Gogh, Gauguin, and the Impressionist Circle.* London: Thames & Hudson, 1970.

Rubin, William, ed. *Cézanne: The Late Work.* New York: MOMA, c. 1977.

Russell, John. *Seurat.* London: Thames & Hudson, 1965.

Schapiro, Meyer. *Paul Cézanne.* Concise ed. London: Thames & Hudson, 1988.

———. *Van Gogh.* Rev. ed. London: Thames & Hudson, 1982.

Shiff, Richard. *Cézanne and the End of Impressionism: A Study of the Theory, Technique, and Critical Evaluation of Modern Art.* Chicago: University of Chicago Press, 1984.

———. *Toulouse-Lautrec.* World of Art. London: Thames & Hudson, 1991.

Shone, Richard. *The Post-Impressionists.* London: Octopus Books, 1980.

Thomson, Belinda. *Gauguin.* World of Art. London: Thames & Hudson, 1987.

Thomson, Richard. *Seurat.* Salem, N.H.: Salem House, 1985.

[Van Gogh] *The Complete Letters of Vincent Van Gogh.* 3 vols. London: Thames & Hudson, 1959.

Verdi, Richard. *Cézanne.* World of Art. London: Thames & Hudson, 1992.

Walther, Ingo F., and Rainer Metzger. *Van Gogh: The Complete Paintings.*
Cologne: Taschen, 1993.

4. THE ORIGINS OF MODERN ARCHITECTURE AND DESIGN

Barthes, Roland. *The Eiffel Tower and Other Mythologies.* Trans. R. Howard. New York: Hill and Wang, 1979.

Charenbhak, Wichit. *Chicago School Architects and Their Critics.* Ann Arbor, Mich.: UMI Research Press, 1984.

Crawford, Alan. *Charles Rennie Mackintosh.* London: Thames & Hudson, 1995.

Hitchcock, Henry-Russell. *The Architecture of H. H. Richardson and His Times.* Rev. ed. Hamden, Conn.: Archon Books, 1961.

Howarth, Thomas. *Charles Rennie Mackintosh and the Modern Movement.* 2nd ed. London: Routledge and Kegan Paul, 1977.

Lethaby, William R. *Philip Webb and His Work.* New ed. London: Raven Oaks Press, c. 1979.

Loyrette, Henri. *Gustave Eiffel.* New York: Rizzoli, 1985.

Manieri-Elia, Mario. *Louis Henry Sullivan, 1856–1924.* Trans. A. Shuggar with C. Creen. New York: Princeton Architectural Press, 1996.

McKean, John. *Crystal Palace: Joseph Paxton and Charles Fox.* Architecture in Detail. London: Phaidon, 1994.

Mead, Christopher Curtis. *Charles Garnier's Paris Opera: Architectural Empathy and the Renaissance of French Classicism.* Cambridge, Mass.: MIT Press, 1991.

Morrison, Hugh. *Louis Sullivan, Prophet of Modern Architecture.* Exh. cat. New York: MOMA, 1935.

Pevsner, Nikolaus. *Charles Rennie Mackintosh.* Milan: Il Balcone, 1950.

Sullivan, Louis H. *Kindergarten Chats and Other Writings.* New York: Dover Publications, 1979.

Szarkowski, John. *The Idea of Louis Sullivan.* Minneapolis: University of Minnesota Press, 1956.

Twombley, Robert, ed. *Louis Sullivan, the Public Papers.* Chicago: University of Chicago Press, 1988.

5. ART NOUVEAU

Adlmann, Jan E. *Vienna Moderne, 1898–1918.* Exh. cat. New York: Cooper-Hewitt Museum, 1979.

Amaya, Mario. *Art Nouveau.* New York: Dutton, 1967.

Arwas, Victor. *Art Deco.* Rev. ed. New York: Harry N. Abrams, 1992.

Chipp, Herschel B. *Viennese Expressionism 1910–1924.* Exh. cat. Berkeley, Calif.: University Museum, University of California, 1963.

Collins, George R. *Antoni Gaudí.* Masters of World Architecture Series. New York: G. Braziller, 1960.

Comini, Alessandra. *Gustav Klimt.* London: Thames & Hudson, 1975.

Duncan, Alastair. *Art Nouveau.* World of Art. London: Thames & Hudson, 1994.

Eggum, Arne. *Edvard Munch: Paintings, Sketches, and Studies.* New York: C. N. Potter, 1985.

Elderfield, John. *The Masterworks of Edvard Munch.* New York: MOMA, 1979.

Farmer, John David. *Ensor.* Exh. cat. New York: G. Braziller, 1976.

Heller, Reinhold. *Munch: His Life and Work.* Chicago: University of Chicago Press, 1984.

Hirsh, Sharon. *Ferdinand Hodler.* New York: G. Braziller, 1982.

Hodin, J.P. *Edvard Munch.* London: Thames & Hudson, 1972.

Kallir, Jane. *Viennese Design and the Wiener Werkstätte.* New York: G. Braziller, 1986.

Kempton, Richard. *Art Nouveau: An Annotated Bibliography.* Los Angeles: Hennessey and Ingalls, 1977.

[Klimt] *Gustav Klimt and Egon Schiele.* Exh. cat. New York: SRGM, 1965. Texts by T. M. Messer and J. Dobai.

Madsen, Stephan T. *Sources of Art Nouveau.* Reprint. New York: Da Capo Press, 1975.

McCully, Marilyn. *El Quatre Gats: Art in Barcelona around 1900.* Exh. cat. Princeton, N.J.: Princeton University Art Museum, 1978.

[Munch] *Edvard Munch: Symbols and Images.* Exh. cat. Washington, D.C.: National Gallery of Art, 1978. Introduction by R. Rosenblum; Texts by A. Eggum, R. Heller, T. Nergaard, R. Stang, B. Torjusen, G. Wohl.

Nebehay, Christian M. *Ver Sacrum, 1898–1903.* Vienna: Edition Tuschi, 1975.

Novotny, Fritz, and Johannes Dobai. *Gustav Klimt.* Salzburg: Verlag Galerie Welz, 1967.

Partsch, Suzanna. *Klimt: Life and Work.* London: Bracken, 1993.

Pevsner, Nikolaus. *Art Nouveau in Britain.* Exh. cat. London: The Arts Council, 1965.

Powell, Nicholas. *The Sacred Spring: The Arts of Vienna, 1898–1918.* Greenwich, Conn.: New York Graphic Society, 1974.

Reade, Brian. *Aubrey Beardsley.* New York: Viking Press, 1967.

Rheims, Maurice. *The Flowering of Art Nouveau.* Trans. P. Evans. New York: Harry N. Abrams, 1966.

Russell, Frank, ed. *Art Nouveau Architecture.* New York: Rizzoli, 1979.

Schmutzler, Robert. *Art Nouveau.* Abridged ed. Trans. E. Rodotti. London: Thames & Hudson, 1978.

Schorske, Carl. *Fin-de-Siècle Vienna: Politics and Culture.* New York: Vintage Books, 1980.

Selz, Peter. *Ferdinand Hodler.* Exh. cat. Berkeley, Calif.: University Art Museum, University of California, 1972. With contributions by J. Brüschweiler et al.

———, and Mildred Constantine, eds. *Art Nouveau: Art and Design at the Turn of the Century.* Rev. ed. New York: MOMA, 1974.

Silverman, Deborah. *Art Nouveau in Fin-de-Siècle France.* Berkeley, Calif.: University of California Press, 1989.

Sotriffer, Kristian. *Modern Austrian Art.* London: Thames & Hudson, 1965.

Stansky, Peter. *Redesigning the World: William Morris, the 1880s, and the Arts and Crafts.* Princeton, N.J.: Princeton University Press, 1985.

Sweeney, James J., and Joseph L. Sert. *Antoni Gaudí.* New York: Praeger, 1960.

[Van de Velde] *Henry van de Velde, 1863–1957.* Exh. cat. Otterlo, The Netherlands: Kröller-Müller Rijksmuseum, 1964.

Varnedoe, Kirk. *Vienna 1900: Art, Architecture, and Design.* Exh. cat. New York: MOMA, 1986.

———, and Elizabeth Streicher. *Graphic Work of Max Klinger.* New York: Dover Publications, 1977.

Vergo, Peter. *Art in Vienna, 1898–1918: Klimt, Kokoschka, Schiele, and Their Contemporaries.* London: Phaidon, 1975.

Waissenberger, Robert. *Vienna Secession.* New York: Rizzoli, 1977.

Whitford, Frank. *Klimt.* London: Thames & Hudson, 1990.

6. THE ORIGINS OF MODERN SCULPTURE

Barr, Margaret S. *Medardo Rosso.* Exh. cat.

New York: MOMA, 1963.

Elsen, Albert, ed. *Rodin Rediscovered.* Exh. cat. Washington, D.C.: National Gallery of Art, 1981.

George, Waldemar. *Aristide Maillol.* London: Cory, Adams & Mackay, 1965.

Gray, Christopher. *Sculpture and Ceramics of Paul Gauguin.* Baltimore: Johns Hopkins Press, 1963.

Jianu, Ionel. *Bourdelle.* Paris: Arted, 1965.

Lampert, Catherine. *Rodin: Sculpture and Drawings.* Exh. cat. London: The Arts Council, 1986.

Longwell, Dennis. *Steichen: The Master Prints, 1895–1914.* New York: MOMA, 1978.

Lorquin, Bertrand. *Maillol.* London: Thames & Hudson, 1995.

Rewald, John, and Leonard von Matt. *Degas: Sculpture, the Complete Works.* London: Thames & Hudson, c. 1957.

Rilke, Rainer Maria. *Rodin.* Salt Lake City: Peregrine Smith, 1979.

7. FAUVISM

Apollonio, Umbro. *Fauves and Cubists.* London: Batsford, 1960.

Barr, Jr., Alfred H. *Matisse: His Art and His Public.* Exh. cat. New York: MOMA, 1951.

Courthion, Pierre. *Georges Rouault.* New York: Harry N. Abrams, 1962.

Crespelle, Jean–Paul. *The Fauves.* Trans. A. Brookner. Greenwich, Conn.: New York Graphic Society, 1962.

Diehl, Gaston. *Derain.* Les maîtres de la peinture moderne. Paris: Flammarion, 1964.

Duthuit, Georges. *The Fauvist Painters.* Trans. R. Mannheim. New York: Wittenborn, Schultz, 1950.

Elderfield, John. *Henri Matisse, A Retrospective.* Exh. cat. New York: MOMA, 1992.

——. *The "Wild Beasts": Fauvism and Its Affinities.* Exh. cat. New York: MOMA, 1976.

Flam, Jack D. *Matisse: The Man and His Art, 1869–1918.* London: Thames & Hudson, 1986.

——, ed. *Matisse: A Retrospective.* New York: Park Lane, 1990.

——, ed. *Matisse On Art.* Berkeley, Calif.: University of California Press, 1994.

Freeman, Judi. *The Fauve Landscape.* Exh. cat. New York: Abbeville, 1990.

Gowing, Lawrence. *Matisse.* London: Thames & Hudson, 1979.

Herbert, James D. *Fauve Painting: The Making of Cultural Politics.* New Haven, Conn.: Yale University Press, 1992.

Leymarie, Jean. *Fauvism, Biographical and Critical Study.* Trans. James Emmons. New York: Skira, 1959.

Rewald, John. *Les Fauves.* Exh. cat. New York: MOMA, 1952.

Rouault, Georges. *Miserere.* Boston: Boston Book and Art Shop, 1963. Introduction A. Blunt; preface by the artist.

Schneider, Pierre. *Matisse.* New York: Rizzoli, 1984.

Selz, Jean. *Vlaminck.* Les maîtres de la peinture moderne. Paris: Flammarion, 1963.

Sutton, Denys. *André Derain.* London: Phaidon, 1959.

Venturi, Lionello. *Rouault: Biographical and Critical Study.* Trans. J. Emmons. Paris: Skira, 1959.

Whitfield, Sarah. *Fauvism.* London: Thames & Hudson, 1996.

8. EXPRESSIONISM IN GERMANY

Barron, Stephanie. *German Expressionist Sculpture.* Exh. cat. Los Angeles: LACMA, 1980.

——, ed. *"Degenerate art": The Fate of the Avant-Garde in Nazi Germany.* Exh. cat. Los Angeles: LACMA, 1991.

Benson, Timothy. *Expressionist Utopias: Paradise, Metropolis, Architectural Fantasy.* Exh. cat. Los Angeles: LACMA, 1993.

Comini, Alessandra. *Egon Schiele, 1890–1918.* London: Thames & Hudson, 1976.

Dabrowski, Magdalena. *Kandinsky Compositions.* Exh. cat. New York: MOMA, 1995.

Dube, Wolf-Dieter, *The Expressionists.* London: Thames & Hudson, 1988.

Expressionism: A German Intuition. Exh. cat. New York: SRGM, 1980.

Gordon, Donald. *Ernst Ludwig Kirchner.* Cambridge, Mass.: Harvard University Press, 1968.

——. *Oskar Kokoschka and the Visionary Tradition.* Bonn: Bouvier, 1981.

Grohmann, Will. *E. L. Kirchner.* Stuttgart: W. Kohlhammer, 1961.

——. *Karl Schmidt-Rottluff.* Stuttgart: W. Kohlhammer, 1956.

Hess, Hans. *Lyonel Feininger.* Stuttgart: W. Kohlhammer, 1959.

Hinz, Renate, ed. *Käthe Kollwitz: Graphics, Posters, Drawings.* Trans. Rita and Robert Kimber. New York: Pantheon Books, 1982. Foreword by Lucy R. Lippard.

[Jawlensky] *Alexej von Jawlensky.* Rotterdam: Museum Boymans - van Beuningen, 1994.

Jordan, Jim. *Paul Klee and Cubism.* Princeton, N.J.: Princeton University Press, 1984.

Kallir, Otto. *Egon Schiele: Oeuvre Catalog of the Paintings.* Wien: Zsolnay, 1966.

Kandinsky in Munich: 1896–1914. Exh. cat. New York: SRGM, 1982.

Klee, Felix, ed. *The Diaries of Paul Klee: 1898–1918.* Berkeley and Los Angeles: University of California Press, 1964.

Klipstein, August. *The Graphic Work of Käthe Kollwitz: Complete Illustrated Catalogue.* New York: Galerie St. Etienne, 1955.

Kokoschka: A Retrospective Exhibition of Paintings, Drawings, Lithographs, Stage Design and Books. Exh. cat. London: Tate Gallery, 1962. Texts by E. H. Gombrich, F. Novotny, H. M. Wingler, et al.

[Kokoschka] *Oskar Kokoschka, 1886–1980.* Exh. cat. New York: SRGM, 1986.

Kollwitz, Hans, ed. *The Diary and Letters of Käthe Kollwitz.* Evanston, Ill.: Northwestern University Press, 1988.

Lanchner, Carolyn, ed. *Paul Klee.* Exh. cat. New York: MOMA, 1987.

Levine, Frederick S. *The Apocalyptic Vision: The Art of Franz Marc as German Expressionism.* New York: Harper & Row, 1979.

Lincoln, Louise, ed. *German Realism of the Twenties: The Artist as Social Critic.* Exh. cat. Minneapolis: Minneapolis Institute of Arts, c.1980.

Lindsay, Kenneth and Peter Vergo. *Kandinsky: Complete Writings on Art.* 2 vols. Boston: G. K. Hall, 1982.

Lloyd, Jill. *German Expressionism: Primitivism and Modernity.* New Haven, Conn.: Yale University Press, 1991.

[Modersohn-Becker] *The Letters and Journals of Paula Modersohn-Becker.* Trans. and ed. J. D. Radycki. Metuchen, N.J.: Scarecrow Press, 1980.

Neue Sachlichkeit and German Realism of the Twenties. Exh. cat. London: Arts Council of Great Britain, 1978.

Painters of the Brücke. Exh. cat. London: Tate Gallery, 1964.

Perry, Gillian. *Paula Modersohn-Becker: Her Life and Work.* New York: Harper & Row, 1979.

Prelinger, Elizabeth. *Käthe Kollwitz.* Exh. cat. Washington, D.C.: National Gallery of Art, 1992.

Richard, Lionel. *Phaidon Encyclopedia of Expressionism.* Trans. S. Tint. Oxford: Phaidon, 1978.

Roethel, Hans. *The Blue Rider.* New York: Praeger, 1971.

——, and Jean Benjamin. *Kandinsky Catalogue Raisonné of the Oil Paintings.* 2 vols. Ithaca, N.Y.: Cornell University Press, 1982.

Sabarsky, Serge. *Egon Schiele.* New York: Rizzoli, 1985.

Samuel, Richard and R. Hinton Thomas. *Expressionism in German Life, Literature, and the Theatre (1910–1924).* Cambridge: W. Heffer & Sons, Ltd., 1939.

Schvey, Henry. *Oskar Kokoschka, the Painter as Playwright.* Detroit: Wayne State University Press, 1982.

Urban, Martin. *Emil Nolde: Catalogue Raisonné of the Oil Paintings.* 2 vols. London: Sotheby's Publications, 1987–1990.

Vergo, Peter. *Art in Vienna, 1898–1918: Klimt, Kokoschka, Schiele, and Their Contemporaries.* London: Phaidon, 1975.

Vogt, Paul. *Erich Heckel.* Recklinghausen, Germany: Bongers, 1965.

——. *Expressionism: German Painting, 1905–1920.* Trans. A. Vivus. New York: Harry N. Abrams, 1980.

Washton, Rose-Carol, ed. *German Expressionism: Documents from the End of the Wilhelmine Empire to the Rise of National Socialism.* Documents of Twentieth Century Art. New York: G. K. Hall, 1993.

Weiler, Clemens. *Alexej Jawlensky.* Cologne: M. Du Mont Schauberg, 1959.

Whitford, Frank. *Expressionist Portraits.* New York: Abbeville, 1987.

——. *Egon Schiele.* London: Thames & Hudson, 1981.

Zigrosser, Carl. *The Expressionists: A Survey of Their Graphic Art.* New York: G. Braziller, 1957.

9. EARLY TWENTIETH-CENTURY SCULPTURE

Chave, Anna. *Constantin Brancusi: Shifting the Bases of Art.* New Haven, Conn.: Yale University Press, 1993.

Elsen, Albert E. *The Sculpture of Henri Matisse.* New York: Harry N. Abrams, 1972.

Geist, Sydney. *Brancusi: A Study of the Sculpture.* Rev. enl. ed. New York: Hacker Art Books, 1983.

Hofmann, Werner. *Wilhelm Lehmbruck.* London: A. Zwemmer, c. 1958.

Maddow, Ben. *Edward Weston: Fifty Years.* Millerton, N.Y.: Aperture, 1973.

Miller, Sanda. *Constantin Brancusi: A Survey of His Work.* Oxford: Clarendon, 1995.

Shanes, Eric. *Constantin Brancusi.* New York: Abbeville, 1989.

Stebbins, Jr., Theodore E. *Weston's Westons: Portraits and Nudes.* Boston: Museum of Fine Arts, 1994.

——, and Karen E. Quinn. *Weston's Westons: California and the West.* Boston: Museum of Fine Arts, 1994.

10. CUBISM

Apollinaire, Guillaume. *The Cubist Painters.* New York: Wittenborn & Co., 1949.

Archipenko, Aleksandr, et al. *Archipenko: Fifty Creative Years, 1908–1958.* New York:

Tekhne, 1960.

Arnason, H. H. *Jacques Lipchitz: Sketches in Bronze.* New York: Praeger, 1969.

Barr, Jr., Alfred H. *Cubism and Abstract Art.* Exh. cat. New York: MOMA, 1936.

——, ed. *Picasso: 75th Anniversary Exhibition.* Exh. cat. New York: MOMA, 1957.

Bois, Yve-Alain. "Kahnweiler's Lesson." *Representations* (Spring 1987): 33–68.

Cogniat, Raymond. *Georges Braque.* New York: Harry N. Abrams, 1980.

——. *Georges Braque: Rétrospective.* Exh. cat. Saint-Paul: Foundation Maeght, 1994.

Cooper, Douglas. *The Cubist Epoch.* Exh. cat. New York and London: LACMA, MMA, and Phaidon, 1970.

——, and Gary Tinterow. *Essential Cubism: Braque, Picasso and Their Friends, 1907–1920.* Exh. cat. London: Tate Gallery, 1983.

——, ed. and trans. *Juan Gris: Catalogue raisonné de l'oeuvre peinte.* Paris: Berggruen, 1977.

——, ed. *Letters of Juan Gris (1913–1927).* London: Lund Humphries, 1956.

Cowling, Elizabeth, and John Golding. *Picasso: Sculptor/Painter.* London: Tate Gallery, 1994.

De Francia, Peter. *Fernand Léger.* New Haven, Conn.: Yale University Press, 1983.

Delevoy, Robert. *Léger: Biographical and Critical Study.* Trans. S. Gilbert. Lausanne: Skira 1962.

Golding, John. *Cubism: A History and an Analysis, 1907–1914.* 3rd ed. Cambridge, Mass.: Belknap Press of Harvard University Press, 1988.

Goldscheider, Cécile. *Laurens.* Cologne: Kiepenheur & Witsch, 1957.

Green, Christopher. *Cubism and Its Enemies.* New Haven, Conn.: Yale University Press, 1987.

——. *Juan Gris.* New Haven, Conn.: Yale University Press, 1992.

——. *Léger and the Avant-Garde.* New Haven, Conn.: Yale University Press, 1976.

Hammacher, A. M. *Jacques Lipchitz.* Trans. J. Brockway. New York: Harry N. Abrams, 1975. Introductory statement by the artist.

Harrison, Charles, Francis Frascina, and Gill Perry. *Primitivism, Cubism, Abstraction: The Early Twentieth Century.* New Haven, Conn.: Yale University Press, 1993.

Hilton, Timothy. *Picasso.* World of Art. London: Thames & Hudson, 1975.

Hofmann, Werner, and Daniel-Henry Kahnweiler. *The Sculpture of Henri Laurens.* New York: Harry N. Abrams, 1970.

Jacques Villon, Raymond Duchamp-Villon, Marcel Duchamp. Exh. cat. New York: SRGM, 1957.

Kahnweiler, Daniel-Henry. *The Rise of Cubism.* Trans. Henry Aronson. New York: Wittenborn, Schultz, 1949.

Karshan, Donald. *Archipenko: The Sculpture and Graphic Art.* Tübingen: Wasmuth, 1974.

[Laurens] *Henri Laurens.* Exh. cat. Lille: Musée d'art moderne, 1992.

[Laurens] *Henri Laurens.* Exh. cat. New York: Curt Valentin Gallery, 1952. With a statement by the artist.

Leighton, Patricia. *Re-Ordering the Universe: Picasso and Anarchism, 1897–1914.* Princeton, N.J.: Princeton University Press, 1989.

Les Desmoiselles d'Avignon. 2 vols. Paris: Musée Picasso, 1988.

Lipchitz, Jacques, with H. H. Arnason. *My Life in Sculpture.* New York: Viking Press, 1972.

Monod-Fontaine, Isabelle, and E. A. Carmean, Jr. *Braque: The Papiers Collés.* Exh. cat. Washington, D.C.: National Gallery, 1982. Text by A. Martin.

Penrose, Roland. *The Sculpture of Picasso.* Exh. cat. New York: MOMA, 1967.

——, and John Golding. *Picasso in Retrospect.* New York and Washington, D.C.: Praeger, 1973.

Picasso: Sixty Years of Graphic Works. Exh. cat. Los Angeles: LACMA, 1966.

Robbins, Daniel. *Albert Gleizes, 1881–1953: A Retrospective Exhibition.* Exh. cat. New York: SRGM, 1964.

Rosenblum, Robert. *Cubism and Twentieth-Century Art.* New York: Harry N. Abrams, 1976.

Rowell, Margit, and Daniel Robbins. *Cubist Drawings: 1907–1929.* Exh. cat. Houston: Janie C. Lee Gallery, 1982.

Rubin, William. "Pablo and Georges and Leo and Bill." *Art in America* 67:2 (March–April 1979): 128ff.

——, ed. *Pablo Picasso: A Retrospective.* Exh. cat. New York: MOMA, 1980.

——, ed. *Picasso and Braque: Pioneering Cubism.* New York: MOMA, 1989.

——, ed. *"Primitivism" in 20th-Century Art: Affinity of the Tribal and the Modern.* 2 vols. New York: MOMA, 1984.

Schiff, Gert, ed. *Picasso in Perspective.* Englewood Cliffs, N.J.: Prentice-Hall, 1976.

Schmalenbach, Werner. *Léger.* New York: Harry N. Abrams, 1976.

Spate, Virginia. *Orphism: The Evolution of Non-Figurative Painting in Paris: 1910–1914.* Exh. cat. Oxford and New York: Clarendon Press and Oxford University Press, 1979.

Steinberg, Leo. "From Narrative to 'Iconic' in Picasso: The Buried Allegory in *Bread and Fruit Dish on a Table* and the Role of *Les Desmoiselles d'Avignon.*" *Art Bulletin* 65:4 (December 1983): 615–649.

——. "The Philosophical Brothel, Part 1." *October* 44 (Spring 1988): 7–74.

——. "The Polemical Part." *Art in America* 67:2 (March–April 1979): 114–127.

——. "Resisting Cézanne: Picasso's 'Three Women.'" *Art in America* (November–December 1978): 114–133.

Vallier, Dora. *Jacques Villon: Oeuvres de 1897 à 1956.* Paris: Éditions cahiers d'art, 1957.

[Villon] *Jacques Villon, Master of Graphic Art (1875–1963).* Exh. cat. Boston: Museum of Fine Arts, 1964. Text by P. A. Wick.

Wallen, Burr, and Donna Stein. *The Cubist Print.* Exh. cat. Santa Barbara, Calif.: University of California Art Museum, 1981.

Wight, Frederick. *Jacques Lipchitz: A Retrospective Selected by the Artist.* Exh. cat. Los Angeles: UCLA Art Galleries, 1963.

Wilkin, Karen. *Georges Braque.* New York: Abbeville, 1991.

Wilkinson, Alan G. *Jacques Lipchitz: A Life in Sculpture.* Exh. cat. Toronto: Art Gallery of Ontario, 1989.

Zelevansky, Lynn, ed. *Picasso and Braque, A Symposium.* New York: MOMA, 1992.

Zurcher, Bernard. *Georges Braque: Life and Work.* Trans. S. Nye. New York: Rizzoli, 1988.

11. TOWARDS ABSTRACTION

Andersen, Troels, ed. *Essays on Art by K. S. Malevich.* 2 vols. Trans. X. Glowacki-Prus and A. McMillan. Copenhagen: Borgen, 1968.

Art into Life: Russian Constructivism, 1914–32. New York: Rizzoli, 1990.

[Balla] *Giacomo Balla.* Exh. cat. Turin: Galleria Civica d'Arte Moderna, 1963. Documentation by E. Crispolti and M. Drudi Gambillo.

Ballo, Giacomo. *Umberto Boccioni: La Vita e l'Opera.* I Maestri della pittura contemporanea. 1st ed. Milan: Il Saggiatore, 1964.

Barr, Jr., Alfred H. *Cubism and Abstract Art.* Exh. cat. New York: MOMA, 1936.

Barron, Stephanie, and Maurice Tuchman. *The Avant-Garde in Russia, 1910–1930: New Perspectives.* Exh. cat. Los Angeles: LACMA, 1980.

Blotkamp, Carel. *Mondrian: The Art of Destruction.* New York: Harry N. Abrams, 1995.

Bowlt, John, ed. and trans. *Russian Art of the Avant-Garde: Theory and Criticism, 1902–1934.* Rev. ed. London: Thames & Hudson, 1989

——, and Rose-Carol Washton Long, eds. *The Life of Vasilli Kandinsky in Russian Art: A Study of "On the Spiritual in Art."* Trans. J. Bowit. Newtonville, Mass.: Oriental Research Partners, 1980.

Bruni, Claudio, and Maria Drudi Gambillo, eds. *After Boccioni, Futurist Paintings and Documents from 1915 to 1919.* Trans. H. G. Heath. Rome: Studio d'arte contemporanea, La Medusa, 1961.

Bulhof, Francis, ed. *Nijhoff, Van Ostaijen, "De Stijl": Modernism in the Netherlands and Belgium in the First Quarter of the Twentieth Century.* The Hague: Nijhoff, 1976.

[Carrà] *Carlo Carrà.* Exh. cat. Paris: Editions Galerie Eric Touchaleaume, 1994.

Carrieri, Raffaele. *Futurism.* Trans. L. van Rensselaer White. Milan: Edizioni del Milione, 1963.

Coen, Ester. *Umberto Boccioni.* Exh. cat. New York: MMA, 1988.

Cork, Richard. *Vorticism and Abstract Art in the First Machine Age.* Berkeley, Calif.: University of California Press, 1976.

Dabrowski, Magdalena. *Contrasts of Form: Geometric Abstract Art, 1910–1980.* Exh. cat. New York: MOMA, 1985.

——. *Liubov Popova.* Exh. cat. New York: MOMA, 1991.

D'Andrea, Jean. *Kazimir Malevich, 1878–1935.* Exh. cat. Los Angeles: Armand Hammer Museum of Art, 1990.

De Stijl. Exh. cat. New York: MOMA, 1961.

d'Harnoncourt, Anne. *Futurism and the International Avant-Garde.* Exh. cat. Philadelphia: Philadelphia Museum of Art, 1980.

Doesburg, Theo van. *New Movement in Painting.* Delft: J. Waltman, 1971.

Elliott, David. *El Lissitzky: 1890–1941.* Exh. cat. Oxford: Museum of Modern Art, 1977.

——. *New Worlds: Russian Art and Society: 1900–1935.* London: Thames & Hudson, 1986.

——, ed. *Rodchenko and the Arts of Revolutionary Russia.* New York: Pantheon Books, 1979.

Goncharova—Larianov. Exh. cat. Paris: Musée d'Art Moderne de la Ville de Paris, 1963.

Goncharova, Larianov, Mansurov. Exh. cat. Bergamo: Galleria Lorenzelli, 1966.

Gray, Camilla, and Marian Burleigh-Motley. *Russian Experiment in Art: 1863–1922.* Rev. ed. London and New York: Thames & Hudson, 1986.

The Great Utopia: The Russian and Soviet Avant-Garde. Exh. cat. New York: SRGM, 1992.

Guerman, Mikhail, ed. *Art of the October Revolution.* Trans. W. Freeman, D. Saunders, and C. Binns. New York: Harry N. Abrams, 1979.

Hahl-Koch, Jelena. *Kandinsky.* New York: Rizzoli, 1993.

Hammacher, A. M. *Mondrian, De Stijl, and Their Impact.* Exh. cat. New York: Marlborough-Gerson Gallery, Inc., 1964.

Hanson, Anne Coffin, ed. *The Futurist Imagination: Word + Image in Italian Futurist Painting, Drawing, Collage, and Free-word Poetry.* New Haven, Conn.: Yale University Art Gallery 1983.

Hoffman, Katherine, ed. *Collage: Critical Views.* Ann Arbor, Mich.: UMI Research Press, 1989.

Hulten, Pontus. *Futurism and Futurisms.* New York: Abbeville, 1986.

———, and Jean-Hubert Martin. *Malevich.* Exh. cat. Paris: Centre National d'Art Contemporain et de Culture Georges Pompidou, 1978.

Jaffe, Hans, et al. *De Stijl, 1917–1931: Visions of Utopia.* Exh. cat. Minneapolis and New York: Walker Art Center and Abbeville, 1982.

Kandinsky, Complete Writings on Art. Ed. K. Lindsay and P. Vergo. Documents of Twentieth Century Art. 2 vols. Boston: G. K. Hall, 1982.

Karginov, German. *Rodchenko.* Trans. E. Hoch. London: Thames & Hudson, 1979.

Khan-Magomedov, Selim. *Rodchenko: The Complete Work.* Cambridge, Mass.: MIT Press, 1986.

Kozloff, Max. *Cubism/Futurism.* New York: Charterhouse, 1973.

Lissitzky-Küppers, Sophie. *El Lissitzky: Life, Letters, Texts.* London: Thames & Hudson, 1980.

Lodder, Christina. *Russian Constructivism.* New Haven, Conn.: Yale University Press, 1983.

[Malevich] *Kazimir Malevich, 1878–1935.* Exh. cat. London: Whitechapel Art Gallery, 1959.

Malevich, Kazimir. *The Non-Objective World.* Chicago: Theobald, 1959.

Mansbach, Steven. *Visions of Totality: László Moholy-Nagy, Theo van Doesberg, and El Lissitzky.* Ann Arbor, Mich.: UMI Research Press, 1980.

Martin, Marianne. *Futurist Art and Theory 1909–1915.* Oxford: Clarendon Press, 1968.

———, and Anne Coffin Hanson, eds. "Futurism." *Art Journal* (Winter 1981) special issue.

Milner, John. *Kazimir Malevich and the Art of Geometry.* New Haven, Conn.: Yale University Press, 1996.

———. *Vladmir Tatlin and the Russian Avant-Garde.* New Haven, Conn.: Yale University Press, 1983.

Mondrian, Piet. *The New Art, the New Life: The Complete Writings.* Ed. and trans. H. Holtzmann and M. James. Documents of Twentieth Century Art. Boston: G. K. Hall, 1986.

[Mondrian] *Piet Mondrian, 1872–1944.* Exh. cat. Toronto: Toronto Art Gallery, 1966. Texts by R. P. Welsh, R. Rosenblum, M. Seuphor.

[Mondrian] *Piet Mondrian, 1872–1944: Centennial Exhibition.* Exh. cat. New York: SRGM, 1971.

Overy, Paul. *De Stijl.* World of Art. London: Thames & Hudson, 1991.

Ozenfant, Amédée. *Foundations of Modern Art.* New American Edition. New York: Dover Publications, 1952.

Poggi, Christine. *In Defiance of Painting: Cubism, Futurism and the Invention of Collage.* New Haven, Conn.: Yale University Press, 1992.

Roman, Gail Harrison, ed. "The Russian Avant-Garde." *Art Journal* XLI:3 (Fall 1981) special issue.

Rowell, Margit. *The Planar Dimension: Europe 1912–1932.* New York: SRGM, 1979.

———, and Angelika Z. Rudenstine. *Art of the Avant-Garde in Russia: Selections from the George Costakis Collection.* Exh. cat. New York: SRGM, 1981.

Rudenstine, Angelika Z., ed. *Piet Mondrian, 1872–1944.* Exh. cat. Boston: Little, Brown, 1994.

———, ed. *Russian Avant-Garde Art: The George Costakis Collection.* Exh. cat. New York: SRGM, 1981.

Taylor, Joshua C. *Futurism: Politics, Painting, and Performance.* Ann Arbor, Mich.: UMI Research Press, 1979.

———. *The Graphic Work of Umberto Boccioni.* Exh. cat. New York: MOMA, 1961.

Tisdall, Caroline, and Angelo Bozzolla. *Futurism.* London: Thames & Hudson, 1978.

Troy, Nancy. *The De Stijl Environment.* Cambridge, Mass.: MIT Press, 1983.

Vantongerloo, Georges. *Paintings, Sculptures, Reflections.* Problems of Contemporary Art, 5. New York: Wittenborn, Schultz, 1948.

Zhadova, Larissa A. *Malevich: Suprematism and Revolution in Russian Art.* Trans. A. Lieven. London: Thames & Hudson, 1982.

12. EARLY TWENTIETH-CENTURY ARCHITECTURE

Graf, Otto Antonia. *Otto Wagner.* Vienna: H. Böhlaus, 1985.

Gravagnuolo, Benedetto. *Adolf Loos: Theory and Works.* Milan: Idea Books Edizioni, Locker Verlang, 1983.

Gutheim, Frederick, ed. *Frank Lloyd Wright on Architecture: Selected Writings, 1894–1940.* New York: Duell, Sloan and Pearce, 1941.

Haiko, Peter, ed. *Architecture of the Early XX Century.* Trans. G. Clough. New York: Rizzoli, 1989.

Hitchcock, Henry-Russell. *In the Nature of Materials: The Buildings of Frank Lloyd Wright, 1887–1941.* New York: Da Capo Press, 1975.

———. *J. J. P. Oud.* Paris: Éditions cahiers d'art, 1931.

Larkin, David, and Bruce Brooks Pfeiffer. *Frank Lloyd Wright: The Masterworks.* New York: Rizzoli, 1993.

Mallgrave, Harry Francis. *Otto Wagner: Reflections on the Raiment of Modernity.* Santa Monica, Calif.: Getty Center for the History of Art and the Humanities, 1993.

Munz, Ludwig. *Adolf Loos.* New York: Praeger, 1966.

Naden, Corinne J. *Frank Lloyd Wright, the Rebel Architect.* New York, 1968.

O'Gorman, James F. *Three American Architects: Richardson, Sullivan, and Wright, 1865–1915.* Chicago: University of Chicago Press, 1991.

Riley, Terrence, and Peter Redd, eds. *Frank Lloyd Wright, Architect.* Exh. cat. New York: MOMA, 1994.

Rogers, Ernesto N. *Auguste Perret.* Milan: Il Balcone, 1955.

Sant'Elia, A. "Manifesto of Futurist Architecture." *Architectural Review* 126 (1959).

Scully, Vincent. *Frank Lloyd Wright.* New York: G. Braziller, 1960.

Veronesi, Giulia. *J. J. Pieter Oud.* Milan: Il Balcone, 1953.

Whittick, Arnold. *Eric Mendelsohn.* 2nd ed. London: L. Hill, 1956.

Windsor, Alan. *Peter Behrens: Architect and Designer, 1879–1940.* New York: Whitney Library of Design, 1981.

Wright, Frank Lloyd. *Frank Lloyd Wright, Collected Writings.* Ed. B.B. Pfeiffer. 5 vols. New York: Rizzoli, 1992–1995.

13. FROM FANTASY TO DADA AND THE NEW OBJECTIVITY

Abrams, Stephan. *Max Beckmann.* New York, 1985.

Ades, Dawn, et al., eds. *In the Mind's Eye: Dada and Surrealism.* New York: Abbeville, 1986.

Andreotti, Margherita. *The Early Sculpture of Jean Arp.* Ann Arbor, Mich.: UMI Research Press, 1989.

Arnason, H. H. *Reality and Fantasy 1900–1954.* Minneapolis: Walker Art Center, 1954.

Arp: 1886–1966. Exh. cat. Minneapolis: Minneapolis Institute of Arts, 1987.

Arp, Hans. *On My Way: Poetry and Essays, 1912–1947.* New York: Wittenborn, Schultz, 1948.

Barr, Jr., Alfred H., ed. *Fantastic Art, Dada, Surrealism.* 2nd ed. rev. Exh. cat. New York: MOMA, 1937. Text by G. Hugnet.

Borràs, Maria Lluïsa. *Picabia.* London: Thames & Hudson, 1985.

Camfield, William A. *Francis Picabia: His Art, Life, and Times.* Princeton, N.J.: Princeton University Press, 1979.

———. *Max Ernst: Dada and the Dawn of Surrealism.* Exh. cat. New York: MOMA, 1993.

Clair, Jean. *Marcel Duchamp: Catalogue Raisonné.* Exh. cat. Paris: Musée National d'Art Moderne, Centre Georges Pompidou, 1977.

Dachy, Marc. *The Dada Movement, 1915–1923.* New York: Skira/Rizzoli, 1990.

d'Harnoncourt, Anne, and Kynaston McShine. *Marcel Duchamp.* Exh. cat. Munich: Prestel and Philadelphia Museum of Art, 1989.

Dietrich, Dorothea. *The Collages of Kurt Schwitters: Tradition and Innovation.* New York: Cambridge University Press, 1993.

[Dix] *Otto Dix, 1891–1969.* Exh. cat. London: Tate Gallery, 1992.

Duchamp, Marcel. *Marcel Duchamp, Notes.* Trans. P. Matisse. Documents of Twentieth Century Art. Boston: G. K. Hall, 1983.

Eberle, Matthias. *World War I and the Weimar Artists: Dix, Grosz, Beckmann, Schlemmer.* New Haven, Conn.: Yale University Press, 1985.

Elderfield, John. *Kurt Schwitters.* Exh. cat. New York: MOMA and Thames & Hudson, 1985.

Foresta, Merry. *Man Ray, 1890–1976: His Complete Works.* Düsseldorf: Edition Stemmle, 1989.

———, et al. *Perpetual Motif: The Art of Man Ray.* Exh. cat. Washington, D.C.: National Museum of American Art, 1988.

Foster, Stephen, and Rudolf Kuenzi. *Dada Spectrum: The Dialectics of Revolt.* Iowa City, Iowa: University of Iowa Press, 1979.

Giedion-Welcker, Carola. *Hans Arp.* Stuttgart: Gerd Hatje, 1957. Documentation by M. Hagenbach.

Grosz, George. *George Grosz: An Autobiography.* Trans. N. Hodges. New York: Macmillan, 1983.

Haftmann, Werner. *Chagall.* Trans. H. Baumann and A. Brown. New York: Harry N. Abrams, 1984.

———. *Marc Chagall: Gouaches, Drawings, Watercolors.* Trans. R. E. Wolf. New York: Harry N. Abrams, 1984.

Hess, Hans. *George Grosz.* London: Studio

Vista, 1974.

Jones, Amelia. *Postmodernism and the En-Gendering of Marcel Duchamp.* Cambridge, England and New York: Cambridge University Press, 1994.

Kuenzli, Rudolf, and Francis M. Naumann. *Marcel Duchamp: Artist of the Century.* Cambridge, Mass.: MIT Press, 1989.

Kuthy, Sandor. *Sophie Taeuber–Hans Arp.* Exh. cat. Bern: Kunstmuseum Bern, 1988.

Lanchner, Carolyn. *Sophie Taeuber-Arp.* Exh. cat. New York: MOMA, 1981.

Lavin, Maud. *Cut with the Kitchen Knife: The Weimar Photomontages of Hannah Höch.* New Haven, Conn.: Yale University Press, 1993.

Lewis, Beth Irwin. *George Grosz: Art and Politics in the Weimar Republic.* Rev. ed. Princeton, N.J.: Princeton University Press, c. 1991.

Man Ray. Exh. cat. Los Angeles: LACMA, 1966. Texts by the artist and his friends.

Man Ray Photographs. Intro. Jean-Hubert Martin. London: Thames & Hudson, 1987.

Man Ray. *Self-Portrait.* Boston: Little, Brown, 1963.

Motherwell, Robert, ed. *The Dada Painters and Poets: An Anthology.* 2nd ed. Boston: G. K. Hall, 1981.

Oesterreicher-Mollwo, Marianne. *Surrealism and Dadaism.* Trans. Stephen Crawshaw. Oxford: Phaidon, 1979.

Paz, Octavio. *Marcel Duchamp: Appearance Stripped Bare.* New York: Viking Press, 1978.

Penrose, Roland. *Man Ray.* Boston: New York Graphic Society, 1975.

Pincus-Witten, Robert. *Giorgio de Chirico: Post Metaphysical and Baroque Paintings, 1920–1970.* Exh. cat. New York: Robert Miller Gallery, 1984.

Rau, Bernd. *Jean Arp: The Reliefs, Catalogue of Complete Works.* New York: Rizzoli, 1981.

Read, Herbert. *The Art of Jean Arp.* New York: Harry N. Abrams, 1968.

Richter, Hans. *Dada, Art, and Anti-Art.* London: Thames & Hudson, 1965.

Rubin, William. *Dada, Surrealism, and Their Heritage.* Exh. cat. New York: MOMA, 1968.

———, ed. *De Chirico.* Exh. cat. New York: MOMA, 1982. Text by F. dell'Arco.

[Sander] *August Sander: Photographer of an Epoch, 1904–1959.* Millerton, N.Y.: Aperture, 1980.

Schmalenbach, Werner. *Kurt Schwitters.* New York: Harry N. Abrams, 1970.

Schulz-Hoffmann, Carla, and Judith Weiss. *Max Beckmann: Retrospective.* Exh. cat. St. Louis and Munich: St. Louis Art Museum and Prestel, 1984.

Schwarz, Arturo. *The Complete Works of Marcel Duchamp.* Rev. ed. London: Thames & Hudson, 1997.

Selz, Peter. *Max Beckmann.* Exh. cat. New York: MOMA, 1964. Texts by H. Joachim and P. T. Rathbone.

Tatar, Maria. Lustmord. *Sexual Murder in Weimar Germany.* Princeton, N.J.: Princeton University Press, 1995.

14. THE SCHOOL OF PARIS BETWEEN THE WARS

Castaing, Marcellin, and Jean Leymarie. *Soutine.* Milan: Silvana editoriale d'arte, 1964.

Chipp, Herschel. *Georges Braque: The Late Paintings, 1940–1963.* Exh. cat. Washington, D.C.: Phillips Collection, 1982.

Ducrot, Nicolaus, ed. *André Kertész: Sixty Years of Photography, 1912–1972.* New York: Grossman Pub., 1972.

Durrell, Lawrence. Exh. cat. *Brassaï.* New York: MOMA, 1968.

Mangin, Nicole S., ed. *Catalogue de l'oeuvre de Georges Braque.* Paris: Maeght, 1959.

Mann, Carol. *Modigliani.* World of Art. London: Thames & Hudson, 1980.

Modigliani, Jeanne. *Modigliani: Man and Myth.* New York: Orion, 1958.

Perez-Tibi, Dora. *Dufy.* Trans. S. Whiteside. London: Thames & Hudson, 1989.

Silver, Kenneth. *Esprit de Corps: The Art of the Parisian Avant-Garde and the First World War, 1914–1925.* London: Thames & Hudson, 1989.

Soby, James T. *Modigliani: Paintings, Drawings, Sculpture.* Exh. cat. 3rd ed. rev. New York: MOMA, 1963.

Sylvester, David. *Chaim Soutine, 1894–1943.* Exh. cat. London: The Arts Council, 1963.

Tuchman, Maurice. *Soutine.* Exh. cat. Los Angeles: LACMA, 1968.

[Utrillo] *Maurice Utrillo.* Exh. cat. Pittsburgh: Museum of Art, Carnegie Institute, 1963.

[Utrillo] *Maurice Utrillo: L'oeuvre complet.* 2 vols. Paris: Paul Pétridès, 1959–62.

Werner, Alfred. *Amedeo Modigliani.* London: Thames & Hudson, 1985.

———. *Chaim Soutine.* London: Thames & Hudson, 1977.

15. SURREALISM

Abbott, Berenice. *The World of Atget.* New York: Horizon Press, 1979.

Ades, Dawn, et al., eds. *In the Mind's Eye: Dada and Surrealism.* New York: Abbeville, 1986.

———. *Dalí.* London: Thames & Hudson, 1995.

Alexandrian, Sarane. *Surrealist Art.* London: Thames & Hudson, 1985.

Arnason, H. H. *Reality and Fantasy, 1900–1954.* Minneapolis: Walker Art Center, 1954.

Barr, Jr., Alfred H., ed. *Fantastic Art, Dada, Surrealism.* 2nd ed. rev. Exh. cat. New York: MOMA, 1937. Texts by G. Hugnet.

Blanc, Giulio V., et al. *Wilfredo Lam and His Contemporaries.* Exh. cat. New York: Studio Museum of Harlem, 1992.

Bonnefoy, Yves. *Giacometti.* New York: Flammarion/Abbeville, 1991.

Breton, André. *Manifestoes of Surrealism.* Ann Arbor, Mich.: University of Michigan Press, 1969.

———. *What Is Surrealism? Selected Writings.* Ed. F. Rosemont. New York: Monad, 1978.

Camfield, William. *Max Ernst: Dada and the Dawn of Surrealism.* Exh. cat. Houston and Munich: The Menil Collection and Prestel, 1993.

[Cartier-Bresson] *Henri Cartier-Bresson: Photographer.* London: Thames & Hudson, 1979.

[Cartier-Bresson] *The World of Henri Cartier-Bresson.* New York: Viking Press, 1968.

Chadwick, Whitney. *Women Artists and the Surrealist Movement.* London: Thames & Hudson, 1991.

Dalí, Salvador. *The Secret Life of Salvador Dalí.* Trans. H. M. Chevalier. New York: Dial Press, 1942.

De Bock, Paul Aloïse. *Paul Delvaux.* Hamburg: J. Asmus, 1965.

Descharnes, Robert. *Salvador Dalí: The Work, the Man.* London: Thames & Hudson, 1984.

Ernst, Max. *Beyond Painting, and Other Writings by the Artist and His Friends.* New York: Wittenborn, Schultz, 1948.

[Ernst] *Max Ernst: A Retrospective.* Exh. cat. New York: SRGM, 1975.

Fer, Briony. *Realism, Rationalism, Surrealism: Art Between the Wars.* New Haven, Conn.: Yale University Press, 1993.

Foster, Hal. *Compulsive Beauty.* Cambridge, Mass.: MIT Press, 1993.

Foucault, Michel. *This Is Not a Pipe.* Trans. and ed. J. Harkness. Berkeley, Calif.: University of California Press, 1983.

Gablik, Suzy. *Magritte.* World of Art. Reprint. London: Thames & Hudson, 1989.

Gerard, Max, ed. *Dalí.* Trans. E. Morse. New York: Harry N. Abrams, 1968.

Hahn, Otto. *Masson.* New York: Harry N. Abrams, 1965.

Haslam, Malcolm. *The Real World of the Surrealists.* New York: Rizzoli, 1978. Introduction by Barbara Rose.

Henning, Edward B. *The Spirit of Surrealism: 1919–1939.* Exh. cat. Cleveland: Cleveland Museum of Art, 1979.

Hohl, Reinhold. *Alberto Giacometti: A Retrospective Exhibition.* Exh. cat. New York: SRGM, 1974.

Krauss, Rosalind. *L'amour fou: Photography and Surrealism.* New York: Abbeville, 1985.

Lanchner, Carolyn. *Joan Miró.* Exh. cat. New York: MOMA, 1993.

Leiris, Michel, and Georges Limbour. *André Masson and His Universe.* London, Geneva, and Paris: Éditions des Trois Collines, 1947.

Lewis, Helena. *The Politics of Surrealism.* New York: Paragon, 1988.

Lippard, Lucy R., ed. *Surrealists on Art.* Englewood Cliffs, N.J.: Prentice-Hall, 1970.

Lord, James. *Giacometti, A Biography.* New York: Farrar, Straus and Giroux, 1985.

Miró, Joan. *Joan Miró: Selected Writings and Interviews.* Ed. M. Rowell. Documents of Twentieth Century Art. Boston: G. K. Hall, 1986.

Nadeau, Maurice. *History of Surrealism.* Cambridge, Mass.: Harvard University Press, 1989.

Oesterreicher-Mollwo, Marianne. *Surrealism and Dadaism.* Trans. Stephen Crawshaw. Oxford: Phaidon, 1979.

Picon, Gaëtan. *Surrealists and Surrealism: 1919–1939.* Trans. J. Emmons. New York: Rizzoli, 1977.

Rowell, Margit. *Julio González: A Retrospective.* Exh. cat. New York: SRGM, 1983.

Rubin, William. *Dada, Surrealism, and Their Heritage.* Exh. cat. New York: MOMA, 1968.

———. *Matta.* Exh. cat. New York: MOMA, 1957.

Selz, Peter. *Alberto Giacometti.* Exh. cat. New York: MOMA, 1965. With autobiographical statement by the artist.

———. *Yves Tanguy.* Exh. cat. New York: MOMA, 1955.

Spies, Werner. *Max Ernst: A Retrospective.* Exh. cat. London: Tate Gallery and Prestel, 1991.

———. *Max Ernst: Loplop, the Artist in the Third Person.* London: Thames & Hudson, 1983.

Stich, Sidra. *Anxious Visions: Surrealist Art.* New York: Abbeville, 1990.

Svendsen, Louise Averill. *Alberto Giacometti: Sculptor and Draftsman.* Exh. cat. New York: The American Federation of Art, 1977.

Sylvester, David. *Magritte: The Silence of the World.* New York: Harry N. Abrams, 1994.

Szarkowski, John, and Maria Hambourg. *The Work of Atget.* 4 vols. New York: MOMA, 1981–84.

[Tanguy] *Yves Tanguy: Un Recueil de ses Oeuvres*. New York: P. Matisse, 1963.

Torczyner, Harry. *Magritte: The True Art of Painting*. New York: Abradale Press, 1985.

Withers, Josephine. *Julio González: Sculpture in Iron*. New York: New York University Press, 1978.

16. MODERN ARCHITECTURE BETWEEN THE WARS

Bauhaus Photography Cambridge, Mass.: MIT Press, 1985.

Bayer, Herbert, Walter Gropius, and Ise Gropius. *Bauhaus, 1919–1928*. New York: MOMA, 1975.

Boesiger, Willy. *Le Corbusier and Pierre Jeanneret: Oeuvre complète*. 6 vols. Zurich: Éditions d'Architecture, 1937–57.

Boulton, Alexander. *Frank Lloyd Wright, Architect: An Illustrated Biography*. New York: Rizzoli, 1993.

Choay, Françoise. *Le Corbusier*. New York: G. Braziller, 1960.

Christ-Janer, Albert. *Eliel Saarinen: Finnish-American Architect and Educator*. Rev. ed. Chicago: University of Chicago Press, 1979.

Curtis, William J. R. *Le Corbusier: Idea and Forms*. New York: Rizzoli, 1986.

Franciscono, Marcel. *Walter Gropius and the Creation of the Bauhaus in Weimar*. Urbana, Ill.: University of Illinois Press, 1971.

———, and Philip Johnson. *The International Style*. 2nd ed. New York: Norton, 1966.

Gardiner, Stephen. *Le Corbusier*. London: Fontana, 1974.

Giedion, Siegfried. *Walter Gropius: Work and Teamwork*. New York: Reinhold, 1954.

Grohmann, Will. *Painters of the Bauhaus*. Exh. cat. London: Marlborough Fine Art Ltd., 1962.

Gropius, Walter. *The New Architecture and the Bauhaus*. Trans. P. M. Shand. London: Faber and Faber, 1935.

———, and László Moholy-Nagy. *Bauhausbücher*. 14 vols. Munich: A. Langen, 1925–28.

Hertz, David. *Frank Lloyd Wright in Word and Form*. New York: G. K. Hall, 1995.

Hines, Thomas. *Richard Neutra and the Search for Modern Architecture: A Biography and History*. New York: Oxford University Press, 1982.

Hitchcock, Henry-Russell, and Philip Johnson. *The International Style*. 2nd ed. New York: Norton, 1966.

Itten, Johannes. *The Art of Colour: The Subjective Experience and Objective Rationale of Colour*. New York: Reinhold, 1961.

———. *Design and Form: The Basic Course at the Bauhaus*. Rev. ed. London: Thames & Hudson, 1975.

Jencks, Charles. *Le Corbusier and the Tragic View of Architecture*. Cambridge, Mass.: Harvard University Press, 1973.

Johnson, Philip. *Mies van der Rohe*. 3rd ed. rev. Greenwich, Conn.: New York Graphic Society, 1978.

Kilham, Walter H., Jr. *Raymond Hood, Architect: Form through Function in the Skyscraper*. New York: Architectural Book Publishing Co., 1973.

Krinsky, Carol Herselle. *Rockefeller Center*. Oxford: Oxford University Press, 1978.

Lane, Barbara Miller. *Architecture and Politics in Germany 1918–1945*. New ed. Cambridge, Mass.: Harvard University Press, 1985.

Le Corbusier. *The City of To-morrow and Its Planning*. London: J. Rodker, 1929.

Le Corbusier, une encyclopédie. Paris: Éditions du Centre Pompidou, 1987.

Le Corbusier. *Towards a New Architecture*. Trans. F. Etchells. London: Architectural Press, 1952.

Levine, Neil. *The Architecture of Frank Lloyd Wright*. Princeton, N.J.: Princeton University Press, 1996.

Papadaki, Stamo, ed. *Le Corbusier: Architect, Painter, Writer*. New York: MacMillan, 1948.

Pfeiffer, Bruce Brooks. *Frank Lloyd Wright: The Masterworks*. Ed. D. Larkin and B. B. Pfeiffer. London: Thames & Hudson, 1993.

Riley, Terrence, and Peter Reed, eds. *Frank Lloyd Wright, Architect*. Exh. cat. New York: MOMA, 1994.

Saarinen, Aline, ed. *Eero Saarinen*. London: Thames & Hudson, 1968.

———, ed. *Eero Saarinen on His Work*. New Haven and London: Yale University Press, 1962. With a statement by the architect.

Schulze, Franz. *Mies van der Rohe: A Critical Biography*. Chicago: University of Chicago Press, 1985.

———, ed. *Mies van der Rohe: Critical Essays*. Cambridge, Mass.: MIT Press, 1990.

Spaeth, David A. *Mies van der Rohe*. New York: Rizzoli, 1985.

Storrer, William Allin. *The Frank Lloyd Wright Companion*. Chicago: University of Chicago Press, 1993.

Tegethoff, Wolf. *Mies van der Rohe: The Villas and Country Houses*. New York: MOMA, 1985.

Whitford, Frank. *Bauhaus*. London: Thames & Hudson, 1984.

Wingler, Hans Maria. *The Bauhaus: Weimar, Dessau, Berlin, Chicago*. Cambridge, Mass: MIT Press, 1969.

Willett, John. *The New Sobriety, 1917-1933*. London: Thames & Hudson, 1978.

Wright, Frank Lloyd. *Frank Lloyd Wright, Collected Writings*. Ed. B. B. Pfeiffer. New York: Rizzoli, 1992. Introduction by K. Frampton.

———. *Letters to Apprentices*. Ed. B. B. Pfeiffer. Fresno, Calif.: Press at Cal State University, 1982.

17. INTERNATIONAL ABSTRACTION BETWEEN THE WARS

Albers, Joseph. *Interaction of Color*. New Haven, Conn.: Yale University Press, 1971.

[Albers] *Josef Albers, A Retrospective*. Exh. cat. New York: SRGM, 1988.

Arnason, H. H., and Ugo Mulas, eds. *Calder*. New York: Viking Press, 1971.

[Baumeister] *Willi Baumeister*. Exh. cat. Stuttgart: Institut für Auslandsbeziehungen, 1981.

[Baumeister] *Willi Baumeister, 1889–1955: Peintures, Dessins*. Exh. cat. Brussels: Musée d'Ixelles, 1982.

Bell, Keith. *Stanley Spencer: A Complete Catalogue of the Paintings*. London: Phaidon, 1992.

Bowness, Alan, ed. *Henry Moore: Complete Sculpture*. 6 vols. London: Lund Humphries, 1988.

[Calder] *Alexander Calder: A Retrospective Exhibition*. Exh. cat. New York: SRGM, 1964.

Calder, Alexander. *Calder: An Autobiography with Pictures*. New York: Pantheon Books, 1966.

Caton, Joseph Harris. *The Utopian Vision of Moholy-Nagy*. Ann Arbor, Mich.: UMI Research Press, 1984.

Curtis, Penelope. *Kandinsky in Paris: 1934–1944*. Exh. cat. New York: SRGM, 1985. Text by V. E. Barnett.

———, and Alan Wilkinson. *Barbara Hepworth: A Retrospective*. Exh. cat. Liverpool: Tate Gallery, 1994.

[Gabo] *Gabo: Constructions, Sculpture, Paintings, Drawings, Engravings*. London: Lund Humphries, 1957. Texts by H. Read and L. Martin.

[Gabo] *Naum Gabo*. Exh. cat. London: Tate Gallery, 1966.

Gabo, Naum. *Of Diverse Arts*. New York: Pantheon Books, 1962.

Grohmann, Will. *Willi Baumeister: Life and Work*. New York: Harry N. Abrams, 1966.

Haus, Andreas. *Moholy-Nagy: Photographs und Photograms*. Trans. F. Samson. New York: Pantheon, 1980.

[Hepworth] *Barbara Hepworth*. Exh. cat. London: Tate Gallery, 1968. Texts by R. Alley, N. Gray, R. W. D. Oxenaar.

Hight, Eleanor. *Picturing Modernism: Moholy-Nagy and Photography in Weimar Germany*. Cambridge, Mass.: MIT Press, 1995.

Hildebrandt, Hans. *Oskar Schlemmer*. Munich: Prestel, 1952.

James, Philip, ed. *Henry Moore on Sculpture*. New York: Viking Press, 1971.

Kandinsky: Russian and Bauhaus Years, 1915–1933. Exh. cat. New York: SRGM, 1983.

Klee, Paul. *On Modern Art*. London: Faber and Faber, 1949.

[Klee] *Paul Klee Centennial: Prints and Transfer Drawings*. Exh. cat. New York: MOMA, 1978.

Klee, Paul. *The Thinking Eye: The Notebooks of Paul Klee*. London: Lund Humphries, 1961.

Kostelanetz, Richard, ed. *Moholy-Nagy*. New York and Washington, D.C.: Praeger, 1970.

Lehman, Arnold and Brenda Richardson, eds. *Oskar Schlemmer*. Exh. cat. Baltimore: Baltimore Museum of Art, 1986.

[Lewis] *Wyndham Lewis and Vorticism*. Exh. cat. London: Tate Gallery, 1956.

Lewison, Jeremy. *Ben Nicholson*. Exh. cat. London: Tate Gallery, 1993.

Lipman, Jean. *Calder Creatures: Great and Small*. New York: E. P. Dutton, 1985.

———. *Calder's Universe*. New York: Viking Press and WMAA, 1977.

Long, Rose-Carol Washton. *Kandinsky: The Development of an Abstract Style*. Oxford and New York: Clarendon Press and Oxford University Press, 1980.

Marter, Joan. *Alexander Calder*. Exh. cat. Cambridge: Cambridge University Press, 1991.

Moholy-Nagy, László. *The New Vision, and 1928: Abstract of an Artist*. New York: Wittenborn Schultz, 1949.

———. *Vision in Motion*. Chicago: P. Theobald, 1947.

Nash, Steven. *Ben Nicholson: Fifty Years of His Art*. Exh. cat. Buffalo, N.Y.: Albright-Knox Art Gallery, 1978.

———, and Jörn Merkert, eds. *Naum Gabo: Sixty Years of Constructivism*. Exh. cat. Munich and New York: Prestel, c. 1985.

Naum Gabo, Antoine Pevsner. Exh. cat. New York: MOMA, 1948. Introduction by H. Read; Texts by R. Olson and A. Chanin.

[Nicholson] *Ben Nicholson: Paintings, Reliefs and Drawings*. London: Lund Humphries, 1948. Introduction by Herbert Read.

Packer, William. *Henry Moore: an Illustrated Biography*. London: Weidenfeld & Nicolson, 1985.

Passuth, Krisztina. *Moholy-Nagy*. London: Thames & Hudson, 1980.

Pevsner, Alexi. *A Biographical Sketch of My Brothers—Naum Gabo and Antoine Pevsner*. Amsterdam: Augustin and Schnooman, 1964.

Roskill, Mark. *Klee, Kandinsky, and the Thought of their Time: A Critical Perspective.* Urbana, Ill.: University of Illinois Press, 1992.

Sabarsky, Serge. *Paul Klee: The Late Years, 1930–1940.* Exh. cat. New York: Sabarsky Gallery, 1977.

Schlemmer, Oskar, László Moholy-Nagy, and Farkas Molnár. *The Theatre of the Bauhaus.* Trans. Arthur S. Wensinger. Middletown, Conn.: Wesleyan University Press, 1961.

Verdi, Richard. *Klee and Nature.* New York: Rizzoli, 1985.

18. AMERICAN ART BETWEEN THE WARS

Abstract Painting and Sculpture in America: 1927–1944. Exh. cat. New York: WMAA, 1983.

[Adams] *Ansel Adams: Images 1924–1974.* Boston: New York Graphic Society, 1974.

Adams, Henry. *Thomas Hart Benton.* New York: Alfred A. Knopf, 1989.

———. *Thomas Hart Benton: Drawing from Life.* Exh. cat. New York: Abbeville, 1990.

Alland, Alexander. *Jacob A. Riis, Photographer and Citizen.* Millerton, N.Y.: Aperture, 1974.

Arnason, H. H. *Theodore Roszak.* Exh. cat. Minneapolis: Walker Art Center, 1956.

Baigell, Matthew. *The American Scene: American Painting of the 1930's.* New York: Praeger, 1974.

———. *Thomas Hart Benton.* New York: Harry N. Abrams, 1974.

Baker, Houston A., Jr. *Modernism and the Harlem Renaissance.* Chicago: Chicago University Press, 1987.

Benton, Thomas Hart. *An American in Art: A Professional and Technical Autobiography.* Lawrence, Kans.: University of Kansas Press, 1969.

Bearden, Romare, and Harry Henderson. *Six Black Masters of American Art.* New York: Zenith Books, 1972.

Brown, Milton. *Story of the Armory Show: The 1913 Exhibition that Changed American Art.* 2nd ed. New York: Abbeville, 1988.

Callahan, Sean, ed. *The Photographs of Margaret Bourke-White.* Boston: New York Graphic Society, 1972.

Castro, Jan. *The Life & Art of Georgia O'Keeffe.* New York: Crown, 1985.

Constantin, Mildred. *Tina Modotti: A Fragile Life.* New York: Rizzoli, 1983.

Corn, Wanda. *Grant Wood: The Regionalist Vision.* Exh. cat. New York: WMAA, 1983.

Curry, Larry. *John Marin/1870–1953.* Exh. cat. Los Angeles, LACMA, 1970.

Curtis, James. *Mind's Eye, Mind's Truth: FSA Photography Reconsidered.* Philadelphia: Temple University Press, 1989.

Davidson, Abraham. *Early American Modernist Painting: 1910–1935.* New York: Harper & Row, 1981.

[Diller] *Burgoyne Diller: Paintings, Sculpture, Drawings* Exh. cat. Minneapolis: Walker Art Center, 1971.

Doherty, Robert J., ed. *The Complete Photographic Work of Jacob A. Riis.* New York: MacMillan, 1981.

Dwight, Edward. *Armory Show 50th Anniversary Exhibition 1913–1963.* Exh. cat. Utica, N.Y.: Munson-Williams-Proctor Institute, 1963.

Eldridge, Charles D. *American Imagination and Symbolist Painting.* Exh. cat. New York: Grey Gallery, New York University, 1980.

———. *Georgia O'Keeffe.* Library of American Art. New York: Harry N. Abrams, 1991.

Fávela, Ramón. *Diego Rivera: The Cubist Years.* Exh. cat. Phoenix: Phoenix Art Museum, 1984.

Fine, Elsa. *The Afro-American Artist: A Search for Identity.* New York: Holt, Rinehart and Winston, 1973.

———. *John Marin.* New York and Washington, D.C.: Abbeville and National Gallery of Art, 1990.

Folgarait, Leonard. *So Far From Heaven: David Alfaro Siqueiros' "The March of Humanity" and Mexican Revolutionary Politics.* New York: Cambridge University Press, 1987.

Glackens, Ira. *William Glackens and the Ashcan Group.* Rev. ed. New York: Horizon Press, 1984.

Goldberg, Vicki. *Margaret Bourke-White: A Biography.* New York: Harper & Row, 1986.

Goodrich, Lloyd. *Max Weber: Retrospective Exhibition.* New York: WMAA, 1949.

Green, Eleanor. *Augustus Vincent Tack: 1870–1949: Twenty-Six Paintings from the Phillips Collection.* Exh. cat. Austin, Tex.: University of Texas, 1972.

Greenough, Sarah, and Juan Hamilton. *Alfred Stieglitz, Photographs and Writings.* Washington, D.C.: National Gallery of Art, 1983.

Haskell, Barbara. *Arthur Dove.* Exh. cat. San Francisco: Museum of Modern Art, 1974.

———. *Joseph Stella.* Exh. cat. New York: WMAA, 1994.

Herrera, Hayden. *Frida Kahlo: The Paintings.* New York: Harper Collins, 1991.

Heyman, Therese, et al. *Dorothea Lange: American Photographs.* San Francisco: San Francisco Museum of Modern Art and Chronicle Books, 1994.

Homer, William Innes. *Alfred Stieglitz and the Photo-Secession.* Exh. cat. Boston: Little, Brown, 1983.

Hurley, F. Jack. *Portrait of a Decade: Roy Stryker and the Development of Documentary Photography in the Thirties.* Baton Rouge, La.: Louisiana State University Press, 1972.

Jaffe, Irma. *Joseph Stella.* Cambridge, Mass.: Harvard University Press, 1970.

Lane, John R. *Stuart Davis: Art and Art Theory.* Exh. cat. New York: Brooklyn Museum, 1978.

———, and Susan C. Larsen. *Abstract Painting and Sculpture in America 1927–1944.* Exh. cat. Pittsburgh: Museum of Art, Carnegie Institute, 1984.

[Lange] *Dorothea Lange.* Exh. cat. New York: MOMA, 1966. With an Introductory Essay by George P. Elliot.

[Lange] *Dorothea Lange: Photographs of a Lifetime.* Millerton, N.Y.: Aperture, 1982.

Levin, Gail. *Edward Hopper: A Catalogue Raisonné.* Exh. cat. New York: WMAA, 1995.

———. *Edward Hopper: The Art and the Artist.* Exh. cat. New York: WMAA, 1981.

———. *Synchromism and American Color Abstraction: 1910–1925.* Exh. cat. New York: WMAA, 1978.

Lorenz, Richard. *Imogen Cunningham: Flora/Photographs.* Boston: Little, Brown, 1996.

Lowe, Sarah M. *Tina Modotti: Photographs.* Exh. cat. Philadelphia and New York: Philadelphia Museum of Art and Harry N. Abrams, 1995.

Lynes, Barbara Buhler. *O'Keeffe, Stieglitz and the Critics, 1916–1929.* Ann Arbor, Mich.: UMI Research Press, 1989.

[Macdonald-Wright] *The Art of Stanton Macdonald-Wright: A Retrospective Exhibition.* Exh. cat. Washington, D.C.: Smithsonian Institution, 1967.

Marter, J. *Beyond the Plane: American Constructions 1930–1965.* Exh. cat. Trenton, N.J.: New Jersey State Museum, 1983.

Meltzer, Milton. *Dorothea Lange, a Photographer's Life.* New York: Farrar, Straus, and Giroux, 1978.

Miller, Dorothy, ed. *The Sculpture of John B. Flannagan.* Exh. cat. New York: MOMA, 1942.

[Nadelman] *The Sculpture and Drawings of Elie Nadelman.* Exh. cat. New York: WMAA, 1975.

Newhall, Nancy. *Ansel Adams: The Eloquent Light.* Millerton, N.Y.: Aperture, 1980.

Nordland, Gerald. *Gaston Lachaise: The Man and His Work.* New York: G. Braziller, 1974.

———. *Gaston Lachaise 1882–1935: Sculpture and Drawings.* Exh. cat. Los Angeles and New York: LACMA and WMAA, 1964.

Norman, Dorothy. *Alfred Stieglitz, an American Seer.* New York: Aperture, 1990.

———, ed. *Selected Writings of John Marin.* New York: Pellegrini & Cudahy, 1949.

[O'Keeffe] *Georgia O'Keeffe.* New York: Viking Press, 1976.

O'Neal, Hank. *Berenice Abbott, American Photographer.* New York: McGraw-Hill, 1982.

Perlman, Bernard B. *Painters of the Ashcan School: The Immortal Eight.* New York: Dover Publications, 1979.

Reich, Sheldon. *John Marin: A Stylistic Analysis and Catalogue Raisonné.* 2 vols. Tucson, Ariz.: University of Arizona Press, 1970.

Ritchie, A. C. *Charles Demuth.* Exh. cat. New York: MOMA, 1950. With a tribute to the artist by Marcel Duchamp.

[Rivera] *Diego Rivera: A Retrospective.* Exh. cat. Detroit: Detroit Institute of Arts, 1986.

Roberts, Brady, et al. *Grant Wood: An American Master Revealed.* Davenport, Iowa and San Francisco: Davenport Museum of Art and Pomegranate Art Books, 1995.

Roots of Abstract Art in America 1910–1930. Exh. cat. Washington, D.C.: Smithsonian Institution, 1965.

Rosenblum, Walter, et al. *America and Lewis Hine: Photographs, 1904–1940.* Millerton, N.Y.: Aperture, 1977.

Schmidt Campbell, Mary. *Memory and Metaphor: The Art of Romare Bearden, 1940–1987.* Exh. cat. New York: Oxford University Press, 1991.

Silverman, Jonathan. *For the World to See: The Life of Margaret Bourke-White.* New York: Viking Press, 1983.

Sims, Lowery Stokes. *Stuart Davis: American Painter.* Exh. cat. New York: MMA, 1991.

Sims, Patterson. *Gaston Lachaise.* Exh. cat. New York: WMAA, 1980.

Smith, Terry. *Making the Modern: Industry, Art, and Design in America.* Chicago: University of Chicago Press, 1993.

Stein, Judith E. *I Tell My Heart: The Art of Horace Pippin.* Exh. cat. Philadelphia: Pennsylvania Academy of Fine Arts, 1994.

[Strand] *Paul Strand: A Retrospective Monograph, the Years 1915–1968.* 2 vols. Millerton, N.Y.: Aperture, 1971.

Szarkowski, John. *Walker Evans.* Exh. cat. New York: MOMA, 1971.

[Tamayo] *Rufino Tamayo: Myth and Magic.* Exh. cat. New York: SRGM, 1979.

Tomkins, Calvin. *Paul Strand: Sixty Years of Photographs.* Millerton, N.Y.: Aperture, 1976.

Troyen, Carol, and Erica Hirschler. *Charles Sheeler: Paintings and Drawings.* Exh. cat. Boston: Museum of Fine Arts, 1987.

Wagner, Anne. *Three Artists (Three Women): Modernism and the Art of Hesse, Krasner, and O'Keeffe*. Berkeley, Calif.: University of California Press, 1996.

Wight, Frederick. *Arthur G. Dove*. Exh. cat. Berkeley and Los Angeles: University of California Press, 1958. Introduction by D. Phillips.

Yglesias, Helen. *Isabel Bishop*. New York: Rizzoli, 1989.

19. ABSTRACT EXPRESSIONISM AND THE NEW AMERICAN SCULPTURE

Alloway, Lawrence. *Barnett Newman: The Stations of the Cross*. Exh. cat. New York: SRGM, 1966.

———. *William Baziotes: A Memorial Exhibition*. Exh. cat. New York: SRGM, 1965. With statements by the artist.

Anfam, David. *Abstract Expressionism*. World of Art. London: Thames & Hudson, 1990.

———. *Franz Kline: Black and White, 1950–1961*. Exh. cat. Houston: The Menil Collection, 1994.

Arnason, H. H. *Abstract Expressionists and Imagists*. Exh. cat. New York: SRGM, 1961.

———. *Robert Motherwell*. Rev. ed. New York: Harry N. Abrams, 1982.

———. *Robert Motherwell: Works on Paper*. Exh. cat. New York: MOMA, 1967.

Ashton, Dore. *About Rothko*. New York: Oxford University Press, 1983.

———. *A Joseph Cornell Album*. New York: Viking Press, 1974.

———. *The New York School: A Cultural Reckoning*. New York: Viking Press, 1973.

———. *Noguchi East and West*. Berkeley, Calif.: University of California Press, 1993.

———. *Yes, but . . . : A Critical Study of Philip Guston*. New York: Viking Press, 1976.

———, and Jack Flam. *Robert Motherwell*. Exh. cat. Buffalo, N.Y. and New York: Albright-Knox Gallery and Abbeville, 1983.

Auping, Michael. *Abstract Expressionism: The Critical Developments*. New York: Harry N. Abrams, 1987.

[Baziotes] *William Baziotes: A Retrospective Exhibition*. Exh. cat. Newport Beach, Calif.: Newport Harbor Art Museum, 1978. Texts by B. Cavaliere and M. Hadler.

[Bourgeois] *The Iconography of Louise Bourgeois*. Exh. cat. New York: Max Hutchinson Gallery, 1980.

Carmean, Jr., E. A., ed. *Robert Motherwell: The Reconciliation Elegy*. Geneva and New York: Skira and Rizzoli, 1980.

———. *David Smith*. Exh. cat. Washington, D.C.: National Gallery of Art, 1982.

Castleman, Riva. *Art of the Forties*. Exh. cat. New York: MOMA, 1991.

Cernuschi, Claude. *Jackson Pollock: Meaning and Significance*. New York: Icon Editions, 1992.

[de Kooning] *Willem de Kooning*. Exh. cat. New York: WMAA, 1983. Texts by P. Cummings, J. Merkert, C. Stoullig.

Doty, Robert, and Diane Waldman. *Adolph Gottlieb*. Exh. cat. New York: SRGM and WMAA, 1968.

Elsen, Albert. *Seymour Lipton*. New York: Harry N. Abrams, 1970.

Frank, Elizabeth. *Jackson Pollock*. New York: Abbeville, 1982.

Fry, E., and J. Ablow. *Philip Guston: The Late Works*. Exh. cat. Sydney, Australia, 1984.

Gaugh, Harry F. *Franz Kline*. Exh. cat. New York: Abbeville, 1994.

———. *The Vital Gesture*. Cincinnati and New York: Cincinnati Art Museum and Abbeville Press, 1985.

———. *Willem de Kooning*. New York:

Abbeville, 1983.

Glimcher, Arnold. *Louise Nevelson*. New York and Washington, D.C.: Dutton, 1972.

Gordon, John. *Franz Kline, 1910–1962*. Exh. cat. New York: WMAA, 1968.

———. *Louise Nevelson*. Exh. cat. New York: WMAA, 1967.

Greenberg, Clement. *Art and Culture: Critical Essays*. Boston: Beacon Press, 1961.

———. *Hans Hofmann*. Paris and London: G. Fall, 1961.

Guilbaut, Serge. *How New York Stole the Idea of Modern Art*. Chicago: University of Chicago Press, 1983.

[Guston] *Philip Guston: Paintings 1969–1980*. London: Whitechapel Art Gallery, 1983.

Hess, Thomas B. *Barnett Newman*. Exh. cat. New York: MOMA, 1971.

Hirsch, Sanford, and Mary Davis MacNaughton, eds. *Adolph Gottlieb: A Retrospective*. Exh. cat. Washington, D.C.: Corcoran Gallery of Art, 1982.

Hobbs, Robert Carleton, and Gail Levin. *Abstract Expressionism: The Formative Years*. Ithaca, N.Y.: Cornell University Press, 1981.

Hofmann, Hans. *Search for the Real and Other Essays*. Ed. S. T. Weeks and B. H. Hayes, Jr. Cambridge, Mass.: MIT Press, 1967.

Hunter, Sam. *Isamu Noguchi*. New York: Abbeville, 1978.

Jordan, Jim M., and Robert Goldwater. *The Paintings of Arshile Gorky: A Critical Catalogue*. New York: New York University Press, 1982.

Kingsley, April. *The Turning Point: The Abstract Expressionists and the Transformation of American Art*. New York: Simon & Schuster, 1992.

[Kline] *Franz Kline Memorial Exhibition*. Exh. cat. Washington, D.C.: Washington Gallery of Modern Art, 1962.

Krauss, Rosalind. *The Sculpture of David Smith: A Catalogue Raisonné*. New York: Garland Pub., 1977.

Lader, Melvin. P. *Gorky*. New York: Abbeville, 1985.

Landau, Ellen. *Jackson Pollock*. New York: Harry N. Abrams, 1989.

Larson, Philip, and Peter Schjeldahl. *De Kooning: Drawings/Sculptures*. Exh. cat. Minneapolis: Walker Art Center, 1974.

Leja, Michael. *Reframing Abstract Expressionism: Subjectivity and Painting in the 1940's*. New Haven, Conn.: Yale University Press, 1993.

Lieberman, William S. *Jackson Pollock: The Last Sketchbook*. New York: Harcourt, Brace, Jovanovich, 1982.

Lippard, Lucy R. *Ad Reinhardt*. New York: Harry N. Abrams, 1981.

Lyons, Nathan, ed. *Aaron Siskind, Photographer*. Rochester, N.Y.: George Eastman House, 1965.

Marcus, Stanley. *David Smith: The Sculptor and His Work*. Ithaca, N.Y.: Cornell University Press, 1983.

McShine, Kynaston, ed. *Joseph Cornell*. Exh. cat. New York: MOMA, 1990.

Motherwell, Robert. *Collected Writings*. New York: Oxford University Press, 1992.

[Motherwell] *Robert Motherwell*. Exh. cat. Buffalo, N.Y.: Albright-Knox Art Gallery, 1983. Essays by J. Flam and D. Ashton.

Nevelson. Exh. cat. New York: Pace Gallery, 1976.

Noguchi, Isamu. *A Sculptor's World*. New York: Harper & Row, 1968.

O'Connor, Francis V., and Eugene V. Thaw, eds. *Jackson Pollock: A Catalogue Raisonné of Paintings, Drawings, and Other Works*.

New Haven, Conn.: Yale University Press, 1978.

Polcari, Stephan. *Abstract Expressionism and the Modern Experience*. New York: Cambridge University Press, 1991.

[Pollock] *Jackson Pollock*. Exh. cat. Paris: Centre Georges Pompidou, Musée National d'Art Moderne, 1982.

Porter, Eliot. *Intimate Landscapes: Photographs*. New York: MMA, 1979.

Prather, Marla. *Willem de Kooning/Paintings*. Exh. cat. Washington: National Gallery of Art, 1994.

Rand, Harry. *Arshile Gorky: The Implications of Symbol*. Montclair, N.J.: Allanheld & Schram, 1980.

[Reinhardt] *Ad Reinhardt*. Exh. cat. New York: Rizzoli, 1991.

Roberts, Colette. *Mark Tobey*. New York: Grove Press, 1960.

Rosand, David, ed. *Robert Motherwell on Paper: Drawings, Prints, Collages*. Exh. cat. New York: Harry N. Abrams, 1997.

Rose, Barbara. *Jackson Pollock: Drawing into Painting*. Exh. cat. New York: MOMA, 1980.

———. *Krasner/Pollock, a Working Relationship*. Exh. cat. New York: Grey Art Gallery, 1981.

———. *Lee Krasner: A Retrospective*. Exh. cat. Houston, Tex.: Museum of Fine Arts, 1983.

———, ed. *Art-as-Art: The Selected Writings of Ad Reinhardt*. New York: Viking Press, 1975.

Rosenberg, Harold. *Barnett Newman*. New York: Harry N. Abrams, 1978.

———. *Isamu Noguchi: The Sculpture of Spaces*. Exh. cat. New York: WMAA, 1980. Foreword by T. Armstrong.

———. *The Tradition of the New*. New York: Horizon Press, 1959.

———. *Willem de Kooning*. New York: Harry N. Abrams, 1974.

Rosenblum, Robert. *Mark Rothko: Notes on Rothko, Surrealist Years*. Exh. cat. New York: Pace Gallery, 1981.

Ross, Clifford. *Abstract Expressionism: Creators and Critics*. New York: Harry N. Abrams, 1990.

Sandler, Irving. *The New York School: The Painters and Sculptors of the Fifties*. New York: Harper & Row, 1978.

———. *The Triumph of American Painting: A History of Abstract Expressionism*. New York: Praeger, 1970.

Schwartz, Constance. *The Abstract Expressionists and Their Precursors*. Nassau County: Nassau County Museum of Fine Art, 1981.

Seitz, William. *Abstract Expressionist Painting in America*. Cambridge, Mass.: Harvard University Press, 1983.

———. *Abstract Expressionist Painting in America*. Exh. cat. Washington, D.C.: National Gallery of Art, 1983.

———. *Hans Hofmann*. Exh. cat. New York: MOMA, 1963. With selected writings by the artist.

———. *Mark Tobey*. Exh. cat. New York: MOMA, 1962.

Shapiro, David, and Cecile Shapiro, eds. *Abstract Expressionism: A Critical Record*. New York: Cambridge University Press, 1990.

Stein, Sally, and Terrence Pitts. *Harry Callahan: Photographs in Color, the Years 1946–1978*. Tucson, Ariz.: Center for Creative Photography, c. 1980.

Sylvester, Julie. *John Chamberlain: A Catalogue Raisonné of the Sculpture, 1954–1985*. New York: Hudson Hills Press,

c. 1986.

Szarkowski, John, ed. *Callahan*. New York: MOMA, 1976.

Tuchman, Maurice, ed. *New York School, The First Generation: Paintings of the 1940s and 1950s*. Los Angeles: LACMA, 1965. With statements by artists and critics.

Wagner, Anne. *Three Artists (Three Women): Modernism and the Art of Hesse, Krasner, and O'Keeffe*. Berkeley, Calif.: University of California Press, 1996.

Waldman, Diane. *Arshile Gorky 1904–1948: A Retrospective*. Exh. cat. New York: SRGM, 1981.

———. *Joseph Cornell*. New York: G. Braziller, 1977.

———. *Mark Rothko, 1903–1970: A Retrospective*. Exh. cat. New York: SRGM, 1978.

———. *Willem de Kooning*. Library of American Art. New York: Harry N. Abrams, 1988.

Wight, Frederick. *Hans Hofmann: Retrospective Exhibition*. Exh. cat. Los Angeles: The University Art Museum, UCLA, 1957.

Wilkin, Karen. *David Smith*. New York: Abbeville, 1984.

Wye, Deborah. *Louise Bourgeois*. Exh. cat. New York: MOMA, 1982.

20. POSTWAR EUROPEAN ART

Ades, Dawn, and Andrew Forge. *Francis Bacon*. London: Thames & Hudson, 1985.

Alloway, Lawrence. *Francis Bacon*. Exh. cat. New York: SRGM, 1963.

Bacon, Francis, et al. *Francis Bacon*. Exh. cat. New York and Washington, D.C.: Thames & Hudson, 1989.

Daries, Hugh, and Sally Yard. *Francis Bacon*. Exh. cat. New York: Abbeville, 1986.

De Staël: A Retrospective Exhibition. Exh. cat. Boston: Museum of Fine Arts, 1965.

Dorival, B. *Soulages*. Exh. cat. Paris: Musée National d'Art Moderne, 1967.

Flomenhaft, Eleanor. *The Roots and Development of CoBrA Art*. Hempstead, N.Y.: Fine Arts Museum of Long Island, c. 1985.

Franzke, Andreas. *Dubuffet*. New York: Harry N. Abrams, 1981.

Freund, Gisèle. *Gisèle Freund, Photographer*. Trans. J. Shepley. New York: Harry N. Abrams, 1985. Introduction by C. Caujolle.

Gimenez, Carmen. *Tàpies*. Exh. cat. New York: SRGM, 1995.

Hughes, Robert. *Lucian Freud Paintings*. London: Thames & Hudson, 1987.

Leiris, Michael. *Francis Bacon: Full Face and in Profile*. New York: Rizzoli, 1983.

Leymarie, Jean. *Art Since Mid-Century: The New Internationalism*. 2 vols. Greenwich, Conn.: New York Graphic Society, 1971.

———. *Balthus*. Exh. cat. New York: Rizzoli, 1990.

Loreau, Max. *Catalogue intégral des travaux de Jean Dubuffet*. 34 vols. Paris: J.-J. Pauvert, 1964–.

Penrose, Roland. *Tàpies*. New York: Rizzoli, 1978.

Permanyer, Lluís. *Tàpies and the New Culture*. Trans. K. Lyons. New York: Rizzoli, 1986.

Putnam, Jacques, ed. *Bram van Velde*. New York: Grove Press, 1959. With statements by S. Beckett and G. Duthuit.

Rewald, John. *Manzù*. New York: Thames & Hudson, 1967.

Rewald, Sabine. *Balthus*. Exh. cat. New York: MMA, 1984.

Ritchie, Andrew. *The New Decade: Twenty-two European Painters and Sculptors*. Exh. cat.

New York: MOMA, 1955.

Russell, John. *Francis Bacon*. London: Thames & Hudson, 1993.

Selz, Peter, and James Sweeney. *Chillida*. New York: Harry N. Abrams, 1986.

Sweeney, James. *Soulages*. Greenwich, Conn.: New York Graphic Society, 1972.

Sylvester, David. *Interviews with Francis Bacon*. 3rd ed. London: Thames & Hudson, 1987.

Trier, Eduard. *The Sculpture of Marino Marini*. London: Praeger, 1961.

[Van Velde] *Bram van Velde*. Exh. cat. New York: Knoedler and Co., 1962. Text by J. van der Marck.

Weiss, Allen S. *Shattered Forms: Art Brut, Phantasms, Modernism*. Albany, N.Y.: SUNY Press, 1992.

Wye, Deborah. *Antoni Tàpies in Print*. Exh. cat. New York: MOMA, 1991.

21. POP ART AND EUROPE'S NEW REALISM

Alloway, Lawrence. *American Pop Art*. Exh. cat. New York and London: WMAA, 1975.

———. *Christo*. New York: Harry N. Abrams, 1969.

———. *Roy Lichtenstein*. New York: Abbeville, 1983.

———, et al. *Modern Dreams: The Rise and Fall and Rise of Pop*. Cambridge: MIT Press, 1988.

Mario. *Pop Art . . . and After*. New York: Viking, 1965.

[Arbus] *Diane Arbus*. New York: Aperture, 1972.

Armstrong, Richard. *Richard Artschwager*. Exh. cat. New York and London: WMAA, 1988.

Ashton, Dore. *Richard Lindner*. New York: Harry N. Abrams, 1970.

Auping, Michael. *Jess: Paste-Ups (and Assemblies), 1951–1983*. Exh. cat. Sarasota, Fla.: The John and Mable Ringling Museum of Art, 1983.

Ball-Teshuva, Jacob. *Christo: The Reichstag and Urban Projects*. Munich and New York: Prestel, 1993.

Bourdon, David. *Christo*. New York: Harry N. Abrams, 1989.

———. *Warhol*. New York: Harry N. Abrams, 1989.

[Broodthaers] *Marcel Broodthaers*. Exh. cat. Minneapolis: Walker Art Center and Rizzoli, 1989.

Calas, Nicolas and Elena. *Icons and Images of the Sixties*. New York: Dutton, 1971.

Castleman, Riva. *Jasper Johns, A Print Retrospective*. Exh. cat. New York: MOMA, 1986.

Cathcart, Linda. *Robert Rauschenberg: Work from Four Series*. Exh. cat. Houston: Contemporary Arts Museum, 1985.

Celant, Germano. *Piero Manzoni*. Milan: Arnoldo Mandadori Arte, 1991.

Crichton, Michael. *Jasper Johns*. Exh. cat. New York: WMAA, 1977.

Ferguson, Russell. *Hand-painted Pop: American Art in Transition, 1955–1962*. Exh. cat. Los Angeles: Museum of Contemporary Art, 1992.

Francis, Richard. *Jasper Johns*. New York: Abbeville, 1984.

[Frank] *Robert Frank: The Americans*. New York, Zurich and Berlin: Scalo, 1993.

Friedman, Martin. *Nevelson: Wood Sculptures*. Exh. cat. Minneapolis: Walker Art Center, 1973.

———. *Oldenburg: Six Themes*. Exh. cat. Minneapolis: Walker Art Center, 1975.

Geldzahler, Henry. *New York Painting and Sculpture: 1940–1970*. Exh. cat. New York:

MMA, 1969.

Glenn, Constance. *Jim Dine Drawings*. New York: Harry N. Abrams, 1985.

———. *Lucas Samaras Photo-Transformations*. Exh. cat. Long Beach, Calif. and New York: Cal State University and Dutton, 1975.

———. *Time Dust: James Rosenquist. Complete Graphics: 1962–1992*. Exh. cat. Long Beach, Calif. and New York: Cal State University and Rizzoli, 1993.

Gordon, John. *Jim Dine*. Exh. cat. New York: WMAA, 1970.

———. *Rosenquist*. New York, 1986.

Greenough, Sarah, and Philip Brookman. *Robert Frank*. Exh. cat. Washington, D.C.: National Gallery of Art, 1994.

Hackett, Pat, ed. *The Andy Warhol Diaries*. New York: Warner Books, 1989.

Harrison, Helen. *Larry Rivers*. New York: Harper & Row, 1984.

Haskell, Barbara. *Blam! The Explosion of Pop, Minimalism, and Performance: 1958–1964*. Exh. cat. New York: WMAA, 1984.

———. *Claes Oldenburg: Object into Monument*. Exh. cat. Pasadena, Calif.: Pasadena Art Museum, 1971.

Henri, Adrian. *Total Art: Environments, Happenings, and Performance*. London: Thames & Hudson, 1979.

Hickey, Dave and Peter Plagens. *The Works of Edward Ruscha*. Exh. cat. New York and San Francisco: Hudson Hills Press and San Francisco Museum of Modern Art, 1982. Foreword H. T. Hopkins.

[Hockney] *David Hockney: A Retrospective*. Los Angeles: LACMA, 1988.

Hockney, David. *That's the Way I See It*. Ed. Nikos Stangos. London: Thames & Hudson, 1993.

Hulten, Pontus. *Jean Tinguely: A Magic Stronger Than Death*. New York: Abbeville, 1987.

———. *Larry Rivers*. New York: Rizzoli, 1991.

———. *The Machine: As Seen at the End of the Mechanical Age*. Exh. cat. New York: MOMA, 1968.

———. *Niki de Saint Phalle*. Exh. cat. Stuttgart: G. Hatje, c. 1992.

Hunter, Sam. *Arman's Apocalypse*. Exh. cat. New York: Marisa del Re Gallery, 1984.

Jimenez, Luis. *Man on Fire/ El Hombre en Llamas*. Exh. cat. Albuquerque, N.Mex.: The Albuquerque Museum, 1994.

Kaprow, Allan. *Assemblage, Environments, and Happenings*. New York: Harry N. Abrams, 1966.

Kienholz: A Retrospective: Edward and Nancy Reddin Kienholz. Exh. cat. New York: WMAA, 1996.

Kirby, Michael. *Happenings: An Illustrated Anthology*. New York: Dutton, 1965.

Kitaj, R. B. *An Artist's Eye*. Exh. cat. London: National Gallery, 1980.

[Kitaj] *R. B. Kitaj*. Exh. cat. New York: Marlborough Gallery, 1979.

Kozloff, Max. *Jasper Johns*. New York: Harry N. Abrams, 1974.

Kuspit, Donald. *Lucas Samaras*. Athens, 1986.

Leggio, James, and Susan Weiley, eds. *American Art of the 1960s*. New York: MOMA, 1991.

Levin, Kim. *Lucas Samaras*. New York: Harry N. Abrams, 1975.

Leymarie, Jean. *Art Since Mid-Century: The New Internationalism*. 2 vols. Greenwich, Conn.: New York Graphic Society, 1971.

Lima Greene, Alison de, and Pierre Restany. *Arman 1955–1991: A Retrospective*. Exh. cat. Houston: The Museum of Fine Arts, 1991.

Lippard, Lucy. *Pop Art*. London: Thames &

Hudson, 1966.

Livingstone, Marco. *David Hockney*. London: Thames & Hudson. Rev. ed. 1996.

———. *Pop Art: A Continuing History*. London: Thames & Hudson, 1990.

———. *R. B. Kitaj*. New York: Rizzoli, 1985.

———, ed. *Pop Art*. London: Royal Academy of Arts, 1991.

Mahsun, Carol Ann, ed. *Pop Art: The Critical Dialogue*. Ann Arbor, Mich.: UMI Research Press, 1989.

McConnell, Gerald. *Assemblage: Three Dimensional Picture Making*. New York: Van Nostrand Reinhold Co., 1976.

McShine, Kynaston, ed. *Andy Warhol, A Retrospective*. Exh. cat. New York: MOMA, 1989.

Morphet, Richard, ed. *R. B. Kitaj*. Exh. cat. London: Tate Gallery, 1994.

Nochlin, Linda. *Andy Warhol Nudes*. New York: Robert Miller Gallery, 1995.

Oldenburg, Claes, and Coosje van Bruggen. *Large-Scale Projects*. New York: Monacelli Press, 1994.

Pellegrini, Aldo. *New Tendencies in Art*. Trans. R. Carson. New York: Crown, 1966.

Popper, Frank. *Origins and Development of Kinetic Art*. Trans. S. Bann. Greenwich, Conn.: New York Graphic Society, 1968.

Ratcliff, Carter. *Red Grooms*. New York: Abbeville, 1984.

Rauschenberg Overseas Cultural Interchange. Washington, D.C.: National Gallery of Art, 1991.

Restany, Pierre. *Botero*. Trans. J. Shepley. New York: Harry N. Abrams, 1984.

———. *Yves Klein*. Trans. J. Shepley. New York: Harry N. Abrams, 1982.

Rose, Barbara. *Claes Oldenburg*. Exh. cat. New York: MOMA, 1970.

Rosenthal, Nan, and Ruth Fine. *The Drawings of Jasper Johns*. Washington, D.C.: The National Gallery of Art, 1990.

Russell, John, and Suzi Gablik. *Pop Art Redefined*. London: Thames & Hudson, 1969.

[Saint Phalle] *Niki de Saint Phalle*. Exh. cat. Paris: Centre Georges Pompidou, 1980.

Sandford, Mariellen. *Happenings and Other Acts*. New York and London: Routledge, 1995.

Sandler, Irving. *American Art of the 1960's*. New York: Harper & Row, 1988.

[Segal] *George Segal: Environments*. Exh. cat. Philadelphia: Institute of Contemporary Art, 1976. Text by J. L. Barrio-Garay.

Seitz, William C. *Segal*. New York: Harry N. Abrams, 1972.

Shanahan, Mary, ed. *Evidence, 1944–1994: Richard Avedon*. Exh. cat. New York: WMAA, 1994.

Shapiro, David. *Jasper Johns Drawings: 1954–1984*. Ed. C. Sweet. New York: Harry N. Abrams, 1984.

———. *Jim Dine: Painting What One Is*. New York: Harry N. Abrams, 1981.

Shirey, David. *Edward and Nancy Reddin Kienholz: Human Scale*. Exh. cat. San Francisco: Museum of Modern Art, 1985. Texts by H. T. Hopkins, L. Weschler, E. Kienholz, R. Glower.

Slemmons, Rod. *Like a One-Eyed Cat: Photographs by Lee Friedlander*. New York and Seattle: Harry N. Abrams and Seattle Art Museum, 1989.

Spies, Werner, ed. *Fernando Botero: Paintings and Drawings*. Munich: Prestel, 1992.

Stich, Sidra. *Yves Klein*. Exh. cat. London: Hayward Gallery and Cantz, 1995.

Szarkowski, John. *Winogrand, Figments from the Real World*. Exh. cat. New York: MOMA, 1988.

Taylor, Paul. *Post-Pop Art*. Cambridge, Mass.: MIT Press, 1989.

Tomkins, Calvin. *Off the Wall: Robert Rauschenberg and the Art World of Our Time*. Garden City, N.Y.: Doubleday, 1980.

Tuchman, Maurice, ed. *American Sculpture of the Sixties*. Exh. cat. Los Angeles: LACMA, 1967.

Tuchman, Phyllis. *George Segal*. New York: Abbeville, 1984.

Tucker, Marcia. *James Rosenquist*. Exh. cat. New York: WMAA, 1972.

Vaizey, Marina. *Christo*. New York: Rizzoli, 1990.

Van der Marck, Jan. *George Segal*. Rev. ed. New York: Harry N. Abrams, 1979.

Varnedoe, Kirk. *Jasper Johns: A Retrospective*. Exh. cat. New York: MOMA, 1996.

Waldman, Diane. *Roy Lichtenstein*. Exh. cat. New York: SRGM, 1993.

Weinhardt, Jr., Carl J. *Robert Indiana*. New York: Harry N. Abrams, 1990.

Whiting, Cécile. *A Taste for Pop: Pop Art, Gender and Consumer Culture*. Cambridge: Cambridge University Press, 1997.

[Winogrand] *Garry Winogrand: Early Work*. Archive 26. Tucson, Ariz.: Center for Creative Photography, 1987.

22. SIXTIES ABSTRACTION

Alloway, Lawrence. *Systemic Painting*. Exh. cat. New York: SRGM, 1966.

Ashton, Dore. *Richard Diebenkorn: Small Paintings from "Ocean Park."* Exh. cat. Lincoln, Nebr.: Sheldon Memorial Art Gallery, 1985.

Baker, Kenneth. *Minimalism: Art of Circumstance*. New York: Abbeville, 1988.

Barker, Ian, ed. *Caro*. Munich: Prestel, 1991.

Baro, Gene. "The Achievement of Helen Frankenthaler," *Art International* 11 (September 1967).

Barrett, Cyril. *Op Art*. London: Studio Vista, 1970.

Bastian, Heiner. *Cy Twombly: Catalogue Raisonné*. 4 vols. Berlin: Schirmer/Mosel, 1992.

Belz, Carl. *Frankenthaler, the 1950s*. Exh. cat. Waltham, Mass.: Rose Art Museum, 1981.

Berger, Maurice. *Labyrinth: Robert Morris, Minimalism, and the 1960s*. New York: Harper & Row, 1989.

Bourdon, David. *Carl Andre: Sculpture, 1959–1977*. New York: J. Rietman, 1978. Foreword by B. Rose.

———, et al. *Jackie Ferrara Sculpture: A Retrospective*. Exh. cat. Sarasota, Fla.: The John and Mable Ringling Museum of Art, 1992.

Buck, Jr., Robert. *Richard Diebenkorn: Paintings and Drawings: 1943–1976*. Exh. cat. Buffalo, N.Y.: Albright-Knox Art Gallery, 1976. Texts by L. Cathcart, G. Nordland, M. Tuchman.

Carmean, E. A. *Helen Frankenthaler, A Paintings Retrospective*. New York: Harry N. Abrams, 1989.

Cathcart, Linda. *Richard Diebenkorn: 38th Venice Biennial, US Pavilion*. New York: International Exhibitions Committee of the American Federation of Arts, 1978.

Colpitt, Frances. *Minimal Art: The Critical Perspective*. Ann Arbor, Mich.: UMI Research Press, 1990.

Coplans, John. *Ellsworth Kelly*. New York: Harry N. Abrams, 1973.

Elderfield, John. *Morris Louis*. Exh. cat. New York: MOMA, 1986.

Fine, Ruth. *Helen Frankenthaler Prints*. Exh. cat. Washington, D.C. and New York:

National Gallery of Art and Harry N. Abrams, 1993.

[Flavin] *Dan Flavin: Three Installations in Fluorescent Light*. Exh. cat. Cologne: Kunsthalle Köln, 1974.

Fried, Michael. *Morris Louis*. New York: Harry N. Abrams, 1971.

———. *Three American Painters: Kenneth Noland, Jules Olitski, Frank Stella*. Exh. cat. Cambridge, Mass.: Fogg Art Museum, 1965.

———. "Art and Objecthood," *Artforum* 5, no. 10 (Summer 1967).

Coenen, E. C. *Ellsworth Kelly*. Exh. cat. New York: MOMA, 1973.

———. *Helen Frankenthaler*. Exh. cat. New York: WMAA and MOMA's International Council, 1969.

Greenberg, Clement. *Art and Culture*. Boston: Beacon Press, 1965.

———. *Post-Painterly Abstraction*. Exh. cat. Los Angeles: LACMA, 1964.

Guberman, Sidney. *Frank Stella: An Illustrated Biography*. New York: Rizzoli, 1995.

Güse, Ernst-Gerhard. *Richard Serra*. New York: Rizzoli, 1988.

Harrison, Charles and Paul Wood, eds. *Art in Theory, 1900–1990*. Oxford: Blackwell Publishers, 1995.

Haskell, Barbara. *Agnes Martin*. New York: WMAA, 1992. Texts by B. Haskell, A. Chave, R. Krauss.

———. *Donald Judd*. Exh. cat. New York: W. W. Norton & Co., 1988.

Judd, Donald. *Complete Writings, 1959–1975*. Halifax and New York: Press of the Nova Scotia College of Art and Design and New York University Press, 1975.

[Judd] *Donald Judd*. Exh. cat. Ottawa: National Gallery of Canada, 1975. Exhibition organized by B. Smith; Text by R. Smith.

[Judd] *Donald Judd: Large Scale Works*. Exh. cat. New York: The Pace Gallery, 1993.

Kertess, Klaus. *Brice Marden: Painting and Drawings*. New York: Harry N. Abrams, 1992.

———. *Joan Mitchell*. New York: Harry N. Abrams, 1997.

Krauss, Rosalind. *Richard Serra/Sculpture*. Exh. cat. New York: MOMA, 1986. Texts by L. Rosenstock and D. Crimp.

Legg, Alicia. *Sol LeWitt*. Exh. cat. New York: MOMA, 1978.

Leggio, James, and Susan Weiley, eds. *American Art of the 1960s*. New York: MOMA, 1991.

Leider, Philip. *Stella Since 1970*. Exh. cat. Fort Worth, Tex.: Fort Worth Art Museum, 1978.

LeWitt, Sol. *Sol LeWitt: Geometric Figures and Color*. New York: Harry N. Abrams, 1979.

Leymarie, Jean. *Art Since Mid-Century: The New Internationalism*. 2 vols. Greenwich, Conn.: New York Graphic Society, 1971.

[Mangold] *Robert Mangold: Paintings 1971–1984*. Exh. cat. Akron, Ohio: Akron Art Museum, 1984.

[Mitchell] *Joan Mitchell*. Exh. cat. Paris: Éditions du Jeu de Paume, 1994.

Moffett, Kenworth. *Jules Olitski*. New York: Harry N. Abrams, 1981.

———. *Kenneth Noland*. New York: Harry N. Abrams, 1977.

[Morris] *Robert Morris: The Mind/Body Problem*. Exh. cat. New York: SRGM, 1994.

Necol, J., and D. Droll. *Abstract Painting: 1960–1969*. Exh. cat. New York: P.S.1, 1983.

Nordland, Gerald. *Fourteen Abstract Painters*.

Exh. cat. Los Angeles: Wright Art Gallery, University of California, 1975.

Popper, Frank. *Agam.* 3rd ed. rev. New York: Harry N. Abrams, 1990.

———. *Origins and Development of Kinetic Art.* Greenwich, Conn.: New York Graphic Society, 1968.

Richardson, Brenda. *Frank Stella: The Black Paintings.* Exh. cat. Baltimore: Baltimore Museum of Art, 1976.

Rickey, George. *Constructivism: Origins and Evolution.* New York: G. Braziller, 1967.

Rose, Barbara. *Frankenthaler.* New York: Harry N. Abrams, 1972.

Rosenblum, Robert. *Frank Stella.* Harmondsworth and Baltimore: Penguin, 1971.

Rosenthal, Mark. *Abstraction in the Twentieth Century: Total Risk, Freedom, Discipline.* Exh. cat. New York: SRGM, 1996.

Rubin, Lawrence. *Frank Stella: Paintings 1958 to 1965: A Catalogue Raisonné.* New York: Stewart, Tabori & Chang, 1986.

Rubin, William. *Anthony Caro.* Exh. cat. New York: MOMA, 1975.

———. *Frank Stella.* Exh. cat. New York: MOMA, 1987.

[Ryman] *Robert Ryman.* Exh. cat. Paris: Musée National d'Art Moderne, Centre Georges Pompidou, 1981.

Sandler, Irving. *Al Held.* New York: Hudson Hills Press, 1984.

———. *American Art of the 1960's.* New York: Harper & Row, 1988.

———. "Mitchell Paints a Picture." *Art News* 56, no. 6 (Oct. 1957): 45.

Selz, Peter. *Sam Francis.* New York: Harry N. Abrams, 1975.

Serota, Nicholas. *Carl Andre: Sculpture, 1959–1978.* Exh. cat. London: Whitechapel Art Gallery, 1978.

Shearer, Linda. *Brice Marden.* Exh. cat. New York: SRGM, 1975.

Sims, Patterson, and Emily Rauh Pulitzer. *Ellsworth Kelly: Sculpture.* Exh. cat. New York: WMAA, 1983.

[Smith] *Tony Smith: Recent Sculpture.* Exh. cat. New York: M. Knoedler and Co., 1971.

Soto: A Retrospective Exhibition. Exh. cat. New York: SRGM, 1974.

Spies, Werner. *Victor Vasarely.* New York: Harry N. Abrams, 1971.

Storr, Robert. *Robert Ryman.* Exh. cat. New York: MOMA, 1993.

Tuchman, Maurice. *American Sculpture of the Sixties.* Exh. cat. Los Angeles: LACMA, 1965.

Tucker, Marcia. *Al Held.* Exh. cat. New York: WMAA, 1974.

———. *Robert Morris.* Exh. cat. New York: WMAA, 1970.

[Twombly] *Cy Twombly: Paintings and Drawings 1954–1977.* Exh. cat. New York: WMAA, 1979.

Upright, Diane. *Morris Louis: The Complete Paintings, a Catalogue Raisonné.* New York: Harry N. Abrams, 1985.

Varnedoe, Kirk. *Cy Twombly, A Retrospective.* Exh. cat. New York: MOMA, 1994.

Vasarely. Neuchâtel: Éditions du Griffon, 1965. Introduction by Marcel Joray.

Waldman, Diane. *Anthony Caro.* New York: Abbeville, 1982.

———. *Carl Andre.* Exh. cat. New York: SRGM, 1970.

———. *Kenneth Noland: A Retrospective.* Exh. cat. New York: SRGM, 1977.

———. *Robert Mangold.* Exh. cat. New York: SRGM, 1971.

———, ed. *Ellsworth Kelly: A Retrospective.* Exh. cat. New York: SRGM, 1996.

Weschler, Lawrence. *Seeing Is Forgetting the Name of the Thing One Sees: A Life of Contemporary Artist Robert Irwin.* Berkeley, Los Angeles, and London: University of California Press, 1982.

Whelan, Richard. *Anthony Caro.* Harmondsworth: Penguin, 1974. Texts by C. Greenberg, M. Fried, J. Russell, P. Tuchman.

Wilkin, Karen. *Frankenthaler: Works on Paper, 1949–84.* Exh. cat. New York: SRGM, 1985.

Wood, Paul, et al. *Modernism in Dispute.* New Haven, Conn.: Yale University Press, 1993.

23. THE SECOND WAVE OF INTERNATIONAL STYLE ARCHITECTURE

Aalto, Alvar. *The Architectural Drawings of Alvar Aalto: 1917–1939.* New York: Garland, 1994.

Applewhite, E. J., ed. *Synergetics Dictionary: The Mind of Buckminster Fuller.* New York: Garland, 1986.

Argan, Giulio C. *Marcel Breuer.* Milan: Görlich, 1957.

Arnell, Peter, and Ted Bickford, eds. *James Stirling: Buildings and Projects.* New York: Rizzoli, 1985. Text by C. Rowe.

Blake, Peter. *Marcel Breuer: Architect and Designer.* Exh. cat. New York: MOMA, 1949.

———. *Philip Johnson.* Basel and Boston: Birkhäuser Verlag, 1996.

Brownlee, David, and David De Long. *Louis I. Kahn: In the Realm of Architecture.* Exh. cat. Los Angeles: LACMA, 1991. Introduction by V. Scully.

Cannell, Michael T. *I. M. Pei: Mandarin of Modernism.* New York: Carol Southern Books, 1995.

Ciucci, Giorgio, ed. *Giuseppe Terragni: Opera Completa.* Milan: Electa, 1996.

Frampton, Kenneth. *Richard Meier, Architect: Buildings and Projects, 1966–1976.* New York: Oxford University Press, 1976. Postscript by J. Hejduk.

Fuller, R. Buckminster. *Ideas and Integrities: A Spontaneous Autobiographical Disclosure.* New York: MacMillan, 1963.

———. *Synergetics: Explorations in the Geometry of Thinking.* New York: MacMillian, 1975.

Giurgola, Romaldo. *Louis I. Kahn.* Boulder, Colo.: Westview Press, 1975.

Hines, Thomas. *Richard Neutra and the Search for Modern Architecture: A Biography and History.* New York: Oxford University Press, 1982.

Hitchcock, Henry-Russell. *Philip Johnson, Architect: 1949–1965.* New York: Holt, Rinehart, and Winston, 1966.

[Johnson] *Philip Johnson: The Glass House.* Ed. D. Whitney and J. Kipnis. New York: Pantheon Books, 1993.

Kahn, Louis I. *Louis I. Kahn: Writings, Lectures, and Interviews.* Ed. A. Latour. New York: Rizzoli, 1991.

[Kiesler] *Frederick Kiesler: Environmental Sculpture.* Exh. cat. New York: SRGM, 1964.

Kultermann, Udo, ed. *Kenzo Tange, 1946–1969: Architecture and Urban Design.* New York: Praeger, 1970.

Lewis, Hilary and John O'Connor. *Philip Johnson: The Architect in His Own Words.* New York: Rizzoli, 1994.

Lobell, John. *Between Silence and Light: Spirit in the Architecture of Louis I. Kahn.* Boulder, Colo.: Shambhala, 1979.

Loud, Patricia. *The Art Museums of Louis I. Kahn.* Durham, N.C.: Duke University Press, 1989.

The Louis I. Kahn Archive: Personal Drawings: The Completely Illustrated Catalogue of the Drawings in the Louis I. Kahn Collection. 7 vols. Philadelphia and New York: University of Pennsylvania, Pennsylvania Historical and Museum Commission, and Garland: 1987.

Murray, Irena Z., ed. *Moshe Safdie: Buildings and Projects, 1967–1992.* Montreal and Buffalo, N.Y.: Canadian Architecture Collection et al., 1996.

Nervi, Pier Luigi. *Aesthetics and Technology in Building.* Trans. R. Einaudi. Cambridge, Mass.: Harvard University Press, 1965.

Neutra, Richard. *Life and Human Habitat.* Stuttgart: A. Koch, 1956.

[Neutra] *Richard Neutra: Promise and Fulfillment, 1919–1932: Selections from the Letters and Diaries of Richard and Dione Neutra.* Ed. and trans. D. Neutra. Carbondale, Ill.: Southern Illinois University Press, 1985.

Newhouse, Victoria. *Wallace K. Harrison, Architect.* New York: Rizzoli, 1989.

Papadaki, Stamo. *Oscar Niemeyer.* New York: Reinhold, 1950.

———. *Oscar Niemeyer: Works in Progress.* New York: Reinhold, 1956.

Pearson, Paul D. *Alvar Aalto and the International Style.* New York: Whitney Library of Design, 1978.

Phillips, Lisa. *Frederick Kiesler.* New York: WMAA, 1989.

Proun, Jules David. *The Architecture of the Yale Center for British Art.* New Haven, Conn.: Yale University Press, 1977. Introduction by E. P. Pillsbury.

Quantrell, Malcolm. *Alvar Aalto: A Critical Study.* New York: Schocken Books, 1983.

Ronner, Heinz, and Sharad Jharen. *Louis I. Kahn: Complete Works, 1935–74.* 2nd ed. rev. and enl. Basel and Boston: Birkhäuser Verlag, 1987.

Rykwert, Joseph. *Richard Meier: 1964/1984.* New York: Rizzoli, 1984.

Schildt, Goran. *Alvar Aalto: The Complete Catalogue of Architecture, Design, and Art.* New York: Rizzoli, 1994.

Schulze, Franz. *Philip Johnson: Life and Work.* New York: Knopf, 1994.

Schumaher, Thomas. *Surface & Symbol: Giuseppe Terragni and the Architecture of Italian Rationalism.* New York: Princeton Architectural Press, 1991.

Scully, Vincent. *Louis I. Kahn.* New York: G. Braziller, 1962.

Snyder, Robert, ed. *Buckminster Fuller: An Autobiographical Monologue/Scenario.* New York: St. Martin's Press, 1980.

[Stirling] *The Architecture of James Stirling: Four Works.* Exh. cat. Minneapolis: Walker Art Center, 1978. Texts by C. Hodgetts and P. Papademetriou.

Temko, Alan. *Eero Saarinen.* New York: G. Braziller, 1962.

Underwood, David. *Oscar Niemeyer and the Architecture of Brazil.* New York: Rizzoli, 1994.

Weston, Richard. *Alvar Aalto.* London: Phaidon, 1995.

Wiseman, Carter. *The Architecture of I. M. Pei: with an Illustrated Catalogue of the Buildings and Projects.* London: Thames & Hudson, 1990.

Zevi, Bruno. *Richard Neutra.* Architetti del movimento moderno, 10. Milan: Il Balcone, 1954.

24. THE PLURALISTIC SEVENTIES

Adriani, Götz, Winfried Konnertz, and Karin Thomas. *Joseph Beuys: Life and Works.* Woodbury, N.Y.: Barron's Educational Series, Inc., 1979.

Anderson, Laurie. *Empty Places*. New York: Harper Perennial, 1991.

———. *Stories from the Nerve Bible: A Retrospective 1972–1992*. New York: Harper Perennial, 1994.

———. *United States*. New York: Harper & Row, 1984.

Armstrong, Elizabeth, and Joan Rothfuss. *In the Spirit of Fluxus*. Exh. cat. Minneapolis: Walker Art Center, 1993.

Ashbery, John, and Kenworth Moffett. *Fairfield Porter: Realist Painter in an Age of Abstraction*. Exh. cat. Boston: Museum of Fine Arts, 1982.

[Aycock] *Alice Aycock Projects: 1979–1981*. Exh. cat. Tampa, Fla.: University of South Florida, 1981. Introduction by E. Fry.

Baker, Kenneth. *Minimalism: Art of Circumstance*. New York: Abbeville, 1988.

Baldessari, John. *This is Not That*. Exh. cat. Manchester: Cornerhouse, 1995.

[Bartlett] *Jennifer Bartlett*. Exh. cat. Minneapolis: Walker Art Center, 1985. Texts by M. Goldwater, R. Smith, C. Tomkins.

Bartman, William, ed. *Vija Celmins*. New York: Art Press, 1992.

Battcock, Gregory, ed. *New Artists Video: A Critical Anthology*. New York: Dutton, 1980.

Beardsley, John. *Earthworks and Beyond: Contemporary Art in the Landscape*. 2nd ed. New York: Abbeville Press, 1984.

———. *Modern Painters at the Corcoran: Sam Gilliam*. Exh. cat. Washington, D.C.: Corcoran Gallery of Art, 1989.

Benezra, Neal. *Ed Paschke*. Exh. cat. Chicago: Art Institute of Chicago, 1990.

Beuys, Joseph. *Where would I have got if I had been intelligent!* Ed. L. Cooke and K. Kelly. New York: Dia Center for the Arts, 1994.

[Bochner] *Mel Bochner 1973–1985*. Exh. cat. Pittsburgh: Carnegie-Mellon University Press, 1985.

Bowman, Russell. *Philip Pearlstein: The Complete Paintings*. New York: Alpine Fine Arts, 1983.

Brenson, Michael, F. Calvo Seraller, and Edward J. Sullivan. *Antonio López Garcia*. New York: Rizzoli, 1990.

Broude, Norma, and Mary D. Garrard, eds. *The Power of Feminist Art: The American Movement of the 1970s, History and Impact*. London: Thames & Hudson, 1994.

Bruggen, Coosje van. *John Baldessari*. New York: Rizzoli, 1990.

Brutvan, Cheryl. *The Paintings of Sylvia Plimack Mangold*. Exh. cat. New York and Buffalo, N.Y.: Hudson Hills Press and Albright-Knox Art Gallery, 1994.

Buchloh, Benjamin. *Hyperrealism*. New York, 1975. Introduction by Salvador Dalí.

[Burden] *Chris Burden: a Twenty-Year Survey*. Exh. cat. Newport Beach, Calif.: Newport Harbor Art Museum, 1988.

Buren, Daniel. *5 Texts*. Exh. cat. New York and London: John Weber and Jack Wendler Galleries, 1973.

[Campus] *Peter Campus: Projected Images*. Exh. cat. New York: WMAA, 1988. Texts by G. Hanhardt and the artist.

Campus, Peter. "Faces." *Aperture* (Spring 1989): 4–5.

Canaday, John. *Richard Estes: The Urban Landscape*. Exh. cat. Boston: Museum of Fine Arts, 1978. Interview by J. Arthur.

Cathcart, Linda. *American Painting of the 1970s*. Exh. cat. Buffalo, N.Y.: Albright-Knox Gallery, 1979.

Celant, Germano. *Arte Povera*. New York: Praeger, 1969.

———. *Mario Merz*. Exh. cat. New York: SRGM, 1989.

Chicago, Judy. *Beyond the Flower: The Autobiography of a Feminist Artist*. New York: Penguin Books, 1996.

———. *The Dinner Party: A Symbol of Our Heritage*. Garden City, N.Y.: Doubleday, 1979.

Christo: Surrounded Islands, Biscayne Bay, Greater Miami, Florida, 1980–1983. Trans. S. Reader. New York: Harry N. Abrams, 1985.

Compton, Michael. *Barry Flanagan: Sculpture*. Exh. cat. London: British Council, 1982.

———. *Some Notes on the Work of Richard Long*. Exh. cat. London: British Council, 1976.

Cooper, Helen, ed. *Eva Hesse: A Retrospective*. Exh. cat. New Haven, Conn.: Yale University Press, 1992.

Corn, Wanda. *The Art of Andrew Wyeth*. Exh. cat. San Francisco: Fine Arts Museums, 1973.

[Darboven] *Hanne Darboven: Venice Biennale, 1982, West German Pavilion*. Hamburg: Sost & Co., 1982.

Deutsche, Rosalyn, et al. *Hans Haacke, Unfinished Business*. Ed. B. Wallis. Cambridge, Mass.: MIT Press, 1986.

Doezema, Marianne, and June Hargrove. *The Public Monument and Its Audience*. Exh. cat. Cleveland: Cleveland Museum of Art, 1977.

Eisenberg, Deborah. *Air: 24 Hours—Jennifer Bartlett*. New York: Harry N. Abrams, 1994.

Enyeart, James. *Jerry N. Uelsmann: Twenty-Five Years - A Retrospective*. Boston: Little, Brown, 1985.

Fairchild, Patricia A. *Primal Acts of Construction/Destruction: The Art of Michael Heizer, 1967–1987*. Ann Arbor: UMI Research Press, 1994.

Fritscher, Jack. *Mapplethorpe: Assault with a Deadly Camera, a Pop Culture Memoir, an Outlaw Reminiscence*. Mamaroneck, N.Y.: Hastings House, 1994.

Fuchs, Rudolf Herman. *Lawrence Weiner: A Selection of Works with Commentary*. Exh. cat. Eindhoven: Van Abbemuseum, 1976.

———. *Richard Long*. London: Thames & Hudson, 1986.

Goldberg, RoseLee. *Performance: From Futurism to the Present*. London: Thames & Hudson, 1988.

Goodyear, Jr., Frank H. *Contemporary American Realism Since 1960*. Boston: New York Graphic Society and the Pennsylvania Academy of the Fine Arts, 1981.

———. *Welliver*. New York: Rizzoli, 1985. Introduction by J. Ashbery.

Gouma-Peterson, Thalia. *Miriam Schapiro: A Retrospective, 1952–1980*. Exh. cat. Ohio: College of Wooster, 1980.

———, ed. *Breaking the Rules: Audrey Flack, A Retrospective 1950–1990*. Exh. cat. New York: Harry N. Abrams, 1992.

Gowing, Lawrence. *Lucien Freud*. London: Thames & Hudson, 1982.

[Haacke] *Hans Haacke Volume I*. Oxford: Museum of Modern Art, 1978.

[Haacke] *Hans Haacke Volume II*. London: Tate Gallery, 1983.

Hanhardt, John. *Nam June Paik*. Exh. cat. New York: WMAA, 1982. Texts by D. Ronte, M. Nyman, J. G. Hanhardt, D. A. Ross.

Haus, M. "Studio Vista: A Little Bit Normal." [Skoglund] *Mirabella* (September 1991): 64–65.

Hendricks, Jon. *Fluxus Codex*. Detroit and New York: The Gilbert and Lila Silverman Fluxus Collection and Harry N. Abrams, 1988.

Henri, Adrian. *Total Art: Environments, Happenings, and Performance*. London: Thames & Hudson, 1974.

Henry, Gerrit. *Janet Fish*. New York: Burton and Skira, 1987.

Hills, Patricia. *Alice Neel*. New York: Harry N. Abrams, 1983.

Hobbs, Robert. *Robert Smithson: Sculpture*. Ithaca, N.Y.: Cornell University Press, 1981. Texts by L. Alloway, J. Coplans, L. Lippard.

Holt, Nancy, ed. *The Writings of Robert Smithson*. New York: New York University Press, 1979.

[Horn] *Rebecca Horn*. Exh. cat. New York: SRGM, 1993.

Huebler, Douglas. *Location Pieces, Site Sculpture, Duration Works, Drawings, Variable Pieces*. Exh. cat. Boston: Museum of Fine Arts, 1972.

Jackie Winsor/Barry Ledoux: Sculpture. Exh. cat. Cambridge, Mass.: MIT Press, 1984.

[Jaudon] *Valerie Jaudon*. Exh. cat. West Berlin: Amerika-Haus, 1983. Introduction by S. Hunter.

Johnson, Ken. "Art and Memory." (Joel Shapiro) *Art in America* (November 1993): 90–99.

Kaiser, W. M. H. *Joseph Kosuth: Artworks and Theories*. Amsterdam, 1977.

Kaprow, Allan. *Essays on the Blurring of Art and Life*. Ed. J. Kelley. Berkeley, Los Angeles, and London: University of California Press, 1993.

Kardon, Janet. *Laurie Anderson: Works from 1969 to 1983*. Philadelphia: Institute of Contemporary Art, 1983.

———. *Robert S. Zakanitch*. Exh. cat. Philadelphia: Institute of Contemporary Art, 1986.

Kirshner, Judith Russi. *Vito Acconci: A Retrospective, 1969–1980*. Exh. cat. Chicago: Museum of Contemporary Art, 1980.

Kismaric, Susan. *Jan Groover*. Exh. cat. New York: MOMA, 1987.

Kozloff, Joyce, and Linda Nochlin. *Patterns of Desire*. New York: Hudson Hills, 1990

Krane, Susan. *Lynda Benglis: Dual Natures*. Exh. cat. Atlanta: High Museum of Art, 1990.

Krauss, Rosalind. *Beverly Pepper: Sculpture in Place*. Exh. cat. New York: Abbeville, 1986.

Lindey, Christine. *Superrealist Painting and Sculpture*. New York: Morrow, 1980.

Linker, Kate. *Mary Miss*. Exh. cat. London: ICA, 1983.

Lippard, Lucy. *Eva Hesse*. New York: New York University Press, 1976.

———. *Get the Message? A Decade of Art for Social Change*. New York: Dutton, 1984.

———. *Six Years: The Dematerialization of the Art Object from 1966 to 1972*. New York: Praeger, 1973.

Livingston, Jane, and Marcia Tucker. *Bruce Nauman: Work from 1965 to 1972*. Exh. cat. Los Angeles: LACMA and WMAA, 1972.

Lucie-Smith, Edward. *Art in the Seventies*. Ithaca, N.Y.: Cornell University Press, 1980.

———. *Super Realism*. Oxford: Phaidon, 1979.

Lyons, Lisa and Martin Friedman. *Close Portraits*. Minneapolis: Walker Art Center, 1980.

———, and Kim Levin. *Wegman's World*. Exh. cat. Minneapolis: Walker Art Center, 1982.

[Mapplethorpe] *Robert Mapplethorpe Photographs*. Exh. cat. Norfolk: Chrysler Museum, 1978.

Marshall, Richard. *Alex Katz*. Exh. cat. New York: WMAA, 1986. Text by R.Rosenblum.

———. *American Art Since 1970: Painting, Sculpture, and Drawings from the Collection of the Whitney Museum of American Art*. Exh. cat. New York: WMAA, 1984.

———. *Developments in Recent Sculpture*. Exh. cat. New York: WMAA, 1981.

———. *New Image Painting*. Exh. cat. New York: WMAA, 1978.

———. *Robert Kushner: Paintings on Paper*. Exh. cat. New York: WMAA, 1984.

———. *Robert Mapplethorpe*. Exh. cat. New York: WMAA, 1988.

McEvilley, Thomas. *Pat Steir*. New York: Harry N. Abrams, 1995.

McShine, Kynaston. *Information*. Exh. cat. New York: MOMA, 1970.

Meisel, Louis K. *Photo-Realism*. New York: Harry N. Abrams, 1989.

———. *Richard Estes: The Complete Paintings 1966–1985*. New York: Harry N. Abrams, 1986. With an Essay by J. Perreault.

[Mendieta] *Ana Mendieta: A Retrospective*. Exh. cat. New York: New Museum of Contemporary Art, 1987.

Meyer, Ursula. *Conceptual Art*. New York: Dutton, 1972.

[Meyerowitz] *Cape Light: Color Photographs by Joel Meyerowitz*. Exh. cat. Boston: Museum of Fine Arts, 1978.

[Meyerowitz] *St. Louis and the Arch: Photographs by Joel Meyerowitz*. Boston, 1980.

Mitchell, W. J. T. *Art and the Public Sphere*. Chicago and London: University of Chicago Press, 1992.

Monte, James K. *Twenty-Two Realists*. Exh. cat. New York: WMAA, 1970.

Morgan, Robert C. *Conceptual Art: An American Perspective*. Jefferson, N.C.: McFarland, 1994.

[Morris] *Robert Morris: Selected Works, 1970–1981*. Exh. cat. Houston: Contemporary Arts Museum, 1981.

[Morris] *Robert Morris: The Mind/Body Problem*. New York: SRGM, 1994.

Morrisroe, Patricia. *Mapplethorpe, a Biography*. New York: Random House, 1995.

Müller, Grégoire, and Gianfranco Gorgoni. *The New Avant-Garde: Issues for the Art of the Seventies*. New York: Praeger, 1972.

Neubert, George. *Public Sculpture/Urban Environment*. Exh. cat. Oakland, Calif.: Oakland Museum, 1974.

Nilson, L. "Seeing the Light." [Joel Meyerowitz] *American Photographer* (September 1981): 40–53.

[Nutt] *Jim Nutt*. Exh. cat. Milwaukee: Milwaukee Art Museum, 1994.

Onorato, J. *Vito Acconci: Domestic Trappings*. Exh. cat. La Jolla, Calif.: La Jolla Museum of Contemporary Art, 1997.

[Pfahl] *Altered Landscapes: Photographs by John Pfahl*. Exh. cat. Carmel, Calif.: Friends of Photography, 1981. Introduction by Peter C. Bunnell.

[Pfahl] *A Distanced Land: The Photographs of John Pfahl*. Exh. cat. Buffalo, N.Y.: Albright-Knox Art Gallery, 1990.

Pincus-Witten, Robert. *Postminimalism: American Art of the Decade*. New York: Out of London Press, 1977.

Princenthal, Nancy. "Vija Celmins: Material Fictions." *Parkett* 44 (1995): 25–8.

Printz, N. *Decoration and Representation*. Exh. cat. Alberta, Canada: Alberta College of Art, 1982.

Raven, Arlene. *Art in the Public Interest*. Ann Arbor, Mich.: UMI Press, 1989.

Real, Really Real, Super Real: Directions in Contemporary American Realism. Exh. cat. San Antonio: San Antonio Museum Association, 1981.

Redstone, Louis, and Ruth Redstone. *Public Art: New Directions*. New York: McGraw-Hill, 1980.

Richardson, Barbara. *Gilbert and George*. Exh. cat. Baltimore: Baltimore Museum of Art, 1984.

Robins, Corinne. *The Pluralist Era: American Art 1968–1981*. New York: Harper & Row, 1984.

Rose, Barbara. *Magdalena Abakanowitz*. New York: Harry N. Abrams, 1994.

Rose, Bernice. *Allegories of Modernism: Contemporary Drawing*. Exh. cat. New York: MOMA, 1992.

Rosenblum, Robert. *Alfred Leslie*. Exh. cat. Boston: Museum of Fine Arts, 1976.

Rosenstock, Laura. *Christopher Wilmarth*. New York: MOMA, 1989.

Rosenthal, Mark. *Neil Jenney: Paintings and Sculpture, 1967–1980*. Exh. cat. Berkeley: University Art Museum, 1981.

———, and Richard Marshall. *Jonathan Borofsky*. Exh. cat. Philadelphia: Museum of Art, 1984.

[Roth] *Dieter Roth*. Exh. cat. New York: David Nolan Gallery, 1989.

Sandler, Irving, ed. "Modernism, Revisionism, Pluralism, and Post-Modernism." *Art Journal* (Winter 1980) special issue.

Sayre, Henry M. *The Object of Performance: The American Avant-Garde Since 1970*. Chicago: University of Chicago Press, 1989.

Senie, Harriet F., and Sally Webster. *Critical Issues in Public Art: Content, Context, and Controversy*. New York: Icon Editions, 1992.

Serra, Richard. *Writings, Interviews*. Chicago and London: University of Chicago Press, 1994.

[Shapiro] *Joel Shapiro: Sculpture and Drawings*. Los Angeles: Pace Wildenstein, 1996.

Shearer, Linda. *Vito Acconci, Public Places*. Exh. cat. New York: MOMA, 1988.

[Singer] *Artworks: Michael Singer*. Exh. cat. Williamstown, Mass.: Williams College Museum of Art, 1990.

Smagula, Howard J. *Currents: Contemporary Directions in the Visual Arts*. Englewood Cliffs, N.J.: Prentice-Hall, 1983.

Smith, Roberta. *Joel Shapiro*. Exh. cat. New York: WMAA, 1982. Text by R. Marshall.

Sondheim, Alan, ed. *Individuals: Post-Movement Art in America*. New York: Dutton, 1976.

Sonfist, Alan, ed. *Art in the Land: A Critical Anthology of Environmental Art*. New York: Dutton, 1983.

[Steir] *Pat Steir: Paintings: Essay/Interview*. New York: Harry N. Abrams, 1986.

Storr, Robert. *Dislocations*. Exh. cat. New York: MOMA, 1991.

Strand, Mark. *William Bailey: Recent Paintings*. Exh. cat. New York: Robert Schoelkopf Gallery, 1982.

Tannenbaum, Judith. *Vija Celmins*. Philadelphia: Institute of Contemporary Art, 1992.

Temkin, Ann. *Thinking Is Form: The Drawings of Joseph Beuys*. Exh. cat. New York: MOMA, 1993.

Thompson, Walter. "Beverly Pepper: Dramas in Space." *Arts Magazine* LXII:10 (Summer 1988): 52–55.

Tisdall, Caroline. *Joseph Beuys*. Exh. cat. New York: SRGM, 1979.

Tomkins, Calvin, and David Bourdon. *Christo's Running Fence*. New York: Harry N. Abrams, 1978. Photography by G. Gorgoni.

Tuchman, Maurice. *Susan Rothenberg*. Los Angeles: LACMA, 1983.

Tucker, Marcia and James Monte. *Anti-Illusion: Procedures/Materials*. Exh. cat. New York: WMAA, 1969.

———, and Robert Pincus-Witten. *John Baldessari*. Exh. cat. New York: New Museum, 1981. Interview by N. Drew.

[Tuttle] *Richard Tuttle: Selected Works, 1964–1994*. Exh. cat. Tokyo: Sezon Museum of Art, 1995.

[Uelsmann] *Jerry Uelsmann: Photo Synthesis*. Gainesville, Fla.: University Press of Florida, 1992.

[Uelsmann] *Jerry Uelsmann: Twenty-five Years —A Retrospective*. Boston: Little, Brown, 1982.

Varnedoe, Kirk. *Duane Hanson*. New York: Harry N. Abrams, 1985.

Wagner, Anne. *Three Artists (Three Women): Modernism and the Art of Hesse, Krasner, and O'Keeffe*. Berkeley, Calif.: University of California Press, 1996.

Waldman, Diane. *Michael Singer*. Exh. cat. New York: SRGM, 1984.

Walker, Barry. *Donald Sultan, a Print Retrospective*. Exh. cat. New York: Rizzoli, 1992.

Weergraf-Serra, Clara, and Martha Buskirk, eds. *The Destruction of "Tilted Arc": Documents*. Cambridge, Mass. and London: MIT Press, 1991.

[Wegman] *William Wegman: Paintings, Drawings, Photographs, Videotapes*. New York: Harry N. Abrams, 1990.

Wegman, William, and Laurence Wieder. *Man's Best Friend*. New York: Harry N. Abrams, 1982.

[Wilmarth] *Christopher Wilmarth: Nine Clearings for a Standing Man*. Exh. cat. Hartford, Conn.: Wadsworth Atheneum, 1975.

[Winsor] *Jackie Winsor*. Exh. cat. New York: MOMA, 1979.

[Wyeth] *Andrew Wyeth*. Exh. cat. Philadelphia: Pennsylvania Academy of the Fine Arts, 1966.

25: THE RETROSPECTIVE EIGHTIES

[Ahearn] *John Ahearn*. Exh. cat. Kansas City, Mo.: Nelson-Atkins Museum, 1990.

[Applebroog] *At the Edge: Ida Applebroog*. Exh. cat. Atlanta: High Museum of Art, 1989.

[Applebroog] *Ida Applebroog: Bilder*. Exh. cat. Ulm, Germany: Ulmer Museum, 1991.

[Applebroog] *Ida Applebroog: Happy Families —A Fifteen Year Survey*. Exh. cat. Houston: Contemporary Arts Museum, 1990.

[Applebroog] *Ida Applebroog: Nostrums (Paintings)/Belladonna (A Film/ Tape by Ida Applebroog and Beth B)*. Exh. cat. New York: Ronald Feldman Fine Arts, 1989.

Audiello, Massimo. "Ross Bleckner: The Joyful and Gloomy Side of Being Alive." *Flash Art* 183 (Summer 1995): 112–115ff.

Auping, Michael. *Francesco Clemente*. Sarasota, Fla.: John and Mable Ringling Museum of Art, 1985.

———. *Judy Pfaff: Installations, Collages, and Drawings*. Exh. cat. Sarasota, Fla.: John and Mable Ringling Museum of Art, 1981.

Benezra, Neal. *Martin Puryear*. Exh. cat. Chicago: Art Institute of Chicago, 1991.

Billeter, Erika, ed. *Eric Fischl: Paintings and*

Drawings. Lausanne: Musée Cantonal des Beaux-Arts, 1990.

[Boltanski] *Christian Boltanski: Lessons of Darkness*. Exh. cat. Chicago: Museum of Contemporary Art, 1988. Texts by L. Gumpert and M. J. Jacob.

Bonito Oliva, Achille. *Trans Avant Garde International*. Moderna Galleria Civica, 1982.

Buchloh, Benjamin, Serge Guilbaut, and David Solkin, eds. *Modernism and Modernity: The Vancouver Conference Papers*. Halifax, Nova Scotia: Press of the Nova Scotia College of Art and Design, 1983.

[Burton] *Scott Burton*. Exh. cat. London: Tate Gallery, 1985.

[Burton] *Scott Burton: Pair Behavior Tableau*. Exh. cat. New York: SRGM, 1976. Text by L. Shearer.

Cahmi, L. "Christian Boltanski: A Conversation." *Print Collector's Newsletter* (January-February 1993): 201–206.

Carmean, Jr., E. A., et al. *The Sculpture of Nancy Graves: A Catalogue Raisonné*. New York: Hudson Hills Press, 1987.

Celant, Germano. *Unexpressionism: Art Beyond the Contemporary*. New York: Rizzoli, 1988.

———, ed. *Keith Haring*. Munich: Prestel, 1992.

[Clemente] *Francesco Clemente: The Fourteen Stations*. Exh. cat. London: Whitechapel Art Gallery, 1983.

Clothier, Peter. *Betye Saar*. Exh. cat. Los Angeles: Museum of Contemporary Art, 1984.

Coe, Sue. *Dead Meat*. New York and London: Four Walls Eight Windows, 1995.

Compton, Michael. *The New Art*. Exh. cat. London: Tate Gallery, 1983.

Contemporary Italian Masters. Exh. cat. Chicago: Council on Fine Arts and the Renaissance Society at the University of Chicago, 1984. Texts by H. Geldzahler and J. R. Kirshner.

Cotter, Holland. "Postmodern Tourist." [Eric Fischl] *Art in America* (April 1991): 154–157, 183.

———. "Taking It Personally: Putting Emotions on Paper." [Joan Snyder] *The New York Times* (April 8, 1994): C26.

Cowart, Jack. *Expressions: New Art from Germany, Georg Baselitz, Jörg Immendorff, Anselm Kiefer, Markus Lüpertz, A. R. Penck*. Exh. cat. St. Louis: St. Louis Art Museum, 1983. Texts by S. Gohr and D. B. Kuspit.

[Cragg] *Tony Cragg*. Exh. cat. London: Hayward Gallery, 1987.

[Cragg] *Tony Cragg*. Exh. cat. London: Tate Gallery, 1989.

[Cragg] *Tony Cragg: Sculpture, 1975–1990*. Newport Beach, Calif.: Newport Harbor Art Museum, 1990.

[Cragg] "Necessary Appetites: An Interview with Tony Cragg." *Border Crossings* (Summer 1992): 45–53. Introduction by Richard Rhodes.

[Cucchi] *Enzo Cucchi*. Exh. cat. Paris: Centre Georges Pompidou, 1986.

Davies, Hugh Marlais. *Martin Puryear*. Exh. cat. Amherst, Mass.: University of Massachusetts, 1984.

Davis, Douglas. *Artculture: Essays on the Postmodern*. New York: Harper & Row, 1979.

Dennison, Lisa, ed. *Ross Bleckner*. Exh. cat. New York: SRGM, 1995.

Endgame: Reference and Simulation in Recent Painting and Sculpture. Boston: Institute of Contemporary Art, 1986.

Feldman, Ronald, et al. *Ida Applebroog*. New York: Ronald Feldman Fine Arts, 1987.

Ferguson, Bruce W. *Eric Fischl*. Exh. cat. Saskatoon, Canada: Mendel Art Gallery, 1985. Texts by J.-C. Ammann, D. B. Kuspit, B. W. Ferguson.

[Fetting] *Rainer Fetting*. Exh. cat. London: Anthony d'Offay, 1982.

[Fetting] *Rainer Fetting*. Exh. cat. New York: Marlborough Gallery, 1984.

[Fischl] *Scenes and Sequences: Recent Monotypes by Eric Fischl*. Exh. cat. Hanover, N.H.: Hood Museum of Art, 1990.

Fox, Howard N. *Avant-Garde in the Eighties*. Los Angeles: LACMA, 1987.

———, ed. *Robert Longo*. Exh. cat. New York: Rizzoli, 1989.

Franzke, Andreas. *Georg Baselitz*. Trans. D. Britt. Exh. cat. New York: Prestel, 1989.

Frazier, I. "Profiles: Partners." [Komar & Melamid] *New Yorker* (December 29, 1986): 33–54.

Freeman, Phyllis, et al., eds. *New Art*. New York: Harry N. Abrams, 1984.

Gablik, Suzy. *Has Modernism Failed?* London: Thames & Hudson, 1984.

Goodman, Cynthia. *Digital Visions: Computers and Art*. New York: Harry N. Abrams, 1987.

Graham-Dixon, Andrew. *Howard Hodgkin*. London: Thames & Hudson, 1994.

[Graves] *Nancy Graves: A Survey 1969/1980*. Exh. cat. Buffalo, N.Y.: Albright-Knox Gallery, 1990.

[Graves] *The Sculpture of Nancy Graves: A Catalogue Raisonné*. New York: Hudson Hills Press and the Fort Worth Art Museum, 1987.

Graze, Sue, and Kathy Halbreich, eds. *Elizabeth Murray: Paintings and Drawings*. Exh. cat. New York: Harry N. Abrams, 1987. With an Essay by R. Smith; Notes by C. Ackley.

[Greenblat] *Rodney A. Greenblat's Reality and Imagination: Two Taste Treats in One!* Exh. cat. University Park, Pa.: Penn State University Museum of Art, 1987.

Gruen, John. *Keith Haring: The Authorized Biography*. London: Thames & Hudson, 1991.

Grundberg, Andy. *Mike and Doug Starn*. New York: Harry N. Abrams, 1990. Introduction by R. Rosenblum.

Hobbs, Robert. *Robert Longo Dis-illusions*. Exh. cat. Iowa City, Iowa: University of Iowa Museum of Art, 1985.

Hockney, David. *Cameraworks*. New York: Knopf, 1984. Text by L. Weschler.

[Hodgkin] *Howard Hodgkin: Paintings*. London: Thames & Hudson, 1995.

Hunter, Sam. *New Directions: Contemporary American Art*. Exh. cat. Princeton, N.J.: University Art Museum, 1981.

Joachimides, Christos. *A New Spirit in Painting*. Exh. cat. London: Royal Academy of Art, 1981. Texts by N. Rosenthal and N. Serota.

[Kiefer] *Anselm Kiefer: The High Priestess*. Exh. cat. New York and London: Harry N. Abrams and Anthony d'Offay, 1989.

Krane, Susan. *Art at the Edge: Alison Saar - Fertile Ground*. Exh. cat. Atlanta: High Museum of Art, 1993.

Kuspit, Donald B. *Leon Golub: Existentialist/Activist Painter*. Rutgers, N.J.: Rutgers University Press, 1985.

LeFalle-Collins, Lizzetta, and R. Barrett. *Betye Saar: Secret Heart*. Exh. cat. Fresno, Calif.: Fresno Art Museum, 1993.

Levick, Melba. *The Big Picture: Murals of Los Angeles*. Boston: Little, Brown, 1988.

Lingwood, James. *Ilya Kabakov: Ten Characters*. London: ICA, 1989.

Linker, Kate. *Love for Sale: The Words and Pictures of Barbara Kruger*. New York: Harry N. Abrams, 1990.

London, Barbara, ed. *Bill Viola*. Exh. cat. New York: MOMA, 1987.

Longo, Robert. *Men in the Cities*. New York: Harry N. Abrams, 1986. Introduction and interview by R. Price.

[Longo] *Robert Longo: Drawings and Reliefs*. Exh. cat. Akron, Ohio: Akron Art Museum, 1984.

Lucie-Smith, Edward. *Art in the Eighties*. Oxford: Phaidon, 1990.

Marmer, Nancy. "Boltanski: The Uses of Contradiction." *Art in America* (October 1989): 169–181, 233–235.

Marshall, Richard. *Jean-Michel Basquiat*. Exh. cat. New York: WMAA, 1992.

McShine, Kynaston. *An International Survey of Recent Painting and Sculpture*. Exh. cat. New York: MOMA, 1984.

[Morris] *Robert Morris: Works of the Eighties*. Exh. cat. Chicago: Museum of Contemporary Art, 1986. Texts by E. F. Fry and D. P. Kuspit.

Neff, Terry. *Gerhard Richter Paintings*. Exh. cat. New York: Thames & Hudson, 1988.

New Figuration in America. Exh. cat. Milwaukee: Milwaukee Art Museum, 1982.

Nixon, Mignon. "You Thrive on Mistaken Identity." [Barbara Kruger] *October* 60 (Spring 1992): 59–81.

Percy, Ann, and Raymond Foye. *Francesco Clemente: Three Worlds*. Exh. cat. Philadelphia: Philadelphia Museum of Art, 1990.

Pincus-Witten, Robert. *Entries (Maximalism)*. New York: Out of London Press, 1983.

———. *Keith Haring*. Exh. cat. New York: Tony Shafrazi Gallery, 1982. Texts by J. Deitch and D. Shapiro.

[Polke] *Sigmar Polke*. Exh. cat. San Francisco: Museum of Modern Art, 1990.

[Polke] *Sigmar Polke: Join the Dots*. Exh. cat. Liverpool: Tate Gallery, 1995.

Princenthal, Nancy. "Reweaving Old Glory." [Donald Lipski] *Art in America* (May 1991): 137–140.

Ratcliff, Carter. *Komar & Melamid*. New York: Abbeville, 1988.

———. *Robert Longo*. New York: Rizzoli, 1985.

Rathbone, Eliza E. *Bill Jensen*. Exh. cat. Washington, D.C.: Phillips Collection, 1987.

[Reed] *David Reed*. Los Angeles: A.R.T. Press, 1990. Essay by C. Hagan.

Richardson, Brenda. *Scott Burton*. Exh. cat. Baltimore: Baltimore Museum of Art, 1986.

[Richter] *Gerhard Richter*. Exh. cat. London: Tate Gallery, 1991.

[Richter] *Gerhard Richter*. Exh. cat. New York: Marian Goodman/Sperone Westwater Gallery, 1985.

Risatti, Howard, ed. *Postmodern Perspectives: Issues in Contemporary Art*. Englewood Cliffs, N.J.: Prentice-Hall, 1990.

Rose, Barbara. *American Painting, the Eighties: A Critical Interpretation*. Exh. cat. New York: Grey Art Gallery and Study Center, New York University, 1979.

Rosenstock, Laura. *Christopher Wilmarth*. New York: MOMA, 1989.

Rosenthal, Mark. *Anselm Kiefer*. Exh. cat. Chicago and Philadelphia: The Art Institute of Chicago and the Philadelphia Museum of Art, 1987.

Schjeldahl, Peter. *Cindy Sherman*. Exh. cat. New York: WMAA, 1984. Texts by P. Schjeldahl and L. Phillips; Afterword by I. M. Danoff.

Schnabel, Julian. *C.V.J.: Nicknames of Maître D's & Other Excerpts from Life.* New York: Random House, 1987.

Schwabsky, Barry. "Memories of Light." [Ross Bleckner] *Art in America* 83 (December 1995): 80–85ff.

[Scully] *Sean Scully.* Pittsburgh: Carnegie Institute, 1985. Exh. cat. Texts by J. Caldwell, D. Carrier, A. Lighthill.

Sculpture for Public Spaces: Maquettes, Models, and Proposals. Exh. cat. New York: Marisa del Re Gallery, 1985.

[Sherman] *Cindy Sherman: History Portraits.* New York: Rizzoli, 1991. Text by A. Danto.

Solomon, Deborah. "Elizabeth Murray: Celebrating Paint." *New York Times Magazine.* (March 31, 1991): 20–25, 40, 46.

[Spero] "Dialogue: An Exchange of Ideas Between Dena Shottenkirk and Nancy Spero." *Arts Magazine* (May 1987): 34–35.

Storr, Robert. "Shape Shifter." [Elizabeth Murray] *Art in America* (April 1989): 210–221, 275.

Transformations: New Sculpture from Britain. [Tony Cragg et al.] Exh. cat. London: British Council, 1983.

Tuchman, Maurice, and Stephanie Barron. *David Hockney, A Retrospective.* Exh. cat. New York: Harry N. Abrams, 1988.

Wakefield, Neville. "Mourning Glory: Ross Bleckner in Retrospect." *Artforum* 33 (February 1995): 69–74ff.

Waldman, Diane. *Enzo Cucchi.* Exh. cat. New York: SRGM, 1986.

———. *Georg Baselitz.* Exh. cat. New York: SRGM, 1995.

———. *Jenny Holzer.* New York: Harry N. Abrams, 1989.

———, ed. *Italian Art Now: An American Perspective.* New York: SRGM, 1982.

Wallach, Amei. *Ilya Kabakov: The Man Who Never Threw Anything Away.* New York: Harry N. Abrams, 1996.

Whitney, David. *David Salle.* New York: Rizzoli, 1994.

Wilmarth, Christopher. *Breath: Inspired by Seven Poems of Stéphane Mallarmé, Translated by Frederick Morgan.* New York: C. Wilmarth, 1982. With an Essay by D. Ashton.

[Wilmarth] *Christopher Wilmarth: Nine Clearings for a Standing Man.* Exh. cat. Hartford, Conn.: Wadsworth Atheneum, 1975.

Yau, John. "Judy Pfaff." *Artforum* (March 1989): 128ff.

26: POSTMODERNISM IN ARCHITECTURE

Ando, Tadao. *The Colors of Light.* London: Phaidon, 1996.

[Ando] *Tadao Ando.* London and New York: Academy Editions and St. Martin's Press, 1990.

Arnell, Peter, and Ted Bickford, eds. *Frank Gehry, Buildings and Projects.* New York: Rizzoli, 1985. Texts by G. Celant and M. Andrews.

Dannatt, Adrian. *The United States Holocaust Memorial Museum: James Ingo Freed.* London: Phaidon, 1995.

Dobney, Stephen, ed. *Eisenman Architects: Selected and Current Works.* Mulgrave, Vic: Images Pub. Group, 1995.

Drew, Philip. *The Architecture of Arata Isozaki.* New York: Harper & Row, 1982.

Duany, Andres, and Elizabeth Plater-Zyberk. "The Second Coming of the American Small Town." *Wilson Quarterly* 16 (Winter 1992): 19–32ff.

[Eisenman] *Re-working Eisenman.* London and Berlin: Academy Editions and Ernst & Sohn, 1993.

Foster Associates: Recent Works. London and New York: Academy Editions and St. Martin's Press, 1992.

[Foster] *Norman Foster.* Ed. I. Yuroukov and E. Kraichkova. Varese, Italy: Arterigere, 1991.

Frampton, Kenneth. *A New Wave of Japanese Architecture.* New York: Institute for Architecture and Urban Studies, 1978.

[Gehry] *The Architecture of Frank Gehry.* Exh. cat. Minneapolis: Walker Art Center, 1986.

Gehry, Frank. *Frank Gehry, Buildings and Projects.* Ed. P. Arnell and T. Bickford. New York: Rizzoli, 1985.

Ghirardo, Diane. *Architecture after Modernism,* London: Thames & Hudson, 1996.

Hadid, Zaha. *The Complete Works.* Intro. Aaron Betsky. London: Thames & Hudson, 1998.

Hays, K. Michael, and Carol Burns, eds. *Thinking the Present: Recent American Architecture.* New York: Princeton Architectural Press, 1990.

Henri, Adrian. *Post-Modernism: The New Classicism in Art and Architecture.* New York: Rizzoli, 1987.

[Hollein] *Architectural Visions for Europe: Hans Hollein.* Braunschweig: Vieweg, 1994.

James, Warren A., ed. *Ricardo Bofill/Taller de Arquitectura: Buildings and Projects: 1960–1985.* New York: Rizzoli, 1986.

Jencks, Charles. *Architecture Today.* Rev. and enl. ed. New York: Harry N. Abrams, 1988.

———. *The Language of Post-Modern Architecture.* Rev. enl. ed. New York: Rizzoli, 1985.

———. *Late Modern Architecture and Other Essays.* London: Academy Editions, 1980.

———. *Post-Modernism: The New Classicism in Art and Architecture.* New York: Rizzoli, 1987.

———, ed. *Frank O. Gehry: Individual Imagination and Cultural Conservatism.* London: Academy Editions, 1995.

Jenson, Robert, and Patricia Conway. *Ornamentalism: The New Decorativeness in Architecture and Design.* New York: C. N. Potter, 1982.

Klotz, Heinrich. *The History of Postmodern Architecture.* Trans. R. Donnell. Cambridge, Mass.: MIT Press, 1988.

[Koolhaas] *Rem Koolhaas: Conversations with Students.* Ed. L. Fitzpatrick and D. Hofius. Houston: Rice University School of Architecture, 1991.

Koolhaas, Rem, and Bruce Mau. *Small, Medium, Large, Extra-large: Office for Metropolitan Architecture.* Ed. J. Sigler. Rotterdam: 010 Publishers, 1995.

Krier, Rob. *Rob Krier on Architecture.* New York: St. Martin's Press, 1982.

Littlejohn, David. *Architect: The Life and Work of Charles W. Moore.* New York: Holt, Rinehart, and Winston, 1984.

Lucan, Jacques. *OMA Rem Koolhaas: Architecture 1970–1990.* New York: Princeton Architectural Press, 1991.

Miller, Nory. *Johnson/Burgee, Architecture: Buildings and Projects.* New York: Random House, 1979.

Norberg-Schulz, Christian. *Ricardo Bofill.* New York: Rizzoli, 1985. Photography by Y. Futagawa.

[Pelli] *Cesar Pelli: Buildings and Projects, 1965–1990.* New York: Rizzoli, 1990.

Portoghesi, Paolo. *Postmodernism: The Architecture of the Post-Industrial Society.* Trans. E. Shapiro. New York: Rizzoli, 1983.

The Presence of the Past: First International Exhibition of Architecture—Venice Biennale 1980. New York: Rizzoli, 1980.

Ray, Keith, ed. *Contextual Architecture: Responding to Existing Style.* New York: McGraw-Hill, 1980.

Rossi, Aldo. *Architecture of the City.* Trans. D. Ghirardo and J. Ockman. Rev. ed. Cambridge, Mass.: MIT Press, 1982.

Rykwert, Joseph. *The Necessity of Artifice.* New York: Rizzoli, 1982.

Smith, C. Ray. *Supermannerism: New Attitudes in Post-Modern Architecture.* New York: Dutton, 1977.

Steele, James. *Schnabel House: Frank Gehry.* London: Phaidon, 1993.

Stern, Robert A. M. *New Directions in American Architecture.* Rev. ed. New York: G. Braziller, 1977.

Tschumi, Bernard. *Event-Cities: Praxis.* Exh. cat. Cambridge, Mass.: MIT Press, 1994.

Venturi, Robert. *Complexity and Contradiction in Architecture.* New York: MOMA, 1966. Introduction by V. Scully.

———, Denise Scott Brown, and Steven Izenour. *Learning from Las Vegas: The Forgotten Symbolism of Architectural Form.* Rev. ed. Cambridge, Mass.: MIT Press, 1977.

Von Moos, Stanislaus. *Venturi, Rauch, and Scott Brown: Buildings and Projects, 1960–1985.* New York: Rizzoli, 1987.

EPILOGUE

1993 Biennial Exhibition. New York: WMAA, 1993.

Always Remember: The NAMES Project AIDS Memorial Quilt. Washington, D.C.: National Museum of American History, 1996.

Bartman, William S., and Miyoshi Barosh, eds. *Mike Kelley.* New York: Art Press, 1992. Interview with the artist by J. Miller.

Blessing, Jennifer, ed. *Rrose is a Rrose is a Rrose: Gender Performance in Photography.* New York: SRGM, 1997.

Blinderman, Barry, ed. *David Wojnarowicz, Tongues of Flame.* Normal, Ill.: University of Illinois State University, 1989.

Bolton, Richard, ed. *Culture Wars: Documents from the Recent Controversies in the Arts.* New York: The New Press, 1992.

Bonami, Francesco. "Back in Five Minutes." [Gabriel Orozco] *Parkett* 48 (1996): 41–47.

Boodro, Michael. "Blood, Spit, and Beauty." *Art News* 93 (March 1994): 126–131.

Bradley, Fiona, ed. *Rachel Whiteread.* Exh. cat. London: Tate Gallery Liverpool, 1996. Text by R. Krauss et al.

Brougher, Kerry. *Jeff Wall.* Exh. cat. Los Angeles: Museum of Contemporary Art, 1997.

"Brush Fires in the Social Landscape." *Aperture* 137 (Fall 1994) special issue on David Wojnarowicz.

Bryson, Norman. "Three Morimura Readings," *Art + Text* 52 (1995): 67–78.

The City Influence: Ross Bleckner, Peter Halley, Jonathan Lasker. Exh. cat. Dayton, Ohio: Dayton Art Institute, 1992.

Conkelton, Sheryl, and Carol S. Eliel. *Annette Messager.* Exh. cat. Los Angeles: LACMA, 1995.

Connor, Kimberly Rae. "To Disembark: The Slave Narrative Tradition." [Glenn Ligon] *African American Review* 30 (Spring 1996): 35–57.

"A Conversation on Recent Feminist Art Practices." *October* 71 (Winter 1995): 48–69.

Cooke, Lynne. "Gary Hill: 'Who am I but a figure of speech?'" *Parkett* 34 (1992): 16–21.

———. "Tony Oursler Altars." *Parkett* 47 (1996): 38–41.

Crone, Rainer, and David Moos. *Jonathan Lasker: Telling the Tales of Painting; About Abstraction at the End of the Millennium.* Stuttgart: Edition Cantz, 1993.

Crow, Thomas. "Profane Illuminations: Social History and the Art of Jeff Wall." *Artforum* (January 1993): 63–69.

Danoff, I. Michael. *Jeff Koons.* Exh. cat. Chicago: Museum of Contemporary Art, 1988.

Danto, Arthur. *Mark Tansey: Visions and Revisions.* Ed. C. Sweet. New York: Harry N. Abrams, 1992. With notes by the artist.

Durant, Alice, Laura Mulvey, and Dirk Snauwaert. *Jimmie Durham.* London: Phaidon, 1995.

Enwezor, Okwui. "Social Grace: The Work of Lorna Simpson." *Third Text* 35 (Summer 1996): 43–58.

Flomenhaft, Eleanor. *Faith Ringgold, A Twenty-Five Year Survey.* Exh. cat. Hempstead, N.Y.: Fine Arts Museum of Long Island, 1990.

Freeman, Judi. *Mark Tansey.* Exh. cat. Los Angeles and San Francisco: LACMA and Chronicle Books, 1993.

[Fritsch] *Katharina Fritsch.* Exh. cat. New York: Dia Center for the Arts, 1993.

Fusco, Coco. *English is Broken Here: Notes on Cultural Fusion in the Americas.* New York: The New Press, 1992.

Gale, Peggy. "Stan Douglas: *Evening* and Others." *Parachute* (July–September 1995): 21–27.

Garrels, Garu, ed. *Amerika: Tim Rollins - K.O.S.* New York: Dia Art Foundation, 1990.

Gilbert-Rolfe, Jeremy. *Roni Horn: Pair Objects 1, 11, 111.* Exh. cat. New York: Galerie Lelong, 1988.

Gillick, Liam, and Rirkrit Tiravanija. "Forget about the Ball and Get on with the Game," *Parkett* 44 (1995).

[Gober] *Robert Gober.* Exh. cat. Rotterdam and Bern: Museum Boymans-Van Beuningen and Kunsthalle Bern, 1990.

Golden, Thelma. *Black Male: Representations of Masculinity in Contemporary American Art.* New York: WMAA, 1994.

Goldin, Nan. *The Ballad of Sexual Dependency.* New York: Aperture, 1996.

Gott, Ted, ed. *Don't Leave Me This Way: Art in the Age of AIDS.* Melbourne: National Gallery of Australia, 1994.

Gumpert, Lynn. "Glamour Girls." *Art in America* 84:7 (July 1996): 62–65.

[Halley] *Peter Halley.* Exh. cat. Madrid: Museo Nacional Renia Sofía, 1991.

[Halley] *Peter Halley: Oeuvres de 1982 a 1991.* Exh. cat. Bordeaux: Musée d'Art Contemporain de Bordeaux, 1991.

[Hamilton] *Ann Hamilton.* Exh. cat. London: Tate Gallery, 1994.

[Hamilton] *Ann Hamilton.* Exh. cat. San Diego: San Diego Museum of Contemporary Art, 1991.

[Hamilton] *Ann Hamilton: Tropos, 1993.* Exh. cat. Ed. L. Cooke and K. Kelly. New York: Dia Center for the Arts, 1995.

[Hamilton] *The Body and the Object: Ann Hamilton, 1984–1996.* Exh. cat. Columbus, Ohio: Wexner Center for the Arts, 1996.

Heartney, Eleanor. "Figuring Absence." [Lorna Simpson] *Art in America* (December 1985).

Hickey, Dave. *In the Dance Hall of the Dead.* New York: Dia Foundation for the Arts, 1992.

[Hill] *Gary Hill.* Exh. cat. Amsterdam: Stedelijk Museum, 1993.

[Hill] *Gary Hill.* Exh. cat. Seattle: Henry Art Gallery, University of Washington, 1994.

[Hill] *Gary Hill: Imagining the Brain Closer Than the Eyes.* Exh. cat. Ostfildern: Cantz and Museum für Gegenwartskunst, 1995. Texts by H. Belting, G. Boehm, the artist et al.

hooks, bell. *Art on My Mind: Visual Politics.* New York: The New Press, 1995.

Horn, Roni. *Earths Grow Thick.* Exh. cat. Columbus, Ohio: Wexner Center for the Arts, 1996. Text by b. hooks.

Johnson, Ken. "Being and Politics." [Adrian Piper] *Art in America* 78 (September 1990): 154–161.

———. "Material Matters." [Roni Horn] *Art in America* (February 1994): 71–78, 117.

Kellein, Thomas. *Mark Tansey.* Exh. cat. Basel: Kunsthalle, 1990.

[Kelley] *Mike Kelley: Three Projects: Half a Man, From My Institution to Yours, Pay for Your Pleasure.* Chicago: University of Chicago Press, 1988.

Kirsch, Andrea, and Susan Fisher Sterling. *Carrie Mae Weems.* Exh. cat. Washington, D.C.: National Museum of Women in the Arts, 1993.

Koons, Jeff. *The Jeff Koons Handbook.* London: Thames & Hudson, 1992. Introduction by R. Rosenblum.

[Lemieux] *Annette Lemieux: The Appearance of Sound.* Sarasota, Fla.: The John and Mable Ringling Museum of Art, 1989.

Linker, Kate. "Went Looking for Africa: Carrie Mae Weems." *Artforum* 31 (February 1993): 79–82.

Lippard, Lucy. "Andres Serrano: The Spirit and the Letter." *Art in America* (April, 1996): 239–245.

———. "Out of the Safety Zone." *Art in America* 78:12 (December 1990): 130–139ff.

Mann, Sally. *At Twelve: Portraits of Young Women.* New York: Aperture, 1988.

———. *Immediate Family.* New York: Aperture, 1992.

———. *Second Sight: The Photographs of Sally Mann.* Boston: D. R. Godine, 1983. Introduction by J. Livingston.

———. *Still Time.* New York: Aperture, 1994.

Masheck, Joseph. "Painting in the Double Negative: Jonathan Lasker's Third Stream Abstraction." *Arts Magazine* 64 (January 1990): 38–43.

Mifflin, Margaret. "An Interview with Karen Finlay." *High Performance* 11 (Spring/ Summer 1988): 86–88.

Murphy, Patrick T., ed. *Andres Serrano: Works 1983–1993.* Exh. cat. Philadelphia: Institute of Contemporary Art, 1994. Texts by R. Hobbs, W. Steiner, M. Tucker.

Nead, Lynda. *The Female Nude: Art, Obscenity and Sexuality.* New York: Routledge, 1992.

Pagel, David. "Still Life: The Tableaux of Ann Hamilton." *Arts Magazine* 64 (May 1990): 56–61.

Piper, Adrian. *Out of Order, Out of Sight.* 2 vols. Cambridge, Mass.: MIT Press, 1996.

———. *Pretend.* Exh. cat. New York: John Weber Gallery, 1990. Texts by M. A. Staniszewski and the artist.

———. *Reflections, 1967–87.* Exh. cat. New York: Alternative Museum, 1987.

Rimanelli, David. "It's My Party, Jeff Koons: A Studio Visit." *Artforum* 35:10 (1997).

[Ringgold] *Faith Ringgold: Twenty Years of Painting, Sculpture, and Performance, 1963–1983.* Exh. cat. New York: The Studio Museum in Harlem, 1984.

Ringgold, Faith. *We Flew Over the Bridge: The Memoirs of Faith Ringgold.* Boston: Little, Brown, 1995.

Rosenblum, Robert. *The Jeff Koons Handbook.* New York: Rizzoli, 1992.

Rubinstein, Raphael. "Counter-Resolution." [Jonathan Lasker] *Art in America* 83 (April 1995): 84–89.

Rugoff, Ralph. "Surfing with Raymond: Nobody Rides for Free." [Raymond Pettibone] *Parkett* 47 (1996): 82–87.

Saltz, Jerry. "The Next Sex." *Art in America* 84:10 (October 1996): 82–91.

Sante, Luc. "The Nude and the Naked." [Sally Mann] *The New Republic* (May 1, 1995): 30–35.

Sensation: Young British Artists from The Saatchi Collection. London: Thames & Hudson, 1997.

Simon, Joan. *Robert Gober.* Exh. cat. Paris: Éditions du Jeu de Paume and Réunion de Musées Nationaux, 1991.

Sims, Patterson. *Mark Tansey: Art and Source.* Exh. cat. Seattle: Seattle Art Museum, 1990.

[Smith] *Kiki Smith.* Exh. cat. Amsterdam: Institute of Contemporary Art, 1990.

Smith, Kiki. "Impossible Liberties." *Art Journal* 54 (Summer 1995): 49–50.

Soutter, Lucy. "By Any Means Necessary: Document and Fiction in the Work of Carrie Mae Weems." *Art and Design* 11 (November/December 1996): 70–75.

Steils, Kristine, and Peter Selz, eds. *Theories and Documents of Contemporary Art: A Sourcebook of Artists' Writings.* Berkeley, Calif.: University of California Press, 1996.

Storr, Robert. "Setting Traps for the Mind and Heart." *Art in America* 84:10 (January 1996): 70–77ff.

Sussman, Elizabeth. *Mike Kelley: Catholic Tastes.* Exh. cat. New York: WMAA, 1993.

Tallman, Susan. "Kiki Smith: Anatomy Lessons." *Art in America* 80 (April 1992): 146–153ff.

Viso, Olga. *Distemper: Dissonant Themes in the Art of the 1990s.* Washington, D.C.: Hirshhorn Museum and Smithsonian Institution, 1996.

Volk, Gregory. "Orozco: Circumstantial Evidence." *World Art* 3 (1995): 55–59.

Wakefield, Neville. "Rachel Whiteread: Separation Anxiety and the Art of Release." *Parkett* 42 (1994): 76–81.

Wallis, Brian. "Questioning Documentary." [Lorna Simpson] *Aperture* 11 (Fall 1988): 60–71.

———, ed. *Body and Soul/Andres Serrano.* New York: Takarajima Books, 1995. Texts by b. hooks, B. Ferguson, A. Arenas.

[Whiteread] *Rachel Whiteread.* Exh. cat. Trans. D. Galloway. Basel and Philadelphia: Kunsthalle and the Institute of Contemporary Art, 1994.

Winzen, Matthias. "Katharina Fritsch." *Journal of Contemporary Art* 7 (Summer 1994): 59–73.

Wojnarowicz, David. *Memories That Smell Like Gasoline.* San Francisco: Artspace Books, 1992.

Wright, Beryl, and Saidiya Hartman. *Lorna Simpson: For the Sake of the Viewer.* Exh. cat. New York and Chicago: Universe Pub. and Museum of Contemporary Art, 1992.

Zelevansky, Lynn. *Sense and Sensibility: Women Artists and Minimalism in the Nineties.* New York: MOMA, 1994.

Index

Credits

The author and publisher wish to thank the libraries, museums, galleries, and private collectors named in the picture captions for permitting the reproduction of works of art in their collections and for supplying the necessary photographs. Photographs from other sources are gratefully acknowledged below. Numbers listed before the sources refer to figure numbers.

Black-and-White Photographs

Chapter 1: 1. Archives Photographiques, Paris; 3. Fototeca Unione, Rome; 5. © A.C.L. Brussels; 6. Alinari, Rome; 10. Alinari, Florence; 13. Bulloz, J.-E., Paris; 18. Réunion des musées nationaux, Paris; 20. Courtesy of Wildenstein Archives; 21. Caisse Nationale des Monuments Historiques, Paris

Chapter 2: 24. Réunion des musées nationaux, Paris; 31. National Portrait Gallery, Smithsonian Institution, Washington, D.C.; 38. Lichtbildwerkstätte Alpenland, Vienna; 39. Réunion des musées nationaux, Paris; 43. Archives Photographiques, Paris; 44. Réunion des musées nationaux, Paris; 53. Réunion des musées nationaux, Paris; 55. Routhier, Studio Lourmel, Paris; 59. Art Resource, New York

Chapter 3: 72. Oliver Baker, New York

Chapter 4: 86. Edwin C. Smith, Essex; 87. Jean Roubier, Paris; 89. Archives Photographiques, Paris; 90. French Government Tourist Office, New York; 91. A.F. Kersting, London; 93. Wayne Andrews, Grosse Pointe, Michigan; 95. T.& R. Annan, Ltd., Glasgow; 96. T.& R. Annan, Ltd., Glasgow; 98. Bettmann Archives, Inc., New York; 100. Wayne Andrews, Grosse Pointe, Michigan; 101. Peter Aaron/Esto; 102. New-York Historical Society

Chapter 5: 103. Derrick E. Witty, Sunbury-on-Thames; 107. Foto Studio Minders, Ghent; 109. Courtesy of the Museum of Modern Art, New York; 111. Courtesy of The Museum of Modern Art, New York; 113. Dr. Franz Stoedtner, Düsseldorf; 114. Felipe Ferré, Paris; 115. Foto Marburg, Marburg/Lahn; 116. Foto Marburg, Marburg/Lahn; 117. Foto Marburg, Marburg/Lahn; 118. Otto Vaering, Oslo; 119. J. Lathion, Nasjonalgalleriet; 123. A.C.L., Brussels; © Spanganberg Verlag, Munich

Chapter 6: 130. Réunion des musées nationaux, Paris; 133. Foto Marburg, Marburg/Lahn; 135. O. E. Nelson

Chapter 7: 147. © 1997 The Museum of Modern Art, New York; 151. Réunion des musées nationaux, Paris

Chapter 8: 154. Foto Witzel, Essen; 157. Kleinhempel, Hamburg; 164. © 1997 The Museum of Modern Art, New York

Chapter 9: 177. Geoffrey Clements, New York; 179. © 1997 The Museum of Modern Art, New York/Peter A. Juley & Son, New York; 181. Fine Arts Associates,

Minneapolis; 191. Courtesy Witkin Gallery, New York

Chapter 10: 199. © 1997 The Museum of Modern Art, New York; 202. Galerie Louise Leiris, Paris; 203. Svetlana; 205. © 1997 The Museum of Modern Art, New York; 211. Courtesy of Galerie Bruno Bischofberger, Zurich; 214. Archives Laurens, Paris; 215. © 1997 The Museum of Modern Art, New York; 219. © 1997 The Museum of Modern Art, New York; 220. © 1997 The Museum of Modern Art, New York

Chapter 11: 234. Charles Uht, New York; 235. Sherwin Greenberg, McGranahan & May, Inc., Buffalo, New York; 236. Courtesy Sotheby's Inc., New York; 238. © 1997 The Museum of Modern Art, New York; 241. Alfred H. Barr, Jr., New York; 251. Alfred H. Barr, Jr., New York; 252. Alfred H. Barr, Jr., New York; 254–255. Irving Blomstrann, New Britain, Connecticut; 256. © 1997 The Museum of Modern Art, New York; 258. © 1997 The Museum of Modern Art, New York; 260. Studio Yves Hervochon, Paris

Chapter 12: 264. Wayne Andrews, Grosse Pointe, Michigan; 265. Hedrich-Blessing, Chicago; 268. Buffalo and Erie County Historical Society; 239. Buffalo and Erie County Society; 270. Wayne Andrews, Grosse Pointe, Michigan; 271. Chicago Architectural Photographing Company; 274. Wayne Andrews/Esto; 280. Benevolo; 284. Doeser Fotos, Lauren; 285. Foto van Ojen, The Hague; 286. Fototechnischer Dienst, Rotterdam; 290. German Information Center, New York; 291. Emil Gmelin, Dornach

Chapter 13: 293. Martin F.J. Coppens, Eindhoven; 295. Courtesy Robert Miller Gallery, New York; 299. Etienne Bertrand Weill, Paris; 300. Elizabeth Sax, Paris; 303. Courtesy Cordier & Ekstrom Inc., New York; 306. eeva-inkeri, New York; 308. Geoffrey Clements; 310. Courtesy Cordier & Ekstrom Inc., New York; 313. Jorg Anders; 315. Landesgalerie, Hanover; 318. Courtesy Dr. Marie-Andreas von Lüttichau; 321. Rheinisches Bildarchiv, Cologne; 323. © 1997 The Museum of Modern Art, New York

Chapter 14: 330. S. Sunami, New York; 332. © 1997 The Museum of Modern Art, New York/Oliver Baker, New York; 340. © 1997 The Museum of Modern Art, New York; 341. Henri Mardyks; 342. © 1997 The Museum of Modern Art, New York; 345. Henri Mardyks; 347. Paul Rosenberg & Co., New York; 349. Geoffrey Clements, New York

Chapter 15: 355. © 1997 The Museum of Modern Art, New York; 356. Etienne Bertrand Weill, Paris; 362. Colten Photos, New York; 380. © Réunion des musées nationaux, Paris; 383. © Réunion des musées nationaux, Paris; 385. © Réunion des musées nationaux, Paris; 393. © 1997 The Museum of Modern Art, New York; 399. New York Public Library; 401. Courtesy of Susan Harder Gallery, New York; © 1997 The Museum of Modern Art,

New York; 406. Courtesy of Magnum Photo Library, New York; 407. Courtesy of Magnum Photos, New York

Chapter 16: 409. © 1997 The Museum of Modern Art, New York; 410. Foto Marburg; 413. Wayne Andrews, Grosse Pointe, Michigan; 415. Lucien Hervé, Paris; 416. Lucien Hervé, Paris; 422. Wayne Andrews, Grosse Pointe, Michigan; 423. Photo Esto/Peter Mauss; William Lescaze, New York; 428. Bill Hedrich, Hedrich-Blessing, Chicago; 429. S.C. Johnson & Johnson, Inc. Racine, Wisconsin; 430. S.C. Johnson & Johnson, Racine, Wisconsin; 431. Wayne Andrews, Grosse Pointe, Michigan

Chapter 17: 435. Fotocollectie, Leiden; 437. L.W. Schmidt, Rotterdam; 440. Photoatelier Gerlach, Vienna; 443. Peter A. Juley & Son, New York; 451. © 1997 The Museum of Modern Art, New York

Chapter 18: 481. David Stansbury, Springfield, Massachusetts; 486. Geoffrey Clements, New York; 487. Geoffrey Clements; 513. Arthur Siegel, Chicago; 514. Helga Photo Studio

Chapter 19: 517. Oliver Baker Associates, New York; 532. Bernard Gotfryd. Courtesy of Amalee Newman; 534. Geoffrey Clements, New York; 537. Eric Sutherland, Minneapolis; 540. Courtesy Marlborough Gallery, New York; 541. Marc Fournier; 543. Rudolph Burckhardt, New York; 544. Rudolph Burckhardt, New York; 551. Magnum Photo Library, New York; 553. © Lee Miller Archives, England; 554. © 1982 by The Trustees of Princeton University. All rights reserved; 556. Courtesy Pace/MacGill Gallery, New York; 558. Courtesy Fraenkel Gallery, San Francisco; 559. © 1997 The Museum of Modern Art, New York

Chapter 20: 561. © Réunion des musées nationaux, Paris; 568. © 1997 The Museum of Modern Art, New York; 578. Squidds & Nunns; 579. Courtesy of the artist; 582. John Schiff, New York; 586. Ernest Nash, Rome; 588. Liselotte Witzel; 594. Courtesy The Witkin Gallery, New York; 595. John Webb FRPS Photography

Chapter 21: 604. Oliver Baker, New York; 608. © David Hockney, Los Angeles; 610. Rudolph Burckhardt; 612. Peter Moore; 613. Courtesy of Leo Castelli Gallery, New York; 614. Rudolph Burckhardt, New York; 616. John Cohen; 618. Robert R. McElroy, Courtesy Sidney Janis Gallery, New York; 619. Geoffrey Clements, New York; 621. Courtesy Leo Castelli Gallery, New York; 622. Eric Pollitzer, New York; 625. Courtesy Allan Stone Gallery, New York; 626. Zindman/Fremont, New York; 628. Oliver Baker, New York; 629. Eric Pollitzer, New York; 630–632. Rudolph Burckhardt, New York; 633. Seymour Rosen; 634. Rudolph Burckhardt, New York; 635. Eric Pollitzer, New York; 636. Eric Pollitzer, New York; 637. Eric Pollitzer, New York; 641. Lee Fatherree; 644. Harry Shunk; 646. David Gahr; 647. © Hans Hammarskiöld Tio

Colorplates

Philadelphia; 168. Bertrand Prevost © Centre G. Pompidou, Paris; 169. Richard Nickel; 170. Geoffrey Clements, New York; 174. Eric Pollitzer, New York; 175. Charles P. Mill, Philadelphia; 180. Courtesy of John Cavaliero; 182. Art Resource, New York; 183. Réunion des musées nationaux, Paris; 185. Galerie Louise Leiris, Paris; 189. Marvin Trachtenberg; 190. Marvin Trachtenberg; 194. Elke Walford; 195. SRGM, New York; 197. SRGM, New York; 199. SRGM, New York; 207. Art Resource, New York; 210. Dimitris Skliris; 218. Bill Jacobson, New York; 219. Dorothy Zeidman; 220. © Cedar Rapids Community School District, Iowa; 222. Edward Owen; 229. © 1997 Estate of Stuart Davis/VAGA New York; 230. Geoffrey Clements, New York; 232. Geoffrey Clements, New York; 233. Edward Owen; 238. Richard Carafelli; 240. Malcolm Varon; 242. Geoffrey Clements, New York; 243. Douglas M. Pater Studio; 244. Eric Pollitzer, New York; 248. Hickey-Robertson, Houston; 249. Hickey-Robertson, Houston; 256. Courtesy the artist; 258. Kevin Noble; 259. Peter Bellamy; 261. Courtesy The Pace Gallery, New York; 262. Jerry L. Thompson, Courtesy Storm King Art Center; 263. David Heald; 270. SRGM, New York; 275. SRGM, New York; 277. Biff Henrich; 279. SRGM, New York; 280. Robert E. Mates, New York; 287. Courtesy James Kirkman, London; 288. © Henry Moore Foundation; 289. Offentliche Kunstmuseum, Basel/Martin Bühler; 290–291. Courtesy the artist; 292. Paul Hester; 295. George Holzer; 300. Courtesy Leo Castelli Gallery, New York; 302. Frank English; 304. Courtesy the artist; 305. Geoffrey Clements, New York; 306. Al Mozell, Courtesy The Pace Gallery, New York; 307. Courtesy Leo Castelli Gallery, New York; 308. Edward Owen; 309. Eric Pollitzer, New York; 312. Courtesy of Leo Castelli Gallery, New York; 313. Geoffrey Clements, New York; 314. Edward Luce, Worcester, Mass; 315. Geoffrey Clements, New York; 316. Statens Kunstmuseum; 317. Lee Stalsworth; 320. © 1997 The Museum of Modern Art, New York; 321. Beth Phillips, Courtesy The Pace Gallery, New York; 322. Sandak, Inc./Division of Macmillan Publishing Company; 323. Courtesy Jeanne-Claude Christo; 324. Wolfgang Volz; 326. Dean Beasom; 329. Geoffrey Clements, New York; 330. Steven Sloman; 331. Cleveland Museum of Art; 332. Geoffrey Clements, New York; 335. Lee Stalsworth; 339. Courtesy the artist; 340. Courtesy of Leo Castelli Gallery, New York; 342. Courtesy the artist; 346. David Heald; 347. Richard Andrews, Larry Tate & Roy Carraher; 349. Eric Pollitzer/SRGM, New York; 351. Steven Sloman; 354. Richard Cheek, Belmont, MA; 355. Robert E. Mates and Paul Katz; 356. © Giorgio Colombo, Milan; 357. Courtesy The Pace Gallery, New York; 359. Courtesy of Mary Boone Gallery, New York; 363. Courtesy the artist; Pedro E. Guerrero, New York; 364. Marvin Trachtenberg, New York; 365. Andrew Garn, New York; 366. Michael Bodycomb; 367. Kazik Ashraf; 368. Esto; 371. Peter Moore; 377. Courtesy Robert Miller Gallery, New York; 378. Courtesy the artist; 381. Courtesy Middendorf Gallery, Washington, D.C.; 383. Lynton Gardiner;

384. Courtesy John Weber Gallery; 386. Susan Zurcher; 389. Philippe Migeat © Centre Georges Pompidou, Paris; 390. Jan Kosmowski; 391. Balthazar Korab, Michigan; 393. Courtesy Laumeier Sculpture Park; 394. © C. Johnson; 396. Bill Jacobson Studio; 397. Courtesy Louis K. Meisel Gallery, NY; 400. Courtesy the artist; 401. George Roos, Courtesy Marlborough Gallery, New York; 402. Eric Pollitzer, Courtesy Robert Miller Gallery, New York; 405. Donald Woodman; 406. D. James Dee; 407. Allan Finkleman; 408. Cymie Payne, Courtesy Barbara Gladstone, New York; 409. Charles Harrison, Courtesy The Willard Gallery, New York; 410. Philipp Scholz Ritterman; 411. Courtesy BlumHelman Gallery, New York; 412. Eric Mitchell; 413. Geoffrey Clements, New York; 418. Robert E. Mates, N.J.; 419. D. James Dee, Courtesy Holly Solomon Gallery, New York; 427. Courtesy Mary Boone Gallery, New York; 423. Anne Gold, Aachen; 424. Courtesy Mary Boone Gallery, New York; 427. David Reynolds; 428. Myles Aronowitz; 429. Jim Frank; 430. Courtesy Sperone Westwater Gallery, New York; 432. © 1997 The Museum of Modern Art, New York; 434. Zindman/Fremont, Courtesy Mary Boone Gallery, New York; 435. Geoffrey Clements, Courtesy Mary Boone Gallery, New York; 436. Sarah Wells, Courtesy Baskerville & Watson Gallery, New York; 437. Courtesy Metro Pictures, New York; 440. William Neetles; 441. Ivan Dalla-Tana, New York; 442. Courtesy Ronald Feldman Fine Arts, New York; 444. Kira Perov/Squidds and Nunns; 445. Geoffrey Clements, New York; 447. Courtesy the Solomon R. Guggenheim Museum, New York; 449. Courtesy M. Knoedler & Co. Inc., New York; 450. Courtesy Marian Goodman Gallery/Sperone Westwater Gallery, New York; 451. Bill Jacobson Studio, New York; 452. D. James Dee, Courtesy David McKee Gallery, New York; 455. Ben Blackwell; 457. Jerry Thompson; 458. Courtesy Walker Art Center, Minneapolis, Minn.; 459. Art Resource, New York; 462. Norman McGrath, New York; 463. Norman McGrath, New York; 464. © Peter Aaron/Esto; 465. Marvin Trachtenberg; 466. Yasuhiro Ishimoto, Tokyo; 467. Timothy Hursley; 468. Timothy Hursley; 469. Shinkenchiku-sha, Tokyo; 471. Timothy Hursley; 472. Shinkenchiku-sha; 473. Attilio Maranzano; 474. Alex S. MacLean; 476. Fred Scrutton; 477. Jim Dow; 478. Courtesy John Weber Gallery; 484. Irma Estwick; 489. Michael O'Brien 490. Michael O'Brien; 495. Bill Johnson/© Dia Center for the Arts; 509. Jean-Marc Meunier

Artist Copyrights

Jacket Cover Photo: © L & M Services B.V. Amsterdam 971103

Black-and-White Photographs

Chapter 8: 155–156. © Dr. Wolfgang and Ingeborg Henz-Ketterer, Wichtrach, Bern; 162. © Nolde-Stiftung Seebüll; 163. © Dr.

Wolfgang and Ingeborg Henz-Ketterer, Wichtrach, Bern
Chapter 9: 182. © Ernst and Hans Barlach Lizenzverwaltung, Ratzeburg, Germany; 191. © 1981 Center for Creative Photography, Arizona Board of Regents
Chapter 10: 230–231. © L & M Services B. V. Amsterdam 971103
Chapter 11: 254–255. © Nina Williams; 256–257. Courtesy of the Mondrian Estate/Holzman Trust; 259. Courtesy of the Mondrian Estate/Holzman Trust
Chapter 15: 400–401. © André Kertész; 402. © Gilberte Brassaï; 404. © Manuel Alvarez Bravo; 406–407. Henri-Cartier Bresson; 408. © Mrs. Noya Brandt, London
Chapter 17: 436–437. © Nina Williams; 450–451. Courtesy of the Mondrian Estate/Holzman Trust; 460–461. © Henry Moore Foundation; 462–463. © Hepworth Estate © Alan Bowness
Chapter 18: 472. Reprinted with permission of Joanna T. Steichen; 488. © Grandma Moses Properties; 496. © Berenice Abbott; 498. © Donna Van der Zee; 501. Instituto Nacional des Bellas Artes, Mexico City; 504. © Frida Kahlo Foundation
Chapter 19: 522. © Hans Namuth, Ltd.; 544–545. © The Isamu Noguchi Foundation; 551. © Robert Capa/Magnum Photo Library; 554. Reproduction courtesy The Minor White Archive, Princeton University, © 1982 by The Trustees of Princeton University. All rights reserved
Chapter 20: 599–601. © Henry Moore Foundation; 602–603. Hepworth Estate © Alan Bowness
Chapter 21: 615. © Alan Kaprow; 618–621. © Claes Oldenburg and © Coosje van Bruggen; 630–631. © Roy Lichtenstein; 661. © 1972 The Estate of Diane Arbus; 662. © Bruce Davidson/Magnum Photos; 663. © 1959 Richard Avedon. All rights reserved. 664. © Associated Press/Worldwide Photos, London
Chapter 22: 668. © Helen Frankenthaler
Chapter 24: 776. © Estate of Eva Hesse; 799. © Claes Oldenburg and © Coosje van Bruggen; 805. © Andrew Wyeth

Colorplates

66. © Nolde-Stiftung Seebüll; 67–68. © Dr. Wolfgang and Ingeborg Henz Ketterer, Wichtrach, Bern; 71. © Nolde-Stiftung Seebüll; 101. © L & M Services B.V. Amsterdam 971103; 104. © L & M Services B.V. Amsterdam 971103; 120–121. Courtesy of the Mondrian Estate/Holzman Trust; 200–202. Courtesy of the Mondrian Estate/Holzman Trust; 227. Instituto Nacional des Bellas Artes, Mexico City; 228. © Frida Kahlo Foundation; 229. © 1997 Estate of Stuart Davis/Vaga, New York; 258. © The Isamu Noguchi Foundation; 290–291. © David Hockney, Los Angeles; 304. © Claes Oldenburg and © Coosje van Bruggen; 307–308. © Roy Lichtenstein; 326. © Robert Frank; 329–330. © Helen Frankenthaler; 377. © Estate of Eva Hesse; 392. © Claes Oldenburg and Coosje van Bruggen